THEORY AND PROBLEMS OF LINEAR ALGEBRA

THEORY AND PROBLEMS OF LINEAR ALGEBRA

DR. R.D. SHARMA
B.Sc. (Hons) (Gold Medallist), M.Sc. (Gold Medallist), Ph.D.
Head of Department of Science & Humanities
Department of Technical Education, Govt. of Delhi
Delhi

DR. RITU JAIN
Associate Professor
Department of Statistics, PGDAV College
University of Delhi

I.K. International Publishing House Pvt. Ltd.

NEW DELHI • BANGALORE

Published by
I.K. International Publishing House Pvt. Ltd.
S-25, Green Park Extension
Uphaar Cinema Market
New Delhi–110 016 (India)
E-mail: info@ikinternational.com
Website: www.ikbooks.com

ISBN 978-93-80578-86-5

© 2011-Author

10   9   8   7   6   5   4   3   2

Reprint 2012

All rights reserved. No part of this publication may be reproduced, stored in a retrieval system, or transmitted in any form or any means: electronic, mechanical, photocopying, recording, or otherwise, without the prior written permission from the publisher.

Published by Krishan Makhijani for I.K. International Publishing House Pvt. Ltd., S-25, Green Park Extension, Uphaar Cinema Market, New Delhi–110 016 and Printed by Rekha Printers Pvt. Ltd., Okhla Industrial Area, Phase II, New Delhi–110 020.

Preface

Linear Algebra has wide range of applications not only in Science and Technology but also in many other fields like Statistics, Economics etc. Recent developments in software technology and information technology have further widened the scope of linear algebra. The knowledge of Linear Algebra is an essential part of mathematical background required by Mathematicians, Computer Scientists, Engineers, Economists and Physicists etc.

This book *Theory and Problems of Linear Algebra* has been designed to cater to the need of students opting Linear Algebra as a subject at undergraduate and postgraduate levels in various Indian universities. The book exhaustively covers the subject matter and its application in various fields.

In order to understand the subject matter covered in the book, reader must be aware of some basic concepts of abstract algebra. These prerequisites have been covered in Chapter-0. The reader is advised to go through this chapter first before switching over to the next chapter. The subject matter has been graded in such a systematic manner that the knowledge of topics covered in each chapter (except Chapter-0) is a prerequisite to understand the topics covered in the chapters to follow.

Chapter-1 deals with modules. A module is an algebraic structure consisting of an Abelian group, a ring and an external mapping, associating an element of the ring and an element of Abelian group to a unique element of the Abelian group, satisfying certain axioms. Emphasis has been given to module morphisms, cyclic module, free modules and notherian modules.

Chapter-2 is an extension of Chapter-1, wherein ring is replaced by a field. The algebraic structure so obtained is known as a vector space. The elements of Abelian group are known as vectors and those of the field are scalars. Linear independence and dependence of vectors have been discussed. Basis and dimension of a vector space have been defined.

Chapter-3, deals with vector space homomorphisms, widely known as linear transformations. Rank of a linear transformation and properties of its dual have also been discussed. Eigenvalues and Eigenvectors of linear transformations have been defined and their properties have been discussed in detail.

Chapter-4 deals with relations between linear transformations between finite dimensional vector spaces and matrices represented by them. Every linear transformation from a finite dimensional vector space to another finite dimensional vector can be represented by a matrix and vice-versa. Various relationships between the properties of matrices and those of linear transformations represented by them have been discussed. Eigenvalues, Eigenvectors of a linear transformations have been discussed.

In Chapter-5, concept of the determinant of a square matrix have been introduced and various properties of determinants have been discussed. Determinant as a multilinear alternating form has been introduced. Eigenvalues and Eigenvectors of a square matrix have been explained.

Inner product spaces, unitary spaces and linear operators on them have been discussed in detail in the next four chapters (Chapter 6 to Chapter 9). Bilinear forms and associated quadratic forms have been explained in the last chapter.

It is hoped that with these features the book will not only serve as a text book for a formal course in Linear Algebra but also as a supplement to standard texts in Linear Algebra and will also be helpful to all readers irrespective of their fields of specification.

I avail this opportunity to thank Vrajesh Makhijani, Director, I.K. International Publishing House Pvt. Ltd. for his sincere efforts in bringing out this book in such an excellent form in very short span of time.

Any suggestion for the improvement of the book will be gratefully acknowledged.

Dr. R. D. Sharma

Contents

0. **Priliminary Concepts** — 1
   - 0.1 INTRODUCTION — 1
   - 0.2 CARTESIAN PRODUCT OF SETS AND RELATIONS — 1
   - 0.3 FUNCTIONS — 2
   - 0.4 BINARY OPERATIONS — 5
   - 0.5 GROUPS — 7
   - 0.6 RINGS AND FIELDS — 9
   - 0.7 MATRICES — 11
   - 0.8 DETERMINANTS — 18
   - 0.9 SYSTEMS OF LINEAR EQUATIONS — 19
   - 0.10 RANK OF A MATRIX — 22

1. **Modules** — 23
   - 1.1 INTRODUCTION — 23
   - 1.2 DEFINITIONS AND EXAMPLES — 23
   - 1.3 ELEMENTARY PROPERTIES OF MODULES — 38
   - 1.4 SUBMODULES — 40
   - 1.5 QUOTIENT MODULES — 54
   - 1.6 R-HOMOMORPHISMS — 58
   - 1.7 ALGEBRAS — 81
   - 1.8 FREE MODULES — 82
   - 1.9 CYCLIC MODULES — 105
   - 1.10 NOETHERIAN AND ARTINIAN MODULES — 109

2. **Vector Spaces** — 115
   - 2.1 INTRODUCTION — 115

2.2	DEFINITIONS AND EXAMPLES	115
2.3	ELEMENTARY PROPERTIES OF VECTOR SPACES	130
2.4	SUB-SPACES	133
2.5	QUOTIENT SPACES	156
2.6	LINEAR COMBINATIONS	160
2.7	LINEAR SPANS	165
2.8	LINEAR INDEPENDENCE AND DEPENDENCE	178
2.9	BASIS AND DIMENSION	192
2.10	COORDINATES	229

3. Linear Transformations — 235

3.1	INTRODUCTION	235
3.2	DEFINITIONS AND EXAMPLES	235
3.3	SOME BASIC PROPERTIES OF LINEAR TRANSFORMATIONS	240
3.4	KERNEL AND IMAGE OF A LINEAR TRANSFORMATION	247
3.5	RANK AND NULLITY OF A LINEAR TRANSFORMATION	264
3.6	ALGEBRA OF LINEAR TRANSFORMATIONS	280
3.7	POLYNOMIALS AND LINEAR OPERATORS	294
3.8	SINGULAR AND NON-SINGULAR LINEAR TRANSFORMATIONS	295
3.9	INVERTIBLE LINEAR TRANSFORMATIONS	296
3.10	DUAL SPACES	308
3.11	ANNIHILATORS	325
3.12	PROJECTIONS	333
3.13	THE DUAL OR THE ADJOINT OF A LINEAR TRANSFORMATION	347

4. Linear Transformations and Matrices — 356

4.1	INTRODUCTION	356
4.2	LINEAR TRANSFORMATIONS AND MATRICES	356
4.3	SOME USEFUL THEOREMS	372
4.4	CHANGE OF BASES	378
4.5	SIMILARITY AND EQUIVALENCY OF MATRICES	395
4.6	RANK OF A MATRIX	402
4.7	INVERTIBLE MATRICES	412
4.8	ELEMENTARY OPERATIONS	414
4.9	ELEMENTARY MATRICES	419
4.10	EIGENVALUES AND EIGENVECTORS OF A LINEAR TRANSFORMATION	429
4.11	EIGENVALUES AND EIGENVECTORS OF A MATRIX	434

5. Determinants — 443
- 5.1 PERMUTATIONS 443
- 5.2 DETERMINANTS 446
- 5.3 PROPERTIES OF DETERMINANTS 447
- 5.4 DETERMINANT OF A LINEAR OPERATOR 459
- 5.5 MULTILINEARITY AND DETERMINANTS 462
- 5.6 DETERMINANT RANK OF A MATRIX 469
- 5.7 COFACTORS AND MINORS 469
- 5.8 CRAMER'S RULE 474
- 5.9 CHARACTERISTIC POLYNOMIAL 477
- 5.10 EIGENVALUES AND EIGENVECTORS OF A SQUARE MATRIX 487
- 5.11 DIAGONALIZING MATRICES 497
- 5.12 THE MINIMAL POLYNOMIAL 521

6. Inner Product Spaces — 531
- 6.1 INTRODUCTION 531
- 6.2 INNER PRODUCT SPACES 531
- 6.3 NORM OR LENGTH OF A VECTOR 542
- 6.4 ANGLE BETWEEN VECTORS 547
- 6.5 ORTHOGONALITY 550
- 6.6 ORTHOGONAL COMPLIMENTS 561
- 6.7 ORTHONORMALITY 567
- 6.8 PROJECTIONS 574
- 6.9 GRAM-SCHMIDT ORTHOGONALIZATION PROCESS 577
- 6.10 INNER PRODUCTS AND MATRICES 588
- 6.11 POSITIVE DEFINITE MATRIX 592
- 6.12 ORTHOGONAL MATRICES 595

7. Linear Operators on Inner Product Spaces — 603
- 7.1 INTRODUCTION 603
- 7.2 INNER PRODUCT SPACE MORPHISMS 603
- 7.3 LINEAR FUNCTIONALS AND INNER PRODUCT SPACES 607
- 7.4 ADJOINT OPERATORS 610
- 7.5 SPECTRAL THEOREM 624
- 7.6 ORTHOGONAL OPERATORS 627
- 7.7 POSITIVE OPERATORS 632

8. Unitary Spaces — 640
- 8.1 INTRODUCTION 640
- 8.2 UNITARY SPACES 640

8.3	NORM OF A VECTOR	647
8.4	ORTHOGONALITY AND ORTHONORMALITY	653
8.5	HERMITIAN INNER PRODUCTS AND MATRICES	665
8.6	UNITARY AND NORMAL MATRICES	668

9. Linear Operators on Unitary Spaces   677

9.1	INTRODUCTION	677
9.2	LINEAR FORMS AND UNITARY SPACES	677
9.3	ADJOINT OPERATORS	680
9.4	POSITIVE OPERATORS	697
9.5	UNITARY OPERATORS	700
9.6	NORMAL OPERATORS	706

10. Bilinear and Quadratic Forms   722

10.1	INTRODUCTION	722
10.2	BILINEAR FORMS	722
10.3	SPACE OF BILINEAR FORMS	731
10.4	BILINEAR FORMS AND MATRICES	739
10.5	SYMMETRIC AND SKEW-SYMMETRIC BILINEAR FORMS	749
10.6	QUADRATIC FORMS	758

Index   767

Chapter 0

Preliminary Concepts

0.1 INTRODUCTION

This Chapter provides a brief review of all basic concepts, definitions, theorems etc which will be used in the subsequent Chapters. Various theorems and results have been stated only without giving their proofs as the same may looked up in any standard text book on Algebra. It has been assumed that the reader is familiar with elementary set theory.

0.2 CARTESIAN PRODUCT OF SETS AND RELATIONS

CARTESIAN PRODUCT OF SETS *Let A and B be any two non-empty sets. The set of all ordered pairs (a, b) such that $a \in A$ and $b \in B$ is called the cartesian product of the sets A and B and is denoted by $A \times B$. That is,*

$$A \times B = \{(a, b) : a \in A \text{ and } b \in B\}$$

If $A = \phi$ or $B = \phi$, then we define $A \times B = \phi$.

RELATION *Let A and B be two sets. A relation R from set A to set B is a subset of $A \times B$.*

If R is a relation from a non-void set A to a non-void set B and if $(a, b) \in R$, then we write $a\,R\,b$ which is read as 'a is related to b by the relation R'. If $(a, b) \notin R$, then we write $a\,\not R\,b$ and we say that a is not related to b by the relation R.

A relation from a set A to itself is called a relation on set A.

INVERSE OF A RELATION *Let A and B be two sets and let R be a relation from a set A to a set B. Then the inverse of R, denoted by R^{-1}, is a relation from B to A and is defined by*

$$R^{-1} = \{(b, a) : (a, b) \in R\}$$

Clearly, $(a, b) \in R \Leftrightarrow (b, a) \in R^{-1}$.

IDENTITY RELATION *The relation I_A on a set is identity relation if every element of A is related to itself only That is,*

$$I_A = \{(a, a) : a \in A\}$$

REFLEXIVE RELATION *A relation R on a set A is said to be reflexive if every element of A is related to itself.*

Thus, R is reflexive on set A iff $(a, a) \in R$ for all $a \in A$.

SYMMETRIC RELATION *A relation R on a set A is said to be a symmetric relation iff*

$$(a, b) \in R \Rightarrow \quad (b, a) \in R \quad \text{for all } a, b \in A.$$

TRANSITIVE RELATION *A relation R on a set A is said to be a transitive relation iff*

$$(a, b) \in R \text{ and } (b, c) \in R \quad \Rightarrow \quad (a, c) \in R \quad \text{for all } a, b, c \in A.$$

EQUIVALENCE RELATION *A relation R on a set A is said to be an equivalence relation iff it is*

  (i) reflexive i.e. $(a, a) \in R$ for all $a \in A$.
  (ii) symmetric i.e. $(a, b) \in R \Rightarrow (b, a) \in R$ for all $a, b \in A$
  (iii) transitive i.e. $(a, b) \in R$ and $(b, c) \in R \Rightarrow (a, c) \in R$ for all $a, b, c \in A$.

If R is an equivalence relation on a non-empty set A and $a \in A$, then the set of all those elements of A which are related to *a* by the relation R is called the equivalence class determined by **a** and is denoted by $[a]$.

Thus, $[a] = \{x \in A : (x, a) \in R\}$

For any $a, b \in A$

  (i) if $b \in [a]$, then $[b] = [a]$
  (ii) $[a] = [b]$ iff $(a, b) \in R$
  (iii) either $[a] = [b]$ or $[a] \cap [b] = \phi$.

0.3 FUNCTIONS

FUNCTION AS A SET OF ORDERED PAIRS *Let A and B be two non-empty sets. A relation f from A to B i.e. a subset of $A \times B$ is called a function (or a mapping or a map) from A to B, if*

  (i) for each $a \in A$ there exists $b \in B$ such that $(a, b) \in f$.
  (ii) $(a, b) \in f$ and $(a, c) \in f \Rightarrow \quad b = c$.

If $(a, b) \in f$, then b is called the image of a under f.

FUNCTION AS A CORRESPONDENCE *Let A and B be two non-empty sets. A function 'f' from set A to set B is a rule relating elements of set A to elements of set B such that*

(i) all elements of set A are associated to elements in set B.

(ii) an element of set A is associated to a unique element in set B.

Terms such as "map" (or "mapping"), "correspondence" are used as synonyms for "function".

If f is a function from a set A to a set B, then we write $f : A \to B$ or, $A \xrightarrow{f} B$, which is read as 'f' is a function from A to B or f maps A to B.

If an element $a \in A$ is associated to an element $b \in B$, then b is called "*the f image of a*" or "*image of a under function f*" or "*the value of the function f at a*". Also, a is called the pre-image of b under the function f and we write $b = f(a)$ or, $f^{-1}(b) = a$.

The set A is known as the domain of f and the set B is known as the co-domain of f. The set of all f-images of elements of A is known as the range of f or image set of A under f and is denoted by $f(A)$.

Thus, $f(A) = \{f(x) : x \in A\}$ = Range of f.

The set of all functions or mappings from a set X to a set A is denoted by A^X.

REMARK. *A rule relating all elements of set A to elements of set B is a function or a well defined rule iff*

$$x = y \Rightarrow f(x) = f(y) \quad \text{for all } x, y \in A.$$

INJECTIVE MAP *A function $f : A \to B$ is said to be an injective map or a one-one function if distinct elements in A have distinct images in B.*

Thus, f is injective if and only if

$$x \neq y \Rightarrow f(x) \neq f(y) \quad \text{for all } x, y \in A.$$
$$\Leftrightarrow \quad f(x) = f(y) \Rightarrow x = y \quad \text{for all } x, y \in A.$$

If $f : A \to B$ is not a one-one function, then it is said to be a *many-one* function. That is, A function $f : A \to B$ is said to be a many-one function if two or more elements of set A have the same image in B.

SURJECTIVE MAP *A function $f : A \to B$ is said to be an onto function or a surjective map if every element of B is the f image of some element of A i.e. if $f(A) = B$ or range of f is the co-domain of f or, $I_m(f) = B$.*

If $f : A \to B$ is not a surjective map or an onto function, then it is said to be an into **function**.

BIJECTIVE MAP *A function $f : A \to B$ is said to be a bijective map if it is injective as well as surjective.*

In other words, a function $f : A \to B$ is a bijective map or a bijection, if

(i) it is injective i.e. $f(x) = f(y) \Rightarrow x = y$ for all $x, y \in A$

(ii) it is surjective i.e. for all $y \in B$ there exists $x \in A$ such that $f(x) = y$.

PERMUTATION *Let A be a non-empty set. A bijective map from A to itself is called a permutation on A.*

If A is a finite set equal to $\{a_1, a_2, \ldots, a_n\}$, then a permutation f on A is written in two row notation as follows:

$$f = \begin{pmatrix} a_1 & a_2 & a_3 & \ldots & a_n \\ f(a_1) & f(a_2) & f(a_3) & \ldots & f(a_n) \end{pmatrix}$$

INVERSE OF A FUNCTION *Let $f : A \to B$ be a bijection. Then a function $g : B \to A$ which associates each element $y \in B$ to a unique element $x \in A$ such that $f(x) = y$ is called the inverse of f and is denoted by f^{-1}.*

Thus, if $f : A \to B$ is a bijection, then $f^{-1} : B \to A$ is such that

$$f(x) = y \Leftrightarrow f^{-1}(y) = x.$$

If f has an inverse, f is said to be invertible. The inverse of an invertible function is unique.

Clearly, f is invertible iff f is a bijection.

COMPOSITION OF FUNCTIONS *Let $f : A \to B$ and $g : C \to D$ be two functions such that $B \subset C$ or range $(f) \subset C$, then the composite h of f and g, denoted by gof, is the mapping $h : A \to D$ defined by*

$$h(x) = g(f(x)) \quad \text{for each } x \in A.$$

The composition of functions is not necessarily commutative i.e. gof $\neq$ fog. But, it is always associative i.e. for any three functions f, g, h, we have

$$(\text{fog})\text{oh} = \text{fo}(\text{goh})$$

provided that fog and goh are defined.

Also, If $f : A \to B$, then

$$\text{fo } I_B = f = I_A \text{ of,}$$

where I_A and I_B are identity functions on A and B respectively.

If f and g are both invertible functions such that gof is defined, then gof is also invertible and $(\text{gof})^{-1} = f^{-1}\text{og}^{-1}$.

Following are some useful results on composition of functions:

(i) If $f : A \to B$ and $g : B \to A$ are two functions such that gof $= I_A$, then f is an injection and g is a surjection.

(ii) if $f : A \to B$ and $g : B \to A$ are two functions such that fog $= I_B$, then f is a surjection and g is an injection.

(iii) If $f : A \to B$ and $g : B \to C$ be two functions, then

    (a) gof: $A \to C$ is onto $\Rightarrow$ $g : B \to C$ is onto

    (b) gof: $A \to C$ is one-one $\Rightarrow$ $f : A \to B$ is one-one

(c) gof : $A \to C$ is onto and $g : B \to C$ is one-one $\Rightarrow$ $f : A \to B$ is onto

(d) gof : $A \to C$ is one-one and $f : A \to B$ is onto $\Rightarrow$ $g : B \to C$ is one-one

LIST Let n denote the set of first n natural numbers i.e. $n = \{1, 2, \ldots, n\}$ Then a function $f : n \to A$ is called a list of elements in A and is written as $(f_1, f_2, f_3, \ldots, f_n)$.

We use the notation $\underline{n}$ to denote the list $(1, 2, 3, \ldots, n)$.

0.4 BINARY OPERATIONS

BINARY OPERATIONS Let S be a non-empty set. A function $f : S \times S \to S$ is called a binary operation (or a binary composition) on the set S.

Thus, a binary operation f on a set S associates each ordered pair $(a, b) \in S \times S$ to a unique element $f(a, b)$ in S. We shall use the notation afb instead of $f(a, b)$ for a binary operation f on a set S. Generally, binary operations are denoted by the symbols like $*$, $\circ$, $\odot$, $\oplus$ etc. Thus, if $*$ is a binary operation on a set S the image of an element $(a, b) \in S \times S$ is written as $a * b$ (instead of the usual notation $*(a, b)$).

Addition $(+)$ and multiplication $(\cdot)$ are binary operations on N but subtraction and division are not binary operations on N. Subtraction is a binary-operation on each of the sets Z, Q, R and C.

COMMUTATIVE BINARY OPERATION A binary operation $*$ on a set S is said to be commutative, if
$$a * b = b * a \quad \text{for all } a, b \in S$$
Addition on R is commutative but subtraction is not commutatively.

ASSOCIATIVE BINARY OPERATION A binary operation $*$ on a set S is said to be associative, if
$$(a * b) * c = a * (b * c) \quad \text{for all } a, b, c \in S.$$

DISTRIBUTIVITY Let S be a non-void set and $*$ and $\odot$ be two binary operations on S. The binary operation $*$ is said to be

(i) left distributive over $\odot$, if
$$a * (b \odot c) = (a * b) \odot (a * c) \quad \text{for all } a, b, c \in S$$

(ii) right distributive over $\odot$, if
$$(b \odot c) * a = (b * a) \odot (c * a) \quad \text{for all } a, b, c \in S$$

The binary operation $*$ is said to be distributive over $\odot$ if it is both left as well as right distributive.

CLOSER PROPERTY Let $*$ be a binary operation on a set S. A subset T of S is said to be closed under $*$ if $a * b \in T$ for all $a, b \in T$.

Clearly, S is closed under $*$ by the definition.

RESTRICTION OF A BINARY OPERATION *Let S and T be two sets such that $T \subset S$. A binary operation $*$ on T is said to be the restriction of a binary operation $\odot$ on S, if $a * b \in T$ for all $a, b \in T$. That is, $*$ and $\odot$ are equal on T.*

If $*$ is restriction of $\odot$ on S, then we also say that the binary operation $*$ is induced by $\odot$ on S.

Usually, we use the same symbol for the binary operation $\odot$ and its restriction $*$ on T.

Addition on Z is restriction of addition on R. Similarly, multiplication on R is restriction of multiplication on C.

LEFT IDENTITY *Let $*$ be a binary operation on a set S. An element $e_1 \in S$ is called a left-identity if*

$$e_1 * a = a \quad \text{for all } a \in S.$$

RIGHT IDENTITY *Let $*$ be a binary operation on a set S. An element $e_2 \in S$ is called a right-identity if*

$$a * e_2 = a \quad \text{for all } a \in S.$$

IDENTITY ELEMENT *Let $*$ be a binary operation on a set S. An element $e \in S$ is called identity element if it is both a left-identity and a right-identity. i.e.*

$$e * a = a = a * e \quad \text{for all } a \in S.$$

The identity element for a binary operation on $*$ on a set S, if it exists, is unique.

LEFT INVERSE *Let $*$ be a binary operation on a set S and $e \in S$ be the identity element for $*$ on S. An element b is a left inverse of $a \in S$ if*

$$b * a = e.$$

RIGHT INVERSE *Let $*$ be a binary operation on a set S and $e \in S$ be the identity element for $*$ on S. An element c is a right inverse of $a \in S$ if*

$$a * c = e$$

INVERSE OF AN ELEMENT *Let $*$ be a binary operation on a set S and $e \in S$ be the identity element for $*$ on S. An element x is an inverse of an element $a \in S$ if x is both a left inverse as well a right inverse of a i.e.*

$$x * a = e = a * x.$$

The inverse of a is usually denoted by a^{-1}. For additive binary operation on a set S, the inverse of a is denoted by $-a$.

An element $a \in S$ is said to be invertible, if it possesses its inverse. The inverse of an invertible element is unique. The identity element is always invertible and is inverse of itself.

ALGEBRAIC STRUCTURE *A non-empty set S equipped with one or more binary operations on it is called an algebraic structure.*

0.5 GROUPS

SEMI-GROUP *An algebraic structure $(G, *)$, consisting of a non-void set G and a binary operation $*$ defined on G, is called a semi-group, if it satisfies the following axiom:*

SG-1 *Associativity*: *The binary operation $*$ is associative on G.*

i.e. $(a*b)*c = a*(b*c)$ for all $a, b, c \in G$.

The algebraic structures $(N, +)$, $(Z, +)$, $(Q, +)$, $(R, +)$, $(C, +)$, $(Z, \times)$, $(Q, \times)$ etc are simi-groups.

Let $P(S)$ be the power set of a set S. Then, $(P(S), \cup)$ and $(P(S), \cap)$ are semi-groups.

MONOID *An algebraic structure $(G, *)$, consisting of a non-void set G and a binary operation $*$ defined on G, is called a monoid, if it satisfies the following axioms:*

M-1 *Associativity*: *The binary operation $*$ is associative on G.*

i.e., $(a*b)*c = a*(b*c)$ for all $a, b, c \in G$.

M-2 *Existence of Identity*: *There exists an element $e \in G$ such that*

$$a * e = a = e * a \quad \text{for all } a \in G.$$

The algebraic structures $(N, \times)$, $(Z, +)$, $(Q, \times)$ are monoids but $(N, +)$ is not a monoid.

GROUP *An algebraic structure $(G, *)$, consisting of a non-void set G and a binary operation $*$ defined on G, is called a group, if it satisfies the following axioms:*

G-1 *Associativity:* *The binary operation $*$ is associative on G.*

i.e. $(a*b)*c = a*(b*c)$ for all $a, b, c \in G$.

G-2 *Existence of identity:* *There exists an element $e \in G$ such that*

$$a * e = a = e * a \quad \text{for all } a \in G$$

G-3 *Existence of inverse:* *For each $a \in G$ there exists an element $a' \in G$ such that*

$$a * a' = e = a' * a.$$

The element a' is called the inverse of a and is denoted by a^{-1}.

The algebraic structures $(Z, +)$, $(Q, +)$, $(R, +)$, $(C, +)$, $(Q_o, \times)$ are groups.

When it is not necessary to indicate the binary operation $*$, the group $(G, *)$ is simply referred to as the group G. If the binary operation on a group G is addition, it is usually called an additive group, its identity element is called zero, written as 0, and the inverse of a is called the negative of a, written as $-a$. When the binary operation on a group G is multiplication, then G is called a multiplicative group and $a \times b$ is written as ab. The identity element is usually denoted by 1 and the inverse of an element a is written as $\dfrac{1}{a}$.

ABELIAN GROUP *A group* $(G, *)$ *is called an abelian group, if* $*$ *is commutative on G.*
i.e. $a * b = b * a$ *for all* $a, b \in G$.
$(Z, +), (Q_o, \times), (R, +), (C, +)$ *etc are abelian groups.*

Following are some useful properties of groups:

(i) The identity element in a group $(G, *)$ is unique.
(ii) The inverse of every element of a group $(G, *)$ is unique.
(iii) The inverse of identity element in group is identity element itself.
(iv) In a group $(G, *)$,
$$(a * b)^{-1} = b^{-1} * a^{-1} \quad \text{for all } a, b \in G$$
(v) Let $(G, *)$ be a group. Then for all $a, b, c \in G$
$$a * b = a * c \Rightarrow b = c \quad \text{(Left cancellation law)}$$
and, $b * a = c * a \Rightarrow b = c$ (Right cancellation law)
(vi) In a group $(G, *)$, $(a^{-1})^{-1} = a$ for all $a \in G$.
(vii) Let $(G, *)$ be a group. Then for any $a, b \in G$, the equation $a * x = b$ and $y * a = b$ have unique solutions in G.

CYCLIC GROUP *A group* $(G, *)$ *is said to be a cyclic if there exists an element* $a \in G$ *such that every element of G is expressible as some integral power of a.*
The element $a \in G$ *is called the generator of G and we write* $G = [a]$.
For example, $(Z, +)$ is a cyclic group generated by 1.
Following are some useful properties of a cyclic group:

(i) Every cyclic group is abelian but an abelian group need not be cyclic.
(ii) If a is a generator of a cyclic group, then a^{-1} is also a generator of G.
(iii) The order of a cyclic group is same as the order of its generator.
(iv) Every infinite cyclic group has two and only two generators.

SUB-GROUP *Let* $(G, *)$ *be a group and H be a non-void subset of G such that*

(i) H is closed for the binary operation $*$ on G.
(ii) H itself is a group for the composition induced by that of G i.e. H itself is a group under the restriction of $*$ on H. Then, we say that $(H, *)$ is a subgroup of $(G, *)$.

Trivially, $\{e\}$ and G itself are subgroups of G.
For the sake of convenience we simply say that H is a subgroup of G, if $(H, *)$ is subgroup of $(G, *)$.

COSETS *Let H be a subgroup of a group G and let $a \in G$. Then the sets*
$$aH = \{ah : h \in H\} \text{ and } Ha = \{ha : h \in H\}$$
are known as left and right cosets respectively of H in G.

Obviously, $aH \subset G$ and $Ha \subset G$ for all $a \in G$.

If the binary operation on G is addition, then
$$a+H = \{a+h : h \in H\} \text{ and } H+a = \{h+a : h \in H\}$$
are respectively the left and right cosets of H in G.

Any two right (left) cosets of a subgroup H of a group G are identical or disjoint.

If H is a subgroup of a group G and $a, b \in G$, then $Ha = Hb \Leftrightarrow ab^{-1} \in H$

If H is a subgroup of an additive group G and $a, b \in G$, then $H+a = H+b \Leftrightarrow a-b \in H$.

NORMAL SUBGROUP *A subgroup N of a group G is said to be a normal subgroup of G if*
$$xax^{-1} \in N \quad \text{for all } x \in G \text{ and all } a \in N.$$
$\Leftrightarrow \quad xNx^{-1} \subset N \quad \text{for all } x \in G$

A subgroup N of a group G is a normal subgroup of G iff $xN = Nx$ for all $x \in G$,
This means that there is no distinction between left and right cosets of a normal subgroup of a group.

If N is a normal subgroup of a group G, then $Na\,Nb = Nab$ for all $a, b \in G$.

QUOTIENT GROUP *Let N be a normal subgroup of a group G. Then, the set $\dfrac{G}{N} = \{Nx : x \in G\}$ of all cosets of N in G is a group under the multiplication of cosets as a binary operation. i.e.*
$$Na\,Nb = Nab \quad \text{for all } a, b \in G.$$
This group is known as the quotient group or factor group of G by N.

0.6 RINGS AND FIELDS

RING *An algebraic structure $(R, +, \cdot)$, consisting of a non-empty set R and two binary operations '+' and '$\cdot$' on R, is called a ring if the following axioms are satisfied:*

R-1 $(R, +)$ is an abelian group.

R-2 $(R, \cdot)$ is a semigroup.

R-3 '$\cdot$' is distributive over '+' i.e. for all $a, b, c \in G$
  (i) $a \cdot (b+c) = a \cdot b + a \cdot c$
  (ii) $(b+c) \cdot a = b \cdot a + c \cdot a$

Clearly, $(Z, +, \times)$, $(Q, +, \times)$, $(R, +, \times)$ are rings.

RING WITH UNITY *A ring $(R, +, \cdot)$ is said to be a ring with unity, if R has the identity element for multiplicative binary operation.*

The identity element for multiplicative binary operation is denoted by 1.

COMMUTATIVE RING *A ring $(R, +, \cdot)$ is said to be a commutative ring if its multiplicative binary operation is commutative.*

i.e. $a \cdot b = b \cdot a$ for all $a, b \in R$.

Following are some useful results in a ring $(R, +, \cdot)$:

(i) $a 0 = 0 a = 0$ for all $a \in R$, where 0 is the zero element i.e. additive identity in R.

(ii) $a(-b) = -(ab) = (-a)b$ for all $a, b \in R$

(iii) $(-a)(-b) = ab$ for all $a, b \in R$

(iv) $a(b - c) = ab - ac$ for all $a, b, c \in R$

(v) $(b - c)a = ba - ca$ for all $a, b, c \in R$.

Let $(R, +, \cdot)$ be a ring and n be a positive integer, then we define

$$na = a + a + a + \cdots + a \text{ (upto } n \text{ terms)}$$

Also, we define $0a = 0$, where 0 on the left hand side is integer 0 and 0 on the right hand side is the zero element (additive identity) of the ring.

CHARACTERISTIC OF A RING *Let $(R, +, \cdot)$ be a ring with zero element 0. If there exists a positive integer n such that $na = 0$ i.e. $a + a + a + \cdots + a$ (n times) $= 0$ (zero of the ring) for all $a \in R$. Then, we say that the ring is of finite characteristic. If n is the smallest positive integer such that $na = 0$ for all $a \in R$, then n is called the characteristic of ring R.*

If there exists no positive integer n such that $na = 0$ for all $a \in R$, then R is said to be of characteristic zero or infinite.

The ring $(Z, +, \times)$ is of characteristic zero whereas $(Z_6, +_6, \times_6)$ is of characteristic 6.

SUBRING *A non-void subset S of a ring $(R, +, \times)$ is a subring of R, iff*

(i) *S is closed with respect to the binary operations of addition and multiplication on R.*

(ii) *S itself is a ring for the induced binary operations.*

The necessary and sufficient conditions for a non-void subset S of a ring R to be a subring of R are (i) $a - b \in S$ (ii) $ab \in S$ for all $a, b \in S$.

FIELD *An algebraic structure $(F, +, \cdot)$, consisting of a non-void set F and two binary operations '+' and '$\cdot$' on F, is called a field if the following axioms are satisfied:*

F-1 $(F, +)$ is an abelian group.

F-2 $(F - \{0\}, \cdot)$ is an abelian group.

F-3 '$\cdot$' is distributive over '+' i.e. for all $a, b, c \in G$

(i) $a \cdot (b+c) = a \cdot b + a \cdot c$
(ii) $(b+c) \cdot a = b \cdot a + c \cdot a$

That is a commutative ring with unity is a field, if its every non-zero element has multiplicative inverse.

$(Q, +, \times)$, $(R, +, \times)$ and $(E, +, \times)$ are fields.

SUB FIELD *A non-void subset K of a field $(F, +, \cdot)$ is a subfield of iff*

(i) *K is closed under the binary operations on F.*
(ii) *K itself is a field for the induced binary operations.*

The necessary and sufficient conditions for a non-void subset K of a field F to be a subfield of F are

(i) $a - b \in K$ for all $a, b \in K$ (ii) $ab^{-1} \in K$ for all $a, 0 \neq b \in K$.

0.7 MATRICES

MATRIX *A matrix over a field F or, simply, a matrix A (when F is implicit) is a rectangular arrangement of scalars.*

If there are mn scalars $a_{ij} \in F$, where $i \in \underline{m}$ and $j \in \underline{n}$. Then, the following arrangement of these mn scalars is a matrix:

$$A = \begin{bmatrix} a_{11} & a_{12} & \cdots & a_{1j} & \cdots & a_{1n} \\ a_{21} & a_{22} & \cdots & a_{2j} & \cdots & a_{2n} \\ \vdots & \vdots & & \vdots & & \vdots \\ a_{i1} & a_{i2} & \cdots & a_{ij} & \cdots & a_{in} \\ \vdots & \vdots & & \vdots & & \vdots \\ a_{m1} & a_{m2} & \cdots & a_{in} & \cdots & a_{mn} \end{bmatrix}$$

The element a_{ij}, called the ij - element or ij - entry, appears in ith row and jth column. Such a matrix is usually denoted by $A = [a_{ij}]_{m \times n}$ or, simply, $A = [a_{ij}]$.

A matrix with m rows and n columns is called an m by n matrix, written as $m \times n$.

The rows of matrix A (given above) are m horizontal lists of scalars:

$(a_{11}, a_{12}, a_{13}, \ldots, a_{1n})$, $(a_{21}, a_{22}, a_{23}, \ldots, a_{2n})$, $\ldots$, $(a_{m1}, a_{m2}, a_{m3}, \ldots, a_{mn})$

If we denote these rows by $A_1, A_2, \ldots, A_m$ respectively, then matrix A can also be written as a vertical list as given below.

$$A = \begin{bmatrix} A_1 \\ A_2 \\ \vdots \\ A_m \end{bmatrix}$$

The columns of matrix A are n vertical lists of scalars

$$\begin{bmatrix} a_{11} \\ a_{21} \\ \vdots \\ a_{m1} \end{bmatrix}, \begin{bmatrix} a_{12} \\ a_{22} \\ \vdots \\ a_{m2} \end{bmatrix}, \ldots, \begin{bmatrix} a_{1n} \\ a_{2n} \\ \vdots \\ a_{mn} \end{bmatrix}$$

These columns are generally denoted by $A^1, A^2, \ldots, A^n$ and the matrix A can be written as list of n columns as given below.

$$A = (A^1, A^2, \ldots, A^n)$$

MATRIX AS A MAPPING Let F be a field and m, n be positive integers. A mapping $A: \underline{m} \times \underline{n} \to F$ associating each ordered pair $(i, j) \in \underline{m} \times \underline{n}$ to the scalar $a_{ij} \in F$ is called an $m \times n$ matrix.

Clearly, a_{ij} is the image of $(i, j) \in \underline{m} \times \underline{n}$ under mapping A i.e. $A(i, j) = a_{ij}$ and is called the ij - entry or ij - element of matrix A. In such a case the matrix A is also written as $A = [a_{ij}]$.

ROW MATRIX A matrix with only one row is called a row matrix or a row vector.

COLUMN MATRIX A matrix with only one column is called a column matrix or a column vector.

As discussed above an $m \times n$ matrix A can be written as a row vector of its columns $A^1, A^2, \ldots, A^n$ and a column vector of its rows $A_1, A_2, \ldots, A_m$.

SQUARE MATRIX An $n \times n$ matrix A is called a square matrix of order n.

If $A = [a_{ij}]$ is a square matrix of order n, then the elements $a_{11}, a_{22}, \ldots, a_{nn}$ are called the diagonal elements and the line along which they lie is called the principal diagonal or leading diagonal of the matrix. The diagonal elements are also written as a_{ii}, $i \in \underline{n}$.

DIAGONAL MATRIX A square matrix $A = [a_{ij}]_{n \times n}$ is called a diagonal matrix if all the elements except those in the leading diagonal are zero. i.e. $a_{ij} = 0$ for all $i \neq j$.

A diagonal matrix of order $n \times n$ having $d_1, d_2, \ldots, d_n$ as diagonal elements is denoted by diag $(d_1, d_2, \ldots, d_n)$. For example the diagonal matrix

$$A = \begin{bmatrix} 1 & 0 & 0 \\ 0 & 2 & 0 \\ 0 & 0 & 3 \end{bmatrix}$$

is written as diag $(1, 2, 3)$.

SCALAR MATRIX A square matrix $A = [a_{ij}]_{n \times n}$ is called a scalar matrix if
(i) $a_{ij} = 0$ for all $i \neq j$ (ii) $a_{ii} = c$ for all i, where $c \neq 0$.

IDENTITY MATRIX A square matrix $A = [a_{ij}]_{n \times n}$ is called an identity matrix if
(i) $a_{ij} = 0$ for all $i \neq j$ (ii) $a_{ii} = 1$ for all i
An identity matrix of order $n \times n$ is generally denoted by I_n.

NULL MATRIX *A matrix whose all elements are zero is called a null matrix or a zero matrix.*

UPPER TRIANGULAR MATRIX *A square matrix $A = [a_{ij}]_{n \times n}$ is called an upper triangular matrix if $a_{ij} = 0$ for all $i > j$.*

All elements below the leading diagonal of an upper triangular matrix are zero.

LOWER TRIANGULAR MATRIX *A square matrix $A = [a_{ij}]_{n \times n}$ is a lower triangular matrix if $a_{ij} = 0$ for all $i < j$.*

All elements above the leading diagonal of a lower triangular matrix are zero.

A triangular matrix $A = [a_{ij}]$ is called strictly triangular iff $a_{ii} = 0$ for all $i \in \underline{n}$.

ECHELON MATRIX *A matrix is called an echelon matrix, or is said to be in echelon form, if either A is the null matrix or A satisfies the following conditions:*

(i) All zero rows, if any, are at the bottom of the matrix. Here, zero row means a row whose all entries are zeros.

(ii) The number of zeros before the first non-zero element in a row is less than the number of such zeros in the next row.

That is, $A = [a_{ij}]$ is an echelon matrix if there exist non-zero entries

$$a_{1j_1}, a_{2j_2}, \ldots, a_{rj_r}, \text{ where } j_1 < j_2 < \cdots < j_r$$

with the property that

$$a_{ij} = 0 \quad \text{for} \quad \begin{cases} i \leq r \text{ and } j < j_i \\ i > r \text{ and } j < j_i \end{cases}$$

The elements $a_{1j_1}, a_{2j_2}, \ldots, a_{rj_r}$, which are the leading non-zero elements in their respective rows, are called the pivots of the echelon matrix.

The matrix A given by

$$A = \begin{bmatrix} 0 & ③ & 2 & 1 \\ 0 & 0 & ② & 5 \\ 0 & 0 & 0 & 0 \end{bmatrix}$$

is an echelon matrix whose pivots have been encircled.

Clearly, each pivot is to the right of the one above.

The following matrix is also an echelon matrix whose pivots have been encircled.

$$\begin{bmatrix} ③ & 4 & 2 & 0 & -1 & -2 & 3 \\ 0 & 0 & ⑤ & 1 & -2 & 3 & 0 \\ 0 & 0 & 0 & 0 & 0 & ⑦ & 3 \\ 0 & 0 & 0 & 0 & 0 & 0 & 0 \end{bmatrix}$$

MATRIX IN ROW CANONICAL FORM *A matrix A is said to be in row canonical form if it is an echelon matrix satisfying following additional conditions:*

(i) Each pivot is equal to 1.

(ii) Each pivot is the only non-zero entry in its column.

It should be noted that in echelon matrix there must be zeros below the pivots but in a matrix in row canonical form each pivot must be equal to 1 and there must also be zeros above the pivots.

The null matrix O and the identity matrix I (of any order) are examples of matrices in row canonical form.

The matrix A given by

$$A = \begin{bmatrix} 0 & 1 & 2 & 0 & 0 & 4 \\ 0 & 0 & 0 & 1 & 0 & 2 \\ 0 & 0 & 0 & 0 & 1 & 3 \end{bmatrix}$$

is in row canonical form but, the matrix B given below

$$B = \begin{bmatrix} -1 & 2 & 3 & 0 & 5 & -46 \\ 0 & 0 & 1 & -2 & 3 & 0 \\ 0 & 0 & 0 & 0 & 6 & 5 \\ 0 & 0 & 0 & 0 & 0 & 0 \end{bmatrix}$$

is not in row canonical form.

EQUALITY OF MATRICES Two matrices $A = [a_{ij}]_{m \times n}$ and $B = [b_{ij}]_{r \times s}$ are equal if

(i) $m = r$ (ii) $n = s$ (iii) $a_{ij} = b_{ij}$ for all i, j

SCALAR MULTIPLICATION Let $A = [a_{ij}]$ be an $m \times n$ matrix and k be a scalar. Then the matrix obtained by multiplying every element of A by k is called the scalar multiple of A by k and is donoted by kA. That is,

$$kA = [k \, a_{ij}]_{m \times n}$$

The negative of an $m \times n$ matrix $A = [a_{ij}]$, written as $-A$, is defined to be the $m \times n$ matrix given by $(-A)_{ij} = -a_{ij}$ for all $i \in \underline{m}, j \in \underline{n}$.

ADDITION OF MATRICES Let $A = [a_{ij}]$ and $B = [b_{ij}]$ be two $m \times n$ matrices over a field F. Then, their sum $A + B$ is also an $m \times n$ matrix over F such that

$$(A+B)_{ij} = a_{ij} + b_{ij} \quad \text{for all } i \in \underline{m}, j \in \underline{n}.$$

Let F be a field and m, n be positive integers. Then, $F^{m \times n}$ denotes the set of all $m \times n$ matrices over field F. It is evident from the above definition that addition of matrices is a binary operation which possesses the following properties:

(i) Matrix addition on $F^{m \times n}$ is commutative.

i.e. $A + B = B + A$ for all $A, B \in F^{m \times n}$

(ii) Matrix addition on $F^{m\times n}$ is associative.

i.e. $(A+B)+C = A+(B+C)$ for all $A, B, C \in F^{m \times n}$

(iii) Null matrix O is the additive identity.

i.e. $A+O = A = O+A$ for all $A \in F^{m \times n}$

(iv) For every matrix $A \in F^{m \times n}$ there exists $-A \in F^{m \times n}$ such that
$$A+(-A) = O = (-A)+A$$

It follows from the above properties that $(F^{m\times n}, +)$ is an abelian group.

Let $A, B, C \in F^{m\times n}$ and $\lambda, \mu \in F$. Then, we also have the following results:
(i) $\lambda(A+B) = \lambda A + \lambda B$ (ii) $(\lambda+\mu)A = \lambda A + \mu A$ (iii) $(\lambda\mu)A = \lambda(\mu A) = \mu(\lambda A)$
(iv) $1A = A$.

0.7.1 MULTIPLICATION OF MATRICES

If $A = [a_1, a_2, \ldots, a_n]$ is a row matrix and $B = \begin{bmatrix} b_1 \\ b_2 \\ \vdots \\ b_n \end{bmatrix}$ is a column matrix, then their product AB is defined to be the scalar (or 1×1 matrix) obtained by multiplying corresponding entries and adding; that is,

$$AB = [a_1, a_2, \ldots, a_n] \begin{bmatrix} b_1 \\ b_2 \\ \vdots \\ b_n \end{bmatrix} = a_1 b_1 + a_2 b_2 + \cdots + a_n b_n = \sum_{r=1}^{n} a_r b_r$$

Note that the product AB is not defined when A and B have different number of elements. Let us now generalize the above definition for arbitrary matrices.

MATRIX MULTIPLICATION Let $A = [a_{ij}]$ and $B = [b_{ij}]$ be two matrices over a filed F such that the number of columns of A is equal to the number of the rows of B; say A is $m \times p$ matrix and B is $p \times n$ matrix. Then, the product AB is $m \times n$ matrix whose ij-entry is obtained by multiplying i^{th} row of A by the jth column of B. That is,

$$(AB)_{ij} = [a_{i1}\ a_{i2} \ldots a_{in}] \begin{bmatrix} b_{1j} \\ b_{2j} \\ \vdots \\ b_{nj} \end{bmatrix} = \sum_{r=1}^{n} a_{ir}\, b_{rj}$$

The multiplication of matrices is not commutative. However, matrix multiplication does satisfy the following properties:

THEOREM *Let A, B, C be three matrices over a field F such that various products and sums are defined. Then,*

(i) $(AB)C = A(BC)$

(ii) $A(B+C) = AB + AC$

(iii) $(B+C)A = BA + CA$

(iv) $k(AB) = (kA)B = A(kB)$, where $k \in F$

(v) $A_{m \times n} O_{n \times p} = O_{m \times p}$ and $O_{p \times m} A_{m \times n} = O_{p \times n}$

(vi) $AI_n = A = I_m A$, where A is an $m \times n$ matrix.

POSITIVE INTEGRAL POWERS OF A SQUARE MATRIX *For any square matrix A, we define (i) $A^1 = A$ and, (ii) $A^{n+1} = A^n A$, where $n \in N$.*
It is evident from the above definition that:

$$A^2 = AA, \ A^3 = A^2 A = AAA \text{ etc.}$$

Also,

(i) $A^m A^n = A^{m+n}$ and, (ii) $(A^m)^n = A^{mn}$ for all $m, n \in N$

MATRIX POLYNOMIAL *Let $f(x) = a_0 x^n + a_1 x^{n-1} + \cdots + a_{n-1} x + a_n$ be a polynomial over a filed F and let A be a square matrix over F. Then,*

$$f(A) = a_0 A^n + a_1 A^{n-1} + \cdots + a_{n-1} A + a_n I$$

is called a matrix polynomial.

Let F be a field and n be a positive integer. Then, the product of two $n \times n$ matrices over F is an $n \times n$ matrix over F. So, the set $F^{n \times n}$ of all $n \times n$ matrices over F is a closed under multiplication of matrices. The foregoing discussion suggests that $(F^{n \times n}, +, \times)$ is a non-commutative ring with unity, if $n > 1$.

TRANSPOSE OF A MATRIX *Let $A = [a_{ij}]$ be an $m \times n$ matrix over a field F. Then, the transpose of A, denoted by A^T or A', is an $n \times m$ matrix such that*

$$(A^T)_{ij} = a_{ji} \text{ for all } i \in \underline{m},\ j \in \underline{n}.$$

Clearly, A^T is obtained from A by interchanging rows and columns of A.

Following theorem lists basic properties of the transpose operation:

THEOREM *Let A and B be two matrices over a field F and λ be a scalar in F. Then, whenever the sum and product are defined:*
(i) $(A^T)^T = A$ (ii) $(A+B)^T = A^T + B^T$ (iii) $(\lambda A)^T = \lambda A^T$ (iv) $(AB)^T = B^T A^T$

SYMMETRIC MATRIX *A square matrix $A = [a_{ij}]$ over a field F is said to be a symmetric matrix if*

$$a_{ij} = a_{ji} \text{ for all } i, j \in \underline{n}$$
$$\Leftrightarrow A = A^T$$

SKEW-SYMMETRIC MATRIX *A square matrix $A = [a_{ij}]$ over a field F is said to be a skew-symmetric matrix if*

$$a_{ij} = -a_{ji} \text{ for all } i, j \in \underline{n}$$
$$\Leftrightarrow A^T = -A.$$

The diagonals entries of a skew-symmetric matrix are all zero and every square matrix A can be uniquely expressed as the sum of a symmetric matrix $\frac{1}{2}(A + A^T)$ and a skew-symmetric matrix $\frac{1}{2}(A - A^T)$.

ORTHOGONAL MATRIX *A square matrix A over R is said to be an orthogonal matrix if $AA^T = A^T A = I$*

NORMAL MATRIX *A square matrix A over R is a normal matrix if it commutes with its transpose A^T. That is,*

$$AA^T = A^T A.$$

Clearly, every symmetric, orthogonal, or skew-symmetric matrix is a normal matrix.

HERMITIAN MATRIX *A matrix A over the filed C of all complex numbers is a Hermitian matrix if its conjugate transpose is equal to A itself. That is,*

$$\overline{A}^T = A \text{ or, } A^* = A, \text{ where } A^* = \overline{A}^T.$$

The diagonal elements of a Hermitian matrix are all real.

SKEW-HERMITIAN MATRIX *A matrix A over the filed C of all complex numbers is a skew-Hermitian matrix if its conjugate transpose is equal to $-A$ i.e. $\overline{A}^T = -A$ or, $A^* = -A$.*

The diagonal elements of a skew-Hermitian matrix are purely imaginary.

UNITARY MATRIX *A square matrix A over C is a unitary matrix if $A^*A = I = AA^*$.*

If A is a square matrix over C, then A is a normal matrix if $AA^* = A^*A$.

Clearly, this definition reduces to the definition of a normal matrix over R, if C is replaced by R.

The conjugate transpose of a square matrix satisfies the following properties:
(i) $(A^*)^* = A$ (ii) $(\lambda A)^* = \overline{\lambda} A^*$, $\lambda \in C$ (iii) $(A+B)^* = A^* + B^*$ (iv) $(AB)^* = B^* A^*$
(v) $(A^*)^{-1} = (A^{-1})^*$

INVERSE OF A MATRIX *A square matrix B is said to be inverse of a square matrix A if* $AB = BA = I$.

If inverse of a square matrix A exists, then it is unique and we say that A is invertible. The inverse of A is denoted by A^{-1}.

Following are properties of inverse of a matrix:

(i) If A is an invertible matrix, then $(A^{-1})^{-1} = A$

(ii) A square matrix is invertible iff it is non-singular.

(iii) Let A and B be invertible matrices. Then, AB is invertible, and $(AB)^{-1} = B^{-1}A^{-1}$.

(iv) Let A, B, C be square matrices of the same order and if A is an invertible matrix. Then,
$$AB = AC \Rightarrow B = C \quad \text{and} \quad BA = CA \Rightarrow B = C.$$

(v) If A is an invertible matrix, then A^T is also invertible and $(A^T)^{-1} = (A^{-1})^T$.

(vi) The inverse of an invertible symmetric matrix is a symmetric matrix.

(vii) The set $F^{n \times n}$ off all invertible $n \times n$ matrices over a field F is a non-abelian group under multiplication of matrices.

0.8 DETERMINANTS

DETERMINANT *Every square matrix over a field F can be associated to a scalar in F which is known as its determinant.*

The determinant of A is denoted by $|A|$.

If $A = \begin{bmatrix} a_{11} & a_{12} \\ a_{21} & a_{22} \end{bmatrix}$ is a square matrix over a filed F, then

$$|A| = \begin{vmatrix} a_{11} & a_{12} \\ a_{21} & a_{22} \end{vmatrix} = a_{11} a_{22} - a_{12} a_{21}$$

If $A = \begin{bmatrix} a_{11} & a_{12} & a_{13} \\ a_{21} & a_{22} & a_{23} \\ a_{31} & a_{32} & a_{33} \end{bmatrix}$ is a square matrix over a field F, then,

$$\begin{aligned} |A| &= \begin{vmatrix} a_{11} & a_{12} & a_{13} \\ a_{21} & a_{22} & a_{23} \\ a_{31} & a_{32} & a_{33} \end{vmatrix} \\ &= a_{11} \begin{vmatrix} a_{22} & a_{23} \\ a_{32} & a_{33} \end{vmatrix} - a_{12} \begin{vmatrix} a_{21} & a_{23} \\ a_{31} & a_{33} \end{vmatrix} + a_{13} \begin{vmatrix} a_{21} & a_{22} \\ a_{31} & a_{32} \end{vmatrix} \\ &= a_{11}(a_{22} a_{33} - a_{23} a_{32}) - a_{12}(a_{21} a_{33} - a_{23} a_{31}) + a_{13}(a_{21} a_{32} - a_{22} a_{31}) \\ &= a_{11} a_{22} a_{33} + a_{12} a_{23} a_{31} + a_{13} a_{32} a_{21} - a_{11} a_{23} a_{32} - a_{12} a_{21} a_{33} - a_{13} a_{22} a_{31} \end{aligned}$$

SINGULAR MATRIX *A square matrix A over a field F is called a singular matrix if $|A| = 0$, otherwise it is a non-singular matrix.*

MINOR *Let $A = [a_{ij}]$ be a square matrix of order n. Then the minor M_{ij} of a_{ij} in A is the determinant of the square submatrix of order $(n-1)$ obtained by leaving i^{th} row and j^{th} column of A.*

COFACTOR *Let $A = [a_{ij}]$ be a square matrix of order n. Then the cofactor C_{ij} of a_{ij} in A is equal to $(-1)^{i+j}$ times the determinant of the submatrix of order $(n-1)$ obtained by leaving i^{th} row and j^{th} column of A.*

Also,

(i) $\sum_{i=1}^{n} a_{ij} C_{ij} = |A|$ and $\sum_{j=1}^{n} a_{ij} C_{ij} = |A|$

(ii) $\sum_{j=1}^{n} a_{ij} C_{kj} = 0$ and $\sum_{i=1}^{n} a_{ij} C_{ik} = 0$.

ADJOINT OF A SQUARE MATRIX *Let $A = [a_{ij}]$ be a square matrix of order n and let C_{ij} be cofactor of a_{ij} in A. Then the transpose of the matrix of cofactors of elements of A is called the adjoint of A and is denoted by adj A. That is,*

$$adj\ A = [C_{ij}]^T$$

Clearly, $(adj\ A)_{ij} = C_{ji}$ for all $i,\ j \in \underline{n}$.

Following are some useful properties of adjoint of a matrix.

(i) For any square matrix A of order n

$$A(adj\ A) = |A|I_n = (adj\ A)A$$

(ii) For any square matrices A and B of order n

$$adj\ AB = adj\ B\ adj\ A$$

(iii) If A is an invertible matrix, then $adj\ A^T = (adj\ A)^T$.

(iv) If A is an invertible matrix of order n, then

$$adj\ (adj\ A) = |A|^{n-2}A$$

and, $|adj.\ (adj\ A)| = |A|^{(n-1)^2}$

(v) If A is a non-singular matrix, then $A^{-1} = \dfrac{1}{|A|}(adj\ A)$

0.9 SYSTEMS OF LINEAR EQUATIONS

A system of m linear equations in n unknowns $x_1,\ x_2,\ \ldots,\ x_n$ can be put in the standard form as follows:

$$a_{11}\ x_1 + a_{12}\ x_2 + \cdots + a_{1n}x_n = b_1$$
$$a_{21}\ x_1 + a_{22}\ x_2 + \cdots + a_{2n}x_n = b_2$$
$$\cdots\ \cdots\ \cdots\ \cdots\ \cdots$$
$$\cdots\ \cdots\ \cdots\ \cdots\ \cdots$$
$$a_{m1}\ x_1 + a_{m2}\ x_2 + \cdots + a_{mn}x_n = b_m$$

where a_{ij} and b_{ij} are constants.

This system of equations is known as $m \times n$ (read as m by n) system and can be written in matrix form as follow:

$$\begin{bmatrix} a_{11} & a_{21} & \cdots & a_{1n} \\ a_{21} & a_{22} & \cdots & a_{2n} \\ \vdots & \vdots & & \\ a_{m1} & a_{m2} & \cdots & a_{mn} \end{bmatrix} \begin{bmatrix} x_1 \\ x_2 \\ \vdots \\ x_n \end{bmatrix} = \begin{bmatrix} b_1 \\ b_2 \\ \vdots \\ b_m \end{bmatrix}$$

Or, $AX = B$, where $A = \begin{bmatrix} a_{11} & a_{12} & \cdots & a_{1n} \\ a_{21} & a_{22} & \cdots & a_{2n} \\ \vdots & \vdots & & \\ a_{m1} & a_{m2} & \cdots & a_{mn} \end{bmatrix}_{m \times n}$, $X = \begin{bmatrix} x_1 \\ x_2 \\ \vdots \\ x_n \end{bmatrix}_{n \times 1}$ and, $B = \begin{bmatrix} b_1 \\ b_2 \\ \vdots \\ b_m \end{bmatrix}_{m \times 1}$

The matrix $A = [a_{ij}]_{m \times n}$ is called the coefficient matrix and the matrix

$$\begin{bmatrix} a_{11} & a_{12} & \cdots & a_{1n} & b_1 \\ a_{21} & a_{22} & \cdots & a_{2n} & b_2 \\ \cdots & \cdots & \cdots & \cdots & \cdots \\ \cdots & \cdots & \cdots & \cdots & \cdots \\ a_{m1} & a_{m2} & \cdots & a_{mn} & b_m \end{bmatrix}$$

is called the **augmented matrix** and is generally denoted by $[A : B]$

A system of equations $AX = B$ is called a homogeneous system if $B = O$. Otherwise the system is said to be non-homogeneous.

A solution of the system of equations $AX = B$ is a list of the values for the unknowns which satisfy each equation of the system. Equivalently, a vector $U \in F^n$ is a solution of the system of equations $AX = B$ iff $AU = B$.

If $AX = B$ is a system of n equations with n unknowns such that $|A| \neq 0$, then the system has unique solution given by $X = A^{-1}B$.

If $|A| \neq 0$, then the system of equations is either inconsistent or it has infinitely many solutions.

A homogeneous system of equations $AX = O$ is always consistent and has trivial solution only if $|A| \neq 0$. If $|A| = 0$, then $AX = O$ has non-trivial solutions also.

0.9.1 SYSTEMS OF EQUATIONS IN TRIANGULAR FORM

Consider the following system of linear equations:

$$3x_1 - 2x_2 + 4x_3 - 3x_4 = 8$$
$$5x_2 - 2x_3 + 3x_4 = 7$$
$$7x_3 - 2x_4 = 3$$
$$3x_4 = 6$$

We observe that the system is square and the first unknown x_1 is the leading unknown in the first equation, the second unknown x_2 is the leading unknown in the second equation, and so on. Such a system is said to be in triangular form.

Thus, a square system of linear equations is said to be in triangular form if each leading unknown is directly to the right of the leading unknown in the preceding equation.

Clearly, a triangular system always has a unique solution which may be obtained by back-substitution.

0.9.2 SYSTEMS OF EQUATIONS IN ECHELON FORM

A system of simultaneous linear equations is said to be in echelon form if the leading unknown in each equation other than the first is to the right of the leading unknown in the preceding equation.

Consider the following system of equations in echelon form:

$$3x_1 + 5x_2 - 2x_3 + 3x_4 - x_5 = 13$$
$$x_3 + 5x_4 + 2x_5 = 7$$
$$2x_4 - 7x_5 = 9$$

Clearly, x_1, x_3 and x_4 are the leading unknowns in this system.

These unknowns are called **pivot** variables and the other unknowns, x_2 and x_5, are called **free variables**.

The solution set of a system of m simultaneous linear equations in n unknowns in echelon form is described in the following theorem.

THEOREM *Let there be a system, of simultaneous linear equations in n unknowns, in echelon form. Then,*

(i) *The system has a unique solution if $m = n$ i.e. the system is in triangular form.*

(ii) *The system has infinite number of solutions, if $m < n$ i.e. there are more variables than the number of equations.*

REMARK. *If the echelon system of simultaneous linear equations contains more variables than equations, then each of the remaining $n - m$ free variables may take any value. So, the system has infinitely many solutions. The general solution of such a system may be obtained in either of the following two equivalent ways:*

(i) *Arbitrarily assign values to the $n - m$ free variables and solve uniquely for the m pivot variables to obtain a solution of the system.*

(ii) *Find the values of m pivot variables in terms of $(n - m)$ free variables to obtain the general solution of the system.*

0.10 RANK OF A MATRIX

The rank of a matrix is defined in many different ways. But, all the definitions lead to the same number.

RANK OF A MATRIX *The rank of a matrix is the order of the highest order non-singular square submatrix.*

It is evident from the above definition that a positive integer r is rank of an $m \times n$ matrix, if

(i) every square submatrix of order $(r+1)$ or more is singular.
(ii) there exists at least one square submatrix of order r which is non-singular.

The rank of a matrix A is written as $\text{rank}(A)$.

Clearly, rank of the identity matrix I_n is n.

If A is an $m \times n$ matrix, then $\text{rank}(A) \leq \min(m, n)$.

The rank of a matrix in echelon form is equal to the number of non-zero rows of the matrix or the number pivots.

THEOREM-1 *The system of linear equations $AX = B$ is consistent iff, the rank of the augmented matrix $[A : B]$ is equal to the rank of the coefficient matrix A.*

THEOREM-2 *Let $AX = B$ be a system of m simultaneous linear equations in n unknowns such that $m \geq n$.*

(i) *If $r(A) = r([A : B]) = n$, the system has a unique solution.*
(ii) *If $r(A) = r([A : B]) = r < n$, the system is consistent and has infinite number of solutions. In fact, in this case $(n - r)$ variables are free-variables.*
(iii) *If $r(A) \neq r([A : B])$, the system is inconsistent i.e. it has no solution.*

Chapter 1

Modules

1.1 INTRODUCTION

In earlier classes, we have dealt with the algebraic structures consisting of a non-void set with one or two binary operations satisfying certain axioms. In this chapter and in the next one, our aim will be to study algebraic structures that consist of two non-void sets, one a ring and other an abelian group, and a mapping associating each ordered pair, consisting of an element of the ring and an element of the abelian group, to a unique element of the abelian group. These structures differ from the previously studied structures (e.g. groups, rings, etc.) in that each pair of an element of the ring and an element of the abelian group is associated with a unique element of the abelian group. This chapter deals with *modules* while *vector space* will be studied in the next chapter.

1.2 DEFINITIONS AND EXAMPLES

LEFT-MODULE. *An algebraic structure* $(M, R, \oplus, \odot)$ *consisting of a non-void set M, a ring R, a binary operation $\oplus$ on M and an external mapping $\odot : R \times M \to M$ associating each $r \in R$, $a \in M$ to a unique element $r \odot a \in M$ is said to be a left R-module or simply a left module over ring R if the following axioms are satisfied:*

M-1 $(M, \oplus)$ *is an abelian group.*

M-2 *For all $a, b \in M$ and $r, s \in R$, we have*

    (i) $r \odot (a \oplus b) = r \odot a \oplus r \odot b$

    (ii) $(r+s) \odot a = r \odot a \oplus s \odot a$

    (iii) $(r \cdot s) \odot a = r \odot (s \odot a)$.

The elements of the ring R are called *scalars* and the mapping $\odot : R \times M \to M$ is called *scalar multiplication*.

If in the above definition, we replace $\odot : R \times M \to M$ by $\odot : M \times R \to M$ such that the analogues of M-2 (i)–(iii) hold, then the algebraic structure $(M, R, \oplus, \odot)$ is called a *right R-module* or simply a *right module* over ring R.

In the above definition, the symbol '+' has been used for addition in the ring R and the symbol '·' has been used for multiplication in R.

REMARK-1 *In modules we will be dealing with two types of zeros (additive identities) (i) zero element of the additive abelian group M and (ii) zero element of the ring R. In order to avoid any confusion, we shall be using the symbol 0 to denote the zero element of ring R and the symbol 0_M to denote the zero element of the abelian group M. The zero element 0 of ring R is also known as the scalar zero.*

REMARK-2 *Since $(M, \oplus)$ is an abelian group, for any $a, b, c \in M$, we have*

(i) $a \oplus b = a \oplus c \Rightarrow b = c$
(ii) $b \oplus a = c \oplus a \Rightarrow b = c$ $\Big\}$ (Cancellation laws)

(iii) $a \oplus b = 0_M \Rightarrow b = -a$ and $a = -b$

(iv) $-(-a) = a$

(v) $a \oplus b = a \Rightarrow b = 0_M$

(vi) $-(a \oplus b) = (-a) \oplus (-b)$

(vii) 0_M is unique

(viii) for each $a \in M$, $-a \in M$ is unique.

If there is no likelihood of any confusion, we shall say that M is a left (right) R-module or M is a left (right) module over ring R, whenever the algebraic structure $(M, R, \oplus, \odot)$ is a left (right) R-module. Thus, whenever we say that M is a left (right) module over ring R or M is a left (right) R-module, it would always mean that $(M, \oplus)$ is an abelian group and $\odot : R \times M \to M (\odot : M \times R \to M)$ is a mapping such that M-2 (i)–(iii) (analogues of M-2 (i)–(iii)) are satisfied.

REMARK-3 *For the sake of convenience in future we shall use the same symbol '+' for addition in the abelian group M and addition in the ring R. But, the context would always make it clear as to which operation is meant. Similarly, the scalar multiplication $\odot$ and the multiplication in R will be denoted by the same symbol '·'.*

We have talked about left and right R-modules. Now a natural question comes. Is it necessary to distinguish between right and left modules? Suppose M is a left R-module. Can't we make M into a right R-module by defining $ar = ra$ for all $a \in M$ and all $r \in R$?. The answer is that we can if R is a commutative ring, as shown in the following lemma. If R is not a commutative ring a left R-module M can't be made into a right R-module.

LEMMA. *Let M be a left R-module over a commutative ring R. Then M is also a right R-module and vice versa.*

PROOF. Since M is a left R-module, for each $a \in M$ and $r \in R$, $r \cdot a$ is a uniquely defined element of M such that for all $a, b \in M$ and all $r, s \in R$.

(i) $r \cdot (a+b) = r \cdot a + r \cdot b$

(ii) $(r+s) \cdot a = r \cdot a + s \cdot a$

(iii) $(r \cdot s) \cdot a = r \cdot (s \cdot a)$

Define $a \cdot r = r \cdot a$ for all $a \in M$ and all $r \in R \ldots$ \hfill (I)

For any $a, b \in M$ and $r, s \in R$ we observe the following properties:

(a) $\quad (a+b) \cdot r = r \cdot (a+b)$ \hfill [By def. (I)]

$\Rightarrow (a+b) \cdot r = r \cdot a + r \cdot b$ \hfill [By (i)]

$\Rightarrow (a+b) \cdot r = a \cdot r + b \cdot r$ \hfill [By def. (I)]

(b) $\quad a \cdot (r+s) = (r+s) \cdot a$ \hfill [By def. (I)]

$\Rightarrow a \cdot (r+s) = r \cdot a + s \cdot a$ \hfill [By (ii)]

$\Rightarrow a \cdot (r+s) = a \cdot r + a \cdot s$ \hfill [By def. (I)]

(c) $\quad a \cdot (r \cdot s) = (r \cdot s) \cdot a$ \hfill [By def. (I)]

$\Rightarrow a \cdot (r \cdot s) = (s \cdot r) \cdot a$ \hfill [By commutativity of multiplication on R]

$\Rightarrow a \cdot (r \cdot s) = s \cdot (r \cdot a)$ \hfill [By (iii)]

$\Rightarrow a \cdot (r \cdot s) = s \cdot (a \cdot r)$ \hfill [By def. (I)]

$\Rightarrow a \cdot (r \cdot s) = (a \cdot r) \cdot s$ \hfill [By def. (I)]

Thus, M is an additive abelian group such that for each $a \in M$ and all $r \in R$, $a \cdot r$ is a uniquely defined element of M and properties $(a) - (c)$ are satisfied.

Hence, M is a right R-module over ring R.

Proceeding in the similar fashion we can show that if M is a right R-module, then it is a left R-module.
\hfill Q.E.D.

REMARK-4 *In future we shall write ar for $a \cdot r$.*

REMARK-5 *As proved in the above lemma that in case of a commutative ring R every left R-module is a right R-module and every right R-module is a left R-module. Thus, in this case, we are justified in referring simply to R-modules and using either left or right notation. However, if R is not a commutative ring, we must be careful to make the distinction between left and right R-modules. Note that left and right R-modules in case of a non-commutative ring are different structures. They cannot be identified or considered the same.*

REMARK-6 *In future unless stated otherwise all R-modules are left R-modules.*

UNITARY MODULE. *Let R be a ring with unity. An R-module M is called a unitary module if $1a = a$ for all $a \in M$.*

Every additive abelian group is a unitary Z-module as Z is a ring with unity.

ILLUSTRATIVE EXAMPLES

EXAMPLE-1 *Every ring is a module over its any subring.*

SOLUTION Let R be a ring, and let S be an arbitrary subring of R. Since R is an additive abelian group. Therefore, M-1 holds.

Taking multiplication in R as scalar multiplication, we find that the axioms M-2(i) to M-2(iii) are respectively the left distributive, right distributive and associative laws.

Hence, R is a module over its subring S.

Since S is an arbitrary subring of ring R. Therefore, every ring is a module over its any subring.

REMARK-7 *The converse of this example is not true, i.e. a subring is not necessarily a module over its over ring. For example, the ring Z of integers is not a module over its over ring Q of rational numbers, because the multiplication of a rational number and an integer is not always an integer.*

EXAMPLE-2 *Every ring is a module over itself.*

SOLUTION Since every ring is a subring of itself, therefore the result follows from Example 1.

REMARK-8 *In order to test whether a given non-void set M forms a module over a ring R, we must proceed as follows:*
  (i) Define a binary operation on M.
  (ii) Define scalar multiplication on M, which associates each scalar in R and each element in M to a unique element in M.
  (iii) Define equality of elements in M.
  (iv) Check whether axioms M-1 and M-2 are satisfied relative to the binary operation on M and scalar multiplication thus defined.

EXAMPLE-3 *Every additive abelian group is a module over the ring Z of integers.*

SOLUTION Let M be an additive abelian group. In order to give a module structure to M over the ring Z of integers, we define scalar multiplication on A as follows:

Scalar Multiplication on M: For any $n \in Z$ and $a \in M$, we define

$$na = \begin{cases} \underbrace{a+a+\cdots+a}_{(n-times)} & , \text{if } n > 0 \\ 0 & , \text{if } n = 0 \\ \underbrace{(-a)+(-a)+\cdots+(-a)}_{(|n|-times)} & , \text{if } n < 0 \end{cases}$$

Clearly $na \in M$ for all $a \in M$ and for all $n \in Z$.

Since A is an additive abelian group. Therefore, addition on M is defined. Thus, addition and scalar multiplication on M are defined. So, it remains now to verify the axioms M-2(i) to M-2(iii).

M-2. For $a, b \in A$ and $m, n \in Z$, we have

(i) If $n > 0$, then

$$n(a+b) = \underbrace{(a+b) + (a+b) + \cdots + (a+b)}_{n-times} \quad \text{[By def. of scalar multiplication]}$$

$$= \underbrace{(a+a+\cdots+a)}_{n-times} + \underbrace{(b+b+\cdots+b)}_{n-times} \quad \text{[By comm. and assoc. of addition on } M\text{]}$$

$$= na + nb \quad \text{[By def. of scalar multiplication]}$$

(ii) If $n = 0$, then

$$n(a+b) = 0 \quad \text{[By def. of scalar multiplication]}$$

$$\Rightarrow n(a+b) = 0 + 0 = na + nb$$

(iii) If $n < 0$, then

$$n(a+b) = \underbrace{\{-(a+b)\} + \{-(a+b)\} + \ldots + \{-(a!+!b)\}}_{|n|-times} \quad \text{[By def. of scalar multiplication.]}$$

$$= \underbrace{(-a-b) + (-a-b) + \cdots + (-a-b)}_{|n|-times} \quad \left[\because +\text{ is commutative on } M\right]$$

$$= \underbrace{\{(-a)+(-a)+\ldots+(-a)\}}_{|n|-times} + \underbrace{\{(-b)+(-b)+\cdots+(-b)\}}_{|n|-times} \quad \left[\begin{array}{l}\text{By commutativity} \\ \text{and associativity} \\ \text{of addition on } M\end{array}\right]$$

$$= |n|(-a) + |n|(-b)$$

$$= (-n)(-a) + (-n)(-b) \quad [\because n < 0 \quad \therefore |n| = -n]$$

$$= na + nb \quad [\text{[By def. of scalar multiplication]}]$$

Thus, $n(a+b) = na + nb$ for all $n \in Z$.

Similarly, we can show that M-2(ii) and M-2(iii) are true.

Hence, M is a module over Z or, M is a Z-module.

REMARK-9 *In the above example, we have proved that every additive abelian group is a Z-module. In fact every abelian group M can be regarded as a Z-module, if we define $n\,a = a^n$ for $a \in M$ and $n \in Z$.*

REMARK-10 *Since every cyclic group is abelian. Therefore, every cyclic group can also be considered as a Z-module.*

EXAMPLE-4 *Let R be a ring and let I be a left ideal of R. Then, I is an R-module.*

SOLUTION Since I is a left ideal of R. Therefore, I is an additive abelian group. We define scalar multiplication on I as follows:

Scalar Multiplication on I: For any $r \in R$ and $a \in I$, let ra be the ordinary product of these elements as elements of R. Since I is a left ideal of R, therefore $ra \in I$ for all $r \in R$ and $a \in I$.

In order to prove that I is an R-module, we have to verify axioms M-2(i) to M-2(iii).

M-2 For any $a, b \in A$ and $r, s \in R$, we have

  (i) $r(a+b) = ra + rb$ by left distributivity of multiplication on R over addition on R.
  (ii) $(r+s)a = ra + sa$ by right distributivity of multiplication on R over addition on R.
  (iii) $(rs)a = r(sa)$ follows from associativity of multiplication on R.

Hence, I is an R-module.

REMARK-11 *Since every ideal is a left ideal, therefore, every ideal of a ring R is an R-module. In fact every ideal is both a left R-module and a right R-module.*

EXAMPLE-5 *Let R be a ring. Then the set R[x] of all polynomials over ring R, in indeterminate x, is an R-module.*

SOLUTION In order to give a module structure to $R[x]$, we define addition, scalar multiplication, and equality in $R[x]$ as follows:

Addition on R[x]: If $f(x) = \sum_i a_i x^i, g(x) = \sum_i b_i x^i \in R[x]$, then we define

$$f(x) + g(x) = \sum_i (a_i + b_i) x^i$$

Clearly, $f(x) + g(x) \in R[x]$, because $a_i + b_i \in R$ for all i.

Scalar Multiplication on R[x]: For any $\lambda \in R$ and $f(x) = \sum_i a_i x^i \in R[x]$, $\lambda f(x)$ is defined as the polynomial $\sum_i (\lambda a_i) x^i$, i.e. $\lambda f(x) = \sum_i (\lambda a_i) x^i$.

Obviously, $\lambda f(x) \in R[x]$, as $\lambda a_i \in R$ for all i.

Equality of two elements of R[x]: For any $f(x) = \sum_i a_i x^i, g(x) = \sum_i b_i x^i \in R[x]$, we define

$$f(x) = g(x) \Leftrightarrow a_i = b_i \text{ for all } i.$$

Now we shall verify axioms M-1 and M-2.

M-1 *R[x] is an abelian group under addition defined above*:

Associativity: If $f(x) = \sum_i a_i x^i, g(x) = \sum_i b_i x^i, h(x) = \sum_i c_i x^i \in R[x]$, then

$$\{f(x) + g(x)\} + h(x) = \left(\sum_i (a_i + b_i) x^i\right) + \sum_i c_i x^i$$
$$= \sum_i \{(a_i + b_i) + c_i\} x^i$$
$$= \sum_i \{a_i + (b_i + c_i)\} x^i \qquad \text{[By associativity of addition on } R\text{]}$$
$$= \sum_i a_i x^i + \sum_i (b_i + c_i) x^i$$
$$= f(x) + \{g(x) + h(x)\}$$

So, addition is associative on $R[x]$.

Commutativity: If $f(x) = \sum_i a_i x^i, g(x) = \sum_i b_i x^i \in R[x]$, then

$$f(x) + g(x) = \sum_i (a_i + b_i) x^i = \sum_i (b_i + a_i) x^i \qquad \text{[By commutativity of addition on } R\text{]}$$
$$\Rightarrow \quad f(x) + g(x) = g(x) + f(x)$$

So, addition is commutative on $R[x]$.

Existence of additive identity: Since $\widehat{0}(x) = \sum_i 0 x^i \in R[x]$ is such that for all $f(x) = \sum_i a_i x^i \in R[x]$

$$\widehat{0}(x) + f(x) = \sum_i (0 + a_i) x^i = \sum_i a_i x^i = f(x) \qquad [\because \ 0 \text{ is the additive identity in } R]$$

Thus, $\widehat{0}(x) + f(x) = f(x) = f(x) + \widehat{0}(x)$ for all $f(x) \in R[x]$

∴ $\widehat{0}(x)$ is the identity element (zero) for addition on $R[x]$.

Existence of additive inverse: Let $f(x) = \sum_i a_i x^i$ be an arbitrary polynomial in $R[x]$. Then, $-f(x) = \sum_i (-a_i) x^i \in R[x]$ such that

$$f(x) + \{-f(x)\} = \sum_i a_i x^i + \sum_i (-a_i) x^i = \sum_i \{a_i + (-a_i)\} x^i = \sum_i 0 x^i = \widehat{0}(x)$$

Thus, for each $f(x) \in R[x]$, there exists $-f(x) \in R[x]$ such that

$$f(x) + \{-f(x)\} = \widehat{0}(x) = \{-f(x)\} + f(x)$$

∴ Each $f(x)$ has its additive inverse in $R[x]$.

30 • *Theory and Problems of Linear Algebra*

Hence, $R[x]$ is an abelian group for addition.

M-2 For any $f(x) = \sum_i a_i x^i, g(x) = \sum_i b_i x^i \in R[x]$ and $\lambda, \mu \in R$, we have

(i) $\lambda\{f(x) + g(x)\} = \lambda\left\{\sum_i (a_i + b_i)x^i\right\}$

$= \sum_i \lambda(a_i + b_i)x^i$

$= \sum_i (\lambda a_i + \lambda b_i)x^i$ [By left distributivity of multiplication over addition]

$= \sum_i (\lambda a_i)x^i + \sum_i (\lambda b_i)x^i$

$= \lambda f(x) + \lambda g(x)$

(ii) $(\lambda + \mu)f(x) = \sum_i (\lambda + \mu)a_i x^i$

$= \sum_i (\lambda a_i + \mu a_i)x^i$ [By right distributivity of multiplication over addition.]

$= \sum_i (\lambda a_i)x^i + \sum_i (\mu a_i)x^i$

$= \lambda f(x) + \mu f(x)$

(iii) $(\lambda \mu)f(x) = \sum_i (\lambda \mu)a_i x^i$

$= \sum_i \lambda(\mu a_i)x^i$ [By associativity of multiplication on R]

$= \lambda \sum_i (\mu a_i)x^i$

$= \lambda\{\mu f(x)\}$

(iv) $1 f(x) = \sum_i (1 a_i)x^i$

$= \sum_i a_i x^i$ [$\because$ 1 is unity in R]

$= f(x)$

Hence, $R[x]$ is an R-module.

REMARK-12 *This example becomes trivial if we consider R as a subring of $R[x]$.*

EXAMPLE-6 *Let R be a ring, and let n be a positive integer. Then,*
$R^n = \{(a_1, a_2, \ldots, a_n) : a_i \in R \text{ for all } i \in \underline{n}\}$ *is an R-module.*

SOLUTION To give a module structure to R^n over ring R, we first define an additive binary operation on R^n, scalar multiplication on R^n and equality of any two elements of R^n as follows:

Addition on R^n: For any $x = (a_1, a_2, \ldots, a_n), y = (b_1, b_2, \ldots, b_n) \in R^n$, we define
$$x + y = (a_1 + b_1, a_2 + b_2, \ldots, a_n + b_n)$$
$\Rightarrow \quad x + y \in R^n \qquad [\because \ a_i + b_i \in R \text{ for all } i \in \underline{n}]$

Thus, addition is a binary operation on R^n.

Scalar multiplication on R^n: For any $x = (a_1, a_2, \ldots, a_n) \in R^n$ and $\lambda \in F$, we define
$$\lambda x = (\lambda a_1, \lambda a_2, \ldots, \lambda a_n).$$
$\Rightarrow \quad \lambda x \in R[x] \qquad [\because \ \lambda a_i \in R \text{ for all } i \in \underline{n}].$

Thus, scalar multiplication is defined on $R[x]$.

Equality of two elements of R^n: For any $x = (a_1, a_2, \ldots, a_n), y = (b_1, b_2, \ldots, b_n) \in R[x]$, we define
$$x = y \Leftrightarrow a_i = b_i \quad \text{for all } i \in \underline{n}.$$

Thus, we have defined addition and scalar multiplication on $R[x]$ and equality of any two elements of $R[x]$. Let us now verify axioms M-1 and M-2.

M-1: R^n is an abelian group under addition

Associativity: For any $x = (a_1, a_2, \ldots, a_n)$, $y = (b_1, b_2, \ldots, b_n)$, $z = (c_1, c_2, \ldots, c_n) \in R^n$, we have

$$\begin{aligned}
(x+y)+z &= (a_1+b_1, a_2+b_2, \ldots, a_n+b_n) + (c_1, c_2, \ldots, c_n) \\
&= ((a_1+b_1)+c_1, (a_2+b_2)+c_2, \ldots, (a_n+b_n)+c_n) \\
&= (a_1+(b_1+c_1), a_2+(b_2+c_2), \ldots, a_n+(b_n+c_n)) \text{ [By associativity of } + \text{ on } R] \\
&= (a_1, a_2, \ldots, a_n) + (b_1+c_1, b_2+c_2, \ldots, b_n+c_n) \\
&= x + (y+z)
\end{aligned}$$

So, addition is associative on R^n.

Commutativity: For any $x = (a_1, a_2, \ldots, a_n), y = (b_1, b_2, \ldots, b_n) \in R^n$, we have
$$x + y = (a_1 + b_1, a_2 + b_2, \ldots, a_n + b_n)$$
$\Rightarrow \quad x + y = (b_1 + a_1, b_2 + a_2, \ldots, b_n + a_n) \qquad$ [By commutativity of $+$ on R]
$\Rightarrow \quad x + y = y + x$

So, addition is commutative on R^n.

Existence of additive identity: Since $0 \in R$, therefore $\underline{0} = (0,0,\ldots,0) \in R^n$ such that for all $x = (a_1, a_2, \ldots, a_n) \in R^n$, we have

$$x + \underline{0} = (a_1, a_2, \ldots, a_n) + (0, 0, \ldots, 0)$$
$$\Rightarrow \quad x + \underline{0} = (a_1 + 0, \ a_2 + 0, \ldots, a_n + 0)$$
$$\Rightarrow \quad x + \underline{0} = (a_1, a_2, \ldots, a_n) = x \qquad [\because \quad a_i + 0 = a_i \text{ for } i \in \underline{n}]$$

$\therefore \quad x + \underline{0} = x = \underline{0} + x$ for all $x \in R^n$ $\qquad [\because$ Addition is commutative on $R^n]$

So, $\underline{0} = (0, 0, \ldots, 0)$ is the identity element for addition on R^n.

Existence of additive inverse: Let $x = (a_1, a_2, \ldots, a_n)$ be an arbitrary element of R^n. Then, $-x = (-a_1, -a_2, \ldots, -a_n) \in R^n$ such that

$$x + (-x) = (a_1 + (-a_1), a_2 + (-a_2), \ldots, a_n + (a_n)) = (0, 0, \ldots, 0) = \underline{0}$$

Thus, for every $x = (a_1, a_2, \ldots, a_n) \in R^n$, there exists $-x = (-a_1, -a_2, \ldots, -a_n) \in R^n$ such that

$$x + (-x) = \underline{0} = (-x) + x$$

So, every $x \in R^n$ has its additive inverse.

Hence, R^n is an abelian group under addition.

M-2 : For any $x = (a_1, a_2, \ldots, a_n), y = (b_1, b_2, \ldots, b_n) \in R^n$ and $\lambda, \mu \in R$, we have

(i) $\lambda(x+y) = \lambda(a_1 + b_1, a_2 + b_2, \ldots, a_n + b_n)$
$\qquad = (\lambda(a_1 + b_1), \lambda(a_2 + b_2), \ldots, \lambda(a_n + b_n))$
$\qquad = (\lambda a_1 + \lambda b_1, \lambda a_2 + \lambda b_2, \ldots, \lambda a_n + \lambda b_n) \quad \left[\begin{array}{l}\text{By left distributivity of} \\ \text{multiplication over addition on } R\end{array}\right]$
$\qquad = (\lambda a_1, \lambda a_2, \ldots, \lambda a_n) + (\lambda b_1, \lambda b_2, \ldots, \lambda b_n)$
$\qquad = \lambda x + \lambda y$

(ii) $(\lambda + \mu)x = ((\lambda + \mu)a_1, (\lambda + \mu)a_2, \ldots, (\lambda + \mu)a_n)$
$\qquad = (\lambda a_1 + \mu a_1, \lambda a_2 + \mu a_2, \ldots, \lambda a_n + \mu a_n) \quad \left[\begin{array}{l}\text{By right distributivity of} \\ \text{multiplication over addition on } R\end{array}\right]$
$\qquad = (\lambda a_1, \lambda a_2, \ldots, \lambda a_n) + (\mu a_1, \mu a_2, \ldots, \mu a_n)$
$\qquad = \lambda x + \mu x$

(iii) $(\lambda \mu)x = ((\lambda \mu)a_1, (\lambda \mu)a_2, \ldots, (\lambda \mu)a_n)$
$\qquad = (\lambda(\mu a_1), \lambda(\mu a_2), \ldots, \lambda(\mu a_n)) \qquad$ [By associativity of multiplication on R]
$\qquad = \lambda(\mu a_1, \mu a_2, \ldots, \mu a_n)$
$\qquad = \lambda(\mu x)$

Hence, R^n is an R-module.

EXAMPLE-7 *Let R be a ring. Then the set $R^{m \times n}$ of all $m \times n$ matrices over R is a module over ring R.*

SOLUTION In order to prove that $R^{m \times n}$ is a module over ring R, we first define addition and scalar multiplication on $R^{m \times n}$ and also the equality of any two elements of $R^{m \times n}$ as follows:

Addition on $R^{m \times n}$: For any $A = [a_{ij}]$, $B = [b_{ij}]$ in $R^{m \times n}$, we define
$A + B = [a_{ij} + b_{ij}]$, i.e. addition on $R^{m \times n}$ is addition of matrices.

Scalar multiplication on $R^{m \times n}$: For any $A = [a_{ij}] \in R^{m \times n}$ and $r \in R$, we define
$rA = [ra_{ij}]$, i.e. scalar multiplication on $R^{m \times n}$ is usually scalar multiplication of the matrix A by the element $r \in R$.

Equality of any two elements of $R^{m \times n}$: For any $A = [a_{ij}], B = [b_{ij}] \in R^{m \times n}$, we define
$A = B \Leftrightarrow a_{ij} = b_{ij}$ for all $i \in \underline{m}, j \in \underline{n}$.

Thus, we have defined addition, scalar multiplication on $R^{m \times n}$ and equality of any two elements of $R^{m \times n}$. Now we have to verify axioms M-1 and M-2.

M-1: *$R^{m \times n}$ is an abelian group under addition of matrices*:

Associativity: For any $A = [a_{ij}], B = [b_{ij}], C = [c_{ij}] \in R^{m \times n}$, we have

$$(A+B)+C = [a_{ij}+b_{ij}] + [c_{ij}]$$
$$= [(a_{ij}+b_{ij})+c_{ij}]$$
$$= [a_{ij}+(b_{ij}+c_{ij})] \qquad \text{[By associativity of + on R]}$$
$$= [a_{ij}] + [b_{ij}+c_{ij}]$$
$$= A + (B+C)$$

So, addition of matrices is associative on $R^{m \times n}$.

Commutativity: For any $A = [a_{ij}], B = [b_{ij}] \in R^{m \times n}$

$$A+B = [a_{ij}+b_{ij}] = [b_{ij}+a_{ij}] \qquad \text{[By commutativity of + on R]}$$
$$= B+A$$

So, addition of matrices is commutative on $R^{m \times n}$.

Existence of additive identity: The null matrix $O = [0]$ is the identity element for addition, because
$$A + O = A = O + A \quad \text{for all} \ A \in R^{m \times n}.$$

Existence of additive inverse: Let $A = [a_{ij}]$ be an arbitrary matrix over ring R. Then, $-A = [-a_{ij}] \in R^{m \times n}$ is such that
$$A + (-A) = [a_{ij} + (-a_{ij})] = [0] = O = (-A) + A.$$

Thus, each $A \in R^{m \times n}$ has its additive inverse $-A \in R^{m \times n}$

So, every element in $R^{m\times n}$ has its additive inverse in $R^{m\times n}$
Hence, $R^{m\times n}$ is an abelian group under matrix addition.

M-2 : For any $A = [a_{ij}]$, $B = [b_{ij}] \in R^{m\times n}$ and $\lambda, \mu \in R$, we have

(i) $\quad \lambda(A+B) = [\lambda(a_{ij}+b_{ij})]$
$\qquad\qquad\quad = [\lambda a_{ij} + \lambda b_{ij}] \qquad$ [By left distributivity of multiplication over addition on R]
$\qquad\qquad\quad = [\lambda a_{ij}] + [\lambda b_{ij}]$
$\qquad\qquad\quad = \lambda A + \lambda B$

(ii) $\quad (\lambda+\mu)A = [(\lambda+\mu)a_{ij}]$
$\qquad\qquad\quad = [\lambda a_{ij} + \mu a_{ij}] \qquad$ [By right distributivity of multiblication over addition on R]
$\qquad\qquad\quad = [\lambda a_{ij}] + [\mu a_{ij}]$
$\qquad\qquad\quad = \lambda A + \mu A$

(iii) $\quad (\lambda\mu)A = [(\lambda\mu)a_{ij}]$
$\qquad\qquad\quad = [\lambda(\mu a_{ij})] \qquad\qquad\qquad$ [By associativity of multiplication on R]
$\qquad\qquad\quad = \lambda(\mu A).$

Hence, $R^{m\times n}$ is a module over ring R.

Particular case: Taking $m = 1$ (or $n = 1$) the set of all $1 \times n$ (or $m \times 1$) matrices, i.e. the set of all n (or m)-tuples, denoted by R^n (or R^m) is a module over ring R.

EXAMPLE-8 *Let X be a non-void set. Let R be a ring and let A be an R-module. Then the set $A^X = \{f : f : X \to A\}$ of all functions from X to A is an R-module under the addition and scalar multiplication on A^X defined by*
$$(f+g)(x) = f(x) + g(x)$$
and, $\qquad (\lambda f)(x) = \lambda f(x) \quad$ *respectively, for all $f, g \in A^X$ and all $\lambda \in R$.*

SOLUTION We define equality of any two elements of A^X as follows:
For any $f, g \in A^X$, we define
$f = g \Leftrightarrow f(x) = g(x)$ for all $x \in X$
Thus addition, scalar multiplication and equality of any two elements of A^X are defined.
We shall now verify axioms M-1 and M-2.

M-1 A^X *is an abelian group under addition defined above:*

Associativity: Let $f, g, h \in A^X$. Then
$\qquad [(f+g)+h](x) = (f+g)(x) + h(x)$
$\Rightarrow \quad [(f+g)+h](x) = [f(x)+g(x)] + h(x)$
$\Rightarrow \quad [(f+g)+h](x) = f(x) + [g(x)+h(x)] \qquad\qquad\qquad$ [By associativity of $+$ on A]

$$\Rightarrow \quad [(f+g)+h](x) = f(x)+(g+h)(x)$$
$$\Rightarrow \quad [(f+g)+h](x) = [f+(g+h)](x) \quad \text{for all } x \in X$$
$$\therefore \quad (f+g)+h = f+(g+h)$$

So, addition is associative on A^X.

Commutativity: For any $f, g \in A^X$
$$(f+g)(x) = f(x)+g(x)$$
$$\Rightarrow \quad (f+g)(x) = g(x)+f(x) \qquad \text{[By commutativity of + on } A]$$
$$\Rightarrow \quad (f+g)(x) = (g+f)(x) \quad \text{for all } x \in X$$
$$\therefore \quad f+g = g+f$$

So, addition is commutative on A^X.

Existence of additive identity: The function $\widehat{0}(x) = 0_A \in A$ for all $x \in X$ is the additive identity, because for any $f \in A^X$
$$(f+\widehat{0})(x) = f(x)+\widehat{0}(x) = f(x)+0_A = f(x) = (\widehat{0}+f)(x) \quad \text{for all } x \in X$$
$$\Rightarrow \quad f+\widehat{0} = f = \widehat{0}+f \quad \text{for all } f \in A^X$$

So, $\widehat{0} : X \to A$ is the additive identity.

Existence of additive inverse: Let f be an arbitrary function in A^X. Then a function $-f$ defined by $(-f)(x) = -f(x)$ for all $x \in X$ is additive inverse of f, because
$$(f+(-f))(x) = f(x)+(-f)(x)$$
$$\Rightarrow \quad (f+(-f))(x) = f(x)-f(x) = 0_A$$
$$\Rightarrow \quad (f+(-f))(x) = \widehat{0}(x) = ((-f)+f)(x) \quad \text{for all } x \in X.$$

Hence, A^X is an abelian group under addition.

M-2: For any $f, g \in A^X$ and $\lambda, \mu \in R$, we have

(i) $\quad [\lambda(f+g)](x) = \lambda(f+g)(x)$
$$\Rightarrow \quad [\lambda(f+g)](x) = \lambda[f(x)+g(x)]$$
$$\Rightarrow \quad [\lambda(f+g)](x) = \lambda f(x) + \lambda g(x) \qquad \text{[By M-2 (i) in } A]$$
$$\Rightarrow \quad [\lambda(f+g)](x) = (\lambda f)(x) + (\lambda g)(x)$$
$$\Rightarrow \quad [\lambda(f+g)](x) = [\lambda f + \lambda g](x) \quad \text{for all } x \in X$$
$$\therefore \quad \lambda(f+g) = \lambda f + \lambda g$$

(ii) $\quad [(\lambda+\mu)f](x) = (\lambda+\mu)f(x)$
$$\Rightarrow \quad [(\lambda+\mu)f](x) = \lambda f(x) + \mu f(x) \qquad \text{[By M-2 (ii) in } A]$$
$$\Rightarrow \quad [(\lambda+\mu)f](x) = (\lambda f)(x) + (\mu f)(x)$$
$$\Rightarrow \quad [(\lambda+\mu)f](x) = [\lambda f + \mu f](x) \quad \text{for all } x \in X$$
$$\therefore \quad (\lambda+\mu)f = \lambda f + \mu f$$

(iii) $\quad [(\lambda\mu)f](x) = (\lambda\mu)f(x)$
$\Rightarrow \quad [(\lambda\mu)f](x) = \lambda(\mu f(x))$ [By M-2 (iii) in A]
$\Rightarrow \quad [(\lambda\mu)f](x) = \lambda[(\mu f)(x)]$
$\Rightarrow \quad [(\lambda\mu)f](x) = [\lambda(\mu f)](x) \quad \text{for all } x \in X$
$\therefore \quad (\lambda\mu)f = \lambda(\mu f)$

Hence, A^X is a module over ring R.

EXAMPLE-9 *Let M_1, and M_2 be two R-modules. Then their cartesian product*
$M_1 \times M_2 = \{(a_1, a_2) : a_1 \in M_1, a_2 \in M_2\}$ *is a module over ring R.*

SOLUTION We define addition, scalar multiplication and equality in $M_1 \times M_2$ as follows:

Addition on $M_1 \times M_2$: If $(a_1, a_2), (b_1, b_2) \in M_1 \times M_2$, then we define
$$(a_1, a_2) + (b_1, b_2) = (a_1 + b_1, a_2 + b_2).$$

Since $a_1 + b_1 \in M_1$ and $a_2 + b_2 \in M_2$, therefore, $(a_1 + b_1, a_2 + b_2) \in M_1 \times M_2$.

So, addition is a binary operation on $M_1 \times M_2$.

Scaler multiplication: For any $(a_1, a_2) \in M_1 \times M_2$ and $\lambda \in R$, we define
$$\lambda(a_1, a_2) = (\lambda a_1, \lambda a_2)$$

Since M_1 and M_2 are modules over ring R, therefore, $\lambda a_1 \in M_1$ and $\lambda a_2 \in M_2$.
Consequently, $\lambda(a_1, a_2) = (\lambda a_1, \lambda a_2) \in M_1 \times M_2$.

Equality of any two elements of $M_1 \times M_2$: (a_1, a_2) and (b_1, b_2) in $M_1 \times M_2$ are defined as equal if $a_1 = b_1$ and $a_2 = b_2$.

Since we have defined addition, scalar multiplication and equality in $M_1 \times M_2$, now it remains to verify axioms M-1 and M-2.

M-1: $M_1 \times M_2$ *is an abelian group under addition*:

Associativity: For any, $(a_1, a_2), (b_1, b_2), (c_1, c_2) \in M_1 \times M_2$, we have

$[(a_1, a_2) + (b_1, b_2)] + (c_1, c_2) = (a_1 + b_1, a_2 + b_2) + (c_1, c_2)$
$\qquad = ((a_1 + b_1) + c_1, (a_2 + b_2) + c_2)$
$\qquad = (a_1 + (b_1 + c_1), a_2 + (b_2 + c_2))$ $\begin{bmatrix}\text{By associativity of addition}\\ \text{on } M_1 \text{ and } M_2\end{bmatrix}$
$\qquad = (a_1, a_2) + (b_1 + c_1, b_2 + c_2)$
$\qquad = (a_1, a_2) + [(b_1, b_2) + (c_1, c_2)]$

So, addition is associative on $M_1 \times M_2$.

Commutativity: For any $(a_1, a_2), (b_1, b_2) \in M_1 \times M_2$, we have
$$(a_1, a_2) + (b_1, b_2) = (a_1 + b_1, a_2 + b_2)$$
$$= (b_1 + a_1, b_2 + a_2) \quad \text{[By commutativity of addition on } M_1 \text{ and } M_2\text{]}$$
$$= (b_1, b_2) + (a_1, a_2)$$
So, addition is commutative on $M_1 \times M_2$.

Existence of additive identity: If 0_{M_1} and 0_{M_2} are zeros in M_1 and M_2 respectively, then $(0_{M_1}, 0_{M_2})$ is zero (additive identity) in $M_1 \times M_2$. Because for any $(a_1, a_2) \in M_1 \times M_2$
$$(a_1, a_2) + (0_{M_1}, 0_{M_2}) = (a_1 + 0_{M_1}, a_2 + 0_{M_2})$$
$$= (a_1, a_2)$$
$$= (0_{M_1}, 0_{M_2}) + (a_1, a_2) \quad \text{[By commutativity of addition on } M_1 \times M_2\text{]}$$

Existence of additive inverse: Let $(a_1, a_2) \in M_1 \times M_2$. Then, $(-a_1, -a_2)$ is its additive inverse, because
$$(a_1, a_2) + (-a_1, -a_2) = (a_1 + (-a_1), a_2 + (-a_2))$$
$$= (0_{M_1}, 0_{M_2})$$
$$= (-a_1, -a_2) + (a_1, a_2) \quad \text{[By commutativity of addition on } M_1 \times M_2\text{]}$$
Thus, each element in $M_1 \times M_2$ has its additive inverse.

Hence, $M_1 \times M_2$ is an abelian group under addition.

M-2: For any $(a_1, a_2) \in M_1 \times M_2$ and $\lambda, \mu \in R$, we have

(i)
$$\lambda[(a_1, a_2) + (b_1, b_2)] = \lambda(a_1 + b_1, a_2 + b_2)$$
$$= (\lambda(a_1 + b_1), \lambda(a_2 + b_2))$$
$$= (\lambda a_1 + \lambda b_1, \lambda a_2 + \lambda b_2) \quad \text{[M-2(i) in } M_1 \text{ and } M_2\text{]}$$
$$= (\lambda a_1, \lambda a_2) + (\lambda b_1, \lambda b_2)$$
$$= \lambda(a_1, a_2) + \lambda(b_1, b_2)$$

(ii)
$$(\lambda + \mu)(a_1, a_2) = ((\lambda + \mu)a_1, (\lambda + \mu)a_2)$$
$$= (\lambda a_1 + \mu a_1, \lambda a_2 + \mu a_2) \quad \text{[By M-2(ii) in } M_1 \text{ and } M_2\text{]}$$
$$= (\lambda a_1, \lambda a_2) + (\mu a_1, \mu a_2)$$
$$= \lambda(a_1, a_2) + \mu(a_1, a_2)$$

(iii)
$$(\lambda \mu)(a_1, a_2) = ((\lambda \mu)a_1, (\lambda \mu)a_2)$$
$$= (\lambda(\mu a_1), \lambda(\mu a_2)) \quad \text{[By M-2(iii) in } M_1 \text{ and } M_2\text{]}$$
$$= \lambda(\mu a_1, \mu a_2)$$
$$= \lambda[\mu(a_1, a_2)]$$

Hence, $M_1 \times M_2$ is a module over ring R.

REMARK. $M_1 \times M_2$ is called the direct product of the R-modules M_1 and M_2.

EXAMPLE-10 Let A be an R-module. Then for any $n \in N$ the set $A^n = \{f : f : \underline{n} \to A\}$ is a module over ring R.

SOLUTION It is a particular case of Example 8, when X is replaced by $\underline{n} = \{1, 2, \ldots, n\}$.

EXERCISE 1.1

1. Let R be a ring and let S be the set of all sequences $<a_i>$, $a_i \in R$. Then show that S is an R-module under the addition and scalar multiplication defined as
$$<a_i> + <b_i> = <a_i + b_i>$$
$$\lambda <a_i> = <\lambda a_i>$$
where $\lambda, a_i, b_i \in R$.

2. Let M be an additive abelian group. Show that there is only one way of making it a Z-module.

3. If R is a ring and M is an R-module, then prove that
$$n(ra) = r(na) \quad \text{for all } n \in Z, r \in R \text{ and } a \in M.$$

4. Let R be a ring and let M be an R-module but not unital. Then show that there exists $a \neq 0$ in M such that $ra = 0_M$ for all $r \in R$.

5. Mark each of the following true or false:
   (i) Every additive abelian group is a Z-module.
   (ii) Every abelian group is a Z-module.
   (iii) Every cyclic group is a Z-module.

ANSWERS

5. (i) T (ii) T (iii) T

1.3 ELEMENTARY PROPERTIES OF MODULES

THEOREM-1 Let M be on R-module. For all $r \in R$ and all $a \in M$, we have

(i) $r0_M = 0_M$
(ii) $0a = 0_M$
(iii) $(-r)a = -(ra) = r(-a)$
(iv) If R is a ring with unity 1 and A is a unitary module over R. Then,
$$(-1)a = -a = 1(-a)$$
(v) If r is a unit and A is a unitary module over ring R. Then,
$$ra = 0_M \Rightarrow a = 0_M$$

PROOF. (i) For any $r \in R$, we have

$$r0_M = r(0_M + 0_M) \qquad [\because 0_M + 0_M = 0_M]$$
$$\Rightarrow r0_M = r0_M + r0_M \qquad \text{[By } M\text{-2 }(i)\text{]}$$
$$\Rightarrow r0_M + 0_M = r0_M + r0_M \qquad [\because 0_M \text{ is the additive identity in } M]$$
$$\Rightarrow r0_M = 0_M \qquad \text{[By left cancellation law in } M\text{]}$$

Hence, $r\, 0_M = 0_M$ for all $r \in R$.

(ii) For any $a \in M$, we have

$$0a = (0+0)a \qquad [\because 0+0 = 0 \text{ in ring } R]$$
$$\Rightarrow 0a = 0a + 0a \qquad \text{[By } M\text{-2}(ii)\text{]}$$
$$\Rightarrow 0a + 0_M = 0a + 0a \qquad [\because 0_M \text{ is the additive identity in } M]$$
$$\Rightarrow 0a = 0_M \qquad \text{[By left cancellation law in } M\text{]}$$

Hence, $0a = 0_M$ for all $a \in M$.

(iii) For any $r \in R$, we have

$$r + (-r) = 0$$
$$\Rightarrow [r + (-r)]a = 0a \qquad \text{for all } a \in M$$
$$\Rightarrow ra + (-r)a = 0a \qquad \text{for all } a \in M \qquad \text{[By } M\text{-2 (ii) in } M\text{]}$$
$$\Rightarrow ra + (-r)a = 0_M \qquad \text{for all } a \in M \qquad \text{[By (ii)]}$$
$$\Rightarrow (-r)a = -(ra) \qquad \text{for all } a \in M$$

Thus, $(-r)a = -(ra)$ for all $r \in R$ and all $a \in M$.

For any $a \in M$, we have

$$a + (-a) = 0_M$$
$$\Rightarrow r[a + (-a)] = r0_M \qquad \text{for all } r \in R$$
$$\Rightarrow ra + r(-a) = r0_M \qquad \text{for all } r \in R \qquad \text{[By } M\text{-2 (i)]}$$
$$\Rightarrow ra + r(-a) = 0_M \qquad \text{for all } r \in R \qquad \text{[By (i)]}$$
$$\Rightarrow r(-a) = -(ra)$$

$\therefore \quad r(-a) = -(ra)$ for all $r \in R$ and all $a \in M$.

Hence, $(-r)a = -(ra) = r(-a)$ for all $r \in R$ and all $a \in M$.

(iv) Putting $r = 1$ in (iii), we get

$$(-1)a = -(1a) = 1(-a)$$
$$\Rightarrow (-1)a = -a = 1(-a) \qquad [\because M \text{ is a unitary module} \therefore 1a = a \text{ for all } a \in M]$$

$\therefore \quad (-1)a = -a = 1(-a)$ for all $a \in M$.

(v) Since r is a unit in R. Therefore, $r^{-1} \in R$ such that $r^{-1}r = 1$.

Now,
$$ra = 0_M$$
$$\Rightarrow \quad r^{-1}(ra) = r^{-1}0_M$$
$$\Rightarrow \quad (r^{-1}r)a = r^{-1}0_M \qquad \text{[By M-2 (iii)]}$$
$$\Rightarrow \quad (1)a = 0_M \qquad \text{[By (i)]}$$
$$\Rightarrow \quad a = 0_M \qquad [\because \ M \text{ is a unitary module}]$$

Thus, if r is a unit and $ra = 0_M$, then $a = 0_M$. Q.E.D.

REMARK. *In future, we shall write $a - b$ for $a + (-b)$ for any $a, b \in M$.*

THEOREM-2 *Let M be an R-module. Then, for all $a, b \in M$ and $r, s \in R$, we have*

(i) $r(a-b) = ra - rb$ (ii) $(r-s)a = ra - sa$.

PROOF. (i) For any $r \in R$ and any $a, b \in M$, we have

$$r(a-b) = r(a + (-b))$$
$$= ra + r(-b) \qquad \text{[By M-2 (i)]}$$
$$= ra + (-(rb)) \qquad \text{[By Theorem 1 (iii)]}$$
$$= ra - rb$$

(ii) For any $r, s \in R$ and $a \in M$, we have

$$(r-s)a = [r + (-s)]a$$
$$= ra + (-s)a \qquad \text{[By M-2 (ii)]}$$
$$= ra + (-(sa)) \qquad \text{[By Theorem 1 (iii)]}$$
$$= ra - sa$$

$\therefore \quad (r-s)a = ra - sa$ for all $r, s \in R$ and all $a \in M$.

Q.E.D.

1.4 SUBMODULES

Continuing the pattern of other algebraic structures, i.e. groups, rings, fields, etc., where we have studied subgroups, subrings, etc, we shall now define submodules.

SUBMODULES *A non-void subset N of an R-module M is called an R-submodule (or simply submodule) of M, if*

(i) $a - b \in N$ *for all $a, b \in N$.*
(ii) $ra \in N$ *for all $a \in N$ and all $r \in R$.*

In other words, an R-submodule of M is a subgroup N of the additive abelian group $(M, +)$ *such that* $ra \in N$ *for all* $r \in R$ *and all* $a \in N$.

Clearly $\{\underline{0}\}$ *and M itself are R-submodules of M. These two submodules are called trivial (improper) R-submodules of M and any other R-submodule of M is called a non-trivial (proper) R-submodule of M.*

REMARK. *Note that if N is an R-submodule of M, then N is also an R-module in its own right.*

IRREDUCIBLE MODULE *An R-module M is said to be an irreducible module if its only submodules are improper submodules.*

ILLUSTRATIVE EXAMPLES

EXAMPLE-1 *If M is an abelian group, then any subgroup of M is a Z-submodule of M.*

SOLUTION Let N be a subgroup of an additive abelian group M. Then,

$$a - b \in N \text{ for all } a, b \in N.$$

Since every subgroup of an abelian group is abelian. Therefore, N is an additive abelian group. But, we know that every abelian group is a Z-module. Therefore, N is a Z-module. Consequently, $na \in N$ for all $n \in Z$ and all $a \in N$.

Hence, N is a Z-submodule of M.

EXAMPLE-2 *If M is an R-module and* $a \in M$, *then the set* $Ra = \{ra : r \in R\}$ *is an R-submodule of M.*

SOLUTION Let x be an arbitrary element of Ra. Then $x = ra$ for some $r \in R$.
Since M is an R-module.
$\therefore \quad r \in R$ and $a \in M \Rightarrow ra \in M \Rightarrow x \in M$
Thus, $x \in Ra \Rightarrow x \in M$.
So, $Ra \subset M$.
Now, $0 \in R \Rightarrow 0a \in Ra \Rightarrow 0_M \in Ra \Rightarrow Ra \neq \phi$
Thus, Ra is a non-void subset of M.
In order to prove that Ra is an R-submodule of M, it remains to prove that for any $x, y \in Ra$, $x - y \in Ra$ and for any $r \in R$, $x \in Ra$, $rx \in Ra$.
Let $x, y \in Ra$. Then there exist $r_1, r_2 \in R$ such that

$$x = r_1 a, \ y = r_2 a$$
$$\Rightarrow x - y = r_1 a - r_2 a = (r_1 - r_2) a \quad \text{[By Theorem 2 on page 40]}$$
$$\Rightarrow x - y \in Ra \quad [\because \ r_1 - r_2 \in R]$$

Thus, $x - y \in Ra$ for all $x, y \in Ra$.

For any $r \in R$, we have

$$rx = r(r_1 a) = (rr_1) a \qquad \text{[By M-2 (iii)]}$$
$$\Rightarrow \quad rx \in Ra \qquad [\because \quad rr_1 \in R]$$

Thus, $rx \in Ra$ for all $r \in R$ and all $x \in Ra$.

Hence, Ra is an R-submodule of M.

EXAMPLE-3 *Let M be an R-module and $a \in M$. Then the set $N = \{ra + na \ : \ r \in R, n \in Z\}$ is a submodule of M.*

SOLUTION Let x be an arbitrary element of N. Then, $x = r_1 a + n_1 a$ for some $r_1 \in R$ and $n_1 \in Z$.
Now,

$$r_1 \in R, a \in M$$
$$\Rightarrow \quad r_1 a \in M \text{ and } n_1 a \in M \qquad [\because \quad M \text{ is an } R\text{-module and } (M, +) \text{ is a group}]$$
$$\Rightarrow \quad r_1 a + n_1 a \in M$$
$$\Rightarrow \quad x \in M$$

Thus, $x \in N \Rightarrow x \in M$
$\therefore \quad N \subset M$.

Since $0 \in R$ and $0 \in Z$, therefore, $0a + 0a = 0_M + 0_M = 0_M \in N$.

$\therefore \quad N$ is a non-void subset of M.

Let x, y be any two elements of N. Then there exist $r_1, r_2 \in R$ and $n_1, n_2 \in Z$ such that

$$x = r_1 a + n_1 a, \ y = r_2 a + n_2 a$$
$$\Rightarrow \quad x - y = (r_1 a + n_1 a) - (r_2 a + n_2 a)$$
$$\Rightarrow \quad x - y = (r_1 - r_2) a + (n_1 - n_2) a$$
$$\Rightarrow \quad x - y = r_3 a + n_3 a, \text{ where } r_3 = r_1 - r_2 \in R \text{ and } n_3 = n_1 - n_2 \in Z$$
$$\Rightarrow \quad x - y \in N.$$

$\therefore \quad x - y \in N$ for all $x, y \in N$.

Let $r \in R$ and $x \in N$. Then,

$$x \in N \Rightarrow x = r_1 a + n_1 a \text{ for some } r_1 \in R \text{ and } n_1 \in Z$$

Case I When $n_1 \geq 0$

In this case, we have

$rx = r(r_1 a + n_1 a)$
$\Rightarrow rx = r(r_1 a) + r(n_1 a)$ $\qquad [\because M$ is an R-module$]$
$\Rightarrow rx = (rr_1)a + r(\underbrace{a + a + \cdots + a}_{n_1\text{-times}})$
$\Rightarrow rx = (rr_1)a + ra + ra + \cdots + ra$ $\qquad [\because M$ is an R-module$]$
$\Rightarrow rx = (rr_1 + \underbrace{r + r + \cdots + r}_{n_1\text{-times}})a$
$\Rightarrow rx = u_1 a$, where $u_1 = rr_1 + \underbrace{r + r + \cdots + r}_{n_1\text{-times}} \in R$
$\Rightarrow rx = u_1 a + 0_M$, where $0_M \in M$
$\Rightarrow rx = u_1 a + 0a$, where $0 \in Z$
$\Rightarrow rx \in N$

Case II When $n_1 < 0$,

In this case, we have

$rx = r(r_1 a + n_1 a)$
$\Rightarrow rx = r(r_1 a) + r(n_1 a)$ $\qquad [\because M$ is an R-module$]$
$\Rightarrow rx = r(r_1 a) + r\{\underbrace{(-a) + (-a) + \cdots + (-a)}_{|n_1|\text{-times}}\}$
$\Rightarrow rx = (rr_1)a + \underbrace{r(-a) + r(-a) + \cdots + r(-a)}_{|n_1|\text{-times}}$ $\qquad [\because M$ is an R-module$]$
$\Rightarrow rx = (rr_1)a + (-r)a + (-r)a + \cdots + (-r)a$
$\Rightarrow rx = \{rr_1 + \underbrace{(-r) + (-r) + \cdots + (-r)}_{|n_1|\text{-times}}\}a$
$\Rightarrow rx = u_2 a$, where $u_2 = rr_1 + (-r) + (-r) + \cdots + (-r) \in R$
$\Rightarrow rx = u_2 a + 0_M$, where $0_M \in M$
$\Rightarrow rx = u_2 a + 0a$, where $0 \in Z$ $\qquad [\because 0_M = 0a]$
$\Rightarrow rx \in N$

$\therefore \quad rx \in N$ for all $r \in R$ and all $x \in N$.

Hence, N is an R-submodule of M.

1.4.1 FINITELY GENERATED SUBMODULE

SUBMODULE GENERATED BY A SUBSET *Let M be an R-module, and let S be a subset of M. An R-submodule N of M is called the R-submodule generated by S, if*

(i) $S \subset N$

(ii) If K is an R-submodule of M such that $K \supset S$, then $K \supset N$.

In other words, the smallest R-submodule of an R-module M containing S is called the R-submodule of M generated by S.

The R-submodule of an R-module M generated by a subset S of M is denoted by $[S]$. If $S = \{a_1, a_2, \ldots, a_n\}$ is a finite set, then the R-submodule of M generated by S is also written as $[a_1, a_2, \ldots, a_n]$.

REMARK. *Since the smallest R-submodule of an R-module M containing the void set is 0_M. Therefore, the R-submodule of M generated by the void set is 0_M.*

FINITELY GENERATED MODULE *An R-module M is called a finitely generated module if there exists a finite subset $\{a_1, a_2, \ldots, a_n\}$ of M such that $M = [a_1, a_2, \ldots, a_n]$*

The elements $a_1, a_2, \ldots, a_n$ are said to generate M.

ILLUSTRATIVE EXAMPLES

EXAMPLE-1 *Let R be a ring with unity 1. If M is an R-module such that $a \in M$, then $Ra = \{ra : r \in R\}$ is the R-submodule of M generated by $\{a\}$. That is, $Ra = [a]$.*

SOLUTION Clearly, Ra is an R-submodule of M (See Example 2 on page 41). In order to prove that Ra is the R-submodule of M generated by $\{a\}$, it is sufficient to prove that Ra is the smallest R-submodule of M containing a.
Since $1 \in R$, therefore, $1a = a \in Ra$.
Let N be an R-submodule of M containing a, and let x be an arbitrary element of Ra. Then, $x = ra$ for some $r \in R$.
Now, $a \in N$ and $r \in R$

$\Rightarrow \quad ra \in N$ \qquad\qquad\qquad\qquad $[\because\ N$ is an R-module in its own$]$
$\Rightarrow \quad x \in N$

Thus, $x \in Ra \Rightarrow x \in N$.
So, $Ra \subset N$.
Thus, every R-submodule of M containing 'a' contains Ra.
Hence, R_a is the smallest submodule of M containing a, i.e. $Ra = [a]$

EXAMPLE-2 *Let M be an R-module and $a \in M$. Then the set $N = \{ra + na : r \in R, n \in Z\}$ is the R-submodule of M generated by a, i.e. $N = [a]$. Further, if R has unity 1, then $N = Ra$.*

SOLUTION By Example 3 on page 42, N is an R-submodule of M.
Now,

$$0 \in R \text{ and } 1 \in Z$$
$\Rightarrow \quad 0a + 1 \cdot a = 0 + a \in N$
$\Rightarrow \quad a \in N$
$\Rightarrow \quad N$ is an R-submodule of M containing a.

Let K be an R-submodule of M containing a. Then, $ra \in K$ for all $r \in R$ and $a \in K$
Also, $na \in K$ for all $n \in Z$ $\qquad [\because \ (K,+)$ is an abelion group$]$
$\therefore \quad ra + na \in K$ for all $r \in R$ and all $n \in Z$
$\Rightarrow \quad N \subset K$.
Thus, N is the smallest-R-submodule of M containing $\{a\}$.
Hence, $N = [a]$.
Now let R be a ring with unity and let x be an arbitrary element of N. Then,
$x = ra + na$ for some $r \in R$ and for some $n \in Z$.
Now two cases arise.

Case I When $n \geq 0$.
In this case, we have
$x = ra + na = ra + n(1a)$
$\Rightarrow \quad x = ra + \underbrace{(1a + 1a + \cdots + 1a)}_{n\text{-}times}$
$\Rightarrow \quad x = (r + \underbrace{1 + 1 + \cdots + 1}_{n\text{-}times})a$
$\Rightarrow \quad x = u_1 a$, where $u_1 = r + \underbrace{1 + 1 + \cdots + 1}_{n\text{-}times}$
$\Rightarrow \quad x \in Ra$.

Case II When $n < 0$.
In this case, we have
$x = ra + na = ra + \underbrace{\big((-a) + (-a) + \cdots + (-a)\big)}_{|n|\text{-}times}$
$\Rightarrow \quad x = ra + \underbrace{(-1)a + (-1)a + \cdots + (-1)a}_{|n|\text{-}times}$
$\Rightarrow \quad x = \big(r + \underbrace{(-1) + (-1) + \cdots + (-1)}_{|n|\text{-}times}\big)a$
$\Rightarrow \quad x = u_2 a$, where $u_2 = r + \underbrace{(-1) + (-1) + \cdots + (-1)}_{|n|\text{-}times} \in R$
$\Rightarrow \quad x \in Ra$.

Thus, in either case, we have $x \in Ra$. So, $N \subset Ra$. Obviously, $Ra \subset N$.
Hence, $N = Ra$.

1.4.2 ALGEBRA OF SUBMODULES

THEOREM-1 *The intersection of any two R-submodules of an R-module M is an R-submodule of M.*

PROOF. Let N_1 and N_2 be two R-submodules of an R-module M. Then,

$0_M \in N_1, 0_M \in N_2$ and $N_1 \subset M, N_2 \subset M_2$

$\Rightarrow \quad 0_M \in N_1 \cap N_2$ and $N_1 \cap N_2 \subset M$.

$\Rightarrow \quad N_1 \cap N_2$ is a non-void subset of M.

Let a, b be any two arbitrary elements of $N_1 \cap N_2$. Then,

$a, b \in N_1 \cap N_2$

$\Rightarrow \quad a, b \in N_1$ and $a, b \in N_2$

$\Rightarrow \quad a - b \in N_1$ and $a - b \in N_2$ $\quad\quad$ [$\because$ N_1 and N_2 are R-submodules of M]

$\Rightarrow \quad a - b \in N_1 \cap N_2$

Thus, $a - b \in N_1 \cap N_2$ for all $a, b \in N_1 \cap N_2$.

Let r be an arbitrary element of R. Then,

$r \in R, a \in N_1 \cap N_2$

$\Rightarrow \quad r \in R, a \in N_1$ and $a \in N_2$

$\Rightarrow \quad r \in R, a \in N_1$ and $r \in R, a \in N_2$

$\Rightarrow \quad ra \in N_1$ and $ra \in N_2$ $\quad\quad$ [$\because$ N_1 and N_2 are R-submodules of M]

$\Rightarrow \quad ra \in N_1 \cap N_2$

Thus, $ra \in N_1 \cap N_2$ for all $r \in R$ and all $a \in N_1 \cap N_2$.

Hence, $N_1 \cap N_2$ is an R-submodule of M.

Q.E.D.

THEOREM-2 *The intersection of an arbitrary family of R-submodules of an R-module M is an R-submodule of M.*

PROOF. Let $\{N_i : i \in I\}$ be an arbitrary family of R-submodules of an R-module M. Here I is the index set such that for each $i \in I$, N_i is an R-submodule of M.

N_i is an R-submodule of M for all $i \in I$

$\Rightarrow \quad 0_M \in N_i$ and $N_i \subset M$ for all $i \in I$

$\Rightarrow \quad 0_M \in \bigcap_{i \in I} N_i$ and $\bigcap_{i \in I} N_i \subset M$

$\Rightarrow \quad \bigcap_{i \in I} N_i$ is a non-void subset of M.

Let a, b be any two arbitrary elements in $\bigcap_{i \in I} N_i$. Then,

$\quad\quad a, b \in \bigcap_{i \in I} N_i$
$\Rightarrow \quad a, b \in N_i$ for all $i \in I$
$\Rightarrow \quad a - b \in N_i$ for all $i \in I$ $\quad\quad\quad\quad\quad$ [$\because\ N_i$ is an R − submodule of M]
$\Rightarrow \quad a - b \in \bigcap_{i \in I} N_i$

Thus, $a - b \in \bigcap_{i \in I} N_i$ for all $a, b \in \bigcap_{i \in I} N_i$

Let r be an arbitrary element of R and $a \in \bigcap_{i \in I} N_i$. Then,

$\quad\quad r \in R,\ a \in \bigcap_{i \in I} N_i$
$\Rightarrow \quad r \in R$ and $a \in N_i$ for all $i \in I$
$\Rightarrow \quad ra \in N_i$ for all $i \in I$ $\quad\quad\quad\quad$ [$\because\ N_i$ is an R-submodules of M for all $i \in I$]
$\Rightarrow \quad ra \in \bigcap_{i \in I} N_i$

Thus, $ra \in \bigcap_{i \in I} N_i$ for all $a \in \bigcap_{i \in I} N_i$ and all $r \in R$

Hence, $\bigcap_{i \in I} N_i$ is an R-submodule of M.

Q.E.D.

THEOREM-3 *The union of any two R-submodules of an R-module M is an R-submodule of M iff one is contained in other.*

PROOF. Let N_1 and N_2 be two R-submodules of an R-module M. We have to prove that $N_1 \cup N_2$ is an R-submodule of M iff either $N_1 \subset N_2$ or $N_2 \subset N_1$.

If $N_1 \subset N_2$, then $N_1 \cup N_2 = N_2$ which is an R-submodule of M.

If $N_2 \subset N_1$, then $N_1 \cup N_2 = N_1$ which is an R-submodule of M.

Hence, in either case $N_1 \cup N_2$ is an R-submodule of M.

Conversely, suppose that $N_1 \cup N_2$ is an R-submodule of M. Then we have to prove that either $N_1 \subset N_2$ or $N_2 \subset N_1$.

If possible, let $N_1 \not\subset N_2$ and $N_2 \not\subset N_1$. Then,

$\quad\quad N_1 \not\subset N_2 \Rightarrow$ There exists $a \in N_1$ such that $a \notin N_2$
and, $\quad N_2 \not\subset N_1 \Rightarrow$ There exists $b \in N_2$ such that $b \notin N_1$.

Now,

$\quad\quad a \in N_1, b \in N_2$
$\Rightarrow \quad a, b \in N_1 \cup N_2$
$\Rightarrow \quad a - b \in N_1 \cup N_2$ $\quad\quad\quad\quad\quad\quad$ [$\because\ N_1 \cup N_2$ is an R-submodule of M]
$\Rightarrow \quad a - b \in N_1$ or, $a - b \in N_2$.

If $a - b \in N_1$, then

$$a - (a - b) \in N_1 \qquad [\because \ a \in N_1 \text{ and } N_1 \text{ is an } R\text{-submodule of } M]$$

$\Rightarrow b \in N_1$, which is a contradiction $\qquad [\because \ b \notin N_1]$

Again, if $a - b \in N_2$, then

$$(a - b) + b \in N_2 \qquad [\because \ b \in N_2 \text{ and } N_2 \text{ is an } R\text{-submodule of } M]$$

$\Rightarrow a \in N_2$, which is a contradiction $\qquad [\because \ a \notin N_2]$

Since the contradictions arise by assuming that $N_1 \not\subset N_2$ and $N_2 \not\subset N_1$.

Hence, either $N_1 \subset N_2$ or, $N_2 \subset N_1$. $\hfill$ Q.E.D.

THEOREM-4 *Let S be a subset of an R-module M. Then the intersection of the family of R-submodules of M containing S is the R-submodule generated by S.*

PROOF. Let $\{N_i : i \in I\}$ be the family of R-submodules of M containing S. Here I is the index set such that for each $i \in I$ there is an R-submodule of M containing S. By Theorem 2, $\bigcap_{i \in I} N_i$ is an R-submodule of M. Since $S \subset N_i$ for all $i \in I$, therefore, $S \subset \bigcap_{i \in I} N_i$. Thus, $\bigcap_{i \in I} N_i$ is an R-submodule of M containing S.

Let K be an R-submodule of M containing S. Then K is one of the members of the family. Consequently, $\bigcap_{i \in I} N_i \subset K$.

Hence, $\bigcap_{i \in I} N_i$ is the smallest R-submodules of M containing S, i.e. $\bigcap_{i \in I} N_i = [S]$.

$\hfill$ Q.E.D.

SUM OF SUBMODULES Let N_1 and N_2 be two R-submodules of an R-module M. Then their sum $N_1 + N_2$ is the set of all elements of the form $a_1 + a_2$, where $a_1 \in N_1$ and $a_2 \in N_2$.

i.e., $\qquad N_1 + N_2 = \{a_1 + a_2 : a_1 \in N_1 \text{ and } a_2 \in N_2\}$.

THEOREM-5 *The sum of two R-submodules of an R-module is an R-submodule of M.*

PROOF. Let N_1 and N_2 be two R-submodules of an R-module M. Then by definition

$$N_1 + N_2 = \{a_1 + a_2 : a_1 \in N_1, a_2 \in N_2\}$$

Since N_1 and N_2 are R-submodules of M. Therefore,

$\qquad 0_M \in N_1, 0_M \in N_2$

$\Rightarrow \qquad 0_M + 0_M \in N_1 + N_2$

$\Rightarrow \qquad 0_M \in N_1 + N_2$

$\Rightarrow \qquad N_1 + N_2 \quad$ is non-void.

Let x be an arbitrary element of $N_1 + N_2$. Then there exist $a_1 \in N_1, a_2 \in N_2$ such that $x = a_1 + a_2$.

Now,
$$a_1 \in N_1 \text{ and } a_2 \in N_2$$
$$\Rightarrow a_1 \in M \text{ and } a_2 \in M \qquad [\because N_1 \subset M, N_2 \subset M]$$
$$\Rightarrow a_1 + a_2 \in M \qquad [\because (M,+) \text{ is a group}]$$
$$\Rightarrow x \in M$$
$$\therefore x \in N_1 + N_2$$
$$\Rightarrow x \in M$$
$$\Rightarrow N_1 + N_2 \subset M.$$

So, $N_1 + N_2$ is a non-void subset of M.

Let $x = a_1 + a_2$, $y = b_1 + b_2 \in N_1 + N_2$. Then, $a_1, b_1 \in N_1$ and $a_2, b_2 \in N_2$.
$$\therefore x - y = (a_1 + a_2) - (b_1 + b_2)$$
$$\Rightarrow x - y = (a_1 - b_1) + (a_2 - b_2) \qquad \text{[Using commutativity and associativity of addition on } M\text{]}$$

Since N_1 and N_2 are R-submodules of M. Therefore,
$$a_1, b_1 \in N_1 \Rightarrow a_1 - b_1 \in N_1$$
and, $$a_2, b_2 \in N_2 \Rightarrow a_2 - b_2 \in N_2.$$
$$\therefore (a_1 - b_1) + (a_2 - b_2) \in N_1 + N_2$$
$$\Rightarrow x - y \in N_1 + N_2.$$

So, $x - y \in N_1 + N_2$ for all $x, y \in N_1 + N_2$.

Let r be an arbitrary element of R. Then,
$$rx = r(a_1 + a_2) = ra_1 + ra_2 \qquad [\text{By M-2 (i)}]$$

Since N_1 and N_2 are R-submodules of M. Therefore,
$$\left. \begin{array}{l} r \in R, a_1 \in N_1 \Rightarrow ra_1 \in N_1 \\ \text{and, } r \in R, a_2 \in N_2 \Rightarrow ra_2 \in N_2 \end{array} \right\} \Rightarrow rx = ra_1 + ra_2 \in N_1 + N_2$$

Thus, $rx \in N_1 + N_2$ for all $r \in R$ and all $x \in N_1 + N_2$.

Hence, $N_1 + N_2$ is an R-submodule of M.

Q.E.D.

THEOREM-6 *Let N_1 and N_2 be R-submodules of an R-module M. Then $N_1 + N_2$ is the smallest R-submodule of M containing $N_1 \cup N_2$. That is, $N_1 + N_2 = [N_1 \cup N_2]$.*

PROOF. Clearly, $N_1 + N_2$ is an R-submodules of M. [see Theorem 5]

Let x be an arbitrary element of $N_1 \cup N_2$. Then, $x \in N_1$ or $x \in N_2$.

If $x \in N_1$, then $x = x + 0_M \in N_1 + N_2$, where $0_M \in N_2$.

If $x \in N_2$, then $x = 0_M + x \in N_1 + N_2$, where $0_M \in N_1$.

Thus, $x \in N_1 \cup N_2 \Rightarrow x \in N_1 + N_2$
$$\therefore N_1 \cup N_2 \subset N_1 + N_2.$$

Let K be an R-submodule of M containing $N_1 \cup N_2$

Let $x = a_1 + a_2$ be an arbitrary element of $N_1 + N_2$. Then,

$a_1 \in N_1$, and $a_2 \in N_2$
$\Rightarrow \quad a_1 \in N_1 \cup N_2$ and $a_2 \in N_1 \cup N_2$
$\Rightarrow \quad a_1 \in K$ and $a_2 \in K$ $\qquad [\because \quad K \supset N_1 \cup N_2]$
$\Rightarrow \quad a_1 + a_2 \in K$ $\qquad [\because \quad (K,+) \text{ is a group}]$
$\Rightarrow \quad x \in K$

Thus, $x \in N_1 + N_2 \Rightarrow x \in K$

$\therefore \quad N_1 + N_2 \subset K$.

Thus, if K is an R-submodule of M containing $N_1 \cup N_2$, then it contains $N_1 + N_2$. Hence, $N_1 + N_2$ is the smallest R-submodule of M containing $N_1 \cup N_2$.

Q.E.D.

THEOREM-7 Let $N_1, N_2, \ldots, N_k$ be R-submodules of an R-module M. Then $N_1 + N_2 + \cdots + N_k = \{a_1 + a_2 + \cdots + a_k : a_i \in N_i, i = 1, \ldots, k\}$ is also an R-submodule of M.

PROOF. It is a trivial generalization of Theorem 6.

For any a in an R-module M, we know that Ra is an R-submodule of M (see Example 2 on page 41). From above theorem it follows that if $(a_1, a_2, \ldots, a_n)$ is a list of n elements in an R-module M, then $Ra_1 + Ra_2 + \cdots + Ra_n$ is an R-submodule of M.

DIRECT SUM OF SUBMODULES An R-module M is said to be direct sum of its two R-submodules N_1 and N_2 if each element $a \in M$ is uniquely expressible as $a = a_1 + a_2$, where $a_1 \in N_1$ and $a_2 \in N_2$. If M is direct sum of N_1 and N_2, then we write $M = N_1 \oplus N_2$.

An R-module M is said to be the direct sum of its R-submodules $N_1, N_2, \ldots, N_k$ if each element $a \in M$ can be expressed uniquely as

$$a = a_1 + a_2 + \cdots + a_k, \text{ where } a_i \in N_i, i = 1, 2, \ldots, k.$$

If M is direct sum of its R-submodules $N_1, N_2, \ldots, N_k$, then we write

$$M = N_1 \oplus N_2 \oplus \cdots \oplus N_k.$$

THEOREM-8 Let $N_1, N_2, \ldots, N_k$ be R-submodules of an R-module M. Then the following are equivalent:

(i) $M = N_1 \oplus N_2 \oplus \cdots \oplus N_k$

(ii) $\sum_{i=1}^{k} a_i = 0_M \Rightarrow a_i = 0_M$ for all $i \in \underline{k}$

(iii) $N_i \cap \sum_{\substack{j=1 \\ j \neq i}}^{k} N_j = \{0_M\}, i \in \underline{k}$.

PROOF. In order to prove the equivalence of three statements, it is sufficient to show that $(i) \Rightarrow (ii) \Rightarrow (iii) \Rightarrow (i)$.

$(i) \Rightarrow (ii)$.

Let M be the direct sum of $N_1, N_2, \ldots, N_k$, and let $\sum_{i=1}^{k} a_i = 0_M$.

Since each element in M has a unique representation in terms of elements of $N_1, N_2, \ldots, N_k$.

$\therefore \quad \sum_{i=1}^{k} a_i = 0_M = 0_M + 0_M + \cdots + 0_M \Rightarrow a_i = 0_M$ for all $i \in \underline{k}$

$(ii) \Rightarrow (iii)$.

Let x be an arbitrary element of $N_i \cap \sum_{\substack{j=1 \\ j \neq i}}^{k} N_j$. Then,

$x \in N_i$ and $x \in \sum_{\substack{j=1 \\ j \neq i}}^{k} N_j$

$\Rightarrow$ There exist elements $a_1 \in A_1, a_2 \in A_2, \ldots, a_i \in A_i, \ldots, a_k \in A_n$ such that

$x = a_1 + a_2 + \cdots + a_{i-1} + a_{i+1} + \cdots + a_k$ and, $x = a_i$.

$\Rightarrow a_i = a_1 + a_2 + \cdots + a_{i-1} + a_{i+1} + \cdots + a_k$

$\Rightarrow a_1 + a_2 + \cdots + a_{i-1} + (-a_i) + a_{i+1} + \cdots + a_k = 0_M$

$\Rightarrow a_1 = a_2 = \cdots = a_{i-1} = a_i = a_{i+1} = \cdots = a_k = 0_M$ [From (ii)]

$\Rightarrow x = 0_M$

Since x is an arbitrary element of $N_i \cap \sum_{\substack{j=1 \\ j \neq i}}^{k} N_j$. Therefore,

$$x \in N_i \cap \sum_{\substack{j=1 \\ j \neq i}}^{k} N_j \Rightarrow x = 0_M \quad \text{for all } i = 1, 2, \ldots, k$$

Hence, $N_i \cap \sum_{\substack{j=1 \\ j \neq i}}^{k} N_j = \{0_M\}$ for all $i = 1, 2, \ldots, k$.

$(iii) \Rightarrow (i)$.

Let $N_i \cap \sum_{\substack{j=1 \\ j \neq i}}^{k} N_j = \{0_M\}$ for all $i = 1, 2, \ldots, k$. Then, we have to show that each element in M has a unique representation in terms of elements in $N_1, N_2, \ldots, N_k$.

Let a be an arbitrary element of M such that

$a = a_1 + a_2 + \cdots + a_k, a = b_1 + b_2 + \cdots + b_k, a_i, b_i \in N_i$ for $i = 1, 2, \ldots, k$.

Then,
$$a_1 + a_2 + \cdots + a_k = b_1 + b_2 + \cdots + b_k$$
$$\Rightarrow (a_1 - b_1) + (a_2 - b_2) + \cdots + (a_k - b_k) = 0_M \quad \left[\begin{array}{l}\text{By commutativity and associativity} \\ \text{of addition on A}\end{array}\right]$$
$$\Rightarrow a_i - b_i = - \sum_{\substack{j=1 \\ j \neq i}}^{k} (a_j - b_j)$$
$$\Rightarrow a_i - b_i = \sum_{\substack{j=1 \\ j \neq i}}^{k} \{-(a_j - b_j)\}$$
$$\Rightarrow a_i - b_i \in \sum_{\substack{j=1 \\ j \neq i}}^{k} N_j \quad \text{for all } i = 1, 2, \ldots, k \qquad [\because \ -(a_j - b_j) \in N_j \quad \text{for all } j]$$
$$\Rightarrow a_i - b_i \in N_i \cap \sum_{\substack{j=1 \\ j \neq i}}^{k} N_j \quad \text{for all } i = 1, 2, \ldots, k \qquad [\because \ a_i - b_i \in N_i \quad \text{for all } i]$$
$$\Rightarrow a_i - b_i = 0_M \quad \text{for all } i = 1, 2, \ldots, k \qquad [\because \ N_i \cap \sum_{\substack{j=1 \\ j \neq i}}^{n} N_j = \{\underline{0}\} \quad \text{for all } i]$$
$$\Rightarrow a_i = b_i \quad \text{for all } i = 1, 2, \ldots, k$$

Thus, $a \in M$ has a unique representation. Hence, M is the direct sum of $N_1, N_2, N_3, \ldots, N_k$.

Q.E.D.

COMPLEMENT OF A SUBMODULE. *Let M be an R-module, and let N_1 be an R-submodule of M. Then, an R-submodule N_2 of M is said to be complement of N_1, if*
$$M = N_1 \oplus N_2$$

INDEPENDENT SUBMODULES. *Let $N_1, N_2, \ldots, N_n$ be R-submodules of an R-module M. Then, $N_1, N_2, \ldots, N_n$ are said to be independent, if*
$$N_i \cap \sum_{\substack{j=1 \\ j \neq i}}^{n} N_j = \{0_M\} \quad \text{for all } i = 1, 2, \ldots, n.$$

THEOREM-9 *Let $N_1, N_2, \ldots, N_n$ be R-submodules of an R-module M such that $M = N_1 \oplus N_2 \oplus \cdots \oplus N_n$. Then the following are equivalent:*

(i) $M = N_1 \oplus N_2 \oplus \ldots \oplus N_n$.
(ii) $N_1, N_2, \ldots, N_n$ are independent.
(iii) For each $a_i \in N_i$, $\sum_{i=1}^{n} a_i = 0_M \Rightarrow a_i = 0_M$.

PROOF. *See Theorem 8 on page 50*

Q.E.D.

THEOREM-10 Let R be a ring with unity 1, and let M be an R-module. If M is generated by a set $\{a_1, a_2, \ldots, a_n\}$, then $M = \{r_1 a_1 + r_2 a_2 + \cdots + r_n a_n : r_i \in R\}$. That is,
$$M = Ra_1 + Ra_2 + \cdots + Ra_n = \sum_{i=1}^{n} Ra_i.$$

PROOF. Since M is generated by the set $\{a_1, a_2, \ldots, a_n\}$. Therefore, M is the smallest R-submodule of itself containing $\{a_1, a_2, \ldots, a_n\}$. In order to prove the theorem, it is sufficient to prove that $\sum_{i=1}^{n} Ra_i$ is an R-submodule of M contain $a_1, a_2, \ldots, a_n$.

Let m_1, m_2 be any two elements of $\sum_{i=1}^{n} Ra_i$. Then,

$$m_1 = r_1 a_1 + \cdots + r_n a_n, \quad m_2 = s_1 a_1 + \cdots + s_n a_n \text{ for some } r_i, s_i \in R.$$
$$\therefore \quad m_1 - m_2 = (r_1 a_1 + \cdots + r_n a_n) - (s_1 a_1 + \cdots + s_n a_n)$$
$$\Rightarrow \quad m_1 - m_2 = (r_1 - s_1) a_1 + \cdots + (r_n - s_n) a_n \in \sum_{i=1}^{n} Ra_i \qquad [\because \; r_i - s_i \in R]$$
$$\Rightarrow \quad m_1 - m_2 \in \sum_{i=1}^{n} Ra_i$$

Thus, $m_1 - m_2 \in \sum_{i=1}^{n} Ra_i$ for all $m_1, m_2 \in \sum_{i=1}^{n} Ra_i$.

Let $m = \sum r_i a_i \in \sum_{i=1}^{n} Ra_i$ and $r \in R$. Then,

$$rm = r \left(\sum_{i=1}^{n} r_i a_i \right) = r(r_1 a_1 + \cdots + r_n a_n)$$
$$\Rightarrow \quad rm = r(r_1 a) + \cdots + r(r_n a) = (rr_1)a + \cdots + (rr_n)a \in \sum_{i=1}^{n} Ra_i$$

Thus, $rm \in \sum_{i=1}^{n} Ra_i$ for all $r \in R$ and all $m \in \sum_{i=1}^{n} Ra_i$.

Also,

$$a_i = 1 a_i = 0 a_1 + 0 a_2 + \cdots + 0 a_{i-1} + 1 a_i + 0 a_{i+1} + \cdots + 0 a_n \in \sum_{i=1}^{n} Ra_i \text{ for all i} \in \underline{n}$$

Hence, $\sum_{i=1}^{n} Ra_i$ is an R-submodule of M containing $\{a_1, a_2, \ldots, a_n\}$.

Consequently, we have $M = \sum_{i=1}^{n} Ra_i$

Q.E.D.

THEOREM-11 *Let $\{N_i : i \in \underline{n}\}$ be a family of R-submodules of an R-module M. Then $\sum_{i=1}^{n} N_i$ is the smallest R-submodule of M containing $\underset{i \in \underline{n}}{\cup} N_i$. That is, $\sum_{i=1}^{n} N_i = [\underset{i \in \underline{n}}{\cup} N_i]$.*

PROOF. It is generalization of Theorem 6 on page 49.

Q.E.D.

1.5 QUOTIENT MODULES

Like for other algebraic structures, we also have *quotient modules* or *factor modules*. In this section, we will develop the structure of a quotient module and study some of its properties.

Let N be an R-submodule of an R-module M. Then N is additive subgroup of the additive abelian group M. Let M/N be the set of all left cosets (Since M is an additive abelian group. Therefore, there is no distinction between left and right cosets of N in M and we may simply call cosets) of N in M. i.e.,
$$M/N = \{a+N : a \in M\}$$
Our main objective in this section is to equip the set M/N with the structure of an R-module. For that purpose, we define addition and scalar multiplication on M/N as follows:

Addition on M/N: For any $a+N, b+N \in M/N$, we define
$$(a+N) + (b+N) = (a+b) + N$$
Scalar multiplication on M/N: If $a+N \in M/N$ and $r \in R$, then we define scalar multiplication on M/N as follows:
$$r(a+N) = ra + N$$
Equality of elements in M/N: Since N is an additive abelian subgroup of the additive abelian group M. Therefore, for any $a+N, b+N \in M/N$
$$a+N = b+N \iff a-b \in N$$
Thus, we have defined addition and scalar multiplication on M/N and also the equality of any two elements of M/N. Now we shall establish that M/N possesses a module structure under addition and scalar multiplication defined above.

THEOREM-1 *Let N be an R-submodule of an R-module M. Then the set*
$$M/N = \{a+N : a \in M\}$$
is an R-module for the addition and scalar multiplication defined as follows:
$$(a+N) + (b+N) = (a+b) + N$$
and,
$$r(a+N) = ra + N \text{ for all } a+N, b+N \in M/N \text{ and for all } r \in R.$$

PROOF. First of all we will show that above rules for addition and scalar multiplication on M/N are well defined rules, i.e. are independent of the particular representative chosen to define a coset.

Let $a+N = a'+N$ and $b+N = b'+N$, where $a,b,a',b' \in M$. Then,
$$a+N = a'+N \Rightarrow a-a' \in N \text{ and } b+N = b'+N \Rightarrow b-b' \in N$$

Now,
$$a-a' \in N, b-b' \in N$$
$$\Rightarrow (a-a')+(b-b') \in N \qquad [\because (N,+) \text{ is a group}]$$
$$\Rightarrow (a+b)-(a'+b') \in N$$
$$\Rightarrow (a+b)+N = (a'+b')+N \qquad [\because (N,+) \text{ is subgroup of } (M,+)]$$
$$\Rightarrow (a+N)+(b+N) = (a'+N)+(b'+N)$$

Thus,
$$a+N = a'+N \text{ and } b+N = b'+N$$
$$\Rightarrow (a+N)+(b+N) = (a'+N)+(b'+N) \text{ for all } a,b,a',b' \in M.$$

So, above defined addition on M/N is well defined.

And,
$$r \in R, a-a' \in N$$
$$\Rightarrow r(a-a') \in N \qquad [\because N \text{ is an } R\text{-submodule}]$$
$$\Rightarrow ra-ra' \in N$$
$$\Rightarrow ra+N = ra'+N \qquad [\because (N,+) \text{ is subgroup of } (M,+)]$$

Therefore, scalar multiplication on M/N is well defined.

M/N is an abelian group under addition:

Associativity: For any $a+N, b+N, c+N \in M/N$, we have
$$\{(a+N)+(b+N)\}+(c+N) = \{(a+b)+N\}+(c+N)$$
$$= \{(a+b)+c\}+N$$
$$= \{a+(b+c)\}+N \qquad \text{[By associativity of addition on } M\text{]}$$
$$= (a+N)+((b+c)+N)$$
$$= (a+N)+\{(b+N)+(c+N)\}$$

So, addition is associative on M/N.

Commutativity: For any $a+N, b+N \in M/N$, we have
$$(a+N)+(b+N) = (a+b)+N$$
$$= (b+a)+N \qquad \text{[By commutativity of addition on } M\text{]}$$
$$= (b+N)+(a+N)$$

So, addition is commutative on M/N.

Existence of additive identity: Since $0_M \in M$, therefore, $0_M + N = N \in M/N$.

Now,
$$(a+N) + (0_M+N) = (a+0_M) + N = (a+N) \quad \text{for all } a+N \in M/N$$
$$\therefore \quad (a+N) + (0_M+N) = a+N = (0_M+N) + (a+N) \quad \text{for all } a+N \in M/N$$

So, $0_M + N = N$ is the identity element for addition on M/N.

Existence of additive inverse: Let $a+N$ be an arbitrary element in M/N. Then,
$$a \in M \Rightarrow -a \in M \Rightarrow (-a) + N \in M/N$$

Thus, for each $a+N \in M/N$ there exists $(-a) + N \in M/N$ such that
$$(a+N) + ((-a)+N) = [a+(-a)] + N = 0_M + N = N$$
$$\Rightarrow \quad (a+N) + ((-a)+N) = 0_M + N = ((-a)+N) + (a+N) \quad \text{[By comm. of add. on } M/N\text{]}$$

So, each $a+N \in M/N$ has its additive inverse $(-a+N) \in M/N$.

Hence, M/N is an additive abelian group.

For any $a+N, b+N \in M/N$ and $r, s \in R$, we have

(i) $\quad r[(a+N) + (b+N)] = r[(a+b) + N]$
$$= [r(a+b)] + N$$
$$= (ra+rb) + N \quad \text{[By M-2 (i) in } M\text{]}$$
$$= (ra+N) + (rb+N) \quad \text{[By def. of add. on } M/N\text{]}$$
$$= r(a+N) + r(b+N) \quad \text{[By def. of scalar mult. on } M/N\text{]}$$

(ii) $\quad (r+s)(a+N) = (r+s)a + N \quad \text{[By def. of scalar mult. on } M/N\text{]}$
$$= (ra+sa) + N \quad \text{[By M-2 (ii) in } M\text{]}$$
$$= (ra+N) + (sa+N) \quad \text{[By def. of add. on } M/N\text{]}$$
$$= r(a+N) + s(a+N) \quad \text{[By def. of scalar mult. on } M/N\text{]}$$

(iii) $\quad rs(a+N) = (rs)a + N$
$$= r(sa) + N \quad \text{[By M-2 (iii) in } M\text{]}$$
$$= r[sa+N]$$
$$= r[s(a+N)] \quad \text{[By def. of scalar mult. on } M/N\text{]}$$

Hence, M/N is an R-module.

Q.E.D.

QUOTIENT MODULE. *Let M be an R-module, and let N be an R-submodule of M. Then the set $M/N = \{a+N : a \in M\}$ is an R-module for the addition and scalar multiplication defined as follows:*

$$(a+N) + (b+N) = (a+b) + N$$
$$r(a+N) = ra+N$$

for all $a+N, b+N \in M/N$ and for all $r \in R$.

This R-module is called a *quotient module* (or a *factor module*) of M by N.

EXERCISE 1.2

1. Mark each of the following as true or false:

    (i) If M is an R-module over a ring R with unity 1, then
    $(-1)a = -a = a(-1)$ for all $a \in M$.

    (ii) Every module has exactly two submodules.

    (iii) Every module has at least two submodules.

    (iv) The union of two submodules of an R-module is a submodule.

2. Let M be an R-module. Then prove that
$$n(ra) = r(na) \quad \text{for all } n \in Z, r \in R \text{ and } a \in M.$$

3. Show that a left (right) ideal I in a ring R is a left (right) R-module and conversely.

4. Let M be an R-module, and let $RM = \left\{ \sum_i r_i a_i : r_i \in R, a_i \in M \right\}$. Then prove that RM is an R-submodule of M.

5. Which of the following are R-submodules of R^3?

    (i) $\{(a_1, a_2, a_3) : a_1 + a_2 + a_3 = 0\}$.

    (ii) $\{(a_1, a_2, a_3) : a_1 + a_2 + a_3 = 1\}$.

    (iii) $\{(a_1, a_2, a_3) : a_1 = 1\}$.

6. Let M be an R-module. Show that the set $\{r \in R : rM = \{0_M\}\}$ is a left ideal of R.

7. For any ring R, prove that an R-submodule of the R-module R is exactly the same thing as a left ideal of R.

8. Let R be a ring with unity, and let M be an R-module (not a unitary module). Then there exists $0_M \neq a \in M$ such that $ra = 0_M$ for all $r \in R$.

9. Let A, B and C be R-submodules of an R-module M such that $A \subset B$. Show that
$A \cap (B+C) = B + A \cap C$

    Give an example of three R-submodules A, B, C of an R-module M such that $A \cap (B+C) \neq A \cap B + A \cap C$.

10. Let $A \oplus B$ and $C \oplus D$ be direct sums of submodules of M such that $A \oplus B = C \oplus D$. Show that $A = C$ does not necessarily imply that $C = D$.

ANSWERS

1. (i) T (ii) F (iii) T (iv) F   5.(i)

1.6 R-HOMOMORPHISMS

After defining submodules and quotient modules, we now define homomorphisms from one module to another module. This is done only for modules over the same ring.

R-HOMOMORPHISM *Let M and N be R-modules. A mapping f from M to N is called an R-homomorphism (or an R-linear mapping or a linear transformation) of M into N, if*

(i) $f(a+b) = f(a) + f(b)$
(ii) $f(ra) = rf(a)$ for all $a, b \in M$ and all $r \in R$.

This definition stipulates that an R-homomorphism is a homomorphism of abelian groups.

The set of all R-homomorphisms of an R-module M into an R-module N is denoted by $Hom_R(M, N)$.

An R-homomorphism f is called an R-monomorphism, R-epimorphism and an R-isomorphism according as f is an injective, surjective and bijective. An R-homomorphism of an R-module M into itself is called an R-endomorphism of M, and the set $Hom_R(M, M)$ is also denoted by $End_R(M)$. An R-isomorphism from an R-module M onto itself is called an R-automorphism, and two R-modules are R-isomorphic if there is an R-isomorphism from one to other. If M and N are two R-isomorphic R-modules, then we write $M \cong N$. It can be easily shown that $\cong$ is an equivalence relation on the set of all R-modules.

EXAMPLE-1 *Let M be an R-module. Then the mapping $I_M : M \to M$ defined by*
$$I_M(a) = a \text{ for all } a \in M$$
is an R-homomorphism of M onto M.

SOLUTION For all $a, b, \in M$ and all $r \in R$, we have
$$I_M(a+b) = a+b = I_M(a) + I_M(b)$$
and,
$$I_M(ra) = ra = rI_M(a)$$

Further, $I_M : M \to M$ is surjective.

Hence, I_M is an R-homomorphism of M onto itself.

REMARK. I_M is called the identity endomorphism of M.

EXAMPLE-2 *Let M be an R-module. Then the mapping $\hat{0} : M \to M$ defined by $\hat{0}(a) = 0_M$ for all $a \in M$ is an R-homomorphism of M into M.*

SOLUTION For all $a, b \in M$ and all $r \in R$, we have
$$\hat{0}(a+b) = 0_M = 0_M + 0_M = \hat{0}(a) + \hat{0}(b)$$
and,
$$\hat{0}(ra) = 0_M = r(0_M) = r(\hat{0}(a)).$$

Hence, $\hat{0} : M \to M$ is an R-homomorphism.

This mapping $\hat{0} : M \to M$ is called the zero endomorphism of M.

EXAMPLE-3 *Let M be an R-module, and let r be some fixed element of R which is a commutative ring. Then the mapping $f : M \to M$ given by*
$$f(a) = ra \text{ for all } a \in M,$$
is an R-homomorphism of M into itself.

SOLUTION Let a, b be any two elements of M. Then,
$$f(a+b) = r(a+b) = ra + rb \qquad \text{[By } M\text{-2(i)]}$$
$$\Rightarrow \qquad f(a+b) = f(a) + f(b)$$
Let $a \in M$ and $s \in R$. Then,
$$f(sa) = r(sa) = (rs)a \qquad \text{[By } M\text{-2(iii)]}$$
$$\Rightarrow \qquad f(sa) = (sr)a \qquad [\because R \text{ is commutative ring}]$$
$$\Rightarrow \qquad f(sa) = s(ra) \qquad \text{[By } M\text{-2(iii)]}$$
$$\Rightarrow \qquad f(sa) = sf(a)$$
Hence, f is an R-homomorphism of M into itself.

THEOREM–1 *Let M and M' be two R-modules, and let $f : M \to M'$ be an R-homomorphism. Then,*

(i) $f(0_M) = 0_{M'}$
(ii) $f(-a) = -f(a)$ for all $a \in M$.
(iii) $f(a-b) = f(a) - f(b)$ for all $a, b \in M$.

PROOF. (i) We have

$$0_M + 0_M = 0_M$$
$$\Rightarrow \qquad f(0_M + 0_M) = f(0_M)$$
$$\Rightarrow \qquad f(0_M) + f(0_M) = f(0_M) \qquad [\because f \text{ is an } R\text{-homomorphism}]$$
$$\Rightarrow \qquad f(0_M) + f(0_M) = f(0_M) + 0_{M'} \qquad [\because 0_{M'} \in M' \text{ is the additive identity}]$$
$$\Rightarrow \qquad f(0_M) = 0_{M'} \qquad \text{[By cancellation laws in } M'\text{]}$$

(ii) For all $a \in M$, we have
$$a + (-a) = 0_M$$
$$\Rightarrow \qquad f(a + (-a)) = f(0_M)$$
$$\Rightarrow \qquad f(a) + f(-a) = 0_{M'}$$
$$\Rightarrow \qquad f(-a) = -f(a)$$
Thus, $f(-a) = -f(a)$ for all $a \in M$

(iii) For all $a, b \in M$, we have

$$f(a-b) = f(a+(-b))$$
$$\Rightarrow f(a-b) = f(a) + f(-b)$$
$$\Rightarrow f(a-b) = f(a) - f(b) \qquad \text{[From(ii)]}$$

Thus, $f(a-b) = f(a) - f(b)$ for all $a, b \in M$.

Q.E.D.

KERNEL OF R-HOMOMORPHISM Let M and M' be R-modules, and let $f : M \to M'$ be an R-homomorphism. Then, the set $K = \{x \in M : f(x) = 0_{M'}\}$ is called the kernel of f and is denoted by $\mathrm{Ker}(f)$.

By Theorem 1, $f(0_M) = 0_{M'}$. Therefore, $0_M \in \mathrm{Ker}(f)$.
Obviously, $\mathrm{Ker}(f) = f^{-1}\{0_{M'}\}$.

HOMOMORPHIC IMAGE Let $f : M \to M'$ be an R-homomorphism of an R-module M into an R-module M'. Then the set $f(M) = \{f(x) : x \in M\}$ is called the homomorphic image of M under f and is denoted by $I_m(f)$.

THEOREM-2 Let $f : M \to M'$ be an R-homomorphism of an R-module M into an R-module M'. Then,

(i) $\mathrm{Ker}(f) = \{a \in M : f(a) = 0_{M'}\}$ is an R-submodule of M,
(ii) $I_m(f) = \{f(a) : a \in M\}$ is an R-submodule of M'.

PROOF. (i) Since $f(0_M) = 0_{M'}$, therefore, $0_M \in \mathrm{Ker}(f)$. Consequently, $\mathrm{Ker}(f)$ is a non-void subset of M. Let a, b be arbitrary elements of $\mathrm{Ker}(f)$. Then,

$$f(a) = 0_{M'}, f(b) = 0_{M'}$$

$$\therefore \quad f(a-b) = f(a) - f(b) = 0_{M'} - 0_{M'} = 0_{M'}$$
$$\Rightarrow \quad a - b \in \mathrm{Ker}(f).$$

Thus, $a - b \in \mathrm{Ker}(f)$ for all $a, b \in \mathrm{Ker}(f)$.
Let $r \in R$ and $a \in \mathrm{Ker}(f)$. Then,

$$f(ra) = rf(a) \qquad [\because f \text{ is an } R\text{-homomorphism}]$$
$$\Rightarrow f(ra) = r0_{M'} \qquad [\because a \in \mathrm{Ker} \Rightarrow f(ra) = 0_{M'}]$$
$$\Rightarrow f(ra) = 0_{M'}$$
$$\Rightarrow ra \in \mathrm{Ker}(f)$$

Thus, $ra \in \mathrm{Ker}(f)$ for all $a \in \mathrm{Ker}(f)$ and all $r \in R$.
Hence, $\mathrm{Ker}(f)$ is an R-submodule of M.

(ii) Since $0_M \in M$, therefore, $f(0_M) = 0_{M'} \in I_m(f)$. Thus, $I_m(f)$ is a non-void subset of M'. Let a', b' be any two elements in $I_m(f)$. Then there exist $a, b \in M$ such that $f(a) = a'$ and $f(b) = b'$.

Now,
$$a' - b' = f(a) - f(b)$$
$\Rightarrow \quad a' - b' = f(a - b) \hfill [\because f \text{ is an } R\text{-homomorphism}]$

$\Rightarrow \quad a' - b' \in I_m(f) \hfill [\because a - b \in M]$

Thus, $a' - b' \in I_m(f)$ for all $a', b' \in I_m(f)$.

Let $r \in R$ and $a' \in I_m(f)$. Then there exists $a \in M$ such that $f(a) = a'$.

Now, $\quad ra' = rf(a) = f(ra) \hfill [\because f \text{ is an } R\text{-homomorphism}]$

$\Rightarrow \quad ra' \in I_m(f) \hfill [\because ra \in M]$

Thus, $ra' \in I_m(f)$ for all $a' \in I_m(f)$ and all $r \in R$.

Hence, $I_m(f)$ is an R-submodule of M'.

Q.E.D.

THEOREM-3 *If $f : M \to M'$ is an R-homomorphism of an R-module M into an R-module M'. Then, f is an R-monomorphism iff $\mathrm{Ker}(f) = \{0_M\}$.*

PROOF. First let f be an R-monomorphism of M into M', and let a be an arbitrary element of $\mathrm{Ker}(f)$. Then,
$$f(a) = 0_{M'}$$
$\Rightarrow \quad f(a) = f(0_M) \hfill [\because f(0_M) = 0_{M'}]$

$\Rightarrow \quad a = 0_M \hfill [\because f \text{ is an injective map}]$

Since a is an arbitrary element of $\mathrm{Ker} f$.

$\therefore \qquad\qquad\qquad a \in \mathrm{Ker}(f) \Rightarrow a = 0_M \quad \text{for all } a \in \mathrm{Ker}(f).$

Hence, $\qquad\qquad\qquad \mathrm{Ker}(f) = \{0_M\}.$

Conversely, let f be an R-homomorphism of an R-module M into an R-module M' such that $\mathrm{Ker}(f) = \{0_M\}$. We have to prove that f is an R-monomorphism.

Let a, b be any two elements of M such that $f(a) = f(b)$. Then,
$$f(a) - f(b) = 0_{M'}$$
$\Rightarrow \quad f(a - b) = 0_{M'} \hfill [\text{By Theorem 1(iii)}]$

$\Rightarrow \quad a - b \in \mathrm{Ker}(f)$

$\Rightarrow \quad a - b = 0_M \hfill [\because \mathrm{Ker} f = \{0_M\}]$

$\Rightarrow \quad a = b.$

Thus, $f(a) = f(b) \Rightarrow a = b$ for all $a, b \in M$.

Hence, f is an R-monomorphism. Q.E.D.

THEOREM-4 Let $f : M \to M'$ be an R-homomorphism. Then, f is an R-epimorphism iff $I_m(f) = M'$.

PROOF. Left as an exercise for the reader.

THEOREM-5 Let M and M' be R-modules, and let $f : M \to M'$ be an R-homomorphism. If N is an R-submodule of M, then $f(N) = \{f(a) : a \in N\}$ is an R-submodule of M'.

PROOF. Since N is an R-submodule of M. Therefore,

$$0_M \in N$$
$$\Rightarrow f(0_M) \in f(N)$$
$$\Rightarrow 0_{M'} \in f(N) \qquad [\because \ f(0_M) = 0_{M'}]$$
$$\Rightarrow f(N) \text{ is non-void.}$$

Also, $f(N) \subset I_m(f) \subset M'$.

Thus, $f(N)$ is a non-void subset of M'.

Let a', b' be arbitrary elements of $f(N)$. Then there exist $a, b \in N$ such that $f(a) = a', f(b) = b'$.

Now,

$$\Rightarrow a' - b' = f(a) - f(b)$$
$$\Rightarrow a' - b' = f(a - b) \qquad \text{[By Theorem 1(iii)]}$$
$$\Rightarrow a' - b' \in f(N) \qquad [\because \ N \text{ is an } R\text{-submodule} \ \therefore \ a - b \in N \Rightarrow f(a - b) \in f(N)]$$

Thus, $a', b' \in f(N) \Rightarrow a' - b' \in f(N)$.

Let r be an arbitrary element of R and let a' be an arbitrary element of $f(N)$. Then there exists $a \in N$ such that $f(a) = a'$.

Since N is an R-submodule of M. Therefore,

$$r \in R, \quad a \in N$$
$$\Rightarrow ra \in N$$
$$\Rightarrow f(ra) \in f(N)$$
$$\Rightarrow rf(a) \in f(N) \qquad [\because \ f \text{ is an } R\text{-homomorphism}]$$
$$\Rightarrow ra' \in f(N)$$

Thus, $r \in R$, $a' \in f(N) \Rightarrow ra' \in f(N)$.
Hence, $f(N)$ is an R-submodule of M'.

Q.E.D.

COROLLARY. *Let M and M' be R-modules, and let $f : M \to M'$ be an R-homomorphism. Then, $I_m(f)$ is an R-submodule of M'.*

THEOREM-6 *Let M and M' be R-modules, and let $f : M \to M'$, $g : M \to M'$ be R-homomorphisms. Then, $f + g : M \to M'$ given by $(f+g)(a) = f(a) + g(a)$ for all $a \in M$, is also an R-homomorphism.*

PROOF. Let a, b be arbitrary elements of M. Then,

$$\begin{aligned}
(f+g)(a+b) &= (f+g)(a) + (f+g)(b) \\
&= [f(a) + g(a)] + [f(b) + g(b)] \\
&= [f(a) + f(b)] + [g(a) + g(b)] \quad \text{[By comm. and assoc. of add. on } M'] \\
&= f(a+b) + g(a+b) \quad \text{[By definition of } f+g]
\end{aligned}$$

Thus, $(f+g)(a+b) = f(a+b) + g(a+b)$ for all $a, b \in M$.

Let r be an arbitrary element R and let a be an arbitrary element of M. Then,

$$\begin{aligned}
(f+g)(ra) &= f(ra) + g(ra) \quad &&\text{[By definition of } f+g] \\
&= rf(a) + rg(a) \quad &&[\because f \text{ and } g \text{ are } R\text{-homomorphisms}] \\
&= r[f(a) + g(a)] \quad &&[\text{By } M'-2(i)] \\
&= r(f+g)(a) \quad &&[\text{By definition of } f+g]
\end{aligned}$$

Thus, $(f+g)(ra) = r(f+g)(a)$ for all $r \in R$ and all $a \in M$.
Hence, $f + g$ is an R-homomorphism of M into M'.

Q.E.D.

SUM OF R-HOMOMORPHISMS *Let M and M' be R-modules and let $f : M \to M'$, $g : M \to M'$ be R-homomorphisms. Then, the R-homomorphism $f + g : M \to M'$ is called the sum of f and g.*

REMARK. *The sum $f + g$ of R-homomorphisms f and g is also called the pointwise sum of R-homomorphisms f and g.*

THEOREM-7 *Let M and M' be R-modules. Then the set $Hom_R(M, M')$ is an abelian group under pointwise sum of R-homomorphisms.*

PROOF. Let $f : M \to M'$, $g : M \to M'$ be any two R-homomorphisms. Then as proved in Theorem 6, $f + g$ is an R-homomorphism. Thus, the pointwise sum of R-homomorphisms is a binary operation on $Hom_R(M, M')$.
We now observe the following properties:

Associativity: Let f, g, h be any three R-homomorphisms in $\text{Hom}_R(M, M')$. Then,

$$[(f+g)+h](a) = (f+g)(a) + h(a) \quad \text{[By definition of pointwise sum]}$$
$$= \{f(a)+g(a)\} + h(a) \quad \text{[By definition of pointwise sum]}$$
$$= f(a) + \{g(a)+h(a)\} \quad \text{[By associativity of addition on } M']$$
$$= f(a) + (g+h)(a) \quad \text{[By definition of pointwise sum]}$$
$$= [f+(g+h)](a) \quad \text{for all } a \in M.$$

Thus, $(f+g)+h = f+(g+h)$ for all $f, g, h \in \text{Hom}_R(M, M')$.

So, pointwise sum is an associative binary operation on $\text{Hom}_R(M, M')$.

Commutativity: Let f, g be any two R-homomorphisms in $\text{Hom}_R(M, M')$. Then for all $a \in M$, we have

$$(f+g)(a) = f(a) + g(a) \quad \text{[By definition of pointwise sum]}$$
$$\Rightarrow \quad (f+g) = g(a) + f(a) \quad \text{[By commutativity of addition on } M']$$
$$\Rightarrow \quad (f+g) = (g+f)(a) \quad \text{[By definition of pointwise sum]}$$
$$\therefore \quad f+g = g+f \quad \text{[Using definition of equality of two functions]}$$

So, pointwise sum is a commutative binary operation on $\text{Hom}_R(M, M')$.

Existence of identity: The mapping $\hat{0} : M \to M'$ given by $\hat{0}(a) = 0_{M'}$ for all $a \in M$ is an R-homomorphism. Also, for any $f \in \text{Hom}_R(M, M')$, we have

$$(\hat{0}+f)(a) = \hat{0}(a) + f(a) \quad \text{[By definition of pointwise sum]}$$
$$\Rightarrow \quad (\hat{0}+f)(a) = 0_{M'} + f(a)$$
$$\Rightarrow \quad (\hat{0}+f)(a) = f(a) \quad \text{for all } a \in M. \quad [\because 0_M \in M \text{ is the additive identity}]$$
$$\therefore \quad \hat{0}+f = f$$
$$\Rightarrow \quad \hat{0}+f = f = f+\hat{0} \quad \text{[By commutativity of pointwise sum]}$$

Thus, $\hat{0} \in \text{Hom}_R(M, M')$ is such that $f + \hat{0} = f = \hat{0} + f$ for all $f \in \text{Hom}_R(M, M')$.
So, $\hat{0}$ is the additive identity.

Existence of inverse: Let f be an arbitrary element of $\text{Hom}_R(M, M')$. Consider the mapping $-f : M \to M'$ defined by $(-f)(a) = -f(a)$ for all $a \in M$. It can be easily seen that $-f \in \text{Hom}_R(M, M')$. Also, for all $a \in M$, we have

$$[f+(-f)](a) = f(a) + (-f)(a) = f(a) - f(a) \quad \text{[By Theorem 1(ii) on page 59]}$$
$$\Rightarrow \quad [f+(-f)](a) = 0_M = \hat{0}(a) \quad \text{for all } a \in M$$
$$\therefore \quad f+(-f) = \hat{0}$$
$$\Rightarrow \quad f+(-f) = \hat{0} = (-f)+f \quad \text{[By commutativity of pointwise sum]}$$

Thus, for each $f \in \text{Hom}_R(M, M')$ there exists $-f \in \text{Hom}_R(M, M')$ such that

$$f+(-f) = \overline{0} = (-f)+f.$$

So, each element in $\text{Hom}_R(M, M')$ possesses its inverse in $\text{Hom}_R(M, M')$.
Hence, $\text{Hom}_R(M, M')$ is an abelian group under the pointwise sum of R-homomorphisms.

Q.E.D.

THEOREM-8 *Let M_1, M_2, M_3 and M_4 be R-modules, and let $f_1 : M_1 \to M_2, f_2 : M_1 \to M_2$, $f_3 : M_2 \to M_3$ and $f_4 : M_4 \to M_1$ be R-homomorphisms. Then,*

(i) $f_3 o (f_1 + f_2) = f_3 o f_1 + f_3 o f_2$

(ii) $(f_1 + f_2) o f_4 = f_1 o f_4 + f_2 o f_4$.

PROOF. (i) Since f_1, f_2, f_3 and f_4 are R-homomorphisms. Therefore, $f_1 + f_2, f_3 o (f_1 + f_2)$, $(f_1 + f_2) o f_4, f_3 o f_1, f_3 o f_2, f_1 o f_4, f_2 o f_4, f_3 o f_1 + f_3 o f_2$ and $f_1 o f_4 + f_2 o f_4$ are also R-homomorphisms.

For all $a \in M_1$, we have

$$[f_3 o (f_1 + f_2)](a) = f_3[(f_1 + f_2)(a)] \quad \text{[By the def. of composition of functions]}$$
$$= f_3[f_1(a) + f_2(a)] \quad \text{[By the def. of pointwise sum]}$$
$$= f_3(f_1(a)) + f_3(f_2(a)) \quad [\because f_3 : M_2 \to M_3 \text{ is an } R\text{-homomorphism}]$$
$$= (f_3 o f_1)(a) + (f_3 o f_2)(a)$$
$$\therefore \quad f_3 o (f_1 + f_2) = f_3 o f_1 + f_3 o f_2 \quad \text{[By the equality of two functions]}$$

(ii) For all $a \in M_4$, we have

$$[(f_1 + f_2) o f_4](a) = (f_1 + f_2)(f_4(a)) \quad \text{[By def. of composition of functions]}$$
$$= f_1(f_4(a)) + f_2(f_4(a)) \quad \text{[By def. of pointwise sum]}$$
$$= (f_1 o f_4)(a) + (f_2 o f_4)(a)$$
$$\therefore \quad (f_1 + f_2) o f_4 = f_1 o f_4 + f_2 o f_4.$$

Q.E.D.

This theorem stipulates that the composition of functions is distributive over the pointwise sum and we know that the composition of functions is an associative operation. Therefore, from the preceding two theorems, we obtain the following theorem.

THEOREM-9 *Let M be an R-module. Then the set $\text{End}_R(M)$ of all endomorphisms of M is a ring with unity under the pointwise sum and composition of mappings as the ring operations.*

Let A be an arbitrary abelian group. Then, A is a Z-module and therefore $\text{End}_Z(A)$ is a ring with unity under the pointwise sum and the composition of functions as the ring operations. Now let R be an arbitrary ring, and let M be an R-module. Then M is an additive abelian group. But, we know that every abelian group is a Z-module. Therefore, M is a Z-module and consequently an R-homomorphism is a Z-homomorphism. Thus, $\text{End}_R(M)$ is a subring of $\text{End}_Z(M)$.

THEOREM-10 *Let R be a commutative ring, and let $f : M \to M'$ be an R-homomorphism from an R-module M to an R-module M'. Then for each $r \in R$ the pointwise multiple rf defined by $(rf)(a) = rf(a)$ for all $a \in M$, is also an R-homomorphism.*

PROOF. Let a be an arbitrary element of M. Then,

$\qquad f(a) \in M'$ $\qquad\qquad\qquad\qquad\qquad$ [$\because$ $f : M \to M'$ is an R-homomorphism]

$\Rightarrow \qquad rf(a) \in M' \quad$ for all $r \in R$ $\qquad\qquad\qquad\qquad$ [$\because$ M' is an R-module]

$\Rightarrow \qquad (rf)(a) \in M' \quad$ for all $r \in R$ $\qquad\qquad\qquad\qquad$ [$\because$ $(rf)(a) = rf(a)$]

Thus, for each $r \in R$, $rf : M \to M'$.

Let a, b be any two elements of M. Then,

$\qquad (rf)(a+b) = r[f(a+b)]$ $\qquad\qquad\qquad\qquad$ [$\because$ $(rf)(x) = r(f(x))$]

$\qquad\qquad\qquad = r[f(a) + f(b)]$ $\qquad\qquad$ [$\because$ $f : M \to M'$ is an R-homomorphism]

$\qquad\qquad\qquad = rf(a) + rf(b)$ $\qquad\qquad\qquad\qquad\qquad$ [By M'-2 (i)]

$\qquad\qquad\qquad = (rf)(a) + (rf)(b)$ $\qquad\qquad\qquad$ [By def. of $rf : M \to M'$]

$\therefore \quad (rf)(a+b) = (rf)(a) + (rf)(b)$ $\qquad\qquad\qquad\qquad\qquad$ for all $a, b \in M$.

Now, let s be an arbitrary element of R and a be an arbitrary element of M. Then,

$\qquad (rf)(sa) = r[f(sa)]$ $\qquad\qquad\qquad\qquad\qquad$ [By def. of $rf : M \to M'$]

$\qquad\qquad\quad = r[sf(a)]$ $\qquad\qquad\qquad\qquad\qquad$ [$\because$ f is an R-homomorphism]

$\qquad\qquad\quad = (rs)f(a)$ $\qquad\qquad\qquad\qquad\qquad\qquad$ [By M'-2 (iii)]

$\qquad\qquad\quad = (sr)f(a)$ $\qquad\qquad\qquad\qquad\qquad\qquad$ [$\because$ R is commutative]

$\qquad\qquad\quad = s[rf(a)]$ $\qquad\qquad\qquad\qquad\qquad\qquad$ [By M'-2 (iii)]

$\qquad\qquad\quad = s[(rf)(a)]$ $\qquad\qquad\qquad\qquad\qquad$ [By def. of $rf : M \to M'$]

Thus, $\quad (rf)(sa) = s[(rf)(a)] \quad$ for all $a \in M$.

Hence, $rf : M \to M'$ is an R-homomorphism for all $r \in R$. $\qquad\qquad\qquad\qquad$ Q.E.D.

REMARK. *This theorem is valid only for commutative rings and fails for ordinary rings.*

In the set $\text{Hom}_R(M, M')$ of all R-homomorphisms of M into M', we have defined *pointwise addition* and *pointwise multiplication* (also known as scalar multiplication) and in Theorem 7 we have proved that $\text{Hom}_R(M, M')$ is an abelian group under pointwise sum. The following theorem proves that $\text{Hom}_R(M, M')$ is an R-module if R is a commutative ring.

THEOREM-11 *Let R be a commutative ring and let M and M' be R-modules. Then the set $\text{Hom}_R(M, M')$ is an R-module under the pointwise sum as addition on $\text{Hom}_R(M, M')$ and pointwise multiplication as scalar multiplication on $\text{Hom}_R(M, M')$.*

PROOF. By Theorem 7, $\text{Hom}_R(M, M')$ is an abelian group under pointwise sum. Therefore, in order to prove that $\text{Hom}_R(M, M')$ is an R-module, it is sufficient to show that for any $f, g \in \text{Hom}_R(M, M')$ and for any $r, s \in R$

(i) $r(f+g) = rf + rg$
(ii) $(r+s)f = rf + sf$
(iii) $(rs)f = r(sf)$

We shall now verify these one by one.

(i) For all $a \in M$, we have

$$\begin{aligned}
[r(f+g)](a) &= r(f+g)(a) && \text{[By def. of scalar multiplication]} \\
&= r(f(a) + g(a)) && \text{[By def. of pointwise sum]} \\
&= rf(a) + rg(a) && \text{[By } M'\text{-2 (i)]} \\
&= (rf)(a) + (rg)(a) && \text{[By def. of scalar multiplication]} \\
&= [rf + rg](a) && \text{[By def. of pointwise sum]}
\end{aligned}$$

$\therefore \quad r(f+g) = rf + rg$ \hfill [By def. of equality of functions]

(ii) Let a be an arbitrary element of M, Then,

$$\begin{aligned}
[(r+s)f](a) &= [(r+s)f(a)] && \text{[By def. of scalar multiplication]} \\
&= [rf(a) + sf(a)] && \text{[By } M'\text{-2(ii)]} \\
&= (rf)(a) + (sf)(a) && \text{[By def. of scalar multiplication]} \\
&= [rf + sf](a)
\end{aligned}$$

$\therefore \quad [(r+s)f](a) = [rf + sf](a) \quad$ for all $a \in M$.

$\Rightarrow \quad (r+s)f = rf + sf$ \hfill [By def. of equality of functions]

(iii) For any $a \in M$, we have

$$\begin{aligned}
[(rs)f](a) &= (rs)f(a) && \text{[By def. of scalar multiplication]} \\
&= r[s(f(a))] && \text{[By } M'\text{-2(iii)]} \\
&= r[(sf)(a)] && \text{[By def. of scalar multiplication]} \\
&= [r(sf)](a) && \text{[By def. of scalar multiplication]}
\end{aligned}$$

$\therefore \quad (rs)f = r(sf)$

Hence $\text{Hom}_R(M, M')$ is an R-module.

Q.E.D.

THEOREM-12 Let R be a commutative ring and M be an R-module. Prove that $\text{End}_R(M)$ (set of all endomorphisms of M) is a subring of $\text{End}(M)$ (set of all endomorphisms of M regarded M as an abelian group).

PROOF. We know that for any abelian group M, the set of all endomorphisms of group M into itself forms a ring under the pointwise sum and composition of functions as ring operations defined as follows:

(i) $(f+g)(x) = f(x) + g(x)$

(ii) $(fg)(x) = f(g(x))$ for all $f, g \in \text{End}(M)$ and for all $x \in M$.

Clearly, $\text{End}_R(M) \subset \text{End}(M)$.

Let $f, g \in \text{End}_R(M), x \in M$ and $r \in R$. Then,

$$(f-g)(rx) = f(rx) - g(rx) = rf(x) - rg(x) = r(f(x) - g(x)) = r\{(f-g)(x)\}$$

and,
$$(fg)(rx) = f(g(rx)) = f(r\,g(x)) = r(f(g(x))) = r\{(fg)(x)\}$$

$\therefore \quad f - g, fg \in \text{End}_R(M)$.

Hence, $\text{End}_R(M)$ is a subring of $\text{End}(M)$.

Q.E.D.

THEOREM-13 *Let R be a commutative ring with unity and let $\text{Hom}_R(R, R)$ denote the ring of endomorphisms of ring R regarded as an R-module. Then, $R \cong \text{Hom}_R(R, R)$ as rings.*

PROOF. Let a be an arbitrary element of R. Consider the mapping $f_a : R \to R$ given by

$$f_a(x) = ax \quad \text{for all } x \in R$$

Let us first show that f_a is an R-homomorphism of R-module R into itself.

For any $x, y, r \in R$, we have

and,
$$f_a(x+y) = a(x+y) = ax + ay = f_a(x) + f_a(y)$$
$$f_a(rx) = a(rx) = (ar)x = (ra)x \quad [\because R \text{ is commutative}]$$
$$= r(ax) = rf_a(x)$$

Thus, f_a is an R-homomorphism of R-module R into itself, i.e. $f_a \in \text{Hom}_R(R, R)$.

Let $a, b \in R$. Then, for any $x \in R$

$$f_{(a+b)}(x) = (a+b)x = ax + bx = f_a(x) + f_b(x) = (f_a + f_b)(x)$$
$$\Rightarrow \quad f_{a+b} = f_a + f_b$$

and,
$$f_{ab}(x) = (ab)x = a(bx) = f_a(bx) = f_a(f_b(x)) = f_a \circ f_b(x)$$
$$\Rightarrow \quad f_{ab} = f_a \circ f_b$$

Consider the mapping $\phi : R \to \text{Hom}_R(R, R)$ given by

$$\phi(a) = f_a \text{ for all } a \in R.$$

We shall now show that ϕ is a ring isomorphism.

ϕ *is one-one*: Let $a, b \in R$. Then,
$$\phi(a) = \phi(b)$$
$\Rightarrow \quad f_a = f_b$
$\Rightarrow \quad f_a(x) = f_b(x) \quad \text{for all } x \in R$
$\Rightarrow \quad ax = bx \quad \text{for all } x \in R$
$\Rightarrow \quad (a-b)x = 0 \quad \text{for all } x \in R$
$\Rightarrow \quad a - b = 0$
$\Rightarrow \quad a = b.$

So, ϕ is one-one.

ϕ *is onto*: Let $g \in \text{Hom}_R(R,R)$. Then, for any $x \in R$, we have
$$g(x) = g(x1) = xg(1) \qquad [\because \; g: R \to R \text{ is an endomorphism of } R\text{-modules}]$$
$\Rightarrow \quad g(x) = xa, \quad \text{where} \quad a = g(1)$
$\Rightarrow \quad g(x) = ax \qquad\qquad\qquad\qquad\qquad\qquad\qquad\qquad [\because \; R \text{ is commutative}]$
$\Rightarrow \quad g(x) = f_a(x)$
$\therefore \quad g = f_a = \phi(a)$

Thus, for each $g \in \text{Hom}_R(R,R)$ there exists $a = g(1) \in R$ such that $\phi(a) = g$.
So, ϕ is onto.

ϕ *is a ringhomomorphism*: Let $a, b \in R$. Then for any $x \in R$, we have
$$\phi(a+b) = f_{a+b} = f_a + f_b = \phi(a) + \phi(b)$$
and,
$$\phi(ab) = f_{ab} = f_a \circ f_b = \phi(a) \circ \phi(b)$$

So, ϕ is a ring homomorphism.

Hence, $R \cong \text{Hom}_R(R, R)$.

THEOREM-14 *Let N be an R-submodule of an R-module M. Then the mapping $p: M \to M/N$ given by $p(a) = a + N$ for all $a \in M$, is an R-epimorphism.*

PROOF. Let a, b be any two elements of M. Then,
$$p(a+b) = (a+b) + N$$
$\Rightarrow \quad p(a+b) = (a+N) + (b+N) \qquad$ [By definition of addition on M/N]
$\Rightarrow \quad p(a+b) = p(a) + p(b)$

Also, for any $r \in R$, we have
$$p(ra) = ra + N = r(a+N) \qquad \text{[By definition of scalar multiplication on } M/N\text{]}$$
$\Rightarrow \quad p(ra) = rp(a)$

70 • *Theory and Problems of Linear Algebra*

Thus, $p : M \to M/N$ is an R-homomorphism.

Obviously, p is surjective, because for each $a + N \in M/N$, there exists $a \in M$ such that $p(a) = a + N$.

Hence, $p : M \to M/N$ is an R-epimorphism.

Q.E.D.

PROJECTION MAPPING *Let N be an R-submodule of an R-module M. Then the mapping $p : M \to M/N$ given by $p(a) = a + N$ for all $a \in M$, is called projection mapping (or natural mapping or canonical mapping).*

THEOREM-15 *Let N be an R-submodule of an R-module M, and let $p : M \to M/N$ be the projection mapping. Then, $\operatorname{Ker}(p) = N$.*

PROOF. Let a be an arbitrary element of $\operatorname{Ker}(p)$. Then,

$$a \in \operatorname{Ker}(p) \Leftrightarrow p(a) = N \Leftrightarrow a + N = N \Leftrightarrow a \in N$$

Hence, $\operatorname{Ker}(p) = N$.

Q.E.D.

THEOREM-16 *(Main Theorem for Quotient Modules). Let M and M' be R-modules, and let N be an R-submodule of M. Then for each R-homomorphism $f : M \to M'$ with $\operatorname{Ker}(f) \supset N$ there exists a unique R-homomorphism $\varphi : M/N \to M'$ such that $\varphi \circ p = f$, where $p : M \to M/N$ is the projection mapping. In other words, the diagram in Fig.1 is commutative.*

Moreover, $I_m(\varphi) = I_m(f)$ and $\operatorname{Ker}(\varphi) = \operatorname{Ker}(f)/N$.

PROOF. Consider the mapping $\varphi : M/N \to M'$ defined by

$$\varphi(a + N) = f(a) \quad \text{for all } a \in M.$$

We observe that φ is well defined, because

$$a+N = b+N$$
$\Rightarrow \quad a-b \in N$
$\Rightarrow \quad a-b \in \text{Ker}(f)$ $[\because \text{Ker}(f) \supset N]$
$\Rightarrow \quad f(a-b) = 0_{M'}$
$\Rightarrow \quad f(a) - f(b) = 0_{M'}$ $[\because f \text{ is an } R\text{-homomorphism}]$
$\Rightarrow \quad f(a) = f(b)$
$\Rightarrow \quad \varphi(a+N) = \varphi(b+N)$

We shall now show that φ is an R-homomorphism.

Let $a+N, b+N$ be any two arbitrary elements of M/N. Then,

$$\varphi[(a+N)+(b+N)] = \varphi[(a+b)+N] \quad \text{[By definition of addition on } M/N\text{]}$$
$$= f(a+b) \quad \text{[By definition of } \varphi\text{]}$$
$$= f(a) + f(b) \quad [\because f \text{ is an } R\text{-homomorphism}]$$
$$= \varphi(a+N) + \varphi(b+N)$$

Now, let $a+N$ be an arbitrary element of M/N and let r be an arbitrary element of R. Then
$$\varphi[r(a+N)] = \varphi(ra+N) \quad \text{[By definition of scalar multiplication on } M/N\text{]}$$
$$= f(ra) \quad \text{[By definition of } \varphi\text{]}$$
$$= rf(a) \quad [\because f \text{ is an } R\text{-homomorphism}]$$
$$= r\varphi(a+N)$$

Thus,
$$\varphi[(a+N)+(b+N)] = \varphi(a+N) + \varphi(b+N)$$
and, $\quad \varphi[r(a+N)] = r\varphi(a+N)$ for all $a+N, b+N \in M/N$ and all $r \in R$.

Hence, $\varphi : M/N \to M'$ is an R-homomorphism.

Since $\varphi : M/N \to M'$ and $p : M \to M/N$. Therefore, $\varphi \circ p : M \to M'$.

Also, $(\varphi \circ p)(a) = \varphi[p(a)] = \varphi(a+N) = f(a)$ for all $a \in M$.

So, $\quad \varphi \circ p = f$ [By the definition of equality of functions]

Uniqueness of φ: If possible, let $\psi : M/N \to M'$ be an R-homomorphism such that $\psi \circ p = f$. Then,
$$\psi \circ p\, (a) = f(a) \quad \text{for all } a \in M$$
$\Rightarrow \quad \psi[p(a)] = f(a)$ for all $a \in M$
$\Rightarrow \quad \psi(a+N) = f(a)$ for all $a \in M$
$\Rightarrow \quad \psi(a+N) = \varphi(a+N)$ for all $a \in M$
$\Rightarrow \quad \psi = \varphi$

Hence, for each R-homomorphism $f : M \to M'$ with $\text{Ker}(f) \supset N$ there exists a unique R-homomorphism $\varphi : M/N \to M'$ such that $\varphi \circ p = f$.

Now, let a' be an arbitrary element of $I_m(\varphi)$. Then,

$a' \in I_m(\varphi) \Leftrightarrow$ There exists $a+N \in M/N$ such that $\varphi(a+N) = a'$.
$\qquad \Leftrightarrow$ There exists $a \in M$ such that $\varphi[p(a)] = a'$ $\qquad$ $[p : M \to M/N$ is onto$]$
$\qquad \Leftrightarrow \varphi \, op(a) = a'$
$\qquad \Leftrightarrow f(a) = a'$ $\qquad\qquad\qquad\qquad\qquad\qquad\qquad\qquad$ $[\because \quad \varphi \, op = f]$
$\qquad \Leftrightarrow a' \in I_m(f)$

$\therefore \quad I_m(\varphi) = I_m(f)$

Since $\text{Ker}(f)$ and N are R-submodules of M such that $\text{Ker}(f) \supset N$. Therefore, the symbol $\text{Ker}(f)/N$ is meaningful.

Now,

$\qquad\qquad a+N \in \text{Ker}(\varphi)$
$\Leftrightarrow \qquad \varphi(a+N) = 0_{M'}$
$\Leftrightarrow \qquad \varphi[p(a)] = 0_{M'}$
$\Leftrightarrow \qquad \varphi \, op(a) = 0_{M'}$
$\Leftrightarrow \qquad f(a) = 0_{M'}$ $\qquad\qquad\qquad\qquad\qquad\qquad\qquad\qquad$ $[\because \quad \varphi \, op = f]$
$\Leftrightarrow \qquad a \in \text{Ker}(f)$
$\Leftrightarrow \qquad a+N \in \text{Ker}(f)/N$

$\therefore \quad \text{Ker}(\varphi) = \text{Ker}(f)/N$.

$\hfill$ Q.E.D.

THEOREM-17 *(Fundamental Theorem of R-homomorphisms) Let f be an R-homomorphism of an R-module M into an R-module M' with kernel N. Then, $M/N \cong I_m(f)(= f(M))$.*

or

Every R-homomorphic image of an R-module M is isomorphic to some quotient module of M.

PROOF. Let f be an R-homomorphism of an R-module M into an R-module M' with Kernel N. Then by Theorem 2, N and $I_m(f)$ are R-submodules of M and M' respectively.

If $a \in M$, then $a+N \in M/N$ and $f(a) \in I_m(f)$.

Consider the mapping $\varphi : M/N \to I_m(f)$ given by

$$\varphi(a+N) = f(a) \text{ for all } a \in M.$$

First of all we will show that φ is well defined, i.e., if $a,b \in M$ such that $a+N = b+N$, then $\varphi(a+N) = \varphi(b+N)$.

Now, $\qquad a+N = b+N$
$\Rightarrow \qquad a-b \in N$ $\qquad\qquad\qquad\qquad\qquad\qquad\qquad\qquad\qquad$ $[\because \quad (N,+)$ is a group$]$

$$\Rightarrow \quad f(a-b) = 0_{M'} \qquad [\because\ N = \operatorname{Ker}(f)]$$
$$\Rightarrow \quad f(a) - f(b) = 0_{M'}$$
$$\Rightarrow \quad f(a) = f(b)$$
$$\Rightarrow \quad \varphi(a+N) = \varphi(b+N).$$

So, φ is well defined.

We shall now show that $\varphi : M/N \to I_m(f)$ is an isomorphism.

φ *is injective:* Let $a+N, b+N$ be any two arbitrary elements of M/N. Then,
$$\varphi(a+N) = \varphi(b+N)$$
$$\Rightarrow \quad f(a) = f(b)$$
$$\Rightarrow \quad f(a) - f(b) = 0_{M'}$$
$$\Rightarrow \quad f(a-b) = 0_{M'} \qquad [\because\ f(a-b) = f(a) - f(b)]$$
$$\Rightarrow \quad a - b \in N \qquad [\because\ N = \operatorname{Ker}(f)]$$
$$\Rightarrow \quad a + N = b + N$$

So, φ is an injective.

φ *is surjective*: Let a' be an arbitrary element of $I_m(f)$. Then there exists $a \in M$ such that $f(a) = a'$. But, $f(a)$ is image of $a+N$ under $\varphi : M/N \to M'$. Thus, for each $a' \in I_m(f)$ there exists $a+N \in M/N$ such that
$$\varphi(a+N) = f(a) = a'$$

Consequently, φ is surjective.

φ *is an R-homomorphism*: Let $a+N$, $b+N$ be any two arbitrary elements of M/N and let r be an arbitrary element of R. Then,
$$\varphi[(a+N) + (b+N)] = \varphi[(a+b) + N] \qquad \text{[By def. of add. on } M/N]$$
$$= f(a+b) \qquad \text{[By def. of } \varphi]$$
$$= f(a) + f(b)$$
$$= \varphi(a+N) + \varphi(b+N)$$

and,
$$\varphi[r(a+N)] = \varphi(ra+N) \qquad \text{[By def. of scalar multiplication on } M/N]$$
$$= f(ra) \qquad \text{[By def. of } \varphi]$$
$$= rf(a) \qquad [\because\ f \text{ is an } R\text{-homomorphism}]$$
$$= r\varphi(a+N)$$

Therefore, φ is an R-homomorphism.

Hence, $\varphi : M/N \to I_m(f)$ is an R-isomorphism. Consequently, $M/N \cong I_m(f)$.

Q.E.D.

THEOREM-18 *(Another form of the fundamental theorem of R-homomorphisms.) Let M and M' be R-modules, and let φ be an R-epimorphism of M onto M' with kernel N. Then there* exists a unique R-isomorphism ψ of M/N onto M' such that $\psi \circ p = \varphi$, where $p : M \to M/N$ is the projection mapping.

PROOF. We have, $\text{Ker}(\varphi) = N$. Therefore, $\varphi(a) = 0_{M'}$ for all $a \in M$.

Let a be an arbitrary element of M. Then, $a + N \in M/N$.

For any $a + x \in a + N$, we have

$$\varphi(a+x) = \varphi(a) + \varphi(x) \qquad [\because \quad \varphi : M \to M' \text{ is an } R\text{-homomorphism}]$$
$$= \varphi(a) + 0_{M'} \qquad [\because \quad x \in N = \text{Ker}(\varphi) \therefore \quad \varphi(x) = 0_{M'}]$$
$$= \varphi(a)$$

Therefore, φ associates each $a + x \in a + N$ to a unique element $\varphi(a) \in I_m(\varphi)$. Thus, $\varphi(a + N) = \varphi(a)$.

Consider a mapping $\psi : M/N \to M'$ given by

$$\psi(a+N) = \varphi(a) \text{ for all } a \in M.$$

ψ is well defined, because for each $a + N \in M/N, \varphi(a)$ is a unique element of M'.

Now,

$$(\psi \circ p)(a) = \psi[p(a)]$$
$$= \psi(a+N) \qquad [\because \quad p(a) = a + N]$$
$$= \varphi(a) \quad \text{for all } a \in M.$$

$\therefore \qquad \psi \circ p = \varphi.$

We shall now show that $\psi : M/N \to M'$ is an R-isomorphism.

ψ *is an R-homomorphism*: For any $a + N, b + N \in M/N$ and $r \in R$, we have

$$\psi[(a+N) + (b+N)] = \psi[(a+b) + N] \qquad \text{[By def. of add. on } M/N]$$
$$= \varphi(a+b) \qquad \text{[By def. of } \psi]$$
$$= \varphi(a) + \varphi(b) \qquad [\varphi : M \to M' \text{ is an } R\text{-homomorphism}]$$
$$= \psi(a+N) + \psi(b+N)$$

and, $\psi[r(a+N)] = \psi(ra+N) \qquad \text{[By def. of scalar multiplication on } M/N]$
$$= \varphi(ra) \qquad \text{[By def. of } \psi]$$
$$= r\varphi(a) \qquad [\because \quad \varphi : M \to M' \text{ is an } R\text{-homomorphism}]$$
$$= r\psi(a+N)$$

So, $\psi : M/N \to M'$ is an R-homomorphism.

ψ *is an injection*: In order to prove this it is sufficient to show that $\text{Ker}(\psi) = N$ (see Theorem 3 on page 61).

Now, $a + N \in \text{Ker}(\psi) \Leftrightarrow \psi(a+N) = 0_{M'} \Leftrightarrow \varphi(a) = 0_{M'} \Leftrightarrow a \in \text{Ker}(\varphi) = N$.

So, $\text{Ker}(\psi) = \text{Ker}(\varphi) = N$.

Thus, $\psi : M/N \to M'$ is an injection.

ψ is a surjection: To prove that ψ is surjective it is sufficient to show that $I_m(\psi) = M'$.

Let a' be an arbitrary element of $I_m(\psi)$. Then,

$a' \in I_m(\psi) \Leftrightarrow$ There exists $a + N \in M/N$ such that $\psi(a+N) = a'$

$\Leftrightarrow \psi(p(a)) = a'$ [∵ $p : M \to M/N$ is an R-epimorphism]

$\Leftrightarrow (\psi \circ p)(a) = a'$

$\Leftrightarrow \varphi(a) = a'$

$\Leftrightarrow a' \in I_m(\varphi) = M'$ [∵ $\varphi : M \to M'$ is surjective]

Thus, $I_m(\psi) = I_m(\varphi)$. So, ψ is a surjection.

Hence, ψ is an R-isomorphism of M/N onto M' such that $\psi \circ p = \varphi$.

Uniqueness of ψ: Let $\psi' : M/N \to M'$ be an isomorphism such that $\psi' \circ p = \varphi$. Then,

$(\psi' \circ p)(a) = \varphi(a)$ for all $a \in M$.

$\Rightarrow \psi'(p(a)) = \varphi(a)$ for all $a \in M$.

$\Rightarrow \psi'(a+N) = \varphi(a)$ for all $a \in M$ [By def. of $p : M \to M/N$]

$\Rightarrow \psi'(a+N) = \psi(a+N)$ for all $a+N \in M/N$ [By def. of $\psi : M/N \to M'$]

Hence, for each R-epimorphism $\varphi : M \to M'$ there exists a unique isomorphism $\psi : M/N \to M'$ such that $\psi \circ p = \varphi$.

Q.E.D.

THEOREM-19 *Let N be an R-submodule of an R-module M. Then submodules of the quotient module M/N are of the form U/N, where U is an R-submodule of M containing N.*

PROOF. Let X be an R-submodule of M/N. Consider the set $U = \{a \in M : p(a) \in X\}$, where $p : M \to M/N$ is the projection mapping.

Clearly, elements of X are the cosets of N in M and N itself is the additive identity in X.

Since $0_M \in M$ and $p(0_M) = 0_M + N = N$. Therefore, $0_M \in U$.

Thus, U is a non-void subset of M.

We shall now show that U is an R-submodule of M.

Let a, b be any two elements of U. Then,

$p(a) \in X, p(b) \in X$.

$\Rightarrow p(a) - p(b) \in X$ [∵ X is an R-submodule of M/N]

$\Rightarrow p(a-b) \in X$ [∵ $p(a-b) = p(a) - p(b)$]

$\Rightarrow a - b \in U$ [∵ $a - b \in M$]

∴ $a - b \in U$ for all $a, b \in U$.

Let a be an arbitrary element of U and $r \in R$. Then,

$p(a) \in X$

$\Rightarrow rp(a) \in X$ $\qquad$ [$\because$ X is an R-submodule of M/N]

$\Rightarrow p(ra) \in X$ $\qquad$ [$\because$ $p : M \to M/N$ is an R-homomorphism]

$\Rightarrow ra \in U$ $\qquad$ [$\because$ M is an R-module $\therefore$ $ra \in M$]

$\therefore$ $ra \in U$ for all $r \in R$ and all $a \in U$.

Hence, U is an R-submodule of M.

Let a be an arbitrary element of M/N. Then,

$p(a) = a + N = N \in X$ $\qquad$ [$\because$ $a \in N$ $\therefore$ $a + N = N$]

$\Rightarrow \quad a \in U$

$\therefore \quad N \subset U$.

Since U and N are R-submodules of M such that $N \subset U$. Therefore, N is an R-submodule of U. Consequently, U/N exists.

Now $p(U) = \{p(a) : a \in U\} = \{a + N : a \in U\} = U/N$.

We shall now show that $X = p(U) = U/N$.

Clearly, $p(U) \subset X$. Let x be an arbitrary element of X. Then as $p : M \to M/N \supset X$ is onto, there exists $a \in M$ such that $p(a) = x$.

Now,

$$p(a) = x \in X \Rightarrow a \in U \text{ (by def. of } U) \Rightarrow x = p(a) \in p(U).$$

Thus, $x \in X \Rightarrow x \in p(U)$.

So, $X \subset p(U)$.

Hence, $X = p(U) = U/N$.

Since X is an arbitrary R-submodule of M/N. Therefore, R-submodules of M/N are of the form U/N, where U is an R-submodule of M containing N.

Q.E.D.

THEOREM-20 *(First Isomorphism Theorem for R-modules) Let N_1 and N_2 be R-submodules of an R-module M. Then,*

(i) $N_1 \cap N_2$ is an R-submodule of M.

(ii) $N_1 + N_2$ is an R-submodule of M.

(iii) $N_1/N_1 \cap N_2 \cong N_1 + N_2/N_2$.

PROOF. For proofs of (i) and (ii) refer Theorems 1 and 5 given on pages 46 and 48 respectively.

(iii) Since $N_1 \cap N_2$, $N_1 + N_2$, N_1 and N_2 are R-submodules of an R-module M such that $N_1 \cap N_2 \subset N_1$ and $N_2 \subset N_1 + N_2$. Therefore, symbols $N_1/N_1 \cap N_2$ and $N_1 + N_2/N_2$ are meaningful.

Consider a mapping $\varphi : N_1 \to N_1 + N_2/N_2$ given by
$$\varphi(a) = a + N_2 \quad \text{for all } a \in N_1$$
Now,
$$a \in N_1$$
$$\Rightarrow a \in N_1 + N_2 \qquad [\because N_1 C N_1 + N_2]$$
$$\Rightarrow \varphi(a) = a + N_2 \in N_1 + N_2/N_2.$$
Therefore, φ is, well defined.

φ *is an R-homomorphism:* For any $a, b \in N$, and any $r \in R$, we have
$$\varphi(a+b) = (a+b) + N_2 \qquad \text{[By definition of } \varphi\text{]}$$
$$= (a + N_2) + (b + N_2) \qquad \text{[By definition of addition on } N_1 + N_2/N_2\text{]}$$
$$= \varphi(a) + \varphi(b)$$
and,
$$\varphi(ra) = ra + N_2 \qquad \text{[By definition of } \varphi\text{]}$$
$$= r(a + N_2) \qquad \text{[By definition of scalar multiplication on } N_1 + N_2/N_2\text{]}$$
$$= r\varphi(a).$$
Thus, φ is an R-homomorphism.

φ *is surjective*: Let $a + N_2$ be an arbitrary element of $N_1 + N_2/N_2$. Then, $a = a_1 + a_2$ for some $a_1 \in N_1$, $a_2 \in N_2$. Thus for each $a + N_2 \in N_1 + N_2/N_2$ there exists $a_1 \in N_1$ such that
$$\varphi(a_1) = a_1 + N_2 = a_1 + (a_2 + N_2) \qquad [\because a_2 \in N_2 \Rightarrow a_2 + N_2 = N_2]$$
$$\Rightarrow \varphi(a_1) = (a_1 + a_2) + N_2 = a + N_2.$$
So, φ is surjective.

Hence, by the fundamental theorem of R-homomorphisms, we have
$$N_1/\text{Ker}(\varphi) \cong N_1 + N_2/N_2$$
Now,
$a \in \text{Ker}(\varphi)$
$\Leftrightarrow \varphi(a) = N_2$ and $a \in N_1$ $\qquad [\because N_2$ is additive identity in $N_1 + N_2/N_2]$
$\Leftrightarrow a + N_2 = N_2$ and $a \in N_1$
$\Leftrightarrow a \in N_2$ and $a \in N_1$
$\Leftrightarrow a \in N_1 \cap N_2$

$\therefore \quad \text{Ker}(\varphi) = N_1 \cap N_2$.

Hence, $N_1/N_1 \cap N_2 \cong N_1 + N_2/N_2$.

Q.E.D.

THEOREM-21 *(Second Isomorphism Theorem for R-modules)* If N_1 and N_2 are R-submodules of an R-module M such that $N_1 \supset N_2$, then
$$(M/N_2)/(N_1/N_2) \cong M/N_1.$$

PROOF. Since N_1 and N_2 are R-submodules of an R-module M. Therefore, M/N_1 and M/N_2 are meaningful.

Consider a mapping $\varphi : M/N_2 \to M/N_1$ given by
$$\varphi(a + N_2) = a + N_1 \quad \text{for all } a + N_2 \in M/N_2.$$

Clearly, φ is well defined.

φ *is an R-homomorphism*: For any $a + N_2, b + N_2 \in M/N_2$ and any $r \in R$, we have

$$\begin{aligned}
\varphi[(a+N_2)+(b+N_2)] &= \varphi[(a+b)+N_2] && \text{[By def. of add. on } M/N_2\text{]}\\
&= (a+b)+N_1 \\
&= (a+N_1)+(b+N_1) && \text{[By def. of add. on } M/N_1\text{]}\\
&= \varphi(a+N_2)+\varphi(b+N_2)
\end{aligned}$$

and,

$$\begin{aligned}
\varphi[r(a+N_2)] &= \varphi(ra+N_2) && \text{[By def. of scalar mult. on } M/N_2\text{]}\\
&= ra + N_1 && \text{[By def. of } \varphi\text{]}\\
&= r(a+N_1) && \text{[By def. of scalar mult. on } M/N_1\text{]}\\
&= r\varphi(a+N_2)
\end{aligned}$$

So, φ is an R-homomorphism.

φ *is surjective*: Let $a + N_1$ be an arbitrary element of M/N_1. Then, $a \in M$.

Now, $a \in M \Rightarrow a + N_2 \in M/N_2$.

Thus, for each $a + N_1 \in M/N_1$ there exists $a + N_2 \in M/N_2$ such that $\varphi(a+N_2) = a + N_1$.

So, φ is surjective.

Hence, by the fundamental theorem of R-homomorphisms, we have
$$(M/N_2)/\text{Ker}(\varphi) \cong M/N_1$$

Let $a + N_2$ be an arbitrary element of $\text{Ker}(\varphi)$. Then,

$a + N_2 \in \text{Ker}(\varphi)$
$\Leftrightarrow \varphi(a+N_2) = N_1$ $\qquad\qquad\qquad\qquad\qquad$ $[\because \ N_1$ is the additive identity in $M/N_1]$
$\Leftrightarrow a + N_1 = N_1$ $\qquad\qquad\qquad\qquad\qquad\qquad\qquad$ [By def. of φ]
$\Leftrightarrow a \in N_1$ $\qquad\qquad\qquad\qquad\qquad\qquad\qquad\qquad$ $[\because \ N_1$ is a subgroup $]$
$\Leftrightarrow a + N_2 \in N_1/N_2$ $\qquad\qquad\qquad\qquad\qquad$ $[\because \ N_1 \supset N_2 \ \therefore \ N_1/N_2$ exists$]$

$\therefore \quad \text{Ker}(\varphi) = N_1/N_2$.
Hence, $(M/N_2)/(N_1/N_2) \cong M/N_1$. Q.E.D.

THEOREM-22 *Let A and B be R-submodules of R-modules M and N respectively. Then,*
$$((M \times N)/A \times B) \cong (M/A) \times (N/B).$$

PROOF. Consider a mapping $\varphi : M \times N \to (M/A) \times (N/B)$ given by
$$\varphi(m,n) = (m+A, \, n+B), m \in M, n \in N.$$

φ *is an R-homomorphism*: Let $(m,n), (a,b)$ be any two elements of $M \times N$ and $r \in R$. Then,
$$\begin{aligned}
\varphi[(m,n) + (a,b)] &= \varphi[(m+a, n+b)] \\
&= ((m+a)+A, \, (n+b)+B) \\
&= ((m+A)+(a+A), \, (n+B)+(b+B)) \quad \text{[By def. of add. on } M/A \text{ and } N/B] \\
&= (m+A, n+B) + (a+A, b+B) \quad \text{[By def. of addition on } M/A \times N/B] \\
&= \varphi(m,n) + \varphi(a,b)
\end{aligned}$$

and,
$$\begin{aligned}
\varphi[r(m,n)] &= \varphi[(rm, rn)] \\
&= (rm+A, rn+B) \\
&= (r(m+A), r(n+B)) \quad \text{[By def. of scalar multiplication on } M/A \text{ and } N/B] \\
&= r(m+A, n+B) \\
&= r\varphi(m,n)
\end{aligned}$$

So, φ is an R-homomorphism.

φ *is surjective*: Let $(m+A, n+B)$ be an arbitrary element of $M/A \times N/B$. Then, $m \in M, n \in N$ and so $(m,n) \in M \times N$. Thus for each $(m+A, n+B) \in M/A \times N/B$ there exists $(m,n) \in M \times N$ such that
$$\varphi(m,n) = (m+A, n+B)$$
So, φ is surjective.

Hence, by the fundamental theorem of R-homomorphisms, we have
$$((M \times N)/\text{Ker}(\varphi)) \cong M/A \times N/B$$

Now,

$(m,n) \in \text{Ker}(\varphi)$
$\Leftrightarrow \varphi(m,n) = (0_M + A, 0_N + B)$
$\Leftrightarrow (m+A, n+B) = (0_M + A, 0_N + B)$
$\Leftrightarrow (m+A, n+B) = (A, B)$

$\Leftrightarrow m \in A, n \in B$

$\Leftrightarrow (m,n) \in A \times B$.

$\therefore \quad Ker(\varphi) = A \times B$.

Hence, $((M \times N)/(A \times B)) \cong (M/A) \times (N/B)$.

Q.E.D.

EXAMPLE-4 *For any ring R, show that the assignment $(a_1, a_2, a_3) \to (a_1, a_2)$ is an R-epimorphism $R^3 \to R^2$ of R-modules. If $D \subset R^3$ is the set of all triples $(0, 0, a_3)$, deduce that D is an R-submodule with $(R^3/D) \cong R^2$.*

SOLUTION Let $\varphi : R^3 \to R^2$ be the given mapping. Then,

$$\varphi(a_1, a_2, a_3) = (a_1, a_2) \quad \text{for all } (a_1, a_2, a_3) \in R^3.$$

φ *is an R-homomorphism*: For any $(a_1, a_2, a_3), (b_1, b_2, b_3) \in R^3$ and any $\lambda, \mu \in R$, we have

$\varphi[\lambda(a_1, a_2, a_3) + \mu(b_1, b_2, b_3)]$
$= \varphi[(\lambda a_1 + \mu b_1, \lambda a_2 + \mu b_2, \lambda a_3 + \mu b_3)]$
$= (\lambda a_1 + \mu b_1, \lambda a_2 + \mu b_2)$ [By def. of φ]
$= \lambda(a_1, a_2) + \mu(b_1, b_2)$
$= \lambda \varphi(a_1, a_2, a_3) + \mu \varphi(b_1, b_2, b_3)$.

$\therefore \quad \varphi$ is an R-homomorphism.

φ *is surjective*. Obviously $\varphi : R^3 \to R^2$ is surjective, because for each $(a_1, a_2) \in R^2$ there exists $(a_1, a_2, a_3) \in R^3$ such that

$$\varphi(a_1, a_2, a_3) = (a_1, a_2)$$

Hence, $\varphi : R^3 \to R^2$ is an epimorphism of R-modules.

We have, $D = \{(0, 0, a_3) : a_3 \in R\}$.

Let (a_1, a_2, a_3) be an arbitrary element of Ker(φ). Then,

$$\varphi(a_1, a_2, a_3) = (0, 0) \Rightarrow (a_1, a_2) = (0, 0) \Rightarrow a_1 = 0 = a_2$$

Thus, $(a_1, a_2, a_3) \in \text{Ker}(\varphi) \Leftrightarrow a_1 = a_2 = 0 \Leftrightarrow (0, 0, a_3) \in \text{Ker}(\varphi)$

$\therefore \quad \text{Ker}(\varphi) = D$.

Hence, D is an R-submodule of R^3 and by the fundamental theorem of R-homomorphisms, we have $R^3/D \cong R^2$.

EXERCISE 1.3

1. Let $f : R^n \to R$ be the mapping defined by
$$f(a_1, a_2, \ldots, a_n) = a_i, \text{ where } i \text{ is fixed}$$
Then show that f is an R-homomorphism of the R-module R^n onto the R-module R. This is called the projection of R^n onto the ith component.

2. Let $f : M \to N$ be an R-homomorphism of an R-module M into an R-module N. If f is a bijection, then show that $f^{-1} : N \to M$ is an R-homomorphism.

3. Prove that $Hom_Z(Q, Q) \cong Q$ as rings.

4. Let M be an R-module and $x \in M$ be such that $rx = 0$, $r \in R$, implies $r = 0$. Then show that $Rx \cong R$ as R-modules.

5. Let M be an R-module and A, B, C, D be submodules of M such that $M = A \oplus B = C \oplus D$. If $A = C$, show that $B \cong D$.

6. Let $f : R \to S$ be a ring homomorphism, and let M be a left S-module. Show that M can be made into a left R-module.

7. Let $N_1, N_2, \ldots, N_k$ be a family of R-submodules of an R-module M such that
$N_i + (N_1 \cap N_2 \cap \cdots \cap N_{i-1} \cap N_{i+1} \cap \cdots \cap N_k) = M$ for all $i = 1, 2, \ldots, k$. Show that
$$\frac{M}{\bigcap_{i=1}^{k} N_i} \cong \left(\frac{M}{N_1} \times \frac{M}{N_2} \times \cdots \times \frac{M}{N_k} \right)$$

1.7 ALGEBRAS

Throughout this section, unless otherwise stated, R is a commutative ring with unity.

R-ALGEBRA. *Let S be a ring. Then S is called an R-algebra (or an algebra over R) if S is a unitary R-module such that for all $r \in R$ and all $a, b \in S$*
$$r(ab) = (ra)b = a(rb)$$

Any field F can be regarded as an algebra over itself, because F is a module over itself and $r(ab) = (ra)b = a(rb)$ for all $r, a, b \in F$.

EXAMPLE-1 *Let R be a commutative ring with unity. Then the set $R[x]$ of all polynomials over ring R is an R-algebra.*

SOLUTION $R[x]$ is a unitary R-module under addition of polynomials and multiplication of a polynomial by a scalar as scalar multiplication. (see Example 5 on page 28).

For any $f(x) = \sum_i a_i x^i, g(x) = \sum_i b_i x^i \in R[x]$ and any $r \in R$, we have

$$r(f(x)g(x)) = r\left\{\sum_{j+k=i}(a_j b_k)x^i\right\} \qquad \text{[By def. of multiplication of polynomials]}$$

$$\Rightarrow r(f(x)g(x)) = \sum_{j+k=i}\{r(a_j b_k)\}x^i \qquad \text{[By def. of scalar multiplication on module } R[x] \text{ over } R\text{]}$$

$$\Rightarrow r(f(x)g(x)) = \sum_{j+k=i}\{(ra_j)b_k\}x^i \qquad \text{[By associativity of multiplication on } R\text{]}$$

$$\Rightarrow r(f(x)g(x)) = (rf(x))g(x)$$

$$\Rightarrow r(f(x)g(x)) = \sum_{j+k=i}\{(a_j r)b_k\}x^i \qquad \text{[By commutativity of multiplication on } R\text{]}$$

$$\Rightarrow r(f(x)g(x)) = \sum_{j+k=i}\{a_j(rb_k)\}x^i \qquad \text{[By associativity of multiplication on } R\text{]}$$

$$\Rightarrow r(f(x)g(x)) = f(x)(rg(x))$$

Hence, $R[x]$ is an R-algebra.

EXAMPLE-2 Let M be a unitary R-module. Then, $\text{End}_R(M)$ is an algebra over R.

SOLUTION By Theorem 11 on page 66, $\text{End}_R(M) = \text{Hom}_R(M, M)$ is an R-module. Also, for any $f \in \text{End}_R(M)$

$$(1f)(a) = 1f(a) = f(a) \quad \text{for all } a \in M \qquad [\because M \text{ is a unitary } R\text{-module}]$$
$$\therefore \quad 1f = f \qquad \text{for all } f \in \text{End}_R(M).$$

Thus, $\text{End}_R(M)$ is a unitary R-module.

For any $f, g \in \text{End}_R(M)$ and any $r \in R$, we have

$$[r(fg)](a) = r(fg)(a) \qquad \text{[By definition of scalar multiplication on } \text{End}_R(M)\text{]}$$
$$\Rightarrow [r(fg)](a) = r(f(a)g(a)) \qquad \text{[By definition of } fg\text{]}$$
$$\Rightarrow [r(fg)](a) = (rf(a))g(a) \qquad \text{[By M-2 (iii)]}$$
$$\Rightarrow [r(fg)](a) = (rf)(a)g(a) \qquad \text{[By definition of scalar multiplication on } \text{End}_R(M)\text{]}$$
$$\Rightarrow [r(fg)](a) = [(rf)g](a) \quad \text{for all } a \in M \qquad \text{[By def. of the product of functions]}$$
$$\therefore \quad r(fg) = (rf)g$$

Similarly, we have $r(fg) = f(rg)$.
Thus, $r(fg) = (rf)g = f(rg)$ for all $f, g \in \text{End}_R(M)$ and all $r \in R$.

Hence, $\text{End}_R(M)$ is an R-algebra.

1.8 FREE MODULES

Throughout this section, unless otherwise stated, R is a non-trivial (non-zero) ring with unity, that is, $1 \neq 0$ and an R-module is a unitary R-module.

LINEAR COMBINATION Let $a_1, a_2, \ldots, a_n$ be elements of an R-module M, and let $\lambda_1, \lambda_2, \ldots, \lambda_n$ be elements of ring R. Then,

$$\lambda_1 a_1 + \lambda_2 a_2 + \cdots + \lambda_n a_n \text{ (or, } \sum_{i=1}^{n} \lambda_i a_i)$$

is called a linear combination of $a_1, a_2, \ldots, a_n$. It is also called a linear combination of the set $S = \{a_1, a_2, \ldots, a_n\}$. Since there are finite number of elements in S, it is also called a finite linear combination of S.

If S is an infinite subset of M, then a linear combination of a finite subset of S is called a finite linear combination of S.

TRIVIAL LINEAR COMBINATION Let $a_1, a_2, \ldots, a_n$ be elements of an R-module M. Then the linear combination

$$\lambda_1 a_1 + \lambda_2 a_2 + \cdots + \lambda_n a_n$$

is called a trivial linear combination of $a_1, a_2, \ldots, a_n$, if $\lambda_1 = \lambda_2 = \cdots = \lambda_n = 0$.

REMARK. *The trivial linear combination of any set of elements of an R-module is always the zero $0_M \in M$, because*

$$0 a_1 + 0 a_2 + \cdots + 0 a_n = 0_M + 0_M + \cdots + 0_M = 0_M$$

NON-TRIVIAL LINEAR COMBINATION Let $a_1, a_2, \ldots, a_n$ be elements of an R-module M. Then a linear combination

$$\lambda_1 a_1 + \lambda_2 a_2 + \cdots + \lambda_n a_n$$

is called a non-trivial linear combination of $a_1, a_2, \ldots, a_n$, if at least one $\lambda_i \neq 0$.

EXAMPLE-1 *If $a_1, a_2, \ldots, a_n$ are elements of an R-module M, then the linear combination $1 a_1 + 0 a_2 + \cdots + 0 a_n$ is a non-trivial linear combination.*

REMARK. *Note that a non-trivial combination of a set of elements in an R-module M may or may not be the zero element of M. For example, $1(2, -1, 0) + (-2)(1, -1, 1) + 0(0, -1, 2)$ is a non-trivial linear combination of $(2, -1, 0), (1, -1, 1)$ and $(0, -1, 2)$ in R^3 and is zero element $(0, 0, 0)$ of R^3 while a non-trivial linear combination $1(2, -1, 0) + 2(1, -1, 1) + 1(0, -1, 2)$ is a non-zero element of R^3.*

LINEAR INDEPENDENCE A list $(a_1, a_2, \ldots, a_n)$ of elements of an R-module M is called linearly independent if for any $\lambda_1, \lambda_2, \ldots, \lambda_n \in R$,

$$\lambda_1 a_1 + \lambda_2 a_2 + \cdots + \lambda_n a_n = 0_M \Rightarrow \lambda_1 = \lambda_2 = \cdots = \lambda_n = 0.$$

Equivalently, a list $(a_1, a_2, \ldots, a_n)$ of elements of an R-module M is linearly independent (*l.i.*) if the only linear combination of $a_1, a_2, \ldots, a_n$ that equals to the zero in M is the trivial linear combination. In other words, a list $(a_1, a_2, \ldots, a_n)$ of elements of an R-module is $l.i.$ if no non-trivial linear combination of $a_1, a_2, \ldots, a_n$ equals to the zero element in M.

A finite subset of an R-module is linearly independent if every list of distinct vectors of it is $l.i.$

A infinite subset of an R-module is $l.i.$ if every finite subset of it is $l.i.$

LINEAR DEPENDENCE *A list $(a_1, a_2, \ldots, a_n)$ of elements in an R-module M is said to be linearly dependent (abbreviated l.d.), if for $\lambda_1, \lambda_2, \ldots, \lambda_n \in R$*

$$\lambda_1 a_1 + \lambda_2 a_2 + \cdots + \lambda_n a_n = 0_M \Rightarrow \text{ at least one } \lambda_i \neq 0.$$

Thus, a list $(a_1, a_2, \ldots, a_n)$ of elements of an R-module M is said to be $l.d.$ if at least one non-trivial linear combination of $a_1, a_2, \ldots, a_n$ equals to the zero element in M.

A subset (finite or infinite) of an R-module M is said to be $l.d.$ if there exists a finite list in it which is $l.d.$

REMARK. *By convention, the void set is considered to be $l.i.$*

REMARK. *Linear independence and dependence are opposite concepts. A set cannot be simultaneously $l.i$ as well as $l.d.$*

THEOREM-1 *Let M be an R-module. Then every subset of M containing 0_M is always $l.d.$*

PROOF. Let $S = \{a_1, a_2, \ldots, a_n\}$ be a subset of M such that $a_i = 0_M, 1 \leq i \leq n$. Then, $0a_1 + 0a_2 + \cdots + 0a_{i-1} + \lambda a_i + 0a_{i+1} + \cdots + 0a_n = 0_M$, where $0 \neq \lambda \in R$.

Hence, S is a linearly dependent (l.d.) set.

Q.E.D.

Consider R as a module over Q. The set $\left\{1, \sqrt{2}\right\}$ is $l.i.$ over Q. To see this, let $\lambda_1, \lambda_2 \in Q$ be such that $\lambda_1 \cdot 1 + \lambda_2 \sqrt{2} = 0$. Then, $\lambda_2 = 0$, because
$\lambda_2 \neq 0 \Rightarrow \sqrt{2} = -\lambda_1/\lambda_2 \Rightarrow \sqrt{2} \in Q$ $\quad [\because \quad \lambda_1, \lambda_2 \in Q \Rightarrow -\lambda_1/\lambda_2 \in Q]$
Thus, $\lambda_2 = 0$, and so $\lambda_1 = 0$ too.
Hence, the set $\left\{1, \sqrt{2}\right\}$ is $l.i.$ in module R over Q. The same set is not $l.i.$ in R as an R-module over itself, because $(\sqrt{2}) \cdot 1 + (-1)\sqrt{2} = 0$. Similarly, it can be seen that the set $\{1, i\}$ is $l.i.$ in module C over R but not in module C over itself.

EXAMPLE-2 *Let R be a ring with unity and $n \in N$. Then in R-module R^n, the set $\left\{e_1^{(n)}, e_2^{(n)}, \ldots, e_n^{(n)}\right\}$ is $l.i.$ Here, $e_i^{(n)}$ is the n-tuple in which all components except the ith are zero and ith component is 1.*

SOLUTION Let $\lambda_1, \lambda_2, \ldots, \lambda_n \in R$ be such that

$$\lambda_1 e_1^{(n)} + \lambda_2 e_2^{(n)} + \cdots + \lambda_n e_n^{(n)} = (0, 0, \ldots, 0)$$
$$\Rightarrow \lambda_1(1, 0, 0, \ldots, 0) + \lambda_2(0, 1, \ldots, 0) + \cdots + \lambda_n(0, 0, \ldots, 1) = (0, 0, \ldots, 0)$$
$$\Rightarrow (\lambda_1, \lambda_2, \ldots, \lambda_n) = (0, 0, \ldots, 0)$$
$$\Rightarrow \lambda_1 = \lambda_2 = \cdots = \lambda_n = 0$$

Hence, the set $\left\{ e^{(n)}, e_2^{(n)}, \ldots, e_n^{(n)} \right\}$ is linearly independent in R^n.

EXAMPLE-3 *Show that the set $S = \{(3, -1, 2), (0, 2, 5), (0, 0, 4)\}$ is l.i. in Z-module Z^3.*

SOLUTION Let $\lambda_1, \lambda_2, \lambda_3 \in Z$ be such that

$$\lambda_1(3, -1, 2) + \lambda_2(0, 2, 5) + \lambda_3(0, 0, 4) = (0, 0, 0)$$
$$\Rightarrow (3\lambda_1, -\lambda_1, 2\lambda_1) + (0, 2\lambda_2, 5\lambda_2) + (0, 0, 4\lambda_3) = (0, 0, 0)$$
$$\Rightarrow (3\lambda_1, -\lambda_1 + 2\lambda_2, 2\lambda_1 + 5\lambda_2 + 4\lambda_3) = (0, 0, 0)$$
$$\Rightarrow 3\lambda_1 = 0, -\lambda_1 + 2\lambda_2 = 0, 2\lambda_1 + 5\lambda_2 + 4\lambda_3 = 0$$
$$\Rightarrow \lambda_1 = \lambda_2 = \lambda_3 = 0.$$

Hence, the set S is a *l.i.* set in Z^3 over Z.

THEOREM-2 *Every subset of a l.i. set in an R-module is l.i.*

PROOF. Let $S = \{a_1, a_2, \ldots, a_m\}$ be a set of elements of an R-module M and let $S' = \{a_1, a_2, \ldots, a_n\}$, $n \leq m$, be an arbitrary finite subset of S. Then, for any $\lambda_1, \lambda_2, \ldots, \lambda_n \in R$

$$\lambda_1 a_1 + \lambda_2 a_2 + \cdots + \lambda_n a_n = 0_M$$
$$\Rightarrow \lambda_1 a_1 + \lambda_2 a_2 + \cdots + \lambda_n a_n + 0 a_{n+1} + \cdots + 0 a_m = 0_M$$
$$\Rightarrow \lambda_1 = \lambda_2 = \cdots = \lambda_n = 0 \qquad [\because \ S \text{ is linearly independent}]$$

Hence, S' is a *l.i.* set.

Since S' is an arbitrary subset of S. Therefore, every subset of S is linearly independent.

Q.E.D.

THEOREM-3 *Every superset of a l.d. set of elements in an R-module is linearly dependent.*

PROOF. Let $S = \{a_1, a_2, \ldots, a_n\}$ be a *l.d.* set of elements of an R-module M and let $S' = \{a_1, a_2, \ldots, a_n, a_{n+1}\}$ be a superset of S. Then,

  S is *l.d.* set

$\Rightarrow$ There exist $\lambda_1, \lambda_2, \ldots, \lambda_n \in R$ not all zero such that $\lambda_1 a_1 + \lambda_2 a_2 + \cdots + \lambda_n a_n = 0_M$
$\Rightarrow$ $\lambda_1 a_1 + \lambda_2 a_2 + \ldots + \lambda_n a_n + 0 a_{n+1} = 0_M$, where $\lambda_1, \ldots, \lambda_n \in R$ such that at least one $\lambda_i \neq 0$.
$\Rightarrow$ S' is *l.d.*

Since S' is an arbitrary superset of S. Therefore, every superset of S is linearly dependent.

Q.E.D.

THEOREM-4 *Let M be an R-module generated by a set $S = \{a_1, a_2, \ldots, a_n\}$. If S is a l.i. set, then each element of M can be written as a linear combination of $a_1, a_2, \ldots, a_n$ in exactly one way.*

PROOF. Since M is generated by the set $\{a_1, a_2, \ldots, a_n\}$. Therefore, by Theorem 10 on page 53 each element of M can be written as a linear combination of $a_1, a_2, \ldots, a_n$. To prove the uniqueness, let $a \in M$ be such that

$$a = \lambda_1 a_1 + \lambda_2 a_2 + \cdots + \lambda_n a_n = \mu_1 a_1 + \mu_2 a_2 + \cdots + \mu_n a_n$$
$$\Rightarrow (\lambda_1 - \mu_1)a_1 + (\lambda_2 - \mu_2)a_2 + \cdots + (\lambda_n - \mu_n)a_n = 0_M \quad \text{[By comm. and assoc. of add on } M\text{]}$$
$$\Rightarrow \lambda_1 - \mu_1 = 0, \lambda_2 - \mu_2 = 0, \ldots, \lambda_n - \mu_n = 0 \quad [\because \{a_1, a_2, \ldots, a_n\} \text{ is linearly independent}]$$
$$\Rightarrow \lambda_1 = \mu_1, \lambda_2 = \mu_2, \ldots, \lambda_n = \mu_n$$

Hence, each element of M can be expressed as a linear combination of $a_1, a_2, \ldots, a_n$ in a unique way.

Q.E.D.

BASIS *A subset B of an R-module M is called a basis for M, if*
  (i) B generates M, i.e. each element of M can be written as a linear combination of elements of B.
  (ii) B is a l.i. set.

FREE MODULE *An R-module M is called a free module, if M has a basis.*
  In other words, an R-module M is a free module if there exists a subset B of M such that
  (i) M is generated by B and, (ii) B is a l.i. set.

REMARK. $\{0_M\}$ *is a free module with basis as the void set.*

EXAMPLE-4 *Show that the set $R^n = \{(a_1, a_2, \ldots, a_n) : a_i \in R\}$ is a free module over ring R.*

SOLUTION By Example 2 page on 84 the set $S = \{e_1^{(n)}, e_2^{(n)}, \ldots, e_n^{(n)}\}$ is a l.i. set in R^n.
Let $a = (a_1, a_2, \ldots, a_n)$ be an arbitrary element of R^n. Then,

$$a = (a_1, 0, \ldots, 0) + (0, a_2, \ldots, 0) + \cdots + (0, 0, \ldots, a_n)$$
$$\Rightarrow \quad a = a_1(1, 0, \ldots, 0) + a_2(0, 1, \ldots, 0) + \cdots + a_n(0, 0, \ldots, 1)$$
$$\Rightarrow \quad a = a_1 e_1^{(n)} + a_2 e_2^{(n)} + \cdots + a_n e_n^{(n)}$$
$$\Rightarrow \quad a \text{ is a linear combination of } e_1^{(n)}, e_2^{(n)}, \ldots, e_n^{(n)}.$$

Thus, each element of R^n is a linear combination of elements in S. Consequently, S generates R^n. Hence, S is a basis for R^n and thus R^n is a free R-module.

REMARK. The basis $\{e_1^{(n)}, e_2^{(n)}, \ldots, e_n^{(n)}\}$ is known as the standard basis for R^n.

EXAMPLE-5 *Show that $R[x]$ is a free R-module.*

SOLUTION Consider the set $B = \{1, x, x^2, x^3, \ldots\}$.

B is *l.i.*: For any $\lambda_0, \lambda_1, \lambda_2, \ldots$ in R

$\lambda_0 1 + \lambda_1 x + \lambda_2 x^2 + \cdots + \lambda_n x^n + \cdots = 0$ (Zero polynomial)

$\Rightarrow \lambda_0 1 + \lambda_1 x + \lambda_2 x^2 + \cdots + \lambda_n x^n + \cdots = 0.1 + 0x + 0x^2 + \cdots + 0x^n + \ldots$

$\Rightarrow \lambda_0 = 0 = \lambda_1 = \lambda_2 = \cdots = \lambda_n = \ldots$ [By the def. of equality of poly.]

$\therefore$ B is a *l.i.* set.

B generates $R[x]$: For any $f(x) = a_0 + a_1 x + a_2 x^2 + \ldots$ in $R[x]$, we have

$f(x) = a_0 1 + a_1 x + a_2 x^2 + \ldots$

$\Rightarrow$ $f(x)$ is a linear combination of $1, x, x^2, \ldots$

$\therefore$ B generates $R[x]$.

Thus, B is a basis for $R[x]$. Hence, $R[x]$ is a free R-module.

REMARK. The set $\{1, 1+x, x^2, \ldots, x^n, \ldots\}$ is also a basis for $R[x]$.

THEOREM-5 *Show that a cyclic group regarded as a Z-module has a basis if it is infinite.*

OR

Show that a cyclic group regarded as a Z-module is free Z-module iff it is infinite.

PROOF. Let $G = [a]$ be an infinite additive cyclic group with identity element $\underline{0}$.

First suppose that G is an infinite cyclic group generated by a. We have to prove that G as a Z-module has a basis. Since $G = [a]$ is infinite. Therefore, $a \neq \underline{0}$ (identity element). Clearly the set $B = \{a\}$ generates G, because each element of G is an integral multiple of a. Since G is an infinite cyclic group generated by a. Therefore, order of a is infinite. Thus, $na \neq \underline{0}$ for any $n \in Z$. In other words, $na = \underline{0}$ only when $n = 0$. So, B is a *l.i.* set. Hence, B is a basis for G.

Conversely, let $G = [a]$ be a Z-module such that it has a basis. Then we have to prove that G is an infinite cyclic group. Since each element of G is an integral multiple of a. Therefore, elements in the basis of G are also integral multiples of a. Let $ma, m \in Z$, be some basis element. Then, $n(ma) = \underline{0}, n \in Z$, must imply $n = 0$. However, if G is a finite group of order k, then

$$k(ma) = m(ka) = m(\underline{0}) \qquad [\because \; 0(G) = k \Rightarrow 0(a) = k \Rightarrow ka = \underline{0}]$$

$\Rightarrow \qquad k(ma) = \underline{0}$, a contradiction

Hence, G must be infinite if it has a basis.

Q.E.D.

It follows from the above discussion that an R-module may or may not have a basis and even if it has, it need not be unique.

THEOREM-6 *Let M be a free R-module with a basis B and let N be an R-module. If $f : B \to N$ is any mapping, then there exists a unique R-homomorphism $\varphi : M \to N$ which is extension of f, that is $\varphi(x) = f(x)$ for all $x \in B$.*

PROOF. We will prove the theorem only in the case in which B is finite, leaving the infinite case as an exercise for the reader. Suppose $O(B) = n$ and $B = \{a_1, a_2, \ldots, a_n\}$. Set $f(a_i) = b_i \in N$, $i = 1, 2, \ldots, n$. Let a be an arbitrary element of M. Since M is free R-module with basis B. Therefore, a can be written uniquely as a linear combination of elements in B. Let

$$a = \lambda_1 a_1 + \lambda_2 a_2 + \cdots + \lambda_n a_n, \quad \text{where } \lambda_i \in R$$

Consider a mapping $\varphi : M \to N$ given by

$$\varphi(a) = \lambda_1 f(a_1) + \lambda_2 f(a_2) + \cdots + \lambda_n f(a_n).$$

Clearly φ exists, because each $a \in M$ is uniquely expressible as a linear combination of elements $a_1, a_2, \ldots, a_n$ and $f(a_1), f(a_2), \ldots, f(a_n)$ exist.

We shall now prove that $\varphi : M \to N$ is an R-homomorphism.

Let a, b be any two elements of M. Then there exist $\lambda_1, \lambda_2, \ldots, \lambda_n, \mu_1, \mu_2, \ldots, \mu_n \in R$ such that

$$a = \lambda_1 a_1 + \lambda_2 a_2 + \cdots + \lambda_n a_n, b = \mu_1 a_1 + \mu_2 a_2 + \cdots + \mu_n a_n \quad [\because B \text{ is a basis for } M]$$

$\therefore \quad \varphi(a+b) = \varphi[\lambda_1 a_1 + \lambda_2 a_2 + \cdots + \lambda_n a_n + \mu_1 a_1 + \mu_2 a_2 + \cdots + \mu_n a_n]$

$\Rightarrow \varphi(a+b) = \varphi[(\lambda_1 + \mu_1)a_1 + (\lambda_2 + \mu_2)a_2 + \cdots + (\lambda_n + \mu_n)a_n]$ $\quad \begin{bmatrix} \text{By comm. and assoc.} \\ \text{of add. on } M \end{bmatrix}$

$\Rightarrow \varphi(a+b) = (\lambda_1 + \mu_1)f(a_1) + (\lambda_2 + \mu_2)f(a_2) + \cdots + (\lambda_n + \mu_n)f(a_n)$ [By def. of φ]

$\Rightarrow \varphi(a+b) = (\lambda_1 + \mu_1)b_1 + (\lambda_2 + \mu_2)b_2 + \cdots + (\lambda_n + \mu_n)b_n \quad [\because f(a_i) = b_i, i = 1, 2, \ldots, n]$

$\Rightarrow \varphi(a+b) = (\lambda_1 b_1 + \mu_1 b_1) + (\lambda_2 b_2 + \mu_2 b_2) + \cdots + (\lambda_n b_n + \mu_n b_n)$ [By M-2(ii)]

$\Rightarrow \varphi(a+b) = [(\lambda_1 b_1 + \lambda_2 b_2 + \ldots + \lambda_n b_n) + (\mu_1 b_1 + \mu_2 b_2 + \ldots + \mu_n b_n)]$ $\quad \begin{bmatrix} \text{By comm. and} \\ \text{assoc. of add. on } N \end{bmatrix}$

$\Rightarrow \varphi(a+b) = \{\lambda_1 f(a_1) + \lambda_2 f(a_2) + \cdots + \lambda_n f(a_n)\}$
$\qquad\qquad\qquad + \{\mu_1 f(a_1) + \mu_2 f(a_2) + \cdots + \mu_n f(a_n)\}$ [By def. of f]

$\Rightarrow \varphi(a+b) = \varphi(a) + \varphi(b)$ [By def. of φ]

For any $r \in R$, we have

$$\begin{aligned}
\varphi(ra) &= \varphi[r(\lambda_1 a_1 + \lambda_2 a_2 + \cdots + \lambda_n a_n)] \\
&= \varphi[r(\lambda_1 a_1) + r(\lambda_2 a_2) + \cdots + r(\lambda_n a_n)] && \text{[By M-2 (i)]} \\
&= \varphi[(r\lambda_1)a_1 + (r\lambda_2)a_2 + \cdots + (r\lambda_n)a_n] && \text{[By M-2 (iii)]} \\
&= (r\lambda_1)f(a_1) + (r\lambda_2)f(a_2) + \cdots + (r\lambda_n)f(a_n) && \text{[By definition of } \varphi\text{]} \\
&= (r\lambda_1)b_1 + (r\lambda_2)b_2 + \cdots + (r\lambda_n)b_n && \text{[By definition of } f\text{]} \\
&= r(\lambda_1 b_1) + r(\lambda_2 b_2) + \cdots + r(\lambda_n b_n) && \text{[By M-2 (iii)]} \\
&= r[\lambda_1 b_1 + \lambda_2 b_2 + \cdots + \lambda_n b_n] && \text{[By M-2 (i)]} \\
&= r[\lambda_1 f(a_1) + \lambda_2 f(a_2) + \cdots + \lambda_n f(a_n)] && \text{[By definition of } f\text{]} \\
&= r\varphi(a)
\end{aligned}$$

Thus, $\varphi : M \to N$ is an R-homomorphism.

For any $a_i \in B$, we have

$$a_i = 0a_1 + 0a_2 + \cdots + 0a_{i-1} + 1a_i + 0a_{i+1} + \cdots + 0a_n$$

$$\therefore \quad \varphi(a_i) = 0f(a_1) + 0f(a_2) + \cdots + 0f(a_{i-1}) + 1f(a_i) + 0f(a_{i+1}) + \cdots + 0f(a_n)$$

$$\Rightarrow \quad \varphi(a_i) = 0_N + 0_N + \cdots + 0_N + f(a_i) + 0_N + \cdots + 0_N = f(a_i)$$

Thus, $\varphi(a_i) = f(a_i)$ for all $a_i \in B$.

Since each $a \in M$ is uniquely expressible as a linear combination of $a_1, a_2, \ldots, a_n$. Therefore, $\varphi : M \to N$ is a unique R-homomorphism such that φ is an extension of f.

Q.E.D.

THEOREM-7 *Let M be an R-module generated by n elements. Then M is isomorphic to a quotient module of R^n.*

PROOF. Let M be generated by the set $\{a_1, a_2, \ldots, a_n\}$ and let $\{e_1^{(n)}, e_2^{(n)}, \ldots, e_n^{(n)}\}$ be the standard basis for R^n.

Consider a mapping $\varphi : R^n \to M$ given by

$$\varphi(\lambda_1, \lambda_2, \ldots, \lambda_n) = \varphi(\lambda_1 e_1^{(n)} + \lambda_2 e_2^{(n)} + \cdots + \lambda_n e_n^{(n)}) = \lambda_1 a_1 + \lambda_2 a_2 + \cdots + \lambda_n a_n$$

We shall now show that φ is an R-homomorphism of R^n onto M.

φ *is an R-homomorphism*: For any $(\lambda_1, \lambda_2, \ldots, \lambda_n), (\mu_1, \mu_2, \ldots, \mu_n) \in R^n$ and $r \in R$, we have

$\varphi[(\lambda_1, \lambda_2, \ldots, \lambda_n) + (\mu_1, \mu_2, \ldots, \mu_n)]$
$= \varphi(\lambda_1 + \mu_1, \lambda_2 + \mu_2, \ldots, \lambda_n + \mu_n)$ [By def. of add. on R^n]
$= (\lambda_1 + \mu_1)a_1 + (\lambda_2 + \mu_2)a_2 + \cdots + (\lambda_n + \mu_n)a_n$ [By def. of φ]
$= (\lambda_1 a + \mu_1 a) + (\lambda_2 a + \mu_2 a) + \cdots + (\lambda_n a + \mu_n a)$ [By M-2 (ii)]
$= (\lambda_1 a + \lambda_2 a + \cdots + \lambda_n a) + (\mu_1 a + \mu_2 a + \cdots + \mu_n a)$ [By comm. and assoc. of add. on M]
$= \varphi(\lambda_1, \lambda_2, \ldots, \lambda_n) + \varphi(\mu_1, \mu_2, \ldots, \mu_n)$

and,

$[r(\lambda_1, \lambda_2, \ldots, \lambda_n)] = \varphi(r\lambda_1, r\lambda_2, \ldots, r\lambda_n)$ [By def. of scalar multiplication on R^n]
$= (r\lambda_1)a_1 + (r\lambda_2)a_2 + \cdots + (r\lambda_n)a_n$ [By def. of φ]
$= r(\lambda_1 a) + r(\lambda_2 a) + \cdots + r(\lambda_n a)$ [By M-2 (iii)]
$= r(\lambda_1 a + \lambda_2 a + \cdots + \lambda_n a)$ [By M-2 (i)]
$= r\varphi(a)$.

So, φ is an R-homomorphism.

φ *is surjective*: Let a be an arbitrary element of M. Then there exist $\lambda_1, \lambda_2, \ldots, \lambda_n \in R$ such that

$a = \lambda_1 a_1 + \lambda_2 a_2 + \cdots + \lambda_n a_n$ [$\because$ $\{a_1, a_2, \ldots, a_n\}$ generates M]
$\Rightarrow a = \varphi(\lambda_1, \lambda_2, \ldots, \lambda_n)$

So, φ is surjective.

Thus, φ is an R-homomorphism of R^n onto M.
Hence, by the fundamental theorem of R-homomorphism $M \cong R^n/\text{Ker}(\varphi)$.

Q.E.D.

COROLLARY. *Let M be a free R-module with a basis $\{a_1, a_2, \ldots, a_n\}$. Then, $M \cong R^n$.*

PROOF. By above theorem, we have
$M \cong R^n/\text{Ker}(\varphi)$, where $\varphi: R^n \to M$ is given by
$\varphi(\lambda_1, \lambda_2, \ldots, \lambda_n) = \lambda_1 a_1 + \lambda_2 a_2 + \cdots + \lambda_n a_n$.
Now,

$(\lambda_1, \lambda_2, \ldots, \lambda_n) \in \text{Ker}(\varphi)$
$\Leftrightarrow \varphi(\lambda_1, \lambda_2, \ldots, \lambda_n) = 0_M$
$\Leftrightarrow \lambda_1 a_1 + \cdots + \lambda_n a_n = 0_M$
$\Leftrightarrow \lambda_1 a_1 + \lambda_2 a_2 + \cdots + \lambda_n a_n = 0 a_1 + 0 a_2 + \cdots + 0 a_n$
$\Leftrightarrow \lambda_1 = \lambda_2 = \cdots = \lambda_n = 0$ [$\because$ $\{a_1, a_2, \ldots, a_n\}$ is a basis for M]
$\Leftrightarrow (\lambda_1, \lambda_2, \ldots, \lambda_n) = (0, 0, \ldots, 0)$

Thus, Ker(φ) is trivial and hence by Theorem 3 on (page 61), φ is injective.
Consequently, $\varphi : R^n \to M$ is an isomorphism.

Hence, $\qquad R^n \cong M \Rightarrow M \cong R^n \qquad\qquad [\because \ \cong \text{ is symmetric}]$.

Q.E.D.

As we have seen that an R-module may or may not have a basis but if it has one it may have many bases. Now a natural question arises in our mind. Do any two bases have something in common? Following theorems answer this question.

THEOREM-8 *Let M be a finitely generated free module over a commutative ring R. Then all bases of M are finite.*

PROOF. Let $S = \{a_1, a_2, \ldots, a_n\}$ be a set of generators of M, and let $B = \{e_i : i \in I\}$ be a basis of M. Here, I is index set such that for each $i \in I, e_i \in B$. Since B is a basis for M, therefore B generates M. Consequently, a_j can be written as

$$a_j = \sum_i \lambda_{ij} \, e_i, \quad \lambda_{ij} \in R,$$

and all but finite number of λ_{ij} are zero. Thus, the set B of these e_i's that occur in expressions of all the a_j's, $j = 1, 2, \ldots, n$, is finite.

Q.E.D.

THEOREM-9 *Let M be a finitely generated free module over a commutative ring R. Then all bases of M have the same number of elements.*

PROOF. Let $(e_1, e_2, \ldots, e_m)$ and $(f_1, f_2, \ldots, f_n)$ be any two ordered bases of R-module. Then we have to prove that $m = n$.
Let $m < n$.
By corollary to Theorem 7 (page 89), $M \cong R^m$ and $M \cong R^n$. Therefore, by symmetry and transitivity of the relation '$\cong$', we have $R^m \cong R^n$.
Let $\varphi : R^m \to R^n$ be an R-isomorphism, and let $\psi = \varphi^{-1}$. Then ψ is also an R-isomorphism of R^n onto R^m such that
$$\psi \circ \varphi = \text{Identity } R\text{-homomorphism on } R^m.$$
Consider the standard bases $(e_1^{(m)}, e_2^{(m)}, \ldots, e_m^{(m)})$ and $(e_1^{(n)}, e_2^{(n)}, \ldots, e_n^{(n)})$ of R^m and R^n respectively.

Since for each $i \in \underline{m}$, $\varphi(e_i^{(m)}) \in R^n$, therefore there exist $a_{1i}, a_{2i}, \ldots, a_{ni}$ in R such that

$$\varphi(e_i^{(m)}) = a_{1i} e_1^{(n)} + a_{2i} e_2^{(n)} + \cdots + a_{ni} e_n^{(n)} \qquad [\because \ (e_1^{(n)}, e_2^{(n)}, \ldots, e_n^{(n)}) \text{ is a basis for } R^n]$$

Similarly, for each $j \in \underline{n}$, there exist $b_{1j}, b_{2j}, \ldots, b_{mj}$ in R such that

$$\psi(e_j^{(n)}) = b_{1j} e_1^{(m)} + b_{2j} e_2^{(m)} + \cdots + b_{mj} e_m^{(m)}. \qquad [\because \ (e_1^{(m)}, e_2^{(m)}, \ldots, e_m^{(m)}) \text{ is a basis for } R^m]$$

Let $A = [a_{ji}]$ and $B = [b_{kj}]$ be $n \times m$ and $m \times n$ matrices. Then

$$\psi o \varphi(e_i^{(m)}) = \psi\left\{\varphi(e_i^{(m)})\right\} = \psi\left(\sum_{j=1}^n a_{ji} e_j^{(n)}\right) \qquad \text{[By def. of } \varphi\text{]}$$

$$\Rightarrow \psi o \varphi(e_i^{(m)}) = \sum_{j=1}^n a_{ji} \psi(e_j^{(n)}) \qquad [\because \psi \text{ is an } R\text{-homomorphism}]$$

$$\Rightarrow \psi o \varphi(e_i^{(m)}) = \sum_{j=1}^n a_{ji} \left(\sum_{k=1}^m b_{kj} e_k^{(m)}\right) \qquad \text{[By def. of } \psi\text{]}$$

$$\Rightarrow \psi o \varphi(e_i^{(m)}) = \sum_{j=1}^n \sum_{k=1}^m (a_{ji} b_{kj}) e_k^{(m)} \qquad [\because a_{ji}, b_{kj} \in R]$$

$$\Rightarrow \psi o \varphi(e_i^{(m)}) = \sum_{k=1}^m \sum_{j=1}^n (b_{kj} a_{ji}) e_k^{(m)} \qquad [\because R \text{ is commutative}]$$

$$\Rightarrow e_i^{(m)} = \sum_{k=1}^m \sum_{j=1}^n (b_{kj} a_{ji}) e_k^{(m)} \text{ for all } i \in \underline{m} \qquad [\because \psi 0 \varphi \text{ is the identity mapping}]$$

$$\Rightarrow \sum_{j=1}^n b_{kj} a_{ji} = \delta_{ki} \qquad [\text{Using linear independence of the } e_i^{(m)}]$$

$\Rightarrow BA = I_m$, where $m \times m$ identity matrix on R

Similarly, $AB = I_n$, where $n \times n$ identity matrix on R.

Let $A' = [A \; O]$ and $B' = \begin{bmatrix} B \\ O \end{bmatrix}$ be $n \times n$ augmented matrices, where each of the O block is a matrix of the appropriate size. Then,

$$A'B' = I_n, \text{ and } B'A' = \begin{bmatrix} I_m & O \\ O & O \end{bmatrix}.$$

$\Rightarrow \quad \text{Det}(A'B') = 1$ and $\text{Det}(B'A') = 0$.

But, A' and B' are $n \times n$ matrices over a commutative ring. So, $\text{Det}(A'B') = \text{Det}(B'A')$, which yields a contradiction. Hence, $m \geqq n$. By symmetry, $n \geqq m$. Hence $m = n$.

Q.E.D.

Above two theorems depict that if an R-module has a basis, then its all bases have the same number of elements.

RANK OF AN R-MODULE. *Let M be a finitely generated free module over a commutative ring R with unity. Then the number of elements in its any basis is called the rank of M, written as rank M.*

REMARK. *The assumption of commutativity in Theorems 8 and 9 is essential, because if R is a non-commutative ring, than R^n may be R-isomorphic to R for more than one value of n.*

THEOREM-10 *If M is an R-module and if F is a free R-module on the set of free generators $a_1, a_2, \ldots, a_n$, then*
$$\text{Hom}_R(F, M) \cong M^n$$

PROOF. Let $B = \{b_1, b_2, \ldots, b_n\}$ be the set of free generators of F. Define a mapping $\varphi: \text{Hom}_R(F, M) \to M^n$ given by
$$\varphi(f) = (f(b_1), f(b_2), \ldots, f(b_n)) \quad \text{for all } f \in \text{Hom}_R(F, M)$$

Clearly, φ is well defined.

φ *is injective*: For any $f, g \in \text{Hom}_R(F, M)$

$\varphi(f) = \varphi(g)$
$\Rightarrow \quad (f(b_1), f(b_2), \ldots, f(b_n)) = (g(b_1), g(b_2), \ldots, g(b_n))$
$\Rightarrow \quad f(b_i) = g(b_i) \quad \text{for all } i \in \underline{n}$
$\Rightarrow \quad f = g$ on B.
$\Rightarrow \quad f = g$ on F \hfill [$\because$ B is a basis for F]
$\therefore \quad \varphi$ is injective.

φ *is surjective*: Let $(a_1, a_2, \ldots, a_n)$ be an arbitrary element of M^n.

Then there exists a unique R-homomorphism $f: F \to M$ given by $f(b_1) = a_1, f(b_2) = a_2, \ldots, f(b_n) = a_n$.
That is, $\varphi(f) = (f(b_1), f(b_2), \ldots, f(b_n)) = (a_1, a_2, \ldots, a_n)$.
Thus, for each $(a_1, a_2, \ldots, a_n) \in M^n$ there exists an R-homomorphism $f \in \text{Hom}_R(F, M)$ such that $\varphi(f) = (a_1, a_2, \ldots, a_n)$.

So, φ is surjective.

φ *is an R-homomorphism*: For any $f, g \in \text{Hom}_R(F, M)$ and any $r \in R$, we have

$\varphi(f+g) = ((f+g)(b_1), (f+g)(b_2), \ldots, (f+g)(b_n))$
$\qquad = (f(b_1) + g(b_1), f(b_2) + g(b_2), \ldots, f(b_n) + g(b_n))$ \hfill [By definition of $f+g$]
$\qquad = (f(b_1), f(b_2), \ldots, f(b_n)) + (g(b_1), g(b_2), \ldots, g(b_n))$
$\qquad = \varphi(f) + \varphi(g).$

and,

$\varphi(rf) = ((rf)(b_1), (rf)(b_2), \ldots, (rf)(b_n))$
$\qquad = (rf(b_1), rf(b_2), \ldots, rf(b_n))$ \hfill [By definition of rf]
$\qquad = r(f(b_1), f(b_2), \ldots, f(b_n))$
$\qquad = r\varphi(f).$

$\therefore$ φ is an R-homomorphism.

Hence, $\varphi : \text{Hom}_R(F, M) \to M^n$ is an isomorphism. Consequently, $\text{Hom}_R(F, M) \cong M^n$.

Q.E.D.

THEOREM-11 *Let M and M' be R-modules, and let F be a free module over ring R. If $p : M \to M'$ is an R-epimorphism, then for each R-homomorphism $f : F \to M'$ there exists an R-homomorphism $g : F \to M$ such that $f = \text{pog}$.*

PROOF. Let B be the set of free generators for R-module F. Let x be an arbitrary element of B. Then,

$f(x) \in M'$ $\hfill [\because \quad f : F \to M']$

$\Rightarrow$ There exists $a_x \in M$ such that $p(a_x) = f(x)$ $\hfill [\because \quad p : M \to M'$ is surjective$]$

Consider a mapping $\varphi : B \to M$ given by

$$\varphi(x) = a_x \quad \text{for all } x \in B.$$

Since F is a free R-module with basis B. Therefore, by Theorem 6 (page 88) there exists an R-homomorphism $g : F \to M$ such that g is extension of φ. That is,

$$g(x) = a_x \quad \text{for all } x \in B.$$

Clearly, for any $x \in B$

$(\text{pog})(x) = p[g(x)] = p(a_x) = f(x)$

$\therefore$ $\text{pog}(x) = f(x)$ for all $x \in B$

$\Rightarrow$ $\text{pog} = f$ on B $\Rightarrow$ $\text{pog} = f$ on F. $\hfill [\because \quad B$ is a basis for $F]$

Q.E.D.

THEOREM-12 *Let F be a free module over a ring R, and let M be an R-module such that $f : M \to F$ is an R-epimorphism. Then M is direct sum of $\text{Ker}(f)$ and a submodule F' of M R-isomorphic to F.*

PROOF. Since F is a free R-module and $f : M \to F$ is an R-epimorphism, therefore, by Theorem 11 corresponding to the identity R-homomorphism I_F there exists an R-homomorphism $\varphi : F \to M$ such that $f \circ \varphi = I_F$.

Let $F' = I_m(\varphi)$. Then F' is an R-submodule of M. Clearly for each $x \in M$, $(\varphi \circ f)(x) \in F'$.

Let $y = x - (\varphi \circ f)(x)$. Then,

$f(y) = f(x - (\varphi \circ f)(x))$

$\Rightarrow f(y) = f(x) - f[\varphi \circ f(x)]$ $\hfill [\because \quad f : M \to F$ is an R-homomorphism$]$

$\Rightarrow f(y) = f(x) - f \circ \varphi(f(x))$
$\Rightarrow f(y) = f(x) - I_F(f(x))$ $\qquad [\because\ f \circ \varphi = I_F]$
$\Rightarrow f(y) = f(x) - f(x) = 0_F$
$\therefore \quad y \in \text{Ker}(f)$

Thus, for each $x \in M$, $\varphi 0 f(x) \in F'$ and $x - (\varphi \circ f)(x) \in \text{Ker}(f)$ such that
$$x = [x - (\varphi \circ f)(x)] + (\varphi \circ f)(x)$$
Therefore, $M = \text{Ker}(f) + F'$

Let $x \in \text{Ker}(f) \cap F'$. Then, $x \in \text{Ker}(f)$ and $x \in F'$.

Now, $x \in F'$
$\Rightarrow \quad$ There exists $a \in F$ such that $\varphi(a) = x$ $\qquad [\because\ F' = I_m(\varphi)]$
$\Rightarrow \qquad f(\varphi(a)) = f(x)$
$\Rightarrow \qquad f \circ \varphi(a) = f(x)$
$\Rightarrow \qquad I_F(a) = f(x)$ $\qquad [\because\ f \circ \varphi = I_F]$
$\Rightarrow \qquad f(x) = a$

and, $x \in \text{Ker}(f) \Rightarrow f(x) = 0_F$
$\therefore \qquad a = 0_F \in F$
$\Rightarrow \qquad \varphi(a) = \varphi(o_F) = 0_M \in M$
$\Rightarrow \qquad x = 0_M$ $\qquad [\because\ x = \varphi(a)]$

Thus, $x \in \text{Ker}(f) \cap F' \Rightarrow x = 0_M$
$\therefore \qquad \text{Ker}(f) \cap F' = \{0_M\}$

Hence, $M = \text{Ker}(f) \oplus F'$.

Since $\varphi : F \to M$ restricts to an R-isomorphism, therefore, $F \cong F'$.

Hence, M is direct sum of $\text{Ker}(f)$ and an R-submodule which is R-isomorphic to F.

EXERCISE 1.4

1. Every finitely generated module is homorphic image of a finitely generated free module. [Hint: See Theorem 7 on page 89]

2. Show that every module is a homomorphic image of a free module.

3. Show that every principal left ideal in an integral domain R with unity is free as a left R-module.

96 • *Theory and Problems of Linear Algebra*

4. Show that every ideal of Z is free as Z-module.
5. Prove that the direct product $M_1 \times M_2 \times \cdots \times M_n$ of free R-modules M_i is again free.
6. Let R be a commutative ring with unity, and let $e \neq 0, 1$ be an idempotent. Prove that Re cannot be a free R-module.
7. Let $\{a_1, a_2, \ldots, a_n\}$ be a basis of a free R-module M. Prove that
$M = Ra_1 \oplus Ra_2 \oplus \cdots \oplus Ra_n$.

1.8.1 FREE MODULES AND MATRICES

Let R be a ring with unity. Consider a linear mapping (an R-homomorphism) $t : R^n \to R^m$ from free R-module R^n to free R-module R^m. Since R^n and R^m are free R-modules with standard bases $(e_1^{(n)}, e_2^{(n)}, \ldots, e_n^{(n)})$ and $(e_1^{(m)}, e_2^{(m)}, \ldots, e_m^{(m)})$ respectively. Therefore, a linear mapping $t : R^n \to R^m$ can be described completely by simply listing the images $t(e_1^{(n)}), t(e_2^{(n)}), \ldots, t(e_n^{(n)})$ of n basis elements of R^n. Each of these n images is an element of R^m, thus a column of m scalars in R. Consequently, the whole list of columns $t(e_1^{(n)}), t(e_2^{(n)}), \ldots, t(e_n^{(n)})$ is an $m \times n$ matrix A over ring R which can be written as a list $(t(e_1^{(n)}), t(e_2^{(n)}), \ldots, t(e_n^{(n)}))$ of n columns $t(e_1^{(n)}), t(e_2^{(n)}), \ldots, t(e_n^{(n)})$, where each $t(e_j^{(n)}), j \in \underline{n}$ is a list of m scalars in R. Thus, each linear mapping $t : R^n \to R^m$ determines an $m \times n$ matrix A over ring R.

Conversely, let $A = [a_{ij}]$ be an $m \times n$ matrix over ring R. Then, A can be considered as a list $(A^1, A^2, \ldots, A^n)$ of n columns $A^1, A^2, \ldots, A^n$, where each column is a list of m scalars in R, that is, an element of free R-module R^m. Thus, the matrix A is a list of n elements (vectors) in R^m. Since R^m is free R-module with $(e_1^{(m)}, e_2^{(m)}, \ldots, e_m^{(m)})$ as a basis. Therefore, for each $j \in \underline{n}, A^j$ can be written as

$$A^j = \sum_{i=1}^{m} a_{ij} e_i^{(m)}, \ a_{ij} \in R \quad \text{for all } i \in \underline{m}.$$

Consider a mapping φ from the set $\left\{e_1^{(n)}, e_2^{(n)}, \ldots, e_n^{(n)}\right\}$ of free generators of R^n to R^m given by the rule

$$\varphi(e_j^{(n)}) = A^j \quad \text{for all } j \in \underline{n}.$$

Since R^n is free module over ring R with basis $\left\{e_1^{(n)}, e_2^{(n)}, \ldots, e_n^{(n)}\right\}$. Therefore, by Theorem 6 (page 88) φ can be extended to a unique linear mapping (or an R-homomorphism) $t_A : R^n \to R^m$ given by the rule

$$t_A(e_j^{(n)}) = A^j$$

$\Rightarrow \quad t_A(e_j^{(n)}) = \sum_{i=1}^{m} a_{ij} e_i^{(m)} \quad \text{for all } j \in \underline{n}.$

Thus, each $m \times n$ matrix $A = [a_{ij}]$ over ring R determines a unique linear mapping $t_A : R^n \to R^m$ given by

$$t_A(e_j^{(n)}) = \sum_{i=1}^{m} a_{ij} e_i^{(m)} \quad \text{for all } j \in \underline{n}.$$

This linear mapping is called *linear transformation* or *linear mapping* or *R-homomorphism* corresponding to matrix A and is denoted by t_A.

It follows from above discussion that the assignment $A \to t_A$ is a bijection from the set $R^{m \times n}$ of all $m \times n$ matrices over ring R to the set $\text{Hom}_R(R^n, R^m)$ of all linear mappings from R module R^n to R-module R^m.

Q.E.D.

NOTE *Throughout this section R will be a ring with unity unless stated otherwise.*

THEOREM-1 *Let A and B be $m \times n$ matrices over a ring R. Then,*

$$t_{A+B} = t_A + t_B.$$

PROOF. Let $A = [a_{ij}]$, $B = [b_{ij}]$ be two $m \times n$ matrices over ring R. Then, $A + B = [a_{ij} + b_{ij}]$. is an $m \times n$ matrix such that $(A+B)_{ij} = a_{ij} + b_{ij}$ for all $i \in \underline{m}, j \in \underline{n}$.

For any $j \in \underline{n}$, we have

$$t_{A+B}(e_j^{(n)}) = \sum_{i=1}^{m} (a_{ij} + b_{ij}) e_i^{(m)}$$

$\Rightarrow \quad t_{A+B}(e_j^{(n)}) = \sum_{i=1}^{m} [a_{ij} e_i^{(m)} + b_{ij} e_i^{(m)}]$ $\quad [\because R^m \text{ is a left } R\text{-module}]$

$\Rightarrow \quad t_{A+B}(e_j^{(n)}) = \sum_{i=1}^{m} a_{ij} e_i^{(m)} + \sum_{i=1}^{m} b_{ij} e_i^{(m)}$

$\Rightarrow \quad t_{A+B}(e_j^{(n)}) = t_A(e_j^{(n)}) + t_B(e_j^{(n)})$

$\Rightarrow \quad t_{A+B}(e_j^{(n)}) = (t_A + t_B)(e_j^{(n)})$

$\therefore \quad t_{A+B} = t_A + t_B$ on the basis $\left\{e_1^{(n)}, e_2^{(n)}, \ldots, e_n^{(n)}\right\}$ of R^n.

Hence, $t_{A+B} = t_A + t_B$ on R^n.

Q.E.D.

THEOREM-2 *Let A be a matrix over a commutative ring R. Then, $t_{(\lambda A)} = \lambda t_A$ for all $\lambda \in R$.*

PROOF. Let $A = [a_{ij}]$ be an $m \times n$ matrix over a commutative ring R. Then, $\lambda A = [\lambda a_{ij}]$ for all $\lambda \in R$.

For any $j \in \underline{n}$, we have

$$t_{(\lambda A)}(e_j^{(n)}) = \sum_{i=1}^{m} (\lambda a_{ij}) e_i^{(m)}$$

98 • *Theory and Problems of Linear Algebra*

$\Rightarrow \quad t_A\left(e_j^{(n)}\right) = \sum_{i=1}^{m} \lambda(a_{ij}e_i^{(m)})$ $\qquad [\because \quad R^m \text{ is a left } R\text{-module}]$

$\Rightarrow \quad t_A\left(e_j^{(n)}\right) = \lambda\left\{\sum_{i=1}^{m} a_{ij}e_i^{(m)}\right\}$ $\qquad [\because \quad R^m \text{ is a left } R\text{-module}]$

$\Rightarrow \quad t_A\left(e_j^{(n)}\right) = \lambda\left\{t_A(e_j^{(n)})\right\}$

Since R is a commutative ring. Therefore, λt_A is an R-homomorphism (or a linear mapping) (see Theorem 10 on page 93) given by

$$(\lambda t_A)(e_j^{(n)}) = \lambda[t_A(e_j^{(n)})]$$

$\therefore \quad t_{\lambda A}(e_j^{(n)}) = (\lambda t_A)(e_j^{(n)}) \quad \text{for all } j \in \underline{n}.$

$\Rightarrow \quad t_{\lambda A} = \lambda t_A \quad \text{on the basis } \left\{e_1^{(n)}, e_2^{(n)}, \ldots, e_n^{(n)}\right\} \text{ of } R^n.$

Hence, $\quad t_{\lambda A} = \lambda t_A \text{ on } R^n.$

Q.E.D.

REMARK. *It should be noted that the above result is true only for a commutative ring.*

THEOREM-3 *Let R be a commutative ring. Then the R-module $R^{m \times n}$ of all $m \times n$ matrices over ring R is R-isomorphic to the R-module $\mathrm{Hom}_R(R^n, R^m)$. That is, $R^{m \times n} \cong \mathrm{Hom}_R(R^n, R^m)$.*

PROOF. Define a mapping $\varphi : R^{m \times n} \to \mathrm{Hom}_R(R^n, R^m)$ given by

$$\varphi(A) = t_A \quad \text{for all } A \in R^{m \times n}$$

Since for each $m \times n$ matrix $A = [a_{ij}]$ over R there exists a unique linear mapping $t_A : R^n \to R^m$ given by

$$t_A(e_j^{(n)}) = \sum_{i=1}^{m} a_{ij}e_i^{(m)} \quad \text{for all } j \in \underline{n}.$$

Therefore, φ is well defined.

φ is injective: For any two matrices $A = [a_{ij}], B = [b_{ij}]$ in $R^{m \times n}$

$\varphi(A) = \varphi(B)$

$\Rightarrow \quad t_A = t_B$

$\Rightarrow \quad t_A(e_j^{(n)}) = t_B(e_j^{(n)}) \quad \text{for all } j \in \underline{n}$

$\Rightarrow \quad \sum_{i=1}^{m} a_{ij}e_i^{(m)} = \sum_{i=1}^{m} b_{ij}e_i^{(m)} \quad \text{for all } j \in \underline{n}$

$$\Rightarrow \quad \sum_{i=1}^{m}(a_{ij}-b_{ij})e_i^{(m)} = \underline{0} \in R^m \quad \text{for all } j \in \underline{n} \quad [\because R^m \text{ is a left } R\text{-module}]$$
$$\Rightarrow \quad a_{ij} - b_{ij} = 0 \quad \text{for all } i \in \underline{m} \text{ and all } j \in \underline{n}$$
$$\Rightarrow \quad a_{ij} = b_{ij} \quad \text{for all } i \in \underline{m}, j \in \underline{n}$$
$$\Rightarrow \quad A = B$$

∴ φ is injective

φ is surjective: For each $t: R^n \to R^m$, we get a list $t(e_1^{(n)}), t(e_2^{(n)}), \ldots, t(e_n^{(n)})$ of n column vectors in R^m. These column vectors define a matrix A whose n columns are $t(e_1^{(n)}), t(e_2^{(n)}), \ldots, t(e_n^{(n)})$ i.e,

$$A = \left(t(e_1^{(n)}), t(e_2^{(n)}), \ldots, t(e_n^{(n)}) \right)$$

Thus, for each R-module homomorphism $t: R^n \to R^m$ there exists a matrix A such that $\varphi(A) = t$. So, φ is surjective.

φ is an R-homomorphism: For any $A, B \in R^{m \times n}$ and any $\lambda \in R$, we have

$$\varphi(A+B) = t_{A+B} = t_A + t_B \quad \text{[By Theorem 1]}$$
$$\Rightarrow \quad \varphi(A+B) = \varphi(A) + \varphi(B)$$
and,
$$\varphi(\lambda A) = t_{(\lambda A)} = \lambda t_A \quad \text{[By Theorem 2]}$$
$$\Rightarrow \quad \varphi(\lambda A) = \lambda \varphi(A)$$

∴ φ is an R-homomorphism.

Thus, φ is an R-isomorphism of $R^{m \times n}$ onto $\text{Hom}_R(R^n, R^m)$.

Hence, $R^{m \times n} \cong \text{Hom}_R(R^n, R^m)$.

Q.E.D.

THEOREM-4 *Let A be an $m \times n$ matrix over a ring R. Then the linear mapping t_A of A is the zero R-homomorphism from R^n to R^m iff A is the null matrix over ring R.*

PROOF. First suppose that $A = [a_{ij}]$ is the null matrix over ring R. Then, $a_{ij} = 0$ for all $i \in \underline{m}, j \in \underline{n}$.

By definition, we have

$$t_A\left(e_j^{(n)}\right) = \sum_{i=1}^{m} a_{ij}\, e_i^{(m)} \quad \text{for all } j \in \underline{n}$$
$$\Rightarrow \quad t_A\left(e_j^{(n)}\right) = \sum_{i=1}^{m} 0\, e_i^m = \underline{0} \in R^m \quad \text{for all } j \in \underline{n}$$

Let a be an arbitrary element of R^n. Then there exist scalars $\lambda_1, \lambda_2, \ldots, \lambda_n \in R$ such that

$$a = \lambda_1 e_1^{(n)} + \lambda_2 e_2^{(n)} + \cdots + \lambda_n e_n^{(n)} \qquad [\because \{e_1^{(n)}, e_2^{(n)} \ldots, e_n^{(n)}\} \text{ is basis for } R^n]$$

$\Rightarrow \quad t_A(a) = t_A\left(\lambda_1 e_1^{(n)} + \lambda_2 e_2^{(n)} + \cdots + \lambda_n e_n^{(n)}\right)$

$\Rightarrow \quad t_A(a) = \lambda_1 t_A(e_1^{(n)}) + \lambda_2 t_A(e_2^{(n)}) + \cdots + \lambda_n t(e_n^{(n)}) \qquad [\because t_A \text{ is an } R\text{-homomorphism}]$

$\Rightarrow \quad t_A(a) = \lambda_1 \underline{0} + \lambda_2 \underline{0} + \cdots + \lambda_n \underline{0}$

$\Rightarrow \quad t_A(a) = \underline{0} \in R^m$

$\therefore \quad t_A(a) = \underline{0} \in R^m \quad$ for all $a \in R^n$.

$\Rightarrow \quad t_A = \widehat{0} \quad$ (zero R-homomorphism).

Conversely, suppose that $t_A : R^n \to R^m$ is the zero R-homomorphism. Then, we have to prove that A is the null matrix over ring R.

Now,

$t_A = \widehat{0}$

$\Rightarrow t_A\left(e_j^{(n)}\right) = \widehat{0}\left(e_j^{(n)}\right) \qquad$ for all $j \in \underline{n}$

$\Rightarrow t_A\left(e_j^{(n)}\right) = \underline{0} \in R^m \qquad$ for all $j \in \underline{n}$

$\Rightarrow \sum_{i=1}^{m} a_{ij} e_i^{(m)} = \underline{0} \in R^m \qquad$ for all $j \in \underline{n}$ \qquad [By def. of t_A]

$\Rightarrow a_{ij} = 0 \qquad$ for all $i \in \underline{m}, j \in \underline{n}$ $[\because \{e_1^{(m)}, \ldots, e_m^{(m)}\}$ is basis for $R^m]$

$\Rightarrow A$ is the null matrix over ring R.

Q.E.D.

THEOREM-5 *Let A be an $n \times n$ matrix over a ring R. Then the linear mapping t_A corresponding to matrix A is the identity homomorphism $I : R^n \to R^n$ iff A is the identity matrix over ring R.*

PROOF. Let $A = [a_{ij}]$ be the identity matrix over ring R. Then, $a_{ij} = \delta_{ij}$ for all $i, j \in \underline{n}$.

By definition, we have

$t_A\left(e_j^{(n)}\right) = \sum_{i=1}^{n} a_{ij} e_i^{(n)} \quad$ for all $j \in \underline{n}$

$\Rightarrow t_A\left(e_j^{(n)}\right) = \sum_{i=1}^{n} \delta_{ij} e_i^{(n)} \quad$ for all $j \in \underline{n}$

$\Rightarrow t_A\left(e_j^{(n)}\right) = e_j^{(n)} \quad$ for all $j \in \underline{n}$ \qquad $[\because \delta_{ij} = 0 \text{ for } i \neq j]$

Let a be an arbitrary element of R^n. Then,

$$a = \lambda_1 e_1^{(n)} + \lambda_2 e_2^{(n)} + \cdots + \lambda_n e_n^{(n)} \qquad [\because \ \{e_1^{(n)}, e_2^{(n)}, \ldots, e_n^{(n)}\} \text{ is basis for } R^n]$$

$\Rightarrow \quad t_A(a) = t_A\left(\lambda_1 e_1^{(n)} + \lambda_2 e_2^{(n)} + \cdots + \lambda_n e_n^{(n)}\right)$

$\Rightarrow \quad t_A(a) = \lambda_1 t_A(e_1^{(n)}) + \lambda_2 t_A(e_2^{(n)}) + \ldots + \lambda_n t_A(e_n^{(n)}) \ [\because \ t_A : R^n \to R^n \text{ is } R\text{-homomorphism}]$

$\Rightarrow \quad t_A(a) = \lambda_1 e_1^{(n)} + \lambda_2 e_2^{(n)} + \cdots + \lambda_n e_n^{(n)} \qquad [t_A(e_j^{(n)}) = e_j^{(n)} \text{ for all } j \in \underline{n}]$

$\Rightarrow \quad t_A(a) = a$

$\therefore \quad t_A(a) = a \quad$ for all $a \in R^n$.

$\Rightarrow \quad t_A = I \quad$ (identity R-homomorphism on R^n)

Conversely, let $t_A : R^n \to R^n$ be the identity R-homomorphism. Then we have to prove that $A = [a_{ij}]$ is the identity matrix over ring R.

Now, $t_A = I$

$\Rightarrow \quad t_A(e_j^{(n)}) = I(e_j^{(n)}) \qquad$ for all $j \in \underline{n}$

$\Rightarrow \quad \sum_{i=1}^{n} a_{ij} e_i^{(n)} = e_j^{(n)} \qquad$ for all $j \in \underline{n}$

$\Rightarrow \quad \sum_{i=1}^{n} a_{ij} e_i^{(n)} = \delta_{ij} e_i^{(n)} \qquad$ for all $i, j \in \underline{n}$

$\Rightarrow \quad \sum_{i=1}^{n} (a_{ij} - \delta_{ij}) e_i^{(n)} = \underline{0} \in R^n \ $ for all $i, j \in \underline{n}$

$\Rightarrow \quad a_{ij} - \delta_{ij} = 0 \qquad$ for all $i, j \in \underline{n} \qquad [\because \ \{e_1^{(n)}, e_2^{(n)}, \ldots, e_n^{(n)}\} \text{ is basis for } R^n]$

$\Rightarrow \quad a_{ij} = \delta_{ij} \qquad$ for all $i, j \in \underline{n}$

$\Rightarrow \quad A = [a_{ij}] = I$

Q.E.D.

In the foregoing discussion, we have seen that the linear transformation corresponding to the sum of two matrices of the same order is equal to the sum of the linear transformations of the two matrices. Also, the linear transformation corresponding to the identity matrix is an identity matrix and the linear transformation corresponding to null matrix is the null transformation or null morphism. Now the question arises, what is the linear transformation of the product of two matrices? Before proceeding to answer this question, let us first recall the definition of product of two matrices. Two matrices $A = [a_{ij}]$ and $B = [b_{ij}]$ are conformable for the product AB, if the number of columns in pre-multiplier A is equal to the number of rows in the post-multiplier B. Thus, if $A = [a_{ij}]_{m \times n}$ and $B = [b_{ij}]_{n \times p}$ are two matrices, then the product AB is an $m \times p$ matrix such that

$$(AB)_{ij} = \sum_{r=1}^{n} a_{ir} b_{rj} \text{ for all } i \in \underline{m} \text{ and } j \in \underline{p}$$

THEOREM-6 Let A and B be two matrices over a commutative ring R, then the composite linear transformation $t_A \circ t_B$ exists only when the matrix product AB exists and in that case $t_{AB} = t_A \circ t_B$.

PROOF. Let $A = [a_{ij}]_{m \times n}$ and $B = [b_{ij}]_{n \times p}$ be matrices over ring R. Then the corresponding linear transformations are $t_A : R^n \to R^m$ and $t_B : R^p \to R^n$ respectively. Consequently, the composite linear transformation $t_A \circ t_B$ maps R^p to R^m.

For any $j \in \underline{m}$, we have

$$(t_A \circ t_B)\left(e_j^{(p)}\right) = t_A\left(t_B(e_j^{(p)})\right)$$

$$\Rightarrow (t_A \circ t_B)\left(e_j^{(p)}\right) = t_A\left(\sum_{r=1}^{n} b_{rj}\, e_r^{(n)}\right) \qquad [\because\ t_B\left(e_j^{(p)}\right) = \sum_{r=1}^{n} b_{rj}\, e_r^{(n)}]$$

$$\Rightarrow (t_A \circ t_B)\left(e_j^{(p)}\right) = \sum_{r=1}^{n} b_{rj}\, t_A\left(e_r^{(n)}\right) \qquad [\because\ t_A : R^n \to R^m \text{ is a linear transformation}]$$

$$\Rightarrow (t_A \circ t_B)\left(e_j^{(p)}\right) = \sum_{r=1}^{n} b_{rj} \left\{\sum_{s=1}^{m} a_{sr}\, e_s^{(m)}\right\}$$

$$\Rightarrow (t_A \circ t_B)\left(e_j^{(p)}\right) = \sum_{r=1}^{n}\sum_{s=1}^{m} (b_{rj}\, a_{sr})\, e_s^{(m)}$$

$$\Rightarrow (t_A \circ t_B)\left(e_j^{(p)}\right) = \sum_{s=1}^{m}\left(\sum_{r=1}^{n} a_{sr}\, b_{rj}\right) e_s^{(m)} \qquad [\because\ R \text{ is a commutative ring}]$$

$$\Rightarrow (t_A \circ t_B)\left(e_j^{(p)}\right) = \sum_{s=1}^{m} (AB)_{sj}\, e_s^{(m)}$$

$$\Rightarrow (t_A \circ t_B)\left(e_j^{(p)}\right) = t_{AB}\left(e_j^{(p)}\right)$$

$\therefore\quad t_A \circ t_B = t_{AB}$ on the set of generators of R^p.

Hence, $\quad t_A \circ t_B = t_{AB}$ on R^p.

Q.E.D.

It has been proved in Theorem 3 that the function $\varphi : R^{m \times n} \to \mathrm{Hom}_R(R^n, R^m)$ associating each $m \times n$ matrix A over a ring R to its morphism (linear transformation) t_A is an isomorphism for addition and scalar multiplication, i.e. R-module isomorphism. The above theorem also establishes that φ is also an isomorphism for multiplication of matrices. We can, therefore, infer that all the properties which are true for module morphisms are also true for matrices and vice versa. For example, multiplication of matrices is associative and distributive over addition, because composition of module morphisms is associative and distributive over addition as proved in Theorem 8 on page 65.

If we restrict ourselves to $n \times n$ square matrices over a commutative ring R, then theorems proved in the foregoing discussion suggest the following theorem.

THEOREM-7 *The ring $R^{n\times n}$ of all $n \times n$ matrices over a commutative ring R is isomorphism to $\text{End}(R^n)$ (ring of all endomorphisms of the free R-module R^n) i.e. $R^{n\times n} \cong \text{End}(R^n)$.*

PROOF. Consider a mapping $\varphi : R^{n\times n} \to \text{End}(R^n)$ given by

$$\varphi(A) = t_A \quad \text{for all } A \in R^{n\times n}$$

Corresponding to each $A = [a_{ij}] \in R^{n\times n}$ there exists a unique linear mapping $t_A : R^n \to R^n$ given by

$$t_A\left(e_j^{(n)}\right) = \sum_{i=1}^{n} a_{ij}\, e_i^{(n)} \quad \text{for all } j \in \underline{n}$$

So, φ is well defined.

We shall now show that φ is an isomorphism.

φ *is injective:* For any two matrices $A = [a_{ij}], B = [b_{ij}] \in R^{n\times n}$

$\varphi(A) = \varphi(B)$
$\Rightarrow t_A = t_B$
$\Rightarrow t_A\left(e_j^{(n)}\right) = t_B\left(e_j^{(n)}\right)$ for all $j \in \underline{n}$
$\Rightarrow \sum_{i=1}^{n} a_{ij}\, e_i^{(n)} = \sum_{i=1}^{n} b_{ij}\, e_i^{(n)}$ for all $j \in \underline{n}$
$\Rightarrow \sum_{i=1}^{n} (a_{ij} - b_{ij})\, e_i^{(n)} = \underline{0} \in R^n$ for all $j \in \underline{n}$ $[\because \ R^n \text{ is an } R\text{-module}]$
$\Rightarrow (a_{ij} - b_{ij}) = 0$ for all $i \in \underline{m}$ and all $j \in \underline{n}$ $\begin{bmatrix} \because \ R^n \text{ is free } R\text{-module} \\ \text{and } e_i^{(n)}, i \in \underline{n} \text{ are in basis} \end{bmatrix}$
$\Rightarrow a_{ij} = b_{ij}$ for all $i \in \underline{m}, j \in \underline{n}$
$\Rightarrow A = B$

So, $\varphi : R^{n\times n} \to \text{End}(R^n)$ is injective.

φ *is surjective:* Let t be an arbitrary endomorphism in $\text{End}(R^n)$. Then, $\left(t(e_1^{(n)}), t(e_2^{(n)}), \ldots, t(e_n^{(n)})\right)$ is a list of n column vectors in R^n. These column vectors define a matrix A having $\left(t(e_1^{(n)}), t(e_2^{(n)}), \ldots, t(e_n^{(n)})\right)$ as a list of its columns. Thus, for each endomorphism $t \in \text{End}(R^n)$ there exists a matrix $A \in R^{n\times n}$ such that $\varphi(A) = t$.

So, φ is surjective.

φ is a homomorphism: Let $A = [a_{ij}]_{n \times n}$ and $b = [b_{ij}]_{n \times n}$ be any two matrices in $R^{n \times n}$. Then,

$$\varphi(A+B) = t_{A+B} = t_A + t_B \qquad \text{[By Theorem 1]}$$

$\Rightarrow \quad \varphi(A+B) = \varphi(A) + \varphi(B)$

and, $\quad \varphi(AB) = t_{AB} = t_A \circ t_B \qquad \text{[By Theorem 6]}$

$\Rightarrow \quad \varphi(AB) = \varphi_A \circ \varphi_B$

So, φ is a ring homomorphism.

Hence, $\quad R^{n \times n} \cong \text{End}(R^n)$.

THEOREM-8 Let $A = [a_{ij}]$ be an $m \times n$ matrix over a commutative ring R and $t_A : R^n \to R^m$ be the corresponding linear transformation. For any column vector (matrix) X in R^n, prove that $t_A(X) = AX$.

PROOF. Since $X \in R^n$ and R^n is free module with $\left(e_1^{(n)}, e_2^{(n)}, \ldots, e_n^{(n)}\right)$ as a list of free generators. Therefore, X can be written as

$$X = \sum_{j=1}^{n} x_j \, e_j^{(n)}, \quad \text{where } X = \begin{bmatrix} x_1 \\ x_2 \\ \vdots \\ x_n \end{bmatrix}, \, x_j \in R \text{ for all } j \in \underline{n}$$

$\Rightarrow \quad t_A(X) = t_A \left(\sum_{j=1}^{n} x_j \, e_j^{(n)} \right)$

$\Rightarrow \quad t_A(X) = \sum_{j=1}^{n} x_j \, t_A \left(e_j^{(n)} \right) \qquad [\because \, t_A : R^n \to R^m \text{ is a linear transformation}]$

$\Rightarrow \quad t_A(X) = \sum_{j=1}^{n} x_j \left(\sum_{i=1}^{m} a_{ij} \, e_i^{(m)} \right)$

$\Rightarrow \quad t_A(X) = \sum_{j=1}^{n} \sum_{i=1}^{m} (x_j \, a_{ij}) e_i^{(m)}$

$\Rightarrow \quad t_A(X) = \sum_{j=1}^{n} \sum_{i=1}^{m} (a_{ij} \, x_j) e_i^{(m)} \qquad [\because \, R \text{ is commutative}]$

$\Rightarrow \quad t_A(X) = \sum_{i=1}^{m} \left(\sum_{j=1}^{n} a_{ij} \, x_j \right) e_i^{(m)}$

$$\therefore \quad i^{th} \text{ coordinate of } t_A(X) = \sum_{j=1}^{n} a_{ij} x_j$$

$$\Rightarrow \quad i^{th} \text{ coordinate of } t_A(X) = [a_{i1}\ a_{i2}\ldots a_{in}] \begin{bmatrix} x_1 \\ x_2 \\ \vdots \\ x_n \end{bmatrix} = (i^{th} \text{ row of } A) X, \quad \text{for all } i \in \underline{n}$$

Hence, $t_A(X) = AX$.

Q.E.D.

EXERCISE 1.5

1. Let $A = [a_{ij}]_{m \times n}$ and $B = [b_{ij}]_{n \times p}$ be two matrices over a commutative ring R. Then prove that

    (i) $t_{(AB)^T} = t_{B^T} o t_{A^T}$

    (ii) $t_{(AB)\lambda} = t_{A(B\lambda)} = t_{(A\lambda)B}$ for some scalar λ

2. Let $A = [a_{ij}]_{m \times n}$, $B = [b_{ij}]_{n \times p}$ and $C = [c_{ij}]_{n \times p}$ be matrices over a commutative ring R. Then,

    $$t_{(AB)C} = t_{A(BC)} = t_A o t_{BC} = t_A o (t_B o t_C).$$

3. Let $A = [a_{ij}]_{m \times n}$, $B = [b_{ij}]_{n \times p}$ and $C = [c_{ij}]_{n \times p}$ be matrices over a commutative ring R. Then, prove that

    $$t_{A(B+C)} = t_A o (t_B + t_C) = t_A o t_B + t_A o t_C$$

4. Let A be an $n \times n$ matrix over a commutative ring R and I be the identity matrix. Then,

    $$t_A = t_A o t_I$$

1.9 CYCLIC MODULES

CYCLIC MODULE *An R-module M is called a cyclic module if it is generated by a single element in it.*

Thus, an R-module M is called a cyclic module if there exists an element $a \in M$ such that $M = [a]$.

Example 2 (page 44) shows that a cyclic module generated by a is $\{ra + na : r \in R, n \in Z\}$, and if R has unity then $Ra = \{ra : r \in R\}$ is a cyclic module generated by a. Also $R = R.1$ shows that R is a cyclic module over itself, and is generated by 1. Theorem 10 (page 53) shows that $Ra_1 + Ra_2 + \cdots + Ra_n = \{r_1 a_1 + r_2 a_2 + \cdots + r_n a_n : r_i \in R\}$ is a module generated by the set $\{a_1, a_2, \ldots, a_n\}$ but it is not cyclic.

THEOREM-1 *Let R be a Euclidean ring. Then any finitely generated R-module is direct sum of finite number of cyclic modules.*

PROOF. Let M be a finitely generated R-module of rank n. We will prove the theorem by induction on n. If $n = 1$, then M is generated by a single element, hence it is cyclic and theorem is proved.

Suppose that the theorem is true for all R-modules of rank $(n-1)$. In other words, assume that each module of rank $(n-1)$ is direct sum of finite number of its cyclic submodules.

We know that an R-module may have many minimal generating sets, so if $\{a_1, a_2, \ldots, a_n\}$ is a minimal generating set of M such that

$$r_1 a_1 + r_2 a_2 + \cdots + r_n a_n = 0_M \Rightarrow r_1 a_1 = r_2 a_2 = \cdots = r_n a_n = 0_M$$

Then, obviously M is direct sum of $N_1, N_2, \ldots, N_n$ where each N_i is a cyclic module generated by a_i. Thus, in this case the theorem is true for M.

Now, let $\{a_1, a_2, \ldots, a_n\}$ be a minimal generating set for M such that

$$r_1 a_1 + r_2 a_2 + \cdots + r_n a_n = 0_M \text{ but not all } r_i a_i \text{ are } 0_M$$

Among all possible such relations for all minimal generating sets, let $s_1 \in R$ be such that $d(s_1)$ is the smallest positive integer (d is Euclidean map). Let the minimal generating set for which it occurs be $\{b_1, b_2, \ldots, b_n\}$. Thus,

$$s_1 b_1 + s_2 b_2 + \cdots + s_n b_n = 0_M \quad \text{for some } s_i \in R \qquad \ldots \text{(i)}$$

We claim that if

$$r_1 b_1 + r_2 b_2 + \cdots + r_n b_n = 0_M \qquad \ldots \text{(ii)}$$

then $s_1 | r_1$.

Since $s_1, r_1 \in R$ and R is an Euclidean ring. Therefore, there exists $p_1, q_1 \in R$ such that $r_1 = p_1 s_1 + q_1$, where either $q_1 = 0$ or $d(q_1) < d(s_1)$.

Multiplying (i) by p_1 and subtracting it from (II), we obtain

$$(r_1 - ps_1) b_1 + (r_2 - ps_2) b_2 + \cdots + (r_n - ps_n) b_n = 0_M$$
$$\Rightarrow q_1 b_1 + (r_2 - p_1 s_2) b_2 + \cdots + (r_n - p_1 s_n) b_n = 0_M \qquad [\because r_1 = p_1 s_1 + q_1] \ldots \text{(iii)}$$

If $q_1 \neq 0$, then $d(q_1) < d(s_1)$ and therefore (iii) contradicts that $d(s_1)$ is the smallest positive integer. Therefore, we must have

$$q_1 = 0 \Rightarrow r_1 = p_1 s_1 \Rightarrow s_1 | r_1.$$

We further claim that $s_1 | s_i$ for $i = 2, 3, \ldots, n$ and to assert we show that $s_1 | s_2$. Since $s_1, s_2 \in R$, so there exist $p_2, q_2 \in R$ such that $s_2 = p_2 s_1 + q_2$ where either $q_2 = 0$ or, $d(q_2) < d(s_1)$. Clearly, the set of elements $\{b_1' = b_1 + p_2 b_2, b_2, b_3, \ldots, b_n\}$ generates M and

$$s_1 b_1' + q_2 b_2 + s_3 b_3 + \cdots + s_n b_n$$
$$= s_1 (b_1 + p_2 b_2) + q_2 b_2 + \cdots + s_n b_n$$
$$= s_1 b_1 + (s_1 p_2 + q_2) b_2 + \cdots + \cdots + s_n b_n$$
$$= s_1 b_1 + s_2 b_2 + \cdots + s_n b_n = 0_M \qquad \ldots \text{(iv)}$$

If $q_2 \neq 0$, then $d(q_2) < d(s_1)$. Therefore,
$$s_1 b_1' + q_2 b_2 + \cdots + s_n b_n = 0_M \text{ contradicts our choice of } s_1.$$
So we must have $q_2 = 0 \Rightarrow s_2 = p_2 s_1 \Rightarrow s_1 | s_2$.

Similarly, it can be shown that $s_1 | s_i, i = 3, \ldots n$ and we can write $s_i = p_i s_1, i = 3, \ldots, n$.

Now the set $\{b_1^* = b_1 + p_2 b_2 + \cdots + p_n b_n, b_2, b_3, \ldots, b_n\}$ is a generating set for R-module M. Let N_1 be the cyclic module generated by b_1^* and N_2 be the R-submodule generated by $b_2, b_3, \ldots, b_n$. Then $M = N_1 + N_2$ since the set $\{b_1^*, b_2, \ldots, b_n\}$ generates M.
If $b \in N_1 \cap N_2$, then $b \in N_1$ and $b \in N_2$.
Now, $b \in N_1 \Rightarrow b = \lambda_1 b_1^*$ for some $\lambda_1 \in R$, and $b \in N_2$
$\Rightarrow b = \lambda_2 b_2 + \cdots + \lambda_n b_n$ for some $\lambda_2, \lambda_3, \ldots, \lambda_n \in R$.

$\therefore \lambda_1 b_1^* = \lambda_2 b_2 + \cdots + \lambda_n b_n$
$\Rightarrow \lambda_1 b_1^* - \lambda_2 b_2 - \cdots - \lambda_n b_n = 0_M$
$\Rightarrow \lambda_1 (b_1 + p_2 b_2 + \cdots + p_n b_n) - \lambda_2 b_2 \cdots - \lambda_n b_n = 0_M$
$\Rightarrow \lambda_1 b_1 + (\lambda_1 p_2 - \lambda_2) b_2 + \cdots + (\lambda_1 p_n - \lambda_n) b_n = 0_M$
$\Rightarrow s_1 | \lambda_1$ [Using (ii)]
$\Rightarrow \lambda_1 = s_1 k_1 \quad$ for some $k_1 \in R$.

Now,
$$b = \lambda_1 b_1^* = (s_1 k_1) b_1^* = k_1 (s_1 b_1^*) \qquad [\because \ R \text{ is commutative}]$$
$\Rightarrow b = k_1 [s_1 (b_1 + p_2 b_2 + \cdots + p_n b_n)]$
$\Rightarrow b = k_1 [s_1 b_1 + s_1 (p_2 b_2) + \cdots + s_1 (p_n b_n)]$
$\Rightarrow b = k_1 [s_1 b_1 + (p_2 s_1) b_2 + \cdots + (p_n s_1) b_n]$
$\Rightarrow b = k_1 [s_1 b_1 + s_2 b_2 + \cdots + s_n b_n] \qquad [\because \ s_i = p_i s_1 \text{ for all } i = 2, 3, \ldots, n]$
$\Rightarrow b = k_1 0_M = 0_M \qquad$ [Using (iv)]

Thus, $b \in N_1 \cap N_2 \Rightarrow b = 0_M$
$\therefore \ N_1 \cap N_2 = \{0_M\}$

Hence, $M = N_1 \oplus N_2$

Since M_2 is generated by $b_2, b_3, \ldots, b_n$. Its rank is $(n-1)$, so by induction assumption N_2 is direct sum of cyclic submodules. Hence, M is direct sum of cyclic modules.

Q.E.D.

COROLLARY. *A finite abelian group is the direct product (sum) of cyclic groups.*

PROOF. It is a simple consequence of the main theorem as a module is an abelian group.

THEOREM-2 Let R be a ring with unity. An R-module M is cyclic iff $M \cong \dfrac{R}{I}$ for some left ideal I of R.

PROOF. First, let M be a cyclic module generated by a. Then, $M = Ra$.

Let $I = \{r \in R : ra = 0_M\}$.

Let us first show that I is a left ideal of R.

Clearly, $0 \in R$ and $0a = 0_M \Rightarrow 0 \in I \Rightarrow I$ is non-void subset of R.

Let $r_1, r_2 \in I$. Then,

$r_1 a = 0_M$ and $r_2 a = 0_M$
$\Rightarrow \quad r_1 a - r_2 a = 0_M$
$\Rightarrow \quad (r_1 - r_2) a = 0_M \qquad$ [$\because$ Multiplication is distributive over addition]
$\Rightarrow \quad r_1 - r_2 \in I$

Let $r \in R$ and $r_1 \in I$. Then,

$r_1 a = 0_M$
$\Rightarrow \quad r(r_1 a) = r\, 0_M$
$\Rightarrow \quad (rr_1) a = 0_M \qquad$ [$\because$ $r\, 0_M = 0_M$ and multiplication is associative on R]
$\Rightarrow \quad r_1 r_1 \in I$

So, I is a left ideal of R.

Consider a mapping $f : R \to M$ defined by $f(r) = ra$ for all $r \in R$.

f is an R-homomorphism: Let $r, s, t \in R$. Then,

$$f(r+s) = (r+s)a = ra + sa = f(r) + f(s)$$
$$f(tr) = (tr)a = t(ra) = tf(r)$$

So, f is an R-homomorphism.

f is onto: Let $x \in M$. Then, $x = ra$ for some $r \in R$.

Thus, for each $x \in M$, there exists $r \in R$ such that $f(r) = ra = x$

So, $f : R \to M$ is onto.

Kernel of f is I: Let $r \in \text{Ker}(f)$. Then,

$$r \in \text{Ker}(f) \Leftrightarrow f(r) = 0_M \in M \Leftrightarrow ra = 0_M \Leftrightarrow r \in I$$

$\therefore \quad \text{Ker}(f) = I.$

Hence, by the fundamental theorem on R-homomorphisms $\dfrac{R}{I} \cong M$.

Conversely, let R be a ring with unity and M be an R-module such that $M \cong \dfrac{R}{I}$ for some left ideal I of R. We have to prove that M is cyclic.

Let $I + a \in \dfrac{R}{I}$. Then,

$$(I+1)(I+a) = I + (1a) \qquad [\because R \text{ is a ring with unity} \therefore 1 \in R]$$
$$\Rightarrow \quad (I+1)(I+a) = I + a$$

Thus, each element $I + a$ of $\dfrac{R}{I}$ is of the form $(I+1)(I+a)$.

So, $\dfrac{R}{I}$ is a left R-module generated by $1 + I$, i.e. $1 + R(1+I) = \dfrac{R}{I}$.

Consequently, $\dfrac{R}{I}$ is a cyclic module generated by $1 + I$

But, $M \cong \dfrac{R}{I}$. Hence, M is a cyclic module.

Q.E.D.

1.10 NOETHERIAN AND ARTINIAN MODULES

ASCENDING CHAIN CONDITION *An R-module M is said to possess ascending (descending) chain condition on R-submodules if for every ascending (descending) sequence of R-submodules of M,*

$$M_1 \subset M_2 \subset M_3 \cdots \quad (M_1 \supset M_2 \supset M_3 \cdots),$$

there exists a positive integer k such that

$$M_k = M_{k+1} = M_{k+2} = \cdots$$

Thus, ascending chain condition (abbreviated Acc) means that every ascending sequence of R-submodules is finite. Similarly, descending chain condition (abbreviated Dcc) means that every descending sequence of R-submodules is finite.

NOETHERIAN MODULE *An R-module M is called noetherian (artinian) if Acc (Dcc) for submodules holds in M.*

If an R-module M is noetherian (artinian), then we also say that M has Acc (Dcc) on submodules, or simply that M has Acc (Dcc).

EXAMPLE-1 *Show that the ring Z of integers as a Z-module is noetherian but not artinian.*

SOLUTION Z is a PID and every ascending chain of ideals (submodules) of a PID is finite. In fact, any ascending chain of ideals in Z starting with n can have at most n distinct terms. This shows that Z as a Z-module is noetherian. But, Z as a Z-module has an infinite properly descending chain
$$[n] \supset [n^2] \supset [n^3] \supset \cdots$$
of ideals (submodules) showing that Z is not artinian Z-module.

EXAMPLE-2 *Let P be the additive group of rational numbers whose denominators are powers of a fixed prime p, i.e. $P = \left\{ \dfrac{m}{p^k} : m \in Z, k = 0, 1, 2, \ldots \right\}$. Then, P is an abelian group. Therefore, P is a Z-module. We have the ascending chain of submodules*
$$Z \subsetneq \left[\frac{1}{p}\right] \subsetneq \left[\frac{1}{p^2}\right] \subsetneq \left[\frac{1}{p^3}\right] \subsetneq \cdots$$
It can be easily seen that the submodules of this chain are the only submodules of P containing Z. Therefore, $M = P|Z$ is artinian but not noetherian. On the other hand, since Z is a submodule of P, therefore P is neither artinian nor noetherian.

EXAMPLE-3 *Any finite abelian group is both artinian and noetherian Z-module.*

EXAMPLE-4 *Every finite dimensional vector space is both noetherian and artinian.*

SOLUTION Let V be an n-dimensional vector space over a field F. Let $S_1 \subsetneq S_2 \subsetneq S_3 \subsetneq \cdots \subsetneq V$ ($S_1 \supsetneq S_2 \supsetneq S_3 \cdots \supseteq$) be an ascending (or a descending) chain of subspaces of V. We know that if S is a proper subspace of V, then $\dim S < \dim V = n$. Thus, any properly ascending (or descending) chain of subspaces cannot have more than $n+1$ terms. Hence, V is both noetherian and artinian.

Before we give more examples, let us prove some theorems which provide us criteria for a module to be noetherian or artinian.

THEOREM-1 *An R-module M is noetherian iff every submodule of M is finitely generated.*

PROOF. Let M be a noetherian R-module. Then we have to show that every R-submodule of M is finitely generated. If possible, let N be an R-submodule of M which is not finitely generated. Then no finite list of its elements can generate N. Let $a_1 \in N$, and let N_1 be the submodule generated by a_1. Clearly $N_1 \subsetneq N$. Choose $a_2 \in N$ such that $a_2 \notin N_1$. Let N_2 be the submodule generated by $\{a_1, a_2\}$. Then, $N_1 \subsetneq N_2$. Again by the same argument there exists an element $a_3 \in N$ such that $a_3 \notin N_2$. Let N_3 be the submodule generated by the set $\{a_1, a_2, a_3\}$. Then, $N_1 \subsetneq N_2 \subsetneq N_3$. Proceeding in this manner we obtain a properly ascending chain
$$N_1 \subsetneq N_2 \subsetneq N_3 \subsetneq \cdots$$
of R-submodules of M which is not finite. This contradicts the fact that M is noetherian. Hence, every R-submodule of M is finitely generated.

Conversely, let M be an R-module such that every R-submodule of M is finitely generated. Then we have to show that M is noetherian.

Let $N_1 \subset N_2 \subset N_3 \subset \cdots$, be an ascending chain of R-submodules of M, and let $N = \bigcup_i N_i$. Then, N is an R-submodule of M. Since each R-submodule of M is finitely generated. Therefore, so is N. Let $N = [a_1, a_2, \ldots, a_n]$. Since each $a_i \in N$ and $N = \bigcup_{i=1} N_i$. Therefore, each a_i is in some module of the sequence, say N_{m_i}. Let $m = max\,(m_1, m_2, \ldots, m_n)$. Then all a_i are in N_m. So $N_m = N$.
Hence, $N_m = N_{m+1} = \cdots = N$
Consequently, M is noetherian.

Q.E.D.

MAXIMAL ELEMENT *Let M be an R-module, and let S be a non-void family of R-submodules of M. Then an element M_0 of S is said to be maximal in S if for each N_0 in S,*

$$N_0 \supset M_0 \Rightarrow N_0 = M_0.$$

In other words, an R-submodule M_0 is maximal in S iff there exists no R-submodule N_0 in S satisfying $M_0 \subset N_0$.

THEOREM-2 *An R-module M is noetherian iff every non-void family S of R-submodules of M has a maximal elements.*

PROOF. Let M be a noetherian R-module, and let S be a non-void family of R-submodules of M. Let N_1 be an element of S. If N_1 is not maximal, then it is properly contained in an R-submodule $N_2 \in S$. If N_2 is maximal, then theorem is proved. If N_2 is not maximal, then it is properly contained in an R-submodule $N_3 \in S$. In case S has no maximal element, we obtain an infinite sequence $N_1 \subset N_2 \subset N_3 \subset \ldots$ of R-submodules of M. This contradicts the fact that M is noetherian. Hence, S has a maximal element.

Conversely, let every non-void family of R-submodules of M has a maximal element. Then, we have to prove that M is noetherian R-module. Let $N_1 \subset N_2 \subset N_3 \subset \ldots$ be an ascending chain of R-submodules of M. Then by hypothesis, the family S of all these submodules has a maximal element, say N_k. But, then $N_k = N_{k+1} = \ldots$. Hence, M is noetherian.

Q.E.D.

Combining Theorems 1 and 2, we obtain the following theorem.

THEOREM-3 *Let M be an R-module. Then the following are equivalent:*

(i) M is noetherian.
(ii) Every submodule of M is finitely generated.
(iii) Every non-void family S of submodules of M has a maximal element.

The following theorem is dual to the above theorem.

THEOREM-4 *Let M be an R-module. Then the following are equivalent:*

(i) *M is artinian.*

(ii) *Every submodule of M is finitely generated.*

(iii) *Every non-void family S of submodules of M has a minimal element (that is, a submodule M_0 in S such that for any submodule N_0 in S with $N_0 \subset M_0$, we have $N_0 = M_0$).*

PROOF. Left as an exercise for the reader.
Let us now discuss some properties of noetherian modules.

Q.E.D.

THEOREM-5 *Any submodule of a noetherian (artinian) module is noetherian.*

PROOF. Let N be an R-submodule of a noetherian R-module M. Since any R-submodule of N is also an R-submodule of M, therefore any ascending chain of R-submodules of N is also of M. But, M is noetherian. Therefore, N is also noetherian.

Q.E.D.

THEOREM-6 *Every R-homomorphic image of a noetherian (artinian) R-module is noetherian.*

PROOF. Let M' be an R-homomorphic image of a noetherian R-module M under an R-homomorphism f. Then by the fundamental theorem of R-homomorphisms

$$M/\mathrm{Ker}(f) \cong M'.$$

We know that each R-submodule of $M/\mathrm{Ker}(f)$ is of the form $N/\mathrm{Ker}(f)$, where N is an R-submodule of M containing $\mathrm{Ker}(f)$.

Since M is noetherian. Therefore, N is finitely generated. Let $\{a_1, a_2, \ldots, a_n\}$ be the set of generators of N.

Then the set $\{a_1 + \mathrm{Ker}(f), a_2 + \mathrm{Ker}(f), \ldots, a_n + \mathrm{Ker}(f)\}$ would generate $N/\mathrm{Ker}(f)$. Therefore, every submodule of $N/\mathrm{Ker}(f)$ is finitely generated. Consequently, every R-submodule $M/\mathrm{Ker}(f)$ is finitely generated and thus $M/\mathrm{Ker}(f)$ is noetherian. Hence, M' is noetherian.

Q.E.D.

THEOREM-7 *Let M be an R-module, and let N be an R-submodule of M. Then M is noetherian (artinian) iff both N and M/N are noetherian (artinian).*

PROOF. Let M be a noetherian R-module. Then by Theorems 5 and 6, N and M/N are noetherian.

Conversely, let N and M/N be noetherian, and let

$$N_1 \subset N_2 \subset N_3 \ldots$$

be an ascending chain of R-submodules of M contained in N. Since N is noetherian. Therefore, there exists a positive integer n such that

$$N_n \cap N = N_{n+k} \cap N \quad \text{for all } k.$$

But, for all k, $N_k + N$ is an R-submodule of M containing N. Therefore,
$$(N_1 + N/N) \subset (N_2 + N/N) \subset (N_3 + N/N) \subset \ldots,$$
is an ascending chain of R-submodules of M/N. As M/N is noetherian there exists a positive integer m such that
$$(N_m + N/N) = (N_{m+k} + N/N) \quad \text{for all } k.$$
If $n_0 = \max(n,m)$, then
$$N_n \cap N = N_{n_0} \cap N \text{ and } (N_n + N/N) = (N_{n_0} + N/N) \quad \text{for all } n \geq n_0.$$
But, then $N_n + N = N_{n_0} + N$ for all $n \geq n_0$.
We claim that $N_n = N_{n_0}$ for all $n \geq n_0$.
To establish our claim, we proceed as follows:
For all $n \geq n_0$, we have

$N_n = N_n \cap (N_n + N) = N_n \cap (N_{n_0} + N)$ $\quad [\because\ N_n + N = N_{n_0} + N \text{ for all } n \geq n_0]$
$\Rightarrow N_n = N_{n_0} + (N_n \cap N)$ $\quad [\because\ N_{n_0} \subset N_n \text{ for all } n \geq n_0]$
$\Rightarrow N_n = N_{n_0} + (N_{n_0} \cap N)$ $\quad [\because\ N_n \cap N = N_{n_0} \cap N \text{ for all } n \geq n_0]$
$\Rightarrow N_n = N_{n_0}$ $\quad [\because\ N_{n_0} \cap N \subset N_{n_0}]$

Thus, $N_n = N_{n_0}$ for all $n \geq n_0$.
Hence, there exists a positive integer n_0 such that $N_{n_0} = N_{n_0+k}$ for all k.
Hence, M is noetherian.

Q.E.D.

EXERCISES 1.6

1. Mark each of the following as true or false.
   (i) An R-homomorphism is a homomorphism of abelian groups.
   (ii) The trivial linear combination of any set of elements of an R-module M is not necessarily the zero in M.
   (iii) Every subset of a linearly independent set is linearly independent.
   (iv) Every superset of a linearly independent set is linearly independent.
   (v) The void set is linearly independent.
   (vi) Every R-module is a free R-module.
   (vii) $\{0_M\}$ is a free R-module.
   (viii) For each $n \in N, R^n$ is a free R-module, where R is a commutative ring with unity.
   (ix) Every R-module has a basis.
   (x) Any two bases for an R-module over a commutative ring R have the same number of elements.

(xi) Every noetherian module is artinian.

(xii) Every artinian module is noetherian.

(xiii) Every finite abelian group is both noetherian and artinian.

(xiv) Any submodule of a noetherian (artinian) module is noetherian (artinian).

(xv) Every R-homomorphic image of a noetherian (artinian) module is noetherian (artinian).

2. If M is an irreducible R-module such that $ra \neq 0_M$ for some $r \in R$ and $a \in M$, prove that $\text{End}(M)$, i.e. the set of all homomorphisms from M to itself is a division ring.
3. Let M and M' be R-modules, and let $f : M \to M'$ be an R-epimorphism with kernel N. Then M is noetherian if M' and N are noetherian.
4. Prove that Q is not a noetherian Z-module.
5. If F is a field, then show that an F-module M is noetherian iff it is finite dimensional.
6. Let R be a commutative ring with unity, and let $e \neq 0, 1$ be an idempotent. Prove that R_e cannot be a free R-module.
7. Prove that the direct product $M_1 \times M_2 \times \cdots \times M_k$ of free R-modules M_i is again a free R-module.
8. Prove that Q is not a free Z-module.

ANSWERS

1. (i) T (ii) F (iii) T (iv) F (v) T (vi) F (vii) T (viii) T (ix) F (x) T (xi) F (xii) F (xiii) T (xiv) T (xv) T

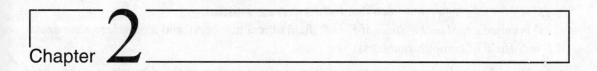

Vector Spaces

2.1 INTRODUCTION

In various practical and theoretical problems, we come across a set V whose elements may be vectors in two or three dimensions or real-valued functions, which can be added and multiplied by a constant (number) in a natural way, the result being again an element of V. Such concrete situations suggest the concept of a vector space. Vector spaces play an important role in many branches of mathematics and physics. In analysis infinite dimensional vector spaces (in fact, normed vector spaces) are more important than finite dimensional vector spaces while in linear algebra finite dimensional vector spaces are used, because they are simple and linear transformations on them can be represented by matrices. This chapter mainly deals with finite dimensional vector spaces.

2.2 DEFINITION AND EXAMPLES

VECTOR SPACE. *An algebraic structure $(V, F, \oplus, \odot)$ consisting of a non-void set V, a field F, a binary operation $\oplus$ on V and an external mapping $\odot : F \times V \to V$ associating each $a \in F, v \in V$ to a unique element $a \odot v \in V$ is said to be a vector space over field F, if the following axioms are satisfied:*

V-1 $(V, \oplus)$ is an abelian group.
V-2 For all $u, v \in V$ and $a, b \in F$, we have
    (i) $a \odot (u \oplus v) = a \odot u \oplus a \odot v$,
    (ii) $(a+b) \odot u = a \odot u \oplus b \odot u$,
    (iii) $(ab) \odot u = a \odot (b \odot u)$,
    (iv) $1 \odot u = u$.

The elements of V are called vectors and those of F are called scalars. The mapping $\odot$ is called scalar multiplication and the binary operation $\oplus$ is termed vector addition.

If there is no danger of any confusion we shall say V is a vector space over a field F, whenever the algebraic structure $(V, F, \oplus, \odot)$ is a vector space. Thus, whenever we say that V is a vector space over a field F, it would always mean that $(V, \oplus)$ is an abelian group and $\odot : F \times V \to V$ is a mapping such that V-2 (i)–(iv) are satisfied.

V is called a *real vector space* if $F = R$ (field of real numbers), and a *complex vector space* if $F = C$ (field of complex numbers).

REMARK-1 *V is called a left or a right vector space according as the elements of a skew-field F are multiplied on the left or right of vectors in V. But, in case of a field these two concepts coincide.*

REMARK-2 *The symbol '+' has been used to denote the addition in the field F. For the sake of convenience, in future, we shall use the same symbol '+' for vector addition $\oplus$ and addition in the field F. But, the context would always make it clear as to which operation is meant. Similarly, multiplication in the field F and scalar multiplication $\odot$ will be denoted by the same symbol '·'.*

REMARK-3 *In this chapter and in future also, we will be dealing with two types of zeros. One will be the zero of the additive abelian group V, which will be known as the vector zero and other will be the zero element of the field F which will be known as scalar zero. We will use the symbol 0_V to denote the zero vector and 0 to denote the zero scalar.*

REMARK-4 *Since $(V, \oplus)$ is an abelian group, therefore, for any $u, v, w \in V$ the following properties will hold:*

(i) $u \oplus v = u \oplus w \Rightarrow v = w$ \} (Cancellation laws)
(ii) $v \oplus u = w \oplus u \Rightarrow v = w$
(iii) $u \oplus v = 0_v \Rightarrow u = -v$ and $v = -u$
(iv) $-(-u) = u$
(v) $u \oplus v = u \Rightarrow v = 0_v$
(vi) $-(u \oplus v) = (-u) \oplus (-v)$
(vii) 0_v is unique
(viii) For each $u \in V$, $-u \in V$ is unique.

REMARK-5 *If V is a vector space over a field F, then we will write $V(F)$.*

ILLUSTRATIVE EXAMPLES

EXAMPLE-1 *Every field is a vector space over its any subfield.*

SOLUTION Let F be a field, and let S be an arbitrary subfield of F. Since F is an additive abelian group, therefore, V-1 holds.

Taking multiplication in F as scalar multiplication, the axioms V-2 (i) to V-2 (iii) are respectively the right distributive law, left distributive law and associative law. Also V-2 (iv) is property of unity in F.

Hence, F is a vector space over S.

Since S is an arbitrary sub field of F, therefore, every field is a vector space over its any subfield.

REMARK. *The converse of the above Example is not true, i.e. a subfield is not necessarily a vector space over its over field. For example, R is not a vector space over C, because multiplication of a real number and a complex number is not necessarily a real number.*

EXAMPLE-2 *R is a vector space over Q, because Q is a subfield of R.*

EXAMPLE-3 *C is a vector space over R, because R is a subfield of C.*

EXAMPLE-4 *Every field is a vector space over itself.*

SOLUTION Since every field is a subfield of itself. Therefore, the result directly follows from Example 1.

REMARK. *In order to know whether a given non-void set V forms a vector space over a field F, we must proceed as follows:*
  (i) Define a binary operation on V and call it vector addition.
  (ii) Define scalar multiplication on V, which associates each scalar in F and each vector in V to a unique vector in V.
  (iii) Define equality of elements (vectors) in V.
  (iv) Check whether V-1 and V-2 are satisfied relative to the vector addition and scalar multiplication thus defined.

EXAMPLE-5 *The set $F^{m \times n}$ of all $m \times n$ matrices over a field F is a vector space over F with respect to the addition of matrices as vector addition and multiplication of a matrix by a scalar as scalar multiplication.*

SOLUTION For any $A, B, C \in F^{m \times n}$, we have
  (i) $(A+B)+C = A+(B+C)$ (Associative law)
  (ii) $A+B = B+A$ (Commutative law)
  (iii) $A+O = O+A = A$ (Existence of Identity)
  (iv) $A+(-A) = O = (-A)+A$ (Existence of Inverse)

Hence, $F^{m \times n}$ is an abelian group under matrix addition.

If $\lambda \in F$ and $A \in F^{m \times n}$, then $\lambda A \in F^{m \times n}$ and for all $A, B \in F^{m \times n}$ and for all $\lambda, \mu \in F$, we have
  (i) $\lambda(A+B) = \lambda A + \lambda B$
  (ii) $(\lambda + \mu)A = \lambda A + \mu A$
  (iii) $\lambda(\mu A) = (\lambda \mu) A$
  (iv) $1A = A$

Hence, $F^{m \times n}$ is a vector space over F.

EXAMPLE-6 *The set $R^{m \times n}$ of all $m \times n$ matrices over the field R of real numbers is a vector space over the field R of real numbers with respect to the addition of matrices as vector addition and multiplication of a matrix by a scalar as scalar multiplication.*

SOLUTION Proceed parallel to Example 5.

REMARK. *The set $Q^{m \times n}$ of all $m \times n$ matrices over the field Q of rational numbers is not a vector space over the field R of real numbers, because multiplication of a matrix in $Q^{m \times n}$ and a real number need not be in $Q^{m \times n}$. For example, if $\sqrt{2} \in R$ and $A \in Q^{m \times n}$, then $\sqrt{2}A \notin Q^{m \times n}$ because the elements of matrix $\sqrt{2}A$ are not rational numbers.*

EXAMPLE-7 *The set of all ordered n-tuples of the elements of any field F is a vector space over F.*

SOLUTION Recall that if $a_1, a_2, \ldots, a_n \in F$, then $(a_1, a_2, \ldots, a_n)$ is called an ordered n-tuple of elements of F. Let $V = \{(a_1, a_2, \ldots, a_n) : a_i \in F \text{ for all } i \in \underline{n}\}$ be the set of all ordered n-tuples of elements of F.

Now, to give a vector space structure to V over the field F, we define a binary operation on V, scalar multiplication on V and equality of any two elements of V as follows:

Vector addition (*addition on V*): For any $u = (a_1, a_2, \ldots, a_n)$, $v = (b_1, b_2, \ldots, b_n) \in V$, we define

$$u + v = (a_1 + b_1, a_2 + b_2, \ldots, a_n + b_n)$$

Since $a_i + b_i \in F$ for all $i \in \underline{n}$. Therefore, $u + v \in V$.

Thus, addition is a binary operation on V.

Scalar multiplication on V: For any $u = (a_1, a_2, \ldots, a_n) \in V$ and $\lambda \in F$, we define

$$\lambda u = (\lambda a_1, \lambda a_2, \ldots, \lambda a_n).$$

Since $\lambda a_i \in F$ for all $i \in \underline{n}$. Therefore, $\lambda u \in V$.

Thus, scalar multiplication is defined on V.

Equality of two elements of V: For any $u = (a_1, a_2, \ldots, a_n), v = (b_1, b_2, \ldots, b_n) \in V$, we define

$$u = v \Leftrightarrow a_i = b_i \quad \text{for all } i \in \underline{n}.$$

Since we have defined vector addition, scalar multiplication on V and equality of any two elements of V. Therefore, it remains to check whether V-1 and V-2(i) to V-2(iv) are satisfied.

V-1. *V is an abelian group under vector addition.*

Associativity: For any $u = (a_1, a_2, \ldots, a_n), v = (b_1, b_2, \ldots, b_n), w = (c_1, c_2, \ldots, c_n) \in V$, we have

$$\begin{aligned}
(u+v)+w &= (a_1+b_1, \ldots, a_n+b_n) + (c_1, c_2, \ldots, c_n) \\
&= ((a_1+b_1)+c_1, \ldots, (a_n+b_n)+c_n) \\
&= (a_1+(b_1+c_1), \ldots, a_n+(b_n+c_n)) \quad \text{[By associativity of addition on } F\text{]} \\
&= (a_1, \ldots, a_n) + (b_1+c_1, \ldots, b_n+c_n) \\
&= u+(v+w)
\end{aligned}$$

So, vector addition is associative on V.

Commutativity: For any $u = (a_1, a_2, \ldots, a_n), v = (b_1, \ldots, b_n) \in V$, we have

$$\begin{aligned}
u+v &= (a_1+b_1, \ldots, a_n+b_n) \\
&= (b_1+a_1, \ldots, b_n+a_n) \quad \text{[By commutativity of addition on } F\text{]} \\
&= v+u
\end{aligned}$$

So, vector addition is commutative on V.

Existence of identity (Vector zero): Since $0 \in F$, therefore, $\underline{0} = (0, 0, \ldots, 0) \in V$ such that

$$u + \underline{0} = (a_1, a_2, \ldots, a_n) + (0, 0, \ldots, 0)$$
$$\Rightarrow \quad u + \underline{0} = (a_1+0, a_2+0, \ldots, a_n+0) = (a_1, a_2, \ldots, a_n) = u \text{ for all } u \in V$$

Thus, $\underline{0} = (0, 0, \ldots, 0)$ is the identity element for vector addition on V.

Existence of inverse: Let $u = (a_1, a_2, \ldots, a_n)$ be an arbitrary element in V. Then, $(-u) = (-a_1, -a_2, \ldots, -a_n) \in V$ such that

$$u + (-u) = \left(a_1 + (-a_1), \ldots, a_n + (-a_n)\right) = (0, 0, \ldots, 0) = \underline{0}.$$

Thus, every element in V has its additive inverse in V.

Hence, V is an abelian group under vector addition.

V-2 For any $u = (a_1, \ldots, a_n), v = (b_1, \ldots, b_n) \in V$ and $\lambda, \mu \in F$, we have

(i) $\begin{aligned}[t]
\lambda(u+v) &= \lambda(a_1+b_1, \ldots, a_n+b_n) \\
&= (\lambda(a_1+b_1), \ldots, \lambda(a_n+b_n)) \\
&= (\lambda a_1 + \lambda b_1, \ldots, \lambda a_n + \lambda b_n) \quad \begin{bmatrix}\text{By left distributivity of multiplication} \\ \text{over addition in } F\end{bmatrix} \\
&= (\lambda a_1, \ldots, \lambda a_n) + (\lambda b_1, \ldots, \lambda b_n) \\
&= \lambda u + \lambda v
\end{aligned}$

(ii) $(\lambda+\mu)u = ((\lambda+\mu)a_1,\ldots,(\lambda+\mu)a_n)$
$= (\lambda a_1+\mu a_1,\ldots,\lambda a_n+\mu a_n)$ [By right dist. of mult. over add. in F]
$= (\lambda a_1,\ldots,\lambda a_n)+(\mu a_1,\ldots,\mu a_n)$
$= \lambda u + \mu u.$

(iii) $(\lambda\mu)u = ((\lambda\mu)a_1,\ldots,(\lambda\mu)a_n)$
$= (\lambda(\mu a_1),\ldots,\lambda(\mu a_n))$ [By associativity of multiplication in F]
$= \lambda(\mu u)$

(iv) $1u = (1a_1,\ldots,1a_n) = (a_1,\ldots,a_n) = u$ [$\because$ 1 is unity in F]

Hence, V is a vector space over F. This vector space is usually denoted by F^n.

REMARK. *If $F=R$, then R^n is generally known as Euclidean space and if $F=C$, then C^n is called the complex Euclidean space.*

EXAMPLE-8 *Show that the set $F[x]$ of all polynomials over a field F is a vector space over F.*

SOLUTION In order to give a vector space structure to $F[x]$, we define vector addition, scalar multiplication, and equality in $F[x]$ as follows:

Vector addition on F[x]: If $f(x) = \sum_i a_i x^i$, $g(x) = \sum_i b_i x^i \in F[x]$, then we define

$$f(x)+g(x) = \sum_i (a_i+b_i)x^i$$

Clearly, $f(x)+g(x) \in F[x]$, because $a_i+b_i \in F$ for all i.

Scalar multiplication on F[x]: For any $\lambda \in F$ and $f(x) = \sum_i a_i x^i \in F[x]$ $\lambda f(x)$ is defined as the polynomial $\sum_i (\lambda a_i)x^i$. i.e,

$$\lambda f(x) = \sum_i (\lambda a_i)x^i$$

Obviously, $\lambda f(x) \in F[x]$, as $\lambda a_i \in F$ for all i.

Equality of two elements of F[x]: For any $f(x) = \sum_i a_i x^i, g(x) = \sum_i b_i x^i$, we define

$$f(x) = g(x) \Leftrightarrow a_i = b_i \quad \text{for all } i.$$

Now we shall check whether V-1 and V-2(i) to V-2(iv) are satisfied.

V-1 $F[x]$ *is an abelian group under vector addition.*

Associativity: If $f(x) = \sum_i a_i x^i$, $g(x) = \sum_i b_i x^i$, $h(x) = \sum_i c_i x^i \in F[x]$, then

$$[f(x) + g(x)] + h(x) = \left(\sum_i (a_i + b_i) x^i\right) + \sum_i c_i x^i$$
$$= \sum_i [(a_i + b_i) + c_i] x^i$$
$$= \sum_i [a_i + (b_i + c_i)] x^i \qquad \text{[By associativity of } + \text{ on } F\text{]}$$
$$= \sum_i a_i x^i + \sum_i (b_i + c_i) x^i$$
$$= f(x) + [g(x) + h(x)]$$

So, vector addition is associative on $F[x]$.

Commutativity: If $f(x) = \sum_i a_i x^i$, $g(x) = \sum_i b_i x^i \in F[x]$, then

$$f(x) + g(x) = \sum_i (a_i + b_i) x^i = \sum_i (b_i + a_i) x^i \qquad \text{[By commutativity of } + \text{ on } F\text{]}$$
$$\Rightarrow \quad f(x) + g(x) = \sum_i b_i x^i + \sum_i a_i x^i = g(x) + f(x)$$

So, vector addition is commutative on $F[x]$.

Existence of zero vector: Since $\hat{0}(x) = \sum_i 0 x^i \in F[x]$ is such that for all $f(x) = \sum_i a_i x^i \in F[x]$

$$\Rightarrow \quad \hat{0}(x) + f(x) = \sum_i 0 x^i + \sum_i a_i x^i = \sum_i (0 + a_i) x^i$$
$$\Rightarrow \quad \hat{0}(x) + f(x) = \sum_i a_i x^i \qquad [\because 0 \in F \text{ is the additive identity}]$$
$$\Rightarrow \quad \hat{0}(x) + f(x) = f(x)$$

Thus, $\hat{0}(x) + f(x) = f(x) = f(x) + \hat{0}(x)$ for all $f(x) \in F[x]$.

So, $\hat{0}(x)$ is the zero vector (additive identity) in $F[x]$.

Existence of additive inverse: Let $f(x) = \sum_i a_i x^i$ be an arbitrary polynomial in $F[x]$. Then, $-f(x) = \sum_i (-a_i) x^i \in F[x]$ such that

$$f(x) + (-f(x)) = \sum_i a_i x^i + \sum_i (-a_i) x^i = \sum_i [a_i + (-a_i)] x^i = \sum_i 0 x^i = \hat{0}(x)$$

Thus, for each $f(x) \in F[x]$, there exists $-f(x) \in F[x]$ such that

$$f(x) + (-f(x)) = \hat{0}(x) = (-f(x)) + f(x)$$

So, each $f(x)$ has its additive inverse in $F[x]$.
Hence, $F[x]$ is an abelian group under vector addition.

V-2 For any $f(x) = \sum_i a_i x^i$, $g(x) = \sum_i b_i x^i \in F[x]$ and $\lambda, \mu \in F$, we have

(i) $\lambda[f(x) + g(x)] = \lambda \left(\sum_i (a_i + b_i) x^i \right) = \sum_i \lambda(a_i + b_i) x^i$

$\qquad\qquad\qquad = \sum_i (\lambda a_i + \lambda b_i) x^i$ [By left distributivity of multiplication over addition]

$\qquad\qquad\qquad = \sum_i (\lambda a_i) x^i + \sum_i (\lambda b_i) x^i$

$\qquad\qquad\qquad = \lambda f(x) + \lambda g(x)$

(ii) $(\lambda + \mu) f(x) = \sum_i (\lambda + \mu) a_i x^i$

$\qquad\qquad\qquad = \sum_i (\lambda a_i + \mu a_i) x^i$ [By right distributivity of multiplication over addition]

$\qquad\qquad\qquad = \sum_i (\lambda a_i) x^i + \sum_i (\mu a_i) x^i$

$\qquad\qquad\qquad = \lambda f(x) + \mu f(x).$

(iii) $(\lambda \mu) f(x) = \sum_i (\lambda \mu) a_i x^i$

$\qquad\qquad\qquad = \sum_i \lambda(\mu a_i) x^i$ [By associativity of multiplication on F]

$\qquad\qquad\qquad = \lambda \sum_i (\mu a_i) x^i$

$\qquad\qquad\qquad = \lambda(\mu f(x))$

(iv) $1 f(x) = \sum_i (1 a_i) x^i$

$\qquad\qquad\qquad = \sum_i a_i x^i$ [$\because$ 1 is unity in F]

$\qquad\qquad\qquad = f(x)$

Hence, $F[x]$ is a vector space over field F.

EXAMPLE-9 *The set V of all real valued continuous (differentiable or integrable) functions defined on the closed interval $[a, b]$ is a real vector space with the vector addition and scalar multiplication defined as follows:*

$$(f + g)(x) = f(x) + g(x)$$
$$(\lambda f)(x) = \lambda f(x)$$

For all $f, g \in V$ and $\lambda \in R$.

SOLUTION Since the sum of two continuous functions is a continuous function. Therefore, vector addition is a binary operation on V. Also, if f is continuous and $\lambda \in R$, then λf is continuous. We know that for any $f, g \in V$.

$$f = g \Leftrightarrow f(x) = g(x) \quad \text{for all } x \in [a,b].$$

Thus, we have defined vector addition, scalar multiplication and equality in V.

It remains to verify V-1 and V-2(i) to V-2(iv).

V-1 V is an abelian group under vector addition.

Associativity: Let f, g, h be any three functions in V. Then,

$$\begin{aligned}
[(f+g)+h](x) &= (f+g)(x) + h(x) = [f(x) + g(x)] + h(x) \\
&= f(x) + [g(x) + h(x)] \quad \text{[By associativity of addition on } R\text{]} \\
&= f(x) + (g+h)(x) \\
&= [f + (g+h)](x) \quad \text{for all } x \in [a, b]
\end{aligned}$$

So, $\quad (f+g) + h = f + (g+h)$

Thus, vector addition is associative on V.

Commutativity: For any $f, g \in V$, we have

$$(f+g)(x) = f(x) + g(x) = g(x) + f(x) \quad \text{[By commutativity of + on } R\text{]}$$
$$\Rightarrow (f+g)(x) = (g+f)(x) \quad \text{for all } x \in [a, b]$$

So, $\quad f + g = g + f$

Thus, vector addition is commutative on V.

Existence of additive identity: The function $\hat{0}(x) = 0$ for all $x \in [a, b]$ is the additive identity, because for any $f \in V$

$$(f + \hat{0})(x) = f(x) + \hat{0}(x) = f(x) + 0 = f(x) = (\hat{0} + f)(x) \quad \text{for all } x \in [a, b].$$

Existence of additive inverse: Let f be an arbitrary function in V. Then, a function $-f$ defined by $(-f)(x) = -f(x)$ for all $x \in [a, b]$ is continuous on $[a, b]$, and

$$[f + (-f)](x) = f(x) + (-f)(x) = f(x) - f(x) = 0 = \hat{0}(x) \quad \text{for all } x \in [a, b].$$

Thus, $\quad f + (-f) = \hat{0} = (-f) + f$

So, each $f \in V$ has its additive inverse $-f \in V$.

Hence, V is an abelian group under vector addition.

V-2. For any $f, g \in V$ and $\lambda, \mu \in R$, we have

(i) $\quad [\lambda(f+g)](x) = \lambda[(f+g)(x)] = \lambda[f(x)+g(x)] = \lambda f(x) + \lambda g(x)$
$\Rightarrow \quad [\lambda(f+g)](x) = (\lambda f)(x) + (\lambda g)(x) \quad$ for all $x \in [a, b]$
$\therefore \quad \lambda(f+g) = \lambda f + \lambda g$

(ii) $\quad [(\lambda+\mu)f](x) = (\lambda+\mu)f(x) = \lambda f(x) + \mu f(x)$
$\Rightarrow \quad [(\lambda+\mu)f](x) = (\lambda f)(x) + (\mu f)(x) = (\lambda f + \mu f)(x) \quad$ for all $x \in [a, b]$
$\therefore \quad (\lambda+\mu)f = \lambda f + \mu f$

(iii) $\quad [(\lambda\mu)f](x) = (\lambda\mu)f(x) = \lambda[\mu f(x)] = \lambda[\mu f](x) \quad$ for all $x \in [a, b]$
$\therefore \quad (\lambda\mu)f = \lambda(\mu f)$

(iv) $\quad (1f)(x) = 1f(x) = f(x) \quad$ for all $x \in [a, b] \qquad [\because 1 \text{ is unity in } R]$
$\therefore \quad 1f = f.$

Hence, V is a vector space over R. Consequently, V is a real vector space.

NOTE *This vector space is denoted by $C[a, b]$.*

EXAMPLE-10 *Let V_1 and V_2 be two vector spaces over a field F. Then their Cartesian product $V_1 \times V_2 = \{(v_1, v_2) : v_1 \in V_1, v_2 \in V_2\}$ is a vector space over field F.*

SOLUTION We define vector addition, scalar multiplication and equality in $V_1 \times V_2$ as follows:

Vector addition on $V_1 \times V_2$: If $(u_1, u_2), (v_1, v_2) \in V_1 \times V_2$, then we define

$$(u_1, u_2) + (v_1, v_2) = (u_1 + v_1, u_2 + v_2)$$

Since $u_1 + v_1 \in V_1$ and $u_2 + v_2 \in V_2$. Therefore, $(u_1, u_2) + (v_1, v_2) \in V_1 \times V_2$.

Thus, vector addition is a binary operation on $V_1 \times V_2$.

Scalar multiplication: For any $(u_1, u_2) \in V_1 \times V_2$ and $\lambda \in F$, we define

$$\lambda(u_1, u_2) = (\lambda u_1, \lambda u_2).$$

Since V_1 and V_2 are vector spaces over F. Therefore, $\lambda u_1 \in V_1$ and $\lambda u_2 \in V_2$.

Consequently, $\lambda(u_1, u_2) = (\lambda u_1, \lambda u_2) \in V_1 \times V_2$.

Equality: (u_1, u_2) and (v_1, v_2) in $V_1 \times V_2$ are defined as equal if $u_1 = v_1$ and $u_2 = v_2$.

Since we have defined vector addition, scalar multiplication, and equality in $V_1 \times V_2$, now it remains to verify V-1 and V-2 (i) to V-2 (iv).

V-1. $V_1 \times V_2$ *is an abelian group under vector addition:*

Associativity: For any $(u_1, u_2), (v_1, v_2), (w_1, w_2) \in V_1 \times V_2$, we have

$$[(u_1, u_2) + (v_1, v_2)] + (w_1, w_2) = ((u_1 + v_1), (u_2 + v_2)) + (w_1, w_2)$$
$$= ((u_1 + v_1) + w_1, (u_2 + v_2) + w_2)$$
$$= (u_1 + (v_1 + w_1), u_2 + (v_2 + w_2)) \quad \begin{bmatrix} \text{By associativity of} \\ \text{vector addition on } V_1 \\ \text{and } V_2 \end{bmatrix}$$
$$= (u_1, u_2) + (v_1 + w_1, v_2 + w_2)$$
$$= (u_1, u_2) + [(v_1, v_2) + (w_1, w_2)]$$

So, vector addition is associative on $V_1 \times V_2$.

Commutativity: For any $(u_1, u_2), (v_1, v_2) \in V_1 \times V_2$, we have

$$(u_1, u_2) + (v_1, v_2) = (u_1 + v_1, u_2 + v_2)$$
$$= (v_1 + u_1, v_2 + u_2) \quad [\text{By commutativity of vector addition on } V_1 \text{ and } V_2]$$
$$= (v_1, v_2) + (u_1, u_2)$$

So, vector addition is commutative on $V_1 \times V_2$.

Existence of additive identity: If 0_{V_1} and 0_{V_2} are zero vectors in V_1 and V_2 respectively, then $(0_{V_1}, 0_{V_2})$ is zero vector (additive identity) in $V_1 \times V_2$. Because for any $(u_1, u_2) \in V_1 \times V_2$

$$(u_1, u_2) + (0_{V_1}, 0_{V_2}) = (u_1 + 0_{V_1}, u_2 + 0_{V_2})$$
$$= (u_1, u_2) = (0_{V_1}, 0_{V_2}) + (u_1, u_2) \quad [\text{By commutativity of vector addition}]$$

Existence of additive Inverse: Let $(u_1, u_2) \in V_1 \times V_2$. Then, $(-u_1, -u_2)$ is its additive inverse. Because,

$$(u_1, u_2) + (-u_1, -u_2) = (u_1 + (-u_1), u_2 + (-u_2))$$
$$= (0_{V_1}, 0_{V_2}) = (-u_1, -u_2) + (u_1, u_2) \quad \begin{bmatrix} \text{By commutativity of vector} \\ \text{addition} \end{bmatrix}$$

Thus, each element of $V_1 \times V_2$ has its additive inverse in $V_1 \times V_2$.

Hence, $V_1 \times V_2$ is an abelian group under vector addition.

V-2. For any $(u_1, u_2), (v_1, v_2) \in V_1 \times V_2$ and $\lambda, \mu \in F$, we have

(i) $\lambda[(u_1, u_2) + (v_1, v_2)] = \lambda(u_1 + v_1, u_2 + v_2)$
$$= (\lambda(u_1 + v_1), \lambda(u_2 + v_2))$$
$$= (\lambda u_1 + \lambda v_1, \lambda u_2 + \lambda v_2) \quad [\text{By } V\text{-2(i) in } V_1 \text{ and } V_2]$$
$$= (\lambda u_1, \lambda u_2) + (\lambda v_1, \lambda v_2)$$
$$= \lambda(u_1, u_2) + \lambda(v_1, v_2)$$

(ii) $\quad (\lambda+\mu)(u_1,u_2) = ((\lambda+\mu)u_1, (\lambda+\mu)u_2)$
$\qquad\qquad\qquad = (\lambda u_1 + \mu u_1, \lambda u_2 + \mu u_2)$ [By V-2(ii) in V_1 and V_2]
$\qquad\qquad\qquad = (\lambda u_1, \lambda u_2) + (\mu u_1, \mu u_2)$
$\qquad\qquad\qquad = \lambda(u_1, u_2) + \mu(u_1, u_2)$

(iii) $\quad (\lambda\mu)(u_1, u_2) = ((\lambda\mu)u_1, (\lambda\mu)u_2)$
$\qquad\qquad\qquad = (\lambda(\mu u_1), \lambda(\mu u_2))$ [By V-2(iii) in V_1 and V_2]
$\qquad\qquad\qquad = \lambda(\mu u_1, \mu u_2)$
$\qquad\qquad\qquad = \lambda\{\mu(u_1, u_2)\}$

(iv) $\quad 1(u_1, u_2) = (1u_1, 1u_2)$
$\qquad\qquad\qquad = (u_1, u_2)$ [By V-2(iv) in V_1 and V_2]

Hence, $V_1 \times V_2$ is a vector space over field F.

EXAMPLE-11 *Let V be a vector space over a field F. Then for any $n \in N$ the function space $V^n = \{f : f : \underline{n} \to V\}$ is a vector space over field F.*

SOLUTION We define vector addition, scalar multiplication and equality in V^n as follows:

Vector addition: Let $f, g \in V^n$, then the vector addition on V^n is defined as

$$(f+g)(x) = f(x) + g(x) \text{ for all } x \in \underline{n}.$$

Scalar multiplication: If $\lambda \in F$ and $f \in V^n$, then λf is defined as

$$(\lambda f)(x) = \lambda(f(x)) \text{ for all } x \in \underline{n}.$$

Equality: $f, g \in V^n$ are defined as equal if $f(x) = g(x)$ for all $x \in \underline{n}$.

Now we shall verify V^n-1 and V^n-2(i) to V^n-2 (iv).

$V^n - 1$. V^n *is an abelian group under vector addition:*

Associativity: For any $f, g, h \in V^n$, we have

$[(f+g)+h](x) = (f+g)(x) + h(x) = [f(x)+g(x)] + h(x) \quad \text{for all } x \in \underline{n}$
$\qquad\qquad = f(x) + [g(x) + h(x)] \quad$ [By associativity of vector addition on V]
$\qquad\qquad = f(x) + (g+h)(x)$
$\qquad\qquad = [f + (g+h)](x) \quad \text{for all } x \in \underline{n}$

$\therefore \quad (f+g) + h = f + (g+h)$

So, vector addition is associative on V^n.

Commutativity: For any $f, g \in V^n$, we have

$$(f+g)(x) = f(x) + g(x) \quad \text{for all } x \in \underline{n}$$
$$\Rightarrow \quad (f+g)(x) = g(x) + f(x) \quad \text{[By commutativity of vector addition on } V\text{]}$$
$$\Rightarrow \quad (f+g)(x) = (g+f)(x) \quad \text{for all } x \in \underline{n}$$
$$\therefore \quad f+g = g+f$$

So, vector addition is commutative on V^n.

Existence of additive identity (zero vector): The function $\widehat{0} : \underline{n} \to V$ defined by $\widehat{0}(x) = 0_v$ for all $x \in \underline{n}$ is the additive identity, because

$$(f+\widehat{0})(x) = f(x) + \widehat{0}(x) = f(x) + 0_V$$
$$\Rightarrow \quad (f+\widehat{0})(x) = f(x) \quad \text{for all } f \in V^n \text{ and for all } x \in \underline{n} \quad [\because 0_V \text{ is zero vector in } V]$$

Existence of additive inverse: Let f be an arbitrary function in V^n. Then $-f : \underline{n} \to V$ defined by

$$(-f)(x) = -f(x) \quad \text{for all } x \in \underline{n}$$

is additive inverse of f, because

$$[f+(-f)](x) = f(x) + (-f)(x) = f(x) - f(x) = 0_V = \widehat{0}(x) \quad \text{for all } x \in \underline{n}.$$

Hence, V^n is an abelian group under vector addition.

V^n-2. For any $f, g \in V^n$ and $\lambda, \mu \in F$, we have

(i)
$$[\lambda(f+g)](x) = \lambda[(f+g)(x)] = \lambda[f(x) + g(x)]$$
$$\Rightarrow \quad [\lambda(f+g)](x) = \lambda f(x) + \lambda g(x) \quad \text{[By } V\text{-2(i) in } V\text{]}$$
$$\Rightarrow \quad [\lambda(f+g)](x) = (\lambda f + \lambda g)(x) \quad \text{for all } x \in \underline{n}$$
$$\therefore \quad \lambda(f+g) = \lambda f + \lambda g.$$

(ii)
$$[(\lambda + \mu)f](x) = (\lambda + \mu)f(x)$$
$$\Rightarrow \quad [(\lambda + \mu)f](x) = \lambda f(x) + \mu f(x) \quad \text{[By } V\text{-2(ii) in } V\text{]}$$
$$\Rightarrow \quad [(\lambda + \mu)f](x) = (\lambda f)(x) + (\mu f)(x)$$
$$\Rightarrow \quad [(\lambda + \mu)f](x) = (\lambda f + \mu f)(x) \quad \text{for all } x \in \underline{n}$$
$$\therefore \quad (\lambda + \mu)f = \lambda f + \mu f$$

(iii)
$$[(\lambda \mu)f](x) = (\lambda \mu)f(x)$$
$$\Rightarrow \quad [(\lambda \mu)f](x) = \lambda(\mu f(x)) \quad \text{[By } V\text{-2(iii) in } V\text{]}$$
$$\Rightarrow \quad [(\lambda \mu)f](x) = \lambda((\mu f)(x))$$

$$\Rightarrow \quad [(\lambda\mu)f](x) = [\lambda(\mu f)](x) \quad \text{for all } x \in \underline{n}$$
$$\therefore \quad (\lambda\mu)f = \lambda(\mu f)$$

(iv) $\quad (1f)(x) = 1f(x)$

$\Rightarrow \quad (1f)(x) = f(x)$ [By V-2(iv) in V]

$\therefore \quad 1f = f$

Hence, V^n is a vector space over field F.

EXERCISE 2.1

1. Let F be a field and let V be the set of all ordered pairs $(a_1, a_2), a_1, a_2 \in F$. Define
$$(a_1, a_2) + (b_1, b_2) = (a_1 + b_1, a_2 + b_2)$$
$$\lambda(a_1, a_2) = (\lambda a_1, a_2)$$
where $\lambda \in F$. Show that the above defined operations do not make V into a vector space over F.

2. Is the set of all non-zero polynomials of degree 2 a vector space?

3. If X is a non-void set and V is a vector space over a field F, then show that $V^X = \{f : f : X \to V\}$ is a vector space over F under the vector addition and scalar multiplication defined below:
$$(f+g)(x) = f(x) + g(x)$$
$$(\lambda f)(x) = \lambda(f(x))$$
for all $f, g \in V^X$ and $\lambda \in F$.

4. Show that the set of all real valued continuous (differentiable or integrable) functions defined in some interval $[0,1]$ is a vector space over R.

5. Prove that the set $P_n(t)$ of all polynomials of degree at most n over the field R of real numbers is a real vector space.

6. Show that the set of all measurable functions is a real vector space.

7. Let F be a field and V be the set of all infinite sequences $(a_1, a_2, a_3, \ldots)$, where $a_i \in F$ for all i. Then show that V is a vector space over F with addition and scalar multiplication defined by
$$(a_1, a_2, a_3, \ldots) + (b_1, b_2, b_3, \ldots) = (a_1 + b_1, a_2 + b_2, \ldots)$$
and,
$$\lambda(a_1, a_2, a_3, \ldots) = (\lambda a_1, \lambda a_2, \lambda a_3, \ldots)$$

8. Let S be a non-empty set and F be an arbitrary field. Let V denote the set of all functions from S to F. For any $f, g \in V$ and $a \in F$, we define $f + g : S \to F, af : S \to F$ such that for any $x \in S$
$$(f+g)(x) = f(x) + g(x)$$
$$(af)(x) = af(x)$$
Show that V is a vector space over F.

9. Let U and W be two vector spaces over a field F. Let V be the set of ordered pairs (u,w) where $u \in U$ and $w \in W$. Show that V is a vector space over F with addition in V and scalar multiplication V defined by
$$(u_1, w_1) + (u_2, w_2) = (u_1 + u_2, w_1 + w_2)$$
and
$$a(u, w) = (au, aw)$$
(This space is called the external direct product of u and w).

10. Let V be a vector space over a field F and $n \in N$. Let $V^n = \{(v_1, v_2, \ldots, v_n) : v_i \in V; i = 1, 2, \ldots, n\}$. Show that V^n is a vector space over field F with addition in V and scalar multiplication on V defined by
$$(u_1, u_2, \ldots, u_n) + (v_1, v_2, \ldots, v_n) = (u_1 + v_1, u_2 + v_2, \ldots, u_n + v_n)$$
and
$$a(v_1, v_2, \ldots, v_n) = (av_1, av_2, \ldots, av_n)$$

11. Let $V = \{(x, y) : x, y \in R\}$. Show that V is not a vector space over R under the addition and scalar multiplication defined by
$$(a_1, b_1) + (a_2, b_2) = (3b_1 + 3b_2, -a_1 - a_2)$$
$$k(a_1, b_1) = (3kb_1, -ka_1)$$
for all $(a_1, b_1), (a_2, b_2) \in V$ and $k \in R$.

12. Let $V = \{(x, y) : x, y \in R\}$. For any $\alpha = (x_1, y_1), \beta = (x_2, y_2) \in V$, $c \in R$, define
$$\alpha \oplus \beta = (x_1 + x_2 + 1, y_1 + y_2 + 1)$$
$$c \odot \alpha = (cx_1, cy_1)$$

   (i) Prove that $(V, \oplus)$ is an abelian group.
   (ii) Verify that $c \odot (\alpha \oplus \beta) \neq c \odot \alpha \oplus c \odot \beta$.
   (iii) Prove that V is not a vector space over R under the above two operations.

13. For any $u = (x_1, x_2, x_3)$, $v = (y_1, y_2, y_3) \in R^3$ and $a \in R$, let us define
$$u \oplus v = (x_1 + y_1 + 1, x_2 + y_2 + 1, x_3 + y_3 + 1)$$
$$a \odot u = (ax_1 + a - 1, ax_2 + a - 1, ax_3 + a - 1)$$
Prove that R^3 is a vector space over R under these two operations.

14. Let V denote the set of all positive real numbers. For any $u, v \in V$ and $a \in R$, define
$$u \oplus v = uv$$
$$a \odot u = u^a$$
Prove that V is a vector space over R under these two operations.

15. Prove that R^3 is not a vector space over R under the vector addition and scalar multiplication defined as follows:
$$(x_1, x_2, x_3) + (y_1, y_2, y_3) = (x_1 + y_1, x_2 + y_2, x_3 + y_3)$$
$$a(x_1, x_2, x_3) = (ax_1, ax_2, ax_3)$$

16. Let $V = \{(x, 1) : x \in R\}$. For any $u = (x, 1)$, $v = (y, 1) \in V$ and $a \in R$, define
$$u \oplus v = (x+y, 1)$$
$$a \odot u = (ax, 1)$$

   Prove that V is a vector space over R under these two operations.

17. Let V be the set of ordered pairs (a, b) of real numbers. Let us define
$$(a, b) \oplus (c, d) = (a+c, b+d)$$
and
$$k(a, b) = (ka, 0)$$

   Show that V is not a vector space over R.

18. Let V be the set of ordered pairs (a, b) of real numbers. Show that V is not a vector space over R with addition and scalar multiplication defined by

   (i) $(a, b) + (c, d) = (a+d, b+c)$ and $k(a, b) = (ka, kb)$

   (ii) $(a, b) + (c, d) = (a+c, b+d)$ and $k(a, b) = (a, b)$

   (iii) $(a, b) + (c, d) = (0, 0)$ and $k(a, b) = (ka, kb)$

   (iv) $(a, b) + (c, d) = (ac, bd)$ and $k(a, b) = (ka, kb)$

19. The set V of all convergent real sequences is a vector space over the field R of all real numbers.

20. Let p be a prime number. The set $Z_p = \{0, 1, 2, \ldots, (p-1)\}$ is a field under addition and multiplication modulo p as binary operations. Let V be the vector space of polynomials of degree at most n over the field Z_p. Find the number of elements in V.

   [Hint: Each polynomial in V is of the form $f(x) = a_0 + a_1 x + \cdots + a_n x^n$, where $a_0, a_1, a_2, \ldots, a_n \in Z_p$. Since each a_i can take any one of the p values $0, 1, 2, \ldots, (p-1)$. So, there are p^{n+1} elements in V.]

2.3 ELEMENTARY PROPERTIES OF VECTOR SPACES

THEOREM-1 Let V be a vector space over a field F. Then

(i) $a \cdot 0_V = 0_V$ for all $a \in F$

(ii) $0 \cdot v = 0_V$ for all $v \in V$

(iii) $a \cdot (-v) = -(a \cdot v) = (-a) \cdot v$ for all $v \in V$ and for all $a \in F$

(iv) $(-1) \cdot v = -v$ for all $v \in F$

(v) $a \cdot v = 0_V \Rightarrow$ either $a = 0$ or $v = 0_V$

PROOF. (i) We have,

$$a \cdot 0_V = a \cdot (0_V + 0_V) \quad \text{for all } a \in F \qquad [\because\ 0_V + 0_V = 0_V]$$
$$\Rightarrow \quad a \cdot 0_V = a \cdot 0_V + a \cdot 0_V \quad \text{for all } a \in F \qquad [\text{By } V\text{-2(i)}]$$
$$\therefore \quad a \cdot 0_V + 0_V = a \cdot 0_V + a \cdot 0_V \quad \text{for all } a \in F \qquad [0_V \text{ is additive identity in } V]$$
$$\Rightarrow \quad 0_V = a \cdot 0_V \quad \text{for all } a \in F \qquad \begin{bmatrix}\text{By left cancellation law for addition} \\ \text{on } V\end{bmatrix}$$

(ii) For any $v \in V$, we have

$$0 \cdot v = (0+0) \cdot v \qquad [\because\ 0 \text{ is additive identity in } F]$$
$$\Rightarrow \quad 0 \cdot v = 0 \cdot v + 0 \cdot v \qquad [\text{By } V\text{-2(ii)}]$$
$$\therefore \quad 0 \cdot v + 0_V = 0 \cdot v + 0 \cdot v \qquad [\because\ 0_V \text{ is additive identity in } V]$$
$$\Rightarrow \quad 0 \cdot v = 0_V \qquad [\text{By left cancellation law for addition in } V]$$

So, $0 \cdot v = 0_V$ for all $v \in V$.

(iii) For any $v \in V$ and $a \in F$, we have

$$a \cdot [v + (-v)] = a \cdot v + a \cdot (-v) \qquad [\text{By } V\text{-2(i)}]$$
$$\Rightarrow \quad a \cdot 0_V = a \cdot v + a \cdot (-v) \qquad [\because\ v + (-v) = 0_V \text{ for all } v \in V]$$
$$\Rightarrow \quad 0_V = a \cdot v + a \cdot (-v) \qquad [\text{By (i)}]$$
$$\Rightarrow \quad a \cdot (-v) = -(a \cdot v) \qquad [\because\ V \text{ is an additive group}]$$

Again,
$$[a + (-a)] \cdot v = a \cdot v + (-a) \cdot v \qquad [\text{By } V\text{-2(ii)}]$$
$$\Rightarrow \quad 0 \cdot v = a \cdot v + (-a) \cdot v \qquad [\because\ a + (-a) = 0]$$
$$\Rightarrow \quad 0_V = a \cdot v + (-a) \cdot v \qquad [\text{By (ii)}]$$
$$\Rightarrow \quad (-a) \cdot v = -(a \cdot v) \qquad [\because\ V \text{ is an additive group}]$$

Hence, $a \cdot (-v) = -(a \cdot v) = (-a) \cdot v \qquad$ for all $a \in F$ and for all $v \in V$.

(iv) Taking $a = 1$ in (iii), we obtain

$$(-1) \cdot v = -v \quad \text{for all } v \in V.$$

(v) Let $a \cdot v = 0_V$, and let $a \neq 0$. Then $a^{-1} \in F$.
Now

$$a \cdot v = 0_V$$
$$\Rightarrow \quad a^{-1} \cdot (a \cdot v) = a^{-1} \cdot 0_V$$
$$\Rightarrow \quad (a^{-1} a) \cdot v = a^{-1} \cdot 0_V \qquad [\text{By } V\text{-2(iii)}]$$
$$\Rightarrow \quad (a^{-1} a) \cdot v = 0_V \qquad [\text{By (i)}]$$
$$\Rightarrow \quad 1 \cdot v = 0_V$$
$$\Rightarrow \quad v = 0_V \qquad [\text{By } V\text{-2(iv)}]$$

Again let $a \cdot v = 0_V$ and $v \neq 0_V$. Then to prove that $a = 0$, suppose $a \neq 0$. Then $a^{-1} \in F$. Now

$\qquad a \cdot v = 0_V$
$\Rightarrow \quad a^{-1} \cdot (a \cdot v) = a^{-1} \cdot 0_V$
$\Rightarrow \quad (a^{-1}a) \cdot v = a^{-1} \cdot 0_V$ [By V-2(iii)]
$\Rightarrow \quad 1 \cdot v = 0_V$ [By (i)]
$\Rightarrow \quad v = 0_V$, which is a contradiction. [By V-2(iv)]

Thus, $a \cdot v = 0_V$ and $v \neq 0_V \Rightarrow a = 0$.
Hence, $a \cdot v = 0_V \Rightarrow$ either $a = 0$ or $v = 0_V$.

Q.E.D.

It follows from this theorem that the multiplication by the zero of V to a scalar in F or by the zero of F to a vector in V always leads to the zero of V.

Since it is convenient to write $u - v$ instead of $u + (-v)$. Therefore, in what follows we shall write $u - v$ for $u + (-v)$.

REMARK. *For the sake of convenience, we drop the scalar multiplication '·' and shall write av in place of $a \cdot v$.*

THEOREM-2 *Let V be a vector space over a field F. Then,*

(i) *If $u, v \in V$ and $0 \neq a \in F$, then*

$$au = av \Rightarrow u = v$$

(ii) *If $a, b \in F$ and $0_V \neq u \in V$, then*

$$au = bu \Rightarrow a = b$$

PROOF. (i) We have,

$\qquad au = av$
$\Rightarrow \quad au + (-(av)) = av + (-(av))$
$\Rightarrow \quad au - av = 0_V$
$\Rightarrow \quad a(u - v) = 0_V$
$\Rightarrow \quad u - v = 0_V$ [By Theorem 1(v)]
$\Rightarrow \quad u = v$

(ii) We have,

$\qquad au = bu$
$\Rightarrow \quad au + (-(bu)) = bu + (-(bu))$
$\Rightarrow \quad au - bu = 0_V$
$\Rightarrow \quad (a - b)u = 0_V$
$\Rightarrow \quad a - b = 0$ [By Theorem 1(v)]
$\Rightarrow \quad a = b$

Q.E.D.

EXERCISE 2.2

1. Let V be a vector space over a field F. Then prove that
$$a(u-v) = au - av \quad \text{for all } a \in F \text{ and } u, v \in V.$$

2. Let V be a vector space over field R and $u, v \in V$. Simplify each of the following:

    (i) $4(5u - 6v) + 2(3u + v)$ (ii) $6(3u + 2v) + 5u - 7v$

    (iii) $5(2u - 3v) + 4(7v + 8)$ (iv) $3(5u + \dfrac{2}{v})$

3. Show that the commutativity of vector addition in a vector space V can be derived from the other axioms in the definition of V.

4. Mark each of the following as true or false.

    (i) Null vector in a vector space is unique.

    (ii) Let $V(F)$ be a vector space. Then scalar multiplication on V is a binary operation on V.

    (iii) Let $V(F)$ be a vector space. Then $au = av \Rightarrow u = v$ for all $u, v \in V$ and for all $a \in F$.

    (iv) Let $V(F)$ be a vector space. Then $au = bu \Rightarrow a = b$ for all $u \in V$ and for all $a, b \in F$.

5. If $V(F)$ is a vector space, then prove that for any integer n, $\lambda \in F$, $v \in V$, prove that $n(\lambda v) = \lambda(nv) = (n\lambda)v$.

ANSWERS

2. (i) $26u - 22v$ (ii) $23u + 5v$

    (iii) The sum $7v + 8$ is not defined, so the given expression is not meaningful.

    (iv) Division by v is not defined, so the given expression is not meaningful.

4. (i) T (ii) F (iii) F (iv) F

2.4 SUBSPACES

SUB-SPACE *Let V be a vector space over a field F. A non-void subset S of V is said to be a subspace of V if S itself is a vector space over F under the operations on V restricted to S.*

If V is a vector space over a field F, then the null (zero) space $\{0_V\}$ and the entire space V are subspaces of V. These two subspaces are called trivial (improper) subspaces of V and any other subspace of V is called a non-trivial (proper) subspace of V.

THEOREM-1 *(Criterion for a non-void subset to be a subspace) Let V be a vector space over a field F. A non-void subset S of V is a subspace of V iff or all $u, v \in S$ and for all $a \in F$*
(i) $u - v \in S$ and, (ii) $au \in S$.

134 • *Theory and Problems of Linear Algebra*

PROOF. First suppose that S is a subspace of vector space V. Then S itself is a vector space over field F under the operations on V restricted to S. Consequently, S is subgroup of the additive abelian group V and is closed under scalar multiplication.

Hence, (i) and (ii) hold.

Conversely, suppose that S is a non-void subset of V such that (i) and (ii) hold. Then,

(i) $\Rightarrow$ S is an additive subgroup of V and therefore S is an abelian group under vector addition.

(ii) $\Rightarrow$ S is closed under scalar multiplication.

Axioms V-2(i) to V-2(iv) hold for all elements in S as they hold for elements in V.

Hence, S is a subspace of V.

Q.E.D.

THEOREM-2 *(Another criterion for a non-void subset to be a subspace) Let V be a vector space over a field F. Then a non-void subset S of V is a subspace of V iff $au + bv \in S$ for all $u, v \in S$ and for all $a, b \in F$.*

PROOF. First let S be a subspace of V. Then,

$$au \in S, bv \in S \quad \text{for all } u, v \in S \text{ and for all } a, b \in F \quad \text{[By Theorem 1]}$$

$\Rightarrow \quad au + bv \in S \quad$ for all $u, v \in S$ and for all $a, b \in F [\because \ S$ is closed under vector addition$]$

Conversely, let S be a non-void subset of V such that $au + bv \in S$ for all $u, v \in S$ and for all $a, b \in F$.

Since $1, -1 \in F$, therefore

$$1u + (-1)v \in S \quad \text{for all } u, v \in S.$$

$\Rightarrow \quad u - v \in S \quad$ for all $u, v \in S$ \hfill (i)

Again, since $0 \in F$. Therefore,

$$au + 0v \in S \quad \text{for all } u \in S \text{ and for all } a \in F.$$

$\Rightarrow \quad au \in S \quad$ for all $u \in S$ and for all $a \in F$ \hfill (ii)

From (i) and (ii), we get

$u - v \in S$ and $au \in S$ for all $u, v \in S$ and for all $a \in F$.

Hence, by Theorem 1, S is a subspace of vector space V.

Q.E.D.

ILLUSTRATIVE EXAMPLES

EXAMPLE-1 Show that $S = \{(0, b, c) : b, c \in R\}$ is a subspace of real vector space R^3.

SOLUTION Obviously, S is a non-void subset of R^3. Let $u = (0, b_1, c_1)$, $v = (0, b_2, c_2) \in S$ and $\lambda, \mu \in R$. Then,

$$\lambda u + \mu v = \lambda(0, b_1, c_1) + \mu(0, b_2, c_2)$$
$$\Rightarrow \quad \lambda u + \mu v = (0, \lambda b_1 + \mu b_2, \lambda c_1 + \mu c_2) \in S.$$

Hence, S is a subspace of R^3.

REMARK. Geometrically-R^3 is three dimensional Euclidean plane and S is yz-plane which is itself a vector space. Hence, S is a sub-space of R^3. Similarly, $\{(a, b, 0) : a, b \in R\}$, i.e. xy-plane, $\{(a, 0, c) : a, c \in R\}$, i.e. xz plane are subspaces of R^3. In fact, every plane through the origin is a subspace of R^3. Also, the coordinate axes $\{(a, 0, 0) : a \in R\}$, i.e. x-axis, $\{(0, a, 0) : a \in R\}$, i.e. y-axis and $\{(0, 0, a) : a \in R\}$, i.e. z-axis are subspaces of R^3.

EXAMPLE-2 Let V denote the vector space R^3, i.e. $V = \{(x, y, z) : x, y, z \in R\}$ and S consists of all vectors in R^3 whose components are equal, i.e. $S = \{(x, x, x) : x \in R\}$. Show that S is a subspace of V.

SOLUTION Clearly, S is a non-empty subset of R^3 as $(0, 0, 0) \in S$.

Let $u = (x, x, x)$ and $v = (y, y, y) \in S$ and $a, b \in R$. Then,

$$au + bv = a(x, x, x) + b(y, y, y)$$
$$\Rightarrow \quad au + av = (ax + by, ax + by, ax + by) \in S.$$

Thus, $au + bv \in S$ for all $u, v \in S$ and $a, b \in R$.
Hence, S is a subspace of V.

REMARK. Geometrically S, in the above example, represents the line passing through the origin O and having direction ratios proportional to $1,1,1$ as shown in the following figure.

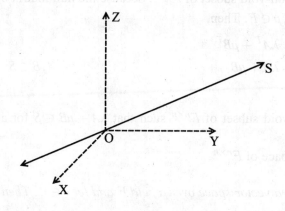

EXAMPLE-3 Let a_1, a_2, a_3 be fixed elements of a field F. Then the set S of all triads (x_1, x_2, x_3) of elements of F, such that $a_1 x_1 + a_2 x_2 + a_3 x_3 = 0$, is a subspace of F^3.

SOLUTION Let $u = (x_1, x_2, x_3), v = (y_1, y_2, y_3)$ be any two elements of S. Then $x_1, x_2, x_3, y_1, y_2, y_3 \in F$ are such that

$$a_1 x_1 + a_2 x_2 + a_3 x_3 = 0 \qquad \text{(i)}$$

and, $$a_1 y_1 + a_2 y_2 + a_3 y_3 = 0 \qquad \text{(ii)}$$

Let a, b be any two elements of F. Then,

$$au + bv = a(x_1, x_2, x_3) + b(y_1, y_2, y_3) = (ax_1 + by_1, ax_2 + by_2, ax_3 + by_3).$$

Now,

$$a_1(ax_1 + by_1) + a_2(ax_2 + by_2) + a_3(ax_3 + by_3)$$
$$= a(a_1 x_1 + a_2 x_2 + a_3 x_3) + b(a_1 y_1 + a_2 y_2 + a_3 y_3)$$
$$= a0 + b0 = 0 \qquad \text{[From (i) and (ii)]}$$

$\therefore \quad au + bv = (ax_1 + by_1, ax_2 + by_2, ax_3 + by_3) \in S$.

Thus, $au + bv \in S$ for all $u, v \in S$ and all $a, b \in F$. Hence, S is a subspace of F^3.

EXAMPLE-4 Show that the set S of all $n \times n$ symmetric matrices over a field F is a subspace of the vector space $F^{n \times n}$ of all $n \times n$ matrices over field F.

SOLUTION Note that a square matrix A is symmetric, if $A^T = A$.

Obviously, S is a non-void subset of $F^{n \times n}$, because the null matrix $0_{n \times n}$ is symmetric.

Let $A, B \in S$ and $\lambda, \mu \in F$. Then,

$$(\lambda A + \mu B)^T = \lambda A^T + \mu B^T$$
$\Rightarrow \quad (\lambda A + \mu B)^T = \lambda A + \mu B \qquad [\because A, B \in S \quad \therefore A^T = A, B^T = B]$
$\Rightarrow \quad \lambda A + \mu B \in S.$

Thus, S is a non-void subset of $F^{n \times n}$ such that $\lambda A + \mu B \in S$ for all $A, B \in S$ and for all $\lambda, \mu \in F$.

Hence, S is a subspace of $F^{n \times n}$.

EXAMPLE-5 Let V be a vector space over a field F and let $v \in V$. Then $F_v = \{av : a \in F\}$ is a subspace of V.

SOLUTION Since $1 \in F$, therefore $v = 1 \, v \in F_v$. Thus, F_v is a non-void subset of V.

Let $\alpha, \beta \in F_v$. Then, $\alpha = a_1 v$, $\beta = a_2 v$ for some $a_1, a_2 \in F$.

Let $\lambda, \mu \in F$. Then,

$$\lambda\alpha + \mu\beta = \lambda(a_1 v) + \mu(a_2 v)$$
$\Rightarrow \quad \lambda\alpha + \mu\beta = (\lambda a_1)v + (\mu a_2)v$ [By V-2(iii)]
$\Rightarrow \quad \lambda\alpha + \mu\beta = (\lambda a_1 + \mu a_2)v$ [By V-2(ii)]
$\Rightarrow \quad \lambda\alpha + \mu\beta \in F_v$ [$\because \lambda a_1 + \mu a_2 \in F$]

Thus, F_v is a non-void subset of V such that $\lambda\alpha + \mu\beta \in F_v$ for all $\alpha, \beta \in F_v$ and for all $\lambda, \mu \in F$.

Hence, F_v is a subspace of V.

EXAMPLE-6 *Let $AX = O$ be homogeneous system of linear equations, where A is an $n \times n$ matrix over R. Let S be the solution set of this system of equations. Show S is a subspace of R^n.*

SOLUTION Every solution U of $AX = O$ may be viewed as a vector in R^n. So, the solution set of $AX = O$ is a subset of R^n. Clearly, $AX = O$. So, the zero vector $O \in S$. Consequently, S is non-empty subset of R^n.

Let $X_1, X_2 \in S$. Then,
X_1 and X_2 are solutions of $AX = O$
$\Rightarrow AX_1 = O$ and $AX_2 = O$ (i)

Let a, b be any two scalars in R. Then,

$$A(aX_1 + bX_2) = a(AX_1) + b(AX_2) = aO + bO = O \quad \text{[Using (i)]}$$
$\Rightarrow aX_1 + bX_2$ is a solution of $AX = O$
$\Rightarrow aX_1 + bX_2 \in S$.

Thus, $aX_1 + bX_2 \in S$ for $X_1, X_2 \in S$ and $a, b \in R$.

Hence, S is a subspace of R^n.

REMARK. *The solution set of a non-homogeneous system $AX = B$ of linear equations in n unknowns is not a subspace of R^n, because zero vector O does not belong to its solution set.*

EXAMPLE-7 *Let V be the vector space of all real valued continuous functions over the field R of all real numbers. Show that the set S of solutions of the differential equation*

$$2\frac{d^2y}{dx^2} - 9\frac{dy}{dx} + 2y = 0$$

is a subspace of V.

SOLUTION We have,

$$S = \left\{ y : 2\frac{d^2y}{dx^2} - 9\frac{dy}{dx} + 2y = 0 \right\}, \quad \text{where } y = f(x).$$

Clearly, $y = 0$ satisfies the given differential equation.

$$\therefore \quad 0 \in S \Rightarrow S \neq \phi$$

Let $y_1, y_2 \in S$. Then,

y_1, y_2 are solutions of $2\dfrac{d^2y}{dx^2} - 9\dfrac{dy}{dx} + 2y = 0$

$\Rightarrow \quad 2\dfrac{d^2y_1}{dx^2} - 9\dfrac{dy_1}{dx} + 2y_1 = 0 \quad$ and, $\quad 2\dfrac{d^2y_2}{dx^2} - 9\dfrac{dy_2}{dx} + 2y_2 = 0 \quad$ (i)

Let $a, b \in R$ and $y = ay_1 + by_2$. Then,

$$2\dfrac{d^2y}{dx^2} - 9\dfrac{dy}{dx} + 2y = 2\dfrac{d^2}{dx^2}(ay_1 + by_2) - 9\dfrac{d}{dx}(ay_1 + by_2) + 2(ay_1 + by_2)$$

$\Rightarrow \quad 2\dfrac{d^2y}{dx^2} - 9\dfrac{dy}{dx} + 2y = 2\left\{a\dfrac{d^2y_1}{dx^2} + b\dfrac{d^2y_2}{dx^2}\right\} - 9\left\{a\dfrac{dy_1}{dx} + b\dfrac{dy_2}{dx}\right\} + 2\{ay_1 + by_2\}$

$\Rightarrow \quad 2\dfrac{d^2y}{dx^2} - 9\dfrac{dy}{dx} + 2y = a\left(2\dfrac{d^2y_1}{dx^2} - 9\dfrac{dy_1}{dx} + 2y_1\right) + b\left(\dfrac{d^2y_2}{dx^2} - 9\dfrac{dy_2}{dx} + 2y_2\right)$

$\Rightarrow \quad 2\dfrac{d^2y}{dx^2} - 9\dfrac{dy}{dx} + 2y = a \times 0 + b \times 0 = 0 \qquad$ [Using (i)]

$\therefore \quad y = ay_1 + by_2$ is a solution of the given differential equation.

$\Rightarrow \quad y \in S$

Thus, $ay_1 + by_2 \in S$ for all $y_1, y_2 \in S$ and $a, b \in R$.

Hence, S is a subspace of V.

EXAMPLE-8 *Let R be the field of real numbers and S be the set of all solutions of the equation $x + y + 2z = 0$. Show that S is a subspace of R^3.*

SOLUTION We have, $S = \{(a, b, c) : a + b + 2c = 0, a, b, c \in R\}$.

Clearly, $1 \times 0 + 1 \times 0 + 2 \times 0 = 0$. So, $(0, 0, 0)$ satisfies the equation $a + b + 2c = 0$.

$\therefore \quad (0, 0, 0) \in S$

$\Rightarrow \quad S$ is non-empty subset of R^3.

Let $u = (a_1, b_1, c_1)$ and $v = (a_2, b_2, c_2)$ be any two elements of S. Then,

$$a_1 + b_1 + 2c_1 = 0 \text{ and } a_2 + b_2 + 2c_2 = 0 \qquad (i)$$

Let a, b be any two elements of R. Then,

$$au + bv = a(a_1, b_1, c_1) + b(a_2, b_2, c_2)$$

$\Rightarrow \quad au + bv = (aa_1 + ba_2, ab_1 + bb_2, ac_1 + bc_2)$

Now,
$$(aa_1+ba_2)+(ab_1+bb_2)+2(ac_1+bc_2)$$
$$=a(a_1+b_1+2c_1)+b(a_2+b_2+2c_2)$$
$$=a\times 0+b\times 0=0 \qquad\text{[Using (i)]}$$
$\therefore \quad au+bv=(aa_1+ba_2,\ ab_1+bb_2,\ ac_1+bc_2)\in S$

Thus, $au+bv\in S$ for all $u,v\in S$ and $a,b\in R$.
Hence, S is a subspace of R^3.

EXAMPLE-9 *Let S be the set of all elements of the form $(x+2y,\ y,\ -x+3y)$ in R^3, where $x,y\in R$. Show that S is a subspace of R^3.*

SOLUTION We have, $S=\{(x+2y,\ y,\ -x+3y):x,y\in R\}$
Let u,v be any two elements of S. Then,
$$u=(x_1+2y_1,y_1,-x_1+3y_1),\ v=(x_2+2y_2,y_2,-x_2+3y_2)$$
where $x_1,y_1,x_2,y_2\in R$.

Let a,b be any two elements of R. In order to prove that S is a subspace of V, we have to prove that $au+bv\in S$, for which we have to show that $au+bv$ is expressible in the form $(\alpha+2\beta,\beta,-\alpha+3\beta)$.

Now,
$$au+bv=a(x_1+2y_1,y_1,-x_1+3y_1)+b(x_2+2y_2,y_2,-x_2+3y_2)$$
$\Rightarrow\quad au+bv=\big((ax_1+bx_2)+2(ay_1+by_2),ay_1+by_2,-(ax_1+bx_2)+3(ay_1+by_2)\big)$
$\Rightarrow\quad au+bv=(\alpha+2\beta,\beta,-\alpha+3\beta),\quad$ where $\alpha=ax_1+bx_2\ $ and $\ \beta=ay_1+by_2$.
$\Rightarrow\quad au+bv\in S$

Thus, $au+bv\in S$ for all $u,v\in S$ and $a,b\in R$.
Hence, S is a subspace of R^3.

EXAMPLE-10 *Let V be the vector space of all 2×2 matrices over the field R of all real numbers. Show that:*

(i) *the set S of all 2×2 singular matrices over R is not a subspace of V.*
(ii) *the set S of all 2×2 matrices A satisfying $A^2=A$ is not a subspace of V.*

SOLUTION (i) We observe that $A=\begin{bmatrix}0 & a\\ 0 & 0\end{bmatrix},\ B=\begin{bmatrix}0 & 0\\ b & 0\end{bmatrix},\ a\neq 0\ b\neq 0$ are elements of S.

But, $A+B=\begin{bmatrix}0 & a\\ b & 0\end{bmatrix}$ is not an element of S as $|A+B|=-ab\neq 0$.

So, S is not a subspace of V.

(ii) We observe that $I = \begin{bmatrix} 1 & 0 \\ 0 & 1 \end{bmatrix} \in S$ as $I^2 = I$. But, $2I = \begin{bmatrix} 2 & 0 \\ 0 & 2 \end{bmatrix} \notin S$, because

$$(2I)^2 = \begin{bmatrix} 2 & 0 \\ 0 & 2 \end{bmatrix} \begin{bmatrix} 2 & 0 \\ 0 & 2 \end{bmatrix} = \begin{bmatrix} 4 & 0 \\ 0 & 4 \end{bmatrix} \neq 2I.$$

So, S is not a subspace of V.

SMALLEST SUBSPACE CONTAINING A GIVEN SUBSET. *Let V be a vector space over a field F, and let W be a subset of V. Then a subspace S of V is called the smallest subspace of V containing W, if*

(i) $W \subset S$

and   (ii) S' *is a subspace of V such that* $W \subset S' \Rightarrow S \subset S'$.

The smallest subspace containing W is also called the subspace generated by W or subspace spanned by W and is denoted by $[W]$. If W is a finite set, then S is called a finitely generated space.

In Example 5, F_v is a finitely generated subspace of V containing $\{v\}$.

EXERCISE 2.3

1. Show that the set of all upper (lower) triangular matrices over field C of all complex numbers is a subspace of the vector space V of all $n \times n$ matrices over C.
2. Let V be the vector space of real-valued functions. Then, show that the set S of all continuous functions and set T of all differentiable functions are subspaces of V.
3. Let V be the vector space of all polynomials in indeterminate x over a field F and S be the set of all polynomials of degree at most n. Show that S is a subspace of V.
4. Let V be the vector space of all $n \times n$ square matrices over a field F. Show that:

    (i) the set S of all symmetric matrices over F is a subspace of V.

    (ii) the set S of all upper triangular matrices over F is a subspace of V.

    (iii) the set S of all diagonal matrices over F is a subspace of V.

    (iv) the set S of all scalar matrices over F is a subspace of V.

5. Let V be the vector space of all functions from the real field R into R. Show that S is a subspace of V where S consists of all:

    (i) bounded functions.

    (ii) even functions.

    (iii) odd functions.

6. Let V be the vector space R^3. Which of the following subsets of V are subspaces of V?

    (i) $S_1 = \{(a,b,c) : a+b = 0\}$

    (ii) $S_2 = \{(a,b,c) : a = 2b+1\}$

(iii) $S_3 = \{(a,b,c) : a \geq 0\}$
(iv) $S_4 = \{(a,b,c) : a^2 = b^2\}$
(v) $S_5 = \{(a,2b,3c) : a,b,c \in R\}$
(vi) $S_6 = \{(a,a,a) : a \in R\}$
(vii) $S_7 = \{(a,b,c) : a,b,c \in Q\}$

7. Let V be the vector space of all real-valued continuous functions defined on the closed interval $[0,1]$ over the field R of real numbers.

   (i) Prove that $S_1 = \{f \in V : f \text{ is differentiable over } [0,1]\}$ is a subspace of V.

   (ii) Prove that $S_2 = \{f \in V : f(\frac{1}{2}) = 0\}$ is a subspace of V.

8. Let V be the set of all 2×3 matrices over field C of all complex number. V is a vector space over C. Determine which of the following subsets of V are subspaces of V?

   (i) $S_1 = \left\{ \begin{bmatrix} a & b & c \\ d & 0 & 0 \end{bmatrix} : a,b,c,d \in C \right\}$

   (ii) $S_2 = \left\{ \begin{bmatrix} a & b & c \\ d & e & f \end{bmatrix} : a+c = e+f \right\}$

   (iii) $S_3 = \left\{ \begin{bmatrix} a & b & c \\ d & e & f \end{bmatrix} : a \in R, b,c,d,e,f \in C \right\}$

   (iv) $S_4 = \left\{ \begin{bmatrix} a & b & c \\ d & e & f \end{bmatrix} : a > 0, b = c \right\}$

9. Let $V = \{a_0 + a_1 x + a_2 x^2 + a_3 x^3 : a_i \in R\}$ be the set of all polynomials of degree less than or equal to 3 over the field R of real numbers. V is a vector space over R. Prove that

   (i) $S_1 = \{a_0 + a_2 x^2 : a_0, a_2 \in R\}$ is a subspace of V.

   (ii) $S_2 = \{a_0 + a_1 x + a_2 x^2 + a_3 x^3 : a_0 = a_2 + a_3\}$ is a subspace of V.

10. Let S be the set of ordered triads $(a_1, a_2, 0)$, where $a_1, a_2 \in R$. Show that S is a subspace of R^3.

11. Let V be the set of infinite sequences $(a_1, a_2, \ldots)$ in a field F. Clearly, V is a vector space over F with addition and scalar multiplication defined by

$$(a_1, a_2, \ldots) + (b_1, b_2, \ldots) = (a_1 + b_1, a_2 + b_2, \ldots)$$

and,
$$k(a_1, a_2, \ldots) = (ka_1, ka_2, \ldots)$$

Show that the set S:

   (i) of all sequences with 0 as first element is a subspace of V.

   (ii) of all sequences with only a finite number of non-zero elements is a subspace of V.

12. Which of the following sets of vectors $\alpha = (a_1, a_2, \ldots, a_n)$ in R^n are subspaces of R^n? ($n \geq 3$).

    (i) all α such that $a_1 \geq o$

    (ii) all α such that $a_1 + 3a_2 = a_3$

    (iii) all α such that $a_2 = a_1^2$

    (iv) all α such that $a_1 a_2 = 0$

    (v) all α such that a_2 is rational.

13. Let V be the vector space of all functions from $R \to R$ over the field R of all real numbers. Show that each of the following subsets of V are subspaces of V.

    (i) $S_1 = \{f : f : R \to R \text{ satisfies } f(0) = f(1)\}$

    (ii) $S_2 = \{f : f : R \to R \text{ is continuous }\}$

    (iii) $S_3 = \{f : f : R \to R \text{ satisfies } f(-1) = 0\}$

14. Let V be the vector space of all functions from R to R over the field R of all real numbers. Show that each of the following subsets of V is not a subspace of V.

    (i) $S_1 = \left\{f : f \text{ satisfies } f(x^2) = \{f(x)\}^2 \text{ for all } x \in R\right\}$

    (ii) $S_2 = \{f : f \text{ satisfies } f(3) - f(-5) = 1\}$

15. Let $R^{n \times n}$ be the vector space of all $n \times n$ real matrices. Prove that the set S consisting of all $n \times n$ real matrices which commutate with a given matrix A in $R^{n \times n}$ form a subspace of $R^{n \times n}$.

16. Let C be the field of all complex numbers and let n be a positive integer such that $n \geq 2$. Let V be the vector space of all $n \times n$ matrices over C. Show that the following subsets of V are not subspaces of V.

    (i) $S = \{A : A \text{ is invertible}\}$ (ii) $S = \{A : A \text{ is not invertible}\}$

17. Prove that the set of vectors $(x_1, x_2, \ldots, x_n) \in R^n$ satisfying the m equations is a subspace of $R^n(R)$:

$$a_{11}x_1 + a_{12}x_2 + \cdots + a_{1n}x_n = 0$$
$$a_{21}x_1 + a_{22}x_2 + \cdots + a_{2n}x_n = 0$$
$$\vdots \quad \vdots \quad \vdots$$
$$a_{m1}x_1 + a_{m2}x_2 + \cdots + a_{mn}x_n = 0,$$

a_{ij}'s are fixed reals.

18. Which of the following sets of vectors $u = (a_1, a_2, \ldots, a_n) \in R^n$ are subspaces of $R^n (n \geq 3)$?

    (i) all $u : a_1 \geq 0$

    (ii) all $u : a_2 = a_1^2$

    (iii) all $u : a_1 a_2 = 0$

    (iv) all $u : a_2$ is rational

    (v) all $u : a_1 + 3a_2 = a_3$.

19. Show that $R^2(R)$ is not a subspace of $R^3(R)$. [Hint : $R^2 \not\subset R^3$].

20. Show that the set of all polynomials of degree less than or equal to n over a field F is a subspace of the vector space of all polynomials over F.

21. Show that the set of all infinite real sequences $(a_1, a_2, \ldots, a_n, \ldots)$ such that $\sum_{i=1}^{\infty} a_i$ converges is a subspace of R^∞.

22. Let $V(F)$ be a vector space, and let v_1, v_2 be fixed vectors in V. Show that the set $S = \{a_1 v_1 + a_2 v_2 : a_1, a_2 \in F\}$ is a subspace of V.

23. Let $R^{2 \times 2}$ be the vector space of all 2×2 matrices over the field R of real numbers. Then the set of all matrices $A \in R^{2 \times 2}$ for which $A^2 = A$ is not a subspace of $R^{2 \times 2}$.

24. Let $V(F)$ be a vector space, and let S be a subspace of V. For any $u, v \in S$ we define $u \cong v \pmod{s}$ iff $u - v \in S$ and for all $a \in F$, $a(u - v) \in S$. Prove that $\cong$ is an equivalence relation on V and for each $u \in S$, $cl(u) = \{v \in S : u \cong v\} = u + S$.

25. If F is a field having more than k elements, then a vector space $V(F)$ cannot be expressed as the set-theoretic union of k proper subspaces.

ANSWERS

6. (i) Yes (ii) No (iii) No (iv) No (v) Yes (vi) Yes (vii) No

12. (i) No (ii) Yes (iii) No (iv) No (v) No

2.4.1 ALGEBRA OF SUBSPACES

THEOREM-1 *Let V be a vector space over a field F. Then the intersection of any two subspace of V is a subspace of V.*

PROOF. Let S_1 and S_2 be any two subspaces of vector space V. Then we have to prove that $S_1 \cap S_2$ is a subset of V.

$$\left. \begin{array}{l} S_1 \text{ is a subspace of } V \Rightarrow S_1 \subset V \text{ and } 0_V \in S_1 \\ S_2 \text{ is a subspace of } V \Rightarrow S_2 \subset V \text{ and } 0_V \in S_1 \end{array} \right\} \Rightarrow 0_V \in S_1 \cap S_2 \subset V.$$

So, $S_1 \cap S_2$ is a non-void subset of V.

Let $u, v \in S_1 \cap S_2$ and $a, b \in F$. Then,

$u, v \in S_1 \cap S_2$ and $a, b \in F$

$\Rightarrow \quad u, v \in S_1,\ u, v \in S_2$ and $a, b \in F$

$\Rightarrow \quad au + bv \in S_1$ and $au + bv \in S_2$ $\qquad [\because\ S_1$ and S_2 are subspaces of V]

$\Rightarrow \quad au + bv \in S_1 \cap S_2$

Thus, $S_1 \cap S_2$ is a non-void subset of V such that $au + bv \in S_1 \cap S_2$ for all $u, v \in S_1 \cap S_2$ and for all $a, b \in F$.

Hence, $S_1 \cap S_2$ is a subspace of V.

Q.E.D.

THEOREM-2 *The intersection of any family of subspaces of a vector space V over a field F is a subspace of V.*

PROOF. Let $\{S_i : i \in I\}$ be a family of subspaces of vector space V, and let $S = \bigcap_{i \in I} S_i$.

Now,

S_i is a subspace of V for all $i \in I$

$\Rightarrow \quad 0_V \in S_i \subset V$ for all $i \in I$

$\Rightarrow \quad 0_V \in \bigcap_{i \in I} S_i \subset V$.

$\Rightarrow \quad 0_V \in S \subset V$.

So, S is a non-void subset of V.

Let $u, v \in S = \bigcap_{i \in I} S_i$ and $a, b \in F$. Then $u, v \in S_i$ for all $i \in I$.

Now,

$u, v \in S_i$ and $a, b \in F$ for all $i \in I$

$\Rightarrow \quad au + bv \in S_i$ for all $i \in I$

$\Rightarrow \quad au + bv \in S = \bigcap_{i \in I} S_i$

Thus, S is a non-void subset of V such that $au + bv \in S$ for all $u, v \in S$ and for all $a, b \in F$.

Hence, S is a subspace of V.

Q.E.D.

THEOREM-3 *The union of two subspaces of a vector space V over a field F is a subspace of V if one is contained in other.*

PROOF. Let S_1 and S_2 be two subspaces of a vector space V over a field F. We have to prove that $S_1 \cup S_2$ is a subspace of V iff either $S_1 \subset S_2$ or $S_2 \subset S_1$.

First suppose that S_1 and S_2 are subspaces of V such that either $S_1 \subset S_2$ or $S_2 \subset S_1$.

If $S_1 \subset S_2$, then $S_1 \cup S_2 = S_2$, which is a subspace of V. Again, if $S_2 \subset S_1$, then $S_1 \cup S_2 = S_1$ which is also a subspace of V.

Hence, in either case $S_1 \cup S_2$ is a subspace of V.

Conversely, suppose that $S_1 \cup S_2$ is a subspace of V. Then we have to prove that either $S_1 \subset S_2$ or $S_2 \subset S_1$.

If possible, let $S_1 \not\subset S_2$ and $S_2 \not\subset S_1$. Then,

$$S_1 \not\subset S_2 \Rightarrow \text{There exists } u \in S_1 \text{ such that } u \notin S_2$$

and, $S_2 \not\subset S_1 \Rightarrow$ There exists $v \in S_2$ such that $v \notin S_1$.

Now,

$u \in S_1, v \in S_2$
$\Rightarrow \quad u, v \in S_1 \cup S_2$
$\Rightarrow \quad u + v \in S_1 \cup S_2$ $\quad [\because S_1 \cup S_2 \text{ is a subspace of } V]$
$\Rightarrow \quad u + v \in S_1 \text{ or, } u + v \in S_2$.

If $\quad u + v \in S_1$
$\Rightarrow \quad (u + v) - u \in S_1$ $\quad [\because u \in S_1 \text{ and } S_1 \text{ is a subspace of } V]$
$\Rightarrow \quad v \in S_1$, which is a contradiction $\quad [\because v \notin S_1]$

If $\quad u + v \in S_2$
$\Rightarrow \quad (u + v) - v \in S_2$ $\quad [\because v \in S_2 \text{ and } S_2 \text{ is a subspace of } V]$
$\Rightarrow \quad u \in S_2$, which is a contradiction $\quad [\because u \notin S_2]$

Since the contradictions arise by assuming that $S_1 \not\subset S_2$ and $S_2 \not\subset S_1$.

Hence, either $S_1 \subset S_2$ or, $S_2 \subset S_1$.

Q.E.D.

REMARK. *In general, the union of two subspaces of a vector space is not necessarily a subspace. For example, $S_1 = \{(a, 0, 0) : a \in R\}$ and $S_2 = \{(0, b, 0) : b \in R\}$ are subspaces of R^3. But, their union $S_1 \cup S_2$ is not a subspace of R^3, because $(2, 0, 0), (0, -1, 0) \in S_1 \cup S_2$, but $(2, 0, 0) + (0, -1, 0) = (2, -1, 0) \notin S_1 \cup S_2$.*

As remarked above the union of two subspaces is not always a subspace. Therefore, unions are not very useful in the study of vector spaces and we shall say no more about them.

THEOREM-4 *The inter section of a family of subspaces of $V(F)$ containing a given subset S of V is the (smallest subspace of V containing S) subspace generated by S.*

PROOF. Let $\{S_i : i \in I\}$ be a family of subspaces of $V(F)$ containing a subset S of V. Let $T = \bigcap_{i \in I} S_i$. Then by Theorem 2, T is a subspace of V. Since $S_i \supset S$ for all i. Therefore, $T = \bigcap_{i \in I} S_i \supset S$.

Hence, T is a subspace of V containing S.

Now it remains to show that T is the smallest subspace of V containing S.

Let W be a subspace of V containing S. Then W is one of the members of the family $\{S_i : i \in I\}$. Consequently, $W \supset \bigcap_{i \in I} S_i = T$.

Thus, every subspace of V that contains S also contains T. Hence, T is the smallest subspace of V containing S.

Q.E.D.

THEOREM-5 *Let $A = \{v_1, v_2, \ldots, v_n\}$ be a non-void finite subset of a vector space $V(F)$. Then the set $S = \left\{ \sum_{i=1}^{n} a_i v_i : a_i \in F \right\}$ is the subspace of V generated by A, i.e. $S = [A]$.*

PROOF. In order to prove that S is a subspace of V generated by A, it is sufficient to show that S is the smallest subspace of V containing A.

We have, $v_i = 0 \cdot v_1 + 0 v_2 + \cdots + 0 v_{i-1} + 1 \cdot v_i + 0 v_{i+1} + \cdots + 0 v_n$

So, $v_i \in S$ for all $i \in \underline{n}$

$\Rightarrow \quad A \subset S$.

Let $u = \sum_{i=1}^{n} a_i v_i$, $v = \sum_{i=1}^{n} b_i v_i \in S$, and let $\lambda, \mu \in F$. Then,

$$\lambda u + \mu v = \lambda \left(\sum_{i=1}^{n} a_i v_i \right) + \mu \left(\sum_{i=1}^{n} b_i v_i \right)$$

$\Rightarrow \quad \lambda u + \mu v = \sum_{i=1}^{n} (\lambda a_i) v_i + \sum_{i=1}^{n} (\mu b_i) v_i$ [By V-2(i) and V-2(iii)]

$\Rightarrow \quad \lambda u + \mu v = \sum_{i=1}^{n} (\lambda a_i + \mu b_i) v_i \in S$ [$\because \lambda a_i + \mu b_i \in F$ for all $i \in \underline{n}$]

Thus, $\lambda u + \mu v \in S$ for all $u, v \in S$ and for all $\lambda, \mu \in F$.

So, S is a subspace of V containing A.

Now, let W be a subspace of V containing A. Then, for each $i \in \underline{n}$

$\qquad a_i \in F, v_i \in A$

$\Rightarrow \quad a_i \in F, v_i \in W$ [$\because A \subset W$]

$\Rightarrow \quad a_i v_i \in W$ [$\because W$ is a subspace of V]

$\Rightarrow \quad \sum_{i=1}^{n} a_i v_i \in W$

Thus, $S \subset W$.

Hence, S is the smallest subspace of V containing A.

Q.E.D.

REMARK. *If φ is the void set, then the smallest subspace of V containing φ is the null space $\{0_V\}$. Therefore, the null space is generated by the void set.*

SUM OF SUBSPACES. *Let S and T be two subspaces of a vector space $V(F)$. Then their sum (linear sum) $S + T$ is the set of all sums $u + v$ such that $u \in S, v \in T$.*

Thus, $S + T = \{u + v : U \in S, v \in T\}$.

THEOREM-6 *Let S and T be two subspaces of a vector space $V(F)$. Then,*

  (i) *$S+T$ is a subspace of V.*
  (ii) *$S+T$ is the smallest subspace of V containing $S \cup T$, i.e. $S+T = [S \cup T]$.*

PROOF. (i) Since S and T are subspaces of V. Therefore,
$$0_V \in S, \ 0_V \in T \Rightarrow 0_V = 0_V + 0_V \in S+T.$$
So, $S+T$ is a non-void subset of V.
Let $u = u_1 + u_2, v = v_1 + v_2 \in S+T$ and let $\lambda, \mu \in F$. Then, $u_1, v_1 \in S$ and $u_2, v_2 \in T$.
Now,
$$\lambda u + \mu v = \lambda(u_1 + u_2) + \mu(v_1 + v_2)$$
$$\Rightarrow \quad \lambda u + \mu v = (\lambda u_1 + \lambda u_2) + (\mu v_1 + \mu v_2) \qquad \text{[By } V\text{-2(i)]}$$
$$\Rightarrow \quad \lambda u + \mu v = (\lambda u_1 + \mu v_1) + (\lambda u_2 + \mu v_2) \qquad \text{[Using comm. and assoc. of vector addition]}$$
Since S and T are subspaces of V. Therefore,
$$u_1, v_1 \in S \text{ and } \lambda, \mu \in F \Rightarrow \lambda u_1 + \mu v_1 \in S$$
and, $\quad u_2, v_2 \in T$ and $\lambda, \mu \in F \Rightarrow \lambda u_2 + \mu v_2 \in T.$
Thus,
$$\lambda u + \mu v = (\lambda u_1 + \mu v_1) + (\lambda u_2 + \mu v_2) \in S+T.$$
Hence, $S+T$ is a non-void subset of V such that $\lambda u + \mu v \in S+T$ for all $u, v \in S+T$ and $\lambda, \mu \in F$. Consequently, $S+T$ is a subspace of V.

 (ii) Let W be a subspace of V such that $W \supset (S \cup T)$.
Let $u+v$ be an arbitrary element of $S+T$. Then $u \in S$ and $v \in T$.
Now,
$$u \in S, v \in T \Rightarrow u \in S \cup T, v \in S \cup T \Rightarrow u, v \in W \qquad [\because W \supset S \cup T]$$
$$\Rightarrow u+v \in W \qquad [\because W \text{ is a subspace of } V]$$
Thus, $u+v \in S+T \Rightarrow u+v \in W$
$$\Rightarrow \quad S+T \subset W.$$
Obviously, $S \cup T \subset S+T$.
Thus, if W is a subspace of V containing $S \cup T$, then it contains $S+T$. Hence, $S+T$ is the smallest subspace of V containing $S \cup T$. Q.E.D.

DIRECT SUM OF SUBSPACES. *A vector space $V(F)$ is said to be direct sum of its two subspaces S and T if every vector u in V is expressible in one and only one as $u = v+w$, where $v \in S$ and $w \in T$.*

If V is direct sum of S and T, then we write $V = S \oplus T$.

THEOREM-7 Let S and T be two subspaces of a vector space $V(F)$. Then V is direct sum of S and T, if

(i) $V = S + T$ (ii) $S \cap T = \{0_V\}$

PROOF. First suppose that V is direct sum of S and T. Then each vector in V has a unique representation as a sum of a vector in S and a vector in T. Thus, $V = S + T$.

Now it remains to prove that $S \cap T = \{0_V\}$. Let $0_V \neq v \in S \cap T$. Then, $v \in S$ and $v \in T$.

Since $v = 0_V + v$ where $0_V \in S$ and $v \in T$, also $v = v + 0_V$ where $v \in S$ and $0_V \in T$. Therefore, $v \in V$ is expressible in two ways as a sum of a vector in S and a vector in T. This contradicts the fact that V is direct sum of S and T. Hence, $v = 0_V$, which implies that $S \cap T = \{0_V\}$.

Conversely, suppose that (i) and (ii) are true. Then for each $u \in V$ there exist $v \in S, w \in T$ such that $u = v + w$. Thus, each $u \in V$ is expressible as a sum of a vector in S and a vector in T. But we have to show that each $u \in V$ is uniquely expressible as a sum of a vector of S and a vector of T. Let, if possible, there be two representations for $u \in V$, i.e. $u = v_1 + w_1$ and $u = v_2 + w_2$ where $v_1, v_2 \in S$ and $w_1, w_2 \in T$. Then,

$$v_1 + w_1 = v_2 + w_2 \Rightarrow v_1 - v_2 = w_1 - w_2$$

Since S and T are subspaces of V. Therefore,

$v_1, v_2 \in S$ and $w_1, w_2 \in T$
$\Rightarrow$ $v_1 - v_2 \in S$ and $w_1 - w_2 \in T$
$\Rightarrow$ $v_1 - v_2 \in S$ and $v_1 - v_2 \in T$ $[\because v_1 - v_2 = w_1 - w_2]$
$\Rightarrow$ $v_1 - v_2 \in S \cap T$
$\Rightarrow$ $v_1 - v_2 = w_1 - w_2 \in S \cap T$
$\Rightarrow$ $v_1 - v_2 = 0_V = w_1 - w_2$
$\Rightarrow$ $v_1 = v_2, w_1 = w_2$.

Therefore, $u \in V$ has a unique representation as a sum of an element of S and an element of T.

Since u is an arbitrary vector in V. Therefore, each vector in V has a unique representation as a sum of a vector in S and a vector in T.

Hence, V is direct sum of S and T.

Q.E.D.

ILLUSTRATIVE EXAMPLES

EXAMPLE-1 The real vector space R^3 is direct sum of its two subspaces $S = \{(a,b,0) : a, b \in R\}$ and $T = \{(0,0,c) : c \in R\}$, because for each $(a,b,c) \in R^3$ there exist unique $(a,b,0) \in S$, $(0,0,c) \in T$ such that $(a,b,c) = (a,b,0) + (0,0,c)$.

EXAMPLE-2 *Show that the vector space $R^{n \times n}$ of all $n \times n$ matrices over real numbers is direct sum of its two subspaces S and T of all symmetric and skew-symmetric matrices respectively*

SOLUTION In Example 4 (Page 136), we have shown that S is a subspace of $R^{n \times n}$. On the similar lines it can be shown that T is a subspace of $R^{n \times n}$.

Let A be an arbitrary matrix in $R^{n \times n}$. Then,
$$A = \tfrac{1}{2}(A+A^T) + \tfrac{1}{2}(A-A^T) = P+Q, \text{ where } P = \tfrac{1}{2}(A+A^T) \text{ and } Q = \tfrac{1}{2}(A-A^T).$$
We have,
$$P^T = \left[\tfrac{1}{2}(A+A^T)\right]^T = \tfrac{1}{2}\left[A^T + (A^T)^T\right] = \tfrac{1}{2}(A^T + A) = P$$
and, $\quad Q^T = \left[\tfrac{1}{2}(A-A^T)\right]^T = \tfrac{1}{2}(A^T - (A^T)^T) = \tfrac{1}{2}(A^T - A) = -Q.$

Therefore, $P \in S$ and $Q \in T$.

Thus, $A = P+Q$, where $P \in S$ and $Q \in T$.

Hence, each matrix in $R^{n \times n}$ is expressible as a sum of a matrix in S and a matrix in T.

To show the uniqueness of the representation, let $A = P' + Q'$ where $P' \in S$ and $Q' \in T$. Then,
$$A^T = (P')^T + (Q')^T = P' - Q' \quad [Q' \in T \Rightarrow Q'^T = -Q' \text{ and}, P' \in S \Rightarrow P'^T = P']$$
$$\therefore \quad P' = \frac{1}{2}(A+A^T) \text{ and } Q' = \frac{1}{2}(A-A^T) \quad [\because \; A = P'+Q' \;\&\; A^T = P' - Q']$$
$$\Rightarrow \quad P' = P \text{ and } Q' = Q.$$

Thus, the representation is unique. Hence, $R^{n \times n} = S \oplus T$.

EXAMPLE-3 *Let $V(R)$ be the vector space of all functions from R to R. Let V_e and V_0 be the subsets of V consisting of all even and odd functions respectively. Prove that:*
  (i) V_e and V_0 are subspaces of V.
  (ii) $V_e + V_0 = V$
  (iii) $V_e \cap V_0 = \{0\}$
  i.e. V is the direct sum of V_e and V_0.

SOLUTION Recall that a function $f : R \to R$ is an even function, if
$$f(-x) = f(x) \quad \text{for all } x \in R$$
$f : R \to R$ is an odd function, if $f(-x) = -f(x)$ for all $x \in R$.

We have,
$$V_e = \{f \in V : f(-x) = f(x) \quad \text{for all } x \in R\}$$
$$V_0 = \{f \in V : f(-x) = -f(x) \quad \text{for all } x \in R\}$$

Let $f, g \in V_e$ and $a, b \in R$. Then, $af + bg : R \to R$ such that
$$(af + bg)(x) = af(x) + bg(x) \quad \text{for all } x \in R.$$

Now,
$$(af+bg)(-x) = af(-x)+bg(-x)$$
$$\Rightarrow (af+bg)(-x) = af(x)+bg(x) \quad [\because f,g \in V_e \therefore f(-x)=f(x) \text{ and } g(-x)=g(x)]$$
$$\Rightarrow (af+bg)(-x) = (af+bg)(x)$$
$$\therefore af+bg: R \to R \text{ is an even function.}$$
$$\Rightarrow af+bg \in V_e$$

Thus, $af+bg \in V_e$ for all $f,g \in V_e$ and $a,b \in R$.

Hence, V_e is a subspace of V.

Similarly, it can be shown that V_0 is also a subspace of V.

(ii) Since V_e and V_0 are subspaces of V. Therefore, V_e+V_0 is also a subspace of V. Hence,
$$V_e+V_0 \subset V \tag{i}$$

Let f be an arbitrary function in V. Then,
$$f(x) = \frac{1}{2}\{f(x)+f(-x)\} + \frac{1}{2}\{f(x)-f(-x)\}$$
$$\Rightarrow f(x) = g(x)+h(x), \text{ where } g(x) = \frac{1}{2}\{f(x)+f(-x)\} \text{ and, } h(x) = \frac{1}{2}\{f(x)-f(-x)\}$$

Now,
$$g(-x) = \frac{1}{2}\{f(-x)+f(x)\} = \frac{1}{2}\{f(x)+f(-x)\} = g(x)$$
and, $$h(-x) = \frac{1}{2}\{f(-x)-f(x)\} = -\frac{1}{2}\{f(x)-f(-x)\} = -h(x)$$

So, $g(x)$ is an even function and $h(x)$ is an odd function.
$$\Rightarrow g \in V_e \text{ and } h \in V_0$$
$$\therefore f = g+h, \text{ where } g \in V_e \text{ and } h \in V_0$$

Thus,
$$f \in V \Rightarrow f \in V_e+V_0$$
$$\Rightarrow V \subset V_e+V_0 \tag{ii}$$

From (i) and (ii), we get
$$V = V_e+V_0$$

(iii) Let $f \in V_e \cap V_0$. Then,
$$f \in V_e \text{ and } f \in V_0$$
$$\Rightarrow f(-x) = f(x) \text{ and } f(-x) = -f(x) \quad \text{for all } x \in R$$
$$\Rightarrow f(x) = -f(x) \quad \text{for all } x \in R$$
$$\Rightarrow f(x) = 0 \quad \text{for all } x \in R$$
$$\Rightarrow f \text{ is zero function on } R$$

Thus, $f \in V_e \cap V_0 \Rightarrow f = \hat{0}$ (zero function)

Hence, $V_e \cap V_0 = \{\hat{0}\}$.

EXAMPLE-4 Consider the vector space $V = R^3$. Let $S = \{(x,y,0) : x,y \in R\}$ and $T = \{(x,0,z) : x,z \in R\}$. Prove that S and T are subspaces of V such that $V = S + T$, but V is not the direct sum of S and T.

SOLUTION Clearly, S and T represent xy-plane and xz-plane respectively. It can be easily checked that S and T are subspaces of V.

Let $v = (x,y,z)$ be an arbitrary point in V. Then,

$$v = (x,y,z) = (x,y,0) + (0,0,z)$$

$\Rightarrow \quad v \in S + T \qquad\qquad\qquad [\because (x,y,0) \in S \text{ and } (0,0,z) \in T]$

$\therefore \quad V \subset S + T$

Also, $S + T \subset V$.

Hence, $V = S + T$ i.e. R^3 is a sum of the xy-plane and the yz-plane.

Clearly, $S \cap T = \{(x,0,0) : x \in R\}$ is the x-axis.

$\therefore \quad S \cap T \neq \{(0,0,0)\}$.

So, V is not the direct sum of S and T.

COMPLEMENT OF A SUBSPACE. Let $V(F)$ be a vector space and let S be a subspace of V. Then a subspace T of V is said to be complement of S, if $V = S \oplus T$.

INDEPENDENT SUBSPACES. Let $S_1, S_2, \ldots, S_n$ be subspaces of a vector space V. Then $S_1, S_2, \ldots, S_n$ are said to be independent, if

$$S_i \cap \sum_{\substack{j=1 \\ j \neq i}}^{n} S_j = \{0_V\} \text{ for all } i = 1, 2, \ldots, n.$$

THEOREM-8 Let $S_1, S_2, \ldots, S_n$ be subspaces of a vector space $V(F)$ such that $V = S_1 + S_2 + \cdots + S_n$. Then, the following are equivalent:

(i) $V = S_1 \oplus S_2 \oplus \cdots \oplus S_n$.
(ii) $S_1, S_2, \ldots, S_n$ are independent.
(iii) For each $v_i \in S_i$, $\sum_{i=1}^{n} v_i = 0_V \Rightarrow v_i = 0_V$.

PROOF. Let us first prove that (i) implies (ii).

Let $u \in S_i \cap \sum_{\substack{j=1 \\ j \neq i}}^{n} S_j$. Then there exist $v_1 \in S_1, v_2 \in S_2, \ldots, v_i \in S_i, \ldots, v_n \in S_n$ such that

$$u = v_i = v_1 + v_2 + \cdots + v_{i-1} + v_{i+1} + \cdots + v_n$$

$\Rightarrow \quad u = v_1 + v_2 + \cdots + v_{i-1} + 0_V + v_{i+1} + \cdots + v_n = 0v_1 + 0v_2 + \cdots + 1v_i + \cdots + 0v_n$

$\Rightarrow \quad v_1 = 0_V = v_2 = \cdots = v_i = \cdots = v_n \qquad\qquad [\because V = S_1 \oplus S_2 \oplus \cdots \oplus S_n]$

$\Rightarrow \quad u = 0_V$

Thus, $u \in S_i \cap \sum_{\substack{j=1 \\ j \neq i}}^{n} S_j \Rightarrow u = 0_V$.

Hence, $S_i \cap \sum_{\substack{j=1 \\ j \neq i}}^{n} S_j = 0_V$. Consequently, $S_1, S_2, \ldots, S_n$ are independent.

(ii) $\Rightarrow$ (iii).

Let $v_1 + v_2 + \cdots + v_n = 0_V$ with $v_i \in S_i, i \in \underline{n}$. Then,

$\Rightarrow \quad v_i = -(v_1 + v_2 + \cdots + v_{i-1} + v_{i+1} + \cdots + v_n)$

$\Rightarrow \quad v_i = - \sum_{\substack{j=1 \\ j \neq i}}^{n} v_j$.

Since $v_i \in S_i$ and $-\sum_{\substack{j=1 \\ j \neq i}}^{n} v_j \in \sum_{\substack{j=1 \\ j \neq i}}^{n} S_j$. Therefore,

$v_i \in S_i \cap \sum_{\substack{j=1 \\ j \neq i}}^{n} S_j \quad$ for all $i \in \underline{n}$.

$\Rightarrow \quad v_i = 0_V \quad$ for all $i \in \underline{n}$ $\qquad [\because S_1, S_2, \ldots, S_n \text{ are independent}]$

(iii) $\Rightarrow$ (i)

Since $V = S_1 + S_2 + \cdots + S_n$ is given. Therefore, it is sufficient to show that each $v \in V$ has a unique representation as a sum of vectors in $S_1, S_2, \ldots, S_n$.

Let, if possible, $v = v_1 + \cdots + v_n$ and $v = w_1 + \cdots + w_n$ be two representations for $v \in V$. Then,

$v_1 + v_2 + \cdots + v_n = w_1 + w_2 + \cdots + w_n$

$\Rightarrow \quad (v_1 - w_1) + \cdots + (v_n - w_n) = 0_V$

$\Rightarrow \quad v_1 - w_1 = 0_V, \ldots, v_n - w_n = 0_V \qquad$ [Using (iii)]

$\Rightarrow \quad v_1 = w_1, \ldots, v_n = w_n$.

Thus, each $v \in V$ has a unique representation. Hence, $V = S_1 \oplus S_2 \oplus \cdots \oplus S_n$.

Consequently, (iii) $\Rightarrow$ (i).

Hence, (i) $\Rightarrow$ (ii) $\Rightarrow$ (iii) $\Rightarrow$ (i) i.e. (i) $\Leftrightarrow$ (ii) $\Leftrightarrow$ (iii)

Q.E.D.

LINEAR VARIETY. *Let S be a subspace of a vector space $V(F)$, and let $v \in V$. Then,*

$$v + S = \{v + u : u \in S\}$$

is called a linear variety of S by v or a translate of S by v or a parallel of S through v.

S is called the base space of the linear variety and v a leader.

THEOREM-9 *Let S be a subspace of a vector space $V(F)$ and let $P = v + S$ be the parallel of S through v. Then,*

(i) *every vector in P can be taken as a leader of P, i.e $u + S = P$ for all $u \in P$.*

(ii) *two vectors $v_1, v_2 \in V$ are in the same parallel of S iff $v_1 - v_2 \in S$*

PROOF. (i) Let u be an arbitrary vector in $P = v + S$. Then there exists $v_1 \in S$ such that
$$u = v + v_1. \Rightarrow v = u - v_1.$$
Let w be an arbitrary vector in P. Then,

$\quad\quad w = v + v_2 \quad$ for some $v_2 \in S$.
$\Rightarrow \quad w = (u - v_1) + v_2 \quad\quad\quad\quad\quad\quad\quad\quad\quad\quad [\because v = u - v_1]$
$\Rightarrow \quad w = u + (v_2 - v_1) \quad\quad\quad\quad\quad\quad\quad\quad\quad [\because v_2 - v_1 \in S]$
$\Rightarrow \quad w \in u + S$

Thus, $P \subset u + S$.
Now let $w' \in u + S$. Then,

$\quad\quad w' = u + v_3 \quad$ for some $v_3 \in S$.
$\Rightarrow \quad w' = (v + v_1) + v_3 \quad\quad\quad\quad\quad\quad\quad\quad\quad [\because v = u - v_1]$
$\Rightarrow \quad w' = v + (v_1 + v_3)$
$\Rightarrow \quad w' \in v + S \quad\quad\quad\quad\quad\quad\quad\quad\quad\quad\quad\quad [\because v_1 + v_3 \in S]$

Thus, $u + S \subset P$. Hence, $u + S = P$
Since u is an arbitrary vector in P. Therefore, $u + S = P$ for all $u \in P$.

(ii) Let v_1, v_2 be in the same parallel of S, say $u + S$. Then there exist $u_1, u_2 \in S$ such that

$\quad\quad v_1 = v + u_1, v_2 = v + u_2$
$\Rightarrow \quad v_1 - v_2 = u_1 - u_2 \Rightarrow v_1 - v_2 \in S \quad\quad\quad [\because u_1, u_2 \in S \Rightarrow u_1 - u_2 \in S]$

Conversely, if $v_1 - v_2 \in S$, then there exists $u \in S$ such that

$\quad\quad v_1 - v_2 = u$
$\Rightarrow \quad v_1 = v_2 + u$
$\Rightarrow \quad v_1 \in v_2 + S. \quad\quad\quad\quad\quad\quad\quad\quad\quad\quad\quad\quad\quad [\because u \in S]$

Also, $v_2 = v_2 + 0_V$ where $0_V \in S$
$\Rightarrow \quad v_2 \in v_2 + S.$

Hence, v_1, v_2 are in the same parallel of S.

Q.E.D.

EXERCISE 2.4

1. Let $V = R^3$, $S = \{(0, y, z) : y, z \in R\}$ and $T = \{(x, 0, z) : x, z \in R\}$
   Show that S and T are subspaces of V such that $V = S + T$. But, V is not the direct sum of S and T.

2. Let $V(R)$ be the vector space of all 2×2 matrices over R. Let S be the set of all symmetric matrices in V and T be the set of all skew-symmetric matrices in V. Show that S and T are subspaces of V such that $V = S \oplus T$.

3. Let $V = R^3$, $S = \{(x, y, 0) : x, y \in R\}$ and $T = \{(0, 0, z) : z \in R\}$.
   Show that S and T are subspaces of V such that $V = S \oplus T$.

4. Let S_1, S_2, S_3 be the following subspaces of vector space $V = R^3$:

   $S_1 = \{(x, y, z) : x = z, x, y, z \in R\}$
   $S_2 = \{(x, y, z) : x + y + z = 0, x, y, z \in R\}$
   $S_3 = \{(0, 0, z) : z \in R\}$

   Show that: (i) $V = S_1 + S_2$ (ii) $V = S_2 + S_3$ (iii) $V = S_1 + S_3$ When is the sum direct?

5. Let S and T be the two-dimensional subspaces of R^3. Show that $S \cap T \neq \{0\}$.

6. Let S, S_1, S_2 be the subspaces of a vector space $V(F)$. Show that

   $$(S \cap S_1) + (S \cap S_2) \subseteq S \cap (S_1 + S_2)$$

   Find the subspaces of R^2 for which equality does not hold.

7. Let V be the vector space of $n \times n$ square matrices. Let S be the subspace of upper triangular matrices, and let T be the subspaces of lower triangular matrices.
   Find (i) $S \cap T$ (ii) $S + T$.

8. Let F be a subfield of the field C of complex numbers and let V be the vector space of all 2×2 matrices over F. Let $S = \left\{ \begin{bmatrix} x & y \\ z & 0 \end{bmatrix} : x, y, z \in F \right\}$ and $T = \left\{ \begin{bmatrix} x & 0 \\ 0 & y \end{bmatrix} : x, y \in F \right\}$.
   Show that S and T are subspaces of V such that $V = S + T$ but $V \neq S \oplus T$.
   $\left[\text{Hint: } S \cap T = \left\{ \begin{bmatrix} x & 0 \\ 0 & 0 \end{bmatrix} : x \in F \right\} \quad \therefore \quad S \cap T \neq \left\{ \begin{bmatrix} 0 & 0 \\ 0 & 0 \end{bmatrix} \right\} \right]$

9. Let S and T be subspaces of a vector space $V(F)$ such that $V = S + T$ and $S \cap T = \{0_V\}$. Prove that each vector v in V there are unique vectors v_1 in S and v_2 in T such that $v = v_1 + v_2$.

10. Let U and W be vector spaces over a field F and let $V = U \times W = \{(u, w) : u \in U$ and $w \in W\}$. Then V is a vector space over field F with addition in V and scalar multiplication on V defined by
    $$(u_1, w_1) + (u_2, w_2) = (u_1 + u_2, w_1 + w_2)$$
    and, $$k(u, w) = (ku, kw).$$
    Let $S = \{(u, 0) : u \in U\}$ and $T = \{(0, w) : w \in W\}$. Show that

(i) S and T are subspaces of V. (ii) $V = S \oplus T$.

11. Let $V(R)$ be the real vector space of all functions from R to itself. Let S be the set of all even functions and T be the set of all odd functions in V. Prove that

    (i) S and T are subspaces of V.
    (ii) $V = S \oplus T$.
    (iii) S and T are complementary subspaces of V.

12. Let S and T be the following subspaces of $V = R^3(R)$:
$$S = \{(a, b, c) : a = b = c, \, a, b, c \in R\} \quad \text{and} \quad T = \{(0, b, c) : b, c \in R\}$$
    Show that $V = S \oplus T$.

13. Let V be the vector space of all polynomials in indeterminate x over the field C of complex numbers, and let S and T be subsets of V consisting of all even and odd polynomials in x. Then prove the following:

    (i) S and T are subspaces of V.
    (ii) $V = S \oplus T$
    (iii) S and T are complimentary subspaces of V.

14. If S_1, S_2, S_3 are subspaces of a vector space $V(F)$ such that $S_2 \supset S_1$. Then prove that
$$S_1 \cap (S_2 + S_3) = S_2 + (S_1 \cap S_3).$$
    Give an example to show that if S_1, S_2, S_3 are subspaces of V, then $S_1 \cap (S_2 + S_3) = S_1 \cap S_2 + S_1 \cap S_3$ is not necessarily true.

15. If S_1, S_2, S_3 are subspaces of a vector space $V(F)$, then show that
$$S_1 + [S_2 \cap (S_1 + S_3)] = (S_1 + S_2) \cap (S_1 + S_3)$$

16. If S and T are subspaces of a vector space $V(F)$, prove that $S + T = S \Leftrightarrow T \subset S$.

17. Show that if in a vector space $V(F)$
$$S_1 \cap (S_2 + S_3) = S_1 \cap S_2 + S_1 \cap S_3$$
    holds for all subspaces S_1, S_2, S_3 of V, then
$$S_1 + (S_2 \cap S_3) = (S_1 + S_2) \cap (S_1 + S_3)$$
    also holds.

18. Prove that a non-void subset S of a vector space $V(F)$ is a subspace *iff* for all $\lambda \in F$, $u, v \in V$, $\lambda u + v \in S$.

19. If S_1, S_2, S_3 are three subspaces of a vector space $V(F)$, prove that
$$(S_1 \cap S_3) + (S_2 \cap S_3) \subset (S_1 + S_2) \cap S_3.$$

20. Construct three subspaces S_1, S_2, S_3 of a vector space $V(F)$, so that
$$V = S_1 \oplus S_2 = S_1 \oplus S_3 \text{ but } S_2 \neq S_3.$$
    [Hint:- Take $V = R^2(R)$, $S_1 = \{(\lambda, 0) : \lambda \in R\}$, $S_2 = \{(0, \lambda) : \lambda \in R\}$ and $S_3 = \{(\lambda, \lambda) : \lambda \in R\}$].

21. Mark each of the following as true or false:
    (i) Every vector space has exactly two subspaces.
    (ii) Every vector space has at least two subspaces.
    (iii) The sum of any two subspaces of a vector space $V(F)$ is a subspace of $V(F)$.
    (iv) The union of any two subspaces of a vector space $V(F)$ is a subspace of $V(F)$.
    (v) The null space is generated by the void set.
    (vi) Every subspace of the real vector space R^3 must always contain in the origin.
    (vii) The set $S = \{(a,b,c) : a,b,c \in R, a > 0\}$ is a subspace of $R^3(R)$.
    (viii) The set $S = \{(a,b,c) : a+b = 0, a,b,c \in R\}$ is a subspace of $R^3(R)$.
    (ix) The set $S = \{(x,y) : x,y \in R, x^2 = y^2\}$ is a subspace of $R^2(R)$.

ANSWERS

4. The sum is direct in (ii) and (iii).
6. In R^2, let S, S_1 and S_2 be respectively the lines $y = x$, the x-axis, the y-axis.
7. (i) Diagonal matrices (ii) V.
21. (i) F (ii) T (iii) T (iv) F (v) T (vi) T (vii) F (viii) T (ix) F

2.5 QUOTIENT SPACES

Let S be a subspace of a vector space $V(F)$. Then S is a subgroup of additive abelian group V. Let V/S be the set of all left cosets (Since V is an additive abelian group, therefore, there is no distinction between left and right cosets of S in V and we may simply call cosets) of S in V, i.e.

$$V/S = \{u + S : v \in V\}$$

Our main objective in this section is to equip the set V/S with the structure of a vector space over field F. For that purpose, we define vector addition, scalar multiplication on V/S as follows:

Vector addition: For any $u + S, v + S \in V/S$, we define

$$(u + S) + (v + S) = (u + v) + S$$

Scalar multiplication: If $u + S \in V/S$ and $a \in F$, then the scalar multiplication on V/S is defined as

$$a(u + S) = au + S$$

Since S is a subgroup of additive abelian group V. Therefore, for any $u+S, v+S \in V/S$

$$u+S = v+S \Leftrightarrow u-v \in S$$

Thus, we have defined vector addition, scalar multiplication on V/S and equality of any two elements of V/S. Now we proceed to prove that V/S is a vector space over field F under the above defined operations of addition and scalar multiplication.

THEOREM-1 *Let V be a vector space over a field F and let S be a subspace of V. Then, the set*

$$V/S = \{u+S : u \in V\}$$

is a vector space over field F for the vector addition and scalar multiplication on V/S defined as follows:

$$(u+S) + (v+S) = (u+v) + S$$

and, $\qquad a(u+S) = au + S \quad$ *for all $u+S, v+S \in V/S$ and for all $a \in F$.*

PROOF. First we shall show that above rules for vector addition and scalar multiplication are well defined, i.e. are independent of the particular representative chosen to denote a coset.

Let $\quad u+S = u'+S$, where $u, u' \in V$ and, $v+S = v'+S$, where $v, v' \in V$.

Then,

$$u+S = u'+S \Rightarrow u - u' \in S \text{ and, } v+S = v'+S \Rightarrow v - v' \in S.$$

Now,

$$u - u' \in S, v - v' \in S$$
$$\Rightarrow \quad (u - u') + (v - v') \in S \qquad\qquad [\because S \text{ is a subspace of } V]$$
$$\Rightarrow \quad (u+v) - (u'+v') \in S$$
$$\Rightarrow \quad (u+v) + S = (u'+v') + S$$
$$\Rightarrow \quad (u+S) + (v+S) = (u'+S)(v'+S)$$

Thus, $u+S = u'+S$ and $v+S = v'+S$
$\Rightarrow (u+S)+(v+S) = (u'+S)+(v'+S)$ for all $u, u', v, v' \in V$.

Therefore, vector addition on V/S is well defined.

Again,

$$\lambda \in F, u - u' \in S \Rightarrow \lambda(u - u') \in S \Rightarrow \lambda u - \lambda u' \in S \Rightarrow \lambda u + S = \lambda u' + S$$

Therefore, scalar multiplication on V/S is well defined.

$V-1$: V/S is an abelian group under vector addition:

Associativity: For any $u+S, v+S, w+S \in V/S$, we have

$$\begin{aligned}
[(u+S)+(v+S)]+(w+S) &= ((u+v)+S)+(w+S) \\
&= \{(u+v)+w\}+S \\
&= \{u+(v+w)\}+S \quad \text{[By associativity of vector addition on } V] \\
&= (u+S)+((v+w)+S) \\
&= (u+S)+[(v+S)+(w+S)]
\end{aligned}$$

So, vector addition is associative on V/S.

Commutativity: For any $u+S, v+S \in V/S$, we have

$$\begin{aligned}
(u+S)+(v+S) &= (u+v)+S \\
&= (v+u)+S \quad \text{[By commutativity of vector addition on } V] \\
&= (v+S)+(u+S)
\end{aligned}$$

So, vector addition is commutative on V/S.

Existence of additive identity: Since $0_V \in V$ therefore, $0_V + S = S \in V/S$.

Now,

$$(u+S)+(0_V+S) = (u+0_V)+S = (u+S) \qquad \text{for all } u+S \in V/S$$
$$\therefore \quad (u+S)+(0_V+S) = (u+S) = (0_V+S)+(u+S) \quad \text{for all } u+S \in V/S.$$

So $0_V + S = S$ is the identity element for vector addition on V/S.

Existence of additive inverse: Let $u+S$ be an arbitrary element of V/S. Then,
$$u \in V \Rightarrow -u \in V \Rightarrow (-u)+S \in V/S$$
Thus, for each $u+S \in V/S$ there exists $(-u)+S \in V/S$ such that

$$(u+S)+((-u)+S) = [u+(-u)]+S = 0_V + S = S$$
$$\Rightarrow (u+S)+((-u)+S) = 0_V + S = (-u)+S+(u+S) \text{ [By commutativity of addition on } V/S]$$

So, each $u+S \in V/S$ has its additive inverse $(-u)+S \in V/S$.

Hence, V/S is an abelian group under vector addition.

$V-2$: *For any* $u+S$, $v+S \in V/S$ *and* $\lambda, \mu \in F$, *we have*

(i) $\quad \lambda[(u+S)+(v+S)] = \lambda[(u+v)+S]$
$\qquad\qquad\qquad\qquad = [\lambda(u+v)]+S$
$\qquad\qquad\qquad\qquad = (\lambda u + \lambda v)+S \qquad\qquad$ [By V-2(i) in V]
$\qquad\qquad\qquad\qquad = (\lambda u + S)+(\lambda v + S) \qquad$ [By definition of addition on V/S]
$\qquad\qquad\qquad\qquad = \lambda(u+S)+\lambda(v+S) \qquad$ [By def. of scalar multiplication on V/S]

(ii) $\quad (\lambda+\mu)(u+S) = (\lambda+\mu)u+S \qquad$ [By def. of scalar multiplication on V/S]
$\qquad\qquad\qquad = (\lambda u + \mu u)+S \qquad\qquad$ [By V-2(ii) in V]
$\qquad\qquad\qquad = (\lambda u + S)+(\mu u + S) \qquad$ [By definition of addition on V/S]
$\qquad\qquad\qquad = \lambda(u+S)+\mu(u+S)$

(iii) $\quad (\lambda\mu)(u+S) = (\lambda\mu)u+S$
$\qquad\qquad\qquad = \lambda(\mu u)+S \qquad\qquad$ [By V-2(iii) in V]
$\qquad\qquad\qquad = \lambda(\mu u + S)$
$\qquad\qquad\qquad = \lambda[\mu(u+S)]$

(iv) $\qquad\qquad 1(u+S) = 1u+S = u+S \qquad\qquad$ [By V-2(iv) in V]

Hence, V/S is a vector space over field F.

Q.E.D.

QUOTIENT SPACE. *Let V be a vector space over a field F and let S be a subspace of V. Then the set $V/S = \{u+S : u \in V\}$ is a vector space over field F for the vector addition and scalar multiplication defined as*

$$(u+S)+(v+S) = (u+v)+S$$
$$a(u+S) = au+S$$

For all $u+S, v+S \in V/S$ and for all $a \in F$.

This vector space is called a quotient space of V by S.

EXERCISE 2.5

1. If S is a subspace of a vector space $V(F)$, then show that there is one-one correspondence between the subspaces of V containing S and subspaces of the quotient space V/S.

2. Define quotient space.

2.6 LINEAR COMBINATIONS

LINEAR COMBINATION. *Let V be a vector space over a field F. Let $v_1, v_2, \ldots, v_n$ be n vectors in V and let $\lambda_1, \lambda_2, \ldots, \lambda_n$ be n scalars in F. Then the vector $\lambda_1 v_1 + \cdots + \lambda_n v_n (\text{or } \sum_{i=1}^{n} \lambda_i v_i)$ is called a linear combination of $v_1, v_2, \ldots, v_n$. It is also called a linear combination of the set $S = \{v_1, v_2, \ldots, v_n\}$. Since there are finite number of vectors in S, it is also called a finite linear combination of S.*

If S is an infinite subset of V, then a linear combination of a finite subset of S is called a finite linear combination of S.

ILLUSTRATIVE EXAMPLES

EXAMPLE-1 *Express $v = (-2, 3)$ in $R^2(R)$ as a linear combination of the vectors $v_1 = (1, 1)$ and $v_2 = (1, 2)$.*

SOLUTION Let x, y be scalars such that

$$v = xv_1 + yv_2$$
$\Rightarrow \quad (-2, 3) = x(1, 1) + y(1, 2)$
$\Rightarrow \quad (-2, 3) = (x + y, x + 2y)$
$\Rightarrow \quad x + y = -2 \text{ and } x + 2y = 3$
$\Rightarrow \quad x = -7, y = 5$

Hence, $v = -7v_1 + 5v_2$.

EXAMPLE-2 *Express $v = (-2, 5)$ in $R^2(R)$ as a linear combination of the vectors $v_1 = (-1, 1)$ and $v_2 = (2, -2)$.*

SOLUTION Let x, y be scalars such that

$$v = xv_1 + yv_2$$
$\Rightarrow \quad (-2, 5) = x(-1, 1) + y(2, -2)$
$\Rightarrow \quad (-2, 5) = (-x + 2y, x - 2y)$
$\Rightarrow \quad -x + 2y = -2 \text{ and } x - 2y = 5$

This is an inconsistent system of equations and so it has no solution.

Hence, v cannot be written as the linear combination of v_1 and v_2.

EXAMPLE-3 *Express $v = (1, -2, 5)$ in R^3 as a linear combination of the following vectors:*

$$v_1 = (1, 1, 1), v_2 = (1, 2, 3), v_3 = (2, -1, 1)$$

SOLUTION Let x, y, z be scalars in R such that

$$v = xv_1 + yv_2 + zv_3$$

$\Rightarrow \quad (1, -2, 5) = x(1, 1, 1) + y(1, 2, 3) + z(2, -1, 1)$

$\Rightarrow \quad (1, -2, 5) = (x + y + 2z, \ x + 2y - z, \ x + 3y + z)$

$\Rightarrow \quad x + y + 2z = 1, x + 2y - z = -2, x + 3y + z = 5$

This system of equations can be written in matrix as follows:

$$\begin{bmatrix} 1 & 1 & 2 \\ 1 & 2 & -1 \\ 1 & 3 & 1 \end{bmatrix} \begin{bmatrix} x \\ y \\ z \end{bmatrix} = \begin{bmatrix} 1 \\ -2 \\ 5 \end{bmatrix}$$

This is equivalent to

$$\begin{bmatrix} 1 & 1 & 2 \\ 0 & 1 & -3 \\ 0 & 2 & -1 \end{bmatrix} \begin{bmatrix} x \\ y \\ z \end{bmatrix} = \begin{bmatrix} 1 \\ -3 \\ 4 \end{bmatrix} \text{ Applying } R_2 \to R_2 - R_1, R_3 \to R_3 - R_1$$

or,

$$\begin{bmatrix} 1 & 1 & 2 \\ 0 & 1 & -3 \\ 0 & 0 & 5 \end{bmatrix} \begin{bmatrix} x \\ y \\ z \end{bmatrix} = \begin{bmatrix} 1 \\ -3 \\ 10 \end{bmatrix} \text{ Applying } R_3 \to R_3 - 2R_2$$

This is equivalent to

$$x + y + 2z = 1, \ y - 3z = -3, \ 5z = 10$$

$\Rightarrow \quad x = -6, \ y = 3, \ z = 2$

Hence, $v = -6v_1 + 3v_2 + 2v_3$.

EXAMPLE-4 Express $v = (2, -5, 3)$ in $R^3(R)$ as a linear combination of the vectors $v_1 = (1, -3, 2), v_2 = (2, -4, -1), v_3 = (1, -5, 7)$.

SOLUTION Let x, y, z be scalars such that

$$v = xv_1 + yv_2 + zv_3$$

$\Rightarrow \quad (2, -5, 3) = x(1, -3, 2) + y(2, -4, -1) + z(1, -5, 7)$

$\Rightarrow \quad (2, -5, 3) = (x + 2y + z, \ -3x - 4y - 5z, \ 2x - y + 7z)$

$\Rightarrow \quad x + 2y + z = 2, \ -3x - 4y - 5z = -5, \ 2x - y + 7z = 3$

This system of equations, in matrix form, can be written as follows:

$$\begin{bmatrix} 1 & 2 & 1 \\ -3 & -4 & -5 \\ 2 & -1 & 7 \end{bmatrix} \begin{bmatrix} x \\ y \\ z \end{bmatrix} = \begin{bmatrix} 2 \\ -5 \\ 3 \end{bmatrix}$$

This is equivalent to

$$\begin{bmatrix} 1 & 2 & 1 \\ 0 & 2 & -2 \\ 0 & -5 & 5 \end{bmatrix} \begin{bmatrix} x \\ y \\ z \end{bmatrix} = \begin{bmatrix} 2 \\ 1 \\ -1 \end{bmatrix} \quad \text{Applying } R_2 \to R_2 + 3R_1, R_3 \to R_3 + (-2)R_1$$

or, $$\begin{bmatrix} 1 & 2 & 1 \\ 0 & 2 & -2 \\ 0 & 0 & 0 \end{bmatrix} \begin{bmatrix} x \\ y \\ z \end{bmatrix} = \begin{bmatrix} 2 \\ 1 \\ 3/2 \end{bmatrix} \quad \text{Applying } R_3 \to R_3 + \frac{5}{2}R_2$$

This is an inconsistent system of equations and so has no solution. Hence, v cannot be written as a linear combination of v_1, v_2 and v_3.

EXAMPLE-5 *Express the polynomial $f(x) = x^2 + 4x - 3$ in the vector space V of all polynomials over R as a linear combination of the polynomials $g(x) = x^2 - 2x + 5, h(x) = 2x^2 - 3x$ and $\phi(x) = x + 3$.*

SOLUTION Let α, β, γ be scalars such that

$$f(x) = u\, g(x) + v\, h(x) + w\, \phi(x) \quad \text{for all } x \in R$$

$\Rightarrow \quad x^2 + 4x - 3 = u(x^2 - 2x + 5) + v(2x^2 - 3x) + w(x + 3) \quad \text{for all } x \in R$

$\Rightarrow \quad x^2 + 4x - 3 = (u + 2v)x^2 + (-2u - 3v + w)x + (5u + 3w) \quad \text{for all } x \in R$ \hfill (i)

$\Rightarrow \quad u + 2v = 1, -2u - 3v + w = 4, 5u + w = -3$ \hfill (ii)

The matrix form of this system of equations is

$$\begin{bmatrix} 1 & 2 & 0 \\ -2 & -3 & 1 \\ 5 & 0 & 3 \end{bmatrix} \begin{bmatrix} u \\ v \\ w \end{bmatrix} = \begin{bmatrix} 1 \\ 4 \\ -3 \end{bmatrix}$$

This system of equations is equivalent to

$$\begin{bmatrix} 1 & 2 & 0 \\ 0 & 1 & 1 \\ 0 & -10 & 3 \end{bmatrix} \begin{bmatrix} u \\ v \\ w \end{bmatrix} = \begin{bmatrix} 1 \\ 6 \\ -8 \end{bmatrix} \quad \text{Applying } R_2 \to R_2 + 2R_1, \ R_3 \to R_3 - 5R_1$$

or, $$\begin{bmatrix} 1 & 2 & 0 \\ 0 & 1 & 1 \\ 0 & 0 & 13 \end{bmatrix} \begin{bmatrix} u \\ v \\ w \end{bmatrix} = \begin{bmatrix} 1 \\ 6 \\ 52 \end{bmatrix} \quad \text{Applying } R_3 \to R_3 + 10R_2$$

Thus, the system of equations obtained in (i) is consistent and is equivalent to

$$u + 2v = 1, v + w = 6 \text{ and } 13w = 52 \Rightarrow \quad u = -3, v = 2, w = 4$$

Hence, $$f(x) = -3g(x) + 2h(x) + 4\phi(x).$$

REMARK. *The equation (i) obtained in the above solution is an identity in x, that is it holds for any value of x. So, the values of u, v and w can be obtained by solving three equations which can be obtained by given any three values to variable x.*

EXAMPLE-6 Let $V = R^{2\times 2}$ be the vector space of all 2×2 matrices (with entries in R) over field R and let

$$M = \begin{bmatrix} 4 & 7 \\ 7 & 9 \end{bmatrix}, A = \begin{bmatrix} 1 & 1 \\ 1 & 1 \end{bmatrix}, B = \begin{bmatrix} 1 & 2 \\ 3 & 4 \end{bmatrix} \text{ and } C = \begin{bmatrix} 1 & 1 \\ 4 & 5 \end{bmatrix}$$

be three matrices in V. Express matrix M as a linear combination of A, B and C.

SOLUTION Let x, y, z be three scalars in R such that

$$M = xA + yB + zC$$

$\Rightarrow \quad \begin{bmatrix} 4 & 7 \\ 7 & 9 \end{bmatrix} = x\begin{bmatrix} 1 & 1 \\ 1 & 1 \end{bmatrix} + y\begin{bmatrix} 1 & 2 \\ 3 & 4 \end{bmatrix} + z\begin{bmatrix} 1 & 1 \\ 4 & 5 \end{bmatrix}$

$\Rightarrow \quad \begin{bmatrix} 4 & 7 \\ 7 & 9 \end{bmatrix} = \begin{bmatrix} x+y+z & x+2y+z \\ x+3y+4z & x+4y+5z \end{bmatrix}$

$\Rightarrow \quad x+y+z = 4, \ x+2y+z = 7, \ x+3y+4z = 7, \ x+4y+5z = 9 \ldots \text{(i)}$

The matrix form of this system of equations is

$$\begin{bmatrix} 1 & 1 & 1 \\ 1 & 2 & 1 \\ 1 & 3 & 4 \\ 1 & 4 & 5 \end{bmatrix} \begin{bmatrix} x \\ y \\ z \end{bmatrix} = \begin{bmatrix} 4 \\ 7 \\ 7 \\ 9 \end{bmatrix}$$

This is equivalent to

$$\begin{bmatrix} 1 & 1 & 1 \\ 0 & 1 & 0 \\ 0 & 2 & 3 \\ 0 & 3 & 4 \end{bmatrix} \begin{bmatrix} x \\ y \\ z \end{bmatrix} = \begin{bmatrix} 4 \\ 3 \\ 3 \\ 5 \end{bmatrix} \quad \text{Applying } R_2 \to R_2 - R_1, \ R_3 \to R_3 - R_1, \ R_4 \to R_4 - R_1$$

or, $\begin{bmatrix} 1 & 1 & 1 \\ 0 & 1 & 0 \\ 0 & 0 & 3 \\ 0 & 0 & 4 \end{bmatrix} \begin{bmatrix} x \\ y \\ z \end{bmatrix} = \begin{bmatrix} 4 \\ 3 \\ -3 \\ -4 \end{bmatrix}$ Applying $R_3 \to R_3 - 2R_2, \ R_4 \to R_4 - 2R_2$

Thus, the system of equations in (i) is equivalent to the following system of equations

$$x+y+z = 4, y = 3, 3z = -3, 4z = -4$$

$\therefore \qquad x = 2, \ y = 3, \ z = -1$

Hence, $M = 2A + 3B - C$

EXAMPLE-7 Consider the vectors $v_1 = (1,2,3)$ and $v_2 = (2,3,1)$ in $R^3(R)$. Find k so that $u = (1, k, 4)$ is a linear combination of v_1 and v_2.

SOLUTION Let x, y be scalars such that
$$u = xv_1 + yv_2$$
$\Rightarrow \quad (1, k, 4) = x(1, 2, 3) + y(2, 3, 1)$
$\Rightarrow \quad (1, k, 4) = (x + 2y, 2x + 3y, 3x + y)$
$\Rightarrow \quad x + 2y = 1, 2x + 3y = k$ and $3x + y = 4$

Solving $x + 2y = 1$ and $3x + y = 4$, we get $x = \dfrac{7}{5}$ and $y = -\dfrac{1}{5}$

Substituting these values in $2x + 3y = k$, we get $k = \dfrac{11}{5}$

EXAMPLE-8 *Consider the vectors $v_1 = (1, 2, 3)$ and $v_2 = (2, 3, 1)$ in $R^3(R)$. Find conditions on a, b, c so that $u = (a, b, c)$ is a linear combination of v_1 and v_2.*

SOLUTION Let x, y be scalars in R such that
$$u = xv_1 + yv_2$$
$\Rightarrow \quad (a, b, c) = x(1, 2, 3) + y(2, 3, 1)$
$\Rightarrow \quad (a, b, c) = (x + 2y, 2x + 3y, 3x + y)$
$\Rightarrow \quad x + 2y = a, 2x + 3y = b$ and $3x + y = c$

Solving first two equations, we get $x = -3a + 2b$ and $y = 2a - b$.

Substituting these values in $3x + y = c$, we get
$3(-3a + 2b) + (2a - b) = c$ or, $7a - 5b + c = 0$ as the required condition.

EXERCISE 2.6

1. Express $v = (3, -2)$ in $R^2(R)$ as a linear combination of the vectors $v_1 = (-1, 1)$ and $v_2 = (2, 1)$.
2. Express $v = (-2, 5)$ in $R^2(R)$ as a linear combination of the vectors $v_1 = (2, -1)$ and $v_2 = (-4, 2)$.
3. Express $v = (3, 7, -4)$ in $R^3(R)$ as a linear combination of the vectors $v_1 = (1, 2, 3)$, $v_2 = (2, 3, 7)$ and $v_3 = (3, 5, 6)$.
4. Express $v = (2, -1, 3)$ in $R^3(R)$ as a linear combination of the vectors $v_1 = (3, 0, 3)$, $v_2 = (-1, 2, -5)$ and $v_3 = (-2, -1, 0)$.
5. Let V be the vector space of all real polynomials over field R of all real numbers. Express the polynomial $p(x) = 3x^2 + 5x - 5$ as a linear combination of the polynomials $f(x) = x^2 + 2x + 1, g(x) = 2x^2 + 5x + 4$ and $h(x) = x^2 + 3x + 6$.
6. Let $V = P_2(t)$ be the vector space of all polynomials of degree less than or equal to 2 and t be the indeterminate. Write the polynomial $f(t) = at^2 + bt + c$ as a linear combination of the polynomials $p_1(t) = (t-1)^2, p_2(t) = t - 1$ and $p_3(t) = 1$.
7. Write the vectors $u = (1, 3, 8)$ in $R^3(R)$ as a linear combination of the vectors $v_1 = (1, 2, 3)$ and $v_2 = (2, 3, 1)$.

8. Write the vector $u = (2,4,5)$ in $R^3(R)$ as a linear combination of the vectors $v_1 = (1,2,3)$ and $v_2 = (2,3,1)$.
9. Consider the vector space $R^{2\times 2}$ of all 2×2 matrices over the field R of real numbers. Express the matrix $M = \begin{bmatrix} 5 & -6 \\ 7 & 8 \end{bmatrix}$ as a linear combination of the matrices

$$E_{11} = \begin{bmatrix} 1 & 0 \\ 0 & 0 \end{bmatrix}, E_{12} = \begin{bmatrix} 0 & 1 \\ 0 & 0 \end{bmatrix}, E_{21} = \begin{bmatrix} 0 & 0 \\ 1 & 0 \end{bmatrix} \text{ and } E_{22} = \begin{bmatrix} 0 & 0 \\ 0 & 1 \end{bmatrix}$$

ANSWERS

1. $v = \dfrac{-7}{3}v_1 + \dfrac{1}{3}v_2$ 2. v is not expressible as a linear combination of v_1 and v_2.
3. $v = 2v_1 - 4v_2 + 3v_3$ 4. v is not expressible as a linear combination of v_1, v_2, v_3.
5. $p(x) = 3f(x) + g(x) - 2h(x)$ 6. $f(t) = ap_1(t) + (2a+b)p_2(t) + (a+b+c)p_3(t)$
7. $u = 3v_1 - v_2$ 8. Not Possible 9. $M = 5E_{11} - 6E_{12} + 7E_{21} + 8E_{22}$

2.7 LINEAR SPANS

LINEAR SPAN OF A SET. *Let V be a vector space over a field F and let S be a subset of V. Then the set of all finite linear combinations of S is called linear span of S and is denoted by $[S]$.*

Thus,
$$[S] = \left\{ \sum_{i=1}^{n} \lambda_i v_i : \lambda_1, \lambda_2, \ldots, \lambda_n \in F,\ n \in N \text{ and } v_1, v_2, \ldots, v_n \in S \right\}.$$

If S is a finite set, say $S = \{v_1, v_2, \ldots, v_n\}$, then
$$[S] = \left\{ \sum_{i=1}^{n} \lambda_i v_i : \lambda_i \in F \right\}.$$

In this case $[S]$ is also written as $[v_1, v_2, \ldots, v_n]$.

If $S = \{v\}$, then $[S] = F_v$ as shown in Example 5 on page 136.

Consider a subset $S = \{(1,0,0),(0,1,0)\}$ of real vector space R^3. Any linear combination of a finite number of elements of S is of the form $\lambda(1,0,0) + \mu(0,1,0) = (\lambda, \mu, 0)$. By definition, the set of all such linear combinations is $[S]$. Thus, $[S] = \{(\lambda, \mu, 0) : \lambda, \mu \in R\}$. Clearly $[S]$ is xy-plane in three dimensional Euclidean space R^3 and is a subspace of R^3. In fact, this is true for every non-void subset of a vector space. The following theorem proves this result.

THEOREM-1 *Let S be a non-void subset of a vector space $V(F)$. Then $[S]$, the linear span of S, is a subspace of V.*

PROOF. Since $0_V = 0v_1 + 0v_2 + \cdots + 0v_n$ for every positive integer n. Therefore, 0_V is a finite linear combination of S and so it is in $[S]$.

Thus, $[S]$ is a non-void subset of V.

Let u and v be any two vectors in $[S]$. Then,
$u = \lambda_1 u_1 + \lambda_2 u_2 + \cdots + \lambda_n u_n$ for some scalars $\lambda_i \in F$, some $u_i^{'s} \in S$, and a positive integer n.
and, $v = \mu_1 v_1 + \mu_2 v_2 + \cdots + \mu_m v_m$ for some $\mu_i^{'s} \in F$, some $v_i^{'s} \in S$, and a positive integer m.

If λ, μ are any two scalars in F, then

$$\lambda u + \mu v = \lambda(\lambda_1 u_1 + \lambda_2 u_2 + \cdots + \lambda_n u_n) + \mu(\mu_1 v_1 + \cdots + \mu_m v_m)$$
$\Rightarrow \quad \lambda u + \mu v = (\lambda\lambda_1)u_1 + \cdots + (\lambda\lambda_n)u_n + (\mu\mu_1)v_1 + \cdots + (\mu\mu_m)v_m$

Clearly, $\lambda u + \mu v$ is a finite linear combination of vectors in S and so it is in $[S]$.

Thus, $\lambda u + \mu v \in [S]$ for all $u, v \in [S]$ and for all $\lambda, \mu \in F$.

Hence, $[S]$ is a subspace of V.

Q.E.D.

THEOREM-2 *Let S be a subspace of a vector space $V(F)$. Then $[S]$ is the smallest subspace of V containing S.*

PROOF. By Theorem 1, $[S]$ is a subspace of V. Let v be an arbitrary vector in S. Then,

$v = 1v$ \hfill [By V-2(iv)]
$\Rightarrow \quad v$ is a finite linear combination of S.
$\Rightarrow \quad v \in [S]$.

Thus, $S \subset [S]$.

Now it remains to prove that $[S]$ is the smallest subspace of V containing S. For this, we shall show that if there exists another subspace T containing S, then it also contains $[S]$. So let T be a subspace of V containing S. Let u be an arbitrary vector in $[S]$. Then,
$u = \lambda_1 v_1 + \cdots + \lambda_n v_n$ for some $\lambda_i^{'s} \in F$, some $v_i^{'s} \in S$ and a positive integer n.

Since $S \subset T$ and $v_i \in S$. Therefore,

$v_i \in T$ for each $i \in \underline{n}$
$\Rightarrow \quad \lambda_1 v_1 + \lambda_2 v_2 + \cdots + \lambda_n v_n \in T$ \hfill [$\because$ T is a subspace of V]
$\Rightarrow \quad u \in T$

Since u is an arbitrary vector in S. Therefore,

$u \in T$ for all $u \in [S]$.
$\Rightarrow \quad [S] \subset T$.

Hence, $[S]$ is the smallest subspace of V containing S.

Q.E.D.

REMARK-1 *Since the null space $\{0_V\}$ is the smallest subspace of V containing the void set φ, Therefore, by convention, we take $[\varphi] = \{0_V\}$.*

REMARK-2 *It is clear from the foregoing discussion that if S is a non-void subset of a vector space $V(F)$, then S itself is smaller than $[S]$. In fact, $[S] \neq \{0_V\}$ always contains an infinite number of vectors whatever be the number of vectors in S.*

THEOREM-3 *Let $S = \{v_1, v_2, \ldots, v_n\}$ be a finite subset of a vector space $V(F)$. Then for any $0 \neq a \in F$*

(i) $[v_1, v_2, \ldots, v_k, \ldots, v_n] = [v_1, v_2, \ldots, av_k, \ldots, v_n]$
(ii) $[v_1, v_2, \ldots, v_k, \ldots, v_n] = [v_1 + av_k, v_2, \ldots, v_k, \ldots, v_n]$
(iii) *If* $v = \sum_{i=1}^{n} a_i v_i$, *where* $a_i \in F$ *for all* $i \in \underline{n}$, *then* $[v_1, v_2, \ldots, v_n] = [v_1, v_2, \ldots, v_n, v]$.

PROOF. (i) Let $u \in [v_1, v_2, \ldots, v_k, \ldots, v_n]$. Then u is a linear combination of vectors $v_1, v_2, \ldots, v_k, \ldots, v_n$.

Consequently, there exist scalars $\lambda_1, \lambda_2, \ldots, \lambda_n$ in F such that

$$u = \lambda_1 v_1 + \cdots + \lambda_k v_k + \cdots + \lambda_n v_n$$

$\Rightarrow \quad u = \lambda_1 v_1 + \cdots + \lambda_k a^{-1}(av_k) + \cdots + \lambda_n v_n \qquad [\because \ 0 \neq a \in F \therefore \ a^{-1} \in F]$

$\Rightarrow \quad u$ is a linear combination of $v_1, v_2, \ldots, av_k, \ldots, v_n$

$\Rightarrow \quad u \in [v_1, v_2, \ldots, av_k, \ldots, v_n]$

Thus, $u \in [v_1, v_2, \ldots, v_k, \ldots, v_n] \Rightarrow u \in [v_1, v_2, \ldots, av_k, \ldots, v_n]$

$\therefore \qquad [v_1, v_2, \ldots, v_k, \ldots, v_n] \subset [v_1, v_2, \ldots, av_k, \ldots, v_n] \qquad\qquad (i)$

Now let u be an arbitrary vector in $[v_1, v_2, \ldots, av_k, \ldots, v_n]$. Then there exist scalar $\mu_1, \mu_2, \ldots, \mu_k, \ldots, \mu_n$ in F such that

$$u = \mu_1 v_1 + \mu_2 v_2 + \cdots + \mu_k(av_k) + \cdots + \mu_n v_n$$

$\Rightarrow \quad u = \mu_1 v_1 + \mu_2 v_2 + \cdots + (\mu_k a) v_k + \cdots + \mu_n v_n$

$\Rightarrow \quad u$ is a linear combination of $v_1, v_2, \ldots, v_k, \ldots, v_n$

$\Rightarrow \quad u \in [v_1, v_2, \ldots, v_k, \ldots, v_n]$

Thus, $u \in [v_1, v_2, \ldots, av_k, \ldots, v_n] \Rightarrow u \in [v_1, v_2, \ldots, v_k, \ldots, v_n]$

$\therefore \qquad [v_1, v_2, \ldots, av_k, \ldots, v_n] \subset [v_1, v_2, \ldots, v_k, \ldots, v_n] \qquad\qquad (ii)$

From (i) and (ii), we obtain

$$[v_1, v_2, \ldots, v_k, \ldots, v_n] = [v_1, v_2, \ldots, av_k, \ldots, v_n]$$

(ii) Let u be an arbitrary vector in $[v_1, v_2, \ldots, v_k, \ldots, v_n]$. Then, there exist scalars $\lambda_1, \lambda_2, \ldots, \lambda_n$, such that

$$u = \sum_{i=1}^{n} \lambda_i v_i$$

$\Rightarrow \quad u = \lambda_1(v_1 + av_k) + \lambda_2 v_2 + \cdots + (\lambda_k - a\lambda_1)v_k + \cdots + \lambda_n v_n$

$\Rightarrow \quad u$ is a linear combination of $v_1 + av_k, v_2, \ldots, v_k, \ldots, v_n$

$\Rightarrow \quad u \in [v_1 + av_k, v_2, \ldots, v_k, \ldots, v_n]$

Thus, $u \in [v_1, v_2, \ldots, v_k, \ldots, v_n] \Rightarrow u \in [v_1 + av_k, v_2, \ldots, v_k, \ldots, v_n]$

$\therefore \quad [v_1, v_2, \ldots, v_k, \ldots, v_n] \subset [v_1 + av_k, v_2, \ldots, v_k, \ldots, v_n]$ \hfill (iii)

Now let

$u \in [v_1 + av_k, v_2, \ldots, v_{k-1}, v_k, \ldots, v_n]$

$\Rightarrow \quad u$ is a linear combination of $v_1 + av_k, v_2, \ldots, v_k, \ldots, v_n$

$\Rightarrow \quad$ There exists scalars $\lambda_1, \lambda_2, \ldots, \lambda_k, \ldots, \lambda_n \in F$ such that

$u = \lambda_1(v_1 + av_k) + \cdots + \lambda_k v_k + \cdots + \lambda_n v_n$

$\Rightarrow \quad u = \lambda_1 v_1 + \cdots + (\lambda_1 a + \lambda_k) v_k + \cdots + \lambda_n v_n$

$\Rightarrow \quad u$ is a linear combination of $v_1, v_2, \ldots, v_k, \ldots, v_n$

$\Rightarrow \quad u \in [v_1, v_2, \ldots, v_k, \ldots, v_n]$

$\therefore \quad [v_1 + av_k, \ldots, v_{k-1}, v_k, \ldots, v_n] \subset [v_1, v_2, \ldots, v_k, \ldots, v_n]$ \hfill (iv)

From (iii) and (iv), we get

$$[v_1, v_2, \ldots, v_k, \ldots, v_n] = [v_1 + av_k, \ldots, \ldots, v_k, \ldots, v_n].$$

(iii) Let u be an arbitrary vector in $[v_1, v_2, \ldots, v_n, v]$. Then u is a linear combination of $v_1, v_2, \ldots, v_n, v$. Consequently, there exist scalars in $\lambda_1, \lambda_2, \ldots, \lambda_n, \lambda$ in F such that

$u = \lambda v_1 + \cdots + \lambda_n v_n + \lambda v$

$\Rightarrow \quad u = \lambda_1 v_1 + \cdots + \lambda_n v_n + \lambda \sum_{i=1}^{n} a_i v_i$

$\Rightarrow \quad u = \sum_{i=1}^{n} (\lambda_i + \lambda a_i) v_i$

$\Rightarrow \quad u$ is a linear combination of $v_1, v_2, \ldots, v_n$.

$\Rightarrow \quad u \in [v_1, v_2, \ldots, v_k, \ldots, v_n]$

$\therefore \quad [v_1, v_2, \ldots, v_k, \ldots, v_n, v] \subset [v_1, v_2, \ldots, v_n]$ \hfill (v)

Again, let u be an arbitrary vector in $[v_1, v_2, \ldots, v_k, \ldots, v_n]$. Then there exist scalars $\lambda_1, \lambda_2, \ldots, \lambda_n$ in F such that

$u = \lambda_1 v_1 + \cdots + \lambda_n v_n$

$\Rightarrow \quad u = \lambda_1 v_1 + \cdots + \lambda_n v_n + 0v$

$\Rightarrow \quad u$ is a linear combination of $v_1, v_2, \ldots, v_n, v$

$\Rightarrow \quad u \in [v_1, v_2, \ldots, v_n, v]$

$\therefore \quad [v_1, v_2, \ldots, v_n] \subset [v_1, v_2, \ldots, v_n, v]$ \hfill (vi)

From (v) and (vi), we get

$$[v_1, v_2, \ldots, v_n] = [v_1, v_2, \ldots, v_n, v].$$

\hfill Q.E.D.

REMARK. *Theorem 3(iii) reveals that the removal of a vector, which is a linear combination of the remaining vectors, from a list of vectors does not effect the span of the list.*

THEOREM-4 *If S and T are non-void subsets of a vector space $V(F)$, then*

(i) $S \subset T \Rightarrow [S] \subset [T]$.

(ii) $[S] = S \Leftrightarrow S$ is a subspace of V.

(iii) $[[S]] = [S]$

(iv) $[S \cup T] = [S] + [T]$

PROOF. (i) Let u be an arbitrary vector in $[S]$. Then,

$$u = \sum_{i=1}^{n} \lambda_i v_i \text{ for some } v_i's \in S, \lambda_i's \in F \text{ and for some positive integer } n.$$

Now, $\quad v_i \in S \Rightarrow v_i \in T \quad$ for all $i \in \underline{n}$ $\hfill [\because S \subset T]$

$\Rightarrow \quad \sum_{i=1}^{n} \lambda_i v_i \in [T]$

$\Rightarrow \quad u \in [T]$

Thus, $\quad u \in [S] \Rightarrow u \in [T]$

$\therefore \quad [S] \subset [T]$

(ii) First, let S be a subspace of V. Then we have to show that $[S] = S$.

Let u be an arbitrary vector in S. Then,

$$u = 1u \hfill \text{[By } V\text{-2(iv)]}$$

$\Rightarrow \quad u$ is a finite linear combination of vectors in S

$\Rightarrow \quad u \in [S]$

$\therefore \quad S \subset [S]$

Let v be an arbitrary vector in $[S]$. Then

$$v = \sum_{i=1}^{n} \lambda_i v_i \text{ for some } v_i's \in S, \lambda_i's \in F, \text{ and for some positive integer } n.$$

$\Rightarrow \quad v \in S \quad \left[\begin{array}{l} \because \ S \text{ is a subspace of } V, \\ \therefore \ S \text{ is closed under vector addition and scalar multiplication} \end{array}\right]$

Thus, $\quad v \in [S] \Rightarrow v \in S$

$\therefore \quad [S] \subset S$.

Hence, $\quad [S] = S$.

Conversely, let us suppose that $[S] = S$. Then we have to show that S is a subspace of V. By Theorem 1, $[S]$ is a subspace of V. Since $[S] = S$, therefore, S is also a subspace of V.

(iii) We know that $[S]$ is a subspace of V. Therefore, by (ii) it follows that $[[S]] = [S]$.

(iv) Let u be an arbitrary vector in $[S \cup T]$. Then,
$u = \lambda_1 u_1 + \lambda_2 u_2 + \cdots + \lambda_m u_m + \mu_1 v_1 + \cdots + \mu_n v_n$ for some $u_i's \in S$, $v_i's \in T$, $\lambda_i's$, $\mu_i's \in F$, and positive integers m and n.

Now, $\quad u_i's \in S$ and $\lambda_i's \in F \Rightarrow \lambda_1 u_1 + \cdots + \lambda_m u_m \in [S]$

and, $\quad v_i's \in T$ and $\mu_i's \in F \Rightarrow \mu_1 v_1 + \cdots + \mu_n v_n \in [T]$.

$\therefore \quad \lambda_1 u_1 + \cdots + \lambda_m u_m + \mu_1 v_1 + \cdots + \mu_n v_n \in [S] + [T]$

$\Rightarrow \quad u \in [S] + [T]$

Thus, $u \in [S \cup T] \Rightarrow u \in [S] + [T]$. Consequently,
$$[S \cup T] \subset [S] + [T].$$

Now, let v be an arbitrary vector in $[S] + [T]$. Then,
$$v = u + w \text{ for some } u \in [S], w \in [T].$$

Now, $\quad u \in [S] \Rightarrow u$ is a finite linear combination of S.

and, $\quad w \in [T] \Rightarrow w$ is a finite linear combination of T.

Therefore, $v = u + w$ is a finite linear combination of $S \cup T$ and so it is in $[S \cup T]$.

Thus, $\quad v \in [S] + [T] \Rightarrow v \in [S \cup T]$

Consequently, $\quad [S] + [T] \subset [S \cup T]$

Hence, $\quad [S \cup T] = [S] + [T]$

Q.E.D.

2.7.1 ROW AND COLUMN SPACES OF A MATRIX

Let $A = [a_{ij}]$ be an arbitrary $m \times n$ matrix over a field F. The rows of A are
$R_1 = (a_{11}, a_{12}, a_{13}, \ldots a_{1n})$, $R_2 = (a_{21}, a_{22}, a_{23}, \ldots a_{2n})$, $\ldots$, $R_m = (a_{m1}, a_{m2}, a_{m3}, \ldots, a_{mn})$
These rows may be viewed as vectors in F^n. So, they span a subspace of F^n, which is known as the row space of matrix A and is denoted by $rowsp(A)$. That is,
$$rowsp(A) = [R_1, R_2, \ldots, R_m]$$
Thus, row space of a matrix is the subspace of F^n spanned the rows of the matrix.

Similarly, the column space of A is the subspace of F^m spanned by the columns
$C_1 = (a_{11}, a_{21}, \ldots, a_{m1})$, $C_2 = (a_{12}, a_{22}, \ldots, a_{m2}) \ldots$, $C_n = (a_{1n}, a_{2n}, \ldots, a_{mn})$ and is denoted by $colsp(A)$.

Clearly, $\quad colsp(A) = rowsp(A^T)$.

THEOREM-1 *Row equivalent matrices have the same row space.*

PROOF. Let A be a matrix and B be the matrix obtained by applying one of the following row operations on matrix A:

(i) Interchange R_i and R_j (ii) Replace R_i by kR_i (iii) Replace R_j by $R_j + kR_i$

Then each row of B is a row of A or a linear combination of rows of A. So, the row space of B is contained in the row space of A. On the other hand, we can apply the inverse elementary row operation on B to obtain A. So, the row space of A is contained in the row space of B. Consequently, A and B have the same row space.

Q.E.D.

THEOREM-2 *Let A and B be row canonical matrices. Then A and B have the same row space if and only if they have the same non-zero rows.*

EXAMPLE Let $v_1 = (1, 2, -1, 3), v_2 = (2, 4, 1, -2), v_3 = (3, 6, 3, -7), v_4 = (1, 2, -4, 11)$ and $v_5 = (2, 4, -5, 14)$ be vectors in $R^4(R)$ such that $S = \{v_1, v_2, v_3\}$ and $T = \{v_4, v_5\}$. Show that $[S] = [T]$.

SOLUTION There are two ways to show that $[S] = [T]$. Show that each of v_1, v_2, v_3 is a linear combination of v_3 and v_4 and show that each of v_4 and v_5 is a linear combination of v_1, v_2, v_3. But, this method is not very convenient. So, let us discuss an alternative method. Let A be the matrix whose rows are v_1, v_2, v_3 and B be the matrix whose rows v_4 and v_5. That is,

$$A = \begin{bmatrix} 1 & 2 & -1 & 3 \\ 2 & 4 & 1 & -2 \\ 3 & 6 & 3 & -7 \end{bmatrix}$$

$\Rightarrow A \sim \begin{bmatrix} 1 & 2 & -1 & 3 \\ 0 & 0 & 3 & -8 \\ 0 & 0 & 6 & -16 \end{bmatrix}$ Applying $R_2 \rightarrow R_2 - 2R_1, R_3 \rightarrow R_3 - 3R_1$

$\Rightarrow A \sim \begin{bmatrix} 1 & 2 & -1 & 3 \\ 0 & 0 & 3 & -8 \\ 0 & 0 & 0 & 0 \end{bmatrix}$ Applying $R_3 \rightarrow R_3 - 2R_2$

$$B = \begin{bmatrix} 1 & 2 & -4 & 11 \\ 2 & 4 & -5 & 14 \end{bmatrix}$$

$\Rightarrow B \sim \begin{bmatrix} 1 & 2 & -4 & 11 \\ 0 & 0 & 3 & -8 \end{bmatrix}$ Applying $R_2 \rightarrow R_2 - 2R_1$

$\Rightarrow B \sim \begin{bmatrix} 1 & 2 & -1 & 3 \\ 0 & 0 & 3 & -8 \end{bmatrix}$ Applying $R_1 \rightarrow R_1 + R_2$

Clearly, non-zero rows of the matrices in row canonical form are identical. Therefore, $rowsp(A) = rowsp(B)$.

Hence, $[S] = [T]$

ILLUSTRATIVE EXAMPLES

EXAMPLE-1 *Prove that any vector $v = (a,b)$ in $R^2(R)$ is a linear combination of $v_1 = (1,1)$ and $v_2 = (-1,1)$.*

SOLUTION Let $v = xv_1 + yv_2$ for some $x, y \in R$. Then,

$$(a,b) = x(1,1) + y(-1,1)$$
$$\Rightarrow (a,b) = (x-y, x+y)$$
$$\Rightarrow x - y = a \text{ and } x + y = b$$

Clearly, this system of equations is consistent with unique solution as the determinant of the coefficient matrix is non-zero. The values of x and y are $\dfrac{a+b}{2}$ and $\dfrac{a-b}{2}$ respectively.

Hence, every vector in R^2 is a linear combination of v_1 and v_2.

EXAMPLE-2 *Show that the vectors $v_1 = (1,1)$ and $v_2 = (1,2)$ span $R^2(R)$.*

SOLUTION In order to show that vectors v_1 and v_2 span $R^2(R)$, it is sufficient to show that any vector in R^2 is a linear combination of v_1 and v_2. Let $v = (a,b)$ be an arbitrary vector in $R^2(R)$. Further, let

$$v = xv_1 + yv_2$$
$$\Rightarrow (a,b) = x(1,1) + y(1,2)$$
$$\Rightarrow (a,b) = (x+y, x+2y)$$
$$\Rightarrow x + y = a \text{ and } x + 2y = b$$

This is a consistent system of equations for all $a, b \in R$. So, every vector in $R^2(R)$ is a linear combination of v_1 and v_2.

EXAMPLE-3 *Let $V = P_n(x)$ be the vector space of all polynomials of degree less than or equal to n over the field R of all real numbers. Show that the polynomials $1, x, x^2, \ldots, x^{n-1}, x^n$ span V.*

SOLUTION Any polynomial $f(x)$ of degree less than or equal to n can be written as

$$f(x) = a_0 + a_1 x + a_2 x^2 + \cdots + a_n x^n, \text{ where } a_0, a_1, a_2, \ldots, a_n \in R$$

$\Rightarrow f(x)$ is a linear combination of $1, x, x^2, \ldots, x^{n-1}, x^n$

Hence, polynomials $1, x, x^2, \ldots, x^{n-1}, x^n$ span V.

EXAMPLE-4 *Prove that the polynomials $1, 1+x, (1+x)^2$ span the vector space $V = P_2(x)$ of all polynomials of degree at most 2 over the field R of real numbers.*

SOLUTION Let $f(x) = a + bx + cx^2$ be an arbitrary polynomial in V, where $a, b, c \in R$. In order to prove that the polynomials $1, 1+x, (1+x)^2$ span V, it is sufficient to show that

$f(x)$ is expressible as a linear combination of $1, 1+x, (1+x)^2$.
Let
$$f(x) = \alpha \times 1 + \beta(1+x) + \gamma(1+x)^2$$
$\Rightarrow \quad a + bx + cx^2 = (\alpha + \beta + \gamma) + (\beta + 2\gamma)x + \gamma x^2$
$\Rightarrow \quad \alpha + \beta + \gamma = a, \quad \beta + 2\gamma = b \quad \text{and} \quad \gamma = c$
$\Rightarrow \quad \alpha = a - b + c \quad \beta = b - 2c \quad \text{and} \quad \gamma = c$

Thus, for a given polynomial $f(x)$, we can find the values of α, β, γ such that $f(x)$ is expressible as a linear combination of $1, 1+x, (1+x)^2$. Hence, the polynomials $1, 1+x, (1+x)^2$ span V.

EXAMPLE-5 *Show that the vector space $V = R[x]$ of real polynomials cannot be spanned by a finite number of polynomials.*

SOLUTION Let S be a finite set of polynomials such that the degree of highest degree polynomial is m. Then, the linear span of S, i.e. $[S]$ cannot contain a polynomial of degree greater than m. Thus, $[S] \neq V$, for any finite set S.

EXAMPLE-6 *Show that the vectors $v_1 = (1,1,1), v_2 = (1,2,3), v_3 = (1,5,8)$ span $R^3(R)$.*

SOLUTION In order to show that vectors v_1, v_2, v_3 span $R^3(R)$, i.e. $[\{v_1, v_2, v_3\}] = R^3$, it is sufficient to show that an arbitrary vector $v = (a,b,c)$ in R^3 is a linear combination of v_1, v_2, v_3.
Let $v = xv_1 + yv_2 + zv_3$, where $x, y, z \in R$.
$\Rightarrow \quad (a,b,c) = x(1,1,1) + y(1,2,3) + z(1,5,8)$
$\Rightarrow \quad (a,b,c) = (x+y+z, x+2y+5z, x+3y+8z)$
$\Rightarrow \quad x+y+z = a, \ x+2y+5z = 6, \ x+3y+8z = c$
The determinant of the coefficient matrix of this system of equations is
$$\Delta = \begin{vmatrix} 1 & 1 & 1 \\ 1 & 2 & 5 \\ 1 & 3 & 8 \end{vmatrix} = \begin{vmatrix} 1 & 1 & 1 \\ 0 & 1 & 4 \\ 0 & 2 & 7 \end{vmatrix} \text{ Applying } R_2 \to R_2 - R_1, R_3 \to R_3 - R_1$$
$\Rightarrow \quad \Delta = -1 \neq 0$
So, the above system of equations is consistent and has unique solution.
Hence, v_1, v_2, v_3 span $R^3(R)$.

EXAMPLE-7 *Find conditions on a, b, c so that $v = (a,b,c)$ in $R^3(R)$ belongs to the subspace spanned by $v_1 = (1,2,0), v_2 = (-1,1,2), v_3 = (3,0,-4)$.*

SOLUTION If $v = (a,b,c)$ belongs to the subspace spanned by v_1, v_2, v_3. Then, there exist scalars $x, y, z \in R$ such that
$$v = xv_1 + yv_2 + zv_3$$
$\Rightarrow \quad (a,b,c) = x(1,2,0) + y(-1,1,2) + z(3,0,-4)$
$\Rightarrow \quad (a,b,c) = (x-y+3z, 2x+y, 2y-4z)$
$\Rightarrow \quad x-y+3z = a, \ 2x+y = b, \ 2y-4z = c$

This system of equations can be written in matrix form as follows:

$$\begin{bmatrix} 1 & -1 & 3 \\ 2 & 1 & 0 \\ 0 & 2 & -4 \end{bmatrix} \begin{bmatrix} x \\ y \\ z \end{bmatrix} = \begin{bmatrix} a \\ b \\ c \end{bmatrix}$$

or, $\begin{bmatrix} 1 & -1 & 3 \\ 0 & 3 & -6 \\ 0 & 2 & -4 \end{bmatrix} \begin{bmatrix} x \\ y \\ z \end{bmatrix} = \begin{bmatrix} a \\ b-2a \\ c \end{bmatrix}$ Applying $R_2 \to R_2 - 2R_1$

or, $\begin{bmatrix} 1 & -1 & 3 \\ 0 & 3 & -6 \\ 0 & 0 & 0 \end{bmatrix} \begin{bmatrix} x \\ y \\ z \end{bmatrix} = \begin{bmatrix} a \\ b-2a \\ c - \frac{2}{3}(b-2a) \end{bmatrix}$ Applying $R_3 \to R_3 - \frac{2}{3}R_2$

Clearly, the above system of equations is consistent, iff $c - \frac{2}{3}(b-2a) = 0$ or, $4a - 2b + 3c = 0$. Hence, $v = (a,b,c)$ belongs to the subspace spanned by v_1, v_2, v_3 iff $4a - 2b + 3c = 0$.

EXAMPLE-8 *Let F be a field with unity 1, and let n be a positive integer such that*
$e_1^{(n)} = (1,0,\ldots,0), e_2^{(n)} = (0,1,0,\ldots,0), e_i^{(n)} = (0,0,\ldots,1,0\ldots 0),\ldots, e_n^{(n)} = (0,0,\ldots,1)$. *Show*
$\qquad\qquad\qquad\qquad\qquad\qquad\qquad\qquad\qquad\qquad$ ith place
that the subset $S = \{e_i^{(n)}, e_2^{(n)}, \ldots, e_n^{(n)}\}$ spans (generates) the entire vector space $F^n(F)$.

SOLUTION Let $(a_1, a_2, \ldots, a_n)$ be an arbitrary vector in F^n. Then,
$$(a_1, a_2, \ldots, a_n) = a_1(1,0,\ldots,0) + a_2(0,1,0,\ldots,0) + \cdots + a_n(0,0,\ldots,1)$$
$$\Rightarrow (a_1, a_2, \ldots, a_n) = a_1 e_1^{(n)} + a_2 e_2^{(n)} + \cdots + a_n e_n^{(n)}$$
Therefore, $(a_1, a_2, \ldots, a_n)$ is a linear combination of S. Since $(a_1, a_2, \ldots, a_n)$ is an arbitrary vector in F^n. Therefore, every vector in F^n is a linear combination of S.
Hence, $F^n \subset [S]$. Also, $[S] \subset F^n$. Consequently, $[S] = F^n$.

Particular Case: Three dimensional Euclidean space R^3 is generated (spanned) by its subset $S = \{e_1^{(3)} = (1,0,0),\ e_2^{(3)} = (0,1,0),\ e_3^{(3)} = (0,0,1)\}$.

EXAMPLE-9 *Is the vector $(2, -5, 3) \in R^3$, in the subspace of R^3 spanned by the vectors $v_1 = (1, -3, 2), v_2 = (2, -4, -1), v_3 = (1, -5, 7)$?*

SOLUTION In order to check whether $v = (2, -5, 3)$ belongs to the subspace spanned by the vectors v_1, v_2, v_3 or not. It is sufficient to check whether v is expressible as a linear combination of v_1, v_2, v_3 or not. Let $\lambda_1, \lambda_2, \lambda_3 \in R$ be such that
$$(2, -5, 3) = \lambda_1 v_1 + \lambda_2 v_2 + \lambda_3 v_3$$
$\Rightarrow (2, -5, 3) = \lambda_1(1, -3, 2) + \lambda_2(2, -4, -1) + \lambda_3(1, -5, 7)$
$\Rightarrow (2, -5, 3) = (\lambda_1 + 2\lambda_2 + \lambda_3, -3\lambda_1 - 4\lambda_2 - 5\lambda_3, 2\lambda_1 - \lambda_2 + 7\lambda_3)$
$\Rightarrow \lambda_1 + 2\lambda_2 + \lambda_3 = 2,\ -3\lambda_1 + 4\lambda_2 - 5\lambda_3 = -5,\ 2\lambda_1 - \lambda_2 + 7\lambda_3 = 3$

This system of equations is inconsistent, because the rank of the coefficient matrix is not equal to the rank of the augmented matrix.

EXAMPLE-10 *Let $v_1 = (-1,2,0), v_2 = (3,2,-1)$ and $v_3 = (1,6,-1)$ be three vectors in $R^3(R)$. Then show that $[v_1, v_2] = [v_1, v_2, v_3]$.*

SOLUTION Clearly,
$$[v_1, v_2] = \{\lambda_1 v_1 + \lambda_2 v_2 : \lambda_1, \lambda_2 \in R\}$$
and, $[v_1, v_2, v_3] = \{\mu_1 v_1 + \mu_2 v_2 + \mu_3 v_3 : \mu_1, \mu_2, \mu_3 \in R\}$.
Let $v_3 = (1,6,-1) = \alpha_1 v_1 + \alpha_2 v_2$. Then,
$$(1,6,-1) = (-\alpha_1 + 3\alpha_2, 2\alpha_1 + 2\alpha_2, -\alpha_2)$$
$$\Rightarrow \quad -\alpha_1 + 3\alpha_2 = 1, 2\alpha_1 + 2\alpha_2 = 6, -\alpha_2 = -1.$$
$$\Rightarrow \quad \alpha_1 = 2, \alpha_2 = 1$$
$$\therefore \quad (1,6,-1) = 2(-1,2,0) + 1(3,2,-1)$$
$$\Rightarrow \quad \mu_3(1,6,-1) = 2\mu_3(-1,2,0) + \mu_3(3,2,-1)$$
$$\Rightarrow \quad \mu_3 v_3 = 2\mu_3 v_1 + \mu_3 v_2$$
Now, $\quad [v_1, v_2, v_3] = \{\mu_1 v_1 + \mu_2 v_2 + 2\mu_3 v_1 + \mu_3 v_2 : \mu_1, \mu_2, \mu_3 \in R\}$
$$\Rightarrow \quad [v_1, v_2, v_3] = \{(\mu_1 + 2\mu_3)v_1 + (\mu_2 + \mu_3)v_2 : \mu_1, \mu_2, \mu_3 \in R\}$$
$$\Rightarrow \quad [v_1, v_2, v_3] = \{\lambda_1 v_1 + \lambda_2 v_2 : \lambda_1, \lambda_2 \in R\}$$
$$\Rightarrow \quad [v_1, v_2, v_3] = [v_1, v_2].$$

EXAMPLE-11 *Let $v_1 = (1,2,-1), v_2 = (2,-3,2), v_3 = (4,1,3)$ and $v_4 = (-3,1,2)$ be four vectors in $R^3(R)$. Then show that $[v_1, v_2] \neq [v_3, v_4]$.*

SOLUTION Suppose $[v_1, v_2] = [v_3, v_4]$. Then for given $\lambda_3, \lambda_4 \in R$ there exist $\lambda_1, \lambda_2 \in R$ such that
$$\lambda_1 v_1 + \lambda_2 v_2 = \lambda_3 v_3 + \lambda_4 v_4$$
$$\Rightarrow \quad \lambda_1(1,2,-1) + \lambda_2(2,-3,2) = \lambda_3(4,1,3) + \lambda_4(-3,1,2)$$
$$\Rightarrow \quad (\lambda_1, 2\lambda_1, -\lambda_1) + (2\lambda_2, -3\lambda_2, 2\lambda_2) = (4\lambda_3, \lambda_3, 3\lambda_3) + (-3\lambda_4, \lambda_4, 2\lambda_4)$$
$$\Rightarrow \quad (\lambda_1 + 2\lambda_2, 2\lambda_1 - 3\lambda_2, -\lambda_1 + 2\lambda_2) = (4\lambda_3 - 3\lambda_4, \lambda_3 + \lambda_4, 3\lambda_3 + 2\lambda_4)$$
$$\Rightarrow \quad \lambda_1 + 2\lambda_2 = 4\lambda_3 - 3\lambda_4 \qquad \text{(i)}$$
$$\quad 2\lambda_1 - 3\lambda_2 = \lambda_3 + \lambda_4 \qquad \text{(ii)}$$
$$\quad -\lambda_1 + 2\lambda_2 = 3\lambda_3 + 2\lambda_4 \qquad \text{(iii)}$$

Solving (i) and (iii), we get $\lambda_1 = \frac{1}{2}(\lambda_3 - 5\lambda_4), \lambda_2 = \frac{1}{4}(7\lambda_3 - \lambda_4)$
These values of λ_1 and λ_2 should satisfy (ii) but they do not satisfy (ii).
Hence, $[v_1, v_2] \neq [v_3, v_4]$.

EXAMPLE-12 Let $R^{2\times 2}$ be the vector space of all 2×2 matrices. Show that the matrices
$E_{11} = \begin{bmatrix} 1 & 0 \\ 0 & 0 \end{bmatrix}, E_{12} = \begin{bmatrix} 0 & 1 \\ 0 & 0 \end{bmatrix}, E_{21} = \begin{bmatrix} 0 & 0 \\ 1 & 0 \end{bmatrix}, E_{22} = \begin{bmatrix} 0 & 0 \\ 0 & 1 \end{bmatrix}$ form a spanning set of $R^{2\times 2}$.

SOLUTION Let $M = \begin{bmatrix} a & b \\ c & d \end{bmatrix}$ be any matrix in $R^{2\times 2}$, where $a,b,c,d \in R$.

Clearly, $M = aE_{11} + bE_{12} + cE_{21} + dE_{22}$

Thus, any matrix M in $R^{2\times 2}$ is expressible as a linear combination of $E_{11}, E_{12}, E_{21}, E_{22}$. Hence, $\{E_{11}, E_{12}, E_{21}, E_{22}\}$ forms a spanning set of $R^{2\times 2}$.

EXAMPLE-13 Find one vector in $R^3(R)$ that spans the intersection of subspaces S and T where S is the xy-plane, i.e. $S = \{(a,b,0) : a, b \in R\}$ and T is the space spanned by the vectors $v_1 = (1,1,1)$ and $v_2 = (1,2,3)$.

SOLUTION Since S and T are subspaces of $R^3(R)$. Therefore, $S \cap T$ is a subspace of $R^3(R)$.

Let $u = (a,b,c)$ be a vector in $S \cap T$. Then,

$u \in S$ and $u \in T$

$\Rightarrow$ $c = 0$ and u is linear combination of v_1 and v_2 $[\because \quad S = \{(a,b,0) : a,b \in R\}]$

$\Rightarrow$ $u = xv_1 + yv_2$ for some $x, y \in R$

$\Rightarrow$ $(a,b,0) = x(1,1,1) + y(1,2,3)$

$\Rightarrow$ $(a,b,0) = (x+y, x+2y, x+3y)$

$\Rightarrow$ $x+y = a, x+2y = b, x+3y = 0$

Solving first two equations, we get $x = 2a - b, y = b - a$

Substituting these values in $x + 3y = 0$, we get $a = 2b$

$\therefore \quad v = (2b, b, 0), b \in R$.

EXAMPLE-14 Is the vector $v = (3, -1, 0, -1)$ in the subspace of $R^4(R)$ spanned by the vectors $v_1 = (2, -1, 3, 2), v_2 = (-1, 1, 1, -3)$ and $v_3 = (1, 1, 9, -5)$?

SOLUTION If v is expressible as a linear combination of v_1, v_2 and v_3, then v is the subspace spanned by v_1, v_2 and v_3, otherwise not. So, let

$v = xv_1 + yv_2 + zv_3$ for some scalars x, y, z.

$\Rightarrow$ $(3, -1, 0, -1) = x(2, -1, 3, 2) + y(-1, 1, 1, -3) + z(1, 1, 9, -5)$

$\Rightarrow$ $(3, -1, 0, -1) = (2x - y + z, -x + y + z, 3x + y + 9z, 2x - 3y - 5z)$

$\Rightarrow$ $2x - y + z = 3, -x + y + z = -1, 3x + y + 9z = 0, 2x - 3y - 5z = -1$

This system of equations can be expressed in matrix form as follows:

$$\begin{bmatrix} 2 & -1 & 1 \\ -1 & 1 & 1 \\ 3 & 1 & 9 \\ 2 & -3 & -5 \end{bmatrix} \begin{bmatrix} x \\ y \\ z \end{bmatrix} = \begin{bmatrix} 3 \\ -1 \\ 0 \\ -1 \end{bmatrix}$$

This is equivalent to

$$\begin{bmatrix} 0 & 1 & 3 \\ -1 & 1 & 1 \\ 0 & 4 & 12 \\ 0 & -1 & -3 \end{bmatrix} \begin{bmatrix} x \\ y \\ z \end{bmatrix} = \begin{bmatrix} 1 \\ -1 \\ -3 \\ -3 \end{bmatrix} \text{ Applying } R_1 \to R_1 + 2R_2,\ R_3 \to R_3 + 3R_2,\ R_4 \to R_4 + 2R_2$$

or,
$$\begin{bmatrix} 0 & 0 & 0 \\ -1 & 0 & -2 \\ 0 & 0 & 0 \\ 0 & -1 & -3 \end{bmatrix} \begin{bmatrix} x \\ y \\ z \end{bmatrix} = \begin{bmatrix} -2 \\ -4 \\ -15 \\ -3 \end{bmatrix} \text{ Applying } R_1 \to R_1 + R_4,\ R_2 \to R_2 + R_4,\ R_3 \to R_3 + 4R_4$$

Clearly, this is an inconsistent system of equations. So, v is not expressible as a linear combination of v_1, v_2, v_3. Hence, $v \notin [v_1, v_2, v_3]$.

EXERCISE 2.7

1. Show that the vectors $e_1^{(3)} = (1,0,0), e_2^{(3)} = (0,1,0)$ and $e_3^{(3)} = (0,0,1)$ form a spanning set of $R^3(R)$.

2. Show that the vectors $v_1 = (1,1,1), v_2 = (1,1,0)$ and $v_3 = (1,0,0)$ form a spanning set of $R^3(R)$.

3. Consider the vector space $V = P_n(t)$ consisting of all polynomials of degree $\leq n$ in indeterminate t. Show that the set of polynomials $1, t, t^2, t^3, \ldots, t^n$ forms a spanning set of V.

4. Show that the vectors $v_1 = (1,2,3), v_2 = (1,3,5)$ and $v_3 = (1,5,9)$ do not spam $R^3(R)$. [Hint: $(2,7,8)$ cannot be written as a linear combination of v_1, v_2, v_3].

5. If v_1, v_2 and v_3 are vectors in a vector space $V(F)$ such that $v_1 + v_2 + v_3 = 0_V$, then show that v_1 and v_2 span the same subspace of V as v_2 and v_3.

   [Hint: Let S_1 and S_2 be subspaces spanned by vectors v_1, v_2 and v_2, v_3 respectively. i.e, $S_1 = [\{v_1, v_2\}]$ and $S_2 = [\{v_2, v_3\}]$. We have to show that $S_1 = S_2$.
   Let v be an arbitrary vector in S_1. Then,

   $v \in S_1$
   $\Rightarrow v = \lambda_1 v_1 + \lambda_2 v_2$ for some $\lambda_1, \lambda_2 \in F$
   $\Rightarrow v = \lambda_1(-v_2 - v_3) + \lambda_2 v_2$ $\quad [\because\ v_1 + v_2 + v_3 = 0_V]$
   $\Rightarrow v = (\lambda_1 - \lambda_2)v_2 + (-\lambda)v_3$
   $\Rightarrow v \in S_2$
   $\therefore S_1 \subset S_2$.

Now, let u be an arbitrary vector in S_2. Then,

$$u \in S_2$$
$$\Rightarrow \quad u = \alpha_2 v_2 + \alpha_3 v_3 \quad \text{for some } \alpha_2, \alpha_3 \in F$$
$$\Rightarrow \quad u = \alpha_2 v_2 + \alpha_3(-v_1 - v_2) \qquad [\because \quad v_1 + v_2 + v_3 = 0_V]$$
$$\Rightarrow \quad u = (\alpha_2 - \alpha_3)v_2 + (-\alpha_3)v_1$$
$$\Rightarrow \quad u \in S_1$$
$$\therefore \quad S_2 \subset S_1$$

Hence, $S_1 = S_2$.]

6. Let $v_1 = (1,2,1), v_2 = (3,1,5)$ and $v_3 = (3,-4,7)$ be three vectors in $R^3(R)$. Show that the subspaces spanned by the sets $\{v_1, v_2\}$ and $\{v_1, v_2, v_3\}$ are the same.

[Hint: Let $S = \{v_1, v_2\}$ and $T = \{v_1, v_2, v_3\}$.
Clearly, $\qquad S \subset T \Rightarrow [S] \subset [T]$
In order to prove that $[T] = [S]$, it is sufficient to show that v_3 is a linear combination of v_1, v_2.

Let $\qquad v_3 = xv_1 + yv_2$. Then,
$$(3, -4, 7) = x(1, 2, 1) + y(3, 1, 5)$$
$$\Rightarrow \qquad x + 3y = 3, 2x + y = -4 \text{ and } x + 5y = 7$$
$$\Rightarrow \qquad x = -3, y = 2$$
$$\therefore \qquad v_3 = -3v_1 + 2v_2$$

Hence, $[v_1, v_2] = [v_1, v_2, v_3]$, i.e. $[S] = [T]$.]

2.8 LINEAR INDEPENDENCE AND DEPENDENCE

In this section, we shall introduce two important related concepts which will be used over and over again. But, we start with the definitions of 'trivial' and 'non-trivial' linear combinations.

TRIVIAL LINEAR COMBINATION *Let* $v_1, v_2, \ldots, v_n$ *be* n *vectors in a vector space* $V(F)$. *Then the linear combination* $\lambda_1 v_1 + \lambda_2 v_2 + \cdots + \lambda_n v_n$ *is called a trivial linear combination if* $\lambda_1 = 0 = \lambda_2 = \cdots = \lambda_n$.

REMARK. *The trivial linear combination of any set of vectors in a vector space is always the null vector, for*
$$0v_1 + 0v_2 + \cdots + 0v_n = 0_V + 0_V + \cdots + 0_V = 0_V$$

NON-TRIVIAL LINEAR COMBINATION *Let* $v_1, v_2, \ldots, v_n$ *be* n *vectors in a vector space* $V(F)$. *Then a linear combination* $\lambda_1 v_1 + \lambda_2 v_2 + \cdots + \lambda_n v_n$ *is called a non-trivial linear combination if at least one* $\lambda_i \neq 0$.

EXAMPLE-1 *If $v_1, v_2, \ldots, v_n$ are n vectors in a vector space $V(F)$, then the combination $1 \cdot v_1 + 0v_2 + \cdots + 0v_n$ is a non-trivial linear combination.*

REMARK. *Note that a non-trivial linear combination of a set of vectors in a vector space may or may not be the null vector. For example, $1(2,-1,0)+(-2)(1,-1,1)+(1)(0,-1,2)$ is a non-trivial linear combination of $(2,-1,0),(1,-1,1)$ and $(0,-1,2)$ and is the null vector while a non-trivial linear combination $1(2,-1,0)+2(1,-1,1)+(1)(0,-1,2)$ is a non-null vector in $R^3(R)$.*

LINEAR INDEPENDENCE *A list $(v_1, v_2, \ldots, v_n)$ of vectors in a vector space $V(F)$ is said to be linearly independent (abbreviated as l.i.) if for $\lambda_1, \lambda_2, \ldots, \lambda_n \in F$*

$$\lambda_1 v_1 + \lambda_2 v_2 + \cdots + \lambda_n v_n = 0_V \quad \Rightarrow \quad \lambda_1 = \lambda_2 = \cdots = \lambda_n = 0.$$

Equivalently, a list $(v_1, v_2, \ldots, v_n)$ of vectors is l.i. if the only linear combination of $v_1, v_2, \ldots, v_n$ that equals to the null vector is the trivial linear combination. In other words, a list $(v_1, v_2, \ldots, v_n)$ of vectors is l.i. if no non-trivial linear combination of $v_1, v_2, \ldots, v_n$ equals to the null vector.

A finite subset of a vector space is said to be linearly independent if every list of distinct vectors of it is linearly independent.

An infinite set of vectors in a vector space is said to be linearly independent if each of its finite subset is linearly independent.

LINEAR DEPENDENCE *A list $(v_1, v_2, \ldots, v_n)$ of vectors in a vector space $V(F)$ is said to be linearly dependent (abbreviated l.d.) if for $\lambda_1, \lambda_2, \ldots, \lambda_n \in F$*

$$\lambda_1 v_1 + \lambda_2 v_2 + \cdots + \lambda_n v_n = 0_V \quad \Rightarrow \quad \text{at least one } \lambda_i \neq 0.$$

Thus, a list $(v_1, v_2, \ldots, v_n)$ of vectors in a vector space $V(F)$ is said to be l.d. if at least one non-trivial linear combination of $v_1, v_2, \ldots, v_n$ equals to the zero (null) vector.

A subset (finite or infinite) of a vector space is said to be linearly dependent (l.d.) if there exists a finite list in it which is linearly dependent.

REMARK-1 *By convention, the void set is considered to be l.i.*

REMARK-2 *Linear independence and linear dependence are opposite concepts. A set cannot be simultaneously l.i. and l.d. It is evident from the definition that two non-zero vectors in a vector space $V(F)$ are linearly dependent iff one is a scalar multiple of other. For example, vectors $v_1 = (2,1,2)$ and $v_2 = (8,4,8)$ form a linearly dependent set of vectors, because $v_2 = 4v_1$.*

REMARK-3 *Let v be a non-zero vector in a vector space $V(F)$. Then, v, by itself, is linearly independent. Because,*

$$\lambda v = 0, v \neq 0_V \Rightarrow \lambda = 0$$

REMARK-4 Note that the set $\{0_V\}$ in a vector space $V(F)$ is always l.d., because $\lambda 0_V = 0_V$ for all non-zero $\lambda \in F$. On the other hand, a singleton set containing a non-zero vector v is always l.i., because $\lambda v = 0_V \Rightarrow \lambda = 0$.

THEOREM-1 *Any list of vectors containing the null vector is always l.d.*

SOLUTION Let $\underline{v} = (v_1, v_2, \ldots, v_n)$ be a list of vectors in a vector space $V(F)$ such that $v_k = 0_V$. Then for any $0 \neq \lambda \in F$

$$0v_1 + 0v_2 + \cdots + \lambda v_k + 0v_{k+1} + \cdots + 0v_n = 0_V + 0_V + \cdots + 0_V = 0_V$$

This shows that a non-trivial linear combination of $v_1, v_2, \ldots, v_n$ equals to the null vector. Hence, $\underline{v}$ is a l.d. list.

Q.E.D.

REMARK. *It follows from the foregoing discussion that the nature of the null vector is so strong that its presence in the set always makes it l.d.*

THEOREM-2 *A list $\underline{v} = (v_1, v_2, \ldots, v_n)$ of non-zero vectors in a vector space $V(F)$ is l.d. iff some one vector v_k of the list is a linear combination (l.c.) of the previous vectors of the list. Also, if for some $j, 1 \leq j \leq n$, the list $(v_1, v_2, \ldots, v_j)$ is a list of linearly independent vectors, then $j < k$.*

PROOF. First suppose that the list $(v_1, v_2, \ldots, v_n)$ of vectors in $V(F)$ is linearly dependent. Then there exist scalars $\lambda_1, \lambda_2, \ldots, \lambda_n \in F$ not all zero such that

$$\lambda_1 v_1 + \lambda_2 v_2 + \cdots + \lambda_n v_n = 0_V$$

Let k be the largest positive integer such that $\lambda_k \neq 0$, so that

$$\lambda_1 v_1 + \lambda_2 v_2 + \cdots + \lambda_k v_k = 0_V$$

We claim that $k > 1$. For, if $k = 1$

$\lambda_1 v_1 = 0_V$ and $\lambda_2 = \lambda_3 = \cdots = \lambda_n = 0$

$\Rightarrow \lambda_1 = 0 = \lambda_2 = \cdots = \lambda_n$ $\qquad [\because v_1 \neq 0_V \quad \therefore \lambda_1 v_1 = 0_V \Rightarrow \lambda_1 = 0]$

which contradicts the fact that $(v_1, v_2, \ldots, v_n)$ is a l.d. list.

Now, $\lambda_1 v_1 + \lambda_2 v_2 + \cdots + \lambda_k v_k = 0_V$

$\Rightarrow \quad v_k = (-\lambda_1 \lambda_k^{-1}) v_1 + \cdots + (-\lambda_{k-1} \lambda_k^{-1}) v_{k-1}$ \hfill (i)

$\Rightarrow \quad v_k$ is a linear combination of the previous vectors of the list.

Hence, some one vector of the list is a linear combination of the previous vectors of the list.

Conversely, suppose that a vector v_k (say) of the list $(v_1, v_2, \ldots, v_k, \ldots, v_n)$ is a linear combination of the previous vectors $v_1, v_2, \ldots, v_{k-1}$. Then there exist scalars $\lambda_1, \lambda_2, \ldots, \lambda_{k-1}$ such that

$v_k = \lambda_1 v_1 + \lambda_2 v_2 + \cdots + \lambda_{k-1} v_{k-1}$

$\Rightarrow \quad \lambda_1 v_1 + \lambda_2 v_2 + \cdots + \lambda_{k-1} v_{k-1} + (-1) v_k = 0_V$

$\Rightarrow \quad \lambda_1 v_1 + \lambda_2 v_2 + \cdots + \lambda_{k-1} v_{k-1} + (-1) v_k + 0 v_{k+1} + \cdots + 0 v_n = 0_V$

$\Rightarrow \quad$ A non-trivial linear combination of the list $(v_1, v_2, \ldots, v_k, \ldots, v_n)$ is zero vector

Hence, the list is linearly dependent.

Q.E.D.

If possible, let the list $(v_1, v_2, \ldots, v_j)$ be a list of independent vectors and $j \geq k$. Then, the list of vectors $v_1, v_2, \ldots, v_k$ is linear independent. But,

$$v_k = (-\lambda_1 \lambda_k^{-1}) v_1 + \cdots + (-\lambda_{k-1} \lambda_k^{-1}) v_{k-1} \qquad \text{[From (i)]}$$

$\Rightarrow$ v_k is a linear combination of $v_1, v_2, \ldots, v_{k-1}$

$\Rightarrow$ $v_1, v_2, \ldots, v_{k-1}, v_k$ are linearly dependent.

This is a contradiction. Hence, $j < k$.

THEOREM-3 *Let $S = \{v_1, v_2, v_3, \ldots v_k\}$ be a linearly independent set of vectors in a vector space $V(F)$ and v be a non-zero vector in $V(F)$. Then, $S_1 = S \cup \{v\}$ is linearly independent iff $v \notin [S]$.*

PROOF. First, let S_1 be linearly independent. Suppose that $v \in [S]$. Then, there exist scalars $\lambda_1, \lambda_2, \ldots, \lambda_k$ in F such that $v = \sum\limits_{i=1}^{k} \lambda_i v_i$

$\Rightarrow$ $S_1 = S \cup \{v\}$ is linearly dependent.

This contradicts the hypothesis that S_1 is linearly independent. Hence, $v \notin [S]$.

Conversely, let $v \notin [S]$. Then, we have to prove that S_1 is linearly independent. If possible, let S_1 be linearly dependent. Then, there exists a finite non-empty subset B (say) of S_1 that is linearly dependent.

Now,

$B \subset S_1 = S \cup \{v\}$ and S is linearly independent

$\Rightarrow$ B contains v and some vectors in $S = \{v_1, v_2, \ldots, v_k\}$

$\Rightarrow$ v is a linear combination of some vectors in S

$\Rightarrow$ v is a linear combination of vectors in S

$\Rightarrow$ $v \in [S]$

This is a contradiction. Hence, S_1 is linearly independent.

Q.E.D.

EXAMPLE-2 *Show that the set $S = \{(1,2,4), (1,0,0), (0,1,0), (0,0,1)\}$ is a linearly dependent set of vectors in $R^3(R)$.*

SOLUTION Clearly,

$$(1,2,4) = 1(1,0,0) + 2(0,1,0) + 4(0,0,1)$$

$\Rightarrow$ $(1,2,4)$ is linear combination of $(1,0,0)$, $(0,1,0)$ and $(0,0,1)$.

Hence, S is a linearly dependent set of vectors.

THEOREM-4 *Each subset of a linearly independent set is linearly independent.*

PROOF. Let $S = \{v_1, v_2, \ldots, v_m\}$ be a linearly independent set of vectors in a vector space $V(F)$. Let $S' = \{v_1, v_2, \ldots, v_n\}$ $n \leq m$, be an arbitrary finite subset of S. Then,

$$\lambda_1 v_1 + \lambda_2 v_2 + \cdots + \lambda_n v_n = 0_V$$
$$\lambda_1 v_1 + \lambda_2 v_2 + \cdots + \lambda_n v_n + 0 v_{n+1} + \cdots + 0 v_m = 0_V$$
$$\Rightarrow \lambda_1 = \lambda_2 = \lambda_3 = \cdots = \lambda_n = 0 \qquad [\because S \text{ is l.i.}]$$

Hence, S' is a linearly independent set.

Q.E.D.

THEOREM-5 *Any super set of linearly dependent set of vectors in a vector space is linearly dependent.*

PROOF. Let $S = \{v_1, v_2, \ldots, v_m\}$ be a linearly dependent set of vectors in a vector space $V(F)$, and let $S' = \{v_1, v_2, \ldots, v_m, v_{m+1}\}$ be a super set of S.

Now,

S is a linearly dependent set

$\Rightarrow$ There exist scalars $\lambda_1, \lambda_2, \ldots, \lambda_m$ not all zero such that $\lambda_1 v_1 + \lambda_2 v_2 + \cdots + \lambda_m v_m = 0_V$

$\Rightarrow \lambda_1 v_1 + \lambda_2 v_2 + \cdots + \lambda_m v_m + \lambda_{m+1} v_{m+1} = 0_V$, where $\lambda_{m+1} = 0$

$\Rightarrow$ A non-trivial linear combination of $v_1, v_2, \ldots, v_m, v_{m+1}$ is zero vector $\quad \left[\because \text{At least one } \lambda_i \neq 0\right]$

$\Rightarrow S'$ is a linearly dependent set.

Q.E.D.

THEOREM-6 *Let S be any linearly independent subset of a vector space $V(F)$ and let $S = S_1 \cup S_2$ such that $S_1 \cap S_2 = \phi$. Prove that (i) $[S] = [S_1] + [S_2]$ (ii) $[S_1] \cap [S_2] = \phi$ i.e. $[S]$ is the direct sum of $[S_1]$ and $[S_2]$.*

PROOF. (i) By Theorem 4(iv) on page 169, we have

$$[S_1 \cup S_2] = [S_1] + [S_2] \text{ i.e, } [S] = [S_1] + [S_2]$$

(ii) Let $S_1 = \{v_1, v_2, \ldots, v_m\}$ and $S_2 = \{u_1, u_2, \ldots, u_n\}$.

Let v be an arbitrary vector in $[S_1] \cap [S_2]$. Then,

$v \in [S_1] \cap [S_2]$

$\Rightarrow v \in [S_1]$ and $v \in [S_2]$

$\Rightarrow v = \sum_{i=1}^{m} \lambda_i v_i$ and $v = \sum_{i=1}^{n} \mu_j u_j$ for some $\lambda_i \in F$ and $\mu_j \in F$

$$\Rightarrow \quad \sum_{i=1}^{m} \lambda_i v_i = \sum_{j=1}^{n} \mu_j u_j$$

$$\Rightarrow \quad \sum_{i=1}^{m} \lambda_i v_i + \sum_{j=1}^{n} (-\mu_j) u_j = 0_V$$

$$\Rightarrow \quad \lambda_i = 0 \text{ and } \mu_j = 0 \quad \text{for all } i \in \underline{m} \text{ and } j \in \underline{n} \qquad \left[\because \begin{array}{l} S = S_1 \cup S_2 = \{v_1, v_2, \ldots, v_m, \\ u_1, u_2, \ldots, u_n\} \text{ is } l.i. \end{array} \right]$$

$$\therefore \quad v = 0_V$$

Since v is an arbitrary vector in $[S_1] \cap [S_2]$. Therefore, $[S_1] \cap [S_2] = \phi$.

Q.E.D.

ILLUSTRATIVE EXAMPLES

EXAMPLE-1 *Let F be a field with unity 1. Show that the vectors $e_1^{(n)} = (1, 0, 0, \ldots, 0)$, $e_2^{(n)} = (0, 1, 0, \ldots, 0), e_3^{(n)} = (0, 0, 1, \ldots, 0) \ldots, e_n^{(n)} = (0, 0, 0, \ldots, 1)$ are linearly independent in the vector space $F^n(F)$.*

SOLUTION Let $\lambda_1, \lambda_2, \ldots, \lambda_n$ be scalars such that

$$\lambda_1 e_1^{(n)} + \lambda_2 e_2^{(n)} + \lambda_3 e_3^{(n)} + \cdots + \lambda_n e_n^{(n)} = 0$$

$$\Rightarrow \quad (\lambda_1, \lambda_2, \lambda_3, \ldots, \lambda_n) = (0, 0, 0, \ldots, 0)$$

$$\Rightarrow \quad \lambda_1 = \lambda_2 = \lambda_3 = \cdots = \lambda_n = 0$$

Hence, $e_1^{(n)}, e_2^{(n)}, \ldots, e_n^{(n)}$ are linearly independent.

Particular Cases:

(i) $e_1^{(2)} = (1, 0), e_2^{(2)} = (0, 1)$ are linearly independent vectors in $R^2(R)$.

(ii) $e_1^{(3)} = (1, 0, 0), e_2^{(3)} = (0, 1, 0), e_3^{(3)} = (0, 0, 1)$ are linearly independent vectors in $R^3(R)$.

EXAMPLE-2 *Show that the vectors $v_1 = (1, 1, 0), v_2 = (1, 3, 2), v_3 = (4, 9, 5)$ are linearly dependent in $R^3(R)$.*

SOLUTION Let x, y, z be scalars i.e., real numbers such that

$$x v_1 + y v_2 + z v_3 = 0$$

$$\Rightarrow \quad x(1, 1, 0) + y(1, 3, 2) + z(4, 9, 5) = (0, 0, 0)$$

$$\Rightarrow \quad (x + y + 4z, x + 3y + 9z, 2y + 5z) = (0, 0, 0)$$

$$\Rightarrow \quad x + y + 4z = 0, \ x + 3y + 9z = 0, \ 0x + 2y + 5z = 0$$

This is a homogeneous system of linear equations. The determinant of the coefficient matrix A is

$$|A| = \begin{vmatrix} 1 & 1 & 4 \\ 1 & 3 & 9 \\ 0 & 2 & 5 \end{vmatrix} = 0$$

So, the system has non-trivial solutions. Hence, given vectors are linearly dependent in $R^3(R)$.

Aliter In order to check the linear independence or dependence of vectors, we may follow the following algorithm:

ALGORITHM

Step I Form a matrix A whose columns are given vectors.
Step II Reduce the matrix in step-I to echelon form.
Step III See whether all columns have pivot elements or not. If all columns have pivot elements, then given vectors are linearly independent. If there is a column not having a pivot element, then the corresponding vector is a linear combination of the preceding vectors and hence linearly dependent.

In Example 2, the matrix A whose columns are v_1, v_2, v_3 is

$$A = \begin{bmatrix} 1 & 1 & 4 \\ 1 & 3 & 9 \\ 0 & 2 & 5 \end{bmatrix}$$

$$\therefore \quad A \sim \begin{bmatrix} 1 & 1 & 4 \\ 0 & 2 & 5 \\ 0 & 2 & 5 \end{bmatrix} \quad \text{Applying } R_2 \to R_2 - R_1$$

$$\Rightarrow \quad A \sim \begin{bmatrix} \boxed{1} & 1 & 4 \\ 0 & \boxed{2} & 5 \\ 0 & 0 & 0 \end{bmatrix} \quad \text{Applying } R_3 \to R_3 - R_2$$

Pivots in the echelon form of matrix A have been encircled. We observe that the third column does not have a pivot. So, the third vector v_3 is a linear combination of the first two vectors v_1 and v_2. Thus, the vectors v_1, v_2, v_3 are linearly dependent.

EXAMPLE-3 *Show that the vectors $v_1 = (1,2,3), v_2 = (2,5,7), v_3 = (1,3,5)$ are linearly independent in $R^3(R)$.*

SOLUTION Let x, y, z be scalars such that

$xv_1 + yv_2 + zv_3 = 0$
$\Rightarrow \quad x(1,2,3) + y(2,5,7) + z(1,3,5) = (0,0,0)$
$\Rightarrow \quad (x+2y+z, 2x+5y+3z, 3x+7y+5z) = (0,0,0)$
$\Rightarrow \quad x+2y+z = 0, \ 2x+5y+3z = 0, \ 3x+7y+5z = 0$

The determinant of the coefficient matrix A of the above homogeneous system of equations is given by

$$|A| = \begin{vmatrix} 1 & 2 & 1 \\ 2 & 5 & 3 \\ 3 & 7 & 5 \end{vmatrix} = 1 \neq 0$$

So, the above system of equations has trivial solution only, i.e. $x = y = z = 0$.

Hence, the given vectors are linearly independent in $R^3(R)$.

EXAMPLE-4 *Show that the vectors* $v_1 = (1,1,2,4), v_2 = (2,-1,-5,2), v_3 = (1,-1,-4,0)$ *and* $v_4 = (2,1,1,6)$ *are linearly dependent in* $R^4(R)$.

SOLUTION Let x, y, z, t be scalars in R such that

$$xv_1 + yv_2 + zv_3 + tv_4 = 0$$
$$\Rightarrow \quad x(1,1,2,4) + y(2,-1,-5,2) + z(1,-1,-4,0) + t(2,1,1,6) = 0$$
$$\Rightarrow \quad (x+2y+z+2t, x-y-z+t, 2x-5y-4z+t, 4x+2y+0z+6t) = (0,0,0,0)$$
$$\Rightarrow \quad x+2y+z+2t = 0, \ x-y-z+t = 0, \ 2x-5y-4z+t = 0, \ 4x+2y+0z+6t = 0$$

The coefficient matrix A of the above homogeneous system of equations is

$$A = \begin{bmatrix} 1 & 2 & 1 & 2 \\ 1 & -1 & -1 & 1 \\ 2 & -5 & -4 & 1 \\ 4 & 2 & 0 & 6 \end{bmatrix}$$

$$\Rightarrow \quad A \sim \begin{bmatrix} 1 & 2 & 1 & 2 \\ 0 & -3 & -2 & -1 \\ 0 & -9 & -6 & -3 \\ 0 & -6 & -4 & -2 \end{bmatrix} \quad \text{Applying } R_2 \to R_2 - R_1, \ R_3 \to R_3 - 2R_1, \ R_4 \to R_4 - 4R_1$$

$$\Rightarrow \quad A \sim \begin{bmatrix} 1 & 2 & 1 & 2 \\ 0 & -3 & -2 & -1 \\ 0 & 0 & 0 & 0 \\ 0 & 0 & 0 & 0 \end{bmatrix} \quad \text{Applying } R_3 \to R_3 - 3R_2, \ R_4 \to R_4 - 2R_2$$

Clearly, rank of $A = 2 <$ Number of unknowns. So, the above system has non-trivial solutions. Hence, given vectors are linearly dependent in $R^4(R)$.

Aliter The matrix A whose columns are v_1, v_2, v_3, v_4 is

$$A = \begin{bmatrix} 1 & 2 & 1 & 2 \\ 1 & -1 & -1 & 1 \\ 2 & -5 & -4 & 1 \\ 4 & 2 & 0 & 6 \end{bmatrix}$$

The echelon form of matrix A is

$$\begin{bmatrix} \boxed{1} & 2 & 1 & 2 \\ 0 & \boxed{-3} & -2 & -1 \\ 0 & 0 & 0 & 0 \\ 0 & 0 & 0 & 0 \end{bmatrix}$$

We observe that the third and fourth columns does not have pivots. So, third and fourth vectors are linear combinations of first two vectors v_1 and v_2. Thus, the vectors are linearly dependent.

EXAMPLE-5 *Determine whether the vectors $f(x) = 2x^3 + x^2 + x + 1, g(x) = x^3 + 3x^2 + x - 2$ and $h(x) = x^3 + 2x^2 - x + 3$ in the vector space $R[x]$ of all polynomials over the real number field are linearly independent or not.*

SOLUTION Let a, b, c be real numbers such that

$$af(x) + bg(x) + ch(x) = 0 \quad \text{for all } x$$

$\Rightarrow \quad a(2x^3 + x^2 + x + 1) + b(x^3 + 3x^2 + x - 2) + c(x^3 + 2x^2 - x + 3) = 0 \quad \text{for all } x$

$\Rightarrow \quad (2a + b + c)x^3 + (a + 3b + 2c)x^2 + (a + b - c)x + (a - 2b + 3c) = 0 \quad \text{for all } x$

$\Rightarrow \quad 2a + b + c = 0,\ a + 3b + 2c = 0,\ a + b - c = 0,\ a - 2b + 3c = 0$

The coefficient matrix A of the above system of equations is

$$A = \begin{bmatrix} 2 & 1 & 1 \\ 1 & 3 & 2 \\ 1 & 1 & -1 \\ 1 & -2 & 3 \end{bmatrix}$$

$\Rightarrow \quad A \sim \begin{bmatrix} 1 & 1 & -1 \\ 1 & 3 & 2 \\ 2 & 1 & 1 \\ 1 & -2 & 3 \end{bmatrix}$ Applying $R_1 \leftrightarrow R_3$

$\Rightarrow \quad A \sim \begin{bmatrix} 1 & 1 & -1 \\ 0 & 2 & 3 \\ 0 & -1 & 3 \\ 0 & -3 & 4 \end{bmatrix}$ Applying $R_2 \to R_2 - R_1, R_3 \to R_3 - 2R_1, R_4 \to R_4 - R_1$

$\Rightarrow \quad A \sim \begin{bmatrix} 1 & 1 & -1 \\ 0 & 0 & 9 \\ 0 & -1 & 3 \\ 0 & 0 & -5 \end{bmatrix}$ Applying $R_2 \to R_2 + 2R_3, R_4 \to R_4 - 3R_3$

$$\Rightarrow \quad A \sim \begin{bmatrix} 1 & 1 & -1 \\ 0 & 0 & 9 \\ 0 & -1 & 3 \\ 0 & 0 & 0 \end{bmatrix} \text{ Applying } R_4 \to R_4 + \frac{5}{9}R_2$$

$$\Rightarrow \quad A \sim \begin{bmatrix} 1 & 1 & -1 \\ 0 & -1 & 3 \\ 0 & 0 & 9 \\ 0 & 0 & 0 \end{bmatrix} \text{ Applying } R_2 \leftrightarrow R_3$$

Clearly, rank of A is equal to 3 which is equal to the number of unknowns a, b, c. So, the system has trivial solution only, i.e. $a = b = c = 0$.

So, $f(x), g(x), h(x)$ are linearly independent in $R[x]$.

EXAMPLE-6 *Let V be the vector space of functions from R into R. Show that the functions $f(t) = \sin t, g(t) = e^t, h(t) = t^2$ are linearly independent in V.*

SOLUTION Let x, y, z be scalars such that

$$xf(t) + y g(t) + z h(t) = 0 \text{ for all } t \in R$$

Putting $t = 0, \pi$ and $\frac{\pi}{2}$, respectively we get

$$xf(0) + y g(0) + zh(0) = 0$$
$$xf(\pi) + y g(\pi) + z h(\pi) = 0$$
$$xf\left(\frac{\pi}{2}\right) + y g\left(\frac{\pi}{2}\right) + z h\left(\frac{\pi}{2}\right) = 0$$

$$\Rightarrow \quad y = 0, \; y \, e^\pi + z \, \pi^2 = 0, x + y \, e^{\frac{\pi}{2}} + h \frac{\pi^2}{4} = 0$$

$$\Rightarrow \quad y = 0, z = 0, x = 0$$

Thus,
$$xf(t) + yg(t) + zh(t) = 0 \Rightarrow x = y = z = 0$$

Hence, $f(t), g(t), h(t)$ are linearly independent.

EXAMPLE-7 *Show that in the vector space $F[x]$ of all polynomials in an indeterminate x, the set $S = \{1, x, x^2, \ldots\}$ is linearly independent.*

PROOF. Let $S' = \{x^{n_1}, x^{n_2}, \ldots, x^{n_k}\}$ be a finite subset of S, where $n_1, n_2, \ldots, n_k$ are distinct non-negative integers.

Let $\lambda_1, \lambda_2, \ldots, \lambda_k \in F$ be such that
$$\lambda_1 x^{n_1} + \lambda_2 x^{n_2} + \cdots + \lambda_k x^{n_k} = 0(x) \qquad \text{[0(x) is zero polynomial over } F\text{]}$$
$$\Rightarrow \quad \lambda_1 = \lambda_2 = \cdots = \lambda_k = 0 \qquad \text{[By the definition of equality of two polynomials]}$$

So, S' is a linearly independent set.

188 • *Theory and Problems of Linear Algebra*

Since S' is an arbitrary finite subset of S. Therefore, every finite subset of S is linearly independent. Consequently, S is linearly independent.

Q.E.D.

EXAMPLE-8 *Show that the vectors (matrices)* $A_1 = \begin{bmatrix} 1 & 1 \\ 1 & 1 \end{bmatrix}$, $A_2 = \begin{bmatrix} 1 & 0 \\ 0 & 1 \end{bmatrix}$, $A_3 = \begin{bmatrix} 1 & 1 \\ 0 & 0 \end{bmatrix}$ *in* $R^{2 \times 2}$ *are linearly independent.*

SOLUTION   Let $\lambda_1, \lambda_2, \lambda_3 \in R$ be such that

$$\lambda_1 A_1 + \lambda_2 A_2 + \lambda_3 A_3 = O_{2 \times 2} \quad \text{(Null matrix)}.$$

$$\Rightarrow \quad \lambda_1 \begin{bmatrix} 1 & 1 \\ 1 & 1 \end{bmatrix} + \lambda_2 \begin{bmatrix} 1 & 0 \\ 0 & 1 \end{bmatrix} + \lambda_3 \begin{bmatrix} 1 & 1 \\ 0 & 0 \end{bmatrix} = \begin{bmatrix} 0 & 0 \\ 0 & 0 \end{bmatrix}$$

$$\Rightarrow \quad \begin{bmatrix} \lambda_1 + \lambda_2 + \lambda_3 & \lambda_1 + \lambda_3 \\ \lambda_1 & \lambda_1 + \lambda_2 \end{bmatrix} = \begin{bmatrix} 0 & 0 \\ 0 & 0 \end{bmatrix}$$

$$\Rightarrow \quad \lambda_1 + \lambda_2 + \lambda_3 = 0, \; \lambda_1 + \lambda_3 = 0, \; \lambda_1 = 0, \; \lambda_1 + \lambda_2 = 0$$

$$\Rightarrow \quad \lambda_1 = \lambda_2 = \lambda_3 = 0.$$

Hence, A_1, A_2, A_3 are linearly independent vectors in $R^{2 \times 2}$.

EXAMPLE-9 *If v_1, v_2 are vectors in a vector space $V(F)$ and $\lambda_1, \lambda_2 \in F$, show that the set $\{v_1, v_2, \lambda_1 v_1 + \lambda_2 v_2\}$ is linearly dependent set.*

SOLUTION   Since the vector $\lambda_1 v_1 + \lambda_2 v_2$ in the set $\{v_1, v_2, \lambda_1 v_1 + \lambda_2 v_2\}$ is a linear combination of other two vectors. Therefore, by Theorem 2 on page 180 the set is linearly dependent.

Aliter. Clearly, $(-\lambda_1) v_1 + (-\lambda_2) v_2 + 1(\lambda_1 v_1 + \lambda_2 v_2) = 0_V$. Therefore, a non-trivial linear combination of $v_1, v_2, \lambda_1 v_1 + \lambda_2 v_2$ equals to the null vector. Hence, the given set is linearly dependent.

2.8.1   LINEAR DEPENDENCE AND ECHELON MATRICES

Consider the following echelon matrix A, whose pivots have been circled:

$$A = \begin{bmatrix} 0 & ② & 3 & 4 & 5 & 6 & 7 \\ 0 & 0 & ④ & 3 & 2 & 3 & 4 \\ 0 & 0 & 0 & 0 & ⑦ & 8 & 9 \\ 0 & 0 & 0 & 0 & 0 & ⑥ & 7 \\ 0 & 0 & 0 & 0 & 0 & 0 & 0 \end{bmatrix}$$

We observe that the rows R_2, R_3, R_4 have 0's in the second column below the non-zero pivot in R_1, and hence any linear combination of R_2, R_3 and R_4 must have 0 as its entry as the second component. Whereas R_1 has a non-zero entry 2 as the second component. Thus, R_1 cannot be a linear combination of the rows below it. Similarly, the rows R_3 and R_4 have 0's in the third column below the non-zero pivot in R_2, and hence R_2 cannot be a linear combination of the rows below it. Finally, R_3 cannot be a multiple of R_4 as R_4 has a 0 in the fifth column below the non-zero pivot in R_3. Thus, if we look at the rows from the bottom and move upward, we find that out of rows R_4, R_3, R_2, R_1 non row is a linear combination of the preceding rows. So, R_1, R_2, R_3, R_4 are linearly independent vectors $R^7(R)$.

The above discussion suggests the following theorem.

THEOREM-1 *The non-zero rows of a matrix in echelon form are linearly independent.*

EXAMPLE-1 *Let v_1, v_2, v_3 be vectors in a vector space $V(F)$, and let $\lambda_1, \lambda_2 \in F$. Show that the set $\{v_1, v_2, v_3\}$ is linearly dependent if the set $\{v_1 + \lambda_1 v_2 + \lambda_2 v_3, v_2, v_3\}$ is linearly dependent.*

SOLUTION If the set $\{v_1 + \lambda_1 v_2 + \lambda_2 v_3, v_2, v_3\}$ is linearly dependent, then there exist scalars $\lambda, \mu, \nu \in F$ (not all zero) such that

$$\lambda(v_1 + \lambda_1 v_2 + \lambda_2 v_3) + \mu v_2 + \nu v_3 = 0_V$$
$$\Rightarrow \quad \lambda v_1 + (\lambda \lambda_1 + \mu) v_2 + (\lambda \lambda_2 + \nu) v_3 = 0_V \tag{i}$$

The set $\{v_1, v_2, v_3\}$ will be linearly dependent if in (i) at least one of the scalar coefficients is non-zero.

If $\lambda \neq 0$, then the set will be linearly dependent whatever may be the values of μ and ν. But, if $\lambda = 0$, then at least one of μ and ν should not be equal to zero and hence at least one of $\lambda \lambda_1 + \mu$ and $\lambda \lambda_2 + \nu$ will not be zero (since $\lambda = 0 \Rightarrow \lambda \lambda_1 + \mu = \mu$ and $\lambda \lambda_2 + \nu = \nu$).

Hence, from (i), we find that the scalars λ, $\lambda \lambda_1 + \mu$, $\lambda \lambda_2 + \nu$ are not all zero. Consequently, the set $\{v_1, v_2, v_3\}$ is linearly dependent.

EXAMPLE-2 *Suppose the vectors u, v, w are linearly independent vectors in a vector space $V(F)$. Show that the vectors $u + v, u - v, u - 2v + w$ are also linearly independent.*

SOLUTION Let $\lambda_1, \lambda_2, \lambda_3$ be scalars such that

$$\lambda_1(u+v) + \lambda_2(u-v) + \lambda_3(u - 2v + w) = 0_V$$
$$\Rightarrow \quad (\lambda_1 + \lambda_2 + \lambda_3)u + (\lambda_1 - \lambda_2 - 2\lambda_3)v + \lambda_3 w = 0_V$$
$$\Rightarrow \quad \lambda_1 + \lambda_2 + \lambda_3 = 0, \ \lambda_1 - \lambda_2 - 2\lambda_3 = 0, \ 0\lambda_1 + 0\lambda_2 + \lambda_3 = 0 \quad \left[\begin{array}{l} \because \ u, v, w \text{ are linearly} \\ \text{independent} \end{array} \right]$$

The coefficient matrix A of the above system of equations is given by

$$A = \begin{bmatrix} 1 & 1 & 1 \\ 1 & -1 & -2 \\ 0 & 0 & 1 \end{bmatrix} \Rightarrow |A| = -2 \neq 0$$

So, the above system of equations has only trivial solution $\lambda_1 = \lambda_2 = \lambda_3 = 0$.

Thus, $u + v$, $u - v$, $u - 2v + w$ are linearly independent vectors.

EXERCISES 2.8

1. Mark each of the following true or false.
   (i) For any subset S of a vector space $V(F)$, $[S+S] = [S]$
   (ii) If S and T are subsets of a vector space $V(F)$, then $S \neq T \Rightarrow [S] \neq [T]$.
   (iii) If S and T are subsets of a vector space $V(F)$, then $S \subset T \Rightarrow [S] \subset [T]$
   (iv) Every subset of a linearly independent set is linearly independent.
   (v) Every superset of a linearly independent set is linearly independent.
   (vi) Every subset of a linearly dependent set is linearly dependent.
   (vii) Every superset of a linearly dependent set is linearly dependent.
   (viii) A subset of a linearly dependent set can never be linearly dependent.
   (ix) The void set is linearly dependent.
   (x) If a non-void subset of a vector space is not linearly dependent, then it is linearly independent.
   (xi) Any set of vectors containing the null vector is linearly dependent.
   (xii) Every singleton set consisting of a non-zero vector is linearly independent.
   (xiii) Intersection of two linearly independent subsets of a vector space is linearly independent.
   (xiv) Union of two linearly independent subsets of a vector space is linearly independent.

2. Which of the following subsets S of R^3 are linearly independent?
   (i) $S = \{(1,2,1), (-1,1,0), (5,-1,2)\}$
   (ii) $S = \{(1,1,0), (0,0,1), (1,5,2)\}$
   (iii) $S = \{(1,0,0), (1,1,1), (0,0,0)\}$
   (iv) $S = \{(1,3,2), (1,-7,8), (2,1,1)\}$
   (v) $S = \{(1,5,2), (1,0,0), (0,1,0)\}$

3. Which of the following subsets S of $R[x]$ are linearly independent?
   (i) $S = \{1, x, x^2, x^3, x^4\}$
   (ii) $S = \{1, x - x^2, x + x^2, 3x\}$
   (iii) $S = \{x^2 - 1, x + 1, x - 1\}$
   (iv) $S = \{x, x - x^3, x^2 + x^4, x + x^2 + x^4 + \frac{1}{2}\}$
   (v) $S = \{1, 1+x, 1+x+x^2, x^4\}$
   (vi) $S = \{1, x, x(1-x)\}$
   (vii) $S = \{1, x, 1+x+x^2\}$

4. Let $f(x), g(x), h(x), k(x) \in R[x]$ be such that $f(x) = 1, g(x) = x, h(x) = x^2$ and $k(x) = 1+x+x^2$ for all $x \in R$. Show that $f(x), g(x), h(x), k(x)$ are linearly dependent but any three of them are linearly independent.
5. Which of the following subsets of the space of all continuous functions on R are linearly independent?

    (i) $S = \{\sin x, \cos x, \sin(x+1)\}$
    (ii) $S = \{xe^x, x^2e^x, (x^2+x-1)e^x\}$
    (iii) $S = \{\sin^2 x, \cos 2x, 1\}$
    (iv) $S = \{x, \sin x, \cos x\}$
    (v) $S = \{x, x^2, e^{2x}\}$

6. Prove that the set $S = \{1, i\}$ is linearly independent in the vector space C of all complex numbers over the field R of real numbers while it is linearly dependent in the vector space C of all complex numbers over the field of complex numbers.
7. If the set $\{v_1, v_2, v_3\}$ is linearly independent in a vector space $V(F)$, then prove that the set $\{v_1+v_2, v_2+v_3, v_3+v_1\}$ is also linearly independent.
8. Prove that the vectors $v_1 = (1, 1, 0, 0)$, $v_2 = (0, 0, 0, 3)$, and $v_3 = (0, 1, -1, 0)$ in $F^4(F)$ are linearly independent if F is a field of characteristic zero and are linearly dependent if F is of characteristic 3.
9. If $\{v_1, v_2, \ldots, v_n\}$ is a linearly independent set of vectors in a vector space $V(F)$ and $\{v_1, v_2, \ldots, v_n, u\}$ is a linearly dependent set in V. Then show that u is a linear combination of $v_1, v_2, \ldots, v_n$.
10. If the set $S = \{v_1, v_2, \ldots, v_n\}$ is a linearly independent set of vectors in a vector space $V(F)$ and $u \in [S]$, then show that representation of u as a linear combination of $v_1, v_2, \ldots, v_n$ is unique.
11. Show that the vectors $v_1 = (1+i, 2i)$ and $v_2 = (1, 1+i)$ are linearly dependent in $C^2(C)$ but linearly independent in $C^2(R)$.
12. If v_1, v_2, v_3 are linearly dependent vectors of $V(F)$ where F is any subfield of the field of complex numbers, then so also are $v_1+v_2, v_2+v_3, v_3+v_1$.
13. Show that the functions $f(t) = \sin t, g(t) = \cos t, h(t) = t$ from R to R are linearly independent.
14. Find a maximal linearly independent subsystem of the system of vectors:

$$v_1 = (2, -2, -4), v_2 = (1, 9, 3), v_3 = (-2, -4, 1), v_4 = (3, 7, -1).$$

ANSWERS

1. (i) F   (ii) F   (iii) T   (iv) T   (v) F   (vi) F   (vii) T   (viii) F   (ix) F   (x) T   (xi) T   (xii) T   (xiii) T   (xiv) F
2. (i), (ii), (iv), (v)   3. (i), (iii), (iv), (v), (vi), (vii)   5. (ii), (iv), (v)
14. $\{v_1, v_2\}$

2.9 BASIS AND DIMENSION

BASIS. *A non-void subset B of a vector space $V(F)$ is said to be a basis for V, if*

 (i) *B spans V, i.e. $[B] = V$.*
and, (ii) *B is linearly independent (l.i.).*

In other words, a non-void subset B of a vector space V is a linearly independent set of vectors in V that spans V.

FINITE DIMENSIONAL VECTOR SPACE. *A vector space $V(F)$ is said to be a finite dimensional vector space if there exists a finite subset of V that spans it.*

A vector space which is not finite dimensional may be called an infinite dimensional vector space.

REMARK. *Note that the null vector cannot be an element of a basis, because any set containing the null vector is always linearly dependent.*

For any field F and a positive integer n, the set $B = \{e_1^{(n)}, e_2^{(n)}, \ldots, e_n^{(n)}\}$ spans the vector space $F^n(F)$ and is linearly independent. Hence, it is a basis for F^n. This basis is called standard basis for F^n.

REMARK. *Since the void set φ is linearly independent and spans the null space $\{0_V\}$. Therefore, the void set φ is the only basis for the null space $\{0_V\}$.*

Consider the subset $B = \{e_1^{(3)}, e_2^{(3)}, e_3^{(3)}\}$, where $e_1^{(3)} = (1, 0, 0)$, $e_2^{(3)} = (0, 1, 0)$, $e_3^{(3)} = (0, 0, 1)$, of the real vector space R^3. The set B spans R^3, because any vector (a, b, c) in R^3 can be written as a linear combination of $e_1^{(3)}$, $e_2^{(3)}$ and $e_3^{(3)}$, namely,

$$(a, b, c) = ae_1^{(3)} + be_2^{(3)} + ce_3^{(3)}$$

Also, B is linearly independent, because

$$\lambda e_1^{(3)} + \mu e_2^{(3)} + \nu e_3^{(3)} = (0, 0, 0)$$
$$\Rightarrow \lambda(1, 0, 0) + \mu(0, 1, 0) + \nu(0, 0, 1) = (0, 0, 0)$$
$$\Rightarrow (\lambda, \mu, \nu) = (0, 0, 0) \Rightarrow \lambda = \mu = \nu = 0$$

Hence, B is a basis for the real vector space R^3.

Now consider the subset $B_1 = \{(1, 1, 0), (1, 0, 1), (0, 1, 1)\}$ of R^3.

This set is also a basis for R^3, because any vector $(a,b,c) \in R^3$ can be written as

$$(a, b, c) = \left(\frac{a+b-c}{2}\right)(1, 1, 0) + \left(\frac{a_1+c-b}{2}\right)(1, 0, 1) + \left(\frac{b+c-a}{2}\right)(0, 1, 1)$$

and, $\lambda(1, 1, 0) + \mu(1, 0, 1) + \nu(0, 1, 1) \Rightarrow \lambda = \mu = \nu = 0$.

REMARK. *It follows from the above discussion that a basis for a vector space need not be unique. In fact, we shall show that corresponding to every non-zero vector in a vector space V one can obtain a basis for V.*

EXAMPLE-1 *Show that the set $B = \{1, x, x^2, \ldots, x^n\}$ is a basis for the vector space of all real polynomials of degree not exceeding n.*

SOLUTION Let $\lambda_0, \lambda_1, \ldots, \lambda_n \in R$ be such that
$$\lambda_0 \cdot 1 + \lambda_1 \cdot x + \cdots + \lambda_n x^n = 0(x) \quad \text{(zero polynomial)}$$
Then,
$$\lambda_0 \cdot 1 + \lambda_1 \cdot x + \cdots + \lambda_n x^n = 0(x) \quad \text{(zero polynomial)}$$
$$\Rightarrow \lambda_0 + \lambda_1 x + \lambda_2 x^2 + \cdots + \lambda_n x^n = 0 + 0x + 0x^2 + \cdots + 0x^n + \ldots$$
$$\Rightarrow \lambda_0 = \lambda_1 = \cdots = \lambda_n = 0$$

Therefore, the set B is linearly independent.

Also, the set B spans the vector space $P_n(x)$ of all real polynomials of degree not exceeding n, because every polynomial of degree less than or equal n is a linear combination of B.

Hence, B is a basis for the vector space $P_n(x)$ of all real polynomials of degree not exceeding n.

EXAMPLE-2 *Show that the infinite set $\{1, x, x^2, \ldots\}$ is a basis for the vector space $R[x]$ of all polynomials over the field R of real numbers.*

SOLUTION The set $B = \{1, x, x^2, \ldots\}$ is linearly independent. Also, the set B spans $R[x]$, because every real polynomial can be expressed as a linear combination of B.

Hence, B is a basis for $R[x]$. Since B is an infinite set. Therefore, $R[x]$ is an infinite dimensional vector space over R.

EXAMPLE-3 *Let $a, b \in R$ such that $a < b$. Then the vector space $C[a, b]$ of all real valued continuous functions on $[a, b]$ is an infinite dimensional vector space.*

EXAMPLE-4 *The field R of real numbers is an infinite dimensional vector space over its subfield Q of rational numbers. But, it is a finite dimensional vector space over itself.*

SOLUTION Since π is a transcendental number, that is, π is not a root of any polynomial over Q. Thus, for any $n \in N$ and $a_i \in Q (i = 0, 1, 2, \ldots, n)$
$$a_0 + a_1 \pi + a_2 \pi^2 + \cdots + a_n \pi^n \neq 0$$
$\Rightarrow 1, \pi, \pi^2, \pi^3, \ldots, \pi^n, \ldots$ are linearly independent over Q.

Hence, R is an infinite dimensional vector space over Q.

EXAMPLE-5 *The set $\{1, i\}$ is a basis for the vector space C of all complex numbers over the field R of real numbers.*

We have defined basis of a vector space and we have seen that basis of a vector space need not be unique. Now a natural question arises, does it always exist?. The answer is in the affirmative as shown in the following theorem.

THEOREM–1 *Every finite dimensional vector space has a basis.*

PROOF. Let V be a finite dimensional vector space over a field F. If V is the null space, then the void set φ is its basis, we are done. So, let V be a non-null space. Since V is finite dimensional. So, there exists a finite subset $S = \{v_1, v_2, \ldots, v_n\}$ of V that spans V. If S is a linearly independent set, we are done. If S is not linearly independent, then by Theorem 2 on page 180, there exists a vector $v_k \in S$ which is linear combination of the previous vectors. Remove v_k from S and let $S_1 = S - \{v_k\}$. By Theorem 3 on page 167, S_1 spans V.

If S_1 is a linearly independent set, the theorem is proved. If not we repeat the above process on S_1 and omit one more vector to obtain $S_2 \subset S_1$. Continuing in this manner, we obtain successively $S_1 \supset S_2 \supset S_3, \ldots$, where each S_i spans V.

Since S is a finite set and each S_i contains one vector less than S_{i-1}. Therefore, we ultimately arrive at a linearly independent set that spans V. Note that this process terminates before we exhaust all vectors in S, because if not earlier, then after $(n-1)$ steps, we shall be left with a singleton set containing a non-zero vector that spans V. This singleton set will form a basis for V because each singleton set containing a non-zero vector in V forms a linearly independent set. Hence, V has a basis.

Q.E.D.

REMARK. *This theorem remains valid if the word finite is omitted. In that case, the proof requires the use of Zorn's lemma. This lemma involves several concepts which are beyond the scope of this book.*

NOTE. *In future unless otherwise mentioned a vector space will mean a finite dimensional vector space.*

THEOREM-2 *The representation of a vector in terms of the basis vectors is unique.*

PROOF. Let V be a vector space over a field F, and let $B = \{b_1, b_2, \ldots, b_n\}$ be a basis for V.

Let v be an arbitrary vector in V. We have to prove that v has a unique representation in terms of vectors in B.

Let $v = \lambda_1 b_1 + \cdots + \lambda_n b_n$ and $v = \mu_1 b_1 + \cdots + \mu_n b_n$ be two representations for $v \in V$ as a linear combination of basis vectors. Then,

$$\lambda_1 b_1 + \cdots + \lambda_n b_n = \mu_1 b_1 + \cdots + \mu_n b_n$$
$$\Rightarrow \quad (\lambda_1 - \mu_1)b_1 + \cdots + (\lambda_n - \mu_n)b_n = 0_V$$
$$\Rightarrow \quad \lambda_1 - \mu_1 = 0, \ldots, \lambda_n - \mu_n = 0 \qquad [\because B \text{ is a linearly independent set}]$$
$$\Rightarrow \quad \lambda_1 = \mu_1, \lambda_2 = \mu_2, \ldots, \lambda_n = \mu_n$$

Hence, v has a unique representation in terms of vectors in B.

Q.E.D.

THEOREM-3 *Let S be a subset of a vector space $V(F)$ such that S spans V. Then S contains a basis for V.*

PROOF. By Theorem 1, the basis for V is obtained by removing those vectors from S, which are linear combinations of previous vectors in S. Hence, S contains a basis for V.

Q.E.D.

THEOREM-4 *A subset B of a vector space $V(F)$ is a basis for V if every vector in V has a unique representation as a linear combination of vectors of B.*

PROOF. First suppose that B is a basis for V. Then B spans V and so every vector in V is a linearly combination of vectors of B. To prove the uniqueness, we consider the following two cases:

Case I When B is finite, say $B = \{b_1, b_2, \ldots, b_n\}$.

Let v be an arbitrary vector in V and suppose that

$$v = \lambda_1 b_1 + \lambda_2 b_2 + \cdots + \lambda_n b_n \quad \text{and, also} \quad v = \mu_1 b_1 + \mu_2 b_2 + \cdots + \mu_n b_n,$$

where $\lambda_i, \mu_i \in F$ for all $i \in \underline{n}$. Then,

$$\lambda_1 b_1 + \lambda_2 b_2 + \cdots + \lambda_n b_n = \mu_1 b_1 + \mu_2 b_2 + \cdots + \mu_n b_n$$

$\Rightarrow \quad (\lambda_1 - \mu_1)b_1 + \cdots + (\lambda_n - \mu_n)b_n = 0_V$

$\Rightarrow \quad \lambda_1 - \mu_1 = 0, \ldots, \lambda_n - \mu_n = 0$ [$\because$ B is linearly independent]

$\Rightarrow \quad \lambda_1 = \mu_1, \lambda_2 = \mu_2, \ldots, \lambda_n = \mu_n.$

Thus, $v \in V$ has a unique representation as a linear combination of vectors of B.

Case II When B is infinite.

Let $v \in V$ and suppose that $v = \sum_{b \in B} \lambda_b\, b$ and, also $v = \sum_{b \in B} \mu_b\, b$

where $\lambda_b, \mu_b \in F$ and each sum contains only finite number of terms (i.e. λ_b, μ_b are zero in all except a finite number of terms). Then,

$$\sum_{b \in B} \lambda_b\, b = \sum_{b \in B} \mu_b\, b$$

$\Rightarrow \quad \sum_{b \in B} (\lambda_b - \mu_b) b = 0_V$

$\Rightarrow \quad \lambda_b - \mu_b = 0$ for all $b \in B$ [$\because$ B is linearly independent]

$\Rightarrow \quad \lambda_b = \mu_b$ for all $b \in B$

This proves that $v \in V$ has a unique representation as a linear combination of vectors of B.

Conversely, suppose that every vector in V has a unique representation as a linear combination of vectors of B. Then, B spans V. Further suppose that

$$\lambda_1 b_1 + \lambda_2 b_2 + \cdots + \lambda_n b_n = 0_V \quad \text{for some } \lambda_1, \lambda_2, \ldots, \lambda_n \in F,\ b_1, b_2, \ldots, b_n \in B.$$

Then,
$$\lambda_1 b_1 + \lambda_2 b_2 + \cdots + \lambda_n b_n = 0 b_1 + 0 b_2 + \cdots + 0 b_n$$
$$\Rightarrow \quad \lambda_1 = \lambda_2 = \cdots = \lambda_n = 0 \quad \left[\begin{array}{l} \because \text{ Every vector in } V \text{ has a unique representation} \\ \text{as a linear combination of vectors in } B \end{array}\right]$$

Thus, the only linear combination of vectors of B that equals to the null vector is the trivial linear combination. Hence, B is a basis for V.

Q.E.D.

The following theorem proves an extremely important result that one cannot have more linearly independent vectors than the number of vectors in a spanning set.

THEOREM-5 *Let V be a vector space over a field F. If V is spanned by the set $\{v_1, v_2, \ldots, v_n\}$ of n vectors in V and if $\{w_1, w_2, \ldots, w_m\}$ is a linearly independent set of vectors in V, then $m \leq n$. Moreover, V can be spanned by a set of n vectors containing the set $\{w_1, w_2, \ldots, w_m\}$.*

PROOF. We shall prove both the results together by induction on m.

First we shall prove the theorem for $m = 1$.

Let $\{w_1\}$ be a linearly independent set. Then $w_1 \neq 0_V$.

Now,

$\quad w_1 \in V$

$\Rightarrow \quad w_1$ is a linear combination of $v_1, v_2, \ldots, v_n$ $\qquad [\because \{v_1, v_2, \ldots, v_n\} \text{ spans } V]$

$\Rightarrow \quad \{w_1, v_1, v_2, \ldots, v_n\}$ is linearly dependent set $\qquad$ [By Theorem 2 on page 180]

$\Rightarrow \quad$ There exist $v_k \in \{w_1, v_1, \ldots, v_n\}$ such that v_k is a linear combination of the preceeding vectors

Let us rearrange the vectors $w_1, v_1, \ldots, v_n$ in such a way that v_n is a linear combination of the previous vectors. Removing v_n from the set $\{w_1, v_1, \ldots, v_n\}$ we obtain the set $\{w_1, v_1, \ldots, v_{n-1}\}$ of n vectors containing w_1 such that it spans V and $n - 1 \geq 0 \Rightarrow n \geq 1 = m$.

Note that the vector removed is one of v's because the set of w's is linearly independent. Hence, the theorem holds for $m = 1$.

Now suppose that the theorem is true for m. This means that for a given set $\{w_1, \ldots, w_m\}$ of m linearly independent vectors in V, we have

(i) $m \leq n$

and, (ii) there exists a set of n vectors $w_1, \ldots, w_m, v_1, \ldots, v_{n-m}$ in V that spans V.

To prove the theorem for $m + 1$, we have to show that for a given set of $(m + 1)$ linearly independent vectors $w_1, w_2, \ldots, w_m, w_{m+1}$ in V, we have

(i) $m + 1 \leq n$

and, (ii) there exists a set of n vectors containing $w_1, \ldots, w_m, w_{m+1}$ that spans V.

Now,

$w_{m+1} \in V$

$\Rightarrow$ w_{m+1} is a linear combination of $w_1, \ldots, w_m, v_1, \ldots, v_{n-m}$

$\Rightarrow$ $\{w_{m+1}, w_m, \ldots, w_1, v_1, \ldots, v_{n-m}\}$ is a linearly dependent set and spans V.

$\Rightarrow$ There exist a vector v_{n-m} (say) such that it is a linear combination of
$w_{m+1}, \ldots, w_1, v_1, v_2, \ldots, v_{n-m-1}$

Removing v_{n-m} from the set $\{w_{m+1}, \ldots, w_1, v_1, \ldots, v_{n-m}\}$ we obtain a set $\{w_{m+1}, \ldots, w_1, v_1, \ldots, v_{n-m-1}\}$ of n vectors that also spans V and $n - m - 1 \geq 0$, i.e. $m + 1 \geq n$.

Hence, the theorem is true for $m + 1$.

Hence, by induction theorem is true for every $m \in N$.

Q.E.D.

COROLLARY-1 *If a basis of a vector space $V(F)$ has n vectors, then every set of p vectors with $p > n$, is linearly dependent set.*

PROOF. Let $B = \{b_1, b_2, \ldots, b_n\}$ be a basis for V, and let $S = \{v_1, v_2, \ldots, v_p\}$ be a set of p vectors in V with $p > n$. If S is linearly independent, then $p \leq n$ (by Theorem 5 on the page 196). Hence, S is linearly dependent.

Q.E.D.

COROLLARY-2 *(Invariance of dimension). Every basis for a finite dimensional vector space has the same number of vectors.*

PROOF. Let $B_1 = \{b_1, b_2, \ldots, b_n\}$ and $B_2 = \{v_1, v_2, \ldots, v_m\}$ be any two bases for a finite dimensional vector space $V(F)$. Then B_1 and B_2 are linearly independent sets both spanning V.

Since B_1 spans V and B_2 is a linearly independent set of vectors in V, by Theorem 5 on the page 196, we get

$$m \leq n.$$

Since B_2 spans V and B_1 is a linearly independent set of vectors in V, by the same theorem, we get

$$n \leq m$$

Hence, $m = n$.

As seen earlier a vector space may have many bases and by the above corollary the number of vectors in every basis is same. Hence, we have the following definition.

DIMENSION. *Let V be a finite dimensional vector space over a field F. The number of elements in a basis of V is called dimension of V and is denoted by $\dim V$.*

If $\dim V = n$, V is said to be an n-dimensional vector space. If V is the null space, its dimension is taken to be zero.

Let R be the field of real numbers, then the dimension of vector space R over itself is one, because 1 forms a basis for R. The real vector space R^2 is 2-dimensional, because $e_1^{(2)} = (1,0)$ and $e_2^{(2)} = (0,1)$ form a basis for R^2. R^3 is 3-dimensional vector space over field R, because $e_1^{(3)} = (1,0,0), e_2^{(3)} = (0,1,0)$ and $e_3^{(3)} = (0,0,1)$ form a basis for R^3. In general, R^n is n-dimensional vector space over R, because $e_1^{(n)} = (1,0,\ldots,0), e_2^{(n)} = (0,1,\ldots,0),\ldots, e_n^{(n)} = (0,0,\ldots,1)$ form a basis for R^n. Thus, the definition of dimension coincides with the familiar practice of regarding a line as one-dimensional, a plane as two-dimensional, and so on. From Examples 1 and 2 (page 193), we observe that the vector space $P_n(x)$ of all polynomials of degree less than or equal to n is an n-dimensional vector space while the vector space $R[x]$ of all real polynomials is an infinite dimensional vector space.

THEOREM-6 *Let $V(F)$ be a vector space of dimension n. Then,*

  (i) *any set of $(n+1)$ or more vectors of V is linearly dependent.*
  (ii) *no set of $(n-1)$ or less vectors of V can span V.*

PROOF. (i) Since $\dim V = n$, therefore, every basis of V has n vectors that span V. By Theorem 5 the number of linearly independent vectors in V is less than or equal to n. Hence, a set of $(n+1)$ or more vectors cannot be linearly independent.
(ii) Suppose a set of $(n-1)$ or less vectors of V spans V. Then the number of linearly independent. vectors in V must be less than or equal to $(n-1)$. This contradicts that a basis of V is a linearly independent set of n vectors of V. Hence, no set of $(n-1)$ or less vectors of V can span V.

Q.E.D.

THEOREM-7 *Let V be a finite dimensional vector space over a field F, and let S be a linearly independent subset of V. Then there is a basis for V which contains S.*

OR

Every linearly independent set of vectors of a finite dimensional vector space $V(F)$ can be extended to form a basis for V.

PROOF. Since V is a finite dimensional vector space. Therefore, there exists a finite subset, say $B = \{b_1, b_2, \ldots, b_n\}$ of V such that B spans V. Since S is a linearly independent set of vectors of V and B spans V. Therefore S has at most n vectors. Let $S = \{v_1, v_2, \ldots, v_m\}, m \leq n$. Since B spans V, therefore, $B' = B \cup S = \{v_1, v_2, \ldots, v_m, b_1, b_2, \ldots, b_n\}$ will also span V. Since B spans V, therefore, each v_i is expressible as a linear combination of vectors of B. Consequently, B' is a linearly dependent set. Therefore, there exists a $b_k \in B'$ which is a linear combination of the preceding vectors. Since S is a linearly independent set, therefore, no v_i is a linear combination of preceding vectors. Remove b_k from B', and let $B'' = \{v_1, \ldots, v_m, b_1, \ldots, b_{k-1}, \ldots, b_n\}$. If B'' is linearly independent, then B'' is a basis for V containing S. If B'' is linearly dependent, we repeat the above process. Continuing in this manner we remove, one by one, every vector which is a linear combination of the preceding vectors, till we obtain a linearly independent set spanning V. Since S is a linearly independent set, v_i is a linear combination of the preceding vectors. Therefore, the vectors removed are some of b_i's. Hence, the reduced linearly independent set consists of all v_i's and some of b_i's. This reduced set $\{v_1, v_2, \ldots, v_m, b_1, b_2, \ldots, b_{n-m}\}$ is a basis for V containing S.

Q.E.D.

This theorem can also be stated as

"Any l.i. set of vectors of a finite dimensional vector space is a part of its basis."

COROLLARY-1 *In a vector space $V(F)$ of dimension n*

(i) any set of n linearly independent vectors is a basis,

and, (ii) any set of n vectors that spans V is a basis.

PROOF. (i) Let $B = \{b_1, b_2, \ldots, b_n\}$ be a set of n linearly independent vectors in vector space V. Then by Theorem 7, B is a part of the basis. But, the basis cannot have more than n vectors. Hence, B itself is a basis for V.

(ii) Let $B = \{b_1, b_2, \ldots, b_n\}$ be a set of n vectors in V that spans V. If B is not linearly independent, then there exists a vector in B which is linear combination of the preceding vectors and by removing it from B we will obtain a set of $(n-1)$ vectors in V that will also span V. But, this is a contradiction to the fact that a set of $(n-1)$ vectors cannot span V. Hence, B is a linearly independent set. Consequently, it is a basis for V.

Q.E.D.

COROLLARY-2 *Let $V(F)$ be a vector space of dimension n, and let S be a subspace of V. Then,*

(i) every basis of S is a part of a basis of V.

(ii) $\dim S \leq \dim V$.

(iii) $\dim S = \dim V \Leftrightarrow S = V$.

and, (iv) $\dim S < \dim V$, if S is a proper subspace of V.

PROOF. (i) Let $B' = \{b_1, b_2, \ldots, b_m\}$ be a basis for S. Since S is a subspace of V. Therefore, B' is a linearly independent set of vectors in V and hence it is a part of a basis for V.
(ii) Since basis for S is a part of the basis for V. Therefore, $\dim S \leq \dim V$.
(iii) Let $\dim S = \dim V = n$. Let $B = \{b_1, b_2, \ldots, b_n\}$ be a basis for V so that $[B] = S$.

Since B is a linearly independent set of n vectors in V and $\dim V = n$. Therefore, B spans V, i.e. $[B] = V$.
Thus, $S = [B] = V \Rightarrow S = V$.
Conversely, let $S = V$. Then,

$S = V$
$\Rightarrow$ S is a subspace of V and V is a subspace of S
$\Rightarrow$ $\dim S \leq \dim V$ and $\dim V \leq \dim S$ [By (ii)]
$\Rightarrow$ $\dim S = \dim V$.

(iv) Let S be a proper subspace of V. Then there exists a vector $v \in V$ such that $v \notin S$. Therefore, v cannot be expressed as a linear combination of vectors in B', the basis for S. That is, $v \notin [S]$. Consequently, the set $\{b_1, b_2, \ldots, b_m, v\}$ forms a linearly independent subset of V. Therefore, the basis for V will contain more than m vectors.

Hence, $\dim S < \dim V$.

Q.E.D.

THEOREM-8 *Let $V(F)$ be a finite dimensional vector space, and let m be a positive integer. Then the function space $V^m = \{f : f : \underline{m} \to V\}$ is also finite dimensional and*

$$\dim(V^m) = m \cdot \dim V$$

PROOF. Let $\dim V = n$, and let $B = \{b_1, b_2, \ldots, b_n\}$ be a basis for V. By Example 11 on page 126, the function space V^m is a vector space over field F. For any $i \in \underline{m}, j \in \underline{n}$ define a mapping

$$\varphi_{ij} : \underline{m} \to V \quad \text{given by}$$
$$\varphi_{ij}(k) = \delta_{ik} b_j \quad \text{for all } k \in \underline{m}.$$

Here δ_{ik} is Kronecker delta

Let $B' = \{\varphi_{ij} : i \in \underline{m}, j \in \underline{n}\}$. Clearly $O(B') = mn$.

We shall now show that B' is a basis for the function space V^m.

B' is $l.i.$: Let $a_{ij} \in F, i \in \underline{m}, j \in \underline{n}$ be such that

$$\sum_{i=1}^{m} \sum_{j=1}^{n} a_{ij} \varphi_{ij} = \widehat{0} \quad \text{(zero function)}$$

Then,

$$\left(\sum_{i=1}^{m}\sum_{j=1}^{n} a_{ij}\varphi_{ij}\right)(k) = \widehat{0}(k) \quad \text{for all } k \in \underline{m}$$

$\Rightarrow \quad \sum_{i=1}^{m}\sum_{j=1}^{n} a_{ij}\varphi_{ij}(k) = 0_V$

$\Rightarrow \quad \sum_{i=1}^{m}\sum_{j=1}^{n} a_{ij}\delta_{ik}b_j = 0_V$

$\Rightarrow \quad \sum_{j=1}^{n}\left(\sum_{i=1}^{m} a_{ij}\delta_{ik}\right) b_j = 0_V$

$\Rightarrow \quad \sum_{j=1}^{n} a_{kj}b_j = 0_V \hfill [\because \ \delta_{ik} = 1 \text{ for } i = k]$

$\Rightarrow \quad a_{kj} = 0 \quad \text{for all } j \in \underline{n} \hfill [\because \ B = \{b_1, \ldots, b_n\} \text{ is a basis for } V]$

Thus, $a_{kj} = 0$ for all $k \in \underline{m}$ and for all $j \in \underline{n}$.

So, $\sum_{i=1}^{m}\sum_{j=1}^{n} a_{ij}\varphi_{ij} = \widehat{0} \Rightarrow a_{ij} = 0$ for all $i \in \underline{m}$ and for all $j \in \underline{n}$.

Therefore, B' is a linearly independent set.

B' spans V^m: Let $f : \underline{m} \to V$ be an arbitrary mapping in V^m. Then for each $k \in \underline{m}$, $f(k) \in V$.

Since B is a basis for V. Therefore, $f(k) \in V$ is expressible as a linear combination of vectors in B. Let

$$f(k) = \sum_{j=1}^{n} \lambda_{kj} b_j \qquad \text{(i)}$$

For any $k \in \underline{m}$, we have

$$\left(\sum_{i=1}^{m}\sum_{j=1}^{n} \lambda_{ij}\varphi_{ij}\right)(k) = \sum_{i=1}^{m}\sum_{j=1}^{n} \lambda_{ij}\varphi_{ij}(k) = \sum_{i=1}^{m}\sum_{j=1}^{n} \lambda_{ij}\delta_{ik}b_j$$

$\Rightarrow \quad \left(\sum_{i=1}^{m}\sum_{j=1}^{n} \lambda_{ij}\varphi_{ij}\right)(k) = \sum_{j=1}^{n}\left(\sum_{i=1}^{m} \lambda_{ij}\delta_{ik}\right) b_j$

$\Rightarrow \quad \left(\sum_{i=1}^{m}\sum_{j=1}^{n} \lambda_{ij}\varphi_{ij}\right)(k) = \sum_{j=1}^{n} \lambda_{kj} b_j = f(k) \hfill \text{[From (i)]}$

$\therefore \quad f = \sum_{i=1}^{m}\sum_{j=1}^{n} \lambda_{ij}\varphi_{ij}$

Thus, $f \in V^m$ is expressible as a linear combination of B'. Since f is an arbitrary function in V^m. Therefore, every function in V^m is a linear combination of B'.

Thus, B' spans V^m. Hence, B' is a basis for V^m.

Consequently, $\dim V^m = O(B') = mn = m \dim V$.

Q.E.D.

THEOREM-9 *If S and T are two subspaces of a finite dimensional vector space $V(F)$, then*

$$\dim(S+T) = \dim S + \dim T - \dim(S \cap T).$$

PROOF. Let $\dim S = p, \dim T = q, \dim(S \cap T) = r$ and, $\dim V = n$. Then by corollary 2 to Theorem 7, $p \leq n, q \leq n, r \leq n$. Let $B = \{b_1, \ldots, b_r\}$ be a basis for $S \cap T$. Then B is a linearly independent set of vectors in $S \cap T$ and therefore in S as well as in T. So it can be extended to a basis $\{b_1, \ldots, b_r, u_1, \ldots, u_{p-r}\}$ for S and to a basis $\{b_1, b_2, \ldots, b_r, v_1, \ldots, v_{q-r}\}$ for T.

We shall now show that the set $B_1 = \{b_1, \ldots, b_r, u_1, \ldots, u_{p-r}, v_1, \ldots, v_{q-r}\}$ is a basis for $S+T$.

To prove that B_1 is a basis for $S+T$, we have to prove that B_1 is a linearly independent subset of $S+T$ and B_1 spans $S+T$.

B_1 is a l.i. set in $S+T$: Let $\lambda_i (i=1,2,\ldots,r), \mu_j(j=1,2,\ldots,p-r), \alpha_k(k=1,2,\ldots,q-r)$ be scalars in F such that

$$\sum_{i=1}^{r} \lambda_i b_i + \sum_{j=1}^{p-r} \mu_j u_j + \sum_{k=1}^{q-r} \alpha_k v_k = 0_V$$

$$\Rightarrow \sum_{i=1}^{r} \lambda_i b_i + \sum_{j=1}^{p-r} \mu_j u_j = -\sum_{k=1}^{q-r} \alpha_k v_k = v, \text{say} \tag{i}$$

Since $\{b_1, \ldots, b_r, u_1, u_2, \ldots, u_{p-r}\}$ is a basis for S. Therefore,

$$v = \sum_{i=1}^{r} \lambda_i b_i + \sum_{j=1}^{p-r} \mu_j u_j \in S$$

Since $\{b_1, b_2, \ldots, b_r, v_1, v_2, \ldots, v_{q-r}\}$ is a basis for T. Therefore,

$$v = -\sum_{k=1}^{q-r} \alpha_k v_k = (-\alpha_1)v_1 + (-\alpha_2)v_2 + \cdots + (-\alpha_{q-r})v_r + 0b_1 + \cdots + 0b_r \in T.$$

Thus, $v \in S \cap T$. Therefore, v can be expressed uniquely as a linear combination of basis vectors $b_1, b_2, \ldots, b_r$ of $S \cap T$. Let

$$v = \sum_{i=1}^{r} \beta_i b_i \quad \text{for some } \beta_i \in F.$$

Then, $\sum_{i=1}^{r} \beta_i b_i = -\sum_{k=1}^{q-r} \alpha_k v_k$ [From (i)]

$\Rightarrow \sum_{i=1}^{r} \beta_i b_i + \sum_{k=1}^{q-r} \alpha_k v_k = 0_V$

$\Rightarrow \beta_i = 0, \alpha_k = 0 \quad \text{for all } i = 1, 2, \ldots, r \, ; k = 1, \ldots, q-r \ [\because \{b_1, \ldots, b_r, v_1, \ldots, v_{q-r}\} \text{ is l.i.}]$

Putting $\alpha_1 = \alpha_2 = \cdots = \alpha_{q-r} = 0$ in (i), we find that

$$\sum_{i=1}^{r} \lambda_i b_i + \sum_{j=1}^{p-r} \mu_j u_j = 0_V$$

$\Rightarrow \lambda_1 = \cdots = \lambda_r = \mu_1 = \cdots = \mu_{p-r} = 0 \quad [\because \{b_1, \ldots, b_r, u_1, \ldots, u_{p-r}\} \text{ is a basis for } S]$

Thus,

$$\sum_{i=1}^{r} \lambda_i b_i + \sum_{j=1}^{p-r} \mu_j u_j + \sum_{k=1}^{q-r} \alpha_k v_k = 0_V$$

$\Rightarrow \lambda_1 = \lambda_2 = \cdots = \lambda_r = \mu_1 = \cdots = \mu_{p-r} = \alpha_1 = \cdots = \alpha_{q-r} = 0.$

Hence, B_1 is a linearly independent set.

B_1 spans $S + T$: Let u be an arbitrary vector in $S + T$. Then there exist $v \in S, w \in T$ such that $u = v + w$.

Now,

$$v \in S \Rightarrow v = \sum_{i=1}^{r} \alpha_i b_i + \sum_{j=1}^{p-r} \beta_j u_j \text{ for some } \alpha_i, \beta_j \in F \quad \left[\begin{array}{l} \because \{b_1, \ldots, b_r, u_1, \ldots, u_{p-r}\} \\ \text{is a basis for } S \end{array}\right]$$

and,

$$w \in T \Rightarrow w = \sum_{k=1}^{r} \gamma_i b_i + \sum_{l=1}^{q-r} \delta_l v_l \text{ for some } \gamma_i, \delta_l \in F \quad \left[\begin{array}{l} \because \{b_1, \ldots, b_r, v_1, \ldots, v_{q-r}\} \\ \text{is a basis for } T \end{array}\right]$$

$\therefore \quad u = v + w = \sum_{i=1}^{r} \alpha_i b_i + \sum_{j=1}^{p-r} \beta_j u_j + \sum_{k=1}^{r} \gamma_i b_i + \sum_{l=1}^{q-r} \delta_l v_l$

$\Rightarrow u = \sum_{i=1}^{r} (\alpha_i + \gamma_i) b_i + \sum_{j=1}^{p-r} \beta_j u_j + \sum_{k=1}^{q-r} \delta_k v_k$

This shows that u is a linear combination of vectors in B_1. Since u is an arbitrary vector in $S + T$. Therefore, every vector in $S + T$ is a linear combination of vectors in B_1.

Thus, B_1 spans $S + T$.

Hence, B_1 is a basis for $S+T$. Consequently, we have
$\dim S + T = $ number of vectors in $B_1 = p+q-r = \dim S + \dim T - \dim(S \cap T)$.

Q.E.D.

COROLLARY. *If S and T are subspaces of a finite-dimensional vector space $V(F)$ such that $S \cap T = \{0_V\}$, then*
$$\dim(S+T) = \dim S + \dim T.$$

PROOF. We have,

$\qquad S \cap T = \{0_V\}$

$\Rightarrow \qquad \dim(S \cap T) = 0$

Thus, $\qquad \dim(S+T) = \dim S + \dim T - \dim(S \cap T)$

$\Rightarrow \qquad \dim(S+T) = \dim S + \dim T.$

Q.E.D.

In three-dimensional Euclidean space R^3, let $S = $ the xy-plane and $T = $ the yz-plane. Clearly, S and T are subspaces of R^3 and $\dim S = 2$, $\dim T = 2$. Clearly, $S \cap T = y$-axis, whose dimension is 1, and $S + T = R^3$. Therefore, $\dim S + T = \dim R^3 = 3$ and $\dim S + \dim T - \dim(S \cap T) = 2 + 2 - 1 = 3$. So, $\dim(S+T) = \dim S + \dim T - \dim(S \cap T)$. This verifies the result of Theorem 9. On the other hand, if $S = $ the xy-plane and $T = z$-axis, then $S \cap T = \{0_V\}$ and $S + T = R^3$. Also, $\dim R^3 = \dim(S+T) = 3 = 2 + 1 = \dim S + \dim T$. This verifies the result of corollary to Theorem 9.

THEOREM-10 *If a finite dimensional vector space $V(F)$ is direct sum of its two subspaces S and T, then*
$$\dim V = \dim S + \dim T.$$

PROOF. Since V is direct sum of S and T, therefore, $V = S + T$ and $S \cap T = \{0_V\}$.

Now, $S \cap T = \{0_V\} \quad \Rightarrow \quad \dim(S \cap T) = 0$

By Theorem 9, we have

$\qquad \dim(S+T) = \dim S + \dim T - \dim(S \cap T)$

$\Rightarrow \qquad \dim(S+T) = \dim S + \dim T.$

Q.E.D.

THEOREM-11 *Every subspace of a finite dimensional vector space has a complement.*

PROOF. Let V be a finite dimensional vector space over a field F, and let S be a subspace of V. Further, let $\dim V = n$ and $\dim S = m$.

Let $B_1 = \{b_1, b_2, \ldots, b_m\}$ be a basis for S. Then B is a linearly independent set of vectors in B. So it can be extended to a basis for V. Let $B = \{b_1, b_2, \ldots, b_m, v_1, v_2, \ldots, v_{n-m}\}$ be extended basis for V. Let T be a subspace of V spanned by the set $B_2 = \{v_1, v_2, \ldots, v_{n-m}\}$

We shall now show that T is the complement of S.
Let v be an arbitrary vector in V. Then,

$$v = \sum_{i=1}^{m} \lambda_i b_i + \sum_{i=1}^{n-m} \mu_i v_i \text{ for some } \lambda_i, u_i \in F \qquad [\because \ B \text{ is a basis for } V]$$

Since B_1 and B_2 are bases for S and T respectively. Therefore,

$$\sum_{i=1}^{m} \lambda_i b_i \in S \text{ and, } \sum_{i=1}^{n-m} \mu_i v_i \in T$$

$$\Rightarrow \quad v = \sum_{i=1}^{m} \lambda_i b_i + \sum_{i=1}^{n-m} \mu_i v_i \in S + T$$

Thus, $v \in V \Rightarrow v \in S + T$
$\therefore \quad V \subset S + T$
Also, $S + T \subset V$.
Hence, $V = S + T$.
Let $u \in S \cap T$. Then, $u \in S$ and $u \in T$.
Now,

$u \in S \Rightarrow$ There exists $\lambda_1, \ldots, \lambda_m \in F$ such that $u = \sum_{i=1}^{m} \lambda_i b_i \qquad [\because \ B_1 \text{ is a basis for } S]$

and,

$u \in T \Rightarrow$ There exists $\mu_1, \ldots, \mu_{n-m} \in F$ such that $u = \sum_{i=1}^{n-m} \mu_i v_i \qquad [\because \ B_2 \text{ is a basis for } T]$

$\therefore \quad \sum_{i=1}^{m} \lambda_i b_i = \sum_{i=1}^{n-m} \mu_i v_i$

$\Rightarrow \quad \lambda_1 b_1 +, \ldots, + \lambda_m b_m + (-\mu_1) v_1 + \cdots + (-\mu_{n-m}) v_{n-m} = 0_V$
$\Rightarrow \quad \lambda_1 = \lambda_2 = \cdots = \lambda_m = 0 = \mu_1 = \cdots = \mu_{n-m} \qquad [\because \ B \text{ is a basis for } V]$
$\Rightarrow \quad u = 0 b_1 + \cdots + 0 b_m = 0_V$

Thus, $u \in S \cap T \Rightarrow u = 0_V$
$\therefore \quad S \cap T = \{0_V\}$
Hence, $V = S \oplus T$. Consequently, T is complement of S.

Q.E.D.

THEOREM-12 *Let $V(F)$ be a finite dimensional vector space, and let S be a subspace of V. Then, $\dim(V/S) = \dim V - \dim S$.*

PROOF. Let $\dim V = n, \dim S = m$. Then $n \geq m$. Let $B_1 = \{v_1, v_2, \ldots, v_m\}$ be a basis for S. Then, B_1 is a linearly independent set of vectors in V. So it can be extended to a basis for V, say $B_2 = \{v_1, \ldots, v_m, b_1, \ldots, b_{n-m}\}$.

We shall now show that the set $B = \{b_1 + S, b_2 + S, \ldots, b_{n-m} + S\}$ is basis for V/S.

To prove that B is a basis for V/S, we have to show that B is a linearly independent set and B spans V/S.

B is a l.i set in V/S: Let $\lambda_1, \ldots, \lambda_{n-m}$ be scalars in F such that

$$\lambda_1(b_1 + S) + \cdots + \lambda_{n-m}(b_{n-m} + S) = S \quad \text{(zero of } V/S\text{)}$$

Then,

$$(\lambda_1 b_1 + S) + \cdots + (\lambda_{n-m} b_{n-m} + S) = S \quad \text{[By def. of scalar multiplication on } V/S\text{]}$$
$$\Rightarrow (\lambda_1 b_1 + \cdots + \lambda_{n-m} b_{n-m}) + S = S$$
$$\Rightarrow \lambda_1 b_1 + \cdots + \lambda_{n-m} b_{n-m} \in S \quad [\because S \text{ is additive subgroup of } (V, +) \therefore S + u = S \Leftrightarrow u \in S]$$
$$\Rightarrow \lambda_1 b_1 + \cdots + \lambda_{n-m} b_{n-m} \text{ is a linear combination of } B_1 = \{v_1, \ldots, v_m\} [\because B_1 \text{ is a basis for } S]$$
$$\Rightarrow \lambda_1 b_1 + \cdots + \lambda_{n-m} b_{n-m} = \mu_1 v_1 + \cdots + \mu_m v_m \quad \text{for some } \mu_i \in F$$
$$\Rightarrow \lambda_1 b_1 + \cdots + \lambda_{n-m} b_{n-m} + (-\mu_1) v_1 + \cdots + (-\mu_m) v_m = 0_V$$
$$\Rightarrow \lambda_1 = \lambda_2 = \cdots = \lambda_{n-m} = 0 = \mu_1 = \cdots = \mu_m \quad [\because B_2 \text{ is linearly independent}]$$

Thus, only the trivial linear combination of vectors of B equals to the null vector of V/S.

Hence, B is a linearly independent set in V/S.

B spans V/S: Let $u + S$ be an arbitrary vector in V/S. Then,

$$u \in V$$
$$\Rightarrow u \text{ is a linear combination of } B_2 \quad [\because B_2 \text{ is a basis for } V]$$
$$\Rightarrow \text{there exist } \lambda_1, \ldots, \lambda_m, \mu_1, \ldots, \mu_{n-m} \text{ in } F \text{ such that}$$

$$u = \lambda_1 v_1 + \cdots + \lambda_m v_m + \mu_1 b_1 + \cdots + \mu_{n-m} b_{n-m}$$
$$\Rightarrow u = v + \mu_1 b_1 + \cdots + \mu_{n-m} b_{n-m}, \quad \text{where } v = \lambda_1 v_1 + \cdots + \lambda_m v_m \in S \, [\because B_1 \text{ is a basis for } S]$$
$$\Rightarrow u = (\mu_1 b_1 + \cdots + \mu_{n-m} b_{n-m}) + v \quad [\because \text{ vector addition is commutative on } V]$$
$$\Rightarrow u + S = (\mu_1 b_1 + \cdots + \mu_{n-m} b_{n-m}) + v + S$$
$$\Rightarrow u + S = (\mu_1 b_1 + \cdots + \mu_{n-m} b_{n-m}) + S \quad [\because v \in S \quad \therefore v + S = S]$$
$$\Rightarrow u + S = (\mu_1 b_1 + S) + \cdots + (\mu_{n-m} b_{n-m} + S)$$
$$\Rightarrow u + S = \mu_1(b_1 + S) + \cdots + \mu_{n-m}(b_{n-m} + S)$$
$$\Rightarrow u + S \text{ is a linear combination of } B.$$

Since $u + S$ is an arbitrary vector in V/S. Therefore, every vector in V/S is a linear combination of B. Consequently, B spans V/S,

Hence, B is a basis for V/S.

$\therefore \dim(V/S) =$ Number of vectors in $B = n - m = \dim V - \dim S$.

Q.E.D.

CO-DIMENSION. *Let $V(F)$ be a finite dimensional vector space and let S be a subspace of V. Then codimension of S is defined to be the dimension of the quotient space V/S.*

Clearly, Codimension $S = \dim V - \dim S$. [By Theorem 12]

MAXIMAL LINEARLY INDEPENDENT SET. *Let $V(F)$ be a vector space. Then a linearly independent set of vectors in V is said to be a maximal linearly independent set if every set of vectors in V containing it as a proper subset is a linearly dependent set.*

THEOREM-13 *If S is a maximal linearly independent set of vectors in a finite dimensional vector space $V(F)$, then S spans V, i.e. $[S] = V$.*

PROOF. Since V is a finite dimensional vector space. Therefore, S is a finite set. Let $S = \{v_1, \ldots, v_n\}$, and let $v = \sum_{i=1}^{n} \lambda_i v_i$ be an arbitrary vector in V. Since S is a maximal linearly independent set. Therefore, the larger set $S' = \{v_1, \ldots, v_n, v\}$ must be linearly dependent. Hence, we can find scalars $\mu_1, \mu_2, \ldots, \mu_n, \mu$ not all zero such that

$$\mu_1 v_1 + \cdots + \mu_n v_n + \mu v = 0_V$$

We claim that $\mu \neq 0$, because

$\mu = 0$

$\Rightarrow \quad \mu_1 v_1 + \cdots + \mu_n v_n = 0_V$ where at least one $\mu_i \neq 0$.

$\Rightarrow \quad S$ is a linearly dependent

$\Rightarrow \quad$ a contradiction.

Now, $\mu_1 v_1 + \cdots + \mu_n v_n + \mu v = 0_V$

$\Rightarrow \quad \mu v = -\sum_{i=1}^{n} \mu_i v_i$

$\Rightarrow \quad v = \sum_{i=1}^{n} (\mu_i \mu^{-1}) v_i \qquad\qquad [\because \mu \neq 0 \quad \therefore \quad \mu^{-1} \in F]$

$\Rightarrow \quad v$ is a linear combination of S

Hence, S spans V.

Q.E.D.

THEOREM-14 *Let $V(F)$ be a finite dimensional vector space. Then a set of n linearly independent vectors in V is maximal iff $n = \dim V$.*

PROOF. Let $S = \{v_1, \ldots, v_n\}$ be a set of n linearly independent vectors in V.

First suppose that S is a maximal linearly independent set. Then, by the previous theorem, S spans V and hence it is a basis for V.

Since S contains n vectors. Therefore, $\dim V = n$.

Conversely, suppose that S is a linearly independent set of n vectors in V such that $n = \dim V$.

Since $n = \dim V$. Therefore, every set of $(n+1)$ or more vectors is a linearly dependent set, i.e. every proper superset of S is linearly dependent. Hence, S is a maximal linearly independent set.

Q.E.D.

THEOREM-15 *Let $V(F)$ be a finite dimensional vector space, and let $B = \{b_1, \ldots, b_n\}$ be a non-void subset of V. Then, B is a basis for V iff it is a maximal linearly independent set of vectors in V.*

PROOF. First, let B be a basis for V. Then B is a linearly independent set of vectors in V. In order to show that B is a maximal linearly independent set, it is sufficient to show that every subset of V containing $n+1$ or more vectors in V is linearly dependent. This is evident from Theorem 6(i) on page 198. Hence, B is a maximal linearly independent set.

Conversely, let B be a maximal linearly independent set. Then by Theorem 13, B spans V. Consequently, B is a basis for V.

Q.E.D.

THEOREM-16 *Let $V(F)$ be a finite dimensional vector space and let S_1, S_2 be subspaces of V such that $V = S_1 + S_2$ and $\dim V = \dim S_1 + \dim S_2$. Then, $V = S_1 \oplus S_2$.*

PROOF. Let $\dim S_1 = m$ and $\dim S_2 = n$. Then, $\dim V = \dim S_1 + \dim S_2 = m + n$.

Let $B_1 = \{v_1, \ldots, v_m\}$ and $B_2 = \{u_1, \ldots, u_n\}$ be bases for S_1 and S_2 respectively.

First we shall show that $B_1 \cup B_2$ is a basis for V.

$B_1 \cup B_2$ spans V: Let v be an arbitrary vector in V. Then,

$$V = S_1 + S_2 \Rightarrow v \in S_1 + S_2 \Rightarrow v = s_1 + s_2 \quad \text{for some } s_1 \in S_1 \text{ and } s_2 \in S_2.$$

Now, $\quad s_1 \in S_1 \Rightarrow s_1$ is a linear combination of vectors in B_1 $\quad [\because B_1$ is a basis for $S_1]$

$s_2 \in S_2 \Rightarrow s_2$ is a linear combination of vectors in B_2 $\quad [\because B_2$ is a basis for $S_2]$

$\therefore \quad v = s_1 + s_2$ is a linear combination of vectors in $B_1 \cup B_2$.

Since v is an arbitrary vector in V. Therefore, every vector in V is a linear combination of vectors in $B_1 \cup B_2$.

Hence, $B_1 \cup B_2$ spans V i.e., $V = [B_1 \cup B_2]$

$B_1 \cup B_2$ is l.i set in V: Since $\dim V = m + n$ and $[B_1 \cup B_2] = V$. Therefore, by Theorem 6(i), the number of distinct vectors in $B_1 \cup B_2$ cannot be less than $m + n$. Thus, $B_1 \cup B_2$ is a spanning set of $m + n$ vectors in V and $\dim V = m + n$. Therefore, by corollary 1 to Theorem 7(page 198), $B_1 \cup B_2$ is a basis for V. Hence, $B_1 \cup B_2$ is a linearly independent set in V.

Hence, $B_1 \cup B_2$ is a basis for V.

Now we shall show that $V = S_1 \oplus S_2$. For this, it is sufficient to show that $S_1 \cap S_2 = \{0_V\}$. Let $v \in S_1 \cap S_2$. Then, $v \in S_1$ and $v \in S_2$.

Since $B_1 = \{v_1, \ldots, v_m\}$ is a basis for S_1. Therefore, there exist scalars $\lambda_1, \ldots, \lambda_m \in F$ such that
$$v = \lambda_1 v_1 + \cdots + \lambda_m v_m.$$

Since $B_2 = \{u_1, \ldots, u_n\}$ is a basis for S_2 and $v \in S_2$. Therefore, there exist scalars $\mu_1, \mu_2, \ldots, \mu_n \in F$ such that

$$v = \mu_1 u_1 + \cdots + \mu_n u_n$$
$\therefore \quad \lambda_1 v_1 + \cdots + \lambda_m v_m = \mu_1 u_1 + \cdots + \mu_n u_n$
$\Rightarrow \quad \lambda_1 v_1 + \cdots + \lambda_m v_m + (-\mu_1) u_1 + \cdots + (-\mu_n) u_n = 0_V$
$\Rightarrow \quad \lambda_1 = \cdots = \lambda_m = 0 = \mu_1 = \cdots = \mu_n \qquad [\because \ B_1 \cup B_2 \text{ is a basis for } V]$
$\Rightarrow \quad v = 0 v_1 + \cdots + 0 v_m = 0_V$

Thus, $\quad v \in S_1 \cap S_2 \Rightarrow v = 0_V$
$\therefore \quad S_1 \cap S_2 = \{0_V\}$
Hence, $V = S_1 \oplus S_2$

Q.E.D

THEOREM-17 *Let S and T be subspaces of a finite dimensional vector space V such that $V = S \oplus T$. Let B_1 and B_2 be bases for S and T respectively. Then $B_1 \cup B_2$ is basis for V and $\dim V = \dim S + \dim T$.*

PROOF. Since V is a finite dimensional vector space over a field F, therefore, S and T are also finite dimensional vector spaces over field F. Let $\dim S = m$ and $\dim T = n$. Further let $B_1 = \{b_1, b_2, \ldots, b_m\}$ be a basis for S and $B_2 = \{b_{m+1}, b_{m+2}, \ldots, b_{m+n}\}$ be a basis for T. Let $B = B_1 \cup B_2 = \{b_1, b_2, \ldots, b_m, b_{m+1}, \ldots, b_{m+n}\}$.

We shall now establish that B is a basis for V.

B is linearly independent: Let $\lambda_1, \lambda_2, \ldots, \lambda_m, \lambda_{m+1}, \ldots, \lambda_{m+n}$ be scalars in F such that

$$\lambda_1 b_1 + \lambda_2 b_2 + \cdots + \lambda_m b_m + \lambda_{m+1} b_{m+1} + \cdots + \lambda_{m+n} b_{m+n} = 0_V$$
$\Rightarrow \quad \lambda_1 b_1 + \cdots + \lambda_m b_m = -(\lambda_{m+1} b_{m+1} + \cdots + \lambda_{m+n} b_{m+n}) \qquad (i)$
$\Rightarrow \quad \lambda_1 b_1 + \cdots + \lambda_m b_m \in T \qquad \left[\because \ -\sum_{i=1}^{n} \lambda_{m+i} b_{m+i} \in T\right]$

$$\Rightarrow \quad \sum_{i=1}^{m} \lambda_i b_i \in S \cap T \qquad \left[\because \sum_{i=1}^{m} \lambda_i b_i \in S\right]$$

$$\Rightarrow \quad \sum_{i=1}^{m} \lambda_i b_i = 0_V$$

$$\Rightarrow \quad \lambda_1 = \lambda_2 = \cdots = \lambda_m = 0 \qquad [\because B_1 \text{ is a basis for } S]$$

$$\Rightarrow \quad \sum_{i=1}^{n} \lambda_{m+i} b_{m+i} = 0_V \qquad [\text{From (i)}]$$

$$\Rightarrow \quad \lambda_{m+1} = \lambda_{m+2} = \cdots = \lambda_{m+n} = 0 \qquad [\because B_2 \text{ is a basis for } T]$$

$$\Rightarrow \quad \lambda_1 = \lambda_2 = \cdots = \lambda_m = \lambda_{m+1} = \cdots = \lambda_{m+n} = 0.$$

So, B is a l.i. set.

B spans V: Let v be an arbitrary vector of V. Then,

$$v \in S \oplus T \qquad [\because V = S \oplus T]$$

$$\Rightarrow \quad v = u + w \quad \text{for some } u \in S, w \in T$$

$$\Rightarrow \quad v = \sum_{i=1}^{m} \lambda_i b_i + \sum_{i=m+1}^{m+n} \lambda_i b_i \text{ for some } \lambda_i \in F$$

$\Rightarrow$ v is a linear combination of vectors in B. [$\because B_1$ and B_2 are bases for S and T respectively]

Thus, B spans V.

Hence, B is a basis for V. Consequently $\dim V = m + n = \dim S + \dim T$. Q.E.D.

THEOREM-18 Let V_1 and V_2 be finite dimensional vector spaces over a field F. Then $V_1 \times V_2$ is also a finite dimensional vector space over field F and $\dim V_1 \times V_2 = \dim V_1 + \dim V_2$.

PROOF. Clearly, $V_1 \times V_2$ is a vector space over field F. [See Example 10 on page 124]

Let $V_1' = V_1 \times \{0_{V_1}\} = \{(v, 0_{V_1}) : v \in V_1\}$ and $V_2' = \{0_{V_2}\} \times V_2 = \{(0_{V_2}, v) : v \in V_2\}$. Then it can be easily seen that $V_1' \cong V_1$ and $V_2' \cong V_2$.

Let v be an arbitrary vector of $V_1 \times V_2$. Then,

$$v = (v_1, v_2) \quad \text{where } v_1 \in V_1, v_2 \in V_2$$

$$\Rightarrow \quad v = (v_1, 0_{V_2}) + (0_{V_1}, v_2) \qquad [\text{By definition of vector addition on } V_1 \times V_2]$$

$$\therefore \quad V_1 \times V_2 = V_1' + V_2'$$

Also, $V_1' \cap V_2' = \{(0_{V_1}, 0_{V_2})\}$

Thus, $V_1 \times V_2 = V_1' \oplus V_2'$

So, $\dim(V_1 \times V_2) = \dim(V_1' \oplus V_2')$

$\Rightarrow \dim(V_1 \times V_2) = \dim V_1' + \dim V_2'$ [By Theorem 10 on page 204]

$\Rightarrow \dim(V_1 \times V_2) = \dim V_1 + \dim V_2$ [$\because V_1' \cong V_1$ and $V_2' \cong V_2$]

Hence, $\dim(V_1 \times V_2) = \dim V_1 + \dim V_2$.

Q.E.D.

Particular case: If V is a finite dimensional vector space over a field F, then so is $V \times V$ and $\dim(V \times V) = 2 \dim V$.

Generalization: If $V_1, V_2, \ldots, V_n$ are finite dimensional vector spaces over a field F, then so is $V_1 \times V_2 \times \cdots \times V_n$, and

$$\dim(V_1 \times V_2 \times \cdots \times V_n) = \dim V_1 + \dim V_2 + \cdots + \dim V_n.$$

THEOREM-19 Let $V(F)$ be a finite dimensional vector space with $\dim V = n \geq 2$. If $B = \{b_1, b_2, \ldots, b_n\}$ is a basis of V and $S_i = \{\lambda b_i : \lambda \in F\}, i = 1, 2, \ldots, n$. Then, each S_i is a subspace of V such that $V = S_1 \oplus S_2 \oplus S_3 \oplus \cdots \oplus S_n$.

PROOF. Clearly, each S_i is one-dimension subspace of V. In fact, $S_i = [b_i]$.

Let v be an arbitrary vector in V. Then,

$v = \lambda_1 b_1 + \lambda_2 b_2 + \cdots + \lambda_n b_n$ for some $\lambda_i \in F$ [$\because$ B is a basis]

$\Rightarrow v \in S_1 + S_2 + \cdots + S_n$ [$\because S_i = \{\lambda b_i : \lambda \in F\}$]

In order to prove that each $v \in V$ is uniquely expressible as the sum of vectors in $S_1, S_2, \ldots, S_n$.

Let $v = \lambda_1' b_1 + \lambda_2' b_2 + \cdots + \lambda_n' b_n$ be another representation of v as the sum of vectors in $S_1, S_2, \ldots, S_n$. Then,

$\lambda_1 b_1 + \lambda_2 b_2 + \cdots + \lambda_n b_n = \lambda_1' b_1 + \lambda_2' b_2 + \cdots + \lambda_n' b_n$

$\Rightarrow \sum_{i=1}^{n} (\lambda_i - \lambda_i') b_i = 0_V$

$\Rightarrow \lambda_i - \lambda_i' = 0$; $i = 1, 2, \ldots, n$ [$\because$ B is linearly independent.]

$\Rightarrow \lambda_i = \lambda_i'$; $i = 1, 2, \ldots, n$

Hence, $V = S_1 \oplus S_2 \oplus \cdots \oplus S_n$

Q.E.D.

THEOREM-20 *A vector space $V(F)$ is the direct sum of its $n(\geq 2)$ subspaces $S_1, S_2, \ldots, S_n$ i.e. $V = S_1 \oplus S_2 \oplus S_3 \oplus \cdots \oplus S_n$ iff*

(i) $V = S_1 + S_2 + \cdots + S_n$ *(ii)* $S_i \cap \sum_{\substack{j=1 \\ j \neq i}}^{n} S_j = \{0_V\}$ *for each* $i = 1, 2, \ldots, n$

PROOF. First suppose that $V = S_1 \oplus S_2 \oplus \cdots \oplus S_n$. Then, $V = S_1 + S_2 + \cdots + S_n$

Let v be an arbitrary vector in $S_i \cap \sum_{\substack{j=1 \\ j \neq i}}^{n} S_j$. Then,

$$v \in S_i \cap \sum_{\substack{j=1 \\ j \neq i}}^{n} S_j \Rightarrow v \in S_i \text{ and } v \in \sum_{\substack{j=1 \\ j \neq i}}^{n} S_j$$

Now, $v \in S_i$

$\Rightarrow \quad v = v_i \quad$ for some $v_i \in S_i$

$\Rightarrow \quad v = 0_V + 0_V + \cdots + v_i + 0_V + \cdots + 0_V \qquad [\because \quad 0_V \in S_j \text{ for all } j]$

and, $v \in \sum_{\substack{j=1 \\ j \neq i}}^{n} S_j$

$\Rightarrow$ There exist vectors $v_j \in S_j$, $j = 1, 2, \ldots, n$ and $j \neq i$ such that

$$v = v_1 + v_2 + \cdots + v_{i-1} + v_{i+1} + \cdots + v_n$$

But, $V = S_1 \oplus S_2 \oplus \cdots \oplus S_n$. So, each vector in V has unique representation as the sum of vectors in $S_1, S_2, \ldots, S_n$.

$\therefore \quad v_1 = 0_V = v_2 = \cdots = v_{i-1} = v_i = v_{i+1} = \cdots = v_n$

$\Rightarrow \quad v = 0_V$

Since v is an arbitrary vector in $S_i \cap \sum_{\substack{j=1 \\ j \neq i}}^{n} S_j$.

$\therefore \quad S_i \cap \sum_{\substack{j=1 \\ j \neq i}}^{n} S_j = \{0_V\}$

Conversely, let $S_1, S_2, \ldots, S_n$ be subspaces of $V(F)$ such that (i) and (ii) hold. We have to prove that V is the direct sum of its subspaces $S_1, S_2, \ldots, S_n$.

Let v be an arbitrary vector in V. Then,

$v \in S_1 + S_2 + \cdots + S_n \qquad\qquad\qquad\qquad\qquad [\because \quad V = S_1 + S_2 + \cdots + S_n]$

$\Rightarrow \quad v = v_1 + v_2 + \cdots + v_n, \quad$ where $v_i \in S_i$; $i = 1, 2, \ldots, n$

If possible, let
$$v = u_1 + u_2 + \cdots + u_n, \quad \text{where } u_i \in S_i$$
be another representation of v as the sum of vectors in $S_1, S_2, \ldots, S_n$.

$\therefore \quad v_1 + v_2 + \cdots + v_n = u_1 + u_2 + \cdots + u_n$

$\Rightarrow \quad \sum_{j=1}^{n} (v_j - u_j) = 0_V$

$\Rightarrow \quad \sum_{\substack{j=1 \\ j \neq i}}^{n} (v_j - u_j) = u_i - v_i \qquad (i)$

Since $S_1, S_2, \ldots, S_n$ are subspaces of V.

$\therefore \quad v_i, u_i \in S_i \Rightarrow v_i - u_i \in S_i, i = 1, 2, \ldots, n$

$\therefore \quad \sum_{\substack{j=1 \\ j \neq i}}^{n} (v_j - u_j) \in \sum_{\substack{j=1 \\ j \neq i}}^{n} S_j$ and $u_i - v_i \in S_i$

$\Rightarrow \quad u_i - v_i \in \sum_{\substack{j=1 \\ j \neq i}}^{n} S_j$ and $u_i - v_i \in S_i \qquad \left[\text{From (i), we have } \sum_{\substack{j=1 \\ j \neq i}}^{n} (v_j - u_j) = \sum_{\substack{j=1 \\ j \neq i}}^{n} u_i - v_i \right]$

$\Rightarrow \quad u_i - v_i \in S_i \cap \sum_{\substack{j=1 \\ j \neq i}}^{n} S_j$ for all $i = 1, 2, \ldots, n$

$\Rightarrow \quad u_i - v_i = 0_V \qquad \text{for all } i \qquad \left[\because S_i \cap \sum_{\substack{j=1 \\ j \neq i}}^{n} S_j = \{0_V\} \right]$

$\Rightarrow \quad u_i = v_i \qquad \text{for all } i$

Thus, each vector in V has a unique representation as the sum of vectors in S_i; $i = 1, 2, \ldots, n$.

Hence, $V = S_1 \oplus S_2 \oplus S_3 \oplus \cdots \oplus S_n$

Q.E.D.

THEOREM-21 *Let $V(F)$ be a finite dimensional vector space, and let $S_1, S_2, \ldots, S_k$ be subspaces of V. Then the following statements are equivalent:*

(i) $V = S_1 \oplus S_2 \oplus \cdots \oplus S_k$, i.e. V is direct sum of $S_1, S_2, \ldots, S_k$.

(ii) If B_i is a basis for S_i, $i = 1, 2, \ldots, k$, then $B = \bigcup_{i=1}^{k} B_i$ is a basis for V.

PROOF. For any $i \in \underline{k}$, let $B_i = \{b_1^i, \ldots, b_{n_i}^i\}$ be a basis for S_i, where $\dim B_i = n_i$. Let $B = \bigcup_{i=1}^{k} B_i$.

(i) $\Rightarrow$ (ii)

Let $V = S_1 \oplus S_2 \oplus \cdots \oplus S_k$.

B spans V: Let v be an arbitrary vector in V. Since V is the direct sum of $S_1,\ldots,S_k$. Therefore, there exist unique $v_1 \in S_1,\ldots,v_k \in S_k$ such that
$$v = v_1 + \cdots + v_k.$$

Since for each $i \in \underline{k}$, B_i is a basis for S_i. Therefore, each v_i can be expressed as a linear combination of vectors in B_i. Consequently, v is expressible as a linear combination of vectors in $\bigcup_{i=1}^{k} B_i = B$. Hence, B spans V.

B is a l.i. set in V: To show that B is a linearly independent set, let $\lambda_1^1,\ldots,\lambda_{n_1}^1,\lambda_1^2,\ldots,\lambda_{n_2}^2,\ldots,\lambda_1^i,\ldots,\lambda_{n_i}^i,\ldots,\lambda_1^k,\ldots,\lambda_{n_k}^k$ be scalars in F such that

$$(\lambda_1^1 b_1^1 + \cdots + \lambda_{n_1}^1 b_{n_1}^1) + \cdots + (\lambda_1^i b_1^i + \cdots + \lambda_{n_i}^i b_{n_i}^i) + \cdots + (\lambda_1^k b_1^k + \cdots + \lambda_{n_k}^k b_{n_k}^k) = 0_V$$
$$\Rightarrow (\lambda_1^1 b_1^1 + \cdots + \lambda_{n_1}^1 b_{n_1}^1) + \cdots + (\lambda_1^k b_1^k + \cdots + \lambda_{n_k}^k b_{n_k}^k) = \underbrace{0_V + 0_V + \cdots + 0_V}_{k-\text{terms}} \quad \text{(i)}$$

Since V is the direct sum of $S_1,\ldots,S_k$. Therefore, each vector in V is uniquely expressible as a sum of vectors one in each S_i. Consequently, (i) implies that

$$\lambda_1^i b_1^i + \cdots + \lambda_{n_i}^i b_{n_i}^i = 0_V \quad \text{for all } i = 1,2,\ldots,k$$
$$\Rightarrow \lambda_1^i = \cdots = \lambda_{n_i}^i = 0 \quad \text{for all } i = 1,2,\ldots,k \quad [\because B_i \text{ is a basis for } S_i]$$

Hence, B is a linearly independent set in V.

Thus, B is a basis for V.

(ii) $\Rightarrow$ (i).

Let $B = \bigcup_{i=1}^{k} B_i$ be a basis for V. Then we have to show that V is the direct sum of $S_1, S_2,\ldots,S_k$.

Let v be an arbitrary vector in V. Then,

$$v = \sum_{i=1}^{k} (\lambda_1^i b_1^i + \cdots + \lambda_{n_i}^i b_{n_i}^i) \qquad [\because B = \bigcup_{i=1}^{k} B_i \text{ is a basis for } V]$$
$$\Rightarrow v = \sum_{i=1}^{k} v_i, \text{ where } v_i = \lambda_1^i b_1^i + \cdots + \lambda_{n_i}^i b_{n_i}^i \in S_i \qquad [\because B_i \text{ is a basis for } S_i]$$

Thus, v is expressible as a sum of vectors one in each S_i. Since v is an arbitrary vector in V. Therefore, each vector in V is expressible as a sum of vectors one in each S_i. Consequently, $V = S_1 + S_2 + \cdots + \cdots + S_k$.

Now we shall show that the expression of each vector in V as a sum of vectors one in each S_i is unique. If possible, let $v \in V$ be such that

$$v = \sum_{i=1}^{k} (\lambda_1^i b_1^i + \cdots + \lambda_{n_i}^i b_{n_i}^i) \text{ and } v = \sum_{i=1}^{k} (\mu_1^i b_1^i + \cdots + \mu_{n_i}^i b_{n_i}^i).$$

Then,
$$\sum_{i=1}^{k}(\lambda_1^i b_1^i + \cdots + \lambda_{n_i}^i b_{n_i}^i) = \sum_{i=1}^{k}(\mu_1^i b_1^i + \cdots + \mu_{n_i}^i b_{n_i}^i)$$

$\Rightarrow \quad \sum_{i=1}^{k}\{(\lambda_1^i - \mu_1^i)b_1^i + (\lambda_2^i - \mu_2^i)b_2^i + \cdots + (\lambda_{n_i}^i - \mu_{n_i}^i)b_{n_i}^i\} = 0_V$

$\Rightarrow \quad \lambda_1^i - \mu_1^i = 0, \lambda_2^i - \mu_2^i = 0, \ldots, \lambda_{n_i}^i - \mu_{n_i}^i = 0 \quad \text{for } i = 1, 2, \ldots, k \quad [\because \ B = \bigcup_{i=1}^{k} B_i \text{ is l.i.}]$

$\Rightarrow \quad \lambda_1^i = \mu_1^i, \ldots, \lambda_{n_i}^i = \mu_{n_i}^i \quad \text{for } i = 1, 2, \ldots, k$

Thus, v is uniquely expressible as a sum of vectors one in each S_i. Therefore, every vector in V is expressible as a sum of vectors one in each S_i.

Hence, V is the direct sum of $S_1, S_2, \ldots, S_k$.

Q.E.D.

THEOREM-22 Let S and T be two subspaces of a vector space $V(F)$ such that $S \cap T = \{0_V\}$. If B_1 and B_2 be bases of S and T respectively, then show that $B = B_1 \cup B_2$ is a basis of $S + T$ and $B_1 \cap B_2 = \varphi$.

PROOF. Let $B_1 = \{b_1, b_2, \ldots, b_m\}$ and $B_2 = \{b_1', b_2', \ldots, b_n'\}$ be bases of S and T respectively. Then, B_1 and B_2 are linearly independent subsets of S and T respectively. Therefore, $B_1 \cap B_2$ is also linearly independent and $B_1 \cap B_2 \subseteq S \cap T$.

$\therefore \quad B_1 \cap B_2 \subseteq \{0_V\}$

$\Rightarrow \quad B_1 \cap B_2 = \varphi \qquad [\because \ B_1 \text{ and } B_2 \text{ are l.i. and any l.i. set cannot have null vector in it}]$

We shall now show that $B = B_1 \cup B_2$ is a basis for $S + T$.

$B = B_1 \cup B_2$ is a l.i. subset of $S + T$: Clearly, S and T are subsets of $S + T$. So, $S \cup T \subseteq S + T$.

$\therefore \quad B_1 \cup B_2 \subseteq S \cup T \subseteq S + T$.

Let $\lambda_i (i = 1, 2, \ldots, m)$ and $\mu_j (j = 1, 2, \ldots, n)$ be scalars in F such that

$$\sum_{i=1}^{m} \lambda_i b_i + \sum_{j=1}^{n} \mu_j b_j' = 0_V$$

$\Rightarrow \quad \sum_{i=1}^{m} \lambda_i b_i = -\sum_{j=1}^{n} \mu_j b_j' = u \text{ (say)}$

Clearly, $\sum_{i=1}^{m} \lambda_i b_i \in S$ and $-\sum_{j=1}^{n} \mu_j b_j' \in T$

$\Rightarrow \quad u \in S \text{ and } u \in T$

$\Rightarrow \quad u \in S \cap T$

$\Rightarrow \quad u = 0_V \qquad\qquad\qquad\qquad\qquad\qquad\qquad [\because \ S \cap T = \{0_V\}]$

$\therefore \quad \sum_{i=1}^{m} \lambda_i b_i = 0_V \text{ and, } -\sum_{j=1}^{n} \mu_j b_j' = 0_V$

$\Rightarrow \quad \lambda_i = 0 (i = 1, 2, \ldots, m), \mu_j = 0 (j = 1, 2, \ldots, n) [\because \ B_1 \text{ and } B_2 \text{ are linearly independent}]$

∴ $B_1 \cup B_2$ is a linearly independent subset of $S+T$.

$B_1 \cup B_2$ *spans* $S+T$: Let v be an arbitrary vector in $S+T$. Then there exist $v \in S$ and $w \in T$ such that

$$u = v + w$$

$\Rightarrow \quad u = \sum_{i=1}^{m} \lambda_i b_i + \sum_{j=1}^{n} \mu_j b'_j \qquad [\because B_1 \text{ and } B_2 \text{ are bases of } S \text{ and } T \text{ respectively}]$

$\Rightarrow \quad u$ is a linear combination of vectors in $B_1 \cup B_2$.

Since u is an arbitrary vector in $S+T$. Therefore, every vector in $S+T$ is a linear combination of vectors in $B_1 \cup B_2$. So, $B_1 \cup B_2$ spans $S+T$.

Hence, $B_1 \cup B_2$ is a basis of $S+T$.

Q.E.D.

THEOREM-23 Let $S_1, S_2, \ldots, S_k (k \geq 1)$ be subspaces of a finitely generated vector space $V(F)$ such that $V = S_1 \oplus S_2 \oplus S_3 \oplus \cdots \oplus S_k$. Further, let $B_i (i = 1, 2, \ldots, k)$ be a basis of S_i. Then,

(i) $B_1, B_2, \ldots, B_k$ are pairwise disjoint and $B = \bigcup_{i=1}^{k} B_i$ is a basis of V.

(ii) $\dim V = \sum_{i=1}^{k} \dim S_i$

PROOF. We shall prove the theorem by using the principle of mathematical induction on k.

For $k = 1$, we have

$$V = S_1 \text{ and } B = B_1$$

So, the result is obvious.

Let $k > 1$ and let the result holds for internal direct sum of $(k-1)$ subspaces of $V(F)$.

We have,

$$V = S_1 \oplus S_2 \oplus S_3 \cdots \oplus S_k$$

$\Rightarrow \quad V = S_1 + S_2 + \cdots + S_k$

$\Rightarrow \quad V = S_1 + \sum_{i=2}^{k} S_i$

$\Rightarrow \quad V = S_1 + S, \quad \text{where } S = \sum_{i=2}^{k} S_i$

Let v be an arbitrary vector in V. Then,

$\quad v = S_1 + S$

$\Rightarrow \quad v \in S_1$ and $v \in S$

$\Rightarrow \quad v = v_1 \quad$ for some $v_1 \in S_1$ and, $v = v_2 + v_3 + \cdots + v_k$ for some $v_i \in S_i (i = 2, \ldots, k)$

$\Rightarrow \quad v = v_1 + 0_V + 0_V + \cdots + 0_V$ and, $v = 0_V + v_2 + v_3 + \cdots + v_k$

But, $V = S_1 \oplus S_2 \oplus S_3 \oplus \cdots \oplus S_k$. So, v must have unique representation as the sum of vectors in $S_1, S_2, \ldots, S_k$.

$\therefore \quad v = v_1 + 0_V + 0_V + \cdots + 0_V$ and $v = 0_V + v_2 + v_3 + \cdots + v_k$

$\Rightarrow \quad v = 0_V = v_2 = \cdots = v_k$

$\Rightarrow \quad v = 0_V$

$\therefore \quad S_1 \cap S_2 = \{0_V\}$

$\Rightarrow \quad V = S_1 \oplus S$ $\hfill [\because \quad v = S_1 + S]$

$\therefore \quad B = B_1 \cup B'$ is a basis for V for any basis B' of S_2 $\hfill$ [See Theorem 17]
Also, $B_1 \cap B' = \varphi$.

$\therefore \quad \dim V = \dim S_1 + \dim S \qquad \ldots \text{(i)}$

Since $S = S_2 \oplus S_3 \oplus \cdots \oplus S_K$ is the internal direct product of $k-1$ subspaces. So, by the induction hypothesis

$$\dim S = \sum_{i=2}^{k} \dim(S_i) \qquad \ldots \text{(ii)}$$

and, $\quad B' = \bigcup_{i=2}^{k} B_i$ is a basis of S.

Also, $B_2, B_3, \ldots, B_k$ are pairwise disjoint.

$\therefore \quad B = B_1 \cup B' = \bigcup_{i=1}^{k} B_i$ is a basis of V such that $B_1, B_2, \ldots, B_k$ are pairwise disjoint.

From (i) and (ii), we get

$$\dim V = \dim S_1 + \sum_{i=2}^{k} \dim S_i \Rightarrow \dim V = \sum_{i=1}^{k} \dim S_i$$

Q.E.D.

ILLUSTRATIVE EXAMPLES

TYPE I **ON CHECKING WHETHER A GIVEN SET OF VECTORS FORMS A BASIS OR NOT**

EXAMPLE-1 *Determine whether or not each of the following sets form a basis of $R^3(R)$:*

(i) $B_1 = \{(1,1,1),(1,0,1)\}$

(ii) $B_2 = \{(1,1,1),(1,2,3),(2,-1,1)\}$

(iii) $B_3 = \{(1,2,3),(1,3,5),(1,0,1),(2,3,0)\}$

(iv) $B_4 = \{(1,1,2),(1,2,5),(5,3,4)\}$

SOLUTION (i) Since $\dim R^3 = 3$. So, a basis of $R^3(R)$ must contain exactly 3 elements. Hence, B_1 is not a basis of $R^3(R)$.

(ii) Since $\dim R^3 = 3$. So, B_2 will form a basis of $R^3(R)$ if and only if vectors in B_2 are linearly independent. To check this, let us form the matrix A whose rows are the given vectors as given below.

$$A = \begin{bmatrix} 1 & 1 & 1 \\ 1 & 2 & 3 \\ 2 & -1 & 1 \end{bmatrix}$$

Since non-zero rows of a matrix in echelon form are linearly independent. So, let us reduce A to echelon form.

$$A \sim \begin{bmatrix} 1 & 1 & 1 \\ 0 & 1 & 2 \\ 0 & -3 & -1 \end{bmatrix} \quad \text{Applying } R_2 \to R_2 - R_1, R_3 \to R_3 - 2R_1$$

$$\Rightarrow \quad A \sim \begin{bmatrix} 1 & 1 & 1 \\ 0 & 1 & 2 \\ 0 & 0 & 5 \end{bmatrix} \quad \text{Applying } R_3 \to R_3 + 3R_2$$

Clearly, the echelon form of A has no zero rows. Hence, the three vectors are linearly independent and so they form a basis of R^3.

(iii) Since $(n+1)$ or more vectors in a vector space of dimension n are linearly dependent. So, B_3 is a linearly dependent set of vectors in $R^3(R)$. Hence, it cannot be a basis of R^3.

(iv) The matrix A whose rows are the vectors in B_4 is given by

$$A = \begin{bmatrix} 1 & 1 & 2 \\ 1 & 2 & 5 \\ 5 & 3 & 4 \end{bmatrix}$$

Since non-zero rows of a matrix in echelon form are linearly independent. So, let us reduce A to echelon form.

$$A \sim \begin{bmatrix} 1 & 1 & 2 \\ 0 & 1 & 3 \\ 0 & -2 & -6 \end{bmatrix} \quad \text{Applying } R_2 \to R_2 - R_1, R_3 \to R_3 - 5R_1$$

$$\Rightarrow \quad A \sim \begin{bmatrix} 1 & 1 & 2 \\ 0 & 1 & 3 \\ 0 & 0 & 0 \end{bmatrix} \quad \text{Applying } R_3 \to R_3 + 2R_2$$

The echelon form of A has a zero row, hence the given vectors are linearly dependent and so B_4 does not form a basis of R^3.

EXAMPLE-2 *Let V be the vector space of all 2×2 matrices over a field F. Prove that V has dimension 4 by finding a basis for V.*

SOLUTION Let $E_{11} = \begin{bmatrix} 1 & 0 \\ 0 & 0 \end{bmatrix}, E_{12} = \begin{bmatrix} 0 & 1 \\ 0 & 0 \end{bmatrix}, E_{21} = \begin{bmatrix} 0 & 0 \\ 1 & 0 \end{bmatrix}$ and $E_{22} = \begin{bmatrix} 0 & 0 \\ 0 & 1 \end{bmatrix}$ be four matrices in V, where 1 is the unity element in F.

Let $B = \{E_{11}, E_{12}, E_{21}, E_{22}\}$.

We shall now show that B forms a basis of V.

B is l.i.: Let x, y, z, t be scalars in F such that

$$xE_{11} + yE_{12} + zE_{21} + tE_{22} = 0$$

$\Rightarrow \begin{bmatrix} x & y \\ z & t \end{bmatrix} = \begin{bmatrix} 0 & 0 \\ 0 & 0 \end{bmatrix} \Rightarrow x = y = z = t = 0$

So, B is l.i.

B spans v: Let $A = \begin{bmatrix} a & b \\ c & d \end{bmatrix}$ be an arbitrary matrix in V. Then,

$$A = aE_{11} + bE_{12} + cE_{21} + dE_{22}$$

$\Rightarrow$ A is expressible as a linear combination of matrices in B.

So, B spans V. Thus, B is a basis of V.

Hence, $\dim V = 4$.

EXAMPLE-3 Let $v_1 = (1, i, 0), v_2 = (2i, 1, 1), v_3 = (0, 1+i, 1-i)$ be three vectors in $C^3(C)$. Show that the set $B = \{v_1, v_2, v_3\}$ is a basis of $C^3(C)$.

SOLUTION We know that $\dim C^3(C) = 3$. Therefore, B will be a basis of $C^3(C)$ if B is a linearly independent set. Let $x, y, z \in C$ such that

$xv_1 + yv_2 + zv_3 = 0$

$\Rightarrow$ $(x + 2iy, xi + y + z(1+i), y + z(1-i)) = (0, 0, 0)$

$\Rightarrow$ $x + 2iy = 0, xi + y + z(1+i) = 0, y + z(1-i) = 0$

$\Rightarrow$ $x = y = z = 0$.

Hence, B is a basis of $C^3(C)$.

EXAMPLE-4 Determine whether $(1,1,1,1), (1,2,3,2), (2,5,6,4), (2,6,8,5)$ form a basis of $R^4(R)$. If not, find the dimension of the subspace they span.

SOLUTION Given four vectors can form a basis of $R^4(R)$ iff they are linearly independent as the dimension of R^4 is 4.

The matrix A having given vectors as its rows is

$$A = \begin{bmatrix} 1 & 1 & 1 & 1 \\ 1 & 2 & 3 & 2 \\ 2 & 5 & 6 & 4 \\ 2 & 6 & 8 & 5 \end{bmatrix}$$

Since non-zero rows of a matrix in echelon form are linearly independent. So, let us reduce A to echelon form

$$A \sim \begin{bmatrix} 1 & 1 & 1 & 1 \\ 0 & 1 & 2 & 1 \\ 0 & 3 & 4 & 2 \\ 0 & 4 & 6 & 3 \end{bmatrix} \quad \text{Applying } R_2 \to R_2 - R_1, R_3 \to R_3 - 2R_2, R_4 \to R_4 - 2R_1$$

$$\Rightarrow \quad A \sim \begin{bmatrix} 1 & 1 & 1 & 1 \\ 0 & 1 & 2 & 1 \\ 0 & 0 & -2 & -1 \\ 0 & 0 & -2 & -1 \end{bmatrix} \quad \text{Applying } R_3 \to R_3 - 3R_2, R_4 \to R_4 - 4R_3$$

$$\Rightarrow \quad A \sim \begin{bmatrix} 1 & 1 & 1 & 1 \\ 0 & 1 & 2 & 1 \\ 0 & 0 & -2 & -1 \\ 0 & 0 & 0 & 0 \end{bmatrix} \quad \text{Applying } R_4 \to R_4 - R_3$$

The echelon matrix has a zero row. So, given vectors are linearly independent and do not form a basis of R^4. Since the echelon matrix has three non-zero rows, so the four vectors span a subspace of dimension 3.

TYPE II ON EXTENDING A GIVEN SET TO FORM A BASIS OF A GIVEN VECTOR SPACE

EXAMPLE-5 *Extend the set* $\{u_1 = (1,1,1,1), u_2 = (2,2,3,4)\}$ *to a basis of* R^4.

SOLUTION Let us first form a matrix A with rows u_1 and u_2, and reduce it to echelon form:

$$A = \begin{bmatrix} 1 & 1 & 1 & 1 \\ 2 & 2 & 3 & 4 \end{bmatrix}$$

$$\Rightarrow \quad A \sim \begin{bmatrix} 1 & 1 & 1 & 1 \\ 0 & 0 & 1 & 2 \end{bmatrix}$$

We observe that the vectors $v_1 = (1,1,1,1)$ and $v_2 = (0,0,1,2)$ span the same space as spanned by the given vectors u_1 and u_2. In order to extend the given set of vectors to a basis of $R^4(R)$, we need two more vectors u_3 and u_4 such that the set of four vectors v_1, v_2, u_3, u_4 is linearly independent. For this, we chose u_3 and u_4 in such a way that the matrix having v_1, v_2, u_3, u_4 as its rows is in echelon form. Thus, if we chose $u_3 = (0, a, 0, 0)$ and $u_4 = (0, 0, 0, b)$, where a, b are non-zero real numbers, then v_1, u_3, v_2, u_4 in the same order form a matrix in echelon form. Thus, they are linearly independent, and they form a basis of R^4. Hence, u_1, u_2, u_3, u_4 also form a basis of R^4.

EXAMPLE-6 *Let* $v_1 = (-1, 1, 0), v_2 = (0, 1, 0)$ *be two vectors in* $R^3(R)$ *and let* $S = \{v_1, v_2\}$. *Extend set S to a basis of* $R^3(R)$.

SOLUTION Let A be the matrix having v_1 and v_2 as its two rows. Then,

$$A = \begin{bmatrix} -1 & 1 & 0 \\ 0 & 1 & 0 \end{bmatrix}$$

Clearly, A is in echelon form. In order to form a basis of $R^3(R)$, we need one more vector such that the matrix having that vector as third row and v_1, v_2 as first and second rows is in echelon form. If we take $v_3 = (0, 0, a)$, where $a (\neq 0) \in R$, then matrix having its three rows as v_1, v_2, v_3 is in echelon form. Thus, v_1, v_2, v_3 are linearly independent and they form a basis of $R^3(R)$.

REMARK. *Sometimes, we are given a list of vectors in the vector space $R^n(R)$ and we want to find a basis for the subspace S of R^n spanned by the given vectors, that is, a basis of $[S]$. The following two algorithms help us for finding such a basis of $[S]$.*

ALGORITHM 1 *(Row space algorithm)*

Step I   Form the matrix A whose rows are the given vectors
Step II Reduce A to echelon form by elementary row operations.
Step III Take the non-zero rows of the echelon form. These rows form a basis of the subspace spanned by the given set of vectors.
         In order to find a basis consisting of vectors from the original list of vectors, we use the following algorithm.

ALGORITHM 2 *(Casting-out algorithm)*

Step I   Form the matrix A whose columns are the given vectors.
Step II Reduce A to echelon form by elementary row operations.
Step III Delete (cast out) those vectors from the given list which correspond to columns without pivots and select the remaining vectors in S which correspond to columns with pivots. Vectors so selected form a basis of $[S]$.

TYPE III ON FINDING THE DIMENSION OF SUBSPACE SPANNED BY A GIVEN SET OF VECTORS

EXAMPLE-7 *Let S be the set consisting of the following vectors in R^5:*

$v_1 = (1, 2, 1, 3, 2), v_2 = (1, 3, 3, 5, 3), v_3 = (3, 8, 7, 13, 8), v_4 = (1, 4, 6, 9, 7), v_5 = (5, 13, 13, 25, 19)$

Find a basis of $[S]$ (i.e the subspace spanned by S) consisting of the original given vectors. Also, find the dimension of $[S]$.

SOLUTION Let A be the matrix whose columns are the given vectors. Then,

$$A = \begin{bmatrix} 1 & 1 & 3 & 1 & 5 \\ 2 & 3 & 8 & 4 & 13 \\ 1 & 3 & 7 & 6 & 13 \\ 3 & 5 & 13 & 9 & 25 \\ 2 & 3 & 8 & 7 & 19 \end{bmatrix}$$

Let us now reduce A to echelon form by using elementary row operations.

$$A \sim \begin{bmatrix} 1 & 1 & 3 & 1 & 5 \\ 0 & 1 & 2 & 2 & 3 \\ 0 & 2 & 4 & 5 & 8 \\ 0 & 2 & 4 & 6 & 10 \\ 0 & 1 & 2 & 5 & 9 \end{bmatrix}$$ Applying $R_2 \to R_2 - 2R_1, R_3 \to R_3 - R_1,$
$R_4 \to R_4 - 3R_1$ and $R_5 \to R_5 - R_1$

$\Rightarrow$ $A \sim \begin{bmatrix} 1 & 1 & 3 & 1 & 5 \\ 0 & 1 & 2 & 2 & 3 \\ 0 & 0 & 0 & 1 & 2 \\ 0 & 0 & 2 & 4 \\ 0 & 0 & 3 & 6 \end{bmatrix}$ Applying $R_3 \to R_3 - 2R_2, R_4 \to R_4 - 2R_2, R_5 \to R_5 - R_1$

$\Rightarrow$ $A \sim \begin{bmatrix} ① & 1 & 3 & 1 & 5 \\ 0 & ① & 2 & 2 & 3 \\ 0 & 0 & 0 & ① & 2 \\ 0 & 0 & 0 & 0 & 0 \\ 0 & 0 & 0 & 0 & 0 \end{bmatrix}$ Applying $R_4 \to R_4 - 2R_3, R_5 \to R_5 - 2R_3$

We observe that pivots (encircled entries) in the echelon form of A appear in the columns C_1, C_2, C_4. So, we "cast out" the vectors u_3 and u_5 from set S and the remaining vectors u_1, u_2, u_4, which correspond to the columns in the echelon matrix with pivots, form a basis of $[S]$. Hence, $\dim[S] = 3$.

EXAMPLE-8 *Let S be the set consisting of following vectors in R^4:*

$$v_1 = (1, -2, 5, -3), v_2 = (2, 3, 1, -4), v_3 = (3, 8, -3, -5).$$

(i) Find a basis and dimension of the subspace spanned by S, i.e. $[S]$.
(ii) Extend the basis of $[S]$ to a basis of R^4.
(iii) Find a basis of $[S]$ consisting of the original given vectors.

SOLUTION (i) Let A be the matrix whose rows are the given vectors. Then,

$$A = \begin{bmatrix} 1 & -2 & 5 & -3 \\ 2 & 3 & 1 & -4 \\ 3 & 8 & -3 & -5 \end{bmatrix}$$

Let us now reduce A to echelon form. The row reduced echelon form of A is as given below.

$$A \sim \begin{bmatrix} 1 & -2 & 5 & -3 \\ 0 & 7 & -9 & 2 \\ 0 & 0 & 0 & 0 \end{bmatrix}$$

The non-zero rows $(1,-2,5,-3)$ and $(0,7,-9,2)$ of the above echelon form of A form a basis of $[S]$. Hence, $\dim[S] = 2$

(ii) In order to extend the basis of $[S]$ to the basis of R^4, we need two more vectors which form a linearly independent set with the vectors in the basis of $[S]$. We observe that the vectors in the basis of $[S]$ with the vectors $(0,0,a,0)$ and $(0,0,0,b)$ (a,b are non-zero real numbers) form an echelon matrix and so they are linearly independent. Hence, vectors $(1,-2,5,-3), (0,7,-9,2), (0,0,a,0), (0,0,0,b)$ is an extension of the basis of $[S]$.

(iii) In order to find a basis of $[S]$ that consists of original given vectors, let us form a matrix A whose columns are the given vector as given below.

$$A = \begin{bmatrix} 1 & 2 & 3 \\ -2 & 3 & 8 \\ 5 & 1 & -3 \\ -3 & -4 & -5 \end{bmatrix}$$

$$\Rightarrow A \sim \begin{bmatrix} 1 & 2 & 3 \\ 0 & 7 & 14 \\ 0 & -9 & -18 \\ 0 & 2 & 4 \end{bmatrix} \quad \text{Applying } R_2 \to R_2 + 2R_1, R_3 \to R_3 - 5R_1, R_4 \to R_4 + 3R_1$$

$$\Rightarrow A \sim \begin{bmatrix} ① & 2 & 3 \\ 0 & ⑦ & 14 \\ 0 & 0 & 0 \\ 0 & 0 & 0 \end{bmatrix} \quad \text{Applying } R_3 \to R_3 + \frac{9}{7}R_2, R_4 \to R_4 + \left(\frac{-2}{7}\right)R_2$$

We observe that pivots in the echelon form of A appear in the first two columns. The corresponding vectors in the sets, i.e. $(1,-2,5,-3), (2,3,1,-4)$ form a basis of $[S]$

EXAMPLE-9 *Let $R^{2\times 2}$ be the vector space of all 2×2 matrices over a field R and let S be the subspace of symmetric matrix. Find a basis of S and also find $\dim S$.*

SOLUTION We have,

$$S = \left\{ \begin{bmatrix} a & b \\ b & d \end{bmatrix} : a, b \in R \right\}$$

Clearly, $E_1 = \begin{bmatrix} 1 & 0 \\ 0 & 0 \end{bmatrix}, E_2 = \begin{bmatrix} 0 & 1 \\ 1 & 0 \end{bmatrix}$ and $E_3 = \begin{bmatrix} 0 & 0 \\ 0 & 1 \end{bmatrix}$ are members of S.

Let $B = \{E_1, E_2, E_3\}$. We shall now show that B is a basis of S.

B spans S: Let $A = \begin{bmatrix} a & b \\ b & d \end{bmatrix}$ be any matrix in S. Then,

$$A = aE_1 + bE_2 + dE_3$$

Thus, any symmetric matrix in S is expressible as a linear combination of matrices in B.
So, B spans S.

B is linearly independent: Let x,y,z be scalars such that
$$xE_1 + yE_2 + zE_3 = O$$
$$\Rightarrow x\begin{bmatrix} 1 & 0 \\ 0 & 0 \end{bmatrix} + y\begin{bmatrix} 0 & 1 \\ 1 & 0 \end{bmatrix} + z\begin{bmatrix} 0 & 0 \\ 0 & 1 \end{bmatrix} = \begin{bmatrix} 0 & 0 \\ 0 & 0 \end{bmatrix}$$
$$\Rightarrow \begin{bmatrix} x & y \\ y & z \end{bmatrix} = \begin{bmatrix} 0 & 0 \\ 0 & 0 \end{bmatrix}$$
$$\Rightarrow x = y = z = 0$$

Thus, B is linearly independent.

Hence, B forms a basis of S.

Clearly, $\dim S =$ Number of matrices in $B = 3$.

EXAMPLE-10 *Find a basis and dimension of the subspace S of $R^3(R)$, where*
  (i) $S = \{(a,b,c) : a+b+c = 0\}$ (ii) $S = \{(a,a,a) : a \in R\}$

SOLUTION (i) We observe that $(1,1,1) \in R^3$ but $(1,1,1) \notin S$ as $1+1+1 \neq 0$. So, $S \neq R^3$
$\therefore \quad \dim S < \dim R^3 \Rightarrow \dim S < 3$

Clearly, $v_1 = (1,0,-1)$ and $v_2 = (0,-1,1)$ are two independent vectors in S.

$\therefore \quad \dim S \geq 2$

Hence, $\dim S = 2$.

(ii) We have $S = \{(a,a,a) : a \in R\}$

$\because \quad (a,a,a) = a(1,1,1) = au$, where $u = (1,1,1)$

$\therefore \quad u = (1,1,1)$ spans S.

Hence, $\dim S = 1$.

EXAMPLE-11 *Let $S = \{u_1, u_2, u_3\}$ and $T = \{v_1, v_2, v_3\}$ be subsets of R^3, where*
$$u_1 = (1,1,-1), u_2 = (2,3,-1), u_3 = (3,1,-5)$$
$$v_1 = (1,-1,-3), v_2 = (3,-2,-8), v_3 = (2,1,-3).$$

Show that $[S] = [T]$, i.e. S and T span the same subspace of R^3.

SOLUTION Let A denote the matrix whose rows are u_1, u_2, u_3. Then,
$$A = \begin{bmatrix} 1 & 1 & -1 \\ 2 & 3 & -1 \\ 3 & 1 & -5 \end{bmatrix}$$
$$\Rightarrow A \sim \begin{bmatrix} 1 & 1 & -1 \\ 0 & 1 & 1 \\ 0 & -2 & -2 \end{bmatrix} \quad \text{Applying } R_2 \to R_2 - 2R_1, R_3 \to R_3 - 3R_1$$
$$\Rightarrow A \sim \begin{bmatrix} 1 & 1 & -1 \\ 0 & 1 & 1 \\ 0 & 0 & 0 \end{bmatrix}$$

Let B denote the matrix whose rows are v_1, v_2, v_3. Then, $B = \begin{bmatrix} 1 & -1 & -3 \\ 3 & -2 & -8 \\ 2 & 1 & -3 \end{bmatrix}$

$\Rightarrow B = \begin{bmatrix} 1 & -1 & -3 \\ 3 & -2 & -8 \\ 2 & 1 & -3 \end{bmatrix}$

$\Rightarrow B \sim \begin{bmatrix} 1 & -1 & -3 \\ 0 & 1 & 1 \\ 0 & 3 & 3 \end{bmatrix}$ Applying $R_2 \to R_2 - 3R_1, R_3 \to R_3 - 3R_1$

$\Rightarrow B \sim \begin{bmatrix} 1 & -1 & -3 \\ 0 & 1 & 1 \\ 0 & 0 & 0 \end{bmatrix}$ Applying $R_3 \to R_3 - 3R_2$

$\Rightarrow B \sim \begin{bmatrix} 1 & 1 & -1 \\ 0 & 1 & 1 \\ 0 & 0 & 0 \end{bmatrix}$ Applying $R_1 \to R_1 + 2R_2$

Clearly, A and B have the same row canonical form. So, row spaces of A and B are equal. Hence, $[S] = [T]$.

EXERCISE 2.9

1. Mark each of the following true or false.
    (i) The vectors in a basis of a vector space are linearly dependent.
    (ii) The null (zero) vector may be a part of a basis.
    (iii) Every vector space has a basis.
    (iv) Every vector space has a finite basis.
    (v) A basis cannot have the null vector.
    (vi) If two bases of a vector space have one common vector, then the two bases are the same.
    (vii) A basis for $R^3(R)$ can be extended to a basis for $R^4(R)$.
    (viii) Any two bases of a finite dimensional vector space have the same number of vectors.
    (ix) Every set of $n+1$ vectors in an n-dimensional vector space is linearly dependent.
    (x) Every set of $n+1$ vectors in an n-dimensional vector space is linearly independent.
    (xi) An n-dimensional vector space can be spanned by a set $n-1$ vectors in it.
    (xii) Every set of n linearly independent vectors in an n-dimensional vector space is a basis.

(xiii) A spanning set of n vectors in an n-dimensional vector space is a basis.

(xiv) $[v]_B$ is independent of B.

(xv) If S and T are subspaces of a vector space $V(F)$, then
$\dim S < \dim T \Rightarrow S \subsetneq T$.

2. Find three bases for the vector space $R^2(R)$ such that no two of which have a vector in common.

3. Determine a basis for each of the following vector spaces:
   (i) $R^2(R)$
   (ii) $R(\sqrt{2}) = \{a + b\sqrt{2} : a, b \in R\}$ over R
   (iii) $Q(\sqrt{2}) = \{a + b\sqrt{2} : a, b \in Q\}$ over Q
   (iv) C over R
   (v) C over itself
   (vi) $Q(i) = \{a + ib : a, b \in Q\}$ over Q

4. Which of the following subsets B form a basis for $R^3(R)$?
   (i) $B = \{(1,0,0), (1,1,0), (1,1,1)\}$
   (ii) $B = \{(0,0,1), (1,0,1)(1,-1,1), (3,0,1)\}$
   (iii) $B = \{(1,2,3), (3,1,0), (-2,1,3)\}$
   (iv) $B = \{(1,1,1), (1,2,3), (-1,0,1)\}$
   (v) $B = \{(1,1,2), (1,2,5), (5,3,4)\}$

5. Which of the following subsets B form a basis for the given vector space V?
   (i) $B = \{(1,0), (i,0), (0,1), (0,i)\}, V = C^2(R)$.
   (ii) $B = \{(1,i,1+i), (1,i,1-i), (i,-i,1)\}, V = C^3(C)$.
   (iii) $B = \{(1,1,1,1), (2,6,4,5), (1,2,1,2), (0,3,2,3)\}, V = R^4(R)$.
   (iv) $B = \{1, \sin x, \sin^2 x, \cos^2 x\}, V = C[-\pi, \pi]$
   (v) $B = \{x - 1, x^2 + x - 1, x^2 - x + 1\}, V = $ Space of all real polynomials of degree at most two.

6. Extend the set $S = \{(1,1,1,1), (1,2,1,2)\}$ to a basis for $R^4(R)$.

7. Extend the set $S = \{(3, -1, 2)\}$ to two different bases for $R^3(R)$.

8. Consider the set S of vectors in $R^3(R)$ consisting of $v_1 = (1,1,0), v_2 = (0,2,3)$, $v_3 = (1,2,3), v_4 = (1,-2,3)$.
   (i) Prove that S is linearly dependent.
   (ii) Find a subset B of S such that B is a basis of $[S]$.
   (iii) Find $\dim[S]$.

9. Let S be the subspace of R^5 spanned by the vectors. $v_1 = (1, 2, -1, 3, 4), v_2 = (2, 4, -2, 6, 8), v_3 = (1, 3, 2, 2, 6), v_4 = (1, 4, 5, 1, 8), v_5 = (2, 7, 3, 3, 9)$. Find a subset of these vectors that form a basis of S.

10. Prove that the set $S = \{a + 2ai : a \in R\}$ is a subspace of the vector space $C(R)$. Find a basis and dimension of S. Let $T = \{a - 2ai : a \in R\}$. Prove that $C(R) = S \oplus T$.

11. Extend each of the following subsets of R^3 to a basis of $R^3(R)$:
    (i) $B_1 = \{(1, 2, 0)\}$ (ii) $B_2 = \{(1, -2, 0), (0, 1, 1)\}$

12. Find a basis of the vector space $R^4(R)$ that contains the vectors $(1, 1, 0, 0), (0, 1, 0, -1)$.

13. Find four matrices A_1, A_2, A_3, A_4 in $R^{2 \times 2}(R)$ such that $A_i^2 = A_i$ for $1 \leq i \leq 4$ and $\{A_1, A_2, A_3, A_4\}$ is a basis of $R^{2 \times 2}$.

14. Let S be the set of all matrices A in $R^{2 \times 2}$ which commute with the matrix $\begin{bmatrix} 1 & 1 \\ 2 & 2 \end{bmatrix}$. Prove that S is a subspace of $R^{2 \times 2}$. Find $\dim S$.

15. Let $V = R^{3 \times 3}$, S be the set of symmetric matrices and T be the set of skew-symmetric matrices in V. Prove that S and T are subspaces of V such that $V = S \oplus T$. Find dimensions of S and T.

16. Express $R^5(R)$ as a direct sum of two subspaces S and T with $\dim S = 2, \dim T = 3$.

17. If S and T be two subspaces of a finite dimensional vector space $V(F)$ such that
$$\dim V = \dim S + \dim T \text{ and } S \cap T = \{0_V\}.$$
Prove that $V = S \oplus T$.

18. Let $V(F)$ be a finite dimensional vector space, and let $S_1, S_2, \ldots, S_k$ be subspaces of V such that
$$V = S_1 + S_2 + \cdots + S_k \text{ and } \dim V = \dim S_1 + \dim S_2 + \cdots + \dim S_k.$$
Prove that $V = S_1 \oplus S_2 \oplus \cdots \oplus S_k$.

19. Show by means of an example that under certain circumstances, it is possible to find three subspaces S_1, S_2, S_3 of a finite dimensional vector space $V(F)$ such that $V = S_1 \oplus S_2 = S_2 \oplus S_3 = S_3 \oplus S_1$. What does this imply about the dimension of V?

20. Give an example of an infinite dimensional vector space $V(F)$ with a subspace S such that V/S is a finite dimensional vector space.

21. Construct three subspaces S_1, S_2, S_3 of a vector space V such that $V = S_1 \oplus S_2 = S_1 \oplus S_3$ but $S_2 \neq S_3$.

22. Give an example of a vector space $V(R)$ having three different non-zero subspaces S_1, S_2, S_3 such that
$$V = S_1 \oplus S_2 = S_2 \oplus S_3 = S_3 \oplus S_1.$$

23. Let $S = \{(a, b, 0) : a, b \in R\}$ be a subspace of R^3. Find its two different complements in R^3.

24. Determine a basis of the subspace spanned by the vectors:

    $v_1 = (1, 2, 3)$, $v_2 = (2, 1, -1)$, $v_3 = (1, -1, -4)$, $v_4 = (4, 2, -2)$

25. Show that the vectors $v_1 = (1, 0, -1)$, $v_2 = (1, 2, 1)$, $v_3 = (0, -3, 2)$ form a basis for R^3. Express each of the standard basis vectors $e_1^{(3)}$, $e_2^{(3)}$, $e_3^{(3)}$ as a linear combination of v_1, v_2, v_3.

ANSWERS

1. (i) F (ii) F (iii) T (iv) F (v) T (vi) F (vii) F (viii) T (ix) T (x) F (xi) F (xii) T (xiii) T (xiv) F (xv) F

2. $B_1 = \{(1,0),(0,1)\}, B_2 = \{(1,-1),(2,3)\}, B_3 = \{(-2,1),(2,-3)\}$.
   Vectors (a_1, b_1) and (a_2, b_2) form a basis of $R^2(R)$ iff $a_1 b_2 \neq a_2 b_1$.

3. (i) $B = \{(1,0),(0,1)\}$ (ii) $B = \{1, \sqrt{2}\}$ (iii) $B = \{1, \sqrt{2}\}$ (iv) $B = \{1, i\}$
   (v) $B = \{1, i\}$ (vi) $B = \{1, i\}$

4. (i)

5. (ii), (iii)

6. $\{(1,1,1,1),(1,2,1,2),(0,0,1,0),(0,0,0,1)\}$

7. $\{(3,-1,2),(0,1,0),(0,0,1)\}, \{(3,-1,2),(0,-1,0),(0,0,2)\}$

8. (ii) $B = \{(1,1,0),(0,2,3)\}$ (ii) $\dim[S] = 2$

9. $\{v_1, v_3, v_5\}$

10. $\{1 + 2i\}$, 1

11. (i) $\{(1,2,0),(0,1,0),(0,0,1)\}$ (ii) $\{(1,-2,0),(0,1,1),(0,0,1)\}$

12. $\{(1,1,0,0),(0,1,0,-1),(0,0,1,0),(0,0,0,1)\}$

13. $A_1 = \begin{bmatrix} 1 & 0 \\ 0 & 0 \end{bmatrix}, A_2 = \begin{bmatrix} 0 & 0 \\ 0 & 1 \end{bmatrix}, A_3 = \begin{bmatrix} 1 & 1 \\ 0 & 0 \end{bmatrix}, A_4 = \begin{bmatrix} 0 & 0 \\ 1 & 1 \end{bmatrix}$

15. $\dim S = \dim T = 3$

16. $S = \{(a,b,0,0,0) : a, b \in R\}, T = \{(0,0,a,b,c) : a,b,c \in R\}$

19. $\dim V$ is an even integer.

21. $V = R^2(R), S_1 = \{(a,0) : a \in R\}, S_2 = \{(0,a) : a \in R\}$ and $S_3 = \{(a,a) : a \in R\}$

23. (i) $S_1 = \{(0,0,a) : a \in R\}$ (ii) $\{(a,a,a) : a \in R\}$

24. $\{v_1, v_2\}$.

25. $e_1^{(3)} = \frac{7}{10}v_1 + \frac{3}{10}v_2 + \frac{1}{5}v_3$, $e_2^{(3)} = -\frac{1}{5}v_1 + \frac{1}{5}v_2 - \frac{1}{5}v_3$, $e_3^{(3)} = \frac{-3}{10}v_1 + \frac{3}{10}v_2 + \frac{1}{5}v_3$

2.10 COORDINATES

ORDERED BASIS. *Let $V(F)$ be a finite dimensional vector space of dimension n, and let $B = \{b_1, \ldots, b_n\}$ be a basis for V. Then a list of length n (ordered n-tuple) of vectors in B is called an ordered basis for V.*

Note that in an ordered basis, the order of arrangements of vectors is very important. If there are n vectors in a basis of a vector space V, then corresponding to various arrangements of vectors we can obtain $n!$ ordered bases for V. For example, $(e_1^{(3)}, e_2^{(3)}, e_3^{(3)})$, $(e_1^{(3)}, e_3^{(3)}, e_2^{(3)})$, $(e_3^{(3)}, e_1^{(3)}, e_2^{(3)})\ldots$ etc. are ordered bases for the real vector space R^3, corresponding to a basis $B = \{e_1^{(3)}, e_2^{(3)}, e_3^{(3)}\}$.

The concept of ordered basis is very useful in the study of matrices.

Let $(b_1, b_2, \ldots, b_n)$ be an ordered basis for a finite dimensional vector space $V(F)$, and let $v \in V$. Then by Theorem 2 on page 194 there exists a unique n-tuple $(\lambda_1, \lambda_2, \ldots, \lambda_n)$ of scalars in F such that

$$v = \lambda_1 b_1 + \lambda_2 b_2 + \cdots + \lambda_n b_n.$$

The n-tuple $(\lambda_1, \lambda_2, \ldots, \lambda_n)$ is called the n-tuple of *coordinates* of V relative to the ordered basis B and the scalar λ_i is called *ith coordinate* of v relative to the ordered basis $(b_1, b_2, \ldots, b_n)$.

The column matrix

$$\begin{bmatrix} \lambda_1 \\ \lambda_2 \\ \vdots \\ \lambda_n \end{bmatrix}_{n \times 1}$$

is called the *coordinate matrix* coordinate vector of v relative to the ordered basis $B = (b_1, b_2, \ldots, b_n)$ and is denoted by $[v]_B$.

ILLUSTRATIVE EXAMPLES

EXAMPLE-1 *Relative to the basis $B = \{v_1, v_2\} = \{(1,1), (2,3)\}$ of R^2, find the coordinate matrix of (i) $v = (4, -3)$ (ii) (a, b).*

SOLUTION Let $x, y \in R$ such that

$$v = xv_1 + yv_2 = x(1,1) + y(2,3) = (x + 2y, x + 3y)$$

(i) If $v = (4, -3)$, then
$$v = xv_1 + yv_2$$
$$\Rightarrow \quad (4, -3) = (x + 2y, x + 3y)$$
$$\Rightarrow \quad x + 2y = 4, x + 3y = -3$$
$$\Rightarrow \quad x = 18, y = -7$$

Hence, coordinate matrix $[v]_B$ of v relative to the basis B is given by $[v_B] = \begin{bmatrix} 18 \\ -7 \end{bmatrix}$

(ii) If $v = (a,b)$, then
$$v = xv_1 + yv_2$$
$\Rightarrow \quad (a,b) = (x+2y, x+3y)$
$\Rightarrow \quad x+2y = a, x+3y = b$
$\Rightarrow \quad x = 3a - 2b, y = -a+b$

Hence, the coordinate matrix $[v]_B$ of v relative to the basis B is given by $[v_B] = \begin{bmatrix} 3a-2b \\ -a+b \end{bmatrix}$

EXAMPLE-2 *Find the coordinate vector of* $v = (1,1,1)$ *relative to the basis* $B = \{v_1 = (1,2,3), v_2 = (-4,5,6), v_3 = (7,-8,9)\}$ *of vector space* R^3.

SOLUTION Let $x, y, z \in R$ be such that
$$v = xv_1 + yv_2 + zv_3$$
$\Rightarrow \quad (1,1,1) = x(1,2,3) + y(-4,5,6) + z(7,-8,9)$
$\Rightarrow \quad (1,1,1) = (x - 4y + 7z, 2x + 5y - 8z, 3x + 6y + 9z)$
$\Rightarrow \quad x - 4y + 7z = 1, 2x + 5y - 8z = 1, 3x + 6y + 9z = 1$
$\Rightarrow \quad x = \dfrac{7}{10}, y = \dfrac{-2}{15}, z = \dfrac{-1}{30}$

Hence $[v]_B = \begin{bmatrix} \dfrac{7}{10} \\ \dfrac{-2}{15} \\ \dfrac{-1}{30} \end{bmatrix}$ is the coordinate vector of v relative to basis B.

EXAMPLE-3 *Find the coordinates of the vector* (a,b,c) *in the real vector space* R^3 *relative to the ordered basis* (b_1, b_2, b_3), *where* $b_1 = (1,0,-1), b_2 = (1,1,1), b_3 = (1,0,0)$.

SOLUTION Let $\lambda, \mu, \gamma \in R$ be such that
$$(a,b,c) = \lambda(1,0,-1) + \mu(1,1,1) + \gamma(1,0,0).$$
$\Rightarrow \quad (a,b,c) = (\lambda + \mu + \gamma, \mu, -\lambda + \mu)$
$\Rightarrow \quad \lambda + \mu + \gamma = a, \mu = b, -\lambda + \mu = c$
$\Rightarrow \quad \lambda = b - c, \mu = b, \gamma = a - 2b + c$.

Hence, the coordinates of vector $(a,b,c) \in R^3$ relative to the given ordered basis are $(b-c, b, a-2b+c)$.

EXAMPLE-4 Let V be the vector space of all real polynomials of degree less than or equal to two. For a fixed $c \in R$, let $f_1(x) = 1, f_2(x) = x+c, f_3(x) = (x+c)^2$. Obtain the coordinates of $c_0 + c_1 x + c_2 x^2$ relative to the ordered basis (f_1, f_2, f_3) of V.

SOLUTION   Let $\lambda, \mu, \gamma \in R$ be such that
$$c_0 + c_1 x + c_2 x^2 = \lambda f_1(x) + \mu f_2(x) + \gamma f_3(x).$$

Then,
$$c_0 + c_1 x + c_2 x^2 = (\lambda + \mu c + \gamma c^2) + (\mu + 2\gamma c)x + \gamma x^2$$
$\Rightarrow \quad \lambda + \mu c + \gamma c^2 = c_0, \ \mu + 2\gamma c = c_1, \ \gamma = c_2$ [Using equality of two polynomials]
$\Rightarrow \quad \lambda = c_0 - c_1 c + c_2 c^2, \ \mu = c_1 - 2c_2 c, \gamma = c_2$.

Hence, the coordinates of $c_0 + c_1 x + c_2 x^2$ relative to the ordered basis (f_1, f_2, f_3) are $(c_0 - c_1 c + c_2 c^2, c_1 - 2c_2 c, c_2)$.

EXAMPLE-5 Consider the vector space $P_3[t]$ of polynomials of degree less than or equal to 3.
(i) Show that $B = \{(t-1)^3, (t-1)^2, (t-1), 1\}$ is a basis of $P_3[t]$.
(ii) Find the coordinate matrix of $f(t) = 3t^3 - 4t^2 + 2t - 5$ relative to basis B.

SOLUTION   (i) Consider the polynomials in B in the following order:
$$(t-1)^3, (t-1)^2, (t-1), 1$$

We see that no polynomial is a linear combination of preceding polynomials. So, the polynomials are linearly independent, and, since $\dim R_3[t] = 4$. Therefore, B is a basis of $P_3[t]$.

(ii) Let x, y, z, s be scalars such that
$$f(t) = x(t-1)^3 + y(t-1)^2 + z(t-1) + s(1)$$
$\Rightarrow \quad f(t) = xt^3 + (-3x+y)t^2 + (3x-2y+z)t + (-x+y-z+s)$
$\Rightarrow \quad 3t^3 - 4t^2 + 2t - 5 = xt^3 + (-3x+y)t^2 + (3x-2y+z)t + (-x+y-z+s)$
$\Rightarrow \quad x = 3, -3x+y = -4, 3x-2y+z = 2, -x+y-z+s = -5$
$\Rightarrow \quad x = 3, y = 13, z = 19, s = 4$

Hence, the coordinate matrix of $f(t)$ relative to the given basis is $[f(t)]_B = \begin{bmatrix} 3 \\ 13 \\ 19 \\ 4 \end{bmatrix}$.

EXAMPLE-6 Let F be a field and n be a positive integer. Find the coordinates of the vector $(a_1, a_2, \ldots, a_n) \in F^n$ relative to the standard basis.

SOLUTION We know that the ordered set $(e_1^{(n)}, e_2^{(n)}, \ldots, e_n^{(n)})$ is the standard basis for $F^n(F)$, where $e_1^{(n)} = (1, 0, \ldots, 0), e_2^{(n)} = (0, 1, 0, \ldots, 0), \ldots, e_n^{(n)} = (0, 0, \ldots, 1)$ and 1 is unity in F.

Let $\lambda_1, \lambda_2, \ldots, \lambda_n$ be scalars in F such that

$$(a_1, a_2, \ldots, a_n) = \lambda_1 e_1^{(n)} + \lambda_2 e_2^{(n)} + \cdots + \lambda_n e_n^{(n)}.$$

Then,

$(a_1, a_2, \ldots, a_n) = \lambda_1(1, 0, \ldots, 0) + \lambda_2(0, 1, \ldots, 0) + \cdots + \lambda_n(0, 0, \ldots, 1)$

$\Rightarrow (a_1, a_2, \ldots, a_n) = (\lambda_1, 0, \ldots, 0) + (0, \lambda_2, \ldots, 0) + \cdots + (0, 0, \ldots, \lambda_n)$

$\Rightarrow (a_1, a_2, \ldots, a_n) = (\lambda_1, \lambda_2, \ldots, \lambda_n)$

$\Rightarrow \lambda_1 = a_1, \lambda_2 = a_2, \ldots, \lambda_n = a_n$.

Hence, the n-tuple $(a_1, a_2, \ldots, a_n)$ is the n-tuple of coordinates of itself.

EXAMPLE-7 Find the coordinate matrix of matrix $A = \begin{bmatrix} 2 & 3 \\ 4 & -7 \end{bmatrix}$ in the real vector space $R^{2 \times 2}(R)$ relative to the basis

(i) $B_1 = \left\{ E_{11} = \begin{bmatrix} 1 & 0 \\ 0 & 0 \end{bmatrix}, E_{12} = \begin{bmatrix} 0 & 1 \\ 0 & 0 \end{bmatrix}, E_{21} = \begin{bmatrix} 0 & 0 \\ 1 & 0 \end{bmatrix}, E_{22} = \begin{bmatrix} 0 & 0 \\ 0 & 1 \end{bmatrix} \right\}$

(ii) $B_2 = \left\{ P = \begin{bmatrix} 1 & 1 \\ 1 & 1 \end{bmatrix}, Q = \begin{bmatrix} 1 & -1 \\ 1 & 0 \end{bmatrix}, R = \begin{bmatrix} 1 & -1 \\ 0 & 0 \end{bmatrix}, S = \begin{bmatrix} 1 & 0 \\ 0 & 0 \end{bmatrix} \right\}$

SOLUTION (i) Let x, y, z, t be scalars such that

$$A = xE_{11} + yE_{12} + zE_{21} + tE_{22}$$

$\Rightarrow \begin{bmatrix} 2 & 3 \\ 4 & -7 \end{bmatrix} = \begin{bmatrix} x & y \\ z & t \end{bmatrix} \Rightarrow x = 2, y = 3, z = 4, t = -7$

Hence, the coordinate matrix of A relative to the usual basis B_1 is $[A]_{B_1} = \begin{bmatrix} 2 \\ 3 \\ 4 \\ -7 \end{bmatrix}$

(ii) Let x, y, z, t be scalars such that

$$A = xP + yQ + zR + tS$$

$\Rightarrow \begin{bmatrix} 2 & 3 \\ 4 & -7 \end{bmatrix} = x \begin{bmatrix} 1 & 1 \\ 1 & 1 \end{bmatrix} + y \begin{bmatrix} 1 & -1 \\ 1 & 0 \end{bmatrix} + z \begin{bmatrix} 1 & -1 \\ 0 & 0 \end{bmatrix} + t \begin{bmatrix} 1 & 0 \\ 0 & 0 \end{bmatrix}$

$$\Rightarrow \begin{bmatrix} 2 & 3 \\ 4 & -7 \end{bmatrix} = \begin{bmatrix} x+y+z+t & x-y-z \\ x+y & x \end{bmatrix}$$

$\Rightarrow \quad x+z+t = 2, x-y-z = 3, x+y = 4, x = -7$

$\Rightarrow \quad x = -7, y = 11, z = -21, t = 30$

Hence, the coordinate matrix of A relative to the basis B_2 is $[A]_{B_2} = \begin{bmatrix} 7 \\ 11 \\ -21 \\ 30 \end{bmatrix}$.

EXERCISE 2.10

1. Find the coordinate matrices of vectors $u = (5,3)$ and $v = (a,b)$ relative to the ordered basis $B = \{(1,-2), (4,-7)\}$ of vector space $R^2(R)$.

2. The vectors $v_1 = (1,2,0), v_2 = (1,3,2), v_3 = (0,1,3)$ form a basis of $R^3(R)$. Find the coordinate matrix of vector v relative to the ordered basis $B = \{v_1, v_2, v_3\}$, where
   (i) $v = (2,7,-4)$ (ii) $v = (a,b,c)$

3. $B = \{t^3 + t^2, t^2 + t, t + 1, 1\}$ is an ordered basis of vector space $P_3[t]$. Find the coordinate matrix $f(t)$ relative to B where:
   (i) $f(t) = 2t^3 + t^2 - 4t + 2$ (ii) $f(t) = at^3 + bt^2 + ct + d$

4. $B = \left\{ \begin{bmatrix} 1 & 1 \\ 1 & 1 \end{bmatrix}, \begin{bmatrix} 1 & -1 \\ 1 & 0 \end{bmatrix}, \begin{bmatrix} 1 & 1 \\ 0 & 0 \end{bmatrix}, \begin{bmatrix} 1 & 0 \\ 0 & 0 \end{bmatrix} \right\}$ is an ordered basis of vector space $R^{2 \times 2}(R)$.
   Find the coordinate matrix of matrix A relative to B where:
   (i) $A = \begin{bmatrix} 3 & -5 \\ 6 & 7 \end{bmatrix}$ (ii) $A = \begin{bmatrix} a & b \\ c & d \end{bmatrix}$

5. Consider the vector space $R_2(t)$ of all polynomials of degree less than or equal to 2. Show that the polynomials $f_1(t) = t - 1, f_2(t) = t - 1, f_3(t) = (t-1)^2$ form a basis of $P_2[t]$. Find the coordinate matrix of polynomial $f(t) = 2t^2 - 5t + 9$ relative to the ordered basis $B = \{f_1(t), f_2(t), f_3(t)\}$.

6. Find the coordinate vector of $v = (a,b,c)$ in $R^3(R)$ relative to the ordered basis $B = \{v_1, v_2, v_3\}$, where $v_1 = (1,1,1), v_2 = (1,1,0), v_3 = (1,0,0)$.

ANSWERS

1. $[u]_B = \begin{bmatrix} -41 \\ 11 \end{bmatrix}, [v]_B = \begin{bmatrix} -7a-4b \\ 2a+b \end{bmatrix}$

2. (i) $[v]_B = \begin{bmatrix} -11 \\ 13 \\ -10 \end{bmatrix}$ (ii) $[v]_B = \begin{bmatrix} 7a-3b+c \\ -6a+3b-c \\ 4a-2b+c \end{bmatrix}$

3. (i) $[f(t)]_B = \begin{bmatrix} 2 \\ -1 \\ -2 \\ 2 \end{bmatrix}$ (ii) $[f(t)]_B = \begin{bmatrix} a \\ b-c \\ a-b+c \\ -a+b-c+d \end{bmatrix}$

4. (i) $[A]_B = \begin{bmatrix} d \\ c-d \\ b+c-2d \\ a-b-2c+2d \end{bmatrix}$ (ii) $[A]_B = \begin{bmatrix} 7 \\ -1 \\ -13 \\ 10 \end{bmatrix}$

5. $[f(t)]_B = \begin{bmatrix} 3 \\ -4 \\ 2 \end{bmatrix}$ 6. $[v]_B = \begin{bmatrix} c \\ b-c \\ a-b \end{bmatrix}$

Chapter 3

Linear Transformations

3.1 INTRODUCTION

Analogous to the concepts of group homomorphism, ring homomorphism and R-homomorphism (module homomorphism), let us introduce the concept of homomorphism among vector spaces. Homomorphisms among vector spaces are generally known as linear transformations, which help us to study the relationships between different vector spaces over the same field.

3.2 DEFINITIONS AND EXAMPLES

LINEAR TRANSFORMATION *Let V and V' be vector spaces over the same field F. A mapping $t : V \to V'$ is called a linear transformation or a vector space homomorphism or simply a linear map, if*

(i) $t(u+v) = t(u) + t(v)$

(ii) $t(au) = at(u)$

for all $u, v \in V$ and all $a \in F$.

Thus, $t : V \to V'$ is a linear transformation if it "preserves" the two basic operations of a vector space, that of vector addition and that of scalar multiplication.

MONOMORPHISM *Let V and V' be two vector spaces over the same field F. Then, a linear transformation $t : V \to V'$ is called a monomorphism, if t is one-one.*

EPIMORPHISM *Let V and V' be two vector spaces over the same field F. Then, a linear transformation $t : V \to V'$ is called an epimorphism, if t is onto.*

ISOMORPHISM *A linear transformation t from a vector space $V(F)$ to a vector space $V'(F)$ is an isomorphism if t is both one-one and onto, i.e. a bijection.*

In such a case, we say that V is isomorphic to V' and we write $V \cong V'$.

Thus, two vector spaces V and V' over the same field F are isomorphic if there exists a bijective linear transformation $t : V \to V'$.

Other definitions like automorphism, endomorphism, etc. are similar to what we have studied in modules.

A linear transformation from a vector space to itself is called a **linear operator**.

ILLUSTRATIVE EXAMPLES

EXAMPLE-1 *Let V be a vector space over a field F. Then the identity mapping I_V on V is a linear transformation. Moreover, it is an automorphism.*

SOLUTION   For any $u, v \in V$ and $a \in F$, we have
$$I_V(u+v) = u+v = I_V(u) + I_V(v),$$
and, $I_V(au) = au = aI_V(u).$

So, I_V is a linear transformation.
We know that $I_V : V \to V$ is bijective. Hence, I_V is an automorphism on V.

EXAMPLE-2 *Let V and V' be vector spaces over the same field F and let t be the mapping that assigns every vector $v \in V$ to the zero vector $0_{V'} \in V'$. Show that t is a linear transformation.*

SOLUTION   For any $u, v \in V$ and $a \in F$, we have
$$t(u+v) = 0_{V'} = 0_{V'} + 0_{V'} = t(u) + t(v)$$
and, $t(au) = 0_{V'} = a0_{V'} = at(u)$

Thus, $t : V \to V'$ is a linear transformation.

REMARK.   *The above transformation is called the zero linear transformation or zero linear mapping and is denoted by $\hat{0}$.*

EXAMPLE-3 *Let F be a field. Then the mapping $t : F^2 \to F^3$ given by*
$$t(a,b) = (a,b,0) \quad \text{for all } (a,b) \in F^2$$
is a linear transformation.

SOLUTION   Let $u = (a_1, b_1), v = (a_2, b_2) \in F^2$ and $\lambda \in F$. Then,
$$t(u+v) = t(a_1+a_2, b_1+b_2) = (a_1+a_2, b_1+b_2, 0)$$
$$\Rightarrow \quad t(u+v) = (a_1, b_1, 0) + (a_2, b_2, 0) = t(u) + t(v)$$
and, $t(\lambda u) = t(\lambda a_1, \lambda b_1) = (\lambda a_1, \lambda b_1, 0) = \lambda(a_1, b_1, 0) = \lambda t(u)$

Hence, $t : F^2 \to F^3$ is a linear transformation.

EXAMPLE-4 Let $t: R^3 \to R^3$ be the mapping defined by
$$t(x,y,z) = (x,y,0) \quad \text{for all } (x,y,z) \in R^3.$$
Show that t is a linear transformation.

SOLUTION For any $u = (a,b,c), v = (a',b',c') \in R^3$ and $\lambda \in R$, we have
$$t(u+v) = t(a+a', b+b', c+c') = (a+a', b+b', 0)$$
$$\Rightarrow \quad t(u+v) = (a,b,0) + (a',b',0) = t(u) + t(v)$$
and, $\quad t(\lambda u) = t(\lambda a, \lambda b, \lambda c) = (\lambda a, \lambda b, 0) = \lambda(a,b,0) = \lambda t(u)$

Thus, t is a linear transformation.
This linear transformation is known as the projection mapping in the xy-plane.

EXAMPLE-5 Let $V = R[x]$ be the vector space of all polynomials over field R and let $D: V \to V$ be the mapping associating each polynomial $f(x)$ to its derivative $\dfrac{d}{dx}(f(x))$. Show that D is a linear transformation.

SOLUTION Let $f(x)$ and $g(x)$ be any two polynomials in $R[x]$ and $a \in R$. Then,
$$D\big(f(x)+g(x)\big) = \frac{d}{dx}\big(f(x)+g(x)\big) = \frac{d}{dx}\big(f(x)\big) + \frac{d}{dx}\big(g(x)\big) = D\big(f(x)\big) + D\big(g(x)\big)$$
and, $\quad D\big(\lambda f(x)\big) = \dfrac{d}{dx}\big(\lambda f(x)\big) = \lambda \dfrac{d}{dx}\big(f(x)\big) = \lambda D\big(f(x)\big)$

Thus, $D: V \to V$ is a linear transformation and is known as the derivative mapping.
Note that D is not a monomorphism, because $D(4x^2 + 3) = D(4x^2)$, but $4x^2 + 3 \neq 4x^2$.

EXAMPLE-6 Let V be the real vector space of all continuous functions from R into itself. Show that a mapping $T: V \to V$ given by
$$T[f(x)] = \int_0^x f(t)\, dt \quad \text{for all } f(x) \in V \text{ and } x \in R$$
is a linear transformation from V into itself.

SOLUTION For any $f(x), g(x) \in V$ and $\lambda \in R$, we have
$$T\big(f(x)+g(x)\big) = \int_0^x \{f(t)+g(t)\}\, dt = \int_0^x f(t)\, dt + \int_0^x g(t)\, dt$$
$$\Rightarrow \quad T\big(f(x)+g(x)\big) = T(f(x)) + T(g(x))$$
and, $\quad T\big(\lambda f(x)\big) = \displaystyle\int_0^x \lambda f(t)\, dt = \lambda \int_0^x f(t)\, dt = \lambda T(f(x))$

Thus, T is a linear mapping and is known as the integral mapping.

EXAMPLE-7 Let A be an $m \times n$ real matrix and let $t_A : R^{n \times 1} \to R^{m \times 1}$ be a mapping defined by
$$t_A(U) = AU \quad \text{for all } U \in R^{n \times 1}$$
Show that t_A is a linear transformation.

SOLUTION For any $U, V \in R^{n \times 1}$ and $a \in R$, we have

$$t_A(U+V) = A(U+V) = AU + AV \quad \left[\begin{array}{l} \text{By distributivity of matrix multiplication} \\ \text{over addition} \end{array} \right]$$

$\Rightarrow \quad t_A(U+V) = t_A(U) + t_A(V)$

and, $\quad t_A(aU) = A(aU) = a(AU) = a\, t_A(U)$

Hence, t_A is a linear transformation.

EXAMPLE-8 Let C be the vector space of all complex numbers over the field of complex numbers, and let $t : C \to C$ be a mapping given by
$$t(x+iy) = x, \quad \text{for all } x+iy \in C.$$
Show that t is not a linear transformation.

SOLUTION Let $z_1 = x_1 + iy_1$, $z_2 = x_2 + iy_2 \in C$. Then,

$$t(z_1 + z_2) = t(x_1 + x_2 + i(y_1 + y_2)) = x_1 + x_2 = t(x_1 + iy_1) + t(x_2 + iy_2)$$

$\Rightarrow \quad t(z_1 + z_2) = t(z_1) + t(z_2)$

So, t satisfies condition (i) in the definition of a linear transformation.
Now, $t\{(3+i)(3-i)\} = t(10) = 10$ but, $(3+i)t(3-i) = (3+i)3 = 9+3i \neq 10$.

$\therefore \quad t[(3+i)(3-i)] \neq (3+i)t(3-i)$.

So, t does not satisfy condition (ii), in the definition of a linear transformation.
Hence, t is not a linear transformation.

EXAMPLE-9 Let C be the field of complex numbers regarded as a vector space over the field R of real numbers. Then show that the mapping $t : C \to R$ given by
$$t(z) = (x^3 + y^3)^{\frac{1}{3}} \quad \text{for all } z = x + iy \in C$$
is not a linear transformation.

SOLUTION For any $z = x + iy \in C$ and $\lambda \in R$

$$t(\lambda z) = t(\lambda x + i\lambda y) = (\lambda^3 x^3 + \lambda^3 y^3)^{\frac{1}{3}} = \lambda (x^3 + y^3)^{\frac{1}{3}} = \lambda\, t(z)$$

So, t satisfies condition (ii) in the definition of a linear transformation.
Now $t(2) = 2, t(3i) = 3$ and $t(2+3i) = (8+27)^{\frac{1}{3}} = (35)^{\frac{1}{3}} \neq t(2) + t(3i)$
So, t does not satisfy condition (ii) in the definition of a linear transformation.
Hence, t is not a linear transformation.

REMARK. *It follows from above two examples that the conditions (i) and (ii) in the definition of a linear transformation, given above, are independent of each other, in the sense that a mapping t from a vector space $V(F)$ into a vector space $U(F)$ may satisfy (i) but not (ii) or it may satisfy (ii) but not (i).*

EXERCISE 3.1

1. Show that the mapping $t : R^2 \to R^2$ defined by
$$t(x,y) = (x+y, x) \quad \text{for all } (x,y) \in R^2$$
is a linear transformation.

2. Show that each of the following mappings is a linear transformation:
    (i) $t : R^2 \to R^2$ defined by $t(x,y) = (ax+by, cx+dy)$, where $a, b, c, d \in R$
    (ii) $t : R^3 \to R^2$ defined by $t(x,y,z) = (x+y+z, 2x-3y+4z)$
    (iii) $t : R^3 \to R^2$ defined by $t(x,y,z) = (x+2y-3z, 4x-5y+6z)$

3. Show that the following mappings are not linear transformations:
    (i) $t : R^2 \to R^2$ defined by $t(x,y) = (x+1, y+2)$
    (ii) $t : R^2 \to R^2$ defined by $t(x,y) = (x^2, y^2)$
    (iii) $t : R^3 \to R^2$ defined by $t(x,y,z) = (x+1, y+z)$
    (iv) $t : R^2 \to R^2$ defined by $t(x,y) = (xy, y)$
    (v) $t : R^3 \to R^2$ defined by $t(x,y,z) = \left(|x|, y+z\right)$

4. Let $V = R^{n \times n}$ vector space of all $n \times n$ real matrices. Let M be an arbitrary but fixed matrix in V. Let $t : V \to V$ be defined by $t(A) = AM + MA$, where A is any matrix in V. Show that t is a linear transformation.

5. Let $V = R^{n \times n}$ be the vector space of all $n \times n$ square matrices, and let M be a fixed non-null matrix in V. Show that the two of the following mapping $t : V \to V$ are linear transformations, but the third is not :
    (i) $t(A) = MA$
    (ii) $t(A) = MA + AM$
    (iii) $t(A) = M + A$

6. Let $V = F^{n \times n}$ be the vector space of all $n \times n$ matrices over the field F, and let M be a fixed $n \times n$ matrix. Then, a mapping $t : V \to V$ defined by
$$t_M(A) = AM - MA \quad \text{for all } A \in V$$
is a linear transformation.

7. Let V and V' be two vector spaces over the same field F and let t_1, t_2 be two linear transformations from V_1 to V_2. If $a, b \in F$, then the mapping $t : V \to V'$ defined by

$$t(v) = at_1(v) + bt_2(v) \quad \text{for all } v \in V$$

is a linear transformation from V to V'.

8. Does there exist a linear transformation such that $t(1,0,0,0) = (1,0,0)$, $t(0,1,0,0) = (1,1,0)$, $t(0,0,1,0) = (1,0,-1)$, $t(1,1,1,0) = (2,1,3)$.

3.3 SOME BASIC PROPERTIES OF LINEAR TRANSFORMATIONS

In this section, we shall study some basic properties of linear transformations which will be very useful in the subsequent discussion.

THEOREM-1 *Let V and V' be vector spaces over the same field F. A mapping $t : V \to V'$ is a linear transformation (abbreviated l.t.) iff $t(au + bv) = at(u) + bt(v)$ for all $u, v \in V$ and all $a, b \in F$.*

PROOF. Let $t : V \to V'$ be a linear transformation. Then, by definition
$t(au) = at(u), t(bv) = bt(v)$ for all $u, v \in V$ and all $a, b \in F$.
$\therefore \quad t(au + bv) = t(au) + t(bv) = at(u) + bt(v)$ for all $u, v \in V$ and all $a, b \in F$.

Conversely, let $t : V \to V'$ be a mapping such that $t(au + bv) = at(u) + bt(v)$ for all $u, v \in V$ and all $a, b \in F$. By taking $a = b = 1$ (unity in F), we get

$$t(u + v) = t(u) + t(v) \quad \text{for all } u, v \in V.$$

Again by taking $b = 0$, we get

$$t(au) = at(u) \quad \text{for all } u \in V \text{ and all } a \in F.$$

Hence, t is a linear transformation.

Q.E.D.

THEOREM-2 *Let t be a linear transformation from a vector space V to a vector space V' (both over the same field F). Then,*

(i) $t(0_V) = 0_{V'}$, *where 0_V and $0_{V'}$ are zero vectors of V and V' respectively.*

(ii) $t(-u) = -t(u)$

(iii) $t(u - v) = t(u) - t(v)$

for all $u, v \in V$.

PROOF. (i) Let $u \in V$. Then $t(u) \in V'$

Now, $\quad t(u+0_V) = t(u) + t(0_V) \qquad [\because t \text{ is a linear transformation}]$

$\Rightarrow \quad t(u) = t(u) + t(0_V) \qquad [\because u+0_V = u]$

$\Rightarrow \quad t(u) + 0_{V'} = t(u) + t(0_V) \qquad [\because 0_{V'} \text{ is zero vector of } V']$

$\Rightarrow \quad t(0_V) = 0_{V'} \qquad [\text{By left cancellation law in } V']$

(ii) For any $u \in V$, we have

$\quad u + (-u) = 0_V$

$\Rightarrow \quad t\big(u+(-u)\big) = t(0_V)$

$\Rightarrow \quad t(u) + t(-u) = 0_{V'} \qquad [\text{From (i)}]$

$\Rightarrow \quad t(-u) = -t(u)$

$\therefore \quad t(-u) = -t(u)$ for all $u \in V$.

(iii) For any $u, v \in V$, we have

$\quad u - v = u + (-v)$

$\Rightarrow \quad t(u-v) = t(u+(-v)) = t(u) + t(-v) \qquad [\because t \text{ is linear}]$

$\Rightarrow \quad t(u-v) = t(u) + \big(-t(v)\big) \qquad [\text{From (ii)}]$

$\Rightarrow \quad t(u-v) = t(u) - t(v)$

$\therefore \quad t(u-v) = t(u) - t(v) \quad$ for all $u, v \in V$.

Q.E.D.

THEOREM-3 *Let t be a linear transformation from a vector space V to a vector space V' (both over the same field F). Then,*

$$t(\lambda_1 v_1 + \lambda_2 v_2 + \cdots + \lambda_n v_n) = \lambda_1 t(v_1) + \lambda_2 t(v_2) + \cdots + \lambda_n t(v_n)$$

for all $v_1, v_2, \ldots, v_n \in V$ and $\lambda_1, \lambda_2, \ldots, \lambda_n \in F$.

PROOF. We will prove the theorem by induction on n i.e. the number of vectors in the linear combination $\lambda_1 v_1 + \lambda_2 v_2 + \cdots + \lambda_n v_n$.

By Theorem 1, the result holds for $n = 2$.

Suppose the result holds for $(n-1)$, i.e.

$$t(\lambda_1 v_1 + \lambda_2 v_2 + \cdots + \lambda_{n-1} v_{n-1}) = \lambda_1 t(v_1) + \lambda_2 t(v_2) + \cdots + \lambda_{n-1} t(v_{n-1})$$

Then,

$t(\lambda_1 v_1 + \lambda_2 v_2 + \cdots + \lambda_{n-1} v_{n-1} + \lambda_n v_n)$
$= t\{(\lambda_1 v_1 + \lambda_2 v_2 + \cdots + \lambda_{n-1} v_{n-1}) + \lambda_n v_n\}$

$$= t(\lambda_1 v_1 + \lambda_2 v_2 + \cdots + \lambda_{n-1} v_{n-1}) + t(\lambda_n v_n)$$
$$= \lambda_1 t(v_1) + \lambda_2 t_2(v_2) + \cdots + \lambda_{n-1} t(v_{n-1}) + \lambda_n t(v_n) \quad \text{[By induction hypothesis and linearity of } t\text{]}$$

Hence, by induction the theorem is true for all n.

Q.E.D.

THEOREM-4 *Let V and V' be vector spaces over the same field F. Let $B = \{b_1, b_2, b_3, \ldots, b_n\}$ be a basis of V and let $b'_1, b'_2, \ldots, b'_n$ be any vectors in V'. Then there exist a unique linear transformation $t : V \to V'$ such that $t(b_i) = b'_i$, $i \in \underline{n}$.*

PROOF. Since B is a basis for V, therefore for each $v \in V$ there exist unique scalars $\lambda_1, \lambda_2, \ldots, \lambda_n \in F$ such that $v = \sum_{i=1}^{n} \lambda_i b_i$.

Consider the mapping $t : V \to V'$ given by the rule

$$t(v) = t\left(\sum_{i=1}^{n} \lambda_i b_i\right) = \sum_{i=1}^{n} \lambda_i b'_i$$

Since $\lambda_1, \lambda_2, \ldots, \lambda_n$ are unique scalars. Therefore, $\sum_{i=1}^{n} \lambda_i b'_i$ is a unique vector in V' i.e. $t(v)$ is unique. So, t is well defined.

We shall now show that $t : V \to V'$ is a linear transformation:

For any $u = \sum_{i=1}^{n} \lambda_i b_i$, $v = \sum_{i=1}^{n} \mu_i b_i$ in V and α, β in F, we have

$$t(\alpha u + \beta v) = t\left\{\alpha\left(\sum_{i=1}^{n} \lambda_i b_i\right) + \beta\left(\sum_{i=1}^{n} \mu_i b_i\right)\right\}$$

$$\Rightarrow \quad t(\alpha u + \beta v) = t\left\{\sum_{i=1}^{n} (\alpha \lambda_i + \beta \mu_i) b_i\right\}$$

$$\Rightarrow \quad t(\alpha u + \beta v) = \sum_{i=1}^{n} (\alpha \lambda_i + \beta \mu_i) b'_i$$

$$\Rightarrow \quad t(\alpha u + \beta v) = \sum_{i=1}^{n} (\alpha \lambda_i b'_i + \beta \mu_i b'_i)$$

$$\Rightarrow \quad t(\alpha u + \beta v) = \sum_{i=1}^{n} \{\alpha (\lambda_i b'_i) + \beta (\mu_i b'_i)\}$$

$$\Rightarrow \quad t(\alpha u + \beta v) = \sum_{i=1}^{n} \alpha (\lambda_i b'_i) + \sum_{i=1}^{n} \beta (\mu_i b'_i)$$

$$\Rightarrow \quad t(\alpha u + \beta v) = \alpha \left(\sum_{i=1}^{n} \lambda_i b'_i\right) + \beta \left(\sum_{i=1}^{n} \mu_i b'_i\right)$$

$$\Rightarrow \quad t(\alpha u + \beta v) = \alpha\, t(u) + \beta\, t(v)$$

Thus, $t(\alpha u + \beta v) = \alpha\, t(u) + \beta t(v)$ for all $u, v \in V$

So, $t : V \to V'$ is a linear transformation.

Now, for any $i \in \underline{n}$, we have

$$b_i = \sum_{r=1}^{n} \lambda_r b_r, \text{ where } \lambda_r = 0 \text{ for } r = 1, 2, \ldots, n, \ r \neq i \text{ and } \lambda_i = 1$$

$\Rightarrow \quad t(b_i) = t\left(\sum_{r=1}^{n} \lambda_r b_r\right)$

$\Rightarrow \quad t(b_i) = \sum_{r=1}^{n} \lambda_r t(b_r)$

$\Rightarrow \quad t(b_i) = \sum_{r=1}^{n} \lambda_r b'_r$

$\Rightarrow \quad t(b_i) = b'_i$

Thus, $t : V \to V'$ is a linear transformation such that $t(b_i) = b'_i$ for all $i \in \underline{n}$.

To prove the uniqueness of t, let, if possible $t' : V \to V'$ be a linear transformation such that $t'(b_i) = b'_i$ for all $i \in \underline{n}$. Then for any $v = \sum_{i=1}^{n} \lambda_i b_i \in V$, we have

$$t'(v) = t'\left(\sum_{i=1}^{n} \lambda_i b_i\right) = \sum_{i=1}^{n} \lambda_i t'(b_i)$$

$\Rightarrow \quad t'(v) = \sum_{i=1}^{n} \lambda_i b'_i = t(v)$

$\therefore \quad t'(v) = t(v) \quad \text{for all } v \in V$

So, $t = t'$.

Hence, $t : V \to V'$ is a unique linear transformation such that $t(b_i) = b'_i$ for all $i \in \underline{n}$.

Q.E.D.

3.3.1 MATRICES AS LINEAR TRANSFORMATIONS

Let A be any $m \times n$ matrix over a field F. Consider the mapping $t_A : F^{n \times 1} \to F^{m \times 1}$ defined by

$$t_A(X) = AX \text{ for every } X = \begin{bmatrix} x_1 \\ x_2 \\ \vdots \\ x_n \end{bmatrix} \in F^{n \times 1}$$

For any $X, Y \in F^{n \times 1}$ and $\lambda \in F$, we have

$\qquad t_A(X+Y) = A(X+Y)$

$\Rightarrow \quad t_A(X+Y) = AX + AY \qquad$ [By distributivity of matrix multiplication over addition]

$\Rightarrow \quad t_A(X+Y) = t_A(X) + t_A(Y)$

and, $\quad t_A(\lambda X) = A(\lambda X) = \lambda(AX) = \lambda \, t_A(X)$

Thus, $t_A : F^{n\times 1} \to F^{m\times 1}$ is a linear transformation.

It follows from the above discussion that every $m \times n$ matrix over a field F may be viewed as a linear transformation from $F^{n\times 1}$ to $F^{m\times 1}$.

Conversely, every linear transformation $t : F^n \to F^m$ may be written in the form $t(X) = AX$, where $X = [x_1, x_2, \ldots, x_n]^T$ and A is an $m \times n$ matrix.

For example, linear transformation $t : R^3 \to R^2$ defined by

$$t(x,y,z) = (x+y+z,\ 2x-3y+4z)$$

may be written in the form

$$t(X) = AX,\ \text{where}\ X = [x,y,z]^T\ \text{and},\ A = \begin{bmatrix} 1 & 1 & 1 \\ 2 & -3 & 4 \end{bmatrix}$$

ILLUSTRATIVE EXAMPLES

EXAMPLE-1 Let $t : R^2 \to R^2$ be a linear transformation for which $t(1,2) = (2,3)$ and $t(0,1) = (1,4)$. Find a formula for t, find $t(x,y)$.

SOLUTION We know that $B = \{(a_1,b_1), (a_2,b_2)\}$ forms a basis of R^2 iff $a_1 b_2 \neq a_2 b_1$. Clearly, $v_1 = (1,2), v_2 = (0,1)$ satisfy this relation. So, $\{v_1, v_2\}$ is a basis of R^2.

Let $v = (x,y)$ be an arbitrary vector in R^2. Then there exist unique scalars $a, b \in R$ such that

$$v = av_1 + bv_2$$
$\Rightarrow\quad (x,y) = a(1,2) + b(0,1)$
$\Rightarrow\quad (x,y) = (a, 2a+b)$
$\Rightarrow\quad a = x,\ b = y - 2x$

As $t : R^2 \to R^2$ is a linear transformation.

$\therefore\qquad v = av_1 + bv_2$
$\Rightarrow\quad t(v) = t(av_1 + bv_2)$
$\Rightarrow\quad t(v) = at(v_1) + bt(v_2)$
$\Rightarrow\quad t(x,y) = at(1,2) + bt(0,1)$
$\Rightarrow\quad t(x,y) = a(2,3) + b(1,4)$
$\Rightarrow\quad t(x,y) = (2a+b, 3a+4b)$
$\Rightarrow\quad t(x,y) = (y, -5x+4y)$

Hence, $t : R^2 \to R^2$ is given by $t(x,y) = (y, -5x+4y)$ for all $(x,y) \in R^2$.

EXAMPLE-2 Let $t : R^2 \to R^2$ be a linear transformation such that $t(1,1) = (1,3)$, $t(-1,1) = (3,1)$. Find a complete formula for t.

SOLUTION Clearly, $B = \{(1,1), (-1,1)\}$ forms a basis for R^2. Let $(a,b) \in R^2$, then there exist scalars $\lambda, \mu \in R$ such that

$$(a,b) = \lambda(1,1) + \mu(-1,1)$$
$\Rightarrow \quad (a,b) = (\lambda - \mu, \lambda + \mu)$
$\Rightarrow \quad \lambda - \mu = a$ and $\lambda + \mu = b \Rightarrow \lambda = \dfrac{a+b}{2}$ and $\mu = \dfrac{b-a}{2}$

Since $t : R^2 \to R^2$ is a linear transformation.

$\therefore \quad (a,b) = \lambda(1,1) + \mu(-1,1)$
$\Rightarrow \quad t(a,b) = \lambda t(1,1) + \mu t(-1,1)$
$\Rightarrow \quad t(a,b) = \lambda(1,3) + \mu(3,1)$
$\Rightarrow \quad t(a,b) = (\lambda + 3\mu, 3\lambda + \mu)$
$\Rightarrow \quad t(a,b) = (2b - a, a + 2b)$

Hence, $t : R^2 \to R^2$ is given by $t(a, b) = (2b - a, a + 2b)$ for all $(a, b) \in R^2$.

EXAMPLE-3 *Let $B = \{(-1,0,1), (0,1,-1), (1,-1,1)\}$ be a basis of $R^3(R)$ and $t : R^3 \to R^3$ be a linear transformation such that $t(-1,0,1) = (1,0,0)$, $t(0,1,-1) = (0,1,0)$, $t(1,-1,1) = (0,0,1)$. Find formula for $t(x,y,z)$ and use it to compute $t(1,-2,3)$.*

SOLUTION Clearly, B forms a basis for R^3. Let $(x,y,z) \in R^3$. Then there exist scalars $a, b, c \in R$ such that

$$(x,y,z) = a(-1,0,1) + b(0,1,-1) + c(1,-1,1)$$
$\Rightarrow \quad (x,y,z) = (-a+c, b-c, a-b+c)$
$\Rightarrow \quad -a+c = x,\ b-c = y,\ a-b+c = z$
$\Rightarrow \quad a = y+z,\ b = x+2y+z,\ c = x+y+z$

Since $t : R^3 \to R^3$ is a linear transformation. Therefore,

$(x,y,z) = a(-1,0,1) + b(0,1,-1) + c(1,-1,1)$
$\Rightarrow \quad t(x,y,z) = a\, t(-1,0,1) + b\, t(0,1,-1) + c\, t(1,-1,1)$
$\Rightarrow \quad t(x,y,z) = a(1,0,0) + b(0,1,0) + c(0,0,1)$
$\Rightarrow \quad t(x,y,z) = (a,b,c)$
$\Rightarrow \quad t(x,y,z) = (y+z,\ x+2y+z,\ x+y+z)$
$\therefore \quad t(1,-2,3) = (1,0,2)$

EXAMPLE-4 *The range of a linear transformation $t : R^3 \to R^3$ has the subspace spanned by the vectors $v_1 = (1,0,-1)$ and $v_2 = (1,2,2)$. Find the transformation explicitly.*

SOLUTION We know that $B = \left\{ e_1^{(3)} = (1,0,0),\ e_2^{(3)} = (0,1,0),\ e_3^{(3)} = (0,0,1) \right\}$ is a basis of $R^3(R)$. Also, $v_1 = (1,0,-1)$ and $v_2 = (1,2,2)$ are two vectors spanning the image of t. Therefore, $v_1 = (1,0,-1), v_2 = (1,2,2)$ and $v_3 = (0,0,0)$ also span $I_m(t)$. By Theorem 4, there exists a unique linear transformation $t: R^3 \to R^3$ such that
$$t(e_1^{(3)}) = v_1,\ t(e_2^{(3)}) = v_2 \text{ and } t(e_3^{(3)}) = v_3$$
Let $v = (a,b,c)$ be an arbitrary vector in R^3 whose basis is B.
Clearly,
$$v = a\, e_1^{(3)} + b\, e_2^{(3)} + c\, e_3^{(3)}$$
$\Rightarrow \quad t(v) = a\, t\left(e_1^{(3)}\right) + b\, t\left(e_2^{(3)}\right) + c\, t\left(e_3^{(3)}\right)$

$\Rightarrow \quad t(v) = a v_1 + b v_2 + c v_3$

$\Rightarrow \quad t(a,b,c) = a(1,0,-1) + b(1,2,2) + c(0,0,0)$

$\Rightarrow \quad t(a,b,c) = (a+b, 2b, -a+2b)$

REMARK. *In the above example, we may choose v_3 as a linear combination of v_1 and v_2, accordingly t will also change. So, t is not unique.*

EXAMPLE-5 *Find a linear transformation $t: R^3 \to R^4$ whose image is spanned by the vectors $v_1 = (1,2,0,-4)$ and $v_2 = (2,0,-1,-3)$.*

SOLUTION We know that $B = \left\{ e_1^{(3)}, e_2^{(3)}, e_3^{(3)} \right\}$ is the standard basis of $R^3(R)$. It is given that $v_1 = (1,2,0,-4), v_2 = (2,0,-1,-3)$ span $I_m(t)$. Therefore, $v_1 = (1,2,0,-4), v_2 = (2,0,-1,-3)$ and $v_3 = (0,0,0,0)$ also span $I_m(t)$. By Theorem 4, there exists a unique linear transformation $t: R^3 \to R^4$ such that

$$t\left(e_1^{(3)}\right) = v_1,\ t\left(e_2^{(3)}\right) = v_2 \text{ and } t\left(e_3^{(3)}\right) = v_3$$

Let $v = (a,b,c)$ be an arbitrary vector in R^3 whose basis is B.
Clearly,
$$v = a\, e_1^{(3)} + b\, e_2^{(3)} + c\, e_3^{(3)}$$
$\Rightarrow \quad t(v) = a\, t\left(e_1^{(3)}\right) + b\, t\left(e_2^{(3)}\right) + c\, t\left(e_3^{(3)}\right)$

$\Rightarrow \quad t(v) = a v_1 + b v_2 + c v_3$

$\Rightarrow \quad t(a,b,c) = a(1,2,0,-4) + b(2,0,-1,-3) + c(0,0,0,0)$

$\Rightarrow \quad t(a,b,c) = (a+2b,\ 2a-b,\ -b,\ -4a-3b)$

REMARK. *In the above example t is not unique. Instead of taking v_3 as the null vector in R^4, we may take $v_3 = v_1$ or, $v_3 = v_2$ or, v_3 as any linear combination of v_1 and v_2. Accordingly linear transformation t changes.*

EXERCISE 3.2

1. Let $t : R^3 \to R^2$ be a linear transformation given by
$$T(x, y, z) = (x+y, y+z).$$
Find a basis and the dimension of (i) the image of t (ii) the kernel of t.

2. Let $t : R^3 \to R^2$ be the linear transformation such that $t(1,2,3) = (1,0,0)$, $t(1,2,0) = (0,1,0)$, $t(1,-1,0) = (0,1,0)$. Find $t(a,b,c)$ for any $(a,b,c) \in R^3$.

3. Let F be a field and $t : F^2 \to F^2$ be a linear transformation such that $t(1,0) = (a,b)$ and $t(0,1) = (c,d)$. Find $t(x,y)$ for any $(x,y) \in F^2$.

4. Describe explicitly the linear transformation $t : R^2 \to R^2$ such that $t(2,3) = (4,5)$ and $t(1,0) = (0,0)$.

5. Find the linear transformation $t : R^2 \to R^2$ such that $t(1, 0) = (1, 1)$, $t(0, 1) = (-1, 2)$. Prove that t maps the square with vertices $(0,0)$, $(1,0)$, $(1,1)$ and $(0,1)$ into a parallelogram.

ANSWERS

1. (i) $\{(1,0), (0,1)\}$, dim $I_m(t) = 2$ (ii) $\{(-1, 1, -1)\}$, dim $\text{Ker}(t) = 1$

2. $t(a,b,c) = \left(\dfrac{c}{3}, \dfrac{3a-c}{3}, 0\right)$ 3. $t(x,y) = (xy+yc, xb+yd)$

4. $t(x, y) = \left(\dfrac{4y}{3}, \dfrac{5y}{3}\right)$

3.4 KERNEL AND IMAGE OF A LINEAR TRANSFORMATION

KERNEL OF A LINEAR TRANSFORMATION Let V and V' be two vector spaces over the same field F, and let $t : V \to V'$ be a linear transformation. Then kernel of t, written as $\text{Ker}(t)$, is defined by
$$\text{Ker}(t) = \{v \in V : t(v) = 0_{V'}\}$$
$\text{Ker}(t)$ is also called the null space of t.

REMARK. Note that $\text{Ker}(t)$ is always non-void, because $t(0_V) = 0_{V'}$.

THEOREM-1 Let V and V' be two vector spaces over the same field F and let $t : V \to V'$ be a linear transformation. Then,

(i) $\text{Ker}(t)$ is a subspace of V.

(ii) $I_m(t) = \{t(u) : u \in V\}$ is a subspace of V'.

PROOF. We have,
   Ker$(t) = \{u \in V : t(u) = 0_{V'}\}$
Since $t(0_V) = 0_{V'}$, therefore $0_V \in$ Ker(t).
Thus, Ker(t) is a non-void subset of V.
Let $u, v \in$ Ker(t) and $a, b \in F$. Then,

$\qquad t(au + bv) = at(u) + bt(v)$ $\qquad\qquad\qquad$ [$\because$ t is a linear transformation]
$\Rightarrow \quad t(au + bv) = a\,0_{V'} + b\,0_{V'}$ $\qquad\qquad$ [$\because$ $u, v \in$ Ker$(t) \Rightarrow t(u) = 0_{V'}, t(v) = 0_{V'}$]
$\Rightarrow \quad t(au + bv) = 0_{V'} + 0_{V'} = 0_{V'}$
$\Rightarrow \quad au + bv \in$ Ker(t)

Thus, $au + bv \in$ Ker(t) for all $u, v \in$ Ker(t) and all $a, b \in F$.
Hence, Ker(t) is a subspace of V.

(ii) Since $0_V \in V$, therefore $0_{V'} = t(0_V) \in I_m(t)$. So, $I_m(t)$ is a non-void subset of V'.

Let $u', v' \in I_m(t)$, then there exist $u, v \in V$ such that $t(u) = u'$ and $t(v) = v'$.

If $a, b \in F$, then $au + bv \in V$.

$\therefore \quad au' + bv' = a\,t(u) + b\,t(v) = t(au + bv)$ $\qquad\qquad\qquad$ [$\because$ t is linear]
$\Rightarrow \quad au' + bv' \in I_m(t)$.

Thus, $au' + bv' \in I_m(t)$ for all $u', v' \in I_m(t)$ and all $a, b \in F$.
Hence, $I_m(t)$ is a subspace of V'.
$\qquad\qquad\qquad\qquad\qquad\qquad\qquad\qquad\qquad\qquad\qquad\qquad\qquad\qquad$ Q.E.D.

THEOREM-2 *A linear transformation t from a vector space $V(F)$ to a vector space $V'(F)$ is a monomorphism iff* Ker$(t) = \{0_V\}$.

PROOF. Let $t : V \to V'$ be a monomorphism. Then t is injective. Let u be an arbitrary vector in Ker(t). Then,

$\qquad t(u) = 0_{V'}$
$\Rightarrow \quad t(u) = t(0_V)$ $\qquad\qquad\qquad\qquad\qquad\qquad\qquad\qquad$ [$\because$ $t(0_V) = 0_{V'}$]
$\Rightarrow \quad u = 0_V$ $\qquad\qquad\qquad\qquad\qquad\qquad\qquad\qquad\qquad$ [$\because$ t is injective]

Since u is an arbitrary vector in Ker(t) such that,

$\qquad u \in$ Ker$(t) \Rightarrow u = 0_V$

Hence, Ker$(t) = \{0_V\}$.

Conversely, let $t : V \to V'$ be a linear transformation such that Ker$(t) = \{0_V\}$. Then we have to show that t is a monomorphism. For this we have to show that t is injective.

Let u, v be any two vectors in V. Then,

$\quad\quad t(u) = t(v)$

$\Rightarrow \quad t(u) - t(v) = 0_{V'}$

$\Rightarrow \quad t(u - v) = 0_{V'}$ \quad\quad\quad\quad\quad\quad\quad\quad\quad\quad [By Theorem 2(iii) on page 240]

$\Rightarrow \quad u - v \in \text{Ker}(t)$

$\Rightarrow \quad u - v = 0_V$ \quad\quad\quad\quad\quad\quad\quad\quad\quad\quad\quad\quad\quad\quad $[\because \text{Ker}(t) = \{0_V\}]$

$\Rightarrow \quad u = v$

Therefore, t is injective and hence a monomorphism.

Q.E.D.

THEOREM-3 *Let t be a linear transformation from a vector space $V(F)$ to a vector space $V'(F)$. If $\{v_1, v_2, \ldots, v_n\}$ spans V, then $\{t(v_1), t(v_2), \ldots, t(v_n)\}$ spans $I_m(t)$.*

PROOF. Since $\{v_1, v_2, \ldots, v_n\}$ spans V, therefore for each $v \in V$ there exist scalars $\lambda_1, \lambda_2, \ldots, \lambda_n \in F$ such that

$\quad\quad v = \lambda_1 v_1 + \lambda_2 v_2 + \cdots + \lambda_n v_n$

$\Rightarrow \quad t(v) = t(\lambda_1 v_1 + \lambda_2 v_2 + \cdots + \lambda_n v_n)$

$\Rightarrow \quad t(v) = \lambda_1 t(v_1) + \lambda_2 t(v_2) + \cdots + \lambda_n t(v_n)$

$\Rightarrow \quad t(v)$ is a linear combination of $t(v_1), t(v_2), \ldots, t(v_n)$.

But, $\quad t(v) \in I_m(t)$

Thus, each vector in $I_m(t)$ is a linear combination of $t(v_1), t(v_2), \ldots, t(v_n)$.

Hence, $\{t(v_1), t(v_2), \ldots, t(v_n)\}$ spans $I_m(t)$.

Q.E.D.

THEOREM-4 *Let V and V' be vector spaces over a field F and let $t : V \to V'$ be a linear transformation. If $(v_1, v_2, \ldots, v_n)$ is a linearly dependent list of vectors in V, then the list $(t(v_1), t(v_2), \ldots, t(v_n))$ is also linearly dependent.*

PROOF. Since $(v_1, v_2, \ldots, v_n)$ is a linearly dependent list of vectors in V. Therefore, there exist scalars $\lambda_1, \lambda_2, \ldots, \lambda_n \in F$ not all zero such that

$\quad\quad \lambda_1 v_1 + \lambda_2 v_2 + \cdots + \lambda_n v_n = 0_V$

$\Rightarrow \quad t(\lambda_1 v_1 + \lambda_2 v_2 + \cdots + \lambda_n v_n) = t(0_V)$

$\Rightarrow \quad \lambda_1 t(v_1) + \lambda_2 t(v_2) + \cdots + \lambda_n t(v_n) = 0_{V'}$

Thus, there exist scalars $\lambda_1, \lambda_2, \ldots, \lambda_n \in F$ not all zero such that

$\quad\quad \lambda_1 t(v_1) + \lambda_2 t(v_2) + \cdots + \lambda_n t(v_n) = 0_{V'}$

Hence, the list $\big(t(v_1), t(v_2), \ldots, t(v_n)\big)$ is linearly dependent.

Q.E.D.

THEOREM-5 *Let $t : V \to V'$ be a linear transformation from a vector space $V(F)$ to a vector space $V'(F)$. If $(v_1, v_2, \ldots, v_n)$ is a list of vectors in V such that the list $(t(v_1), t(v_2), \ldots, t(v_n))$ is linearly independent in V', then the list $(v_1, v_2, \ldots, v_n)$ is linearly independent in V.*

PROOF. Let $\lambda_1, \lambda_2, \ldots, \lambda_n \in F$ be such that

$$\lambda_1 v_1 + \lambda_2 v_2 + \cdots + \lambda_n v_n = 0_V$$
$$\Rightarrow \quad t(\lambda_1 v_1 + \lambda_2 v_2 + \cdots + \lambda_n v_n) = t(0_V)$$
$$\Rightarrow \quad \lambda_1 t(v_1) + \lambda_2 t(v_2) + \cdots + \lambda_n t(v_n) = 0_{V'}$$
$$\Rightarrow \quad \lambda_1 = \lambda_2 = \cdots = \lambda_n = 0 \qquad [\because (t(v_1), t(v_2), \ldots, t(v_n)) \text{ is linearly independent}]$$

Hence, $(v_1, v_2, \ldots, v_n)$ is a linearly independent list in V.

Q.E.D.

REMARK. *These two theorems depict that the image of a linearly dependent list of vectors is linearly dependent and the pre-image of a linearly independent list is linearly independent. However, the image of a linearly independent list need not be linearly independent, because basis of the kernel of a linear transformation is linearly independent and its image set $\{0_{V'}\}$ is linearly dependent.*

THEOREM-6 *Let V and V' be two vector spaces over the same field F and let $t : V \to V'$ be a linear transformation.*

(i) *For any subspace S of V, prove that $t(S) = \{t(v) : v \in S\}$ is a subspace of V'.*

(ii) *For any subspace S' of V', prove that $t^{-1}(S') = \{v \in V' : t(v) \in S'\}$ is a subspace of V containing the $\mathrm{Ker}(t)$.*

PROOF. (i) As $0_V \in S$. Therefore,
$$t(0_V) \in t(S) \Rightarrow 0_{V'} \in t(S). \qquad [\because t(0_V) = 0_{V'}]$$
So, $t(S)$ is a non-empty subset of V'.
Let $u', v' \in t(S)$. Then there exist $u, v \in S$ such that $t(u) = u'$ and $t(v) = v'$.
Let $a, b \in F$. Then,

$$au' + bv' = at(u) + bt(v) = t(au + bv) \qquad [\because t : V \to V' \text{ is a linear transformation}]$$
$$\Rightarrow \quad au' + bv' \in t(S) \qquad [\because S \text{ is a subspace of } V \quad \therefore \quad au + bv \in S \Rightarrow t(au + bv) \in t(S)]$$

Thus, $t(S)$ is a non-void subset of V' such that $au' + bv' \in t(S)$ for all $u', v' \in t(S)$ and $a, b \in F$.

(ii) Since S' is a subspace of V'.

$$\therefore \quad 0_{V'} \in S'$$
$$\Rightarrow \quad t(0_V) \in S' \Rightarrow 0_V \in t^{-1}(S') \Rightarrow t^{-1}(S') \neq \phi$$

Thus, $t^{-1}(S')$ is a non-empty subset of V'.

Let $u, v \in t^{-1}(S')$ and $a, b \in F$. Then,

$$t(u) \in S', t(v) \in S'.$$
$$\Rightarrow \quad at(u) + bt(v) \in S' \quad\quad\quad [\because S' \text{ is a subspace of } V']$$
$$\Rightarrow \quad t(au + bv) \in S' \quad\quad\quad [\because t \text{ is a linear transformation}]$$
$$\Rightarrow \quad au + bv \in t^{-1}(S')$$

Thus, $au + bv \in t^{-1}(S')$ for all $u, v \in t^{-1}(S')$ and $a, b \in F$.
So, $t^{-1}(S')$ is a subspace of V.
Let $u \in \text{Ker}(t)$. Then,

$$t(u) = 0_{V'} \Rightarrow u \in t^{-1}(S')$$

Thus, $\quad u \in \text{Ker}(t) \Rightarrow u \in t^{-1}(S')$

$\therefore \quad\quad \text{Ker}(t) \subset t^{-1}(S')$

Hence, $t^{-1}(S')$ is a subspace of V containing $\text{Ker}(t)$. Q.E.D.

REMARK. *The above theorem gives that range $(t) = t(V)$ is a subspace of V'. A linear transformation $t : V \to V'$ for which range $(t) \neq \{0_{V'}\}$ is called a non-zero linear transformation.*

EXAMPLE *Let t be a linear transformation from a vector space $V(F)$ to a vector space $V'(F)$ such that $\text{Ker}(t) \neq \{0_V\}$. Show that there exist vectors $v_1, v_2 \in V$ such that $v_1 \neq v_2$ and $t(v_1) = t(v_2)$.*

SOLUTION Let $v_1 = 0_V$. Then, $t(v_1) = 0_{V'}$. Since $\text{Ker}(t) \neq \{0_V\}$, therefore, there exists a non-zero vector, say v_2, in V such that $t(v_2) = 0_{V'}$.
Thus, v_1 and v_2 are vectors in V such that $v_1 \neq v_2$ and $t(v_1) = 0_{V'} = t(v_2)$.

ISOMORPHIC VECTOR SPACES *A vector space $V(F)$ is said to be isomorphic to a vector space $V'(F)$ if there exists a bijective linear transformation of V onto V'.*

If V is isomorphic to V', then we write $V \cong V'$.

It can be easily seen that the relation '$\cong$' is an equivalence relation on the set of all vector spaces over a field F.

REMARK. *In future unless otherwise mentioned, a vector space will always mean a finite dimensional vector space.*

THEOREM-7 *Every n-dimensional vector space over a field F is isomorphic to F^n.*

PROOF. Let V be an n-dimensional vector space over a field F, and let $B = \{b_1, b_2, \ldots, b_n\}$ be an ordered basis for V. Then for each $v \in V$, there exists a unique list $(\lambda_1, \lambda_2, \ldots, \lambda_n)$ of scalars in F such that

$$v = \lambda_1 b_1 + \lambda_2 b_2 + \cdots + \lambda_n b_n$$

Consider a mapping $t : V \to F^n$ given by

$$t(v) = t\left(\sum_{i=1}^{n} \lambda_i b_i\right) = (\lambda_1, \lambda_2, \ldots, \lambda_n)$$

Since for each $v \in V$ there exists a unique list $(\lambda_1, \lambda_2, \ldots, \lambda_n) \in F_n$. Therefore, t is well defined.

t is injective: Let u, v be any two vectors in V. Then there exist scalars $\lambda_1, \lambda_2, \ldots, \lambda_n$, $\mu_1, \mu_2, \ldots, \mu_n \in F$ such that

$$u = \sum_{i=1}^{n} \lambda_i b_i \text{ and } v = \sum_{i=1}^{n} \mu_i b_i. \qquad [\because B \text{ is an ordered basis for } V]$$

$\therefore \qquad t(u) = t(v)$

$\Rightarrow \qquad t\left(\sum_{i=1}^{n} \lambda_i b_i\right) = t\left(\sum_{i=1}^{n} \mu_i b_i\right)$

$\Rightarrow \qquad (\lambda_1, \lambda_2, \ldots, \lambda_n) = (\mu_1, \mu_2, \ldots, \mu_n)$

$\Rightarrow \qquad \lambda_i = \mu_i \quad \text{for all } i \in \underline{n}.$

$\Rightarrow \qquad \sum_{i=1}^{n} \lambda_i b_i = \sum_{i=1}^{n} \mu_i b_i \Rightarrow u = v$

Thus, $t(u) = t(v) \Rightarrow u = v$ for all $u, v \in V$. So, t is injective.

t is surjective: Let $(\lambda_1, \lambda_2, \ldots, \lambda_n)$ be an arbitrary element of F^n. Then there exists a vector $v = \lambda_1 b_1 + \lambda_2 b_2 + \cdots + \lambda_n b_n \in V$ such that

$$t(v) = t\left(\sum_{i=1}^{n} \lambda_i b_i\right) = (\lambda_1, \lambda_2, \ldots, \lambda_n).$$

Thus, for each $(\lambda_1, \lambda_2, \ldots, \lambda_n) \in F^n$ there exists a vector $v \in V$ such that $t(v) = (\lambda_1, \lambda_2, \ldots, \lambda_n)$.
So, t is surjective.

t is a linear transformation: For any $u = \sum_{i=1}^{n} \lambda_i b_i, v = \sum_{i=1}^{n} \mu_i b_i \in V$ and $\lambda, \mu \in F$, we have

$$t(\lambda u + \mu v) = t\left\{\lambda\left(\sum_{i=1}^{n} \lambda_i b_i\right) + \mu\left(\sum_{i=1}^{n} \mu_i b_i\right)\right\}$$

$\Rightarrow \qquad t(\lambda u + \mu v) = t\left\{\sum_{i=1}^{n} (\lambda \lambda_i + \mu \mu_i) b_i\right\}$

$\Rightarrow \qquad t(\lambda u + \mu v) = (\lambda \lambda_1 + \mu \mu_1, \lambda \lambda_2 + \mu \mu_2, \ldots, \lambda \lambda_n + \mu \mu_n)$

$\Rightarrow \qquad t(\lambda u + \mu v) = (\lambda \lambda_1, \lambda \lambda_2, \ldots, \lambda \lambda_n) + (\mu \mu_1, \mu \mu_2, \ldots, \mu \mu_n)$

$\Rightarrow \qquad t(\lambda u + \mu v) = \lambda(\lambda_1, \lambda_2, \ldots, \lambda_n) + \mu(\mu_1, \mu_2, \ldots, \mu_n)$

$\Rightarrow \qquad t(\lambda u + \mu v) = \lambda\, t(u) + \mu t(v)$

So, t is a linear transformation.

Thus, $t: V \to F^n$ is an isomorphism of vector spaces. Hence, $V \cong F^n$.

Q.E.D.

THEOREM-8 *Two finite dimensional vector spaces over the same field are isomorphic iff they are of the same dimension.*

PROOF. Let V and V' be two finite dimensional vector spaces over the same field F such that $\dim V = \dim V' = n$. Then we have to show that $V \cong V'$.

Let $B = \{b_1, b_2, \ldots, b_n\}$ and $B' = \{b'_1, b'_2, \ldots, b'_n\}$ be bases for V and V' respectively.

Let $v \in V$. Then there exist unique scalars $\lambda_1, \lambda_2, \ldots, \lambda_n$ such that $v = \sum_{i=1}^{n} \lambda_i b_i$.

Consider a mapping $t: V \to V'$ given by

$$t(v) = t\left(\sum_{i=1}^{n} \lambda_i b_i\right) = \sum_{i=1}^{n} \lambda_i b'_i.$$

Clearly, $t: V \to V'$ is well defined.

t is injective: Let u, v be any two vectors in V. Then $u = \sum_{i=1}^{n} \lambda_i b_i, v = \sum_{i=1}^{n} \mu_i b_i$ for some $\lambda_i, \mu_i \in F$.

$\therefore \quad t(u) = t(v)$

$\Rightarrow \quad t\left(\sum_{i=1}^{n} \lambda_i b_i\right) = t\left(\sum_{i=1}^{n} \mu_i b_i\right)$

$\Rightarrow \quad \sum_{i=1}^{n} \lambda_i b'_i = \sum_{i=1}^{n} \mu_i b'_i$

$\Rightarrow \quad \sum_{i=1}^{n} (\lambda_i - \mu_i) b'_i = 0_{V'}$

$\Rightarrow \quad \lambda_i - \mu_i = 0 \qquad \text{for all } i \in \underline{n} \qquad [\because B' \text{ is basis for } V']$

$\Rightarrow \quad \lambda_i = \mu_i \qquad \text{for all } i \in \underline{n}$

$\Rightarrow \quad \sum_{i=1}^{n} \lambda_i b_i = \sum_{i=1}^{n} \mu_i b_i$

$\Rightarrow \quad u = v$

Thus, $t(u) = t(v) \Rightarrow u = v$. for all $u, v \in V$.

So, t is injective.

t is surjective: Let v' be an arbitrary vector in V'. Then there exist scalars $\lambda_1, \lambda_2, \ldots, \lambda_n \in F$ such that $v' = \sum_{i=1}^{n} \lambda_i b'_i$, as B' is basis for V'. Consequently, there exists a vector $v = \sum_{i=1}^{n} \lambda_i b_i \in V$ such that $t(v) = t\left(\sum_{i=1}^{n} \lambda_i b_i\right) = \sum_{i=1}^{n} \lambda_i b'_i = v'$.

So, t is surjective.

t is a linear transformation: Let $u = \sum_{i=1}^{n} \lambda_i b_i, v = \sum_{i=1}^{n} \mu_i b_i$ be any two vectors in V and $\lambda, \mu \in F$. Then,

$$t(\lambda u + \mu v) = t\left\{\lambda\left(\sum_{i=1}^{n} \lambda_i b_i\right) + \mu\left(\sum_{i=1}^{n} \mu_i b_i\right)\right\}$$

$\Rightarrow \quad t(\lambda u + \mu v) = t\left\{\sum_{i=1}^{n} \lambda(\lambda_i b_i) + \sum_{i=1}^{n} \mu(\mu_i b_i)\right\}$

$\Rightarrow \quad t(\lambda u + \mu v) = t\left\{\sum_{i=1}^{n} (\lambda \lambda_i + \mu \mu_i) b_i\right\}$

$\Rightarrow \quad t(\lambda u + \mu v) = \left\{\sum_{i=1}^{n} (\lambda \lambda_i + \mu \mu_i) b'_i\right\}$ [By definition of t]

$\Rightarrow \quad t(\lambda u + \mu v) = \sum_{i=1}^{n} (\lambda \lambda_i) b'_i + \sum_{i=1}^{n} (\mu \mu_i) b'_i$

$\Rightarrow \quad t(\lambda u + \mu v) = \sum_{i=1}^{n} \lambda(\lambda_i b'_i) + \sum_{i=1}^{n} \mu(\mu_i b'_i)$

$\Rightarrow \quad t(\lambda u + \mu v) = \lambda\left(\sum_{i=1}^{n} \lambda_i b'_i\right) + \mu\left(\sum_{i=1}^{n} \mu_i b'_i\right)$

$\Rightarrow \quad t(\lambda u + \mu v) = \lambda t(u) + \mu t(v)$

Thus, $t(\lambda u + \mu v) = \lambda t(u) + \mu t(v)$ for all $u, v \in V$ and all $\lambda, \mu \in F$.

So, t is a linear transformation.

Hence, $t : V \to V'$ is an isomorphism. Consequently, $V \cong V'$.

Conversely, let V and V' be two finite dimensional vector space over the same field such that $V \cong V'$. Then, we have to show that $\dim V = \dim V'$.

Since V is isomorphic to V'. Therefore, there exists an isomorphism t of V onto V'.

Let $B = \{b_1, b_2, \ldots, b_n\}$ be a basis for V. Then, $\dim V = n$.

We shall now show that $B' = \{t(b_1), t(b_2), \ldots, t(b_n)\}$ is a basis for V'.

B' is l.i.: Let $\lambda_1, \lambda_2, \ldots, \lambda_n$ be scalars in F such that

$\lambda_1 t(b_1) + \lambda_2 t(b_2) + \cdots + \lambda_n t(b_n) = 0_{V'}$

$\Rightarrow \quad t(\lambda_1 b_1 + \lambda_2 b_2 + \cdots + \lambda_n b_n) = t(0_V)$

$\Rightarrow \quad \lambda_1 b_1 + \lambda_2 b_2 + \cdots + \lambda_n b_n = 0_V$ [$\because t$ is a monomorphism]

$\Rightarrow \quad \lambda_1 = \lambda_2 = \cdots = \lambda_n = 0$ [$\because B$ is a basis for V]

So, B' is linearly independent.

B' spans V': Since $t : V \to V'$ is surjective. Therefore, for each $v' \in V'$ there exists $v \in V$ such that $t(v) = v'$.

Now, $v \in V$

$\Rightarrow$ There exist scalars $\lambda_1, \lambda_2, \ldots, \lambda_n \in F$ such that $v = \lambda_1 b_1 + \lambda_2 b_2 + \cdots + \lambda_n b_n$ $\left[\begin{array}{c} \because B \text{ is a} \\ \text{basis for } V \end{array} \right]$

$\Rightarrow v' = t(v) = t(\lambda_1 b_1 + \lambda_2 b_2 + \cdots + \lambda_n b_n)$

$\Rightarrow v' = \lambda_1 t(b_1) + \lambda_2 t(b_2) + \cdots + \lambda_n t(b_n)$

$\Rightarrow v'$ is a linear combination of $t(b_1), t(b_2), \ldots, t(b_n)$

Thus, each vector in V' is a linear combination of vectors in B'. So, B' spans V'.

Hence, B' is a basis for V'. Consequently, $\dim V' = n$.

Hence, $\dim V = \dim V'$.

Q.E.D.

THEOREM-9 *Let V be a finite dimensional vector space over a field F, and let $\underline{v} = (v_1, v_2, \ldots, v_n)$ be a list of n vectors in V. Then the mapping $L_{\underline{v}} : F^n \to V$ given by*

$$L_{\underline{v}}(a_1, a_2, \ldots, a_n) = a_1 v_1 + a_2 v_2 + \cdots + a_n v_n$$

is a linear transformation. Further

(i) $L_{\underline{v}}$ is a monomorphism $\Leftrightarrow \underline{v}$ is linearly independent list.

(ii) $L_{\underline{v}}$ is an epimorphism $\Leftrightarrow \underline{v}$ spans V.

(iii) $L_{\underline{v}}$ is an isomorphism $\Leftrightarrow \underline{v}$ forms a basis for V.

PROOF. Let $x = (\lambda_1, \lambda_2, \ldots, \lambda_n), y = (\mu_1, \mu_2, \ldots, \mu_n)$ be any two elements of F^n and $\lambda, \mu \in F$. Then,

$L_{\underline{v}}(\lambda x + \mu y) = L_{\underline{v}}(\lambda \lambda_1 + \mu \mu_1, \lambda \lambda_2 + \mu \mu_2, \ldots, \lambda \lambda_n + \mu \mu_n)$

$\Rightarrow L_{\underline{v}}(\lambda x + \mu y) = (\lambda \lambda_1 + \mu \mu_1) v_1 + (\lambda \lambda_2 + \mu \mu_2) v_2 + \cdots + (\lambda \lambda_n + \mu \mu_n) v_n$

$\Rightarrow L_{\underline{v}}(\lambda x + \mu y) = \{(\lambda \lambda_1) v_1 + \cdots + (\lambda \lambda_n) v_n\} + \{(\mu \mu_1) v_1 + (\mu \mu_2) v_2 + \cdots + (\mu \mu_n) v_n\}$

$\Rightarrow L_{\underline{v}}(\lambda x + \mu y) = \lambda(\lambda_1 v_1 + \cdots + \lambda_n v_n) + \mu(\mu_1 v_1 + \mu_2 v_2 + \cdots + \mu_n v_n)$

$\Rightarrow L_{\underline{v}}(\lambda x + \mu y) = \lambda L_{\underline{v}}(x) + \mu L_{\underline{v}}(y)$

So, $L_{\underline{v}} : F^n \to V$ is a linear transformation.

(i) Let $L_{\underline{v}} : F^n \to V$ be a monomorphism. Then, we have to prove that the list $\underline{v} = (v_1, v_2, \ldots, v_n)$ is linearly independent.

Let $\lambda_1, \lambda_2, \ldots, \lambda_n$ be scalars in F such that

$\lambda_1 v_1 + \lambda_2 v_2 + \cdots + \lambda_n v_n = 0_V$

$\Rightarrow L_{\underline{v}}(\lambda_1, \lambda_2, \ldots, \lambda_n) = 0_V$ [By definition of $L_{\underline{v}}$]

$\Rightarrow L_{\underline{v}}(\lambda_1, \lambda_2, \ldots, \lambda_n) = L_{\underline{v}}(0, 0, \ldots, 0)$

$\Rightarrow \quad (\lambda_1, \lambda_2, \ldots, \lambda_n) = (0, 0, \ldots, 0)$ $\qquad [\because L_{\underline{v}} \text{ is injective}]$

$\Rightarrow \quad \lambda_1 = \lambda_2 = \cdots = \lambda_n = 0$

So, the list $\underline{v} = (v_1, v_2, \ldots, v_n)$ is linearly independent.

Conversely, let $\underline{v} = (v_1, v_2, \ldots, v_n)$ be a linearly independent list of vectors in V. Then we have to prove that $L_{\underline{v}}$ is a monomorphism. For which it is sufficient to prove that $L_{\underline{v}}$ is injective.

Let $x = (\lambda_1, \lambda_2, \ldots, \lambda_n), y = (\mu_1, \mu_2, \ldots, \mu_n)$ be any two elements of F^n. Then,

$\qquad L_{\underline{v}}(x) = L_{\underline{v}}(y)$

$\Rightarrow \quad L_{\underline{v}}(\lambda_1, \lambda_2, \ldots, \lambda_n) = L_{\underline{v}}(\mu_1, \mu_2, \ldots, \mu_n)$

$\Rightarrow \quad \lambda_1 v_1 + \lambda_2 v_2 + \cdots + \lambda_n v_n = \mu_1 v_1 + \mu_2 v_2 + \cdots + \mu_n v_n$

$\Rightarrow \quad (\lambda_1 - \mu_1) v_1 + (\lambda_2 - \mu_2) v_2 + \cdots + (\lambda_n - \mu_n) v_n = 0_V$

$\Rightarrow \quad \lambda_1 - \mu_1 = 0, \lambda_2 - \mu_2 = 0, \ldots, \lambda_n - \mu_n = 0$ $\qquad [\because \underline{v} \text{ is linearly independent list}]$

$\Rightarrow \quad \lambda_1 = \mu_1, \lambda_2 = \mu_2, \ldots, \lambda_n = \mu_n$

$\Rightarrow \quad (\lambda_1, \lambda_2, \ldots, \lambda_n) = (\mu_1, \mu_2, \ldots, \mu_n)$

$\Rightarrow \quad x = y$

So, $L_{\underline{v}}$ is injective and hence a monomorphism.

(ii) Let $L_{\underline{v}} : F^n \to V$ be an epimorphism. Then, we have to prove that the list $\underline{v} = (v_1, v_2, \ldots, v_n)$ spans V.

$\qquad L_{\underline{v}} : F^n \to V$ is an epimorphism

$\Rightarrow \quad$ For each $v \in V$ there exists $(\lambda_1, \lambda_2, \ldots, \lambda_n) \in F^n$ such that

$\qquad L_{\underline{v}}(\lambda_1, \lambda_2, \ldots, \lambda_n) = v$

$\Rightarrow \quad \lambda_1 v_1 + \lambda_2 v_2 + \cdots + \lambda_n v_n = v$

$\Rightarrow \quad v$ is a linear combination of vectors in the list $\underline{v}$.

$\Rightarrow \quad$ each vector in V is a linear combination of vectors in the list $\underline{v}$.

$\Rightarrow \quad \underline{v}$ spans V.

Conversely, let the list $\underline{v} = (v_1, v_2, \ldots, v_n)$ spans V. Then we have to prove that $L_{\underline{v}} : F^n \to V$ is an epimorphism. For this it is sufficient to show that $L_{\underline{v}} : F^n \to V$ is surjective.

Let v be an arbitrary vector in V. Then there exist scalars $\lambda_1, \lambda_2, \ldots, \lambda_n \in F$ such that

$\qquad v = \lambda_1 v_1 + \lambda_2 v_2 + \cdots + \lambda_n v_n$ $\qquad [\because \text{The list } \underline{v} = (v_1, v_2, \ldots, v_n) \text{ spans } V]$

$\Rightarrow \quad v = L_{\underline{v}}(\lambda_1, \lambda_2, \ldots, \lambda_n)$

Thus, for each $v \in V$ there exists $(\lambda_1, \lambda_2, \ldots, \lambda_n) \in F^n$ such that $L_{\underline{v}}(\lambda_1, \lambda_2, \ldots, \lambda_n) = v$.

So, $L_{\underline{v}} : F^n \to V$ is surjective.

Hence, $L_{\underline{v}}$ is an epimorphism.

(iii) Combining (i) and (ii), we get

$L_{\underline{v}}$ is an isomorphism $\Leftrightarrow$ List $\underline{v} = (v_1, v_2, \ldots, v_n)$ is a basis for V.

Q.E.D.

COROLLARY-1 *Every n-dimension vector space $V(F)$ is isomorphic to F^n.*

PROOF. Let the list $B = (b_1, b_2, \ldots, b_n)$ be an ordered basis for V. Then by the main theorem $L_B : F^n \to V$ is an isomorphism. Hence, $F^n \cong V$.

Q.E.D.

COROLLARY-2 *Any two vector spaces (over the same field) of the same dimension are isomorphic to each other.*

PROOF. Let V and V' be two vector spaces over a field F, and let $\dim V = \dim V' = n$. Then, by the main theorem

$\quad\quad F^n \cong V$ and $F^n \cong V'$

$\Rightarrow \quad V \cong F^n$ and $F^n \cong V'$ \hfill $[\because \cong$ is symmetric$]$

$\Rightarrow \quad V \cong V'$ \hfill $[\because \cong$ is transitive$]$

Q.E.D.

THEOREM-10 *Let S be a subspace of a vector space $V(F)$. Then the quotient space V/S is homomorphic image of V with kernel S.*

PROOF. Consider a mapping $\varphi : V \to V/S$ given by the rule

$$\varphi(v) = v + S \text{ for all } v \in V$$

φ is a linear transformation: Let u, v be any two vectors in V and $\lambda, \mu \in F$. Then,

$\quad\quad \varphi(\lambda u + \mu v) = (\lambda u + \mu v) + S$

$\Rightarrow \quad \varphi(\lambda u + \mu v) = (\lambda u + S) + (\mu v + S)$ \hfill [By definition of addition on V/S]

$\Rightarrow \quad \varphi(\lambda u + \mu v) = \lambda(u + S) + \mu(v + S)$ \hfill [By definition of scalar multiplication on V/S]

$\Rightarrow \quad \varphi(\lambda u + \mu v) = \lambda \varphi(u) + \mu \varphi(v)$

So, φ is a linear transformation.

φ is surjective: Let $v + S$ be an arbitrary element of V/S. Then there exists $v \in V$ such that $\varphi(v) = v + S$.

So, φ is surjective.

Hence, $\varphi : V \to V/S$ is an epimorphism. Consequently V/S is homomorphic image of V.

Now, let $v \in V$ such that $v \in \text{Ker}(\varphi)$. Then,

$v \in \text{Ker}(\varphi)$

$\Leftrightarrow \quad \varphi(v) = S$ \hfill [$\because$ S is zero of V/S]

$\Leftrightarrow \quad v + S = S$ \hfill [By definition of φ]

$\Leftrightarrow \quad v \in S$

So, $\text{Ker } \varphi = S$.

Hence, V/S is homomorphic image of V with kernel S. \hfill Q.E.D.

PROJECTION MAPPING. *Let S be a subspace of a vector space $V(F)$. Then the mapping $p : V \to V/S$ given by $p(v) = v + S$ for all $v \in V$, is an epimorphism. This mapping is called the natural mapping or the canonical mapping or the projection mapping.*

REMARK. *The kernel of the natural mapping $p : V \to V/S$ is S.*

THEOREM-11 (*Main Theorem for Quotient Spaces*) *Let V and V' be two vector spaces over the same field F and let S be a subspace of V. Then for each linear transformation $t : V \to V'$ with $\text{Ker}(t) \supset S$ there exists a unique linear transformation $\varphi : V/S \to V'$ such that $\varphi \circ p = t$, where $p : V \to V/S$ is the projection mapping. In other words, the diagram in Fig. 1 is commutative.*

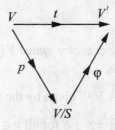

Fig. 1.

Moreover, $I_m(\varphi) = I_m(t)$ and $\text{Ker}(\varphi) = \text{Ker}(t)/S$.

PROOF. Analogous to the proof of main Theorem 16 on page 70 of Chapter 1 on modules in Chapter 1.

THEOREM-12 (*Fundamental Theorem of homomorphisms for vector spaces*). *Let V and V' be vector spaces over the same field F. Let $t : V \to V'$ be a linear transformation. Then, $V/\text{Ker}(t) \cong I_m(t)$.*

OR

Every homomorphic image of a vector space $V(F)$ is isomorphic to some quotient space of V.

PROOF. Let $\text{Ker}(t) = S$. Define the mapping $\varphi : V/S \to V'$ by the rule

$$\varphi(v+S) = t(v) \quad \text{for all } v \in V.$$

First of all, we will show that φ is well defined, i.e. if $u, v \in V$ such that $u + S = v + S$, then $\varphi(u+S) = \varphi(v+S)$.

Now,

$$u + S = v + S$$
$\Rightarrow \quad u - v \in S$ $[\because (S, +) \text{ is an abelian group}]$
$\Rightarrow \quad t(u-v) = 0_{V'}$ $[\because \text{Ker}(t) = S]$
$\Rightarrow \quad t(u) - t(v) = 0_{V'}$
$\Rightarrow \quad t(u) = t(v)$
$\Rightarrow \quad \varphi(u+S) = \varphi(v+S)$

So, φ is well defined.

We shall now show that $\varphi : V/S \to I_m(t)$ is an isomorphism.

φ *is injective*: Let $u+S, v+S$ be arbitrary elements of V/S. Then,

$$\varphi(u+S) = \varphi(v+S)$$
$\Rightarrow \quad t(u) = t(v)$
$\Rightarrow \quad t(u) - t(v) = 0_{V'}$
$\Rightarrow \quad t(u-v) = 0_{V'}$
$\Rightarrow \quad u - v \in S$ $[\because S = \text{Ker}(t)]$
$\Rightarrow \quad u + S = v + S$

So, φ is injective.

φ *is surjective*: Let v' be an arbitrary vector in $I_m(t) \in V'$. Then there exists $v \in V$ such that $t(v) = v'$. But, $t(v)$ is image of $v + S$ under $\varphi : V/S \to V'$. Thus, for each $v' \in I_m(t)$, there exists $v + S \in V/S$ such that $\varphi(v+S) = t(v) = v'$.

So, φ is surjective.

φ *is a linear transformation*: Let $u+S, v+S$ be arbitrary elements of V/S, and let λ, μ be arbitrary scalars in F. Then,

$$\varphi[\lambda(u+S) + \mu(v+S)] = \varphi[(\lambda u + S) + (\mu v + S)] \quad \text{[By definition of scalar multiplication on } V/S]$$
$$= \varphi[(\lambda u + \mu v) + S] \quad \text{[By definition of addition on } V/S]$$
$$= t(\lambda u + \mu v) \quad \text{[By definition of } \varphi]$$
$$= \lambda t(u) + \mu t(v) \quad [\because t \text{ is a linear map}]$$
$$= \lambda \varphi(u+S) + \mu \varphi(v+S)$$

So, φ is a linear transformation.

Hence, $\varphi : V/S \to I_m(t)$ is an isomorphism. Consequently, $V/\text{Ker}(t) \cong I_m(t)$.

Q.E.D.

THEOREM-13 (*Alternative form of the fundamental theorem of homomorphisms*) *Let V and V' be vector spaces over a field F, and let $t : V \to V'$ be an epimorphism with kernel S. Then there exists a unique isomorphism $\varphi : V/S \to V'$ such that $\varphi 0 p = t$, where $p : V \to V/S$ is the projection mapping.*

PROOF. Analogous to the proof of Theorem 18 on page 73 of Chapter 1 on modules.

THEOREM-14 *Let V and V' be vector spaces over the same field F and let $t : V \to V'$ be an epimorphism with kernel S. Then there is a one-to-one correspondence between the set of subspaces of V containing S and the set of subspaces of V'.*

PROOF. We first establish that if W is a subspace of V containing S, then $t^{-1}(t(W)) = W$.

Obviously, $t^{-1}(t(W)) \supset W$.

Now let u be an arbitrary vector $t^{-1}(t(W))$. Then,

$$t(u) \in t(W)$$
$$\Rightarrow \quad t(u) = t(v) \quad \text{for some } v \in W$$
$$\Rightarrow \quad t(u) - t(v) = 0_{V'}$$
$$\Rightarrow \quad t(u-v) = 0_{V'}$$
$$\Rightarrow \quad u - v \in \text{Ker}(t)$$
$$\Rightarrow \quad u - v \in W \qquad [\because \text{Ker}(t) = S \subset W]$$
$$\Rightarrow \quad (u-v) + v \in W \qquad [\because v \in W]$$
$$\Rightarrow \quad u \in W$$

Thus, $u \in t^{-1}(t(W)) \Rightarrow u \in W$.

$\therefore \quad t^{-1}(t(W)) \subset W$.

Hence, $t^{-1}(t(W)) = W$ \hfill (i)

Let A and B be the family of subspaces of V containing S and the family of subspaces of V'.

We now consider the mapping $\varphi : A \to B$ given by

$$\varphi(W) = t(W) \quad \text{for all } W \in A.$$

φ is injective: Let W_1, W_2 be any two members of A such that $\varphi(W_1) = \varphi(W_2)$. Then,

$$t(W_1) = t(W_2)$$
$$\Rightarrow \quad t^{-1}(t(W_1)) = t^{-1}(t(W_2))$$
$$\Rightarrow \quad W_1 = W_2 \qquad \text{[Using (i)]}$$

So, φ is injective.

φ *is surjective*: Let S' be an arbitrary subspaces of V'. Then it can be easily seen that the set

$$t^{-1}(S) = \{v \in V : t(v) \in S'\}$$

is a subspace of V containing S.

Obviously, $t(t^{-1}(S)) = S$.

So, t is surjective.

Hence, t is a bijective map between the family of subspaces of V containing S and the subspaces of V'.

Q.E.D.

THEOREM-15 *Let S be a subspace of a vector space $V(F)$. Then every subspace of V/S is of the form W/S where W is a subspace of V containing S.*

PROOF. Let X be a subspace of V/S. We know that the projection mapping $p : V \to V/S$ is an epimorphism with kernel S. Therefore, by Theorem 14, there exists a subspace W of V containing S such that $X = p(W) = \{v + S : v \in W\} = W/S$.

Q.E.D.

THEOREM-16 *(First Isomorphism Theorem for vector space homomorphisms). Let S and T be subspaces of a vector space $V(F)$. Then,*

(i) $S \cap T$ is a subspace of V.
(ii) $S + T$ is a subspace of V.
(iii) $S/S \cap T \cong S + T/T$.

PROOF. Analogous to the proof of Theorem 20 on page 76 on modules.

Q.E.D.

THEOREM-17 *(Second Isomorphism Theorem for vector space homomorphisms). If S and T are subspaces of a vector space $V(F)$ such that $S \supset T$, then*

$$(V/T)/(S/T) \cong V/S.$$

PROOF. Analogous to the proof of Theorem 21 on page 78 on modules.

Q.E.D.

THEOREM-18 *Let S_1 and S_2 be subspaces of vector spaces $V_1(F)$ and $V_2(F)$ respectively. Then*

$$((V_1 \times V_2)/(S_1 \times S_2)) \cong (V_1/S_1) \times (V_2/S_2).$$

PROOF. Analogous to the proof of Theorem 22 on page 79 on modules.

Q.E.D.

THEOREM-19 Let V be a finite dimensional vector space over a field F and let $t : V \to V$ be a monomorphism. Then t is an epimorphism, and so, an isomorphism.

PROOF. Let $B = \{b_1, b_2, \ldots, b_n\}$ be a basis for V. First we will establish that $B' = \{t(b_1), t(b_2), \ldots, t(b_n)\}$ is also a basis for V.

B' is l.i.: Let $\lambda_1, \lambda_2, \ldots, \lambda_n \in F$ be such that $\lambda_1 t(b_1) + \lambda_2 t(b_2) + \cdots + \lambda_n t(b_n) = 0_V$. Then,

$$t(\lambda_1 b_1 + \lambda_2 b_2 + \cdots + \lambda_n b_n) = t(0_V) \qquad [\because t : V \to V \text{ is linear}]$$
$$\Rightarrow \quad \lambda_1 b_1 + \lambda_2 b_2 + \cdots + \lambda_n b_n = 0_V \qquad [\because t \text{ is injective}]$$
$$\Rightarrow \quad \lambda_1 = \lambda_2 = \cdots = \lambda_n = 0 \qquad [\because B \text{ is a basis for } V]$$

So, B' is linearly independent.

Since V is of dimension n and B' is a linearly independent set of n vectors in V. Therefore B' is a basis for V.

Now let v be an arbitrary vector in V, then there exist scalars $\lambda_1, \lambda_2, \ldots, \lambda_n \in F$ such that

$$v = \lambda_1 t(b_1) + \lambda_2 t(b_2) + \cdots + \lambda_n t(b_n) \qquad [\because B' \text{ is a basis for } V]$$
$$\Rightarrow \quad v = t(\lambda_1 b_1 + \lambda_2 b_2 + \cdots + \lambda_n b_n) \qquad [\because t \text{ is linear}]$$
$$\Rightarrow \quad v = t(u), \text{ where } u = \sum_{i=1}^{n} \lambda_i b_i \in V \qquad [\because B \text{ is a basis for } V]$$

Thus, for each $v \in V$, there exists $u \in V$ such that $t(u) = v$. Therefore, t is surjective.

Hence, t is an isomorphism.

Q.E.D.

THEOREM-20 Let $V(F)$ and $V'(F)$ be two finite dimensional vector spaces such that $\dim V = \dim V'$. If $t : V \to V'$ is a monomorphism, then t is an isomorphism.

PROOF. Proceed as in Theorem 19.

Q.E.D.

THEOREM-21 Let $V(F)$ be a finite dimensional vector space and let $t : V \to V$ be an epimorphism. Then t is a monomorphism, and so, an isomorphism.

PROOF. In order to prove that t is a monomorphism it is sufficient to show that t is injective.

Let $B = \{b_1, b_2, \ldots, b_n\}$ be a basis for V, and let $B' = \{t(b_1), t(b_2), \ldots, t(b_n)\}$.

First we shall show that B' spans V.

Let v be an arbitrary vector in V. Then there exists $u \in V$ such that $t(u) = v$, since $t : V \to V$ is surjective.

Now,

$u \in V$

$\Rightarrow \quad u = \lambda_1 b_1 + \lambda_2 b_2 + \cdots + \lambda_n b_n$ $\qquad [\because B \text{ is a basis for } V]$

$\Rightarrow \quad t(u) = t(\lambda_1 b_1 + \lambda_2 b_2 + \cdots + \lambda_n b_n)$

$\Rightarrow \quad t(u) = \lambda_1 t(b_1) + \lambda_2 t(b_2) + \cdots + \lambda_n t(b_n)$

$\Rightarrow \quad v = \lambda_1 t(b_1) + \lambda_2 t(b_2) + \cdots + \lambda_n t(b_n)$

$\Rightarrow \quad v$ is a linear combination of vectors in B'.

Thus, each $v \in V$ is a linear combination of vectors in B'. So, B' spans V.

Since $\dim(V) = n$ and B' is a set of n vectors in V such that B' spans V. Therefore, B' is a linearly independent set and so, a basis for V.

Let u, v be any two vectors in V. Then,

$$u = \sum_{i=1}^{n} \lambda_i b_i, \; v = \sum_{i=1}^{n} \mu_i b_i \quad \text{for some } \lambda_i, \mu_i \in F$$

$\therefore \quad t(u) = t(v)$

$\Rightarrow \quad t\left(\sum_{i=1}^{n} \lambda_i b_i\right) = t\left(\sum_{i=1}^{n} \mu_i b_i\right)$

$\Rightarrow \quad \sum_{i=1}^{n} \lambda_i t(b_i) = \sum_{i=1}^{n} \mu_i t(b_i)$ $\qquad [\because t \text{ is linear}]$

$\Rightarrow \quad \sum_{i=1}^{n} (\lambda_i - \mu_i) t(b_i) = 0_V$

$\Rightarrow \quad \lambda_i - \mu_i = 0 \quad \text{for all } i \in \underline{n}$ $\qquad [\because B' \text{ is linearly independent}]$

$\Rightarrow \quad \lambda_i = \mu_i \quad \text{for all } i \in \underline{n}$

$\Rightarrow \quad \sum_{i=1}^{n} \lambda_i b_i = \sum_{i=1}^{n} \mu_i b_i \Rightarrow u = v$

$\therefore \quad t$ is injective and hence an isomorphism.

Q.E.D.

REMARK. *In the above two examples, we have shown that every monomorphism from a finite dimensional vector to itself is an epimorphism and every epimorphism is a monomorphism. However, if V is not finite dimensional, then an epimorphism (or a monomorphism) from V to itself need not be monomorphism (epimorphism). For example, the differential operator D in Example 5 (page 237) is an epimorphism but not a monomorphism.*

THEOREM-22 *Let V be a finite dimensional vector space over a field F, and let $t : V \to V$ be a linear transformation which is not epimorphism. Then there exists $v(\neq 0_V)$ in V such that $t(v) = 0_V$.*

PROOF. Suppose on the contrary that there is no non-zero vector $v \in V$ such that $t(v) = 0_V$. Then for any $u, v \in V$

$$t(u) = t(v) \Rightarrow t(u) - t(v) = 0_V \Rightarrow t(u-v) = 0_V \Rightarrow u - v = 0_V \Rightarrow u = v$$

$\therefore \quad t : V \to V$ is injective

$\Rightarrow \quad t : V \to V$ is surjective [By Theorem 19]

$\Rightarrow \quad t$ is an epimorphism, a contradiction to the hypothesis.

Therefore, our supposition is wrong.

Hence, there exists a non-zero vector $v \in V$ such that $t(v) = 0_V$.

Q.E.D.

3.4.1 KERNEL AND IMAGE OF A MATRIX AS A LINEAR TRANSFORMATION

Let A be an $m \times n$ matrix over a field F. As discussed in Section 3.3.1, matrix A may be viewed as a linear transformation t_A from $F^{n \times 1}$ to $F^{m \times 1}$ such that

$$t_A(X) = AX \quad \text{for all } X \in F^{n \times 1}$$

Consider the usual basis $e_1^{(n)} = [1, 0, 0, \ldots, 0]^T$, $e_2^{(n)} = [0, 1, 0, \ldots, 0]^T, \ldots, e_n^{(n)} = [0, 0, 0, \ldots, 1]^n$ of $F^{n \times 1}$.

In section 1.8.1, we have learnt that, $t_A\left(e_1^{(n)}\right)$, $t_A\left(e_2^{(n)}\right), \ldots, t_A\left(e_n^{(n)}\right)$, i.e. the columns of matrix A span the image of t_A. Thus, the image of t_A is the column space of A.

The kernel of t_A consists of all vectors X in $F^{n \times 1}$ for which

$$t_A(X) = O \quad \text{i.e.} \quad AX = O$$

Thus, the kernel of t_A is the solution space of the homogeneous system $AX = O$, which is also known as the null space of matrix A.

If follows from the above discussion that every matrix may be viewed as a linear transformation whose image is its column space and kernel is the solution space of the homogeneous system $AX = O$.

3.5 RANK AND NULLITY OF A LINEAR TRANSFORMATION

RANK Let V and V' be vector spaces over the same field F, and let $t : V \to V'$ be a linear transformation. Then $I_m(t)$ is a subspace of V'. If $I_m(t)$ is finite dimensional, then its dimension is called rank of t, written as rank (t).

Thus, rank $(t) = \dim(I_m(t))$.

NULLITY Let $V(F)$ and $V'(F)$ be two vector spaces, and let $t : V \to V'$ be a linear transformation. If $\text{Ker}(t)$ is finite dimensional, then its dimension is called nullity of t.

Thus, nullity $(t) = \dim(\text{Ker}(t))$.

THEOREM-1 *Let V and V' be vector spaces over the same field F such that V is spanned by a finite subset $S = \{v_1, v_2, \ldots, v_m\}$ of V. Let $t : V \to V'$ be a linear transformation. Then, $I_m(t)$ is spanned by $t(S) = \{t(v_1), t(v_2), \ldots, t(v_m)\}$ and $\operatorname{rank}(t) \leq \dim V$.*

PROOF. If $S = \phi$, then $[S] = \{0_V\}$, i.e. $V = \{0_V\}$.

$\therefore \quad I_m(t) = \{t(0_V)\} = \{0_{V'}\}$, which is spanned by $\phi = t(\phi)$.

Also, $\operatorname{rank}(t) = \dim(I_m(t)) = 0 = \dim V$.

So, the theorem is proved in this case.

Let us now assume that $S \neq \phi$.

Let v be an arbitrary vector in V. As S spans V. So, there exist scalars $\lambda_1, \lambda_2, \ldots, \lambda_m$ in F such that

$$v = \sum_{i=1}^{m} \lambda_i v_i$$

$\Rightarrow \quad t(v) = \sum_{i=1}^{m} \lambda_i t(v_i)$

$\Rightarrow \quad t(v)$ is expressible as a linear combination of $t(v_1), t(v_2), \ldots, t(v_m)$.

Thus, each vector in $I_m(t)$ is expressible as a linear combination of $t(v_1), t(v_2), \ldots, t(v_m)$. So, $t(S) = \{t(v_1), t(v_2), \ldots, t(v_m)\}$ spans $I_m(t)$.

Let $B = \{b_1, b_2, \ldots, b_n\}$ be a basis of V. Then, B spans V and hence $t(B)$ spans $I_m(t)$ (as proved above).

$\therefore \quad \operatorname{rank}(t) = \dim\{I_m(t)\} \leq n$

$\Rightarrow \quad \operatorname{rank}(t) \leq \dim V.$

Q.E.D.

THEOREM-2 *Let V and V' be two vector spaces over the same field F, and let $t : V \to V'$ be a linear transformation. If S is a finite dimensional subspace of V such that $S \cap \operatorname{Ker}(t) = \{0_V\}$. Then, the following hold:*

(i) *If $B = \{b_1, b_2, \ldots, b_m\}$ is a basis of S, then $t(B) = \{t(b_1), t(b_2), \ldots, t(b_m)\}$ is a basis of $t(S)$.*

(ii) *$\dim S = \dim t(S)$*

(iii) *If t is one-one and V is finite dimensional, then $\operatorname{rank}(t) = \dim V$.*

PROOF. (i) By Theorem 1, $t(B)$ spans $t(S)$. Let $\lambda_1, \lambda_2, \ldots, \lambda_m$ be scalars in F such that

$$\sum_{i=1}^{m} \lambda_i t(b_i) = 0_{V'}$$

$\Rightarrow \quad t\left(\sum_{i=1}^{m} \lambda_i b_i\right) = 0_{V'}$

$\Rightarrow \quad \sum_{i=1}^{m} \lambda_i b_i \in \operatorname{Ker}(t)$

Also, B spans S. Therefore, $\sum_{i=1}^{m} \lambda_i b_i \in S$.

$\therefore \quad \sum_{i=1}^{m} \lambda_i b_i \in S \cap \text{Ker}(t)$

$\Rightarrow \quad \sum_{i=1}^{m} \lambda_i b_i = 0_V$ $\quad\quad\quad\quad\quad\quad\quad\quad\quad\quad\quad\quad\quad\quad$ [$\because \quad S \cap \text{Ker}(t) = \{0_V\}$]

$\Rightarrow \quad \lambda_i = 0, \quad \text{for all } i \in \underline{m}$ $\quad\quad\quad\quad\quad\quad\quad\quad\quad\quad\quad$ [$\because \quad B$ is a basis of S]

Thus, $\sum_{i=1}^{m} \lambda_i t(b_i) = 0_{V'} \Rightarrow \lambda_i = 0, \quad \text{for all } i \in \underline{m}$

$\therefore \quad t(b_1), t(b_2), \ldots, t(b_m)$ are linearly independent.

Hence, $t(B)$ is a basis of $t(S)$.

(ii) Since B and $t(B)$ are bases of S and $t(S)$ respectively.

$\therefore \quad \dim S = \dim t(S)$ $\quad\quad\quad\quad\quad\quad\quad\quad$ [B and $t(B)$ have same number of vectors]

(iii) Let V be finite dimensional of dimension n and let t be one-one. Then,

$\text{Ker}(t) = \{0_V\}$

$\therefore \quad V \cap \text{Ker}(t) = \{0_V\}$

So, taking $S = V$ in (ii), we get

$\dim V = \dim\left(t(V)\right) \Rightarrow \dim V = \text{rank}(t)$

Q.E.D.

Let V and V' be finite dimensional vector spaces over a field F. Then each linear transformation $t : V \to V'$ provides us two subspaces, namely, $\text{Ker}(t)$ and $I_m(t)$. Dimensions of these two subspaces are related with the dimension of V. The following theorem establishes that relation.

THEOREM-3 (*Sylvester's Law of Nullity*). *Let $V(F)$ and $V'(F)$ be finite dimensional vector spaces. If $t : V \to V'$ is a linear transformation, then*

$$\dim(I_m(t)) + \dim(\text{Ker}(t)) = \dim V \quad \text{i.e.} \quad \text{rank}(t) + \text{nullity}(t) = \dim V.$$

PROOF. Since V and V' are finite dimensional vector spaces. Therefore, $\text{Ker}(t)$ and $I_m(t)$ being subspaces of V and V', respectively are also finite dimensional.

Let $\dim V = n$ and $\dim(\text{Ker}(t)) = \text{nullity}(t) = m$. Then, $m \leq n$.

Let $\{b_1, b_2, \ldots, b_m\}$ be a basis for $\text{Ker}(t)$. Then it is a linearly independent set of vectors in V and hence it can be extended to form a basis for V. Let $B = \{b_1, b_2, \ldots, b_m, v_1, v_2, \ldots, v_{n-m}\}$ be the basis for V. Then vectors $t(b_1), t(b_2), \ldots, t(b_m), t(v_1), t(v_2), \ldots, t(v_{n-m})$ are in $I_m(t)$.

Let $B' = \{t(v_1), t(v_2), \ldots, t(v_{n-m})\}$. Then, $B' \subset I_m(t)$.

We shall now establish that B' is a basis for $I_m(t)$.

B' is l.i.: Let $\lambda_1, \lambda_2, \ldots, \lambda_{n-m}$ be scalars in F such that

$$\lambda_1 t(v_1) + \lambda_2 t(v_2) + \cdots + \lambda_{n-m} t(v_{n-m}) = 0_{V'}$$
$\Rightarrow \quad t(\lambda_1 v_1 + \lambda_2 v_2 + \cdots + \lambda_{n-m} v_{n-m}) = 0_{V'}$
$\Rightarrow \quad \lambda_1 v_1 + \lambda_2 v_2 + \cdots + \lambda_{n-m} v_{n-m} \in \text{Ker}(t)$
$\Rightarrow \quad$ There exist scalars $\mu_1, \mu_2, \ldots, \mu_m \in F$ such that

$$\lambda_1 v_1 + \lambda_2 v_2 + \cdots + \lambda_{n-m} v_{n-m} = \mu_1 b_1 + \mu_2 b_2 + \cdots + \mu_m b_m \quad \left[\begin{array}{l} \because \{b_1, b_2, \ldots, b_m\} \text{ is a} \\ \text{basis for Ker}(t). \end{array}\right]$$

$\Rightarrow \quad \lambda_1 v_1 + \lambda_2 v_2 + \cdots + \lambda_{n-m} v_{n-m} + (-\mu_1) b_1 + (-\mu_2) b_2 + \cdots + (-\mu_m) b_m = 0_V$
$\Rightarrow \quad \lambda_1 = \lambda_2 = \cdots = \lambda_{n-m} = 0 = \mu_1 = \mu_2 = \cdots = \mu_m \qquad [\because B \text{ is a basis for } V]$

So, B' is a linearly independent set.

B' spans $I_m(t)$: Let v' be an arbitrary vector in $I_m(t)$. Then there exists $v \in V$ such that $t(v) = v'$.

Now,
$v \in V \Rightarrow$ There exist scalars $\lambda_1, \lambda_2, \ldots, \lambda_n \in F$ such that

$$v = \lambda_1 b_1 + \lambda_2 b_2 + \cdots + \lambda_m b_m + \lambda_{m+1} v_1 + \cdots + \lambda_n v_{n-m} \qquad [\because B \text{ is a basis for } V]$$

$\Rightarrow \quad t(v) = t(\lambda_1 b_1 + \lambda_2 b_2 + \cdots + \lambda_m b_m + \lambda_{m+1} v_1 + \cdots + \lambda_n v_{n-m})$
$\Rightarrow \quad t(v) = \lambda_1 t(b_1) + \lambda_2 t(b_2) + \cdots + \lambda_m t(b_m) + \lambda_{m+1} t(v_1) + \cdots + \lambda_n t(v_{n-m}) \qquad [\because t \text{ is linear}]$
$\Rightarrow \quad t(v) = \lambda_1 0_V + \lambda_2 0_V + \cdots + \lambda_m 0_V + \lambda_{m+1} t(v_1) + \cdots + \lambda_n t(v_{n-m}) \qquad \left[\begin{array}{l} \because b_i \in \text{Ker}(t) \\ \text{for all } i \in \underline{m} \end{array}\right]$
$\Rightarrow \quad t(v) = \lambda_{m+1} t(v_1) + \lambda_{m+2} t(v_2) + \cdots + \lambda_n t(v_{n-m})$
$\Rightarrow \quad v' = \lambda_{m+1} t(v_1) + \lambda_{m+2} t(v_2) + \cdots + \lambda_n t(v_{n-m}) \qquad [\because t(u) = v']$
$\Rightarrow \quad v'$ is a linear combination of vectors in B'.

Thus, $v' \in I_m(t) \Rightarrow v'$ is a linear combination of vectors in B'.

So, B' spans $I_m(t)$.

Hence, B' is a basis for $I_m(t)$.

Consequently,

$\text{rank}(t) = \dim(I_m(t)) = $ Number of vectors in $B' = n - m$
$\Rightarrow \quad \text{rank}(t) = \dim V - \dim(\text{Ker}(t)) = \dim V - \text{nullity}(t)$

Hence, rank (t) + nullity $(t) = \dim V$.

Q.E.D.

Aliter By the fundamental theorem of homomorphisms for vector spaces, we have

$$I_m(t) \cong V/\text{Ker}(t)$$
$\Rightarrow \quad \dim(I_m(t)) = \dim(V/\text{Ker}(t))$ [By Theorem 8 on page 253]
$\Rightarrow \quad \dim(I_m(t)) = \dim V - \dim(\text{Ker}(t))$ $\quad [\because \dim(V/S) = \dim V - \dim S]$
$\Rightarrow \quad \dim(I_m(t)) + \dim(\text{Ker}(t)) = \dim V$
$\Rightarrow \quad \text{rank}(t) + \text{nullity}(t) = \dim V.$

Q.E.D.

COROLLARY-1 *Let $V(F)$ and $V'(F)$ be two vector spaces of dimensions m and n respectively, and let $t : V \to V'$ be a linear transformation of rank r. Then*

(i) $r \leq \min(m, n)$

and, (ii) *there exist bases $B = \{b_1, b_2, \ldots, b_m\}$ and $B' = \{b'_1, b'_2, \ldots, b'_n\}$ of V and V' respectively such that $t(b_1) = b'_1, t(b_2) = b'_2, \ldots, t(b_r) = b'_r, t(b_{r+1}) = 0_{V'}, t(b_{r+2}) = 0_{V'}, \ldots, t(b_m) = 0_{V'}.$*

PROOF. (i) By the main theorem, we have

$$\dim V = \text{rank}(t) + \text{nullity}(t)$$
$\Rightarrow \quad \text{rank}(t) \leq \dim V \Rightarrow r \leq m$ \hfill (i)

Since $I_m(t)$ is a subspace of V'. Therefore,

$$\dim(I_m(t)) \leq \dim V'$$
$\Rightarrow \quad \text{rank}(t) \leq \dim V'$
$\Rightarrow \quad r \leq n$ \hfill (ii)

From (i) and (ii), we get $r \leq \min(m, n)$

(ii) By the main theorem, we have

$$\dim V = \text{rank}(t) + \text{nullity}(t)$$
$\Rightarrow \quad \text{nullity}(t) = \dim V - \text{rank}(t)$
$\Rightarrow \quad \dim(\text{Ker}(t)) = m - r$

Let $B_1 = \{b_{r+1}, b_{r+2}, \ldots, b_m\}$ be a basis for $\text{Ker}(t)$. Then B_1 is a linearly independent set of vectors in V and so it can be extended to form a basis $B = \{b_1, b_2, \ldots, b_r, b_{r+1}, \ldots, b_m\}$ of V. Let $B_2 = \{b_1, b_2, \ldots, b_r\}$. Then B_2 being subset of a linearly independent set, is a linearly independent set.

We will now show that the set $\{t(b_1), t(b_2), \ldots, t(b_r)\}$ is a linearly independent set.

Let $\lambda_1, \lambda_2, \ldots, \lambda_r$ be scalars in F such that
$$\lambda_1 t(b_1) + \lambda_2 t(b_2) + \cdots + \lambda_r t(b_r) = 0_{V'}. \text{Then,}$$
$$t(\lambda_1 b_1 + \lambda_2 b_2 + \cdots + \lambda_r b_r) = 0_{V'} \qquad [\because t \text{ is linear}]$$
$\Rightarrow \quad \lambda_1 b_1 + \lambda_2 b_2 + \cdots + \lambda_r b_r \in \text{Ker}(t)$
$\Rightarrow \quad \lambda_1 b_1 + \lambda_2 b_2 + \cdots + \lambda_r b_r = \lambda_{r+1} b_{r+1} + \lambda_{r+2} b_{r+2} + \cdots + \lambda_m b_m \quad [\because B_1 \text{ is a basis for Ker}(t)]$
$\Rightarrow \quad \lambda_1 b_1 + \lambda_2 b_2 + \cdots + \lambda_r b_r + (-\lambda_{r+1}) b_{r+1} + \cdots + (-\lambda_m) b_m = 0_V$
$\Rightarrow \quad \lambda_1 = \lambda_2 = \cdots = \lambda_r = \lambda_{r+1} = \cdots = \lambda_m = 0 \qquad [\because B \text{ is a basis for } V]$

So, the set $\{t(b_1), t(b_2), \ldots, t(b_r)\}$ is a linearly independent set of vectors in V'. Therefore, it is a part of the basis of V'. Consequently, there exist vectors $b'_1, b'_2, \ldots, b'_r$ in the basis B' of V' such that $t(b_1) = b'_1$, $t(b_2) = b'_2, \ldots, t(b_r) = b'_r$. Since vectors $b_{r+1}, b_{r+2}, \ldots, b_m$ are in Ker(t). Therefore, $t(b_{r+1}) = 0_{V'}, \ldots, t(b_m) = 0_{V'}$. Thus, there exist a basis $B = \{b_1, b_2, \ldots, b_m\}$ of V and a basis $B' = \{b'_1, b'_2, \ldots, b'_n\}$ of V' such that $t(b_1) = b'_1, t(b_2) = b'_2, \ldots, t(b_r) = b'_r$, $t(b_{r+1}) = 0_{V'}, t(b_{r+2}) = 0_{V'}, \ldots, t(b_m) = 0_{V'}$.

Q.E.D.

COROLLARY-2 *Let $V(F)$ and $V'(F)$ be two vector spaces of the same dimension. Then any epimorphism $t : V \to V'$ and also any monomorphism $t : V \to V'$ is an isomorphism.*

PROOF. First let $t : V \to V'$ be an epimorphism. Then $I_m(t) = V'$.
By Sylvester's law of nullity, we have
$$\text{rank}(t) + \text{nullity}(t) = \dim V$$
$\Rightarrow \quad \dim V' + \text{nullity}(t) = \dim V \qquad [\because I_m(t) = V']$
$\Rightarrow \quad \text{nullity}(t) = 0 \qquad [\because \dim V = \dim V']$
$\Rightarrow \quad \text{Ker}(t) = \{0_V\}$
$\Rightarrow \quad t : V \to V' \text{ is a monomorphism.} \qquad \text{[By Theorem 2 on page 248]}$
Hence, $t : V \to V'$ is an isomorphism. $\qquad [\because t : V \to V' \text{ is an epimorphism}]$

Now let $t : V \to V'$ be a monomorphism, then by Theorem 2 on the page 248, Ker$(t) = \{0_V\}$. This implies that
$$\dim(\text{Ker}(t)) = 0$$
$\Rightarrow \quad \text{nullity}(t) = 0$
$\Rightarrow \quad \dim V = \text{rank}(t) \qquad [\because \text{rank}(t) + \text{nullity}(t) = \dim V]$
$\Rightarrow \quad \dim V' = \dim(I_m(t)) \qquad [\because \dim V = \dim V']$
$\Rightarrow \quad t$ is an epimorphism
$\Rightarrow \quad t$ is an isomorphism $\qquad [\because t \text{ is a monomorphism}]$

Q.E.D.

COROLLARY-3 Let S be a subspace of a finite dimensional vector space V. Then,
$$\dim(V/S) = \dim V - \dim S.$$

PROOF. Since the projection mapping $p : V \to V/S$, given by $p(v) = v + S$ for all $v \in V$, is an epimorphism with kernel S. Therefore, by Sylvester's law of nullity, we have

$$\dim V = \dim(I_m(p)) + \dim S$$
$$\Rightarrow \quad \dim V = \dim(V/S) + \dim S \qquad [\because I_m(p) = V/S]$$
$$\Rightarrow \quad \dim(V/S) = \dim V - \dim S$$

Q.E.D.

COROLLARY-4 Let S be a subspace of a finite dimensional vector space $V(F)$. Then there exists a subspace T of V such that (i) $V = S \oplus T$ (ii) $V/S \cong T$

PROOF. (i) Let $\dim V = n$, $\dim S = m$, and let $B_1 = \{b_1, b_2, \ldots, b_m\}$ be a basis of S. Then B_1 is a linearly independent set of vectors in V. Therefore, it can be extended to form a basis $B = \{b_1, b_2, \ldots, b_m, b_{m+1}, \ldots, b_n\}$ of V. Let T be a subspace of V spanned by the set $B_2 = \{b_{m+1}, b_{m+2}, \ldots, b_n\}$.

We shall now show that T is the required subspace of V.

In order to prove that $V = S \oplus T$, we have to prove that every vector in V is expressible as the sum of a vector in S and a vector in T and $S \cap T = \{0_V\}$.

Let v be an arbitrary vector in V. Then there exist scalars $\lambda_1, \lambda_2, \ldots, \lambda_m, \lambda_{m+1}, \ldots, \lambda_n$ in F such that

$$v = \lambda_1 v_1 + \lambda_2 v_2 + \cdots + \lambda_m v_m + \lambda_{m+1} v_{m+1} + \cdots + \lambda_n v_n \qquad [\because B \text{ is basis for } V]$$
$$\Rightarrow v = (\lambda_1 v_1 + \lambda_2 v_2 + \cdots + \lambda_m v_m) + (\lambda_{m+1} v_{m+1} + \cdots + \lambda_n v_n)$$
$$\Rightarrow v = u + w, \text{ where } u = \sum_{i=1}^{m} \lambda_i v_i \in S \text{ and } w = \sum_{i=m+1}^{n} \lambda_i v_i \in T.$$

Thus, each vector in V is expressible as the sum of a vector in S and a vector in T.
So, $V = S + T$.
Let v be an arbitrary vector in $S \cap T$. Then, $v \in S$ and $v \in T$.
Now,

$$v \in S \Rightarrow v = \lambda_1 b_1 + \lambda_2 b_2 + \cdots + \lambda_m b_m \quad \text{for some } \lambda_i \in F \qquad [\because B_1 \text{ is basis for } S]$$
$$\text{and, } v \in T \Rightarrow v = \mu_1 b_{m+1} + \mu_2 b_{m+2} + \cdots + \mu_{n-m} b_n \text{ for some } \mu_i \in F. \qquad [\because B_2 \text{ is basis for } T]$$

$$\therefore \quad \lambda_1 b_1 + \lambda_2 b_2 + \cdots + \lambda_m b_m = \mu_1 b_{m+1} + \mu_2 b_{m+2} + \cdots + \mu_{n-m} b_n$$
$$\Rightarrow \quad \lambda_1 b_1 + \lambda_2 b_2 + \cdots + \lambda_m b_m + (-\mu_1) b_{m+1} + (-\mu_2) b_{m+2} + \cdots + (-\mu_{n-m}) b_n = 0_V$$
$$\Rightarrow \quad \lambda_1 = \lambda_2 = \cdots = \lambda_m = 0 = \mu_1 = \mu_2 = \cdots = \mu_{n-m} \qquad [\because B \text{ is basis for } V]$$
$$\Rightarrow \quad v = 0_V.$$

Thus, $v \in S \cap T \Rightarrow v = 0_V$.
So, $S \cap T = \{0_V\}$.
Hence, $V = S \oplus T$.
(ii) By Corollary 3 (page 270), we have
$$\dim(V/S) = \dim V - \dim S = n - m = \dim T.$$
Therefore, by Theorem 8 (page 253), $V/S \cong T$. \hfill Q.E.D.

THEOREM-4 *Let $V(F)$ and $V'(F)$ be finite dimensional vector spaces, and let $t : V \to V'$ be a linear transformation. Then for any subspace S of V, $\dim(t(S)) \geq \dim S - \text{nullity}(t)$.*

PROOF. Let t_1 be the restriction of t to S. Then $t_1 : S \to V'$ is a linear transformation. Therefore, by Sylvester's law of nullity, we have
$$\dim S = \text{rank}(t_1) + \text{nullity}(t_1) \qquad (i)$$
Since t_1 is restriction of t to S. Therefore,
$$t(S) = t_1(S)$$
$$\Rightarrow \quad \dim(t(S)) = \dim t_1(S)$$
$$\Rightarrow \quad \dim(t(S)) = \text{rank}(t_1)$$
$$\Rightarrow \quad \dim S = \dim(t(S)) + \text{nullity}(t_1) \qquad \text{[Using (i)]} \quad (ii)$$
Since $\text{Ker}(t_1) = \text{Ker}(t) \cap S$. Therefore,
$$\dim \text{Ker}(t_1) \leq \dim \text{Ker}(t)$$
$$\Rightarrow \quad \text{nullity}(t_1) \leq \text{nullity}(t) \qquad (iii)$$
From (ii) and (iii), we have
$$\dim S \leq \dim(t(S)) + \text{nullity}(t)$$
$$\Rightarrow \quad \dim(t(S)) \geq \dim S - \text{nullity}(t).$$
\hfill Q.E.D.

ILLUSTRATIVE EXAMPLES

EXAMPLE-1 *Show that the mapping $t : R^2 \to R^3$ given by*
$$t(a,b) = (a+b, a-b, b) \quad \text{for all } (a,b) \in R^2,$$
is a linear transformation. Find the range, rank, kernel and nullity of t.

SOLUTION For any $x = (a,b)$, $y = (c,d) \in R^2$ and any $\lambda, \mu \in R$, we have

$$t(\lambda x + \mu y) = t(\lambda a + \mu c, \lambda b + \mu d)$$
$$\Rightarrow t(\lambda x + \mu y) = (\lambda a + \mu c + \lambda b + \mu d, \lambda a + \mu c - \lambda b - \mu d, \lambda b + \mu d)$$
$$\Rightarrow t(\lambda x + \mu y) = (\lambda a + \lambda b + \mu c + \mu d, \lambda a - \lambda b + \mu c - \mu d, \lambda b + \mu d)$$
$$\Rightarrow t(\lambda x + \mu y) = (\lambda a + \lambda b, \lambda a - \lambda b, \lambda b) + (\mu c + \mu d, \mu c - \mu d, \mu d)$$
$$\Rightarrow t(\lambda x + \mu y) = \lambda (a+b, a-b, b) + \mu (c+d, c-d, d)$$
$$\Rightarrow t(\lambda x + \mu y) = \lambda t(x) + \mu t(y)$$

$\therefore \quad t : R^2 \to R^3$ is a linear transformation.

Since $B = \{(1,0), (0,1)\}$ is a basis for R^2 and $t(1,0) = (1,1,0), t(0,1) = (1,-1,1)$. Therefore, $B' = \{t(1,0), t(0,1)\} = \{(1,1,0), (1,-1,1)\}$ spans range $= I_m(t)$ (See Theorem 3 on page 249). Moreover, B' is a linearly independent set, because

$$\lambda(1,1,0) + \mu(1,-1,1) = (0,0,0) \Rightarrow \lambda = \mu = 0$$

Hence, B' forms a basis for range(t).

$\therefore \qquad \text{rank}(t) = \dim(I_m(t)) = \dim(\text{range}(t)) = 2$
and, $\qquad \text{nullity}(t) = \dim R^2 - \text{rank}(t) = 2 - 2 = 0$.
Thus, $\qquad \text{Ker}(t) = \text{Null Space} = \{(0,0)\}$.

EXAMPLE-2 Let C be the field of complex numbers, and let $t : C^3 \to C^3$ be a mapping given by

$$t(a,b,c) = (a - b + 2c, 2a + b - c, -a - 2b).$$

Show that t is a linear transformation and find its kernel.

SOLUTION It can be easily checked that

$$t[\lambda(a,b,c) + \mu(x,y,z)] = \lambda t(a,b,c) + \mu t(x,y,z) \text{ for all } (a,b,c), (x,y,z) \in C^3 \text{ and all } \lambda, \mu \in C.$$

Therefore, $t : C^3 \to C^3$ is a linear transformation.

$\qquad \text{Ker}(t) = \{(a,b,c) : t(a,b,c) = (0,0,0)\}$
$\therefore \quad (a,b,c) \in \text{Ker}(t)$
$\Leftrightarrow \quad t(a,b,c) = (0,0,0)$
$\Leftrightarrow \quad (a-b+2c, 2a+b-c, -a-2b) = (0,0,0)$
$\Leftrightarrow \quad a-b+2c = 0, 2a+b-c = 0, -a-2b = 0$

This is a homogeneous system of equations whose solution space is the kernel of t.

The coefficient matrix A given by

$$A = \begin{bmatrix} 1 & -1 & 2 \\ 2 & 1 & -1 \\ -1 & -2 & 0 \end{bmatrix}$$

$$\Rightarrow A \sim \begin{bmatrix} 1 & -1 & 2 \\ 0 & 3 & -5 \\ 0 & -3 & 2 \end{bmatrix} \text{ Applying } R_2 \to R_2 - 2R_1, R_3 \to R_3 + R_1$$

$$\Rightarrow A \sim \begin{bmatrix} 1 & -1 & 2 \\ 0 & 3 & -5 \\ 0 & 0 & -3 \end{bmatrix} \text{ Applying } R_3 \to R_3 + R_2$$

Clearly, rank$(A) = 3 =$ Number of unknowns

So, it has only the trivial solution $a = b = c = 0$.

Hence, Ker$(t) = \{(0,0,0)\} =$ Null space.

EXAMPLE-3 *Let* $t : R^4 \to R^3$ *be the linear transformation defined by*

$$t(x,y,z,t) = (x-y+z+t,\ x+2z-t,\ x+y+3z-3t)$$

Find a basis and dimension of (i) the image of t (ii) the kernel of t.

SOLUTION (i) We know that if vectors $v_1, v_2, \ldots, v_n$ span a vector space $V(F)$ and $t : V \to V'$ is a linear transformation, then $t(v_1), t(v_2), \ldots, t(v_n)$ span $I_m(t)$. Consider vectors $e_1^{(4)} = (1,0,0,0)$, $e_2^{(4)} = (0,1,0,0)$, $e_3^{(4)} = (0,0,1,0)$ and $e_4^{(4)} = (0,0,0,1)$ forming standard basis of R^4. Then, $t(e_1^{(4)})$, $t(e_2^{(4)})$, $t(e_3^{(4)})$, $t(e_4^{(4)})$ span $I_m(t)$.

We have, $t(e_1^4) = (1,1,1)$, $t(e_2^{(4)}) = (-1,0,1)$, $t(e_3^{(4)}) = (1,2,3)$, $t(e_4^{(4)}) = (1,-1,-3)$

In order to find the basis and dimension of $I_m(t)$, let us form a matrix A whose rows are images of $e_1^{(4)}, e_2^{(4)}, e_3^{(4)}, e_4^{(4)}$.

We have,

$$A = \begin{bmatrix} 1 & 1 & 1 \\ -1 & 0 & 1 \\ 1 & 2 & 3 \\ 1 & -1 & -3 \end{bmatrix}$$

Let us now reduce A to echelon form by using elementary row operations.

$$A \sim \begin{bmatrix} 1 & 1 & 1 \\ 0 & 1 & 2 \\ 0 & 1 & 2 \\ 0 & -2 & -4 \end{bmatrix} \text{ Applying } R_2 \to R_2 + R_1, R_3 \to R_3 - R_1, R_4 \to R_4 - R_1$$

$$\Rightarrow \quad A \sim \begin{bmatrix} 1 & 1 & 1 \\ 0 & 1 & 2 \\ 0 & 0 & 0 \\ 0 & 0 & 0 \end{bmatrix} \text{ Applying } R_3 \to R_3 - R_2, R_4 \to R_4 + 2R_2$$

Clearly, there are two non-zero rows. So, $v_1 = (1,1,1)$, $v_2 = (0,1,2)$ form a basis of $I_m(t)$, hence $\dim(I_m(t)) = 2$.

(ii) Let $v = (x,y,z,t)$ be an arbitrary element in Ker(t). Then,

$t(v) = (0,0,0)$
$\Rightarrow \quad (x-y+z+t,\ x+2z-t,\ x+y+3z-3t) = (0,0,0)$
$\Rightarrow \quad x-y+z+t = 0,\ x+2z-t = 0,\ x+y+3z-3t = 0$

Ker(t) is the solution space of the above homogeneous system of equations which can be written in matrix form as follows:

$$\begin{bmatrix} 1 & -1 & 1 & 1 \\ 1 & 0 & 2 & -1 \\ 1 & 1 & 3 & -3 \end{bmatrix} \begin{bmatrix} x \\ y \\ z \\ t \end{bmatrix} = \begin{bmatrix} 0 \\ 0 \\ 0 \end{bmatrix}$$

$$\Rightarrow \begin{bmatrix} 1 & -1 & 1 & 1 \\ 0 & 1 & 1 & -2 \\ 0 & 2 & 2 & -4 \end{bmatrix} \begin{bmatrix} x \\ y \\ z \\ t \end{bmatrix} = \begin{bmatrix} 0 \\ 0 \\ 0 \end{bmatrix} \text{ Applying } R_2 \to R_2 - R_1, R_3 \to R_3 - R_1.$$

$$\Rightarrow \begin{bmatrix} 1 & -1 & 1 & 1 \\ 0 & 1 & 1 & -2 \\ 0 & 0 & 0 & 0 \end{bmatrix} \begin{bmatrix} x \\ y \\ z \\ t \end{bmatrix} = \begin{bmatrix} 0 \\ 0 \\ 0 \end{bmatrix} \text{ Applying } R_3 \to R_3 + (-2)R_2$$

We observe that there are two non-zero rows in the coefficient matrix, so its rank is 2. So, there are two (= Number of variables − rank) free variables.

The homogeneous system is equivalent to

$$x - y + z + t = 0,\ y + z - 2t = 0$$

Taking z and t as free variables, we obtain the following solutions:

(i) $z = -1,\ t = 0,\ y = 1,\ x = 2$ (ii) $z = 0,\ t = 1,\ y = 2,\ x = 1$

Thus, $(2,1,-1,0)$ and $(1,2,0,1)$ form a basis of Ker(t)

EXAMPLE-4 *Let* $t : R^3 \to R^3$ *be the linear transformation defined by*
$$t(x, y, z) = (x+2y-z, y+z, x+y-2z)$$
Find a basis and the dimension of (i) the image of t (ii) the kernel of t.

SOLUTION (i) The standard basis of R^3 is $\{e_1^{(3)}, e_2^{(3)}, e_3^{(3)}\}$, where $e_1^{(3)} = (1,0,0)$, $e_2^{(3)} = (0,1,0)$ and $e_3^{(3)} = (0,0,1)$.

$\therefore \quad t\left(e_1^{(3)}\right) = (1,0,1),\ t\left(e_2^{(3)}\right) = (2,1,1),\ t\left(e_3^{(3)}\right) = (-1,1,-2)$

Since $t\left(e_1^{(3)}\right), t\left(e_2^{(3)}\right), t\left(e_3^{(3)}\right)$ span $I_m(t)$. So, to find a basis and the dimension of $I_m(t)$, we form a matrix A whose rows are $t\left(e_1^{(3)}\right), t\left(e_2^{(3)}\right), t\left(e_3^{(3)}\right)$ as given below.

$$A = \begin{bmatrix} 1 & 0 & 1 \\ 2 & 1 & 1 \\ -1 & 1 & -2 \end{bmatrix}$$

Now we reduce A to echelon form as shown below:

$$A \sim \begin{bmatrix} 1 & 0 & 1 \\ 0 & 1 & -1 \\ 0 & 1 & -1 \end{bmatrix} \quad \text{Applying } R_2 \to R_2 - 2R_1,\ R_3 \to R_3 + R_1$$

$$\Rightarrow \quad A \sim \begin{bmatrix} 1 & 0 & 1 \\ 0 & 1 & -1 \\ 0 & 0 & 0 \end{bmatrix} \quad \text{Applying } R_3 \to R_3 - R_2$$

Thus, $v_1 = (1,0,1)$ and $v_2 = (0,1,-1)$ form a basis of $I_m(t)$ and hence $\dim(I_m(t)) = 2$.

(ii) Let $v = (x,y,z) \in \text{Ker}(t)$. Then,

$t(v) = (0, 0, 0)$

$\Rightarrow \quad (x+2y-z, y+z, x+y-2z) = (0, 0, 0)$

$\Rightarrow \quad x+2y-z = 0,\ y+z = 0,\ x+y-2z = 0$

Kernel of t is the solution space of this homogeneous system of equations which can be written in matrix form as follows:

$$\begin{bmatrix} 1 & 2 & -1 \\ 0 & 1 & 1 \\ 1 & 1 & -2 \end{bmatrix} \begin{bmatrix} x \\ y \\ z \end{bmatrix} = \begin{bmatrix} 0 \\ 0 \\ 0 \end{bmatrix}$$

or, $\begin{bmatrix} 1 & 2 & -1 \\ 0 & 1 & 1 \\ 0 & -1 & -1 \end{bmatrix} \begin{bmatrix} x \\ y \\ z \end{bmatrix} = \begin{bmatrix} 0 \\ 0 \\ 0 \end{bmatrix}$ Applying $R_3 \to R_3 - R_1$

or, $\begin{bmatrix} 1 & 2 & -1 \\ 0 & 1 & 1 \\ 0 & 0 & 0 \end{bmatrix} \begin{bmatrix} x \\ y \\ z \end{bmatrix} = \begin{bmatrix} 0 \\ 0 \\ 0 \end{bmatrix}$

or, $x+2y-z=0$, $y+z=0$

Clearly, rank of the coefficient matrix is 2. So, there is one free variable and hence $\dim(\text{Ker}(t)) = 1$. Taking z as the free variable and setting $z = 1$, we get $y = -1$ and $x = 3$. Hence, $(3, -1, 1)$ forms a basis of Ker(t).

EXAMPLE-5 *Let $t : R^4 \to R^3$ be a linear transformation defined by*
$$t(a,b,c,d) = (a-b+c+d,\ a+2c-d,\ a+b+3c-3d).$$
Then find the basis and dimension of the (i) $I_m(t)$, (ii) Ker(t).

SOLUTION (i) By definition, $I_m(t) = \{t(v) : v \in R^4\}$. Since the set $\{e_1^{(4)}, e_2^{(4)}, e_3^{(4)}, e_4^{(4)}\}$, being standard basis, spans R^4. Therefore, $t(e_1^{(4)}) = (1,1,1)$, $t(e_2^{(4)}) = (-1,0,1)$, $t(e_3^{(4)}) = (1,2,3)$ and $t(e_4^{(4)}) = (1,-1,-3)$ span $I_m(t)$.

To find the basis of $I_m(t)$, we will now find linearly independent vectors from the set $\{t(e_1^{(4)}), t(e_2^{(4)}), t(e_3^{(4)}), t(e_4^{(4)})\}$, which spans $I_m(t)$. For this, form a matrix A whose rows are $t(e_1^{(4)}), t(e_2^{(4)}), t(e_3^{(4)}), t(e_4^{(4)})$, i.e.

$$A = \begin{bmatrix} 1 & 1 & 1 \\ -1 & 0 & 1 \\ 1 & 2 & 3 \\ 1 & -1 & -3 \end{bmatrix}.$$

Applying elementary transformations this matrix reduces to the following echelon matrix

$$A \sim \begin{bmatrix} 1 & 1 & 1 \\ 0 & 1 & 2 \\ 0 & 0 & 0 \\ 0 & 0 & 0 \end{bmatrix}.$$

The non-zero rows of the above echelon matrix form a basis of $I_m(t)$. Hence, $\{(1,1,1), (0,1,2)\}$ is a basis of $I_m(t)$. Hence, $\dim(I_m(t)) = \text{rank}(t) = 2$.

(ii) By definition, Ker$(t) = \{v \in R^4 : t(v) = 0 \in R^3\}$.
If $v \in \text{Ker}(t) = 0 \in R^3$ for all $v \in R^4$.

$\therefore \quad t(a,b,c,d) = (0,0,0)$ for all $a,b,c,d \in R$

$\Rightarrow \quad a-b+c+d = 0,\ a+2c-d = 0,\ a+b+3c-3d = 0.$ \hfill (i)

Forming the matrix of coefficients of the above system of equations and reducing it to echelon form, we obtain

$$\begin{bmatrix} 1 & -1 & 1 & 1 \\ 0 & 1 & 1 & 2 \\ 0 & 0 & 0 & 0 \end{bmatrix}.$$

Therefore, the system of equations given in (i) is equivalent to the system

$$\left.\begin{matrix} a-b+c+d=0 \\ b+c+d=0 \end{matrix}\right\} \Rightarrow \left.\begin{matrix} b=2d-c \\ a=-2c+d \end{matrix}\right\} \qquad \text{(ii)}$$

From (ii), we observe that the values of a and b depend upon c and d and hence c and d are two free variables. Therefore,

$\text{Ker}(t) = \{(-2c+d,\ 2d-c,\ c,d) : c,d \in R\}$

$\Rightarrow$ Nullity$(t) = \dim(\text{Ker}(t)) = $ Number of free variables $= 2$.

To find a basis of Ker(t), we give arbitrary values to c and d. For example, for $c=0,\ d=1$, we have $a=1,\ b=2$ and for $c=1,\ d=0$, we have $a=-2,\ b=-1$. Therefore, $\{(1,2,0,1),\ (-2,-1,1,0)\}$ is a basis for Ker(t).

EXAMPLE-6 Let $t_A : R^{3\times 1} \to R^{3\times 1}$ be a linear transformation given by $t_A(X) = AX$, where $A = \begin{bmatrix} 1 & 1 & 2 \\ 1 & 0 & 1 \\ 2 & 1 & 3 \end{bmatrix}$. Find

(i) basis for range of t_A.

(ii) basis for kernel of t_A.

(iii) rank and nullity of t_A.

SOLUTION (i) As discussed in section 3.4.1 the range of t_A is the column space of A. To find the same, we reduce A^T to echelon form.

We have,

$$A^T = \begin{bmatrix} 1 & 1 & 2 \\ 1 & 0 & 1 \\ 2 & 1 & 3 \end{bmatrix}$$

$\Rightarrow \qquad A^T \sim \begin{bmatrix} 1 & 1 & 2 \\ 0 & -1 & -1 \\ 0 & -1 & -1 \end{bmatrix} \qquad$ Applying $R_2 \to R_2 - R_1,\ R_3 \to R_3 - 2R_1$

$\Rightarrow \qquad A^T \sim \begin{bmatrix} 1 & 1 & 2 \\ 0 & -1 & -1 \\ 0 & 0 & 0 \end{bmatrix} \qquad$ Applying $R_3 \to R_3 - R_2$

There are two non-zero rows in A^T. So, $\{(1, 1, 2), (0, -1, -1)\}$ is a basis of range of t_A and hence rank $(t_A) = 2$.

(ii) The kernel of t_A is the solution of $t_A(X) = O$ i.e. $AX = O$. Let $x = [x, y, z]^T$ be in kernel of t_A. Then,

$t_A(X) = O$

$\Rightarrow \quad AX = O$

$$\Rightarrow \begin{bmatrix} 1 & 1 & 2 \\ 1 & 0 & 1 \\ 2 & 1 & 3 \end{bmatrix} \begin{bmatrix} x \\ y \\ z \end{bmatrix} = \begin{bmatrix} 0 \\ 0 \\ 0 \end{bmatrix}$$

$$\Rightarrow \begin{bmatrix} 1 & 1 & 2 \\ 0 & -1 & -1 \\ 0 & -1 & -1 \end{bmatrix} \begin{bmatrix} x \\ y \\ z \end{bmatrix} = \begin{bmatrix} 0 \\ 0 \\ 0 \end{bmatrix} \quad \text{Applying } R_2 \to R_2 - R_1, R_3 \to R_3 - 2R_1$$

$$\Rightarrow \begin{bmatrix} 1 & 1 & 2 \\ 0 & -1 & -1 \\ 0 & 0 & 0 \end{bmatrix} \begin{bmatrix} x \\ y \\ z \end{bmatrix} = \begin{bmatrix} 0 \\ 0 \\ 0 \end{bmatrix} \quad \text{Applying } R_3 \to R_3 - R_2$$

$$\Rightarrow x+y+2z=0, \; -y-z=0$$

This system of equations has only one free variable say, z. Putting $z = 1$ in $-y - z = 0$, we get $y = -1$. Substituting $y = -1$, $z = 1$ in the first equation, we get $x = -1$. Thus, $\{(-1, -1, 1)\}$ is a basis of Ker (t_A).

(iii) Clearly, rank (t_A) = Dim $(I_m(t_A))$ = 2 and, Nullity (t_A) = Dim (Ker (t_A)) = 1.

EXERCISE–3.3

1. For each of the following linear transformations, find a basis and the dimension of the kernel and the image :

    (i) $t : R^3 \to R^3$ defined by $t(x, y, z) = (x+2y-3z, 2x+5y-4z, x+4y+z)$

    (ii) $t : R^4 \to R^3$ defined by
    $$t(x, y, z, u) = (x+2y+3z+2u, 2x+4y+7z+5u, x+2y+6z+5u)$$

    (iii) $t : R^3 \to R^2$ defined by $t(x, y, z) = (x+y+z, 2x+2y+2z)$

2. Find a linear transformation $t : R^3 \to R^3$ whose image is spanned by $(1,2,3)$ and $(4,5,6)$.
3. Find a linear mapping $t : R^4 \to R^3$ whose kernel is spanned by $(1,2,3,4)$ and $(0,1,1,1)$.
4. Let V and V' be finite dimensional vector spaces over the same field F, and let $t : V \to V'$ be a linear transformation such that t is onto, then prove that $\dim V' \leq \dim V$. Determine all linear transformations $t : R^3 \to R^4$ that are onto.
5. Consider the zero linear transformation $\hat{0} : V \to V'$ defined by $\hat{0}(v) = 0_{V'}$ for all $v \in V$. Find the kernel and the image of $\hat{0}$.
6. Let $V = P_{10}(t)$ be the vector space of polynomials of degree ≤ 10. Consider the linear transformation $D^4 : V \to V$, where D^4 denotes the fourth derivative $\dfrac{d^4 f}{dt^4}$. Find a basis and the dimension of (i) the image of D^4 (ii) the kernel of D^4.
7. Let V and V' be finite dimensional vector spaces of same dimension over the same field F. Then, for a linear transformation $t : V \to V'$, the following are equivalent:

    (i) t is an isomorphism (ii) t is onto (iii) t is one-one.

8. Let $t : R^5 \to R^2$ be a non-zero linear transformation such that t is not onto. Find rank (t) and nullity (t).

9. Let $D : P_3(t) \to P_3(t)$ be given by
$$D(a_0 + a_1t + a_2t^2 + a_3t^3) = a_1 + 2a_2t + 3a_3t^2$$

   Prove that:

   (i) D is a linear transformation.

   (ii) Find kernel (D) and range (D)

   (iii) Verify that rank (D) + nullity$(D) = 4$.

10. Let $t : R^{4 \times 1} \to R^{3 \times 1}$ be defined by $t \begin{bmatrix} x \\ y \\ z \\ u \end{bmatrix} = \begin{bmatrix} 1 & 0 & -1 & 3 \\ 2 & 0 & -1 & 5 \\ 3 & 0 & -2 & 8 \end{bmatrix} \begin{bmatrix} x \\ y \\ z \\ u \end{bmatrix}$

   (i) Does $\begin{bmatrix} 1 \\ 0 \\ 1 \\ 0 \end{bmatrix} \in \text{Ker}(t)$?   (ii) Does $\begin{bmatrix} 3 \\ 5 \\ 8 \end{bmatrix} \in \text{Range}(t)$?

   (iii) What are nullity (t) and rank (t)?

11. Let $V_1, V_2, \ldots, V_n$ be any $n \geq 2$, finite dimensional vector spaces over a field F. Let $t_i : V_i \to V_{i+1} (1 \leq i \leq n-1)$ be any linear transformation, such that
   (i) $\text{Ker}(t_1) = \{0\}$ (ii) $\text{Ker}(t_{i+1}) = \text{Range}(t_i)$ for $1 \leq i \leq n-2$ (iii) Range $(t_{n-1}) = V_n$.
   Prove that $\sum_{i=1}^{n} (-1)^i \dim V_i = 0$

ANSWERS

1. (i) $\dim(\text{Ker}(t)) = 1, \text{Ker}(t) = \{(7, -2, 1)\}, \dim(I_m(t)) = 2, I_m(t) = \{(1,2,1), (0,1,2)\}$

   (ii) $\dim(\text{Ker}(t)) = 2, \text{Ker}(t) = \{(-2,1,0,0), (1,0,-1,1)\}, \dim(I_m(t)) = 2,$
   $I_m(t) = \{(1,2,1), (0,1,3)\}$

   (iii) $\dim(\text{Ker}(t)) = 2, \text{Ker}(t) = \{(1,0,-1), (1,-1,0)\}, \dim(I_m(t)) = 1, I_m(t) = \{(1,2)\}$

2. $t(x,y,z) = (x+4y, 2x+5y, 3x+6y)$ 3. $t(x,y,z,u) = (x+y-z, 2x+y-u, 0)$

4. None, As $\dim R^4 > \dim R^3$ 5. $\text{Ker}(\hat{0}) = V, I_m(\hat{0}) = \{0_{V'}\}$

6. (i) $\{1, t, t^2, \ldots, t^6\}$ (ii) $\{1, t, t^2, t^3\}$

3.6 ALGEBRA OF LINEAR TRANSFORMATIONS

Let V and V' be vector spaces over the same field F. The set of all linear transformations from V to V' is denoted by $\text{Hom}_F(V,V')$ or simply by $\text{Hom}(V,V')$. We want to impose a vector space structure on the set $\text{Hom}(V,V')$ over the field F. For this purpose, we define addition and scalar multiplication on $\text{Hom}(V,V')$ as given below.

ADDITION. *Let $t_1, t_2 \in \text{Hom}(V,V')$. Then the sum $t_1 + t_2$ of t_1 and t_2 is a mapping from V to V' given by*
$$(t_1 + t_2)(v) = t_1(v) + t_2(v) \quad \text{for all } v \in V.$$

SCALAR MULTIPLICATION. *For any $t \in \text{Hom}(V,V')$ and $\lambda \in F$, the scalar multiplication λt is defined as*
$$(\lambda t)(v) = \lambda t(v) \quad \text{for all } v \in V.$$

THEOREM-1 *Let V and V' be two vector spaces over the same field F. Then for any $t_1, t_2 \in \text{Hom}(V,V')$, the sum $t_1 + t_2 \in \text{Hom}(V,V')$.*

PROOF. For any $u, v \in V$ and any $\lambda, \mu \in F$, we have

$(t_1 + t_2)(\lambda u + \mu v) = t_1(\lambda u + \mu v) + t_2(\lambda u + \mu v)$

$\Rightarrow (t_1 + t_2)(\lambda u + \mu v) = [\lambda t_1(u) + \mu t_1(v)] + [\lambda t_2(u) + \mu t_2(v)]$ $\quad [\because t_1, t_2 \text{ are linear}]$

$\Rightarrow (t_1 + t_2)(\lambda u + \mu v) = [\lambda t_1(u) + \lambda t_2(u)] + [\mu t_1(v) + \mu t_2(v)]$ $\quad \left[\begin{array}{l}\text{By commutativity and} \\ \text{associativity of addition on } V'\end{array}\right]$

$\Rightarrow (t_1 + t_2)(\lambda u + \mu v) = \lambda[t_1(u) + t_2(u)] + \mu[t_1(v) + t_2(v)]$ $\quad [\text{By } V'\text{-2(i)}]$

$\Rightarrow (t_1 + t_2)(\lambda u + \mu v) = \lambda(t_1 + t_2)(u) + \mu(t_1 + t_2)(v)$ $\quad [\text{By definition of } t_1 + t_2]$

So, $t_1 + t_2$ is a linear transformation from V to V'.
Hence, $t_1 + t_2 \in \text{Hom}(V,V')$.

Q.E.D.

THEOREM-2 *Let V and V' be vector spaces over the same field F. Then for any $t \in \text{Hom}(V,V')$ and $\lambda \in F$, $\lambda t \in \text{Hom}(V,V')$.*

PROOF. For any $u, v \in V$ and any $\alpha, \beta \in F$, we have

$(\lambda t)(\alpha u + \beta v) = \lambda[t(\alpha u + \beta v)]$ $\quad [\text{By definition of } \lambda t]$

$\Rightarrow (\lambda t)(\alpha u + \beta v) = \lambda[\alpha t(u) + \beta t(v)]$ $\quad [\because t : V \to V' \text{ is linear}]$

$\Rightarrow (\lambda t)(\alpha u + \beta v) = \lambda(\alpha t(u)) + \lambda(\beta t(v))$ $\quad [\text{By } V'\text{-2(i)}]$

$\Rightarrow (\lambda t)(\alpha u + \beta v) = (\lambda \alpha)t(u) + (\lambda \beta)t(v)$ $\quad [\text{By } V'\text{-2(iii)}]$

$\Rightarrow (\lambda t)(\alpha u + \beta v) = (\alpha \lambda)t(u) + (\beta \lambda)t(v)$ $\quad [\text{By commutativity of multiplication on } F]$

$\Rightarrow \quad (\lambda t)(\alpha u + \beta v) = \alpha(\lambda t(u)) + \beta(\lambda t(v))$ [By V'–2(iii)]

$\Rightarrow \quad (\lambda t)(\alpha u + \beta v) = \alpha(\lambda t)(u) + \beta(\lambda t)(v)$ [By definition of λt]

Therefore, $\lambda t : V \to V'$ is a linear transformation.

Hence, $\lambda t \in \mathrm{Hom}(V, V')$ for all $t \in \mathrm{Hom}(V, V')$ and all $\lambda \in F$.

Q.E.D.

THEOREM-3 *Let V and V' be vector spaces over a field F. Then the set $\mathrm{Hom}(V, V')$ is a vector space over field F.*

PROOF. In order to give a vector space structure to $\mathrm{Hom}(V, V')$, we define vector addition, scalar multiplication and equality as follows:

Vector addition on $\mathrm{Hom}(V, V')$: For any $t_1, t_2 \in \mathrm{Hom}(V, V')$, we define

$$(t_1 + t_2)(v) = t_1(v) + t_2(v) \quad \text{for all } v \in V$$

Clearly, $t_1 + t_2 \in \mathrm{Hom}(V, V')$ [By Theorem 1]

So, vector addition is well defined.

Vector addition on $\mathrm{Hom}(V, V')$ is also called pointwise addition.

Scalar multiplication on $\mathrm{Hom}(V, V')$: For any $t \in \mathrm{Hom}(V, V')$ and $\lambda \in F$, we define

$$(\lambda t)(v) = \lambda t(v) \text{ for all } v \in V.$$

Clearly, $\lambda t \in \mathrm{Hom}(V, V')$ [By Theorem 2]

Thus, scalar multiplication is defined on $\mathrm{Hom}(V, V')$.

Equality: For any $t_1, t_2 \in \mathrm{Hom}(V, V')$, we define

$$t_1 = t_2 \Leftrightarrow t_1(v) = t_2(v) \text{ for all } v \in V$$

Thus, we have defined vector addition, scalar multiplication, and equality in $\mathrm{Hom}(V, V')$. We shall now verify vector space axioms.

$\mathrm{Hom}(V, V')$ is an abelian group under above defined pointwise addition:

Associativity: For any $t_1, t_2, t_3 \in \mathrm{Hom}(V, V')$, we have

$[(t_1 + t_2) + t_3](v) = (t_1 + t_2)(v) + t_3(v)$ [By definition of pointwise addition on Hom (V, V')]

$\Rightarrow [(t_1 + t_2) + t_3](v) = [t_1(v) + t_2(v)] + t_3(v)$ [By definition of pointwise addition on Hom (V, V')]

$\Rightarrow [(t_1 + t_2) + t_3](v) = t_1(v) + [t_2(v) + t_3(v)]$ [By associativity of addition on V']

$\Rightarrow \quad [(t_1+t_2)+t_3](v) = t_1(v) + (t_2+t_3)(v)$ $\left[\begin{array}{l}\text{By definition of pointwise}\\ \text{addition on } \text{Hom}(V,V')\end{array}\right]$

$\Rightarrow \quad [(t_1+t_2)+t_3](v) = [t_1+(t_2+t_3)](v) \quad \text{for all } v \in V$ $\left[\begin{array}{l}\text{By definition of pointwise}\\ \text{addition}\end{array}\right]$

$\therefore \quad (t_1+t_2)+t_3 = t_1+(t_2+t_3)$ [By definition of equality]

Thus, $(t_1+t_2)+t_3 = t_1+(t_2+t_3)$ for all $t_1, t_2, t_3 \in \text{Hom}(V,V')$.

So, pointwise addition is associative on $\text{Hom}(V,V')$.

Commutativity: For any $t_1, t_2 \in \text{Hom}(V,V')$, we have

$\quad (t_1+t_2)(v) = t_1(v) + t_2(v)$ [By definition of pointwise addition]

$\Rightarrow \quad (t_1+t_2)(v) = t_2(v) + t_1(v)$ [By commutativity of addition on V']

$\Rightarrow \quad (t_1+t_2)(v) = (t_2+t_1)(v)$ for all $v \in V$ [By definition of pointwise addition]

$\therefore \quad t_1+t_2 = t_2+t_1$ [By definition of equality]

Thus, $t_1+t_2 = t_2+t_1$ for all $t_1, t_2 \in \text{Hom}(V,V')$.

So, pointwise addition is commutative on $\text{Hom}(V,V')$.

Existence of additive identity in $\text{Hom}(V,V')$: Let $\hat{0}$ be a mapping from V to V' given by

$$\hat{0}(v) = 0_{V'} \text{ for all } v \in V.$$

Then, $\hat{0} \in \text{Hom}(V,V')$.

For any $t \in \text{Hom}(V,V')$

$\quad (t+\hat{0})(v) = t(v) + \hat{0}(v)$ [By definition of pointwise addition]

$\Rightarrow \quad (t+\hat{0})(v) = t(v) + 0_{V'}$ $[\because \hat{0}(v) = 0_{V'}]$

$\Rightarrow \quad (t+\hat{0})(v) = t(v) \quad \text{for all } v \in V.$ [By definition of $\hat{0}$]

$\Rightarrow \quad t+\hat{0} = t$ [By definition of equality]

$\Rightarrow \quad t+\hat{0} = t = \hat{0}+t$ [By commutativity of addition on $\text{Hom}(V,V')$]

Thus, $t+\hat{0} = t = \hat{0}+t$ for all $t \in \text{Hom}(V,V')$.

So, $\hat{0}$ is the additive identity in $\text{Hom}(V,V')$.

Existence of additive inverse of each element of $\text{Hom}(V,V')$: Let t be an arbitrary element of $\text{Hom}(V,V')$. Let us define $-t$ by the rule

$$(-t)(v) = -t(v) \quad \text{for all } v \in V$$

It can be easily seen that $-t \in \text{Hom}(V,V')$.

Now

$[t+(-t)](v) = t(v) + (-t)(v)$ [By definition of pointwise addition]

$\Rightarrow \quad [t+(-t)](v) = t(v) - t(v)$

$\Rightarrow \quad [t+(-t)](v) = 0_{V'} = \hat{0}(v) \quad$ for all $v \in V$

$\therefore \quad t+(-t) = \hat{0}$

$\Rightarrow \quad t+(-t) = \hat{0} = (-t) + t$ [By commutativity of pointwise addition]

Thus, for each $t \in \text{Hom}(V, V')$ there exists $-t \in \text{Hom}(V, V')$ such that $-t$ is additive inverse of t.

So, each element of $\text{Hom}(V, V')$ possesses additive inverse.

Hence, $\text{Hom}(V, V')$ is an abelian group under pointwise addition.

Further for any $t_1, t_2 \in \text{Hom}(V, V')$ and any $\lambda, \mu \in F$ we make the following observations:

(i) $[\lambda(t_1+t_2)](v) = \lambda[(t_1+t_2)(v)]$ [By definition of scalar multiplication on $\text{Hom}(V, V')$]

$\qquad\qquad = \lambda[t_1(v) + t_2(v)]$ [By definition of pointwise addition]

$\qquad\qquad = \lambda t_1(v) + \lambda t_2(v)$ [By V'-2(i)]

$\qquad\qquad = (\lambda t_1)(v) + (\lambda t_2)(v)$ [By definition of scalar multiplication on $\text{Hom}(V, V')$]

$\qquad\qquad = (\lambda t_1 + \lambda t_2)(v) \quad$ for all $v \in V$ [By definition of pointwise addition]

$\therefore \quad \lambda(t_1+t_2) = \lambda t_1 + \lambda t_2$

(ii) $[(\lambda+\mu)t_1](v) = (\lambda+\mu)t_1(v)$ [By definition of scalar multiplication]

$\qquad\qquad = \lambda t_1(v) + \mu t_2(v)$ [By V'-2(ii)]

$\qquad\qquad = (\lambda t_1)(v) + (\mu t_2)(v)$ [By definition of scalar multiplication]

$\qquad\qquad = (\lambda t_1 + \mu t_2)(v) \quad$ for all $v \in V$ [By definition of pointwise addition]

$\therefore \quad (\lambda+\mu)t_1 = \lambda t_1 + \mu t_2$

(iii) $[(\lambda\mu)t_1](v) = (\lambda\mu)t_1(v)$ [By definition of scalar multiplication on $\text{Hom}(V, V')$]

$\qquad\qquad = \lambda[\mu t_1(v)]$ [By V'-2(iii)]

$\qquad\qquad = \lambda[(\mu t_1)(v)]$ [By definition of scalar multiplication]

$\qquad\qquad = [\lambda(\mu t_1)](v) \quad$ for all $v \in V$ [By definition of scalar multiplication]

$\therefore \quad (\lambda\mu)t_1 = \lambda(\mu t_1)$

(iv) $(1t_1)(v) = 1t_1(v)$ [By definition of scalar multiplication]

$\Rightarrow \quad (1t_1)(v) = t_1(v) \quad$ for all $v \in V$ [By V'-2(iv)]

$\therefore \quad 1t_1 = t_1$

Hence, $\text{Hom}(V, V')$ is a vector space over field F. Q.E.D.

REMARK. *In what follows the set $Hom(V, V')$ will denote the vector space of all linear transformations from V to V'.*

THEOREM-4 *Let $V(F)$ be a finite dimensional vector space with basis $B = \{b_1, b_2, \ldots, b_n\}$. Then for any vector space $V'(F)$, a mapping $f : B \to V'$ can be extended uniquely to a linear transformation $t : V \to V'$. Further two linear transformations t and t' from V to V' are equal iff $t(b_i) = t'(b_i)$ for all $i \in \underline{n}$.*

PROOF. Since B is a basis for V. Therefore, each $v \in V$ can be uniquely expressed as

$$v = \sum_{i=1}^{n} \lambda_i b_i \text{ for some } \lambda_i \in F.$$

Define $t : V \to V'$ by the rule $t(v) = \sum_{i=1}^{n} \lambda_i f(b_i)$.

First we will show that $t : V \to V'$ is a linear transformation.

For any $u = \sum_{i=1}^{n} \lambda_i b_i, v = \sum_{i=1}^{n} \mu_i b_i \in V$ and any $\alpha, \beta \in F$

$$t(\alpha u + \beta v) = t\left\{\alpha \left(\sum_{i=1}^{n} \lambda_i b_i\right) + \beta \left(\sum_{i=1}^{n} \mu_i b_i\right)\right\}$$

$\Rightarrow \quad t(\alpha u + \beta v) = t\left\{\sum_{i=1}^{n} (\alpha \lambda_i + \beta \mu_i) b_i\right\}$

$\Rightarrow \quad t(\alpha u + \beta v) = \sum_{i=1}^{n} (\alpha \lambda_i + \beta \mu_i) f(b_i)$ [By definition of t]

$\Rightarrow \quad t(\alpha u + \beta v) = \sum_{i=1}^{n} \{\alpha(\lambda_i f(b_i)) + \beta(\mu_i f(b_i))\}$ [By V'–2(i) and V'–2(iii)]

$\Rightarrow \quad t(\alpha u + \beta v) = \sum_{i=1}^{n} \alpha(\lambda_i f(b_i)) + \sum_{i=1}^{n} \beta(\mu_i f(b_i))$

$\Rightarrow \quad t(\alpha u + \beta v) = \alpha \left(\sum_{i=1}^{n} \lambda_i f(b_i)\right) + \beta \left(\sum_{i=1}^{n} \mu_i f(b_i)\right)$

$\Rightarrow \quad t(\alpha u + \beta v) = \alpha\, t(u) + \beta\, t(v)$

$\therefore \quad t : V \to V'$ is a linear transformation.

Now for any $b_i \in B$, we have

$b_i = 0b_1 + \cdots + 0b_{i-1} + 1b_i + 0b_{i+1} + \cdots + 0b_n = \sum_{j=1}^{n} \lambda_j b_j,$ where $\lambda_j = 1$ for $j = i$, $\lambda_j = 0$ for $j \neq i$

$\Rightarrow \quad t(b_i) = \sum_{j=1}^{n} \lambda_j f(bj)$ [By definition of t]

$\Rightarrow \quad t(b_i) = f(b_i)$ [$\because \lambda_j = 1$ for $j = i, \lambda_j = 0$ for $j \neq i$]

Thus, $\quad t(b_i) = f(b_i) \quad$ for all $i \in \underline{n}$.

So, t is an extension of f.

For the second part, we have to show that if t and t' agree on B, then they are equal. Let v be an arbitrary vector in V. Then,

$$v = \sum_{i=1}^{n} \lambda_i b_i \quad \text{for some } \lambda_i \in F \qquad [\because B \text{ is basis for } V]$$

$\Rightarrow \quad t(v) = \sum_{i=1}^{n} \lambda_i t(b_i) \qquad [\because t \text{ is linear}]$

$\Rightarrow \quad t(v) = \sum_{i=1}^{n} \lambda_i t'(b_i) \qquad [\because t(b_i) = t'(b_i) \text{ for all } i \in \underline{n}]$

$\Rightarrow \quad t(v) = t'\left(\sum_{i=1}^{n} \lambda_i b_i\right) \qquad [\because t' \text{ is linear}]$

$\Rightarrow \quad t(v) = t'(v)$

Thus, $t(v) = t'(v)$ for all $v \in V$.
Hence, $t = t'$ \hfill Q.E.D.

THEOREM-5 *Let V and V' be finite dimensional vector spaces over a field F, and let $B = \{b_1, b_2, \ldots, b_n\}$ be an ordered basis for V. Then for any list $(b'_1, b'_2, \ldots, b'_n)$ of n vectors in V', there exists a unique linear transformation $t : V \to V'$ such that $t(b_1) = b'_1, t(b_2) = b'_2, \ldots, t(b_n) = b'_n$. Further, t is an isomorphism iff the list $(b'_1, b'_2, \ldots, b'_n)$ is a basis for V'.*

PROOF. Since B is a basis for V, therefore, for each $v \in V$ there exists scalars $\lambda_1, \lambda_2, \ldots, \lambda_n$ in F such that $v = \lambda_1 b_1 + \lambda_2 b_2 + \cdots + \lambda_n b_n$.

Consider a mapping $t : V \to V'$ given by the rule

$$t(v) = t\left(\sum_{i=1}^{n} \lambda_i b_i\right) = \sum_{i=1}^{n} \lambda_i b'_i$$

$t : V \to V'$ **is a linear transformation:** For any $u = \sum_{i=1}^{n} \lambda_i b_i, v = \sum_{i=1}^{n} \mu_i b_i$ in V and any $\alpha, \beta \in F$

$$t(\alpha u + \beta v) = t\left\{\alpha\left(\sum_{i=1}^{n} \lambda_i b_i\right) + \beta\left(\sum_{i=1}^{n} \mu_i b_i\right)\right\}$$

$\Rightarrow \quad t(\alpha u + \beta v) = t\left\{\sum_{i=1}^{n}(\alpha\lambda_i + \beta\mu_i)b_i\right\} \qquad$ [Using V–2(i)–(iii)]

$\Rightarrow \quad t(\alpha u + \beta v) = \sum_{i=1}^{n}(\alpha\lambda_i + \beta\mu_i)b'_i \qquad$ [By definition of t]

$\Rightarrow \quad t(\alpha u + \beta v) = \sum_{i=1}^{n}[(\alpha\lambda_i)b'_i + (\beta\mu_i)b'_i] \qquad$ [By V'–2(ii)]

$\Rightarrow \quad t(\alpha u + \beta v) = \sum_{i=1}^{n}[\alpha(\lambda_i b'_i) + \beta(\mu_i b'_i)] \qquad$ [By V'–2(iv)]

$\Rightarrow \quad t(\alpha u + \beta v) = \sum_{i=1}^{n} \alpha\left(\lambda_i b_i'\right) + \sum_{i=1}^{n} \beta\left(\mu_i b_i'\right)$

$\Rightarrow \quad t(\alpha u + \beta v) = \alpha\left(\sum_{i=1}^{n} \lambda_i b_i'\right) + \beta\left(\sum_{i=1}^{n} \mu_i b_i'\right)$ [By V'-2(i)]

$\Rightarrow \quad t(\alpha u + \beta v) = \alpha t(u) + \beta t(v)$

$\therefore \quad t(\alpha u + \beta v) = \alpha t(u) + \beta t(v)$ for all $u, v \in V$ and all $\alpha, \beta \in F$.

So, $t : V \to V'$ is a linear transformation.

Now, for any $b_i \in B$, we have

$$b_i = 0b_1 + \cdots + 0b_{i-1} + 1b_i + 0b_{i+1} + \cdots + 0b_n$$

$\Rightarrow \quad t(b_i) = t(0b_1 + \cdots + 0b_{i-1} + 1b_i + 0b_{i+1} + \cdots + 0b_n)$

$\Rightarrow \quad t(b_i) = 0b_1' + \cdots + 0b_{i-1}' + 1b_i' + 0b_{i+1}' + \cdots + 0b_n'$

$\Rightarrow \quad t(b_i) = 0_{V'} + \cdots + 0_{V'} + b_i' + 0_{V'} + \cdots + 0_{V'} = b_i'$

Thus, $t : V \to V'$ is a linear transformation such that $t(b_i) = b_i'$ for all $i \in \underline{n}$.

To prove the uniqueness of t, let, if possible $t' : V \to V'$ be a linear transformation such that $t'(b_i) = b_i'$ for all $i \in \underline{n}$. Then for any $v = \sum_{i=1}^{n} \lambda_i b_i \in V$, we have

$t'(v) = t'\left(\sum_{i=1}^{n} \lambda_i b_i\right) = \sum_{i=1}^{n} \lambda_i t'(b_i)$ $\qquad [\because\ t' : V \to V'$ is linear$]$

$\Rightarrow \quad t'(v) = \sum_{i=1}^{n} \lambda_i b_i' = t(v)$ $\qquad [\because\ t'(b_i) = b_i'$ for all $i \in \underline{n}]$

Thus, $t'(v) = t(v)$ for all $v \in V$

$\therefore \quad t' = t$

Hence, $t : V \to V'$ is a unique linear transformation such that $t(b_i) = b_i'$ for all $i \in \underline{n}$.

Now we shall show that $t : V \to V'$ is an isomorphism iff the list $B' = (b_1', b_2', \ldots, b_n')$ is a basis for V'.

First let $t : V \to V'$ be an isomorphism such that $t(b_i) = b_i'$ for all $i \in \underline{n}$. Then we have to show that the list $B' = (b_1', b_2', \ldots, b_n')$ is a basis for V'.

B' is l.i.: Let $\lambda_1, \lambda_2, \ldots, \lambda_n$ be scalars in F such that

$\lambda_1 b_1' + \lambda_2 b_2' + \cdots + \lambda_n b_n' = 0_{V'}$

$\Rightarrow \quad \lambda_1 t(b_1) + \lambda_2 t(b_2) + \cdots + \lambda_n t(b_n) = 0_{V'}$ $\qquad [\because\ t(b_i) = b_i'$ for all $i \in \underline{n}]$

$\Rightarrow \quad t(\lambda_1 b_1 + \lambda_2 b_2 + \cdots + \lambda_n b_n) = 0_{V'}$

$\Rightarrow \quad \lambda_1 b_1 + \lambda_2 b_2 + \cdots + \lambda_n b_n \in \mathrm{Ker}(t)$

$\Rightarrow \quad \lambda_1 b_1 + \lambda_2 b_2 + \cdots + \lambda_n b_n = 0_V$ $\qquad [\because\ \mathrm{Ker}(t) = \{0_V\}$ as t is a monomorphism$]$

$\Rightarrow \quad \lambda_1 = \lambda_2 = \cdots = \lambda_n = 0$ $\qquad [\because\ B$ is a basis for $V]$

So, B' is a linear independent list.

B' *spans* V': Let v' be an arbitrary vector in V'. As $t : V \to V'$ is an isomorphism, so there exists a vector $v = \sum_{i=1}^{n} \lambda_i b_i$ in V such that $t(v) = v'$.

Now,

$$t(v) = v'$$

$$\Rightarrow t\left(\sum_{i=1}^{n} \lambda_i b_i\right) = v'$$

$$\Rightarrow \sum_{i=1}^{n} \lambda_i t(b_i) = v' \qquad [\because t : V \to V' \text{ is linear}]$$

$$\Rightarrow \sum_{i=1}^{n} \lambda_i b_i' = v'$$

$\Rightarrow v'$ is a linear combination of $b_1', b_2', \ldots, b_n'$.

Thus, $v' \in V' \Rightarrow v'$ is a linear combination of vectors in B'.

So, B' spans V'.

Hence, B' is a basis for V'.

Conversely, let the list B' be a basis for V'. Then we have to prove that the linear transformation $t : V \to V'$ is an isomorphism. For this, it is sufficient to show that $t : V \to V'$ is an epimorphism with $\text{Ker}(t) = \{0_V\}$.

t *is surjective*: Let v' be an arbitrary vector in V'. Then,

$$v' = \sum_{i=1}^{n} \lambda_i b_i' \quad \text{for some } \lambda_i \in F \qquad [\because B' \text{ is a basis for } V']$$

$$\Rightarrow v' = \sum_{i=1}^{n} \lambda_i t(b_i) \qquad [\because t(b_i) = b_i' \text{ for all } i \in \underline{n}]$$

$$\Rightarrow v' = t\left(\sum_{i=1}^{n} \lambda_i b_i\right) \qquad [\because t : V \to V' \text{ is linear}]$$

$$\Rightarrow v' = t(v), \text{ where } v = \sum_{i=1}^{n} \lambda_i b_i \in V \qquad [\because B \text{ is basis for } V]$$

Thus, for each $v' \in V'$ there exists $v \in V$ such that $t(v) = v'$.

So, $t : V \to V'$ is surjective.

$Ker(t) = \{0_V\}$: Let $v = \sum_{i=1}^{n} \lambda_i b_i$ be a vector in V such that $v \in Ker(t)$. Then,

$$t(v) = 0_{V'}$$

$$\Leftrightarrow t\left(\sum_{i=1}^{n} \lambda_i b_i\right) = 0_{V'}$$

$$\Leftrightarrow \sum_{i=1}^{n} \lambda_i t(b_i) = 0_{V'}$$

$$\Leftrightarrow \sum_{i=1}^{n} \lambda_i b'_i = 0_{V'} \qquad\qquad [\because \quad t(b_i) = b'_i]$$

$$\Leftrightarrow \lambda_1 = \lambda_2 = \cdots = \lambda_n = 0 \qquad\qquad [\because \quad B' \text{ is a basis for } V']$$

$$\Leftrightarrow v = 0_V$$

Thus, $v \in Ker(t) \Leftrightarrow v = 0_V$
So, $Ker(t) = \{0_V\}$.
Hence, $t : V \to V'$ is an isomorphism.

Q.E.D.

REMARK. *Since each vector in a vector space is uniquely expressible in terms of basis vectors. Therefore, a linear transformation t from a vector space V to a vector space V' is completely defined if we mention under t the images of basis vectors. Moreover, if $t_1 : V \to V'$ and $t_2 : V \to V'$ are linear transformations, then t_1 and t_2 become equal if they agree on a basis of V, i.e. $t_1(b_i) = t_2(b_i)$ for all b_i in basis of V.*

THEOREM-6 *Let V and V' be finite dimensional vector spaces over the same field F. Then the vector space $\mathrm{Hom}(V, V')$ of all linear transformations from V to V' is finite dimensional and that*

$$\dim \left[\mathrm{Hom}(V, V')\right] = \dim V \dim V'$$

PROOF. Let $\dim V = m$ and $\dim V' = n$, and let $B = \{b_1, b_2, \ldots, b_m\}$ be an ordered basis for V.

Consider a mapping $\varphi : \mathrm{Hom}(V, V') \to V'^m$ given by

$$\varphi(t) = \Big(t(b_1), t(b_2), \ldots, t(b_m)\Big) \quad \text{for all } t \in \mathrm{Hom}(V, V')$$

φ *is a linear transformation*: For any $t_1, t_2 \in \mathrm{Hom}(V, V')$, and any $\lambda, \mu \in F$

$$\varphi(\lambda t_1 + \mu t_2) = \Big((\lambda t_1 + \mu t_2)(b_1), (\lambda t_1 + \mu\varphi(\lambda t_1 + \mu t_2)t_2)(b_2), \ldots, (\lambda t_1 + \mu t_2)(b_m)\Big)$$

$$\Rightarrow \varphi(\lambda t_1 + \mu t_2) = \Big((\lambda t_1)(b_1) + (\mu t_2)(b_1), (\lambda t_1)(b_2) + (\mu t_2)(b_2), \ldots, (\lambda t_1)(b_m) + (\mu t_2)(b_m)\Big)$$

$$\Rightarrow \varphi(\lambda t_1 + \mu t_2) = \Big(\lambda t_1(b_1) + \mu t_2(b_1), \lambda t_1(b_2) + \mu t_2(b_2), \ldots, \lambda t_1(b_m) + (\mu t_2)(b_m)\Big)$$

Linear Transformations • **289**

$\Rightarrow \quad \varphi(\lambda t_1 + \mu t_2) = \Big(\lambda t_1(b_1), \lambda t_1(b_2), \ldots, \lambda t_1(b_m)\Big) + \Big(\mu t_2(b_1), \mu t_2(b_2), \ldots, \mu t_2(b_m)\Big)$

$\Rightarrow \quad \varphi(\lambda t_1 + \mu t_2) = \lambda\Big(t_1(b_1), t_1(b_2), \ldots, t_1(b_m)\Big) + \mu\Big(t_2(b_1), t_2(b_2), \ldots, t_2(b_m)\Big)$

$\Rightarrow \quad \varphi(\lambda t_1 + \mu t_2) = \lambda \varphi(t_1) + \mu \varphi(t_2)$

φ *is injective*: For any $t_1, t_2 \in \text{Hom}(V, V')$

$\varphi(t_1) = \varphi(t_2)$

$\Rightarrow \quad \Big(t_1(b_1), t_1(b_2), \ldots, t_1(b_m)\Big) = \Big(t_2(b_1), t_2(b_2), \ldots, t_2(b_m)\Big)$

$\Rightarrow \quad t_1(b_i) = t_2(b_i) \quad$ for all $i \in \underline{m}$.

$\Rightarrow \quad t_1 = t_2 \quad$ on B

$\Rightarrow \quad t_1 = t_2 \qquad\qquad\qquad\qquad\qquad\qquad\qquad\qquad\qquad$ [$\because$ B is a basis for V]

So, φ is injective.

φ *is surjective*: Let $(v'_1, v'_2, \ldots, v'_m)$ be an arbitrary list in V'^m. Then, by Theorem 5(page 285) there exists a unique linear transformation $t : V \to V'$ such that $t(b_i) = v'_i$ for all $i \in \underline{m}$.

$\therefore \qquad \varphi(t) = \Big(t(b_1), t(b_2), \ldots, t(b_m)\Big) = (v'_1, v'_2, \ldots, v'_m)$

Thus, for each $(v'_1, v'_2, \ldots, v'_m) \in V'^m$ there exists a $t \in \text{Hom}(V, V')$ such that
$$\varphi(t) = (v'_1, v'_2, \ldots, v'_m).$$
So, φ is surjective.

Hence, $\varphi : \text{Hom}(V, V') \to V'^m$ is an isomorphism. Consequently, by Theorem 8(page 253)

$\qquad \dim[\text{Hom}(V, V')] = \dim V'^m$

$\Rightarrow \quad \dim[\text{Hom}(V, V')] = m \dim V' \qquad$ [By Theorem 8 on page 200 of Chapter 2]

$\Rightarrow \quad \dim[\text{Hom}(V, V')] = \dim V \dim V'$.

$\hfill$ Q.E.D.

Aliter: Let $\dim V = m, \dim V' = n$, and let $B = \{b_1, b_2, \ldots, b_m\}$ and $B' = \{b'_1, b'_2, \ldots, b'_n\}$ be bases for V and V' respectively. Then by Theorem 5(page 285) for each $i \in \underline{m}$ and $j \in \underline{n}$, there exists a unique linear transformation. $\varphi_{ij} : V \to V'$ given by the rule

$$\varphi_{ij}(b_k) = \delta_{ik} b'_j \quad \text{for all } k \in \underline{m}.$$

Here δ_{ik} is *Kronecker* delta and is such that $\delta_{ik} = 1$ for $i = k$ and $\delta_{ik} = 0$ for $i \neq k$.
Let $B_1 = \{\varphi_{ij} : i \in \underline{m}, j \in \underline{n}\}$. Then B_1 consists of mn linear transformations from V to V'. We will now show that B_1 is a basis for $\text{Hom}(V, V')$.

B_1 *is linearly independent*: Let $\lambda_{ij} \in F, i \in \underline{m}, j \in \underline{n}$ be scalars in F such that

$$\sum_{i=1}^{m} \sum_{j=1}^{n} \lambda_{ij} \varphi_{ij} = \hat{0}. \qquad [\hat{0} \text{ is zero map in } \text{Hom}(V, V')]$$

Then,

$$\left\{\sum_{i=1}^{m}\sum_{j=1}^{n}\lambda_{ij}\,\varphi_{ij}\right\}(b_k) = \hat{0}(b_k) \quad \text{for all } k \in \underline{m}.$$

$$\Rightarrow \sum_{i=1}^{m}\sum_{j=1}^{n}\lambda_{ij}\,\varphi_{ij}(b_k) = 0_{V'} \quad \text{for all } k \in \underline{m}.$$

$$\Rightarrow \sum_{i=1}^{m}\sum_{j=1}^{n}\lambda_{ij}\,\delta_{ik}\,b'_j = 0_{V'}$$

$$\Rightarrow \sum_{j=1}^{n}\lambda_{kj}\,b'_j = 0_{V'} \Rightarrow \lambda_{kj} = 0 \text{ for all } k \in \underline{m}, j \in \underline{n} \qquad [\because B' \text{ is linearly independent}]$$

So, B_1 is a linearly independent set.

B_1 spans $\text{Hom}(V, V')$: Let t be an arbitrary linear transformation in $\text{Hom}(V,V')$. Then, $t(b_i) \in V'$ for all $i \in \underline{m}$.
Now,

$$t(b_i) \in V' \quad \text{for all } i \in \underline{m},$$

$$\Rightarrow t(b_i) = \sum_{j=1}^{n}\lambda_{ij}\,b'_j \text{ for some } \lambda_{ij} \in F \qquad [\because B' \text{ is basis for } V']$$

Now,

$$\sum_{k=1}^{m}\sum_{j=1}^{n}\lambda_{kj}\,\varphi_{kj}(b_i) = \sum_{k=1}^{m}\sum_{j=1}^{n}\lambda_{kj}\,\delta_{ki}\,b'_j = \sum_{j=1}^{n}\lambda_{ij}\,b'_j = t(b_i) \quad \text{for all } i \in \underline{m}.$$

$$\therefore \quad t = \sum_{k=1}^{m}\sum_{j=1}^{n}\lambda_{kj}\,\varphi_{kj} = \sum_{i=1}^{m}\sum_{j=1}^{n}\lambda_{ij}\,\varphi_{ij}$$

Thus, each $t \in \text{Hom}(V,V')$ is expressible as a linear combination elements in B_1.

So, B_1 spans $\text{Hom}(V,V')$.

Hence, B_1 is a basis for $\text{Hom}(V,V')$. Consequently, we have

$$\dim[\text{Hom}(V,V')] = \text{Number of elements in } B_1 = mn = \dim V \dim V'.$$

EXAMPLE-1 *Find dimensions of each of the following:*
  (i) $\text{Hom}(R^3, R^4)$ (ii) $\text{Hom}(R^5, R^3)$ (iii) $\text{Hom}(P_3(t), R^2)$ (iv) $\text{Hom}(M_{2\times 3}, R^4)$

SOLUTION We have,
$$\dim[\text{Hom}(V,V')] = \dim V \times \dim V'$$
Therefore,
  (i) $\dim[\text{Hom}(R^3, R^4)] = \dim(R^3) \times \dim(R^4) = 3 \times 4 = 12$
  (ii) $\dim[\text{Hom}(R^5, R^3)] = \dim(R^5) \times \dim(R^3) = 5 \times 3 = 15$
  (iii) $\dim[\text{Hom}(P_3(t), R^2)] = \dim P_3(t) \times \dim R^2 = 4 \times 2 = 8$
  (iv) $\dim[\text{Hom}(M_{2\times 3}, R^4)] = \dim(M_{2\times 3}) \times \dim R^4 = (2 \times 3) \times 4 = 24$

EXAMPLE-2 Consider the linear transformations $t_1 : R^3 \to R^2$, $t_2 : R^3 \to R^2$, $t_3 : R^3 \to R^2$ defined by $t_1(x,y,z) = (x+y+z, x+y)$, $t_2(x,y,z) = (2x+z, x+y)$ $t_3(x,y,z) = (2y, x)$. show that t_1, t_2, t_3 are linearly independent as elements of $\text{Hom}(R^3, R^2)$.

SOLUTION Let a, b, c be scalars in R such that
$$at_1 + bt_2 + ct_3 = \hat{0}$$
$\Rightarrow \quad (at_1 + bt_2 + ct_3)(e_1^{(3)}) = \hat{O}(e_1^{(3)})$
$ (at_1 + bt_2 + ct_3)(e_2^{(3)}) = \hat{O}(e_2^{(3)})$
$ (at_1 + bt_2 + ct_3)(e_3^{(3)}) = \hat{O}(e_3^{(3)})$
$\Rightarrow \quad at_1(e_1^{(3)}) + bt_2(e_1^{(3)}) + ct_3(e_1^{(3)}) = \hat{O}(e_1^{(3)})$
$ at_1(e_2^{(3)}) + bt_2(e_2^{(3)}) + ct_3(e_2^{(3)}) = \hat{O}(e_2^{(3)})$
$ at_1(e_3^{(3)}) + bt_2(e_3^{(3)}) + ct_3(e_3^{(3)}) = \hat{O}(e_3^{(3)})$
$\Rightarrow \quad (a+2b, a+b+c) = (0,0)$
$ (a+2c, a+b) = (0,0)$
$ (a+b, 0) = (0,0)$
$\Rightarrow \quad a+2b = 0,\ a+b+c = 0,\ a+2c = 0,\ a+b = 0$
$\Rightarrow \quad a = b = c = 0.$

Hence t_1, t_2, t_3 are linearly independent vectors in $\text{Hom}(R^3, R^2)$

THEOREM-7 Let V, V' and V'' be vector spaces over the same field F. Let $t : V \to V'$ and $s : V' \to V''$ be linear transformations. Then the composite map $\text{sot} : V \to V''$ given by $(\text{sot})(v) = s(t(v))$ for all $v \in V$, is a linear transformation.

PROOF. Let u, v be any two vectors in V and λ, μ be any two scalars in F. Then,
$$(\text{sot})(\lambda u + \mu v) = s\big(t(\lambda u + \mu v)\big) \qquad \text{[By definition of the composite map]}$$
$\Rightarrow \quad (\text{sot})(\lambda u + \mu v) = s\big(\lambda t(u) + \mu t(v)\big) \qquad [\because t : V \to V' \text{ is linear}]$
$\Rightarrow \quad (\text{sot})(\lambda u + \mu v) = \lambda s(t(u)) + \mu s(t(v)) \qquad [\because s : V' \to V'' \text{ is linear}]$
$\Rightarrow \quad (\text{sot})(\lambda u + \mu v) = \lambda(\text{sot})(u) + \mu(\text{sot})(v) \qquad \text{[By definition of sot]}$

Hence, $\text{sot} : V \to V''$ is a linear transformation.

Q.E.D.

REMARK. If $s : V \to V$ and $t : V \to V$ are linear transformations, then $\text{sot}: V \to V$ and $\text{tos}: V \to V$ both are linear transformations. But, sot need not be equal to tos, as the composition of mappings is not commutative.

EXAMPLE-3 Let t_1 and t_2 be linear operators R^2 defined by

$$t_1(a, b) = (b, a) \text{ and } t_2(a, b) = (a, 0) \text{ for all } (a, b) \in R^2.$$

Show that $t_1 o t_2 \neq t_2 o t_1$.

PROOF. For any $(a, b) \in R^2$

$$(t_1 o t_2)(a, b) = t_1\Big(t_2(a, b)\Big) = t_1(a, 0) = (0, a)$$

and, $(t_2 o t_1)(a, b) = t_2\Big(t_1(a, b)\Big) = t_2(b, a) = (b, 0)$

$\therefore \quad (t_1 o t_2)(a, b) \neq (t_2 o t_1)(a, b)$

Hence, $t_1 o t_2 \neq t_2 o t_1$.

Q.E.D.

EXAMPLE-4 Let D and T be two linear operators, defined on the vector space $R[x]$ of all polynomials over R, defined by

$$D[f(x)] = \frac{d}{dx}(f(x)) \text{ and, } T(f(x)) = xf(x) \text{ for all } f(x) \in R[x]$$

Show that $DoT \neq ToD$ i.e. $DT \neq TD$ and $DT - TD = I$

SOLUTION We have,

$$[DT](f(x)) = D\{T(f(x))\} = D\{xf(x)\} = \frac{d}{dx}(xf(x)) = f(x) + xf'(x)$$

and, $[TD](f(x)) = T\{D(f(x))\} = T\left[\frac{d}{dx}\{f(x)\}\right] = x\frac{d}{dx}\{f(x)\} = xf'(x)$

Clearly, $(DT)f(x) \neq (TD)f(x)$ for every non-zero polynomial.

$\therefore \quad DT \neq TD$

Also, $(DT)f(x) - (TD)f(x) = f(x) \quad$ for all $f(x) \in R[x]$

$\Rightarrow \quad (DT - TD)f(x) = f(x) \quad$ for all $f(x) \in R[x]$

$\Rightarrow \quad DT - TD = I$

EXAMPLE-5 Let D and T be two operators on the vector space $R[x]$ of all polynomials over R defined by

$$D\{f(x)\} = \frac{d}{dx}\{f(x)\} \text{ and, } T\{f(x)\} = \int_0^x f(x)dx$$

for every $f(x)$. Then show that $DT = I$ and $TD \neq I$

SOLUTION Let $f(x) = a_0 + a_1x + a_2x^2 + \ldots$ be an arbitrary polynomial in $R[x]$. Then,

$$(DT)(f(x)) = D[T(f(x))]$$
$$\Rightarrow (DT)(f(x)) = D\left\{\int_0^x (a_0 + a_1x + a_2x^2 + \ldots)dx\right\}$$
$$\Rightarrow (DT)(f(x)) = D\left(a_0x + a_1\frac{x^2}{2} + a_2\frac{x^3}{3} + \ldots\right)$$
$$\Rightarrow (DT)(f(x)) = a_0 + a_1x + a_2x^2 + \cdots = f(x)$$
$$\therefore (DT)(f(x)) = f(x) \quad \text{for all } f(x) \in R[x]$$
$$\Rightarrow DT = I$$

Now,
$$(TD)(f(x)) = T[D(f(x))]$$
$$\Rightarrow (TD)(f(x)) = T\left\{\frac{d}{dx}\left(a_0 + a_1x + a_2x^2 + \ldots\right)\right\}$$
$$\Rightarrow (TD)(f(x)) = T\left(a_1 + 2a_2x + \ldots\right)$$
$$\Rightarrow (TD)(f(x)) = a_1x + a_2x^2 + \cdots \neq f(x) \quad \text{Unless} \quad a_0 = 0$$
$$\therefore (TD)f(x) \neq f(x) \quad \text{for some } f(x) \in R[x]$$
$$\Rightarrow TD \neq I$$

Hence, $TD \neq DT$.

THEOREM-8 Let $V(F)$ be a vector space, and let $t_1 : V \to V, t_2 : V \to V, t_3 : V \to V$ be linear transformations. Then,

(i) $t_1 o \hat{0} = \hat{0} o t_1 = \hat{0}, \hat{0}$ is the zero transformation.

(ii) $t_1 o I_V = I_V o t_1 = t_1$

(iii) $t_1 o (t_2 o t_3) = (t_1 o t_2) o t_3$

(iv) $t_1 o (t_2 + t_3) = t_1 o t_2 + t_1 o t_3$

(v) $(t_2 + t_3) o t_1 = t_2 o t_1 + t_3 o t_1$

(vi) $\lambda(t_1 o t_2) = (\lambda t_1) o t_2 = t_1 o (\lambda t_2)$ for all $\lambda \in F$.

PROOF. Left as an exercise for the reader.

It follows from Theorems 3 and 8 that the set $\text{Hom}(V,V)$ is a ring with unity under pointwise addition and composition of mappings as two ring operations.

ALGEBRA. *A ring R is said to be an algebra over a field F if the additive group $(R,+)$ of R is a vector space over F and for any $u, v \in R, \lambda \in F$, we have $\lambda(uv) = (\lambda u)v = u(\lambda v)$.*

THEOREM-9 *For any vector space $V(F)$. $\mathrm{Hom}(V, V)$ is an algebra over field F.*

PROOF. By Theorems 3 and 8, we see that $\mathrm{Hom}(V, V)$ is a ring with unity and $\mathrm{Hom}(V, V)$ is a vector space over field F.

For any $t_1, t_2 \in \mathrm{Hom}(V, V')$ and any $\lambda \in F$

$$[t_1 o(\lambda t_2)](v) = t_1\Big((\lambda\, t_2)(v)\Big) \qquad \text{[By definition of composition map]}$$

$\Rightarrow \quad [t_1 o(\lambda t_2)](v) = t_1\Big(\lambda\, t_2(v)\Big) \qquad$ [By definition of scalar multiplication on $\mathrm{Hom}(V_1 V')$]

$\Rightarrow \quad [t_1 o(\lambda t_2)](v) = \lambda[t_1(t_2(v))] \qquad [\because t_1 \text{ is linear}]$

$\Rightarrow \quad [t_1 o(\lambda t_2)](v) = \lambda[(t_1 o t_2)(v)]$

$\Rightarrow \quad [t_1 o(\lambda t_2)](v) = [\lambda(t_1 o t_2)](v) \quad \text{for all } v \in V.$

$\therefore \qquad t_1 o(\lambda t_2) = \lambda(t_1 o t_2)$

Similarly, $(\lambda t_1) o t_2 = \lambda(t_1 o t_2)$.

So, $(\lambda t_1) o t_2 = \lambda(t_1 o t_2) = t_1 o(\lambda t_2)$

Hence, $\mathrm{Hom}(V, V')$ is an algebra over field F.

Q.E.D.

THEOREM-10 *For any vector space $V(F)$ of dimension n, $\mathrm{Hom}(V, V)$ is an algebra of dimension n^2 over field F.*

PROOF. It is a direct consequence of Theorems 6 and 8.

REMARK. *If $V(F)$ is a vector space, then a linear transformation from V into itself is also called a linear operator on V. The vector space of all linear operators on V i.e. $\mathrm{Hom}(V, V)$ is also denoted by $A(V)$.*

In $A(V)$, we sometimes write $t_1 t_2$ instead of $t_1 o t_2$.

3.7 POLYNOMIALS AND LINEAR OPERATORS

Let V be a vector space over field F, and let $A(V)$ be the algebra of linear operators on V. Clearly, the identity operator $I \in A(V)$. Also, for any $t \in A(V)$, we have

$toI = Iot = t$, which is also written as $tI = It = t$.

We define powers of t as follows:

$t^0 = I,\ t^2 = tot$ which is also written as $t^2 = tt$

$t^3 = t^2 ot = totot$ also written as $t^3 = ttt$

$t^4 = t^3 ot = totototot$ also written as $t^4 = tttt$

and so on.

If $m, n \in N$, then it can be easily shown that
$$t^m t^n = t^{m+n} \text{ and } (t^m)^n = t^{mn} = (t^n)^m$$
Furthermore, for any polynomial $f(x) = a_o + a_1 x + a_2 x^2 + \cdots + a_n x^n$ over field F, we can form the linear operator $f(t)$ defined by
$$f(t) = a_o I + a_1 t + a_2 t^2 + \cdots + a_n t^n$$
The polynomials in a linear operator behave like ordinary polynomials.

3.8 SINGULAR AND NON-SINGULAR LINEAR TRANSFORMATIONS

SINGULAR *A linear transformation t from a vector space $V(F)$ to a vector space $V'(F)$ is said to be non-singular, if $\text{Ker}(t) = \{0_V\}$. That is,*
$$v \in V \text{ and } t(v) = 0_{V'} \Rightarrow v = 0_V$$

If $\text{Ker}(t) \neq \{0_V\}$, that is, there exists a non-null vector $v \in V$ such that $t(v) = 0_{V'}$, then t is said to be *a singular transformation*.

It follows from Theorem 2 on page 248, that a linear transformation $t : V \to V'$ is non-singular iff t is a monomorphism.

In Example 1 on page 271, the mapping $t : R^2 \to R^3$ given by
$$t(a,b) = (a+b,\ a-b, b) \text{ for all } a,b \in R^2$$
is a non-singular linear transformation, because $\text{Ker}(t) = \{(0,0)\}$.

THEOREM-1 *A linear transformation t from a vector space $V(F)$ to a vector space $V'(F)$ is non-singular iff t carries each linearly independent subset of V onto a linearly independent subset of V'.*

PROOF. *First suppose that $t : V \to V'$ is a non-singular linear transformation.*
Let $B = \{b_1, b_2, \ldots, b_n\}$ be a linearly independent subset of V. Then we have to prove that $B' = \{t(b_1), t(b_2), \ldots, t(b_n)\}$ is also linearly independent
Let $\lambda_1, \lambda_2, \ldots, \lambda_n$ be scalars in F such that
$$\lambda_1 t(b_1) + \lambda_2 t(b_2) + \cdots + \lambda_n t(b_n) = 0_{V'}$$
Then, $\quad t(\lambda_1 b_1 + \lambda_2 b_2 + \cdots + \lambda_n b_n) = 0_{V'}$ $\qquad [\because t : V \to V' \text{ is linear}]$
$\Rightarrow \quad \lambda_1 b_1 + \lambda_2 b_2 + \cdots + \lambda_n b_n = 0_V$ $\qquad [\because t \text{ is non-singular}]$
$\Rightarrow \quad \lambda_1 = \lambda_2 = \cdots = \lambda_n = 0$ $\qquad [\because B \text{ is linearly independent set}]$
$\therefore \quad B'$ is a linearly independent set.

Hence, image of a linearly independent set under a non-singular linear transformation is a linearly independent set.

Conversely, suppose that t carries linearly independent subsets of V onto linearly independent subsets of V'. Then we have to prove that t is non-singular. That is, if $0_V \neq v \in V$, then $t(v) \neq 0_{V'}$.

Let $v(\neq 0_V) \in V$. Then the set $S = \{v\}$ consisting of a single non-zero vector is linearly independent

Let $S' = \{t(v)\}$. Then, by hypothesis S' is linearly independent Therefore, $t(v) \neq 0_{V'}$.

Thus,
$$v \neq 0_V \Rightarrow t(v) \neq 0_{V'} \Rightarrow \text{Ker}(t) = \{0_V\}.$$

Hence, t is a non-singular linear transformation.

Q.E.D.

3.9 INVERTIBLE LINEAR TRANSFORMATION

Let us begin with the following definition.

INVERTIBLE LINEAR TRANSFORMATION. *A linear transformation t from a vector space $V(F)$ to a vector space $V'(F)$ is said to be invertible if it is bijective.*

THEOREM-1 *Let V and V' be vector spaces over a field F, and let $t : V \to V'$ be a linear transformation. If $t : V \to V'$ is bijective, then the inverse function $t^{-1} : V' \to V$ is a linear transformation.*

PROOF. Let u', v' be any two vectors in V' and $\lambda, \mu \in F$. Since $t : V \to V'$ is bijective. Therefore, there exist $u, v \in V$ such that $t(u) = u'$, $t(v) = v'$. By definition of t^{-1}, we have $t^{-1}(u') = u, t^{-1}(v') = v$. Now, $\lambda u + \mu v \in V$ is such that

$\qquad t(\lambda u + \mu v) = \lambda\, t(u) + \mu t(v)$ $\qquad\qquad$ [$\because t : V \to V'$ is linear]

$\Rightarrow \qquad t(\lambda u + \mu v) = \lambda u' + \mu v'$

$\Rightarrow \qquad t^{-1}(\lambda u' + \mu v') = \lambda u + \mu v$ $\qquad\qquad$ [By definition of t^{-1}]

$\Rightarrow \qquad t^{-1}(\lambda u' + \mu v') = \lambda t^{-1}(u') + \mu t^{-1}(v')$ $\qquad\qquad$ [$\because u = t^{-1}(u'), v = t^{-1}(v')$]

So, $t^{-1} : V' \to V$ is a linear transformation.

Q.E.D.

THEOREM-2 *Let $V(F)$ be a finite dimensional vector space. Then a linear transformation $t : V \to V$ is invertible iff t is non-singular.*

PROOF. First suppose that $t : V \to V$ is invertible. Then t is an isomorphism. Therefore, by Theorem 2 on page 248, $\text{Ker}(t) = \{0_V\}$ and so t is non-singular.

Conversely, suppose that $t : V \to V$ is non-singular. Then we have to prove that t is invertible, i.e. t is bijective.

Let v_1, v_2 be any two vectors in V. Then,

$$\begin{aligned}
& t(v_1) = t(v_2) \\
\Rightarrow\ & t(v_1) - t(v_2) = 0_V \\
\Rightarrow\ & t(v_1 - v_2) = 0_V \\
\Rightarrow\ & v_1 - v_2 = 0_V \qquad [\because t \text{ is non-singular}] \\
\Rightarrow\ & v_1 = v_2
\end{aligned}$$

$\therefore\ t : V \to V$ is injective.

By Theorem 19 on page 262, t is surjective.

Thus, $t : V \to V$ is bijective. Hence, $t : V \to V$ is invertible.

Q.E.D.

THEOREM-3 *Let $V(F)$ be a finite dimensional vector space. Then a linear transformation $t : V \to V$ is invertible iff t is an epimorphism.*

PROOF. Let $t : V \to V$ be an invertible linear transformation. Then, t is an epimorphism.

Conversely, let $t : V \to V$ be an epimorphism. Then we have to show that t is invertible. For this it is sufficient to show that t is an isomorphism, which follows from Theorem 20 on page 262. Hence, $t : V \to V$ is invertible.

Q.E.D.

THEOREM-4 *Let $V(F)$ be a finite dimensional vector space. Then a linear transformation $t : V \to V$ is invertible iff t is a monomorphism.*

PROOF. *Left as an exercise for the reader.*

REMARK. *In the above theorems, we have shown that a monomorphism from a finite dimensional vector space to itself is an epimorphism and vice-versa. But, this result fails if V is not finite dimensional which is clear from the following.*

Consider the linear transformation D from the vector space $F[x]$, of all polynomials in x over field F to itself given by

$$D[f(x)] = f'(x)$$

Clearly, D is surjective, i.e. an epimorphism but it is not injective, i.e. a monomorphism. Because, $D(2x+3) = D(2x-3)$ but $2x+3 \neq 2x-3$.

Now, let $t : F[x] \to F[x]$ be a linear transformation given by

$$t\big(f(x)\big) = xf(x) \quad \text{for all } f(x) \in F[x].$$

Then, t is a monomorphism. But, t is not an epimorphism, because a non-zero constant polynomial in $F[x]$ is not image of any polynomial in $F[x]$ under t.

THEOREM-5 *Let V and V' be finite dimensional vector spaces over a field F such that $\dim V = \dim V'$, and let $t : V \to V'$ be a linear transformation. Then the following are equivalent:*

(i) t is invertible.

(ii) t is non-singular.

(iii) t is an epimorphism, i.e. $I_m(t) = V'$.

(iv) If $\{b_1, b_2, \ldots, b_n\}$ is a basis for V, then $\{t(b_1), t(b_2), \ldots, t(b_n)\}$ is a basis for V'.

(v) There is some basis $\{b_1, b_2, \ldots, b_n\}$ for V such that $\{t(b_1), t(b_2), \ldots, t(b_n)\}$ is a basis for V'.

PROOF. First we shall show that (i) implies (ii)

$t : V \to V'$ is invertible

$\Rightarrow \quad t : V \to V'$ is an isomorphism

$\Rightarrow \quad t : V \to V'$ is a monomorphism

$\Rightarrow \quad \text{Ker}(t) = \{0_V\}$ [By Theorem 2 on page 248]

$\Rightarrow \quad t$ is non-singular.

Thus, (i) implies (ii).

Now, we shall show that (ii) implies (iii)

Let $\{b_1, b_2, \ldots, b_n\}$ be a basis for V. Since $t : V \to V'$ is non-singular. Therefore, by Theorem 1 on page 295, $\{t(b_1), t(b_2), \ldots, t(b_n)\}$ is a linearly independent set of n vectors in V'. But, $\dim V' = n$. Therefore, $\{t(b_1), t(b_2), \ldots, t(b_n)\}$ is a basis for V'.

Now, let v' be an arbitrary vector in V'. Then there exist scalars $\lambda_1, \lambda_2, \ldots, \lambda_n \in F$ such that

$v' = \lambda_1 t(b_1) + \lambda_2 t(b_2) + \cdots + \lambda_n t(b_n)$ $[\because \{t(b_1), \ldots, t(b_n)\}$ is basis for $B']$

$\Rightarrow \quad v' = t(\lambda_1 b_1 + \lambda_2 b_2 + \cdots + \lambda_n b_n)$

$\Rightarrow \quad v' \in I_m(t)$

Thus, every vector in V' is in $I_m(t)$. So, $V' \subset I_m(t)$.

Also, $I_m(t) \subset V'$. Hence, $I_m(t) = V'$.

Thus, (ii) implies (iii)

Let us now show that (iii) implies (iv)

Now,

(iii) holds good.

$\Rightarrow$ $t : V \to V'$ is an epimorphism

$\Rightarrow$ $I_m(t) = V'$

$\Rightarrow$ $\dim(I_m(t)) = \dim V' = n$

$\Rightarrow$ $\{t(b_1), t(b_2), \ldots, t(b_n)\}$ spans V' of dimension equal to n

$\Rightarrow$ $\{t(b_1), t(b_2), \ldots, t(b_n)\}$ is a basis for V'.

Thus, (iii) implies (iv)

Now, suppose (iv) holds good.

Since V is finite dimensional. Therefore, there exists a basis for V. Let $\{b_1, b_2, \ldots, b_n\}$ be a basis for V. Then $\{t(b_1), t(b_2), \ldots, t(b_n)\}$ is a basis for V' as it is given in (iv). So, (iv) implies (v).

Finally, we shall show that (v) implies (i).

Let $\{b_1, b_2, \ldots, b_n\}$ be a basis for V such that $\{t(b_1), t(b_2), \ldots, t(b_n)\}$ is a basis for V'. Then we have to prove that t is invertible.

The set $\{t(b_1), t(b_2), \ldots, t(b_n)\}$ spans $I_m(t)$. Also it spans V'. Therefore,

$I_m(t) = V' \Rightarrow t$ is surjective.

Now, let v be an arbitrary vector in V. Then,

$v = \sum_{i=1}^{n} \lambda_i b_i$ for some $\lambda_i \in F$ $\qquad [\because \{b_1, b_2, \ldots, b_n\}$ is basis for $V]$

$\Rightarrow$ $t(v) = t\left(\sum_{i=1}^{n} \lambda_i b_i\right) = \sum_{i=1}^{n} \lambda_i t(b_i)$ $\qquad [\because t$ is linear$]$

Now,

$v \in \text{Ker}(t)$

$\Rightarrow$ $t(v) = 0_{V'}$

$\Rightarrow$ $\sum_{i=1}^{n} \lambda_i t(b_i) = 0_{V'}$

$\Rightarrow$ $\lambda_i = 0$ for all $i \in \underline{n}$ $\qquad [\{t(b_1), t(b_2), \ldots, t(b_n)\}$ is linearly independent$]$

$\Rightarrow$ $v = 0_V$

Thus, $v \in \text{Ker}(t) \Rightarrow v = 0_V$

$\therefore$ $\text{Ker}(t) = \{0_V\}$

$\Rightarrow$ $t : V \to V'$ is monomorphism $\qquad$ [By Theorem 2 on page 248]

$\Rightarrow$ t is injective.

$\Rightarrow$ $t : V \to V'$ is invertible.

Thus, (vi) ⇒ (i).
Hence, (i) ⇔ (ii) ⇔ (iii) ⇔ (iv) ⇔ (v).

Q.E.D.

EXAMPLE-2 Let $t : R^3(R) \to R^3(R)$ be a linear transformation given by

$$t(a,b,c) = (3a, a-b, 2a+b+c) \quad \text{for all } (a,b,c) \in R^3(R).$$

Is t invertible? If so, find a rule for t^{-1} like one which defines t.

SOLUTION We have to see whether $t : R^3 \to R^3$ is bijective or not.
Let $u = (a_1, b_1, c_1), v = (a_2, b_2, c_2)$ be any two vectors in R^3. Then,

$t(u) = t(v)$
⇒ $(3a_1, a_1 - b_1, 2a_1 + b_1 + c_1) = (3a_2, a_2 - b_2, 2a_2 + b_2 + c_2)$
⇒ $3a_1 = 3a_2, a_1 - b_1 = a_2 - b_2, 2a_1 + b_1 + c_1 = 2a_2 + b_2 + c_2$
⇒ $a_1 = a_2, b_1 = b_2, c_1 = c_2$
⇒ $(a_1, b_1, c_1) = (a_2, b_2, c_2)$

So, $t : R^3 \to R^3$ is injective.

Since t is a linear transformation from a finite dimensional vector space to itself and it is injective. Therefore, by Theorem t is surjective.

Hence, $t : R^3 \to R^3$ is invertible.

To find rule for t^{-1}, let $(a,b,c) \in R^3$ be such that $t(a,b,c) = (\alpha, \beta, \gamma)$. Then by definition of t^{-1}, we have,

$t^{-1}(\alpha, \beta, \gamma) = (a, b, c)$
$t(a,b,c) = (\alpha, \beta, \gamma)$
⇒ $(3a, a-b, 2a+b+c) = (\alpha, \beta, \gamma)$
⇒ $3a = \alpha, a - b = \beta, 2a + b + c = \gamma$
⇒ $a = \dfrac{\alpha}{3}, b = \dfrac{\alpha}{3} - \beta, c = \gamma - \alpha + \beta$.
∴ $t^{-1}(\alpha, \beta, \gamma) = \left(\dfrac{\alpha}{3}, \dfrac{\alpha}{3} - \beta, \gamma - \alpha + \beta\right)$ for all $(\alpha, \beta, \gamma) \in R^3$ is the rule for t^{-1}.

EXAMPLE-3 Let t be a linear operator on R^2 defined by $t(x,y) = (2x+y, 3x+2y)$. Show that t is invertible and find t^{-1}.

SOLUTION In order to show that t is invertible, it is sufficient to show that t is non-singular i.e. $t(x,y) = (0,0)$ iff $(x,y) = (0,0)$.

Now,

$$t(x,y) = (0,0)$$
$$\Leftrightarrow (2x+y, 3x+2y) = (0,0)$$
$$\Leftrightarrow 2x+y = 0, \; 3x+2y = 0$$
$$\Leftrightarrow x = y = 0$$
$$\Leftrightarrow (x,y) = (0,0)$$

Thus, $t(x,y) = (0,0)$ iff $(x,y) = (0,0)$.

So, t is non-singular and hence invertible.

To find the formula for t^{-1}, let $t(x,y) = (a,b)$. Then, $t^{-1}(a,b) = (x,y)$

Now,

$$t(x,y) = (a,b)$$
$$\Rightarrow (2x+y, 3x+2y) = (a,b)$$
$$\Rightarrow 2x+y = a \text{ and } 3x+2y = b$$
$$\Rightarrow x = 2a-b, \; y = -3a+2b$$
$$\therefore t^{-1}(a,b) = (2a-b, -3a+2b)$$

THEOREM-6 *Let V and V' be vector spaces over a field F. Then for any $t_1, t_2 \in \text{Hom}(V, V')$*

(i) $\text{rank}(\lambda t_1) = \text{rank}(t_1)$ *for all* $0 \neq \lambda \in F$

(ii) $|\text{rank}(t_1) - \text{rank}(t_2)| \leq \text{rank}(t_1 + t_2) \leq \text{rank}(t_1) + \text{rank}(t_2)$.

PROOF. (i) Since $I_m(t_1) = t_1(V)$ is a subspace of V'. Therefore,
$$I_m(\lambda t_1) = (\lambda t_1)(V) = \lambda \, t_1(V) \subset t_1(V) = I_m(t_1)$$

Similarly, $\lambda^{-1} t_1(V) \subset t_1(V)$

$$\Rightarrow \lambda[\lambda^{-1} t_1(V)] \subset (\lambda t_1)(V)$$
$$\Rightarrow t_1(V) \subset (\lambda t_1) V$$
$$\Rightarrow I_m(t_1) \subset I_m(\lambda t_1)$$
$$\therefore I_m(t_1) = I_m(\lambda t_1)$$
$$\Rightarrow \dim I_m(t_1) = \dim I_m(\lambda t_1)$$
$$\Rightarrow \text{rank}(t_1) = \text{rank}(\lambda t_1)$$

(ii) For any $v \in V$, we have

$$(t_1 + t_2)(v) = t_1(v) + t_2(v) \text{ for all } v \in V \quad \text{[By definition of pointwise addition]}$$

$\therefore \quad (t_1 + t_2)(V) \subset t_1(V) + t_2(V)$

$\Rightarrow \quad \dim(t_1 + t_2)(V) \leq \dim\{t_1(V) + t_2(V)\} \quad \begin{bmatrix} \because \dim(S+T) = \dim S + \dim T - \dim(S \cap T) \\ \therefore \dim(S+T) \leq \dim S + \dim T \end{bmatrix}$

$\Rightarrow \quad \dim(t_1 + t_2)(V) \leq \dim t_1(V) + \dim t_2(V)$

$\Rightarrow \quad \text{rank }(t_1 + t_2) \leq \text{rank } t_1 + \text{rank } t_2 \quad \text{(i)}$

Now, $\quad t_1 = (t_1 + t_2) + (-t_2),$

$\Rightarrow \quad \text{rank}(t_1) \leq \text{rank}(t_1 + t_2) + \text{rank}(-t_2) \quad \text{[Using (i)]}$

$\Rightarrow \quad \text{rank}(t_1) \leq \text{rank}(t_1 + t_2) + \text{rank}(t_2) \quad \text{[From (i), rank }(-t_2) = \text{rank}(t_2)]$

$\Rightarrow \quad \text{rank}(t_1) - \text{rank}(t_2) \leq \text{rank}(t_1 + t_2) \quad \text{(ii)}$

Similarly, we have

$$\text{rank}(t_2) - \text{rank}(t_1) \leq \text{rank}(t_1 + t_2) \quad \text{(iii)}$$

From (i), (ii) and (iii), we obtain

$$|\text{rank }(t_1) - \text{rank }(t_2)| \leq \text{rank }(t_1 + t_2) \leq \text{rank }(t_1) + \text{rank }(t_2).$$

Q.E.D.

THEOREM-7 Let V, V' and V'' be finite dimensional vector space over the same field F, and let $t_1 \in \text{Hom}(V, V'), t_2 \in \text{Hom}(V'', V)$. Then,

$$\text{rank }(t_1) + \text{rank }(t_2) - n \leq \text{rank }(t_1 o t_2) \leq \min\left[\text{rank}(t_1), \text{rank}(t_2)\right], \quad \text{where } n = \dim V.$$

PROOF. Since $t_1 : V \to V'$ and $t_2 : V'' \to V$ are linear transformations. Therefore, $t_1 o t_2 : V'' \to V'$ is a linear transformation.

Now,

$\text{rank}(t_1 o t_2) = \dim\{I_m(t_1 o t_2)\}$

$\Rightarrow \quad \text{rank}(t_1 o t_2) = \dim\left[(t_1 o t_2)(V'')\right] = \dim\left[t_1(t_2(V''))\right]$

$\Rightarrow \quad \text{rank}(t_1 o t_2) \leq \min\left[\dim t_2(V''), \dim V'\right] \quad \begin{bmatrix} \because \text{ For any } t : V_1 \to V_2, \\ \text{rank}(t) \leq \min(\dim V_1, \dim V_2) \end{bmatrix}$

$\therefore \quad \text{rank}(t_1 o t_2) \leq \dim t_2(V'') \quad \text{(i)}$

Since $(t_1 o t_2)(V'') \subset t_1(V)$

$\therefore \quad \dim\left[(t_1 o t_2)(V'')\right] \leq \dim t_1(V)$

i.e., $\quad \text{rank}(t_1 o t_2) \leq \text{rank } t_1 \quad \text{(ii)}$

From (i) and (ii), we get

$$\text{rank}(t_1 o t_2) \leq \min\left[\text{rank}(t_1), \text{rank}(t_2)\right] \quad \text{(iii)}$$

By Sylvester's law of nullity, we have

$$\operatorname{rank}(t_1) + \operatorname{nullity}(t_1) = \dim V = n$$

$\therefore \qquad \operatorname{nullity}(t_1) = n - \operatorname{rank}(t_1) \qquad$ (iv)

Let t_1' be the restriction of t_1 to $t_2(V'')$. Then,

$$\operatorname{Ker}(t_1') = \operatorname{Ker}(t_1) \cap t_2(V'')$$

$\Rightarrow \qquad \operatorname{nullity}(t_1') \leqq \operatorname{nullity}(t_1) \qquad$ (v)

From (iv) and (v), we get

$\operatorname{nullity}(t_1') \leqq n - \operatorname{rank}(t_1) \qquad$ (vi)

Now,

$$t_1'[t_2(V'')] = (t_1 o\, t_2)(V'')$$

$\Rightarrow \qquad \operatorname{rank}(t_1') = \operatorname{rank}(t_1 o\, t_2)$

Since $\quad \dim t_2(V'') = \operatorname{rank}(t_1') + \operatorname{nullity}(t_1')$

$\Rightarrow \qquad \dim t_2(V'') = \operatorname{rank}(t_1 o t_2) + \operatorname{nullity}(t_1')$

$\therefore \qquad \operatorname{rank}(t_2) = \operatorname{rank}(t_1 o t_2) + \operatorname{nullity}(t_1')$

$\Rightarrow \qquad \operatorname{rank}(t_2) \leqq \operatorname{rank}(t_1 o t_2) + n - \operatorname{rank}(t_1) \qquad$ [Using (vi)]

$\Rightarrow \qquad \operatorname{rank}(t_1) + \operatorname{rank}(t_2) - n \leqq \operatorname{rank}(t_1 o t_2) \qquad$ (vii)

From (iii) and (vii), we get

$$\operatorname{rank}(t_1) + \operatorname{rank}(t_2) - n \leqq \operatorname{rank}(t_1 o t_2) \leqq \min\bigl[\operatorname{rank}(t_1), \operatorname{rank}(t_2)\bigr]$$

Q.E.D.

COROLLARY. *Let $t_1 \in \operatorname{Hom}(V, V'')$, $t_2 \in \operatorname{Hom}(V'', V)$. Then,*

(i) $\operatorname{rank}(t_1 o t_2) = \operatorname{rank}(t_2)$, *if t_1 is an isomorphism,*

(ii) $\operatorname{rank}(t_1 o t_2) = \operatorname{rank}(t_1)$, *if t_2 is an isomorphism.*

PROOF. (i) Since $t_1 : V \to V''$ is an isomorphism. Therefore, t_1 is invertible and so t_1^{-1} exists. Now, $\quad t_2 = t_1^{-1} o (t_1 o t_2)$

$\Rightarrow \qquad \operatorname{rank}(t_2) = \operatorname{rank}\bigl[t_1^{-1} o (t_1 o t_2)\bigr]$

$\Rightarrow \qquad \operatorname{rank}(t_2) \leqq \min\bigl[\operatorname{rank}(t_1^{-1}), \operatorname{rank}(t_1 o t_2)\bigr]$

$\Rightarrow \qquad \operatorname{rank}(t_2) \leqq \operatorname{rank}(t_1 o t_2)$

By the above theorem, we have

$$\operatorname{rank}(t_1 o t_2) \leqq \operatorname{rank}(t_2)$$

Hence, $\quad \operatorname{rank}(t_1 o t_2) = \operatorname{rank}(t_2)$

(ii) Can be proved on the same lines.

Q.E.D.

ILLUSTRATIVE EXAMPLES

EXAMPLE-1 *Determine whether or not each of the following linear transformations is non-singular. If not, find a non-zero vector v whose image is 0.*

(i) $t : R^2 \to R^2$ defined by $t(x,y) = (x-y, x-2y)$
(ii) $t : R^2 \to R^2$ defined by $t(x,y) = (2x-4y, 3x-6y)$

SOLUTION We know that a linear transformation is non-singular iff its kernel is the null space. So, let us find the kernel of t in each case.

(i) Let $v = (x,y) \in R^2$ be such that $t(v) = 0$. Then,

$$t(v) = 0 \Rightarrow t(x,y) = (0,0) \Rightarrow (x-y, x-2y) = (0,0) \Rightarrow x-y = 0 \text{ and } x-2y = 0$$

Clearly, $x = 0, y = 0$ is the only solution of the above system of equations. Therefore, $\text{Ker}(t) = \{(0,0)\}$ and hence t is non-singular.

(ii) Let $v = (x,y) \in R^2$ be such that $t(v) = 0$. Then,

$t(v) = 0$

$\Rightarrow t(x,y) = (0,0)$

$\Rightarrow (2x-4y, 3x-6y) = (0,0) \Rightarrow 2x-4y = 0 \text{ and } 3x-6y = 0$

This system of homogeneous equations is equivalent to $x - 2y = 0$, which has non-zero solutions. Hence, t is singular. Putting $y = 1$, we get $x = 2$. Thus, $v = (2,1)$ is a non-zero vector in R^2 such that $t(v) = 0$.

EXAMPLE-2 *Show that the linear transformation $t : R^2 \to R^2$ defined by $t(x,y) = (x+y, 2x-y)$ is non-singular. Also, find a formula for t^{-1}.*

SOLUTION In order to prove that t is non-singular, we will have to show that kernel of t is the null space.

Let $v = (x,y) \in \text{Ker}(t)$. Then,

$$t(x,y) = (0,0) \Rightarrow (x+y, 2x-y) = (0,0) \Rightarrow x+y = 0 \text{ and, } 2x-y = 0$$

This is a homogeneous system of equations having only the trivial solution $x = 0, y = 0$. Hence, t is invertible.

In order to find t^{-1}, let $(x,y) \in R^2$ such that

$t(x,y) = (a,b)$

$\Rightarrow (x+y, 2x-y) = (a,b)$

$\Rightarrow x+y = a, 2x-y = b \Rightarrow x = \dfrac{a+b}{3}, y = \dfrac{2a-b}{3}$

Now, $t(x,y) = (a,b)$

$\Rightarrow t^{-1}(a,b) = (x,y)$

$\Rightarrow t^{-1}(a,b) = \left(\dfrac{a+b}{3}, \dfrac{2a-b}{3}\right)$ or, $t^{-1}(x,y) = \left(\dfrac{x+y}{3}, \dfrac{2x-y}{3}\right)$

EXAMPLE-3 Let $t : R^2 \to R^3$ be defined by $t(x,y) = (x+y, x-2y, 3x+y)$. Show that t is non-singular. Also, find a formula for t^{-1}.

SOLUTION In order to show that t is non-singular, we need to find kernel of t.
So, let $(x,y) \in \text{Ker}(t)$. Then,
$$t(x,y) = (0,0,0)$$
$\Rightarrow \quad (x+y, x-2y, 3x+y) = (0,0,0)$
$\Rightarrow \quad x+y = 0, x-2y = 0, 3x+y = 0$

This system of equations has $x=0, y=0$ as the only solution.
$\therefore \qquad \text{Ker}(t) = \{(0,0)\}$

Hence, t is non-singular.
Since R^2 and R^3 have different dimensions. So, t^{-1} does not exist.

EXAMPLE-4 Let t be a linear operator on R^3 defined by
$$t(x,y,z) = (3x, x-y, 2x+y+z)$$
Is t non-singular (invertible)? If so, find a rule for t^{-1}.

SOLUTION Let $(x,y,z) \in R^3$ be such that $(x,y,z) \in \text{Ker}(t)$. Then,
$$t(x,y,z) = (0,0,0)$$
$\Rightarrow \quad (3x, x-y, 2x+y+z) = (0,0,0)$
$\Rightarrow \quad 3x = 0, x-y = 0, 2x+y+z = 0$

This system of homogeneous equations has only the trivial solution $x = y = z = 0$.
$\therefore \quad (x,y,z) \in \text{Ker}(t) \Leftrightarrow (x,y,z) = (0,0,0)$
$\Rightarrow \quad \text{Ker}(t) = \{(0,0,0)\}$

Hence, t is invertible.
In order to find a rule for t^{-1}. Let $t(x,y,z) = (a,b,c)$, then $t^{-1}(a,b,c) = (x,y,z)$.
Now,
$$t(x,y,z) = (a,b,c)$$
$\Rightarrow \quad (3x, x-y, 2x+y+z) = (a,b,c)$
$\Rightarrow \quad 3x = a, x-y = b, 2x+y+z = c$
$\Rightarrow \quad x = \dfrac{a}{3}, y = \dfrac{a-3b}{3}, z = -a+b+c$
$\therefore \quad t^{-1}(a,b,c) = \left(\dfrac{a}{3}, \dfrac{a-3b}{3}, -a+b+c\right)$

or $\quad t^{-1}(x,y,z) = \left(\dfrac{x}{3}, \dfrac{x-3y}{3}, -x+y+z\right)$

EXAMPLE-5 Let $f : R^3 \to R_2[t]$ be a linear transformation defined by
$$f(x,y,z) = (x+y)t^2 + (x+2y+2z)t + y + z$$
Show that f is non-singular. Also, find a formula for f^{-1}.

SOLUTION Let (x,y,z) be an arbitrary vector in kernel f. Then,

$(x,y,z) \in \text{Ker}(f)$

$\Leftrightarrow \quad (x+y)t^2 + (x+2y+2z)t + y + z = 0(t) \quad$ for all t [$0(t)$ is zero polynomial]

$\Leftrightarrow \quad x+y = 0, x+2y+2z = 0, y+z = 0$

This system of homogeneous equations has only the trivial solution $x = y = z = 0$.

$\therefore \quad \text{Ker}(f) = \{(0,0,0)\}$ and hence f is non-singular.

In order to find f^{-1}, let $f(x,y,z) = at^2 + bt + c$. Then, $f^{-1}(at^2 + bt + c) = (x,y,z)$

Now, $f(x,y,z) = at^2 + bt + c$

$\Rightarrow \quad (x+y)t^2 + (x+2y+2z)t + y + z = at^2 + bt + c \quad$ for all t

$\Rightarrow \quad x+y = a, x+2y+2z = b, y+z = c$

$\Rightarrow \quad x = b - 2c, y = a - b + 2c, z = -a + b - c$

$\therefore \quad f^{-1}(at^2 + bt + c) = (b - 2c, a - b + 2c, -a + b - c)$

EXAMPLE-6 Let t_1 and t_2 be linear transformations from a finite dimensional vector space V to itself such that $t_1 o t_2 = I_V$. Then show that t_1 and t_2 are both invertible and $t_1^{-1} = t_2$. Also, show that this is false when V is not finite dimensional.

SOLUTION Recall that a linear transformation is invertible iff it is a bijection.

First we shall show that t_2 is invertible.

t_2 *is injective*: Let $v_1, v_2 \in V$. Then,

$t_2(v_1) = t_2(v_2)$

$\Rightarrow \quad t_1(t_2(v_1)) = t_1(t_2(v_2))$

$\Rightarrow \quad (t_1 o t_2)(v_1) = (t_1 o t_2)(v_2)$

$\Rightarrow \quad I_V(v_1) = I_V(v_2) \Rightarrow v_1 = v_2$

$\therefore \quad t_2$ is injective.

Since $t_2 : V \to V$ is an injective map and V is finite dimensional. Therefore, t_2 is surjective. Consequently, t_2 is invertible.

Now $t_1 o t_2 = I_V \Rightarrow (t_1 o t_2) o t_2^{-1} = I_V o t_2^{-1} \Rightarrow t_1 o (t_2 o t_2^{-1}) = t_2^{-1} \Rightarrow t_1 o I_V = t_2^{-1} \Rightarrow t_1 = t_2^{-1}$

But, $t_2^{-1} o t_2 = I_V = t_2 o t_2^{-1}$

$\Rightarrow \quad t_1 o t_2 = I_V = t_2 o t_1$

$\Rightarrow \quad t_1$ is invertible and $t_1^{-1} = t_2$.

EXERCISE 3.4

1. Determine whether or not each of the following linear transformations is non-singular. If not find a non-zero vector v whose image is the null vector; otherwise find a formula for the inverse linear transformation.

   (i) $t : R^2 \to R^2$ defined by $t(x,y) = (x-2y, x-y)$

   (ii) $t : R^2 \to R^2$ defined by $t(x,y) = (2x-3y, 4x-6y)$

   (iii) $t : R^2 \to R^3$ defined by $t(x,y) = (x-y, 2x+y, x+3y)$

   (iv) $t : R^3 \to R^3$ defined by $t(x,y,z) = (x+z, x-z, y)$

2. A linear transformation $t : R^2 \to R^2$ is defined by $t(x,y) = (\alpha x + \beta y, \gamma x + \delta y)$
   Prove that t is non-singular if $\alpha\delta \neq \beta\gamma$.

3. Let t be the (unique) linear operator on $C^3(C)$ for which $t(1,0,0) = (1,0,i), t(0,1,0) = (0,1,1), t(0,0,1) = (i,1,0)$. Is t non-singular (invertible)?

   [Hint: We observe that $e_1^{(3)} = (1,0,0), e_2^{(3)} = (0,1,0)$ and $e_3^{(3)} = (0,0,1)$ are linearly independent but $t\left(e_1^{(3)}\right), t\left(e_2^{(3)}\right), t\left(e_3^{(3)}\right)$ are not so. Hence, t is non-singular.]

4. Determine whether or not each of the following linear transformations is non-singular. If not, find a non-zero vector v whose image is the zero vector ; otherwise find a formula for the inverse linear transformation :

   (i) $t : R^3 \to R^3$ defined by $t(x,y,z) = (x+y+z, 2x+3y+5z, x+3y+7z)$

   (ii) $f : R^2 \to P_2(t)$ defined by $f(x,y) = (x+2y)t^2 + (x-y)t + x + y$

5. Let V and V' be finite dimensional vector spaces such that $\dim V = \dim V'$. Suppose $t : V \to V'$ is a linear transformation. Then t is an isomorphism iff t is non-singular.

6. Let V be a finite dimensional vector space and V' be another vector space over the same field. Then $\dim V = \dim(I_m(t))$ iff t is non-singular.
   [Hint: $\dim V = \dim(I_m(t)) + \dim(\text{Ker}(t)) \Rightarrow \dim(\text{Ker}(t)) = 0 \Rightarrow \text{Ker}(t) = \{0_V\}$]

7. Is there any non-singular linear transformation $t : R^4 \to R^3$.
   [Hint: $\dim(I_m(t)) < \dim R^4$. So, there is no non-singular linear transformation $t : R^4 \to R^3$]

ANSWERS

1 (i) Non-singular, $t^{-1}(x,y) = (2y-x, y-x)$ (ii) Singular

(iii) Non-singular, t^{-1} does not exist (iv) Non-singular, $t^{-1}(x,y,z) = \left(\dfrac{x+y}{2}, z, \dfrac{x-y}{2}\right)$

4 (i) Singular, $v = (2,-3,1)$ (ii) Non-singular, but not invertible as $\dim P_2(t) > \dim R^2$

3.10 DUAL SPACES

In the previous section, we have seen that the set of all linear transformations from a vector space to another vector space forms a vector space. If we take the second vector space to be just the field of scalars, we get a new vector space which is closely related to one we started with. This new vector space is very useful indeed and is called dual of the original space.

LINEAR FUNCTIONAL OR LINEAR FORM *Let V be a vector space over a field F. A linear functional (or a linear form) on V is a linear transformation from V to F.*

In other words, a linear form f on a vector space $V(F)$ is a mapping $f : V \to F$ such that

$$f(\lambda u + \mu v) = \lambda f(u) + \mu f(v) \quad \text{for all } u, v \in V \text{ and all } \lambda, \mu \in F.$$

Since every field is a vector space over itself. Therefore, by Theorem 3 on page 281, the set $\text{Hom}(V, F)$ of all linear forms on V is a vector space over field F.

DUAL SPACE. *Let V be a vector space over a field F. Then the set of all linear forms on V is a vector space over field F. This vector space is called the dual space of V and is denoted by V^*.*

The dual space of V is also called the conjugate space or the adjoint space.

As we know that every field is a vector space of dimension one over itself. Therefore, by Theorem 6 on page 288, the dual space V^* of a finite dimensional vector space $V(F)$ is finite dimensional vector space with dimension equal to that of V.

THEOREM-1 *Let V be a vector space of dimension n over a field F, and let $B = \{b_1, b_2, \ldots, b_n\}$ be an ordered basis for V. Then for each list $(\lambda_1, \lambda_2, \ldots, \lambda_n)$ of n scalars in F there exists a unique linear form f on V such that $f(b_i) = \lambda_i$ for all $i \in \underline{n}$. This linear form is defined by the formula*

$$f\left(\sum_{i=1}^{n} \alpha_i b_i\right) = \sum_{i=1}^{n} \alpha_i \lambda_i.$$

PROOF. Since B is a basis for V. Therefore, for each $v \in V$ there exist unique scalars $\alpha_1, \alpha_2, \ldots, \alpha_n$ in F such that $v = \sum_{i=1}^{n} \alpha_i b_i$.

Consider a mapping $f : V \to F$ given by

$$f(v) = \sum_{i=1}^{n} \alpha_i \lambda_i.$$

Obviously, for each $v \in V$, $f(v)$ is a unique element of F. Therefore, $f : V \to F$ is well defined.

We shall now establish that $f : V \to F$ is a linear form on V.

Let $u = \sum_{i=1}^{n} \alpha_i b_i, v = \sum_{i=1}^{n} \beta_i b_i \in V$ and $a, b \in F$. Then,

$$f(au+bv) = f\left\{a\left(\sum_{i=1}^{n} \alpha_i b_i\right) + b\left(\sum_{i=1}^{n} \beta_i b_i\right)\right\}$$

$\Rightarrow \quad f(au+bv) = f\left\{\sum_{i=1}^{n} (a\alpha_i + b\beta_i) b_i\right\}$

$\Rightarrow \quad f(au+bv) = \sum_{i=1}^{n} (a\alpha_i + b\beta_i) \lambda_i$ [By definition of f]

$\Rightarrow \quad f(au+bv) = \sum_{i=1}^{n} (a\alpha_i)\lambda_i + \sum_{i=1}^{n} (b\beta_i)\lambda_i$

$\Rightarrow \quad f(au+bv) = a\left(\sum_{i=1}^{n} \alpha_i \lambda_i\right) + b\left(\sum_{i=1}^{n} \beta_i \lambda_i\right)$

$\Rightarrow \quad f(au+bv) = af(u) + bf(v)$

$\therefore$ f is a linear form on V.
Now for each $b_i \in B$, we have
$b_i = \sum_{j=1}^{n} \alpha_j b_j$, where $\alpha_j = 1$ for $j = i$ and $\alpha_j = 0$ for all $j \neq i$.

$\Rightarrow \quad f(b_i) = f\left(\sum_{j=1}^{n} \alpha_j b_j\right) = \sum_{j=1}^{n} \alpha_j \lambda_j$ [By definition of f]

$\Rightarrow \quad f(b_i) = \lambda_i$ [$\alpha_j = 0$ for all $j \neq i$]

Thus, f is a linear form on V such that $f(b_i) = \lambda_i$ for all $i \in \underline{n}$.
Let g be a linear form on V such that $g(b_i) = \lambda_i$ for all $i \in \underline{n}$.
Then for any $v = \sum_{i=1}^{n} \alpha_i b_i \in V$, we have

$$g(v) = g\left(\sum_{i=1}^{n} \alpha_i b_i\right) = \sum_{i=1}^{n} \alpha_i g(b_i)$$ [$\because$ g is linear]

$\Rightarrow \quad g(v) = \sum_{i=1}^{n} \alpha_i \lambda_i$ [By definition of g]

$\Rightarrow \quad g(v) = f(v)$ [By definition of f]

$\therefore \quad g(v) = f(v)$ for all $v \in V$.
Thus, $g = f$.
Hence, there exists a unique linear form $f : V \rightarrow F$ such that $f(b_i) = \lambda_i$ for all $i \in \underline{n}$.
Q.E.D.

THEOREM-2 *Let V be a vector space of dimension n over a field F. Then $V^* \cong F^n$.*

PROOF. Let $B = \{b_1, b_2, \ldots, b_n\}$ be an ordered basis for V.
Consider a mapping $\varphi : V^* \rightarrow F^n$ given by

$$\varphi(f) = \Big(f(b_1), f(b_2), \ldots, f(b_n)\Big) \quad \text{for all } f \in V^*.$$

φ is a linear transformation: For any $f, g \in V^*$ and $\alpha, \beta \in F$, we have

$$\varphi(\alpha f + \beta g) = \big((\alpha f + \beta g)(b_1), (\alpha f + \beta g)(b_2), \ldots, (\alpha f + \beta g)(b_n)\big)$$

$\Rightarrow \quad \varphi(\alpha f + \beta g) = ((\alpha f)(b_1) + (\beta g)(b_1), (\alpha f)(b_2) + (\beta g)(b_2), \ldots, (\alpha f)(b_n) + (\beta g)(b_n))$

$\Rightarrow \quad \varphi(\alpha f + \beta g) = \big(\alpha f(b_1) + \beta g(b_1), \alpha f(b_2) + \beta g(b_2), \ldots, \alpha f(b_n) + \beta g(b_n)\big)$

$\Rightarrow \quad \varphi(\alpha f + \beta g) = \big(\alpha f(b_1), \alpha f(b_2), \ldots, \alpha f(b_n)\big) + \big(\beta g(b_1), \beta g(b_2), \ldots, \beta g(b_n)\big)$

$\Rightarrow \quad \varphi(\alpha f + \beta g) = \alpha \big(f(b_1), f(b_2), \ldots, f(b_n)\big) + \beta \big(g(b_1), g(b_2), \ldots, g(b_n)\big)$

$\Rightarrow \quad \varphi(\alpha f + \beta g) = \alpha \varphi(f) + \beta \varphi(g)$

$\therefore \quad \varphi : V^* \to F^n$ is a linear transformation.

φ is injective: For any $f, g \in V^*$

$\varphi(f) = \varphi(g)$

$\Rightarrow \quad \big(f(b_1), f(b_2), \ldots, f(b_n)\big) = \big(g(b_1), g(b_2), \ldots, g(b_n)\big)$

$\Rightarrow \quad f(b_i) = g(b_i) \quad \text{for all } i \in \underline{n}.$

$\Rightarrow \quad f = g$ on B

$\Rightarrow \quad f = g$

$\therefore \quad \varphi$ is injective.

φ is surjective: By Theorem 1 for each list $(\lambda_1, \lambda_2, \ldots, \lambda_n)$ of n scalars in F^n there exists a unique linear form f on V such that $f(b_i) = \lambda_i$ for all $i \in \underline{n}$. Therefore, for each $(\lambda_1, \lambda_2, \ldots, \lambda_n) \in F^n$ there exists $f \in V^*$ such that

$$\varphi(f) = \big(f(b_1), f(b_2), \ldots, f(b_n)\big) = (\lambda_1, \lambda_2, \ldots, \lambda_n)$$

$\therefore \quad \varphi$ is surjective.

Hence, $\varphi : V^* \to F^n$ is an isomorphism. Consequently, $V^* \cong F^n$. Q.E.D.

COROLLARY. *Any finite dimensional vector space V has the same dimension as its dual space V^*.*

PROOF. Let $V(F)$ be a vector space of dimension n. Then by the main Theorem $V^* \cong F^n$.

$\therefore \quad \dim V^* = \dim F^n$

$\Rightarrow \quad \dim V^* = n = \dim V.$ $\quad [\because \dim F^n = n]$

As we have seen in Theorem 9 on page 255 that if B is an ordered basis of an n-dimensional vector space $V(F)$, then $L_B : F^n \to V$ is an isomorphism. That is, $F^n \cong V$. Theorem 2 shows that $V^* \cong F^n$. Therefore,

$$V^* \cong F^n \text{ and } F^n \cong V \Rightarrow V^* \cong V \qquad \text{[By transitivity of } \cong \text{]}$$

Thus, every finite dimensional vector space is isomorphic to its dual space.

Q.E.D.

THEOREM-3 *Let $V(F)$ be an n-dimensional vector space, and let $B = \{b_1, b_2, \ldots, b_n\}$ be an ordered basis for V. Then there exists a basis $X = \{x_1, x_2, \ldots, x_n\}$ for V^* such that $x_i(b_j) = \delta_{ij}$ for all $i, j \in \underline{n}$.*

PROOF. Since for each $i \in \underline{n}$, $e_i^{(n)} = (0, 0, \ldots, 1, 0, \ldots, 0)$ is a list of n scalars in F. Therefore, by Theorem 5 on page 285 for each $i \in \underline{n}$ there exists a unique linear form $x_i : V \to F$ such that

$$x_i(b_j) = \begin{cases} 0 \text{ if } i \neq j \\ 1 \text{ if } i = j \end{cases}$$

i.e, $\quad x_i(b_j) = \delta_{ij}, \ j \in \underline{n}$.

Let $X = \{x_1, x_2, \ldots, x_n\}$. Then, X is an ordered subset of V^* containing n distinct elements of V^*.

We will now show that X is basis for V^*.

X is $l.i.$: Let $\lambda_1, \lambda_2, \ldots, \lambda_n$ be n scalars in F such that $\sum\limits_{i=1}^{n} \lambda_i x_i = \hat{0} \in V^*$. Then,

$$\left(\sum_{i=1}^{n} \lambda_i x_i\right)(v) = \hat{0}(v) \quad \text{for all } v \in V$$

$\Rightarrow \quad \sum\limits_{i=1}^{n} \lambda_i x_i(v) = 0 \quad \text{for all } v \in V \qquad [\because \hat{0}(v) = 0]$

$\Rightarrow \quad \sum\limits_{i=1}^{n} \lambda_i x_i(b_j) = 0 \quad \text{for all } j \in \underline{n}$

$\Rightarrow \quad \sum\limits_{i=1}^{n} \lambda_i \delta_{ij} = 0 \quad \text{for all } j \in \underline{n}$

$\Rightarrow \quad \lambda_j = 0 \quad \text{for all } j \in \underline{n}$

Thus, $\sum\limits_{i=1}^{n} \lambda_i x_i = \hat{0} \Rightarrow \lambda_i = 0 \quad \text{for all } i \in \underline{n}$

So, X is a linearly independent set.

X spans V^*: Let f be an arbitrary linear form in V^*. The linear form f will be completely defined if we define it on a basis for V. So, let

$$f(b_i) = \lambda_i \quad \text{for all } i \in \underline{n}.$$

We shall now show that $f = \sum\limits_{i=1}^{n} \lambda_i x_i$, i.e. f is a linear combination of elements in X.

Now for any $j \in \underline{n}$, we have

$$\left(\sum_{i=1}^{n} \lambda_i x_i\right)(b_j) = \sum_{i=1}^{n} \lambda_i x_i(b_j)$$

$$\Rightarrow \left(\sum_{i=1}^{n} \lambda_i x_i\right)(b_j) = \sum_{i=1}^{n} \lambda_i \delta_{ij} = \lambda_j = f(b_j) \qquad [\because f(b_i) = \lambda_i \text{ for all } i \in \underline{n}]$$

$$\therefore \sum_{i=1}^{n} \lambda_i x_i = f \text{ on } B$$

$$\Rightarrow \sum_{i=1}^{n} \lambda_i x_i = f \text{ on } V \qquad [\because B \text{ is a basis for } V]$$

Thus, each $f \in V^*$ is expressible as $f = \sum_{i=1}^{n} \lambda_i x_i$. So, X spans V^*.
Hence, X is a basis for V^*.

Q.E.D.

COROLLARY. *If $V(F)$ is a finite dimensional vector space, then $\dim V = \dim V^*$.*

DUAL BASIS. *Let $V(F)$ be an n-dimensional vector space, and let $B = \{b_1, b_2, \ldots, b_n\}$ be an ordered basis for V. Then there exists a basis $X = \{x_1, x_2, \ldots, x_n\}$ of V^* such that*

$$x_i(b_j) = \delta_{ij} \quad \text{for all } i, j \in \underline{n}.$$

This basis X is called the basis dual to B or dual basis. The pair (X, B) is called dual basis pair for V.

BIDUAL. *Let $V(F)$ be a vector space. Then the dual V^{**} of the dual space V^* is called the bidual (or double dual) of V.*

If V is a finite dimensional vector space, then by corollary to Theorem 2, we have $\dim V = \dim V^* = \dim V^{**}$.

Now we present two basic facts about linear forms.

THEOREM-4 *Let $V(F)$ be a vector space. Then to each non-zero linear form f on V there is at least one vector $v \in V$ such that $f(v) = 1$.*

PROOF. Since $f : V \to F$ is a non-zero linear form on V. Therefore, there exists a vector $u \in V$ such that $f(u) \neq 0$.

Let $\lambda = f(u)$. Then $0 \neq \lambda \in F$. Let $\mu = \lambda^{-1} \in F$. Further, let $v = \mu u$.

Now, $f(v) = f(\mu u) = \mu f(u)$ \qquad [$\because f$ is linear]

$\Rightarrow f(v) = \lambda^{-1} f(u)$ \qquad [$\because \mu = \lambda^{-1}$]

$\Rightarrow f(v) = [f(u)]^{-1} f(u)$ \qquad [$\because \lambda = f(u)$]

$\Rightarrow f(v) = 1$

Hence, there exists a vector $v \in V$ such that $f(v) = 1$.

Q.E.D.

THEOREM-5 *Let $V(F)$ be a finite dimensional vector space. Then for every non-zero vector $v \in V$ there exists at least one linear form f with $f(v) = 1$.*

PROOF. Let $\dim V = n$. Since $v \neq 0_V$, so $\{v\}$ is a linearly independent subset of V and hence it can be extended to form a basis $B = \{b_1, b_2, \ldots, b_n\}$ for V, where $b_i = v$. Let $X = \{x_1, x_2, \ldots, x_n\}$ be basis dual to B. Then,

$$x_i(b_j) = \delta_{ij} \quad \text{for all } i, j \in \underline{n}$$

$\Rightarrow \quad x_i(b_i) = 1 \Rightarrow x_i(v) = 1 \Rightarrow f(v) = 1$, where $f = x_i$

Thus, corresponding to a non-zero vector $v \in V$ there exists a linear form $f \in V^*$ such that $f(v) = 1$.

Q.E.D.

THEOREM-6 *Let $V(F)$ be an n-dimensional vector space, and let $B = \{b_1, b_2, \ldots, b_n\}$ be a basis for V. Let $X = \{x_1, x_2, \ldots, x_n\}$ be the basis dual to B. Then for each linear form f on V, we have $f = \sum_{i=1}^{n} f(b_i) x_i$ and for each vector $v \in V$, we have $v = \sum_{i=1}^{n} x_i(v) b_i$.*

PROOF. Since X is the basis dual to basis B. Therefore,

$$x_i(b_j) = \delta_{ij} \quad \text{for all } i, j \in \underline{n}$$

Let f be an arbitrary linear form on V. Then there exist scalars $\lambda_1, \lambda_2, \ldots, \lambda_n \in F$ such that

$$f = \sum_{i=1}^{n} \lambda_i x_i \qquad [\because X \text{ is basis for } V^*]$$

$\Rightarrow \quad f(b_j) = \left(\sum_{i=1}^{n} \lambda_i x_i \right)(b_j) = \sum_{i=1}^{n} \lambda_i x_i(b_j)$

$\Rightarrow \quad f(b_j) = \sum_{i=1}^{n} \lambda_i \delta_{ij} \qquad [\because x_i(b_j) = \delta_{ij}]$

$\Rightarrow \quad f(b_j) = \lambda_j \qquad \text{for all } j \in \underline{n}.$

$\therefore \quad f = \sum_{i=1}^{n} f(b_i) x_i.$

Now let v be an arbitrary vector in V. Then,

$$v = \sum_{j=1}^{n} \mu_j b_j \qquad \text{for some } \mu_j \in F \qquad [\because B \text{ is a basis for } V]$$

$\Rightarrow \quad x_i(v) = x_i \left(\sum_{j=1}^{n} \mu_j b_j \right) = \sum_{j=1}^{n} \mu_j x_i(b_j) \qquad [\because x_i \text{ is linear}]$

$\Rightarrow \quad x_i(v) = \sum_{j=1}^{n} \mu_j \delta_{ij} = \mu_i \quad \text{for all } i \in \underline{n} \qquad [\because x_i(b_j) = \delta_{ij}]$

$\therefore \quad v = \sum_{i=1}^{n} \mu_i b_i = \sum_{i=1}^{n} x_i(v) b_i.$

Q.E.D.

REMARK-1 *It follows from the second part of this theorem that $v = \sum_{i=1}^{n} x_i(v) b_i$ for all $v \in V$. If we define $\left(\sum_{i=1}^{n} x_i b_i \right)(v) = \sum_{i=1}^{n} x_i(v) b_i$, then this result can be written as $I_V(v) = v = \left(\sum_{i=1}^{n} x_i b_i \right)(v)$ for all $v \in V$ Thus, we may write $\sum_{i=1}^{n} x_i b_i = I_V$.*

REMARK-2 *It should be noted that if $B = \{b_1, b_2, \ldots, b_n\}$ is an ordered basis of an n-dimensional vector space V and $X = \{x_1, x_2, \ldots, x_n\}$ is the basis dual to B, then the linear form x_i on V associates each vector v in V with the ith coordinate of v relative to the basis B. That is, $\left(x_1(v), x_2(v), \ldots, x_i(v), \ldots, x_n(v) \right)$ is the coordinate vector of $v \in V$.*

THEOREM-7 *Let $V(F)$ be an n-dimensional vector space. Then for any non-zero vector $v \in V$ there exists a linear form f on V such that $f(v) \neq 0$.*

PROOF. Since $v \neq 0_V$, therefore $\{v\}$ is a linearly independent set of vectors in V. Consequently it can be extended to form a basis for V. Let $B = \{b_1, b_2, \ldots, b_n\}$ be extended basis for V such that $b_i = v$, and let $X = \{x_1, x_2, \ldots, x_n\}$ be basis dual to B. Then,

$$x_i(b_j) = \delta_{ij} \text{ for all } i, j \in \underline{n}.$$

Since B is a basis for V. Therefore, there exist scalars $\lambda_1, \lambda_2, \lambda_3, \ldots, \lambda_n \in F$ such that

$$v = \sum_{i=1}^{n} \lambda_i b_i$$

$$\Rightarrow \quad x_i(v) = x_i \left(\sum_{j=1}^{n} \lambda_j b_j \right) = \sum_{j=1}^{n} \lambda_j x_i(b_j)$$

$$\Rightarrow \quad x_i(v) = \sum_{j=1}^{n} \lambda_j \delta_{ij} = \lambda_i \quad \text{for all } i \in \underline{n}.$$

Now, if $f(v) = 0$ for all $f \in V^*$, then as a particular case, we have $x_1(v) = 0, x_2(v) = 0, \ldots, x_n(v) = 0$;

i.e. $x_i(v) = 0$ for all $i \in \underline{n}$

$\Rightarrow \quad \lambda_i = 0$ for all $i \in \underline{n}$ $[\because \; x_i(v) = \lambda_i]$

$\Rightarrow \quad v = \sum_{i=1}^{n} \lambda_i v_i = 0_V$, which is a contradiction as $v \neq \underline{0}$.

Thus, if $v \neq \underline{0}$. Then for at least one $f \in V^*$, we have $f(v) \neq 0$.   Q.E.D.

COROLLARY-1 *If u and v are two distinct vectors in a finite dimensional vector space $V(F)$, then there exists a linear form f on V such that $f(u) \neq f(v)$.*

PROOF. Let $w = u - v$. Then, $w \neq \underline{0}$. Therefore, by above theorem there exists a linear form f on V such that

$$f(w) \neq 0 \Rightarrow f(u-v) \neq 0 \Rightarrow f(u) \neq f(v)$$

Q.E.D.

COROLLARY-2 Let $V(F)$ be a finite dimensional vector space such that $f(v) = 0$ for all $f \in V^*$. Then, $v = 0_V$.

PROOF. Suppose $V \neq 0_V$. Then by above theorem there exists a linear form $f \in V^*$ such that $f(v) \neq 0$. This contradicts the hypothesis that $f(v) = 0$ for all $f \in V^*$. Hence, $v = 0_V$.

Q.E.D.

THEOREM-8 (Principle of Duality). Let $V(F)$ be a finite dimensional vector space. Then for each $v \in V$, the function $\eta_v : V^* \to F$ given by

$$\eta_v(f) = f(v) \quad \text{for all } f \in V^*$$

is a linear form on V^*, i.e. $\eta_v \in V^{**}$.

Also, V is isomorphic to its bidual V^{**} under the canonical isomorphism $\varphi : V \to V^{**}$ which sends each vector $v \in V$ into the linear form $\eta_v : V^* \to F$.

PROOF. For any $v \in V$ and $f \in V^*$, $f(v)$ is a unique element of F. Therefore, the mapping $\eta_v : V^* \to F$ is well defined.

First we shall show that for each $v \in V$, n_v is a linear form on V^*.

η_v is a linear form on V^*: Let $f, g \in V^*$ and $\lambda, \mu \in F$. Then,

$$\begin{aligned}
\eta_v(\lambda f + \mu g) &= (\lambda f + \mu g)(v) && \text{[By definition of } \eta_v\text{]} \\
&= (\lambda f)(v) + (\mu g)(v) && \text{[By definition of pointwise sum]} \\
&= \lambda f(v) + \mu g(v) && \text{[By definition of scalar multiplication on Hom}(V, F)\text{]} \\
&= \lambda \eta_v(f) + \mu \eta_v(g)
\end{aligned}$$

$\therefore \quad \eta_v(\lambda f + \mu g) = \lambda \eta_v(f) + \mu \eta_v(g)$ for all $f, g \in V^*$ and $a, b \in F$.

So, $\eta_v : V^* \to F$ is a linear form and thus $\eta_v \in V^{**}$.

Now consider the mapping $\varphi : V \to V^{**}$ given by

$$\varphi(v) = \eta_v \quad \text{for all } v \in V.$$

φ *is injective*: For any $u, v \in V$

$\quad \varphi(u) = \varphi(v)$

$\Rightarrow \quad \eta_u = \eta_v$

$\Rightarrow \quad \eta_u(f) = \eta_v(f)$ for all $f \in V^*$

$\Rightarrow \quad f(u) = f(v)$ for all $f \in V^*$

$\Rightarrow \quad f(u) - f(v) = 0$ for all $f \in V^*$

$\Rightarrow \quad f(u-v) = 0_V$ for all $f \in V^*$ $\qquad\qquad\qquad\qquad\qquad$ [$\because$ f is linear]

$\Rightarrow \quad u - v = 0_V$ $\qquad\qquad\qquad\qquad\qquad\qquad\qquad\qquad\qquad$ [See Corollary 2 to Theorem 7]

$\Rightarrow \quad u = v.$

$\therefore \quad \varphi : V \to V^{**}$ is injective.

φ *is a linear transformation*: For any $u, v \in V$ and $a, b \in F$

$\quad \varphi(au+bv) = \eta_{au+bv}$ $\qquad\qquad\qquad\qquad\qquad\qquad\qquad\qquad\qquad$ ...(i)

For any $f \in V^*$, we have

$\quad \eta_{au+bv}(f) = f(au+bv) = af(u) + bf(v)$ $\qquad\qquad\qquad\qquad$ [$\because$ f is linear]

$\Rightarrow \quad \eta_{au+bv}(f) = a\,\eta_u(f) + b\eta_v(f)$ $\qquad\qquad\qquad\qquad\qquad$ [By definition of η_v]

$\Rightarrow \quad \eta_{au+bv}(f) = (a\,\eta_u + b\eta_v)(f)$

$\therefore \quad \eta_{au+bv} = a\,\eta_u + b\eta_v$ $\qquad\qquad\qquad\qquad\qquad\qquad\qquad\qquad$ ...(ii)

From (i) and (ii), we get

$\quad \varphi(au+bv) = \eta_{au+bv} = a\eta_u + b\eta_v = a\varphi(u) + b\varphi(v).$

$\therefore \quad \varphi : V \to V^{**}$ is a linear transformation.

Hence, $\varphi : V \to V^{**}$ is a monomorphism. But, $\dim V = \dim V^{**}$. Therefore, by Theorem 20 on page 262 $\varphi : V \to V^{**}$ is an isomorphism. Hence, $V \cong V^{**}$. $\qquad\qquad$ Q.E.D.

REMARK. *Note that in the above theorem $\varphi : V \to V^{**}$ does not depend upon any particular choice of basis of V, that is why it is called canonical isomorphism.*

COROLLARY. *Let $V(F)$ be a finite dimensional vector space. If η is a linear form on the dual space V^* of V, then there exists a unique vector $v \in V$ such that*

$$\eta(f) = f(v) \text{ for all } f \in V^*.$$

PROOF. By above theorem, $\varphi : V \to V^{**}$ is an isomorphism. Therefore, for each $\eta \in V^{**}$, i.e. for each linear form η on V^*, there exists a unique vector $v \in V$ such that

$$\varphi(v) = \eta$$
$$\Rightarrow \quad \eta_v = \eta$$
$$\Rightarrow \quad \eta_v(f) = \eta(f) \quad \text{for all } f \in V^*$$
$$\Rightarrow \quad \eta(f) = f(v) \quad \text{for all } f \in V^* \qquad \text{[By definition of } \eta_v : V^* \to F\text{]}$$

Q.E.D.

THEOREM-9 *Let $V(F)$ be a finite dimensional vector space. Then each basis for the dual space V^* is the dual of some basis for V.*

PROOF. Let $X = \{x_1, x_2, \ldots, x_n\}$ be an arbitrary basis for the dual space V^*. Then there exists a dual basis $Y = \{y_1, y_2, \ldots, y_n\}$ for V^{**} such that

$$y_i(x_j) = \delta_{ij} \quad \text{for all } i, j \in \underline{n}$$

By corollary to Theorem 8, we know that for each $y_i \in V^{**}$ there exists a unique vector $b_i \in V$ such that

$$\varphi(b_i) = y_i.$$
$$\Rightarrow \quad \eta_{b_i} = y_i$$
$$\Rightarrow \quad \eta_{b_i}(f) = y_i(f) \text{ for all } f \in V^* \qquad (i)$$

Since the correspondence $v \leftrightarrow \eta_v$ is an isomorphism from V to V^{**} and under an isomorphism a basis is mapped to a basis. Therefore, $B = \{b_1, b_2, \ldots, b_n\}$ is a basis for V.

Putting $f = x_j$ in (i), we get

$$\eta_{b_i}(x_j) = y_i(x_j)$$
$$\Rightarrow \quad x_j(b_i) = \delta_{ij} \qquad [\because y_i(x_j) = \delta_{ij}]$$
$$\Rightarrow \quad X = \{x_1, x_2, \ldots, x_n\} \text{ is the dual of the basis } B.$$

Q.E.D.

Theorems 3 and 9 stipulate that each basis for a finite dimensional vector space V determines a basis for the dual space V^* and vice - versa.

ILLUSTRATIVE EXAMPLES

EXAMPLE-1 *Consider the basis $B = \{b_1 = (2,1), b_2 = (3,1)\}$ of R^2. Find the dual basis of B.*

SOLUTION Let $X = \{x_1, x_2\}$ be the basis dual to B. Then,

$$x_1(b_1) = 1, x_1(b_2) = 0, \ x_2(b_1) = 0, x_2(b_2) = 1$$

Let (a,b) be an arbitrary element of R^2. since B is the basis of R^2.

$\therefore \quad (a,b) = \lambda_1 b_1 + \lambda_2 b_2$

$\Rightarrow \quad x_1(a,b) = \lambda_1$ and $x_2(a,b) = \lambda_2$

Now, $(a,b) = \lambda_1 b_1 + \lambda_2 b_2$

$\Rightarrow \quad (a,b) = \lambda_1(2,1) + \lambda_2(3,1)$

$\Rightarrow \quad (a,b) = (2\lambda_1 + 3\lambda_2, \lambda_1 + \lambda_2)$

$\Rightarrow \quad 2\lambda_1 + 3\lambda_2 = a, \lambda_1 + \lambda_2 = b$

$\Rightarrow \quad \lambda_1 = -a + 3b, \lambda_2 = a - 2b$

$\therefore \quad x_1(a,b) = -a + 3b$ and $x_2(a,b) = a - 2b$.

EXAMPLE-2 Find the dual basis of the basis $B = \{(1,-1,3),(0,1,-1),(0,3,-2)\}$ of $R^3(R)$.

SOLUTION Let $b_1 = (1,-1,3), b_2 = (0,1,-1), b_3 = (0,3,-2)$. Then, $B = \{b_1, b_2, b_3\}$.

Let $X = \{x_1, x_2, x_3\}$ be the basis dual to B. Then,

$$x_1(b_1) = 1, \ x_1(b_2) = 0, \ x_1(b_3) = 0,$$

$$x_2(b_1) = 0, \ x_2(b_2) = 1, \ x_2(b_3) = 0,$$

and, $\quad x_3(b_1) = 0, \ x_3(b_2) = 0, \ x_3(b_3) = 1.$

To find explicit expressions for x_1, x_2, x_3. Let (a,b,c) be an arbitrary element of R^3. Since B is basis for R^3. Therefore,

$$(a,b,c) = \lambda_1 b_1 + \lambda_2 b_2 + \lambda_3 b_3 \text{ for some } \lambda_1, \lambda_2, \lambda_3 \in R.$$

$\Rightarrow \quad x_1(a,b,c) = \lambda_1, x_2(a,b,c) = \lambda_2, x_3(a,b,c) = \lambda_3$

Now,

$(a,b,c) = \lambda_1 b_1 + \lambda_2 b_2 + \lambda_3 b_3$

$\Rightarrow \quad (a,b,c) = \lambda_1(1,-1,3) + \lambda_2(0,1,-1) + \lambda_3(0,3,-2)$

$\Rightarrow \quad \lambda_1 = a, -\lambda_1 + \lambda_2 + 3\lambda_3 = b, 3\lambda_1 - \lambda_2 - 2\lambda_3 = c$

$\Rightarrow \quad \lambda_1 = a, \lambda_2 = 7a - 2b - 3c, \lambda_3 = -2a + b + c.$

$\therefore \quad x_1(a,b,c) = a, x_2(a,b,c) = 7a - 2b - 3c, x_3(a,b,c) = -2a + b + c.$

Therefore, $X = \{x_1, x_2, x_3\}$ is the basis dual to B where x_1, x_2, x_3 are as defined above.

EXAMPLE-3 Consider the basis $B = \{b_1 = (1,1,1), b_2 = (1,1,-1), b_3 = (1,-1,-1)\}$ of $R^3(R)$ and a vector $v = (0,1,0) \in R^3$. If $X = \{x_1, x_2, x_3\}$ is the basis dual to B, find $x_1(v), x_2(v)$ and $x_3(v)$.

SOLUTION Since X is the basis dual to B. Therefore,

$$x_1(b_1) = 1, \; x_1(b_2) = 0, \; x_1(b_3) = 0$$
$$x_2(b_1) = 0, \; x_2(b_2) = 1, \; x_2(b_3) = 0$$
$$x_3(b_1) = 0, \; x_3(b_2) = 0, \; x_3(b_3) = 1$$

Let $v = \lambda_1 b_1 + \lambda_2 b_2 + \lambda_3 b_3$. Then,

$$x_1(v) = \lambda_1, \; x_2(v) = \lambda_2, \; x_3(v) = \lambda_3$$

Now, $v = \lambda_1 b_1 + \lambda_2 b_2 + \lambda_3 b_3$

$\Rightarrow \quad (0,1,0) = \lambda_1(1,1,1) + \lambda_2(1,1,-1) + \lambda_3(1,-1,-1)$

$\Rightarrow \quad \lambda_1 + \lambda_2 + \lambda_3 = 0, \; \lambda_1 + \lambda_2 - \lambda_3 = 1, \; \lambda_1 - \lambda_2 - \lambda_3 = 0$

$\Rightarrow \quad \lambda_1 = 0, \; \lambda_2 = \dfrac{1}{2}, \; \lambda_3 = -\dfrac{1}{2}$

$\Rightarrow \quad x_1(v) = 0, \; x_2(v) = \dfrac{1}{2}, \; x_3(v) = -\dfrac{1}{2}$

EXAMPLE-4 Let $V = P_1(t) = \{a + bt : a,b \in R\}$ be the vector space of real polynomials of degree at most one. Find the basis $\{b_1, b_2\}$ of V that is dual to the basis $\{x_1, x_2\}$ of V^* defined by

$$x_1(f(t)) = \int_0^1 f(t)dt \text{ and } x_2(f(t)) = \int_0^2 f(t)dt$$

SOLUTION Let $b_1 = \alpha + \beta t$ and $b_2 = p + qt$. By definition of the dual basis, we have

$$x_1(b_1) = 1, \; x_1(b_2) = 0, \; x_2(b_1) = 0 \text{ and } x_2(b_2) = 1$$

$\Rightarrow \quad \int_0^1 (\alpha + \beta t)dt = 1, \; \int_0^1 (p + qt)dt = 0, \; \int_0^2 (\alpha + \beta t)dt = 0, \; \int_0^2 (p + qt)dt = 0$

$\Rightarrow \quad \alpha + \dfrac{\beta}{2} = 1, \; p + \dfrac{q}{2} = 0, \; \alpha + \beta = 0, \; p + q = \dfrac{1}{2}$

$\Rightarrow \quad \alpha = 2, \beta = -2, \; p = -\dfrac{1}{2}, q = 1$

$\Rightarrow \quad b_1 = 2 - 2t, \; b_2 = -\dfrac{1}{2} + t$

EXAMPLE-5 Let $V = P_2(t)$ be the vector space of all real polynomials of degree at most two. For any $p(t) \in V$, define

$$f_1(p(t)) = \int_0^1 p(t)dt, \quad f_2(p(t)) = \int_0^2 p(t)dt, \quad f_3(p(t)) = \int_0^{-1} p(t)dt$$

Show that f_1, f_2, f_3 is a basis of the dual space V^*. Find a basis of V such that it is dual to basis $\{f_1, f_2, f_3\}$ of V^*.

SOLUTION Clearly, $B = \{1, t, t^2\}$ is a basis of V. Let $\lambda_1, \lambda_2, \lambda_3$ be scalars such that

$$\lambda_1 f_1 + \lambda_2 f_2 + \lambda_3 f_3 = \widehat{0}$$
$$\Rightarrow \quad \lambda_1 f_1(p(t)) + \lambda_2 f_2(p(t)) + \lambda_3 f_3(p(t)) = \widehat{0}(p(t)) \quad \text{for all } p(t) \in V$$
$$\Rightarrow \quad \lambda_1 \int_0^1 p(t)dt + \lambda_2 \int_0^2 p(t)dt + \lambda_3 \int_0^{-1} p(t)dt = 0 \quad \text{(i)}$$

Substituting $p(t) = 1, t, t^2$ successively in (i), we obtain

$$\lambda_1 + 2\lambda_2 - \lambda_3 = 0$$
$$\frac{\lambda_1}{2} + 2\lambda_2 + \frac{\lambda_3}{2} = 0$$
$$\frac{\lambda_1}{3} + \frac{8}{3}\lambda_2 - \frac{\lambda_3}{3} = 0$$

This is a homogeneous system of equations such that the determinant of the coefficient matrix is non-zero.

$$\therefore \quad \lambda_1 = \lambda_2 = \lambda_3 = 0$$

Thus, $\{f_1, f_2, f_3\}$ is a linearly independent subset of V^*.

Also, $\dim V^* = \dim V = 3$ $\quad\quad\quad\quad [\because \dim V = 3]$

Hence, $\{f_1, f_2, f_3\}$ is a basis of V^*.

Let $\{\varphi_1, \varphi_2, \varphi_3\}$ be an ordered basis of V such that it is dual to the ordered basis $\{f_1, f_2, f_3\}$ of V^*. Then, $f_i(\varphi_j) = \delta_{ij}$ for all $i, j = 1, 2, 3$.

Let $\varphi_1(t) = a_0 + a_1 t + a_2 t^2$. Then,

$$f_1(\varphi_1) = 1, \; f_2(\varphi_1) = 0, \; f_3(\varphi_1) = 0$$

$$\Rightarrow \int_0^1 (a_0 + a_1 t + a_2 t^2) dt = 1, \; \int_0^2 (a_0 + a_1 t + a_2 t^2) dt = 0, \; \int_0^{-1} (a_0 + a_1 t + a_2 t^2) dt = 0.$$

$$\Rightarrow a_0 + \frac{a_1}{2} + \frac{a_2}{3} = 1, \; 2a_0 + 2a_1 + \frac{8}{3} a_2 = 0, \; -a_0 + \frac{a_1}{2} - \frac{a_2}{3} = 0$$

$$\Rightarrow a_0 = 1, \; a_1 = 1, \; a_2 = -\frac{3}{2}$$

$$\therefore \varphi_1(t) = 1 + t - \frac{3}{2} t^2$$

Let $\varphi_2(t) = b_0 + b_1 t + b_2 t^2$. Then,

$$f_1(\varphi_2) = 0, \; f_2(\varphi_2) = 1, \; f_2(\varphi_2) = 0$$

$$\Rightarrow \int_0^1 (b_0 + b_1 t + b_2 t^2) dt = 0, \; \int_0^2 (b_0 + b_1 t + b_2 t^2) dt = 1, \; \int_0^{-1} (b_0 + b_1 t + b_2 t^2) dt = 0$$

$$\Rightarrow b_0 + \frac{b_1}{2} + \frac{b_2}{3} = 0, \; 2b_0 + 2b_1 + \frac{8}{3} b_2 = 1, \; -b_0 + \frac{b_1}{2} - \frac{b_2}{3} = 0$$

$$\Rightarrow b_0 = -\frac{1}{6}, \; b_1 = 0, \; b_2 = \frac{1}{2}$$

$$\therefore \varphi_2(t) = -\frac{1}{6} + \frac{1}{2} t^2$$

Finally, let $\varphi_3(t) = c_0 + c_1 t + c_2 t^2$. Then,

$$f_1(\varphi_3) = 0, \; f_2(\varphi_3) = 0, \; f_3(\varphi_3) = 1$$

$$\Rightarrow \int_0^1 (c_0 + c_1 t + c_2 t^2) dt = 0, \; \int_0^2 (c_0 + c_1 t + c_2 t^2) dt = 0, \; \int_0^{-1} (c_0 + c_1 t + c_2 t^2) dt = 1$$

$$\Rightarrow c_0 + \frac{c_1}{2} + \frac{c_2}{3} = 0, \; 2c_0 + 2c_1 + \frac{8}{3} c_2 = 0, \; -c_0 + \frac{c_1}{2} - \frac{c_2}{3} = 1$$

$$\Rightarrow c_0 = -\frac{1}{3}, \; c_1 = 1, \; c_2 = -\frac{1}{2}$$

$$\therefore \varphi_3(t) = -\frac{1}{3} + t - \frac{1}{2} t^2$$

Hence, $\left\{ 1 + t - \frac{3}{2} t^2, \; -\frac{1}{6} + \frac{1}{2} t^2, \; -\frac{1}{3} + t - \frac{1}{2} t^2 \right\}$ is a basis dual to basis $\{f_1, f_2, f_3\}$ of V^*.

EXAMPLE-6 Let $V = P_2(t)$ be the vector space of all polynomials of degree at most 2 over a field F. Let $a, b, c \in F$ be distinct scalars. Let $\varphi_a, \varphi_b, \varphi_c$ be linear functionals defined by

$$\varphi_a(f(t)) = f(a), \quad \varphi_b(f(t)) = f(b), \quad \varphi_c(f(t)) = f(c) \quad \text{for all } f(t) \in P_2(t)$$

Show that $\{\varphi_a, \varphi_b, \varphi_c\}$ is basis for V^*, and find the basis $\{f_1(t), f_2(t), f_3(t)\}$ of V that is its dual.

SOLUTION Clearly, V is a vector space of dimension 3 such that $\{1, t, t^2\}$ is a basis for V.

Since $\dim V = \dim V^*$ and $\dim V = 3$. Therefore, $\dim V^* = 3$. In order to prove that $X = \{\varphi_a, \varphi_b, \varphi_c\}$ is a basis for V^* it is sufficient to show that X is linearly independent set.

Let $\lambda_1, \lambda_2, \lambda_3$ be scalars in F such that

$$\lambda_1 \varphi_a + \lambda_2 \varphi_b + \lambda_3 \varphi_c = \hat{0}$$

$\Rightarrow \quad \lambda_1 \varphi_a(p(t)) + \lambda_2 \varphi_b(p(t)) + \lambda_3 \varphi_c(p(t)) = \hat{0}(p(t)) \quad \text{for all } p(t) \in V.$

Replacing $p(t)$ successively by $1, t$ and t^2, we get

$$\lambda_1 + \lambda_2 + \lambda_3 = 0$$
$$\lambda_1 a + \lambda_2 b + \lambda_3 c = 0$$
$$\lambda_1 a^2 + \lambda_2 b^2 + \lambda_3 c^2 = 0$$

The determinant of the coefficient matrix is equal to $(a-b)(b-c)(c-a) \neq 0$ as a, b, c are distinct scalars in F. Therefore, $\lambda_1 = \lambda_2 = \lambda_3 = 0$.

Hence, X is a basis for the dual space V^*.

Let $B = \{f_1, f_2, f_3\}$ be the basis for $V = P_2(t)$ dual to basis X of the dual space V^*. Then,

$$\varphi_a(f_1(t)) = 1, \quad \varphi_b(f_1(t)) = 0, \quad \varphi_c(f_1(t)) = 0 \quad \text{for all } t \in R$$

Let $f_1(t) = a_0 + a_1 t + a_2 t^2$. Then,

$\varphi_a(f_1(t)) = 1, \quad \varphi_b(f_1(t)) = 0, \quad \varphi_c(f_1(t)) = 0 \quad \text{for all } t$

$\Rightarrow \quad f_1(a) = 1, \; f_1(b) = 0, \; f_1(c) = 0$

$\Rightarrow \quad a_0 + a_1 a + a_2 a^2 = 1$

$a_0 + a_1 b + a_2 b^2 = 0$
$a_0 + a_1 c + a_2 c^2 = 0$

Solving these equations, we obtain

$$a_0 = \frac{bc}{(a-b)(a-c)}, \quad a_1 = \frac{-(b+c)}{(a-b)(a-c)}, \quad a_2 = \frac{1}{(a-b)(a-c)}$$

$\therefore \quad f_1(t) = \frac{t^2 - (b+c)t + bc}{(a-b)(a-c)} = \frac{(t-b)(t-c)}{(a-b)(a-c)}$

Similarly,

$$\varphi_a(f_2(t)) = 0, \quad \varphi_b(f_2(t)) = 1, \quad \varphi_c(f_2(t)) = 0$$

and, $\quad \varphi_a(f_3(t)) = 0, \quad \varphi_b(f_3(t)) = 0, \quad \varphi_c(f_3(t)) = 1$

provide us
$$f_2(t) = \frac{(t-a)(t-c)}{(b-a)(b-c)}, \quad f_3(t) = \frac{(c-a)(c-b)}{(t-a)(t-b)}$$
Hence, $B = \{f_1(t), f_2(t), f_3(t)\}$ is the basis of V dual to the basis $X = \{\varphi_a, \varphi_b, \varphi_c\}$ of V^*.

EXAMPLE-7 *Let $V(F)$ be a finite dimensional vector space, and let for any two linear forms f and g on V, $f(v) = 0 \Rightarrow g(v) = 0$. Prove that $g = \lambda f$ for some $\lambda \in F$.*

SOLUTION Let v be an arbitrary vector in $\text{Ker}(f)$. Then,
$$f(v) = 0$$
$\Rightarrow \quad g(v) = 0$
$\Rightarrow \quad v \in \text{Ker}(g)$
$\therefore \quad \text{Ker}(f) \subset \text{Ker}(g)$
Now two cases arise.

Case I. When f is the zero linear form:
If f is the zero linear form, then $\text{Ker}(f) = \{v \in V : f(v) = 0\} = V$.
Since $\text{Ker}(f) \subset \text{Ker}(g)$, therefore $\text{Ker}(g) = V$. So g is also zero linear form on V.
Hence, we have
$$g = \lambda f \quad \text{for all } \lambda \in F.$$

Case II. When f is a non-zero linear form:
Since f is a non-zero linear form on V. Therefore, there exists a non-zero vector $v_0 \in V$ such that $f(v_0) = \alpha$, where α is a non-zero scalar in F.
Let $\lambda = g(v_0)/f(v_0)$.
Let v be an arbitrary vector in V, and let $f(v) = \alpha, a \in F$. Then,
$f(v) = a$
$\Rightarrow \quad f(v) = (a\alpha^{-1})\alpha \qquad\qquad [\because 0 \neq \alpha \in F \Rightarrow \alpha^{-1} \in F]$
$\Rightarrow \quad f(v) = \beta\alpha, \text{ where } \beta = a\alpha^{-1} \in F$
$\Rightarrow \quad f(v) = \beta f(v_0) \qquad\qquad [\because \alpha = f(v_0)]$
$\Rightarrow \quad f(v) = f(\beta v_0) \qquad\qquad [\because f : V \to F \text{ is linear}]$
$\Rightarrow \quad f(v - \beta v_0) = 0 \qquad\qquad [\because f : V \to F \text{ is linear}]$
$\Rightarrow \quad v - \beta v_0 \in \text{Ker}(f)$
$\Rightarrow \quad v - \beta v_0 = v_1 \text{ for some } v_1 \in \text{Ker}(f)$
$\Rightarrow \quad v = \beta v_0 + v_1 \text{ for some } \beta \in F \text{ and some } v_1 \in \text{Ker}(f)$
$\Rightarrow \quad g(v) = g(\beta v_0 + v_1)$
$\Rightarrow \quad g(v) = \beta g(v_0) + g(v_1) \qquad\qquad [\because g \text{ is linear}]$
$\Rightarrow \quad g(v) = \beta g(v_0) \qquad (i) \quad [\because v_1 \in \text{Ker}(f) \Rightarrow v_1 \in \text{Ker}(g) \Rightarrow g(v_1) = 0]$

We have,

$$(\lambda f)(v) = \lambda f(v) = \lambda f(\beta v_0 + v_1) \qquad [\because v = \beta v_0 + v_1]$$
$$\Rightarrow (\lambda f)(v) = \lambda[\beta f(v_0) + f(v_1)] \qquad [\because f \text{ is linear}]$$
$$\Rightarrow (\lambda f)(v) = \lambda \beta f(v_0) \qquad [\because f(v_1) = 0]$$
$$\Rightarrow (\lambda f)(v) = \frac{g(v_0)}{f(v_0)} \beta f(v_0) \qquad [\because \lambda = g(v_0)/f(v_0)]$$
$$\Rightarrow (\lambda f)(v) = \beta g(v_0) \qquad \text{(ii)}$$

From (i) and (ii), we obtain

$$g(v) = \lambda f(v) \quad \text{for all } v \in V.$$
$$\Rightarrow g = \lambda f$$

Hence, in either case, we obtain $g = \lambda f$ for some $\lambda \in F$.

EXERCISE 3.5

1. Find the basis $X = \{x_1, x_2, x_3\}$ that is dual to the standard basis $B = \{e_1^{(3)}, e_2^{(3)}, e_3^{(3)}\}$ of $R^3(R)$.

2. Find the dual basis of the basis $B = \{b_1 = (1, -2, 3), b_2 = (1, -1, 1), b_3 = (2, -4, 7)\}$ of $R^3(R)$.

3. Let $V = P_2(t)$ be the vector space of all polynomials of degree at most 2. Find the basis $B = \{f_1(t), f_2(t), f_3(t)\}$ of V that is dual to the basis $X = \{x_1, x_2, x_3\}$ of V^* defined by

$$x_1\Big(f(t)\Big) = \int_0^1 f(t)dt, \; x_2\Big(f(t)\Big) = f'(1), \; x_3\Big(f(t)\Big) = f(0)$$

4. Let $V(F)$ be a vector space and $u, v \in V$ such that $f(u) = 0 \;\Rightarrow\; f(v) = 0$ for all $f \in V^*$. Show that $v = \lambda u$ for some $\lambda \in F$.

5. Let $V(F)$ be a vector space and $x_1, x_2 \in V^*$ such that $x_1(v) = 0 \;\Rightarrow\; x_2(v) = 0$ for all $v \in V$. Show that $x_2 = \lambda x_1$ for some $\lambda \in F$.

6. Let V be the vector space of polynomials over F. For $a \in F$, define $\phi_a : V \to F$ by $\phi_a\Big(f(t)\Big) = f(a)$. Show that: (i) ϕ_a is linear (ii) if $a \neq b$, then $\phi_a \neq \phi_b$.

7. Let V be a vector space over R. Let $x_1, x_2 \in V^*$ and subspace $f : V \to R$ be defined by $f(v) = x_1(v) x_2(v)$ also belongs to V^*. Show that either $x_1 = \hat{0}$ or $x_2 = \hat{0}$.

8. Let F be a field and $n \in N$. Let $B = \{e_1^{(n)}, e_2^{(n)}, \ldots, e_n^{(n)}\}$ be the standard basis of vector space $F^n(F)$. Show that the dual basis is $X = \{x_1, x_2, \ldots, x_n\}$, where $x_i : F^n \to F$ is given by $x_i(a_1, a_2, \ldots, a_n) = a_i$, $i \in \underline{n}$.

ANSWERS

1. $x_1(a, b, c) = a$, $x_2(a, b, c) = b$, $x_3(a, b, c) = c$
2. $x_1(a, b, c) = -3a - 5b - 2c$, $x_2(a, b, c) = 2a + b$, $x_3(a, b, c) = a + 2b + c$
3. $f_1(t) = 3t - \frac{3}{2}t^2$, $f_2(t) = -\frac{t}{2} + \frac{3}{4}t^2$, $f_3(t) = 1 - 3t + \frac{3}{2}t^2$

3.11 ANNIHILATORS

ANNIHILATOR *Let $V(F)$ be a vector space, and let S be a subset of V. Then the set $\{x \in V^* : x(v) = 0 \text{ for all } v \in S\}$ is called the annihilator of S in V^* and is denoted by* Annih(S).

Thus, Annih(S) = $\{x \in V^* : x(v) = 0 \text{ for all } v \in S\}$.

Clearly, annihilator of a subset of V is the set of all those linear forms of V which map vectors in S to the additive identity in field F.

Similarly, if T is a subset of V^*, then annihilator of T is the set $\{v \in V : x(v) = 0 \text{ for all } x \in T\}$.

REMARK. *Note that in the definition of annihilator S should be simply a set. It should not necessarily be a subspace of V.*

If $S = \{0_V\}$, then $x(0_V) = 0$ for all $x \in V^*$. Therefore,
Annih(0_V) = $\{x : x \in V^*\} = V^*$.

Further, if $S = V$, then $\hat{0} \in V^*$ is such that $\hat{0}(v) = 0$ for all $v \in S = V$. So, Annih(V) = $\{\hat{0}\}$.

If $V(F)$ is a finite dimensional vector space, then by Theorem 7 on page 314 for each non-zero vector $v \in V$ there exists a linear form $x \in V^*$ such that $x(v) \neq 0$. Thus, for each $\{0_V\} \neq S \subset V$ there exists $x \in V^*$ such that $x \notin$ Annih(S). Therefore, Annih(S) $\neq V^*$.

THEOREM-1 *Let S be a subset of a vector space $V(F)$. Then* Annih(S) *is a subspace of the dual space V^*.*

PROOF. We have,
$$\text{Annih}(S) = \{x \in V^* : x(v) = 0 \text{ for all } v \in S\}.$$
Since $\hat{0}(v) = 0$ for all $v \in S$. Therefore, $\hat{0} \in$ Annih(S).

Thus, Annih(S) is a non-void subset of V^*.

Let x, y be any two linear forms in Annih(S) and a, b be any two scalars in F. Then $ax + by \in V^*$ is such that

$(ax + by)(v) = (ax)(v) + (by)(v)$
$\Rightarrow \quad (ax + by)(v) = ax(v) + by(v)$ [By definition of scalar multiplication on V^*]
$\Rightarrow \quad (ax + by)(v) = a0 + b0 = 0 \quad$ for all $v \in S$ $\quad [\because x, y \in \text{Annih}(S)]$
$\therefore \quad ax + by \in$ Annih(S).

Thus, $ax + by \in \text{Annih}(S)$ for all $x, y \in \text{Annih}(S)$ and all $a, b \in F$.
Hence, $\text{Annih}(S)$ is a subspace of V^*.

Q.E.D.

THEOREM-2 *Let $V(F)$ be a finite dimensional vector space, and let S be a subspace of V. Then,*

$$\dim S + \dim(\text{Annih}(S)) = \dim V.$$

PROOF. Let $\dim V = n$, and $\dim S = m$. Then, $m \leq n$.

Let $\{b_1, b_2, \ldots, b_m\}$ be a basis for S, then it can be extended to form a basis $\{b_1, b_2, \ldots, b_m, b_{m+1}, \ldots, b_n\}$ of V.

Let $X = \{x_1, x_2, \ldots, x_m, x_{m+1}, \ldots, x_n\}$ be the basis dual to the basis $\{b_1, b_2, \ldots, b_m, \ldots, b_n\}$ of V. Then,

$$x_i(b_j) = \delta_{ij} \quad \text{for all } i, j \in \underline{n}.$$

By Theorem 1, $\text{Annih}(S)$ is a subspace of V^*. We shall now establish that $X' = \{x_{m+1}, \ldots, x_n\}$ is a basis for $\text{Annih}(S)$.

X' *is a l.i. set*: Since $\{x_{m+1}, \ldots, x_n\}$ is a part of the basis $\{x_1, x_2, \ldots, x_n\}$ of V^*. Therefore, X' is a linearly independent set.

X' *spans* $\text{Annih}(S)$: Let f be an arbitrary linear form in $\text{Annih}(S)$. Then,

$f \in V^*$

$\Rightarrow \quad f = \sum\limits_{i=1}^{n} \lambda_i x_i \quad \text{for some } \lambda_i \in F \hfill [\because X \text{ is a basis for } V^*]$

$\Rightarrow \quad f(v) = \left(\sum\limits_{i=1}^{n} \lambda_i x_i\right)(v) \quad \text{for all } v \in S$

$\Rightarrow \quad f(v) = \sum\limits_{i=1}^{n} \lambda_i x_i(v) \quad \text{for all } v \in S$

$\Rightarrow \quad f(b_j) = \sum\limits_{i=1}^{n} \lambda_i (x_i(b_j)) \quad \text{for all } b_j, j = 1, 2, \ldots, m$

$\Rightarrow \quad f(b_j) = \sum\limits_{i=1}^{n} \lambda_i \delta_{ij} \hfill [\because x_i(b_j) = \delta_{ij}]$

$\Rightarrow \quad f(b_j) = \lambda_j \quad \text{for all } j = 1, 2, \ldots, m$

But, $\quad f \in \text{Annih}(S) \Rightarrow f(b_j) = 0 \quad \text{for all } j = 1, 2, \ldots, m.$

$\therefore \quad \lambda_j = 0 \quad \text{for all } j \in \underline{m}.$

$\Rightarrow \quad f = \sum\limits_{i=m+1}^{n} \lambda_i x_i \hfill \left[\because f = \sum\limits_{i=1}^{n} \lambda_i x_i\right]$

$\Rightarrow \quad f$ is a linear combination of elements in X'.

$\Rightarrow \quad X'$ spans $\text{Annih}(S)$.

Hence, X' is a basis for Annih(S). Consequently

$$\dim(\text{Annih}(S)) = \text{Number of linear forms in } X'$$
$$\Rightarrow \quad \dim(\text{Annih}(S)) = n - m = \dim V - \dim S$$
Hence, $\quad \dim S + \dim(\text{Annih}(S)) = \dim V.$

Q.E.D.

COROLLARY. *If S is a subspace of a finite dimensional vector space $V(F)$, then S^* (the dual space of S) is isomorphic to $V^*/\text{Annih}(S)$.*

PROOF. Let $\dim V = n$ and $\dim S = m$. Then $m \leq n$. Since S^* is the space dual to S. Therefore, $\dim S^* = m$.

Now,

$$\dim\left(V^*/\text{Annih}(S)\right) = \dim V^* - \dim(\text{Annih}(S)) \quad \left[\because \dim\left(\frac{V}{S}\right) = \dim V - \dim S\right]$$
$$\Rightarrow \quad \dim\left(V^*/\text{Annih}(S)\right) = \dim V - \dim(\text{Annih}(S)) \quad [\because \dim V = \dim V^*]$$
$$\Rightarrow \quad \dim\left(V^*/\text{Annih}(S)\right) = \dim S \quad \text{[By Theorem 2]}$$
$$\Rightarrow \quad \dim\left(V^*/\text{Annih}(S)\right) = \dim S^* \quad [\because \dim S = \dim S^*]$$

Thus, S^* and $V^*/\text{Annih}(S)$ are vector spaces of the same dimension over the same field F.
Hence, $\quad S^* \cong V^*/\text{Annih}(S)$.

Q.E.D.

Let $V(F)$ be a finite dimensional vector space, and let S be a subset of V. Then Annih(S) is a subspace of V^*. Now, again by the definition of an annihilator, we have

$$\text{Annih}(\text{Annih}(S)) = \{\alpha \in V^{**} : \alpha(x) = 0 \quad \text{for all } x \in \text{Annih}(S)\}$$

By Theorem 1, Annih (Annih(S)) is a subspace of V^{**}. But, if V is a finite dimensional vector space, then we can identify V^{**} with V through the canonical homomorphism $v \leftrightarrow \eta_v$. Therefore, we may regard Annih (Annih(S)) as a subspace of V. Thus,

$$\text{Annih}(\text{Annih}(S)) = \{v \in V : x(v) = 0 \quad \text{for all } x \in \text{Annih}(S)\}$$

THEOREM-3 *Let $V(F)$ be a finite dimensional vector space, and let S be a subspace of V. Then,* Annih(Annih(S)) = S.

PROOF. We have,

$$\text{Annih}(S) = \{x \in V^* : x(v) = 0 \quad \text{for all } v \in S\},$$
and, $\quad \text{Annih}(\text{Annih}(S)) = \{v \in V : x(v) = 0 \quad \text{for all } x \in \text{Annih}(S)\}$

Let v be an arbitrary vector in Annih(S). Then, by definition

$$x(v) = 0 \quad \text{for all } x \in \text{Annih}(S)$$

$\Rightarrow \quad v \in \text{Annih (Annih}(S))$.

$\therefore \quad v \in S \Rightarrow v \in \text{Annih (Annih}(S))$.

Thus, $\quad S \subset \text{Annih (Annih}(S))$.

Since S and Annih (Annih(S)) are subspaces of V such that $S \subset$ Annih (Annih(S)). Therefore, S is a subspace of Annih (Annih(S)).

Now,

$$\dim S + \dim \text{Annih}(S) = \dim V \qquad \text{[By Theorem 2]}$$

and, $\quad \dim \text{Annih}(S) + \dim \Big(\text{Annih}(\text{Annih}(S))\Big) = \dim V^* = \dim V.$

$\therefore \quad \dim(S) = \dim \Big(\text{Annih (Annih}(S))\Big)$

Hence, $\quad S = \text{Annih (Annih}(S))$.

Q.E.D.

ILLUSTRATIVE EXAMPLES

EXAMPLE-1 *Let S_1 and S_2 be subspaces of a finite dimensional vector space $V(F)$. Then $S_1 = S_2$ iff* Annih$(S_1) =$ Annih(S_2).

SOLUTION   Obviously, if $S_1 = S_2$ then, Annih$(S_1) =$ Annih(S_2).

Conversely, let Annih$(S_1) =$ Annih(S_2). Then,

$$\text{Annih (Annih}(S_1)) = \text{Annih (Annih}(S_2))$$

$\Rightarrow \qquad\qquad\qquad S_1 = S_2$

EXAMPLE-2 *If S_1 and S_2 are two subsets of a vector space $V(F)$ such that $S_1 \subset S_2$, then show that* Annih$(S_2) \subset$ Annih(S_1).

SOLUTION   Let x be an arbitrary linear form in Annih(S_2). Then,

$\qquad x \in \text{Annih}(S_2)$

$\Rightarrow \quad x(v) = 0 \quad$ for all $v \in S_2$

$\Rightarrow \quad x(v) = 0 \quad$ for all $v \in S_1$ $\qquad\qquad\qquad\qquad [\because \; S_1 \subset S_2]$

$\Rightarrow \quad x \in \text{Annih}(S_1)$

$\therefore \qquad \text{Annih}(S_2) \subset \text{Annih}(S_1)$

EXAMPLE-3 Let S_1 and S_2 be two subspaces of a finite dimensional vector space $V(F)$. Then,
(i) $\text{Annih}(S_1 + S_2) = \text{Annih}(S_1) \cap \text{Annih}(S_2)$
(ii) $\text{Annih}(S_1 \cap S_2) = \text{Annih}(S_1) + \text{Annih}(S_2)$.

SOLUTION (i) Clearly, $S_1 \subset S_1 + S_2$ and $S_2 \subset S_1 + S_2$. Therefore, by the above example
$\text{Annih}(S_1 + S_2) \subset \text{Annih}(S_1)$ and, $\text{Annih}(S_1 + S_2) \subset \text{Annih}(S_2)$
$\Rightarrow \quad \text{Annih}(S_1 + S_2) \subset \text{Annih}(S_1) \cap \text{Annih}(S_2)$ \hfill (i)

Let $x \in \text{Annih}(S_1) \cap \text{Annih}(S_2)$. Then,
$x \in \text{Annih}(S_1)$ and $x \in \text{Annih}(S_2)$.
Let $v \in S_1 + S_2$. Then, $v = v_1 + v_2$ where $v_1 \in S_1, v_2 \in S_2$.
$\therefore \quad x(v) = x(v_1 + v_2) = x(v_1) + x(v_2) = 0 + 0 = 0$ \hfill $[\because v_1 \in S_1, v_2 \in S_2]$
$\Rightarrow \quad x(v) = 0$ for all $v \in S_1 + S_2$.
$\therefore \quad x \in \text{Annih}(S_1 + S_2)$

Thus, $x \in \text{Annih}(S_1) \cap \text{Annih}(S_2) \Rightarrow x \in \text{Annih}(S_1 + S_2)$
$\therefore \quad \text{Annih}(S_1) \cap \text{Annih}(S_2) \subset \text{Annih}(S_1 + S_2)$.
Hence, $\text{Annih}(S_1 + S_2) = \text{Annih}(S_1) \cap \text{Annih}(S_2)$.

(ii) Replacing S_1 by $\text{Annih}(S_1)$ and S_2 by $\text{Annih}(S_2)$ in (i), we get
$\text{Annih}(\text{Annih}(S_1) + \text{Annih}(S_2)) = \text{Annih}(\text{Annih}(S_1)) - \text{Annih}(\text{Annih}(S_2))$
$\Rightarrow \quad \text{Annih}(\text{Annih}(S_1) + \text{Annih}(S_2)) = S_1 \cap S_2$.
$\Rightarrow \quad \text{Annih}(S_1) + \text{Annih}(S_2) = \text{Annih}(S_1 \cap S_2)$ \hfill [Taking Annih. of both sides]

EXAMPLE-4 Let S be a subset of a vector space $V(F)$. Then $\text{Annih}(S) = \text{Annih}([S])$.

SOLUTION We have, $S \subset [S]$.
$\therefore \quad \text{Annih}([S]) \subset \text{Annih}(S)$.
Now, let x be an arbitrary linear form in $\text{Annih}(S)$. Then $x(v) = 0$ for all $v \in S$.
Let $u \in [S]$. Then,
$u = \Sigma \lambda_i v_i$ where $v_i \in S$ and $\lambda_i \in F$.
$\Rightarrow \quad x(u) = x(\Sigma \lambda_i v_i)$
$\Rightarrow \quad x(u) = \Sigma \lambda_i x(v_i)$ \hfill $[\because x \text{ is linear}]$
$\Rightarrow \quad x(u) = \Sigma \lambda_i 0 = 0$ \hfill $[\because x(v_i) = 0 \text{ for each } i]$
$\therefore \quad x(u) = 0$ for all $u \in [S]$.
So, $x \in \text{Annih}([S])$.
Thus, $x \in \text{Annih}(S) \Rightarrow x \in \text{Annih}([S])$
$\therefore \quad \text{Annih}(S) \subset \text{Annih}([S])$
Hence, $\text{Annih}(S) = \text{Annih}([S])$.

EXAMPLE-5 Let S be the subspace spanned by $(1,2,3)$ and $(0,4,-1)$ in the real vector space R^3. Determine the annihilator of S, i.e. Annih(S).

SOLUTION Let $v_1 = (1,2,3)$ and $v_2 = (0,4,-1)$. Clearly, v_1, v_2 are linearly independent vectors in R^3. Therefore, $\dim S = 2$.

Now,
$$\dim S + \dim\left(\text{Annih}(S)\right) = \dim V$$
$$\Rightarrow \quad 2 + \dim\left(\text{Annih}(S)\right) = 3$$
$$\Rightarrow \quad \dim\left(\text{Annih}(S)\right) = 1$$

We have,
$$\text{Annih}(S) = \{\varphi \in V^* : \varphi(v) = 0 \text{ for all } v \in S\}$$

In order to find Annih(S), we have to find a linear functional $\varphi : R^3 : R(\because \dim(\text{Annih}(S)) = 1)$ such that $\varphi(v_1) = 0$ and $\varphi(v_2) = 0$.

Let $\varphi(x,y,z) = ax + by + cz$. Then,
$$\varphi(v_1) = 0 \quad \text{and} \quad \varphi(v_2) = 0$$
$$\Rightarrow \quad a + 2b + 3c = 0 \quad \text{and} \quad 4b - c = 0$$

This system of equations has only one free variable say b. Putting $b = 1$, we get $c = 4$ and $a = -14$.
$$\therefore \quad \varphi(x,y,z) = -14x + y + 4z$$

Hence, Annih$(S) = \{\varphi(x,y,z)\}$ such that $\varphi(x,y,z) = -14x + y + 4z$

EXAMPLE-6 Find a basis of annihilator of the subspace S of R^4 spanned by the vectors $v_1 = (1,2,-3,4)$ and $v_2 = (0,1,4,-1)$.

SOLUTION Clearly, v_1, v_2 are linearly independent. So, $\dim S = 2$ and hence $\dim(\text{Annih}(S)) = 2$. Thus, we have to find two linear functional from R^4 to R such that $\varphi(v_1) = 0$ and $\varphi(v_2) = 0$. Let $\varphi(x,y,z,t) = ax + by + cz + dt$. Then,
$$\varphi(v_1) = 0 \quad \text{and} \quad \varphi(v_2) = 0$$
$$\Rightarrow \quad \varphi(1,2,-3,4) = a + 2b - 3c + 4d = 0 \quad \text{and} \quad \varphi(0,1,4,-1) = b + 4c - d = 0$$

This is a system of two equations in four unknowns a,b,c,d. If we form the coefficient matrix, it is in echelon form with two zero rows. That is,
$$\begin{bmatrix} 1 & 2 & -3 & 4 \\ 0 & 1 & 4 & -1 \\ 0 & 0 & 0 & 0 \\ 0 & 0 & 0 & 0 \end{bmatrix}$$

So, there are two free variables say, c and d.

For $c = 1$, $d = 0$, we get $a = 11$, $b = -4$, $c = 1$, $d = 0$ as a solution.
For $c = 0$, $d = 1$, we get $a = 6$, $b = -1$, $c = 0$, $d = 1$ as a solution.
Thus, we obtain two linear functionals φ_1 and φ_2 given by
$\varphi_1(x,y,z,t) = 11x - 4y + z$ and $\varphi_2(x,y,z,t) = 6x - y + t$ forming a basis of Annih(S).

EXAMPLE-7 Let S_1 and S_2 be subspaces of a vector space $V(F)$ such that $V = S_1 \oplus S_2$. Then $V^* = \text{Annih}(S_1) \oplus \text{Annih}(S_2)$.

SOLUTION Let f be an arbitrary linear form in Annih$(S_1) \cap$ Annih(S_2). Then,

$$f \in \text{Annih}(S_1) \text{ and } f \in \text{Annih}(S_2).$$

Since $V = S_1 \oplus S_2$, therefore, each vector $v \in V$ can be written as $v = v_1 + v_2$ where $v_1 \in S_1$, $v_2 \in S_2$.

We have, $f(v) = f(v_1 + v_2)$
$\Rightarrow \quad f(v) = f(v_1) + f(v_2)$ $\qquad [\because f \text{ is a linear form on } V]$
$\Rightarrow \quad f(v) = 0 + 0$ $\qquad [\because f \in \text{Annih}(S_1) \text{ and } v_1 \in S_1, v_2 \in S_2 \Rightarrow f(v_1) = 0, f(v_2) = 0]$
$\Rightarrow \quad f(v) = 0$

Thus, $f(v) = 0$ for all $v \in V$.
$\therefore \quad f = \hat{0}$ (zero linear form on V)
$\therefore \quad$ Annih$(S_1) \cap$ Annih$(S_2) = \{\hat{0}\}$.

Let f be an arbitrary linear form in V^*. If $v \in V$, then
$$V = S_1 \oplus S_2$$
$\Rightarrow v$ can be uniquely written as $v = v_1 + v_2$ where $v_1 \in S_1, v_2 \in S_2$.
For each $f \in V^*$, we define two functions f_1 and f_2 from V into F such that

$$f_1(v) = f_1(v_1 + v_2) = f(v_2) \text{ and, } f_2(v) = f_2(v_1 + v_2) = f(v_1).$$

f_1 *is a linear form on* V: Let $u = u_1 + u_2, v = v_1 + v_2 \in$ where $u_1, v_1 \in S_1, u_2, v_2 \in S_2$, and let $a, b \in F$. Then,

$\qquad f_1(au + bv) = f_1(au_1 + bv_1 + au_2 + bv_2)$
$\Rightarrow \quad f_1(au + bv) = f(au_2 + bv_2)$ $\qquad [\because au_1 + bv_1 \in S_1,\ au_2 + bv_2 \in S_2]$
$\Rightarrow \quad f_1(au + bv) = af(u_2) + bf(v_2)$ $\qquad [\because f \text{ is linear form}]$
$\Rightarrow \quad f_1(au + bv) = af_1(u) + bf_1(v)$ $\qquad [\text{By definition of } f_1]$

$\therefore \quad f_1$ is a linear form on V, i.e. $f_1 \in V^*$.
Now we shall show that $f_1 \in$ Annih(S_1).
Let v_1 be an arbitrary vector in S_1. Then, $v_1 \in V$.

We have,
$$v_1 = v_1 + 0_V \text{ where } v_1 \in S_1, \ 0_V \in S_2.$$
$$\therefore \quad f_1(v_1) = f_1(v_1 + 0_V) = f(0_V) \qquad \text{[By definition of } f_1\text{]}$$
$$\Rightarrow \quad f_1(v_1) = 0.$$

Thus, $f_1(v_1) = 0$ for all $v_1 \in S_1$.

So, $f_1 \in \text{Annih}(S_1)$.

Similarly, we can show that f_2 is a linear form on V and $f_2 \in \text{Annih}(S_2)$.

Let v be an arbitrary vector in V. Then,

$$v = v_1 + v_2 \text{ where } v_1 \in S_1 \text{ and } v_2 \in S_2 \qquad [\because V = S_1 \oplus S_2]$$
$$\Rightarrow \quad (f_1 + f_2)(v) = f_1(v) + f_2(v) \qquad \text{[By definition of pointwise sum]}$$
$$\Rightarrow \quad (f_1 + f_2)(v) = f(v_2) + f(v_1) \qquad \text{[By definition of } f_1, f_2\text{]}$$
$$\Rightarrow \quad (f_1 + f_2)(v) = f(v_1) + f(v_2)$$
$$\Rightarrow \quad (f_1 + f_2)(v) = f(v_1 + v_2) \qquad [\because f \text{ is linear}]$$
$$\Rightarrow \quad (f_1 + f_2)(v) = f(v)$$

Thus, $(f_1 + f_2)(v) = f(v)$ for all $v \in V$.
$$\therefore \quad f = f_1 + f_2.$$
Thus, $f \in V^* \Rightarrow f = f_1 + f_2$ where $f_1 \in \text{Annih}(S_1), f_2 \in \text{Annih}(S_2)$.
$$\therefore \quad V^* = \text{Annih}(S_1) + \text{Annih}(S_2).$$
Hence, $V^* = \text{Annih}(S_1) \oplus \text{Annih}(S_2)$.

EXAMPLE-8 *If S_1 and S_2 are subspaces of a finite dimensional vector space V such that $V = S_1 \oplus S_2$, then*   *(i) $S_1^* \cong \text{Annih}(S_2)$*   *(ii) $S_2^* \cong \text{Annih}(S_1)$.*

SOLUTION Let $\dim V = n$, $\dim S_1 = m$. Then,
$$V = S_1 \oplus S_2 \Rightarrow \dim S_2 = \dim V - \dim S_1 = n - m.$$

We have,

$\dim S_1^* = \dim S_1 = m$ and, $\dim (\text{Annih}(S_2)) = \dim V - \dim S_2 = n - (n - m) = m$.
$$\therefore \quad S_1^* \cong \text{Annih}(S_2).$$

Similarly, we have
$$S_2^* \cong \text{Annih}(S_1).$$

EXERCISE 3.6

1. Let S be the subspace of R^3 spanned by $(1,1,0)$ and $(0,1,1)$. Find a basis of the annihilator of S.

2. Let S be the subspace of R^4 spanned by the vectors $(1,2,-3,4)$, $(1,3,-2,6)$, $(1,4,-1,8)$. Find a basis of the annihilator of S.

3. Let S be a subset of a vector space V, show that $[S] = \text{Annih}(\text{Annih}(S))$.

4. Let $V(F)$ be a vector space such that $x \in V^*$ annihilates a subset of S of V. Show that φ annihilates $[S]$.

5. Let S be a subspace of R^4 spanned by the vectors $(0,0,1,-1), (2,1,1,0), (2,1,1,-1)$. Find a basis of annihilator of S.

ANSWERS

1. $\{\varphi(x,y,z) = x-y+z\}$ 2. $\{\varphi_1(x,y,z,t) = 5x-y+z,\ \varphi_2(x,y,z,t) = 2y-t\}$

5. $\{\varphi(x,y,z,t) = x-2y\}$

3.12 PROJECTIONS

In the previous Chapter, we have seen that the two dimensional real vector space R^2 (xy-plane) is the direct sum of its two subspaces $S_1 = \{(a,0) : a \in R\}$ i.e. (x-axis) and $S_2 = \{(0,b) : b \in R\}$ i.e. (y-axis). Therefore, corresponding to each point $P(a,b)$ in R^2 there exist unique points $(a,0)$ and $(0,b)$ in S_1 and S_2 respectively such that $(a,b) = (a,0) + (0,b)$. Point $(a,0)$ in S_1 (or on x-axis) is called the projection of the point $P(a,b)$ on S_1 (or on x-axis) along S_2 (y-axis). Similarly, $(0,b) \in S_2$ is called the projection of the point $P(a,b)$ on S_2 (or on y-axis) along S_1 (i.e. along x-axis). We now formalize this concept in the following definition.

PROJECTIONS. *Let $V(F)$ be a vector space, and let S_1 and S_2 be subspaces of V such that $V = S_1 \oplus S_2$. Then every vector v in V can be uniquely written as $v = v_1 + v_2$ where $v_1 \in S_1$ and $v_2 \in S_2$. The projection on S_1 along S_2 is the linear transformation $t : V \to V$ such that $t(v) = v_1$.*

Similarly, the projection on S_2 along S_1 is the linear transformation $t' : V \to V'$ given by $t'(v) = v_2$.

In order to make the definition meaningful, we must show that t, as defined above, is a linear transformation.

Since V is the direct sum of S_1 and S_2, therefore, each v in V uniquely determines a vector v_1 in S_1 and a vector $v_2 \in S_2$ such that $v = v_1 + v_2$. Hence, t is a mapping from V into itself.

To prove that t is a linear transformation, let $u = u_1 + u_2$, $v = v_1 + v_2 \in V$ and $\lambda, \mu \in F$, where $u_1, v_1 \in S_1$ and $u_2, v_2 \in S_2$. Then $t(u) = u_1, t(v) = v_1$. Also, since S_1 and S_2 are subspaces of V, therefore, $\lambda u_1 + \mu v_1 \in S_1$ and $\lambda u_2 + \mu v_2 \in S_2$.

Now, $t(\lambda u + \mu v) = t[\lambda(u_1 + u_2) + \mu(v_1 + v_2)]$

$\Rightarrow \quad t(\lambda u + \mu v) = t[(\lambda u_1 + \mu v_1) + (\lambda u_2 + \mu v_2)]$

$\Rightarrow \quad t(\lambda u + \mu v) = \lambda u_1 + \mu v_1$ [By definition of t, since $\lambda u_1 + \mu v_1 \in S_1$ and $\lambda u_2 + \mu v_2 \in S_2$]

$\Rightarrow \quad t(\lambda u + \mu v) = \lambda t(u) + \mu t(v)$ $\quad\quad [\because t(u) = u_1, t(v) = v_1]$

$\therefore \; t: V \to V$ is a linear transformation.

Similarly, it can be shown that $t': V \to V$ is a linear transformation.

THEOREM-1 *Let $V(F)$ be a vector space. A linear transformation $t: V \to V$ is a projection on some subspace of V iff it is idempotent, i.e. $t^2 = t$.*

PROOF. First, let S_1 and S_2 be subspaces of V such that $V = S_1 \oplus S_2$ and let t be the projection on S_1 along S_2. Then we have to prove that $t^2 = t$.

Let v be an arbitrary vector in V. Then there exist vectors $v_1 \in S_1, v_2 \in S_2$ such that

$v = v_1 + v_2$ $\quad\quad [\because V = S_1 \oplus S_2]$

$\Rightarrow \quad t(v) = v_1$ $\quad\quad$ [By definition of t]

$\therefore \quad t^2(v) = t(t(v)) = t(v_1) = t(v_1 + 0_V),$ where $v_1 \in S_1$ and $0_V \in S_2$

$\Rightarrow \quad t^2(v) = v_1$ $\quad\quad$ [By definition of t]

$\Rightarrow \quad t^2(v) = t(v)$ $\quad\quad [\because t(v) = v_1]$

Thus, $t^2(v) = t(v)$ for all $v \in V$.

So, $t^2 = t$.

Conversely, let $t: V \to V$ be a linear transformation such that $t^2 = t$. Then we have to prove that t is a projection on some subspace of V.

Let $S_1 = \{v \in V : t(v) = v\}$ and $S_2 = \{v \in V : t(v) = 0_V\} = \text{Ker}(t)$. Then S_2 is a subspace of V. Also, S_1 is a subspace as shown below.

Let $u_1, v_1 \in S_1$ and $\lambda, \mu \in F$. Then, $t(u_1) = u_1, t(v_1) = v_1$ and

$t(\lambda u_1 + \mu v_1) = \lambda t(u_1) + \mu t(v_1)$ $\quad\quad [\because t$ is linear$]$

$\Rightarrow \quad t(\lambda u_1 + \mu v_1) = \lambda u_1 + \mu v_1$ $\quad\quad$ [By definiton of S_1]

$\therefore \quad \lambda u_1 + \mu v_1 \in S_1$

Hence, S_1 is also a subspace of V.

Now to prove that t is projection on S_1 along S_2 we have to show that $V = S_1 \oplus S_2$.

Let v be an arbitrary vector in V. Then v can be written as $v = t(v) + [v - t(v)]$.

Let $v_1 = t(v)$ and $v_2 = v - t(v)$. Then,

$$t(v_1) = t(t(v)) = t^2(v) = t(v) \qquad [\because t^2 = t]$$
$$\Rightarrow \quad t(v_1) = v_1.$$
$$\Rightarrow \quad v_1 \in S_1.$$

Also, $\quad t(v_2) = t(v - t(v))$
$$\Rightarrow \quad t(v_2) = t(v) - t(t(v))$$
$$\Rightarrow \quad t(v_2) = t(v) - t^2(v)$$
$$\Rightarrow \quad t(v_2) = t(v) - t(v) = 0_V \qquad [\because t^2 = t]$$
$$\therefore \quad v_2 \in S_2.$$

Thus, each $v \in V$ can be expressed as $v = v_1 + v_2$ where $v_1 \in S_1, v_2 \in S_2$.
So, $V = S_1 + S_2$.
Let v be an arbitrary vector in $S_1 \cap S_2$. Then $v \in S_1$ and $v \in S_2$.
Now, $v \in S_1 \Rightarrow t(v) = v$ and, $v \in S_2 \Rightarrow t(v) = 0_V$.
Therefore $v = t(v) = 0_V$.
So, $\quad S_1 \cap S_2 = 0_V$.
Thus, $\quad V = S_1 \oplus S_2$.
Now let $v \in V$. Then,

$$v = v_1 + v_2 \text{ for some } v_1 \in S_1, v_2 \in S_2 \qquad [\because V = S_1 \oplus S_2]$$
$$\Rightarrow \quad t(v) = t(v_1 + v_2)$$
$$\Rightarrow \quad t(v) = t(v_1) + t(v_2) \qquad [\because t \text{ is linear}]$$
$$\Rightarrow \quad t(v) = v_1 + 0_V \qquad [\because v_1 \in S_1 \Rightarrow t(v_1) = v_1 \text{ and, } v_2 \in S_2 \Rightarrow t(v_2) = 0_V]$$
$$\Rightarrow \quad t(v) = v_1.$$

Thus, S_1 and S_2 are subspaces of vector space V such that $V = S_1 \oplus S_2$ and $t : V \to V$ is a linear transformation such that for all $v = v_1 + v_2$ where $v_1 \in S_1, v_2 \in S_2$ such that $t(v) = v_1$. So, t is the projection on S_1 along S_2. Q.E.D.

THEOREM-2 *Let $V(F)$ be a vector space, and let S_1 and S_2 be subspaces of V such that $V = S_1 \oplus S_2$. If t is the projection on S_1 along S_2, then*
(i) $I_m(t) = S_1$, and (ii) Ker$(t) = S_2$.

PROOF. (i) Let v be an arbitrary vector in S_1. Then $v = v + 0_V$ where $v \in S_1$ and $0_V \in S_2$.
Therefore, by the definition of t, we have

$$t(v) = v, \quad \Rightarrow v \in I_m(t).$$

Thus, $\quad v \in S_1 \Rightarrow v \in I_m(t)$.
$\therefore \quad S_1 \subset I_m(t).$ \hfill (i)

Now let u be an arbitrary vector in $I_m(t)$. Then there exists a vector $v \in V$ such that $t(v) = u$. This implies that

$$t(t(v)) = t(u)$$
$\Rightarrow \quad t^2(v) = t(u)$
$\Rightarrow \quad t(v) = t(u) \qquad [\because t^2 = t]$
$\Rightarrow \quad t(u) = u \qquad [\because t(v) = u]$
$\Rightarrow \quad u \in S_1$
$\therefore \quad I_m(t) \subset S_1. \qquad \qquad \qquad \qquad \qquad \qquad \qquad \text{(ii)}$
Hence, $\quad I_m(t) = S_1. \qquad \qquad \qquad \qquad \qquad \text{[From (i) and (ii)]}$

(ii) Let v be an arbitrary vector in S_2. Then $v \in V$ can be written as $v = 0_V + v$, where $0_V \in S_1, v \in S_2$. Therefore, by definition of projection t, we have

$$t(v) = 0_V$$
$\Rightarrow \quad v \in \text{Ker}(t)$
$\therefore \quad S_2 \subset \text{Ker}(t).$

Now let v be an arbitrary vector in $\text{Ker}(t)$. Then $t(v) = 0_V$. Let $v = v_1 + v_2$ where $v_1 \in S_1, v_2 \in S_2$.

We have, $\quad t(v) = v_1 \qquad \qquad \qquad \qquad \qquad \text{[By definition of } t\text{]}$
$\Rightarrow \quad v_1 = 0_V \qquad \qquad \qquad \qquad \qquad [\because t(v) = 0_V]$
$\Rightarrow \quad v = v_2$
$\Rightarrow \quad v \in S_2 \qquad \qquad \qquad \qquad \qquad \qquad [\because v_2 \in S_2]$

$\therefore \quad \text{Ker}(t) \subset S_2.$
Hence, $\quad \text{Ker}(t) = S_2.$

Q.E.D.

THEOREM-3 Let S_1 and S_2 be two subspaces of a vector space $V(F)$ such that $V = S_1 \oplus S_2$. If t is the projection on S_1 along S_2, then

$$S_1 = \{v \in V : t(v) = v\} \text{ and, } S_2 = \text{Ker}(t).$$

PROOF. Let $W = \{v \in V : t(v) = v\}$. Then we have to prove that $W = S_1$.

Let $v \in S_1$. Then $v \in V$ can be written as $v = v + 0_V$ where $v \in S_1, 0_V \in S_2$. Therefore, by the definition of t, we have

$$t(v) = v \Rightarrow v \in W$$
$\therefore \quad S_1 \subset W.$

Now, let $v \in W$. Then, $t(v) = v$.

$\Rightarrow \quad v \in W$

$\Rightarrow \quad v \in V$

$\Rightarrow \quad v = v_1 + v_2$ where $v_1 \in S_1, v_2 \in S_2$

$\Rightarrow \quad t(v) = v_1$ [By definition of t]

$\Rightarrow \quad v = v_1 \in S_1$ $[\because t(v) = v]$

$\therefore \quad W \subset S_1$.

Hence $\quad W = S_1$.

For $\quad S_2 = \text{Ker}(t)$, see Theorem 2(ii).

Q.E.D.

THEOREM-4 *Let S_1 and S_2 be subspaces of a vector space $V(F)$. Then a linear transformation $t : V \rightarrow V$ is a projection on S_1 along S_2 iff $I_v - t$ is the projection on S_2 along S_1.*

PROOF. First, let t be a projection on S_1 along S_2. Then, $V = S_1 \oplus S_2$.
Let v be an arbitrary vector in V. Then, $v = v_1 + v_2$ for some $v_1 \in S_1$, $v_2 \in S_2$.
Since t is the projection on S_1 along S_2. Therefore, $t(v) = v_1$.
Now, $(I_v - t)(v) = I_v(v) - t(v) = v - v_1 = v_2$ $[\because v = v_1 + v_2]$
$\therefore$ $I_v - t$ is the projection on S_2 along S_1.

Conversely, let $I_v - t$ be the projection on S_2 along S_1. Then we have to show that t is the projection on S_1 along S_2.

Let v be an arbitrary vector in V. Then,

$v = v_1 + v_2$ where $v_1 \in S_1, v_2 \in S_2$ $[\because V = S_1 \oplus S_2]$

$\Rightarrow (I_v - t)(v) = v_2$ $[\because I_v - t$ is the projection on S_2 along $S_1]$

$\Rightarrow I_v(v) - t(v) = v_2$

$\Rightarrow v - t(v) = v_2$

$\Rightarrow (v_1 + v_2) - t(v) = v_2$

$\Rightarrow t(v) = v_1$.

$\therefore$ $t : V \rightarrow V$ is the projection on S_1 along S_2.

Q.E.D.

THEOREM-5 *Let $S_1, S_2, \ldots, S_k$ be subspaces of a vector space $V(F)$ such that $V = S_1 \oplus S_2 \oplus \cdots \oplus S_k$. Then there exist k linear transformations $t_1, t_2, \ldots, t_k$ on V such that*

(i) *each t_i is a projection.*
(ii) *$t_i o t_j = \hat{0}$ if $i \neq j$.*
(iii) *$t_1 + t_2 + \cdots + t_k = I_v$.*

(iv) $I_m(t_i) = S_i$ for all $i \in \underline{k}$.

Conversely, if $t_1, t_2, \ldots, t_k$ are k linear transformations on V such that (i)–(iv) hold, then $V = S_1 \oplus S_2 \oplus \cdots \oplus S_k$ where $S_i = I_m(t_i)$, $i = 1, 2, \ldots, k$.

PROOF. (i) Since V is the direct sum of $S_1, S_2, \ldots, S_k$. Therefore, each $v \in V$ can be uniquely written as

$$v = v_1 + v_2 + \cdots + v_k, \text{ where each } v_i \in S_i, i = 1, 2, \ldots, k.$$

Let t_i be a rule associating $v \in V$ to $v_i \in S_i \subset V$. Then $t_i : V \to V$ is a mapping for all $i = 1, 2, \ldots, k$.

t_i *is a linear transformation*: Let $u, v \in V$ and $a, b \in F$. Then,
$u = u_1 + u_2 + \cdots + u_k$, $u_i \in S_i$; $v = v_1 + v_2 + \cdots + v_k$, $v_i \in S_i$.

$\therefore \quad t_i(au + bv) = t_i\left\{a(u_1 + u_2 + \cdots + u_k) + b(v_1 + v_2 + \cdots + v_k)\right\}$

$\Rightarrow \quad t_i(au + bv) = t_i\left\{(au_1 + bv_1) + \cdots + (au_k + bv_k)\right\}$

$\Rightarrow \quad t_i(au + bv) = au_i + bv_i$ [By definition of t_i]

$\Rightarrow \quad t_i(au + bv) = at_i(u) + bt_i(v)$

So, t_i is a linear transformation on V.

Now,

$\Rightarrow \quad t_i^2(v) = t_i\big(t_i(v)\big)$

$\Rightarrow \quad t_i^2(v) = t_i(v_i)$ [By definition of t_i]

$\Rightarrow \quad t_i^2(v) = t_i(0_V + 0_V + \cdots + v_i + 0_V + \cdots + 0_V)$

$\Rightarrow \quad t_i^2(v) = v_i$ [By definition of t_i]

$\Rightarrow \quad t_i^2(v) = t_i(v)$

$\therefore \quad t_i^2(v) = t_i(v) \quad \text{for all } v \in V.$

Thus, $\quad t_i^2 = t_i \quad \text{for all } i = 1, 2, \ldots, k.$

Hence, each t_i is a projection. (ii) Let v be an arbitrary vector in V. Then, $v = v_1 + v_2 + \cdots + v_k$, where $v_i \in S_i$.

For any $i \neq j$, we have

$(t_i o t_j)(v) = t_i[t_j(v)]$

$\Rightarrow \quad (t_i o t_j)(v) = t_i(v_j)$ [By definition of t_j]

$\Rightarrow \quad (t_i o t_j)(v) = t_i(0_V + \cdots + 0_V + v_j + 0_V + \cdots + 0_V)$

$\Rightarrow \quad (t_i o t_j)(v) = 0_V$

$\Rightarrow \quad (t_i o t_j)(v) = \hat{0}(v).$

$\therefore \quad t_i o t_j = \hat{0} \quad \text{for all } i \neq j.$

(iii) Let $v \in V$. Then, $v = v_1 + v_2 + \cdots + v_k$ where each $v_i \in S_i$.

We have,

$$(t_1+t_2+\cdots+t_k)(v) = t_1(v)+t_2(v)+\cdots+t_k(v)$$
$$\Rightarrow (t_1+t_2+\cdots+t_k)(v) = v_1+v_2+\cdots+v_k \qquad \text{[By definition of } t_i,\ t_i(v)=v_i\text{]}$$
$$\Rightarrow (t_1+t_2+\cdots+t_k)(v) = v$$
$$\Rightarrow (t_1+t_2+\cdots+t_k)(v) = I_v(v)$$
$$\therefore \quad t_1+t_2+\cdots+t_k = I_v$$

(iv) Let $v \in I_m(t_i)$. Then there exists $u \in V$ such that $t_i(u) = v$. Since $V = S_1 \oplus S_2 \oplus \cdots \oplus S_k$.

$$\therefore \quad u = u_1+u_2+\cdots+u_k, \text{ where } u_i \in S_i.$$
$$\Rightarrow t_i(u) = u_i \quad \text{for all } i=1,2,\ldots,k.$$
$$\Rightarrow v = u_i \qquad\qquad\qquad\qquad\qquad [\because t_i(u)=v]$$
$$\Rightarrow v \in S_i \qquad\qquad\qquad\qquad\qquad [\because u_i \in S_i]$$

Thus, $v \in I_m(t_i) \Rightarrow v \in S_i$

So, $I_m(t_i) \subset S_i$.

Now, let $v \in S_i$. Then, $t_i(v) = v$. Therefore, $v \in I_m(t_i)$.

Thus, $v \in S_i \Rightarrow v \in I_m(t_i)$.

So, $S_i \subset I_m(t_i)$.

Hence, $I_m(t_i) = S_i$ for all $i=1,2,\ldots,k$.

Conversely, let $t_1, t_2, \ldots, t_k$ are linear transformations on V such that (i) to (iv) hold and let $S_i = I_m(t_i)$. Then we have to prove that V is the direct sum of $S_1, S_2, \ldots, S_k$.

From (iii), we have

$$I_V = t_1+t_2+\cdots+t_k$$
$$\Rightarrow I_V(v) = (t_1+t_2+\ldots+t_k)(v) \quad \text{for all } v \in V.$$
$$\Rightarrow v = t_1(v)+t_2(v)+\ldots+t_k(v) \quad \text{for all } v \in V.$$
$$\Rightarrow v = v_1+v_2+\cdots+v_k \quad \text{for all } v \in V, \text{ where } v_i = t_i(v) \in I_m(t_i) = S_i \text{ for all } i=1,2,\ldots,k.$$

Thus, each $v \in V$ can be written as $v = v_1+v_2+\cdots+v_k$, where $v_i \in S_i$.

$$\therefore \quad V = S_1+S_2+\cdots+S_k.$$

Now we shall show that each vector in V is uniquely expressible as the sum of the vectors in $S_1, S_2, \ldots, S_k$.

Let $v = v_1+v_2+\cdots+v_k$ where $v_i \in S_i$.

Since $S_i = I_m(t_i)$, therefore, $v_i = t_i(u_i)$ where $u_i \in V, 1 \leq i \leq k$.

We have,
$$t_j(v) = t_j(v_1 + v_2 + \cdots + v_k) = t_j(v_1) + t_j(v_2) + \cdots + t_j(v_k)$$
$$\Rightarrow \quad t_j(v) = \sum_{i=1}^{k} t_j(v_i) = \sum_{i=1}^{k} t_j\big(t_i(u_i)\big) \qquad [\because v_i = t_i(u_i)]$$
$$\Rightarrow \quad t_j(v) = \sum_{i=1}^{k} (t_j o t_i)(u_i)$$
$$\Rightarrow \quad t_j(v) = t_j o t_j(u_j) \qquad [\because t_i o t_j = \widehat{0} \text{ if } i \neq j]$$
$$\Rightarrow \quad t_j(v) = t_j^2(u_j)$$
$$\Rightarrow \quad t_j(v) = t_j(u_j) \qquad [\because t_j^2 = t_j]$$
$$\Rightarrow \quad t_j(v) = v_j$$

Hence, the representation of v as the sum of vectors in $S_1, S_2, \ldots, S_k$ is unique. Consequently, $V = S_1 \oplus S_2 \oplus \cdots \oplus S_k$.

Q.E.D.

ILLUSTRATIVE EXAMPLES

EXAMPLE-1 *Let $V(F)$ be a vector space, and let $t : V \to V$ be an idempotent linear transformation, i.e. $t^2 = t$. Then,*
  (i) $v \in I_m(t) \Leftrightarrow t(v) = v$
  (ii) $V = I_m(t) \oplus \mathrm{Ker}(t)$
  (iii) t is the projection on $I_m(t)$ along $\mathrm{Ker}(t)$.

SOLUTION (i) Let $v \in I_m(t)$. Then there exists $u \in V$ such that $t(u) = v$.
Now, $\quad t(u) = v$
$$\Rightarrow \quad t(t(u)) = t(v)$$
$$\Rightarrow \quad t^2(u) = t(v)$$
$$\Rightarrow \quad t(u) = t(v) \qquad [\because t^2 = t]$$
$$\Rightarrow \quad v = t(v) \qquad [\because t(u) = v]$$

Let $v \in V$ be such that $t(v) = v$. Then we have to show that $v \in I_m(t)$. Since v is the image of $v \in V$ under t. Therefore, $v \in I_m(t)$.

Thus, $v \in I_m(t) \Leftrightarrow t(v) = v$.

(ii) Let v be an arbitrary vector in V. Then,
$$v = t(v) + [v - t(v)] = v_1 + v_2 \text{ where } v_1 = t(v) \in I_m(t) \text{ and } v_2 = v - t(v).$$
Now, $v_2 = v - t(v)$
$$\Rightarrow \quad t(v_2) = t(v - t(v)) = t(v) - t(t(v))$$
$$\Rightarrow \quad t(v_2) = t(v) - t^2(v) = t(v) - t(v) \qquad [\because t^2 = t]$$
$$\Rightarrow \quad t(v_2) = 0_V$$

Therefore, $v_2 \in \text{Ker}(t)$.
Thus, $v = v_1 + v_2$ where $v_1 \in I_m(t)$ and $v_2 \in \text{Ker}(t)$.
Hence, $V = I_m(t) + \text{Ker}(t)$.
Now, let $v \in I_m(t) \cap \text{Ker}(t)$. Then,

$$v \in I_m(t) \text{ and } v \in \text{Ker}(t)$$
$\Rightarrow \quad t(v) = v \text{ and } t(v) = 0_V$ [By (i)]
$\Rightarrow \quad v = 0_V$

Thus, $v \in I_m(t) \cap \text{Ker}(t) \Rightarrow v = 0_V$
So, $I_m(t) \cap \text{Ker}(t) = \{0_V\}$.
Hence, $V = I_m(t) \oplus \text{Ker}(t)$.
(iii) Let v be an arbitrary vector in V. Then by (ii), we have
$v = v_1 + v_2$ where $v_1 \in I_m(t)$ and $v_2 \in \text{Ker}(t)$.
$\Rightarrow \quad t(v) = t(v_1 + v_2) = t(v_1) + t(v_2) = t(v_1) + 0_V$ $\quad [\because v_2 \in \text{Ker}(t)]$
$\Rightarrow \quad t(v) = t(v_1) = v_1$ $\quad [\because v_1 \in I_m(t)]$
$\therefore$ t is the projection on $I_m(t)$ along $\text{Ker}(t)$.

EXAMPLE-2 *Let $V(F)$ be a vector space, and let S_1 and S_2 be two subspaces of V such that $V = S_1 \oplus S_2$. If t_1 is the projection on S_1 along S_2 and t_2 is the projection on S_2 along S_1, prove that* (i) $t_1 + t_2 = I_V$ *and* (ii) $t_1 o t_2 = \widehat{0}, t_2 o t_1 = \widehat{0}$.

SOLUTION Let v be an arbitrary vector in V. Since $V = S_1 \oplus S_2$, therefore, there exist $v_1 \in S_1, v_2 \in S_2$ such that $v = v_1 + v_2$. Since t_1 is the projection on S_1 along S_2 and t_2 is the projection on S_2 along S_1. Therefore, $t_1(v) = v_1$ and $t_2(v) = v_2$.
Now, $(t_1 + t_2)(v) = t_1(v) + t_2(v) = v_1 + v_2 = v = I_V(v)$
$\therefore \quad t_1 + t_2 = I_V$
(ii) We have

$t_1 o t_2 = t_1 o (I_V - t_1)$ $\quad [\because t_1 + t_2 = I_V]$
$\Rightarrow \quad t_1 o t_2 = t_1 o I_V - t_1 o t_1$
$\Rightarrow \quad t_1 o t_2 = t_1 - t_1^2$
$\Rightarrow \quad t_1 o t_2 = t_1 - t_1$ $\quad [\because t_1 \text{ is a projection} \Rightarrow t_1^2 = t_1]$
$\Rightarrow \quad t_1 o t_2 = \widehat{0}$

Similarly, we have
$$t_2 o t_1 = t_2 o (I_V - t_2) = t_2 o I_V - t_2 o t_2 = t_2 - t_2 = \widehat{0}.$$

EXAMPLE-3 *Let $t_1, t_2, \ldots, t_n$ be linear transformations from a vector space $V(F)$ into itself such that $t_1 + t_2 + \cdots + t_n = I_V$. Prove that if $t_i o t_j = \widehat{0}$ for $i \neq j$, then $t_i^2 = t_i$ for each $i \in \underline{n}$.*

SOLUTION We have,

$$t_i^2 = t_i o t_i$$
$$\Rightarrow t_i^2 = t_i o(I_V - t_1 - t_2 - \cdots - t_{i-1} - t_{i+1} - \cdots - t_n)$$
$$\Rightarrow t_i^2 = t_i o I_V - t_i o t_1 - t_i o t_2 - \cdots - t_i o t_{i-1} - t_i o t_{i+1} \cdots - t_i o t_n$$
$$\Rightarrow t_i^2 = t_i \qquad [\because t_i o t_j = \widehat{0} \text{ for } i \neq j]$$

THEOREM-4 *Let V be a vector space over a field F of characteristic not equal to 2, and let $S_1, T_1; S_2, T_2$ be respectively complementary subspaces of V. If t_1 and t_2 are projections on S_1 and S_2 along T_1 and T_2 respectively, then the following hold:*

(i) $t_1 + t_2$ *is a projection* $\Leftrightarrow t_1 o t_2 = t_2 o t_1 = \widehat{0}$ *and if this condition holds then $t = t_1 + t_2$ is a projection on $S = S_1 \oplus S_2$ along $T = T_1 \cap T_2$.*

(ii) $t = t_1 - t_2$ *is a projection on $S' = S_1 \cap T_2$ along $T' = S_2 \oplus T_1$ iff $t_1 o t_2 = t_2 o t_1 = t_2$.*

(iii) *If $t_1 o t_2 = t = t_2 o t_1$, then t is the projection on $S'' = S_1 \cap S_2$ along $T'' = T_1 + T_2$.*

PROOF. (i) Let $t_1 + t_2$ be a projection. Then,

$$(t_1 + t_2)^2 = t_1 + t_2$$
$$\Rightarrow t_1^2 + t_1 o t_2 + t_2 o t_1 + t_2^2 = t_1 + t_2$$
$$\Rightarrow t_1 + t_1 o t_2 + t_2 o t_1 + t_2 = t_1 + t_2 \qquad [\because t_1, t_2 \text{ are projections, } \therefore t_1^2 = t_1, t_2^2 = t_2]$$
$$\Rightarrow t_1 o t_2 + t_2 o t_1 = \widehat{0} \qquad (i)$$
$$\Rightarrow t_1 o(t_1 o t_2 + t_2 o t_1) = t_1 o \widehat{0}$$
$$\Rightarrow t_1 o(t_1 o t_2) + t_1 o(t_2 o t_1) = \widehat{0} \qquad [\because t_1 o \widehat{0} = \widehat{0}]$$
$$\Rightarrow (t_1 o t_1) o t_2 + t_1 o(t_2 o t_1) = \widehat{0}$$
$$\Rightarrow t_1 o t_2 + t_1 o(t_2 o t_1) = \widehat{0} \qquad [\because t_1^2 = t_1 o t_1 = t_1] \quad (ii)$$

Again, $t_1 o t_2 + t_2 o t_1 = \widehat{0}$

$$\Rightarrow (t_1 o t_2 + t_2 o t_1) o t_1 = \widehat{0} o t_1$$
$$\Rightarrow (t_1 o t_2) o t_1 + (t_2 o t_1) o t_1 = \widehat{0}$$
$$\Rightarrow (t_1 o t_2) o t_1 + t_2 o(t_1 o t_1) = \widehat{0}$$
$$\Rightarrow (t_1 o t_2) o t_1 + t_2 o t_1 = \widehat{0} \qquad [\because t_1^2 = t_1 o t_1 = t_1] \quad (iii)$$

From (ii) and (iii), we get

$$t_1 o t_2 - t_2 o t_1 = \widehat{0} \qquad (iv)$$

From (i) and (iv), we get
$$t_1 o t_2 + t_1 o t_2 = \hat{0}$$
$\Rightarrow \quad (1+1)(t_1 o t_2) = \hat{0}$
$\Rightarrow \quad t_1 o t_2 = \hat{0}$ \hfill [$\because$ characteristic $F \neq 2, \therefore 1+1 \neq 0$]
$\Rightarrow \quad t_1 o t_2 = \hat{0} = t_2 o t_1$ \hfill [$\because t_1 o t_2 = t_2 o t_1$]

Conversely, let $t_1 o t_2 = t_2 o t_1 = \hat{0}$. Then,
$$(t_1 + t_2)^2 = (t_1 + t_2) o (t_1 + t_2)$$
$\Rightarrow \quad (t_1 + t_2)^2 = t_1^2 + t_1 o t_1 + t_2 o t_1 + t_2^2$
$\Rightarrow \quad (t_1 + t_2)^2 = t_1 + \hat{0} + \hat{0} + t_2 = t_1 + t_2$
$\therefore \quad t_1 + t_2$ is a projection.

Now let $t = t_1 + t_2$ be a projection with $t_1 o t_2 = t_2 o t_1 = \hat{0}$. Then we have to prove that $t = t_1 + t_2$ is a projection on $S = S_1 \oplus S_2$ along $T_1 \oplus T_2 = T$.

Now if t is a projection on S along T, then S and T are respectively the solutions of the equations $t(v) = v$ and $t(v) = 0_V$ for all $v = u + w \in V$, where $u \in S$ and $w \in T$
i.e, $S = \{v \in V : t(v) = v\}$ and $T = \{v \in V : t(v) = 0_V\}$.

We wish to show that (a) $S = S_1 \oplus S_2$, (b) $T = T_1 \cap T_2$.

(a) To show that $S = S_1 \oplus S_2$.
Now, $v = v_1 + u_1$ where $v_1 \in S_1, u_1 \in S_2$
$\Rightarrow \quad t_1(v) = v_1, (I_V - t_1)(v) = u_1 \in T_1$
and, $v = v_2 + u_2$ where $v_2 \in S_1, u_2 \in S_2$
$\Rightarrow \quad t_2(v) = v_2, (I_V - t_2)(v) = u_2 \in T_2$.

To show that $S = S_1 + S_2$.
$v \in S$
$\Rightarrow \quad t(v) = v$ \hfill [By definition of S]
$\Rightarrow \quad (t_1 + t_2)(v) = v$, i.e. $t_1 + t_2 = I_V$ \hfill [$\because t = t_1 + t_2$]
$\Rightarrow \quad t_1(v) = t_2(v) = v$ \hfill [$\because t_1 + t_2 = I_V$]
$\Rightarrow \quad v = t_1(v_2 + u_2) + t_2(v_1 + u_1)$
$\Rightarrow \quad v = t_1(v_2) + t_1(u_2) + t_2(v_1) + t_2(u_1)$
$\Rightarrow \quad v = t_1(t_2(v)) + t_1(u_2) + t_2(t_1(v)) + t_2(u_1)$
$\Rightarrow \quad v = t_1(u_2) + t_2(u_1)$ \hfill [$\because t_2 o t_1 = t_1 o t_2 = \hat{0}$]
$\Rightarrow \quad v \in S_1 + S_2$ [$\because t_1(u_2) = t_1^2(u_2) = t_1(t_1(u_2)) \in S_1$ and, $t_2(u_1) = t_2^2(u_1) = t_2(t_2(u_1)) \in S_2$]
Thus, $v \in S \Rightarrow v \in S_1 + S_2$.
So, $S \subset S_1 + S_2$.

Again, let $v \in S_1 + S_2$. Then,

$$v = v_1 + v_2 \quad \text{where } v_1 \in S_1, v_2 \in S_2$$
$\Rightarrow \quad (t_1 + t_2)(v) = (t_1 + t_2)(v_1 + v_2)$
$\Rightarrow \quad (t_1 + t_2)(v) = t_1(v_1) + t_1(v_2) + t_2(v_1) + t_2(v_2)$
$\Rightarrow \quad (t_1 + t_2)(v) = v_1 + 0_V + 0_V + v_2 \quad \left[\begin{array}{l} \because t_1 \text{ and } t_2 \text{ are projections on } S_1 \text{ and } S_2 \text{ along} \\ T_1 \text{ and } T_2 \text{ respectively and } v_1 \in S_1, v_2 \in S_2 \end{array}\right]$
$\Rightarrow \quad t(v) = v$
$\Rightarrow \quad v \in S$
$\therefore \quad S_1 + S_2 \subset S.$

Hence, we have $S = S_1 + S_2$

To prove that $S_1 \cap S_2 = \{0_V\}$.

Let $v \in S_1 \cap S_2$ Then,

$$v \in S_1 \text{ and } v \in S_2$$
$\Rightarrow \quad t_1(v) = v \text{ and } t_2(v) = v$
$\Rightarrow \quad v = t_1(v) = t_1(t_2(v)) = (t_1 o t_2)(v) = \hat{0}(v) = 0_V \qquad [\because t_1 o t_2 = \hat{0}]$

Thus, $\quad S_1 \cap S_2 = \{0_V\}$

Hence, $\quad S = S_1 \oplus S_2$.

(b) To show that $T = T_1 \cap T_2$.

Let $v \in T$. Then,

$$t(v) = 0_V$$
$\Rightarrow \quad (t_1 + t_2)(v) = 0_V$
$\Rightarrow \quad t_1(v) + t_2(v) = 0_V$
$\Rightarrow \quad t_1(t_1(v)) + t_1(t_2(v)) = t_1(0_V) \text{ and } t_2(t_1(v)) + t_2(t_2(v)) = t_2(0_V)$
$\Rightarrow \quad t_1^2(v) + (t_1 o t_2)(v) = 0_V \text{ and } (t_2 o t_1)(v) + t_2^2(v) = 0_V$
$\Rightarrow \quad t_1(v) + (t_1 o t_2)(v) = 0_V \text{ and } (t_2 o t_1)(v) + t_2(v) = 0_V \qquad [\because t_1^2 = t_1, t_2^2 = t_2]$
$\Rightarrow \quad t_1(v) = 0_V = t_2(v) \qquad [\because t_1 o t_2 = t_2 o t_1 = \hat{0}]$
$\Rightarrow \quad v \in T_1 \text{ and } v \in T_2$
$\Rightarrow \quad v \in T_1 \cap T_2$
$\therefore \quad T \subset T_1 \cap T_2.$

Again, if $v \in T_1 \cap T_2$, then

$\qquad v \in T_1$ and $v \in T_2$
$\Rightarrow \qquad t_1(v) = 0_V$ and $t_2(v) = 0_V$
$\Rightarrow \qquad t_1(v) + t_2(v) = 0_V$
$\Rightarrow \qquad (t_1 + t_2)(v) = 0_V$
$\Rightarrow \qquad t(v) = 0_V \qquad\qquad\qquad\qquad\qquad\qquad [\because t = t_1 + t_2]$
$\Rightarrow \qquad v \in T$
$\therefore \qquad T_1 \cap T_2 \subset T$
Hence, $\qquad T = T_1 \cap T_2$.

(ii) Since t_1 and t_2 are respectively projections on S_1 and S_2 along T_1 and T_2 respectively. Therefore, $I_V - t_1$ and $I_V - t_2$ are projections on T_1 and T_2 along S_1 and S_2 respectively. Hence, by (i) $(I_V - t_1) + t_2 = I_V - (t_1 - t_2)$ is a projection on $T' = S_2 \oplus T_1$ along $S' = S_1 \cap T_2$ and so $I_V - \Big((I_V - t_1) + t_2\Big) = t_1 - t_2$ is a projection on $S' = S_1 \cap T_2$ along $T' = S_2 \oplus T_1$.

Now, $\quad t_1 - t_2$ is a projection
$\Leftrightarrow \quad I_V - (t_1 - t_2)$ is a projection
$\Leftrightarrow \quad (I_V - t_1) + t_2$ is a projection
$\Leftrightarrow \quad (I_V - t_1)ot_2 = t_2 o(I_V - t_1) = \hat{0}$
$\Leftrightarrow \quad t_2 - t_1 o t_2 = t_2 - t_2 o t_1 = \hat{0}$
$\Leftrightarrow \quad t_1 o t_2 = t_2 o t_1 = t_2$

Hence, $t = t_1 - t_2$ is a projection on $S' = S_1 \cap T_2$ along $T' = S_2 \oplus T_1$ iff $t_1 o t_2 = t_2 o t_1 = t_2$.

(iii)] We have, $t = t_1 o t_2 = t_2 o t_1$
$\therefore \qquad t^2 = (t_1 o t_2) o (t_2 o t_1)$
$\Rightarrow \qquad t^2 = (t_1 o t_2^2) o t_1 = (t_1 o t_2) o t_1 \qquad\qquad [\because t_2^2 = t_2]$
$\Rightarrow \qquad t^2 = t_1 o (t_2 o t_1)$
$\Rightarrow \qquad t^2 = t_1 o (t_1 o t_2) \qquad\qquad\qquad\qquad [\because t_1 o t_2 = t_2 o t_1]$
$\Rightarrow \qquad t^2 = t_1^2 o t_2 = t_1 o t_2 \qquad\qquad\qquad\qquad [\because t_1^2 = t_1]$
$\Rightarrow \qquad t^2 = t$

$\therefore \quad t$ is a projection.

Now, if t is projection on $S'' = S_1 \cap S_2$ along $T'' = T_1 + T_2$, then $S'' = \{v \in V : t(v) = v\}$ and $T'' = \{v \in V : t(v) = 0_V\}$.

(a) To show that $S'' = S_1 \cap S_2$.

Let $v \in S''$. Then, $t(v) = v$

$\Rightarrow \quad t_1(v) = t_1(t(v))$ and $t_2(t(v)) = t_2(v)$

$\Rightarrow \quad t_1(v) = t_1\Big((t_1 o t_2)(v)\Big)$ and $t_2(v) = t_2 o\Big((t_2 o t_1)(v)\Big)$

$\Rightarrow \quad t_1(v) = (t_1^2 o t_2)(v)$ and $t_2(v) = (t_2^2 o t_1)(v)$

$\Rightarrow \quad t_1(v) = (t_1 o t_2)(v)$ and $t_2(v) = (t_2 o t_1)(v) \qquad [\because \; t_1^2 = t_1, t_2^2 = t_2]$

$\Rightarrow \quad t_1(v) = t(v)$ and $t_2(v) = t(v) \qquad\qquad\qquad\qquad [\because \; t_1 o t_2 = t_2 o t_1 = t]$

$\Rightarrow \quad t_1(v) = v$ and $t_2(v) = v \qquad\qquad\qquad\qquad\qquad [\because \; t(v) = v]$

$\Rightarrow \quad v \in S_1$ and $v \in S_2$.

$\Rightarrow \quad v \in S_1 \cap S_2$

$\therefore \quad S'' \subset S_1 \cap S_2$

Again, if $v \in S_1 \cap S_2$, then

$\qquad v \in S_1$ and $v \in S_2$

$\Rightarrow \quad t_1(v) = v$ and $t_2(v) = v$

$\Rightarrow \quad t(v) = (t_1 o t_2)(v) = t_1(t_2(v)) = t_1(v) = v$

$\Rightarrow \quad v \in S''$

$\therefore \quad S_1 \cap S_2 \subset S''$.

Hence, $\quad S'' = S_1 \cap S_2$

(b) To show that $\quad T'' = T_1 + T_2$

Let $v \in T''$. Then,

$\qquad t(v) = 0_V$

$\Rightarrow \quad (t_1 o t_2)(v) = 0_V$

$\Rightarrow \quad t_1(t_2(v)) = 0_V$

$\Rightarrow \quad t_2(v) \in T_1$

Similarly, $v \in T'' \Rightarrow t_1(v) \in T_2$

Since $v = t_2(v) + (I_V - t_2)(v)$, where $t_2(v) \in T_1$ and $(I_V - t_2)(v) \in T_2$

$\therefore \quad T'' \subset T_1 + T_2$

Also, if

$\qquad v \in T_1 + T_2$

$\Rightarrow \quad v = v_1 + v_2$ where $v_1 \in T_1, v_2 \in T_2$

$\Rightarrow \quad t(v) = t(v_1 + v_2) = t(v_1) + t(v_2)$

$\Rightarrow \quad t(v) = (t_2 o t_1)(v_1) + (t_1 o t_2)(v_2)$

$\Rightarrow \quad t(v) = t_2\Big(t_1(v_1)\Big) + t_1\Big(t_2(v_2)\Big)$

$\Rightarrow \quad t(v) = t_2(0_V) + t_1(0_V)$

$\Rightarrow \quad t(v) = 0_V$

$\Rightarrow \quad v \in T''$

$\therefore \quad T_1 + T_2 \subset T''$.

Hence, $T'' = T_1 + T_2$. Q.E.D.

3.13 THE DUAL OR THE ADJOINT OF A LINEAR TRANSFORMATION

In Section 3.10, we have studied dual spaces and their properties. As we have seen that there is a close relation between a space and its dual. Naturally the question arises, is there any relation between a linear transformation between two given spaces and a linear transformation between their dual spaces? In this section, we shall establish the relationship between a linear transformation and its dual.

Let V and V' be vector spaces over the same field F, and let V^* and V'^* be the spaces dual to V and V' respectively. If $t : V \to V'$ is a linear transformation and f' is an arbitrary linear form in V'^*, then $f' o t$ is a mapping from V to F. Since f' and t are linear, therefore, $f' o t : V \to F$ is a linear form, i.e. $f' o t \in V^*$. Thus, for each $t \in \text{Hom}(V, V')$ and for each $f' \in V'^*$ there exists a linear form $f' o t \in V^*$. This provides us a function $t^* : V'^* \to V^*$ which associates each $f' \in V'^*$ to $f' o t \in V^*$. Let us now show that $t^* : V'^* \to V^*$ is a linear transformation.

Let $f', g' \in V'^*$ and $\lambda, \mu \in F$. Then,

$$t^*(\lambda f' + \mu g') = (\lambda f' + \mu g') o t \qquad \text{[By definition of } t^*\text{]}$$

For any $v \in V$, we have

$$\{(\lambda f' + \mu g') o t\}(v) = \Big(\lambda f' + \mu g'\Big)\Big(t(v)\Big) \quad \text{[By definition of composition of functions]}$$

$\Rightarrow \quad \{(\lambda f' + \mu g') o t\}(v) = (\lambda f')\Big(t(v)\Big) + (\mu g')\Big(t(v)\Big)$

$\Rightarrow \quad \{(\lambda f' + \mu g') o t\}(v) = \lambda \{f'(t(v))\} + \mu \{g'(t(v))\}$

$\Rightarrow \quad \{(\lambda f' + \mu g') o t\}(v) = \lambda \{(f' o t)(v)\} + \mu \{(g' o t)(v)\}$

$\therefore \quad (\lambda f' + \mu g') o t = \lambda(f' o t) + \mu(g' o t)$

Thus, $\quad t^*(\lambda f' + \mu g') = \lambda(f' o t) + \mu(g' o t)$

$\Rightarrow \quad t^*(\lambda f' + \mu g') = \lambda\Big(t^*(f')\Big) + \mu\Big(t^*(g')\Big) \qquad \text{[By definition of } t^*\text{]}$

Hence, $t^* : V'^* \to V^*$ is a linear transformation.

It follows from the above discussion that for each linear transformation $t : V \to V'$ there exists a linear transformation $t^* : V'^* \to V^*$ given by

$$t^*(f') = f'ot \quad \text{for all } f' \in V'^*$$

Let us now suppose that $T^* : V'^* \to V^*$ is another linear transformation given by

$$T^*(f') = f'ot \quad \text{for all } f' \in V'^*$$

Then, $\quad t^*(f') = T^*(f') \quad \text{for all } f' \in V'^*$

$\Rightarrow \quad\quad\quad t^* = T^*$

Thus, for each $t : V \to V'$ there exists a unique linear transformation $t^* : V'^* \to V^*$ given by

$$t^*(f') = f'ot \quad \text{for all } f' \in V'^*$$

The above discussion suggests the following.

DUAL OR ADJOINT OF A LINEAR TRANSFORMATION *Let V and V' be vector spaces over the same field F, and let $t : V \to V'$ be a linear transformation. Then a linear transformation $t^* : V'^* \to V^*$ given by*

$$t^*(f') = f'ot \quad \text{for all } f' \in V'^*$$

is called the dual of t.

The dual of t is also called the *adjoint* or the *transpose* of t. We shall use the symbol t^* to denote the dual of t.

EXAMPLE *Let f be a linear functional on F^2 (F is a field) defined by $f(x,y) = ax + by$. For each of the following linear operators t on F^2, find $t^*((f))(x,y)$.*

(i) $t(x,y) = (x,0)$ (ii) $t(x,y) = (-y,x)$ (iii) $t(x,y) = (x-y,x+y)$

SOLUTION   We have, $t^*(f) = fot$

(i) If $t(x,y) = (x,0)$, then

$$\left(t^*(f)\right)(x,y) = fot(x,y) = f\left(t(x,y)\right) = f(x,0) = ax$$

(ii) If $t(x,y) = (-y,x)$, then

$$\left(t^*(f)\right)(x,y) = fot(x,y) = f\left(t(x,y)\right) = f(-y,x) = -ay + bx$$

(iii) If $t(x,y) = (x-y,x+y)$, then

$$\left(t^*(f)\right)(x,y) = fot(x,y)$$

$\Rightarrow \quad \left(t^*(f)\right)(x,y) = f\left(t(x,y)\right)$

$\Rightarrow \quad \left(t^*(f)\right)(x,y) = f(x-y,x+y) = a(x-y) + b(x+y) = (a+b)x + (b-a)y$

THEOREM-1 *Let $V(F)$ be a vector space. Then,*

(i) $\hat{0}^* = \hat{0}_D$ *i.e, the dual of the null map* $\hat{0} : V \to V$ *is the null map* $\hat{0}_D : V^* \to V^*$.

(ii) $I_V^* = I_{V^*}$

PROOF. (i) By the definition of the dual of a linear transformation, we have

$$\hat{0}^*(f) = (f o \hat{0}) \quad \text{for all } f \in V^*$$
$$(f o \hat{0})(v) = f[\hat{0}(v)] \quad \text{for all } v \in V$$
$$\Rightarrow (f o \hat{0})(v) = f(0_V) \quad \text{for all } v \in V$$
$$\Rightarrow (f o \hat{0})(v) = 0 \quad \text{for all } v \in V$$
$$\Rightarrow (f o \hat{0})(v) = \alpha(v) \quad \text{for all } v \in V, \text{ where } \alpha \text{ is the zero linear form on } V.$$
$$\therefore \quad f o \hat{0} = \alpha, \quad \text{where } \alpha : V \to F \text{ is the zero linear form}$$
$$\Rightarrow \hat{0}^*(f) = f o \hat{0} = \alpha$$
$$\Rightarrow \hat{0}^*(f) = \hat{0}_D(f) \quad \text{for all } f \in V^* \quad [\text{Here } \hat{0}_D : V^* \to V^* \text{ is zero map}]$$
$$\Rightarrow \hat{0}^* = \hat{0}_D.$$

(ii) By the definition of the dual of a linear transformation, we have

$$I_V^*(f) = f o I_V \quad \text{for all } f \in V^*$$

Now, $(f o I_V)(v) = f(I_V(v)) = f(v)$ for all $v \in V$

$$\therefore \quad f o I_V = f$$
$$\Rightarrow I_V^*(f) = f \quad \text{for all } f \in V^*$$
$$\Rightarrow I_V^*(f) = I_{V^*}(f) \quad \text{for all } f \in V^*$$

Hence, $I_V^* = I_{V^*}$

Q.E.D.

THEOREM-2 *Let V be a vector space over a field F, and let t_1 and t_2 be linear transformations from V to itself. Then,*

(i) $(t_1 + t_2)^* = t_1^* + t_2^*$ (ii) $(t_1 o t_2)^* = t_2^* o t_1^*$

PROOF. (i) By the definition of the dual of a linear transformation, we have

$$(t_1 + t_2)^*(f) = f o (t_1 + t_2) \quad \text{for all } f \in V^*$$
$$\Rightarrow (t_1 + t_2)^*(f) = f o t_1 + f o t_2 \quad \text{for all } f \in V^* \qquad [\text{By Theorem 8 on page 293}]$$
$$\Rightarrow (t_1 + t_2)^*(f) = t_1^*(f) + t_2^*(f) \quad \text{for all } f \in V^* \qquad [\text{By definition of the dual}]$$
$$\Rightarrow (t_1 + t_2)^*(f) = (t_1^* + t_2^*)(f) \quad \text{for all } f \in V^* \qquad \left[\begin{array}{l}\text{By definition of the sum of} \\ \text{linear transformations}\end{array}\right]$$
$$\therefore \quad (t_1 + t_2)^* = t_1^* + t_2^*.$$

(ii) By the definition of the dual of a linear transformation, we have

$$(t_1 o t_2)^*(f) = f o (t_1 o t_2) \quad \text{for all } f \in V^* \qquad \text{[By definition of the adjoint map]}$$
$$\Rightarrow (t_1 o t_2)^*(f) = (f o t_1) o t_2 \quad \text{for all } f \in V^* \qquad \text{[By associativity]}$$
$$\Rightarrow (t_1 o t_2)^*(f) = g o t_2, \quad \text{where } g = f 0 t_1 \in V^*$$
$$\Rightarrow (t_1 o t_2)^*(f) = t_2^*(g) \qquad \text{[By the definition of adjoint]}$$
$$\Rightarrow (t_1 o t_2)^*(f) = t_2^*(f o t_1) \quad \text{for all } f \in V^*$$
$$\Rightarrow (t_1 o t_2)^*(f) = t_2^*(t_1^*(f)) \quad \text{for all } f \in V^*$$
$$\Rightarrow (t_1 o t_2)^*(f) = (t_2^* o t_1^*)(f) \quad \text{for all } f \in V^*$$
$$\therefore (t_1 o t_2)^* = t_2^* o t_1^*$$

Q.E.D.

THEOREM-3 Let $V(F)$ be a vector space, and let $t : V \to V$ be an invertible linear transformation. Then $(t^{-1})^* = (t^*)^{-1}$.

PROOF. Since t is invertible linear transformation. Therefore, if t^{-1} is inverse of t, then

$$t o t^{-1} = I_V = t^{-1} o t$$
$$\Rightarrow (t o t^{-1})^* = I_V^* = (t^{-1} o t)^*$$
$$\Rightarrow (t^{-1})^* o t^* = I_{V^*} = t^* o (t^{-1})^* \qquad \text{[By Theorem 2 (ii)]}$$
$$\Rightarrow t^* \text{ is invertible and } (t^*)^{-1} = (t^{-1})^*$$

Q.E.D.

THEOREM-4 Let $V(F)$ be a vector space, and let $t : V \to V$ be a linear transformation. Then $(at)^* = at^*$ for all $a \in F$.

PROOF. Let $a \in F$. Then, $at : V \to V$ is a linear transformation. By the definition of the adjoint, we have

$$(at)^*(f) = f o (at) \quad \text{for all } f \in V^*$$
$$\Rightarrow (at)^*(f) = a(f o t) \quad \text{for all } f \in V^*$$
$$\Rightarrow (at)^*(f) = a\, t^*(f) \quad \text{for all } f \in V^*$$
$$\therefore (at)^* = at^*$$

Q.E.D.

THEOREM-5 If t is a linear transformation from a vector space $V(F)$ to a vector space $V'(F)$, then the annihilator of the range of t is equal to the kernel of t^*, i.e. $\text{Annih}(I_m(t)) = \text{Ker}(t^*)$.

PROOF. Let $f' \in \text{Ker}(t^*)$. Then,

$\qquad t^*(f') = \hat{0} \in V^*$

$\Rightarrow \qquad f'ot = \hat{0} \in V^*$ [By the definition of t^*]

$\Rightarrow \qquad (f'ot)(v) = \hat{0}(v) \quad$ for all $v \in V$

$\Rightarrow \qquad f'(t(v)) = 0 \qquad$ for all $v \in V$

$\Rightarrow \qquad f'(v') = 0 \qquad$ for all $v' = t(v) \in I_m(t)$

$\Rightarrow \qquad f' \in \text{Annih}(I_m(t))$

$\therefore \qquad \text{Ker}(t^*) \subset \text{Annih}(I_m(t))$

Now let $f' \in \text{Annih}(I_m(t))$. Then

$\qquad f'(v') = 0 \qquad$ for all $v' \in I_m(t)$

$\Rightarrow \qquad f'(t(v)) = 0 \qquad$ for all $v \in V \qquad [\because \quad v \in V \Rightarrow \quad v' = t(v) \in I_m(t)]$

$\Rightarrow \qquad (f'ot)(v) = \hat{0}(v) \qquad$ for all $v \in V \qquad [\hat{0}$ is the zero linear form on $V]$

$\Rightarrow \qquad f'ot = \hat{0} \in V^*$

$\Rightarrow \qquad t^*(f') = \hat{0}$

$\Rightarrow \qquad f' \in \text{Ker}(t^*)$

$\therefore \qquad \text{Annih}(I_m(t)) \subset \text{Ker}(t^*)$.

Hence, Annih $(I_m(t)) = \text{Ker}(t^*)$.

Q.E.D.

THEOREM-6 *Let $V(F)$ and $V'(F)$ be finite dimensional vector spaces, and let $t : V \to V'$ be a linear transformation. Then*

(i) $\text{rank}(t) = \text{rank}(t^*)$

(ii) *the range of t^* is the annihilator of the* $\text{Ker}(t)$, *i.e.* $I_m(t^*) = \text{Annih}(\text{Ker}(t))$.

PROOF. (i) Let $\dim V = n, \dim V' = m$, and let $\text{rank}(t) = r$. Since $I_m(t)$ is a subspace of V'. Therefore, by Theorem 2 on page 326, we have

$\qquad \dim(I_m(t)) + \dim(\text{Annih}(I_m(t))) = \dim V'$

$\Rightarrow \qquad \dim(\text{Annih}(I_m(t))) = m - r$

$\Rightarrow \qquad \dim(\text{Ker}(t^*)) = m - r \qquad [\because \quad \text{Annih}(I_m(t)) = \text{Ker}(t^*) \text{ by Theorem 5}] \qquad$ (i)

Since $t^* : V'^* \to V^*$ is a linear transformation, therefore by the Sylvester's law of nullity, we have

$$\text{rank}(t^*) + \text{nullity}(t^*) = \dim V'^*$$
$$\Rightarrow \quad \text{rank}(t^*) = \dim V'^* - \text{nullity}(t^*)$$
$$\Rightarrow \quad \text{rank}(t^*) = \dim V' - \dim(\text{Ker}(t^*)) \qquad [\because \dim V' = \dim V'^*]$$
$$\Rightarrow \quad \text{rank}(t^*) = m - (m - r) \qquad \text{[From (i)]}$$
$$\Rightarrow \quad \text{rank}(t^*) = \text{rank}(t) = r$$

Aliter: Let $\dim V = n$, $\dim V' = m$, and let $\text{rank}(t) = r$. In order to prove that $\text{rank}(t^*) = r$ it is sufficient to show that there exist ordered bases $X' = \{x'_1, x'_2, \ldots, x'_m\}$ and $X = \{x_1, x_2, \ldots, x_n\}$ of V'^* and V^* respectively such that $t^*(x'_1) = x_1, t^*(x'_2) = x_2, \ldots, t^*(x'_r) = x_r$, $t^*(x'_{r+1}) = \widehat{0} = \cdots = t^*(x'_m)$. Here, $\widehat{0}$ is the zero linear form on V.

Since $\text{rank}(t) = r$, therefore, by corollary 1 of Sylvester's law of nullity, there exist ordered bases $B = \{b_1, b_2, \ldots, b_n\}$ and $B' = \{b'_1, b'_2, \ldots, b'_m\}$ of V and V' respectively such that

$$t(b_1) = b'_1, t(b_2) = b'_2, \ldots, t(b_r) = b'_r, t(b_{r+1}) = t(b_{r+2}) = \cdots = t(b_n) = 0_{V'}.$$

Let $X = \{x_1, x_2, \ldots, x_n\}$ and $X' = \{x'_1, x'_2, \ldots, x'_m\}$ be bases dual to B and B' respectively. Then,

$$x_i(b_j) = \delta_{ij} \quad \text{for all } i, j \in \underline{n} \text{ and } x'_p(b'_q) = \delta_{pq} \text{ for all } p, q \in \underline{m}.$$

Now,
$$(x'_i o t)(b_j) = x'_i[t(b_j)]$$

$$\Rightarrow \quad (x'_i o t)(b_j) = \begin{cases} x'_i(b'_j) & \text{if } 1 \leq j \leq r, \\ x'_i(0_{V'}) & \text{if } j > r \end{cases} \qquad \left[\because \begin{array}{l} t(b_j) = b'_j \text{ for } 1 \leq j \leq r \\ t(b_j) = 0_{V'} \text{ for } j > r \end{array}\right]$$

$$\Rightarrow \quad (x'_i o t)(b_j) = \begin{cases} \delta_{ij} & \text{if } 1 \leq i \leq r, 1 \leq j \leq r \\ 0 & \text{if } i, j > r \end{cases} \qquad [\because x'_p(b'_q) = \delta_{pq} \text{ for all } p, q \in \underline{m}]$$

$$\Rightarrow \quad (x'_i o t)(b_j) = \begin{cases} x_i(b_j) & \text{if } 1 \leq i \leq r, 1 \leq j \leq r \\ 0 & \text{Otherwise} \end{cases} \qquad [\because x_i(b_j) = \delta_{ij} \text{ for all } i, j \in \underline{n}]$$

$$\Rightarrow \quad (x'_i o t)(b_j) = \begin{cases} x_i(b_j) & \text{if } 1 \leq i \leq r, 1 \leq j \leq r \\ \widehat{0}(b_j) & \text{Otherwise, where } \widehat{0} \text{ is the zero linear form on } V \end{cases}$$

$$\therefore \quad x'_i o t = \begin{cases} x_i & \text{for } i = 1, 2, \ldots, r \\ \widehat{0} & \text{for } i = r+1, \ldots, m. \end{cases}$$

$$\Rightarrow \quad t^*(x'_i) = \begin{cases} x_i & \text{for } i = 1, 2, \ldots, r \\ \widehat{0} & \text{for } i = r+1, r+2, \ldots, m. \end{cases}$$

Hence, $\text{rank}(t) = \text{rank}(t^*) = r$.

(ii) Since $t^* : V'^* \to V^*$ is a linear transformation, therefore, $I_m(t^*)$ is a subspace of V^*. Also, $\text{Annih}(\text{Ker}(t))$ is a subspace of V^*.

Let f be an arbitrary element of $I_m(t^*)$. Then, $f = t^*(g')$ for some $g' \in V'^*$.
If v is any vector in $\text{Ker}(t)$, then

$$f(v) = [t^*(g')](v)$$
$$\Rightarrow \quad f(v) = (g' o t)(v) \qquad \text{[By definition of } t^*]$$
$$\Rightarrow \quad f(v) = g'(t(v))$$
$$\Rightarrow \quad f(v) = g'(0_{V'}) \qquad [\because \ v \in \text{Ker}(t) \Rightarrow t(v) = 0_{V'}]$$
$$\Rightarrow \quad f(v) = 0.$$

Thus, $f(v) = 0$ for all $v \in \text{Ker}(t)$.
$\therefore \ f \in \text{Annih}(\text{Ker}(t))$
Thus, $f \in I_m(t^*) \Rightarrow f \in \text{Annih}(\text{Ker}(t))$.
So, $I_m(t^*) \subset \text{Annih}(\text{Ker}(t))$.
$\Rightarrow \ I_m(t^*)$ is a subspace of $\text{Annih}(\text{Ker}(t))$.
Now,
$\dim(\text{Ker}(t)) + \dim(\text{Annih}(\text{Ker}(t))) = \dim V$ \qquad [By Theorem 2 on page 326]
$\therefore \quad \dim(\text{Annih}(\text{Ker}(t))) = \dim V - \dim(\text{Ker}(t))$
$\Rightarrow \quad \dim(\text{Annih}(\text{Ker}(t))) = \dim(I_m(t))$ \qquad [By Sylvester's law]
$\Rightarrow \quad \dim(\text{Annih}(\text{Ker}(t))) = \text{rank}(t)$
$\Rightarrow \quad \dim(\text{Annih}(\text{Ker}(t))) = \text{rank}(t^*)$
$\Rightarrow \quad \dim(\text{Annih}(\text{Ker}(t))) = \dim(I_m(t^*))$.

Thus, $I_m(t^*)$ and $\text{Annih}(\text{Ker}(t))$ are subspaces of V^* such that $I_m(t^*) \subset \text{Annih}(\text{Ker}(t))$ and $\dim(I_m(t^*)) = \dim(\text{Annih}(\text{Ker}(t)))$.
Hence, $I_m(t^*) = \text{Annih}(\text{Ker}(t))$.

Q.E.D.

Aliter: Obviously, $I_m(t^*)$ and $\text{Annih}(\text{Ker}(t))$ are subspaces of V^*.
Let f be an arbitrary linear form on V such that $f \in I_m(t^*)$.
Then there exists $f' \in V'^*$ such that

$$t^*(f') = f$$
$$\Rightarrow \quad f' o t = f \qquad \text{[By the definition of } t^*]$$
$$\Rightarrow \quad (f' o t)(v) = f(v) \quad \text{for all } v \in V$$
$$\Rightarrow \quad (f' o t)(v) = f(v) \quad \text{for all } v \in \text{Ker}(t)$$
$$\Rightarrow \quad f'\big(t(v)\big) = f(v) \quad \text{for all } v \in \text{Ker}(t)$$
$$\Rightarrow \quad f'(0_{V'}) = f(v) \quad \text{for all } v \in \text{Ker}(t) \qquad [\because \ v \in \text{Ker}(t) \Rightarrow t(v) = 0_V]$$
$$\Rightarrow \quad f(v) = 0 \quad \text{for all } v \in \text{Ker}(t)$$
$$\Rightarrow \quad f \in \text{Annih}(\text{Ker}(t)) \qquad \text{[By the definition of annihilator]}$$

Thus, $f \in I_m(t^*) \Rightarrow f \in \text{Annih}(\text{Ker}(t))$.
So, $I_m(t^*) \subset \text{Annih}(\text{Ker}(t))$.
$\Rightarrow I_m(t^*)$ is a subspace of $\text{Annih}(\text{Ker}(t))$.
By Theorem 2 on page 326, we have

$$\dim(\text{Ker}(t)) + \dim(\text{Annih}(\text{Ker}(t))) = \dim V$$

$\therefore \quad \dim(\text{Annih}(\text{Ker}(t))) = \dim V - \dim(\text{Ker}(t))$
$\Rightarrow \quad \dim(\text{Annih}(\text{Ker}(t))) = \dim V - \text{nullity}(t)$
$\Rightarrow \quad \dim(\text{Annih}(\text{Ker}(t))) = \dim(I_m(t)) \qquad [\because \dim(I_m(t)) + \text{nullity}(t) = \dim V]$
$\Rightarrow \quad \dim(\text{Annih}(\text{Ker}(t))) = \text{rank}(t)$
$\Rightarrow \quad \dim(\text{Annih}(\text{Ker}(t))) = \text{rank}(t^*) \qquad [\because \text{rank}(t) = \text{rank}(t^*)]$
$\Rightarrow \quad \dim(\text{Annih}(\text{Ker}(t))) = \dim(I_m(t^*))$.

Thus, $\dim(I_m(t^*)) = \dim(\text{Annih}(\text{Ker}(t)))$ and $I_m(t^*) \subset \text{Annih}(\text{Ker}(t))$.

$\therefore \quad I_m(t^*) = \text{Annih}(\text{Ker}(t))$.

EXERCISE 3.7

1. Let f be the linear function on R^2 defined by $f(x,y) = 3x - 2y$. For each of the following linear mappings $t : R^3 \to R^2$, find $(t^*(f))(x,y,z)$:
   (i) $t(x,y,z) = (x+y, y+z)$ (ii) $t(x,y,z) = (x+y+z, 2x-y)$
   [Hint: Use $t^*(f) = f \circ t$]

2. Let $V(F)$ be a finite dimensional vector space and t be any linear operator on V. Show that the mapping $\varphi : \text{Hom}(V,V) \to \text{Hom}(V^*,V^*)$ defined by $\varphi(t) = t^*$ is an isomorphism.

3. Let $V(F)$ be a vector space, and let S_1 and S_2 be subspaces of V. If t is the projection on S_1 along S_2, then t^* is the projection on $\text{Annih}(S_2)$ along $\text{Annih}(S_1)$.

4. Let $V(F)$ be a finite dimensional vector space and t be a linear operator on V. Prove that $(t^*)^* = t^{**} = t$.

5. Let V be the vector space of all polynomial functions over the field of real numbers. Let a and b be fixed real numbers and let f be the linear functional on V defined by

$$f(p) = \int_a^b p(x)\,dx$$

If D is the differential operator on V, find $D^*(f)$.

[Hint: $(D^*(f))\,p(x) = (f \circ D)p(x) = f\{D(p(x))\}$
$= f\left\{\dfrac{d}{dx}(p(x))\right\} = \int_a^b \dfrac{d}{dx}\{p(x)\}\,dx = p(b) - p(a)$]

6. Let V be the vector space of all $n \times n$ matrices over a field F and let B be a fixed $n \times n$ matrix. If t is a linear operator on V defined by $t(A) = AB - BA$, and if f is the linear functional on V defined by $f(A) = tr(A)$. Find $t^*(f)$.

[Hint: $\left(t^*(f)\right)(A) = f o t(A) = f\left(t(A)\right) = f(AB - BA)$

$\Rightarrow \quad \left(t^*(f)\right)(A) = tr(AB - BA) = tr(AB) - tr(BA) = 0$

$\therefore \quad (t^*(f))(A) = 0$ for all $A \in V$.]

7. Let V be a finite-dimensional vector space over the field F and let t be a linear operator on V. Let α be a scalar and suppose there is a non-zero vector α in V such that $t(v) = \alpha v$. Prove that there is a non-zero linear functional f on V such that $t^*(f) = \alpha f$.

ANSWERS

1. (i) $3x + y - 2z$ (ii) $-x + 5y + 6z$ 5. $p(b) - p(a)$ 6. 0

Chapter 4

Linear Transformations and Matrices

4.1 INTRODUCTION

In the previous two chapters, we have learnt about vector spaces and linear transformations between two vector spaces $V(F)$ and $V'(F)$ defined over the same field F. In this chapter, we shall see that each linear transformation from an n-dimensional vector space $V(F)$ to an m-dimensional vector space $V'(F)$ corresponds to an $m \times n$ matrix over field F which depends upon the bases of vector spaces V and V'. Conversely, every $m \times n$ matrix over field F determines a linear transformation $t : V \to V'$, depending upon the bases of V and V'. This linear transformation is generally known as the map of matrix A. We shall also see how the matrix representation of a linear transformation changes with the changes of bases of the given vector spaces.

4.2 LINEAR TRANSFORMATIONS AND MATRICES

In this section, we shall show that for each linear mapping from a vector space $V(F)$ to a vector space $V'(F)$ there corresponds a matrix A with respect to bases B and B' of V and V', respectively. Conversely, any $m \times n$ matrix $A = [a_{ij}]$ over F determines a unique linear transformation from V to V' whose matrix is A.

Let $B = \{b_1, b_2, \ldots, b_n\}$ and $B' = \{b'_1, b'_2, \ldots, b'_m\}$ be ordered bases for vector spaces $V(F)$ and $V'(F)$ respectively. Let $t : V \to V'$ be a linear transformation. Then t is determined completely by n vectors $t(b_1), t(b_2), \ldots, t(b_n)$. Since B' is an ordered basis for V', therefore each $t(b_j)$ can be written as

$$t(b_j) = \sum_{i=1}^{m} a_{ij} b'_i \quad \text{for some } a_{ij} \in F \tag{i}$$

Thus, each $t(b_j)$ is determined by m scalars $a_{1j}, a_{2j}, \ldots, a_{mj}$ which are known as coordinates of $t(b_j)$ relative to the basis B'. Consequently, the linear transformation $t : V \to V'$ is determined by mn scalars a_{ij}; $i \in \underline{m}, j \in \underline{n}$. These mn scalars define a matrix $A = [a_{ij}]$ whose j^{th} column is $t(b_j)$; $j = 1, 2, \ldots, n$. That is,

$$A = (t(b_1), t(b_2), \ldots, t(b_n))$$

Let $X = \{x_1, x_2, \ldots, x_n\}$ and $X' = \{x'_1, x'_2, \ldots, x'_m\}$ be bases dual to B and B' respectively. Then, for any $i \in \underline{m}$ and $j \in \underline{n}$, we have

$$x'_i(t(b_j)) = x'_i \left(\sum_{r=1}^{m} a_{rj} b'_r \right)$$

$\Rightarrow \quad x'_i(t(b_j)) = \sum_{r=1}^{m} a_{rj} x'_i(b'_r) \qquad\qquad [\because \; x'_i : V' \to F \text{ is linear}]$

$\Rightarrow \quad x'_i(t(b_j)) = \sum_{r=1}^{m} a_{rj} \delta_{ir} \qquad\qquad [\because \; X' \text{ is basis dual to } B]$

$\Rightarrow \quad x'_i(t(b_j)) = a_{ij}$

Hence, $t : V \to V'$ is determined by mn scalars $a_{ij} = x'_i \big(t(b_j) \big)$; $i \in \underline{m}, j \in \underline{n}$. These mn scalars form a matrix $A = [a_{ij}]$ over F. This matrix is called the matrix of t relative to bases B and B' and is denoted by $M_B^{B'}(t)$. More explicitly, the matrix of linear transformation $t : V \to V'$ is a mapping $M_B^{B'}(t) : \underline{m} \times \underline{n} \to F$ given by

$$\left(M_B^{B'}(t) \right)((i,j)) = x'_i(t(b_j)) \quad \text{for all } i \in \underline{m}, j \in \underline{n}.$$

That is,

$$M_B^{B'}(t) = \begin{bmatrix} x'_1(t(b_1)) & \ldots & x'_1(t(b_j)) & \ldots & x'_1(t(b_n)) \\ \vdots & & \vdots & & \vdots \\ x'_i(t(b_1)) & \ldots & x'_i(t(b_j)) & \ldots & x'_i(t(b_n)) \\ \vdots & & \vdots & & \vdots \\ x'_m(t(b_1)) & \ldots & x'_m(t(b_j)) & \ldots & x'_m(t(b_n)) \end{bmatrix}$$

Thus, each linear transformation $t : V \to V'$ is determined by an $m \times n$ matrix $M_B^{B'}(t)$ with respect to bases B and B' of V and V' respectively.

Conversely, let $A = [a_{ij}]$ be an $m \times n$ matrix over a field F, and let $B = \{b_1, b_2, \ldots, b_n\}$ and $B' = \{b'_1, b'_2, \ldots, b'_m\}$ be ordered bases for V and V' respectively. Then for each $r \in \underline{m}$ and $s \in \underline{n}$, $b'_r : F \to V'$ and $x_s : V \to F$ are linear transformations. So the composite $b'_r x_s$ is a linear transformation from V to V' for all $r \in \underline{m}$ and $s \in \underline{n}$. Let

$$t = \sum_{r=1}^{m} \sum_{s=1}^{n} a_{rs} b'_r x_s$$

Then, $t : V \to V'$ is a linear transformation and

$$\left[M_B^{B'}(t) \right]_{ij} = x'_i \big(t(b_j) \big)$$

$$\Rightarrow \quad \left[M_B^{B'}(t)\right]_{ij} = x_i'\left[\left(\sum_{r=1}^{m}\sum_{s=1}^{n} a_{rs}\, b_r'\, x_s\right)(b_j)\right]$$

$$\Rightarrow \quad \left[M_B^{B'}(t)\right]_{ij} = x_i'\left[\sum_{r=1}^{m}\sum_{s=1}^{n} a_{rs}\, b_r'\, x_s\,(b_j)\right]$$

$$\Rightarrow \quad \left[M_B^{B'}(t)\right]_{ij} = x_i'\left[\sum_{r=1}^{m}\sum_{s=1}^{n} a_{rs}\, b_r'\, \delta_{sj}\right]$$

$$\Rightarrow \quad \left[M_B^{B'}(t)\right]_{ij} = x_i'\left[\sum_{r=1}^{m} a_{rj}\, b_r'\right]$$

$$\Rightarrow \quad \left[M_B^{B'}(t)\right]_{ij} = \sum_{r=1}^{m} a_{rj}\, x_i'(b_r')$$

$$\Rightarrow \quad \left[M_B^{B'}(t)\right]_{ij} = \sum_{r=1}^{m} a_{rj}\, \delta_{ir} = a_{ij} \quad \text{for all } i \in \underline{m}, j \in \underline{n}.$$

$$\therefore \quad M_B^{B'}(t) = A.$$

Thus, each $m \times n$ matrix $A = [a_{ij}]$ over field F with bases $B = \{b_1, b_2, \ldots, b_n\}$ and $B' = \{b_1', b_2', \ldots, b_m'\}$ of V and V' respectively determines a linear transformation $t : V \to V'$ given by

$$t = \sum_{r=1}^{m}\sum_{s=1}^{n} a_{rs}\, b_r'\, x_s$$

such that matrix of t relative to bases B and B' is matrix A itself. This linear transformation is sometimes denoted by $M^{-1}(A)$ and is called the map of matrix A.

Let v be a vector of a vector space $V(F)$, and let $B = \{b_1, b_2, \ldots, b_n\}$ be an ordered basis for V. Then $v \in V$ can be considered as a linear transformation from F to V, associating each $\lambda \in F$ to λv in V. Therefore, one can speak of the matrix of a vector $v \in V$ relative to some ordered basis $B = \{b_1, b_2, \ldots, b_n\}$ of V. Choosing the unity $1 \in F$ as the basis for F, the matrix of $v \in V$ relative to basis B is given by

$$\left[M_B^B(v)\right]_{i1} = x_i(v(1))$$

$$\Rightarrow \quad \left[M_B^B(v)\right]_{i1} = x_i(1v) \qquad\qquad \text{[By definition of } v : F \to V\text{]}$$

$$\Rightarrow \quad \left[M_B^B(v)\right]_{i1} = x_i(v) = i\text{th coordinate of } v \text{ relative to } B$$

Thus, matrix of $v \in V$ relative to an ordered basis $B = \{b_1, b_2, \ldots, b_n\}$ of V is an $n \times 1$ matrix

$$M_B^B(v) = \begin{bmatrix} x_1(v) \\ x_2(v) \\ \vdots \\ x_n(v) \end{bmatrix}$$

This matrix is known as the coordinate matrix of vector v relative to basis B.

Similarly, for each $t(v) \in V'$, $M_{B'}^{B'}(t(v))$ is the coordinate matrix of $t(v)$ relative to basis B' of V' and is given by

$$M_{B'}^{B'}(t(v)) = \begin{bmatrix} x_1'(t(v)) \\ x_2'(t(v)) \\ \vdots \\ x_m'(t(v)) \end{bmatrix}.$$

where X' is basis dual to B'.

THEOREM-1 *Let $B = \{b_1, b_2, \ldots, b_n\}$ and $B' = \{b_1', b_2', \ldots, b_m'\}$ be bases for vector spaces $V(F)$ and $V'(F)$ respectively and let $t : V \to V'$ be a linear transformation. Then,*

$$M_B^{B'}(t) M_B^B(v) = M_{B'}^{B'}(t(v)) \quad \text{for all } v \in V.$$

PROOF. Clearly, $M_B^{B'}(t)$, $M_B^B(v)$ and $M_{B'}^{B'}(t(v))$ are $m \times n$, $n \times 1$ and $m \times 1$ matrices over field F. Therefore, $M_B^{B'}(t)$ and $M_B^B(v)$ are conformable for the product $M_B^{B'}(t) \, M_B^B(v)$ and order of $M_B^{B'}(t) \, M_B^B(v)$ is same as the order of $M_{B'}^{B'}(v)$.

Let $X = \{x_1, x_2, \ldots, x_n\}$ and $X' = \{x_1', x_2', \ldots, x_m'\}$ be bases dual to B and B' respectively. Then,

$$\left[M_B^{B'}(t)\right]_{ij} = x_i'(t(b_j)) \quad \text{for all } i \in \underline{m}, j \in \underline{n}.$$

Now, let v be an arbitrary vector in V. Then,

$$v = \sum_{r=1}^{n} \lambda_r b_r \quad \text{for some } \lambda_r \in F$$

$\Rightarrow \quad t(v) = t\left(\sum_{r=1}^{n} \lambda_r b_r\right) = \sum_{r=1}^{n} \lambda_r t(b_r)$ and $x_r(v) = \lambda_r$ for all $r \in \underline{n}$.

Now,

$$\left[M_B^{B'}(t) \, M_B^B(v)\right]_{i1} = \sum_{r=1}^{n} \left(M_B^{B'}(t)\right)_{ir} \left(M_B^B(v)\right)_{r1}$$

$\Rightarrow \quad \left[M_B^{B'}(t) \, M_B^B(v)\right]_{i1} = \sum_{r=1}^{n} x_i'(t(b_r)) \, x_r(v)$

$\Rightarrow \quad \left[M_B^{B'}(t) \, M_B^B(v)\right]_{i1} = \sum_{r=1}^{n} x_i'(t(b_r)) \, \lambda_r$

$\Rightarrow \quad \left[M_B^{B'}(t) \, M_B^B(v)\right]_{i1} = x_i'\left(\sum_{r=1}^{n} \lambda_r t(b_r)\right)$

$\Rightarrow \quad \left[M_B^{B'}(t) \, M_B^B(v)\right]_{i1} = x_i'(t(v)) = \left[M_{B'}^{B'}(t(v))\right]_{i1} \quad \text{for all } i \in \underline{m}.$

Hence, $\quad M_B^{B'}(t) \, M_B^B(v) = M_{B'}^{B'}(t(v))$

Q.E.D.

THEOREM-2 *Let $A = [a_{ij}]$ be an $m \times n$ matrix over a field F. Then for given ordered bases $B = \{b_1, b_2, \ldots, b_n\}$ and $B' = \{b'_1, b'_2, \ldots, b'_m\}$ of vector spaces $V(F)$ and $V'(F)$ respectively, the map of A relative to the bases B and B' is that linear transformation $t : V \to V'$ which gives its effect on the basis vectors $b_1, b_2, \ldots, b_n$ of V as*

$$t(b_j) = \sum_{i=1}^{m} a_{ij} b'_i \quad \text{for all } j \in \underline{n}.$$

This map t is also determined by giving its composite with each coordinate form $x'_i : V' \to F$ of the dual basis of V' as

$$x'_i \circ t = \sum_{j=1}^{n} a_{ij} x_j \quad \text{for all } i \in \underline{m}$$

where $X = \{x_1, x_2, \ldots, x_n\}$ and $X' = \{x'_1, x'_2, \ldots, x'_m\}$ are bases dual to B and B' respectively.

PROOF. The map of matrix A relative to the bases B and B' is given by

$$t = \sum_{r=1}^{m} \sum_{s=1}^{n} a_{rs} b'_r x_s$$

$\Rightarrow \quad t(b_j) = \left(\sum_{r=1}^{m} \sum_{s=1}^{n} a_{rs} b'_r x_s \right)(b_j)$

$\Rightarrow \quad t(b_j) = \sum_{r=1}^{m} \sum_{s=1}^{n} a_{rs} b'_r x_s(b_j)$

$\Rightarrow \quad t(b_j) = \sum_{r=1}^{m} \sum_{s=1}^{n} a_{rs} b'_r \delta_{sj} \hfill [\because \; x_s(b_j) = \delta_{sj}]$

$\Rightarrow \quad t(b_j) = \sum_{r=1}^{m} a_{rj} b'_r = \sum_{r=1}^{m} a_{rj} b'_r = \sum_{i=1}^{m} a_{ij} b'_i \quad \text{for all } j \in \underline{n}.$

To prove the second part, let v be an arbitrary vector in V. Then,

$$t(v) = \left(\sum_{r=1}^{m} \sum_{s=1}^{n} a_{rs} b'_r x_s \right)(v) = \sum_{r=1}^{m} \sum_{s=1}^{n} a_{rs} b'_r x_s(v)$$

Therefore, for any $i \in \underline{m}$, we have

$$x'_i(t(v)) = x'_i \left[\sum_{r=1}^{m} \sum_{s=1}^{n} a_{rs} b'_r x_s(v) \right]$$

$\Rightarrow \quad x'_i(t(v)) = x'_i \left[\sum_{r=1}^{m} \left(\sum_{s=1}^{n} a_{rs} x_s(v) \right) b'_r \right]$

$\Rightarrow \quad x'_i(t(v)) = \sum_{r=1}^{m} \left(\sum_{s=1}^{n} a_{rs} x_s(v) \right) x'_i b'_r$

$\Rightarrow \quad x'_i(t(v)) = \sum_{r=1}^{m} \left(\sum_{s=1}^{n} a_{rs} x_s(v) \right) \delta_{ir}$

$\Rightarrow \quad x'_i(t(v)) = \sum_{s=1}^{n} a_{is} x_s(v)$

$\Rightarrow \quad x'_i(t(v)) = \left(\sum_{j=1}^{n} a_{ij} x_j \right)(v) \quad \text{for all } v \in V$

$\Rightarrow \quad (x'_i \circ t)(v) = \left(\sum_{j=1}^{n} a_{ij} x_j \right)(v) \quad \text{for all } v \in V$

$\Rightarrow \quad x'_i \circ t = \sum_{j=1}^{n} a_{ij} x_j \quad \text{for all } i \in \underline{m}.$

Q.E.D.

REMARK. *In the above theorem, the equations*

$$t(b_j) = \sum_{i=1}^{m} a_{ij} b'_i \quad \text{for all } j \in \underline{n},$$

relating the vector s of basis B', describe t in terms of bases while equations

$$x'_i \circ t = \sum_{j=1}^{n} a_{ij} x_j \quad \text{for all } i \in \underline{m},$$

relating the linear forms on V, describe t in terms of the coordinates. Any one of these two systems can be used for finding the matrix of a given linear transformation t with respect to bases B and B'.

To find the matrix of a linear transformation $t : V \to V'$ relative to bases B and B' we first write the vectors $t(b_1), t(b_2), \ldots, t(b_n)$ as linear expressions of basis vectors in B'. In these linear expressions the transpose of the matrix of the coefficients is the matrix A.

ILLUSTRATIVE EXAMPLES

EXAMPLE-1 *Let $t : R^2 \to R^2$ be a linear transformation given by $t(a, b) = (2a - 3b, a + b)$ for all $(a, b) \in R^2$. Find the matrix of t relative to bases $B = \{b_1 = (1, 0), b_2 = (0, 1)\}$ and $B' = \{b'_1 = (2, 3), b'_2 = (1, 2)\}$*

SOLUTION In order to find the matrix of t relative to bases B and B', we have to express $t(b_1)$ and $t(b_2)$ as a linear combination of b'_1 and b'_2. For this, we first find the coordinates of an arbitrary vector $(a, b) \in R^2$ with respect to basis B'.

$$(a,b) = xb'_1 + yb'_2$$
$\Rightarrow \quad (a,b) = x(2,3) + y(1,2)$
$\Rightarrow \quad (a,b) = (2x+y, 3x+2y)$
$\Rightarrow \quad 2x+y = a$ and $3x+2y = b$
$\Rightarrow \quad x = 2a-b$ and $y = -3a+2b$
$\therefore \quad (a,b) = (2a-b)b'_1 + (-3a+2b)b'_2$ \hfill (i)

Now, $t(a,b) = (2a-3b, a+b)$

$\Rightarrow \quad t(b_1) = t(1,0) = (2,1)$ and $t(b_2) = t(0,1) = (-3,1)$ [Putting $a=2$ and $b=1$ in (i)]
$\Rightarrow \quad t(b_1) = 3b'_1 - 4b'_2$
and, $\quad t(b_2) = -7b'_1 + 11b'_2$ \hfill [Putting $a=-3$ and $b=1$]

$$\therefore \quad M_B^{B'}(t) = \begin{bmatrix} 3 & -4 \\ -7 & 11 \end{bmatrix}^T = \begin{bmatrix} 3 & -7 \\ -4 & 11 \end{bmatrix}$$

EXAMPLE-2 *Find the matrix of the linear transformation $t : R^3 \to R^3$ given by*

$$t(a,b,c) = (2b+c, a-4b, 3a)$$

relative to ordered basis $B = \{b_1 = (1,1,1), b_2 = (1,1,0), b_3 = (1,0,0)\}$. Also, verify that $M_B^B(t)\, M_B^B(v) = M_B^B(t(v))$ for any vector v in R^3.

SOLUTION   First we find the coordinates of an arbitrary vector $(a,b,c) \in R^3$ relative to basis B. For this, we write (a,b,c) as a linear combination of b_1, b_2, b_3 as given below.

$$(a,b,c) = xb_1 + yb_2 + zb_3$$
$\Rightarrow \quad (a,b,c) = x(1,1,1) + y(1,1,0) + z(1,0,0) = (x+y+z, x+y, x)$
$\Rightarrow \quad x+y+z = a,\ x+y = b$ and $x = c$
$\Rightarrow \quad x = c,\ y = b-c$, and $z = a-b$
$\therefore \quad (a,b,c) = cb_1 + (b-c)b_2 + (a-b)b_3$ \hfill (i)

Now, $t(a,b,c) = (2b+c,\ a-4b,\ 3a)$ \hfill (ii)
$\therefore \quad t(b_1) = t(1,1,1) = (3,-3,3)$ \hfill [Putting $a=b=c=1$ in (ii)]
$\quad\ t(b_2) = t(1,1,0) = (2,-3,3)$ \hfill [Putting $a=b=1,\ c=0$ in (ii)]
and, $\quad t(b_3) = t(1,0,0) = (0,1,3)$ \hfill [Putting $a=1,\ b=c=0$ in (ii)]

$\therefore \quad t(b_1) = (3,-3,3) = 3b_1 + (-6)b_2 + 6b_3$ [Putting $a = 3$, $b = -3$, $c = 3$ in (i)]

$t(b_2) = (2,-3,3) = 3b_1 + (-6)b_2 + 5b_3$ [Putting $a = 2$, $b = -3$, $c = 3$ in (i)]

$t(b_3) = (0,1,3) = 3b_1 + (-2)b_2 + (-1)b_3$ [Putting $a = 0$, $b = 1$, $c = 3$ in (i)]

$\therefore \quad M_B^B(t) = \begin{bmatrix} 3 & -6 & 6 \\ 3 & -6 & 5 \\ 3 & -2 & -1 \end{bmatrix}^T = \begin{bmatrix} 3 & 3 & 3 \\ -6 & -6 & -2 \\ 6 & 5 & -1 \end{bmatrix}$

Let $v = (a,b,c)$ be an arbitrary vector in R^3. Then,

$(a,b,c) = cb_1 + (b-c)b_2 + (a-b)b_3$ [From (i)]

$\Rightarrow \quad M_B^B(v) = \begin{bmatrix} c \\ b-c \\ a-b \end{bmatrix}$

Now, we shall express $t(v)$ as a linear combination of b_1, b_2, b_3.
Let

$t(v) = t(a,b,c) = (2b+c, a-4b, 3a) = xb_1 + yb_2 + zb_3$

$\Rightarrow \quad (2b+c, a-4b, 3a) = x(1, 1, 1) + y(1, 1, 0) + z(1, 0, 0)$

$\Rightarrow \quad (2b+c, a-4b, 3a) = (x+y+z, x+y, x)$

$\Rightarrow \quad x = 3a, \ x+y = a-4b, \ x+y+z = 2b+c$

$\Rightarrow \quad x = 3a, \ y = -2a-4b, \ z = -a+6b+c$

$\therefore \quad t(v) = 3ab_1 + (-2a-4b)b_2 + (-a+6b+c)b_3$

$\Rightarrow \quad M_B^B(t(v)) = \begin{bmatrix} 3a \\ -2a-4b \\ -a+6b+c \end{bmatrix}$

$\therefore \quad M_B^B(t) \, M_B^B(v) = \begin{bmatrix} 3 & 3 & 3 \\ -6 & -6 & -2 \\ 6 & 5 & -1 \end{bmatrix} \begin{bmatrix} c \\ b-c \\ a-b \end{bmatrix}$

$\Rightarrow \quad M_B^B(t) \, M_B^B(v) = \begin{bmatrix} 3a \\ -2a-4b \\ -a+6b+c \end{bmatrix} = M_B^B(t(v))$

EXAMPLE-3 Let $t : R^3 \to R^2$ be the linear transformation defined by

$$t(x,y,z) = (3x+2y-4z, \ x-5y+3z)$$

(i) Find the matrix of t relative to bases $B = \{b_1 = (1,1,1), b_2 = (1,1,0), b_3 = (1,0,0)\}$ and $B' = \{b'_1 = (1,3), b'_2 = (2,5)\}$.

(ii) Verify that $M_B^{B'}(t) M_B^B(v) = M_{B'}^{B'}(t(v))$.

SOLUTION (i) First of all, we shall express any vector $(a,b) \in R^2$ as a linear combination of basis vectors in B'. Let

$$(a,b) = xb'_1 + yb'_2$$
$$\Rightarrow \quad (a,b) = x\,(1,3) + y\,(2,5)$$
$$\Rightarrow \quad x + 2y = a \text{ and } 3x + 5y = b$$
$$\Rightarrow \quad x = -5a + 2b \text{ and } y = 3a - b$$
$$\therefore \quad (a,b) = (-5a + 2b)b'_1 + (3a - b)b'_2 \qquad (i)$$

Now, $t(x,y,z) = (3x + 2y - 4z,\ x - 5y + 3z)$

$$\Rightarrow \quad t(b_1) = t(1,1,1) = (1,-1) = -7b'_1 + 4b'_2 \qquad \text{[Putting } a=1, b=-1 \text{ in (i)]}$$
$$t(b_2) = t(1,1,0) = (5,-4) = -33b'_1 + 19b'_2 \qquad \text{[Putting } a=5, b=-4 \text{ in (i)]}$$
$$t(b_3) = t(1,0,0) = (3,1) = -13b'_1 + 8b'_2 \qquad \text{[Putting } a=3, b=1 \text{ in (i)]}$$

$$\therefore \quad M_B^{B'}(t) = \begin{bmatrix} -7 & 4 \\ -33 & 19 \\ -13 & 8 \end{bmatrix}^T = \begin{bmatrix} -7 & -33 & -13 \\ 4 & 19 & 8 \end{bmatrix}$$

(ii) Let $v = (x,y,z)$ be any vector in R^3. Then,

$$v = lb_1 + mb_2 + nb_3$$
$$\Rightarrow \quad (x,y,z) = l(1,1,1) + m(1,1,0) + n(1,0,0)$$
$$\Rightarrow \quad (x,y,z) = (l+m+n,\ l+m,\ l)$$
$$\Rightarrow \quad l+m+n = x,\ l+m = y,\ l = z$$
$$\Rightarrow \quad l = z,\ m = y-z,\ n = x-y$$

$$\therefore \quad M_B^B(v) = \begin{bmatrix} l \\ m \\ n \end{bmatrix} = \begin{bmatrix} z \\ y-z \\ x-y \end{bmatrix}$$

Now, let

$$t(v) = t(x,y,z) = pb'_1 + qb'_2$$
$$\Rightarrow \quad (3x+2y-4z,\ x-5y+3z) = p(1,3) + q(2,5)$$
$$\Rightarrow \quad p + 2q = 3x+2y-4z,\ 3p+5q = x-5y+3z$$
$$\Rightarrow \quad p = -13x - 20y + 26z,\ q = 8x + 11y - 15z$$
$$\therefore \quad t(v) = (-13x - 20y + 26z)b'_1 + (8x + 11y - 15z)b'_2$$
$$\Rightarrow \quad M_{B'}^{B'}(t(v)) = \begin{bmatrix} -13x - 20y + 26z \\ 8x + 11y - 15z \end{bmatrix}$$

Thus,

$$M_B^{B'}(t)M_B^B(v) = \begin{bmatrix} -7 & -33 & -13 \\ 4 & 19 & 8 \end{bmatrix} \begin{bmatrix} z \\ y-z \\ x-y \end{bmatrix} = \begin{bmatrix} -13x - 20y + 26z \\ 8x + 11y - 15z \end{bmatrix} = M_{B'}^{B'}(t(v))$$

EXAMPLE-4 *The set $B = \{e^{3t}, te^{3t}, t^2 e^{3t}\}$ is an ordered basis of the vector space V of all functions $f : R \to R$. Let D be the differential operator on V, that is, $D(f) = \dfrac{df}{dt}$. Find the matrix representation of D relative to basis B.*

SOLUTION In order to find the matrix representation of D relative to basis B, we will have to express images of basis vectors as a linear combination of the basis vectors in B.

We have
$$D(e^{3t}) = 3e^{3t} = 3(e^{3t}) + 0(te^{3t}) + 0(t^2 e^{3t})$$
$$D(te^{3t}) = 3te^{3t} + e^{3t} = 1(e^{3t}) + 3(te^{3t}) + 0(t^2 e^{3t})$$
$$D(t^2 e^{3t}) = 2te^{3t} + 3t^2 e^{3t} = 0(e^{3t}) + 2(te^{3t}) + 3(t^2 e^{3t})$$

$$\therefore \quad M_B^B(D) = \begin{bmatrix} 3 & 0 & 0 \\ 1 & 3 & 0 \\ 0 & 2 & 3 \end{bmatrix}^T = \begin{bmatrix} 3 & 1 & 0 \\ 0 & 3 & 2 \\ 0 & 0 & 3 \end{bmatrix}$$

EXAMPLE-5 *Let $t : R^2 \to R^2$ be a linear transformation given by $t(1,1) = (3,7)$ and $t(1,2) = (5,-4)$. Find the matrix of t relative to the standard basis of R^2.*

SOLUTION The standard basis of R^2 is $B = \{b_1 = (1,0), b_2 = (0,1)\}$
For any $(a,b) \in R^2$, let
$$t(a,b) = (la + mb, pa + qb) \qquad (i)$$

Then,
$$t(1,1) = (3,7) \Rightarrow (l+m, p+q) = (3,7) \Rightarrow l+m = 3, p+q = 7$$
$$t(1,2) = (5,-4) \Rightarrow (l+2m, p+2q) = (5,-4) \Rightarrow l+2m = 5, p+2q = -4$$

Solving these equations, we get
$$l = 1, m = 2, p = 18 \text{ and } q = -11$$

$$\therefore \quad t(a,b) = (a+2b, 18a - 11b)$$
$$\Rightarrow \quad t(1,0) = (1,18) \text{ and } t(0,1) = (2,-11)$$

Clearly,
$$t(1,0) = (1,18) = 1(1,0) + 18(0,1) = 1b_1 + 18b_2$$
$$t(0,1) = (2,-11) = 2(1,0) + (-11)(0,1) = 2b_1 + (-11)b_2$$

$$\therefore \quad M_B^B(t) = \begin{bmatrix} 1 & 18 \\ 2 & -11 \end{bmatrix}^T = \begin{bmatrix} 1 & 2 \\ 18 & -11 \end{bmatrix}$$

EXAMPLE-6 Let $A = \begin{bmatrix} 3 & -2 \\ 4 & -5 \end{bmatrix}$ be a matrix in $R^{2\times 2}$ which defines a linear transformation $t_A : R^2 \to R^2$ by the rule $t_A(X) = AX$ for all $X \in R^2$. Find the matrix of t_A relative to the basis $B = \left\{ \begin{bmatrix} 1 \\ 2 \end{bmatrix}, \begin{bmatrix} 2 \\ 5 \end{bmatrix} \right\}$ for R^2.

SOLUTION Let us first find the coordinates of an arbitrary vector $\begin{bmatrix} a \\ b \end{bmatrix}$ in R^2 with respect to basis B. Let

$$\begin{bmatrix} a \\ b \end{bmatrix} = x \begin{bmatrix} 1 \\ 2 \end{bmatrix} + y \begin{bmatrix} 2 \\ 5 \end{bmatrix}$$

$\Rightarrow \quad x + 2y = a$ and $2x + 5y = b$
$\Rightarrow \quad x = 5a - 2b, \; y = -2a + b$.

$\therefore \quad \begin{bmatrix} a \\ b \end{bmatrix} = (5a - 2b) \begin{bmatrix} 1 \\ 2 \end{bmatrix} + (-2a + b) \begin{bmatrix} 2 \\ 5 \end{bmatrix}$ \hfill (i)

Let $B = \{X_1, X_2\}$, where $X_1 = \begin{bmatrix} 1 \\ 2 \end{bmatrix}$ and $X_2 = \begin{bmatrix} 2 \\ 5 \end{bmatrix}$. Then,

$$t_A(X_1) = AX_1 = \begin{bmatrix} 3 & -2 \\ 4 & -5 \end{bmatrix} \begin{bmatrix} 1 \\ 2 \end{bmatrix} = \begin{bmatrix} -1 \\ -6 \end{bmatrix}$$

$$t_A(X_2) = AX_2 = \begin{bmatrix} 3 & -2 \\ 4 & -5 \end{bmatrix} \begin{bmatrix} 2 \\ 5 \end{bmatrix} = \begin{bmatrix} -4 \\ -17 \end{bmatrix}$$

Now,

$t_A(X_1) = \begin{bmatrix} -1 \\ -6 \end{bmatrix} = 7X_1 - 4X_2$ \hfill [Putting $a = -1$, $b = -6$ in (i)]

$t_A(X_2) = \begin{bmatrix} -4 \\ -17 \end{bmatrix} = 14X_1 - 9X_2$ \hfill [Putting $a = -4$, $b = -17$ in (i)]

$\therefore \quad M_B^B(t_A) = \begin{bmatrix} 7 & -4 \\ 14 & -9 \end{bmatrix}^T = \begin{bmatrix} 7 & 14 \\ -4 & -9 \end{bmatrix}$

EXAMPLE-7 The matrix $A = \begin{bmatrix} 1 & -2 & 1 \\ 3 & -1 & 0 \\ 1 & 4 & -2 \end{bmatrix}$ on R defines a linear transformation $t_A : R^3 \to R^3$ by the rule $t_A(X) = AX$. Find the matrix representing t_A relative to the basis $B = \{X_1, X_2, X_3\}$, where $X_1 = \begin{bmatrix} 1 \\ 1 \\ 1 \end{bmatrix}$, $X_2 = \begin{bmatrix} 0 \\ 1 \\ 1 \end{bmatrix}$, $X_3 = \begin{bmatrix} 1 \\ 2 \\ 3 \end{bmatrix}$

Linear Transformations and Matrices • 367

SOLUTION First of all we find the coordinates of an arbitrary vector, $X = \begin{bmatrix} a \\ b \\ c \end{bmatrix}$ in R^3. So, let

$$X = xX_1 + yX_2 + zX_3$$

$$\Rightarrow \quad \begin{bmatrix} a \\ b \\ c \end{bmatrix} = x \begin{bmatrix} 1 \\ 1 \\ 1 \end{bmatrix} + y \begin{bmatrix} 0 \\ 1 \\ 1 \end{bmatrix} + z \begin{bmatrix} 1 \\ 2 \\ 3 \end{bmatrix}$$

$$\Rightarrow \quad x + 0y + z = a$$
$$x + y + 2z = b$$
$$x + y + 3z = c$$

Solving these equations, we get

$$x = a + b - c, \ y = -a + 2b - c, \ z = c - b$$

$$\therefore \quad \begin{bmatrix} a \\ b \\ c \end{bmatrix} = (a+b-c)\begin{bmatrix} 1 \\ 1 \\ 1 \end{bmatrix} + (-a+2b-c)\begin{bmatrix} 0 \\ 1 \\ 1 \end{bmatrix} + (c-b)\begin{bmatrix} 1 \\ 2 \\ 3 \end{bmatrix} \qquad \text{(i)}$$

Now,

$$t_A(X_1) = AX_1 = \begin{bmatrix} 1 & -2 & 1 \\ 3 & -1 & 0 \\ 1 & 4 & -2 \end{bmatrix} \begin{bmatrix} 1 \\ 1 \\ 1 \end{bmatrix} = \begin{bmatrix} 0 \\ 2 \\ 3 \end{bmatrix}$$

$$t_A(X_2) = AX_2 = \begin{bmatrix} 1 & -2 & 1 \\ 3 & -1 & 0 \\ 1 & 4 & -2 \end{bmatrix} \begin{bmatrix} 0 \\ 1 \\ 1 \end{bmatrix} = \begin{bmatrix} -1 \\ -1 \\ 2 \end{bmatrix}$$

$$t_A(X_3) = AX_3 = \begin{bmatrix} 1 & -2 & 1 \\ 3 & -1 & 0 \\ 1 & 4 & -2 \end{bmatrix} \begin{bmatrix} 1 \\ 2 \\ 3 \end{bmatrix} = \begin{bmatrix} 0 \\ 1 \\ 3 \end{bmatrix}$$

$$\therefore \quad t_A(X_1) = -X_1 + X_2 + X_3 \qquad \text{[Putting } a = 0, \ b = 2, \ c = 3 \text{ in (i)]}$$
$$t_A(X_2) = -4X_1 - 3X_2 + 3X_3 \qquad \text{[Putting } a = -1, \ b = -2, \ c = 2 \text{ in (i)]}$$
$$t_A(X_3) = -2X_1 - X_2 + 2X_3 \qquad \text{[Putting } a = 1, \ b = 2, \ c = 3 \text{ in (i)]}$$

$$\therefore \quad M_B^B(t_A) = \begin{bmatrix} -1 & 1 & -1 \\ -4 & -3 & 3 \\ -2 & -1 & 2 \end{bmatrix}^T = \begin{bmatrix} -1 & -4 & -2 \\ 1 & -3 & -1 \\ -1 & 3 & 2 \end{bmatrix}$$

REMARK. *The matrix representation of $t_A : F^n \to F^n$ relative to the standard basis of F^n is matrix A itself. Therefore, matrix A can also be considered as a linear transformation from $F^{n \times 1}$ to $F^{n \times 1}$ associating each $X \in F^{n \times 1}$ to AX in $F^{n \times 1}$.*

EXAMPLE-8 Let $V = R^{2\times 2}$ denote the vector space of all 2×2 matrices over R and $M = \begin{bmatrix} a & b \\ c & d \end{bmatrix}$ be the matrix in $R^{2\times 2}$. Let $t : V \to V$ be a linear transformation given by $t(A) = MA$ for all $A \in V$. Find the matrix of t relative to the standard basis

$$B = \left\{ E_{11} = \begin{bmatrix} 1 & 0 \\ 0 & 0 \end{bmatrix}, E_{12} = \begin{bmatrix} 0 & 1 \\ 0 & 0 \end{bmatrix}, E_{21} = \begin{bmatrix} 0 & 0 \\ 1 & 0 \end{bmatrix}, E_{22} = \begin{bmatrix} 0 & 0 \\ 0 & 1 \end{bmatrix} \right\}.$$

SOLUTION We have,

$$t(A) = MA, \text{ where } M = \begin{bmatrix} a & b \\ c & d \end{bmatrix}$$

$$\therefore \quad t(E_{11}) = ME_{11} = \begin{bmatrix} a & 0 \\ c & 0 \end{bmatrix} = aE_{11} + 0E_{12} + cE_{21} + 0E_{22}$$

$$t(E_{12}) = ME_{12} = \begin{bmatrix} 0 & a \\ 0 & c \end{bmatrix} = 0E_{11} + aE_{12} + 0E_{21} + cE_{22}$$

$$t(E_{21}) = ME_{21} = \begin{bmatrix} b & 0 \\ d & 0 \end{bmatrix} = bE_{11} + 0E_{12} + dE_{21} + 0E_{22}$$

$$t(E_{22}) = ME_{22} = \begin{bmatrix} 0 & b \\ 0 & d \end{bmatrix} = 0E_{11} + bE_{12} + 0E_{21} + dE_{22}$$

$$\therefore \quad M_B^B(t) = \begin{bmatrix} a & 0 & c & 0 \\ 0 & a & 0 & c \\ b & 0 & d & 0 \\ 0 & b & 0 & d \end{bmatrix}^T = \begin{bmatrix} a & 0 & b & 0 \\ 0 & a & 0 & b \\ c & 0 & d & 0 \\ 0 & c & 0 & d \end{bmatrix}$$

EXERCISE 4.1

1. Let $t : R^2 \to R^2$ be a linear transformation defined by $t(x,y) = (2x + 7y, x - 3y)$.

    (i) Find the matrix A representing t relative to the bases $B = \{(1, 1), (1, 2)\}$ and $B' = \{(1, 4), (1, 5)\}$

    (ii) Find the matrix B representing t relative to the bases B' and B.

2. Find the matrix representation of each of the following linear transformation relative to the standard basis of R^n:

    (i) $t : R^3 \to R^2$ defined by $t(x, y, z) = (2x - 4y + 9z, 5x + 3y - 2z)$
    (ii) $t : R^2 \to R^4$ defined by $t(x, y) = (3x + 4y, 5x - 2y, x + 7y, 4x)$
    (iii) $t : R^4 \to R$ defined by $t(x, y, z, u) = 2x + 3y - 7z - u$

3. Let $t : R^3 \to R^2$ be a linear transformation defined by

$$t(x,y,z) = (2x + y - z, 3x - 2y + 4z)$$

(i) Find the matrix A representing t relative to the ordered bases $B = \{b_1 = (1, 1, 1), b_2 = (1, 1, 0), b_3 = (1, 0, 0)\}$ and $B' = \{b'_1 = (1, 3), b'_2 = (1, 4)\}$ of R^3 and R^2 respectively.

(ii) For any $v = (a, b, c) \in R^3$ verify that
$$M_B^{B'}(t) \, M_B^B(v) = M_{B'}^{B'}(t(v))$$

4. Let $t : R^3 \to R^2$ be a linear transformation defined by $t(x, y, z) = (2x+3y-z, 4x-y+2z)$.

   (i) Find the matrix A of t relative to the bases
   $B = \{b_1 = (1, 1, 0), b_2 = (1, 2, 3), b_3 = (1, 3, 5)\}$ and $B' = \{b'_1 = (1, 2), b'_2 = (2, 3)\}$

   (ii) For any $v = (a, b, c)$ in R^3, find $[v]_B^B$ and $[t(v)]_{B'}^{B'}$.

   (iii) Verify that $A[v]_B^B = [t(v)]_{B'}^{B'}$.

5. Let t be the linear operator on R^2 defined by $t(x, y) = (4x - 2y, 2x + y)$
   Find the matrix of t relative to the basis $B = \{b_1 = (1, 1), b_2 = (-1, 0)\}$.

6. Let t be the linear operator on R^2 defined by $t(x, y) = (2y, 3x - y)$. Find the matrix of t relative to the basis $B = \{b_1 = (1, 3), b_2 = (2, 5)\}$.

7. Let t be the linear operator on R^3 defined by $t(x, y, z) = (3x+z, -2x+y, -x+2y+4z)$.
   Find the matrix of t relative to the ordered basis $B = \{b_1, b_2, b_3\}$, where $b_1 = (1, 0, 1)$, $b_2 = (-1, 2, 1)$, $b_3 = (2, 1, 1)$.

8. Let V be the vector space of functions $f : R \to R$ with the basis $B = \{\sin t, \cos t, e^{3t}\}$ and let $D : V \to V$ be the differential operator defined by $D(f(t)) = \dfrac{d}{dt}(f(t))$. Find the matrix of t relative to the basis B.

9. Consider the linear transformation $t : R^2 \to R^2$ defined by $t(x, y) = (3x + 4y, 2x - 5y)$ and the following bases of R^2: $B = \{b_1, b_2\} = \{(1, 0), (0, 1)\}$ and $B' = \{b'_1, b'_2\} = \{(1, 2), (2, 3)\}$

   (i) Find the matrix A representing t relative to the basis B.

   (ii) Find the matrix B representing t relative to the basis B'.

10. Consider the linear transformation $t : R^2 \to R^2$ defined by
    $$t(x, y) = (2x - 7y, 4x + 3y).$$

    (i) Find the matrix of t relative to the basis $B = \{b_1 = (1, 3), b_2 = (2, 5)\}$

    (ii) Verify that $M_B^B(t) \, M_B^B(v) = M_B^B(t(v))$ for the vector $v = (4, -3)$ in R^2.

11. Let $t : R^3 \to R^3$ be the linear transformation defined by $t(1, 0, 0) = (1, 1, 1)$, $t(0, 1, 0) = (1, 3, 5)$, $t(0, 0, 1) = (2, 2, 2)$. Find the matrix representing t relative to the standard basis of R^3.

12. For each linear transformation $t : R^2 \to R^2$, find the matrix representing t relative to the standard basis of R^2:

(i) t is defined by $t(1, 0) = (3, 5)$ and $t(0, 1) = (7, -2)$

(ii) t is the reflection in R^2 about the line $y = x$

(iii) t is the reflection in R^2 about the line $y = -x$

(iv) t is the rotation in R^2 counterclockwise by $90°$.

(v) t is the rotation in R^2 counterclockwise by $45°$.

13. Let $V = R^{2 \times 2}$ be the vector space of all 2×2 matrices over R and $M = \begin{bmatrix} a & b \\ c & d \end{bmatrix}$ be a matrix in V. Let $B = \{E_{11}, E_{12}, E_{21}, E_{22}\}$ be the usual basis of V. Find the matrix representing each of the following linear transformations relative to the ordered basis B.

    (i) $t : V \to V$ defined by $t(A) = AM$

    (ii) $t : V \to V$ defined by $t(A) = MA - AM$

14. Let V denote the vector space of all functions $f : R \to R$ and let D be the differential operator on V, that is, $D(f) = \dfrac{df}{dt}$. Find the matrix representing D relative to each of the following bases:

    (i) $B = \{e^t, e^{2t}, te^{2t}\}$ (ii) $\{1, t, \sin 3t, \cos 3t\}$ (iii) $\{e^{5t}, te^{5t}, t^2 e^{5t}\}$

15. Let t be a linear transformation from R^3 to R^2 defined by $t(x, y, z) = (x + y, 2z - x)$.

    (i) If B and B' are standard ordered bases of R^3 and R^2 respectively, what is the matrix of t relative to the pair of bases B, B'?

    (ii) If $B = \{b_1 = (1, 0, -1), b_2 = (1, 1, 1), b_3 = (1, 0, 0)\}$ and $B' = \{b'_1 = (0, 1), b'_2 = (1, 0)\}$, what is the matrix of t relative to the pair B, B'?

16. Let t be the linear operation on R^3 defined by
$$t(x, y, z) = (3x + z, -2x + y, -x + 2y + 4z)$$

    (i) What is matrix of t relative to the standard ordered basis for R^3?

    (ii) What is the matrix of t relative to the ordered basis $B = \{b_1, b_2, b_3\}$, where $b_1 = (1, 0, 1), b_2 = (-1, 2, 1), b_3 = (2, 1, 1)$?

17. The matrix $A = \begin{bmatrix} 5 & -1 \\ 2 & 4 \end{bmatrix}$ defines a linear transformation $t_A : R^2 \to R^2$ by the rule
$$t_A(X) = AX \quad \text{for all } X \in R^2.$$

    (i) Find the matrix of t_A relative to the basis $B = \{X_1, X_2\}$ of R^2, where $X_1 = [1, 3]^T$ and $X_2 = [2, 8]^T$.

    (ii) For any $X = [a, b]^T$ in R^2, find $[X]_B^B$ and $[t_A(X)]_B^B$

18. Consider the matrix $A = \begin{bmatrix} 1 & 3 & 1 \\ 2 & 7 & 4 \\ 1 & 4 & 3 \end{bmatrix}$ over field R. Find the matrix representing the linear operator $t_A : R^3 \to R^3$, defined by $t_A(X) = AX$ for all $X \in R^3$, relative to the basis $B = \{[1,1,1]^T, [0,1,1]^T, [1,2,3]^T\}$.

19. Let $\begin{bmatrix} 1 & 3 & 1 \\ 2 & 5 & -4 \\ 1 & -2 & 2 \end{bmatrix}$. Find the matrix that represents the linear operator given by A relative to the basis $B = \{[1,1,0]^T, [0,1,1]^T, [1,2,2]^T\}$.

ANSWERS

1. (i) $\begin{bmatrix} 47 & 85 \\ -38 & 69 \end{bmatrix}$ (ii) $\begin{bmatrix} 71 & 88 \\ -41 & -51 \end{bmatrix}$

2. (i) $\begin{bmatrix} 2 & -4 & 9 \\ 5 & 3 & -2 \end{bmatrix}$ (ii) $\begin{bmatrix} 3 & 4 \\ 5 & -2 \\ 1 & 7 \\ 4 & 0 \end{bmatrix}$ (iii) $[2 \quad 1 \quad -7 \quad -1]$

3. (i) $A = \begin{bmatrix} 3 & 11 & 5 \\ -1 & -8 & -3 \end{bmatrix}$

4. (i) $\begin{bmatrix} -9 & 1 & 4 \\ 7 & 2 & 1 \end{bmatrix}$ (ii) $[v]_B^B = \begin{bmatrix} -a+2b-c \\ 5a-5b+2c \\ -3a+3b-c \end{bmatrix}$ (iii) $[t(v)]_{B'}^{B'} = \begin{bmatrix} 2a-11b+7c \\ 7b-4c \end{bmatrix}$

5. $\begin{bmatrix} 3 & -2 \\ 1 & 2 \end{bmatrix}$ 6. $\begin{bmatrix} -30 & -48 \\ 18 & 29 \end{bmatrix}$ 7. $\begin{bmatrix} \dfrac{17}{4} & \dfrac{35}{4} & \dfrac{11}{2} \\ \dfrac{-3}{4} & \dfrac{15}{4} & \dfrac{-3}{2} \\ \dfrac{1}{2} & \dfrac{-7}{2} & 0 \end{bmatrix}$

8. $\begin{bmatrix} 0 & -1 & 0 \\ 1 & 0 & 0 \\ 0 & 0 & 3 \end{bmatrix}$ 9. (i) $A = \begin{bmatrix} 3 & 4 \\ 2 & -5 \end{bmatrix}$ (ii) $\begin{bmatrix} -49 & -76 \\ 30 & 47 \end{bmatrix}$

10. (i) $\begin{bmatrix} 121 & 201 \\ -70 & -116 \end{bmatrix}$ 11. $\begin{bmatrix} 1 & 1 & 2 \\ 1 & 3 & 2 \\ 1 & 5 & 2 \end{bmatrix}$

12. (i) $\begin{bmatrix} 3 & 7 \\ 5 & -2 \end{bmatrix}$ (ii) $\begin{bmatrix} 0 & 1 \\ 1 & 0 \end{bmatrix}$ (iii) $\begin{bmatrix} 0 & -1 \\ -1 & 0 \end{bmatrix}$ (iv) $\begin{bmatrix} 0 & -1 \\ 1 & 0 \end{bmatrix}$ (v) $\begin{bmatrix} \sqrt{2} & -\sqrt{2} \\ \sqrt{2} & \sqrt{2} \end{bmatrix}$

13. (i) $\begin{bmatrix} a & c & 0 & 0 \\ b & d & 0 & 0 \\ 0 & 0 & a & c \\ 0 & 0 & b & d \end{bmatrix}$ (ii) $\begin{bmatrix} 0 & -c & b & 0 \\ -b & a-d & 0 & b \\ c & 0 & d-a & -c \\ 0 & c & -b & 0 \end{bmatrix}$

14. (i) $\begin{bmatrix} 1 & 0 & 0 \\ 0 & 2 & 1 \\ 0 & 0 & 2 \end{bmatrix}$ (ii) $\begin{bmatrix} 0 & 1 & 0 & 0 \\ 0 & 0 & 0 & 0 \\ 0 & 0 & 0 & -3 \\ 0 & 0 & 3 & 0 \end{bmatrix}$ (iii) $\begin{bmatrix} 5 & 1 & 0 \\ 0 & 5 & 2 \\ 0 & 0 & 5 \end{bmatrix}$

17. (i) $\begin{bmatrix} -6 & 4 \\ -28 & 15 \end{bmatrix}$ (ii) $[X]_B^B = \begin{bmatrix} 4a - b \\ -\dfrac{3}{2}a + \dfrac{1}{2}b \end{bmatrix}$ and $[t_A(X)]_B^B = \begin{bmatrix} 18a - 8b \\ \dfrac{1}{2}(-13a + 7b) \end{bmatrix}$

18. $\begin{bmatrix} 10 & 13 & -5 \\ 8 & 11 & -4 \\ 20 & 28 & -10 \end{bmatrix}$ 19. $\begin{bmatrix} 8 & 1 & 3 \\ 7 & -6 & -11 \\ -5 & 3 & 6 \end{bmatrix}$

4.3 SOME USEFUL THEOREMS

In this section, we will study some useful results on matrices associated to a linear transformation relative to different bases of given vector spaces. These results have been stated and proved in the following theorems.

THEOREM-1 Let $V(F)$ and $V'(F)$ be vector spaces with bases B and B' respectively, and let A be an $m \times n$ matrix over field F. Then the map $M^{-1}(A)$ (relative to bases B and B') of matrix A can be expressed as

$$M^{-1}(A) = L_{B'} o t_A o L_B^{-1}$$

PROOF. Let $B = \{b_1, b_2, \ldots, b_n\}$, $B' = \{b'_1, b'_2, \ldots, b'_m\}$, and let $t = M^{-1}(A)$. Then

$$M_{B'}^{B'}t(v) = M_B^{B'}(t) M_B^B(v) \quad \text{for all } v \in V \qquad \text{[By Theorem 1 on page 359]}$$

$\Rightarrow \quad \begin{bmatrix} \text{Coordinate matrix} \\ \text{of } t(v) \text{ relative to } B' \end{bmatrix} = [\text{Matrix } A] \quad [\text{Coordinate matrix of } v \text{ relative to } B]$

$\Rightarrow \quad L_{B'}^{-1}(t(v)) = t_A(L_B^{-1}(v)) \quad \text{for all } v \in V$

$\Rightarrow \quad (L_{B'}^{-1} o t)(v) = (t_A o L_B^{-1})(v) \quad \text{for all } v \in V$

$\Rightarrow \quad L_{B'}^{-1} o t = t_A o L_B^{-1}$

$\Rightarrow \quad t = L_{B'} o t_A o L_B^{-1}$

$\Rightarrow \quad t = L_{B'} o t_A o L_B^{-1}$

$\Rightarrow \quad M^{-1}(A) = L_{B'} o t_A o L_B^{-1}$

Q.E.D.

THEOREM-2 Let $B = \{b_1, b_2, \ldots, b_n\}$ be a basis for an n dimensional vector space $V(F)$. Then,

(i) $M_B^B(I_V) = I_n$, I_V is the identity map on V.

(ii) $M_B^B(\hat{0}) = O_{n \times n}$, $\hat{0} : V \to V$ is the zero map.

PROOF. (i) Let $X = \{x_1, x_2, \ldots, x_n\}$ be basis dual to B. Then,

$$x_i(b_j) = \delta_{ij} \quad \text{for all } i, j \in \underline{n}.$$

Now,

$$[M_B^B(I_V)]_{ij} = x_i(I_V(b_j)) = x_i(b_j) = \delta_{ij} \quad \text{for all } i, j \in \underline{n}.$$

$\therefore \quad M_B^B(I_V) = I_n$

(ii) $[M_B^B(\hat{0})]_{ij} = x_i(\hat{0}(b_j)) = x_i(0_V) = 0 \quad \text{for all } i, j \in \underline{n}.$

$\therefore \quad M_B^B(\hat{0}) = O_{n \times n}$

Q.E.D.

THEOREM-3 Let V and V' be n and m dimensional vector spaces over a field F. Then for given bases B and B' of V and V' respectively, the function assigning to each linear transformation $t : V \to V'$ its matrix $M_B^{B'}(t)$ is an isomorphism between the vector spaces $\text{Hom}(V, V')$ and the space $F^{m \times n}$ of all $m \times n$ matrices over F. That is, $\text{Hom}(V, V') \cong F^{m \times n}$.

PROOF. Let $B = \{b_1, b_2, \ldots, b_n\}$, $B' = \{b'_1, b'_2, \ldots, b'_m\}$, and let $X = \{x_1, x_2, \ldots, x_n\}$ and $X' = \{x'_1, x'_2, \ldots, x'_m\}$ be bases dual to B and B' respectively.

Consider a mapping $\varphi : \mathrm{Hom}(V, V') \to F^{m \times n}$ given by

$$\varphi(t) = M_B^{B'}(t) \quad \text{for all } t \in \mathrm{Hom}(V, V').$$

Since for each $t \in \mathrm{Hom}(V, V')$ there exists a unique matrix with respect to bases B and B'. Therefore, φ is well defined.

φ *is injective:* Let t_1, t_2 be any two linear transformations. Then,

$$\varphi(t_1) = \varphi(t_2)$$
$\Rightarrow \quad M_B^{B'}(t_1) = M_B^{B'}(t_2)$
$\Rightarrow \quad \left[M_B^{B'}(t_1)\right]_{ij} = \left[M_B^{B'}(t_2)\right]_{ij} \quad \text{for all } i \in \underline{m}, j \in \underline{n}$
$\Rightarrow \quad a_{ij} = b_{ij} \quad \text{for all } i \in \underline{m}, j \in \underline{n}.$
$\Rightarrow \quad a_{ij}\, b_i' = b_{ij}\, b_i' \quad \text{for all } i \in \underline{m}, j \in \underline{n}.$
$\Rightarrow \quad \sum_{i=1}^{m} a_{ij}\, b_i' = \sum_{i=1}^{m} b_{ij}\, b_i' \quad \text{for all } j \in \underline{n}.$
$\Rightarrow \quad t_1(b_j) = t_2(b_j) \quad \text{for all } j \in \underline{n}.$
$\Rightarrow \quad t_1 = t_2 \text{ on } B$
$\Rightarrow \quad t_1 = t_2 \text{ on } V.$ [By Theorem 2 on page 360]

Thus, $\varphi(t_1) = \varphi(t_2) \Rightarrow t_1 = t_2$ for all $t_1, t_2 \in \mathrm{Hom}((V, V'))$.

So, φ is injective.

φ *is surjective:* Let $A = [a_{ij}]$ be an $m \times n$ matrix in $F^{m \times n}$. Then there exists a linear transformation $t : V \to V'$ given by

$$t = \sum_{r=1}^{m} \sum_{s=1}^{n} a_{rs}\, b_r'\, x_s$$

such that $M_B^{B'}(t) = A$, i.e. $\varphi(t) = A$.

Thus, for each $t \in \mathrm{Hom}(V, V')$ there exists $A = M_B^{B'}(t) \in F^{m \times n}$ such that $\varphi(t) = A$.

So, φ is surjective.

φ *is a linear transformation:* Let $t_1, t_2 \in \mathrm{Hom}(V, V')$, and let $\lambda, \mu \in F$. Then, $\lambda t_1 + \mu t_2 \in \mathrm{Hom}(V, V')$ is such that

$$\left[M_B^{B'}(\lambda t_1 + \mu t_2)\right]_{ij} = x_i'[(\lambda t_1 + \mu t_2)(b_j)]$$

$\Rightarrow \quad \left[M_B^{B'}(\lambda t_1 + \mu t_2)\right]_{ij} = x_i'[\lambda t_1(b_j) + \mu t_2(b_j)]$

$$\Rightarrow \quad \left[M_B^{B'}(\lambda t_1 + \mu t_2)\right]_{ij} = x_i'[\lambda t_1(b_j)] + x_i'[\mu t_2(b_j)]$$

$$\Rightarrow \quad \left[M_B^{B'}(\lambda t_1 + \mu t_2)\right]_{ij} = \lambda x_i'[t_1(b_j)] + \mu x_i'[t_2(b_j)]$$

$$\Rightarrow \quad \left[M_B^{B'}(\lambda t_1 + \mu t_2)\right]_{ij} = \lambda \left[M_B^{B'}(t_1)\right]_{ij} + \mu \left[M_B^{B'}(t_2)\right]_{ij}.$$

$$\Rightarrow \quad \left[M_B^{B'}(\lambda t_1 + \mu t_2)\right]_{ij} = \left[\lambda M_B^{B'}(t_1) + \mu M_B^{B'}(t_2)\right]_{ij} \quad \text{for all } i \in \underline{m}, j \in \underline{n}.$$

$$\therefore \quad M_B^{B'}(\lambda t_1 + \mu t_2) = \lambda M_B^{B'}(t_1) + \mu M_B^{B'}(t_2)$$

Thus, for any $t_1, t_2 \in \text{Hom}(V, V')$ and $\lambda, \mu \in F$, we have

$$\varphi(\lambda t_1 + \mu t_2) = M_B^{B'}(\lambda t_1 + \mu t_2) = \lambda M_B^{B'}(t_1) + \mu M_B^{B'}(t_2) = \lambda \varphi(t_1) + \mu \varphi(t_2).$$

So, φ is a linear transformation.

Hence, $\varphi : \text{Hom}(V, V') \to F^{m \times n}$ is an isomorphism. Consequently, we obtain

$$\text{Hom}(V, V') \cong F^{m \times n}$$

Q.E.D.

THEOREM-4 Let V, V' and V'' be finite dimensional vector spaces over the same field F, and let B, B' and B'' be their respective bases. Then for any linear transformations $t : V \to V'$ and $s : V' \to V''$

$$M_B^{B''}(sot) = M_{B'}^{B''}(s) \, M_B^{B'}(t)$$

PROOF. Let $B = \{b_1, b_2, \ldots, b_n\}, B' = \{b_1', b_2', \ldots, b_p'\}$ and $B'' = \{b_1'', b_2'', \ldots, b_m''\}$ be bases for V, V' and V'' respectively. Further, let $X = \{x_1, x_2, \ldots, x_n\}, X' = \{x_1', x_2', \ldots, x_p'\}$ and $X'' = \{x_1'', x_2'', \ldots, x_m''\}$ be bases dual to B, B' and B'' respectively.

Let $M_{B'}^{B''}(s) = P$, $M_B^{B'}(t) = Q$ and $M_B^{B''}(sot) = R$. Then, P, Q and R are $m \times p$, $p \times n$ and $m \times n$ matrices. Clearly, P and Q are conformable for the product PQ and PQ is of order $m \times n$. Thus, PQ and R are of the same order.

Now,

$$[PQ]_{ij} = \sum_{r=1}^{p} [P]_{ir} [Q]_{rj}$$

$$\Rightarrow \quad [PQ]_{ij} = \sum_{r=1}^{p} x_i''(s(b_r')) \cdot x_r'(t(b_j))$$

$$\Rightarrow \quad [PQ]_{ij} = x_i'' \left[\sum_{r=1}^{p} s(b_r') \cdot x_r'(t(b_j))\right]$$

$$\Rightarrow \quad [PQ]_{ij} = x_i'' \left[s\left\{\left(\sum_{r=1}^{p} b_r' \, x_r'\right) t(b_j)\right\}\right]$$

$\Rightarrow \quad [PQ]_{ij} = x_i'' [s [I_{V'}(t(b_j))]] \qquad \qquad [\because \sum_{r=1}^{p} b_r' x_r' = I_{V'}]$

$\Rightarrow \quad [PQ]_{ij} = x_i'' [s(t(b_j))]$

$\Rightarrow \quad [PQ]_{ij} = x_i'' [(sot)(b_j)]$

$\Rightarrow \quad [PQ]_{ij} = \left[M_B^{B''}(sot) \right]_{ij} \quad \text{for all } i \in \underline{m}, j \in \underline{n}.$

$\Rightarrow \quad [PQ]_{ij} = [R]_{ij} \quad \text{for all } i \in \underline{m}, j \in \underline{n}.$

$\therefore \quad PQ = R$

Hence, $M_B^{B''}(sot) = M_{B'}^{B''}(s) \, M_B^{B'}(t)$

Q.E.D.

THEOREM-5 Let V and V' be finite dimensional vector spaces over the same field F. Let $t: V \to V'$ be a linear transformation with the dual $t^*: V'^* \to V^*$. Then,

$$M(t^*) = [M(t)]^T$$

PROOF. Let $B = \{b_1, b_2, \ldots, b_n\}$ and $B' = \{b_1', b_2', \ldots, b_m'\}$ be bases for V and V' respectively. Further, let $X = \{x_1, x_2, \ldots, x_n\}$ and $X' = \{x_1', x_2', \ldots, x_m'\}$ be bases dual to B and B' respectively. Let $M_B^{B'}(t) = [a_{ij}]$. Then,

$$t(b_j) = \sum_{i=1}^{m} a_{ij} b_i' \quad \text{for all } j \in \underline{n} \qquad \qquad (i)$$

and $\quad x_i' ot = \sum_{j=1}^{n} a_{ij} x_j \quad \text{for all } i \in \underline{m}.$

The dual map $t^*: V'^* \to V^*$ is given by

$t^*(f') = f' ot \quad \text{for all } f' \in V'^*$

$\therefore \quad x_i' ot = \sum_{j=1}^{n} a_{ij} x_j \quad \text{for all } i \in \underline{m}.$

$\Rightarrow \quad t^*(x_i') = \sum_{j=1}^{n} a_{ij} x_j \quad \text{for all } i \in \underline{m} \qquad \qquad (ii)$

From (i) and (ii), we observe that the effect on the basis vectors in bases B and X' of V and V'^* respectively. Therefore,

$[M_B^{B'}(t)] = a_{ij} \quad \text{and}, \, [M_{X'}^{X}(t^*)]_{ji} = a_{ij} \quad \text{for all } i \in \underline{m} \text{ and all } j \in \underline{n}.$

$\Rightarrow \quad [M_{X'}^{X}(t^*)]_{ij} = a_{ji} = \left([M_B^{B'}(t)]^T \right)_{ij} \quad \text{for all } i \in \underline{m}, j \in \underline{n}.$

Hence, $\quad M(t^*) = [M(t)]^T$

Q.E.D.

THEOREM-6 Let $V(F)$ be an n-dimensional vector space and B_1, B_2 ordered bases of V. If $t : V \to V$ is an isomorphism, then

$$M_{B_2}^{B_1}(t^{-1}) = \left(M_{B_1}^{B_2}(t)\right)^{-1}$$

PROOF. Since $t : V \to V$ is an isomorphism. Therefore, it is a bijection and hence $t^{-1} : V \to V$ exists such that

$$t o t^{-1} = I_V$$

Now,

$$M_{B_1}^{B_2}(t) \, M_{B_2}^{B_1}(t^{-1}) = M_{B_2}^{B_2}(t o t^{-1}) \qquad \text{[By Theorem 4]}$$

$$\Rightarrow \quad M_{B_1}^{B_2}(t) \, M_{B_2}^{B_1}(t^{-1}) = M_{B_2}^{B_2}(I_V)$$

$$\Rightarrow \quad M_{B_1}^{B_2}(t) \, M_{B_2}^{B_1}(t^{-1}) = I \qquad \text{[By Theorem 2(i)]}$$

$$\Rightarrow \quad \left(M_{B_1}^{B_2}(t)\right)^{-1} = M_{B_1}^{B_2}(t^{-1})$$

Q.E.D.

THEOREM-7 Let V ba a finite dimensional vector space over field F. If $t : V \to V$ is an endomorphism and $B = \{b_1, b_2, \ldots, b_n\}$ is any basis of V, show that $M_B^B(t)$ is a scalar matrix if and only if t is the morphism of "scalar multiple of some scalar $\lambda \in F$".

PROOF. First, let $t : V \to V$ is an endomorphism of "scalar multiple of some scalar $\lambda \in F$" i.e. $t(v) = \lambda v$ for all $v \in V$.

Let $X = \{x_1, x_2, \ldots, x_n\}$ be the basis dual of B. Then,

$$\left(M_B^B(t)\right)_{ij} = x_i(t(b_j))$$

$$\Rightarrow \quad \left(M_B^B(t)\right)_{ij} = x_i(\lambda b_j) = \lambda \, x_i(b_j) = \lambda \delta_{ij} = \begin{cases} \lambda, & \text{for } i = j \\ 0, & \text{for } i \neq j \end{cases}$$

$\therefore \quad M_B^B(t)$ is a scalar matrix.

Conversely, let $A = [a_{ij}]$ be $n \times n$ scalar matrix such that

$$a_{ij} = \begin{cases} 0, & \text{for } i \neq j \\ \lambda, & \text{for } i = j \end{cases}$$

i.e. $a_{ij} = \lambda \delta_{ij}$ for $i, j \in \underline{n}$

Let t be the linear morphism representing A. Then,

$$t = M^{-1}(A) = \sum_{r=1}^{n} \sum_{s=1}^{n} a_{rs} \, b_r \, x_s$$

$\Rightarrow \quad t = \sum_{r=1}^{n} \sum_{s=1}^{n} \lambda \delta_{rs} \, b_r \, x_s$

$\Rightarrow \quad t = \lambda \sum_{r=1}^{n} b_r \, x_r$

$\Rightarrow \quad t = \lambda I_V \qquad\qquad\qquad\qquad\qquad\qquad\qquad \left[\because \ \sum_{r=1}^{n} b_r \, x_r = I_V\right]$

$\Rightarrow \quad t(v) = \lambda I_V(v) = \lambda v \quad \text{for all } v \in V.$

Hence, t is the morphism of "scalar multiple of some scalar λ".

Q.E.D.

4.4 CHANGE OF BASIS

Let V be an n dimensional vector space over a field F and let $B = \{b_1, b_2, \ldots, b_n\}$ be an ordered basis of V. Let $X = \{x_1, x_2, \ldots, x_n\}$ be basis dual to B. Then every vector $v \in V$ can be represented by means of $n \times 1$ matrix $[x_1(v), x_2(v), \ldots, x_n(v)]^T$ which is known as the coordinate vector of v relative to basis B. Also, every linear transformation $t : V \to V$ can be represented by an $n \times n$ matrix $A = M_B^B(t)$ over field F such that

$$a_{ij} = x_i(t(b_j)), \text{ where } A = [a_{ij}] \text{ and } i, j \in \underline{n}$$

In this section, we shall see how the matrix of a linear transformation depends on the choice of bases. Let us now study the effect of change of bases on the matrix of a linear transformation.

Let $B = \{b_1, b_2, \ldots, b_n\}$ and $C = \{c_1, c_2, \ldots, c_n\}$ be ordered bases of vector space V, and let $X = \{x_1, x_2, \ldots, x_n\}$ and $Y = \{y_1, y_2, \ldots, b_n\}$ be bases dual to B and C respectively. Then, $x_i(b_j) = y_i(c_j) = \delta_{ij}$ for all $i, j \in \underline{n}$. Also, $x_i(v)$ and $y_i(v)$ are the ith coordinates of v relative to the bases B and C respectively. We define the change-of-basis matrix $P = [p_{ij}]$ from basis B to basis C as an $n \times n$ matrix over F such that

$p_{ij} = x_i(c_j)$

i.e. $\quad p_{ij} = i$th old coordinate of the jth new basis vector c_j

$\Rightarrow \quad p_{ij} = \left(M_C^B(I_V)(c_j)\right)_{ij}$

$\Rightarrow \quad P = M_C^B(I_V)$

Thus, the change-of-basis matrix P from basis B to basis C is the matrix of the identity map $I_V : V \to V$ relative to these two bases.

THEOREM-1 *Let V be an n dimensional vector space over a field F. Let $B = \{b_1, b_2, \ldots, b_n\}$ and $C = \{c_1, c_2, \ldots, c_n\}$ be ordered bases of V and $X = \{x_1, x_2, \ldots, x_n\}$ and $Y = \{y_1, y_2, \ldots, y_n\}$ be bases dual to B and C respectively. If $P = [p_{ij}]$ is change-of-basis matrix from old basis B to new basis C, then*

$$x_i = \sum_{j=1}^{n} p_{ij} \, y_j, \quad i \in \underline{n} \text{ and, } c_j = \sum_{i=1}^{n} p_{ij} \, b_i, \quad j \in \underline{n}$$

PROOF. Since X and Y are bases dual to B and C respectively. Therefore,

$$\sum_{i=1}^{n} b_i x_i = I_V \text{ and, } \sum_{j=1}^{n} c_j y_j = I_V$$

For any $i \in \underline{n}$, we have

$$x_i(v) = x_i(I_V(v)) \quad \text{for all } v \in V$$

$$\Rightarrow \quad x_i(v) = \left\{ x_i \left(\sum_{j=1}^{n} c_j y_j \right) \right\}(v)$$

$$\Rightarrow \quad x_i(v) = \left\{ \sum_{j=1}^{n} x_i(c_j) y_j \right\}(v)$$

$$\Rightarrow \quad x_i(v) = \left\{ \sum_{j=1}^{n} p_{ij} y_j \right\}(v) \qquad [\because \ x_i(c_j) = p_{ij}]$$

$$\therefore \quad x_i = \sum_{j=1}^{n} p_{ij} y_j \quad \text{for all } i \in \underline{n}$$

Similarly, for any $j \in \underline{n}$, we have

$$c_j = I_V(c_j)$$

$$\Rightarrow \quad c_j = \left(\sum_{i=1}^{n} b_i x_i \right)(c_j) \qquad \left[\because \ I_V = \sum_{i=1}^{n} b_i x_i \right]$$

$$\Rightarrow \quad c_j = \sum_{i=1}^{n} x_i(c_j) b_i$$

$$\Rightarrow \quad c_j = \sum_{i=1}^{n} p_{ij} b_i$$

$$\therefore \quad c_j = \sum_{i=1}^{n} p_{ij} b_i, \quad j \in \underline{n}$$

Q.E.D.

The above theorem suggests the following algorithm for finding out the change matrix from an old basis B of a vector space V to a new basis C.

ALGORITHM

Step I Obtain the old and new bases of the given vector space.

Step II Express each vector in the new basis as a linear combination of the old basis vectors.

Step II Find the transpose of the matrix of coefficients in step II.

The matrix so obtained is the required change-of-basis matrix.

REMARK-1 *The change-of-basis matrix P obtained by above algorithm may be viewed as the matrix whose columns are, respectively, the coordinate column vectors of the new basis vectors*

c_j relative to the old basis B, i.e.
$$P = \left[[c_1]_B^B, [c_2]_B^B, \ldots, [c_n]_B^B \right]$$

or, $$P = \begin{bmatrix} x_1(c_1) & x_1(c_2) & x_1(c_3) & \ldots & x_n(c_1) \\ x_2(c_1) & x_2(c_2) & x_2(c_3) & \ldots & x_n(c_2) \\ x_3(c_1) & x_3(c_2) & x_3(c_3) & \ldots & x_n(c_3) \\ \vdots & \vdots & \vdots & & \\ x_n(c_1) & x_n(c_2) & x_n(c_3) & \ldots & x_n(c_n) \end{bmatrix} = [x_i(c_j)]$$

REMARK-2 Similar to the change-of-basis matrix P from an old basis B to new basis C, we may obtain the change-of-basis matrix Q from new basis C to the old basis B whose columns are, respectively, the coordinate column vectors of the old basis vectors b_j relative to the new basis C, i.e.
$$Q = \left[[b_1]_C^C, [b_2]_C^C, \ldots, [b_n]_C^C \right]$$

REMARK-3 Since bases B and C are linearly independent sets of vectors. Therefore, P and Q are invertible such that $P^{-1} = Q$ and $Q^{-1} = P$.

THEOREM-2 Let $V(F)$ be a vector space and let P be the change-of-basis matrix from a basis B to a basis C. Then, for any vector $v \in V$
$$M_C^B(I_V) \, M_C^C(v) = M_B^B(v)$$

i.e. P (coordinate vector of v relative to C) = (coordinate vector of v relative to B)

PROOF. By Theorem 1 on page 359, for any linear transformation $t : V \to V$ and for any $v \in V$, we have
$$M_B^{B'}(t) \, M_B^B(v) = M_{B'}^{B'} t(v)$$
$\Rightarrow \quad M_C^B(I_V) \, M_C^C(v) = M_B^B(I_V(v)) \quad \text{for all } v \in V$
$\Rightarrow \quad M_C^B(I_V) \, M_C^C(v) = M_B^B(v) \quad \text{for all } v \in V$
$\Rightarrow \quad P \, M_C^C(v) = M_B^B(v) \quad \text{for all } v \in V$

Q.E.D.

REMARK-1 This theorem helps us in finding the coordinate vector of any vector relative to some basis when the coordinate vector of the same vector relative to another basis is given. In fact, it states,

[Coordinate vector of a vector relative to some basis B]
= (Change-of-basis matrix from B to some basis C) (Coordinate vector of the same vector relative to C)

REMARK-2 $P\, M_C^C(v) = M_B^B(v)$

$\Rightarrow \qquad M_C^C(v) = P^{-1} M_B^B(v)$

$\Rightarrow \qquad M_C^C(v) = $ (Change-of-basis matrix from basis C to B) $M_B^B(v)$

ILLUSTRATIVE EXAMPLES

EXAMPLE-1 *Consider the following bases of R^2:*

$$B = \{e_1^{(2)} = (1,0),\ e_2^{(2)} = (0,1)\} \quad \text{and}\ C = \{c_1 = (1,3),\ c_2 = (1,4)\}$$

(i) Find the change-of-basis matrix P from the standard basis B to basis C.

(ii) Find the change-of-basis matrix Q from the basis C to basis B.

(iii) Find the coordinate vectors of $v = (5,-3)$ relative to bases B and C.

SOLUTION (i) To find P, we first express each one of the new basis vectors in terms of old basis vectors. Clearly,

$$c_1 = (1,3) = 1e_1^{(2)} + 3e_2^{(2)}$$
$$c_2 = (1,4) = 1e_1^{(2)} + 4e_2^{(2)}$$

$$\therefore\ P = \begin{bmatrix} 1 & 3 \\ 1 & 4 \end{bmatrix}^T = \begin{bmatrix} 1 & 1 \\ 3 & 4 \end{bmatrix}$$

(ii) In order to find Q, we have to express each one of the old basis vectors as a linear combination of the new basis vectors. For this, let us find the coordinates of an arbitrary vector (a,b) relative to C. Let

$(a,b) = x(1,3) + y(1,4)$

$\Rightarrow\ (a,b) = (x+y,\ 3x+4y)$

$\Rightarrow\ x+y = a,\ 3x+4y = b \quad \Rightarrow \quad x = 4a-b,\ y = -3a+b$

$\therefore\ (a,b) = (4a-b)c_1 + (-3a+b)c_2$ \hfill (i)

$\Rightarrow\ (1,0) = 4c_1 + (-3)c_2$ \hfill [Putting $a=1$ and $b=0$ in (i)]

$(0,1) = (-1)c_1 + (1)c_2$ \hfill [Putting $a=0$ and $b=1$ in (i)]

$$\therefore\ Q = \begin{bmatrix} 4 & -3 \\ -1 & 1 \end{bmatrix}^T = \begin{bmatrix} 4 & -1 \\ -3 & 1 \end{bmatrix}$$

(iii) Since B is the standard basis of R^2. So, coordinate matrix (or coordinate vector) of $v = (5, -3)$ relative to B is $\begin{bmatrix} 5 \\ -3 \end{bmatrix}$.

Now,
$$M_C^C(v) = M_B^C(I_V) \, M_B^B(v)$$

$\Rightarrow \quad M_C^C(v) = \begin{bmatrix} 4 & -1 \\ -3 & 1 \end{bmatrix} \begin{bmatrix} 5 \\ -3 \end{bmatrix} \qquad \left[\because M_B^C(I_V) = Q \text{ and } M_B^B(v) = \begin{bmatrix} 5 \\ -3 \end{bmatrix} \right]$

$\Rightarrow \quad M_C^C(v) = \begin{bmatrix} 23 \\ -18 \end{bmatrix}$

Aliter The coordinate vector of $v = (5, -3)$ relative to basis C can also be obtained by expressing v as a linear combination of c_1 and c_2.

REMARK-3 It is evident from Example 1(i) that the change-of-basis matrix from the standard basis B of F^n to any basis C of F^n is the matrix P whose columns are, respectively the basis vectors of C.

EXAMPLE-2 Consider the following bases of R^2 :
$B = \{b_1 = (1,2), \, b_2 = (3,5)\}, \, C = \{c_1 = (1,-1), \, c_2 = (1,-2)\}$

(i) Find the change-of-basis matrix P from basis B to basis C.

(ii) Find the change-of-basis matrix Q from basis C to basis B.

(iii) Find the coordinate vectors of vector $v = (a,b) \in R^2$ relative to base B and C respectively.

SOLUTION (i) To find the change-of-basis matrix P from basis B to basis C. We have to express each vector in C as a linear combination of basis vectors in B. For this, let us find the coordinate vectors of an arbitrary vector (a,b) in R^2 relative to the old basis B as follows.

Let $\qquad (a,b) = xb_1 + yb_2 \qquad\qquad\qquad\qquad\qquad\qquad\qquad\qquad\qquad\qquad$ (i)

Then,
$\qquad (a,b) = x(1,2) + y(3,5)$

$\Rightarrow \qquad (a,b) = (x+3y, 2x+5y)$

$\Rightarrow \qquad x+3y = a, \, 2x+5y = b$

$\Rightarrow \qquad x = -5a+3b, y = 2a-b$

$\therefore \qquad (a,b) = (-5a+3b)b_1 + (2a-b)b_2 \qquad\qquad$ [Putting the values of x and y in (i)]

$\Rightarrow \qquad c_1 = (1,-1) = -8b_1 + 3b_2$ [Putting $a = 1$, $b = -1$ in (ii)]

and, $\quad c_2 = (1,-2) = -11b_1 + 4b_2$

$\therefore \qquad P = \begin{bmatrix} -8 & 3 \\ -11 & 4 \end{bmatrix}^T = \begin{bmatrix} -8 & -11 \\ 3 & 4 \end{bmatrix}$

(ii) In order to find the change-of-basis matrix Q from basis C to basis B. We have to express each vector in B as a linear combination of basis vectors in C. For this, let us find the coordinate vectors of an arbitrary vector (a,b) in R^2 relative to C as follows:

Let $\quad (a,b) = xc_1 + yc_2$

$\Rightarrow \quad (a,b) = x(1,-1) + y(1,-2)$

$\Rightarrow \quad (a,b) = (x+y, -x-2y)$

$\Rightarrow \quad x+y = a \quad \text{and} \quad -x-2y = b$

$\Rightarrow \quad x = 2a + b \quad y = -a - b$

$\therefore \quad (a,b) = (2a+b)c_1 + (-a-b)c_2$

$\Rightarrow \quad b_1 = (1,2) = 4c_1 - 3c_2$

$\quad b_2 = (3,5) = 11c_1 - 8c_2$

$\therefore \quad Q = \begin{bmatrix} 4 & -3 \\ 11 & -8 \end{bmatrix}^T = \begin{bmatrix} 4 & 11 \\ -3 & -8 \end{bmatrix}$

(iii) From (i) and (ii) the coordinate vectors of $v = (a,b)$ relative to basis B and C is

$M_B^B(v) = \begin{bmatrix} -5a+3b \\ 2a-b \end{bmatrix}$ and, $M_C^C(v) = \begin{bmatrix} 2a+b \\ -a-b \end{bmatrix}$ respectively.

We can also obtain one of the coordinate vectors, if we are known the other coordinate vector and the change-of-basis matrix as given below.

$M_C^C(v) = M_B^C(I_V) \, M_B^B(v)$

$\Rightarrow \quad M_C^C(v) = Q \, M_B^B(v)$ $\qquad \left[\because M_B^C(I_V) = Q \right]$

$\Rightarrow \quad M_C^C(v) = \begin{bmatrix} 4 & 11 \\ -3 & -8 \end{bmatrix} \begin{bmatrix} -5a+3b \\ 2a-b \end{bmatrix} \begin{bmatrix} 2a+b \\ -a-b \end{bmatrix}$

EXAMPLE-3 *Consider the following bases of R^3:*

$B = \{e_1^{(3)}, e_2^{(3)}, e_3^{(3)}\}$ and $C = \{c_1 = (1,0,1), c_2 = (2,1,2), c_3 = (1,2,2)\}$

(i) Find the change-of-basis matrix P from basis B to the basis C.

(ii) Find the change-of-basis matrix Q from basis C to the basis B.

SOLUTION (i) Clearly,

$$c_1 = 1 \times e_1^{(3)} + 0 \times e_2^{(3)} + 1 \times e_3^{(3)}$$

$$c_2 = 2 \times e_1^{(3)} + 1 \times e_2^{(3)} + 2 \times e_3^{(3)}$$

$$c_3 = 1 \times e_1^{(3)} + 2 \times e_2^{(3)} + 2 \times e_3^{(3)}$$

$$\therefore \quad P = \begin{bmatrix} 1 & 0 & 1 \\ 2 & 1 & 2 \\ 1 & 2 & 2 \end{bmatrix}^T = \begin{bmatrix} 1 & 2 & 1 \\ 0 & 1 & 2 \\ 1 & 2 & 2 \end{bmatrix}$$

(ii) Let (a,b,c) in R^3 such that

$$(a,b,c) = x\, c_1 + y\, c_2 + z\, c_3$$
$\Rightarrow \quad (a,b,c) = x(1,0,1) + y(2,1,2) + z(1,2,2)$
$\Rightarrow \quad (a,b,c) = (x+2y+z,\ 0x+y+2z,\ x+2y+2z)$
$\Rightarrow \quad x+2y+z = a,\ y+2z = b,\ x+2y+2z = c$
$\Rightarrow \quad x = -2a-2b+3c,\ y = 2a+b-2c,\ z = c-a$
$\therefore \quad (a,b,c) = (-2a-2b+3c)c_1 + (2a+b-2c)c_2 + (c-a)c_3$ \hfill (i)

$\Rightarrow \quad e_1^{(3)} = -2c_1 + 2c_2 - c_3$ \hfill [Putting $a=1,\ b=0,\ c=0$ in (i)]

$\qquad e_2^{(3)} = -2c_1 + 2c_2 + 0c_3$ \hfill [Putting $a=0,\ b=1,\ c=0$ in (i)]

$\qquad e_3^{(3)} = 3c_1 - 2c_2 + c_3$ \hfill [Putting $a=0,\ b=0,\ c=1$ in (i)]

$$\therefore \quad Q = \begin{bmatrix} -2 & 2 & -1 \\ -2 & 1 & 0 \\ 3 & -2 & 1 \end{bmatrix}^T = \begin{bmatrix} -2 & -2 & 3 \\ 2 & 1 & -2 \\ -1 & 0 & 1 \end{bmatrix}$$

EXAMPLE-4 *The vectors $b_1 = (1,1,0)$, $b_2 = (0,1,1)$, $b_3 = (1,2,2)$ form a basis of R^3. Find the coordinates of an arbitrary vector $v = (a,b,c)$ relative to the ordered basis $B' = \{b_1, b_2, b_3\}$.*

SOLUTION Consider the standard basis $B = \{e_1^{(3)}, e_2^{(3)}, e_3^{(3)}\}$ of R^3.
We have,
$$M_{B'}^{B'}(v) = M_B^{B'}(I_V)\, M_B^B(v)$$
i.e. Coordinate vector of v relative to B' = (Change-of-basis matrix from B' to B) (Coordinate vector of v relative to B)

Clearly, $\qquad M_B^B(v) = \begin{bmatrix} a \\ b \\ c \end{bmatrix} \qquad [\because (a,b,c) = a\, e_1^{(3)} + b\, e_2^{(3)} + c\, e_3^{(3)}]$

Also, $\qquad M_B^{B'}(I_V) = \left(M_{B'}^{B}(I_V)\right)^{-1}$

In order to find the change-of-basis matrix from B to B', we have to express vectors in B' in terms of vectors in B. Clearly,

$$b_1 = 1e_1^{(3)} + 1e_2^{(3)} + 0e_3^{(3)}$$
$$b_2 = 0e_1^{(3)} + 1e_2^{(3)} + 1e_3^{(3)}$$
$$b_3 = 1e_1^{(3)} + 2e_2^{(3)} + 2e_3^{(3)}$$

$$\therefore \quad M_{B'}^{B}(I_V) = \begin{bmatrix} 1 & 1 & 0 \\ 0 & 1 & 1 \\ 1 & 2 & 2 \end{bmatrix}^T = \begin{bmatrix} 1 & 0 & 1 \\ 1 & 1 & 2 \\ 0 & 1 & 2 \end{bmatrix}$$

$$\Rightarrow \quad M_{B}^{B'}(I_v) = \left(M_{B'}^{B}(I_V)\right)^{-1} = \begin{bmatrix} 0 & 1 & -1 \\ -2 & 2 & -1 \\ 1 & -1 & 1 \end{bmatrix}$$

$$\therefore \quad M_{B'}^{B'}(v) = M_{B}^{B'}(I_V) M_{B}^{B}(v)$$

$$\Rightarrow \quad M_{B'}^{B'}(v) = \begin{bmatrix} 0 & 1 & -1 \\ -2 & 2 & -1 \\ 1 & -1 & 1 \end{bmatrix} \begin{bmatrix} a \\ b \\ c \end{bmatrix} = \begin{bmatrix} b-c \\ -2a+2b-c \\ a-b+c \end{bmatrix}$$

Hence, the coordinate vector of $v = (a, b, c)$ relative to basis B' is $\begin{bmatrix} b-c \\ -2a+2b-c \\ a-b+c \end{bmatrix}$.

REMARK. *The coordinates of v can also be obtained by expressing v as a linear combination of b_1, b_2, b_3 as follows:*

$$(a, b, c) = x(1, 1, 0) + y(0, 1, 1) + z(1, 2, 2)$$

This yields a system of simultaneous linear equations, which can be solved by back-substitution or any other method.

EXAMPLE-5 *Show that the vectors $b_1 = (1, 0, -1), b_2 = (1, 2, 1), b_3 = (0, -3, 2)$ form a basis of R^3. Express each of the standard basis vectors $e_1^{(3)}, e_2^{(3)}, e_3^{(3)}$ as a linear combination of b_1, b_2, b_3.*

SOLUTION Let x, y, z be scalars such that

$$xb_1 + yb_2 + zb_3 = O$$

$$\Rightarrow \quad x + y + z = 0, \; 0x + 2y - 3z = 0, \; -x + y + 2z = 0$$

This is a homogeneous system of equations having trivial solution only as the coefficient matrix is non-singular.

$$\therefore \quad x = y = z = 0$$

So, b_1, b_2, b_3 are linearly independent vectors.

Let $C = \{b_1, b_2, b_3\}$ be the given ordered basis and $B = \{e_1^{(3)}, e_2^{(3)}, e_3^{(3)}\}$ be the standard basis of R^3. Then,
$$M_B^C(I_V) \, M_B^B(v) = M_C^C(v) \text{ for all } v \in R^3 \tag{i}$$
We have,
$$M_B^C(I_V) = \left(M_C^B(I_V)\right)^{-1}$$

Now, $\quad M_C^B(I_V) = \begin{bmatrix} 1 & 1 & 0 \\ 0 & 2 & -3 \\ -1 & 1 & 2 \end{bmatrix}$

$\Rightarrow \quad \left(M_C^B(I_V)\right)^{-1} = \dfrac{1}{10} \begin{bmatrix} 7 & -2 & -3 \\ 3 & 2 & 3 \\ 2 & -2 & 2 \end{bmatrix}$

$\Rightarrow \quad M_B^C(I_V) = \dfrac{1}{10} \begin{bmatrix} 7 & -2 & -3 \\ 3 & 2 & 3 \\ 2 & -2 & 2 \end{bmatrix}$

From (i), we get
$$\dfrac{1}{10} \begin{bmatrix} 7 & -2 & -3 \\ 3 & 2 & 3 \\ 2 & -2 & 2 \end{bmatrix} M_B^B(v) = M_C^C(v) \text{ for all } v \in R^3 \tag{ii}$$

Replacing v by $e_1^{(3)}$, we get
$$\dfrac{1}{10} \begin{bmatrix} 7 & -2 & -3 \\ 3 & 2 & 3 \\ 2 & -2 & 2 \end{bmatrix} M_B^B(e_1^{(3)}) = M_C^C(e_1^{(3)})$$

$\Rightarrow \quad M_C^C\left(e_1^{(3)}\right) = \dfrac{1}{10} \begin{bmatrix} 7 & -2 & -3 \\ 3 & 2 & 3 \\ 2 & -2 & 2 \end{bmatrix} \begin{bmatrix} 1 \\ 0 \\ 0 \end{bmatrix} = \begin{bmatrix} \frac{7}{10} \\ \frac{3}{10} \\ \frac{1}{5} \end{bmatrix}$

$\Rightarrow \quad e_1^{(3)} = \dfrac{7}{10} b_1 + \dfrac{3}{10} b_2 + \dfrac{1}{5} b_2$

Similarly, we have
$$M_C^C\left(e_2^{(3)}\right) = \begin{bmatrix} -1/5 \\ 1/5 \\ -1/5 \end{bmatrix} \text{ and } M_C^C\left(e_3^{(3)}\right) = \begin{bmatrix} -3/10 \\ 3/10 \\ 1/5 \end{bmatrix}$$

$\Rightarrow \quad e_2^{(3)} = -\dfrac{1}{5} b_1 + \dfrac{1}{5} b_2 - \dfrac{1}{5} b_3 \quad \text{and} \quad e_3^{(3)} = -\dfrac{3}{10} b_1 + \dfrac{3}{10} b_2 + \dfrac{1}{5} b_3$

4.4.1 MORE THEOREMS ON CHANGE-OF-BASIS

THEOREM-1 *If B_1, B_2 and B_3 are three different bases for a finite dimensional vector space V and if the change-of-basis matrix from a basis B_1 to a basis B_2 is P, that from a basis B_2 to a basis B_3 is Q and that from a basis B_1 to a basis B_3 is R, then $PQ = R$.*

PROOF. Since change-of-basis matrix from one basis to another basis of a vector space $V(F)$ is the matrix of identity map I_V relative to the two bases. Therefore,

$$P = M_{B_2}^{B_1}(I_V), Q = M_{B_3}^{B_2}(I_V) \text{ and } R = M_{B_3}^{B_1}(I_V)$$

$\Rightarrow \quad PQ = M_{B_2}^{B_1}(I_V) \, M_{B_3}^{B_2}(I_V)$

$\Rightarrow \quad PQ = M_{B_3}^{B_1}(I_V \circ I_V)$ [By Theorem 4 on page 375]

$\Rightarrow \quad PQ = M_{B_3}^{B_1}(I_V)$

$\Rightarrow \quad PQ = R.$ Q.E.D.

COROLLARY. *Let B and C be two bases of a finite dimensional vector space $V(F)$. Then the change-of-basis matrix P from basis B to basis C is invertible and its inverse P^{-1} is change-of-basis matrix from C to B.*

PROOF. Let Q be the change-of-basis matrix from basis C to basis B. Then, $Q = M_B^C(I_V)$
It is given that $P = M_C^B(I_V)$.

$\therefore \quad QP = M_B^C(I_V) \, M_C^B(I_V)$ and, $PQ = M_C^B(I_V) \, M_B^C(I_V)$

$\Rightarrow \quad QP = M_C^C(I_V)$ and, $PQ = M_B^B(I_V)$ [By Theorem 4 on page 375]

$\Rightarrow \quad QP = I$ and, $PQ = I$

$\Rightarrow \quad PQ = I = QP$

$\Rightarrow \quad P$ is invertible and $P^{-1} = Q = M_B^C(I_V)$.

Aliter Since a change-of-basis matrix is the matrix of the identity map I_V relative to two different bases of V and every identity map is invertible. Therefore, by Theorem 6 on page 377, P is invertible and $P^{-1} = M_B^C(I_V)$.

It is evident from the above discussion that each pair of bases of a finite dimensional vector space determines a change-of-basis matrix which is invertible. In the following theorem, we will prove that for a given basis B of a finite dimensional vector space V and an invertible matrix P, there exists a new basis C of V such that P is the change-of-basis matrix from B to C. In other words, for a given basis every invertible matrix can be considered as a change-of-basis matrix.

THEOREM-2 *Let $B = \{b_1, b_2, \ldots, b_n\}$ be a basis of an n dimensional vector space $V(F)$ and $P = [p_{ij}]$ be any invertible matrix over F. Then there exists exactly one basis C of V such that matrix P is the change-of-basis matrix from basis B to basis C.*

PROOF. Let $C = \{c_1, c_2, c_3, \ldots, c_n\}$ be such that

$$c_j = \sum_{i=1}^{n} p_{ij} b_i \quad \text{for all } j \in \underline{n} \qquad (i)$$

Clearly, C is a set consisting of n vectors and V is an n dimensional vector space. Therefore, to prove that C is a basis of V, it is sufficient to show that it is linearly independent set of vectors. So, let $\lambda_j, j \in \underline{n}$ be scalars in F such that

$$\sum_{j=1}^{n} \lambda_j c_j = 0$$

$$\Rightarrow \quad \sum_{j=1}^{n} \lambda_j \left(\sum_{i=1}^{n} p_{ij} b_i \right) = 0 \qquad \text{[From (i)]}$$

$$\Rightarrow \quad \sum_{i=1}^{n} \left(\sum_{j=1}^{n} p_{ij} \lambda_j \right) b_i = 0$$

$$\Rightarrow \quad \sum_{j=1}^{n} p_{ij} \lambda_j = 0 \text{ for all } i \in \underline{n} \qquad [\because B \text{ is linearly independent }] \quad (ii)$$

It is given that P is an invertible matrix. So, let $P^{-1} = [q_{ij}]$. Multiplying both sides of (ii) by q_{ri}, we get

$$q_{ri} \left(\sum_{j=1}^{n} p_{ij} \lambda_j \right) = q_{ri} 0 \quad \text{for all } i, r \in \underline{n}$$

$$\Rightarrow \quad \sum_{i=1}^{n} q_{ri} \left(\sum_{j=1}^{n} p_{ij} \lambda_j \right) = 0 \quad \text{for all } r \in \underline{n}$$

$$\Rightarrow \quad \sum_{j=1}^{n} \left\{ \sum_{i=1}^{n} (q_{ri} p_{ij}) \lambda_j \right\} = 0 \quad \text{for all } r \in \underline{n}$$

$$\Rightarrow \quad \sum_{j=1}^{n} \delta_{rj} \lambda_j = 0 \quad \text{for all } r \in \underline{n} \qquad \left[\because QP = I \Rightarrow \sum_{i=1}^{n} q_{ri} p_{ij} = \delta_{rj} \right]$$

$$\Rightarrow \quad \lambda_r = 0 \quad \text{for all } r \in \underline{n}$$

Thus, $\quad \sum_{j=1}^{n} \lambda_j c_j = 0$

$$\Rightarrow \quad \lambda_1 = \lambda_2 = \cdots = \lambda_n = 0$$

$\Rightarrow \quad C$ is a linearly independent set of vectors in V

$\Rightarrow \quad C$ is a basis of V.

Let us now show that P is the change-of-basis matrix from basis B to basis C.

Let $X = \{x_1, x_2, \ldots, x_n\}$ be the basis dual to basis B.

For any $i, j \in \underline{n}$, we have

$$\left(M_C^B(I_V) \right)_{ij} = x_i (I_V(c_j))$$

$$\Rightarrow \quad \left(M_C^B(I_V) \right)_{ij} = x_i(c_j)$$

$$\Rightarrow \quad (M_C^B(I_V))_{ij} = x_i \left(\sum_{r=1}^n p_{rj} b_r \right) \quad \quad \text{[Using (i)]}$$

$$\Rightarrow \quad (M_C^B(I_V))_{ij} = \sum_{r=1}^n p_{rj} x_i(b_r)$$

$$\Rightarrow \quad (M_C^B(I_V))_{ij} = \sum_{r=1}^n p_{rj} \delta_{ir} = p_{ij} = (P)_{ij}$$

$$\therefore \quad M_C^B(I_v) = P$$

Hence, P is the change-of-basis matrix from basis B to basis C.

Q.E.D.

REMARK. *It follows from the above theorem that corresponding to every basis B of a finite dimensional vector space $V(F)$ and an invertible matrix P, there exists a basic C of vector space V such that $M_C^B(I_V) = P$. Thus, if B is a given basis of V, then every invertible matrix $P = [p_{ij}]$ determines a basis $C = \{c_1, c_2, \ldots, c_n\}$, where $c_j = \sum_{i=1}^n p_{ij} b_j$, of V. In other words, there is one to one correspondence, i.e. a bijection between the set of all bases of V and the set of all $n \times n$ invertible matrices over F.*

THEOREM-3 *Let V and V' be vector spaces over a field F with dimensions n and m respectively. Let A be the matrix of a linear transformation $t : V \to V'$ with respect to a given pair of bases B and B' of V and V' respectively. Then, the matrix A' of t with respect to a new pair of bases C and C' of V and V' respectively is given by $A' = Q^{-1}AP$, where P is the change-of-basis matrix from basis B to basis C and Q is that from B' to C'.*

PROOF. It is given that P and Q are change-of-basis matrices from bases B to C and B' to C' respectively. Therefore,

$$P = M_C^B(I_V) \quad \text{and} \quad Q = M_{C'}^{B'}(I_{V'})$$

Also, we have

$$M_B^{B'}(t) = A \quad \text{and} \quad M_C^{C'}(t) = A'$$

$$\therefore \quad Q^{-1}AP = M_{B'}^{C'}(I_{V'}) \, M_B^{B'}(t) \, M_C^B(I_V) \quad \quad [\because Q = M_{C'}^{B'}(I_{V'}) \Rightarrow Q^{-1} = M_{B'}^{C'}(I_{V'})]$$

$$\Rightarrow \quad Q^{-1}AP = M_{B'}^{C'}(I_{V'}) \left(M_B^{B'}(t) \, M_C^B(I_V) \right) \quad \quad \text{[By associativity of matrix multiplication]}$$

$$\Rightarrow \quad Q^{-1}AP = M_{B'}^{C'}(I_{V'}) \, M_C^{B'}(to \, I_V)$$

$$\Rightarrow \quad Q^{-1}AP = M_{B'}^{C'}(I_{V'}) \, M_C^{B'}(t)$$

$$\Rightarrow \quad Q^{-1}AP = M_{B'}^{C'}(I_{V'}) \, M_C^{B'}(t)$$

$$\Rightarrow \quad Q^{-1}AP = M_C^{C'}(I_{V'} o t)$$

$$\Rightarrow \quad Q^{-1}AP = M_C^{C'}(t)$$

$$\Rightarrow \quad Q^{-1}AP = A' \quad \text{or}, A' = Q^{-1}AP$$

Q.E.D.

COROLLARY. *Let V be a finite dimensional vector space over a field F and $t: V \to V$ be an endomorphism. Then any two matrices A and A' of t relative to two different bases B and B' respectively of V are related by $A' = P^{-1}AP$ where P is the change-of-basis matrix from basis B to basis B'.*

PROOF. We have,

$$A = M_B^B(t), \; A' = M_{B'}^{B'}(t), \quad \text{and} \quad P = M_{B'}^B(I_V)$$

$\therefore \quad P^{-1}AP = M_B^{B'}(I_V) M_B^B(t) \, M_{B'}^B(I_V)$

$\Rightarrow \quad P^{-1}AP = M_B^{B'}(I_V) \left(M_B^B(t) \, M_{B'}^B(I_V) \right)$ [By associativity of matrix multiplication]

$\Rightarrow \quad P^{-1}AP = M_B^{B'}(I_V) \, M_{B'}^B(to \, I_V)$

$\Rightarrow \quad P^{-1}AP = M_B^{B'}(I_V) \, M_{B'}^B(t)$

$\Rightarrow \quad P^{-1}AP = M_{B'}^{B'}(I_V \, ot)$

$\Rightarrow \quad P^{-1}AP = M_{B'}^{B'}(t)$

$\Rightarrow \quad P^{-1}AP = A'$

Q.E.D.

ILLUSTRATIVE EXAMPLES

EXAMPLE-1 *Consider the vector space R^2. Let $t: R^2 \to R^2$ be a linear transformation such that the matrix of t relative to ordered basis $B = \left\{ e_1^{(2)}, e_2^{(2)} \right\}$ is $A = \begin{bmatrix} 1 & 1 \\ 1 & 1 \end{bmatrix}$. Find the matrix of t relative to the ordered basis $C = \{ c_1 = (1,1), c_2 = (1,-1) \}$.*

SOLUTION Let P be the change-of-basis matrix from basis B to basis C and A' be the matrix of t relative to ordered basis C. Then,

$$A' = P^{-1}AP$$

In order to find P, we express each vector in basis C as a linear combination of basis vectors in B.

Clearly,

$$c_1 = (1, 1) = 1 e_1^{(2)} + 1 e_2^{(2)}$$

and

$$c_2 = (1, -1) = 1 e_1^{(2)} + (-1) e_2^{(2)}$$

$\therefore \quad P = \begin{bmatrix} 1 & 1 \\ 1 & -1 \end{bmatrix}^T = \begin{bmatrix} 1 & 1 \\ 1 & -1 \end{bmatrix}$

$$\Rightarrow \qquad P^{-1} = -\frac{1}{2}\begin{bmatrix} -1 & -1 \\ -1 & 1 \end{bmatrix} = \frac{1}{2}\begin{bmatrix} 1 & 1 \\ 1 & -1 \end{bmatrix}$$

Hence, $\qquad A' = P^{-1}AP = \frac{1}{2}\begin{bmatrix} 1 & 1 \\ 1 & -1 \end{bmatrix}\begin{bmatrix} 1 & 1 \\ 1 & 1 \end{bmatrix}\begin{bmatrix} 1 & 1 \\ 1 & -1 \end{bmatrix} = \frac{1}{2}\begin{bmatrix} 1 & 1 \\ 1 & -1 \end{bmatrix}\begin{bmatrix} 2 & 0 \\ 2 & 0 \end{bmatrix} = \begin{bmatrix} 2 & 0 \\ 0 & 0 \end{bmatrix}$

EXAMPLE-2 Consider the linear transformation $t : R^2 \to R^2$ defined by $t(x,y) = (5x-y, 2x+y)$ and the following bases of R^2:

$$B = \{e_1^{(2)} = (1,0),\ e_2^{(2)} = (0,1)\} \text{ and } C = \{c_1 = (1,4),\ c_2 = (2,7)\}$$

(i) Find the change-of-basis matrix P from B to C and the change-of-basis matrix Q from C to B.
(ii) Find the matrix A representing t relative to basis B.
(iii) Find the matrix A' representing t relative to basis C.

SOLUTION (i) To find the change-of-basis matrix from B to C, we first express each basis vector in C as a linear combination of basis vectors in B.

Clearly, $\quad c_1 = (1,4) = 1e_1^{(2)} + 4e_2^{(2)}$ and, $c_2 = (2,7) = 2e_1^{(2)} + 7e_2^{(2)}$

$$\therefore \qquad P = \begin{bmatrix} 1 & 4 \\ 2 & 7 \end{bmatrix}^T = \begin{bmatrix} 1 & 2 \\ 4 & 7 \end{bmatrix}$$

We know that $\qquad Q = P^{-1}$

$$\therefore \qquad Q = \frac{1}{-1}\begin{bmatrix} 7 & -2 \\ -4 & 1 \end{bmatrix} = \begin{bmatrix} -7 & 2 \\ 4 & -1 \end{bmatrix}$$

(ii) In order to find $A = M_B^B(t)$, we express $t(e_1^{(2)})$ and $t(e_2^{(2)})$ as a linear combination of basis vectors in B. We have,

$t(x,y) = (5x-y,\ 2x+y)$

$\therefore \quad t(e_1^{(2)}) = t(1,0) = (5,2) = 5e_1^{(2)} + 2e_2^{(2)}$

$t(e_2^{(2)}) = t(0,1) = (-1,1) = (-1)e_1^{(2)} + 1e_2^{(2)}$

$$\therefore \qquad A = M_B^B(t) = \begin{bmatrix} 5 & 2 \\ -1 & 1 \end{bmatrix}^T = \begin{bmatrix} 5 & -1 \\ 2 & 1 \end{bmatrix}$$

(iii) By above corollary, we have

$A' = P^{-1}AP$

$$\Rightarrow \qquad A' = \begin{bmatrix} -7 & 2 \\ 4 & -1 \end{bmatrix}\begin{bmatrix} 5 & -1 \\ 2 & 1 \end{bmatrix}\begin{bmatrix} 1 & 2 \\ 4 & 7 \end{bmatrix} = \begin{bmatrix} 5 & 1 \\ -2 & 1 \end{bmatrix}$$

REMARK. A' can also be obtained by expressing $t(c_1)$ and $t(c_2)$ as a linear combination of basis vectors in C and then taking the transpose of the coefficient matrix.

EXAMPLE-3 Consider the following bases of R^3:

$$B = \{e_1^{(3)}, e_2^{(3)}, e_3^{(3)}\} \text{ and } C = \{c_1 = (1,0,1),\ c_2 = (-1,2,1),\ c_3 = (2,1,1)\}.$$

Let $t: R^3 \to R^3$ be the linear transformation defined by $t(x,y,z) = (3x+z,\ -2x+y,\ -x+2y+4z)$

(i) Find the change-of-basis matrix P from basis B to basis C.
(ii) Find the matrix A representing t relative to basis B.
(iii) Find the matrix A' representing t relative to basis C.

SOLUTION (i) To find the change-of-basis matrix from B to C, we have to express vectors in C in terms of vectors in B as follows:

$$c_1 = (1,0,1) = 1e_1^{(3)} + 0e_2^{(3)} + 1e_3^{(3)}$$
$$c_2 = (-1,2,1) = (-1)e_1^{(3)} + 2e_2^{(3)} + 1e_3^{(3)}$$
$$c_3 = (2,1,1) = 2e_1^{(3)} + 1e_2^{(3)} + 1e_3^{(3)}$$

$$\therefore \quad P = \begin{bmatrix} 1 & 0 & 1 \\ -1 & 2 & 1 \\ 2 & 1 & 1 \end{bmatrix}^T = \begin{bmatrix} 1 & -1 & 2 \\ 0 & 2 & 1 \\ 1 & 1 & 1 \end{bmatrix}$$

(ii) We have,

$$t(e_1^{(3)}) = t(1,0,0) = (3,-2,-1) = 3e_1^{(3)} + (-2)e_2^{(3)} + (-1)e_3^{(3)}$$
$$t(e_2^{(3)}) = t(0,1,0) = (0,1,2) = 0e_1^{(3)} + 1e_2^{(3)} + 2e_3^{(3)}$$
$$t(e_3^{(3)}) = t(0,0,1) = (1,0,4) = 1e_1^{(3)} + 0e_2^{(3)} + 4e_3^{(3)}$$

$$\therefore \quad A = \begin{bmatrix} 3 & -2 & -1 \\ 0 & 1 & 2 \\ 1 & 0 & 4 \end{bmatrix}^T = \begin{bmatrix} 3 & 0 & 1 \\ -2 & 1 & 0 \\ -1 & 2 & 4 \end{bmatrix}$$

(iii) Clearly,
$$A' = P^{-1}AP$$

$$\Rightarrow \quad A' = -\frac{1}{4} \begin{bmatrix} 1 & 3 & -5 \\ 1 & -1 & -1 \\ -2 & -2 & 2 \end{bmatrix} \begin{bmatrix} 3 & 0 & 1 \\ -2 & 1 & 0 \\ -1 & 2 & 4 \end{bmatrix} \begin{bmatrix} 1 & -1 & 2 \\ 0 & 2 & 1 \\ 1 & 1 & 1 \end{bmatrix}$$

$$\Rightarrow \quad A' = -\frac{1}{4} \begin{bmatrix} 1 & 3 & -5 \\ 1 & -1 & -1 \\ -2 & -2 & 2 \end{bmatrix} \begin{bmatrix} 4 & -2 & 7 \\ -2 & 4 & -3 \\ 3 & 9 & 4 \end{bmatrix} = \begin{bmatrix} \dfrac{17}{4} & \dfrac{35}{4} & \dfrac{11}{2} \\ -\dfrac{3}{4} & \dfrac{15}{4} & -\dfrac{3}{2} \\ -\dfrac{1}{2} & -\dfrac{7}{2} & 0 \end{bmatrix}$$

Aliter Matrix A' can also be obtained by expressing $t(c_1), t(c_2), t(c_3)$ in terms of basis vectors c_1, c_2, c_3.

EXERCISE 4.2

1. Consider the bases $B = \{(1,2),(2,3)\}$ and $C = \{(1,3),(1,4)\}$ of R^2. Find the change-of-basis matrix (i) P from B to C (ii) Q from C to B. Also, show that $PQ = I_2$.
2. Consider the bases $B = \{b_1 = (1,-2), b_2 = (3,-4)\}$ and $C = \{c_1 = (1,3), c_2 = (3,8)\}$ of R^2.
   - (i) Find the coordinate vector of $v = (a,b)$ relative to the basis B.
   - (ii) Find the change-of-basis matrix P from B to C.
   - (iii) Find the coordinate vector of $v = (a,b)$ relative to the basis C.
   - (iv) Find the change-of-basis matrix Q from C to B.
   - (v) Verify that $Q = P^{-1}$.
   - (vi) For any vector $v = (a,b)$ in R^2, show that $P^{-1} M_B^B(v) = M_C^C(v)$.
3. Find the change-of-basis matrix P from the standard basis B of R^2 to a basis C, the change-of-basis matrix Q from C back to B, and the coordinates of $v = (a,b)$ relative to C, for each of the following bases C:
   - (i) $C = \{c_1 = (1,2), c_2 = (3,5)\}$
   - (ii) $C = \{c_1 = (2,5), c_2 = (3,7)\}$
   - (iii) $C = \{(1,-3),(3,-8)\}$
   - (iv) $C = \{(2,3),(4,5)\}$
4. Consider the linear transformation $t : R^2 \to R^2$ defined by $t(x,y) = (5x+y, 3x-2y)$ and the following bases of R^2:
   $$B = \{(1,2),\ (2,3)\} \text{ and } C = \{(1,3),(1,4)\}$$
   - (i) Find the matrix A representing t relative to the basis B.
   - (ii) Find the matrix B representing t relative to the basis C.
   - (iii) Find the change-of-basis matrix P from B to C.
   - (iv) How are A and B related?
5. Find the change-of-basis matrix from the standard basis B of R^3 to a basis C, the change-of-basis matrix Q from C to B, and the coordinates of $v = (a,b,c)$ relative to C, where C is the ordered basis consisting of the vectors:
   - (i) $c_1 = (1,1,0),\ c_2 = (0,1,2),\ c_3 = (0,1,1)$
   - (ii) $c_1 = (1,0,1),\ c_2 = (1,1,2),\ c_3 = (1,2,4)$
   - (iii) $c_1 = (1,2,1),\ c_2 = (1,3,4),\ c_3 = (2,5,6)$
6. Let V and V' be finite dimensional vector spaces over field F and $t : V \to V'$ be a linear transformation. Let P be the change-of-basis matrix from a basis B to a basis C in V and let Q be the change-of-basis matrix from a basis B' to a basis C' in V'. Then,
$$M_C^{C'}(t) = Q^{-1} M_B^{B'}(t) P$$

7. The vectors $v_1 = (1,2,0)$, $v_2 = (1,3,2), v_3 = (0,1,3)$ form a basis B of R^3. Find:

   (i) the change-of-basis matrix P from the standard basis $B = \{e_1^{(3)}, e_2^{(3)}, e_3^{(3)}\}$ to basis $B' = \{v_1, v_2, v_3\}$.

   (ii) the change-of-basis matrix Q from B' back to B.

8. Let V be a finite dimensional vector space and let P be the change-of-basis matrix from a basis B to a basis C for V. Prove that for any linear operator t on V
$$M_C^C(t) = P^{-1} M_B^B(t) P.$$

9. Let $V(F)$ be a finite dimensional vector space and let P be the change-of-basis matrix from a basis B to a basis C for V. Then, $P^{-1} M_B^B(v) = M_C^C(v)$ for all $v \in V$.

ANSWERS

1. $P = \begin{bmatrix} 3 & 5 \\ -1 & -2 \end{bmatrix}$, $Q = \begin{bmatrix} 2 & 5 \\ -1 & -3 \end{bmatrix}$

2. (i) $\begin{bmatrix} -2a - \dfrac{3}{2}b \\ a + \dfrac{1}{2}b \end{bmatrix}$ (ii) $P = \begin{bmatrix} -\dfrac{13}{2} & -18 \\ \dfrac{5}{2} & 7 \end{bmatrix}$ (iii) $\begin{bmatrix} -8a + 3b \\ 3a - b \end{bmatrix}$ (iv) $\begin{bmatrix} -14 & -36 \\ 5 & 13 \end{bmatrix}$

3. (i) $\begin{bmatrix} 1 & 3 \\ 2 & 5 \end{bmatrix}$, $\begin{bmatrix} -5 & 3 \\ 2 & -1 \end{bmatrix}$, $\begin{bmatrix} -5a + 3b \\ 2a - b \end{bmatrix}$

   (ii) $\begin{bmatrix} 1 & 3 \\ -3 & -8 \end{bmatrix}$, $\begin{bmatrix} -8 & -3 \\ 3 & 1 \end{bmatrix}$, $\begin{bmatrix} -8a - 3b \\ 3a + b \end{bmatrix}$

   (iii) $\begin{bmatrix} 2 & 3 \\ 5 & 7 \end{bmatrix}$, $\begin{bmatrix} -7 & 5 \\ 3 & -2 \end{bmatrix}$, $\begin{bmatrix} -7a + 3b \\ 5a - 2b \end{bmatrix}$

   (iv) $\begin{bmatrix} 2 & 4 \\ 3 & 5 \end{bmatrix}$, $\begin{bmatrix} -\dfrac{5}{2} & 2 \\ \dfrac{3}{2} & -1 \end{bmatrix}$, $\begin{bmatrix} -\dfrac{5}{2}a + 2b \\ \dfrac{3}{2}a - b \end{bmatrix}$

4. (i) $\begin{bmatrix} -23 & -39 \\ 15 & 26 \end{bmatrix}$ (ii) $\begin{bmatrix} 35 & 41 \\ -27 & -32 \end{bmatrix}$ (iii) $\begin{bmatrix} 3 & 5 \\ -1 & -2 \end{bmatrix}$ (iv) $B = P^{-1} A P$

5. P is the matrix whose columns are c_1, c_2, c_3, $Q = P^{-1}$, $M_C^C(v) = Q \begin{bmatrix} a \\ b \\ c \end{bmatrix}$

(i) $Q = \begin{bmatrix} 1 & 0 & 0 \\ 1 & -1 & 1 \\ -2 & 2 & -1 \end{bmatrix}$, $M_C^C(v) = \begin{bmatrix} a \\ a-b+c \\ -2a+2b-c \end{bmatrix}$

(ii) $Q = \begin{bmatrix} 0 & -2 & 1 \\ 2 & 3 & -2 \\ -1 & -1 & 1 \end{bmatrix}$, $M_C^C(v) = \begin{bmatrix} -2b+c \\ 2a+3b-2c \\ -a-b+c \end{bmatrix}$

(iii) $Q = \begin{bmatrix} -2 & 2 & -1 \\ -7 & 4 & -1 \\ 5 & -3 & 1 \end{bmatrix}$, $M_C^C(v) = \begin{bmatrix} -2a+2b-c \\ -7a+4b-c \\ 5a-3b+c \end{bmatrix}$

7. (i) $P = \begin{bmatrix} 1 & 1 & 0 \\ 2 & 3 & 1 \\ 0 & 2 & 3 \end{bmatrix}$ (ii) $Q = \begin{bmatrix} 7 & -3 & 1 \\ -6 & 3 & -1 \\ 4 & -2 & 1 \end{bmatrix}$

4.5 SIMILARITY AND EQUIVALENCY OF MATRICES

SIMILAR MATRICES *Let A and B be square matrices of order n over the same field F. Then B is said to be similar to A over field F if there is an $n \times n$ invertible matrix P over F such that $B = P^{-1}AP$.*

If matrix B is similar to matrix A, then we also say that B is obtained from A by a similarity transformation.

THEOREM-1 *The relation of similarity is an equivalence relation in the set of all $n \times n$ square matrices over a field F.*

PROOF. We observe the following properties of the relation of similarity:
Reflexivity: Let A be any $n \times n$ square matrix over field F. Then,
$\quad A = I^{-1}AI$, where I is $n \times n$ identity matrix over F.
$\therefore \quad A$ is similar to itself as I is invertible.
So, the relation of similarity is reflexive.
Symmetry: Let A and B be two $n \times n$ square matrices over F such that A is similar to B. Then, there exists an invertible matrix P over F such that
$\quad\quad A = P^{-1}BP$
$\Rightarrow \quad PAP^{-1} = P(P^{-1}BP)P^{-1}$
$\Rightarrow \quad PAP^{-1} = IBI$
$\Rightarrow \quad B = (P^{-1})^{-1} AP^{-1}$
$\Rightarrow \quad B$ is similar to A $\quad\quad\quad\quad$ [$\because$ P is invertible $\Rightarrow P^{-1}$ is invertible]
So, the relation of similarity is symmetric.

Transitivity: Let A, B, C be $n \times n$ square matrices over F such that A is similar to B and B is similar to C. Then, there exist $n \times n$ invertible matrices P and Q over F such that

$$A = P^{-1}BP \text{ and } B = Q^{-1}CQ$$
$\Rightarrow \quad A = P^{-1}(Q^{-1}CQ)P$
$\Rightarrow \quad A = (P^{-1}Q^{-1})C(QP)$
$\Rightarrow \quad A = (QP)^{-1}C(QP) \qquad\qquad [\because \ (QP)^{-1} = P^{-1}Q^{-1}]$
$\Rightarrow \quad A$ is similar to C

So, the relation of similarity is transitive.

Hence, the relation of similarity is an equivalence relation on the set of all $n \times n$ matrices over F.

Q.E.D.

THEOREM-2 *The only matrix similar to the identity matrix I over a field F is I itself.*

PROOF. The identity matrix over a field is invertible such that it is inverse of itself i.e. $I^{-1} = I$.
Also,
$$I = I^{-1} I I$$
$\Rightarrow \quad I$ is similar to itself.

If possible, let A be a matrix over F such that A is similar to I. Then, there exists an invertible matrix P such that

$$A = P^{-1} I P \Rightarrow A = P^{-1} P \Rightarrow A = I$$

Hence, the only matrix similar to I is I itself.

Q.E.D.

THEOREM-3 *Let V be a finite dimensional vector space over a field F and $t : V \to V$ be an endomorphism. Then t is represented by matrices A and A', relative to possibly different bases of V iff A' is similar to A.*

OR

Let V be a vector space over a field F. Two matrices over F are similar iff they are represented by the same linear transformation relative to possibly different bases of V.

PROOF. Let matrices A and A' be represented by a linear transformation $t : V \to V$ relative to bases B and B' respectively, i.e. $A = M_B^B(t)$ and $A' = M_{B'}^{B'}(t)$

We have to prove that A and A' are similar matrices.

Let P be the change-of-basis matrix from basis B to B'. Then,

$$P = M_{B'}^B(I_V) \text{ and } P^{-1} = M_B^{B'}(I_V).$$

$$\therefore \quad P^{-1}AP = M_B^{B'}(I_V)\, M_B^B(t)\, M_{B'}^B(I_V)$$
$$\Rightarrow \quad P^{-1}AP = M_B^{B'}(I_V)\, M_{B'}^B(toI_V)$$
$$\Rightarrow \quad P^{-1}AP = M_B^{B'}(I_V)\, M_{B'}^B(t)$$
$$\Rightarrow \quad P^{-1}AP = M_{B'}^{B'}(I_V ot) = M_{B'}^{B'}(t) = A'$$

Thus, if A and A' be represented by a linear transformation $t : V \to V$ relative to bases B and B' respectively, then there exists an invertible matrix $P = M_{B'}^B(I_V)$ such that $A' = P^{-1}AP$

$\therefore \quad A' = M_{B'}^{B'}(t)$ and $A = M_B^B(t)$ are similar matrices.

Conversely, A and A' be two square matrices of the same order over field F such that A' is similar to A. Then we have to prove that A and A' are represented by the same linear transformation. Let A be the matrix of t relative to some basis B of vector space V.

Now,

A' is similar to $A \Rightarrow$ There exists an invertible matrix P such that $A' = P^{-1}AP$.

Since, B is a basis of vector space V and P is an invertible matrix.

So, by Theorem 2 on page 387, there exists exactly one basis C of V such that P is the change-of-basis matrix from B to C, i.e. $P = M_C^B(I_V)$.

Now,

$$A' = P^{-1}AP$$
$$\Rightarrow \quad A' = M_B^C(I_V)\, M_B^B(t)\, M_C^B(I_V) \qquad [\because\ P = M_C^B(I_V) \Rightarrow P^{-1} = M_B^C(I_V)]$$
$$\Rightarrow \quad A' = M_B^C(I_V)\, M_C^B(toI_V)$$
$$\Rightarrow \quad A' = M_B^C(I_V)\, M_C^B(t)$$
$$\Rightarrow \quad A' = M_C^C(I_V ot)$$
$$\Rightarrow \quad A' = M_C^C(t)$$
$$\Rightarrow \quad A' \text{ is the matrix of } t \text{ relative to some different basis } C.$$

Q.E.D.

EQUIVALENT MATRICES Let A and A' be two $m \times n$ matrices over a field F. Then matrix A' is said to be equivalent to matrix A if there exist invertible square matrices P and Q over F such that $A' = Q^{-1}AP$

Clearly, Q is an $m \times m$ matrix and P is an $n \times n$ matrix over F.

If matrix A' is equivalent to matrix A, then we also say A' is obtained from A by equivalency transformation.

THEOREM-4 *The relation of equivalency of matrices over a field F is an equivalence relation.*

PROOF. We observe the following properties of the relation of equivalency on the set of all matrices over a field F.

Reflexivity: Let A be any $m \times n$ matrix over field F, and I_m and I_n be $m \times m$ and $n \times n$ identity matrices over F. Then,

$$A = I_m A I_n$$
$\Rightarrow \quad A = (I_m)^{-1} A I_n \qquad [\because (I_m)^{-1} = I_m]$
$\Rightarrow \quad A$ is equivalent to itself $\qquad [\because I_m$ and I_n are invertible$]$

So, the relation of equivalency is a reflexive relation.

Symmetry: Let A and B be two $m \times n$ matrices over F such that A is equivalent to B. Then, there exist invertible square matrices P and Q over F such that

$$A = Q^{-1}BP$$
$\Rightarrow \quad QAP^{-1} = Q(Q^{-1}BP)P^{-1}$
$\Rightarrow \quad B = QAP^{-1}$
$\Rightarrow \quad B = (Q^{-1})^{-1} AP^{-1}$
$\Rightarrow \quad B = (Q')^{-1} AP'$, where $Q' = Q^{-1}$ and $P' = P^{-1}$
$\Rightarrow \quad B$ is equivalent to A

So, the relation of equivalency is symmetric relation.

Transitivity: Let A, B, C be $m \times n$ matrices over field F such that A is equivalent to B and B is equivalent to C. Then there exist invertible matrices P, Q, R and S such that

$$A = Q^{-1}BP \text{ and } B = S^{-1}CR$$
$\Rightarrow \quad A = Q^{-1}(S^{-1}CR)P$
$\Rightarrow \quad A = (Q^{-1}S^{-1})C(RP)$
$\Rightarrow \quad A = (SQ)^{-1}C(RP)$
$\Rightarrow \quad A$ is equivalent to C

So, the relation of equivalency is transitive relation.

Hence, the relation of equivalency is an equivalence relation.

Q.E.D.

THEOREM-5 *Let V and V' be finite dimensional vector spaces over the same field F. A linear transformation $t: V \to V'$ is represented by matrices A and A' relative to possibly different bases of V and V' if and only if A and A' are equivalent.*

PROOF. Let B, C be two bases of V and B', C' be bases of V' such that

$$A = M_B^{B'}(t) \quad \text{and} \quad A' = M_C^{C'}(t)$$

Let P be the change-of-basis matrix from B to C and Q be the change-of-basis matrix from B' to C'. Then,

$$P = M_C^B(I_V) \text{ and } Q = M_{C'}^{B'}(I_{V'})$$

$\therefore \quad Q^{-1}AP = M_{B'}^{C'}(I_{V'}) \, M_B^{B'}(t) \, M_C^B(I_V)$

$\Rightarrow \quad Q^{-1}AP = M_{B'}^{C'}(I_{V'}) \, M_C^{B'}(toI_V)$

$\Rightarrow \quad Q^{-1}AP = M_{B'}^{C'}(I_{V'}) \, M_C^{B'}(t)$

$\Rightarrow \quad Q^{-1}AP = M_C^{C'}(I_{V'}ot)$

$\Rightarrow \quad Q^{-1}AP = M_C^{C'}(t) = A'$

Hence, A' is equivalent to A.

Conversely, let A and A' be equivalent matrices. Then, there exist invertible matrices P and Q such that $A' = Q^{-1}AP$.

Let B and B' be bases of V and V' respectively such that $A = M_B^{B'}(t)$. Then, we have to prove that there exist bases C and C' of V and V' respectively such that $A' = M_C^{C'}(t)$.

Since B is a basis for V and P is an invertible matrix. Therefore, by Theorem 2 on page 387, there exists unique basis C for V such that P is the change-of-basis matrix from B to C, i.e. $P = M_C^B(I_V)$. Similarly, B' is a basis for V' and Q is an invertible matrix. Therefore, there exist unique basis C' for V' such that Q is the change-of-basis matrix from B' to C', i.e. $Q = M_{C'}^{B'}(I_{V'})$.

Now,

$$A' = Q^{-1}AP$$

$\Rightarrow \quad A' = M_{B'}^{C'}(I_{V'}) \, M_B^{B'}(t) \, M_C^B(I_V)$

$\Rightarrow \quad A' = M_{B'}^{C'}(I_{V'}) \, M_C^{B'}(toI_V)$

$\Rightarrow \quad A' = M_{B'}^{C'}(I_{V'}) \, M_C^{B'}(t)$

$\Rightarrow \quad A' = M_C^{C'}(I_{V'}ot) = M_C^{C'}(t)$

Hence, A' is the matrix of t relative to bases C and C' of V and V' respectively.

Q.E.D.

THEOREM-6 *Let $V(F)$ be a finite dimensional vector space with $B = \{u_1, u_2, \ldots, u_n\}$ and $C = \{v_1, v_2, \ldots, v_n\}$ as its two bases and let $X = \{x_1, x_2, \ldots, x_n\}$ and $Y = \{y_1, y_2, \ldots, y_n\}$ be bases dual to B and C respectively. If P is the change-of-basis matrix from B to C, then $(P^{-1})^T$ is the change-of-basis matrix from X to Y.*

PROOF. Let $P = [a_{ij}]$ and $Q = [b_{ij}]$. Then,

$$v_i = \sum_{r=1}^{n} a_{ir} \, u_r \text{ and } y_i = \sum_{s=1}^{n} b_{is} \, x_s$$

For any $i, j \in \underline{n}$, we have

$$y_i(v_j) = \left(\sum_{s=1}^{n} b_{is} x_s\right)\left(\sum_{r=1}^{n} a_{jr} u_r\right)$$

$\Rightarrow \quad \delta_{ij} = \sum_{r=1}^{n}\sum_{s=1}^{n} b_{is} a_{jr} x_s(u_r)$

$\Rightarrow \quad \delta_{ij} = \sum_{r=1}^{n}\sum_{s=1}^{n} b_{is} a_{jr} \delta_{sr}$

$\Rightarrow \quad \delta_{ij} = \sum_{r=1}^{n} b_{ir} a_{jr}$

$\Rightarrow \quad \delta_{ij} = \sum_{r=1}^{n} (Q)_{ir} (P^T)_{rj}$

$\Rightarrow \quad \delta_{ij} = (Q P^T)_{ij}$

$\Rightarrow \quad Q P^T = I$

$\Rightarrow \quad Q = (P^T)^{-1} = (P^{-1})^T \qquad \left[\because (P^T)^{-1} = (P^{-1})^T\right]$

Q.E.D.

TRACE OF A MATRIX *Let $A = [a_{ij}]$ be a square matrix over a field F. Then the sum of the diagonal elements of A is called the trace of A and generally written as $tr(A)$.*

Thus, $tr(A) = \sum_{i=1}^{n} a_{ii}$

THEOREM-7 *Similar matrices have the same trace.*

PROOF. Let $A = [a_{ij}]_{n \times n}$ and $B = [b_{ij}]_{n \times n}$ be two matrices over a field F such that B is similar to A. Then there exists an invertible matrix P such that

$\qquad B = P^{-1} A P$

$\Rightarrow \quad tr(B) = tr(P^{-1} A P)$

$\Rightarrow \quad tr(B) = tr(P^{-1}(AP))$ \hfill [By associativity of multiplication of matrices]

$\Rightarrow \quad tr(B) = tr((AP)P^{-1})$ \hfill $[\because tr(AB) = tr(BA)]$

$\Rightarrow \quad tr(B) = tr(A(PP^{-1}))$ \hfill [By associativity of matrix multiplication]

$\Rightarrow \quad tr(B) = tr(AI)$

$\Rightarrow \quad tr(B) = tr(A)$

Q.E.D.

Let V be a finite dimensional vector space over a field F and $t : V \to V$ be a linear transformation. If B and B' are bases of V, then matrices $A = M_B^B(t)$ and $A' = M_{B'}^{B'}(t)$ are similar. Therefore, they have the same trace. Thus, we define the trace of a linear transformation as follows.

TRACE OF A LINEAR TRANSFORMATION *Let $V(F)$ be a finite dimensional vector space and $t : V \to V$ be a linear transformation. Then, the trace of t is defined as the trace of the matrix represented by t relative to any ordered basis of V.*

DETERMINAT OF A LINEAR TRANSFORMATION *Let $V(F)$ be a finite dimensional vector space and $t : V \to V$ be a linear operator. Then, the determinant of t is defined as the determinant of the matrix represented by t relative to any ordered basis of V.*

EXERCISE 4.3

1. Let A and B be two square matrices over a field F. Then, prove that $tr(AB) = tr(BA)$
2. Let A be an $n \times n$ matrix over R. Prove that $A = O$ iff $tr(A^T A) = 0$.
3. If A and B are two matrices over field F, such that B is similar to A, then show that B^2 is similar to A^2.
4. If A and B are two invertible matrices such that B is similar to A, then prove that B^{-1} is similar to A^{-1}.
5. Show that the only matrix similar to the null matrix over a field F is the null matrix itself.
6. If A and B are two matrices over a field F such that B is similar to A, then prove that B' is similar to A'.
7. Consider the linear transformation $t : R^2 \to R^2$ defined by $t(x,y) = (2x+3y, 4x-5y)$ and bases $B_1 = \{e_1^{(2)} = (1,0), e_2^{(2)} = (0,1)\}$ and $B_2 = \{(1,2),(2,5)\}$ of R^2. Find the matrix representations of t relative to the bases B_1 and B_2 respectively. Also, find the trace of t.
8. If A and B are two matrices over field F such that B is similar to A, then show that B^n is similar to A^n for all $n \in N$.
9. Find the trace of each of the following linear transformations relative to standard order basis:
   (i) $t : R^3 \to R^3$ defined by $t(x,y,z) = (x+2y-3z, 4x-5y-6z, 7x+8y+9z)$.
   (ii) $t : R^3 \to R^3$ defined by $t(e_1^{(3)}) = (1,3,5), t(e_2^{(3)}) = (2,4,6), t(e_3^{(3)}) = (7,7,7)$.
10. Find the trace and determinant of each of the following linear operators on R^2 :
    (i) $t_1(x, y) = (2x-3y, 5x+4y)$ (ii) $t_2(x, y) = (ax+by, cx+dy)$
11. Find the trace and determinant of each of the following linear operators on R^3:
    (i) $t_1(x, y, z) = (x+3y, 3x-2z, x-4y-3z)$
    (ii) $t_2(x, y, z) = (y+3z, 2x-4z, 5x+7y)$

12. Let $B = \{v_1, v_2\}$ be a basis of V, and $t : V \to V$ be defined by $t(v_1) = 3v_1 - 2v_2$ and $t(v_2) = v_1 + 4v_2$. Suppose $B' = \{u_1, u_2\}$ is a basis of V such that $u_1 = v_1 + v_2$ and $u_2 = 2v_1 + 3v_2$.

   (i) Find the matrices A and B representing t relative to the bases B and B' respectively.

   (ii) Find the matrix P such that $B = P^{-1}AP$.

ANSWERS

9. (i) 5   (ii) 12

10. (i) $tr(t_1) = 6$, $\text{Det}(t_1) = 23$ (ii) $tr(t_2) = a + d$, $\text{Det}(t_2) = ad - bc$

11. (i) $tr(t_1) = -2$, $\text{Det}(t_1) = 13$ (ii) $tr(t_2) = 0$, $\text{Det}(t_2) = 22$

12. (i) $A = \begin{bmatrix} 3 & -2 \\ 1 & 4 \end{bmatrix}$, $B = \begin{bmatrix} 8 & -2 \\ 11 & -1 \end{bmatrix}$ (ii) $P = \begin{bmatrix} 1 & 1 \\ 2 & 3 \end{bmatrix}$

4.6 RANK OF A MATRIX

In Chapter 0, we have learnt that a positive integer r is rank of a matrix over field R, if

   (i) every square submatrix (if any) of order $(r+1)$ or more is singular.

   (ii) there exists at least one non-singular, square submatrix of order r.

We have also learnt that the rank of a matrix in row-reduced echelon form is equal to the number of non-zero rows in it. Here non-zero row means a row having at least one non-zero element. There are several equivalent definitions of the rank of a matrix depending upon the concept employed to define them. There are three ways to introduce the concept of the rank of a matrix.

   (i) Using linear transformations

   (ii) Using the concept of linear independence and dependence

   (iii) Using determinants.

In this section, we will study first two only. The third approach which makes use of determinant function will be discussed in the next chapter.

4.6.1 RANK OF A MATRIX THROUGH LINEAR TRANSFORMATIONS

In the previous Chapter, we have introduced the concept of rank and nullity of a linear transformation. In Chapter 1, we have studied that each $m \times n$ matrix A over a free R-module M defines a linear transformation $t_A : R^n \to R^m$ which is known as the linear transformation of matrix A. We know that every vector space $V(F)$ is a free module over its basis. Therefore, if A is an $m \times n$ matrix over a field F then there exists a linear transformation $t_A : F^n \to F^m$ representing matrix A.

In this section, we will use the concept of the rank and nullity of a linear transformation to define the rank and nullity of a matrix defined over a field F.

RANK OF A MATRIX *Let A be an $m \times n$ matrix over a field F. The rank of A is defined as the rank of the linear transformation $t_A : F^n \to F^m$.*

NULLITY OF A MATRIX *Let A be an $m \times n$ matrix over a field F. The nullity of A is defined to be the nullity of the linear transformation $t_A : F^n \to F^m$.*

Let $V(F)$ and $V'(F)$ be vector spaces of dimensions n and m respectively and let $t : V \to V'$ be a linear transformation. We shall now show that the rank and nullity of any $m \times n$ matrix A over F are same as that of linear transformation $t : V \to V'$ representing matrix A relative to some pair of bases B and B' of V and V' respectively.

THEOREM-1 *Let $V(F)$ and $V'(F)$ be two finite dimensional vector spaces and $t : V \to V'$ be a linear transformation. The rank of t is same as that of any matrix representing t relative to some bases of V and V'.*

PROOF. Let $B = \{b_1, b_2, \ldots, b_n\}$ and $B' = \{b'_1, b'_2, \ldots, b'_m\}$ be bases of V and V' respectively such that $M_B^{B'}(t) = A$. Let $t_A : F^n \to F^m$ be the linear map corresponding to A. Then,

$$t = M^{-1}(A) = L_{B'} o t_A o L_B^{-1}$$ [See Theorem 1 on page 372]

$\Rightarrow \quad \text{rank}(t) = \text{rank}\left(L_{B'} o t_A o L_B^{-1}\right)$

$\Rightarrow \quad \text{rank}(t) = \text{rank}(L_{B'} o t_A)$ $\quad \left[\because L_B^{-1} : V \to F^n \text{ is invertible and } \text{rank}(t_1 o t_2) = \text{rank}(t_1), \text{ if } t_2 \text{ is invertible}\right]$

$\Rightarrow \quad \text{rank}(t) = \text{rank}(t_A)$ $\quad [\because L_{B'} : F^n - V' \text{ is a bijection}]$

$\Rightarrow \quad \text{rank}(t) = \text{rank}(A)$ $\quad$ [By definition of rank of A]

Q.E.D.

COROLLARY. *Let $V(F)$ and $V'(F)$ be two finite dimensional vector spaces of dimensions n and m respectively. Let $t : V \to V'$ be a linear transformation. Then, the nullity of t is same as the nullity of any matrix representing t relative to some bases of V and V'.*

PROOF. Since $t : V \to V'$ is a linear transformation.

$\therefore \quad \text{rank}(t) + \text{nullity}(t) = \dim V$

$\Rightarrow \quad \text{rank}(t) + \text{nullity}(t) = n$ $\hspace{5cm}$ (i)

Also, $t_A : F^n \to F^m$ is the linear transformation representing matrix A.

$\therefore \quad \text{rank}(t_A) + \text{nullity}(t_A) = \dim(F^n)$

$\Rightarrow \quad \text{rank}(t_A) + \text{nullity}(t_A) = n$

$\Rightarrow \quad \text{rank}(t) + \text{nullity}(t_A) = n \qquad [\because \text{rank}(t_A) = \text{rank}(t)] \quad \text{(ii)}$

From (i) and (ii), we have

$\text{rank}(t) + \text{nullity}(t) = \text{rank}(t) + \text{nullity}(t_A)$

$\Rightarrow \quad \text{nullity}(t) = \text{nullity}(t_A)$

Q.E.D.

THEOREM-2 *For any matrix A over a field F, prove that rank (A) = rank (A^T).*

PROOF. Let $V(F)$ and $V'(F)$ be two finite dimensional vector spaces, and let $B = \{b_1, b_2, \ldots, b_n\}$ and $B' = \{b'_1, b'_2, \ldots, b'_m\}$ be bases of V and V' respectively.

Let $t : V \to V'$ be a linear transformation such that $M_B^{B'}(t) = A$. Then,

$\text{rank}(t) = \text{rank}(A) \qquad\qquad\qquad\qquad\qquad\qquad [\text{By Theorem 1}] \quad \text{(i)}$

Let $\quad t^* : V'^* \to V^*$ be the map dual to t. Then,

$\text{rank}(t) = \text{rank}(t^*) \qquad\qquad\qquad\qquad\qquad [\text{By Theorem 6 on page 351}] \quad \text{(ii)}$

From (i) and (ii), we get

$\text{rank}(A) = \text{rank}(t^*) \qquad\qquad\qquad\qquad\qquad\qquad\qquad\qquad\qquad\qquad \text{(iii)}$

By Theorem 5 on page 376, we have

$M(t^*) = \{M(t)\}^T = A^T \qquad\qquad [\because A = M(t) = M_B^{B'}(t)] \quad \text{(iv)}$

But,

$\text{rank}(t^*) = \text{rank}(M(t^*)) \qquad\qquad\qquad\qquad [\text{By Theorem 1 on page 403}]$

$\Rightarrow \quad \text{rank}(t^*) = \text{rank}(A^T) \qquad\qquad\qquad\qquad\qquad\qquad [\text{Using (iv)}] \quad \text{(v)}$

From (iii) and (v), we get

$\text{rank}(A) = \text{rank}(A^T)$

Q.E.D.

THEOREM-3 *Let A and B be two $m \times n$ matrices over field F. Then,*

$\text{rank}(A+B) \leq \text{rank}(A) + \text{rank}(B)$

PROOF. Clearly, $A + B$ is an $m \times n$ matrix over F. Therefore, $t_A : F^n \to F^m$, $t_B : F^n \to F^m$ and $t_{A+B} : F^n \to F^m$ are linear maps corresponding to matrices A, B and $A+B$ respectively. Clearly, $\text{rank}(A) = \text{rank}(t_A)$, $\text{rank}(B) = \text{rank}(t_B)$ and $\text{rank}(A+B) = \text{rank}(t_{A+B})$.

Now,
$$t_{A+B} = t_A + t_B$$
$\Rightarrow \quad \text{rank}(t_{A+B}) = \text{rank}(t_A + t_B)$
$\Rightarrow \quad \text{rank}(t_{A+B}) \leq \text{rank}(t_A) + \text{rank}(t_B) \qquad [\because \text{rank}(t_1 + t_2) \leq \text{rank}(t_1) + \text{rank}(t_2)]$
$\Rightarrow \quad \text{rank}(A + B) \leq \text{rank}(A) + \text{rank}(B)$

Q.E.D.

THEOREM-4 *The rank of the product of two matrices never exceeds the rank of either matrix.*

PROOF. Let A and B be two $m \times n$ and $n \times p$ matrices respectively over a field F. Further, let $t_A : F^n \to F^m$ and $T_B : F^p \to F^n$ be the linear transformations corresponding to matrices A and B respectively. Then,
$$\text{rank}(A) = \text{rank}(t_A) \quad \text{and} \quad \text{rank}(B) = \text{rank}(t_B)$$
Clearly, AB is an $m \times p$ matrix over F and $t_{AB} : F^p \to F^m$ is the corresponding linear transformation such that
$$t_{AB} = t_A o t_B$$
$\Rightarrow \quad \text{rank}(t_{AB}) = \text{rank}(t_A o t_B)$
$\Rightarrow \quad \text{rank}(t_{AB}) \leq \min\{\text{rank}(t_A), \text{rank}(t_B)\} \qquad [\because \text{rank } t_1 o t_2 \leq \min\{\text{rank}(t_1), \text{rank}(t_2)\}]$
$\Rightarrow \quad \text{rank}(AB) \leq \min\{\text{rank}(A), \text{rank}(B)\}$
$\Rightarrow \quad \text{rank}(AB) \leq \text{rank}(A) \text{ and, } \text{rank}(AB) \leq \text{rank}(B).$

Q.E.D.

THEOREM-5 *Let A and B be $n \times n$ matrices of ranks r and s, prove that the rank of AB is never less than $r + s - n$.*

PROOF. Let A and B be two $n \times n$ matrices over a field F such that rank $(A) = r$ and rank$(B) = s$. Let $t_A : F^n \to F^n$ and $t_B : F^n \to F^n$ be the corresponding linear transformations. Then, rank $(t_A) = r$ and rank $(t_B) = s$. Also, AB is an $n \times n$ matrix over F with $t_{AB} : F^n \to F^n$ as corresponding linear map.
Clearly, $t_A o t_B : F^n \to F^n$ such that
$$t_{AB} = t_A o t_B$$
$\Rightarrow \quad \text{rank}(t_A o t_B) = \text{rank}(t_{AB})$
But, $\quad \text{rank}(t_A o t_B) \geq \text{rank}(t_A) + \text{rank}(t_B) - n \qquad$ [See Theorem 7 on page 302]
$\Rightarrow \quad \text{rank}(t_{AB}) \geq \text{rank}(t_A) + \text{rank}(t_B) - n$
$\Rightarrow \quad \text{rank}(AB) \geq \text{rank}(A) + \text{rank}(B) - n$
$\Rightarrow \quad \text{rank}(AB) \geq r + s - n$

Q.E.D.

THEOREM-6 Let A and B be $n \times n$ matrices over a field F such that A is invertible. Then, rank $(AB) = $ rank (B).

PROOF. Let $t_A : F^n \to F^n$ and $t_B : F^n \to F^n$ be the linear maps corresponding to matrices A and B respectively.
Now,

A is invertible matrix

$\Rightarrow \quad t_A$ is invertible

$\Rightarrow \quad$ rank $(t_A o t_B) = $ rank (t_B) [See Corollary to Theorem 7 on page 302]

$\Rightarrow \quad$ rank $(t_{AB}) = $ rank (t_B)

$\Rightarrow \quad$ rank $(AB) = $ rank (B)

                                Q.E.D.

THEOREM-7 *(Sylvester's law of nullity)* Let A and B be two matrices over a field F such that the product AB exists. The nullity of the product AB never exceeds the sum of the nullities of the factors A and B and is never less than the nullity of either factor, if B is a square matrix.

PROOF. Let A and B be $m \times n$ and $n \times p$ matrices over a field F. Then, AB is $m \times p$ matrix. Let $t_A : F^n \to F^m$, $t_B : F^p \to F^n$ and $t_{AB} : F^p \to F^m$ be linear transformations corresponding to the matrices A, B and AB respectively.
Now,

$$t_{AB} = t_A o t_B$$

$\Rightarrow \quad$ rank $(t_{AB}) = $ rank $(t_A o t_B)$

$\Rightarrow \quad p - $ rank $(t_{AB}) = p - $ rank $(t_A o t_B)$

$\Rightarrow \quad$ nullity $(t_{AB}) = $ nullity $(t_A o t_B)$ (i)

By Theorem 7 on page 302 we have

$$\text{rank } (t_A) + \text{rank } (t_B) - n \leq \text{rank } (t_A o t_B) \quad \quad \quad \text{(ii)}$$

Also, we have

$$\text{rank } (t_A) + \text{nullity } (t_A) = n$$

$$\text{rank } (t_B) + \text{nullity } (t_B) = p$$

and,   rank $(t_A o t_B) + $ nullity $(t_A o t_B) = p$

Using these results in (ii), we get

$$n - \text{nullity } (t_A) + p - \text{nullity } (t_B) - n \leq p - \text{nullity } (t_A o t_B)$$

$\Rightarrow \quad$ nullity $(t_A) + $ nullity $(t_B) \geq $ nullity $(t_A o t_B)$

$\Rightarrow \quad$ nullity $(t_A) + $ nullity $(t_B) \geq $ nullity (t_{AB})

$\Rightarrow \quad$ nullity $(A) + $ nullity $(B) \geq $ nullity (AB)

Hence,   nullity $(AB) \leq $ nullity $(A) + $ nullity (B)

Let us now prove the second part of the theorem. Let B be a square matrix of order $n \times n$. Then, $t_A : F^n \to F^m$, $t_B : F^n \to F^n$ and $t_A o t_B : F^n \to F^m$ are corresponding linear transformations. Then,

$$\text{rank } (t_A o t_B) \leq \text{rank}(t_A)$$
$$\Rightarrow \quad n - \text{nullity } (t_A o t_B) \leq n - \text{nullity } (t_A)$$
$$\Rightarrow \quad \text{nullity } (t_A) \leq \text{nullity } (t_A o t_B)$$
$$\Rightarrow \quad \text{nullity } (A) \leq \text{nullity } (AB)$$
$$\Rightarrow \quad \text{nullity } (AB) \geq \text{nullity } (A)$$

Similarly, we have

$$\text{nullity}(AB) \geq \text{nullity } (B)$$

Q.E.D.

4.6.2 RANK OF A MATRIX BASED UPON THE CONCEPT OF LINEAR INDEPENDENCE AND DEPENDENCE OF VECTORS

Let A be an $m \times n$ matrix over a field F. Then each row of A can be viewed as an element of vector space F^n. If A_i denote the ith row of A, then matrix A determines the list $(A_1, A_2, \ldots A_n)$ of m vectors in F^n. Similarly, each column of A is an element of vector space F^m and if A^j denote the jth column of A, then A can be considered as the list $(A^1, A^2, \ldots, A^n)$ of n vectors in F^m. Thus, the matrix A simultaneously determines a list of m vectors $(A_1, A_2, \ldots, A_m)$ in F^n and a list of n vectors $(A^1, A^2, \ldots, A^n)$ in F^m. This will help us in defining row and column ranks of a matrix as defined below.

ROW RANK *Let A be an $m \times n$ matrix over a field F. The maximum number of linearly independent rows of A is called the row rank of A.*

COLUMN RANK *Let A be an $m \times n$ matrix over a field F. The maximum number of linearly independent columns of A is called the column rank of A.*

We shall now show that the row and column ranks of a matrix are equal to the rank of the matrix.

THEOREM-1 *Let $A = [a_{ij}]$ be an $m \times n$ matrix over a field F. Then, column rank of A is same as the rank of matrix A.*

PROOF. Let $t_A : F^n \to F^m$ be the linear transformation corresponding to the matrix A. Then,

$$\text{rank } (A) = \text{rank } (t_A)$$

Let S_1 be the subspace of F^m spanned by the vectors $t(e_1^{(n)}), t(e_2^{(n)}), \ldots, t(e_n^{(n)})$ and S_2 be the subspace of F^m spanned by the column vectors $A^1, A^2, \ldots, A^n$ of matrix A, i.e.

$$S_1 = \left\{ \sum_{j=1}^{n} \alpha_j t_A(e_j^{(n)}) : \alpha_j \in F \right\} \text{ and, } S_2 = \left\{ \sum_{j=1}^{n} \beta_j A^j : \beta_j \in F \right\}$$

We observe that S_1 is the space spanned by the images of basis vectors in F^n under t_A.

$$\therefore \qquad \dim S_1 = \dim(I_m(t_A)) = \text{rank }(t_A) = \text{rank }(A) \qquad \text{(i)}$$

Clearly, S_2 is the space spanned by the columns of matrix A.

$$\therefore \qquad \dim S_2 = \text{Column rank of } A \qquad \text{(ii)}$$

In order to prove that the column rank of A is same as the rank of matrix A, it is sufficient to show that $\dim S_1 = \dim S_2$. For this, let us consider the mapping $\phi : S_1 \to S_2$ defined by

$$\phi\left(\sum_{j=1}^{n} \alpha_j \, t_A\left(e_j^{(n)}\right)\right) = \sum_{j=1}^{n} A^j \alpha_j, \quad \text{where } \alpha_j \in F \quad \text{for all } j \in \underline{n}$$

ϕ *is well defined:* For any $\sum_{j=1}^{n} \alpha_j \, t_A\left(e_j^{(n)}\right)$ and $\sum_{j=1}^{n} \beta_j \, t_A\left(e_j^{(n)}\right)$ in S_1, we observe that

$$\sum_{j=1}^{n} \alpha_j \, t_A\left(e_j^{(n)}\right) = \sum_{j=1}^{n} \beta_j \, t_A\left(e_j^{(n)}\right)$$

$\Rightarrow \quad \sum_{j=1}^{n} \alpha_j \sum_{i=1}^{m} a_{ij} e_i^{(m)} = \sum_{j=1}^{n} \beta_j \sum_{i=1}^{m} a_{ij} e_i^{(m)} \qquad \left[\because t_A\left(e_j^{(n)}\right) = \sum_{i=1}^{m} a_{ij} e_i^{(m)}\right]$

$\Rightarrow \quad \sum_{i=1}^{m} \sum_{j=1}^{n} a_{ij} \alpha_j e_i^{(m)} = \sum_{i=1}^{m} \sum_{j=1}^{n} a_{ij} \beta_j e_i^{(m)}$

$\Rightarrow \quad \sum_{i=1}^{m} \left(\sum_{j=1}^{n} a_{ij} \alpha_j - a_{ij} \beta_j\right) e_i^{(m)} = 0$

$\Rightarrow \quad \sum_{i=1}^{m} \left\{\sum_{j=1}^{n} a_{ij}(\alpha_j - \beta_j)\right\} e_i^{(m)} = 0$

$\Rightarrow \quad \sum_{j=1}^{n} a_{ij}(\alpha_j - \beta_j) = 0 \quad \text{for all } i \in \underline{m} \qquad \left[\because e_1^{(m)}, e_2^{(m)}, \ldots, e_m^{(m)} \text{ are l.i.}\right]$

$\Rightarrow \quad \sum_{j=1}^{n} a_{ij} \alpha_j = \sum_{j=1}^{n} a_{ij} \beta_j \quad \text{for all } i \in \underline{m}$

$\Rightarrow \quad \sum_{j=1}^{n} A^j \alpha_j = \sum_{j=1}^{n} A^j \beta_j$

$\Rightarrow \quad \phi\left(\sum_{j=1}^{n} \alpha_j \, t_A\left(e_j^{(n)}\right)\right) = \phi\left(\sum_{j=1}^{n} \beta_j \, t_A\left(e_j^{(n)}\right)\right)$

So, ϕ is well defined.

We shall now show that ϕ is a monomorphism. For this, it is sufficient to show that $\text{Ker}(\phi) = \{0\}$.

ϕ *is a linear transformation:* For any $\sum_{j=1}^{n} \alpha_j \, t_A\left(e_j^{(n)}\right), \sum_{j=1}^{n} \beta_j \, t_A\left(e_j^{(n)}\right)$ is S_1 and $\lambda, \mu \in F$, we have

$$\phi\left\{\lambda\left(\sum_{j=1}^{n} \alpha_j \, t_A\left(e_j^{(n)}\right)\right) + \mu\left(\sum_{j=1}^{n} \beta_j \, t_A\left(e_j^{(n)}\right)\right)\right\} = \phi\left\{\sum_{j=1}^{n} (\lambda\alpha_j + \mu\beta_j) t_A\left(e_j^{(n)}\right)\right\}$$

$$= \sum_{j=1}^{n} A^j (\lambda \alpha_j + \mu \beta_j)$$

$$= \sum_{j=1}^{n} \left(A^j \lambda \alpha_j + A^j \mu \beta_j \right)$$

$$= \lambda \sum_{j=1}^{n} \left(A^j \alpha_j \right) + \mu \sum_{j=1}^{n} \left(A^j \beta_j \right)$$

$$= \lambda \phi \left(t_A \left(e_j^{(n)} \right) \right) + \mu \phi \left(t_A \left(e_j^{(n)} \right) \right)$$

So, ϕ is a linear transformation from S_1 to S_2.

ϕ *is a monomorphism:* We observe that

$$\sum_{j=1}^{n} \alpha_j t_A \left(e_j^{(n)} \right) = 0$$

$$\Leftrightarrow \quad \sum_{j=1}^{n} \alpha_j \left(\sum_{i=1}^{m} a_{ij} e_i^{(m)} \right) = 0$$

$$\Leftrightarrow \quad \sum_{i=1}^{m} \left(\sum_{j=1}^{n} (a_{ij} \alpha_j) \right) e_i^{(m)} = 0$$

$$\Leftrightarrow \quad \sum_{j=1}^{n} a_{ij} \alpha_j = 0 \quad \text{for all } i \in \underline{m} \qquad [\because e_1^{(m)}, e_2^{(m)}, \ldots, e_m^{(m)} \text{ are linearly independent}]$$

$$\Leftrightarrow \quad \sum_{j=1}^{n} A^j \alpha_j = 0$$

$$\therefore \quad \phi \left(\sum_{j=1}^{n} \alpha_j t_A \left(e_j^{(n)} \right) \right) = 0 \text{ iff } \sum_{j=1}^{n} A^j \alpha_j = 0$$

So, $\text{Ker}(\phi) = \{0\}$

Hence, ϕ is a monomorphism.

Thus, $\phi : S_1 \to S_2$ is a linear transformation such that $\text{Ker}(\phi) = \{0\}$. Therefore,

$$\phi(S_1) \cong \frac{S_1}{\text{Ker}(\phi)} \qquad \text{[By Fundamental Theorem on Homomorphisms]}$$

$\Rightarrow \quad \phi(S_1) \cong S_1$

$\Rightarrow \quad \dim(\phi(S_1)) = \dim(S_1)$

But, $\phi(S_1)$ is a subspace of S_2

$\therefore \quad \dim \left(\phi(S_1) \right) \leq \dim S_2$

$\Rightarrow \quad \dim S_1 \leq \dim S_2$

Similarly, the mapping $\psi : S_2 \to S_1$ given by

$$\psi \left(\sum_{j=1}^{n} A^j \alpha_j \right) = \sum_{j=1}^{n} \alpha_j t_A \left(e_j^{(n)} \right)$$

is a monomorphism of S_2 onto S_1 with $\text{Ker}(\psi) = \{0\}$.

$\therefore \qquad \dim S_2 \leq \dim S_1$

Hence, $\qquad \dim S_1 = \dim S_2$

$\Rightarrow \qquad$ rank $(A) =$ Column rank of A \hfill [From (i) and (ii)]

\hfill Q.E.D.

COROLLARY. *Let A be an $m \times n$ matrix over a field F. Then, row rank of A is same as its rank.*

PROOF. As the rows of matrix A are columns of A^T.

$\therefore \quad$ Row rank of $A =$ Column rank of A^T

$\qquad\qquad\qquad = \text{rank } (A^T)$ \hfill [By the above Theorem]

$\qquad\qquad\qquad = \text{rank } (A)$ \hfill [$\because$ rank $(A^T) =$ rank (A)]

The equality of row rank and column rank of a matrix suggests us the following definition of the rank of a matrix.

\hfill Q.E.D.

RANK OF A MATRIX *Let A be an $m \times n$ matrix over a field F. The maximum number of linearly independent rows or columns of A is defined as the rank of A.*

It is evident from the above definition that:

(i) If A is a non-null matrix, then rank $(A) \geq 1$

(ii) If A is an $m \times n$ matrix, then rank $(A) \leq \min(m,n)$.

REMARK-1 *The rank of a null matrix is defined to be zero.*

REMARK-2 *Let I_n be the identity matrix of order n over a field F. Then, rows and columns of I_n are unit vectors $e_1^{(n)}, e_2^{(n)}, \ldots, e_n^{(n)}$ which are linearly independent. So, row and column ranks of I_n are each equal to n.*

THEOREM-2 *Let $V(F)$ and $V'(F)$ be vector spaces of dimensions n and m respectively and $t : V \to V'$ be a linear transformation. Then any $m \times n$ matrix of rank r over field F is equivalent to the matrix D (over the same field) with all entries zero except for the first r entries unity along the main diagonal, i.e.*

$$D_{ii} = 1, \quad i = 1, 2, \ldots, r$$
$$D_{ij} = 0 \quad \text{for} \quad i \neq j \text{ or } i = j > r.$$

PROOF. Let B and B' be bases of V and V' respectively such that $M_B^{B'}(t) = A$ and let rank $(A) = r$. Then, rank $(t) = r$. Therefore, there exist bases $C = \{c_1, c_2, \ldots, c_n\}$ and $C' = \{c'_1, c'_2, \ldots, c'_m\}$ of V and V' respectively such that

$$t(c_i) = \begin{cases} c_i' & \text{for } i = 1, 2, \ldots, r \\ 0, & \text{for } i = r+1, \ldots, n \end{cases}$$

Let $D = M_C^{C'}(t)$.

Since A and D are matrices of the same linear transformation. Therefore, they are equivalent to each other.

Let $X' = \{x_1', x_2', \ldots, x_m'\}$ be the basis dual to basis C'. Then, for any $i \in \underline{m}, j \in \underline{n}$, we have

$$D_{ij} = \left(M_C^{C'}(t)\right)_{ij} = x_i'(t(c_j))$$

$\Rightarrow \quad D_{ij} = \begin{cases} x_i'(c_j'); & i, j = 1, 2, \ldots, r \\ x_i'(0); & j > r \end{cases}$

$\Rightarrow \quad D_{ij} = \begin{cases} \delta_{ij}; & i, j = 1, 2, \ldots, r \\ 0; & j > r \end{cases}$

$\Rightarrow \quad D_{ij} = \delta_{ii} = 1 \quad \text{for} \quad i = 1, 2, \ldots, r$

and, $\quad D_{ij} = 0 \quad \text{for } i \neq j \quad \text{or}, i = j > r$

Q.E.D.

REMARK. *In the above theorem matrix D can also be written in the form*

$$D = \begin{bmatrix} I_r & O_{r \times n-r} \\ O_{m-r \times r} & O_{m-r \times n-r} \end{bmatrix}$$

where I_r is identity matrix of order $r \times r$ and $O_{p \times q}$ denote null matrix of order $p \times q$.

COROLLARY. *Two matrices of the same order and over the same field are equivalent iff they have the same rank.*

PROOF. Let A and B be two $m \times n$ matrices over the same field F.

First, let A and B be equivalent matrices. Then, they are matrices of the same linear transformation $t : V \to V'$, where V and V' are vector spaces of dimensions n and m respectively over the same field F.

$\therefore \quad \text{rank}(A) = \text{rank}(t) \text{ and } \text{rank}(B) = \text{rank}(t)$

$\Rightarrow \quad \text{rank}(A) = \text{rank}(B)$

Conversely, let A and B be two matrices over field F such that rank (A) = rank (B). Then, by Theorem 2, A and B are equivalent to a matrix D such that

$$D_{ii} = \begin{cases} 1 & \text{for} \quad i, j = 1, 2, \ldots, r \\ 0 & \text{for} \quad i \neq j \quad \text{or} \quad i = j > r \end{cases}$$

Since the relation of equivalency is an equivalence relation, therefore, A and B are equivalent matrices.

Q.E.D.

4.7 INVERTIBLE MATRICES

INVERTIBLE MATRIX. Let A be an $n \times n$ matrix over a field F. If there exists an $n \times n$ matrix B such that $AB = BA = I_n$, then A is said to be an invertible matrix and B is called inverse of A.

In other words, an $n \times n$ matrix A over a field F is invertible if it has both a left and a right inverse and these two are equal.

THEOREM-1 A square matrix A over a field F is invertible iff t_A is invertible.

PROOF. Let A be an $n \times n$ square matrix over F. Then,

A is invertible.
$\Leftrightarrow$ There exists an $n \times n$ matrix B over F such that $AB = I_n = BA$
$\Leftrightarrow$ $t_{AB} = t_{I_n} = t_{BA}$
$\Leftrightarrow$ $t_A o t_B = I = t_B o t_A$ \hspace{2em} $[t_A o t_B = t_{AB}$ and $t_{I_n} = I]$
$\Leftrightarrow$ t_A is invertible

Q.E.D.

THEOREM-2 Let B be a basis of an n dimensional vector space V over a field F. Then a linear transformation $t : V \to V$ is invertible iff matrix of t relative to B is invertible.

PROOF. First, we assume that $t : V \to V$ is an invertible linear transformation. Then, $t^{-1} : V \to V$ exists such that
$$t o t^{-1} = I = t^{-1} o t$$
$\Rightarrow$ $M_B^B(t o t^{-1}) = M_B^B(I) = M_B^B(t^{-1} o t)$
$\Rightarrow$ $M_B^B(t) M_B^B(t^{-1}) = I = M_B^B(t^{-1}) M_B^B(t)$
$\Rightarrow$ $M_B^B(t)$ is invertible and $\left[M_B^B(t)\right]^{-1} = M_B^B(t^{-1})$.

Conversely, let $M_B^B(t)$ be an invertible matrix. Let $[M_B^B(t)]^{-1}$ be the inverse of $M_B^B(t)$. Since $[M_B^B(t)]^{-1} \in F^{n \times n}$. Therefore, there exists a linear transformation $t_1 : V \to V$ such that
$$M_B^B(t_1) = [M_B^B(t)]^{-1}$$
$\Rightarrow$ $M_B^B(t_1) M_B^B(t) = M_B^B(t) M_B^B(t_1) = I_n$
$\Rightarrow$ $M_B^B(t_1 o t) = M_B^B(t o t_1) = I_n = M_B^B(I_n)$
$\Rightarrow$ $t_1 o t = t o t_1 = I$
$\Rightarrow$ t is invertible and $t^{-1} = t_1$.

Q.E.D.

THEOREM-3 Let A be a square matrix of order n over a field F. Then, A is invertible iff rank $(A) = n$. Or equivalently nullity $(A) = 0$.

PROOF. First, let us assume that A is an invertible matrix. Then, there exists an $n \times n$ matrix B over F such that

$$AB = I = BA$$

Now, $AB = I$

$\Rightarrow \quad t_{AB} = t_I$

$\Rightarrow \quad t_A o t_B = I$ $\quad\quad\quad [\because\ t_A o t_B = t_{AB}$ and, $t_I = I]$

$\Rightarrow \quad t_A : F^n \to F^n$ has right inverse t_B

$\Rightarrow \quad t_A : F^n \to F^n$ is an epimorphism

$\Rightarrow \quad$ rank $(t_A) = n$

$\Rightarrow \quad$ rank $(A) = n$

Conversely, let rank $(A) = n$. Then,

$\quad\quad$ rank$(t_A) = n$

$\Rightarrow \quad t_A : F^n \to F^n$ is an epimorphism

$\Rightarrow \quad t_A$ has a right inverse

$\Rightarrow \quad A$ has a right inverse $\quad\quad\quad\quad\quad\quad\quad\quad\quad\quad\quad\quad\quad\quad\quad\quad$ (i)

Now, $t_A : F^n \to F^n$ is an epimorphism.

$\Rightarrow \quad t_A : F^n \to F^n$ is monomorphism. $\quad\quad [\because t_A$ is from F^n to itself$]$

$\Rightarrow \quad t_A$ has a left inverse

$\Rightarrow \quad A$ has a left inverse $\quad\quad\quad\quad\quad\quad\quad\quad\quad\quad\quad\quad\quad\quad\quad\quad\quad$ (ii)

From (i) and (ii), we obtain that A is invertible.

$\quad\quad\quad\quad\quad\quad\quad\quad\quad\quad\quad\quad\quad\quad\quad\quad\quad\quad\quad\quad\quad\quad\quad\quad\quad\quad\quad\quad\quad\quad$ Q.E.D.

THEOREM-4 *Let $A \in F^{n \times n}$ (ring of all $n \times n$ matrices over a field F). Then the following statements are equivalent:*

  (i) A has a left inverse in $F^{n \times n}$
  (ii) nullity $(A) = 0$
  (iii) rank $(A) = n$
  (iv) A has a right inverse in $F^{n \times n}$
  (v) A has two sided inverse in $F^{n \times n}$.

PROOF. Let $t_A : F^n \to F^n$ be the linear transformation corresponding to the matrix A. For the sake of convenience, we transfer these properties into the properties of transformation t_A. Property (i) in terms of t_A becomes

$\quad\quad t_A$ has a left inverse $\quad\quad\quad\quad\quad\quad\quad\quad\quad\quad\quad\quad\quad\quad\quad\quad\quad\quad\quad$ (a)

$\Rightarrow \quad t_A$ is a monomorphism

$\Rightarrow \quad$ Ker $(t_A) =$ null space

$\Rightarrow \quad$ nullity $(t_A) = 0$

$\Rightarrow$ nullity $(A) = 0$ (b)
$\Rightarrow$ rank $(A) = n$ (c)
$\Rightarrow$ $t_A : F^n \to F^n$ is an epimorphism
$\Rightarrow$ t_A is an isomorphism
$\Rightarrow$ t_A is invertible transformation
$\Rightarrow$ A is invertible matrix
$\Rightarrow$ A has a right inverse (d)
$\Rightarrow$ t_A has a right inverse
$\Rightarrow$ t_A is an epimorphism
$\Rightarrow$ t_A is an isomorphism
$\Rightarrow$ t_A has two sided inverse
$\Rightarrow$ A has two sided inverse in $F^{n \times n}$ (e)
$\Rightarrow$ A has a left inverse (f)

From $(a), (b), (c), (d), (e)$ and (f), we find that

$$(i) \Rightarrow (ii) \Rightarrow (iii) \Rightarrow (iv) \Rightarrow (v) \Rightarrow (i)$$

Hence, all statements are equivalent.

Q.E.D.

REMARK. Let A be an invertible matrix and A^{-1} be its inverse. Then,

$$AA^{-1} = I = A^{-1}A$$
$\Rightarrow$ $(AA^{-1})^T = I = (A^{-1}A)^T$
$\Rightarrow$ $(A^{-1})^T A^T = I = A^T (A^{-1})^T$
$\Rightarrow$ A^T is invertible and $(A^T)^{-1} = (A^{-1})^T$

Thus, if A is invertible matrix, then so is A^T and $(A^T)^{-1} = (A^{-1})^T$

4.8 ELEMENTARY OPERATIONS

Consider a list $\underline{v} = (v_1, \ldots, v_n)$ of n vectors in a vector space V over a field F. An elementary operation on v is any one of the following three types.

(i) Interchange of any two vectors v_i and v_j of the list,
(ii) Multiplication of one vector v_i (say) of the list by a non-zero scalar $\lambda \in F$,
(iii) Addition of a scalar multiple λv_j of one vector v_j of the list to another vector v_i (say).

These operations on $\underline{v}$ are denoted by $e_{(i,j)}, e_{\lambda(i)}$ and $e_{(i)+\lambda(j)}$ respectively. The list obtained by performing an elementary operation e on $\underline{v}$ is denoted by $e(\underline{v})$.

The effects of elementary operations $e_{(i,j)}, e_{\lambda(i)}$ and $e_{(i)+\lambda(j)}$ on the list $\underline{v} = (v_1, \ldots, v_n)$ of vectors in vector space V may be exhibited as

$$e_{(i,j)}(\underline{v}) = (v_1, \ldots, v_j, v_{i+1}, \ldots, v_{j-1}, v_i, v_n)$$
$$e_{\lambda(i)}(\underline{v}) = (v_1, \ldots, v_{i-1}, \lambda v_i, v_{i+1}, \ldots, v_n)$$
$$e_{(i)+\lambda(j)}(\underline{v}) = (v_1, \ldots, v_{i-1}, v_i + \lambda v_j, v_{i+1}, \ldots, v_n)$$

Since a list $\underline{v} = (v_1, \ldots, v_n)$ of n vectors in V is an element of the function space $V^n = \{f : f : \underline{n} \to V\}$. Therefore, each elementary operation on $\underline{v}$ may be regarded as a function $e : V^n \to V^n$; in fact as a linear transformation. It can be easily seen that inverse of an elementary operation is an elementary operation, that is, if $e(\underline{v}) = \underline{v}'$ then $e^{-1}(\underline{v}') = (\underline{v})$, where e^{-1} is an elementary operation. One can easily see that

$$e_{(i,j)}^{-1} = e_{(j,i)} \qquad\qquad [\because\ e_{(i,j)}(e_{(j,i)}(\underline{v})) = \underline{v}]$$
$$e_{i\lambda}^{-1} = e_{(i)\frac{1}{\lambda}} \qquad\qquad [e_{1/\lambda}(e_{\lambda(i)}(\underline{v})) = \underline{v}]$$
$$e_{(i)+\lambda(j)}^{-1} = e_{(i)-\lambda(j)} \qquad\qquad [e_{(i)-\lambda(j)}(e_{(i)+\lambda(j)}(\underline{v})) = \underline{v}]$$

Two lists $\underline{v}$ and $\underline{v}'$ of n vectors in a vector space V are said to be equivalent iff one can be obtained from other by applying some elementary operations.

THEOREM-1 *Let e be an elementary operation. Then a list $\underline{v}$ of vectors in V is linearly independent iff $e(\underline{v})$ is linearly independent.*

PROOF. Let $\underline{v} = (v_1, \ldots, v_n)$. If $e = e_{(i,j)}$ or $e = e_{\lambda(i)}$, then the result is obvious. Let $e = e_{(i)+\lambda(j)}$. Then,

$$e(\underline{v}) = (v_1, \ldots, v_{i-1}, v_i + \lambda v_j, v_{i+1}, \quad v_j, \ldots, v_n)$$

Let $\underline{v}$ be a linearly independent list and $\alpha_1, \ldots, \alpha_n \in F$ such that

$$\alpha_1 v_1 + \cdots + \alpha_i(v_i + \lambda v_j) + \cdots + \alpha_j v_j + \quad + \alpha_n v_n = 0$$
$\Rightarrow \quad \alpha_1 v_1 + \cdots + \alpha_i v_i + \cdots + (\lambda \alpha_i + \alpha_j) v_j + \cdots + \alpha_n \lambda_n = 0$
$\Rightarrow \quad \alpha_1 = \alpha_2 = \cdots = \alpha_n = 0 \qquad\qquad [\because\ \text{list } \underline{v} \text{ is linearly independent}]$
$\Rightarrow \quad e(\underline{v})$ is linearly independent.

The converse follows from the fact that if $e(\underline{v}) = \underline{v}'$, then $e^{-1}(\underline{v}') = \underline{v}$, where e^{-1} is an elementary operation.

Q.E.D.

THEOREM-2 *If $\underline{v}$ is a list of vectors in a vector space V and e is an elementary operation, then span of $\underline{v}$ is same as the span of $e(\underline{v})$.*

PROOF. The proof is an easy consequence of Theorem 3 on page 167 of Chapter 2.
Combining the above two theorems, we can state.

THEOREM-3 *Let e be an elementary operation. Then a list $\underline{v}$ of vectors in a vector space V is a basis of V iff $e(\underline{v})$ is a basis for V.*

We shall now show that three elementary operations on a list $\underline{v} = (v_1, \ldots, v_n)$ of vectors in a vector space $V(F)$ become operations on the columns of a matrix. To see this, we choose a basis $B = \{b_1, \ldots, b_m\}$ of vector space V and express each vector v_j of the list in terms of the vectors in B as follows:

$$v_j = \sum_{i=1}^{m} a_{ij} b_i, \quad \text{where} \quad a_{ij} \in F, \, j \in \underline{m}$$

Replacing each v_j in the list $\underline{v}$ by $\sum_{i=1}^{m} a_{ij} b_i$ we find that the list $\underline{v} = (v_1, \ldots, v_n)$ can be replaced by the matrix

$$A = \begin{bmatrix} a_{11} \ldots a_{1j} \ldots a_{1n} \\ \vdots \quad\quad \vdots \\ a_{i1} \ldots a_{ij} \ldots a_{in} \\ \vdots \quad\quad \vdots \\ a_{m1} \ldots a_{mj} \ldots a_{mn} \end{bmatrix} = [a_{ij}]_{m \times n}$$

over field F. If we express the matrix A as a list of its columns, we find that each column $A^j \in F^m$ of A is the list of coordinates of vector v_j relative to basis B. Thus, we can replace vector space V by F^m and each vector v_j by column vector A^j. Hence, the three elementary operations on the list v of vectors in V become operations on the columns of the matrix A.

The corresponding three elementary operations on the columns of matrix A are:

(i) Interchange any two columns A^i and A^j.

(ii) Multiplication of one column A^i (say) by a non-zero scalar $\lambda \in F$.

(iii) Addition of a scalar multiple λA^j of one column A^j to another column A^i (say).

An elementary operation on the list of columns of A is called **elementary column operation**. Similarly, an elementary operation on the list of rows of A is called **elementary row operation** on A.

The following symbols will be used to denote the six elementary operations.

(i) $e^{(i,j)}$ for the interchange of ith and jth columns. Also denoted by $R_i \leftrightarrow R_j$.

(ii) $e^{(\lambda)i}$ for multiplication of the ith column by a non-zero scalar $\lambda \in F$. It is popularly denoted by $C_i \to \lambda C_i$.

(iii) $e^{(i)+\lambda(j)}$ for addition to the ith column, the product of jth column by scalar $\lambda \in F$, which is also denoted by $C_i \to C_i + \lambda C_j$.

The corresponding row operations will be denoted by $e_{(i,j)}(R_i \leftrightarrow R_j)$, $e_{\lambda(i)}(R_i \to \lambda R_i)$ and $e_{(i)+\lambda(j)}(R_i \to R_i + \lambda R_j)$ respectively.

The matrix obtained by applying an elementary column (row) operation e on a matrix A is denoted by $e(A)$.

It is evident from Theorem 2 that A and $e(A)$ have the same column rank. But, what is even more interesting is that A and $e(A)$ have the same row rank, too. In order to prove this, we have to first consider the effect of an elementary operation on the product of two matrices.

THEOREM-4 *Let $A = [a_{ij}]$ and $B = [b_{ij}]$ be $m \times n$ and $n \times p$ matrices respectively over the same field F. Let e' be an elementary column operation and e be an elementary row operation. Then,*

(i) $e(AB) = e(A)B$. (ii) $e'(AB) = Ae'(B)$

PROOF. (i) We shall prove the results separately for each of the three kinds of elementary operations.

Case I. When $e' = e^{(s,t)}$. Let $(i,j) \in \underline{m} \times \underline{p}$.

If $j \neq s$ and $j \neq t$, then

$$(e'(AB))_{ij} = (AB)_{ij} = \sum_{r=1}^{n} a_{ir} b_{rj}$$
$$\Rightarrow (e'(AB))_{ij} = \sum_{r=1}^{n} a_{ir} (e'(B))_{rj} = (Ae'(B))_{ij}.$$

If $j = s$, then

$$(e'(AB))_{is} = (AB)_{it} = \sum_{r=1}^{n} a_{ir} b_{rt}$$
$$\Rightarrow (e'(AB))_{is} = \sum_{r=1}^{n} a_{ir} (e'(B))_{rs} = (A(e'(B)))_{is}$$

Finally, if $j = t$, then

$$(e'(AB))_{it} = (AB)_{is} = \sum_{r=1}^{n} a_{ir} b_{rs} = \sum_{r=1}^{n} a_{ir} (e'(B))_{rt} = (A(e'(B)))_{it}$$

Thus, for all $(i,j) \in \underline{m} \times \underline{p}$, we have

$$(e'(AB))_{ij} = (A(e'(B)))_{ij}$$

Hence, $e'(AB) = Ae'(B)$.

Case II. When $e' = e^{\lambda(s)}$

If $j \neq s$, then for all $(i, j) \in \underline{m} \times \underline{p}$, we have

$$(e'(AB))_{ij} = (AB)_{ij} = \sum_{r=1}^{n} a_{ir}b_{rj} = \sum_{r=1}^{n} a_{ir}(e'(B))_{rj} = (A(e'(B)))_{ij}$$

If $j = s$, then

$$(e'(AB))_{is} = \lambda(AB)_{is} = \lambda\left(\sum_{r=1}^{n} a_{ir}b_{rs}\right)$$

$$\Rightarrow \quad (e'(AB))_{is} = \sum_{r=1}^{n} a_{ir}\lambda b_{rs} = \sum_{r=1}^{n} a_{ir}(e'(B))_{rs} = (A(e'(B)))_{is}$$

Hence, $e'(AB) = Ae'(B)$

Case III. When $e' = e^{(s)+\lambda(t)}$

If $j \neq s$, then

$$e'(AB)_{ij} = (AB)_{ij} = \sum_{r=1}^{n} a_{ir}b_{rj} = \sum_{r=1}^{n} a_{ir}(e'(B))_{rj} = (A(e'(B)))_{ij}$$

If $j = s$, then

$$e'(AB)_{is} = (AB)_{is} + \lambda(AB)_{it}$$

$$\Rightarrow \quad e'(AB)_{is} = \sum_{r=1}^{n} a_{ir}b_{rs} + \lambda\left(\sum_{r=1}^{n} a_{ir}b_{rt}\right)$$

$$\Rightarrow \quad e'(AB)_{is} = \sum_{r=1}^{n} a_{ir}(b_{rs} + \lambda b_{rt})$$

$$\Rightarrow \quad e'(AB)_{is} = \sum_{r=1}^{n} a_{ir}(e'(B))_{rs} = A(e'(B))_{is}$$

Hence, $e'(AB) = Ae'(B)$

Thus, in all the three cases, we have $e'(AB) = Ae'(B)$.

Part (b) can be proved on similar lines.

Q.E.D.

THEOREM-5 *Let A be an $m \times n$ matrix over a field F and e' be an elementary column operation such that $e'(A) = B$. Then,*

$$\alpha_1 A_1 + \alpha_2 A_2 + \cdots + \alpha_m A_m = 0 \quad \text{holds iff} \quad \alpha_1 B_1 + \alpha_2 B_2 + \cdots + \alpha_m B_m = 0$$

where $\alpha_1, \ldots, \alpha_m \in F$.

PROOF. Let $\alpha_1 A_1 + \cdots + \alpha_m A_m = 0$

$$\Rightarrow \quad [\alpha_1, \ldots, \alpha_m]A = 0 \qquad\qquad [\because \ A = [A_1 \ A_2 \ \ldots \ A_m]^T]$$

$$\Rightarrow \quad e'([\alpha_1, \ldots, \alpha_m]A) = 0$$

$$\Rightarrow \quad [\alpha_1, \ldots, \alpha_m]e'(A) = 0 \qquad\qquad \text{[By Theorem 4]}$$

$$\Rightarrow \quad [\alpha_1, \ldots, \alpha_m]B = 0$$

$$\Rightarrow \quad \alpha_1 B_1 + \cdots + \alpha_m B_m = 0 \qquad\qquad [\because \ B = [B_1 \ B_2 \ \ldots \ B_m]^T]$$

The converse follows from the fact that every elementary operation is invertible.

Q.E.D.

The above theorem proves that the maximum number of linearly independent rows of A is same as the maximum number of linearly independent rows of B. Thus, if e' is an elementary column operation, then A and $e'(A)$ have the same row rank. We have already shown that an elementary column operation does not alter the column rank. Hence, if e' is an elementary column operation, then A and $e'(A)$ have the same rank. Similarly, it can be shown that for any row operation e, matrix A and $e(A)$ have the same rank. Thus, we conclude that an elementary column (row) operation does not alter the rank of a matrix.

COLUMN EQUIVALENT MATRICES *Let A, B be $m \times n$ matrices over a field F. Then A is said to be column equivalent to B, if B can be obtained by applying successively a number of elementary column operations on A. If A is column equivalent to B, then we write as*

$$A \overset{C}{\sim} B$$

ROW EQUIVALENT MATRICES *Let A, B be $m \times n$ matrices over a field F. If B can be obtained by applying successively a number of elementary row operations on A, then A is row equivalent to B, written as*

$$A \overset{R}{\sim} B$$

If matrix B is obtained by applying successively a number of row and column operations on A, then A is said to be row-column equivalent to B, written as

$$A \overset{RC}{\sim} B$$

or simply equivalent to B, written as $A \sim B$.

It can be easily shown that row equivalence, column equivalence and row-column equivalence are equivalence relations in the set $F^{m \times n}$. We have seen that elementary column (row) operations do not alter the rank of a matrix. Therefore, two equivalent (row equivalent or column equivalent or row-column equivalent) matrices have the same rank.

4.9 ELEMENTARY MATRICES

ELEMENTARY MATRIX. *A square matrix obtained by applying one of the elementary column or row operation on an identity matrix is called an elementary matrix.*

In other words, if e is an elementary column or row operation, then the matrix $e(I)$ is called an elementary matrix. It may be noted that there is no distinction between the elementary matrices obtained by using an elementary row equivalent or column equivalent or row-column equivalent operation. It is easy to show that

$$e_{(i,j)}(I) = e^{(i,j)}(I), \ e_{\lambda(i)}(I) = e^{\lambda(i)}(I), \ e_{(i)+\lambda(j)}(I) = e^{(i)+\lambda(j)}(I)$$

We shall use the notations $E_{(i,j)}, E_{\lambda(i)}, E_{(i)+\lambda(j)}$ respectively to denote the elementary matrices described above.

Theory and Problems of Linear Algebra

THEOREM-1 *If an elementary column operation carries an $n \times n$ identity matrix I to the elementary matrix E, then it carries any $m \times n$ matrix A to the matrix product AE.*

PROOF. Let e' be an elementary column operation. Then, $e'(I) = E$.

Now, $A = AI$
$\Rightarrow \quad e'(A) = e'(AI)$
$\Rightarrow \quad e'(A) = Ae'(I)$ [By Theorem 4 on page 417]
$\Rightarrow \quad e'(A) = AE$

Q.E.D.

THEOREM-2 *If an elementary row operation carries an $m \times m$ identity matrix I to the elementary matrix E, then it carries any $m \times n$ matrix A to the matrix product EA.*

PROOF. The proof of this theorem is analogous to the proof of earlier theorem. So, it is left as an exercise for the reader.

Above theorems prove that an elementary column (row) operation on a matrix A is equivalent to multiplying A on the right (left) by an elementary matrix obtained by applying the same elementary operation on an identity matrix.

THEOREM-3 *Every elementary matrix is invertible.*

PROOF. Let E be an elementary matrix of order n. Since an elementary operation does not change the rank of a matrix and an elementary matrix E is obtained by such an operation upon the identity matrix of order n. Thus, rank$(E) = n$. Hence, by Theorem 4 on the page 413, E is invertible.

Let e' be an elementary column operation and E be an elementary matrix such that $e'(I) = E$. Since every elementary column operation is invertible. Therefore, e' is invertible. Let $(e')^{-1}$ be the inverse of e' and $(e')^{-1}(I) = E'$. Then,

$$EE' = E\left((e')^{-1}(I)\right)$$
$\Rightarrow \quad EE' = (e')^{-1}(EI)$
$\Rightarrow \quad EE' = (e')^{-1}(E) = (e')^{-1}(e'(I)) = I$

Similarly, we have $E'E = I$

$\therefore \quad EE' = I = E'E \Rightarrow E^{-1} = E'$

Hence, inverse of an elementary matrix is same as the elementary matrix obtained by the inverse elementary transformation. Consequently, we have

$$E^{-1}{}_{(i,j)} = \left\{e^{(i,j)}\right\}^{-1}(I) = e^{(i,j)}(I) = E_{(i,j)}$$

$$E^{-1}{}_{\lambda(i)} = \left\{e^{\lambda(i)}\right\}^{-1}(I) = e^{1/\lambda(i)}(I) = E_{1/\lambda(i)}$$

$$E^{-1}{}_{(i)+\lambda(j)} = \left\{e^{(i)+\lambda(j)}\right\}^{-1}(I) = e^{(i)-\lambda(j)}(I) = E_{(i)-\lambda(j)}.$$

Q.E.D.

THEOREM-4 *A square matrix A over a field F is invertible iff it can be written as a product of elementary matrices.*

PROOF. First, let A be an invertible matrix. Then by applying a sequence $\phi_1, \phi_2, \ldots, \phi_k$ of elementary column operations, it can be reduced to the identity matrix I. If $E_1, E_2, \ldots, E_k$ are corresponding elementary matrices, then

$AE_1E_2\ldots E_k = I$ [By Theorem 1]

$\Rightarrow \quad A = IE^{-1}{}_k E^{-1}{}_{k-1} \ldots E_1^{-1}$ [$\because$ Every elementary matrix is invertible]

$\Rightarrow \quad A = E^{-1}{}_k E^{-1}{}_{k-1} \ldots E_1^{-1}$

Since inverse of an elementary matrix is an elementary matrix. Hence, A is expressible as a product of elementary matrices.

Conversely, suppose $A = E_1 E_2 \ldots E_k$, where $E_1, \ldots, E_k$ are elementary matrices. Since every elementary matrix is invertible and product of a number of invertible matrices is an invertible matrix. Therefore, A is an invertible matrix.

Q.E.D.

THEOREM-5 *If a square matrix A over a field F is reduced to identity matrix I by a sequence of elementary column operations, then A is invertible and the same sequence of column operations applied on the identity matrix I will yield A^{-1}.*

PROOF. Let $e'_1, \ldots, e'_k$ be a sequence of elementary column operations, which reduces A to the identity matrix I. Then,

$$e'_k \ldots e'_1(A) = I \qquad (i)$$

Let $E_1, \ldots E_k$ be the elementary matrices corresponding to the elementary operations $e'_1, \ldots e'_k$. Then, (i) gives

$AE_1 \ldots E_k = I$ [By Theorem 1]

$\Rightarrow \quad A = E_k^{-1} E_{k-1}^{-1} \ldots E_2^{-1} E_1^{-1}$

$\Rightarrow \quad A$ is invertible. [$\because$ Each E_i is invertible]

This proves the first part of the theorem.

Let A^{-1} be the inverse of A. Then,

$$A^{-1}A = I \qquad (ii)$$

Performing column operations $e'_1, \ldots, e'_k$ on both the sides of (ii), we get

$$e'_k \ldots e'_1(A^{-1}A) = e'_k \ldots e'_1(I)$$
$$\Rightarrow \quad A^{-1}(e'_k \ldots e'_1(A)) = e'_k \ldots e'_1(I) \qquad \text{[By Theorem 4 on page 417]}$$
$$\Rightarrow \quad A^{-1}I = e'_k \ldots e'_1(I) \qquad \text{[Using i]}$$
$$\Rightarrow \quad A^{-1} = e'_k \ldots e'_1(I).$$

This proves the second part of the theorem. Q.E.D.

THEOREM-6 *Let A, B be two $m \times n$ matrices over a field F. Then,*

(i) A and B are column equivalent iff there exists an invertible matrix P such that $AP = B$.

(ii) A and B are row equivalent iff there exists an invertible matrix Q such that $QA = B$.

(iii) A is equivalent to B iff there exist invertible matrices P and Q such that $B = QAP$.

PROOF. (i) Suppose A is column equivalent to B. Then B is obtained by applying a sequence $e'_1, \ldots, e'_k$ of elementary column operations on A.

i.e,
$$e'_k \ldots e'_1(A) = B \qquad (i)$$

Let $E_1, \ldots, E_k$ be the elementary matrices corresponding to the elementary column operations $e'_1, \ldots, e'_k$. Then from (i) we get

$$B = AE_1 \ldots E_k = AP$$

where $P = E_1 \ldots E_k$ is the product of elementary matrices and hence invertible.

Conversely, let $B = AP$, where P is an invertible $n \times n$ matrix. By Theorem 4, P is product of elementary matrices, say

$$P = E_1 \ldots E_k \quad \Rightarrow \quad B = AE_1 \ldots E_k \qquad (ii)$$

Let $e'_1, \ldots, e'_k$ be the elementary column operations corresponding to the elementary matrices $E_1, \ldots, E_k$. Then (ii) can be written as

$$B = e'_1 \ldots e'_k(A) \quad \Rightarrow \quad A \text{ is column equivalent to } B$$

Similarly, (ii) and (iii) can be proved. Q.E.D.

This theorem suggests the following algorithm to find the inverse of a square matrix by using elementary operations.

ALGORITHM 1

Step I Obtain the square matrix A.

Step II Write A in the following form: $A = AI$

Linear Transformations and Matrices • 423

Step III Apply a sequence of elementary column operations on A on the left-hand side to reduce it to the identity matrix and perform the same sequence of elementary column operations on I on the right-hand side to get $I = AB$. The matrix B so obtained is the inverse of A.

REMARK. *In the above algorithm while performing column operations if all entries of a column reduce to zero. In other words, A on LHS does not reduce to identity matrix, then we say that A is not invertible.*

ALGORITHM 2

Step I Obtain the square matrix A.

Step II Write A in the following form: $A = IA$

Step III Apply a sequence of elementary row operations on A on the left-hand side to reduce it to the identity matrix and perform the same sequence of elementary row operations on I on the right-hand side so that $A = IA$ reduces to the form $I = BA$. The matrix B so obtained is the required inverse of A.

ILLUSTRATIVE EXAMPLES

EXAMPLE-1 Find the inverse of the matrix $A = \begin{bmatrix} 1 & 2 & -4 \\ -1 & -1 & 5 \\ 2 & 7 & -3 \end{bmatrix}$ by using elementary column operations.

SOLUTION We have,

$$A = AI$$

or, $\begin{bmatrix} 1 & 2 & -4 \\ -1 & -1 & 5 \\ 2 & 7 & -3 \end{bmatrix} = A \begin{bmatrix} 1 & 0 & 0 \\ 0 & 1 & 0 \\ 0 & 0 & 1 \end{bmatrix}$

$\Rightarrow \begin{bmatrix} 1 & 0 & 0 \\ -1 & 1 & 1 \\ 2 & 3 & 5 \end{bmatrix} = A \begin{bmatrix} 1 & -2 & 4 \\ 0 & 1 & 0 \\ 0 & 0 & 1 \end{bmatrix}$ Applying $C_2 \rightarrow C_2 + (-2)C_1$, $C_3 \rightarrow C_3 + 4C_1$

$\Rightarrow \begin{bmatrix} 1 & 0 & 0 \\ 0 & 1 & 0 \\ 5 & 3 & 2 \end{bmatrix} = A \begin{bmatrix} -1 & -2 & 6 \\ 1 & 1 & -1 \\ 0 & 0 & 1 \end{bmatrix}$ Applying $C_1 \rightarrow C_1 + C_2$, $C_3 \rightarrow C_3 - C_2$

$$\Rightarrow \begin{bmatrix} 1 & 0 & 0 \\ 0 & 1 & 0 \\ 0 & 0 & 2 \end{bmatrix} = A \begin{bmatrix} -16 & -11 & 6 \\ \dfrac{7}{2} & \dfrac{5}{2} & -1 \\ \dfrac{-5}{2} & \dfrac{-3}{2} & 1 \end{bmatrix} \quad \text{Applying } C_1 \to C_1 + \left(\dfrac{-5}{2}\right)C_3,$$
$$C_2 \to C_2 + \left(\dfrac{-3}{2}\right)C_1$$

$$\Rightarrow \begin{bmatrix} 1 & 0 & 0 \\ 0 & 1 & 0 \\ 0 & 0 & 1 \end{bmatrix} = A \begin{bmatrix} -16 & -11 & 3 \\ \dfrac{7}{2} & \dfrac{5}{2} & \dfrac{-1}{2} \\ \dfrac{-5}{2} & \dfrac{-3}{2} & \dfrac{1}{2} \end{bmatrix} \quad \text{Applying } C_3 \to C_3 \left(\dfrac{1}{2}\right)$$

$$\therefore \quad A^{-1} = \begin{bmatrix} -16 & -11 & 3 \\ \dfrac{7}{2} & \dfrac{5}{2} & \dfrac{-1}{2} \\ \dfrac{-5}{2} & \dfrac{-3}{2} & \dfrac{1}{2} \end{bmatrix}$$

EXAMPLE-2 *Find the inverse of the matrix* $A = \begin{bmatrix} 1 & -2 & -1 \\ 2 & -3 & 1 \\ 3 & -4 & 4 \end{bmatrix}$ *by using elementary row operations.*

SOLUTION We have,

$$A = IA$$

or, $\begin{bmatrix} 1 & -2 & -1 \\ 2 & -3 & 1 \\ 3 & -4 & 4 \end{bmatrix} = \begin{bmatrix} 1 & 0 & 0 \\ 0 & 1 & 0 \\ 0 & 0 & 1 \end{bmatrix} A$

$\Rightarrow \begin{bmatrix} 1 & -2 & -1 \\ 0 & 1 & 3 \\ 0 & 2 & 7 \end{bmatrix} = \begin{bmatrix} 1 & 0 & 0 \\ -2 & 1 & 0 \\ -3 & 0 & 1 \end{bmatrix} A \quad \text{Applying } R_2 \to R_2 - 2R_1, R_3 \to R_3 - 3R_1$

$\Rightarrow \begin{bmatrix} 1 & 0 & 5 \\ 0 & 1 & 3 \\ 0 & 0 & 1 \end{bmatrix} = \begin{bmatrix} -3 & 2 & 0 \\ -2 & 1 & 0 \\ 1 & -2 & 1 \end{bmatrix} A \quad \text{Applying } R_1 \to R_1 + 2R_2, R_3 \to R_3 + (-2)R_2$

$\Rightarrow \begin{bmatrix} 1 & 0 & 0 \\ 0 & 1 & 0 \\ 0 & 0 & 1 \end{bmatrix} = \begin{bmatrix} -8 & 12 & -5 \\ -5 & 7 & -3 \\ 1 & -2 & 1 \end{bmatrix} A \quad \text{Applying } R_1 \to R_1 + (-5)R_3, R_2 \to R_2 + (-3)R_3$

Hence, $A^{-1} = \begin{bmatrix} -8 & 12 & -5 \\ -5 & 7 & -3 \\ 1 & -2 & 1 \end{bmatrix}$

Let A be an invertible matrix. Then by applying a sequence $e_1, e_2, \ldots, e_k$ of elementary row operations, it can be reduced to the identity matrix I. If $E_1, E_2, \ldots, E_k$ are corresponding elementary matrices, then

$$I = E_1 E_2 \ldots E_k A$$

or $\quad E_1 E_2 E_3 \ldots E_k A = I$

$\Rightarrow \quad A = E_k^{-1} E_{k-1}^{-1} \ldots E_3^{-1} E_2^{-1} E_1^{-1} I \Rightarrow A = E_k^{-1} E_{k-1}^{-1} \ldots E_3^{-1} E_2^{-1} E_1^{-1}$

Since inverse of an elementary matrix is an elementary matrix. Therefore, the above result and Theorem 4 suggest the following algorithm for expressing a square matrix over a field F as a product of elementary matrices.

ALGORITHM

Step I   Obtain the matrix A (say).
Step II   Reduce A to the identity matrix I by applying a sequence of elementary row operations, keeping the track of the elementary row operations.
Step III   Write the inverse row operations.
Step IV   Write an elementary matrix corresponding to each inverse row operation in the same sequence.
Step V   Obtain the product of elementary matrices obtained in step IV in the same order. The product of elementary matrices so obtained is the required representation of A as the elementary matrices.

ILLUSTRATIVE EXAMPLES

EXAMPLE-1   *Express the matrix* $A = \begin{bmatrix} 1 & -3 \\ -2 & 4 \end{bmatrix}$ *as the product of elementary matrices.*

SOLUTION   We have,

$$A = \begin{bmatrix} 1 & -3 \\ -2 & 4 \end{bmatrix}$$

$\Rightarrow \quad A \sim \begin{bmatrix} 1 & -3 \\ 0 & -2 \end{bmatrix}, \quad$ Applying $R_2 \to R_2 + 2R_1$

$\Rightarrow \quad A \sim \begin{bmatrix} 1 & 1 \\ 0 & -2 \end{bmatrix}, \quad$ Applying $R_1 \to R_1 + (-2)R_2$

$\Rightarrow \quad A \sim \begin{bmatrix} 1 & 1 \\ 0 & 1 \end{bmatrix}, \quad$ Applying $R_2 \to R_2 \left(-\dfrac{1}{2}\right)$

$\Rightarrow \quad A \sim \begin{bmatrix} 1 & 0 \\ 0 & 1 \end{bmatrix}, \quad$ Applying $R_1 \to R_1 - R_2$

The sequence of elementary operations is:

$$R_2 \to R_2 + 2R_1; \quad R_1 \to R_1 + (-2R_2); \quad R_2 \to R_2 \left(-\dfrac{1}{2}\right); \quad R_1 \to R_1 - R_2$$

The sequence of inverse operations is:
$$R_2 \to R_2 - 2R_1; \quad R_1 \to R_1 + 2R_2; \quad R_2 \to (-2)R_2; \quad R_1 \to R_1 + R_2$$

The elementary matrices corresponding to these operations are:

$$E_1 = \begin{bmatrix} 1 & 0 \\ -2 & 1 \end{bmatrix}, E_2 = \begin{bmatrix} 1 & 2 \\ 0 & 1 \end{bmatrix}, E_3 = \begin{bmatrix} 1 & 0 \\ 0 & -2 \end{bmatrix}, E_4 = \begin{bmatrix} 1 & 1 \\ 0 & 1 \end{bmatrix}$$

$\therefore \quad A = E_1 E_2 E_3 E_4$

or, $\quad A = \begin{bmatrix} 1 & 0 \\ -2 & 1 \end{bmatrix} \begin{bmatrix} 1 & 2 \\ 0 & 1 \end{bmatrix} \begin{bmatrix} 1 & 0 \\ 0 & -2 \end{bmatrix} \begin{bmatrix} 1 & 1 \\ 0 & 1 \end{bmatrix}$

EXAMPLE-2 Write the matrix $A = \begin{bmatrix} 1 & 2 & 3 \\ 0 & 1 & 4 \\ 0 & 0 & 1 \end{bmatrix}$ as a product of elementary matrices.

SOLUTION We have,

$$A = \begin{bmatrix} 1 & 2 & 3 \\ 0 & 1 & 4 \\ 0 & 0 & 1 \end{bmatrix}$$

$\Rightarrow \quad A = \begin{bmatrix} 1 & 2 & 3 \\ 0 & 1 & 0 \\ 0 & 0 & 1 \end{bmatrix} \quad$ Applying $R_2 \to R_2 + (-4R_3)$

$\Rightarrow \quad A = \begin{bmatrix} 1 & 2 & 0 \\ 0 & 1 & 0 \\ 0 & 0 & 1 \end{bmatrix} \quad$ Applying $R_1 \to R_1 + (-3)R_3$

$\Rightarrow \quad A = \begin{bmatrix} 1 & 0 & 0 \\ 0 & 1 & 0 \\ 0 & 0 & 1 \end{bmatrix} \quad$ Applying $R_1 \to R_1 + (-2)R_2$

The sequence of elementary operations is:
$$R_2 \to R_2 + (-4)R_3; \quad R_1 \to R_1 + (-3)R_3; \quad R_1 \to R_1 + (-2)R_2$$

The sequence of inverse elementary operations is:
$$R_2 \to R_2 + 4R_3; \quad R_1 \to R_1 + 3R_3; \quad R_1 \to R_1 + 2R_2$$

The corresponding elementary matrices are:

$$E_1 = \begin{bmatrix} 1 & 0 & 0 \\ 0 & 1 & 4 \\ 0 & 0 & 1 \end{bmatrix}, \quad E_2 = \begin{bmatrix} 1 & 0 & 3 \\ 0 & 1 & 0 \\ 0 & 0 & 1 \end{bmatrix}, \quad E_3 = \begin{bmatrix} 1 & 2 & 0 \\ 0 & 1 & 0 \\ 0 & 0 & 1 \end{bmatrix},$$

$$\therefore \quad A = E_1 \, E_2 \, E_3$$

$$\Rightarrow \quad A = \begin{bmatrix} 1 & 0 & 0 \\ 0 & 1 & 4 \\ 0 & 0 & 1 \end{bmatrix} \begin{bmatrix} 1 & 0 & 3 \\ 0 & 1 & 0 \\ 0 & 0 & 1 \end{bmatrix} \begin{bmatrix} 1 & 2 & 0 \\ 0 & 1 & 0 \\ 0 & 0 & 1 \end{bmatrix}$$

EXAMPLE-3 Show that the matrix $A = \begin{bmatrix} 1 & 1 & 2 \\ 2 & 3 & 8 \\ -3 & -1 & 2 \end{bmatrix}$ cannot be expressed as the product of elementary matrices.

SOLUTION We have,

$$A = \begin{bmatrix} 1 & 1 & 2 \\ 2 & 3 & 8 \\ -3 & -1 & 2 \end{bmatrix}$$

$$\Rightarrow \quad A \sim \begin{bmatrix} 1 & 1 & 2 \\ 0 & 1 & 4 \\ 0 & 2 & 8 \end{bmatrix} \quad \text{Applying } R_1 \to R_1 - 2R_2, R_2 \to R_3 + 3R_1$$

$$\Rightarrow \quad A \sim \begin{bmatrix} 1 & 1 & 2 \\ 0 & 1 & 4 \\ 0 & 0 & 0 \end{bmatrix} \quad \text{Applying } R_3 \to R_3 + (-2)R_2$$

We find that in echelon form, matrix A has a zero row. So, the matrix A cannot be reduced to the identity matrix and hence it cannot be expressed as a product of elementary matrices.

Note: It should be noted here that $|A| = 0$. So, A is not invertible.

REMARK. *Any elementary row (column) operation on a matrix A is equivalent to multiplying A on the left (right) by an elementary matrix obtained by applying the same operation on an identity matrix. Following example illustrates the application of this result.*

EXAMPLE-4 If $A = \begin{bmatrix} 1 & 1 & 2 \\ 1 & 2 & 3 \\ 0 & -1 & 1 \end{bmatrix}$, find two non-singular matrices P and Q such that $PAQ = I$.

SOLUTION Clearly,

$$A = IAI$$

or, $\begin{bmatrix} 1 & 1 & 2 \\ 1 & 2 & 3 \\ 0 & -1 & 1 \end{bmatrix} = \begin{bmatrix} 1 & 0 & 0 \\ 0 & 1 & 0 \\ 0 & 0 & 1 \end{bmatrix} A \begin{bmatrix} 1 & 0 & 0 \\ 0 & 1 & 0 \\ 0 & 0 & 1 \end{bmatrix}$

$\Rightarrow \quad \begin{bmatrix} 1 & 1 & 2 \\ 0 & 1 & 1 \\ 0 & -1 & 1 \end{bmatrix} = \begin{bmatrix} 1 & 0 & 0 \\ -1 & 1 & 0 \\ 0 & 0 & 1 \end{bmatrix} A \begin{bmatrix} 1 & 0 & 0 \\ 0 & 1 & 0 \\ 0 & 0 & 1 \end{bmatrix} \quad \text{Applying } R_2 \to R_2 - R_1$

$$\Rightarrow \begin{bmatrix} 1 & 1 & 2 \\ 0 & 1 & 1 \\ 0 & 0 & 2 \end{bmatrix} = \begin{bmatrix} 1 & 0 & 0 \\ -1 & 1 & 0 \\ -1 & 1 & 1 \end{bmatrix} A \begin{bmatrix} 1 & 0 & 0 \\ 0 & 1 & 0 \\ 0 & 0 & 1 \end{bmatrix} \quad \text{Applying } R_3 \to R_3 + R_2$$

$$\Rightarrow \begin{bmatrix} 1 & 0 & 0 \\ 0 & 1 & 1 \\ 0 & 0 & 2 \end{bmatrix} = \begin{bmatrix} 1 & 0 & 0 \\ -1 & 1 & 0 \\ -1 & 1 & 1 \end{bmatrix} A \begin{bmatrix} 1 & -1 & -2 \\ 0 & 1 & 0 \\ 0 & 0 & 1 \end{bmatrix} \quad \text{Applying } C_2 \to C_2 - C_1, \, C_3 \to C_3 - 2C_1$$

$$\Rightarrow \begin{bmatrix} 1 & 0 & 0 \\ 0 & 1 & 0 \\ 0 & 0 & 2 \end{bmatrix} = \begin{bmatrix} 1 & 0 & 0 \\ -1 & 1 & 0 \\ -1 & 1 & 1 \end{bmatrix} A \begin{bmatrix} 1 & -1 & -1 \\ 0 & 1 & -1 \\ 0 & 0 & 1 \end{bmatrix} \quad \text{Applying } C_3 \to C_3 - C_2$$

$$\Rightarrow \begin{bmatrix} 1 & 0 & 0 \\ 0 & 1 & 0 \\ 0 & 0 & 1 \end{bmatrix} = \begin{bmatrix} 1 & 0 & 0 \\ -1 & 1 & 0 \\ -1 & 1 & 1 \end{bmatrix} A \begin{bmatrix} 1 & -1 & -\frac{1}{2} \\ 0 & 1 & -\frac{1}{2} \\ 0 & 0 & \frac{1}{2} \end{bmatrix} \quad \text{Applying } C_3 \to \frac{1}{2}C_3$$

EXERCISE 4.4

1. Let e_1, e_2, e_3 denote, respectively, the following elementary row operations on I_3:

    e_1 : "Interchange of R_2 and R_3, i.e. $R_2 \leftrightarrow R_3$"
    e_2 : "Replace R_2 by $3R_2$, i.e. $R_2 \to R_2(3)$"
    e_3 : "Replace R_1 by $R_1 + 2R_3$, i.e. $R_1 \to R_1 + 2R_3$"

    (i) Find the corresponding elementary matrices E_1, E_2, E_3.
    (ii) Find the inverse elementary operations $e_1^{-1}, e_2^{-1}, e_3^{-1}$ and their corresponding elementary matrices E_1', E_2', E_3' and the relationship between them and E_1, E_2, E_3.
    (iii) Describe the corresponding elementary column operations e_1', e_2', e_3'.
    (iv) Find elementary matrices F_1, F_2, F_3 corresponding to e_1', e_2', e_3' and the relationship between them and E_1, E_2, E_3.

2. Find the inverse of each of the following matrices (if it exists):

$$A = \begin{bmatrix} 1 & -2 & -1 \\ 2 & -3 & 1 \\ 3 & -4 & 4 \end{bmatrix}, \, B = \begin{bmatrix} 1 & 2 & 3 \\ 2 & 6 & 1 \\ 3 & 10 & -1 \end{bmatrix}, \, C = \begin{bmatrix} 1 & 3 & -2 \\ 2 & 8 & -3 \\ 1 & 7 & 1 \end{bmatrix}, \, D = \begin{bmatrix} 2 & 1 & -1 \\ 5 & 2 & -3 \\ 0 & 2 & 1 \end{bmatrix}$$

3. Express each of the following matrices as a product of elementary matrices:

$$A = \begin{bmatrix} 1 & 2 \\ 3 & 4 \end{bmatrix}, \, B = \begin{bmatrix} 3 & -6 \\ -2 & 4 \end{bmatrix}, \, C = \begin{bmatrix} 2 & 6 \\ -3 & -7 \end{bmatrix}, \, D = \begin{bmatrix} 1 & 2 & 0 \\ 0 & 1 & 3 \\ 3 & 8 & 7 \end{bmatrix}$$

ANSWERS

1. (i) $E_1 = \begin{bmatrix} 1 & 0 & 0 \\ 0 & 0 & 1 \\ 0 & 1 & 0 \end{bmatrix}$, $E_2 = \begin{bmatrix} 1 & 0 & 0 \\ 0 & 3 & 0 \\ 0 & 0 & 1 \end{bmatrix}$, $E_3 = \begin{bmatrix} 1 & 0 & 2 \\ 0 & 1 & 0 \\ 0 & 0 & 1 \end{bmatrix}$

   (ii) $R_2 \leftrightarrow R_3$, $R_2 \to R_2\left(\dfrac{1}{3}\right)$, $R_1 \to R_1 - 2R_3$, $E_i' = E_i^{-1}$; $i = 1,2,3$.

   (iii) $C_2 \leftrightarrow C_3$, $C_2 \to C_2(3)$, $C_1 \to C_1 + 2C_3$

   (iv) $F_i = E_i^T$; $i = 1, 2, 3$

2. $A^{-1} = \begin{bmatrix} -8 & 12 & -5 \\ 5 & 7 & -3 \\ 1 & -2 & 1 \end{bmatrix}$, B is not invertible,

   $C^{-1} = \begin{bmatrix} \dfrac{29}{2} & \dfrac{-17}{2} & \dfrac{7}{2} \\ \dfrac{-5}{2} & \dfrac{3}{2} & \dfrac{-1}{2} \\ 3 & -2 & 1 \end{bmatrix}$, $D^{-1} = \begin{bmatrix} 8 & -3 & -1 \\ -5 & 2 & 1 \\ 10 & -4 & -1 \end{bmatrix}$

3. $A = \begin{bmatrix} 1 & 0 \\ 3 & 1 \end{bmatrix} \begin{bmatrix} 1 & 0 \\ 0 & -2 \end{bmatrix} \begin{bmatrix} 1 & 2 \\ 0 & 1 \end{bmatrix}$, B is not invertible

   $C = \begin{bmatrix} 1 & 0 \\ -\dfrac{3}{2} & 1 \end{bmatrix} \begin{bmatrix} 1 & 0 \\ 0 & 2 \end{bmatrix} \begin{bmatrix} 1 & 6 \\ 0 & 1 \end{bmatrix} \begin{bmatrix} 2 & 0 \\ 0 & 1 \end{bmatrix}$

   $D = \begin{bmatrix} 1 & 0 & 0 \\ 0 & 1 & 0 \\ 3 & 0 & 1 \end{bmatrix} \begin{bmatrix} 1 & 0 & 0 \\ 0 & 1 & 0 \\ 0 & 2 & 1 \end{bmatrix} \begin{bmatrix} 1 & 0 & 0 \\ 0 & 1 & 3 \\ 0 & 0 & 1 \end{bmatrix} \begin{bmatrix} 1 & 2 & 0 \\ 0 & 1 & 0 \\ 0 & 0 & 1 \end{bmatrix}$

4.10 EIGENVALUES AND EIGENVECTORS OF A LINEAR TRANSFORMATION

For a variety of physical problems, interest centres around a given linear transformation t and those scalars λ and vectors v which satisfy the equation $t(v) = \lambda v, v \neq 0$

EIGNEVECTOR *Let V be a finite dimensional vector space over a field F. A non-zero vector $v \in V$ is an eigenvector of a linear transformation $t : V \to V$, if there exist a scalar $\lambda \in F$ such that $t(v) = \lambda v$.*

Consider the identity map $I : V \to V$ given by

$I(v) = v \quad$ for all $v \in V$

$\Rightarrow \quad I(v) = 1v \quad$ for all $v \in V$

$\Rightarrow$ For every non-zero vector $v \in V$ there is scalar $\lambda = 1 \in F$ such that $I(v) = \lambda v$ for all $v \in V$
$\Rightarrow$ Every non-zero vector in V is an eigenvector of $I : V \to V$.

Hence, the set of eigenvectors of $I : V \to V$ is $V - \{0\}$.

EIGENVALUE *Let V be a finite dimensional vector space over a field F. A scalar $\lambda \in F$ is called an eigenvalue of a linear transformation $t : V \to V$ if there exists a non-zero vector v such that $t(v) = \lambda v$*

Clearly, every vector satisfying the above definition is called an eigenvector of t belonging to the eigenvalue λ. It also follows from the above definitions that each eigenvector v determines an eigenvalue λ, while each eigenvalue must arise in this way from at least one eigenvector.

REMARK. *The null vector 0_V is not an eigenvector of any linear transformation $t : V \to V$.*

Let v be an eigenvector corresponding to an eigenvalue λ of a linear transformation $t : V \to V$. Then,

$$t(v) = \lambda v$$

Let $a \neq 0$ be a non-zero scalar in F. Then, for any vector $av \in F$, we have

$$t(av) = at(v) = a(\lambda v) = \lambda(av) = \lambda\, t(av)$$

$\Rightarrow$ av is an eigenvector corresponding to eigenvalue λ.

Thus, corresponding to an eigenvalue, there are infinitely many eigenvectors.

The eigenvectors and eigenvalues are often called characteristic values and characteristic vectors or proper values and proper vectors.

The set of all eigenvalues of a linear transformation $t : V \to V$ is called the spectrum of t.

THEOREM-1 *Let $V(F)$ be a finite dimensional vector space. If v is an eigenvector of a linear transformation $t : V \to V$, then v cannot correspond to more than one eigenvalue of t.*

PROOF. Let v be an eigenvector of t corresponding to two distinct eigenvalues λ_1 and λ_2 of t. Then,

$$t(v) = \lambda_1 v \text{ and } t(v) = \lambda_2 v$$
$\Rightarrow \quad \lambda_1 v = \lambda_2 v$
$\Rightarrow \quad (\lambda_1 - \lambda_2) v = 0_V$
$\Rightarrow \quad \lambda_1 - \lambda_2 = 0 \qquad\qquad [\because v \neq 0_V]$
$\Rightarrow \quad \lambda_1 = \lambda_2$

Hence, v cannot correspond to more than one eigenvalue of t.

Q.E.D.

THEOREM-2 *Let V be a finite dimensional vector space over a field F. The scalar $0 \in F$ is an eigenvalue of a linear transformation $t : V \to V$ iff t is singular.*

PROOF. First suppose that 0 is an eigenvalue of t. Then there exists a non-zero vector $v \in V$ such that
$$t(v) = 0v \Rightarrow \quad t(v) = 0 \Rightarrow t \text{ is singular}$$

Conversely, let $t : V \to V$ be a singular linear transformation. Then, there exists a non-zero vector $v \in V$ such that

$\qquad t(v) = 0_V$
$\Rightarrow \qquad t(v) = 0v$
$\Rightarrow \qquad 0 \in F$ is an eigenvalue of t.

Q.E.D.

THEOREM-3 *Let $V(F)$ be a finite dimensional vector space and let $t : V \to V$ be an invertible linear transformation. If λ is an eigenvalue of t, then show that λ^{-1} is an eigenvalue of t^{-1}.*

PROOF. Since t is invertible. So, it is non-singular. Therefore, $\lambda \neq 0$ and so $\lambda^{-1} \in F$.
Now, λ is an eigenvalue of t.

$\Rightarrow \quad$ There exists a non-zero vector $v \in V$ such that $t(v) = \lambda v$
$\Rightarrow \quad t^{-1}(t(v)) = t^{-1}(\lambda v)$
$\Rightarrow \quad (t^{-1} o t)(v) = \lambda\, t^{-1}(v) \qquad\qquad [\because\ t^{-1} : V \to V \text{ is a linear transformation}]$
$\Rightarrow \quad I(v) = \lambda\, t^{-1}(v) \qquad\qquad\qquad\qquad\qquad [\because\ t^{-1} o t = I]$
$\Rightarrow \quad v = \lambda\, t^{-1}(v)$
$\Rightarrow \quad t^{-1}(v) = \lambda^{-1} v$

Hence, λ^{-1} is a value of t^{-1}.

Q.E.D.

THEOREM-4 *Let $V(F)$ be a finite dimensional vector space and let λ be an eigenvalue of a linear transformation $t : V \to V$. Then the set of all eigenvectors associated with an eigenvalue λ, by adjoining zero vector to it, is a subspace of V.*

PROOF. Let S_λ denote the set of all eigenvectors of t associated with eigenvalue λ. Then,
$$S_\lambda = \{v \in V : t(v) = \lambda v\}$$

Let $S'_\lambda = S_\lambda \cup \{0_V\}$.
Clearly, $0_V \in S'_\lambda$. So, S'_λ is a non-empty subset of V.
Let $u, v \in S'_\lambda$. Then, $t(u) = \lambda u$ and $t(v) = \lambda v$.

For any $a, b \in F$ and $u, v \in S'_\lambda$, we have

$$\begin{aligned}
t(au+bv) &= at(u) + bt(v) && [\because \; t:V \to V \text{ is a linear transformation}] \\
&= a(\lambda u) + b(\lambda v) && [\because \; t(u) = \lambda u \text{ and } t(v) = \lambda v] \\
&= (a\lambda)u + (b\lambda)v \\
&= (\lambda a)u + (\lambda b)v \\
&= \lambda(au) + \lambda(bv) \\
&= \lambda(au+bv)
\end{aligned}$$

$\therefore \quad au + bv \in S'_\lambda$ for all $u, v \in S'_\lambda$ and $a, b \in F$

Hence, S'_λ is a subspace of V.

Q.E.D.

REMARK. S'_λ is called the eigenspace of λ or the space of eigenvectors of t associated with eigenvalue λ.

THEOREM-5 *Let $V(F)$ be a finite dimensional vector space. Then a scalar λ is an eigenvalue of a linear transformation $t : V \to V$ iff the linear transformation $t - \lambda I$ is singular.*

SOLUTION λ is the eigenvalue of t

$\Leftrightarrow$ There exists a non-zero vector $v \in V$ such that $t(v) = \lambda v$

$\Leftrightarrow t(v) = \lambda I(v)$ $\qquad [\because \; I(v) = v \text{ for all } v \in V]$

$\Leftrightarrow t(v) = (\lambda I)(v)$

$\Leftrightarrow t(v) - (\lambda I)(v) = 0$

$\Leftrightarrow (t - \lambda I)(v) = 0$

$\Leftrightarrow t - \lambda I$ is singular $\qquad [\because \; (t - \lambda I)(v) = 0 \text{ for some non-zero vector } v \in V]$

THEOREM-6 *Let $V(F)$ be a finite dimensional vector space. Then distinct non-zero eigenvectors of a linear transformation $t : V \to V$ corresponding to distinct eigenvalues of t are linearly independent.*

PROOF. Let $v_1, v_2, v_3, \ldots, v_n$ be n distinct eigenvectors of t corresponding to n distinct eigenvalues $\lambda_1, \lambda_2, \ldots, \lambda_n$ respectively. Then,

$$t(v_i) = \lambda_i v_i, \quad i = 1, 2, \ldots, n$$

In order to prove that $v_1, v_2, v_3, \ldots, v_n$ are linearly independent vectors in V, we will use induction on n.

If $n = 1$, then v_1 being a non-zero vector is linearly independent.

Now suppose that the vectors $v_1, v_2, \ldots, v_k$ where $k < n$ are linearly independent.

Consider now the vectors $v_1, v_2, \ldots, v_k, v_{k+1}$. We shall now show that these vectors are linearly independent.

Let $\alpha_1, \alpha_2, \alpha_3, \ldots, \alpha_k, \alpha_{k+1}$ be scalars in F such that

$$\alpha_1 v_1 + \alpha_2 v_2 + \cdots + \alpha_k v_k + \alpha_{k+1} v_{k+1} = 0_V \qquad (i)$$
$$\Rightarrow t(\alpha_1 v_1 + \alpha_2 v_2 + \cdots + \alpha_k v_k + \alpha_{k+1} v_{k+1}) = t(0_V)$$
$$\Rightarrow \alpha_1 t(v_1) + \alpha_2 t(v_2) + \cdots + \alpha_k t(v_k) + \alpha_{k+1} t(v_{k+1}) = 0_V$$
$$\Rightarrow \alpha_1 (\lambda_1 v_1) + \alpha_2 (\lambda_2 v_2) + \cdots + \alpha_k (\lambda_k v_k) + \alpha_{k+1} (\lambda_{k+1} v_{k+1}) = 0_V$$
$$\Rightarrow (\lambda_1 \alpha_1) v_1 + (\lambda_2 \alpha_2) v_2 + \cdots + (\lambda_k \alpha_k) v_k + (\lambda_{k+1} \alpha_{k+1}) v_{k+1} = 0_V \qquad (ii)$$

Multiplying (i) by λ_{k+1} and subtracting from (ii), we get

$$(\lambda_1 - \lambda_{k+1}) \alpha_1 v_1 + (\lambda_2 - \lambda_{k+1}) \alpha_2 v_2 + \cdots + (\lambda_k - \lambda_{k+1}) \alpha_k v_k = 0_V$$
$$\Rightarrow (\lambda_i - \lambda_{k+1}) \alpha_i = 0; \quad i = 1, 2, \ldots, k \qquad [\because v_1, v_2, \ldots, v_k \text{ are } l.i.]$$
$$\Rightarrow \alpha_i = 0; \quad i = 1, 2, \ldots, k \qquad [\because \lambda_i \neq \lambda_{k+1}, i = 1, 2, \ldots, k]$$

Substituting $\alpha_1 = \alpha_2 = \cdots = \alpha_k = 0$ in (i), we get

$$\alpha_{k+1} v_{k+1} = 0_V \Rightarrow \alpha_{k+1} = 0 \qquad [\because v_{k+1} \neq 0_V]$$

Thus,

$$\alpha_1 v_1 + \alpha_2 v_2 + \cdots + \alpha_k v_k + \alpha_{k+1} v_{k+1} = 0_V$$
$$\Rightarrow \alpha_1 = \alpha_2 = \cdots = \alpha_k = \alpha_{k+1} = 0$$
$$\Rightarrow v_1, v_2, \ldots, v_k, v_{k+1} \text{ are linearly independent vectors in } V.$$

Hence, by induction $v_1, v_2, \ldots, v_n$ are linearly independent vectors in V.

Q.E.D.

COROLLARY. *Let $V(F)$ be a finite dimensional vector space of dimension n. If $t : V \to V$ is a linear transformation, then t cannot have more than n distinct eigenvalues in F.*

PROOF. It is given that $\dim V = n$. So, t cannot have more than n linearly independent vectors. Since distinct eigenvalues of t give rise distinct linearly independent vectors. So, t cannot have more than n distinct eigenvalues in F.

Q.E.D.

COROLLARY. *Let V be a finite dimensional vector space of dimension n. Let $t : V \to V$ be a linear transformation. If t has n distinct eigenvalues, then there is a basis of V which consists of eigenvalues of t.*

PROOF. By Theorem 6, n eigenvectors of t corresponding to n eigenvalues are linearly independent. Since $\dim V = n$. Therefore, these n linearly independent vectors in V forms a basis for V. Hence, the list of n eigenvectors of t is a basis for V.

Q.E.D.

EXERCISE 4.5

1. Let $V(F)$ be a finite dimensional vector space. Let $t_1 : V \to V$ and $t_2 : V \to V$ be linear transformations. If v is an eigenvector of t_1 and t_2, show that v is also an eigenvector of $\lambda t_1 + \mu t_2$, where $\lambda, \mu \in F$.

2. Let $V(F)$ be a finite dimensional vector space and $t : V \to V$ be a linear transformation. If v is an eigenvector of t belonging to the eigenvalue λ, prove that

    (i) For $n > 0$, v is an eigenvector of t^n belonging to λ^n, where $t^2 = tot$, $t^3 = totot$, etc.

    (ii) $f(\lambda)$ is an eigenvalue of $f(t)$ for any polynomial $f(t)$.

3. Let $t_1 : V \to V$ and $t_2 : V \to V$ be linear transformations such that $\lambda \neq 0$ is an eigenvalue of $t_1 o t_2$. Show that λ is also an eigenvalue of $t_2 o t_1$.

4. Let $V(F)$ be a finite dimensional vector space and $t : V \to V$ be a linear transformation with λ as an eigenvalue. Show that the eigenspace of $t - \lambda I$ is the null space.

5. Let $V(F)$ be a finite dimensional vector space and $t : V \to V$ be a linear transformation. Show that 0_V is a eigenvector of t iff t is not invertible.

4.11 EIGENVALUES AND EIGENVECTORS OF A MATRIX

The concept of eigenvalues and eigenvectors can be discussed from two point of view, viz.(i) linear transformation point of view (ii) matrix point of view. In the previous section, we have studied about the first one. In this section, we shall discuss the second one. In fact, these two concepts are essentially the same as an $n \times n$ square matrix A over a field F can be viewed as a linear transformation $t_A : F^n \to F^n$ given by $t_A(X) = AX$. So, we define eigenvalues and eigenvectors of a matrix as follows.

EIGENVALUE OF A MATRIX Let A be an $n \times n$ matrix over a field F. A scalar $\lambda \in F$ is called an eigenvalue of A iff it is an eigenvalue of linear transformation $t_A : F^n \to F^n$ representing matrix A.

EIGENVECTOR OF A MATRIX Let A be an $n \times n$ matrix over a field F. A non-zero vector $X \in F^n$ is an eigenvector of A iff it is an eigenvector of the linear transformation $t_A : F^n \to F^n$.

THEOREM-1 Let A be $n \times n$ matrix over a field F. A non-zero vector $X \in F^n$ (or a column vector $X \in F^n$) is an eigenvector of $t_A : F^n \to F^n$ iff there exists a scalar $\lambda \in F$ such that $AX = \lambda X$.

PROOF. First suppose that $O \neq X \in F^n$ is an eigenvector of matrix A. Then, X is an eigenvector of $t_A : F^n \to F^n$. Therefore, there exists a scalar $\lambda \in F$ such that

$$t_A(X) = \lambda X$$
$$\Rightarrow \quad AX = \lambda X \qquad [\because\ t_A : F^n \to F^n \text{ is defined as } t_A(X) = AX]$$

Conversely, let X be a non-zero vector in F^n such that
$$AX = \lambda X \quad \text{for some scalar } \lambda \in F$$
$\Rightarrow \quad t_A(X) = \lambda X \hfill [\because \ t_A(X) = AX]$
$\Rightarrow \quad X$ is an eigenvector of t_A
$\Rightarrow \quad X$ is an eigenvector of A.

Q.E.D.

ILLUSTRATIVE EXAMPLES

EXAMPLE-1 *Show that a square matrix A has 0 as eigenvalue iff A is not invertible.*

SOLUTION Let A be an $n \times n$ matrix over a field F. First, let 0 be an eigenvalue of matrix A and $X(\neq O) \in F^n$ be the corresponding eigenvector. Then,
$$AX = 0X$$
$\Rightarrow \quad AX = O$

If possible, let A be an invertible matrix. Then,
$$AX = O$$
$\Rightarrow \quad A^{-1}(AX) = A^{-1}O \hfill [A^{-1} \text{ is the inverse of } A]$
$\Rightarrow \quad (A^{-1}A)X = O$
$\Rightarrow \quad IX = O$
$\Rightarrow \quad X = O$

But $X \neq O$. So, we arrive at a contradiction.

Hence, A must be a non-invertible matrix.

Conversely, let A be a non-invertible matrix. Then, A is singular. So, the system of equations $AX = O$ has non-trivial solutions. So, there exists a non-zero vector $X \in F^n$ such that $AX = O \Rightarrow AX = 0X \Rightarrow 0$ is an eigenvalue of A.

EXAMPLE-2 *If λ is an eigenvalue of an invertible matrix A over a field F, then λ^{-1} is an eigenvalue of A^{-1}.*

SOLUTION Since A is an invertible matrix with an eigenvalue λ. Therefore, by above example, $\lambda \neq 0$. So, $\lambda^{-1} \in F$.

Now,

λ is an eigenvalue of matrix A
$\Rightarrow \quad$ There exists a non-zero vector $X \in F^n$ such that $AX = \lambda X$
$\Rightarrow \quad A^{-1}(AX) = A^{-1}(\lambda X) \hfill [\because \ A \text{ is invertible} \therefore A^{-1} \text{ exists}]$
$\Rightarrow \quad (A^{-1}A)X = \lambda(A^{-1}X)$
$\Rightarrow \quad IX = \lambda(A^{-1}X)$

436 • *Theory and Problems of Linear Algebra*

$\Rightarrow \quad X = \lambda(A^{-1}X)$
$\Rightarrow \quad \lambda^{-1}X = A^{-1}X$
$\Rightarrow \quad \lambda^{-1}$ is the eigenvalue of A^{-1}.

EXAMPLE-3 *Let A be $n \times n$ matrix over a field F such that the sum of the entries of each row is 1. Prove that 1 is an eigenvalue of A.*

SOLUTION Let $A = [a_{ij}]$ be $n \times n$ matrix such that $\sum_{j=1}^{n} a_{ij} = 1$ for all $i \in \underline{n}$

or,
$$a_{11} + a_{12} + \cdots + a_{1n} = 1$$
$$a_{21} + a_{22} + \cdots + a_{2n} = 1$$
$$\vdots$$
$$a_{n1} + a_{n2} + \cdots + a_{nn} = 1$$

$\Rightarrow \quad \begin{bmatrix} a_{11} & a_{12} & \cdots & a_{1n} \\ a_{21} & a_{22} & \cdots & a_{2n} \\ \vdots & \vdots & & \vdots \\ a_{n1} & a_{n2} & \cdots & a_{nn} \end{bmatrix} \begin{bmatrix} 1 \\ 1 \\ \vdots \\ 1 \end{bmatrix} = \begin{bmatrix} 1 \\ 1 \\ \vdots \\ 1 \end{bmatrix}$

$\Rightarrow \quad AX = X$, where $X = \begin{bmatrix} 1 \\ 1 \\ \vdots \\ 1 \end{bmatrix}$

$\Rightarrow \quad AX = 1X$
$\Rightarrow \quad 1$ is an eigenvalue of A.

EXAMPLE-4 *Let A be an $n \times n$ matrix over a field F. If λ is an eigenvalue of A, then λ^2 is an eigenvalue of A^2.*

SOLUTION It is given that λ is an eigenvalue of A. So, there exists a non-zero vector $X \in F^n$ such that $AX = \lambda X$

or, $\quad AX - \lambda X = O$
$\Rightarrow \quad (A - \lambda I)X = O$ \hfill (i)
$\Rightarrow \quad A(A - \lambda I)X = AO$
$\Rightarrow \quad (A^2 - \lambda AI)X = O$

$\Rightarrow \quad (A^2 - \lambda A)X = O$

$\Rightarrow \quad (A^2 - \lambda(A - \lambda I) - \lambda^2 I)X = O$

$\Rightarrow \quad \{(A^2 - \lambda^2 I) - \lambda(A - \lambda I)\}X = O$

$\Rightarrow \quad (A^2 - \lambda^2 I)X - \lambda(A - \lambda I)X = O$

$\Rightarrow \quad (A^2 - \lambda^2 I)X - \lambda O = O$ [Using (i)]

$\Rightarrow \quad (A^2 - \lambda^2 I)X = O$

Thus, λ^2 is an eigenvalue of A^2.

4.12.1 MORE THEOREMS ON EIGENVALUES AND EIGENVECTORS OF A MATRIX

THEOREM-1 *Let $V(F)$ be a finite dimensional vector space and $t : V \to V$ be a linear transformation. If A is the matrix of t relative to some basis B of vector space V, then the eigenvalues of A are exactly same as those of t.*

PROOF. Let $A = [a_{ij}]$ be an $n \times n$ matrix of linear transformation $t : V \to V$ relative to basis $B = \{b_1, b_2, \ldots, b_n\}$ of vector space V. Then,

$$t(b_j) = \sum_{i=1}^{n} a_{ij} b_i \quad \text{(i)}$$

Let v be any vector in V such that $v = \sum_{j=1}^{n} \alpha_j b_j$, where $\alpha_j \in F$. Then,

$$t(v) = t\left(\sum_{j=1}^{n} \alpha_j b_j\right)$$

$\Rightarrow \quad t(v) = \sum_{j=1}^{n} \alpha_j t(b_j)$

$\Rightarrow \quad t\left(\sum_{i=1}^{n} \alpha_i b_i\right) = \sum_{j=1}^{n} \alpha_j \left(\sum_{i=1}^{n} a_{ij} b_i\right)$ [Using (i)]

$\Rightarrow \quad t\left(\sum_{i=1}^{n} \alpha_i b_i\right) = \sum_{i=1}^{n} \left(\sum_{j=1}^{n} a_{ij} \alpha_j\right) b_i$

$\Rightarrow \quad t\left(\sum_{i=1}^{n} \alpha_i b_i\right) = \sum_{i=1}^{n} \lambda \alpha_i b_i \quad \text{iff} \quad \sum_{j=1}^{n} a_{ij} \alpha_j = \lambda \alpha_i \quad \text{for all } i \in \underline{n}$

$\Rightarrow \quad t\left(\sum_{i=1}^{n} \alpha_i b_i\right) = \lambda \left(\sum_{i=1}^{n} \alpha_i b_i\right) \quad \text{iff} \quad AX = \lambda X, \text{ where } X = \begin{bmatrix} \alpha_1 \\ \alpha_2 \\ \vdots \\ \alpha_n \end{bmatrix}$

$\Rightarrow \quad t(v) = \lambda v \quad \text{iff} \quad AX = \lambda X$

$\Rightarrow \quad \lambda$ is an eigenvalue of t iff it is an eigenvalue of A.

Q.E.D.

THEOREM-2 *Let $V(F)$ be a finite dimensional vector space and $t : V \to V$ be a linear transformation. If A is the matrix of t relative to some basis B of vector space V, then a vector $v \in V$ is an eigenvector of t corresponding to its eigenvalue λ iff its coordinate matrix relative to the basis B is an eigenvector of A corresponding to its eigenvalue λ.*

PROOF. We have, $M_B^B(t) = A$

$\therefore \quad M_B^B(t - \lambda I_V) = M_B^B(t) - M_B^B(\lambda I_V) = M_B^B(t) - \lambda M_B^B(I_V) = A - \lambda I$

Let X be the coordinate vector of v relative to basis B. Then, $X = M_B^B(v)$.

For any non-zero vector $v \in V$, we have

$$M_B^B((t - \lambda I_V)(v)) = M_B^B(t - \lambda I_V) M_B^B(v) \qquad [\because \quad M_B^B(t(v)) = M_B^B(t)\, M_B^B(v)]$$

$\Rightarrow \quad M_B^B(t(v) - \lambda v) = (A - \lambda I) X$

$\Rightarrow \quad M_B^B(t(v) - \lambda v) = AX - \lambda X \qquad \qquad \qquad \qquad \qquad \qquad \qquad \qquad \qquad \qquad \text{(i)}$

Now,

v is an eigenvector of $t : V \to V$ corresponding to eigenvalue λ

$\Leftrightarrow \quad t(v) = \lambda v$

$\Leftrightarrow \quad t(v) - \lambda v = 0_V$

$\Leftrightarrow \quad M_B^B(t(v) - \lambda v) = O$ (Null matrix)

$\Leftrightarrow \quad AX - \lambda X = O \qquad\qquad\qquad\qquad\qquad\qquad\qquad\qquad\qquad\qquad\qquad$ [Using (i)]

$\Leftrightarrow \quad AX = \lambda X$

Thus, v is an eigenvector of t iff $X \left(= M_B^B(v)\right)$ is an eigenvector of A and the linear transformation t and the matrix A have the common eigenvalue λ.

Q.E.D.

REMARK. *It is evident from the above theorem that each eigenvalue of t is an eigenvalue of A, i.e. the matrix of t relative to some basis of V.*

THEOREM-3 *Let A be an $n \times n$ matrix over a field F. Then a scalar λ is an eigenvalue of A iff the matrix $A - \lambda I$ is singular.*

PROOF. The scalar λ is an eigenvalue of A.

$\Leftrightarrow \quad$ There exists a non-zero vector $X \in F^n$ such that $AX = \lambda X$

$\Leftrightarrow \quad AX = \lambda(IX)$

$\Leftrightarrow \quad AX - (\lambda I)X = 0$

$\Leftrightarrow \quad (A - \lambda I)X = 0, \quad$ where X is a non-zero vector in F^n

$\Leftrightarrow \quad |A - \lambda I| = 0 \qquad\qquad\qquad\qquad\qquad\qquad [\because \quad AX = 0$ has non-trivial solutions iff $|A| = 0]$

$\Leftrightarrow \quad A - \lambda I$ is singular.

Q.E.D.

THEOREM-4 *A linear transformation $t : V \to V$ from an n dimensional vector space V into itself has a diagonal matrix A relative to some basis B of V iff the eigenvectors of t span V. When this is the case, the eigenvalues of t are exactly the diagonal entries of A and each occurs on the diagonal a number of times equal to the dimension of its eigenspace.*

PROOF. First suppose that V is spanned by the n-eigenvectors $b_1, b_2, \ldots, b_n$ of $t : V \to V$. By Theorem 6 on page 432, $b_1, b_2, \ldots, b_n$ are linearly independent vectors in V and hence $B = \{b_1, b_2, \ldots, b_n\}$ is a basis for V. Let $\lambda_1, \lambda_2, \ldots, \lambda_n$ be the eigenvalues of t corresponding to the eigenvectors $b_1, b_2, \ldots, b_n$ respectively. Then,

$$t(b_i) = \lambda_i b_i \; ; \; i = 1, 2, \ldots, n.$$

Let $X = \{x_1, x_2, \ldots, x_n\}$ be the basis dual to B. Then,

$$\left(M_B^B(t)\right)_{ij} = x_i\left(t(b_j)\right)$$

$\Rightarrow \quad \left(M_B^B(t)\right)_{ij} = x_i(\lambda_j b_j) = \lambda_j x_i(b_j) = \lambda_j \delta_{ij} \qquad [\because \; x_i(b_j) = \delta_{ij}]$

$\Rightarrow \quad \left(M_B^B(t)\right)_{ij} = \begin{cases} \lambda_i & \text{for } i = j \in \underline{n} \\ 0 & \text{for } i \neq j, \; i, j \in \underline{n} \end{cases}$

$\Rightarrow \quad M_B^B(t)$ is a diagonal matrix with diagonal entries $\lambda_1, \lambda_2, \ldots, \lambda_n$.

Conversely, let us assume that $t : V \to V$ is a linear transformation such that the matrix of t relative to some basis $B = \{b_1, b_2, \ldots, b_n\}$ of V is a diagonal matrix $A = [a_{ij}]$ such that

$$a_{ij} = \begin{cases} 0, & \text{if } i \neq j, \; i, j \in \underline{n} \\ \lambda_i, & \text{if } i = j \in \underline{n} \end{cases}$$

Now,

$A = [a_{ij}] = M_B^B(t)$

$\Rightarrow \quad t(b_i) = \sum_{r=1}^{n} a_{ri} b_r, \quad i \in \underline{n}$

$\Rightarrow \quad t(b_i) = a_{ii} b_i, \quad i \in \underline{n} \qquad [\because \; a_{ij} = 0 \text{ for } i \neq j]$

$\Rightarrow \quad t(b_i) = \lambda_i b_i, \quad i \in \underline{n} \qquad [\because \; a_{ii} = \lambda_i \text{ for all } i \in \underline{n}]$

$\Rightarrow \quad b_1, b_2, \ldots, b_n$ are eigenvectors of t corresponding to the eigenvectors $\lambda_1, \lambda_2, \ldots, \lambda_n$ respectively.

Now, let the matrix of t be diagonal matrix relative to the basis $B = \{b_1, b_2, \ldots, b_n\}$ with diagonal entries $\lambda_1, \lambda_2, \ldots, \lambda_n$ the eigenvalues of t.

Thus, if matrix of $t : V \to V$ is a diagonal matrix relative to some basis, then the basis vectors are the eigenvectors and their corresponding eigenvalues are diagonal entries of the matrix. Since eigenvectors form a basis of V. So, they span V.

Then we have to prove that the eigenvalues t are exactly the diagonal entries of A and an eigenvalue occurs on the diagonal a number of times equal to the dimension of its eigenspace.

Let v be an eigenvector of t with eigenvalue λ. Then, $t(v) = \lambda v$

Since B is a basis of V. Therefore, v can be written as

$$v = \sum_{i=1}^{n} \alpha_i b_i, \quad \alpha_i \in F, \quad i \in \underline{n}$$

$\Rightarrow \quad t(v) = t\left(\sum_{i=1}^{n} \alpha_i b_i\right)$

$\Rightarrow \quad t(v) = \sum_{i=1}^{n} \alpha_i t(b_i)$ \hfill $[\because \; t \text{ is a linear transformation}]$

$\Rightarrow \quad \lambda v = \sum_{i=1}^{n} \alpha_i (\lambda_i b_i)$ \hfill $[\because \; t(b_i) = \lambda_i b_i, i \in \underline{n}]$

$\Rightarrow \quad \lambda\left(\sum_{i=1}^{n} \alpha_i b_i\right) = \sum_{i=1}^{n} (\alpha_i \lambda_i) b_i$

$\Rightarrow \quad \sum_{i=1}^{n} (\lambda \alpha_i - \alpha_i \lambda_i) b_i = 0_V$

$\Rightarrow \quad \sum_{i=1}^{n} (\lambda - \lambda_i) \alpha_i b_i = 0_V$

$\Rightarrow \quad (\lambda - \lambda_i) \alpha_i = 0_V$ \hfill $[\because \; B \text{ is a basis of } V]$

$\Rightarrow \quad \lambda - \lambda_i = 0 \quad \text{for some } i$ \hfill $[\because \; v \neq 0_V \; \therefore \; \alpha_i \neq 0 \text{ for some } i]$

$\Rightarrow \quad \lambda = \lambda_i \quad \text{for some } i$

Thus, any eigenvalue λ of t must be one of the given diagonal entries $\lambda_1, \lambda_2, \ldots, \lambda_n$.

Let us now suppose that an eigenvalue λ appears m times in the diagonal. Let the first m diagonal entries be equal to λ, i.e. $\lambda_1 = \lambda_2 = \cdots = \lambda_m = \lambda$. Then we have to prove that the dimension of eigenspace of λ is m.

Let $S_\lambda = \{v : t(v) = \lambda v\}$ be the eigenspace of λ and $S'_\lambda = S_\lambda \cup \{0_V\}$.

Consider the set $S = \left\{\sum_{i=1}^{m} \alpha_i b_i : \alpha_i \in F; \quad i = 1, 2, \ldots, m\right\}$.

Clearly, S is a subspace of V such that $b_i \in S$ for $i = 1, 2, \ldots, m$.

$\therefore \quad \{b_1, b_2, \ldots, b_m\}$ is a basis of S.

$\Rightarrow \quad \dim S = m$.

We shall now show that $S = S'_\lambda$.

Let $v = \sum_{i=1}^{m} \alpha_i b_i$ be an arbitrary vector in S. Then,

$$t(v) = t\left(\sum_{i=1}^{m} \alpha_i b_i\right)$$

$\Rightarrow \quad t(v) = \sum_{i=1}^{m} \alpha_i t(b_i)$

$\Rightarrow \quad t(v) = \sum_{i=1}^{m} \alpha_i (\lambda_i b_i)$ $\quad\quad\quad\quad\quad\quad\quad\quad\quad\quad\quad [\because t(b_i) = \lambda_i b_i; i = 1, 2, \ldots, n]$

$\Rightarrow \quad t(v) = \sum_{i=1}^{m} (\alpha_i \lambda_i) b_i$

$\Rightarrow \quad t(v) = \lambda \left(\sum_{i=1}^{m} \alpha_i b_i\right)$ $\quad\quad\quad\quad\quad\quad\quad\quad\quad [\because \lambda_1 = \lambda_2 = \cdots = \lambda_m = \lambda]$

$\Rightarrow \quad t(v) = \lambda v$

$\Rightarrow \quad v \in S_\lambda$

$\therefore \quad v \in S \Rightarrow v \in S_\lambda$

$\Rightarrow \quad S \subset S_\lambda \Rightarrow \quad S \subset S'_\lambda$ $\quad\quad\quad\quad\quad\quad\quad\quad\quad\quad\quad\quad\quad\quad\quad\quad\quad\quad\quad$ (i)

Now, let v be an arbitrary element of S_λ. Then,

$$t(v) = \lambda v$$

$\Rightarrow \quad t\left(\sum_{i=1}^{n} \beta_i b_i\right) = \lambda \left(\sum_{i=1}^{n} \beta_i b_i\right)$ $\quad\quad\quad \left[\because B \text{ is a basis of } V \text{ and } v \in V \therefore v = \sum_{i=1}^{n} \beta_i b_i\right]$

$\Rightarrow \quad \sum_{i=1}^{n} \beta_i t(b_i) = \sum_{i=1}^{n} (\lambda \beta_i) b_i$

$\Rightarrow \quad \sum_{i=1}^{n} \beta_i \lambda_i b_i = \sum_{i=1}^{n} (\lambda \beta_i) b_i$ $\quad\quad\quad\quad\quad\quad\quad\quad\quad\quad\quad\quad [\because t(b_i) = \lambda_i b_i]$

$\Rightarrow \quad \sum_{i=1}^{n} (\lambda_i - \lambda) \beta_i b_i = 0_V$

$\Rightarrow \quad (\lambda_i - \lambda) \beta_i = 0_V$ for $i = 1, 2, \ldots, n$ $\quad\quad\quad\quad\quad\quad\quad\quad\quad [\because B \text{ is a basis of } V]$

$\Rightarrow \quad \beta_i = 0_V$ for $i = m+1, m+2, \ldots, n$ $\quad\quad\quad \left[\begin{array}{l}\because \lambda_i = \lambda \text{ for } i = 1, 2, \ldots, m \text{ and,} \\ \lambda_i \neq \lambda \text{ for } i = m+1, \ldots, n\end{array}\right]$

$\therefore \quad v = \sum_{i=1}^{m} \beta_i b_i \quad \Rightarrow \quad v \in S$

Thus, $\quad v \in S_\lambda \Rightarrow v \in S$

$\therefore \quad S_\lambda \subset S \Rightarrow \quad S'_\lambda \subset S$ $\quad\quad\quad\quad\quad\quad\quad\quad\quad\quad\quad\quad\quad\quad\quad\quad\quad\quad\quad\quad\quad$ (ii)

From (i) and (ii), we get $\quad S = S'_\lambda$

$\therefore \quad \dim S'_\lambda = \dim S$

$\Rightarrow \quad \dim S'_\lambda = m$.

$\Rightarrow \quad \dim S'_\lambda = $ Number of times an eigenvalue occurs in the leading diagonal.

Q.E.D.

EXERCISE 4.6

1. Let A be a square matrix over a field F. If λ is an eigenvalue of A, prove that
   (i) $\lambda + k$ is an eigenvalue of $A + kI$.
   (ii) $k\lambda$ is an eigenvalue of kA

2. Show that λ is an eigenvalue of a square matrix A over a field F, then λ^k is an eigenvalue of A^k, where $k \in N$. Deduce the following:
   (i) If A is nilpotent, then 0 is the only eigenvalue of A.
   (ii) If A is a non-zero idempotent, then 0 and 1 are the only eigenvalues of A.

3. Prove that the eigenvalues of a triangular matrix are the diagonal entries of the matrix.

4. If an $n \times n$ matrix over a field F has n distinct eigenvalues in F, show that the corresponding eigenvectors are linearly independent and hence span F^n.

5. Let A, B be $n \times n$ matrices, each having n distinct eigenvalues. Show that $AB = BA$ iff they have the same eigenvectors and in this case there exists an invertible matrix P such that $P^{-1}AP$ and $P^{-1}BP$ are both diagonal matrices.

Chapter 5

Determinants

5.1 PERMUTATIONS

Various concepts involved in this chapter require a good knowledge of permutations. Let us, therefore, review some properties of permutations.

PERMUTATION *A bijection, i.e. a one-one onto mapping from a set A to itself is called a permutation of A.*

If $A = \{a_1, a_2, a_3, \ldots, a_n\}$ is a finite set with n elements and f is a permutation of A, then $f(a_1), f(a_2), \ldots, f(a_n)$ are n distinct elements of set A written possibly in a different order. Instead of using the usual notation of writing a function, we use a two row notation to represent a permutation. In the first row, we list all elements of set A and in the second row, we write the corresponding images. Thus, a permutation f of set A is written as

$$f = \begin{pmatrix} a_1 & a_2 & a_3 & \ldots & a_n \\ f(a_1) & f(a_2) & f(a_3) & \ldots & f(a_n) \end{pmatrix}$$

In this notation, rearrangement of columns is immaterial. For example, if g is a permutation of set $A = \{1, 2, 3, 4\}$ such that $g(1) = 4$, $g(2) = 3$, $g(3) = 1$ and $g(4) = 2$, then g can be written as

$$g = \begin{pmatrix} 1 & 2 & 3 & 4 \\ 4 & 3 & 1 & 2 \end{pmatrix}$$

Also, $\begin{pmatrix} 2 & 3 & 1 & 4 \\ 3 & 1 & 4 & 2 \end{pmatrix}$, $\begin{pmatrix} 3 & 4 & 1 & 2 \\ 1 & 2 & 4 & 3 \end{pmatrix}$ etc. all denote the same permutation g.

Let $A = \{a_1, a_2, a_3, \ldots, a_n\}$ be a finite set consisting of n distinct elements. If we keep the order of elements fixed, then in two row notation, we observe that each arrangement of elements in second row provides a permutation of set A. Thus, there are $n!$ permutations of set A. The set of all these $n!$ permutations of set A is a group under the composition of mappings as a binary operation. For any positive integer n, S_n denotes the group of all permutations of $\{1, 2, 3, \ldots, n\}$ and is called the symmetric group of degree n. The order of S_n is $n!$.

For example, S_2 is the symmetric group of all permutations of set $\{1,2\}$ and it consists of the following permutations

$$I = \begin{pmatrix} 1 & 2 \\ 1 & 2 \end{pmatrix}, \sigma = \begin{pmatrix} 1 & 2 \\ 2 & 1 \end{pmatrix}$$

Similarly, S_3 is the symmetric group of all permutations of set $\{1,2,3\}$. It consists of following six permutations:

$$\sigma_1 = \begin{pmatrix} 1 & 2 & 3 \\ 1 & 2 & 3 \end{pmatrix}, \sigma_2 = \begin{pmatrix} 1 & 2 & 3 \\ 2 & 3 & 1 \end{pmatrix}, \sigma_3 = \begin{pmatrix} 1 & 2 & 3 \\ 3 & 1 & 2 \end{pmatrix}$$

$$\sigma_4 = \begin{pmatrix} 1 & 2 & 3 \\ 2 & 1 & 3 \end{pmatrix}, \sigma_5 = \begin{pmatrix} 1 & 2 & 3 \\ 3 & 2 & 1 \end{pmatrix}, \sigma_6 = \begin{pmatrix} 1 & 2 & 3 \\ 1 & 3 & 2 \end{pmatrix}$$

CYCLIC PERMUTATION *A permutation σ of a set A is a cyclic permutation or a cycle if there exists a finite subset $\{a_1, a_2, \ldots, a_r\}$ of A such that*

$$\sigma(a_1) = a_2, \; \sigma(a_2) = a_3, \ldots, \sigma(a_r) = a_1$$

and, $\sigma(x) = x$ if $x \in A$ but $x \notin \{a_1, a_2, \ldots, a_r\}$

If σ is such a permutation, then it is written as $\sigma = (a_1, a_2, \ldots, a_r)$

The number of elements that appear in such a representation of a cyclic permutation is called its length. Thus, length of σ defined above is r.

Clearly, $\sigma = (2,4,5) \in S_5$ is a cyclic permutation of length 3 such that

$$\sigma(2) = 4, \; \sigma(4) = 5 \text{ and } \sigma(5) = 2$$

Thus, $\sigma = (2,4,5) = \begin{pmatrix} 1 & 2 & 3 & 4 & 5 \\ 1 & 4 & 3 & 5 & 2 \end{pmatrix}$

It should be noted that the order of a cycle is its length.

DISJOINT PERMUTATIONS *Two permutations σ and τ on a set A are said to be disjoint permutations if $\sigma(x) \neq x$, then $\tau(x) = x$ and if $\tau(x) \neq x$, then $\sigma(x) = x$.*

In general, product of permutations is not commutative as the composition of functions is not commutative. However, any two disjoint permutations always commute.

Two cycles $(a_1, a_2, \ldots, a_r)$ and $(b_1, b_2, \ldots, b_s)$ in S_n are disjoint cycles iff

$$\{a_1, a_2, \ldots, a_r\} \cap \{b_1, b_2, \ldots, b_s\} = \phi.$$

For example, in S_5 cycles $(1,2,4)$ and $(3,5)$ are disjoint cycles whereas cycles $(1, 2, 4)$ and $(2, 5)$ are not disjoint cycles.

TRANSPOSITION *A cyclic permutation of length two is called a transposition.*

THEOREM-1 *Any non-identity permutation of a finite set is expressible as a product of pairwise disjoint cyclic permutations each of length at least two.*

THEOREM-2 *Every cyclic permutation of length r can be expressed as a product of $(r-1)$ transpositions.*

It should be noted that the representation of a cyclic permutation as a product of transposition is not unique.

It follows from the above two theorems that any permutation σ of a finite set is expressible as a product of transpositions. But, the representation of σ as a product of transpositions is not unique. However, if an expression of σ as a product of transpositions has even number of terms, then any other expression of σ as a product of transpositions also has even number of terms.

REMARK. *Identity permutation is treated as a product of zero number of transpositions.*

EVEN PERMUTATION *A permutation σ of a set A is called an even permutation if it is a product of even number of transpositions.*

A permutation is an odd permutation if it is not an even permutation.

The identity permutation is an even permutation.

As mentioned earlier that a cycle of length $r \geq 1$ is a product of $(r-1)$ transpositions. Therefore, a cycle of length r is even or odd according as r is odd or even. Then, cycles (1 2), (4 1 3 5), (2 4 7 3 1 5), being of even length are odd permutations whereas cycles (2 3 4), (1 3 5 2 7) are even permutations.

THEOREM-3 *If σ and τ are permutations of a finite set A, then*

(i) $\tau \circ \sigma$ is even if either τ and σ both are even or both are odd.

(ii) $\tau \circ \sigma$ is odd if one of τ and σ is odd and other is even.

For any permutation σ of a finite set A Sgn(parity), denoted by $(-1)^\sigma$, is defined as

$$(-1)^\sigma = \begin{cases} 1, & \text{if } \sigma \text{ is even} \\ -1, & \text{if } \sigma \text{ is odd} \end{cases}$$

If σ and τ are two permutations of set $A = \{1, 2, 3, \ldots, n\}$, then

$$\text{Sgn}(\sigma \circ \tau) = (\text{Sgn}\sigma)(\text{Sgn}\tau)$$

Also, $\quad \text{Sgn}\sigma^{-1} = \text{Sgn}\sigma$

For $n \geq 2$, the set A_n of all even permutations in S_n is a subgroup of S_n such that

$$O(A_n) = \frac{1}{2}n! = \frac{1}{2}O(S_n).$$

5.2 DETERMINANTS

Consider an $n \times n$ matrix $A = [a_{ij}]$ over a field F and a product of its n elements such that it contains one and only one element from each row and one and only one element from each column. Such a product can be written in the following from:

$$a_{1\sigma(1)} \, a_{2\sigma(2)} \, a_{3\sigma(3)} \cdots a_{n\sigma(n)}$$

In forming this product factors are taken from successive rows of matrix A that is why the first subscripts are in the natural order $1, 2, \ldots, n$. As the factors are taken from different columns, so the sequence of second subscripts forms a permutation σ in S_n. Conversely, each permutation $\sigma \in S_n$ determines a product of the above form. Thus, elements of matrix $A = [a_{ij}]$ determine $n!$ such products. If each such product is multiplied by $(-1)^\sigma$, where σ is the corresponding permutation in S_n, then their sum is known as the determinant of A which is formally defined as follows.

DETERMINANT *Let $A = [a_{ij}]$ be a square matrix over a field F. The determinant of A, denoted by Det (A), is defined by*

$$\text{Det}(A) = \sum_{\sigma \in S_n} (-1)^\sigma \, a_{1\sigma(1)} \, a_{2\sigma(2)} \cdots a_{n\sigma(n)}$$

The determinant of A is also denoted by $|A|$ or,
$$\begin{vmatrix} a_{11} & a_{12} & \cdots & a_{1n} \\ a_{21} & a_{22} & \cdots & a_{2n} \\ \vdots & \vdots & & \vdots \\ a_{n1} & a_{n2} & \cdots & a_{nn} \end{vmatrix}$$

Particular Cases:

(i) Let $A = [a_{11}]$ be a 1×1 matrix. Since S_1 has only one permutation, which is even.

$$\therefore \quad \text{Det}(A) = a_{11}$$

(ii) Let $A = [a_{ij}]$ be a 2×2 matrix. In this case, we have

$$S_2 = \left\{ I = \begin{pmatrix} 1 & 2 \\ 1 & 2 \end{pmatrix}, \sigma = \begin{pmatrix} 1 & 2 \\ 2 & 1 \end{pmatrix} \right\}$$

Clearly, I is even and σ is an odd permutation.

$$\therefore \quad (-1)^I = 1 \quad \text{and} \quad (-1)^\sigma = -1$$

Hence, $\quad \text{Det}(A) = \sum_{\sigma \in S_2} (-1)^\sigma \, a_{1\sigma(1)} \, a_{2\sigma(2)}$

$\Rightarrow \quad \text{Det}(A) = (-1)^I \, a_{1I(1)} \, a_{2I(2)} + (-1)^\sigma \, a_{1\sigma(1)} \, a_{2\sigma(2)}$

$\Rightarrow \quad \text{Det}(A) = a_{11} \, a_{22} - a_{12} \, a_{21}$

or, $\quad \begin{vmatrix} a_{11} & a_{12} \\ a_{21} & a_{22} \end{vmatrix} = a_{11} \, a_{22} - a_{12} \, a_{21}$

(iii) Let $A = [a_{ij}]$ be a 3×3 matrix. In this case, we have

$$S_3 = \left\{ I = \begin{pmatrix} 1 & 2 & 3 \\ 1 & 2 & 3 \end{pmatrix}, \sigma_1 = \begin{pmatrix} 1 & 2 & 3 \\ 2 & 3 & 1 \end{pmatrix}, \sigma_2 = \begin{pmatrix} 1 & 2 & 3 \\ 3 & 1 & 2 \end{pmatrix}, \sigma_3 = \begin{pmatrix} 1 & 2 & 3 \\ 1 & 3 & 2 \end{pmatrix}, \right.$$

$$\left. \sigma_4 = \begin{pmatrix} 1 & 2 & 3 \\ 3 & 2 & 1 \end{pmatrix}, \sigma_5 = \begin{pmatrix} 1 & 2 & 3 \\ 2 & 1 & 3 \end{pmatrix} \right\}$$

Clearly, I, σ_1 and σ_2 are even permutations and $\sigma_3, \sigma_4, \sigma_5$ are odd permutations.

$\therefore \qquad (-1)^I = 1, (-1)^{\sigma_1} = (-1)^{\sigma_2} = 1, (-1)^{\sigma_3} = (-1)^{\sigma_4} = (-1)^{\sigma_5} = -1$

Hence, $\quad \text{Det}(A) = \sum_{\sigma \in S_n} (-1)^\sigma a_{1\sigma(1)} a_{2\sigma(2)} a_{3\sigma(3)}$

$\Rightarrow \quad \text{Det}(A) = (-1)^I a_{1I(1)} a_{2I(2)} a_{3I(3)} + (-1)^{\sigma_1} a_{1\sigma_1(1)} a_{2\sigma_1(2)} a_{3\sigma_1(3)}$
$\qquad \qquad + (-1)^{\sigma_2} a_{1\sigma_2(1)} a_{2\sigma_2(2)} a_{3\sigma_2(3)} + (-1)^{\sigma_3} a_{1\sigma_3(1)} a_{2\sigma_3(2)} a_{3\sigma_3(3)}$
$\qquad \qquad + (-1)^{\sigma_4} a_{1\sigma_4(1)} a_{2\sigma_4(2)} a_{3\sigma_4(3)} + (-1)^{\sigma_5} a_{1\sigma_5(1)} a_{2\sigma_5(2)} a_{3\sigma_5(3)}$

$\Rightarrow \quad \text{Det}(A) = a_{11}a_{22}a_{33} + a_{12}a_{23}a_{31} + a_{13}a_{21}a_{32} - a_{11}a_{23}a_{32} - a_{13}a_{22}a_{31} - a_{12}a_{21}a_{33}$

or, $\begin{vmatrix} a_{11} & a_{12} & a_{13} \\ a_{21} & a_{22} & a_{23} \\ a_{31} & a_{32} & a_{33} \end{vmatrix} = a_{11}a_{22}a_{33} + a_{12}a_{23}a_{31} + a_{13}a_{21}a_{32} - a_{11}a_{23}a_{32} - a_{13}a_{22}a_{31} - a_{12}a_{21}a_{33}$

In the definition of the determinant of a square matrix of order n, there are $n!$ terms. Therefore, if $n > 3$, the number of terms in the determinant becomes astronomical. Therefore, evaluation of determinants directly from definition is quite cumbersome and time-consuming. So, we prove a number of properties about determinants which will help us to shorten the computation considerably. In the following section, we will discuss various properties of determinants.

5.3 PROPERTIES OF DETERMINANTS

In this section, we shall discuss some elementary properties of determinants as stated and proved in the following theorems.

THEOREM-1 *Let I_n be the identity matrix over a field F. Then, $\text{Det}(I_n) = 1$.*

PROOF. We have,

$$I_n = [\delta_{ij}], \text{ where } \quad \delta_{ij} = \begin{cases} 1, & \text{if } i = j \\ 0, & \text{if } i \neq j \end{cases}$$

$\therefore \quad \text{Det}(I_n) = \sum_{\sigma \in S_n} (-1)^\sigma \delta_{1\sigma(1)} \delta_{2\sigma(2)} \ldots \delta_{n\sigma(n)}$

$\Rightarrow \quad \text{Det}(I_n) = (-1)^I \delta_{1I(1)} \delta_{2I(2)} \ldots \delta_{nI(n)} + \sum_{\sigma \in S_n - \{I\}} (-1)^\sigma \delta_{1\sigma(1)} \delta_{2\sigma(2)} \ldots \delta_{n\sigma(n)}$

For any $\sigma \in S_n - \{I\}$, there exists k such that $\sigma(k) \neq k$.

$\therefore \qquad \delta_{k\sigma(k)} = 0$

$\Rightarrow \qquad \sum_{\sigma \in S_n - \{I\}} \delta_{1\sigma(1)} \delta_{2\sigma(2)} \ldots \delta_{n\sigma(n)} = 0$

Hence, $\quad \text{Det}(I_n) = \delta_{11}\delta_{22}\ldots\delta_{nn} + 0 = 1$

Q.E.D.

THEOREM-2 *Let A be a square matrix over a field F such that all the elements of a row (column) are zero. Then, $\text{Det}(A) = 0$.*

PROOF. Let $A = [a_{ij}]$ be an $n \times n$ matrix over field F. Then,

$$\text{Det}(A) = \sum_{\sigma \in S_n} (-1)^\sigma a_{1\sigma(1)} a_{2\sigma(2)} \ldots a_{n\sigma(n)}$$

The expression on *RHS* contains $n!$ terms such that each term contains a factor from every row and also from every column. So, each term on *RHS* contains a factor from the row whose all elements are zeros.

$\therefore \qquad \text{Det}(A) = 0.$

Q.E.D.

SINGULAR MATRIX *A square matrix A over a field F is singular if $\text{Det}(A) = 0$, otherwise A is said to be a non-singular matrix.*

THEOREM-3 *Let A be a square matrix over a field F. If B is a matrix obtained from A by multiplying each element of a row (column) by a scalar λ, then*

$$\text{Det}(B) = \lambda \, \text{Det}(A)$$

PROOF. Let $A = [a_{ij}]$ be $n \times n$ matrix over field F and $B = [b_{ij}]$ be matrix obtained from A such that kth row of A is multiplied by λ and all other rows remain same. Then,

$\qquad a_{ij} = b_{ij} \quad \text{for all} \quad i,j \in \underline{n}, i \neq k \text{ and, } b_{kj} = \lambda a_{kj} \quad \text{for all} \quad j \in \underline{n}$

$\therefore \qquad \text{Det}(B) = \sum_{\sigma \in S_n} (-1)^\sigma b_{1\sigma(1)} b_{2\sigma(2)} \ldots b_{k\sigma(k)} \ldots b_{n\sigma(n)}$

$\Rightarrow \qquad \text{Det}(B) = \sum_{\sigma \in S_n} (-1)^\sigma a_{1\sigma(1)} a_{2\sigma(2)} \ldots \lambda a_{k\sigma(k)} \ldots a_{n\sigma(n)}$

$\Rightarrow \qquad \text{Det}(B) = \lambda \sum_{\sigma \in S_n} (-1)^\sigma a_{1\sigma(1)} a_{2\sigma(2)} \ldots a_{k\sigma(k)} \ldots a_{n\sigma(n)}$

$\Rightarrow \qquad \text{Det}(B) = \lambda \, \text{Det}(A)$

Q.E.D.

COROLLARY. Let $A = [a_{ij}]$ be an $n \times n$ matrix over a field F and $\lambda \in F$. Then, $Det(\lambda A) = \lambda^n Det(A)$.

PROOF. Let $B = [b_{ij}]$ be $n \times n$ matrix such that $B = \lambda A$. Then,
$$b_{ij} = \lambda\, a_{ij} \quad \text{for all } i,j \in \underline{n}$$

$\therefore \quad Det(B) = \sum_{\sigma \in S_n} (-1)^\sigma b_{1\sigma(1)} b_{2\sigma(2)} \ldots b_{n\sigma(n)}$

$\Rightarrow \quad Det(B) = \sum_{\sigma \in S_n} (-1)^\sigma (\lambda a_{1\sigma(1)})(\lambda a_{2\sigma(2)}) \ldots (\lambda a_{n\sigma(n)})$

$\Rightarrow \quad Det(B) = \lambda^n \sum_{\sigma \in S_n} (-1)^\sigma a_{1\sigma(1)} a_{2\sigma(2)} \ldots a_{n\sigma(n)}$

$\Rightarrow \quad Det(B) = \lambda^n Det(A)$

Hence, $\quad Det(\lambda A) = \lambda^n Det(A)$.

Q.E.D.

THEOREM-4 Let $A = [a_{ij}]$ be an $n \times n$ triangular matrix over a field F. Then,
$$Det(A) = a_{11}\, a_{22}\, a_{33} \ldots a_{nn}$$

PROOF. Let $A = [a_{ij}]$ be an $n \times n$ lower triangular matrix over a field F. Then,

$$a_{ij} = 0 \quad \text{for all } i < j \qquad \text{(i)}$$

$\therefore \quad Det(A) = \sum_{\sigma \in S_n} (-1)^\sigma a_{1\sigma(1)}\, a_{2\sigma(2)} \ldots a_{n\sigma(n)}$

$\Rightarrow \quad Det(A) = (-1)^I a_{1I(1)}\, a_{2I(2)} \ldots a_{nI(n)} + \sum_{\sigma \in S_n - \{I\}} (-1)^\sigma a_{1\sigma(1)}\, a_{2\sigma(2)} \ldots a_{n\sigma(n)}$

$\Rightarrow \quad Det(A) = a_{11}\, a_{22} \ldots a_{nn} + \sum_{\sigma \in S_n - \{I\}} (-1)^\sigma a_{1\sigma(1)}\, a_{2\sigma(2)} \ldots a_{n\sigma(n)} \qquad \text{(ii)}$

Consider the expression
$$\sum_{\sigma \in S_n - \{I\}} (-1)^\sigma a_{1\sigma(1)}\, a_{2\sigma(2)} \ldots a_{n\sigma(n)} \qquad \text{(iii)}$$

Clearly, this expression is the sum of $(n! - 1)$ terms. Out of these terms, let us consider a term
$$t = (-1)^\sigma a_{1\sigma(1)}\, a_{2\sigma(2)} \ldots a_{n\sigma(n)}$$

Suppose $\sigma(1) \neq 1$. Then, $\sigma(1) > 1$.

$\therefore \quad a_{1\sigma(1)} = 0 \qquad \text{[From (i)]}$

$\Rightarrow \quad t = 0$

So, each term for which $\sigma(1) > 1$ in (iii) is zero.

Now, suppose $\sigma(1) = 1$ but $\sigma(2) \neq 2$. Then, $\sigma(2) > 2$.

$\therefore \quad a_{2\sigma(2)} = 0$ [From (i)]

$\Rightarrow \quad t = 0$

Thus, each term in (iii) for which $\sigma(1) \neq 1$ or $\sigma(2) \neq 2$ is zero.

Similarly, we find that each term in (iii) is zero.

$\therefore \quad \text{Det}(A) = a_{11}\, a_{22}\, a_{33} \ldots a_{nn} \quad \left[\text{Putting } \sum_{\sigma \in S_n - \{I\}} (-1)^\sigma a_{1\sigma(1)} a_{2\sigma(2)} \ldots a_{n\sigma(n)} = 0 \text{ in (ii)}\right]$

Let $A = [a_{ij}]$ be an upper triangular matrix. Then, $A^T = [b_{ij}]$ is a lower triangular matrix.

$\therefore \quad \text{Det}(A^T) = b_{11}\, b_{22} \ldots b_{nn}$

$\Rightarrow \quad \text{Det}(A^T) = a_{11}\, a_{22} \ldots a_{nn} \qquad [\because \quad b_{ij} = a_{ji} \text{ for all } i, j]$

$\Rightarrow \quad \text{Det}(A) = a_{11}\, a_{22} \ldots a_{nn}$

Q.E.D.

COROLLARY. *The determinant of a diagonal matrix is the product of its diagonal entries.*

THEOREM-5 *Let $A = [a_{ij}]$ be any $n \times n$ matrix over a field F such that each element of its kth row is sum of two elements, then $\text{Det}(A)$ can be written as the sum of two determinants having all rows identical to the rows of A except kth rows, i.e. if $A = [a_{ij}], B = [b_{ij}]$ and $C = [c_{ij}]$ are $n \times n$ matrices such that*

$$a_{kj} = b_{kj} + c_{kj} \quad \text{for all } j \in \underline{n}$$
$$a_{ij} = b_{ij} = c_{ij} \quad \text{for all } i, j \in \underline{n}, i \neq k$$

Then, $\text{Det}(A) = \text{Det}(B) + \text{Det}(C)$

PROOF. We have,

$\text{Det}(A) = \sum_{\sigma \in S_n} (-1)^\sigma a_{1\sigma(1)}\, a_{2\sigma(2)} \ldots a_{k\sigma(k)} \ldots a_{n\sigma(n)}$

$\Rightarrow \quad \text{Det}(A) = \sum_{\sigma \in S_n} (-1)^\sigma a_{1\sigma(1)}\, a_{2\sigma(2)} \ldots \left(b_{k\sigma(k)} + c_{k\sigma(k)}\right) \ldots a_{n\sigma(n)}$

$\Rightarrow \quad \text{Det}(A) = \sum_{\sigma \in S_n} (-1)^\sigma a_{1\sigma(1)}\, a_{2\sigma(2)} \ldots b_{k\sigma(k)} \ldots a_{n\sigma(n)}$

$\qquad\qquad + \sum_{\sigma \in S_n} (-1)^\sigma a_{1\sigma(1)}\, a_{2\sigma(2)} \ldots c_{k\sigma(k)} \ldots a_{n\sigma(n)}$

$\Rightarrow \quad \text{Det}(A) = \sum_{\sigma \in S_n} (-1)^\sigma b_{1\sigma(1)}\, b_{2\sigma(2)} \ldots b_{k\sigma(k)} \ldots b_{n\sigma(n)}$

$\qquad\qquad + \sum_{\sigma \in S_n} (-1)^\sigma c_{1\sigma(1)}\, c_{2\sigma(2)} \ldots c_{k\sigma(k)} \ldots c_{n\sigma(n)} \quad \left[\begin{array}{l} \because \quad a_{ij} = b_{ij} = c_{ij} \\ \text{for all } i, j \in \underline{n}, i \neq k \end{array}\right]$

$\Rightarrow \quad \text{Det}(A) = \text{Det}(B) + \text{Det}(C)$

Q.E.D.

THEOREM-6 *If A is a square matrix over a field F, then the determinant of matrix A and its transpose are equal. That is,* $\text{Det}(A) = \text{Det}(A^T)$.

PROOF. Let $A = [a_{ij}]$ be a square matrix of order n. Then,

$$\text{Det}(A) = \sum_{\sigma \in S_n} (-1)^\sigma a_{1\sigma(1)} a_{2\sigma(2)} a_{3\sigma(3)} \cdots a_{n\sigma(n)} \qquad \text{(i)}$$

Let $A^T = [b_{ij}]$. Then, $b_{ij} = a_{ji}$ for all $i, j \in \underline{n}$

$\therefore \quad \text{Det}(A^T) = \sum_{\sigma \in S_n} (-1)^\sigma b_{1\sigma(1)} b_{2\sigma(2)} b_{3\sigma(3)} \cdots b_{n\sigma(n)}$

$$\Rightarrow \text{Det}(A^T) = \sum_{\sigma \in S_n} (-1)^\sigma a_{\sigma(1)1} a_{\sigma(2)2} a_{\sigma(3)3} \cdots a_{\sigma(n)n} \qquad \text{(ii)}$$

Clearly, $\sigma(1), \sigma(2), \ldots, \sigma(n)$ take all the values $1, 2, 3, \ldots, n$.

Also, $\quad \sigma(i) = j \quad \text{iff } \sigma^{-1}(j) = i$

$\therefore \quad a_{\sigma(i)\,i} = a_{j\,\sigma^{-1}(j)}$

Consequently, by re-arranging terms in (ii), we get

$$\text{Det}(A^T) = \sum_{\sigma \in S_n} (-1)^\sigma a_{1\sigma^{-1}(1)} a_{2\sigma^{-1}(2)} a_{3\sigma^{-1}(3)} \cdots a_{n\sigma^{-1}(n)}$$

$$\Rightarrow \text{Det}(A^T) = \sum_{\sigma^{-1} \in S_n} (-1)^{\sigma^{-1}} a_{1\sigma^{-1}(1)} a_{2\sigma^{-1}(2)} a_{3\sigma^{-1}(3)} \cdots a_{n\sigma^{-1}(n)} \quad [\because (-1)^\sigma = (-1)^{\sigma^{-1}}] \quad \text{(iii)}$$

As σ runs through all the permutations of S_n, so $\tau = \sigma^{-1}$ also runs through all the permutations of S_n. Therefore, (iii) becomes

$$\text{Det}(A^T) = \sum_{\tau \in S_n} (-1)^\tau a_{1\tau(1)} a_{2\tau(2)} a_{3\tau(3)} \cdots a_{n\tau(n)}$$

$\Rightarrow \text{Det}(A^T) = \text{Det}(A)$ [From (i)]

Q.E.D.

THEOREM-7 *Let $A = [a_{ij}]$ be any $n \times n$ matrix over a field F and let B be a matrix obtained from A by interchanging two distinct rows (columns). Then,* $\text{Det}(B) = -\text{Det}(A)$.

PROOF. Let $B = [b_{ij}]$ be obtained from A by interchanging kth and lth columns. Then,

$$a_{ij} = b_{ij} \quad \text{for all } i, j \in \underline{n}, j \neq k, l$$

$$a_{ik} = b_{il} \text{ and } a_{il} = b_{ik} \quad \text{for all } i \in \underline{n}$$

Let $\tau = (k, l)$ be a transposition. Then, $\tau(k) = l, \tau(l) = k$ and $\tau(i) = i$ for $i \neq k, l$. Therefore,

$$b_{ij} = a_{i\tau(j)} \quad \text{for } j \neq k, l \text{ and for all } i \in \underline{n}$$

$$b_{ik} = a_{il} \Rightarrow b_{ik} = a_{i\tau(k)} \quad \text{for all } i \in \underline{n}$$

$$b_{il} = a_{ik} \Rightarrow b_{il} = a_{i\tau(l)} \quad \text{for all } i \in \underline{n}$$

Let σ be a permutation in S_n such that $\sigma(p) = k$ and $\sigma(q) = l$. Then, for any $i \neq p, q$

$$b_{i\sigma(i)} = a_{i\ \tau\circ\sigma(i)}$$
$$b_{pk} = a_{pl} = a_{p\tau\circ\sigma(p)}$$
$$b_{ql} = a_{qk} = a_{q\tau\circ\sigma(q)}$$

Thus, for any permutation σ in S_n, we have

$$b_{1\sigma(1)}\, b_{2\sigma(2)} \ldots b_{n\sigma(n)} = a_{1\tau\circ\sigma(1)}\, a_{2\tau\circ\sigma(2)} \ldots a_{n\tau\circ\sigma(n)} \qquad (i)$$

Hence,
$$\text{Det}(B) = \sum_{\sigma \in S_n} (-1)^{\sigma}\, b_{1\sigma(1)}\, b_{2\sigma(2)} \ldots b_{n\sigma(n)}$$
$$\Rightarrow \quad \text{Det}(B) = \sum_{\sigma \in S_n} (-1)^{\sigma}\, a_{1\tau\circ\sigma(1)}\, a_{2\tau\circ\sigma(2)} \ldots a_{n\tau\circ\sigma(n)} \qquad \text{[Using (i)]}$$

Since the transposition τ is an odd permutation.

$\therefore \quad \text{Sgn}(\tau\circ\sigma) = \text{Sgn}(\tau)\,\text{Sgn}(\sigma) = -\text{Sgn}(\sigma)$ $\quad [\because \text{Sgn}(\tau) = -1]$

$\Rightarrow \quad \text{Sgn}(\sigma) = -\text{Sgn}(\tau\circ\sigma)$

$\Rightarrow \quad (-1)^{\sigma} = -(-1)^{\tau\circ\sigma}$

So,
$$\text{Det}(B) = -\sum_{\tau\circ\sigma \in S_n} (-1)^{\tau\circ\sigma}\, a_{1\tau\circ\sigma(1)}\, a_{2\tau\circ\sigma(2)} \ldots a_{n\tau\circ\sigma(n)} \qquad (ii)$$

But, as σ runs through all the permutations in S_n, $\tau\circ\sigma$ also runs through all the permutations in S_n.

$\therefore \quad \text{Det}(A) = \sum_{\tau\circ\sigma \in S_n} (-1)^{\tau\circ\sigma}\, a_{1\tau\circ\sigma(1)}\, a_{2\tau\circ\sigma(2)} \ldots a_{n\tau\circ\sigma(n)} \qquad (iii)$

$\Rightarrow \quad \text{Det}(B) = -\text{Det}(A)$ $\qquad$ [From (ii) and (iii)]

Q.E.D.

THEOREM-8 *Let A be an $n \times n$ matrix over a field F such that it has two identical rows (columns), then $\text{Det}(A) = 0$.*

PROOF. Let $A = [a_{ij}]$ be the given matrix whose kth and lth columns are identical. Let B be the matrix obtained from A by interchanging kth and lth columns of A. Then,

$$\text{Det}(B) = -\text{Det}(A) \qquad \text{[By Theorem 7]} \qquad (i)$$

But, kth and lth columns A are identical. Therefore,

$$B = A$$
$$\Rightarrow \quad \text{Det}(B) = \text{Det}(A) \qquad (ii)$$

From (i) and (ii), we get
$$\text{Det}(A) = -\text{Det}(A)$$
$$\Rightarrow \quad \text{Det}(A) + \text{Det}(A) = 0$$

$\Rightarrow \quad 2\,\text{Det}(A) = 0$

$\Rightarrow \quad \text{Det}(A) = 0$

Q.E.D.

THEOREM-9 *Let A be an $n \times n$ matrix over a field F and $\sigma \in S_n$. For any $i = 1, 2, \ldots, n$, if $\sigma(A)$ be an $n \times n$ matrix whose ith column is the $\sigma(i)$th column of A. Then,*

$$\text{Det}(\sigma(A)) = (-1)^\sigma \text{Det}(A)$$

PROOF. Let $\tau = (k\ l)$ be a transposition in S_n. Then $\tau(A)$ is the matrix whose kth column is the lth column of A and the lth column is the kth column of A; other columns of $\tau(A)$ are the same as the corresponding column of A.

So, by Theorem 7, we have

$$\text{Det}(\tau(A)) = -\text{Det}(A) \qquad \text{(i)}$$

Since every permutation in S_n is expressible as the product of transpositions. Therefore, for $\sigma \in S_n$ there exist transpositions $\tau_1, \tau_2 \ldots \tau_r$ such that

$$\sigma = \tau_1, \tau_2 \ldots \tau_r$$

$\therefore \quad \text{Det}(\sigma(A)) = \text{Det}(\tau_1, \tau_2 \ldots \tau_r(A))$

$\Rightarrow \quad \text{Det}(\sigma(A)) = (-1)^r \text{Det}(A) \qquad$ [Using repeated application of (i)]

$\Rightarrow \quad \text{Det}(\sigma(A)) = (-1)^\sigma \text{Det}(A) \qquad [\because \ (-1)^r = (-1)^\sigma]$

Q.E.D.

THEOREM-10 *Let A be a square matrix of order n over a field F and B be a matrix obtained from A by adding a multiple of a row (column) of A to another row (column) of A, then $\text{Det}(B) = \text{Det}(A)$.*

PROOF. Let $A = [a_{ij}]$ be an $n \times n$ matrix over a field F and $B = [b_{ij}]$ be a matrix obtained from A by adding λ times the kth row to the lth row of A. Then,

$b_{ij} = a_{ij} \quad$ for all $i, j \in \underline{n}, i \neq l$

$b_{lj} = a_{lj} + \lambda a_{kj} \quad$ for all $j \in \underline{n}$

$\therefore \quad \text{Det}(B) = \sum\limits_{\sigma \in S_n} (-1)^\sigma\, b_{1\sigma(1)}\, b_{2\sigma(2)} \ldots b_{k\sigma(k)} \ldots b_{l\sigma(l)} \ldots b_{n\sigma(n)}$

$\Rightarrow \quad \text{Det}(B) = \sum\limits_{\sigma \in S_n} (-1)^\sigma\, a_{1\sigma(1)}\, a_{2\sigma(2)} \ldots a_{k\sigma(k)} \ldots (a_{l\sigma(l)} + \lambda a_{k\sigma(k)}) \ldots a_{n\sigma(n)}$

$\Rightarrow \quad \text{Det}(B) = \sum\limits_{\sigma \in S_n} (-1)^\sigma\, a_{1\sigma(1)}\, a_{2\sigma(2)} \ldots a_{k\sigma(k)} \ldots a_{l\sigma(l)} \ldots a_{n\sigma(n)}$

$\qquad \qquad + \lambda \sum\limits_{\sigma \in S_n} (-1)^\sigma\, a_{1\sigma(1)}\, a_{2\sigma(2)} \ldots a_{k\sigma(k)} \ldots a_{k\sigma(k)} \ldots a_{n\sigma(n)}$

$\Rightarrow \quad \text{Det}(B) = \text{Det}(A) + \lambda \times 0 \qquad \left[\begin{array}{l} \because \ k\text{th and }l\text{th rows of the matrix,} \\ \text{representing second sum, are identical} \end{array}\right]$

$\Rightarrow \quad \text{Det}(B) = \text{Det}(A)$

Q.E.D.

THEOREM-11 *Let A be a matrix over a field F and E be an elementary matrix. Then,*
$$\mathrm{Det}(EA) = \mathrm{Det}(E) \cdot \mathrm{Det}(A)$$

PROOF. There are three types of elementary operation on a matrix:

(i) Multiplying all elements of a row (column) by a non-zero scalar λ.

(ii) Interchange of any two rows (columns).

(iii) Adding a multiple of one row (column) to another.

So, the elementary matrix can be a matrix obtained from the identity matrix by applying any one of the above elementary operations.

Let E_1 be the elementary matrix obtained by applying (i) elementary operation on I. Then,
$$\mathrm{Det}(E_1) = \lambda\, \mathrm{Det}(I) = \lambda \qquad \text{[By Theorem 3]}$$

Let E_2 be the elementary matrix obtained by applying (ii) elementary operation on I. Then,
$$\mathrm{Det}(E_2) = -\mathrm{Det}(I) = -1 \qquad \text{[By Theorem 7]}$$

Let E_3 be the elementary matrix obtained by applying (iii) elementary operation on I. Then,
$$\mathrm{Det}(E_3) = \mathrm{Det}(I) = 1 \qquad \text{[By Theorem 10]}$$

We know that any elementary row operation on a matrix A is equivalent to the premultiplication of A by an elementary matrix obtained by applying the same transformation on I.

Let B_1 be the matrix obtained from A by applying (i) elementary operation. Then, by Theorem 3, we have
$$\mathrm{Det}(B_1) = \lambda\, \mathrm{Det}(A) \qquad (i)$$
But, $\quad B_1 = E_1 A$
$\Rightarrow \quad \mathrm{Det}(B_1) = \mathrm{Det}(E_1 A)$
$\Rightarrow \quad \lambda\, \mathrm{Det}(A) = \mathrm{Det}(E_1 A) \qquad \text{[Using (i)]}$
$\Rightarrow \quad \mathrm{Det}(E_1 A) = \lambda\, \mathrm{Det}(A)$
$\Rightarrow \quad \mathrm{Det}(E_1 A) = \mathrm{Det}(E_1)\mathrm{Det}(A) \qquad [\because \mathrm{Det}(E_1) = \lambda]$

Let B_2 be the matrix obtained from A by applying (ii) elementary operation. Then, by Theorem 7, we have
$$\mathrm{Det}(B_2) = -\mathrm{Det}(A)$$
But, $\quad B_2 = E_2 A$
$\Rightarrow \quad \mathrm{Det}(B_2) = \mathrm{Det}(E_2 A)$
$\Rightarrow \quad -\mathrm{Det}(A) = \mathrm{Det}(E_2 A)$
$\Rightarrow \quad \mathrm{Det}(E_2 A) = \mathrm{Det}(E_2)\mathrm{Det}(A) \qquad [\because \mathrm{Det}(E_2) = -1]$

Let B_3 be the matrix obtained from A by applying elementary operation (iii). Then, by Theorem 10, we have

$$\text{Det}(B_3) = \text{Det}(A)$$

But, $\quad B_3 = E_3 A$

$\Rightarrow \quad \text{Det}(B_3) = \text{Det}(E_3 A)$

$\Rightarrow \quad \text{Det}(A) = \text{Det}(E_3 A)$

$\Rightarrow \quad \text{Det}(E_3 A) = \text{Det}(E_3)\text{Det}(A) \qquad\qquad\qquad\qquad [\because \text{Det}(E_3) = 1]$

Hence, in all the three cases, we have

$$\text{Det}(EA) = \text{Det}(E)\text{Det}(A).$$

Q.E.D.

THEOREM-12 *Let A and B be two $n \times n$ matrices over a field F. Then,*

$$\text{Det}(AB) = \text{Det}(A)\,\text{Det}(B)$$

PROOF. Following cases arise.

Case I When A is a singular matrix, i.e. $\text{Det}(A) = 0$.

If A is a singular matrix, then AB is also a singular matrix.

$\therefore \quad \text{Det}(AB) = 0$

Also, $\quad \text{Det}(A)\,\text{Det}(B) = 0 \times \text{Det}(B) = 0 \qquad\qquad\qquad [\because \text{Det}(A) = 0]$

Hence, $\quad \text{Det}(AB) = \text{Det}(A)\,\text{Det}(B)$

Case II When A is a non-singular matrix, i.e. $\text{Det}(A) \neq 0$.

If $\text{Det}(A) \neq 0$, then A is invertible.

Since every invertible matrix is expressible as the product of elementary matrices. So, there exists a sequence of elementary matrices $E_1, E_2, \ldots, E_n$ such that

$$A = E_n E_{n-1} \ldots E_2 E_1$$

Let $P(n)$ denote the statement: The determinant of the product of n elementary matrices and some given matrix is equal to the products of determinants of the given elementary matrices and the given matrix.

We shall first prove this statement by the principle of mathematical induction on n.

Step I By Theorem 11, we have

$$\text{Det}(E_1 B) = \text{Det}(E_1)\,\text{Det}(B) \qquad\qquad\qquad\qquad\qquad\qquad\qquad (i)$$

So, the statement is true for $n = 1$.

Step II Let the statement be true for $(n-1)$. Then,

$$|E_{n-1}E_{n-2}\ldots E_2E_1B| = |E_{n-1}||E_{n-2}|\ldots|E_2||E_1||B|$$

Now,

$|E_nE_{n-1}E_{n-2}\ldots E_2E_1B| = |E_n(E_{n-1}E_{n-2}\ldots E_2E_1B)|$

$\Rightarrow |E_nE_{n-1}E_{n-2}\ldots E_2E_1B| = |E_n||E_{n-1}E_{n-2}\ldots E_2E_1B|$ [Using (i)]

$\Rightarrow |E_nE_{n-1}E_{n-2}\ldots E_2E_1B| = |E_n||E_{n-1}||E_{n-2}|\ldots|E_2||E_1||B|$ [By induction assumption]

So, the statement is true for n.

Hence, by the principle of mathematical induction, the result is true for all $n \in N$,

i.e. $|E_nE_{n-1}\ldots E_2E_1B| = |E_n||E_{n-1}|\ldots|E_2||E_1||B|$ for all $n \in N$ (ii)

Now,

$A = E_nE_{n-1}\ldots E_2E_1$

$\Rightarrow A = E_nE_{n-1}\ldots E_2E_1 I$

$\Rightarrow |A| = |E_n||E_{n-1}|\ldots|E_2||E_1||I|$ [As proved above]

$\Rightarrow |A| = |E_n||E_{n-1}|\ldots|E_2||E_1|$ $[\because |I| = 1]$ (iii)

From (ii) and (iii), we get

$$|AB| = |A||B|$$

Hence, in either case, we have

$$|AB| = |A||B| \quad \text{i.e. } \text{Det}(AB) = \text{Det}(A)\,\text{Det}(B)$$

Q.E.D.

GENERALIZATION If $A_1, A_2, \ldots, A_n$ are square matrices of the same order over a field F, then
$$|A_1A_2\ldots A_n| = |A_1||A_2|\ldots|A_n|$$

COROLLARY. 1 *If A is an invertible matrix over a field F, then $\text{Det}(A) \neq 0$ and $\text{Det}(A^{-1}) = \dfrac{1}{\text{Det}(A)}$.*

PROOF. Since A is an invertible matrix. Therefore, A^{-1} exists and $AA^{-1} = I$

Now, $AA^{-1} = I$

$\Rightarrow \text{Det}(AA^{-1}) = \text{Det}(I)$

$\Rightarrow \text{Det}(A)\,\text{Det}(A^{-1}) = 1$

$\Rightarrow \text{Det}(A^{-1}) = \dfrac{1}{\text{Det}(A)}$

Q.E.D.

COROLLARY. 2 *If A and B are similar matrices, then $|A| = |B|$, i.e. similar matrices have the same determinant.*

PROOF. Let A and B be similar matrices over a field F. Then there exists an invertible matrix P such that
$$B = P^{-1}AP$$
$\Rightarrow \quad \text{Det}(B) = \text{Det}(P^{-1}AP)$

$\Rightarrow \quad \text{Det}(B) = \text{Det}(P^{-1})\,\text{Det}(A)\,\text{Det}(P)$

$\Rightarrow \quad \text{Det}(B) = \dfrac{1}{\text{Det}(P)} \times \text{Det}(A) \times \text{Det}(P) = \text{Det}(A)$

Q.E.D.

THEOREM-13 *A square matrix A over a field F is invertible iff $\text{Det}(A) \neq 0$, i.e. A is non-singular.*

PROOF. First suppose that A is an invertible matrix over F. Then, A^{-1} exists and

Now, $\quad AA^{-1} = I$

$\Rightarrow \quad \text{Det}(A)\,\text{Det}(A^{-1}) = \text{Det}(I)$

$\Rightarrow \quad \text{Det}(A)\,\text{Det}(A^{-1}) = 1$

$\Rightarrow \quad \text{Det}(A) \neq 0$

Conversely, let $\text{Det}(A) \neq 0$. Then, we have to prove that A is an invertible matrix. In order to prove this it is sufficient to show that $\text{rank}(A) = n$. If possible, let $\text{rank}(A) < n$. Then,

$\quad \text{rank}(A) < n$

$\Rightarrow \quad A^1, A^2, \ldots, A^n$ are linearly dependent columns in F^n.

$\Rightarrow \quad$ One of these n columns A^k (say) is either $O \in F^n$

or, it is a linear combination of the preceding columns in F^n.

If $\quad A^k = O$, then $\text{Det}(A) = 0$ \hfill [By Theorem 2]

If A^k is a linear combination of $A^1, A^2, \ldots, A^{k-1}$, then there exist scalars $\lambda_1, \lambda_2, \ldots, \lambda_{k-1}$ in F such that
$$A^k = \sum_{r=1}^{k-1} \lambda_r A^r$$

$\therefore \quad \text{Det}(A) = \text{Det}(A^1, A^2, \ldots, A^k, \ldots A^n)$

$\Rightarrow \quad \text{Det}(A) = \text{Det}\left(A^1, A^2, \ldots, \sum_{r=1}^{k-1} \lambda_r A^r, \ldots, A^n\right)$

$\Rightarrow \quad \text{Det}(A) = \sum_{r=1}^{k-1} \lambda_r \text{Det}\left(A^1, A^2, \ldots, A^{k-1}, A^r, \ldots A^n\right)$ \hfill [By Theorem 5]

$\Rightarrow \quad \text{Det}(A) = 0$ \hfill [By Theorem 8]

Thus, in both the cases, we obtain $\text{Det}(A) = 0$, which is a contradiction to $\text{Det}(A) \neq 0$. Therefore, columns of A are linearly independent vectors in F^n. Hence, $\text{rank}(A) = n$ and so A is invertible.

Q.E.D.

THEOREM-14 *Let A be an $n \times n$ matrix over a field F. Then the following are equivalent:*

(i) A is invertible
(ii) $AX = O$ has only the zero solution
(iii) $\text{Det}(A) \neq 0$.

PROOF. First we shall prove that (i) $\Rightarrow$ (ii).
Let A be an invertible matrix. Then, A^{-1} exists.
$\therefore \quad AX = O$
$\Rightarrow \quad A^{-1}(AX) = A^{-1}O$
$\Rightarrow \quad (A^{-1}A)X = O$
$\Rightarrow \quad IX = O$
$\Rightarrow \quad X = O$

So, $X = O$ is the only solution of $AX = O$.
So, (i) implies (ii).
Now, we shall show that (ii) implies (iii).
Suppose $AX = O$ has only the zero solution $X = O$. Then,
A is row equivalent to I
$\Rightarrow \quad A$ is invertible
$\Rightarrow \quad \text{Det}(A) \neq 0$

Thus, (ii) $\Leftrightarrow$ (iii).
Finally, we shall show that (iii) implies (i).
We have,
$\text{Det}(A) \neq 0$
$\Rightarrow \quad \text{rank}(A) = n$
$\Rightarrow \quad A$ is invertible.

So, (iii) implies (i).
Hence, (i), (ii) and (iii) are equivalent.

Q.E.D.

EXERCISE 5.1

1. Let A be an $n \times n$ matrix over a field F and $\lambda \in F$. Then, prove that $\text{Det}(\lambda A) = \lambda^n \text{Det}(A)$.
2. Let A be an $n \times n$ matrix over a field F. Then, prove that

$$\text{Det}(-A) = \begin{cases} \text{Det}(A), & \text{if } n \text{ is even} \\ -\text{Det}(A), & \text{if } n \text{ is odd}. \end{cases}$$

3. A square matrix is an orthogonal matrix if $A^T A = I$. Prove that the determinant of an orthogonal matrix over R is ± 1.
4. If B is row equivalent to a square matrix A, then prove that $|B| = 0$ iff $|A| = 0$.

5.4 DETERMINANT OF A LINEAR OPERATOR

Let $V(F)$ be a finite dimensional vector space and let t be a linear operator on V, i.e. a linear transformation from V to itself. Let P be the matrix representation of t relative to some basis B of V, i.e. $P = M_B^B(t)$. Then, Det(P) is defined as the determinant of t relative to basis B.

Let Q be the matrix representation of t relative to another basis B' of V, i.e. $M_{B'}^{B'}(t) = Q$. Then, P and Q are similar matrices (see Theorem 3 on page 389) and hence Det(P) = Det(Q) (See Corollary 2 of Theorem 12 on page 455). Thus, the determinant of t is independent of the choice of basis of V. So, we define the determinant of a linear operator as follows:

DETERMINANT OF A LINEAR OPERATOR *Let t be a linear operator on a vector space $V(F)$. The determinant of any matrix representation of t relative to any basis of V is defined as the determinant of t and is written as Det(t).*

EXAMPLE-1 Find the determinant of $t : R^3 \to R^3$ defined by
$$t(x,y,z) = (x+3y-4z,\; 2y+7z,\; x+5y-3z)$$

SOLUTION Let $B = \left\{e_1^{(3)}, e_2^{(3)}, e_3^{(3)}\right\}$ be the standard basis of R^3. Then,

$$t(e_1^{(3)}) = t(1,0,0) = (1,0,1) = 1e_1^{(3)} + 0e_2^{(3)} + 1e_3^{(3)}$$
$$t(e_2^{(3)}) = t(0,1,0) = (3,2,5) = 3e_1^{(3)} + 2e_2^{(3)} + 5e_3^{(3)}$$
$$t(e_3^{(3)}) = t(0,0,1) = (-4,7,-3) = -4e_1^{(3)} + 7e_2^{(3)} + (-3)e_3^{(3)}$$

$\therefore \quad P = M_B^B(t) = \begin{bmatrix} 1 & 0 & 1 \\ 3 & 2 & 5 \\ -4 & 7 & -3 \end{bmatrix}^T = \begin{bmatrix} 1 & 3 & -4 \\ 0 & 2 & 7 \\ 1 & 5 & -3 \end{bmatrix}$

$\therefore \quad \text{Det}(t) = |P| = \begin{vmatrix} 1 & 3 & -4 \\ 0 & 2 & 7 \\ 1 & 5 & -3 \end{vmatrix} = -6 + 21 + 0 + 8 - 35 + 0 = -12$

REMARK. *The matrix of t relative to the standard basis is a matrix whose rows, respectively, consist of the coefficient of x, y, z. For example, if $t : R^3 \to R^3$ is defined by*
$$t(x,y,z) = (3x+2z,\; 5y+7z,\; x-y+z)$$

Then, $\quad \text{Det}(t) = \begin{vmatrix} 3 & 0 & 2 \\ 0 & 5 & 7 \\ 1 & -1 & 1 \end{vmatrix}.$

EXAMPLE-2 Let V be the vector space of all functions with $B = \{e^t, e^{2t}, e^{3t}\}$ as a basis and let $D : V \to V$ be the differential operator, that is, $D(f(t)) = \dfrac{df}{dt}$. Find Det(D).

SOLUTION We have,

$$D(e^t) = \frac{d}{dt}(e^t) = e^t \Rightarrow D(e^t) = 1e^t + 0e^{2t} + 0e^{3t}$$

$$D(e^{2t}) = \frac{d}{dt}(e^{2t}) = 2e^{2t} \Rightarrow D(e^{2t}) = 0e^t + 2e^{2t} + 0e^{3t}$$

$$D(e^{3t}) = \frac{d}{dt}(e^{3t}) = 3e^{3t} \Rightarrow D(e^{3t}) = 0e^t + 0e^{2t} + 3e^{3t}$$

Let A be the matrix of $D : V \to V$ relative to the given basis. Then,

$$A = \begin{bmatrix} 1 & 0 & 0 \\ 0 & 2 & 0 \\ 0 & 0 & 3 \end{bmatrix}^T = \begin{bmatrix} 1 & 0 & 0 \\ 0 & 2 & 0 \\ 0 & 0 & 3 \end{bmatrix}$$

$$\therefore \quad \text{Det}(D) = \begin{vmatrix} 1 & 0 & 0 \\ 0 & 2 & 0 \\ 0 & 0 & 3 \end{vmatrix} = 1 \times 2 \times 3 = 6$$

THEOREM-1 *Let t_1 and t_2 be linear operators on a finite dimensional vector space $V(F)$. Then, prove that*

$$\text{Det}(t_1 \circ t_2) = \text{Det}(t_1)\,\text{Det}(t_2)$$

PROOF. Let B be any ordered basis of V. Then,

$$M_B^B(t_1 \circ t_2) = M_B^B(t_1)\,M_B^B(t_2)$$

$\Rightarrow \quad R = PQ$, where $R = M_B^B(t_1 \circ t_2), P = M_B^B(t_1)$ and $Q = M_B^B(t_2)$

$\Rightarrow \quad \text{Det}(R) = \text{Det}(PQ)$

$\Rightarrow \quad \text{Det}(R) = \text{Det}(P)\,\text{Det}(Q)$

$\Rightarrow \text{Det}(t_1 \circ t_2) = \text{Det}(t_1)\,\text{Det}(t_2)\;[\because\;\text{Det}(R) = \text{Det}(t_1 \circ t_2), \text{Det}(P) = \text{Det}(t_1), \text{Det}(Q) = \text{Det}(t_2)]$

Q.E.D.

THEOREM-2 *Let t be a linear operator on a finite dimensional vector space $V(F)$. Then, t is invertible iff $\text{Det}(t) \neq 0$.*

PROOF. Let B be an ordered basis for V.

First, let t be invertible. Then there exists $t^{-1} : V \to V$ such that

$$t \circ t^{-1} = I_V = t^{-1} \circ t$$

$\Rightarrow \quad \text{Det}(t \circ t^{-1}) = \text{Det}(I_V)$

$\Rightarrow \quad \text{Det}(t)\,\text{Det}(t^{-1}) = 1 \qquad\qquad [\because\;\text{Det}(I_V) = \text{Det}(M_B^B(I_V)) = \text{Det}(I) = 1]$

Since $\text{Det}(t)$ and $\text{Det}(t^{-1})$ are elements of F and in a field the product of two elements can be 0 iff at least one of them is zero. i.e. a field is without zero divisors.

$\therefore \quad \text{Det}(t)\text{Det}(t^{-1}) = 1 \Rightarrow \text{Det}(t) \neq 0$

Conversely, suppose that t is a linear operator on V such that $\text{Det}(t) \neq 0$.
Let A be the matrix of t relative to any ordered basis for V. Then,

$\text{Det}(A) = \text{Det}(t) \Rightarrow \text{Det}(A) \neq 0 \Rightarrow A$ is invertible $\Rightarrow t$ is invertible

Q.E.D.

EXAMPLE-3 *Let t_1 and t_2 be linear operators on a finite dimensional vector space $V(F)$ such that $t_1 o t_2 = \hat{0}, t_1 \neq \hat{0}$ and, $t_2 \neq \hat{0}$, then prove that $\text{Det}(t_1) = \text{Det}(t_2) = 0$.*

PROOF. If possible, let $\text{Det}(t_1) \neq 0$. Then,

$\text{Det}(t_1) \neq 0 \Rightarrow t_1$ is invertible and t_1^{-1} exists

$\therefore \quad t_1 o t_2 = \hat{0}$

$\Rightarrow \quad t_1^{-1} o (t_1 o t_2) = t_1^{-1} o \hat{0}$

$\Rightarrow \quad (t_1^{-1} o t_1) o t_2 = \hat{0}$

$\Rightarrow \quad I_V o t_2 = \hat{0}$

$\Rightarrow \quad t_2 = \hat{0}$

This is contrary to the hypothesis that $t_2 \neq \hat{0}$.

$\therefore \quad \text{Det}(t_1) = 0$

Similarly, it can be shown that $\text{Det}(t_2) = 0$.

EXERCISE 5.2

1. Let $V(F)$ be a finite dimensional vector space and I_V be identity operator on V. Prove that $\text{Det}(I_V) = 1$.
2. Let t be the linear operator on R^3 defined by

   $t(x,y,z) = (2x - 4y + z, \ x - 2y + 3z, \ 5x + y - z)$ for all $x, y, z \in R$

   Find $\text{Det}(t)$.
3. Find the determinant of each of the following linear transformations:
   (i) $t : R^2 \to R^2$ defined by $t(x, y) = (2x - 9y, \ 3x - 5y)$
   (ii) $t : R^3 \to R^3$ defined by $t(x, y, z) = (3x - 2z, \ 5y + 7z, \ x + y + z)$
   (iii) $t : R^3 \to R^2$ defined by $t(x, y, z) = (2x + 7y - 4z, \ 4x - 6y + 2z)$

4. Let V be the vector space of functions and $D : V \to V$ be the differential operator, that is, $Df(t) = \dfrac{df}{dt}$. Find Det (D) when the basis of V is (i) $B = \{1, t, t^2, \ldots, t^5\}$ (ii) $B = \{\sin t, \cos t\}$

5. Let $t : V \to V$ be an invertible linear transformation. Then, prove that $Det(t^{-1}) = [Det(t)]^{-1}$.

ANSWERS

2. -55 3. (i) 17   (ii) 4   (iii) not defined   4. (i) 0   (ii) 1

5.5 MULTILINEARITY AND DETERMINANTS

In this section, we shall show that the determinant of a square matrix A over a field F can be considered as a function which is linear in each row of matrix A. Due to this reason, the determinant of a square matrix A is said to be a multilinear function of rows of A. This determinant also turns out to be the same multilinear function of columns of matrix A.

MULTILINEAR FUNCTION Let V be a vector space over a field F and $n \in N$. Then a function $D : V^n \to F$ is said to be a multilinear function if it is linear in each component, i.e.

(i) $D(v_1, v_2, \ldots, v_i + u_i, \ldots, v_n) = D(v_1, v_2, \ldots, v_i, \ldots, v_n) + D(v_1, v_2, \ldots, u_i, \ldots, v_n)$

(ii) $D(v_1, v_2, \ldots, \lambda v_i, \ldots, v_n) = \lambda D(v_1, v_2, \ldots, v_i, \ldots, v_n)$ for all $i \in \underline{n}$.

where $v_1, v_2, \ldots, v_i, u_i, \ldots, v_n \in V$ and $\lambda \in F$.

More explicitly, the function $D : V^n \to F$ is said to be a multilinear function if for each index $i (i = 1, 2, \ldots, n)$, the function $\phi : V \to F$ defined by

$\phi(v) = D(v_1, v_2, \ldots v_{i-1}, v, v_{i+1}, \ldots, v_n)$ for all $v \in V$ is a linear transformation

The multilinear function $D : V^n \to F$ is also called an n-linear form.

ALTERNATING FUNCTION A multilinear function $D : V^n \to F$ is said to be an alternating function, if

$D(v_1, v_2, \ldots, v_i, \ldots, v_n) = 0$ whenever $v_i = v_j$, $i \neq j$, $i, j \in \underline{n}$.

THEOREM-1 Let $V(F)$ be an n dimensional vector space and $D : V^n \to F$ be an alternating function. Then the value of $D(v_1, v_2, \ldots, v_n)$ is multiplied by (-1) when any two arguments v_i and $v_j (i \neq j)$ are interchanged.

PROOF. Using multilinearity of $D: V^n \to F$, we have

$D(v_1, v_2, \ldots, v_{i-1}, v_i + v_j, v_{i+1}, \ldots, v_{j-1}, v_i + v_j, v_{j+1}, \ldots, v_n)$
$= D(v_1, v_2, \ldots, v_{i-1}, v_i, v_{i+1}, \ldots, v_{j-1}, v_i + v_j, v_{j+1}, \ldots, v_n)$
$\quad + D(v_1, v_2, \ldots, v_{i-1}, v_j, v_{i+1}, \ldots, v_{j-1}, v_i + v_j, v_{j+1}, \ldots, v_n)$
$= D(v_1, v_2, \ldots, v_{i-1}, v_i, v_{i+1}, \ldots, v_{j-1}, v_i, v_{j+1}, \ldots, v_n)$
$\quad + D(v_1, v_2, \ldots, v_{i-1}, v_i, v_{i+1}, \ldots, v_{j-1}, v_j, v_{j+1}, \ldots, v_n)$
$\quad + D(v_1, v_2, \ldots, v_{i-1}, v_j, v_{i+1}, \ldots, v_{j-1}, v_i, v_{j+1}, \ldots, v_n)$
$\quad + D(v_1, v_2, \ldots, v_{i-1}, v_j, v_{i+1}, \ldots, v_{j-1}, v_j, v_{j+1}, \ldots, v_n)$

Since $D: V^n \to F$ is an alternating function. Therefore, in the above equality LHS $= 0$ and first and fourth terms on RHS are each equal to zero.

$\therefore \quad 0 = D(v_1, v_2, \ldots, v_{i-1}, v_i, v_{i+1}, \ldots, v_{j-1}, v_j, v_{j+1}, \ldots, v_n)$
$\quad + D(v_1, v_2, \ldots, v_{i-1}, v_j, v_{i+1}, \ldots, v_{j-1}, v_i, v_{j+1}, \ldots, v_n)$
$\Rightarrow \quad D(v_1, v_2, \ldots, v_{i-1}, v_i, v_{i+1}, \ldots, v_{j-1}, v_j, v_{j+1}, \ldots, v_n)$
$\quad = -D(v_1, v_2, \ldots, v_{i-1}, v_j, v_{i+1}, \ldots, v_{j-1}, v_i, v_{j+1}, \ldots, v_n)$

Q.E.D.

COROLLARY. *If $D: V^n \to F$ is an alternating function, then for any $\sigma \in S_n$*

$$D(v_{\sigma(1)}, v_{\sigma(2)}, \ldots, v_{\sigma(n)}) = (-1)^\sigma D(v_1, v_2, \ldots, v_n)$$

PROOF. The permutation σ can be expressed as the product of transpositions. Let $\sigma = \tau_1 \tau_2 \ldots \tau_r$. From the above theorem, we find that each transposition of the arguments of D changes the sign.

$\therefore \quad D(v_{\sigma(1)}, v_{\sigma(2)}, \ldots, v_{\sigma(n)}) = (-1)^r D(v_1, v_2, \ldots, v_n) \qquad (i)$

The permutation σ is even or odd according as r is even or odd. Therefore, the sign on RHS of (i) is positive if σ is even and negative if σ is odd.

Hence, $\quad D(v_{\sigma(1)}, v_{\sigma(2)}, \ldots, v_{\sigma(n)}) = (-1)^\sigma D(v_1, v_2, \ldots, v_n)$.

Q.E.D.

THEOREM-2 *Let $D: V^n \to F$ be an alternating function and $\lambda \in F$. Then,*

$$D(v_1, v_2, \ldots, v_i, \ldots, v_j, \ldots, v_n) = D(v_1, v_2, \ldots, v_i + \lambda v_j, v_{i+1}, \ldots, v_j, \ldots, v_n)$$

PROOF. Using multilinearity of D, we have

$D(v_1, v_2, \ldots, v_i + \lambda v_j, v_{i+1}, \ldots, v_j, \ldots, v_n)$
$= D(v_1, v_2, \ldots, v_i, v_{i+1}, \ldots, v_j, \ldots, v_n) + D(v_1, v_2, \ldots, v_{i-1}, \lambda v_j, v_{i+1}, \ldots, v_j, \ldots, v_n)$

$$= D(v_1, v_2, \ldots, v_i, v_{i+1}, \ldots, v_j, \ldots, v_n) + \lambda\, D(v_1, v_2, \ldots, v_{i-1}, v_j, v_{i+1}, \ldots, v_j, \ldots, v_n)$$
$$= D(v_1, v_2, \ldots, v_i, v_{i+1}, \ldots, v_j, \ldots, v_n) + \lambda\, 0$$
$$= D(v_1, v_2, \ldots, v_i, v_{i+1}, \ldots, v_j, \ldots, v_n)$$

Q.E.D.

Let M be the set of all $n \times n$ square matrices over a field F and let A_i denote its ith row, $i \in \underline{n}$. The matrix A may be considered as an n-tuple consisting of its row vectors $A_1, A_2, \ldots, A_n$. That is, we may view A in the form $A = (A_1, A_2, \ldots, A_n)$. We shall now show that the determinant can be considered as a function from M to F. That is why determinant is also viewed as a function and we call it the determinant function.

The following theorem characterizes the determinant function.

THEOREM-3 Let M be the set of all $n \times n$ square matrices over a field F. Prove that there exists a unique function $D : M \to F$ such that:

(i) D is multilinear function
(ii) D is an alternating function
(iii) $D(I) = 1$.

PROOF. Let D be the determinant function associating each $n \times n$ matrix A in M to its determinant $|A|$.

Let us first show that D satisfies (i), (ii) and (iii).

(i) *D is multilinear function:* Let $A = [a_{ij}]$ be a matrix in M such that the ith row of A is of the form $(b_{i1} + c_{i1}, b_{i2} + c_{i2}, \ldots, b_{in} + c_{in})$, i.e. each element in ith row of A is sum of two elements. In other words $A_i = B_i + C_i$ or $a_{ij} = b_{ij} + c_{ij}$ for all $j \in \underline{n}$. Then, by Theorem 5 on page 450, we have

$$\text{Det}(A) = \text{Det}(A_1, A_2, \ldots, B_i + C_i, \ldots, A_n)$$
$$\Rightarrow \quad \text{Det}(A) = \text{Det}(A_1, A_2, \ldots, B_i, \ldots, A_n) + \text{Det}(A_1, A_2, \ldots, C_i, \ldots, A_n)$$

i.e. $D(A_1, A_2, \ldots, B_i + C_i, \ldots, A_n) = D(A_1, A_2, \ldots, B_i, \ldots, A_n) + D(A_1, A_2, \ldots, C_i, \ldots, A_n)$

Also, by Theorem 3 on page 448, we have

$$\text{Det}(A_1, A_2, \ldots, kA_i, \ldots, A_n) = k\, \text{Det}(A_1, A_2, \ldots, A_i, \ldots, A_n)$$

i.e. $D(A_1, A_2, \ldots, kA_i, \ldots, A_n) = k\, D(A_1, A_2, \ldots, A_i, \ldots, A_n)$

Thus, $D : M \to F$ is a multilinear function.

(ii) *D is an alternating function:* Let $A = (A_1, A_2, \ldots, A_n) \in M$ such that ith and jth rows are identical. Then, by Theorem 8 on page 452, we have

$$\text{Det}(A) = 0.$$

i.e. $D(A_1, A_2, \ldots, A_i, \ldots, A_j, \ldots, A_n) = 0.$

So, $D : M \to F$ is an alternating function.

(iii) Let I be the identity matrix in M. Then, Det $(I) = 1$ [See Theorem 1 on page 447]

$\therefore \quad D : M \to K$ is such that $D(I) = 1$.

Thus, $D : M \to K$ satisfies all the three conditions stated in the theorem.

We shall now prove the uniqueness of D, that is, D is the only function satisfying these properties.

Let $d : M \to F$ be a function satisfying the above properties. As $\{e_1^{(n)}, e_2^{(n)}, \ldots, e_n^{(n)}\}$ is the standard basis of F^n such that

$$(e_1^{(n)}, e_2^{(n)}, \ldots, e_n^{(n)}) = I$$

$\Rightarrow \quad d(e_1^{(n)}, e_2^{(n)}, \ldots, e_n^{(n)}) = d(I)$

$\Rightarrow \quad d(e_1^{(n)}, e_2^{(n)}, \ldots, e_n^{(n)}) = 1 \qquad$ [Using (iii) in the statement]... (a)

Since $d : M \to F$ is an alternating function. Therefore, for any permutation $\sigma \in S$, we have

$$d\left(e_{\sigma(1)}^{(n)}, e_{\sigma(2)}^{(n)}, \ldots, e_{\sigma(n)}^{(n)}\right) = (-1)^\sigma d\left(e_1^{(n)}, e_2^{(n)}, \ldots, e_n^{(n)}\right) \quad \begin{bmatrix} \text{From Corollary 1 of} \\ \text{Theorem 1 on page 462} \end{bmatrix}$$

$\Rightarrow \quad d\left(e_{\sigma(1)}^{(n)}, e_{\sigma(2)}^{(n)}, \ldots, e_{\sigma(n)}^{(n)}\right) = (-1)^\sigma \qquad$ [From (a)]

Since F^n is a vector space with $\{e_1^{(n)}, e_2^{(n)}, \ldots, e_n^{(n)}\}$ as standard basis. Therefore, if $A = (A_1, A_2, \ldots, A_n)$ is any matrix in M, then

$$A_i = \sum_{j=1}^{n} a_{ij} e_j^{(n)} \quad \text{for all} \quad i \in \underline{n}$$

$\therefore \quad d(A) = d\left(\sum_{r_1=1}^{n} a_{1r_1} e_{r_1}^{(n)}, \ldots, \sum_{r_1=1}^{n} a_{nr_n} e_{r_n}^{(n)}\right)$

$\Rightarrow \quad d(A) = \sum_{r_1=1}^{n} \sum_{r_1=2}^{n} \cdots \sum_{r_n=1}^{n} a_{1r_1}, a_{2r_2}, \ldots, a_{nr_n} d\left(e_{r_1}^{(n)}, e_{r_2}^{(n)}, \ldots, e_{r_n}^{(n)}\right)$ [Using multilinearity of d]

In this multiple sum $r_1, r_2, \ldots, r_n$ run independently from 1 to n. However, if any two r_p and r_q are equal, then

$$e_{r_p}^{(n)} = e_{r_q}^{(n)}$$

$\Rightarrow \quad d\left(e_{r_1}^{(n)}, e_{r_2}^{(n)}, \ldots, e_{r_p}^{(n)}, \ldots, e_{r_q}^{(n)}, \ldots, e_{r_n}^{(n)}\right) = 0 \qquad [\because d \text{ is an alternating function}]$

Thus, the terms in the sum which may give rise a non-zero contribution are those for which all of $r_1, r_2, \ldots, r_n$ are distinct, that is, for which

$$\sigma = \begin{pmatrix} 1 & 2 & \cdots & n \\ \sigma(1) = r_1 & \sigma(2) = r_2 & \cdots & \sigma(n) = r_n \end{pmatrix}$$

is a permutation in S_n.

$\therefore \quad d(A) = \sum_{\sigma \in S_n} a_{1\sigma(1)} a_{2\sigma(2)} \cdots a_{n\sigma(n)} d\left(e_{\sigma(1)}^{(n)}, e_{\sigma(2)}^{(n)}, \ldots, e_{\sigma(n)}^{(n)}\right)$

$\Rightarrow \quad d(A) = \sum_{\sigma \in S_n} (-1)^\sigma a_{1\sigma(1)} a_{2\sigma(2)} \cdots a_{n\sigma(n)}$

$\Rightarrow \quad d(A) = D(A)$, i.e. d is the determinant function.

So, D is unique.

It follows from the above theorem that there exists one and only one multilinear, alternating function $D : M \to F$ satisfying $D(I) = 1$ such that for any matrix $A = [a_{ij}]$ in M

$$D(A) = \sum_{\sigma \in S_n} (-1)^{\sigma} a_{1\sigma(1)}, a_{2\sigma(2)}, \ldots, a_{n\sigma(n)}$$

This function is called the determinant function and we write $\text{Det}(A) = |A|$.

Q.E.D.

REMARK. *All properties of determinants discussed and proved in Section 5.3 can also be proved by considering D as a multilinear alternating function with $D(I) = 1$.*

THEOREM-4 *Let F be a field and $n \in \mathbb{N}$. If $h : F^n \to F$ is an alternating multilinear form, then for any $n \times n$ square matrix $A = [a_{ij}]$ over F, $h(A) = h(I) \text{Det}(A)$*

PROOF. Let A_i denote the ith row of A. Then, $A = (A_1, A_2, \ldots, A_n)$. Since $\{e_1^{(n)}, e_2^{(n)}, \ldots, e_n^{(n)}\}$ is standard basis of F^n. Therefore, for any $i \in \underline{n}$, we have

$$A_i = \sum_{r_i=1}^{n} a_{ir_i} e_{r_i}^{(n)}$$

$$\therefore \quad h(A) = h(A_1, A_2, \ldots, A_n)$$

$$\Rightarrow \quad h(A) = h\left(\sum_{r_1=1}^{n} a_{1r_1} e_{r_1}^{(n)}, \sum_{r_2=1}^{n} a_{2r_2} e_{r_2}^{(n)}, \ldots \sum_{r_n=1}^{n} a_{nr_n} e_{r_n}^{(n)} \right)$$

$$\Rightarrow \quad h(A) = \sum_{r_1=1}^{n} \sum_{r_2=2}^{n} \ldots \sum_{r_n=1}^{n} a_{1r_1} a_{2r_2} \ldots a_{nr_n} h\left(e_{r_1}^{(n)}, e_{r_2}^{(n)}, \ldots, e_{r_n}^{(n)} \right) \quad \text{(i)}$$

In the sum on RHS each one of the indices run independently from 1 to n. So, it contains n^n terms. If $r_p = r_q$, then

$$h\left(e_{r_1}^{(n)}, e_{r_2}^{(n)}, \ldots, e_{r_p}^{(n)}, \ldots, e_{r_q}^{(n)}, \ldots, e_{r_n}^{(n)} \right) = 0 \qquad [\because h \text{ is an alternating function}]$$

Therefore, the only terms in the above sum that give a non-zero contribution to the sum on RHS of (i) are those for which $r_1, r_2, \ldots, r_n$ are distinct, that is, for which the mapping σ given by

$$\sigma = \begin{pmatrix} 1 & 2 & \ldots & n \\ \sigma(1) = r_1 & \sigma(2) = r_2 & \ldots & \sigma(n) = r_n \end{pmatrix}$$

is a permutation.

$$\therefore \quad h(A) = \sum_{\sigma \in S_n} a_{1\sigma(1)} a_{2\sigma(2)} \ldots a_{n\sigma(n)} h\left(e_{\sigma(1)}^{(n)}, e_{\sigma(2)}^{(n)}, \ldots, e_{\sigma(n)}^{(n)} \right) \qquad \text{[From (i)]}$$

$$\Rightarrow \quad h(A) = \sum_{\sigma \in S_n} (-1)^{\sigma} a_{1\sigma(1)} a_{2\sigma(2)} \ldots a_{n\sigma(n)} h\left(e_1^{(n)}, e_2^{(n)}, \ldots, e_n^{(n)} \right)$$

$$\Rightarrow \quad h(A) = h\left(e_1^{(n)}, e_2^{(n)}, \ldots, e_n^{(n)}\right) \left\{ \sum_{\sigma \in S_n} (-1)^\sigma a_{1\sigma(1)} a_{2\sigma(2)} \cdots a_{n\sigma(n)} \right\}$$

$$\Rightarrow \quad h(A) = h(I) \operatorname{Det}(A)$$

Let us now introduce a notation which will be very helpful in proving properties of determinants particularly when we consider determinant function D as a multilinear, alternating function with $D(I) = 1$.

Q.E.D.

NOTATION Let $\underline{n}^{\underline{n}}$ denote the set of all functions from $\underline{n} = (1,2,3,\ldots,n)$ to itself. Then, for any $\phi \in \underline{n}^{\underline{n}}$, we define

$$\in_\phi = \begin{cases} (-1)^\phi = 1, & \text{if } \phi \text{ is an even permutation in } S_n \\ (-1)^\phi = -1, & \text{if } \phi \text{ is an odd permutation in } S_n \\ 0, & \text{if } \phi \text{ is not a permutation} \end{cases}$$

For any $\phi, \psi \in \underline{n}^{\underline{n}}$, it can be easily shown that $\in_{\phi \circ \psi} = \in_\phi \in_\psi$ and, if $\sigma \in S_n$, then $\in_\sigma = \in_{\sigma^{-1}}$.

Using the above notation, we obtain

$$\operatorname{Det}(A) = \sum_{\sigma \in S_n} \in_\sigma a_{1\sigma(1)} a_{2\sigma(2)} \cdots a_{n\sigma(n)}$$

THEOREM-5 Let $A = [a_{ij}]$ be an $n \times n$ matrix over a field F and $\phi, \psi \in \underline{n}^{\underline{n}}$. Then,

$$\sum_{\sigma \in S_n} \in_\sigma a_{\phi(1)\sigma(1)} a_{\phi(2)\sigma(2)} \cdots a_{\phi(n)\sigma(n)} = \in_\phi \operatorname{Det}(A)$$

PROOF. *Case I When $\phi \in S_n$, i.e. ϕ is a permutation on $\underline{n}$.*

$$\phi \in S_n \Rightarrow \phi^{-1} \in S_n \Rightarrow \sigma \circ \phi^{-1} \in S_n, \text{ where } \sigma \in S_n$$

Let $\quad \rho = \sigma \circ \phi^{-1}$. Then, $\sigma = \rho \circ \phi$.

Now,

$$\in_\sigma a_{\phi(1)\sigma(1)} a_{\phi(2)\sigma(2)} \cdots a_{\phi(n)\sigma(n)}$$
$$= \in_{\rho \circ \phi} a_{\phi(1)\rho(\phi(1))} a_{\phi(2)\rho(\phi(2))} \cdots a_{\phi(n)\rho(\phi(n))} = \in_\rho \in_\phi a_{1\rho(1)} a_{2\rho(2)} \cdots a_{n\rho(n)}$$

Since σ ranges over S_n. Therefore, ρ also ranges over S_n.

Hence,

$$\sum_{\sigma \in S_n} \in_\sigma a_{\phi(1)\sigma(1)} a_{\phi(2)\sigma(2)} \cdots a_{\phi(n)\sigma(n)}$$

$$= \sum_{\rho \in S_n} \in_\rho \in_\phi a_{1\rho(1)} a_{2\rho(2)} \cdots a_{n\rho(n)} = \in_\phi \sum_{\rho \in S_n} \in_\rho a_{1\rho(1)} a_{2\rho(2)} \cdots a_{n\rho(n)} = \in_\phi \text{Det}(A)$$

Case II When $\phi \notin S_n$

In this case, we have

$$\in_\phi = 0.$$

$\Rightarrow \quad \in_\phi \text{Det}(A) = 0.$ \hfill (i)

Now, $\quad \phi \notin S_n$

$\Rightarrow \quad \phi$ is not a bijection on $\underline{n}$

$\Rightarrow \quad$ There exist $i, j \in \underline{n}$ such that $i \neq j$ but $\phi(i) = \phi(j)$.

Let $\tau = (i, j)$ be a transposition and $\sigma \in S_n$ such that $\rho = \sigma \circ \tau$. Then,

$$\rho(i) = \sigma \circ \tau(i) = \sigma(\tau(i)) = \sigma(j)$$
$$\rho(j) = \sigma \circ \tau(j) = \sigma(\tau(j)) = \sigma(i)$$

and, $\quad \rho(r) = \sigma \circ \tau(r) = \sigma(\tau(r)) = \sigma(r)$ for all $r \in \underline{n}, r \neq i, j$

Since τ is an odd permutation. Therefore,

$$\in_\rho = \in_{\sigma \circ \tau} = \in_\sigma \in_\tau = - \in_\sigma$$

Now, consider the sum

$$\sum_{\sigma \in S_n} \in_\sigma a_{\phi(1)\sigma(1)} a_{\phi(2)\sigma(2)} \cdots a_{\phi(n)\sigma(n)} \hfill (ii)$$

The contribution to this sum due to permutations σ and ρ is

$\in_\sigma a_{\phi(1)\sigma(1)} a_{\phi(2)\sigma(2)} a_{\phi(i)\sigma(i)} \cdots a_{\phi(j)\sigma(j)} \cdots a_{\phi(n)\sigma(n)} + \in_\rho a_{\phi(1)\rho(1)} a_{\phi(2)\rho(2)} \cdots a_{\phi(i)\rho(i)} \cdots a_{\phi(j)\rho(j)} \cdots a_{\phi(n)\rho(n)}$

$= \in_\sigma a_{\phi(1)\sigma(1)} a_{\phi(2)\sigma(2)} \cdots a_{\phi(n)\sigma(n)} - \in_\sigma a_{\phi(1)\sigma(1)} \cdots a_{\phi(i)\sigma(j)} \cdots a_{\phi(j)\sigma(i)} \cdots a_{\phi(n)\sigma(n)}$

$= 0$ \hfill $[\because \phi(i) = \phi(j)]$

Since S_n can be partitioned into pairs of the type (σ, ρ) and the contribution to the sum given in (ii) from each such pair is zero.

$$\therefore \sum_{\sigma \in S_n} \in_\sigma a_{\phi(1)\sigma(1)} a_{\phi(2)\sigma(2)} \cdots a_{\phi(n)\sigma(n)} = 0 \hfill (iii)$$

From (ii) and (iii), we get

$$\sum_{\sigma \in S_n} \epsilon_\sigma \, a_{\phi(1)\sigma(1)} \, a_{\phi(2)\sigma(2)} \cdots a_{\phi(n)\sigma(n)} = \epsilon_\phi \, \text{Det}(A)$$

Hence, $\quad \sum_{\sigma \in S_n} \epsilon_\sigma \, a_{\phi(1)\sigma(1)} \, a_{\phi(2)\sigma(2)} \cdots a_{\phi(n)\sigma(n)} = \epsilon_\phi \, \text{Det}(A)$

Q.E.D.

5.6 DETERMINANT RANK OF A MATRIX

SUBMATRIX Let $A = [a_{ij}]$ be an $m \times n$ matrix over a field F. A matrix obtained by dropping some rows and columns from A is called a submatrix of A.

For example, if $A = \begin{bmatrix} 2 & -1 & 3 & 4 \\ -3 & 5 & 0 & 1 \\ 7 & -4 & -2 & 9 \end{bmatrix}$, then $\begin{bmatrix} 2 & -1 \\ -3 & 5 \end{bmatrix}, \begin{bmatrix} 2 & -1 & 3 \\ -3 & 5 & 0 \end{bmatrix}, \begin{bmatrix} -1 & 3 & 4 \\ 5 & 0 & 1 \\ 4 & -2 & 9 \end{bmatrix}$

etc. are submatrices of A.

A matrix A is a submatrix of itself, because it is obtained from A by dropping no rows or columns.

RANK OF A MATRIX Let A be an $m \times n$ non-null matrix over a field F. Then a positive integer r is said to be the rank of A if it possesses the following properties:

(i) there is at least one $r \times r$ submatrix M (say) of A such that $\text{Det}(M) \neq 0$.

(ii) if there is any square submatrix N of order $(r+1)$ or more, then $\text{Det}(N) = 0$.

It is evident from the above definition that $\text{rank}(I_n) = n$ as $\text{Det}(I_n) = 1 \neq 0$.

The rank of a square null matrix is defined to be zero.

Also, if $A = [a_{ij}]$ is an $m \times n$ matrix, then $\text{rank}(A) \leq \min(m, n)$.

If $A = \begin{bmatrix} 1 & 1 & 2 \\ 1 & 2 & 5 \\ 5 & 3 & 4 \end{bmatrix}$, then $\text{Det}(A) = 0$ and $M = \begin{bmatrix} 1 & 2 \\ 2 & 5 \end{bmatrix}$ is a square submatrix such that $\text{Det}(M) \neq 0$. So, $\text{rank}(A) = 2$.

5.7 COFACTORS AND MINORS

Let $A = [a_{ij}]$ be an $n \times n$ matrix over a field F. Then,

$$\text{Det}(A) = \sum_{\sigma \in S_n} (-1)^\sigma a_{1\sigma(1)} \, a_{2\sigma(2)} \cdots a_{i\sigma(i)} \cdots a_{n\sigma(n)} \qquad (i)$$

Let us consider $\text{Det}(A)$ as a function of elements of ith row of matrix A. There are $n!$ terms in the expression on the RHS such that each term contains exactly one of element $a_{i1}, a_{i2}, a_{i3}, \ldots, a_{in}$ as a factor. The terms which contain a given element a_{ij} (say) as a factor are given by the permutations $\sigma \in S_n$ which satisfy the condition $\sigma(i) = j$. Clearly, there $(n-1)!$ such permutations.

The sum of all such terms on the RHS of (i) is

$$\sum_{\substack{\sigma \in S_n \\ \sigma(i)=j}} (-1)^\sigma a_{1\sigma(1)} a_{2\sigma(2)} \cdots a_{i\sigma(i)} \cdots a_{n\sigma(n)}$$

$$= a_{ij} \left\{ \sum_{\substack{\sigma \in S_n \\ \sigma(i)=j}} (-1)^\sigma a_{1\sigma(1)} a_{2\sigma(2)} \cdots a_{i-1\sigma(i-1)} a_{i+1\sigma(i+1)} \cdots a_{n\sigma(n)} \right\}$$

$= a_{ij} A_{ij}$, where A_{ij} stands for the expression within brackets given in the previous step.

$$A_{ij} = \sum_{\substack{\sigma \in S_n \\ \sigma(i)=j}} (-1)^\sigma a_{1\sigma(1)} a_{2\sigma(2)} \cdots a_{i-1\sigma(i-1)} a_{i+1\sigma(i+1)} \cdots a_{n\sigma(n)}$$

is known as the cofactor of a_{ij} in $\mathrm{Det}(A)$.

Let us now find an explicit formula for A_{ij}.

THEOREM-1 Let $A = [a_{ij}]$ be an $n \times n$ matrix over a field F. Then the cofactor A_{ij} of a_{ij} in $\mathrm{Det}(A)$ is the scalar given by

$$A_{ij} = (-1)^{i+j} \mathrm{Det}(A^{(i,j)})$$

where $A^{(i,j)}$ is the submatrix of A obtained by deleting the ith row and jth column of A.

PROOF. We have,

$$A_{ij} = \sum_{\substack{\sigma \in S_n \\ \sigma(i)=j}} (-1)^\sigma a_{1\sigma(1)} a_{2\sigma(2)} \cdots a_{i-1\sigma(i-1)} a_{i+1\sigma(i+1)} \cdots a_{n\sigma(n)} \qquad (\mathrm{i})$$

$$\therefore \quad A_{11} = \sum_{\substack{\sigma \in S_n \\ \sigma(1)=1}} (-1)^\sigma a_{2\sigma(2)} a_{3\sigma(3)} \cdots a_{(i)\sigma(i)} \cdots a_{n\sigma(n)} \qquad (\mathrm{ii})$$

In the summation on the RHS of (ii) the summation is taken over all permutations $\sigma \in S_n$ with $\sigma(1) = 1$.

Now,

$\sigma \in S_n$ and $\sigma(1) = 1$

$\Rightarrow \quad \sigma \in S_{\{2,3,\ldots,n\}}$, where $S_{\{2,3,\ldots,n\}}$ denotes the set of all permutations on $\{2, 3, \ldots, n\}$.

$$\therefore \quad A_{11} = \sum_{\sigma \in S_{\{2,3,\ldots,n\}}} (-1)^\sigma a_{2\sigma(2)} a_{3\sigma(3)} \cdots a_{i\sigma(i)} \cdots a_{n\sigma(n)}$$

$\Rightarrow \quad A_{11} = \mathrm{Det}(A^{(1,1)}) \qquad (\mathrm{iii})$

Now consider the cofactor A_{ij} for any i and j. Interchange the ith row with each previous row in succession till ith row becomes the first row, and then similarly move jth column to first position. In this process, the relative position of rows and columns of the submatrix $A^{(i,j)}$ remain unchanged. Therefore, these operations do not change $\mathrm{Det}(A^{(i,j)})$ but they do change the sign of $\mathrm{Det}(A)$ and hence the sign of A_{ij} by $(i-1) + (j-1) = i+j-2$ times.

$\therefore \quad A_{ij} = (-1)^{i+j-2} \mathrm{Det}(A^{(i,j)}) \qquad$ [Using (iii)]

$\Rightarrow \quad A_{ij} = (-1)^{i+j} \mathrm{Det}(A^{(i,j)})$

Q.E.D.

Above Theorem suggests an alternative definition of cofactors of elements of a square matrix as given below.

COFACTOR Let $A = [a_{ij}]$ be an $n \times n$ matrix over a field F and let B_{ij} denote the $(n-1) \times (n-1)$ square submatrix of A obtained by deleting its ith row and jth column. Then, the cofactor of a_{ij} in A is $(-1)^{i+j}$ times $|B_{ij}|$, i.e $A_{ij} = (-1)^{i+j}|B_{ij}|$.

The determinant $|B_{ij}|$, in the above definition, is called the *minor* of the element a_{ij} of A and is denoted by M_{ij}.

Clearly, $A_{ij} = (-1)^{i+j} M_{ij}$

It is evident from the above definition that A_{ij} is the signed minor and the sign depends upon the sum $i + j$, i.e.

$$A_{ij} = \begin{cases} M_{ij}, & \text{if } i+j \text{ is even} \\ -M_{ij}, & \text{if } i+j \text{ is odd.} \end{cases}$$

REMARK. *A minor is principal minor if row and column indices are the same, or equivalently, if the diagonal elements of the minor come from the diagonal of the matrix.*

EXAMPLE-1 If $A = \begin{bmatrix} 1 & 2 & 3 \\ 4 & 5 & 6 \\ 7 & 8 & 9 \end{bmatrix}$, find the following minors and cofactors:

(i) M_{23} and A_{23} (ii) M_{13} and A_{13}

SOLUTION Using the definitions of minors and cofactors as given in the above discussion, we have

(i) $M_{23} = \begin{vmatrix} 1 & 2 \\ 7 & 8 \end{vmatrix} = 8 - 14 = -6$

$\therefore \quad A_{23} = (-1)^{2+3} M_{23} = -M_{23} = -(-6) = 6$

(ii) $M_{13} = \begin{vmatrix} 4 & 5 \\ 7 & 8 \end{vmatrix} = 32 - 35 = -3$

$\therefore \quad A_{13} = (-1)^{1+3} M_{13} = M_{13} = -3$

THEOREM-2 Let $A = [a_{ij}]$ be an $n \times n$ matrix over a field F. Then,

(i) $\sum_{r=1}^{n} a_{ir} A_{jr} = \delta_{ij} \text{Det}(A)$ (ii) $\sum_{r=1}^{n} a_{ri} A_{rj} = \delta_{ij} \text{Det}(A)$

where A_{ij} is cofactor of a_{ij} in matrix $A = [a_{ij}]$.

PROOF. (i) We have,

$$\text{Det}(A) = \sum_{\sigma \in S_n} (-1)^\sigma a_{1\sigma(1)} a_{2\sigma(2)} \cdots a_{i-1\sigma(i-1)} a_{i\sigma(i)} a_{i+1\sigma(i+1)} \cdots a_{n\sigma(n)}$$

$$\Rightarrow \quad \text{Det}(A) = \sum_{j=1}^{n} a_{ij} \left\{ \sum_{\substack{\sigma \in S_n \\ \sigma(i)=j}} (-1)^\sigma a_{1\sigma(1)} a_{2\sigma(2)} \cdots a_{i-1\sigma(i-1)} a_{i+1\sigma(i+1)} \cdots a_{n\sigma(n)} \right\}$$

$$\Rightarrow \quad \text{Det}(A) = \sum_{j=1}^{n} a_{ij} A_{ij} \tag{i}$$

Now,

$$\sum_{r=1}^{n} a_{ir} A_{jr} = \sum_{r=1}^{n} a_{ir} \left\{ \sum_{\substack{\sigma \in S_n \\ \sigma(j)=r}} (-1)^\sigma a_{1\sigma(1)} a_{2\sigma(2)} \cdots a_{j-1\sigma(j-1)} a_{j+1\sigma(j+1)} \cdots a_{n\sigma(n)} \right\}$$

$$\Rightarrow \quad \sum_{r=1}^{n} a_{ir} A_{jr} = \sum_{\sigma \in S_n} (-1)^\sigma a_{1\sigma(1)} a_{2\sigma(2)} \cdots a_{i\sigma(j)} a_{j+1\sigma(j+1)} \cdots a_{n\sigma(n)}$$

$$\Rightarrow \quad \sum_{r=1}^{n} a_{ir} A_{jr} = 0 \quad [\because \text{ RHS is the determinant of a matrix whose two rows are identical.}]$$

$$\Rightarrow \quad \sum_{r=1}^{n} a_{ir} A_{jr} = \delta_{ij}, i \neq j \tag{ii}$$

Combining (i) and (ii), we get

$$\sum_{r=1}^{n} a_{ir} A_{jr} = \delta_{ij} \text{Det}(A)$$

Similarly, we can prove that

$$\sum_{r=1}^{n} a_{ri} A_{rj} = \delta_{ij} \text{Det}(A).$$

Q.E.D.

REMARK. *The above theorem states that the determinant of a square matrix $A = [a_{ij}]$ is equal to the sum of the products obtained by multiplying the elements of any row (column) by their respective cofactors, i.e.*

$$\sum_{j=1}^{n} a_{ij} A_{ij} = \text{Det}(A), \quad \sum_{i=1}^{n} a_{ij} A_{ij} = \text{Det}(A) \tag{i}$$

It also states that the sum of the products obtained by multiplying the elements of any row (column) by the cofactors of the corresponding elements of some other row (column) is zero, i.e.

$$\sum_{j=1}^{n} a_{rj} A_{ij} = 0, \quad \sum_{i=1}^{n} a_{ir} A_{is} = 0$$

Formulas for $\text{Det}(A)$ given in (i) are called the Laplace expansions of the determinant of A by the ith row and jth column.

ADJOINT OF A MATRIX Let $A = [a_{ij}]$ be an $n \times n$ matrix over a field F and let A_{ij} denote the cofactor of a_{ij} in A. The adjoint of A, denoted by adj A, is the transpose of the matrix of cofactors of A, i.e.

Clearly, adj $A = [A_{ij}]^T$
(adj $A)_{ij} = A_{ji}$ = Cofactor of a_{ji} in A.

EXAMPLE-2 Let $A = \begin{bmatrix} 0 & -1 & 1 \\ 2 & -4 & 3 \\ -1 & 1 & 5 \end{bmatrix}$. Find adj A and verify that $a_{11}A_{11} + a_{12}A_{12} + a_{13}A_{13} = |A|$

SOLUTION Clearly,

$A_{11} = \begin{vmatrix} -4 & 3 \\ 1 & 5 \end{vmatrix} = -23$, $A_{12} = (-1)^{1+2} \begin{vmatrix} 2 & 3 \\ -1 & 5 \end{vmatrix} = -13$, $A_{13} = (-1)^{1+3} \begin{vmatrix} 2 & -4 \\ -1 & 1 \end{vmatrix} = -2$,

$A_{21} = (-1)^{2+1} \begin{vmatrix} -1 & 1 \\ 1 & 5 \end{vmatrix} = 6$, $A_{22} = (-1)^{2+2} \begin{vmatrix} 0 & 1 \\ -1 & 5 \end{vmatrix} = 1$, $A_{23} = (-1)^{2+3} \begin{vmatrix} 0 & -1 \\ -1 & 1 \end{vmatrix} = 1$

$A_{31} = (-1)^{3+1} \begin{vmatrix} -1 & 1 \\ -4 & 3 \end{vmatrix} = 1$, $A_{32} = (-1)^{3+2} \begin{vmatrix} 0 & 1 \\ 2 & 3 \end{vmatrix} = 2$, $A_{33} = (-1)^{3+3} \begin{vmatrix} 0 & -1 \\ 2 & -4 \end{vmatrix} = 2$

$\therefore$ adj $A = \begin{bmatrix} A_{11} & A_{12} & A_{13} \\ A_{21} & A_{22} & A_{23} \\ A_{31} & A_{32} & A_{33} \end{bmatrix}^T = \begin{bmatrix} -23 & -13 & -2 \\ 6 & 1 & 1 \\ 1 & 2 & 2 \end{bmatrix}^T = \begin{bmatrix} -23 & 6 & 1 \\ -13 & 1 & 2 \\ -2 & 1 & 2 \end{bmatrix}$

Now, $|A| = \begin{vmatrix} 0 & -1 & 1 \\ 2 & -4 & 3 \\ -1 & 1 & 5 \end{vmatrix} = 0 \begin{vmatrix} -4 & 3 \\ 1 & 5 \end{vmatrix} - (-1) \begin{vmatrix} 2 & 3 \\ -1 & 5 \end{vmatrix} + 1 \begin{vmatrix} 2 & -4 \\ -1 & 1 \end{vmatrix}$

$\Rightarrow \quad |A| = 0(-20 - 3) + 1(10 + 3) + 1(2 - 4) = 11$

and, $a_{11} A_{11} + a_{12} A_{12} + a_{13} A_{13} = 0 \times -23 + (-1)(-13) + 1 \times -2 = 11$

$\therefore \quad a_{11} A_{11} + a_{12} A_{12} + a_{13} A_{13} = |A|$

THEOREM-3 *Let A be an $n \times n$ matrix over a field F. Then, prove that*
$$A(\text{adj } A) = \text{Det}(A)I = (\text{adj } A)A$$

PROOF. Let $A = [a_{ij}]$ and let A_{ij} denote cofactor of a_{ij} in A. Then,

$(A(\text{adj } A))_{ij} = \sum_{r=1}^{n} a_{ir} (\text{adj } A)_{rj}$

$\Rightarrow \quad (A(\text{adj } A))_{ij} = \sum_{r=1}^{n} a_{ir} A_{jr}$

$\Rightarrow \quad (A(\text{adj } A))_{ij} = \delta_{ij} \text{Det}(A)$

$\Rightarrow \quad (A(\text{adj } A))_{ij} = \begin{cases} \text{Det}(A), & \text{if } i = j \\ 0, & \text{if } i \neq j \end{cases}$

Thus, $A(\text{adj } A)$ is a diagonal matrix whose each diagonal element is equal to Det(A).

$\therefore \quad (A(\text{adj } A)) = \text{Det}(A) I$

Similarly, we have

$$((\text{adj } A)A) = \text{Det}(A) \, I$$

Hence, $A(\text{adj } A) = \text{Det}(A) \, I = (\text{adj } A)A$

Q.E.D.

THEOREM-4 *Let A be an $n \times n$ invertible matrix over a field F such that $\text{Det}(A)$ is a non-zero element of F. Then, prove that $A^{-1} = \bigl(\text{Det}(A)\bigr)^{-1} \text{adj } A$*

PROOF. If $\text{Det}(A)$ is a non-zero element of F. Then, $\bigl(\text{Det}(A)\bigr)^{-1}$ exists.

$\therefore \quad A(\text{adj } A) = \text{Det}(A)I = (\text{adj } A)A$

$\Rightarrow \quad \bigl(\text{Det}(A)\bigr)^{-1}\bigl(A(\text{adj } A)\bigr) = \bigl(\text{Det}(A)\bigr)^{-1}\bigl(\text{Det}(A) \, I\bigr) = \bigl(\text{Det}(A)\bigr)^{-1}\bigl((\text{adj } A)A\bigr)$

$\Rightarrow \quad A\bigl\{\bigl(\text{Det}(A)\bigr)^{-1}(\text{adj } A)\bigr\} = I = \bigl\{\bigl(\text{Det}(A)\bigr)^{-1}(\text{adj } A)\bigr\}A$

$\Rightarrow \quad A^{-1} = \bigl(\text{Det}(A)\bigr)^{-1}(\text{adj } A)$

Q.E.D.

REMARK. *If A is a matrix over R, then we also write*

$$A^{-1} = \frac{1}{\text{Det}(A)} \, \text{adj } A$$

If A is the matrix in Example 2, then

$$A^{-1} = \frac{1}{\text{Det}(A)} \, \text{adj } A \quad \Rightarrow \quad A^{-1} = \frac{1}{11} \begin{bmatrix} -23 & 6 & 1 \\ -13 & 1 & 2 \\ 2 & 1 & 2 \end{bmatrix}$$

5.8 CRAMER'S RULE

Consider a system of n linear equations in n unknowns as given below:

$$\begin{aligned}
a_{11} x_1 + a_{12} x_2 + \cdots + a_{1j} x_j + \cdots + a_{1n} x_n &= b_1 \\
a_{21} x_1 + a_{22} x_2 + \cdots + a_{2j} x_j + \cdots + a_{2n} x_n &= b_2 \\
\vdots \quad \vdots \quad \vdots \quad \vdots \quad \vdots& \\
a_{i1} x_1 + a_{i2} x_2 + \cdots + a_{ij} x_j + \cdots + a_{in} x_n &= b_j \\
\vdots \quad \vdots \quad \vdots \quad \vdots \quad \vdots& \\
a_{n1} x_1 + a_{n2} x_2 + \cdots + a_{nj} x_j + \cdots + a_{nn} x_n &= b_n
\end{aligned}$$

where $a_{ij}, b_j \in F$ for all $i, j \in \underline{n}$

This system of equations can be written in matrix form as follows:
$$AX = B,$$
where $A = [a_{ij}]$ is $n \times n$ matrix of coefficients and $B = [b_j]$ is the column vector of constants. Let A_i be the matrix obtained from A by replacing the ith column of A by the column vector $B = [b_j]$. Furthermore, let
$$D = \text{Det}(A), \quad N_i = \text{Det}(A_i), \quad i \in \underline{n}$$
The relationship between these determinants and the solution of the system $AX = B$ is given by Cramer's rule as given below.

THEOREM-1 *(Cramer's Rule) The system of equations $AX = B$ has a unique solution iff $D = \text{Det}(A) \neq 0$. In this case,*
$$x_i = \frac{N_i}{D} \quad \text{for each } i \in \underline{n}.$$

PROOF. First, let A be an invertible matrix. Then, A^{-1} exists.
Now,
$$A(A^{-1}B) = (AA^{-1})B = IB = B$$
$\Rightarrow \quad U = A^{-1}B$ is a solution of $AX = B$

In order to prove that the uniqueness of the solution, let V be any other solution of $AX = B$. Then, $AV = B$.
Now,
$$V = IV = (A^{-1}A)V = A^{-1}(AV) = A^{-1}B \qquad [\because \ AV = B]$$
Hence, $A^{-1}B$ is the unique solution of $AX = B$.

Thus, if A is invertible, $A^{-1}B$ is the unique solution of $AX = B$. Conversely, let $V = [v_1, v_2, \ldots, v_n]^T$ be the unique solution of the system of equations $AX = B$. Then,
$$AV = B$$
$\Rightarrow \quad \text{adj } A(AV) = (\text{adj } A)B$
$\Rightarrow \quad ((\text{adj } A)A)V = (\text{adj } A)B$
$\Rightarrow \quad |A|V = (\text{adj } A)B$
$\Rightarrow \quad V = \dfrac{1}{|A|}((\text{adj } A)B)$

Clearly, V as the unique solution of $AX = B$ exists if $|A| \neq 0$, that is, if A is invertible.
Hence, $AX = B$ has a unique solution iff $|A| \neq 0$.
Now, $\quad AX = B$
$\Rightarrow \quad X = A^{-1}B$
$\Rightarrow \quad X = \left(\dfrac{1}{|A|} \text{adj } A\right)B$

$$\Rightarrow \qquad X = \frac{1}{|A|}(\operatorname{adj} A)B$$

$$\Rightarrow \qquad x_i = \frac{1}{|A|}\sum_{j=1}^{n}(\operatorname{adj} A)_{ij}(B)_{ji} \quad \text{for all } i \in \underline{n}$$

$$\Rightarrow \qquad x_i = \frac{1}{|A|}\sum_{j=1}^{n} A_{ji}\, b_j \quad \text{for all } i \in \underline{n}$$

$$\Rightarrow \qquad x_i = \frac{1}{|A|}(b_1 A_{1i} + b_2 A_{2i} + b_3 A_{3i} + \cdots + b_n A_{ni}) \quad \text{for all } i \in \underline{n}$$

Clearly, $b_1 A_{1i} + b_2 A_{2i} + \cdots + b_n A_{ni}$ is the determinant of the matrix obtained by replacing ith column of A by the column vector B.

$$\therefore \qquad x_i = \left(\frac{1}{|A|}\right) N_i = \frac{N_i}{D} \quad \text{for all } i \in \underline{n}$$

Q.E.D.

EXAMPLE *Solve the following system of equations by using Cramer's rule:*

$$2x + 3y - z = 1$$
$$3x + 5y + 2z = 8$$
$$x - 2y - 3z = -1.$$

SOLUTION We have,

$$D = \begin{vmatrix} 2 & 3 & -1 \\ 3 & 5 & 2 \\ 1 & -2 & -3 \end{vmatrix} = 22 \neq 0$$

So, the system has a unique solution given by

$$x = \frac{N_1}{D},\ y = \frac{N_2}{D} \text{ and } z = \frac{N_3}{D}, \text{ where}$$

$$N_1 = \begin{vmatrix} 1 & 3 & -1 \\ 8 & 5 & 2 \\ -1 & -2 & -3 \end{vmatrix} = 66,\ N_2 = \begin{vmatrix} 2 & 1 & -1 \\ 3 & 8 & 2 \\ 1 & -1 & -3 \end{vmatrix} = -22,\ N_3 = \begin{vmatrix} 2 & 3 & 1 \\ 3 & 5 & 8 \\ 1 & -2 & -1 \end{vmatrix} = 44$$

$$\therefore \qquad x = \frac{N_1}{D} = \frac{66}{22} = 3,\ y = \frac{N_2}{D} = \frac{-22}{22} = -1,\ z = \frac{N_3}{D} = \frac{44}{22} = 2$$

EXERCISE 5.3

1. If A is an $n \times n$ matrix over a field F, show that
$$\operatorname{Det}(\operatorname{adj} A) = (\operatorname{Det}(A))^{n-1},\ n \geq 2$$
2. Show that the adjoint of a diagonal matrix is a diagonal matrix.
3. Show that the adjoint of a triangular matrix is a triangular matrix.

4. Let $A = [a_{ij}]$ be a triangular matrix. Show that
   (i) A is invertible iff each diagonal element $a_{ii} \neq 0$.
   (ii) The diagonal elements of A^{-1} (if it exists) are a_{ii}^{-1}, the reciprocals of the diagonal elements of A.

5. Let A be an $n \times n$ matrix over a field F. Then,
   (i) $\operatorname{rank}(A) < (n-1) \Rightarrow \operatorname{rank}(\operatorname{adj} A) = 0$
   (ii) $\operatorname{rank}(A) = (n-1) \Rightarrow \operatorname{rank}(\operatorname{adj} A) = 1$
   (iii) $\operatorname{rank}(A) = n \Rightarrow \operatorname{rank}(\operatorname{adj} A) = n$.

6. If A is an $n \times n$ matrix over a field F, prove that for all $n \geq 2$
   (i) $\operatorname{adj}(\operatorname{adj} A) = |A|^{n-2} A$ (ii) $|\operatorname{adj}(\operatorname{adj} A)| = |A|^{(n-1)^2}$

7. Show that if A is an $n \times n$ skew-symmetric matrix and n is odd, then $|A| = 0$.

8. Show that if A is Hermitian, then $|A|$ is a real number. What happens when A is skew-Hermitian?

9. If A and B are $n \times n$ matrices, show that
$$\operatorname{adj}(AB) = (\operatorname{adj} B)(\operatorname{adj} A)$$

10. Show that if A is $n \times n$ skew-symmetric matrix, then $\operatorname{adj} A$ is symmetric or skew-symmetric according as n is odd or even.

5.9 CHARACTERISTIC POLYNOMIAL

Let $V(F)$ be a finite dimensional vector space and $t : V \to V$ be a linear operator. We have learnt that t can be represented by a matrix A relative to some basis and, when a second basis is chosen, t is represented by the matrix $B = P^{-1}AP$, where P is the change-of-basis matrix such that $\operatorname{Det}(B) = \operatorname{Det}(A)$. We have also learnt that every square matrix A over a field F may be viewed as a linear operator $t_A : F^n \to F^n$ defined by

$$t_A(X) = AX \quad \text{for all } X \in F^n$$

Therefore, various concepts in this section will be discussed from two points of view: one in terms of matrices and other in terms of linear operators.

5.9.1 POLYNOMIALS OF MATRICES

Let A be an $n \times n$ matrix over a field F, we define powers of matrix A as follows:

$$A^1 = A, \, A^2 = AA, \, A^3 = A^2 A, \ldots, A^{n+1} = A^n A, \ldots \text{ and } A^0 = I.$$

Consider a polynomial $f(t) = a_n t^n + a_{n-1} t^{n-1} + \cdots + a_1 t + a_0$ over a field F, where a_i are scalars in F and t is the variable (generally called indeterminate). For any square matrix A over field F, we define

$f(A) = a_n A^n + a_{n-1} A^{n-1} + \cdots + a_1 A + a_0 I$, where I is the identity matrix of order same as that of A.

$f(A)$ is known as the polynomial of matrix A. Note that $f(A)$ is obtained from $f(t)$ by substituting the matrix A for the variable t and substituting the scalar matrix $a_0 I$ for the scalar a_0.

If $f(A)$ is the null matrix, then A is called a *zero or root* of $f(t)$.

Let $A = \begin{bmatrix} 1 & 2 \\ 3 & -4 \end{bmatrix}$. Then,

$$A^2 = AA = \begin{bmatrix} 1 & 2 \\ 3 & -4 \end{bmatrix} \begin{bmatrix} 1 & 2 \\ 3 & -4 \end{bmatrix} = \begin{bmatrix} 7 & -6 \\ -9 & 22 \end{bmatrix}.$$

Let $f(t) = 3t^2 - 2t + 4$ and $g(t) = t^2 + 3t - 10$ be two polynomials. Then,

$$f(A) = 3A^2 - 2A + 4I$$

$$\Rightarrow f(A) = 3 \begin{bmatrix} 7 & -6 \\ -9 & 22 \end{bmatrix} - 2 \begin{bmatrix} 1 & 2 \\ 3 & -4 \end{bmatrix} + 4 \begin{bmatrix} 1 & 0 \\ 0 & 1 \end{bmatrix}$$

$$\Rightarrow f(A) = \begin{bmatrix} 21 & -18 \\ -27 & 66 \end{bmatrix} + \begin{bmatrix} -2 & -4 \\ -6 & 8 \end{bmatrix} + \begin{bmatrix} 4 & 0 \\ 0 & 4 \end{bmatrix}$$

$$\Rightarrow f(A) = \begin{bmatrix} 23 & -22 \\ -33 & 78 \end{bmatrix}$$

and,

$$g(A) = A^2 + 3A - 10I$$

$$\Rightarrow g(A) = \begin{bmatrix} 7 & -6 \\ -9 & 22 \end{bmatrix} + 3 \begin{bmatrix} 1 & 2 \\ 3 & -4 \end{bmatrix} - 10 \begin{bmatrix} 1 & 0 \\ 0 & 1 \end{bmatrix}$$

$$\Rightarrow g(A) = \begin{bmatrix} 7 & -6 \\ -9 & 22 \end{bmatrix} + \begin{bmatrix} 3 & 6 \\ 9 & -12 \end{bmatrix} + \begin{bmatrix} -10 & 0 \\ 0 & -10 \end{bmatrix}$$

$$\Rightarrow g(A) = \begin{bmatrix} 0 & 0 \\ 0 & 0 \end{bmatrix}$$

We observe that $f(A) \neq O$ but, $g(A) = O$. So, A is a zero of $g(t)$.

REMARK. *If f and g are polynomials over a field F and A is any square matrix, then*
   (i) $(f+g)(A) = f(A) + g(A)$ (ii) $(\lambda f)(A) = \lambda f(A), \lambda \in F$
   (iii) $(fg)(A) = f(A)g(A)$ (iv) $f(A)g(A) = g(A)f(A)$

5.9.2 POLYNOMIALS OF LINEAR OPERATORS

Let $V(F)$ be a vector space and let $t : V \to V$ be a linear operator on V. The powers of t are defined by the composition operation as follows:

$$t^2 = t \circ t, \ t^3 = t^2 \circ t, \ldots, \ t^{n+1} = t^n \circ t \ldots$$

Let $f(x) = a_n x^n + a_{n-1} x^{n-1} + a_{n-2} x^{n-2} + \cdots + a_1 x + a_0$ be a polynomial over F. Then, we define

$$f(t) = a_n t^n + a_{n-1} t^{n-1} + \cdots + a_1 t + a_0 I, \text{ where } I : V \to V \text{ is the identity operator.}$$

We also say that t is a zero or root of $f(x)$ if $f(t) = 0$, the zero operator.

REMARK. *Let A be the matrix of $t : V \to V$ relative to some basis of V. Then, $f(A)$ is the matrix representation of $f(t)$, and in particular, $f(t) = 0$ if and only if $f(A) = O$.*

5.9.3 CHARACTERISTIC POLYNOMIAL OF A MATRIX

Consider the matrix $A - \lambda I_n$, where $A = [a_{ij}]$ is a square matrix of order n, I_n is the identity matrix and λ is an indeterminate. Clearly, this matrix is obtained from matrix A by subtracting λ down the diagonal. For examples, if

$$A = \begin{bmatrix} 1 & 2 \\ 3 & 4 \end{bmatrix}, \quad \text{then} \quad A - \lambda I = \begin{bmatrix} 1 & 2 \\ 3 & 4 \end{bmatrix} - \lambda \begin{bmatrix} 1 & 0 \\ 0 & 1 \end{bmatrix} = \begin{bmatrix} 1-\lambda & 2 \\ 3 & 4-\lambda \end{bmatrix}$$

Similarly, if

$$A = \begin{bmatrix} 2 & 1 & 1 \\ 0 & 2 & 3 \\ -1 & 1 & 3 \end{bmatrix}, \quad \text{then} \quad A - \lambda I = \begin{bmatrix} 2-\lambda & 1 & 1 \\ 0 & 2-\lambda & 3 \\ -1 & 1 & 3-\lambda \end{bmatrix}$$

Let us find $|A - \lambda I|$ in these two cases:

If $A = \begin{bmatrix} 1 & 2 \\ 3 & 4 \end{bmatrix}$, then $|A - \lambda I| = \begin{vmatrix} 1-\lambda & 2 \\ 3 & 4-\lambda \end{vmatrix} = \lambda^2 - 5\lambda - 2$

If $A = \begin{bmatrix} 2 & 1 & 3 \\ 0 & 2 & 3 \\ -1 & 1 & 3 \end{bmatrix}$, then $|A - \lambda I| = \begin{vmatrix} 2-\lambda & 1 & 1 \\ 0 & 2-\lambda & 3 \\ -1 & 1 & 3-\lambda \end{vmatrix} = -\lambda^3 + 7\lambda^2 - 14\lambda + 5$

We observe that in each case $|A - \lambda I|$ is a polynomial, in indeterminate λ, whose degree is same as that of the order of A. So, we define characteristic polynomial of a square matrix as follows:

CHARACTERISTIC POLYNOMIAL *Let $A = [a_{ij}]$ be a square matrix of order n over a field F and λ be an indeterminate. Then, $\Delta_A(\lambda) = |A - \lambda I_n|$ is called the characteristic polynomial of A.*

There are simple formulas for the characteristic polynomials of square matrices of orders 2 and 3 as discussed below.

(i) Let $A = \begin{bmatrix} a_{11} & a_{12} \\ a_{21} & a_{22} \end{bmatrix}$. Then,

$$\Delta_A(\lambda) = \begin{vmatrix} a_{11} - \lambda & a_{12} \\ a_{21} & a_{22} - \lambda \end{vmatrix}$$

$\Rightarrow \quad \Delta_A(\lambda) = (-1)^2 \{\lambda^2 - (a_{11} + a_{22})\lambda + |A|\} = (-1)^2 \{\lambda^2 - tr(A) + |A|\}$

(ii) Let $A = \begin{bmatrix} a_{11} & a_{12} & a_{13} \\ a_{21} & a_{22} & a_{23} \\ a_{31} & a_{32} & a_{33} \end{bmatrix}$. Then,

$$\Delta_A(\lambda) = \begin{vmatrix} a_{11} - \lambda & a_{12} & a_{13} \\ a_{21} & a_{22} - \lambda & a_{23} \\ a_{31} & a_{32} & a_{33} - \lambda \end{vmatrix}$$

$\Rightarrow \quad \Delta_A(\lambda) = (-1)^3 \{\lambda^3 - tr(A)\lambda^2 + (A_{11} + A_{22} + A_{33})\lambda - |A|\}$

Hence, A_{11}, A_{22}, A_{33} denote, respectively, the cofactors of a_{11}, a_{22}, a_{33}.

EXAMPLE-1 *Find the characteristic polynomial of each of the following matrices:*

(i) $A = \begin{bmatrix} 5 & -1 \\ 3 & 2 \end{bmatrix}$ (ii) $A = \begin{bmatrix} 1 & 1 & 2 \\ 0 & 3 & 2 \\ 1 & 3 & 9 \end{bmatrix}$

SOLUTION (i) We have,

$tr(A) = 5 + 2 = 7$ and $|A| = 10 + 3 = 13$

$\therefore \quad \Delta_A(\lambda) = \lambda^2 - 7\lambda + 13$

(ii) We have,

$tr(A) = 1 + 3 + 9 = 13$, $A_{11} = \begin{vmatrix} 3 & 2 \\ 3 & 9 \end{vmatrix} = 21$, $A_{22} = \begin{vmatrix} 1 & 2 \\ 1 & 9 \end{vmatrix} = 7$, $A_{33} = \begin{vmatrix} 1 & 1 \\ 0 & 3 \end{vmatrix} = 3$.

and, $|A| = \begin{vmatrix} 1 & 1 & 2 \\ 0 & 3 & 2 \\ 1 & 3 & 9 \end{vmatrix} = 17$

$\therefore \quad \Delta_A(\lambda) = (-1)^3 \{\lambda^3 - tr(A)\lambda^2 + (A_{11} + A_{12} + A_{33})\lambda - |A|\}$

$\Rightarrow \quad \Delta_A(\lambda) = -(\lambda^3 - 13\lambda^2 + 31\lambda + 17)$

$\Rightarrow \quad \Delta_A(\lambda) = -\lambda^3 + 13\lambda^2 - 31\lambda + 17$

REMARK. *If $\Delta_A(\lambda)$ is the characteristic polynomial of a matrix A, then $\Delta_A(\lambda) = 0$ is known as its characteristic equation.*

EXAMPLE-2 *If $A = [a_{ij}]$ is $n \times n$ triangular matrix over a field F, then the characteristic polynomial of A is*
$$\Delta_A(\lambda) = (a_{11} - \lambda)(a_{22} - \lambda) \dots (a_{nn} - \lambda)$$

SOLUTION Since A is a triangular matrix, therefore, $|A - \lambda I|$ is also a triangular matrix with diagonal elements $a_{11} - \lambda, a_{22} - \lambda, \dots, a_{nn} - \lambda$.

$\therefore \quad |A - \lambda I| = (a_{11} - \lambda)(a_{22} - \lambda) \dots (a_{nn} - \lambda) \begin{bmatrix} \because \text{Determinant of a triangular matrix is} \\ \text{equal to the product of diagonal elements} \end{bmatrix}$

$\Rightarrow \quad \Delta_A(\lambda) = (a_{11} - \lambda)(a_{22 - \lambda}) \dots (a_{nn} - \lambda)$

THEOREM-1 *Similar matrices have the same characteristic polynomial.*

PROOF. Let A and B be two similar matrices over a field F. Then, there exist an invertible matrix P such that $B = P^{-1}AP$.

$\therefore \quad \Delta_A(\lambda) = |A - \lambda I| \quad \text{and} \quad \Delta_B(\lambda) = |B - \lambda I|$

Now,

$\quad \Delta_B(\lambda) = |B - \lambda I|$

$\Rightarrow \quad \Delta_B(\lambda) = |P^{-1}AP - \lambda I| \hfill [\because B = P^{-1}AP]$

$\Rightarrow \quad \Delta_B(\lambda) = |P^{-1}AP - P^{-1}\lambda IP|$

$\Rightarrow \quad \Delta_B(\lambda) = |P^{-1}(A - \lambda I)P|$

$\Rightarrow \quad \Delta_B(\lambda) = |P^{-1}| \; |A - \lambda I| \; |P|$

$\Rightarrow \quad \Delta_B(\lambda) = |A - \lambda I| \; |P^{-1}| \; |P| \hfill [\text{By commutativity of multiplication in } F]$

$\Rightarrow \quad \Delta_B(\lambda) = |A - \lambda I| \hfill [\because |P^{-1}| = |P|^{-1}]$

$\Rightarrow \quad \Delta_B(\lambda) = \Delta_A(\lambda)$

\hfill Q.E.D.

EXAMPLE-3 *Let A be a square matrix over a field F. Prove that A and A^T have the same characteristic polynomial.*

SOLUTION We have,

$\quad \Delta_A(\lambda) = |A - \lambda I|$

$\Rightarrow \quad \Delta_A(\lambda) = |(A - \lambda I)^T| \hfill [\because |A| = |A^T|]$

$\Rightarrow \quad \Delta_A(\lambda) = |A^T - (\lambda I)^T| \hfill [\because (A - B)^T = A^T - B^T]$

$\Rightarrow \quad \Delta_A(\lambda) = |A^T - \lambda I| \hfill [\because (\lambda I)^T = \lambda I^T = \lambda I]$

$\Rightarrow \quad \Delta_A(\lambda) = \Delta_{A^T}(\lambda)$

Hence, A and A^T have the same characteristic polynomial.

5.9.4 CHARACTERISTIC POLYNOMIAL OF A LINEAR OPERATOR

Let $V(F)$ be a finite dimensional vector space and $t : V \to V$ be a linear operator. Let A and B be the matrix representations of t relative to two different bases of V. Then, A and B are similar matrices and so they have the same characteristic polynomial. So, we define the characteristic polynomial of linear operator t as follows.

CHARACTERISTIC POLYNOMIAL OF A LINEAR OPERATOR *Let $V(F)$ be a finite dimensional vector space and let $t : V \to V$ be a linear operator. Then the characteristic polynomial of t is defined as the characteristic polynomial of any matrix representation of t.*

EXAMPLE-1 Find the characteristic polynomial of the linear operator $t : R^2 \to R^2$ defined by $t(x,y) = (3x + 5y, 2x - 7y)$.

SOLUTION The characteristic polynomial of a linear operator is same as the characteristic polynomial of any matrix A that represents the linear operator.
Consider the standard basis $B = \{e_1^{(2)}, e_2^{(2)}\}$ of R^2.
We have,

$$t(e_1^{(2)}) = t(1,0) = (3,2) = 3e_1^{(2)} + 2e_2^{(2)}$$
$$t(e_2^{(2)}) = t(0,1) = (5,-7) = 5e_1^{(2)} + (-7)e_2^{(2)}$$

Let A be the matrix of t relative to the standard basis B. Then,

$$A = \begin{bmatrix} 3 & 2 \\ 5 & -7 \end{bmatrix}^T = \begin{bmatrix} 3 & 5 \\ 2 & -7 \end{bmatrix}$$

Let $\Delta_t(\lambda)$ be the characteristic polynomial of t. Then,

$$\Delta_t(\lambda) = \Delta_A(\lambda) = (-1)^2\{\lambda^2 - t_r(A)\lambda + |A|\} = \lambda^2 + 4\lambda - 31$$

EXAMPLE-2 Let V be the vector space of all functions and let $D : V \to V$ be the differential linear operator, i.e. $Df = \dfrac{df}{dt}$. Find the characteristic polynomial $A(\lambda)$ of D, if it is given that $\{\sin t, \cos t\}$ is a basis of V.

SOLUTION We have,

$$D(\sin t) = \cos t = 0(\sin t) + 1(\cos t)$$
$$D(\cos t) = -\sin t = (-1)(\sin t) + 0(\cos t)$$

The matrix A representing the differential operator D relative to the given basis is

$$A = \begin{bmatrix} 0 & 1 \\ -1 & 0 \end{bmatrix}^T = \begin{bmatrix} 0 & -1 \\ 1 & 0 \end{bmatrix}$$

$$\Delta_\lambda(D) = (-1)^2\{\lambda^2 - tr(A)\lambda + |A|\} = \lambda^2 + 1$$

THEOREM-1 *(Caley-Hamilton) Every square matrix over a field F is a root of its characteristic polynomial.*

PROOF. Let A be $n \times n$ matrix over a field F and let $\Delta_A(\lambda)$ be the characteristic polynomial of A. Then,
$$\Delta_A(\lambda) = (-1)^n(\lambda^n + a_{n-1}\lambda^{n-1} + a_{n-2}\lambda^{n-2} + \cdots + a_1\lambda + a_0)$$
Consider the adjoint of matrix $A - \lambda I$. Since each entry of matrix $A - \lambda I$ is linear in λ or a constant. Therefore, each cofactor of $A - \lambda I$, being determinant of square submatrix of order $(n-1)$, is a polynomial of degree $(n-1)$ in λ. Thus, elements of adj $(A - \lambda I)$ are polynomials of degree $(n-1)$ in λ. So, adj$(A - \lambda I)$ can be expressed as
$$\text{adj}(A - \lambda I) = B_{n-1}\lambda^{n-1} + B_{n-2}\lambda^{n-2} + \cdots + B_1\lambda + B_0$$
where $B_{n-1}, \ldots, B_0$ are $n \times n$ square matrices over field F.

Now,
$$(A - \lambda I) \text{ adj } (A - \lambda I) = |A - \lambda I|I \qquad [\because A(\text{adj}A) = |A|I]$$
$$\Rightarrow (A - \lambda I) \text{ adj } (A - \lambda I) = \Delta_A(\lambda)I \qquad [\because \Delta_A(\lambda) = |A - \lambda I|]$$
$$\Rightarrow (A - \lambda I)(B_{n-1}\lambda^{n-1} + B_{n-2}\lambda^{n-2} + \ldots + B_1\lambda + B_0) = (-1)^n(\lambda^n + a_{n-1}\lambda^{n-1} + \ldots + a_1\lambda + a_0)I$$

Equating the coefficients of the corresponding powers of λ on both sides, we get
$$-B_{n-1} = (-1)^n I$$
$$AB_{n-1} - B_{n-2} = (-1)^n a_{n-1} I$$
$$AB_{n-2} - B_{n-3} = (-1)^n a_{n-2} I$$
$$\cdots \quad \cdots \quad \cdots$$
$$AB_1 - B_0 = (-1)^n a_1 I$$
$$AB_0 = (-1)^n a_0 I$$

Multiplying the above equations by $A^n, A^{n-1}, A^{n-2}, \ldots, A$ and I respectively on the left-hand side and adding, we get
$$O = (-1)^n(A^n + a_{n-1}A^{n-1} + a_{n-2}A^{n-2} + \cdots + a_1 A + a_0 I)$$
$$\Rightarrow \Delta_A(A) = O$$
$$\Rightarrow A \text{ is a root of } \Delta_A(\lambda).$$

Q.E.D.

REMARK. *Since every linear operator on a finite dimensional vector space $V(F)$ represents a matrix relative to some basis of V and every matrix over F is represented by some linear transformation. So, we have the following analogous Cayley-Hamilton theorem for linear operators.*

THEOREM-2 *(Cayley-Hamilton Theorem for Linear Operators)* *Every linear operator on a finite dimensional vector space is a root or zero of its characteristic polynomial.*

COROLLARY. If $\Delta_A(\lambda) = (-1)^n(\lambda^n + a_{n-1}\lambda^{n-1} + \cdots + a_1\lambda + a_0)$ is the characteristic polynomial of an $n \times n$ matrix A such that $|A| \neq 0$, then prove that

$$A^{-1} = -\frac{1}{a_0}\{a_1 I + a_2 A + \cdots + a_{n-1} A^{n-2} + A^{n-1}\}$$

PROOF. By Cayley-Hamilton theorem, we have

$\Delta_A(A) = O$
$\Rightarrow (-1)^n (A^n + a_{n-1}A^{n-1} + \cdots + a_1 A + a_0 I) = O$
$\Rightarrow A^n + a_{n-1}A^{n-1} + a_{n-2}A^{n-2} + \cdots + a_1 A + a_0 I = O$

Multiplying both sides by A^{-1}, we get

$A^{n-1} + a_1 A^{n-2} + \cdots + a_1 I + a_0\, A^{-1} = O$
$\Rightarrow a_0\, A^{-1} = -\{a_1 I + a_2 A + \cdots + a_1 A^{n-2} + A^{n-1}\}$
$\Rightarrow A^{-1} = -\dfrac{1}{a_0}\{a_1 I + a_2 A + \cdots + a_{n-1} A^{n-2} + A^{n-1}\}$

Q.E.D.

CHARACTERISTIC EQUATION Let A be a square matrix over a field F and $\Delta_A(\lambda)$ be its characteristic polynomial. Then, $\Delta_A(\lambda) = 0$ is known as the characteristic equation of A.

EXAMPLE-3 Verify Cayley-Hamilton theorem for the matrix $A = \begin{bmatrix} 0 & 0 & 1 \\ 3 & 1 & 0 \\ 2 & 1 & 4 \end{bmatrix}$ and hence find A^{-1}.

SOLUTION The characteristic polynomial $C_A(\lambda)$ of A is given by

$$C_A(\lambda) = (-1)^3 \{\lambda^3 - tr(A)\lambda^2 + (A_{11} + A_{22} + A_{33})\lambda - |A|\}$$

We have,

$tr(A) = 0 + 1 + 4 = 5$

$A_{11} = \begin{vmatrix} 1 & 0 \\ 1 & 4 \end{vmatrix} = 4,\ A_{22} = \begin{bmatrix} 0 & 1 \\ 2 & 4 \end{bmatrix} = -2,\ A_{33} = \begin{vmatrix} 0 & 0 \\ 3 & 1 \end{vmatrix} = 0$ and, $|A| = 1$.

$\therefore\quad C_A(\lambda) = (-1)^3(\lambda^3 - 5\lambda^2 + 2\lambda - 1)$

or, $\quad C_A(\lambda) = -\lambda^3 + 5\lambda^2 - 2\lambda + 1$

Now, $\quad A^2 = AA = \begin{bmatrix} 0 & 0 & 1 \\ 3 & 1 & 0 \\ 2 & 1 & 4 \end{bmatrix}\begin{bmatrix} 0 & 0 & 1 \\ 3 & 1 & 0 \\ 2 & 1 & 4 \end{bmatrix} = \begin{bmatrix} 2 & 1 & 4 \\ 3 & 1 & 3 \\ 11 & 5 & 18 \end{bmatrix}$

$$A^3 = A^2 A = \begin{bmatrix} 2 & 1 & 4 \\ 3 & 1 & 3 \\ 11 & 5 & 18 \end{bmatrix} \begin{bmatrix} 0 & 0 & 1 \\ 3 & 1 & 0 \\ 2 & 1 & 4 \end{bmatrix} = \begin{bmatrix} 11 & 5 & 18 \\ 9 & 4 & 15 \\ 51 & 23 & 83 \end{bmatrix}$$

$\therefore \quad C_A(A) = -A^3 + 5A^2 - 2A + I$

$$\Rightarrow \quad C_A(A) = -\begin{bmatrix} 11 & 5 & 18 \\ 9 & 4 & 15 \\ 51 & 23 & 83 \end{bmatrix} + 5 \begin{bmatrix} 2 & 1 & 4 \\ 3 & 1 & 3 \\ 11 & 5 & 18 \end{bmatrix} - 2 \begin{bmatrix} 0 & 0 & 1 \\ 3 & 1 & 0 \\ 2 & 1 & 4 \end{bmatrix} + \begin{bmatrix} 1 & 0 & 0 \\ 0 & 1 & 0 \\ 0 & 0 & 1 \end{bmatrix}$$

$$\Rightarrow \quad C_A(A) = \begin{bmatrix} -11+10+0+1 & -5+5+0+0 & -18+20-2+0 \\ -9+15-6+0 & -4+5-2+1 & -15+15+0+0 \\ 51+55-63+0 & -23+25-2+0 & -83+90-8+1 \end{bmatrix}$$

$$\Rightarrow \quad C_A(A) = \begin{bmatrix} 0 & 0 & 0 \\ 0 & 0 & 0 \\ 0 & 0 & 0 \end{bmatrix} = O_{3 \times 3}$$

So, Cayley-Hamilton theorem is verified.

Now,

$C_A(A) = O$

$\Rightarrow \quad -A^3 + 5A^2 - 2A + I = O$

$\Rightarrow \quad A^3 - 5A^2 + 2A = I$

$\Rightarrow \quad A(A^2 - 5A + 2I) = I$

$\Rightarrow \quad A^{-1} = A^2 - 5A + 2I$

$$\Rightarrow \quad A^{-1} = \begin{bmatrix} 2 & 1 & 4 \\ 3 & 1 & 3 \\ 11 & 5 & 18 \end{bmatrix} + \begin{bmatrix} 0 & 0 & -5 \\ -15 & -5 & 0 \\ -10 & -5 & -20 \end{bmatrix} + \begin{bmatrix} 2 & 0 & 0 \\ 0 & 2 & 0 \\ 0 & 0 & 2 \end{bmatrix} = \begin{bmatrix} 4 & 1 & -1 \\ -12 & -2 & 3 \\ 1 & 0 & 0 \end{bmatrix}$$

EXAMPLE-4 *If A is an invertible matrix with characteristic polynomial $\Delta_A(\lambda) = c_0 + c_1 \lambda + c_2 \lambda^2 + \cdots + (-1)^n \lambda^n$ of degree n, show that the characteristic polynomial of A^{-1}, is*

$$(-1)^n \left\{ \lambda'^n + \frac{c_1}{|A|} \lambda'^{n-1} + \frac{c_2}{|A|} \lambda'^{n-2} + \cdots + \frac{(-1)^n}{|A|} \right\}, \text{ where } \lambda' = \frac{1}{\lambda}.$$

SOLUTION We have,

$$|A - \lambda I| = c_0 + c_1 \lambda + c_2 \lambda^2 + \cdots + (-1)^n \lambda^n \quad \text{(i)}$$

Putting $\lambda = 0$, we get $c_0 = |A|$

Now,

$|A - \lambda I| = c_0 + c_1 \lambda + c_2 \lambda^2 + \cdots + (-1)^n \lambda^n$

$\Rightarrow \quad |A - \lambda(AA^{-1})| = |A| + c_1 \lambda + c_2 \lambda^2 + \cdots + (-1)^n \lambda^n$

$\Rightarrow \quad |A||I - \lambda A^{-1}| = |A| + c_1\lambda + c_2\lambda^2 + \cdots + (-1)^n\lambda^n$

$\Rightarrow \quad |A|(\lambda)^n \left|\dfrac{1}{\lambda}I - A^{-1}\right| = |A| + c_1\lambda + c_2\lambda^2 + \cdots + (-1)^n\lambda^n$

$\Rightarrow \quad |A|(-1)^n\lambda^n|A^{-1} - \lambda^{-1}I| = |A| + c_1\lambda + c_2\lambda^2 + \cdots + (-1)^n\lambda^n$

$\Rightarrow \quad |A^{-1} - \lambda^{-1}I| = (-1)^n \left\{ \dfrac{1}{\lambda^n} + \dfrac{c_1}{|A|}\dfrac{1}{\lambda^{n-1}} + \dfrac{c_2}{|A|\lambda^{n-2}} + \cdots + \dfrac{(-1)^n}{|A|} \right\}$

$\Rightarrow \quad |A^{-1} - \lambda'I| = (-1)^n \left\{ \lambda'^n + \dfrac{c_1}{|A|}\lambda'^{n-1} + \dfrac{c_2}{|A|}\lambda'^{n-2} + \cdots + \dfrac{(-1)^n}{|A|} \right\}$, where $\lambda' = \dfrac{1}{\lambda}$

EXERCISE 5.4

1. Find the characteristic polynomial $\Delta_A(\lambda)$ of each of the following matrices:

   (i) $A = \begin{bmatrix} 1 & 3 \\ 4 & 5 \end{bmatrix}$ (ii) $A = \begin{bmatrix} 5 & 3 \\ 2 & 10 \end{bmatrix}$ (iii) $A = \begin{bmatrix} 7 & -1 \\ 6 & 2 \end{bmatrix}$ (iv) $A = \begin{bmatrix} 5 & -2 \\ 4 & -4 \end{bmatrix}$

2. Find the characteristic polynomial $\Delta_A(\lambda)$ of each of the following matrices:

   (i) $A = \begin{bmatrix} 1 & 2 & 3 \\ 3 & 0 & 4 \\ 6 & 4 & 5 \end{bmatrix}$ (ii) $A = \begin{bmatrix} 1 & 6 & -2 \\ -3 & 2 & 0 \\ 0 & 3 & -4 \end{bmatrix}$

3. Find the characteristic polynomial of each of the following linear operators:
   (i) $t: R^2 \to R^2$ defined by $t(x,y) = (2x + 5y, x - 3y)$
   (ii) $t: R^2 \to R^2$ defined by $t(x,y) = (x - y, x + y)$

4. Let V be the vector space of all functions from R to itself and D be the differential operator on V, that is, $D(f) = \dfrac{df}{dt}$. Find the characteristic polynomial of D relative to the basis $B = \{e^{3t}, te^{3t}, t^2e^{3t}\}$ of V.

5. Find the characteristic polynomial of the matrix

$$A = \begin{bmatrix} 2 & -1 & 1 \\ -1 & 2 & -1 \\ 1 & -1 & 2 \end{bmatrix}$$

   and verify that A is a root of it.

6. Verify Cayley-Hamilton theorem for the matrix $A = \begin{bmatrix} 1 & -1 & 1 \\ 4 & 1 & 0 \\ 8 & 1 & 1 \end{bmatrix}$ and hence find A^{-1}.

7. If A is an invertible matrix with characteristic polynomial $\Delta_A(\lambda) = c_0 + c_1\lambda + \cdots + (-1)^n \lambda^n$, show that the transposed matrix of cofactors of A is
$$-\left(c_1 I + c_2 A + \cdots + c_{n-1} A^{n-2} + (-1)^{n-1} A^{n-1}\right).$$

ANSWERS

1. (i) $\Delta_A(\lambda) = \lambda^2 - 6\lambda - 7$ (ii) $\Delta_A(\lambda) = \lambda^2 - 15\lambda + 44$ (iii) $\Delta_A(\lambda) = \lambda^2 - 9\lambda + 20$
   (iv) $\Delta_A(\lambda) = \lambda^2 - \lambda - 12$

2. (i) $\Delta_A(\lambda) = \lambda^3 - 6\lambda^2 - 35\lambda - 38$ (ii) $\Delta_A(\lambda) = \lambda^3 + \lambda^2 - 8\lambda + 62$

3. (i) $\Delta(t) = \lambda^2 + \lambda - 11$ (ii) $\Delta(\lambda) = \lambda^2 + 2$

4. $\Delta(\lambda) = -\lambda^3 + 9\lambda^2 - 27\lambda + 27$ 6. $A^{-1} = \begin{bmatrix} 1 & 2 & -1 \\ -4 & -7 & 4 \\ -4 & -9 & 5 \end{bmatrix}$

5.10 EIGENVALUES AND EIGENVECTORS OF A SQUARE MATRIX

Let A be an $n \times n$ matrix over a field F. In Chapter 4, we have learnt that a non-zero vector $X \in F^n$ is an eigenvector of A iff it is an eigenvector of $t_A : F^n \to F^n$ defined as $t_A(X) = AX$. We have also learnt that a non-zero vector $X \in F^n$ is an eigenvector of $t_A : F^n \to F^n$ iff there exists a scalar $\lambda \in F$ such that $AX = \lambda X$. It follows from the above discussion that a non-zero vector $X \in F^n$ is an eigenvector of matrix A iff there exists a scalar $\lambda \in F$ such that $AX = \lambda X$.

In this section, we shall discuss methods of finding the eigenvalues and eigenvectors of a square matrix defined over a field F. Following theorem proves that the eigenvalues of a square matrix are the roots of its characteristic equation.

THEOREM-3 *Let A be an $n \times n$ matrix over a field F. A scalar λ is an eigenvalue of A iff $|A - \lambda I| = 0$, i.e. $\Delta_A(\lambda) = 0$*

PROOF. First suppose that λ is an eigenvalue of matrix A. Then there exists a non-zero vector $X \in F^n$ such that
$$AX = \lambda X$$
$\Rightarrow \quad AX - \lambda X = O$
$\Rightarrow \quad AX - \lambda I X = O$ $\quad [\because X = IX]$
$\Rightarrow \quad (A - \lambda I)X = O$
$\Rightarrow \quad t_{A - \lambda I}(X) = O$

$\Rightarrow$ $\quad X \in \text{Ker }(t_{A-\lambda I})$

$\Rightarrow$ $\quad \text{Ker }(t_{A-\lambda I}) \neq \{O\}$ $\hfill [\because \quad X \neq O]$

$\Rightarrow$ $\quad \text{Nullity } t_{A-\lambda I} > 0$

$\Rightarrow$ $\quad \text{Rank }(t_{A-\lambda I}) < n$ $\hfill [\because \quad \text{Rank }(t_{A-\lambda I}) + \text{Nullity }(t_{A-\lambda I}) = n]$

$\Rightarrow$ $\quad \text{Rank }(A - \lambda I) < n$ $\hfill [\because \quad \text{Rank}(A) = \text{Rank}(t_A)]$

$\Rightarrow$ $\quad A - \lambda I$ is not invertible

$\Rightarrow$ $\quad |A - \lambda I| = 0$, i.e. $\Delta_A(\lambda) = 0$

Conversely, let A be an $n \times n$ matrix over a field F such that $|A - \lambda I| = 0$ for some scalar λ. We have to prove that scalar λ is an eigenvalue of matrix A.

Now,

$\quad |A - \lambda I| = 0$

$\Rightarrow$ $\quad \text{Rank }(A - \lambda I) < n$

$\Rightarrow$ $\quad \text{Nullity }(A - \lambda I) > 0$

$\Rightarrow$ $\quad \text{Nullity }(t_{A-\lambda I}) > 0$ $\hfill [\because \quad \text{Nullity }(A - \lambda I) = \text{Nullity } t_{A-\lambda I}]$

$\Rightarrow$ $\quad \text{Kernel }(t_{A-\lambda I}) \neq \{O\}$

$\Rightarrow$ $\quad$ There exists a non-zero vector $X \in F^n$ such that

$$t_{A-\lambda I}(X) = O$$

$\Rightarrow$ $\quad (A - \lambda I)X = O$

$\Rightarrow$ $\quad AX - \lambda IX = O$

$\Rightarrow$ $\quad AX = \lambda X$

$\Rightarrow$ $\quad \lambda$ is an eigenvalue of matrix A.

Q.E.D.

REMARK. *The characteristic equation $\Delta_A(\lambda) = 0$ of an $n \times n$ matrix A over a field F is an nth degree equation in λ. If the field F is algebraically closed, i.e. every polynomial over F possesses a zero in F, then the matrix A will definitely have at least one eigenvalue. If F is not algebraically closed, then matrix A may or may not have an eigenvalue according as the characteristic equation $\Delta_A(\lambda)$ has or has not a root in F. If A is a matrix over the field C of all complex numbers, then by the fundamental theorem of algebra matrix A will definitely have at least one eigenvalue. Since the field R of real numbers is not algebraically closed, therefore, a matrix A over R may or may not have an eigenvalue. For example, the matrix $A = \begin{bmatrix} 0 & 3 \\ -7 & 0 \end{bmatrix}$ over R has no eigenvalue as its characteristic equation $\Delta_A(\lambda) = \lambda^2 + 21$ has no root in R.*

The above discussion suggests us the following algorithm for computing eigenvalues and eigenvectors for a given square matrix and also for finding the basis vectors for the eigenspaces corresponding to each eigenvalue.

ALGORITHM

Step I   Obtain the square matrix A (say).

Step II   Find the characteristic polynomial $\Delta_A(\lambda)$ of matrix A.

Step III   Solve the characteristic equation $\Delta_A(\lambda) = 0$ to obtain the eigenvalues of matrix A.

Step IV   For each eigenvalue λ, find the solution space of the homogeneous system $(A - \lambda I) X = O$. These solution spaces, for distinct values of λ, are known as the eigenspaces for the corresponding eigenvalues.

REMARK. *The eigenvalues and eigenvectors of a linear operator t on a finite dimensional vector space $V(F)$ are the eigenvalues and eigenvectors of a matrix of t relative to some ordered basis of V.*

ILLUSTRATIVE EXAMPLES

EXAMPLE-1 *Find all eigenvalues and corresponding eigenvectors of the following matrices:*

(i) $A = \begin{bmatrix} 4 & -5 \\ 1 & -2 \end{bmatrix}$ (ii) $A = \begin{bmatrix} 0 & 1 \\ 0 & 0 \end{bmatrix}$

SOLUTION   (i) The characteristic polynomial $\Delta_A(\lambda)$ of A is

$$\Delta_A(\lambda) = (-1)^2 \{\lambda^2 - tr(A) + |A|\} = \lambda^2 - 2\lambda - 3$$

The eigenvalues of matrix A are given by

$$\Delta_A(\lambda) = 0 \Rightarrow \lambda^2 - 2\lambda - 3 = 0 \Rightarrow (\lambda - 3)(\lambda + 1) = 0 \Rightarrow \lambda = -1, 3$$

Thus, $\lambda = -1$ and $\lambda = 3$ are eigenvalues of given matrix.

Eigenvectors corresponding to $\lambda = -1$: Let $X = \begin{bmatrix} x \\ y \end{bmatrix}$ be an eigenvector corresponding to the eigenvalue $\lambda = -1$. Then,

$(A+I)X = O$ [Putting $\lambda = -1$ in $(A - \lambda I)X = O$]

$\Rightarrow \begin{bmatrix} 5 & -5 \\ 1 & -1 \end{bmatrix} \begin{bmatrix} x \\ y \end{bmatrix} = \begin{bmatrix} 0 \\ 0 \end{bmatrix}$

$\Rightarrow \quad 5x - 5y = 0$ and $x - y = 0$

$\Rightarrow \quad x - y = 0$.

$\Rightarrow \quad x = y = k$

Thus, $E_{-1} = \left\{ \begin{bmatrix} k \\ k \end{bmatrix} : k \in C \right\}$ is the eigenspace corresponding to eigenvalue $\lambda = -1$. Clearly, this space is spanned by $X_1 = \begin{bmatrix} 1 \\ 1 \end{bmatrix}$. So, $X_1 = \begin{bmatrix} 1 \\ 1 \end{bmatrix}$ is an eigenvector belonging to and spanning the eigenspace corresponding to $\lambda = -1$.

Eigenvectors corresponding to $\lambda = 3$. Let $X = \begin{bmatrix} x \\ y \end{bmatrix}$ be an eigenvector corresponding to the eigenvalue $\lambda = 3$. Then,

$$(A - 3I)X = O$$

$$\Rightarrow \begin{bmatrix} 1 & -5 \\ 1 & -5 \end{bmatrix} \begin{bmatrix} x \\ y \end{bmatrix} = \begin{bmatrix} 0 \\ 0 \end{bmatrix}$$

$$\Rightarrow x - 5y = 0 \Rightarrow x = 5k, \, y = k$$

Thus, $E_2 = \left\{ \begin{bmatrix} 5k \\ k \end{bmatrix} : k \in C \right\}$ is the eigenspace corresponding to the eigenvalue $\lambda = 3$. This is spanned by $X_2 = \begin{bmatrix} 5 \\ 1 \end{bmatrix}$. So, $X_2 = \begin{bmatrix} 5 \\ 1 \end{bmatrix}$ is an eigenvector belonging to and spanning the eigenspace corresponding to $\lambda = 3$.

(ii) The characteristic polynomial $\Delta_A(\lambda)$ of A is

$$\Delta_A(\lambda) = (-1)^2 \{\lambda^2 - tr(A) + |A|\} = \lambda^2$$

So, the eigenvalues of matrix A are given by

$$\Delta_A(\lambda) = 0 \Rightarrow \lambda^2 = 0 \Rightarrow \lambda = 0.$$

Thus, $\lambda = 0$ is the only eigenvalue of matrix A.

Eigenvectors corresponding to $\lambda = 0$: Let $X = \begin{bmatrix} x \\ y \end{bmatrix}$ be an eigenvector of matrix A corresponding to the eigenvalue $\lambda = 0$. Then,

$$(A - 0I)X = O$$

$$\Rightarrow AX = O$$

$$\Rightarrow \begin{bmatrix} 0 & 1 \\ 0 & 0 \end{bmatrix} \begin{bmatrix} x \\ y \end{bmatrix} = \begin{bmatrix} 0 \\ 0 \end{bmatrix}$$

$$\Rightarrow \begin{bmatrix} y \\ 0 \end{bmatrix} = \begin{bmatrix} 0 \\ 0 \end{bmatrix} \Rightarrow y = 0$$

Thus, $y = 0$, $x = k$, where k is any complex number.

$\therefore \quad E = \left\{ \begin{bmatrix} k \\ 0 \end{bmatrix} : k \in C \right\}$ is the eigenspace of A corresponding to $\lambda = 0$.

Clearly, it is spanned by the vector $\begin{bmatrix} 1 \\ 0 \end{bmatrix}$.

EXAMPLE-2 *Find all eigenvalues and eigenvectors of the following matrices:*

(i) $A = \begin{bmatrix} 1 & 0 \\ 0 & i \end{bmatrix}$ (ii) $\begin{bmatrix} 1 & 1 \\ 0 & i \end{bmatrix}$

SOLUTION (i) We have,
$$A = \begin{bmatrix} 1 & 0 \\ 0 & i \end{bmatrix}$$

The characteristic equation of A is
$$\Delta_A(\lambda) = (-1)^2 \{\lambda^2 - \lambda\, tr(A) + |A|\} = \lambda^2 - (1+i)\lambda + i$$

The eigenvalues of A are roots of the equation

$\Delta_A(\lambda) = 0$

$\Rightarrow \quad \lambda^2 - (1+i)\lambda + i = 0$

$\Rightarrow \quad (\lambda^2 - \lambda) - i(\lambda - 1) = 0$

$\Rightarrow \quad \lambda(\lambda - 1) - i(\lambda - 1) = 0$

$\Rightarrow \quad (\lambda - 1)(\lambda - i) = 0$

$\Rightarrow \quad \lambda = 1, i$

Thus, $\lambda = 1$ and $\lambda = i$ are eigenvalues of matrix A.

Eigenvectors corresponding to $\lambda = 1$: Let $X = \begin{bmatrix} x \\ y \end{bmatrix}$ be an eigenvector of matrix A corresponding to the eigenvalue $\lambda = 1$. Then,

$(A - 1I)X = O$

$\Rightarrow \quad (A - I)X = O$

$\Rightarrow \quad \begin{bmatrix} 0 & 0 \\ 0 & i-1 \end{bmatrix} \begin{bmatrix} x \\ y \end{bmatrix} = \begin{bmatrix} 0 \\ 0 \end{bmatrix}$

$\Rightarrow \quad \begin{bmatrix} 0 \\ y(1-i) \end{bmatrix} = \begin{bmatrix} 0 \\ 0 \end{bmatrix} \Rightarrow y = 0$

$\therefore \quad x = k, y = 0$, where k is any non-zero complex number.

Thus, $E_1 = \left\{ X : X = \begin{bmatrix} k \\ 0 \end{bmatrix}, k \in C, k \neq 0 \right\}$ is the eigenspace of A corresponding to the eigenvalue $\lambda = 1$. Clearly, it is spanned by $\begin{bmatrix} 1 \\ 0 \end{bmatrix}$.

Eigenvectors corresponding to $\lambda = i$: Let $X = \begin{bmatrix} x \\ y \end{bmatrix}$ be an eigenvector of matrix A corresponding to the eigenvalue $\lambda = i$. Then,

$(A - iI)X = O$

$\Rightarrow \quad \begin{bmatrix} 1-i & 0 \\ 0 & 0 \end{bmatrix} \begin{bmatrix} x \\ y \end{bmatrix} = \begin{bmatrix} 0 \\ 0 \end{bmatrix}$

$\Rightarrow \quad x(1-i) = 0 \Rightarrow x = 0$

$\therefore \quad x = 0, y = k$, where k is any non-zero complex number.

Thus, $E_i = \left\{ X = \begin{bmatrix} 0 \\ k \end{bmatrix} : k \in C, k \neq 0 \right\}$ is the eigenspace of A corresponding to $\lambda = i$.

Clearly, E_i is spanned by the vector $\begin{bmatrix} 0 \\ 1 \end{bmatrix}$.

(ii) We have,
$$A = \begin{bmatrix} 1 & 1 \\ 0 & i \end{bmatrix}$$

The characteristic polynomial of matrix A is given by
$$\Delta_A(\lambda) = (-1)^2 \{\lambda^2 - \lambda \, tr\,(A) + |A|\} = \lambda^2 - (1+i)\lambda + i$$

The eigenvalues of matrix A are roots of the equation
$$\Delta_A(\lambda) = 0 \Rightarrow \lambda^2 - (1+i)\lambda + i = 0 \Rightarrow (\lambda - 1)(\lambda - i) = 0 \Rightarrow \lambda = 1, i$$

Thus, $\lambda = 1$ and $\lambda = i$ are eigenvalues of matrix A.

Eigenvectors corresponding to $\lambda = 1$: Let $X = \begin{bmatrix} x \\ y \end{bmatrix}$ be an eigenvector of A corresponding to the eigenvalue $\lambda = 1$. Then,

$(A - I)X = O$ [Putting $\lambda = 1$ in $(A - \lambda I)X = 0$]

$\Rightarrow \begin{bmatrix} 0 & 1 \\ 0 & i-1 \end{bmatrix} \begin{bmatrix} x \\ y \end{bmatrix} = \begin{bmatrix} 0 \\ 0 \end{bmatrix}$

$\Rightarrow \begin{bmatrix} y \\ (i-1)y \end{bmatrix} = \begin{bmatrix} 0 \\ 0 \end{bmatrix}$

$\Rightarrow \quad y = 0$

Thus, $x = k$, $k \neq 0$ and $y = 0$.

So, $\left\{ X : X = \begin{bmatrix} k \\ 0 \end{bmatrix}, k \neq 0 \right\}$ is the eigenspace of A corresponding to the eigenvalue $\lambda = 1$.

This space is spanned by the vector $\begin{bmatrix} 1 \\ 0 \end{bmatrix}$.

Eigenvectors corresponding to $\lambda = i$: Let $X = \begin{bmatrix} x \\ y \end{bmatrix}$ be an eigenvector of matrix A corresponding to the eigenvalue $\lambda = i$. Then,

$(A - iI)X = O$

$\Rightarrow \begin{bmatrix} 1-i & 1 \\ 0 & 0 \end{bmatrix} \begin{bmatrix} x \\ y \end{bmatrix} = \begin{bmatrix} 0 \\ 0 \end{bmatrix}$

$\Rightarrow \quad x(1-i) + y = 0$

This system has only one free variable. Let $x = c$. Then, $y = (i-1)c$, $c \neq 0$.

Thus, $\left\{ X : X = \begin{bmatrix} c \\ (i-1)c \end{bmatrix}, c \neq 0 \right\}$ is the eigenspace of A corresponding to the eigenvalue $\lambda = i$. Clearly, this space is spanned by the vector $\begin{bmatrix} 1 \\ i-1 \end{bmatrix}$.

EXAMPLE-3 *For each of the following linear operators $t : R^2 \to R^2$, find all eigenvalues and basis for each eigenspace:*

(i) $t(x, y) = (3x + 3y, x + 5y)$ (ii) $t(x, y) = (3x - 13y, x - 3y)$

SOLUTION (i) The matrix of t relative to the standard basis of R^2 is $A = \begin{bmatrix} 3 & 3 \\ 1 & 5 \end{bmatrix}$

The characteristic polynomial of matrix A is

$$\Delta_A(\lambda) = \lambda^2 - \lambda\, tr\,(A) + |A| = \lambda^2 - 8\lambda + 12$$

The eigenvalues of A are roots of the characteristic equation

$$\Delta_A(\lambda) = 0 \Rightarrow \lambda^2 - 8\lambda + 12 = 0 \Rightarrow \lambda = 2, 6.$$

Thus, eigenvalues of $t : R^2 \to R^2$ are 2 and 6.

Eigenvectors corresponding to $\lambda = 2$: Let $X = \begin{bmatrix} x \\ y \end{bmatrix}$ be an eigenvector of A corresponding to the eigenvalue $\lambda = 2$. Then,

$$(A - 2I)X = O$$

$\Rightarrow \quad \begin{bmatrix} 1 & 3 \\ 1 & 3 \end{bmatrix} \begin{bmatrix} x \\ y \end{bmatrix} = \begin{bmatrix} 0 \\ 0 \end{bmatrix}$

$\Rightarrow \quad x + 3y = 0$

Let $y = k$. Then, $x = -3k, k \neq 0$.

Thus, $\left\{ X : X = \begin{bmatrix} -3k \\ k \end{bmatrix}, k \neq 0 \right\}$ is the eigenspace of t corresponding to $\lambda = 2$. Clearly, it is spanned by $\begin{bmatrix} -3 \\ 1 \end{bmatrix}$. So, basis for this eigenspace is $\begin{bmatrix} -3 \\ 1 \end{bmatrix}$.

Eigenvectors corresponding to $\lambda = 6$: Let $X = \begin{bmatrix} x \\ y \end{bmatrix}$ be an eigenvector of A corresponding to the eigenvalue $\lambda = 6$. Then,

$$(A - 6I)X = 0$$
$$\Rightarrow \begin{bmatrix} -3 & 3 \\ 1 & -1 \end{bmatrix} \begin{bmatrix} x \\ y \end{bmatrix} = \begin{bmatrix} 0 \\ 0 \end{bmatrix}$$
$$\Rightarrow -3x + 3y = 0 \text{ and } x - y = 0$$
$$\Rightarrow x = y$$
$$\Rightarrow x = y = k, \ k \neq 0$$

Thus, $\left\{ X : X = \begin{bmatrix} k \\ k \end{bmatrix}, k \neq 0 \right\}$ is eigenspace of t corresponding to $\lambda = 6$. Clearly, it is spanned by the vector $\begin{bmatrix} 1 \\ 1 \end{bmatrix}$ and hence the basis is $\begin{bmatrix} 1 \\ 1 \end{bmatrix}$.

(ii) We have,

$$t(x, y) = (3x - 13y, \ x - 3y)$$

The matrix of t relative to the standard basis is $A = \begin{bmatrix} 3 & -13 \\ 1 & -3 \end{bmatrix}$

The characteristic polynomial of matrix A is

$$\Delta_A(\lambda) = (-1)^2 \left\{ \lambda^2 - \lambda \ tr(A) + |A| \right\} = \lambda^2 + 4$$

The characteristic equation of A does not have real roots. Thus, A, a real matrix representing $t : R^2 \to R^2$, has no eigenvalues and no eigenvectors.

EXAMPLE-4 *Find the eigenvalues and eigenvectors of the following matrix:*

$$A = \begin{bmatrix} 2 & 1 & 1 \\ 2 & 3 & 2 \\ 3 & 3 & 4 \end{bmatrix}$$

SOLUTION We have,

$$tr(A) = 2 + 3 + 4 = 9, \ A_{11} + A_{22} + A_{33} = 6 + 5 + 4 = 15 \text{ and } |A| = 7$$

So, the characteristic polynomial of A is

$$C_A(\lambda) = (-1)^3 \left\{ \lambda^3 - \lambda^2 \ tr(A) + \lambda(A_{11} + A_{22} + A_{33}) - |A| \right\}$$
$$\Rightarrow C_A(\lambda) = -(\lambda^3 - 9\lambda^2 + 15\lambda - 7)$$

The eigenvalues of matrix A are roots of the characteristic polynomial. So, eigenvalues are given by

$$C_A(\lambda) = 0$$
$$\Rightarrow \quad \lambda^3 - 9\lambda^2 + 15\lambda - 7 = 0 \quad \Rightarrow \quad (\lambda - 1)^2(\lambda - 7) = 0 \quad \Rightarrow \quad \lambda = 1, 1, 7$$

So, the eigenvalues of matrix A are $\lambda_1 = 1, \lambda_2 = 1$ and $\lambda_3 = 7$.

Eigenvectors corresponding to $\lambda_1 = 1$: Let $X = \begin{bmatrix} x \\ y \\ z \end{bmatrix}$ be an eigenvector corresponding to $\lambda_1 = 1$. Then,

$$(A - I)X = 0$$
$$\Rightarrow \quad \begin{bmatrix} 1 & 1 & 1 \\ 2 & 2 & 2 \\ 3 & 3 & 3 \end{bmatrix} \begin{bmatrix} x \\ y \\ z \end{bmatrix} = \begin{bmatrix} 0 \\ 0 \\ 0 \end{bmatrix}$$
$$\Rightarrow \quad x + y + z = 0$$

Thus, there are two free variables, say, y and z. Let $y = 0$. Then,

$$x + z = 0 \Rightarrow x = k_1, z = -k_1$$

$$\therefore \quad X_1 = \begin{bmatrix} k_1 \\ 0 \\ -k_1 \end{bmatrix} = k_1 \begin{bmatrix} 1 \\ 0 \\ -1 \end{bmatrix}, \quad k_1 \neq 0 \text{ is an eigenvector of matrix } A \text{ corresponding to } \lambda = 1.$$

If $z = 0$, then

$$x + y + z = 0 \Rightarrow \quad x + y = 0 \Rightarrow \quad x = -k_2, y = k_2$$

$$\therefore \quad X_2 = \begin{bmatrix} -k_2 \\ k_2 \\ 0 \end{bmatrix} = k_2 \begin{bmatrix} -1 \\ 1 \\ 0 \end{bmatrix} \text{ is another eigenvector corresponding to } \lambda = 1.$$

Thus, X_1 and X_2 are eigenvectors of A corresponding to the eigenvalue $\lambda = 1$. The eigenspace of this eigenvalue is the subspace W spanned by X_1 and X_2. Any non-zero vector in W will be an eigenvector corresponding to the eigenvalue $\lambda = 1$.

Eigenvectors corresponding to $\lambda_3 = 7$: Let $X = \begin{bmatrix} x \\ y \\ z \end{bmatrix}$ be an eigenvector of matrix A corresponding to eigenvalue $\lambda_3 = 7$. Then,

$$(A - 7I)X = O$$

$$\Rightarrow \begin{bmatrix} -5 & 1 & 1 \\ 2 & -4 & 2 \\ 3 & 3 & -3 \end{bmatrix} \begin{bmatrix} x \\ y \\ z \end{bmatrix} = \begin{bmatrix} 0 \\ 0 \\ 0 \end{bmatrix}$$

$$\Rightarrow \quad -5x + y + z = 0$$
$$2x - 4y + 2z = 0$$
$$3x + 3y - 3z = 0$$

Solving last two equations, we get

$$\frac{x}{1} = \frac{y}{2} = \frac{z}{3} \Rightarrow x = k, y = 2k, z = 3k$$

These values also satisfy the first equation. So, $X_3 = \begin{bmatrix} k \\ 2k \\ 3k \end{bmatrix} = k \begin{bmatrix} 1 \\ 2 \\ 3 \end{bmatrix}$ is an eigenvector corresponding to $\lambda_3 = 7$.

Thus, the eigenvectors of matrix A are

$$X_1 = \begin{bmatrix} 1 \\ 0 \\ -1 \end{bmatrix}, X_2 = \begin{bmatrix} -1 \\ 1 \\ 0 \end{bmatrix}, X_3 = \begin{bmatrix} 1 \\ 2 \\ 3 \end{bmatrix}$$

Clearly, X_1, X_2, X_3 are linearly independent vectors in R^3.

EXERCISE 5.5

1. Find all eigenvalues and eigenvectors of each of the following matrices:

    (i) $\begin{bmatrix} 2 & 4 \\ 3 & 13 \end{bmatrix}$ (ii) $\begin{bmatrix} 2 & -3 \\ 2 & -5 \end{bmatrix}$ (iii) $\begin{bmatrix} 5 & 6 \\ -2 & -2 \end{bmatrix}$ (iv) $\begin{bmatrix} 3 & -2 \\ 2 & 1 \end{bmatrix}$

2. Find all eigenvalues and eigenvectors of each of the following matrices:

    (i) $\begin{bmatrix} 3 & 2 & 4 \\ 2 & 0 & 2 \\ 4 & 2 & 3 \end{bmatrix}$ (ii) $\begin{bmatrix} 1 & 1 & 1 \\ 1 & 1 & 1 \\ 1 & 1 & 1 \end{bmatrix}$ (iii) $\begin{bmatrix} 2 & 1 & 0 \\ 0 & 2 & 1 \\ 0 & 0 & 2 \end{bmatrix}$

3. Prove that similar matrices have the same eigenvalues.

4. Determine the eigenvalues and eigenvectors of the identity matrix.

ANSWERS

1. (i) $\lambda_1 = 1$, $X_1 = \begin{bmatrix} 4 \\ -1 \end{bmatrix}$; $\lambda_2 = 14$, $X_2 = \begin{bmatrix} 1 \\ 3 \end{bmatrix}$

   (ii) $\lambda_1 = 1$, $X_1 = \begin{bmatrix} 1 \\ 1 \end{bmatrix}$; $\lambda_2 = 4$, $X_2 = \begin{bmatrix} 1 \\ -2 \end{bmatrix}$

   (iii) $\lambda_1 = 1$, $X_1 = \begin{bmatrix} 3 \\ -2 \end{bmatrix}$; $\lambda_2 = 2$, $X_2 = \begin{bmatrix} 2 \\ -1 \end{bmatrix}$

   (iv) $\lambda_1 = 2 + i\sqrt{3}$, $X_1 = \begin{bmatrix} 2 \\ 1 - i\sqrt{3} \end{bmatrix}$; $\lambda_2 = 2 - i\sqrt{3}$, $X_2 = \begin{bmatrix} 2 \\ 1 + i\sqrt{3} \end{bmatrix}$

2. (i) $\lambda_1 = 8$, $\lambda_2 = -1$, $\lambda_3 = -1$, $X_1 = \begin{bmatrix} 2 \\ 1 \\ 2 \end{bmatrix}$, $X_2 = \begin{bmatrix} 0 \\ 2 \\ -1 \end{bmatrix}$, $X_3 = \begin{bmatrix} 1 \\ 0 \\ -1 \end{bmatrix}$

   (ii) $\lambda_1 = 0$, $\lambda_2 = 0$, $\lambda_3 = 3$, $X_1 = \begin{bmatrix} 1 \\ 0 \\ -1 \end{bmatrix}$, $X_2 = \begin{bmatrix} 0 \\ 1 \\ -1 \end{bmatrix}$, $X_3 = \begin{bmatrix} 1 \\ 1 \\ 1 \end{bmatrix}$.

   (iii) $\lambda = 2, 2, 2$. The eigenvectors are given by $\begin{bmatrix} k \\ 0 \\ 0 \end{bmatrix}$, $k \neq 0$.

4. All eigenvalues are equal to 1. Every non-zero vector is an eigenvector.

5.11 DIAGONALIZING MATRICES

In Chapter 4 (see Theorem 6 on page 432), we have proved that distinct non-zero vectors of a linear transformation corresponding to distinct eigenvalues are linearly independent. Also, a linear transformation $t : V \to V$ has a diagonal matrix A relative to some basis of V iff eigenvectors of t span V and in such a case the eigenvalues of t are exactly the diagonal entries of A. and we say that t is diagonalizable.

Following theorem is analogous to Theorem 6 (on page 432 of Chapter 4) for matrices.

THEOREM-1 *Let A be an $n \times n$ matrix over a field F. If A has n distinct eigenvalues, then the corresponding eigenvectors of A are linearly independent.*

PROOF. Let $\lambda_1, \lambda_2, \ldots, \lambda_n$ be n distinct eigenvalues of A and let $X_1, X_2, \ldots, X_n$ respectively be corresponding eigenvectors. Then,

$$AX_i = \lambda X_i \, ; \, i \in \underline{n}$$

We have to prove that $X_1, X_2, \ldots, X_n$ are linearly independent. If possible, let $X_1, X_2, \ldots, X_n$ be linearly dependent. Then there exists $r, 1 \leq r \leq n$ such that $X_1, X_2, \ldots, X_r$ are linearly independent but $X_1, X_2, X_3, \ldots, X_{r+1}$ are linearly dependent. So, we can choose scalars $\alpha_1, \alpha_2, \ldots, \alpha_{r+1}$ not all zero such that

$$\alpha_1 X_1 + \alpha_2 X_2 + \cdots + \alpha_r X_r + \alpha_{r+1} X_{r+1} = O \tag{i}$$

$$\Rightarrow A(\alpha_1 X_1 + \alpha_2 X_2 + \cdots + \alpha_r X_r + \alpha_{r+1} X_{r+1}) = AO$$

$$\Rightarrow \alpha_1(AX_1) + \alpha_2(AX_2) + \cdots + \alpha_r(AX_r) + \alpha_{r+1}(AX_{r+1}) = O$$

$$\Rightarrow \alpha_1 \lambda_1 X_1 + \alpha_2 \lambda_2 X_2 + \cdots + \alpha_r \lambda_r X_r + \cdots + \alpha_{r+1} \lambda_{r+1} X_{r+1} = O \tag{ii}$$

Multiplying (i) by λ_{r+1} and subtracting from (ii), we get

$$\alpha_1(\lambda_1 - \lambda_{r+1})X_1 + \alpha_2(\lambda_2 - \lambda_{r+1})X_2 + \cdots + \alpha_r(\lambda_r - \lambda_{r+1})X_r = O$$

$$\Rightarrow \alpha_1 = \alpha_2 = \cdots = \alpha_r = 0 \quad [\because X_1, X_2, \ldots, X_r \text{ are l.i. and } \lambda_1, \lambda_2, \ldots, \lambda_n \text{ are distinct}]$$

Putting $\alpha_1 = \alpha_2 = \cdots = \alpha_r = 0$ in (i), we get

$$\alpha_{r+1} X_{r+1} = O$$

$$\Rightarrow \alpha_{r+1} = 0 \qquad [\because X_{r+1} \neq O]$$

$$\therefore \alpha_1 = \alpha_2 = \cdots = \alpha_{r+1} = 0.$$

This is a contradiction to the fact that $\alpha_1, \alpha_2, \ldots, \alpha_{r+1}$ are not all zero. So, our supposition is wrong. Hence, $X_1, X_2, \ldots, X_n$ are linearly independent eigenvectors of A.

Q.E.D.

THEOREM-2 *Let A be an $n \times n$ matrix over a field F. If A has n distinct eigenvalues $\lambda_1, \lambda_2, \ldots, \lambda_n$, then there exists an invertible matrix P such that $P^{-1}AP = \text{diag}(\lambda_1, \lambda_2, \ldots, \lambda_n)$.*

PROOF. Let $X_1, X_2, \ldots, X_n \in F^n$ be the eigenvectors of A corresponding to n distinct eigenvalues $\lambda_1, \lambda_2, \ldots, \lambda_n$ respectively. Then,

$$AX_i = \lambda_i X_i \, ; \, i = 1, 2, \ldots, n.$$

Let P be $n \times n$ matrix over F having $X_1, X_2, \ldots, X_n$ as its columns, i.e. $P = (X_1, X_2, \ldots, X_n)$. By Theorem 6 (on page 432 of Chapter 4) $X_1, X_2, X_3, \ldots, X_n$ are linearly independent vectors in F^n. Therefore, $\text{rank}(P) = n$ and hence P is an invertible matrix.

Now, for any $r = 1, 2, 3, \ldots, n$

rth column of $AP = A(r\text{th column of } P) = AX_r$

$$\therefore AP = (AX_1, AX_2, \ldots, AX_n)$$

$$\Rightarrow AP = (\lambda_1 X_1, \lambda_2 X_2, \ldots, \lambda_n X_n)$$

$$\Rightarrow \quad AP = [X_1, X_2 \ldots X_n] \begin{bmatrix} \lambda_1 & 0 & 0 & \ldots & 0 \\ 0 & \lambda_2 & 0 & \ldots & 0 \\ 0 & 0 & \lambda_3 & \ldots & 0 \\ \vdots & \vdots & \vdots & & \\ 0 & 0 & 0 & \ldots & \lambda_n \end{bmatrix}$$

$$\Rightarrow \quad AP = P \operatorname{diag}(\lambda_1, \lambda_2, \ldots, \lambda_n)$$

$$\Rightarrow \quad P^{-1}AP = \operatorname{diag}(\lambda_1, \lambda_2 \ldots, \lambda_n)$$

Q.E.D.

REMARK. *If a matrix A is diagonalizable as above, i.e.*

$$P^{-1}AP = \operatorname{diag}(\lambda_1, \lambda_2, \ldots, \lambda_n)$$

or, $\quad P^{-1}AP = D, \quad$ where $D = \operatorname{diag}(\lambda_1, \lambda_2, \ldots, \lambda_n)$

Then, A has the extremely useful diagonal factorization $A = PDP^{-1}$
Using this factorization, the algebra of A reduces to the algebra of the diagonal matrix D, which can be easily calculated. For example,

$$A^m = (PDP^{-1})^m = PD^m P^{-1} = P \operatorname{diag}(\lambda_1^m, \lambda_2^m, \ldots, \lambda_n^m) P^{-1}$$

More generally, for any polynomial $f(t)$, we have

$$f(A) = f(PDP^{-1}) = Pf(D)P^{-1} = P \operatorname{diag}(f(k_1), f(k_2), \ldots, f(k_n)) P^{-1}$$

Furthermore, if $\lambda_i > 0$ for all $i = 1, 2, \ldots, n$ and B is a matrix such that $B^2 = A$, then

$$B = P \operatorname{diag}\left(\sqrt{\lambda_1}, \sqrt{\lambda_2}, \ldots, \sqrt{\lambda_n}\right) P^{-1}$$

The matrix B is known as a non-negative square root of A. The eigenvalues of B are non-negative. Similarly, the non-negative cube root of A is given by

$$B = P \operatorname{diag}\left(\lambda_1^{\frac{1}{3}}, \lambda_2^{\frac{1}{3}}, \ldots, \lambda_n^{\frac{1}{3}}\right) P^{-1}, \text{ and so on.}$$

THEOREM-3 *Let A be an $n \times n$ square matrix over a field F. If the characteristic polynomial $\Delta_A(\lambda)$ of matrix A is product of n distinct factors, say, $\Delta_A(\lambda) = (\lambda - a_1)(\lambda - a_2) \ldots (\lambda - a_n)$. Then A is similar to the diagonal matrix $D = \operatorname{diag}(\lambda_1, \lambda_2, \ldots, \lambda_n)$.*

PROOF. Let $X_1, X_2, \ldots, X_n$ be non-zero eigenvectors of A corresponding to the eigenvalues $\lambda_1, \lambda_2, \ldots, \lambda_n$. Then, by Theorem 2, $X_1, X_2, \ldots, X_n$ are linearly independent and hence form a basis of F^n. By Theorem 2, there exists an invertible matrix P such that $P^{-1}AP = \operatorname{diag}(\lambda_1, \lambda_2, \ldots, \lambda_n) = D$, i.e. A is similar to the diagonal matrix D.

Q.E.D.

DIAGONALIZABLE MATRIX *A square matrix A over a field F is said to be diagonalizable if there exists an invertible matrix P such that $P^{-1}AP$ is a diagonal matrix.*

Clearly, A is diagonalizable iff A is similar to a diagonal matrix.

DIAGONALIZABLE LINEAR OPERATOR *A linear operator t on a finite dimensional vector space $V(F)$ is said to be a diagonalizable operator if matrix of t relative to some ordered basis B is diagonalizable.*

THEOREM-4 *If an $n \times n$ square matrix A over a field F is diagonalizable, then the characteristic polynomial $\Delta_A(\lambda)$ of A is of the form $(-1)^n(\lambda - \lambda_1)(\lambda - \lambda_2)\ldots(\lambda - \lambda_n)$ for some $\lambda_i \in F$.*

PROOF. Since A is a diagonalizable matrix, therefore, there exists an invertible matrix P such that $P^{-1}AP$ is a diagonal, i.e. A is similar to a diagonal matrix $D = \text{diag}(\lambda_1, \lambda_2, \ldots, \lambda_n)$.

Now,
$$D - \lambda I_n = \text{diag}(\lambda_1 - \lambda, \lambda_2 - \lambda, \lambda_3 - \lambda, \ldots, \lambda_n - \lambda)$$
$$\Rightarrow \quad \Delta_D(\lambda) = (-1)^n(\lambda - \lambda_1), (\lambda - \lambda_2), \ldots, (\lambda - \lambda_n) \qquad [\because \Delta_D(\lambda) = |D - \lambda I_n|]$$
$$\Rightarrow \quad \Delta_A(\lambda) = (-1)^n(\lambda - \lambda_1), (\lambda - \lambda_2), \ldots, (\lambda - \lambda_n) \qquad \begin{bmatrix} \because \text{Similar matrices have the} \\ \text{same characteristic polynomial} \end{bmatrix}$$

Q.E.D.

The above theorem shows that if a square matrix A over a field F is such that its characteristic polynomial is not factorizable into distinct linear factors over F, then A is not diagonalizable.

The above discussion suggests us the following algorithm to check whether a matrix is diagonalizable or not.

ALGORITHM *(Diagonalization Algorithm)*

Step I   Obtain the square matrix A.
Step II   Find the characteristic polynomial $\Delta_A(\lambda)$ of A.
Step III   Find the eigenvalues of A by solving the characteristic equation $\Delta_A(\lambda) = 0$.
Step IV   For each eigenvalue λ of A:
Solve the homogeneous system $(A - \lambda I)X = O$ and find a basis for the eigenspace. Let $\{X_1, X_2, \ldots, X_n\}$ be the set of basis vectors of different eigenspaces. These basis vectors are linearly independent eigenvectors of A for different eigenvalues of A.
Step V   Consider the set $S = \{X_1, X_2, \ldots, X_m\}$ of all eigenvectors obtained in step IV.

(i) If $m \neq n$, then matrix A is not diagonalizable.

(ii) If $m = n$, then matrix A is diagonalizable.

In order to find the diagonal matrix D similar to matrix A, proceed as follows:
Step VI   Construct a matrix P whose columns are the eigenvectors $X_1, X_2, \ldots, X_n$.
Step VII   Find P^{-1}.
Step VIII   Find $D = P^{-1}AP$. Matrix D will be equal to $\text{diag}(\lambda_1, \lambda_2, \ldots, \lambda_n)$, where λ_i is the eigenvalue of A corresponding to the eigenvector X_i.

ILLUSTRATIVE EXAMPLES

EXAMPLE-1 Let $A = \begin{bmatrix} 4 & 2 \\ 3 & -1 \end{bmatrix}$.

(i) Find all eigenvalues and corresponding eigenvectors.

(ii) Find matrices P and D such that P is non-singular and $D = P^{-1}AP$ is diagonal.

SOLUTION (i) The characteristic polynomial $\Delta_A(\lambda)$ of A is given by

$$\Delta_A(\lambda) = (-1)^2 \{\lambda^2 - tr\,(A)\lambda + |A|\} = \lambda^2 - 3\lambda - 10 = (\lambda - 5)(\lambda + 2)$$

The eigenvalues of A are roots of the characteristic equation $\Delta_A(\lambda) = 0$
So, eigenvalues of A are given by

$$(\lambda - 5)(\lambda + 2) = 0 \Rightarrow \lambda = 5, \lambda = -2$$

Thus, the eigenvalues of A are $\lambda_1 = 5$ and $\lambda_2 = -2$.

Eigenvector corresponding to $\lambda_1 = 5$: Let $X = \begin{bmatrix} x \\ y \end{bmatrix}$ be an eigenvector corresponding to eigenvalue $\lambda_1 = 5$. Then, X is given by the homogeneous system of equations

$$(A - 5I)X = O$$

$$\Rightarrow \begin{bmatrix} -1 & 2 \\ 3 & -6 \end{bmatrix} \begin{bmatrix} x \\ y \end{bmatrix} = \begin{bmatrix} 0 \\ 0 \end{bmatrix}$$

$$\Rightarrow \quad -x + 2y = 0,\ 3x - 6y = 0$$

$$\Rightarrow \quad x - 2y = 0$$

The system has only one free variable. Putting $y = 1$, we get $x = 2$.

So, $X_1 = \begin{bmatrix} 2 \\ 1 \end{bmatrix}$ is an eigenvector that spans the eigenspace of $\lambda_1 = 5$. The eigenspace of $\lambda_1 = 5$ is given by $E_1 = \left\{ X : X = k\begin{bmatrix} 2 \\ 1 \end{bmatrix}, k \neq 0 \right\}$

Eigenvector corresponding to $\lambda_2 = -2$: Let $X = \begin{bmatrix} x \\ y \end{bmatrix}$ be an eigenvector corresponding to the eigenvalue $\lambda_2 = -2$. Then, X is given by the homogeneous system of equations.

$$(A + 2I)X = O$$

$$\Rightarrow \begin{bmatrix} 6 & 2 \\ 3 & 1 \end{bmatrix} \begin{bmatrix} x \\ y \end{bmatrix} = \begin{bmatrix} 0 \\ 0 \end{bmatrix}$$

$$\Rightarrow \quad 6x + 2y = 0 \quad \text{and} \quad 3x + y = 0$$

$$\Rightarrow \quad 3x + y = 0$$

Clearly, there is only one free variable. Putting $x = -1$, we get $y = 3$.

So, $x = -1$ and $y = 3$ is a non-zero solution of the system.

Thus, $X_2 = \begin{bmatrix} -1 \\ 3 \end{bmatrix}$ is an eigenvector that spans the eigenspace of $\lambda_2 = -2$.

The eigenspace is given by $E_2 = \left\{ X : X = k \begin{bmatrix} -1 \\ 3 \end{bmatrix}, k \neq 0 \right\}$.

(ii) Let P be the matrix whose columns are eigenvectors X_1 and X_2. Then,

$$P = \begin{bmatrix} 2 & -1 \\ 1 & 3 \end{bmatrix} \quad \text{and so} \quad P^{-1} = \begin{bmatrix} \frac{3}{7} & \frac{1}{7} \\ -\frac{1}{7} & \frac{2}{7} \end{bmatrix}$$

$$\therefore \quad D = P^{-1}AP = \begin{bmatrix} \frac{3}{7} & \frac{1}{7} \\ -\frac{1}{7} & \frac{2}{7} \end{bmatrix} \begin{bmatrix} 4 & 2 \\ 3 & -1 \end{bmatrix} \begin{bmatrix} 2 & -1 \\ 1 & 3 \end{bmatrix} = \begin{bmatrix} 5 & 0 \\ 0 & -2 \end{bmatrix}$$

Note that the diagonal entries of D are the eigenvalues of A corresponding to the eigenvectors appearing in P.

REMARK. $D = P^{-1}AP$ implies that D is similar to A such that P is the change-of-basis matrix from the standard basis of R^2 to the new basis $B = \{X_1, X_2\}$ and D is the matrix (the matrix function) that represents A relative to the new basis B.

EXAMPLE-2 Show that the matrix $A = \begin{bmatrix} 5 & -1 \\ 1 & 3 \end{bmatrix}$ is not diagonalizable.

SOLUTION We have,
$$tr(A) = 5 + 3 = 8, \quad |A| = 16$$

So, the characteristic polynomial $\Delta_A(\lambda)$ of matrix A is

$$\Delta_A(\lambda) = (-1)^2 \{\lambda^2 - \lambda \, tr(A) + |A|\} = \lambda^2 - 8\lambda + 16 = (\lambda - 4)^2$$

The eigenvalues of A are roots of $\Delta_A(\lambda) = 0$. So, $\lambda = 4$ is the only eigenvalue of A.

Let $X = \begin{bmatrix} x \\ y \end{bmatrix}$ be an eigenvector of A corresponding to the eigenvalue $\lambda = 4$. Then, X is given by the homogeneous system

$$(A - 4I)X = O$$

$$\Rightarrow \quad \begin{bmatrix} 1 & -1 \\ 1 & -1 \end{bmatrix} \begin{bmatrix} x \\ y \end{bmatrix} = 0$$

$$\Rightarrow \quad x - y = 0 \quad \text{and} \quad x - y = 0 \quad \Rightarrow \quad x - y = 0$$

Clearly, the system has only one independent solution, for example, $x = 1, y = 1$. Thus, $X = \begin{bmatrix} 1 \\ 1 \end{bmatrix}$ and its multiples $k \begin{bmatrix} 1 \\ 1 \end{bmatrix}, k \neq 0$ are eigenvectors of A. So, number of independent eigenvectors is not equal to 2. Hence, matrix A is not diagonalizable.

EXAMPLE-3 *Show that the matrix $A = \begin{bmatrix} 3 & -5 \\ 2 & -3 \end{bmatrix}$ is not diagonalizable over the field R. However, it is diagonalizable over the field C of all complex numbers.*

SOLUTION   We have,
$$tr(A) = 0 \quad \text{and} \quad |A| = 1$$
So, characteristic equation of A is
$$\lambda^2 - 0\lambda + 1 = 0 \qquad [\because \Delta_A(\lambda) = (-1)^2 \{\lambda^2 - tr(A)\lambda + |A|\}]$$
$$\Rightarrow \quad \lambda^2 + 1 = 0 \Rightarrow \lambda = \pm i$$

Thus, $\Delta_A(\lambda) = 0$ has no real roots, i.e. $\Delta_A(\lambda) = 0$ is not solvable in R. So, A has no eigenvalues and no eigenvectors. Hence, A is not diagonalizable over the field R.

Let us now consider matrix A over the field C. In this case, the characteristic equation has i and $-i$ as its roots. Thus, A has two distinct eigenvalues i and $-i$, and hence A has two independent eigenvectors. Consequently, there exists a non-singular matrix P over the field C of all complex numbers such that
$$P^{-1}AP = \begin{bmatrix} i & 0 \\ 0 & -i \end{bmatrix}$$
Hence, A is diagonalizable over C.

EXAMPLE-4 *Let $A = \begin{bmatrix} 1 & 1 & 2 \\ -1 & 2 & 1 \\ 0 & 1 & 3 \end{bmatrix}$. Find a matrix P such that $P^{-1}AP = \text{diag}(1, 2, 3)$.*

SOLUTION   We have,
$$tr(A) = 1 + 2 + 3 = 6, \quad A_{11} + A_{22} + A_{33} = 5 + 3 + 3 = 11 \text{ and } |A| = 6$$
So, the characteristic polynomial of A is
$$\Delta_A(\lambda) = (-1)^3 \{\lambda^3 - \lambda^2 tr(A) + \lambda(A_{11} + A_{22} + A_{33}) - |A|\}$$
or, $\quad \Delta_A(\lambda) = -(\lambda^3 - 6\lambda^2 + 11\lambda - 6)$

The characteristic equation of A is
$$\Delta_A(\lambda) = 0$$
$$\Rightarrow \quad \lambda^3 - 6\lambda^2 + 11\lambda - 6 = 0$$
$$\Rightarrow \quad (\lambda - 1)(\lambda - 2)(\lambda - 3) = 0 \quad \Rightarrow \quad \lambda = 1, 2, 3$$

Thus, the eigenvalues of A are $\lambda_1 = 1$, $\lambda_2 = 2$ and $\lambda_3 = 3$.
Let us now find eigenvectors corresponding to these eigenvalues.

Eigenvector corresponding to $\lambda_1 = 1$: Let $X = \begin{bmatrix} x \\ y \\ z \end{bmatrix}$ be an eigenvector corresponding to eigenvalue $\lambda_1 = 1$. Then,

$$(A - \lambda_1 I)X = O$$
$$\Rightarrow \quad (A - I)X = O$$
$$\Rightarrow \quad \begin{bmatrix} 0 & 1 & 2 \\ -1 & 1 & 1 \\ 0 & 1 & 2 \end{bmatrix} \begin{bmatrix} x \\ y \\ z \end{bmatrix} = \begin{bmatrix} 0 \\ 0 \\ 0 \end{bmatrix}$$
$$\Rightarrow \quad y + 2z = 0, \ -x + y + z = 0, \ y + 2z = 0$$
$$\Rightarrow \quad -x + y + z = 0 \quad \text{and} \quad y + 2z = 0$$
$$\Rightarrow \quad x = -k, \ y = -2k, \ z = k$$
$$\Rightarrow \quad X = \begin{bmatrix} x \\ y \\ z \end{bmatrix} = \begin{bmatrix} -k \\ -2k \\ k \end{bmatrix} = k \begin{bmatrix} -1 \\ -2 \\ 1 \end{bmatrix}$$

Thus, $E_1 = \left\{ X : X = k \begin{bmatrix} -1 \\ -2 \\ 1 \end{bmatrix}, k \neq 0 \right\}$ is the eigenspace of $\lambda_1 = 1$ and $X_1 = \begin{bmatrix} -1 \\ -2 \\ 1 \end{bmatrix}$ is an eigenvector for $\lambda_1 = 1$.

Eigenvector corresponding to $\lambda_2 = 2$: Let $X = \begin{bmatrix} x \\ y \\ z \end{bmatrix}$ be an eigenvector corresponding to eigenvalue of $\lambda_2 = 2$. Then, X is a solution of the homogeneous system

$$(A - 2I)X = O$$
$$\Rightarrow \quad \begin{bmatrix} -1 & 1 & 2 \\ -1 & 0 & 1 \\ 0 & 1 & 1 \end{bmatrix} \begin{bmatrix} x \\ y \\ z \end{bmatrix} = \begin{bmatrix} 0 \\ 0 \\ 0 \end{bmatrix}$$
$$\Rightarrow \quad -x + y + 2z = 0, \ -x + z = 0, \ y + z = 0$$
$$\Rightarrow \quad x = k, \ y = -k, \ z = k, \ k \neq 0$$
$$\therefore \quad X = k \begin{bmatrix} 1 \\ -1 \\ 1 \end{bmatrix}, \ k \neq 0 \quad \text{are eigenvectors of } \lambda_2 = 2.$$

So, $E_2 = \left\{ X : k \begin{bmatrix} 1 \\ -1 \\ 1 \end{bmatrix}, k \neq 0 \right\}$ is the eigenspace of $\lambda_2 = 2$.

$\therefore \quad X_2 = \begin{bmatrix} 1 \\ -1 \\ 1 \end{bmatrix}$ is an eigenvector of $\lambda_2 = 2$.

Eigenvector corresponding to $\lambda_3 = 3$: Let $X = \begin{bmatrix} x \\ y \\ z \end{bmatrix}$ be an eigenvector of $\lambda_3 = 3$. Then, X is a solution of the homogeneous system

$(A - 3I)X = O$

$\Rightarrow \quad \begin{bmatrix} -2 & 1 & 2 \\ -1 & -1 & 1 \\ 0 & 1 & 0 \end{bmatrix} \begin{bmatrix} x \\ y \\ z \end{bmatrix} = \begin{bmatrix} 0 \\ 0 \\ 0 \end{bmatrix}$

$\Rightarrow \quad -2x + y + 2z = 0, \ -x - y + z = 0, \ y = 0$

$\Rightarrow \quad x = k, \ y = 0, \ z = k, \ k \neq 0$

Thus, $E_3 = \left\{ X : X = k \begin{bmatrix} 1 \\ 0 \\ 1 \end{bmatrix}, k \neq 0 \right\}$ is the eigenspace of $\lambda_3 = 3$. An eigenvector of $\lambda_3 = 3$ is $X_3 = \begin{bmatrix} 1 \\ 0 \\ 1 \end{bmatrix}$.

Clearly, eigenvectors X_1, X_2 and X_3 are linearly independent. Let P be the matrix whose columns are eigenvectors X_1, X_2 and X_3, i.e.

$$P = \begin{bmatrix} -1 & 1 & 1 \\ -2 & -1 & 0 \\ 1 & 1 & 1 \end{bmatrix} \Rightarrow P^{-1} = \frac{1}{2} \begin{bmatrix} -1 & 0 & 1 \\ 2 & -2 & -2 \\ -1 & 2 & 3 \end{bmatrix}$$

$\therefore \quad P^{-1}AP = \frac{1}{2} \begin{bmatrix} -1 & 0 & 1 \\ 2 & -2 & -2 \\ -1 & 2 & 3 \end{bmatrix} \begin{bmatrix} 1 & 1 & 2 \\ -1 & 2 & 1 \\ 0 & 1 & 3 \end{bmatrix} \begin{bmatrix} -1 & 1 & 1 \\ -2 & -1 & 0 \\ 1 & 1 & 1 \end{bmatrix}$

$$\Rightarrow \quad P^{-1}AP = \frac{1}{2}\begin{bmatrix} -1 & 0 & 1 \\ 2 & -2 & -2 \\ -1 & 2 & 3 \end{bmatrix}\begin{bmatrix} -1 & 2 & 3 \\ -2 & -2 & 0 \\ 1 & 2 & 3 \end{bmatrix}$$

$$\Rightarrow \quad P^{-1}AP = \frac{1}{2}\begin{bmatrix} 2 & 0 & 0 \\ 0 & 4 & 0 \\ 0 & 0 & 6 \end{bmatrix} = \begin{bmatrix} 1 & 0 & 0 \\ 0 & 2 & 0 \\ 0 & 0 & 3 \end{bmatrix} = \text{diag}\,(1,2,3)$$

EXAMPLE-5 Let $t : R^3 \to R^3$ be a linear transformation defined by

$$t(x,y,z) = (2x+y-2z,\ 2x+3y-4z,\ x+y-z)$$

(i) Find all eigenvalues of t, and a basis of each eigenspace.

(ii) Is t diagonalizable? If so, find the basis B of R^3 that diagonalize t, and find its diagonal representation.

SOLUTION (i) Let us first find the matrix representation of t relative to standard basis $B = \left\{e_1^{(3)}, e_2^{(3)}, e_3^{(3)}\right\}$ of R^3 by writing down the coefficients of x, y, z as rows as given below.

$$A = M_B^B(t) = \begin{bmatrix} 2 & 1 & -2 \\ 2 & 3 & -4 \\ 1 & 1 & -1 \end{bmatrix}$$

We have, $tr\,(A) = 4, |A| = 2$ and $A_{11} + A_{22} + A_{33} = 1 + 0 + 4 = 5$. The characteristic polynomial of matrix A is

$$\Delta_A(\lambda) = (-1)^3\{\lambda^3 - 4\lambda^2 + 5\lambda - 2\}$$

The eigenvalues of t are roots of the equation

$$\Delta_A(\lambda) = 0$$
or, $\quad \lambda^3 - 4\lambda^2 + 5\lambda - 2 = 0$
$\Rightarrow \quad (\lambda - 1)^2(\lambda - 2) = 0$
$\Rightarrow \quad \lambda = 1,\ \lambda = 2$

Thus, the eigenvalues of t are $\lambda_1 = 1$ and $\lambda_2 = 2$.

(ii) Let us now find eigenvectors corresponding to each eigenvalue of t.

Eigenvectors corresponding to eigenvalue $\lambda_1 = 1$. Let $X = \begin{bmatrix} x \\ y \\ z \end{bmatrix}$ be an eigenvector for $\lambda_1 = 1$. Then, X is a solution of the homogenous system

$(A - I)X = O$

$\Rightarrow \begin{bmatrix} 1 & 1 & -2 \\ 2 & 2 & -4 \\ 1 & 1 & -2 \end{bmatrix} \begin{bmatrix} x \\ y \\ z \end{bmatrix} = \begin{bmatrix} 0 \\ 0 \\ 0 \end{bmatrix}$

$\Rightarrow \quad x + y - 2z = 0, \ 2x + 2y - 4z = 0, \ x + y - 2z = 0$

$\Rightarrow \quad x + y - 2z = 0$

Here, we have two free variables. So, there are two independent solutions.

$$x = 1, \ y = -1, \ z = 0 \quad \text{and} \quad x = 2, \ y = 0, \ z = 1$$

Thus, $X_1 = \begin{bmatrix} 1 \\ -1 \\ 0 \end{bmatrix}$ and $X_2 = \begin{bmatrix} 2 \\ 0 \\ 1 \end{bmatrix}$ are two independent eigenvectors of t corresponding to eigenvalue $\lambda_1 = 1$. The eigenspace spanned by X_1 and X_2 is $\{X : aX_1 + bX_2, a, b \in R\}$ and has $\{X_1, X_2\}$ as its basis.

Eigenvectors corresponding to eigenvalue $\lambda_2 = 2$: Let $X = \begin{bmatrix} x \\ y \\ z \end{bmatrix}$ be an eigenvector of $\lambda_2 = 2$. Then, X is a solution of homogeneous system

$(A - 2I)X = O$

$\Rightarrow \begin{bmatrix} 0 & 1 & -2 \\ 2 & 1 & -4 \\ 1 & 1 & -3 \end{bmatrix} \begin{bmatrix} x \\ y \\ z \end{bmatrix} = \begin{bmatrix} 0 \\ 0 \\ 0 \end{bmatrix}$

$\Rightarrow \quad y - 2z = 0, \ 2x + y - 4z = 0, \ x + y - 3z = 0$

$\Rightarrow \quad x = z, \ y = 2z$

Thus, there is only one free variable z. So, $X_3 = \begin{bmatrix} 1 \\ 2 \\ 1 \end{bmatrix}$ (taking $z = 1$) is an eigenvector of $\lambda_2 = 2$.

We observe that X_1, X_2, X_3 are linearly independent vectors (as the matrix having X_1, X_2, X_3 as its columns is non-singular). So, t is diagonalizable as it has three linearly independent eigenvectors.

Taking $B = \{X_1, X_2, X_3\}$ as a basis, t is represented by the diagonal matrix D given by $D = \text{diag}(1, 1, 2)$.

EXAMPLE-6 Let $A = \begin{bmatrix} 3 & -1 & 1 \\ 7 & -5 & 1 \\ 6 & -6 & 2 \end{bmatrix}$.

(i) Find all eigenvalues of A.

(ii) Find a maximum set B of linearly independent vectors of A.

(iii) Is A diagonalizable? If yes, find P such that $D = P^{-1}AP$ is a diagonal matrix.

SOLUTION (i) We have,

$$tr(A) = 3 - 5 + 2 = 0, \quad A_{11} + A_{22} + A_{33} = -4 + 0 - 8 = -12 \quad \text{and} \quad |A| = -16$$

So, the characteristic equation of A is

$$\Delta_A(\lambda) = 0$$
$$\Rightarrow (-1)^3 (\lambda^3 - 0\lambda^2 - 12\lambda + 16) = 0$$
$$\Rightarrow \lambda^3 - 12\lambda + 16 = 0$$
$$\Rightarrow (\lambda - 2)^2 (\lambda + 4) = 0$$
$$\Rightarrow \lambda = 2, \quad \lambda = -4$$

Thus, the eigenvalues of A are $\lambda_1 = 2$, $\lambda_2 = -4$.

(ii) *Eigenvector corresponding to* $\lambda_1 = 2$: Let $X = \begin{bmatrix} x \\ y \\ z \end{bmatrix}$ be an eigenvector of $\lambda_1 = 2$.

Then, X is a solution of the homogeneous system

$$(A - 2I)X = O$$
$$\Rightarrow \begin{bmatrix} 1 & -1 & 1 \\ 7 & -7 & 1 \\ 6 & -6 & 0 \end{bmatrix} \begin{bmatrix} x \\ y \\ z \end{bmatrix} = \begin{bmatrix} 0 \\ 0 \\ 0 \end{bmatrix}$$
$$\Rightarrow x - y + z = 0, \quad 7x - 7y + z = 0, \quad 6x - 6y = 0$$
$$\Rightarrow x - y + z = 0, \quad z = 0$$
$$\Rightarrow x - y = 0, \quad z = 0$$

The system has one free variable. So, it has one independent solution. The general solution is $x = y = k$ (say), $z = 0$. So, $\left\{ X : X = k \begin{bmatrix} 1 \\ 1 \\ 0 \end{bmatrix}, k \neq 0 \right\}$ is the eigenspace of $\lambda_1 = 2$ and it has $X_1 = \begin{bmatrix} 1 \\ 1 \\ 0 \end{bmatrix}$ as the basis.

Eigenvector corresponding to $\lambda_2 = -4$: Let $X = \begin{bmatrix} x \\ y \\ z \end{bmatrix}$ be an eigenvector of $\lambda_2 = -4$.

Then, X is a solution of the homogeneous system

$$(A + 4I)X = O$$

$$\Rightarrow \begin{bmatrix} 7 & -1 & 1 \\ 7 & -1 & 1 \\ 6 & -6 & 6 \end{bmatrix} \begin{bmatrix} x \\ y \\ z \end{bmatrix} = \begin{bmatrix} 0 \\ 0 \\ 0 \end{bmatrix}$$

$\Rightarrow \quad 7x - y + z = 0, \ 7x - y + z = 0, \ 6x - 6y + 6z = 0$

$\Rightarrow \quad x - y + z = 0, \ 7x - y + z = 0$

$\Rightarrow \quad x - y + z = 0, \ 6y - 6z = 0$.

$\Rightarrow \quad x - y + z = 0, \ y = z$

$\Rightarrow \quad x = 0, \ y = z.$

The system has only one free variable $y = z = k$(say). So, the eigenspace of $\lambda_2 = -4$ is

$$\left\{ X : X = k \begin{bmatrix} 0 \\ 1 \\ 1 \end{bmatrix}, \ k \neq 0 \right\} \text{ which has } X_2 = \begin{bmatrix} 0 \\ 1 \\ 1 \end{bmatrix} \text{ as the basis.}$$

(iii) Since A has at most two linearly independent eigenvectors. So, A is not diagonalizable.

EXAMPLE-7 Let $A = \begin{bmatrix} 2 & 2 \\ 1 & 3 \end{bmatrix}$. Find

(i) all eigenvalues and corresponding eigenvectors.

(ii) a non-singular matrix P such that $D = P^{-1}AP$ is a diagonal matrix.

(iii) A^6 and $f(A)$, where $f(t) = t^4 - 3t^3 - 6t^2 + 7t + 3$

(iv) a real cube root of A, that is, a matrix B such that $B^3 = A$ and B has real eigenvalues.

SOLUTION (i) We have,

$$tr(A) = 5 \text{ and } |A| = 4$$

The characteristic polynomial of A is

$$\Delta_A(\lambda) = (-1)^2(\lambda^2 - 5\lambda + 4) = (\lambda - 1)(\lambda - 4)$$

The eigenvalues of A are roots of the equation

$$\Delta_A(\lambda) = 0 \Rightarrow \lambda = 1, 4.$$

Thus, $\lambda_1 = 1$ and $\lambda_2 = 4$ are eigenvalues of A.

Eigenvectors corresponding to $\lambda_1 = 1$: Let $X = \begin{bmatrix} x \\ y \end{bmatrix}$ be an eigenvector of A corresponding to $\lambda_1 = 1$. Then, X is a solution of the homogeneous system

$$(A - I)X = O$$

$$\Rightarrow \begin{bmatrix} 1 & 2 \\ 1 & 2 \end{bmatrix} \begin{bmatrix} x \\ y \end{bmatrix} = \begin{bmatrix} 0 \\ 0 \end{bmatrix}$$

$$\Rightarrow x + 2y = 0, \ x + 2y = 0 \Rightarrow x + 2y = 0$$

Clearly, the system has only one free variable. Putting $y = -1$ in $x + 2y = 0$, we get $x = 2$. So, $X_1 = \begin{bmatrix} 2 \\ -1 \end{bmatrix}$ is an eigenvector of $\lambda_1 = 1$. The eigenspace of $\lambda_1 = 1$ is $\left\{ X : X = k \begin{bmatrix} 2 \\ -1 \end{bmatrix}, k \neq 0 \right\}$.

Eigenvectors corresponding to $\lambda_2 = 4$: Let $X = \begin{bmatrix} x \\ y \end{bmatrix}$ be an eigenvector of A corresponding to $\lambda_2 = 4$. Then, X is a solution of the homogeneous system.

$$(A - 4I)X = O$$

$$\Rightarrow \begin{bmatrix} -2 & 2 \\ 1 & -1 \end{bmatrix} \begin{bmatrix} x \\ y \end{bmatrix} = \begin{bmatrix} 0 \\ 0 \end{bmatrix}$$

$$\Rightarrow -2x + 2y = 0 \quad \text{and} \quad x - y = 0 \Rightarrow x - y = 0$$

The system has only one independent variable. Putting $x = 1$ in $x - y = 0$, we get $y = 1$. So, $X_2 = \begin{bmatrix} 1 \\ 1 \end{bmatrix}$ is an eigenvector of $\lambda_2 = 4$. The eigenspace of $\lambda_2 = 4$ is $\left\{ X : X = k \begin{bmatrix} 1 \\ 1 \end{bmatrix}, k \neq 0 \right\}$.

(ii) Since A has distinct eigenvalues. So, X_1 and X_2 are linearly independent and matrix P has X_1 and X_2 as its columns, i.e.

$$P = \begin{bmatrix} 2 & 1 \\ -1 & 1 \end{bmatrix}$$

$$\therefore \quad P^{-1} = \frac{1}{3} \begin{bmatrix} 1 & -1 \\ 1 & 2 \end{bmatrix} \text{ and, } D = P^{-1}AP = \frac{1}{3} \begin{bmatrix} 1 & -1 \\ 1 & 2 \end{bmatrix} \begin{bmatrix} 2 & 2 \\ 1 & 3 \end{bmatrix} \begin{bmatrix} 2 & 1 \\ -1 & 1 \end{bmatrix} = \begin{bmatrix} 1 & 0 \\ 0 & 4 \end{bmatrix}$$

(iii) We have,

$$D = P^{-1}AP$$

$$\Rightarrow A = PDP^{-1}$$

$$\Rightarrow A^6 = PD^6P^{-1}$$

$$\Rightarrow A^6 = \begin{bmatrix} 2 & 1 \\ -1 & 1 \end{bmatrix} \begin{bmatrix} 1^6 & 0 \\ 0 & 4^6 \end{bmatrix} \frac{1}{3} \begin{bmatrix} 1 & -1 \\ 1 & 2 \end{bmatrix} \ [\because D = \text{diag}(1, 4) \therefore D^6 = \text{diag}(1^6, 4^6)]$$

$$\Rightarrow A^6 = \begin{bmatrix} 2 & 1 \\ -1 & 1 \end{bmatrix} \begin{bmatrix} 1 & 0 \\ 0 & 4096 \end{bmatrix} \frac{1}{3} \begin{bmatrix} 1 & -1 \\ 1 & 2 \end{bmatrix}$$

$$\Rightarrow A^6 = \begin{bmatrix} 2 & 4096 \\ -1 & 4096 \end{bmatrix} \frac{1}{3} \begin{bmatrix} 1 & -1 \\ 1 & 2 \end{bmatrix}$$

$$\Rightarrow A^6 = \frac{1}{3} \begin{bmatrix} 4098 & 8190 \\ 4095 & 8193 \end{bmatrix} = \begin{bmatrix} 1366 & 2730 \\ 1365 & 2731 \end{bmatrix}$$

Now,

$$f(t) = t^4 - 3t^3 - 6t^2 + 7t + 3 \text{ and } A = PDP^{-1}$$

$$\Rightarrow f(A) = P f(D) P^{-1}$$

$$\Rightarrow f(A) = \begin{bmatrix} 2 & 1 \\ -1 & 1 \end{bmatrix} \begin{bmatrix} 2 & 0 \\ 0 & -1 \end{bmatrix} \frac{1}{3} \begin{bmatrix} 1 & -1 \\ 1 & 2 \end{bmatrix} \quad [\because f(D) = \text{diag}(f(1), f(4)) = (2, -1)]$$

$$\Rightarrow f(A) = \begin{bmatrix} 1 & 2 \\ -1 & 0 \end{bmatrix}$$

(iv) We have, $D = \begin{bmatrix} 1 & 0 \\ 0 & 4 \end{bmatrix}$

$\therefore \begin{bmatrix} 1 & 0 \\ 0 & \sqrt[3]{4} \end{bmatrix}$ is the real cube root of D.

Hence, the real cube root of A is given by

$$B = P \sqrt[3]{D} P^{-1} = \begin{bmatrix} 2 & 1 \\ -1 & 1 \end{bmatrix} \begin{bmatrix} 1 & 0 \\ 0 & \sqrt[3]{4} \end{bmatrix} \frac{1}{3} \begin{bmatrix} 1 & -1 \\ 1 & 2 \end{bmatrix} = \frac{1}{3} \begin{bmatrix} 2 + \sqrt[3]{4} & -2 + 2\sqrt[3]{4} \\ -1 + \sqrt[3]{4} & 1 + 2\sqrt[3]{4} \end{bmatrix}$$

5.11.1 DIAGONALIZING REAL SYMMETRIC MATRICES

In the above discussion, we have seen that there are many real matrices A (matrices over R) that are not diagonalizable. In fact, we have seen that there are matrices over R not having real eigenvalues. However, these problems do not exist for a symmetric matrix over R as stated and proved below.

THEOREM-1 *Let A be a real symmetric matrix. Then each root of its characteristic polynomial is real.*

OR

Every eigenvalue of a real symmetric matrix is real.

PROOF. Let A be a real symmetric matrix and λ be an eigenvalue of A. Further, let X be the eigenvector corresponding to the eigenvalue λ. Then,

$$AX = \lambda X$$
$$\Rightarrow (AX)^T = (\lambda X)^T$$
$$\Rightarrow \overline{X}^T \overline{A}^T = \overline{\lambda}\, \overline{X}^T$$
$$\Rightarrow \overline{X}^T A = \overline{\lambda}\, \overline{X}^T \qquad [\because A \text{ is real symmetric} \therefore (\overline{A})^T = A] \quad \text{(i)}$$
$$\Rightarrow (\overline{X}^T A)X = (\overline{\lambda}\, \overline{X}^T)X \qquad \text{[Post-multiplying both sides by } X]$$
$$\Rightarrow \overline{X}^T(AX) = \overline{\lambda}(\overline{X}^T X) \qquad \text{(ii)}$$

Also, $\overline{X}^T(AX) = (\overline{X}^T A)X$ [From (i)]
$$\Rightarrow \overline{X}^T(AX) = (\lambda \overline{X}^T)X \qquad [\because AX = \lambda X]$$
$$\Rightarrow \overline{X}^T(AX) = \lambda(\overline{X}^T X) \qquad \text{(iii)}$$

From (ii) and (iii), we get

$$\overline{\lambda}(\overline{X}^T X) = \lambda(\overline{X}^T X)$$
$$\Rightarrow (\overline{\lambda} - \lambda)(\overline{X}^T X) = 0$$
$$\Rightarrow (\overline{\lambda} - \lambda)\left(\sum_{i=1}^{n} x_i^2\right) = 0 \qquad \left[\because X = \begin{bmatrix} x_1 \\ x_2 \\ \vdots \\ x_n \end{bmatrix} \therefore \overline{X}^T X = \sum_{i=1}^{n} x_i^2\right]$$

$$\Rightarrow \overline{\lambda} - \lambda = 0 \qquad \left[\because \sum_{i=1}^{n} x_i^2 \neq 0 \text{ as } X \neq 0\right]$$
$$\Rightarrow \overline{\lambda} = \lambda$$
$$\Rightarrow \lambda \text{ is a real number.}$$

Since λ is an arbitrarily chosen eigenvalue of A. Hence, all eigenvalues of A are real.

Q.E.D.

THEOREM-2 Let A be a real symmetric matrix. If X_1 and X_2 are eigenvectors of A belonging to distinct eigenvalues λ_1 and λ_2 respectively. Then, X_1 and X_2 are orthogonal, i.e. $X_1^T X_2 = 0$ or $X_1 X_2^T = 0$.

PROOF. It is given that X_1 and X_2 are eigenvectors of A corresponding to the eigenvalues λ_1 and λ_2 respectively.

$$\therefore \quad AX_1 = \lambda_1 X_1 \quad \text{and} \quad AX_2 = \lambda_2 X_2$$

Now,
$$AX_1 = \lambda_1 X_1$$
$$\Rightarrow (AX_1)^T = (\lambda_1 X_1)^T$$
$$\Rightarrow X_1^T A^T = \lambda_1 X_1^T$$
$$\Rightarrow X_1^T A = \lambda_1 X_1^T \qquad [\because A \text{ is symmetric} \therefore A^T = A]$$
$$\Rightarrow (X_1^T A) X_2 = (\lambda_1 X_1^T) X_2$$
$$\Rightarrow X_1^T (AX_2) = \lambda_1 (X_1^T X_2)$$
$$\Rightarrow X_1^T (\lambda_2 X_2) = \lambda_1 (X_1^T X_2)$$
$$\Rightarrow \lambda_2 X_1^T X_2 = \lambda_1 X_1^T X_2$$
$$\Rightarrow (\lambda_2 - \lambda_1) X_1^T X_2 = 0$$
$$\Rightarrow X_1^T X_2 = 0 \qquad [\because \lambda_1 \neq \lambda_2]$$

Similarly, we can prove that $X_1 X_2^T = 0$.

Q.E.D.

The above two theorems and theorems studied in earlier section give us the following fundamental result stated as a theorem.

THEOREM-3 *Let A be a real symmetric matrix. Then there exists an orthogonal matrix P such that $D = P^{-1}AP$ is a diagonal matrix.*

ORTHOGONAL DIAGONALIZATION *Let A be a real symmetric matrix. The process of obtaining an orthogonal matrix P such that $P^{-1}AP$ is a diagonal matrix is known as the orthogonal diagonalization of A and we say that A is orthogonally diagonalizable.*

In the above discussion, the orthogonal matrix P is obtained by normalizing each eigenvector of A. It should be noted that if $X = \begin{bmatrix} x_1 \\ x_2 \\ \vdots \\ x_n \end{bmatrix}$ is an eigenvector corresponding to an eigenvalue λ, then $\hat{X} = \begin{bmatrix} x_1' \\ x_2' \\ \vdots \\ x_n' \end{bmatrix}$, where $\hat{x}_i = \dfrac{x_i}{||X||}$, where $||X|| = \sqrt{x_1^2 + x_2^2 + \cdots + x_n^2}$ is normalized form of X.

The procedure of orthogonal diagonalization of a real symmetric matrix is given in the following algorithm.

ORTHOGONAL DIAGONALIZATION ALGORITHM

Step I Obtain the real symmetric matrix A of order $n \times n$.
Step II Find the characteristic polynomial $\Delta_A(\lambda)$ of A.

Step III Find the eigenvalues of A by solving the characteristic equation $\Delta_A(\lambda) = 0$.

Step IV For each eigenvalue λ of A in step III, find an orthogonal basis vector of its eigenspace.

Step V Normalize all eigenvectors obtained in step IV. These vectors form an orthonormal basis of R^n.

Step VI Construct matrix P whose columns are the normalized eigenvectors in step V.

Step VII Find P^{-1} and $D = P^{-1}AP$. The diagonal entries of D are the eigenvalues to the columns of P.

Following examples illustrates the above algorithm.

ILLUSTRATIVE EXAMPLES

EXAMPLE-1 Let $A = \begin{bmatrix} 2 & -2 \\ -2 & 5 \end{bmatrix}$ be a real symmetric matrix. Find an orthogonal matrix P such that $P^{-1}AP$ is a diagonal matrix.

SOLUTION We have,
$$tr(A) = 7 \text{ and } |A| = 10 - 4 = 6$$

So, the characteristic polynomial $\Delta_A(\lambda)$ of A is given by
$$\Delta_A(\lambda) = (-1)^2(\lambda^2 - 7\lambda + 6) = (\lambda - 1)(\lambda - 6)$$

The eigenvalues of A are given by
$$\Delta_A(\lambda) = 0 \Rightarrow (\lambda - 1)(\lambda - 6) = 0 \Rightarrow \lambda = 1, 6$$

Eigenvectors corresponding to $\lambda_1 = 6$: Let $X = \begin{bmatrix} x \\ y \end{bmatrix}$ be the eigenvectors corresponding to $\lambda_1 = 6$. Then, X is a solution of the homogeneous system

$(A - 6I)X = O$

$\Rightarrow \begin{bmatrix} -4 & -2 \\ -2 & -1 \end{bmatrix} \begin{bmatrix} x \\ y \end{bmatrix} = \begin{bmatrix} 0 \\ 0 \end{bmatrix}$

$\Rightarrow -4x - 2y = 0 \quad \text{and} \quad -2x - y = 0$

$\Rightarrow y + 2x = 0$

Clearly, $x = 1, y = -2$ is a solution of this equation.

Therefore, $X_1 = \begin{bmatrix} 1 \\ -2 \end{bmatrix}$ is a non-zero solution of $(A - 6I)X = O$. The eigenspace of $\lambda_1 = 6$ is $\left\{ X : X = k \begin{bmatrix} 1 \\ -2 \end{bmatrix}, k \neq 0 \right\}$ whose basis vector is $X_1 = \begin{bmatrix} 1 \\ -2 \end{bmatrix}$.

Eigenvectors corresponding to $\lambda_2 = 1$: Let $X = \begin{bmatrix} x \\ y \end{bmatrix}$ be the eigenvector corresponding to $\lambda_2 = 1$. Then, X is a solution of the homogeneous system

$$(A - I)X = O$$

$$\Rightarrow \begin{bmatrix} 1 & -2 \\ -2 & 4 \end{bmatrix} \begin{bmatrix} x \\ y \end{bmatrix} = \begin{bmatrix} 0 \\ 0 \end{bmatrix}$$

$$\Rightarrow x - 2y = 0 \quad \text{and} \quad -2x + 4y = 0$$

$$\Rightarrow x - 2y = 0$$

Clearly, $x = 2$, $y = 1$ is a solution of the above homogeneous system. The eigenspace of $\lambda_2 = 1$ is $\left\{ X : X = k \begin{bmatrix} 2 \\ 1 \end{bmatrix}, k \neq 0 \right\}$ whose basis is the vector $X_2 = \begin{bmatrix} 2 \\ 1 \end{bmatrix}$.

Clearly, $X_1 X_2^T = \begin{bmatrix} 1 \\ -2 \end{bmatrix} \begin{bmatrix} 2 & 1 \end{bmatrix} = 0$ and, $X_1^T X_2 = \begin{bmatrix} 1 & -2 \end{bmatrix} \begin{bmatrix} 2 \\ 1 \end{bmatrix} = 0$

So, X_1 and X_2 are orthogonal vectors in R^2. Let us now normalize eigenvectors X_1 and X_2.

We have, $\|X_1\| = \sqrt{1^2 + (-2)^2} = \sqrt{5}$ and $\|X_2\| = \sqrt{2^2 + 1^2} = \sqrt{5}$

$$\therefore \quad \widehat{X}_1 = \begin{bmatrix} \frac{1}{\sqrt{5}} \\ \frac{-2}{\sqrt{5}} \end{bmatrix} \text{ and, } \widehat{X}_2 = \begin{bmatrix} \frac{2}{\sqrt{5}} \\ \frac{1}{\sqrt{5}} \end{bmatrix} \text{ are orthonormal basis vectors in } R^2.$$

Let P be the matrix whose columns are $\widehat{X}_1$ and $\widehat{X}_2$ respectively. Then,

$$P = \begin{bmatrix} \frac{1}{\sqrt{5}} & \frac{2}{\sqrt{5}} \\ \frac{-2}{\sqrt{5}} & \frac{1}{\sqrt{5}} \end{bmatrix} \Rightarrow P^{-1} = \begin{bmatrix} \frac{1}{\sqrt{5}} & \frac{-2}{\sqrt{5}} \\ \frac{2}{\sqrt{5}} & \frac{1}{\sqrt{5}} \end{bmatrix}$$

$$\therefore \quad D = P^{-1} A P = \begin{bmatrix} \frac{1}{\sqrt{5}} & \frac{-2}{\sqrt{5}} \\ \frac{2}{\sqrt{5}} & \frac{1}{\sqrt{5}} \end{bmatrix} \begin{bmatrix} 2 & -2 \\ -2 & 5 \end{bmatrix} \begin{bmatrix} \frac{1}{\sqrt{5}} & \frac{2}{\sqrt{5}} \\ \frac{-2}{\sqrt{5}} & \frac{1}{\sqrt{5}} \end{bmatrix} = \begin{bmatrix} 6 & 0 \\ 0 & 1 \end{bmatrix}$$

EXAMPLE-2 Let $A = \begin{bmatrix} 11 & -8 & 4 \\ -8 & -1 & -2 \\ 4 & -2 & -4 \end{bmatrix}$ be a real summetric matrix. Find:

(i) *all eigenvalues of A.*
(ii) *a maximal set S of non-zero orthogonal eigenvectors of A.*
(iii) *an orthogonal matrix P such that $D = P^{-1}AP$ is a diagonal matrix.*

SOLUTION (i) We have,

$$tr(A) = 11 - 1 - 4 = 6, \ |A| = 400, \ A_{11} = 0, \ A_{22} = -60, \ A_{33} = -75$$

So, characteristic polynomial $\Delta_A(\lambda)$ of matrix A is

$$\Delta_A(\lambda) = (-1)^3 \{\lambda^3 - 6\lambda^2 + (0 - 60 - 75)\lambda - 400\}$$
$$\Rightarrow \quad \Delta_A(\lambda) = -(\lambda^3 - 6\lambda^2 - 135\lambda - 400)$$
$$\Rightarrow \quad \Delta_A(\lambda) = -(\lambda + 5)^2(\lambda - 16)$$

The eigenvalues of A are given by

$$\Delta_A(\lambda) = 0 \Rightarrow \lambda = -5 \text{(twice)}, \lambda = 16.$$

(ii) *Eigenvectors corresponding to $\lambda_1 = -5$*: Let $X = \begin{bmatrix} x \\ y \\ z \end{bmatrix}$ be an eigenvector corresponding to $\lambda_1 = -5$. Then, X is a solution of homogeneous system

$$(A + 5I)X = O$$

$$\Rightarrow \quad \begin{bmatrix} 16 & -8 & 4 \\ -8 & 4 & -2 \\ 4 & -2 & 1 \end{bmatrix} \begin{bmatrix} x \\ y \\ z \end{bmatrix} = \begin{bmatrix} 0 \\ 0 \\ 0 \end{bmatrix}$$

$$\Rightarrow \quad 16x - 8y + 4z = 0, \ -8x + 4y - 2z = 0, \ 4x - 2y + z = 0$$
$$\Rightarrow \quad 4x - 2y + z = 0$$

Clearly, $x = 0, y = 1, z = 2$ is a solution of the above equation, i.e. $X_1 = \begin{bmatrix} 0 \\ 1 \\ 2 \end{bmatrix}$ is an eigenvector of $\lambda_1 = -5$. Now, we wish to find a solution $X_2 = \begin{bmatrix} a \\ b \\ c \end{bmatrix}$ of $4x - 2y + z = 0$ which is orthogonal to X_1, that is

$$X_2^T X_1 = 0 \Rightarrow [0 \ 1 \ 2] \begin{bmatrix} a \\ b \\ c \end{bmatrix} = 0 \Rightarrow b + 2c = 0$$

and, $\quad 4a - 2b + c = 0 \quad [\because \ x = a, y = b, z = c \text{ is a solution of } 4x - 2y + z = 0]$

Clearly, $a = -5, \ b = -8, \ c = 4$ is a solution of $4a - 2b + c = 0$ and $b + 2c = 0$

Thus, $X_2 = \begin{bmatrix} -5 \\ -8 \\ 4 \end{bmatrix}$ is orthogonal to X_1.

Eigenvectors corresponding to $\lambda_2 = 16$: Let $X = \begin{bmatrix} x \\ y \\ z \end{bmatrix}$ be an eigenvector corresponding to $\lambda_2 = 16$. Then, X is a solution of the homogeneous system

$$(A - 16I)X = O$$

$$\Rightarrow \begin{bmatrix} -5 & -8 & 4 \\ -8 & -17 & -2 \\ 4 & -2 & -20 \end{bmatrix} \begin{bmatrix} x \\ y \\ z \end{bmatrix} = \begin{bmatrix} 0 \\ 0 \\ 0 \end{bmatrix}$$

$$\Rightarrow -5x - 8y + 4z = 0, \ -8x - 17y - 2z = 0, \ 4x - 2y - 20z = 0$$

Clearly, $x = 4, y = -2, z = 1$ is a solution of the above system of equations.

Thus, $X_3 = \begin{bmatrix} 4 \\ -2 \\ 1 \end{bmatrix}$ is an eigenvector for $\lambda_2 = 16$.

As eigenvectors corresponding to distinct eigenvalues of a real symmetric matrix are orthogonal. So, X_3 is orthogonal to both X_1 and X_2.

Thus, X_1, X_2, X_3 form a maximal system of non-zero orthogonal vectors of A.

(iii) Let us now normalize X_1, X_2, X_3 to obtain the orthonormal basis.

We have,

$$\|X_1\| = \sqrt{0+1+4} = \sqrt{5}, \ \|X_2\| = \sqrt{25+64+16} = \sqrt{105},$$
$$\|X_3\| = \sqrt{16+4+1} = \sqrt{21}$$

$$\therefore \quad \widehat{X}_1 = \frac{X_1}{\sqrt{5}}, \ \widehat{X}_2 = \frac{X_2}{\sqrt{105}}, \ \widehat{X}_3 = \frac{X_3}{\sqrt{21}}$$

The matrix P is the matrix whose columns are $\widehat{X}_1, \widehat{X}_2, \widehat{X}_3$.

$$\therefore \quad P = \begin{bmatrix} 0 & \frac{-5}{\sqrt{105}} & \frac{4}{\sqrt{21}} \\ \frac{1}{\sqrt{5}} & \frac{-8}{\sqrt{105}} & \frac{-2}{\sqrt{21}} \\ \frac{2}{\sqrt{5}} & \frac{4}{\sqrt{105}} & \frac{1}{\sqrt{21}} \end{bmatrix} \quad \text{and,} \quad D = P^{-1}AP = \begin{bmatrix} -5 & 0 & 0 \\ 0 & -5 & 0 \\ 0 & 0 & 16 \end{bmatrix}$$

EXERCISE 5.6

1. Let $A = \begin{bmatrix} 3 & -4 \\ 2 & -6 \end{bmatrix}$. Find:

    (i) all eigenvalues and corresponding eigenvectors.

    (ii) matrices P and D such that P is non-singular and $D = P^{-1}AP$ is a diagonal matrix.

2. For each of the following matrices, find all eigenvalues and corresponding linearly independent eigenvectors:

(i) $\begin{bmatrix} 2 & -3 \\ 2 & -5 \end{bmatrix}$ (ii) $\begin{bmatrix} 2 & 4 \\ -1 & 6 \end{bmatrix}$ (iii) $\begin{bmatrix} 1 & -4 \\ 3 & -7 \end{bmatrix}$

When possible, find the non-singular matrix P that diagonalizes the matrix P.

3. For each of the following matrices, find all eigenvalues and a maximum set B of linearly independent eigenvectors:

(i) $\begin{bmatrix} 1 & -3 & 3 \\ 3 & -5 & 3 \\ 6 & -6 & 4 \end{bmatrix}$ (ii) $\begin{bmatrix} 3 & -1 & 1 \\ 7 & -5 & 1 \\ 6 & -6 & 2 \end{bmatrix}$ (iii) $\begin{bmatrix} 1 & 2 & 2 \\ 1 & 2 & -1 \\ -1 & 1 & 4 \end{bmatrix}$

Which matrices can be diagonalized, and why?

4. Let $A = \begin{bmatrix} 7 & 3 \\ 3 & -1 \end{bmatrix}$. Find an orthogonal matrix P such that $D = P^{-1}AP$ is diagonal.

5. For each of the following real symmetric matrices A, find an orthogonal matrix P and a diagonal matrix D such that $D = P^{-1}AP$.

(i) $A = \begin{bmatrix} 5 & 4 \\ 4 & -1 \end{bmatrix}$ (ii) $A = \begin{bmatrix} 4 & -1 \\ -1 & 4 \end{bmatrix}$ (iii) $A = \begin{bmatrix} 7 & 3 \\ 3 & -1 \end{bmatrix}$

6. Let $A = \begin{bmatrix} 4 & 1 & -1 \\ 2 & 5 & -2 \\ 1 & 1 & 2 \end{bmatrix}$.

(i) Find all eigenvalues of A.

(ii) Find a maximum set B of linearly independent eigenvectors of A.

(iii) Is A diagonalizable? If yes, find P such that $D = P^{-1}AP$ diagonal.

7. For each of the following symmetric matrices A, find its eigenvalues, a maximal orthogonal set S of eigenvectors, and an orthogonal matrix P such that $D = P^{-1}AP$ is diagonal:

(i) $A = \begin{bmatrix} 0 & 1 & 1 \\ 1 & 0 & 1 \\ 1 & 1 & 0 \end{bmatrix}$ (ii) $A = \begin{bmatrix} 2 & 2 & 4 \\ 2 & 5 & 8 \\ 4 & 8 & 17 \end{bmatrix}$

8. Let $A = \begin{bmatrix} 2 & -1 \\ -2 & 3 \end{bmatrix}$. Find:

(i) all eigenvalues and corresponding eigenvectors.

(ii) a non-singular matrix P such that $D = P^{-1}AP$ is a diagonal matrix.

(iii) A^8 and $f(A)$, where $f(t) = t^4 - 5t^3 + 7t^2 - 2t + 5$

(iv) Find a matrix B such that $B^2 = A$.

9. Let $t : R^3 \to R^3$ be a linear operator represented in the standard basis by the matrix

$$A = \begin{bmatrix} -9 & 4 & 4 \\ -8 & 3 & 4 \\ -16 & 8 & 7 \end{bmatrix}$$

Prove that t is diagonalizable. Also, find the eigenvalues and corresponding eigenvectors.

10. Prove that the matrix $A = \begin{bmatrix} 1 & 2 \\ 0 & 1 \end{bmatrix}$ is not diagonalizable.

11. Is the matrix $A = \begin{bmatrix} 3 & 1 & -1 \\ 2 & 2 & -1 \\ 2 & 2 & 0 \end{bmatrix}$ similar over the field R to a diagonal matrix? Is A similar, over the field C, to a diagonal matrix?

12. Is the matrix $A = \begin{bmatrix} 6 & -3 & -2 \\ 4 & -1 & -2 \\ 10 & -5 & -3 \end{bmatrix}$ diagonalizable over R? Is A diagonalizable over C?

ANSWERS

1. (i) $2, -5$, (ii) $P = \begin{bmatrix} 4 & 1 \\ 1 & 2 \end{bmatrix}$, $D = \begin{bmatrix} 2 & 0 \\ 0 & -5 \end{bmatrix}$

2. (i) $\lambda_1 = 1$, $X_1 = \begin{bmatrix} 3 \\ 1 \end{bmatrix}$; $\lambda_2 = -4$, $X_2 = \begin{bmatrix} 1 \\ 2 \end{bmatrix}$, $P = \begin{bmatrix} 3 & 1 \\ 1 & 2 \end{bmatrix}$

   (ii) $\lambda = 4$, $X = \begin{bmatrix} 2 \\ 1 \end{bmatrix}$, P does not exist

   (iii) $\lambda_1 = -1$, $X_1 = \begin{bmatrix} 2 \\ 1 \end{bmatrix}$; $\lambda_2 = -5$, $X_2 = \begin{bmatrix} 2 \\ 3 \end{bmatrix}$, $P = \begin{bmatrix} 2 & 2 \\ 1 & 3 \end{bmatrix}$

3. (i) $\lambda_1 = -2$, $X_1 = \begin{bmatrix} 1 \\ 1 \\ 0 \end{bmatrix}$, $X_2 = \begin{bmatrix} 1 \\ 0 \\ -1 \end{bmatrix}$; $\lambda_3 = 4$, $X_3 = \begin{bmatrix} 1 \\ 1 \\ 2 \end{bmatrix}$, $P = \begin{bmatrix} 1 & 1 & 1 \\ 1 & 0 & 1 \\ 0 & -1 & 2 \end{bmatrix}$

   (ii) $\lambda_1 = 2$, $X_1 = \begin{bmatrix} 1 \\ 1 \\ 0 \end{bmatrix}$; $\lambda_2 = -4$, $X_2 = \begin{bmatrix} 0 \\ 1 \\ 1 \end{bmatrix}$

(iii) $\lambda_1 = 3$, $X_1 = \begin{bmatrix} 1 \\ 1 \\ 0 \end{bmatrix}$, $X_2 = \begin{bmatrix} 1 \\ 0 \\ 1 \end{bmatrix}$; $\lambda_2 = 1$, $X_3 = \begin{bmatrix} 2 \\ -1 \\ 1 \end{bmatrix}$, $P = \begin{bmatrix} 1 & 1 & 2 \\ 1 & 0 & -1 \\ 0 & 1 & 1 \end{bmatrix}$

4. $P = \begin{bmatrix} \dfrac{3}{\sqrt{10}} & \dfrac{1}{\sqrt{10}} \\ \dfrac{1}{\sqrt{10}} & \dfrac{-3}{\sqrt{10}} \end{bmatrix}$

5. (i) $P = \begin{bmatrix} \dfrac{2}{\sqrt{5}} & \dfrac{-1}{\sqrt{5}} \\ \dfrac{1}{\sqrt{5}} & \dfrac{2}{\sqrt{5}} \end{bmatrix}$, $D = \begin{bmatrix} 7 & 0 \\ 0 & 3 \end{bmatrix}$ (ii) $P = \begin{bmatrix} \dfrac{1}{\sqrt{2}} & \dfrac{1}{\sqrt{2}} \\ \dfrac{1}{\sqrt{2}} & \dfrac{-1}{\sqrt{2}} \end{bmatrix}$, $D = \begin{bmatrix} 3 & 0 \\ 0 & 5 \end{bmatrix}$

(iii) $P = \begin{bmatrix} \dfrac{3}{\sqrt{10}} & \dfrac{-1}{\sqrt{10}} \\ \dfrac{1}{\sqrt{10}} & \dfrac{3}{\sqrt{10}} \end{bmatrix}$, $D = \begin{bmatrix} 8 & 0 \\ 0 & 2 \end{bmatrix}$

6. (i) $\lambda_1 = 3$, $\lambda_2 = 5$ (ii) $B = \{(1, -1, 0),\ (1, 0, 1),\ (1, 2, 1)\}$

(iii) Yes, $P = \begin{bmatrix} 1 & 1 & 1 \\ -1 & 0 & 2 \\ 0 & 1 & 1 \end{bmatrix}$

7. (i) $\lambda_1 = -1$, $X_1 = \begin{bmatrix} 1 \\ -1 \\ 0 \end{bmatrix}$, $X_2 = \begin{bmatrix} 1 \\ 1 \\ -2 \end{bmatrix}$; $\lambda_2 = 2$, $X_3 = \begin{bmatrix} 1 \\ 1 \\ 1 \end{bmatrix}$, $P = \begin{bmatrix} \dfrac{1}{\sqrt{2}} & \dfrac{1}{\sqrt{6}} & \dfrac{1}{\sqrt{3}} \\ \dfrac{-1}{\sqrt{2}} & \dfrac{1}{\sqrt{6}} & \dfrac{1}{\sqrt{3}} \\ 0 & \dfrac{-2}{\sqrt{6}} & \dfrac{1}{\sqrt{3}} \end{bmatrix}$

(ii) $\lambda_1 = 1$, $X_1 = \begin{bmatrix} 2 \\ 1 \\ -1 \end{bmatrix}$, $X_2 = \begin{bmatrix} 2 \\ -3 \\ 1 \end{bmatrix}$; $\lambda_2 = 22$, $X_3 = \begin{bmatrix} 1 \\ 2 \\ 4 \end{bmatrix}$, $P = \begin{bmatrix} \dfrac{2}{\sqrt{6}} & \dfrac{2}{\sqrt{14}} & \dfrac{1}{\sqrt{21}} \\ \dfrac{1}{\sqrt{6}} & \dfrac{-3}{\sqrt{14}} & \dfrac{2}{\sqrt{21}} \\ \dfrac{-1}{\sqrt{6}} & \dfrac{1}{\sqrt{14}} & \dfrac{4}{\sqrt{21}} \end{bmatrix}$

8. (i) $\lambda_1 = 1$, $X_1 = \begin{bmatrix} 1 \\ 1 \end{bmatrix}$; $\lambda_2 = 4, X_2 = \begin{bmatrix} 1 \\ -2 \end{bmatrix}$ (ii) $P = \begin{bmatrix} 1 & 1 \\ 1 & -2 \end{bmatrix}$

(iii) $A^8 = \begin{bmatrix} 21846 & -43690 \\ -21845 & 43691 \end{bmatrix}$, $f(A) = \begin{bmatrix} 3 & 2 \\ 1 & 1 \end{bmatrix}$ (iv) $B = \begin{bmatrix} \frac{4}{3} & \frac{-2}{3} \\ \frac{-1}{3} & \frac{5}{3} \end{bmatrix}$

9. $\lambda_1 = -1$, $X_1 = \begin{bmatrix} 1 \\ 1 \\ 1 \end{bmatrix}$, $X_2 = \begin{bmatrix} 0 \\ 1 \\ -1 \end{bmatrix}$; $\lambda_2 = 3$, $X_3 = \begin{bmatrix} 1 \\ 1 \\ 2 \end{bmatrix}$

11. No, No   12. No, Yes

5.12 THE MINIMAL POLYNOMIAL

Let A be any square matrix over a field F and $f(\lambda)$ be a polynomial, in indeterminate λ, over F. If $f(A) = O$ (null matrix), then we say that the polynomial $f(\lambda)$ annihilates the matrix A. Let S_A denote the collection of all polynomials $f(\lambda)$, over F, which annihilate the matrix A. Clearly, the set S_A is non-empty as every square matrix A over F is a zero of its characteristic polynomial $\Delta_A(\lambda)$. So, $\Delta_A(\lambda)$ belongs to S_A. A polynomial $f(\lambda)$ over F is called a monic polynomial if the coefficient of highest degree term in $f(\lambda)$ is unity, i.e. leading coefficient is unity. Among all non-zero polynomials over F which annihilate a matrix A, the polynomial which is of lowest degree and is monic is of special interest. It is called the minimal polynomial as defined below.

MINIMAL POLYNOMIAL *Let A be a square matrix over a field F. A monic polynomial $m(\lambda)$ of the lowest degree over field F that annihilates A is called minimal polynomial of matrix A.*

The equation $m_A(A) = O$ is called minimial polynomial equation of matrix A.

REMARK. *If A is $n \times n$ square matrix over field F, then its minimal polynomial $m_A(\lambda)$ and characteristic polynomial $\Delta_A(\lambda)$ both annihilate A, i.e. $m_A(A) = O$ and $\Delta_A(A) = O$. But, $m_A(\lambda)$ is of lowest degree polynomial annihilating A.*

$\therefore$ $\qquad\qquad\qquad$ Degree $M_A(\lambda) \leq$ Degree $\Delta_A(\lambda)$

THEOREM-1 *The minimal polynomial of a square matrix is unique.*

PROOF. Let A be a square matrix over a field F and let $m_A(\lambda)$ be its minimal polynomial of degree n. Then, any non-zero polynomial of degree less than n cannot annihilate matrix A.

Let $\quad f_A(\lambda) = a_0 + a_1\lambda + a_2\lambda^2 + \cdots + a_{n-1}\lambda^{n-1} + \lambda^n$
and, $\quad g_A(\lambda) = b_0 + b_1\lambda + b_2\lambda^2 + \cdots + b_{n-1}\lambda^{n-1} + \lambda^n$

be two minimal polynomials of A. Then, $f_A(\lambda)$ and $g_A(\lambda)$ both annihilate matrix A.

$\therefore \quad f_A(A) = O$ and $g_A(A) = O$

$\Rightarrow \quad a_0 I + a_1 A + a_2 A^2 + \cdots + a_{n-1} A^{n-1} + A^n = O$ \hfill (i)

and, $\quad b_0 I + b_1 A + b_2 A^2 + \cdots + b_{n-1} A^{n-1} + A^n = O$ \hfill (ii)

Subtracting (ii) from (i), we get

$(a_0 - b_0)I + (a_1 - b_1)A + \cdots + (a_{n-1} - b_{n-1})A^{n-1} = O$

$\Rightarrow$ Polynomial $\phi(\lambda) = (a_0 - b_0)I + (a_1 - b_1)\lambda + \cdots + (a_{n-1} - b_{n-1})\lambda^{n-1}$ annihilates matrix A.

But, degree of $\phi(\lambda)$ is $(n-1)$ and no polynomial of degree less than n can annihilate A as $m_A(\lambda)$ is minimal polynomial of degree n. Therefore, $\phi(\lambda)$ must be a zero polynomial.

$\therefore \quad a_0 = b_0, a_1 = b_1, \ldots, a_{n-1} = b_{n-1}$

$\Rightarrow \quad f_A(\lambda) = g_A(\lambda)$

Hence the minimal polynomial of matrix A is unique.

Q.E.D.

THEOREM-2 *Similar matrices have the same minimal polynomial.*

PROOF. Let A and B be two similar matrices. Then, there exists an invertible matrix P such that $B = P^{-1}AP$.

Let $f(t)$ be any polynomial. Then,

$$f(B) = P^{-1}f(A)P \hfill (i)$$

Let $m(\lambda)$ be minimal polynomial of matrix A. Then, $m(\lambda)$ is the monic polynomial of lowest degree such that $m(A) = 0$, i.e. $m(\lambda)$ annihilates matrix A.

From (i), we have

$m(B) = P^{-1}m(A)P$

$\Rightarrow \quad m(B) = P^{-1}OP = O$ \hfill $[\because \quad m(A) = O]$

$\Rightarrow \quad m(\lambda)$ annihilates matrix B

Thus, $m(\lambda)$ is the monic polynomial of lowest degree annihilating matrix B. So, $m(\lambda)$ is the monic polynomial of matrix B.

Hence, A and B have the same minimal polynomial.

Q.E.D.

THEOREM-3 *The minimal polynomial of a matrix A is a divisor of every polynomial that annihilates the matrix A.*

PROOF. Let $m_A(\lambda)$ be the minimal polynomial of matrix A and $f(\lambda)$ be a polynomial that annihilates A.

By division algorithm there exist two polynomials $g(\lambda)$ and $h(\lambda)$ such that
$$f(\lambda) = m_A(\lambda) g(\lambda) + h(\lambda) \qquad (i)$$
where either $\deg(h(\lambda)) < \deg(m_A(\lambda))$ or $h(\lambda) = 0$.

From (i), we have
$$f(A) = m_A(A) g(A) + h(A)$$
$\Rightarrow \quad O = 0 \times g(A) + h(A) \quad [\because m_A(\lambda) \text{ and } f(\lambda) \text{ both annihilate } A. \therefore m_A(A) = O, f(A) = O]$
$\Rightarrow \quad h(A) = O$
$\Rightarrow \quad h(\lambda)$ annihilates matrix A.

If $h(\lambda) \neq 0$, then it is a non-zero polynomial of degree less then the degree of $m_A(\lambda)$ and thus we arrive at a contradiction that $m_A(\lambda)$ is the minimal polynomial of A. Therefore, $h(\lambda) = 0$. Substituting $h(\lambda) = 0$ in (i), we get
$$f(\lambda) = m_A(\lambda) g(\lambda) \quad \Rightarrow \quad m_A(\lambda) \text{ is a divisor of } f(\lambda).$$
Q.E.D.

COROLLARY. *The minimal polynomial of a matrix is a divisor of its characteristic polynomial.*

PROOF. Let $m_A(\lambda)$ and $\Delta_A(\lambda)$ be minimal and characteristic polynomials respectively of a matrix A. By Cayley-Hamilton theorem, we have
$$\Delta_A(A) = 0$$
$\Rightarrow \quad \Delta_A(\lambda)$ annihilates matrix A
$\Rightarrow \quad m_A(\lambda)$ is a divisor of $\Delta_A(\lambda)$ \hfill [By Theorem 3]
\hfill Q.E.D.

THEOREM-4 *Let $m_A(\lambda)$ be the minimal polynomial of an $n \times n$ matrix A over a field F and $\Delta_A(\lambda)$ be the characteristic polynomial of A, then $\Delta_A(\lambda)$ divides $\{m_A(\lambda)\}^n$.*

PROOF. Let $m_A(\lambda) = \lambda^r + a_1 \lambda^{r-1} + a_2 \lambda^{r-2} + \cdots + a_{r-1} \lambda + a_r$ be the minimal polynomial of matrix A and the characteristic polynomial of A is $\Delta_A(\lambda) = |A - \lambda I|$.

Let us define matrices $B_0, B_1, \ldots, B_{r-1}$ as follows:

$B_0 = I$ \qquad\qquad\qquad\qquad so, \qquad $I = B_0$

$B_1 = A + a_1 I$ \qquad\qquad\quad so, \qquad $a_1 I = B_1 - A = B_1 - AI = B_1 - AB_0$

$B_2 = A^2 + a_1 A + a_2 I$ \qquad so, \qquad $a_2 I = B_2 - A(A + a_1 I) = B_2 - AB_1$

$B_3 = A^3 + a_1 A^2 + a_2 A + a_3 I$ \qquad so, \qquad $a_3 I = B_3 - A(A^2 + a_1 A + a_2 I) = B_3 - AB_2$

$\vdots \quad \vdots \quad \vdots \quad \vdots \quad \vdots$ \hfill $\vdots$

$B_{r-1} = A^{r-1} + a_1 A^{r-2} + a_2 A^{r-3}$
$\qquad + \cdots + a_{r-2} A + a_{r-1} I$ \qquad so, \qquad $a_{r-1} I = B_{r-1} - AB_{r-2}$

Now,
$$B_{r-1} = A^{r-1} + a_1 A^{r-2} + a_2 A^{r-3} + \cdots + a_{r-2} A + a_{r-1} I$$
$\Rightarrow \quad AB_{r-1} = A^r + a_1 A^{r-1} + a_2 A^{r-2} + \cdots + a_{r-2} A^2 + a_{r-1} A$

$\Rightarrow \quad -AB_{r-1} = a_r I - \left(A^r + a_1 A^{r-1} + a_2 A^{r-2} + \cdots + a_{r-2} A^2 + a_{r-1} A + a_r I\right)$

$\Rightarrow \quad -AB_{r-1} = a_r I - m_A(A)$

$\Rightarrow \quad -AB_{r-1} = a_r I - O \hfill [\because m_A(A) = O]$

$\Rightarrow \quad a_r I = -AB_{r-1}$

Let $B(\lambda) = \lambda^{r-1} B_0 + \lambda^{r-2} B_1 + \cdots + \lambda B_{r-2} + B_{r-1}$. Then,

$(A - \lambda I) B(\lambda) = (A - \lambda I)(\lambda^{r-1} B_0 + \lambda^{r-2} B_1 + \cdots + \lambda B_{r-2} + B_{r-1})$

$\Rightarrow (A - \lambda I) B(\lambda) = (\lambda^{r-1} A B_0 + \lambda^{r-2} A B_1 + \cdots + \lambda A B_{r-2} + A B_{r-1})$
$\qquad \qquad \qquad \qquad - (\lambda^r B_0 + \lambda^{r-1} B_1 + \cdots + \lambda^2 B_{r-2} + \lambda B_{r-1})$

$\Rightarrow (A - \lambda I) B(\lambda) = -\lambda^r B_0 - \lambda^{r-1}(B_1 - A B_0) - \lambda^{r-2}(B_2 - A B_1) - \cdots - \lambda(B_{r-1} - A B_{r-2}) - A B_{r-1}$

$\Rightarrow (A - \lambda I) B(\lambda) = -\lambda^r I - \lambda^{r-1} a_1 I - \lambda^{r-2} a_2 I - \cdots - \lambda a_{r-1} I - a_r I$

$\Rightarrow (A - \lambda I) B(\lambda) = -(\lambda^r + a_1 \lambda^{r-1} + a_2 \lambda^{r-2} + \cdots + a_{r-1} \lambda + a_r) I$

$\Rightarrow (A - \lambda I) B(\lambda) = -m_A(\lambda) I$

$\Rightarrow |A - \lambda I| \ |B(\lambda)| = (-1)^n \{m_A(\lambda)\}^n$

$\Rightarrow \Delta_A(\lambda) \ |B(\lambda)| = (-1)^n \{m_A(\lambda)\}^n$

$\Rightarrow \{m_A(\lambda)\}^n = \Delta_A(\lambda)(-1)^n |B(\lambda)|$

$\Rightarrow \Delta_A(\lambda)$ is a divisor of $\{m_A(\lambda)\}^n$

Q.E.D.

THEOREM-5 *The characteristic polynomial $\Delta_A(\lambda)$ and the minimal polynomial $m_A(\lambda)$ of a matrix A have the same irreducible factors.*

PROOF. Let A be $n \times n$ matrix over a field F and let $f(\lambda)$ be an irreducible polynomial. If $f(\lambda)$ is an irreducible factor of the minimal polynomial $m_A(\lambda)$. Then, $f(\lambda)$ divides the minimal polynomial $m_A(\lambda)$.

$\Rightarrow \quad f(\lambda)$ divides $\Delta_A(\lambda) \hfill [\because \ m_A(\lambda) \text{ divides } \Delta_A(\lambda)]$

On the other hand, if $f(\lambda)$ is an irreducible factor of characteristic polynomial $\Delta_A(\lambda)$, then

$\quad f(\lambda)$ divides $\{m_A(\lambda)\}^n \hfill$ [By Theorem 4]

$\Rightarrow \quad f(\lambda)$ divides $m_A(\lambda) \hfill [\because f(\lambda) \text{ is irreducible}]$

Thus, the minimal polynomial $m_A(\lambda)$ and the characteristic polynomial $\Delta_A(\lambda)$ have the same irreducible factors.

Q.E.D.

REMARK-1 *The above theorem does not say that for a square matrix A, $m_A(\lambda) = \Delta_A(\lambda)$. It only means that any irreducible factor of one must divides the other. As a linear factor is irreducible, therefore $m_A(\lambda)$ and $\Delta_A(\lambda)$ have the same linear factors. Consequently, $m_A(\lambda)$ and $\Delta_A(\lambda)$ have the same roots and hence a scalar λ is an eigenvalue of A iff it is a root of the minimal polynomial $m_A(\lambda)$.*

REMARK-2 *If the characteristic polynomial $\Delta_A(\lambda)$ of an $n \times n$ matrix over a field F has n distinct linear factors, then its minimimal polynomial of A has same n distinct linear factors. In fact, $m_A(\lambda) = (-1)^n \Delta_A(\lambda)$.*

THEOREM-6 *Let $m_A(\lambda)$ be the minimal polynomial of an $n \times n$ matrix A over field F. Then, A is invertible iff constant term in $m_A(\lambda)$ is non-zero.*

SOLUTION Let $m_A(\lambda) = a_0 + a_1\lambda + a_2\lambda^2 + \cdots + a_{r-1}\lambda^{r-1} + \lambda^r$ be the minimal polynomial of matrix A. Then,

$$m_A(A) = O \qquad [\because m_A(\lambda) \text{ annihilates matrix } A]$$
$$\Rightarrow a_0 I + a_1 A + a_2 A^2 + \cdots + a_{r-1} A^{r-1} + A^r = O$$

First, let $a_0 \neq 0$. Then,
$$a_0 I + a_1 A + a_2 A^2 + \cdots + a_{r-1} A^{r-1} + A^r = O$$
$$\Rightarrow a_0 I = -(a_1 A + a_2 A^2 + \cdots + a_{r-1} A^{r-1} + A^r)$$
$$\Rightarrow a_0 I = -(a_1 I + a_2 A + \cdots + a_{r-1} A^{r-2} + A^{r-1}) A$$
$$\Rightarrow -\frac{1}{a_0}(a_1 I + a_2 A + \cdots + a_{r-1} A^{r-2} + A^{r-1}) A = I$$
$$\Rightarrow A \text{ is invertible and } A^{-1} = -\frac{1}{a_0}(a_1 I + a_2 A + \cdots + a_{r-1} A^{r-2} + A^{r-1})$$

Conversely, let A be invertible matrix. Then, A^{-1} exists.
Now, $m_A(A) = O$
$$\Rightarrow a_0 I + a_1 A + a_2 A^2 + \cdots + a_{r-1} A^{r-1} + A^r = O$$
$$\Rightarrow (a_0 I + a_1 A + a_2 A^2 + \cdots + a_{r-1} A^{r-1} + A^r) A^{-1} = O A^{-1}$$
$$\Rightarrow a_0 A^{-1} + a_1 I + a_2 A + \cdots + a_{r-1} A^{r-2} + A^{r-1} = O$$
$$\Rightarrow a_0 A^{-1} = -(a_1 I + a_2 A + \cdots + a_{r-1} A^{r-2} + A^{r-1})$$
$$\Rightarrow A^{-1} = -\frac{1}{a_0}(a_1 I + a_2 A + \cdots + a_{r-1} A^{r-2} + A^{r-1})$$
$$\Rightarrow a_0 \neq 0 \qquad [\because A^{-1} \text{ exists}]$$

Q.E.D.

5.12.1 MINIMAL POLYNOMIAL OF A LINEAR OPERATOR

Let $V(F)$ be a vector space and $t : V \to V$ be a linear operator on V. The minimal polynomial $m(\lambda)$ of t is defined to be the monic polynomial of lowest degree which annihilates t, that is, t is a root of $m(t)$.

If A is any matrix representation of t, then for any polynomial f, we have
$$f(t) = 0 \quad \text{iff} \quad f(A) = 0$$
Therefore, t and A have the same minimal polynomials.

It follows from the above discussion that all theorems on minimal polynomials of a matrix (discussed in section 5.12) also hold for the minimal polynomial of a linear operator. That is, we have the following theorem on linear operators.

THEOREM-1 *The minimal polynomial of a linear operator is unique.*

THEOREM-2 *The minimal polynomial of a linear operator is a divisor of every polynomial that annihilates t.*

THEOREM-3 *The minimal polynomial of a linear operator is a divisor of its characteristic polynomial.*

THEOREM-4 *The characteristic and minimal polynomials of linear operator have the same irreducible factors.*

THEOREM-5 *A scalar λ is an eigenvalue of a linear operator t iff λ is a root of the minimal polynomial of t.*

THEOREM-6 *Let $V(F)$ be a finite dimensional vector space and t be a diagonalizable linear operator on V with $\lambda_1, \lambda_2, \ldots, \lambda_k$ as distinct eigenvalues. Then, the minimal polynomial $m(\lambda)$ of λ is the polynomial $m(\lambda) = (\lambda - \lambda_1)(\lambda - \lambda_2) \ldots (\lambda - \lambda_k)$.*

PROOF. We know that each eigenvalue of a linear operator t is a root of its minimal polynomial. Therefore, $\lambda - \lambda_1, \lambda - \lambda_2, \ldots, \lambda - \lambda_k$ are factors of the minimal polynomial of t.

Let $m(\lambda) = (\lambda - \lambda_1)(\lambda - \lambda_2) \ldots (\lambda - \lambda_k)$

We shall now show that $m(\lambda)$ is the minimal polynomial of t. For this it is sufficient to show that $m(\lambda)$ annihilates t, i.e. $m(t) = 0$.

Let $v_1, v_2, \ldots, v_k$ be eigenvectors of t corresponding to eigenvalues $\lambda_1, \lambda_2, \ldots, \lambda_k$ respectively. Then,
$$t(v_i) = \lambda_i v_i \quad \text{for all} \quad i \in \underline{k} \tag{i}$$

Now, $\quad m(t) = (t - \lambda_1)(t - \lambda_2) \ldots (t - \lambda_k)$

$\Rightarrow \quad (m(t))(v_i) = ((t - \lambda_1)(t - \lambda_2) \ldots (t - \lambda_k))(v_i)$

$\Rightarrow \quad (m(t))(v_i) = ((t - \lambda_1)(t - \lambda_2) \ldots (t - \lambda_k))(t - \lambda_i)(v_i)$

$\Rightarrow \quad (m(t))(v_i) = ((t - \lambda_1)(t - \lambda_2) \ldots (t - \lambda_k))(t(v_i) - \lambda_i v_i)$

$\Rightarrow \quad (m(t))(v_i) = ((t - \lambda_1)(t - \lambda_2) \ldots (t - \lambda_k))(0_V) \quad$ [Using (i)]

$\Rightarrow \quad (m(t))(v_i) = 0_V \quad \text{for all } i \in \underline{k}$

We know that distinct eigenvectors corresponding to distinct eigenvalues of a linear operator are linearly independent. Therefore, $v_1, v_2, \ldots, v_k$ are linearly independent and hence form a part of a basis of V. So, any vector $v \in V$ can be written as a linear combination of eigenvectors $v_1, v_2, \ldots, v_k$. But,

$$(m(t))(v_i) = 0_V \quad \text{for all eigenvectors } v_i$$
$$\therefore \quad (m(t))(v) = 0_V \quad \text{for all } v \in V$$
$$\Rightarrow \quad m(t) = 0$$
$$\Rightarrow \quad m(\lambda) \text{ annihilates } t.$$

Hence, $m(\lambda)$ is the minimal polynomial of t.

Q.E.D.

COROLLARY. *Let $V(F)$ be a finite dimensional vector space and t be a linear operator on V such that the characteristic equation of t has distinct root $\lambda_1, \lambda_2, \ldots, \lambda_n$ then $m(\lambda) = (\lambda - \lambda_1)(\lambda - \lambda_2) \ldots (\lambda - \lambda_n)$ is the minimal polynomial of t.*

PROOF. Since the characteristic equation of t has all distinct roots. Therefore, t has distinct eigenvalues $\lambda_1, \lambda_2, \ldots, \lambda_n$. Consequently, it is diagonalizable and hence

$$m(\lambda) = (\lambda - \lambda_1)(\lambda - \lambda_2) \ldots (\lambda - \lambda_n)$$

Q.E.D.

ILLUSTRATIVE EXAMPLES

EXAMPLE-1 *Find the minimal polynomial $m_A(\lambda)$ of the matrix $A = \begin{bmatrix} 5 & 1 \\ 3 & 7 \end{bmatrix}$.*

SOLUTION We have,

$$tr(A) = 5 + 7 = 12 \quad \text{and} \quad |A| = 35 - 3 = 32.$$

So, the characteristic polynomial of A is

$$\Delta_A(\lambda) = (-1)^2(\lambda^2 - 12\lambda + 32) = (\lambda - 4)(\lambda - 8)$$

Since $\Delta_A(\lambda)$ has distinct linear factors. Therefore,

$$m_A(\lambda) = (-1)^2 \Delta_A(\lambda) = \Delta_A(\lambda) = (\lambda - 4)(\lambda - 8)$$

EXAMPLE-2 *Find the minimal polynomial $m_A(\lambda)$ of the matrix $A = \begin{bmatrix} 1 & 0 & 0 \\ 2 & 2 & 0 \\ 3 & 3 & 3 \end{bmatrix}$*

SOLUTION Clearly, A is a triangular matrix with diagonal elements 1, 2 and 3. So, its characteristic polynomial $\Delta_A(\lambda)$ is given by

$$\Delta_A(\lambda) = (-1)^3(\lambda-1)(\lambda-2)(\lambda-3) = -(\lambda-1)(\lambda-2)(\lambda-3)$$

Since $\Delta_A(\lambda)$ has distinct linear factors.

$\therefore \qquad m_A(\lambda) = (-1)^3 \Delta_A(\lambda) = -\Delta_A(\lambda) = (\lambda-1)(\lambda-2)(\lambda-3)$

EXAMPLE-3 Find the minimal polynomial $m_A(\lambda)$ of the matrix $A = \begin{bmatrix} 2 & 1 \\ -1 & 4 \end{bmatrix}$.

SOLUTION We have,

$$tr(A) = 6 \text{ and } |A| = 9$$

So, the characteristic polynomial of A is given by

$$\Delta_A(\lambda) = (-1)^2(\lambda^2 - 6\lambda + 9) = (\lambda-3)^2$$

Since $m_A(\lambda)$ is a divisor of $\Delta_A(\lambda)$. Therefore, the minimal polynomial of A is

$$f(\lambda) = (\lambda-3) \text{ or, } g(\lambda) = (\lambda-3)^2$$

We shall now see which one of these polynomials annihilates A.

Clearly, $g(\lambda) = \Delta_A(\lambda)$

$\therefore \qquad g(A) = \Delta_A(A) = O$ $\qquad [\because \Delta_A(A) = O]$

$\Rightarrow \qquad g$ annihilates A

But, $\quad f(A) = A - 3I = \begin{bmatrix} 2 & 1 \\ -1 & 4 \end{bmatrix} - \begin{bmatrix} 3 & 0 \\ 0 & 3 \end{bmatrix} = \begin{bmatrix} -1 & 1 \\ -1 & 1 \end{bmatrix} \neq O$

$\therefore \qquad m_A(\lambda) = (-1)^2 g(\lambda) = \Delta_A(\lambda) = (\lambda-3)^2$

EXAMPLE-4 Find the minimal polynomial $m_A(\lambda)$ of the matrix $A = \begin{bmatrix} 2 & 2 & -5 \\ 3 & 7 & -15 \\ 1 & 2 & -4 \end{bmatrix}$.

SOLUTION We have,

$$tr(A) = 2 + 7 - 4 = 5, \; A_{11} + A_{22} + A_{33} = 2 - 3 + 8 = 7 \text{ and } |A| = 3.$$

So, the characteristic polynomial of A is

$$\Delta_A(\lambda) = (-1)^3(\lambda^3 - 5\lambda^2 + 7\lambda - 3)$$

$\Rightarrow \qquad \Delta_A(\lambda) = -(\lambda-1)^2(\lambda-3)$

The minimal polynomial $m_A(\lambda)$ is a divisor of characteristic polynomial $\Delta_A(\lambda)$ and each irreducible factor of $\Delta_A(\lambda)$ must also be a factor of $m_A(\lambda)$. Thus, $m_A(\lambda)$ is exactly one of the following:

$f(\lambda) = (\lambda-1)(\lambda-3) \text{ or, } g(\lambda) = (\lambda-1)(\lambda-3)^2$ $\quad \begin{bmatrix} \text{The minus sign is avoided to make the} \\ \text{coefficient of highest degree term unity} \end{bmatrix}$

Let us see which one of the above polynomials annihilates A.

Clearly, $g(\lambda) = -\Delta_A(\lambda)$

$\Rightarrow \quad g(A) = -\Delta_A(A) = O$ [By Cayley-Hamilton Theorem]

Now, $f(A) = (A-I)(A-3I) = \begin{bmatrix} 1 & 2 & -5 \\ 3 & 6 & -15 \\ 1 & 2 & -5 \end{bmatrix} \begin{bmatrix} 1 & 2 & -5 \\ 3 & 4 & -15 \\ 1 & 2 & -7 \end{bmatrix} = \begin{bmatrix} 0 & 0 & 0 \\ 0 & 0 & 0 \\ 0 & 0 & 0 \end{bmatrix}$

Clearly, $f(\lambda)$ and $g(\lambda)$ both annihilate A but $f(\lambda)$ is of lower degree.

Hence, $m_A(\lambda) = f(\lambda) = (\lambda - 1)(\lambda - 3)$.

EXAMPLE-5 Find the minimal polynomial $m_A(\lambda)$ of the matrix $A = \begin{bmatrix} 3 & -2 & 2 \\ 4 & -4 & 6 \\ 2 & -3 & 5 \end{bmatrix}$

SOLUTION We have,

$tr(A) = 3 - 4 + 5 = 4$, $A_{11} + A_{22} + A_{33} = -2 + 11 - 4 = 5$ and $|A| = 2$

So, the characteristic polynomial of A is

$\Delta_A(\lambda) = (-1)^3(\lambda^3 - 4\lambda^2 + 5\lambda - 2) = -(\lambda - 2)(\lambda - 1)^2$

The minimal polynomial $m_A(\lambda)$ must divide $\Delta_A(\lambda)$ and each irreducible factor of $\Delta_A(\lambda)$ must be a factor of $m_A(\lambda)$. Also, coefficient of highest degree term in $m_A(\lambda)$ must be unity. Thus, $m_A(\lambda)$ must be exactly one of the following

$f(\lambda) = (\lambda - 2)(\lambda - 1)$ or $g(\lambda) = (\lambda - 2)(\lambda - 1)^2$

Clearly, $g(A) = -\Delta_A(A) = O$ [By Cayley-Hamilton Theorem]

But,

$f(A) = (A - 2I)(A - I) = \begin{bmatrix} 1 & -2 & 2 \\ 4 & -6 & 6 \\ 2 & -3 & 3 \end{bmatrix} \begin{bmatrix} 2 & -2 & 2 \\ 4 & -5 & 6 \\ 2 & -3 & 4 \end{bmatrix} = \begin{bmatrix} -2 & 2 & -2 \\ -4 & 4 & -4 \\ -2 & 2 & -2 \end{bmatrix} \neq O$

$\therefore \quad m_A(\lambda) \neq f(\lambda)$.

Hence, $m_A(\lambda) = g(\lambda) = (\lambda - 2)(\lambda - 1)^2$

EXERCISE 5.7

1. Find the minimal polynomials of the identity matrix and null matrix.

2. Find the minimal polynomial of the matrix $A = \begin{bmatrix} 1 & -2 \\ 0 & 4 \end{bmatrix}$.

3. Find the minimal polynomial of the matrix $A = \begin{bmatrix} -3 & 1 \\ 7 & 3 \end{bmatrix}$.

4. Find the minimal polynomial of each of the following matrices:

   (i) $A = \begin{bmatrix} 1 & -1 \\ 1 & 3 \end{bmatrix}$ (ii) $A = \begin{bmatrix} 1 & 2 & -1 \\ 0 & 2 & 5 \\ 0 & 0 & 3 \end{bmatrix}$ (iii) $A = \begin{bmatrix} 4 & -2 & 2 \\ 6 & -3 & 4 \\ 3 & -2 & 3 \end{bmatrix}$

5. Let $A = \begin{bmatrix} 1 & 1 & 0 \\ 0 & 2 & 0 \\ 0 & 0 & 1 \end{bmatrix}$ and $B = \begin{bmatrix} 2 & 0 & 0 \\ 0 & 2 & 2 \\ 0 & 0 & 1 \end{bmatrix}$. Show that A and B have different characteristic polynomials, but have the same minimal polynomial.

6. Find the characteristic and minimal polynomials of each of the following matrices:

   (i) $A = \begin{bmatrix} 3 & 1 & -1 \\ 2 & 4 & -2 \\ -1 & -1 & 3 \end{bmatrix}$ (ii) $B = \begin{bmatrix} 3 & 2 & -1 \\ 3 & 8 & -3 \\ 3 & 6 & -1 \end{bmatrix}$

7. Show that a matrix A and its transpose A^T have the same minimal polynomial.

8. Show that A is a scalar matrix kI iff the minimal polynomial of A is $m_A(\lambda) = \lambda - k$.

9. If $f(\lambda)$ is an irreducible monic polynomial such that $f(A) = O$ for a matrix A, show that $f(\lambda)$ is the minimal polynomial of A.

10. Let $V(F)$ be a finite dimensional vector space. Find the minimal polynomials for the identity operator and zero operator on V.

ANSWERS

1. $m_I(\lambda) = \lambda - 1$, $m_O(\lambda) = O$ 2. $m_A(\lambda) = \lambda^2 - 5\lambda + 6$ 3. $m_A(\lambda) = \lambda^2 - 16$

4. (i) $m_A(\lambda) = (\lambda - 2)^2$,   (ii) $m_A(\lambda) = \lambda^3 - 6\lambda^2 + 11\lambda - 6$,   (iii) $m_A(\lambda) = \lambda^2 - 3\lambda + 2$

6. (i) $\Delta_A(\lambda) = m_A(\lambda) = (\lambda - 2)^2(\lambda - 6)$,
   (ii) $\Delta_A(\lambda) = (\lambda - 2)^2(\lambda - 6)$, $m_A(\lambda) = (\lambda - 2)(\lambda - 6)$

10. $\lambda - 1$, 0

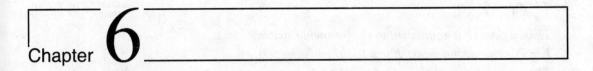

Inner Product Spaces

6.1 INTRODUCTION

Uptill now we have been studying vector spaces over an arbitrary field F. In this chapter, we restrict F to be the real field R and in such a case V is called a real vector space. In the study of arbitrary vector spaces the concepts of length of a vector, angle between two vectors, orthogonality of vectors did not appear. In this chapter, we shall do the same by introducing an additional structure on a real vector space V to obtain an inner product space so that these concepts are defined. In fact, we shall do this by defining a certain type of scalar-valued function from $V \times V$ to R, which will be known as inner product on V. An inner product on a vector space V is a function with properties similar to those of the dot product in R^3 which helps us to know the relationship between linear algebra and geometry.

6.2 INNER PRODUCT SPACES

Let us begin with the definition of inner product.

INNER PRODUCT *Let V be a real vector space. An inner product on V is a function $\langle , \rangle : V \times V \to R$ which assigns each ordered pair $(u,v) \in V \times V$ to a real number $\langle u, v \rangle$ in such a way that following axioms hold:*

I1 *Linearity:* $\langle au_1 + bu_2, v \rangle = a\langle u_1, v \rangle + b\langle u_2, v \rangle$ *for all $u_1, u_2, v \in V$ and $a, b, \in R$.*
I2 *Symmetry:* $\langle u, v \rangle = \langle v, u \rangle$ *for all $u, v \in V$.*
I3 *Positive Definiteness:* $\langle u, u \rangle \geq 0$ *for all $u \in V$ and $\langle u, u \rangle = 0$ if and only if $u = 0_V$*

INNER PRODUCT SPACE *A vector space equipped with an inner product is called an inner product space.*

REMARK. *Using axiom I1, we have*

$$\langle 0_V, 0_V \rangle = \langle 0u, 0_V \rangle = 0\langle u, 0_V \rangle \quad \text{for all } u \in V$$
$$\Rightarrow \quad \langle 0_V, 0_V \rangle = 0 \qquad\qquad [\because 0\langle u, 0_V \rangle = 0]$$

Thus, axiom I3 is equivalent to the following axiom:
I'3: Positive definiteness: If $u \neq 0_V$, then $\langle u, u \rangle > 0$.
Thus, a function satisfying $I1, I2$ and $I'3$ is an inner product.

In the definition of inner product axiom $I1$ states that an inner product function is linear in first position. Using axioms $I1$ and $I2$, we have

$$\langle u, cv_1 + dv_2 \rangle = \langle cv_1 + dv_2, u \rangle \qquad\qquad \text{[Using } I2\text{]}$$
$$\Rightarrow \quad \langle u, cv_1 + dv_2 \rangle = c\langle v_1, u \rangle + d\langle v_2, u \rangle \qquad\qquad \text{[Using } I1\text{]}$$

This means that the inner product function is linear in its second position.

Using the linearity (in both positions) of inner product function and the principle of induction, we obtain the following general formula:

$$\left\langle \sum_i a_i u_i, \sum_j b_j v_j \right\rangle = \sum_i \sum_j a_i b_j \langle u_i, v_j \rangle$$

That is, an inner product of linear combinations of vectors is equal to a linear combination of the inner product of the vectors.

ILLUSTRATIVE EXAMPLES

EXAMPLE-1 *Consider the vector space R^n. Prove that R^n is an inner product space with inner product defined by*
$$\langle u, v \rangle = a_1 b_1 + a_2 b_2 + \cdots + a_n b_n,$$
where $u = (a_1, a_2, \ldots, a_n)$ and $v = (b_1, b_2, \ldots, b_n)$.

SOLUTION We observe the following properties of the product defined above.

I1 *Linearity:* Let $u = (a_1, a_2, \ldots, a_n), v = (b_1, b_2, \ldots, b_n), w = (c_1, c_2, \ldots, c_n) \in R^n$ and $\alpha, \beta \in R$. Then,

$$\alpha u + \beta v = (\alpha a_1 + \beta b_1, \alpha a_2 + \beta b_2, \ldots, \alpha a_n + \beta b_n)$$
$$\therefore \quad \langle \alpha u + \beta v, w \rangle = (\alpha a_1 + \beta b_1) c_1 + (\alpha a_2 + \beta b_2) c_2 + \cdots + (\alpha a_n + \beta b_n) c_n$$
$$\Rightarrow \quad \langle \alpha u + \beta v, w \rangle = \{(\alpha a_1) c_1 + (\alpha a_2) c_2 + \ldots (\alpha a_n) c_n\} + \{(\beta b_1) c_1 + (\beta b_2) c_2 + \ldots + (\beta b_n) c_n\}$$
$$\Rightarrow \quad \langle \alpha u + \beta v, w \rangle = \alpha (a_1 c_1 + a_2 c_2 + \cdots + a_n c_n) + \beta (b_1 c_1 + b_2 c_2 + \cdots + b_n c_n)$$
$$\Rightarrow \quad \langle \alpha u + \beta v, w \rangle = \alpha \langle u, w \rangle + \beta \langle v, w \rangle$$

I2 *Symmetry:* Let $u = (a_1, a_2, \ldots, a_n), v = (b_1, b_2, \ldots, b_n) \in R^n$. Then,

$$\langle u, v \rangle = a_1 b_1 + a_2 b_2 + \cdots + a_n b_n$$

$\Rightarrow \quad \langle u, v \rangle = b_1 a_1 + b_2 a_2 + \cdots + b_n a_n$

$\Rightarrow \quad \langle u, v \rangle = \langle v, u \rangle$

I3 *Positive Definiteness:* For any $u = (a_1, a_2, \ldots, a_n) \in R^n, u \neq 0$, we have

$$\langle u, u \rangle = a_1^2 + a_2^2 + \cdots + a_n^2$$

$\Rightarrow \quad \langle u, u \rangle > 0 \qquad\qquad [\because \ u \neq 0 \ \therefore \ a_i \neq 0 \text{ for atleast one } a_i]$

Thus, the given function is an inner product on R^n.

Hence, R^n is an inner product space for the defined product.

REMARK-1 *The inner product given in the above example is called the standard inner product on R^n and the resulting inner product space is called a Euclidean n-space.*

REMARK-2 *Generally vectors in R^n are represented by column vectors, that is, by $n \times 1$ column matrices. In such a case, the standard inner product is defined by the formula:*

$$\langle u, v \rangle = u^T v.$$

EXAMPLE-2 *Prove that R^2 is an inner product space with an inner product defined by*

$$\langle u, v \rangle = a_1 b_1 - a_2 b_1 - a_1 b_2 + 2 a_2 b_2, \text{ where } u = (a_1, a_2), \quad v = (b_1, b_2) \in R^2$$

SOLUTION Clearly, R^2 is a vector space over R.

We observe the following properties of the function defined above.

I1 *Linearity:* For any $u = (a_1, a_2), v = (b_1, b_2), w = (c_1, c_2) \in R^2$ and $\alpha, \beta \in R$, we have

$\alpha u + \beta v = (\alpha a_1 + \beta b_1, \alpha a_2 + \beta b_2)$

$\langle \alpha u + \beta v, w \rangle = (\alpha a_1 + \beta b_1) c_1 - (\alpha a_2 + \beta b_2) c_1 - (\alpha a_1 + \beta b_1) c_2 + 2(\alpha a_2 + \beta b_2) c_2$

$\Rightarrow \quad \langle \alpha u + \beta v, w \rangle = \alpha(a_1 c_1 - a_2 c_1 - a_1 c_2 + 2 a_2 c_2) + \beta(b_1 c_1 - b_2 c_1 - b_1 c_2 + 2 b_2 c_2)$

$\Rightarrow \quad \langle \alpha u + \beta v, w \rangle = \alpha \langle u, w \rangle + \beta \langle v, w \rangle$

I2 *Symmetry:* For any $u = (a_1, a_2), v = (b_1, b_2) \in R^2$, we have

$\langle u, v \rangle = a_1 b_1 - a_2 b_1 - a_1 b_2 + 2 a_2 b_2$

$\Rightarrow \quad \langle u, v \rangle = b_1 a_1 - b_2 a_1 - b_1 a_2 + 2 b_2 a_2$

$\Rightarrow \quad \langle u, v \rangle = \langle v, u \rangle$

I3 *Positive Definiteness:* For any non-zero $u = (a_1, a_2) \in R^2$, we have

$\langle u, u \rangle = a_1 a_1 - a_2 a_1 - a_1 a_2 + 2 a_2 a_2$

$\Rightarrow \quad \langle u, u \rangle = a_1^2 - 2 a_1 a_2 + 2 a_2^2 = (a_1 - a_2)^2 + a_2^2 > 0 \qquad [\because u \neq 0 \Rightarrow a_1 \neq 0 \text{ or } a_2 \neq 0]$

Thus, the function defined above is an inner product on R^2.

Hence, R^2 is an inner product space for the defined product.

EXAMPLE-3 *Consider the vector space $C[a,b]$ of all continuous functions defined on the closed interval $[a,b]$. Prove that the following defines an inner product on $C[a,b]$*

$$\langle f,g \rangle = \int_a^b f(t)g(t), \text{where } f(t) \text{ and } g(t) \text{ are functions in } C[a,b].$$

SOLUTION We observe the following properties of the function defined above.

I1 *Linearity:* For any $f, g, h \in C[a,b]$ and $\alpha, \beta \in R$, we have

$$\langle \alpha f + \beta g, h \rangle = \int_a^b (\alpha f + \beta g)(t) h(t) dt$$

$$\Rightarrow \quad \langle \alpha f + \beta g, h \rangle = \int_a^b \{\alpha f(t) + \beta g(t)\} h(t) dt$$

$$\Rightarrow \quad \langle \alpha f + \beta g, h \rangle = \int_a^b \{\alpha f(t)h(t) + \beta g(t)h(t)\} dt$$

$$\Rightarrow \quad \langle \alpha f + \beta g, h \rangle = \alpha \int_a^b f(t)h(t) dt + \beta \int_a^b g(t)h(t) dt$$

$$\Rightarrow \quad \langle \alpha f + \beta g, h \rangle = \alpha \langle f, h \rangle + \beta \langle g, h \rangle$$

I2 *Symmetry:* For any $f, g \in C[a,b]$, we have

$$\langle f, g \rangle = \int_a^b f(t)g(t)dt = \int_a^b g(t)f(t)dt = \langle g, f \rangle$$

I3 *Positive Definiteness:* For any non-zero function $f \in C[a,b]$, we have

$$\langle f, f \rangle = \int_a^b f(t)f(t)dt = \int_a^b \{f(t)\}^2 dt > 0 \quad [\because f(t) \neq 0 \text{ for some } t \in [a,b]]$$

Hence, the given function is an inner product on $C[a,b]$.

EXAMPLE-4 *Prove that the vector space $R^{m \times n}$ of all $m \times n$ matrices over R is an inner product space with an inner product defined on it by*

$$\langle A, B \rangle = tr(B^T A) \quad \text{for all} \quad A, B \in R^{m \times n}$$

SOLUTION We observe the following properties of the function defined above:

I1 *Linearity:* For any $A, B, C \in R^{m \times n}$ and $a, b \in R$, we have

$$\langle aA + bB, C \rangle = tr\{C^T(aA + bB)\}$$
$$\Rightarrow \quad \langle aA + bB, C \rangle = tr(aC^TA + bC^TB)$$
$$\Rightarrow \quad \langle aA + bB, C \rangle = tr(aC^TA) + b\,tr(C^TB) \qquad [\because \; tr(A+B) = tr(A) + tr(B)]$$
$$\Rightarrow \quad \langle aA + bB, C \rangle = a\,tr(C^TA) + b\,tr(C^TB) \qquad [\because \; tr(\lambda A) = \lambda tr(A)]$$
$$\Rightarrow \quad \langle aA + bB, C \rangle = a\langle A, C \rangle + b\langle B, C \rangle$$

I2 *Symmetry:* For any $A, B \in R^{m \times n}$, we have

$$\langle A, B \rangle = tr(B^TA)$$
$$\Rightarrow \quad \langle A, B \rangle = tr\{(B^TA)^T\} \qquad [\because \; tr(A) = tr(A^T)]$$
$$\Rightarrow \quad \langle A, B \rangle = tr(A^TB) \qquad [\because \; (B^TA)^T = A^T(B^T)^T = A^TB]$$
$$\Rightarrow \quad \langle A, B \rangle = \langle B, A \rangle$$

I3 *Positive Definiteness:* For any non-null matrix $A = [a_{ij}] \in R^{m \times n}$, we have

$$\langle A, A \rangle = tr(A^TA)$$
$$\Rightarrow \quad \langle A, A \rangle = \sum_{i=1}^{n} (A^TA)_{ii}$$
$$\Rightarrow \quad \langle A, A \rangle = \sum_{i=1}^{n} \left(\sum_{r=1}^{m} (A^T)_{ir} (A)_{ri} \right)$$
$$\Rightarrow \quad \langle A, A \rangle = \sum_{i=1}^{n} \left(\sum_{r=1}^{m} a_{ri} a_{ri} \right)$$
$$\Rightarrow \quad \langle A, A \rangle = \sum_{i=1}^{n} \left(\sum_{r=1}^{m} (a_{ri})^2 \right) > 0 \qquad [\because \; A \neq 0 \; \therefore \; \text{At least one } a_{ij} \neq 0]$$

Thus, the given function is an inner production $R^{m \times n}$ and hence $R^{m \times n}$ is an inner product space.

EXAMPLE-5 *Let V be a real vector space. Show that the sum of two inner products on V is an inner product on V. Is the difference of two inner products on V an inner product on V? Show that a positive multiple of an inner product on V is an inner product on V.*

SOLUTION Let f and g be two inner products on real vector space V. Then, $f, g: V \times V \to R$ such that for all $u, v, w \in V$ and $a, b \in R$

(i) $f(u,v) = f(v,u), g(u,v) = g(v,u)$
(ii) $f(au+bv, w) = af(u,v) + bf(v,w), g(au+bv, w) = ag(u,w) + bg(v,w)$
(iii) $f(u,u) \geq 0$ and $g(u,u) \geq 0$

Let us define the sum $f+g$ of the inner products f and g as follows:

$$(f+g)(u,v) = f(u,v) + g(u,v) \quad \text{for all } u, v \in V$$

We observe the following properties of $f+g$:

I1 *Linearity:* For any $u, v, w \in V$ and $a, b \in R$, we have

$$(f+g)(au+bv, w) = f(au+bv,w) + g(au+bv,w) \quad \text{[By definition of } f+g\text{]}$$
$$= af(u,w) + bf(v,w) + ag(u,w) + bg(v,w) \text{ [By linearity of } f \text{ and } g\text{]}$$
$$= a\{f(u,w) + g(u, w)\} + b\{f(v,w) + g(v,w)\}$$
$$= a(f+g)(u,w) + b(f+g)(u,w)$$

I2 *Symmetry:* For any $u, v \in V$, we have

$$(f+g)(u,v) = f(u,v) + g(u,v) \quad \text{[By definition of } f+g\text{]}$$
$$= f(v,u) + g(v,u) \quad \text{[By symmetry of } f \text{ and } g\text{]}$$
$$= (f+g)(v,u)$$

I3 *Positive Definition :* For any non-zero vector $u \in V$, we have

$$(f+g)(u,u) = f(u,u) + g(u,u) > 0 \quad [\because f(u,u) > 0 \text{ and } g(u,u) > 0]$$

Hence, $f+g$ is an inner product on V.
If we define $f-g$ by the following rule:

$$(f-g)(u,v) = f(u,v) - g(u,v) \quad \text{for all } u, v \in V$$

Then, the sign of $(f-g)(u,u) = f(u,u) - g(u,u)$ is not definite.

So, $f-g$ is not necessarily an inner product on V as the axiom of positive definiteness may not hold good. As $f+g$ is an inner product on V. Therefore, by taking $g = f+f$ is an inner product on V. Also, $2f+f = 3f$ is a inner product on V. Continuing like this, we find that nf is an inner product on V for all $n \in N$.

EXAMPLE-6 *Find the value of k so that the following is an inner product on R^2:*

$$\langle u, v \rangle = x_1y_1 - 3x_1y_2 - 3x_2y_1 + kx_2y_2, \text{ where } u = (x_1,x_2), v = (y_1,y_2) \in R^2.$$

SOLUTION It is given that the above function is an inner produce on R^2. Therefore,

$\langle u, u \rangle > 0$ for all $u = (x_1, x_2) \in R^2, u \neq 0$

$\Rightarrow \quad x_1^2 - 3x_1 x_2 - 3x_2 x_1 + k x_2^2 > 0$ for all $x_1, x_2 \in R$

$\Rightarrow \quad x_2^2 \left\{ \left(\dfrac{x_1}{x_2}\right)^2 - 6\left(\dfrac{x_1}{x_2}\right) + k \right\} > 0$ for all $x_1, x_2 \in R$

$\Rightarrow \quad \left(\dfrac{x_1}{x_2}\right)^2 - 6\left(\dfrac{x_1}{x_2}\right) + k > 0$ for all $x_1, x_2 \in R$

$\Rightarrow \quad a^2 - 6a + k > 0$ for all $a \in R, a \neq 0$

$\Rightarrow \quad 36 - 4k < 0 \qquad [\because ax^2 + bx + c > 0$ for all $x \in R \Rightarrow a > 0$ and Disc $< 0]$

$\Rightarrow \quad k > 9$

EXAMPLE-7 Let V be an inner product space and $u, v \in V$. Simplify $\langle 2u - 5v, 4u + 6v \rangle$

SOLUTION Using linearity of the inner product, we have

$\langle 2u - 5v, 4u + 6v \rangle = 2\langle u, 4u + 6v \rangle - 5\langle v, 4u + 6v \rangle$

$\Rightarrow \quad \langle 2u - 5v, 4u + 6v \rangle = 2\{\langle u, 4u \rangle + \langle u, 6v \rangle\} - 5\{\langle v, 4u \rangle + \langle v, 6v \rangle\}$

$\Rightarrow \quad \langle 2u - 5v, 4u + 6v \rangle = 2\{4\langle u, u \rangle + 6\langle u, v \rangle\} - 5\{4\langle v, u \rangle + 6\langle v, v \rangle\}$

$\Rightarrow \quad \langle 2u - 5v, 4u + 6v \rangle = 8\langle u, u \rangle + 12\langle u, v \rangle - 20\langle u, v \rangle - 30\langle v, v \rangle$

$\Rightarrow \quad \langle 2u - 5v, 4u + 6v \rangle = 8\langle u, u \rangle - 8\langle u, v \rangle - 30\langle v, v \rangle$

EXAMPLE-8 Let $\langle , \rangle$ be the standard inner product on R^2. If $u = (1, 2)$, $v = (-1, 1) \in R^2$, then find $w \in R^2$ satisfying $\langle u, w \rangle = -1$ and $\langle v, w \rangle = 3$.

SOLUTION Let $w = (a, b) \in R^2$. Then,

$\langle u, w \rangle = -1$ and $\langle v, w \rangle = 3$

$\Rightarrow \quad a + 2b = -1$ and $-a + b = 3$

$\Rightarrow \quad a = -\dfrac{7}{3}, b = \dfrac{2}{3}$

$\Rightarrow \quad w = \left(-\dfrac{7}{3}, \dfrac{2}{3}\right).$

EXAMPLE-9 Let $\langle , \rangle$ denote the standard inner production R^2. Show that for any $u \in R^2$, we have

$$u = \langle u, e_1^{(2)} \rangle e_1^{(2)} + \langle u, e_2^{(2)} \rangle e_2^{(2)}$$

SOLUTION We have, $e_1^{(2)} = (1, 0)$ and $e_2^{(2)} = (0, 1)$

Let $u = (a, b) \in R^2$. Then,

$$\langle u, e_1^{(2)} \rangle = a \times 1 + b \times 0 = a \quad \text{and,} \quad \langle u, e_2^{(2)} \rangle = a \times 0 + b \times 1 = b$$

$$\therefore \quad \langle u, e_1^{(2)} \rangle e_1^{(2)} + \langle u, e_2^{(2)} \rangle e_2^{(2)} = ae_1^{(2)} + be_2^{(2)} = a(1,0) + b(0,1) = (a,b) = u$$

EXAMPLE-10 Let V be an inner product space with inner product $\langle , \rangle$. Then show that

(i) $\langle 0_V, v \rangle = 0$ for all $v \in V$ (ii) if $\langle u, v \rangle = 0$ for all $v \in V$, then $u = 0_V$

SOLUTION (i) For any $v \in V$, we have

$$\langle 0_V, v \rangle = \langle 0 0_V, v \rangle$$
$$= 0 \langle 0_V, v \rangle \qquad\qquad [\because \quad \langle au, v \rangle = a \langle u, v \rangle]$$
$$= 0$$

Hence, $\langle 0_V, v \rangle = 0$ for all $v \in V$.

(ii) Let $\langle u, v \rangle = 0$ for all $v \in V$.

$\Rightarrow \quad \langle u, u \rangle = 0$ \hfill [Taking $v = u$]

$\Rightarrow \quad u = 0_V$

EXAMPLE-11 Let V be an inner product space with $\langle , \rangle$ as an inner product and $u, v \in V$. Then, $u = v$ iff $\langle u, w \rangle = \langle v, w \rangle$ for all $w \in V$.

SOLUTION First, let $u = v$. Then,

$$\langle u, w \rangle = \langle v, w \rangle \quad \text{for all } w \in V$$

Conversely, let $u, v \in V$ be such that

$\langle u, w \rangle = \langle v, w \rangle \quad$ for all $w \in V$

$\Rightarrow \quad \langle u, w \rangle - \langle v, w \rangle = 0 \quad$ for all $w \in V$

$\Rightarrow \quad \langle u - v, w \rangle = 0 \quad$ for all $w \in V$

$\Rightarrow \quad \langle u - v, u - v \rangle = 0 \hfill$ [Taking $w = u - v$]

$\Rightarrow \quad u - v = 0_V \hfill [\because \quad \langle u, u \rangle = 0 \Leftrightarrow u = 0_V]$

$\Rightarrow \quad u = v$

EXAMPLE-12 Let $A = [a_{ij}]$ be a 2×2 matrix with real entries. For X, Y in $R^{2 \times 1}$ let

$$f_A(X,Y) = Y^T A X$$

Show that f_A is an inner product on $R^{2 \times 1}$ if and only if $A^T = A$, $a_{11} > 0$, $a_{22} > 0$ and $|A| > 0$.

SOLUTION First, let $A = [a_{ij}]$ be such that $A^T = A$, $a_{11} > 0$, $a_{22} > 0$ and $|A| > 0$. Then, we have to show that f_A is an inner product on $R^{2 \times 1}$.
We observe the following properties of f_A:

I1 *Linearity:* Let $X, Y, Z \in R^{2 \times 1}$ and $a, b \in R$. Then,

$$\begin{aligned} f_A(aX + bY, Z) &= Z^T A(aX + bY) \\ &= Z^T(aAX + bAY) \\ &= a(Z^T A X) + b(Z^T A Y) \\ &= a f_A(X, Z) + b f_A(Y, Z) \end{aligned}$$

I2 *Symmetry:* For any $X, Y \in R^{2 \times 1}$, we have

$$\begin{aligned} f_A(X,Y) = Y^T A X &= Y^T A^T (X^T)^T && [\because A^T = A] \\ &= (X^T A Y)^T \\ &= X^T A Y && [\because X^T A Y \text{ is a scalar}] \\ &= f_A(Y, X) \end{aligned}$$

I3 *Positive Definiteness:* For any $O \neq X = \begin{bmatrix} x_1 \\ x_2 \end{bmatrix} \in R^{2 \times 1}$, we have

$$\begin{aligned} f_A(X,X) &= X^T A X \\ \Rightarrow f_A(X,X) &= a_{11} x_1^2 + a_{12} x_1 x_2 + a_{21} x_1 x_2 + a_{22} x_2^2 \\ \Rightarrow f_A(X,X) &= a_{11} x_1^2 + 2 a_{12} x_1 x_2 + a_{22} x_2^2 && [\because A^T = A \quad \therefore a_{12} = a_{21}] \end{aligned}$$

It is given that $a_{11} > 0$, $a_{22} > 0$ and $|A| < 0$.

$$\begin{aligned} \therefore \quad & a_{11} a_{22} - a_{12} a_{21} > 0 \\ \Rightarrow \quad & a_{11} a_{22} - (a_{12})^2 > 0 \\ \Rightarrow \quad & 4\{(a_{12})^2 - a_{11} a_{22}\} < 0 \\ \Rightarrow \quad & a_{11} x_1^2 + 2 a_{12} x_1 x_2 + a_{22} x_2^2 > 0 && [\because a_{11} > 0 \text{ and Disc} < 0] \\ \Rightarrow \quad & f_A(X, X) > 0 \end{aligned}$$

Hence, f_A is an inner production $R^{2 \times 1}$.

Conversely, let f_A be an inner product on $R^{2\times 1}$. Then,

$f_A(X,Y) = f_A(Y,X)$ for all $X, Y \in R^{2\times 1}$ [By symmetry]

$\Rightarrow Y^T A X = X^T A Y$ for all $X, Y \in R^{2\times 1}$

$\Rightarrow [y_1 \ y_2]\begin{bmatrix} a_{11} & a_{12} \\ a_{21} & a_{22} \end{bmatrix}\begin{bmatrix} x_1 \\ x_2 \end{bmatrix} = [x_1 \ x_2]\begin{bmatrix} a_{11} & a_{12} \\ a_{21} & a_{22} \end{bmatrix}\begin{bmatrix} y_1 \\ y_2 \end{bmatrix}$, where $X = \begin{bmatrix} x_1 \\ x_2 \end{bmatrix}, y = \begin{bmatrix} y_1 \\ y_2 \end{bmatrix}$

$\Rightarrow a_{11}x_1y_1 + a_{12}x_2y_1 + a_{21}x_1y_2 + a_{22}x_2y_2 = a_{11}x_1y_1 + a_{12}x_1y_2 + a_{21}x_2y_1 + a_{22}x_2y_2$

$\Rightarrow (x_1y_2 - x_2y_1)(a_{21} - a_{12}) = 0$ for all $x_1, x_2, y_1, y_2 \in R$

$\Rightarrow a_{21} - a_{12} = 0 \Rightarrow a_{12} = a_{21} \Rightarrow A^T = A$

For any $X = \begin{bmatrix} x_1 \\ x_2 \end{bmatrix} \in R^{2\times 1}$, we have

$f_A(X,X) > 0$ [Positive Definiteness]

$\Rightarrow X^T A X > 0$

$\Rightarrow a_{11}x_1^2 + a_{12}x_1x_2 + a_{21}x_2x_1 + a_{22}x_2^2 > 0$

$\Rightarrow a_{11}x_1^2 + 2a_{12}x_1x_2 + a_{22}x_2^2 > 0$ for all $x_1, x_2 \in R$ $[\because a_{12} = a_{21}]$

$\Rightarrow a_{11} > 0, a_{22} > 0$ and $4a_{12}^2 - 4a_{11}a_{22} < 0$

$\Rightarrow a_{11} > 0, a_{22} > 0$ and $a_{11}a_{22} - a_{12}a_{21} > 0$ $[\because a_{12} = a_{21}]$

$\Rightarrow a_{11} > 0, a_{22} > 0$ and $|A| > 0$.

EXERCISE 6.1

1. Prove that R^2 is an inner product space with an inner product defined by

    $\langle u, v \rangle = x_1y_1 - 2x_1y_2 - 2x_2y_1 + 5x_2y_2$, where $u = (x_1, x_2), \ v = (y_1, y_2) \in R^2$.

2. Prove that the following defines an inner product on R^2:

    $\langle u, v \rangle = x_1y_1 - x_1y_2 - x_2y_1 + 3x_2y_2$, where $u = (x_1, x_2), v = (y_1, y_2) \in R^2$.

3. Find the value of λ so that the following is an inner product on R^2:

    $\langle u, v \rangle = x_1y_1 - 2x_1y_2 - 2x_2y_1 + \lambda x_2y_2$, where $u = (x_1, x_2), v = (y_1, y_2)$

4. Which of the following define inner products on R^2?

    (i) $\langle u, v \rangle = x_1y_1 + 2x_1y_2 + 2x_2y_2 + 5x_2y_2$

    (ii) $\langle u, v \rangle = x_1^2 - 2x_1y_2 - 2x_2y_1 + y_1^2$

(iii) $\langle u, v \rangle = 2x_1y_1 + 5x_2y_2$

(iv) $\langle u, v \rangle = x_1y_1 - 2x_1y_2 - 2x_2y_1 + 4x_2y_2$, where $u = (x_1, x_2), v = (y_1, y_2) \in R^2$.

5. Show that each of the following is not an inner product on R^3:

   (i) $\langle u, v \rangle = x_1y_1 + x_2y_2$

   (ii) $\langle u, v \rangle = x_1y_2x_3 + y_1x_2y_3$, where $u = (x_1, x_2, x_3)$ and $v = (y_1, y_2, y_3) \in R^3$.

6. Let $f(u,v)$ and $g(u,v)$ be inner products on a vector space V. Prove that:

   (i) The sum $f+g$ is an inner product on V, where $(f+g)(u,v) = f(u,v) + g(u,v)$.

   (ii) The scalar product af, for $a>0$, is an inner product on V, where $(af)(u,v) = af(u,v)$.

7. Let $\langle , \rangle$ be the standard inner product on R^2. If $u = (1,3), v = (2,1) \in R^2$ be such that
$$\langle w, u \rangle = 3 \langle w, v \rangle = -1$$
Find w.

8. Compute $\langle u, v \rangle$ with respect to the standard inner product on R^2, if $u = (1, -3)$ and $v = (2, 5)$.

9. If an inner product on R^2 is defined by
$$\langle u, v \rangle = x_1y_1 - 2x_1y_2 - 2x_2y_1 + 5x_2y_2, \quad \text{where} \quad u = (x_1, x_2) \text{ and } v = (y_1, y_2).$$
Find $\langle u, v \rangle$, if $u = (1, -3)$ and $v = (2, 5)$.

10. Let V be the vector space of all infinite sequences of real numbers $(a_1, a_2, a_3, \ldots)$ satisfying $\sum_{i=1}^{\infty} a_i^2 = a_1^2 + a_2^2 + \cdots < \infty$, i.e. the sum converges. If addition and scalar multiplication on V are defined component wise, i.e. if $u = (a_1, a_2, a_3, \ldots)$ and $v = (v_1, v_2, v_3, \ldots)$, then $u + v = (a_1 + b_1, a_2 + b_2, \ldots)$ and $ku = (ka_1, ka_2, \ldots)$. Then show that V is an inner product space with inner product defined by
$$\langle u, v \rangle = a_1b_1 + a_2b_2 + \ldots$$

11. Let $\langle , \rangle$ be the standard inner product on R^2. If $u = (1, 3), v = (2, 5) \in R^2$ be such that $\langle u, w \rangle = 0$ and $\langle v, w \rangle = 3$, find w.

12. Let A be any 2×2 symmetric matrix over R. For any $u, v \in R^{2 \times 1}$ define $\langle u, v \rangle = u^T A v$. Prove that this is an inner product on R^2 iff $\det(A) > 0$ and the diagonal entries of A are positive.

13. For any $u = (x_1, y_1), v = (x_2, y_2)$ in R^2, we defined
$$\langle u, v \rangle = [x_1, y_1] \begin{bmatrix} 1 & 2 \\ 3 & 8 \end{bmatrix} \begin{bmatrix} x_2 \\ y_2 \end{bmatrix} = x_1x_2 + 2x_1y_1 + 3x_2y_1 + 8y_1y_2$$

Verify that R^2 is an inner product space with the above defined inner product.

ANSWERS

3. $\lambda > 4$ 4. (i), (iii)   7. $w = (0, -1/3)$ 8. -13 9. -17 11. $(9, -3)$

6.3 NORM OR LENGTH OF A VECTOR

In this section, we will introduce the concept of length (norm) of a vector in an inner product space. We will also learn about various properties related to the norm of a vector.

NORM OF A VECTOR *Let V be an inner product space and $u \in V$. The non-negative square root of $<u,u>$, i.e. $\sqrt{\langle u,u \rangle}$ is called the norm or length of u and is denoted by $\|u\|$.*

Thus, $\|u\| = \sqrt{\langle u,u \rangle}$

$\Rightarrow$ $\|u\|^2 = \langle u,u \rangle$

If $\|u\| = 1$ or, equivalently $\langle u,u \rangle = 1$, then u is called a **unit vector** and is said to be **normalized**.

REMARK. *Every non-zero vector u in an inner product space V can be normalized by multiplying it by the reciprocal of its length, i.e.*

$$\widehat{u} = \frac{1}{\|u\|} u$$

Clearly, $\widehat{u}$ is a positive multiple of u. The process of getting unit vector $\widehat{u}$ from a vector u is called normalizing vector u.

THEOREM-1 *If V is an inner product space and $v \in V$, then*

(i) $\|u\| \geq 0$ and $\|u\| = 0$ iff $u = 0_V$ (ii) $\|au\| = |a|\,\|u\|$ for all $a \in R$.

PROOF. (i) If $u \neq 0_V$, then $\langle u, u \rangle > 0$.

$\therefore \quad \|u\| = \sqrt{\langle u,u \rangle} \Rightarrow \|u\|^2 = \langle u,u \rangle \Rightarrow \|u\|^2 > 0$

If $u = 0_V$, then

$\langle u, u \rangle = \langle 0_V, 0_V \rangle = 0 \Rightarrow \|u\|^2 = 0$ iff $u = 0_V$.

(ii) We have,

$\|au\|^2 = \langle au, au \rangle = a^2 \langle u,u \rangle = a^2 \|u\|^2$

$\Rightarrow \quad \|au\| = |a|\,\|u\|$ [Taking square root of both sides]

Q.E.D.

THEOREM-2 *(Cauchy-Schwarz Inequality) For any two vectors u and v in an inner product space V*

$$\langle u,v \rangle^2 \leq \langle u,u \rangle \langle v,v \rangle \quad \text{or,} \quad |\langle u,v \rangle| \leq ||u|| \, ||v||$$

PROOF. If $u = 0_V$, then both $\langle u,v \rangle = 0$ and $||u|| \, ||v|| = 0$.
So, theorem holds good.
Let $u \neq 0_V$. Then for any real number t, we have

$\langle tu+v, tu+v \rangle \geq 0$ [By positive definiteness]

$\Rightarrow \quad \langle tu,tu \rangle + \langle tu,v \rangle + \langle v,tu \rangle + \langle v,v \rangle \geq 0$

$\Rightarrow \quad t^2 \langle u,u \rangle + t \langle u,v \rangle + t \langle v,u \rangle + \langle v,v \rangle \geq 0$

$\Rightarrow \quad t^2 ||u||^2 + t \langle u,v \rangle + t \langle u,v \rangle + ||v||^2 \geq 0$

$\Rightarrow \quad t^2 ||u||^2 + 2t \langle u,v \rangle + ||v||^2 \geq 0$

$\Rightarrow \quad 4\{\langle u,v \rangle\}^2 - 4||u||^2 ||v||^2 \leq 0 \qquad \begin{bmatrix} \because & ax^2 + bx + c > 0 \text{ for all } x \text{ and } a > 0 \\ \Rightarrow & b^2 - 4ac \leq 0 \end{bmatrix}$

$\Rightarrow \quad \{\langle u,v \rangle\}^2 \leq ||u||^2 ||v||^2$

$\Rightarrow \quad |\langle u,v \rangle| \leq ||u|| \, ||v|| \qquad\qquad\qquad [\because \ x^2 \leq a^2 \ \Rightarrow \ |x| \leq a]$

Q.E.D.

COROLLARY. *For any real numbers x_1, x_2, y_1 and y_2, prove that*

$$|x_1 y_1 + x_2 y_2| \leq (x_1^2 + x_2^2)^{1/2} (y_1^2 + y_2^2)^{1/2}$$

PROOF. Let $u = (x_1, x_2), v = (y_1, y_2) \in R^2$.
Since R^2 is an inner product space with respect to the standard inner product. Therefore, by Cauchy-schwarz inequality, we have

$|\langle u,v \rangle| \leq ||u|| \, ||v||$

$\Rightarrow \quad |x_1 y_1 + x_2 y_2| \leq (x_1^2 + x_2^2)^{1/2} (y_1^2 + y_2^2)^{1/2}$

Q.E.D.

COROLLARY. *For any two vectors u and v in an inner product space V, prove that*

$$||u+v|| \leq ||u|| + ||v|| \qquad \text{(Triangle inequality)}$$

PROOF. Using the definition of norm of a vector, we have

$||u+v||^2 = \langle u+v, u+v \rangle$

$\Rightarrow \quad ||u+v||^2 = \langle u, u+v \rangle + \langle v, u+v \rangle$

$\Rightarrow \quad ||u+v||^2 = \langle u,u \rangle + \langle u,v \rangle + \langle u,v \rangle + \langle v,v \rangle$

$$\Rightarrow \quad ||u+v||^2 = \langle u,u \rangle + \langle u,v \rangle + \langle u,v \rangle + \langle v,v \rangle \qquad [\because \langle u,v \rangle = \langle v,u \rangle]$$
$$\Rightarrow \quad ||u+v||^2 = ||u||^2 + 2\langle u,v \rangle + ||v||^2$$

By Cauchy–Schwarz inequality, we have

$$\langle u,v \rangle \leq ||u||\,||v||$$
$$\Rightarrow \quad 2\langle u,v \rangle \leq 2||u||\,||v||$$
$$\Rightarrow \quad ||u||^2 + 2\langle u,v \rangle + ||v||^2 \leq ||u||^2 + 2||u||\,||v|| + ||v||^2$$
$$\Rightarrow \quad ||u+v||^2 \leq \Big(||u|| + ||v||\Big)^2$$
$$\Rightarrow \quad ||u+v|| \leq ||u|| + ||v|| \qquad [\because x^2 \leq a^2 \Rightarrow |x| \leq a]$$

Q.E.D.

REMARK. *The above corollary is called the triangle inequality, because if we view $u+v$ as the side of the triangle formed with the sides u and v, then the corollary states that the length of one side of a triangle cannot be greater than the sum of the lengths of the other two sides.*

THEOREM-3 *Let V be an inner product space and $u,v \in V$. Then,*

(i) $||u+v||^2 - ||u-v||^2 = 4\langle u,v \rangle$
(ii) $||u+v||^2 + ||u-v||^2 = 2\Big(||u||^2 + ||v||^2\Big)$ *(Parallelogram law)*

PROOF. By using the definition of norm of a vector, we have

$$||u+v||^2 = \langle u+v, u+v \rangle$$
$$\Rightarrow \quad ||u+v||^2 = \langle u, u+v \rangle + \langle v, u+v \rangle$$
$$\Rightarrow \quad ||u+v||^2 = \langle u,u \rangle + \langle u,v \rangle + \langle v,u \rangle + \langle v,v \rangle$$
$$\Rightarrow \quad ||u+v||^2 = \langle u,u \rangle + 2\langle u,v \rangle + \langle v,v \rangle \qquad [\because \langle u,v \rangle = \langle v,u \rangle]$$
$$\Rightarrow \quad ||u+v||^2 = ||u||^2 + 2\langle u,v \rangle + ||v||^2 \qquad \text{(i)}$$

and,

$$||u-v||^2 = \langle u-v, u-v \rangle$$
$$\Rightarrow \quad ||u-v||^2 = \langle u, u-v \rangle + \langle -v, u-v \rangle$$
$$\Rightarrow \quad ||u-v||^2 = \langle u,u \rangle + \langle u,-v \rangle + \langle -v,u \rangle + \langle -v,-v \rangle$$
$$\Rightarrow \quad ||u-v||^2 = \langle u,u \rangle - \langle u,v \rangle - \langle v,u \rangle + (-1)^2 \langle v,v \rangle$$
$$\Rightarrow \quad ||u-v||^2 = \langle u,u \rangle - 2\langle u,v \rangle + \langle v,v \rangle$$
$$\Rightarrow \quad ||u-v||^2 = ||u||^2 - 2\langle u,v \rangle + ||v||^2 \qquad \text{(ii)}$$

On subtracting (ii) from (i), we get

$$||u+v||^2 - ||u-v||^2 = 4\langle u,v \rangle$$

On adding (i) and (ii), we get

$$||u+v||^2 + ||u-v||^2 = 2\left(||v||^2 + ||v||^2\right)$$

Q.E.D.

REMARK. *The result (ii) in the above theorem is known as Parallelogram law, because if u and v denote the sides of a parallelogram, then its diagonals are $u+v$ and $u-v$ and the sum of the squares of diagonals of a parallelogram, is equal to the sum of the squares of its sides.*

THEOREM-4 *If u and v are vectors in an inner product space V, then*

$$\big| \, ||u|| - ||v|| \, \big| \leq ||u-v||$$

PROOF. We have,

$$\big| \, ||u|| - ||v|| \, \big|^2 = ||u||^2 + ||v||^2 - 2||u|| \, ||v||$$

$$\Rightarrow \quad \big| \, ||u|| - ||v|| \, \big|^2 = \langle u,u \rangle + \langle v,v \rangle - 2||u|| \, ||v|| \qquad \text{(i)}$$

By Cauchy–Schwarz inequality, we have

$$|\langle u,v \rangle| \leq ||u|| \, ||v||$$

$$\Rightarrow \quad -||u|| \, ||v|| \leq \langle u,v \rangle \leq ||u|| \, ||v|| \qquad [\because \ |x| \leq a \Leftrightarrow -a \leq x \leq a]$$

$$\Rightarrow \quad \langle u,v \rangle \leq ||u|| \, ||v||$$

$$\Rightarrow \quad -2||u|| \, ||v|| \leq -2\langle u,v \rangle$$

$$\Rightarrow \quad \langle u,u \rangle + \langle v,v \rangle - 2||u|| \, ||v|| \leq \langle u,u \rangle + \langle v,v \rangle - 2\langle u,v \rangle$$

$$\Rightarrow \quad \big| \, ||u|| - ||v|| \, \big|^2 \leq \langle u,u \rangle + \langle -v,-v \rangle + \langle u,-v \rangle + \langle -v,u \rangle$$

$$\Rightarrow \quad \big| \, ||u|| - ||v|| \, \big|^2 \leq \langle u-v, u-v \rangle$$

$$\Rightarrow \quad \big| \, ||u|| - ||v|| \, \big|^2 \leq ||u-v||^2$$

$$\Rightarrow \quad \big| \, ||u|| - ||v|| \, \big| \leq ||u-v||$$

Q.E.D.

ILLUSTRATIVE EXAMPLES

EXAMPLE-1 *Let u, v be vectors in an inner product space V such that $|\langle u,v \rangle| = ||u|| \, ||v||$, i.e. Cauchy–Schwarz inequality reduces to an equality. Show that u and v are linearly dependent vectors.*

SOLUTION If one of the two vectors u and v is zero vector, then, $|\langle u,v \rangle| = ||u|| \, ||v||$ holds good and u and v from a linearly dependent set of vectors as any set containing zero vector is linearly dependent. So, let us assume that at least one of u and v is a non-zero vector. Let $u \neq 0_V$. Then, $||u|| > 0$.

Consider the vector

$$w = v - \frac{\langle u,v \rangle u}{||u||^2} = v - \lambda u, \quad \text{where } \lambda = \frac{\langle u,v \rangle}{||u||^2}$$

$\Rightarrow \quad \langle w,w \rangle = \langle v-\lambda u, v-\lambda u \rangle$

$\Rightarrow \quad \langle w,w \rangle = \langle v,v \rangle + \langle v,-\lambda u \rangle + \langle -\lambda u, v \rangle + \langle -\lambda u, -\lambda u \rangle$

$\Rightarrow \quad \langle w,w \rangle = \langle v,v \rangle - \lambda \langle v,u \rangle - \lambda \langle u,v \rangle + \lambda^2 \langle u,u \rangle$

$\Rightarrow \quad \langle w,w \rangle = ||v||^2 - 2\lambda \langle u,v \rangle + \lambda^2 ||u||^2$

$\Rightarrow \quad \langle w,w \rangle = ||v||^2 - \frac{2\{\langle u,v \rangle\}^2}{||u||^2} + \frac{\{\langle u,v \rangle\}^2}{||u||^2}$

$\Rightarrow \quad \langle w,w \rangle = ||v||^2 - \frac{\{\langle u,v \rangle\}^2}{||u||^2}$

$\Rightarrow \quad \langle w,w \rangle = ||v||^2 - ||v||^2 = 0 \quad \left[\because \ |\langle u,v \rangle| = ||u|| \ ||v|| \Rightarrow \{\langle u,v \rangle\}^2 = ||u||^2 \ ||v||^2 \right]$

$\Rightarrow \quad w = 0_V \quad\quad\quad\quad\quad\quad\quad\quad\quad\quad\quad\quad\quad [\because \ \langle u,u \rangle = 0 \Rightarrow u = 0_V]$

$\Rightarrow \quad v - \frac{\langle u,v \rangle}{||u||^2} u = 0_V$

$\Rightarrow \quad v = \frac{\langle u,v \rangle}{||u||^2} u$

$\Rightarrow \quad v$ is a scalar multiple of u

$\Rightarrow \quad u$ and v are linearly dependent vectors.

EXAMPLE-2 *Let u and v be two vectors in an inner product space v such that $||u+v|| = ||u|| + ||v||$. Prove that u and v are linear dependent vectors. Give an example to show that the convers of this statement is not true.*

SOLUTION We have,

$||u+v|| = ||u|| + ||v||$

$\Rightarrow \quad ||u+v||^2 = \left(||u|| + ||v|| \right)^2$

$\Rightarrow \quad \langle u+v, u+v \rangle = ||u||^2 + ||v||^2 + 2||u|| \ ||v||$

$\Rightarrow \quad \langle u,u \rangle + 2 \langle u,v \rangle + \langle v,v \rangle = ||u||^2 + ||v||^2 + 2||u|| \ ||v||$

$\Rightarrow \quad \langle u,v \rangle = ||u|| \ ||v||$

$\Rightarrow \quad u$ and v are linearly dependent vectors $\quad\quad\quad\quad\quad\quad$ [See Example 1]

The converse is not true, because vectors $u = (-1, 0, 1)$ and $v = (2, 0, -2)$ in R^3 are linearly dependent as $v = 2u$. But, $||u+v|| \neq ||u|| + ||v||$.

EXERCISE 6.2

1. If u and v are vectors in an innerproduct space V, prove that
$$||2u - 3v||^2 = 4||u||^2 - 12\langle u, v \rangle + 9||v||^2$$
2. If $u = (1, 5)$ and $v = (3, 4) \in R^2$, using standard inner product in R^2, find $||u||$ and $||v||$.
3. Consider the inner product in R^2 defined by
$$\langle u, v \rangle = x_1 y_1 - x_1 y_2 - x_2 y_1 + 3 x_2 y_2,$$
where $u = (x_1, x_2)$, $v = (y_1, y_2)$.
If $u = (-2, 3)$ and $v = (3, -4)$, find $||u||$, $||v||$ and $||u-v||$.
4. Let $P(t)$ denote the inner product space of all polynomials with the inner product defined as
$$\langle f, g \rangle = \int_0^1 f(t)g(t)dt$$
If $f(t) = t + 2$ and $g(t) = 3t - 2$, find $||f||$ and $||g||$. Also, normalize f and g.
5. If u, v be two linearly dependent vectors in an inner product space V. Then prove that
$$|\langle u, v \rangle| = ||u|| \, ||v||.$$
6. Let V be an inner product space and $u, v \in V$ such that $||u|| = ||v|| = 1$. Then, show that
$$|\langle u, v \rangle| \leq 1.$$

ANSWERS

2. $||u|| = \sqrt{26}$, $||v|| = 5$ 3. $||u|| = \sqrt{43}$, $||v|| = 9$, $||u - v|| = \sqrt{242}$

4. $||f|| = \sqrt{\dfrac{19}{3}}$, $||g|| = 1$ $\hat{f} = \sqrt{\dfrac{3}{19}}(t+2)$, $\hat{g} = 3t - 2$

6.4 ANGLE BETWEEN VECTORS

ANGLE BETWEEN TWO VECTORS *Let V be an inner product space. Let u and v be two non-zero vectors in V. The angle between u and v is defined to be the angle θ such that $0 \leq \theta \leq \pi$ and*
$$\cos \theta = \frac{\langle u, v \rangle}{||u|| \, ||v||}$$

By the Cauchy–Scwarz inequality, we have

$$-||u||\,||v|| \le \langle u, v \rangle \le ||u||\,||v||$$

$$\Rightarrow \quad -1 \le \frac{\langle u, v \rangle}{||u||\,||v||} \le 1$$

$$\Rightarrow \quad -1 \le \cos\theta \le 1$$

So, $\cos\theta = \dfrac{\langle u, v \rangle}{||u||\,||v||}$ is meaningful and gives unique value of θ.

ILLUSTRATIVE EXAMPLES

EXAMPLE-1 R^3 *is an inner product space with respect to standard inner product. Find the angle between the vectors* $u = (1, 1, 2)$ *and* $v = (2, -1, 1)$.

SOLUTION Let θ be the angle between u and v. Then,

$$\cos\theta = \frac{\langle u, v \rangle}{||u||\,||v||}$$

We have, $u = (1, 1, 2)$ and $v = (2, -1, 1)$.

$\therefore \langle u, v \rangle = 1 \times 2 + 1 \times -1 + 2 \times 1 = 3$, $||u|| = \sqrt{1+1+4} = \sqrt{6}$ and $||v|| = \sqrt{4+1+1} = \sqrt{6}$

$$\therefore \quad \cos\theta = \frac{3}{\sqrt{6} \times \sqrt{6}} = \frac{1}{2} \Rightarrow \theta = \frac{\pi}{3}$$

EXAMPLE-2 *In the inner product space* R^4 *with respect to standard inner product, find the angle between the vectors* $u = (2, 1, 2, -3)$ *and* $v = (-2, 2, 1, 3)$.

SOLUTION We have,

$$u = (2, 1, 2, -3) \text{ and } v = (-2, 2, 1, 3)$$

$\therefore \quad \langle u, v \rangle = 2 \times -2 + 1 \times 2 + 2 \times 1 + (-3) \times 3 = -9$

$$||u|| = \sqrt{4+1+4+9} = 3\sqrt{2}, \quad ||v|| = \sqrt{4+4+1+9} = 3\sqrt{2}$$

Let θ be the angle between u and v. Then,

$$\cos\theta = \frac{\langle u, v \rangle}{||u||\,||v||} = \frac{-9}{3\sqrt{2} \times 3\sqrt{2}} = \frac{-1}{2} \Rightarrow \theta = \frac{2\pi}{3}$$

EXAMPLE-3 *Find the angle between* $f(t) = t - 1$ *and* $g(t) = t$ *in the polynomial space* $P(t)$ *with inner product* $\langle f, g \rangle = \displaystyle\int_0^1 f(t)g(t)dt$.

SOLUTION We have,

$$f(t) = t - 1 \text{ and } g(t) = t$$

$$\therefore \quad ||f||^2 = \langle f, f \rangle = \int_0^1 f^2(t)dt = \int_0^1 (t-1)^2 dt = \left[\frac{(t-1)^3}{3}\right]_0^1 = \frac{1}{3}$$

$$||g||^2 = \langle g, g \rangle = \int_0^1 g^2(t)dt = \int_0^1 t^2 dt = \frac{1}{3}$$

and,

$$\langle f, g \rangle = \int_0^1 f(t)g(t)dt = \int_0^1 (t-1)t\, dt = \left[\frac{t^3}{3} - \frac{t^2}{2}\right]_0^1 = -\frac{1}{6}$$

Let θ be the angle between f and g. Then,

$$\cos\theta = \frac{\langle f, g \rangle}{||f||\,||g||} = \frac{-\frac{1}{6}}{\sqrt{1/3}\sqrt{1/3}} = -\frac{1}{2} \Rightarrow \theta = \frac{2\pi}{3}$$

EXAMPLE-4 *Consider the inner product space $R^{2\times 3}$ of all 2×3 matrices over R with the inner product defined by*

$$\langle A, B \rangle = tr\,(B^T A) \quad \text{for all } A, B \in R^{2\times 3}$$

Find the angle between $A = \begin{bmatrix} 9 & 8 & 7 \\ 6 & 5 & 4 \end{bmatrix}$ *and* $B = \begin{bmatrix} 1 & 2 & 3 \\ 4 & 5 & 6 \end{bmatrix}$.

SOLUTION We have,

$$\langle A, B \rangle = tr\,(B^T A) = \sum_{i=1}^{3}(B^T A)_{ii} = \sum_{i=1}^{3}\left(\sum_{r=1}^{3} b_{ri}\, a_{ri}\right)$$

$\Rightarrow \quad \langle A, B \rangle = $ Sum of the products of corresponding elements of A and B

$\Rightarrow \quad \langle A, B \rangle = 9 \times 1 + 8 \times 2 + 7 \times 3 + 6 \times 4 + 5 \times 5 + 4 \times 6 = 119$

$$||A||^2 = \langle A, A \rangle = tr\,(A^T A) = \sum_{i=1}^{3}(A^T A)_{ii} = \sum_{i=1}^{3}\left(\sum_{r=1}^{3} a_{ri}^2\right)$$

$\Rightarrow \quad ||A||^2 = $ Sum of the squares of elements of $A = 81 + 64 + 49 + 36 + 25 + 16 = 271$

$\Rightarrow \quad ||A|| = \sqrt{271}$

and,

$\Rightarrow \quad ||B||^2 = $ Sum of the squares of elements of $B = 1 + 4 + 9 + 16 + 25 + 36 = 91$

$\Rightarrow \quad ||B|| = \sqrt{91}$

Let θ be the angle between A and B. Then,

$$\cos\theta = \frac{\langle A, B\rangle}{\|A\|\,\|B\|} = \frac{119}{\sqrt{271}\sqrt{91}}$$

EXERCISE 6.3

1. Consider the inner product space R^3 with the standard inner product. If $u = (2, 3, 5)$ and $v = (1, -4, 3) \in R^3$ and θ is the angle between u and v, find $\cos\theta$.
2. Find $\cos\theta$, where θ is the angle between $u = (1, 3, -5, 4)$ and $v = (2, -3, 4, 1)$ in inner product space R^4 with standard inner product.
3. Let $P(t)$ be the inner product space of all polynomials with inner product defined by

$$\langle f, g\rangle = \int_0^1 f(t)g(t)dt \quad \text{for all } f, g \in P(t). \text{ If } f(t) = 3t - 5 \text{ and } g(t) = t^2. \text{ Then, find}$$

the angle between f and g.

ANSWERS

1. $\dfrac{5}{\sqrt{26}\sqrt{38}}$ 2. $\dfrac{-23}{\sqrt{1530}}$ 3. $\cos^{-1}\dfrac{-55}{12\sqrt{65}}$

6.5 ORTHOGONALITY

In this section, we shall discuss the concept of orthogonality and orthonormality of vectors in an inner product space. The same will be extended to orthogonal and orthonormal bases of an inner product space.

ORTHOGONAL VECTORS *Let V be an inner product space and vectors $u, v \in V$. The vector u is said to be orthogonal to vector v if $\langle u, v\rangle = 0$.*

If u is orthogonal to v, then we write $u \perp v$.

Let u, v be vectors in an inner product space V such that

$$u \perp v \Rightarrow \langle u, v\rangle = 0 \Rightarrow \langle v, u\rangle = 0 \qquad \text{[By symmetry]}$$
$$\Rightarrow v \perp u$$

Thus, the relation 'is orthogonal to' on an inner product space is a symmetric relation.

Let u be an arbitrary vector in an inner product space V. Then $\langle u, 0_V\rangle = 0$. So, every vector in V is orthogonal to the null vector.

Let $u, v \in V$ such that $u \perp v$ and a be any scalar in R. Then,

$$\langle au, v\rangle = a\langle u, v\rangle = a0 = 0 \qquad [\because\ u \perp v \ \therefore\ \langle u, v\rangle = 0]$$
$$\Rightarrow \qquad au \perp v$$

Thus, if $u \perp v$, then every scalar multiple of u is also orthogonal to v.

Since, $\langle u, u \rangle > 0$ for all $u \neq 0_V$ and $\langle u, u \rangle = 0$ iff $u = 0_V$. So, the null vector is the only vector which is orthogonal to itself.

Now, let a vector $u \in V$ be such that u is orthogonal to every vector in V. Then,

$$\langle u, v \rangle = 0 \quad \text{for all } v \in V$$
$$\Rightarrow \quad \langle u, u \rangle = 0 \qquad \qquad \text{[Taking } v = u\text{]}$$
$$\Rightarrow \quad u = 0_V$$

Thus, null vector is the only vector which is orthogonal to every vector in V.

If u and v are two non-zero vectors in V. Then,

$$u \perp v \Leftrightarrow \langle u, v \rangle = 0 \Leftrightarrow \frac{\langle u, v \rangle}{\|u\| \|v\|} = 0 \Leftrightarrow \cos\theta = 0 \Leftrightarrow \theta = \frac{\pi}{2}$$

Thus, two orthogonal vectors are always perpendicular to each other.

A vector u is said to be perpendicular or orthogonal to a subspace S of inner product space V if it is orthogonal to every vector in S, i.e. $\langle u, v \rangle = 0$ for all $v \in S$.

Consider the inner product space R^3 with standard inner product.

Let $u = (1, 2, -3)$, $v = (1, 1, 1)$ and $w = (-1, 4, -3) \in R^3$. Then,

$$\langle u, v \rangle = 1 \times 1 + 2 \times 1 + (-3) \times 1 = 0$$
$$\langle v, w \rangle = 1 \times -1 + 1 \times 4 + 1 \times -3 = 0$$

and $\qquad \langle u, w \rangle = -1 + 8 + 9 = 16$

Thus, v is orthogonal to u and w but u and w are not orthogonal.

ILLUSTRATIVE EXAMPLES

EXAMPLE-1 *Let $C[-\pi,\pi]$ be the inner product space of all continuous functions defined on $[-\pi, \pi]$ with the inner product defined by*

$$\langle f, g \rangle = \int_{-\pi}^{\pi} f(t)g(t)dt$$

Prove that $\sin t$ *and* $\cos t$ *are orthogonal functions in* $C[-\pi,\pi]$.

SOLUTION We have,

$$\langle f, g \rangle = \int_{-\pi}^{\pi} f(t)g(t)dt$$

$$\therefore \quad \langle \sin t, \cos t \rangle = \int_{-\pi}^{\pi} \sin t \cos t \, dt = \frac{1}{2}\int_{-\pi}^{\pi} \sin 2t \, dt = -\frac{1}{4}\Big[\cos 2t\Big]_{-\pi}^{\pi} = -\frac{1}{4}(1-1) = 0$$

Thus, $\sin t$ and $\cos t$ are orthogonal functions in the inner product space $C[-\pi,\pi]$.

EXAMPLE-2 Consider the inner product space R^4 with the standard inner product. If vectors $u = (3, 2, k, -5)$ and $v = (1, k, 7, 3)$ are orthogonal, find the value of k.

SOLUTION It is given that u and v are orthogonal vectors in R^4.

$\therefore \quad \langle u, v \rangle = 0$

$\Rightarrow \quad 3 \times 1 + 2 \times k + k \times 7 + (-5) \times 3 = 0 \Rightarrow 9k - 12 = 0 \Rightarrow k = \dfrac{4}{3}$

EXAMPLE-3 Consider the vector space $R^{2 \times 3}$ of all 2×3 matrices with the inner product defined by
$$\langle A, B \rangle = tr\,(B^T A)$$
If $A = \begin{bmatrix} 2 & a & 3 \\ 1 & 0 & -2 \end{bmatrix}$ and $B = \begin{bmatrix} -4 & 3 & a \\ 5 & 7 & 2 \end{bmatrix}$ are matrices in $R^{2 \times 3}$ such that A is orthogonal to B. Find a.

SOLUTION It is given that A is orthogonal to B.

$\therefore \quad \langle A, B \rangle = 0$

$\Rightarrow \quad tr\,(B^T A) = 0$

$\Rightarrow \quad \sum\limits_{i=1}^{3} (B^T A)_{ii} = 0$

$\Rightarrow \quad \sum\limits_{i=1}^{3} \left(\sum\limits_{r=1}^{3} (B^T)_{ir} (A)_{ri} \right) = 0$

$\Rightarrow \quad \sum\limits_{i=1}^{3} \left(\sum\limits_{r=1}^{3} b_{ri}\, a_{ri} \right) = 0$

$\Rightarrow \quad$ Sum of the products of corresponding elements of A and $B = 0$

$\Rightarrow \quad -8 + 3a + 3a + 5 + 0 - 4 = 0 \Rightarrow 6a - 7 = 0 \Rightarrow a = \dfrac{7}{6}$

EXAMPLE-4 Two vectors u and v in an inner product space are orthogonal iff
$$||u + v||^2 = ||u||^2 + ||v||^2.$$
Interpret the result geometrically.

SOLUTION We have,

$||u + v||^2 = ||u||^2 + ||v||^2$

$\Leftrightarrow \quad \langle u + v, u + v \rangle = \langle u, u \rangle + \langle v, v \rangle$

$\Leftrightarrow \quad \langle u, u \rangle + \langle u, v \rangle + \langle v, u \rangle + \langle v, v \rangle = \langle u, u \rangle + \langle v, v \rangle$

⇔ $2\langle u, v\rangle = 0$
⇔ $\langle u, v\rangle = 0$
⇔ u is orthogonal to v.

Geometrical interpretation If we view vectors u and v as sides AB and AC of a triangle ABC, then vector $u+v$ represents its third side BC.

∴ $$u \perp v \Leftrightarrow ||u+v||^2 = ||u||^2 + ||v||^2$$

can be interpreted geometrically as follows:

$\triangle ABC$ is right angled at $A \Leftrightarrow BC^2 = AB^2 + AC^2$.

EXAMPLE-5 *If u and v are vectors in an inner product space, then $u+v$ is orthogonal to $u-v$, if and only if $||u|| = ||v||$. Interpret the result geometrically.*

SOLUTION $u+v$ is orthogonal to $u-v$.
⇔ $\langle u+v, u-v\rangle = 0$
⇔ $\langle u, u\rangle + \langle u, -v\rangle + \langle v, u\rangle + \langle v, -v\rangle = 0$
⇔ $\langle u, u\rangle - \langle u, v\rangle + \langle v, u\rangle - \langle v, v\rangle = 0$
⇔ $||u||^2 - ||v||^2 = 0$
⇔ $||u|| = ||v||$

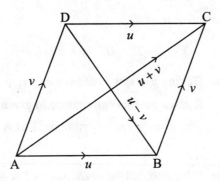

If we view vectors u and v as adjacent sides AB and AC respectively of a parallelogram $ABCD$, then vectors $u+v$ and $u-v$ represent its diagonals AC and BD respectively.

∴ $\langle u+v, u-v\rangle = 0 \Leftrightarrow ||u|| = ||u||$
⇒ $AC \perp BD \Leftrightarrow AB = AD$

i.e. if diagonals of a parallelogram are perpendicular, then it is a rhombus.

EXAMPLE-6 *Find a non-zero vector u in R^3 that is orthogonal to the vectors $v_1 = (1, 1, 2)$, $v_2 = (2, 1, 3)$ and $v_3 = (1, 2, 3)$ in R^3 with standard inner product.*

SOLUTION Let $u = (x, y, z)$ be the required vector. Then,

$$\langle u, v_1 \rangle = 0, \langle u, v_2 \rangle \quad \text{and} \quad \langle u, v_3 \rangle = 0$$

$\Rightarrow \quad x + y + 2z = 0$

$2x + y + 3z = 0$

$x + 2y + 3z = 0$

This is a homogeneous system of equation having non-trivial solutions as the coefficient matrix is singular.

In matrix from the system can be written as

$$\begin{bmatrix} 1 & 1 & 2 \\ 2 & 1 & 3 \\ 1 & 2 & 3 \end{bmatrix} \begin{bmatrix} x \\ y \\ z \end{bmatrix} = \begin{bmatrix} 0 \\ 0 \\ 0 \end{bmatrix}$$

$\Rightarrow \quad \begin{bmatrix} 1 & 1 & 2 \\ 0 & -1 & -1 \\ 0 & 1 & 1 \end{bmatrix} \begin{bmatrix} x \\ y \\ z \end{bmatrix} = \begin{bmatrix} 0 \\ 0 \\ 0 \end{bmatrix}$ Applying $R_2 \to R_2 - 2R_1, R_3 \to R_3 - R_1$

$\Rightarrow \quad \begin{bmatrix} 1 & 1 & 2 \\ 0 & -1 & -1 \\ 0 & 0 & 0 \end{bmatrix} \begin{bmatrix} x \\ y \\ z \end{bmatrix} = \begin{bmatrix} 0 \\ 0 \\ 0 \end{bmatrix}$

Thus, the given system is equivalent to

$$x + y + 2z = 0, \quad -y - z = 0$$

Here, only one variable is free. Taking $z = t$, we obtain $x = -t$, $y = -t$, $z = t$.

Thus, $u = (-t, -t, t)$, $t \in R$ gives vectors orthogonal to given vectors.

Putting $t = 1$, we get $(-1, -1, 1)$ as a vector orthogonal to v_1, v_2 and v_3

EXAMPLE-7 *Find a non-zero vector orthogonal to the vectors $v_1 = (1, 2, 1)$ and $v_2 = (4, 5, 2)$ in inner product space R^3 with standard inner product.*

SOLUTION Let $u = (x, y, z) \in R^3$ be the desired vector. Then,

$\langle u, v_1 \rangle = 0 \quad \text{and} \quad \langle u, v_2 \rangle = 0$

$\Rightarrow \quad x + 2y + z = 0 \quad \text{and} \quad 4x + 5y + 2z = 0$

$\Rightarrow \quad x + 2y + z = 0 \quad \text{and} \quad 2x + y = 0$ $\begin{bmatrix} \text{Mutiplying first equation by 2 and} \\ \text{subtracting from second equation} \end{bmatrix}$

This system has only one free variable (x or y). Putting $x = 1$ in second equation, we get $y = -2$. Substituting these values in first equation, we get $z = 3$.

Thus, $u = (1, -2, 3)$ is a desired vector orthogonal to v_1 and v_2.

6.5.1 ORTHOGONAL SETS

In this section, we shall extend the concept of orthogonality of vectors to orthogonality of sets.

ORTHOGONAL SET *Let V be an inner product space. A set S of non-zero vectors in V is called an orthogonal set if each pair of vectors in S is a pair of orthogonal vectors,*

i.e., $\qquad \langle u, v \rangle = 0 \quad$ for all $u, v \in S, u \neq v.$

THEOREM-1 *Any orthogonal set of non-zero vectors in an inner product space is linearly independent.*

PROOF. Let V be an inner product space and S be an orthogonal set of non-zero vectors in V. Let $S' = \{v_1, v_2, \ldots, v_n\}$ be a finite subset of set S. Then,

$$\sum_{i=1}^{n} \lambda_i v_i = 0_V$$

$\Rightarrow \quad \left\langle \sum_{i=1}^{n} \lambda_i v_i, v_j \right\rangle = \langle 0_V, v_j \rangle \quad$ for all $j \in \underline{n}$

$\Rightarrow \quad \sum_{i=1}^{n} \lambda_i \langle v_i, v_j \rangle = 0 \quad$ for all $j \in \underline{n}$

$\Rightarrow \quad \lambda_j \langle v_j, v_j \rangle = 0 \quad$ for all $j \in \underline{n} \qquad\qquad [\because \langle v_i, v_j \rangle = 0$ for all $i \neq j]$

$\Rightarrow \quad \lambda_j \|v_j\|^2 = 0 \quad$ for all $j \in \underline{n}$

$\Rightarrow \quad \lambda_j = 0 \quad$ for all $j \in \underline{n} \qquad\qquad [\because \|v_j\|^2 \neq 0$ for all $j \in \underline{n}]$

$\Rightarrow \quad \lambda_1 = \lambda_2 = \cdots = \lambda = 0$

Thus, $\sum_{i=1}^{n} \lambda_i v_i = 0_V \quad \Rightarrow \lambda_1 = \lambda_2 = \cdots = \lambda_n = 0$

So, S' is a linearly independent set.

Thus, every finite subset of S is linearly independent. Hence, S is linearly independent.

Q.E.D.

THEOREM-2 *(Pythagoras) Let V be an inner product space and $\{v_1, v_2, \ldots, v_n\}$ be an orthogonal set of vectors in V. Then,*

$$\|v_1 + v_2 + \cdots + v_n\|^2 = \|v_1\|^2 + \|v_2\|^2 + \cdots + \|v_n\|^2$$

PROOF. We have,

$$\|v_1 + v_2 + \cdots + v_n\|^2 = \langle v_1 + v_2 + \cdots + v_n, v_1 + v_2 + \cdots + v_n \rangle$$

$$\Rightarrow \quad \|v_1 + v_2 + \cdots + v_n\|^2 = \langle v_1, v_1 \rangle + \langle v_2, v_2 \rangle + \cdots + \langle v_n, v_n \rangle + 2 \sum_{i=1}^{n} \sum_{\substack{j=1 \\ i<j}}^{n} \langle v_i, v_j \rangle$$

$$\Rightarrow \quad \|v_1 + v_2 + \cdots + v_n\|^2 = \|v_1\|^2 + \|v_2\|^2 + \cdots + \|v_n\|^2 + 2 \times 0 \quad [\because \langle v_i, v_j \rangle = 0 \text{ for } i \neq j]$$

$$\Rightarrow \quad \|v_1 + v_2 + \cdots + v_n\|^2 = \|v_1\|^2 + \|v_2\|^2 + \cdots + \|v_n\|^2$$

Q.E.D.

THEOREM-3 *Let $S = \{v_1, v_2, v_3, \ldots, v_n\}$ be an orthogonal set of non-zero vectors in an inner product space V. If $u \in [S]$, then $u = \sum_{i=1}^{n} \frac{\langle u, v_i \rangle}{\|v_i\|^2} v_i$*

PROOF. Since $u \in [S]$, therefore there exist scalars $\lambda_1, \lambda_2, \ldots, \lambda_n$ such that $u = \sum_{i=1}^{n} \lambda_i v_i$

For any $j \in \underline{n}$, we have

$$\langle u, v_j \rangle = \left\langle \sum_{i=1}^{n} \lambda_i v_i, v_j \right\rangle$$

$$\Rightarrow \quad \langle u, v_j \rangle = \sum_{i=1}^{n} \lambda_i \langle v_i, v_j \rangle$$

$$\Rightarrow \quad \langle u, v_j \rangle = \lambda_j \langle v_j, v_j \rangle \qquad [\because \langle v_i, v_j \rangle = 0 \text{ for } i \neq j]$$

$$\Rightarrow \quad \langle u, v_j \rangle = \lambda_j \|v_j\|^2$$

$$\Rightarrow \quad \lambda_j = \frac{\langle u, v_j \rangle}{\|v_j\|^2}$$

$$\therefore \quad u = \sum_{i=1}^{n} \lambda_i v_i = \sum_{i=1}^{n} \frac{\langle u, v_i \rangle}{\|v_i\|^2} v_i$$

Q.E.D.

ORTHOGONAL BASIS *Let V be an inner product space. An orthogonal set of vectors in V is called an orthogonal basis of V if it is a basis of V.*

An orthogonal set of vectors in an inner product space is linearly independent. Therefore, if V is an n dimensional inner product space and S is an orthogonal set consisting of n non-zero vectors of V, then S forms an orthogonal basis of V.

THEOREM-4 *Let V be an inner product space and $\{v_1, v_2, \ldots, v_n\}$ be an orthogonal basis for V. Then, for any $u \in V$*

$$u = \frac{\langle u, v_1 \rangle}{\langle v_1, v_1 \rangle} v_1 + \frac{\langle u, v_2 \rangle}{\langle v_2, v_2 \rangle} v_2 + \cdots + \frac{\langle u, v_n \rangle}{\langle v_n, v_n \rangle} v_n$$

i.e. $\quad u = \sum_{i=1}^{n} \frac{\langle u, v_i \rangle}{||v_i||^2} v_i$

PROOF. Since $\{v_1, v_2, \ldots, v_n\}$ is a basis for V and $u \in V$. Therefore, there exist scalars $\lambda_1, \lambda_2, \ldots, \lambda_n \in R$ such that $u = \sum_{i=1}^{n} \lambda_i v_i$

Now, proceed as in the previous theorem.

Q.E.D.

REMARK. *In the above theorem* $\dfrac{\langle u, v_1 \rangle}{||v_1||^2}, \dfrac{\langle u, v_2 \rangle}{||v_2||^2}, \ldots, \dfrac{\langle u, v_n \rangle}{||v_n||^2}$ *are known as the coordinates of u relative to basis S.*

ILLUSTRATIVE EXAMPLES

EXAMPLE-1 *Let S be the set consisting of vectors $v_1 = (1, 2, 1)$, $v_2 = (2, 1, -4)$ and $v_3 = (3, -2, 1)$ in the inner product space R^3 with standard inner product.*

(i) *Show that S is orthogonal and a basis of R^3.*

(ii) *Find the coordinates of vector $(7, 1, 9)$ relative to the basis S.*

SOLUTION (i) We have,

$$\langle v_1, v_2 \rangle = 1 \times 2 + 2 \times 1 + 1 \times -4 = 0, \quad \langle v_2, v_3 \rangle = 2 \times 3 + 1 \times -2 + -4 \times 1 = 0$$

and, $\quad \langle v_1, v_3 \rangle = 1 \times 3 + 2 \times -2 + 1 \times 1 = 0$

Thus, S is orthogonal and hence linearly independent. Since $\dim R^3 = 3$. Therefore, every linearly independent set of three vectors in R^3 forms a basis of R^3. Hence, S is a basis of R^3.

(ii) Let $u = (7, 1, 9)$. Then, coordinates of u relative to $S = \{v_1, v_2, v_3\}$ are $\dfrac{\langle u, v_1 \rangle}{||v_1||^2}, \dfrac{\langle u, v_2 \rangle}{||v_2||^2}, \dfrac{\langle u, v_3 \rangle}{||v_3||^2}$

We have,

$\langle u, v_1 \rangle = 7 \times 1 + 1 \times 2 + 1 \times 9 = 18, \quad \langle u, v_2 \rangle = 7 \times 2 + 1 \times 1 + 9 \times -4 = -21$

$\langle u, v_3 \rangle = 7 \times 3 + 1 \times -2 + 9 \times 1 = 28, \quad ||v_1||^2 = \langle v_1, v_1 \rangle = 1 + 4 + 1 = 6$

$||v_2||^2 = \langle v_2, v_2 \rangle = 4 + 1 + 16 = 21, \quad ||v_3||^2 = \langle v_3, v_3 \rangle = 9 + 4 + 1 = 14$

Thus, the coordinates of u relative to S are $\dfrac{18}{6}, \dfrac{-21}{21}, \dfrac{28}{14}$ or $3, -1, 2$.
Hence, $(7, 1, 9) = 3v_1 - v_2 + 2v_3$

EXAMPLE-2 *Consider inner product space R^4 with standard inner product. Let $S = \{v_1, v_2, v_3, v_4\}$ be a subset of R^4, where $v_1 = (1, 1, 0, -1)$, $v_2 = (1, 2, 1, 3)$, $v_3 = (1, 1, -9, 2)$ and $v_4 = (16, -13, 1, 3)$.*

(i) Show that S is orthogonal and a basis of R^4.
(ii) Find the coordinates of an arbitrary vector $u = (a, b, c, d)$ in R^4 relative to the basis S.

SOLUTION (i) We have,

$$\langle v_1, v_2 \rangle = 1 \times 1 + 1 \times 2 + 0 \times 1 + (-1) \times 3 = 0$$
$$\langle v_1, v_3 \rangle = 1 \times 1 + 1 \times 1 + 0 \times 9 + (-1) \times 2 = 0$$
$$\langle v_1, v_4 \rangle = 1 \times 16 + 1 \times -13 + 0 \times 1 + (-1) \times 3 = 0$$
$$\langle v_2, v_3 \rangle = 1 \times 1 + 2 \times 1 + 1 \times -9 + 3 \times 2 = 0$$
$$\langle v_2, v_4 \rangle = 1 \times 16 + 2 \times -13 + 1 \times 1 + 3 \times 3 = 0$$
$$\langle v_3, v_4 \rangle = 1 \times 16 + 1 \times -13 + (-9) \times 1 + 2 \times 3 = 0$$

Thus, S is orthogonal and hence linearly independent. Consequently, S is a basis of R^4 as any four linearly independent vectors from a basis of R^4.

(ii) The coordinates of u relative to the basis S are

$$\dfrac{\langle u, v_1 \rangle}{\|v_1\|^2} = \dfrac{a+b-d}{3}, \quad \dfrac{\langle u, v_2 \rangle}{\|v_2\|^2} = \dfrac{a+2b+c+3d}{15}$$

$$\dfrac{\langle u, v_3 \rangle}{\|v_3\|^2} = \dfrac{a+b-9c+2d}{87}, \quad \dfrac{\langle u, v_4 \rangle}{\|v_4\|^2} = \dfrac{16a-13b+c+3d}{435}$$

EXAMPLE-3 *Let $v_1 = (1, -2, 3, 4)$ and $v_2 = (3, -5, 7, 8)$ be two vectors in the inner product space R^4. Find a basis of the subspace S of R^4 that is orthogonal to v_1 and v_2.*

SOLUTION We have,

$$S = \{u = (x, y, z, t) \in R^4 : \langle u, v_1 \rangle = 0 \text{ and } \langle u, v_2 \rangle = 0\}$$

Now,

$\langle u, v_1 \rangle = 0$ and $\langle u, v_2 \rangle = 0$
$\Rightarrow \quad x - 2y + 3z + 4t = 0$ and $3x - 5y + 7z + 8t = 0$
$\Rightarrow \quad x - 2y + 3z + 4t = 0$ and $x - y + z = 0$ [On eliminating t]

This system of equations has two free variables.
Putting $x = 1, y = 2$ in $x - y + z = 0$, we get $z = 1$

Substituting $x = 1, y = 2, z = 1$ in $x - 2y + 3z + 4t = 0$, we get $t = 0$.

∴ $u_1 = (1, 2, 1, 0) \in S$

Putting $x = 4$, $y = 4$ in $x - y + z = 0$, we get $z = 0$.

Substituting these values in $x - 2y + 3z + 4t = 0$, we get $t = 1$

∴ $u_2 = (4, 4, 0, 1) \in S$

Thus, $u_1 = (1, 2, 1, 0)$ and $u_2 = (4, 4, 0, 1)$ from a basis of S.

EXAMPLE-4 *Let S be the subspace of inner product space R^4 orthogonal to the vectors $v_1 = (1, 1, 2, 2)$ and $v_2 = (0, 1, 2, -1)$. Find an orthogonal basis of S.*

SOLUTION We have,
$$S = \{u \in R^4 : \langle u, v_1 \rangle = 0, \langle u, v_2 \rangle = 0\}$$
Let $u = (x, y, z, t) \in S$ Then,

$\langle u, v_1 \rangle = 0$ and $\langle u, v_2 \rangle = 0$

$\Rightarrow \quad x + y + 2z + 2t = 0$ and $y + 2z - t = 0$

$\Rightarrow \quad x + y + 2z + 2t = 0$ and $x + 3t = 0$ [On subtracting second equation from first]

Clearly, there are two free variables. Second equation has one free variable, say, t. Consequently, first equation also has one free variable, say z.

Let us take z and t as free variables.

Now,

$z = 1, t = 0 \quad \Rightarrow \quad x = 0$ and $y = -2$

∴ $u_1 = (0, -2, 1, 0) \in S$

Now, we have to find one more vector u_2 (say) in S which is orthogonal to u_1 and u_1, u_2 are linearly independent.

Let $u_2 = (a, b, c, d)$. Then,

$\langle u_2, v_1 \rangle = 0, \langle u_2, v_2 \rangle = 0$ and $\langle u_2, u_1 \rangle = 0$

Now, $\langle u_2, v_1 \rangle = 0$ and $\langle u_2, v_2 \rangle = 0$

$\Rightarrow \quad a + b + 2c + 2d = 0$ and $a + 3d = 0$ [Proceeding as above in $\langle u, v_1 \rangle = 0, \langle u, v_2 \rangle = 0$]

$\langle u_2, v_1 \rangle = 0 \quad \Rightarrow \quad -2b + c = 0$

Thus, we have

$a + b + 2c + 2d = 0$, $a + 3d = 0$ and $-2b + c = 0$

$\Rightarrow \quad a = -3d, b = \dfrac{d}{5}, c = \dfrac{2d}{5}$

$\Rightarrow \quad a = -15, b = 1, c = 2, d = 5$ [Putting $d = 5$]

∴ $u_2 = (-15, 1, 2, 5)$

Hence, u_1, u_2 form an orthogonal basis of S.

EXERCISE 6.4

1. Let V be an inner product space and $S = \{v_1, v_2, \ldots, v_k\}$ be an orthogonal set of vectors in V. Show that $\{\lambda_1 v_1, \lambda_2 v_2, \ldots, \lambda_k v_k\}$ is also an orthogonal set for any scalars $\lambda_1, \lambda_2, \ldots, \lambda_k$.

2. Let $V = C[-\pi, \pi]$ be the vector space of continuous functions on the interval $[-\pi, \pi]$ with inner product defined by
$$<f,g> = \int_{-\pi}^{\pi} f(t)\, g(t)\, dt$$
Show that the set $S = \{1, \cos t, \cos 2t, \cos 3t, \ldots, \sin t, \sin 2t, \sin 3t, \ldots\}$ is an orthogonal set.

3. Consider the inner product space R^4 with standard inner product. Let $S = \{v_1, v_2, v_3, v_4\}$, where $v_1 = (1, 1, 1, 1)$, $v_2 = (1, 1, -1, -1)$, $v_3 = (1, -1, 1, -1)$ and $v_4 = (1, -1, -1, 1)$.

    (i) Show that S is orthogonal and a basis of R^4.

    (ii) Express $u = (1, 3, -5, 6)$ as a linear combination of vectors in S.

    (iii) Find the coordinates of an arbitrary vector $v = (a, b, c, d)$ in R^4 relative to the basis S.

4. Let V be an inner product space. Show that

    (i) $u + v$ is orthogonal to $u - v$ iff $||u|| = ||v||$.

    (ii) u is orthogonal to v iff $||u + v||^2 = ||u||^2 + ||v||^2$

5. Let $v_1 = (1, 1, 3, 4, 1)$ and $v_2 = (1, 2, 1, 2, 1)$ be two vectors in the inner product space R^5 and S be a subspace of R^5 orthogonal to v_1 and v_2. Find a basis of S.

6. Let V be the vector space of all polynomials over R of degree ≤ 2 with inner product defined by $<f,g> = \int_0^1 f(t)\, g(t)\, dt$. Find a basis of the subspace S orthogonal to the polynomial $\phi(t) = 2t + 1$.

7. Find a vector of unit length in the inner product space R^3 which is orthogonal to the vector $v = (2, -1, 6)$.

8. In the inner product space R^3 with the standard inner product find two mutually orthogonal vectors each of which is orthogonal to the vector $v = (4, 2, 3)$.

9. Suppose w_1 and w_2 are non-zero orthogonal vectors in an inner product space V. Let v be any vector in V. Find c_1 and c_2 so that v' is orthogonal to w_1 and w_2, where $v' = v - c_1 w_1 - c_2 w_2$

10. Let $w_1, w_2, \ldots, w_k$ be an orthogonal set of non-zero vectors in an inner product space V and $v \in V$. Let $v' = v - \sum_{r=1}^{k} c_r w_r$, where $c_r = \dfrac{\langle v, w_r \rangle}{\langle w_r, w_r \rangle}$. Then prove that v' is orthogonal to $w_1, w_2, \ldots, w_k$.

ANSWERS

3. (i) $u = \dfrac{5}{4}v_1 + \dfrac{3}{4}v_2 - \dfrac{13}{4}v_3 + \dfrac{9}{4}v_4$

   (ii) $\dfrac{a+b+c+d}{4}, \dfrac{a+b-c-d}{4}, \dfrac{a-b+c-d}{4}, \dfrac{a-b-c+d}{4}$

5. $\{(-1, 0, 0, 0, 1), (-6, 2, 0, 1, 0), (-5, 2, 1, 0, 0)\}$

6. $\{f(t) = 7t^2 - 5t, g(t) = 12t^2 - 5\}$ 7. $\left(\dfrac{2}{3}, \dfrac{-2}{3}, \dfrac{-1}{3}\right)$

8. $(3, -3, -2), (5, 17, -18)$ 9. $c_1 = \dfrac{\langle v, w_1 \rangle}{\langle w_1, w_1 \rangle}, c_2 = \dfrac{\langle v, w_2 \rangle}{\langle w_2, w_2 \rangle}$

6.6 ORTHOGONAL COMPLIMENTS

ORTHOGONAL COMPLIMENT OF A VECTOR *Let V be an inner product space and $u \in V$. Then the set of all vectors orthogonal to u is called the orthogonal compliment of u and is denoted by $u^\perp$.*

Thus, $u^\perp = \{v \in V : \langle v, u \rangle = 0\}$

The symbol "$u^\perp$" is usually read as "u perpendicular",.

THEOREM-1 *Let V be an inner product space and u be a vector in V. Then the orthogonal complement $u^\perp$ of vector u is a subspace of V.*

PROOF. We have,
$$u^\perp = \{u \in V : \langle v, u \rangle = 0\}$$

Clearly, $u^\perp$ is a subset of V.

Since $0_V \in V$ such that $\langle 0_V, u \rangle = 0$.

$\therefore \quad 0_V \in u^\perp$

Thus, $u^\perp$ is a non-void subset of V.

Let $v_1, v_2 \in u^\perp$ and $a, b \in R$. Then, $av_1 + bv_2 \in V$ such that

$\qquad \langle av_1 + bv_2, u \rangle = a\langle v_1, u \rangle + b\langle v_2, u \rangle$

$\Rightarrow \quad \langle av_1 + bv_2, u \rangle = a \times 0 + b \times 0 = 0 \qquad [\because \ v_1, v_2 \in u^\perp \Rightarrow \langle v_1, u \rangle = 0, \langle v_2, u \rangle = 0]$

$\therefore \qquad av_1 + bv_2 \in u^\perp$

Thus, $u^\perp$ is a non-empty subset of V such that $av_1 + bv_2 \in u^\perp$ for all $v_1, v_2 \in u^\perp$ and $a, b \in R$.

Hence, $u^\perp$ is a subspace of V.

Q.E.D.

ILLUSTRATIVE EXAMPLES

EXAMPLE-1 *Let $u = (-1, 4, -3)$ be a vector in the inner product space with the standard inner product. Find a basis of the subspace $u^\perp$ of R^3.*

SOLUTION We have,
$$u^\perp = \{v \in R^3 : \langle u, v \rangle = 0\}$$
or, $$u^\perp = \{v = (x, y, z) \in R^3 : -x + 4y - 3z = 0\}$$

Thus, $u^\perp$ consists of all vectors $v = (x, y, z)$ such that $-x + 4y - 3z = 0$. In this equation there are only two free variables. Taking y and z as free variable, we find that
$$y = 1, z = 1 \Rightarrow x = 1; \quad y = 0, z = 1 \Rightarrow x = -3$$

Thus, $v_1 = (1, 1, 1)$ and $v_2 = (-3, 0, 1)$ are two independent solutions of $-x + 4y - 3z = 0$.
Hence, $\{v_1 = (1, 1, 1), v_2 = (-3, 0, 1)\}$ form a basis of for $u^\perp$.

EXAMPLE-2 *Let $u = (1, 2, 3)$ be a vector in R^3. Find an orthogonal basis for $u^\perp$.*

SOLUTION We have,
$$u^\perp = \{v = (x, y, z) \in R^3 : \langle u, v \rangle = 0\}$$
or, $$u^\perp = \{v = (x, y, z) \in R^3 : x + 2y + 3z = 0\}$$

The equation $x + 2y + 3z = 0$ has two free variables.

In order to find an orthogonal basis for $u^\perp$, we have to find non-zero solutions of $x + 2y + 3z = 0$ which are orthogonal to each other. For this, let us find a non-zero solution of $x + 2y + 3z = 0$ by putting $y = 1$ and $z = -1$. The solution so obtained is $v_1 = (1, 1, -1)$.

Now, we have to find a solution $v_2 = (a, b, c)$ of $x + 2y + 3z = 0$ such that $\langle v_1, v_2 \rangle = 0$. That is, we have to find a, b, c satisfying $a + 2b + 3c = 0$ and $a + b - c = 0$
or, $a + 2b + 3c = 0$ and $b + 4c = 0$ [Subtracting second equation from first]
Putting $c = -1$ in $b + 4c = 0$, we get $b = 4$

Substituting these values in any one of the two equations $a + 2b + 3c = 0$ and $a + b - c = 0$, we get $a = -5$.

$\therefore \quad v_2 = (-5, 4, -1)$

Hence, v_1, v_2 form an orthogonal basis for $u^\perp$.

6.6.1 ORTHOGONAL COMPLEMENT OF A SET

Let us now extend the concept of orthogonal complement of a vector to the orthogonal complement of a set.

ORTHOGONAL COMPLEMENT OF A SET Let S be a subset of an inner product space V. The orthogonal complement of S, denoted by $S^\perp$ (read as "S perpendicular"), is the set of all vectors in V that are orthogonal to every vector in S. That is,

$$S^\perp = \{v \in V : \langle v, u \rangle = 0 \quad \text{for all } u \in S\}$$

THEOREM-1 Let S be a subset of an inner product space V. Then, $S^\perp$ is a subspace of V.

PROOF. We have,
$$S^\perp = \{v \in V : \langle v, u \rangle = 0 \quad \text{for all } u \in S\}$$
Clearly, $0_V \in V$ such that $\langle 0_V, u \rangle = 0$ for all $u \in S$.
$\therefore \qquad 0_V \in S^\perp$

Thus, $S^\perp$ is a non-void subset of V.
Let $v_1, v_2 \in S^\perp$. Then, $\langle v_1, u \rangle = 0$ and $\langle v_2, u \rangle = 0$ for all $u \in S$.
For any scalars $a, b \in R$ and any vector $u \in S$, we have

$$\langle av_1 + bv_2, u \rangle = a\langle v_1, u \rangle + b\langle v_2, u \rangle = a \times 0 + b \times 0 = 0$$
$\Rightarrow \qquad av_1 + bv_2 \in S^\perp$

Thus, $av_1 + bv_2 \in S^\perp$ for all $v_1, v_2 \in S^\perp$ and $a, b \in R$.
Hence, $S^\perp$ is a subspace of V.

Q.E.D.

Note: The orthogonal complement of V is the null space $\{0_V\}$ and that of $\{0_V\}$ is V itself.

EXAMPLE Let V be the real inner product space consisting of the space of real-valued continuous functions on the interval $[-1, 1]$ with the inner product defined by

$$\langle f, g \rangle = \int_{-1}^{1} f(t)g(t)dt.$$

Let S be the subspace of all odd functions. Find the orthogonal complement of S.

SOLUTION We have,
$$S^\perp = \{f \in V : \langle f, g \rangle = 0 \text{ for all } g \in S\}$$
Now,
$\langle f, g \rangle = 0$ for all $g \in S$

$\Rightarrow \quad \int_{-1}^{1} f(t)g(t)dt = 0 \quad$ for all real-valued odd functions $g(t)$ on the interval $[-1, 1]$

$\Rightarrow \quad f(t)$ is any even function on the interval $[-1, 1]$

$\therefore \quad S^\perp = \{f \in V : f(-t) = f(t) \text{ for all } t \in [-1, 1]\}$

ORTHOGONAL COMPLEMENT OF AN ORTHOGONAL COMPLEMENT Let S be a subset of an inner product space V. Then, $S^\perp$ is a subspace of V. The set of all vectors in V which are orthogonal to each vector in $S^\perp$ is called orthogonal complement of an orthogonal complement and is denoted by $S^{\perp\perp}$. That is,

$$S^{\perp\perp} = \{v \in V : \langle v, u \rangle = 0 \text{ for all } u \in S^\perp\}$$

By Theorem 1, $S^{\perp\perp}$ is a subspace of V.

THEOREM-2 Let V be an inner product space and S, S_1, S_2 be subsets of V. Then,

(i) $S \subset S^{\perp\perp}$
(ii) $S_1 \subset S_2 \Rightarrow S_2^\perp \subset S_1^\perp$
(iii) $S^\perp = [S]^\perp$
(iv) $[S] \subseteq S^{\perp\perp}$
(v) $S^\perp = S^{\perp\perp\perp}$

PROOF. (i) Let v be an arbitrary vector in S. Then,

$$\langle u, v \rangle = 0 \quad \text{for all } u \in S^\perp$$

$\Rightarrow \quad v \in S^{\perp\perp}$

Thus, $\quad v \in S \Rightarrow v \in S^{\perp\perp}$.

Hence, $\quad S \subset S^{\perp\perp}$.

(ii) Let $v \in S_2^\perp$. Then,

$$\langle v, u \rangle = 0 \quad \text{for all } u \in S_2$$

$\Rightarrow \quad \langle v, u \rangle = 0 \quad \text{for all } u \in S_1$ $\quad\quad\quad [\because S_1 \subset S_2]$

$\Rightarrow \quad v \in S_1^\perp$.

Thus, $\quad v \in S_2^\perp \Rightarrow v \in S_1^\perp$

Hence, $\quad S_2^\perp \subset S_1^\perp$.

(iii) We have,

$$S \subseteq [S]$$

$\Rightarrow \quad [S]^\perp \subseteq S^\perp$ $\quad\quad\quad\quad\quad\quad\quad\quad\quad\quad\quad\quad\quad\quad\quad$ [Using (ii)]

Now, let $u \in S^\perp$. Then, u is orthogonal to every vector in S.

Let v be an arbitrary vector in $[S]$. Then, there exist $v_1, v_2, \ldots, v_k$ in S such that

$$v = \lambda_1 v_1 + \lambda_2 v_2 + \cdots + \lambda_k v_k \quad \text{for } \lambda_i \in R$$

$\therefore \quad \langle u, v \rangle = \langle u, \lambda_1 v_1 + \lambda_2 v_2 + \cdots + \lambda_k v_k \rangle$

$\Rightarrow \quad \langle u, v \rangle = \lambda_1 \langle u, v_1 \rangle + \lambda_2 \langle u, v_2 \rangle + \cdots + \lambda_k \langle u, v_k \rangle$

$\Rightarrow \quad \langle u, v \rangle = \lambda_1(0) + \lambda_2(0) + \cdots + \lambda_k(0) = 0 \qquad$ [u is orthogonal to each $v_i \in S$]

$\Rightarrow \quad \langle u, v \rangle = 0$

$\Rightarrow \quad u \in [S]^\perp$

Thus, $\quad u \in S^\perp \Rightarrow u \in [S]^\perp$

$\therefore \quad S^\perp \subseteq [S]^\perp$

Hence, $\quad [S]^\perp = S^\perp$

(iv) Let $v \in [S]$. Then, there exist vectors $v_1, v_2, \ldots, v_k$ in S such that

$$v = \sum_{i=1}^{k} \lambda_i v_i \quad \text{for some } \lambda_i \in R.$$

Let u be an arbitrary vector in $S^\perp$. Then,

$\Rightarrow \quad \langle u, v \rangle = \left\langle u, \sum_{i=1}^{k} \lambda_i v_i \right\rangle$

$\Rightarrow \quad \langle u, v \rangle = \sum_{i=1}^{k} \lambda_i \langle u, v_i \rangle$

$\Rightarrow \quad \langle u, v \rangle = \sum_{i=1}^{k} \lambda_i \qquad\qquad$ [$\because \quad v_i \in S$ and $u \in S^\perp \therefore \quad \langle u, v_i \rangle = 0$]

$\Rightarrow \quad \langle u, v \rangle = 0$

$\therefore \quad \langle u, v \rangle = 0 \quad$ for all $u \in S^\perp$

$\Rightarrow \quad v \in \left(S^\perp\right)^\perp$

Thus, $\quad v \in [S] \Rightarrow v \in \left(S^\perp\right)^\perp = S^{\perp\perp}$

$\therefore \quad [S] \subset S^{\perp\perp}$

(v) From (i), we have $S \subseteq S^{\perp\perp}$

$\therefore \quad S^\perp \subseteq \left(S^\perp\right)^{\perp\perp} \qquad\qquad$ [Replacing S by $S^\perp$]

$\Rightarrow \quad S^\perp \subseteq S^{\perp\perp\perp}$

Again, $\quad S \subseteq S^{\perp\perp}$

$\Rightarrow \quad \left(S^{\perp\perp}\right)^\perp \subseteq S^\perp \qquad\qquad$ [Using (ii)]

$\Rightarrow \quad S^{\perp\perp\perp} \subset S^\perp$

Hence, $\quad S^\perp = S^{\perp\perp\perp}$

Q.E.D.

ILLUSTRATIVE EXAMPLE

EXAMPLE-1 *Let S be the subspace of the inner product space R^3 (with standard inner product) spanned by $u = (1, 0, 1)$ and $v = (1, 2, -2)$. Find a basis of the orthogonal complement $S^\perp$ of S.*

SOLUTION We have,

$$S^\perp = \{w = (x, y, z) \in R^3 : \langle w, u \rangle = 0 \quad \text{for all } u \in S\}.$$

$\therefore \quad \langle w, u \rangle = 0 \quad \text{and} \quad \langle w, v \rangle = 0 \qquad [\because \quad u, v \in S]$

$\Rightarrow \quad x + z = 0 \quad \text{and} \quad x + 2y - 2z = 0$

$\Rightarrow \quad x = -z, \; y = \dfrac{3z}{2}$

$\therefore \quad S^\perp = \left\{ \left(-z, \dfrac{3z}{2}, z \right) : z \in R \right\}$

Clearly, $S^\perp$ is spanned by $\left(-1, \dfrac{3}{2}, 1 \right)$. Hence, $\left(-1, \dfrac{3}{2}, 1 \right)$ forms a basis for $S^\perp$.

EXAMPLE-2 *Consider the inner product space R^5 with standard inner product space. Let S be the subspace of R^5 spanned by the vectors $v_1 = (1, 2, 3, -1, 2)$ and $v_2 = (2, 4, 7, 2, -1)$. Find a basis of orthogonal complement $S^\perp$ for S.*

SOLUTION Clearly,

$$S^\perp = \{w \in R^5 : \langle w, u \rangle = 0 \quad \text{for all } u \in S\}$$

In order to find $S^\perp$, we need to find all $w = (x, y, z, s, t) \in R^5$ such that

$\langle w, v_1 \rangle = 0 \quad \text{and} \quad \langle w, v_2 \rangle = 0$

$\Rightarrow \quad x + 2y + 3z - s + 2t = 0 \quad \text{and} \quad 2x + 4y + 7z + 2s - t = 0$

These are two equations in five variables. So, there are three free variables. On eliminating x from the second equation, above system reduces to

$$x + 2y + 3z - s + 2t = 0$$
$$z + 4s - 5t = 0$$

In the second equation there are two free variables.

So, taking y, s and t as free variables, we obtain as follows:

(i) Taking $y = -1$, $s = 0$, $t = 0$, we obtain the solution $w_1 = (2, -1, 0, 0, 0)$

(ii) Taking $y = 0$, $s = 1$, $t = 0$, we obtain the solution $w_2 = (13, 0, -4, 1, 0)$

(iii) Taking $y = 0$, $s = 0$, $t = 1$, we obtain the solution $w_3 = (-17, 0, 5, 0, 1)$.

Thus, $\{w_1, w_2, w_3\}$ is a basis for $S^\perp$.

EXERCISE 6.5

1. Let $u = (-1, 2, 1)$ be a vector in the inner product space R^3 with the standard inner product. Find a basis of the subspace $u^\perp$ of R^3.
2. Let $v = (2, 1, -1)$ be a vector in R^3. Find an orthogonal basis for $v^\perp$.
3. Let $w = (1, 2, 3, 1)$ be a vector in R^4. Find an orthogonal basis of $w^\perp$.
4. Let S be the subspace of the inner product space R^3 spanned by the vectors $v_1 = (0, 1, 1)$ and $v_2 = (3, -1, 2)$. Find a basis of the orthogonal complement $S^\perp$ of S.
5. Let $u = (1, -2, -1, 3)$ be a vector in R^4. Find an orthogonal basis for $u^\perp$.
6. Let $R^{2\times 2}$ be the inner product space of all 2×2 matrices with the inner product defined by $\langle A, B \rangle = tr\,(B^T A)$. Find an orthogonal basis for the orthogonal complement of
   (i) diagonal matrices   (ii) symmetric matrices.

ANSWERS

1. $\{(1, 0, 1), (3, 1, 1)\}$ 2. $\{(1, -1, 1), (0, 1, 1)\}$
3. $\{(0, 0, 1, -3), (0, -5, 3, 1), (-14, 2, 3, 1)\}$ 4. $(-1, -1, 1)$
5. $\{(0, 0, 3, 1), (0, 5, -1, 3), (-14, -2, -1, 3)\}$
6. (i) $\left\{ \begin{bmatrix} 0 & 1 \\ 0 & 0 \end{bmatrix}, \begin{bmatrix} 0 & 0 \\ 1 & 0 \end{bmatrix} \right\}$ (ii) $\left\{ \begin{bmatrix} 0 & -1 \\ 1 & 0 \end{bmatrix} \right\}$

6.7 ORTHONORMALITY

ORTHONORMAL SET *Let V be an inner product space. A set S of vectors in V is called an orthonormal set, if*
(i) $\langle u, v \rangle = 0$ *for all $u, v \in S$, $u \neq v$*   (ii) $\langle u, u \rangle = 1$ *for all $u \in S$.*

That is, an orthogonal set of vectors in an inner product space V is an orthonormal set if each vector is of unit length.

We have, $\langle 0_V, 0_V \rangle = 0 \neq 1$

Therefore, 0_V cannot be a vector in an orthonormal set.

Thus, a list $(v_1, v_2, \ldots, v_n)$ of vectors in an inner product space is an orthonormal list, if

$$\langle v_i, v_j \rangle = \delta_{ij} \text{ for all } i, j \in \underline{n}.$$

REMARK. *An orthogonal set of vectors in an inner product space can be transformed into an orthonormal set by multiplying each vector in S by the reciprocal of its length. Thus, if $S = \{v_1, v_2, \ldots, v_k\}$ is an orthogonal set in V, then $\left\{ \dfrac{v_1}{\|v_1\|}, \dfrac{v_2}{\|v_2\|}, \ldots, \dfrac{v_k}{\|v_k\|} \right\}$ is an orthonormal set in V.*

Clearly, every orthonormal set of vectors in an inner product space is an orthogonal set but the converse is not true. Therefore, all properties possessed by an orthogonal set also hold good for an orthonormal set. For example, every orthonormal set of vectors in an inner product space is linearly independent (See Theorem 1 on page 568) and if $S = \{v_1, v_2, \ldots, v_k\}$ is an orthonormal set of non-zero vectors in V, then any $u \in [S]$ is expressible as $u = \sum\limits_{i=1}^{k} \langle u, v_i \rangle v_i$ (See Theorem 2 on page 569).

ORTHONORMAL BASIS *Let V be an inner product space. An orthonormal set of vectors in V which is also a basis for V is called an orthonormal basis.*

Consider the inner product space R^3 with the standard inner product.

Clearly, $S = \left\{ e_1^{(3)}, e_2^{(3)}, e_3^{(3)} \right\}$ is a basis of R^3.

Also, $\left\langle e_1^{(3)}, e_2^{(3)} \right\rangle = 0, \left\langle e_2^{(3)}, e_3^{(3)} \right\rangle = 0$ and $\left\langle e_1^{(3)}, e_3^{(3)} \right\rangle = 0$

and, $\left\langle e_1^{(3)}, e_1^{(3)} \right\rangle = 1 = \left\langle e_2^{(3)}, e_2^{(3)} \right\rangle = \left\langle e_3^{(3)}, e_3^{(3)} \right\rangle.$

So, S is an orthonormal basis of R^3. More generally, the standard basis of R^n is the orthonormal basis for every n.

THEOREM-1 *Let V be an inner product space. Any orthonormal set of vectors in V is linearly independent.*

PROOF. Let S be an orthonormal set of vectors in V and $S_1 = \{v_1, v_2, \ldots, v_n\}$ be a finite subset of S. Further, let $\lambda_1, \lambda_2, \ldots, \lambda_n \in R$ be such that

$$\sum_{i=1}^{n} \lambda_i v_i = 0_V$$

$\Rightarrow \quad \left\langle \sum\limits_{i=1}^{n} \lambda_i v_i, v_j \right\rangle = \langle 0_V, v_j \rangle \quad \text{for all } j = 1, 2, \ldots, n$

$\Rightarrow \quad \sum\limits_{i=1}^{n} \lambda_i \langle v_i, v_j \rangle = 0 \quad \text{for all } j = 1, 2, \ldots, n$

$\Rightarrow \quad \sum\limits_{i=1}^{n} \lambda_i \delta_{ij} = 0 \quad \text{for all } j = 1, 2, \ldots, n$

$\Rightarrow \quad \lambda_j = 0 \quad \text{for all } j = 1, 2, \ldots, n$

$\Rightarrow \quad \lambda_1 = \lambda_2 = \cdots = \lambda_n = 0$

Therefore, S_1 is a linearly independent set. Since S_1 is an arbitrary subset of S. Therefore, every finite subset of S is linearly independent. Hence, S is a linearly independent set.

Q.E.D.

COROLLARY-1 *An orthonormal set of vectors in an inner product space V is a basis for V iff it spans V.*

PROOF. Let S be an orthonormal set of vectors in V such that $[S] = V$. As every orthonormal set of vectors in V is linearly independent. Therefore, S is linearly independent. It is given that S spans V. So, S is a basis for V.

Conversely let S be a basis for V. Then, S spans V.

Hence, S is a basis for V iff it spans V.

Q.E.D.

COROLLARY-2 *A set of n orthonormal vectors in an inner product space V is a basis for V iff $\dim V = n$.*

PROOF. First, let V be an inner product space such that $\dim V = n$ and S be an orthonormal set of n vectors in V. Then, S is linearly independent. Thus, S is a linearly independent set of n vectors in V such that $\dim V = n$. So, S is a basis for V.

Conversely, let S be an orthonormal set of n vectors in V such that S is a basis for V. Then, $\dim V = n$.

Q.E.D.

THEOREM-2 *Let $S = \{v_1, v_2, \ldots, v_n\}$ be an orthonormal basis of an inner product space V and $u \in V$. Then the coordinates of u relative to S are $\langle u, v_i \rangle, i = 1, 2, \ldots, n$. Thus, $u = \sum_{i=1}^{n} \langle u, v_i \rangle v_i$. Moreover, $\|u\|^2 = \sum_{i=1}^{n} |\langle u, v_i \rangle|^2$.*

PROOF. Let $\lambda_1, \lambda_2, \ldots, \lambda_n$ be the coordinates of u relative to the basis S. Then,

$$u = \sum_{j=1}^{n} \lambda_j v_j$$

$\Rightarrow \quad \langle u, v_i \rangle = \left\langle \sum_{j=1}^{n} \lambda_j v_j, v_i \right\rangle \quad \text{for all } i = 1, 2, \ldots, n$

$\Rightarrow \quad \langle u, v_i \rangle = \sum_{j=1}^{n} \lambda_j \langle v_j, v_i \rangle \quad \text{for all } i = 1, 2, \ldots, n$

$\Rightarrow \quad \langle u, v_i \rangle = \sum_{j=1}^{n} \lambda_j \delta_{ij} = \lambda_i \quad \text{for all } i = 1, 2, \ldots, n \qquad (i)$

$\therefore \quad u = \sum_{i=1}^{n} \lambda_i v_i = \sum_{i=1}^{n} \langle u, v_i \rangle v_i$

Now,

$$\|u\|^2 = \langle u, u \rangle = \left\langle \sum_{i=1}^{n} \lambda_i v_i, \sum_{j=1}^{n} \lambda_j v_j \right\rangle$$

$\Rightarrow \quad \|u\|^2 = \sum_{i=1}^{n} \sum_{j=1}^{n} \lambda_i \lambda_j \langle v_i, v_j \rangle$

$\Rightarrow \quad \|u\|^2 = \sum_{i=1}^{n} \sum_{j=1}^{n} \lambda_i \lambda_j \delta_{ij} \qquad\qquad [\because \ S \text{ is orthonormal basis for } V]$

$\Rightarrow \quad ||u||^2 = \sum_{i=1}^{n} \lambda_i^2$

$\Rightarrow \quad ||u||^2 = \sum_{i=1}^{n} |\langle u, v_i \rangle|^2$ [Using (i)]

Q.E.D.

THEOREM-3 *If $S = \{v_1, v_2, \ldots, v_n\}$ is an orthonormal set of vectors in an inner product space V and if $u \in V$, then the vector $w = u - \sum_{i=1}^{n} \langle u, v_i \rangle v_i$ is orthogonal to each of the vectors $v_1, v_2, \ldots, v_n$, i.e. $w \in S^\perp$ and consequently, to the subspace spanned by S.*

PROOF. Clearly,

$$\langle w, v_j \rangle = \left\langle u - \sum_{i=1}^{n} \langle u, v_i \rangle v_i, v_j \right\rangle \quad \text{for all } j = 1, 2, \ldots, n$$

$\Rightarrow \quad \langle w, v_j \rangle = \langle u, v_j \rangle - \left\langle \sum_{i=1}^{n} \langle u, v_i \rangle v_i, v_j \right\rangle \quad \text{for all } j = 1, 2, \ldots, n$

$\Rightarrow \quad \langle w, v_j \rangle = \langle u, v_j \rangle - \sum_{i=1}^{n} \langle u, v_i \rangle \langle v_i, v_j \rangle \quad \text{for all } j = 1, 2, \ldots, n$

$\Rightarrow \quad \langle w, v_j \rangle = \langle u, v_j \rangle - \sum_{i=1}^{n} \langle u, v_i \rangle \delta_{ij} \quad \text{for all } j = 1, 2, \ldots, n$

$\Rightarrow \quad \langle w, v_j \rangle = \langle u, v_j \rangle - \langle u, v_j \rangle = 0 \quad \text{for all } j = 1, 2, \ldots, n$ (i)

$\Rightarrow \quad w$ is orthogonal to each of the vectors $v_1, v_2, \ldots, v_n$.

Now, we shall show that w is orthogonal to the subspace spanned by S i.e. $[S]$.
Let w' be an arbitrary vector in $[S]$. Then, there exist scalars $\lambda_1, \lambda_2, \ldots, \lambda_n$ in R such that

$$w' = \sum_{i=1}^{n} \lambda_i v_i$$

$\therefore \quad \langle w, w' \rangle = \left\langle w, \sum_{i=1}^{n} \lambda_i v_i \right\rangle$

$\Rightarrow \quad \langle w, w' \rangle = \sum_{i=1}^{n} \lambda_i \langle w, v_i \rangle$

$\Rightarrow \quad \langle w, w' \rangle = \sum_{i=1}^{n} \lambda_i 0 = 0$ [Using (i)]

Since w is orthogonal to an arbitrary vector in $[S]$.
Hence, w is orthogonal to $[S]$.

Q.E.D.

THEOREM-4 *(Bessel's inequality) If $\{v_1, v_2, v_3, \ldots, v_n\}$ is an orthonormal set in an inner product space V, and $v \in V$. Then,*

$$\sum_{i=1}^{n} |\langle v, v_i \rangle|^2 \leq \|v\|^2$$

Furthermore, equality holds iff v is in the subspace spanned by $\{v_1, v_2, \ldots, v_n\}$.

PROOF. Consider the vector u given by

$$u = v - \sum_{i=1}^{n} \langle v, v_i \rangle v_i$$

$\Rightarrow \quad \|u\|^2 = \langle u, u \rangle$

$\Rightarrow \quad \|u\|^2 = \left\langle v - \sum_{i=1}^{n} \langle v, v_i \rangle v_i, \ v - \sum_{j=1}^{n} \langle v, v_j \rangle v_j \right\rangle$

$\Rightarrow \quad \|u\|^2 = \langle v, v \rangle - \left\langle v, \sum_{j=1}^{n} \langle v, v_j \rangle v_j \right\rangle - \left\langle \sum_{i=1}^{n} \langle v, v_i \rangle v_i, v \right\rangle + \left\langle \sum_{i=1}^{n} \langle v, v_i \rangle v_i, \sum_{j=1}^{n} \langle v, v_j \rangle v_j \right\rangle$

$\Rightarrow \quad \|u\|^2 = \langle v, v \rangle - \sum_{j=1}^{n} \langle v, v_j \rangle \langle v, v_j \rangle - \sum_{i=1}^{n} \langle v, v_i \rangle \langle v_i, v \rangle + \sum_{i=1}^{n} \sum_{j=1}^{n} \langle v, v_i \rangle \langle v, v_j \rangle \langle v_i, v_j \rangle$

$\Rightarrow \quad \|u\|^2 = \langle v, v \rangle - 2 \sum_{i=1}^{n} \{\langle v, v_i \rangle\}^2 + \sum_{i=1}^{n} \sum_{j=1}^{n} \langle v, v_i \rangle \langle v, v_j \rangle \delta_{ij}$

$\Rightarrow \quad \|u\|^2 = \|v\|^2 - 2 \sum_{i=1}^{n} |\langle v, v_i \rangle|^2 + \sum_{i=1}^{n} |\langle v, v_i \rangle|^2$

$\Rightarrow \quad \|u\|^2 = \|v\|^2 - \sum_{i=1}^{n} |\langle v, v_i \rangle|^2 \qquad (i)$

$\Rightarrow \quad \|v\|^2 - \sum_{i=1}^{n} |\langle v, v_i \rangle|^2 \geq 0 \qquad [\because \ \|u\|^2 \geq 0]$

$\Rightarrow \quad \sum_{i=1}^{n} |\langle v, v_i \rangle|^2 \leq \|v\|^2$

This proves the first part of the theorem.

Now, $\quad \sum_{i=1}^{n} |\langle v, v_i \rangle|^2 = \|v\|^2$

$\Rightarrow \quad \|u\|^2 = 0$

$\Rightarrow \quad u = 0_V$

$\Rightarrow \quad v - \sum_{i=1}^{n} \langle v, v_i \rangle v_i = 0_V$

$\Rightarrow \quad v = \sum_{i=1}^{n} \langle v, v_i \rangle v_i$

$\Rightarrow$ v is the linear combination of $v_1, v_2, \ldots, v_n$.

$\Rightarrow$ $v \in$ subspace spanned by the set $\{v_1, v_2, \ldots, v_n\}$.

Conversely, let v be linear combination of $v_1, v_2, \ldots, v_n$. Then,

$$v = \sum_{i=1}^{n} \langle v, v_i \rangle v_i \qquad \text{[See Theorem 2]}$$

$\Rightarrow$ $u = 0_V$

$\Rightarrow$ $||u||^2 = 0$

$\Rightarrow$ $||v||^2 = \sum_{i=1}^{n} |\langle v, v_i \rangle|^2$

Q.E.D.

THEOREM-5 *(Another form of Bessel's Inequality) If $\{v_1, v_2, \ldots, v_n\}$ is an orthogonal set of non-zero vectors in an inner product space V and $v \in V$ then $\sum_{i=1}^{n} \frac{|\langle v, v_i \rangle|^2}{||v_i||^2} \leq ||v||^2$*

PROOF. Let $w_i = \frac{v_i}{||v_i||}, i = 1, 2, \ldots, n$. Then $\{w_1, w_2, \ldots, w_n\}$ is an orthonormal set of non-zero vectors in V.

Proceeding as in Theorem 4, we get

$$\sum_{i=1}^{n} |\langle v, w_i \rangle|^2 \leq ||v||^2 \qquad \text{(i)}$$

Now,

$$\langle v, w_i \rangle = \left\langle v, \frac{v_i}{||v_i||} \right\rangle = \frac{1}{||v_i||} \langle v, v_i \rangle \text{ for all } i = 1, 2, \ldots, n \qquad \text{(ii)}$$

From (i) and (ii), we get

$$\sum_{i=1}^{n} \frac{|\langle v, v_i \rangle|^2}{||v_i||^2} \leq ||v||^2$$

Q.E.D.

EXAMPLE *Consider the inner product space $R^{2 \times 2}$ of all 2×2 matrices with the inner product defined by $\langle A, B \rangle = tr(B^T A)$. Show that $S = \left\{ \begin{bmatrix} 1 & 0 \\ 0 & 0 \end{bmatrix}, \begin{bmatrix} 0 & 1 \\ 0 & 0 \end{bmatrix}, \begin{bmatrix} 0 & 0 \\ 1 & 0 \end{bmatrix}, \begin{bmatrix} 0 & 0 \\ 0 & 1 \end{bmatrix} \right\}$ is an orthonormal basis for $R^{2 \times 2}$.*

SOLUTION Let $E_{11} = \begin{bmatrix} 1 & 0 \\ 0 & 0 \end{bmatrix}, E_{12} = \begin{bmatrix} 0 & 1 \\ 0 & 0 \end{bmatrix}, E_{21} = \begin{bmatrix} 0 & 0 \\ 1 & 0 \end{bmatrix}$ and $E_{22} = \begin{bmatrix} 0 & 0 \\ 0 & 1 \end{bmatrix}$.

Let $A = \begin{bmatrix} a & b \\ c & d \end{bmatrix}$ be any matrix in $R^{2 \times 2}$. Then,

$$A = aE_{11} + bE_{12} + cE_{21} + dE_{22}$$

Thus, any matrix in $R^{2\times 2}$ is expressible as a linear combination of matrices in S. So, S spans $R^{2\times 2}$.

Now,
$$\lambda_1 E_{11} + \lambda_2 E_{12} + \lambda_3 E_{21} + \lambda_4 E_{22} = 0$$

$$\Rightarrow \begin{bmatrix} \lambda_1 & \lambda_2 \\ \lambda_3 & \lambda_4 \end{bmatrix} = \begin{bmatrix} 0 & 0 \\ 0 & 0 \end{bmatrix} \Rightarrow \lambda_1 = \lambda_2 = \lambda_3 = \lambda_4 = 0.$$

So, S is linearly independent.

Hence, S is a basis of $R^{2\times 2}$.

Now,

$$\langle E_{11}, E_{12} \rangle = tr\left(E_{12}^T E_{11}\right) = tr(E_{21} E_{11}) = tr\begin{bmatrix} 0 & 0 \\ 1 & 0 \end{bmatrix} = 0$$

$$\langle E_{11}, E_{21} \rangle = tr\left(E_{21}^T E_{11}\right) = tr(E_{12} E_{11}) = tr\begin{bmatrix} 0 & 1 \\ 0 & 0 \end{bmatrix} = 0$$

$$\langle E_{11}, E_{22} \rangle = tr\left(E_{22}^T E_{11}\right) = tr(E_{22} E_{11}) = tr\begin{bmatrix} 0 & 0 \\ 0 & 0 \end{bmatrix} = 0$$

$$\langle E_{12}, E_{21} \rangle = tr\left(E_{21}^T E_{12}\right) = tr(E_{12} E_{12}) = tr\begin{bmatrix} 0 & 0 \\ 0 & 0 \end{bmatrix} = 0$$

$$\langle E_{12}, E_{22} \rangle = tr\left(E_{22}^T E_{12}\right) = tr(E_{22} E_{12}) = tr\begin{bmatrix} 0 & 1 \\ 0 & 0 \end{bmatrix} = 0$$

$$\langle E_{21}, E_{22} \rangle = tr\left(E_{22}^T E_{21}\right) = tr(E_{22} E_{21}) = tr\begin{bmatrix} 0 & 0 \\ 0 & 0 \end{bmatrix} = 0$$

$$\langle E_{11}, E_{11} \rangle = tr\left(E_{11}^T E_{11}\right) = tr(E_{11} E_{11}) = tr\begin{bmatrix} 1 & 0 \\ 0 & 0 \end{bmatrix} = 1$$

$$\langle E_{12}, E_{12} \rangle = tr\left(E_{12}^T E_{12}\right) = tr(E_{21} E_{12}) = tr\begin{bmatrix} 0 & 0 \\ 0 & 1 \end{bmatrix} = 1$$

$$\langle E_{21}, E_{21} \rangle = tr\left(E_{21}^T E_{21}\right) = tr(E_{12} E_{21}) = tr\begin{bmatrix} 1 & 0 \\ 0 & 0 \end{bmatrix} = 1$$

$$\langle E_{22}, E_{22} \rangle = tr\left(E_{22}^T E_{22}\right) = tr(E_{22} E_{22}) = tr\begin{bmatrix} 0 & 0 \\ 0 & 1 \end{bmatrix} = 1$$

Hence, S is an orthonormal basis of R.

EXERCISE 6.6

1. Let $B = \{v_1, v_2, \ldots, v_n\}$ be an orthonormal basis of an inner product space V. Prove that:

    (i) $\langle a_1v_1 + a_2v_2 + \cdots + a_nv_n, b_1v_1 + b_2v_2 + \cdots + b_nv_n \rangle = \sum_{i=1}^{n} a_i b_i$, where a_i, b_i are scalars in R.

    (ii) $\langle u, v \rangle = \langle u, v_1 \rangle \langle v, v_1 \rangle + \langle u, v_2 \rangle \langle v, v_2 \rangle + \cdots + \langle u, v_n \rangle \langle v, v_n \rangle$ for all $u, v \in V$.

2. Show that the vectors $v_1 = \left(\frac{1}{2}, \frac{1}{2}, \frac{1}{2}, \frac{1}{2}\right)$, $v_2 = \left(\frac{1}{2}, \frac{1}{2}, \frac{-1}{2}, \frac{-1}{2}\right)$, $v_3 = \left(\frac{1}{2}, \frac{-1}{2}, \frac{1}{2}, \frac{-1}{2}\right)$, $v_4 = \left(\frac{1}{2}, \frac{-1}{2}, \frac{-1}{2}, \frac{1}{2}\right)$ form an orthonormal basis for R^4.

3. Let S be the subspace of R^4 orthogonal to the vectors $v_1 = (1, 1, 2, 2)$ and $v_2 = (0, 1, 2, -1)$. Find an orthonormal basis of S.

ANSWERS

3. $\left\{\left(0, \frac{2}{\sqrt{5}}, \frac{-1}{\sqrt{5}}, 0\right), \left(\frac{-15}{\sqrt{255}}, \frac{1}{\sqrt{255}}, \frac{2}{\sqrt{255}}, \frac{5}{\sqrt{255}}\right)\right\}$

6.8 PROJECTIONS

As mentioned in the beginning the chapter that inner product spaces help us in understanding relationship between linear algebra and coordinate geometry. For example, the concepts of length of a vector, angle between vectors and orthogonality vectors are an analogous to the corresponding concepts in geometry. In this section, we intend to discuss the concept of projection similar to what we have learnt in geometry.

PROJECTION OF A VECTOR *Let V be an inner product space. The projection of a vector v along a non-zero vector w is the scalar multiple cw of w such that $v - cw$ is orthogonal to w.*

Now,

cw is the projection of vector v on vector w

$\Rightarrow \quad v - cw$ is orthogonal to w

$\Rightarrow \quad \langle v - cw, w \rangle = 0$

$\Rightarrow \quad \langle v, w \rangle - c \langle w, w \rangle = 0$

$\Rightarrow \quad c = \dfrac{\langle v, w \rangle}{\langle w, w \rangle}$

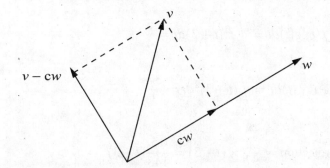

The projection of vector v along a non-zero vector w is denoted by proj (v,w).

Thus, $$\text{proj}(v,w) = cw = \frac{\langle v,w \rangle}{\langle w,w \rangle} w$$

The scalar $c = \dfrac{\langle v,w \rangle}{\langle w,w \rangle}$ is called the *Component of v along w* or the *Fourier coefficient of v with respect to w*.

REMARK. *In three dimension geometry the projecton vector of vector $\vec{v}$ along a vector $\vec{w}$ is defined as $\dfrac{\vec{v} \cdot \vec{w}}{|\vec{w}|^2} \vec{w}$ which is same as what we have defined above.*

EXAMPLE-1 *Find the projection of vector $v = (1,-2,3,-4)$ along the vector $w = (1,2,1,2)$ in the inner product space R^4 with standard inner product.*

SOLUTION We have,

Projection of vector v along vector $w = \dfrac{\langle v,w \rangle}{\langle w,w \rangle} w$

It is given that $v = (1,-2,3,-4)$ and $w = (1,2,1,2)$.

$\therefore \quad \langle v,w \rangle = 1 - 4 + 3 - 8 = -8$ and, $\langle w,w \rangle = 1 + 4 + 1 + 4 = 10$

$\therefore \quad$ Projection of v along $w = \dfrac{\langle v,w \rangle}{\langle w,w \rangle} w = -\dfrac{8}{10}(1,2,1,2) = \left(-\dfrac{4}{5}, -\dfrac{8}{5}, -\dfrac{4}{5}, -\dfrac{8}{5}\right)$

EXAMPLE-2 *Let $V = P_2(t)$ be the inner product space of all polynomials of degree less than or equal to 2 with inner product defined by $\langle f,g \rangle = \displaystyle\int_0^t f(t)\,g(t)\,dt$. If $f(t) = t^2$ and $g(t) = t + 3$ are two polynomials in $P_2(t)$, find the projection $f(t)$ along $g(t)$.*

SOLUTION Required projection $= \dfrac{\langle f,g \rangle}{\langle g,g \rangle} g$

Now,
$$\langle f,g \rangle = \int_0^1 f(t)\,g(t)\,dt = \int_0^t t^2(t+3)\,dt = \frac{5}{4}$$

$$\langle g,g \rangle = \int_0^1 g(t)\,g(t)\,dt = \int_0^1 (t+3)^2\,dt = \frac{37}{3}$$

$\therefore \quad$ Required projection $= \dfrac{\frac{5}{4}}{\frac{37}{3}} \times (t+3) = \dfrac{15}{148}(t+3)$

EXAMPLE-3 *In the inner product space $R^{2\times 2}$ of all 2×2 matrices over R with the inner product defined by $\langle A,B \rangle = tr(B^T A)$, find the projection of $A = \begin{bmatrix} 1 & 2 \\ 3 & 4 \end{bmatrix}$ along $B = \begin{bmatrix} 1 & 1 \\ 5 & 5 \end{bmatrix}$.*

SOLUTION Required projection $= \dfrac{\langle A,B \rangle}{\langle B,B \rangle} B$

Now, $\langle A,B \rangle = tr(B^T A) = tr\left(\begin{bmatrix} 1 & 5 \\ 1 & 5 \end{bmatrix} \begin{bmatrix} 1 & 2 \\ 3 & 4 \end{bmatrix}\right) = tr \begin{bmatrix} 16 & 22 \\ 16 & 22 \end{bmatrix} = 38$

$\langle B,B \rangle = tr(B^T B) = tr\left(\begin{bmatrix} 1 & 5 \\ 1 & 5 \end{bmatrix} \begin{bmatrix} 1 & 1 \\ 5 & 5 \end{bmatrix}\right) = tr \begin{bmatrix} 26 & 26 \\ 26 & 26 \end{bmatrix} = 52$

$\therefore \quad$ Required projection $= \dfrac{\langle A,B \rangle}{\langle B,B \rangle} B = \dfrac{38}{52}\begin{bmatrix} 1 & 1 \\ 5 & 5 \end{bmatrix} = \dfrac{19}{26}\begin{bmatrix} 1 & 1 \\ 5 & 5 \end{bmatrix}$

6.8.1 PROJECTION OF A VECTOR ALONG A SUBSPACE

Let V be an inner product space, v be a vector in V and S be a subspace of V. Then,

$\qquad V = S \oplus S^\perp \qquad$ [See Theorem 3 on page 583]

$\therefore \quad v \in V$

$\Rightarrow \quad v \in S \oplus S^\perp$

$\Rightarrow \quad v$ is uniquely expressible in the form $v = u + u'$, where $u \in S$ and $u' \in S^\perp$

The projection of v along S is defined as the vector u and is denoted proj (v,s).

Thus, proj$(v,S) = u$

If $\{v_1, v_2, \ldots, v_k\}$ is an orthogonal basis of S, then

$$\text{proj}\,(v,S) = \frac{\langle v,v_1 \rangle}{\langle v_1,v_1 \rangle}v_1 + \frac{\langle v,v_2 \rangle}{\langle v_2,v_2 \rangle}v_2 + \cdots + \frac{\langle v,v_k \rangle}{\langle v_k,v_k \rangle}v_k$$

i.e, proj(v,S) = Sum of the projections of v along the vectors $v_1, v_2, \ldots, v_k$.

EXAMPLE-1 Let S be the subspace of the inner product space R^4 spanned by the vectors $v_1 = (1, 1, 1, 1)$ and $v_2 = (-1, 0, 1, 0)$. Find the projection of vector $v = (1, 3, 5, 7)$ onto S.

SOLUTION Clearly, v_1, v_2 are orthogonal vectors and so they form an orthogonal basis for S.

$$\text{proj}(v,S) = \frac{\langle v, v_1 \rangle}{\langle v_1, v_1 \rangle} v_1 + \frac{\langle v, v_2 \rangle}{\langle v_2, v_2 \rangle} v_2$$

$\Rightarrow \quad \text{proj}(v,S) = \frac{16}{4} v_1 + \frac{4}{2} v_2 = 4v_1 + 2v_2$

$\Rightarrow \quad \text{proj}(v,S) = (2, 4, 6, 4)$

EXAMPLE-2 Consider the inner product space $P(t)$ of all polynomials with the inner product $<f,g> = \int_0^1 f(t)g(t)dt$. Let S be a subspace of $P(t)$ spanned by the set $\{1, 2t-1, 6t^2 - 6t + 1\}$ Find the projection of $f(t) = t^3$ onto S.

SOLUTION Clearly, $\{1, 2t-1, 6t^2 - 6t + 1\}$ is an orthogonal set. Let $f_1 = 1, f_2 = 2t - 1$, $f_3 = 6t^2 - 6t + 1$. Then,

$$\text{proj}(f,S) = \frac{\langle f, f_1 \rangle}{\langle f_1, f_1 \rangle} f_1 + \frac{\langle f, f_2 \rangle}{\langle f_2, f_2 \rangle} f_2 + \frac{\langle f, f_3 \rangle}{\langle f_3, f_3 \rangle} f_3$$

Now,

$\langle f_1, f_1 \rangle = \int_0^1 1 dt = 1, \langle f_2, f_2 \rangle = \int_0^1 (2t-1)^2 dt = \frac{1}{3}, \langle f_3, f_3 \rangle = \int_0^1 (6t^2 - 6t + 1)^2 dt = \frac{1}{5}$

$\langle f, f_1 \rangle = \int_0^1 t^3 dt = \frac{1}{4}, \langle f, f_2 \rangle = \int_0^1 t^3 (2t-1) dt = \frac{3}{20}, \langle f, f_3 \rangle = \int_0^1 t^3 (6t^2 - 6t + 1) dt = \frac{1}{20}$

$\therefore \quad \text{proj}(f,S) = \frac{1}{4} + \frac{9}{20}(2t-1) + \frac{5}{20}(6t^2 - 6t + 1) = \frac{3}{2}t^2 - \frac{3}{5}t + \frac{1}{20}$

6.9 GRAM-SCHMIDT ORTHOGONALIZATION PROCESS

Let V be a finite dimensional inner product space. Then, V being a finite dimensional vector space, has a basis $B = \{v_1, v_2, \ldots, v_n\}$ (say). This basis can be used to construct an orthonormal basis of inner product space V. The process of constructing an orthonormal basis of V from an ordinary basis for V is known as the Gram-Schmidt orthogonalization process. The following theorem explains the construction process.

THEOREM-1 *(Gram-Schmidt Orthogonalization process) Every finite dimensional inner product space has an orthonormal basis.*

PROOF. Let V be an n-dimensional inner product space. Then V, being a finite dimensional vector space, has a basis. Let $B = \{v_1, v_2, v_3, \ldots, v_n\}$ be an ordinary basis for V. We shall now use this basis to construct an orthonormal basis for V as follows:
Let $w_1 = v_1$

$$w_2 = v_2 - \frac{\langle v_2, w_1 \rangle}{\langle w_1, w_1 \rangle} w_1$$

$$w_3 = v_3 - \frac{\langle v_3, w_1 \rangle}{\langle w_1, w_1 \rangle} w_1 - \frac{\langle v_3, w_2 \rangle}{\langle w_2, w_2 \rangle} w_2$$

$$w_4 = v_4 - \frac{\langle v_4, w_1 \rangle}{\langle w_1, w_1 \rangle} w_1 - \frac{\langle v_4, w_2 \rangle}{\langle w_2, w_2 \rangle} w_2 - \frac{\langle v_4, w_3 \rangle}{\langle w_3, w_3 \rangle} w_3$$

$$\vdots \quad \vdots \quad \vdots \quad \vdots \quad \vdots \quad \vdots$$

$$w_n = v_n - \frac{\langle v_n, w_1 \rangle}{\langle w_1, w_1 \rangle} w_1 - \frac{\langle v_n, w_2 \rangle}{\langle w_2, w_2 \rangle} w_2 - \cdots - \frac{\langle v_n, w_{n-1} \rangle}{\langle w_{n-1}, w_{n-1} \rangle} w_{n-1}$$

Let $C_{ki} = \dfrac{\langle v_k, w_i \rangle}{\langle w_i, w_i \rangle}, k = 2, 3, \ldots, n$

Then, $w_2, w_3, \ldots, w_n$ are given by

$$w_k = v_k - C_{k_1} w_1 - C_{k_2} w_2 - C_{k_3} w_3 - \cdots - C_{k k-1} w_{k-1} \text{ for } k = 2, 3, \ldots, n.$$

We shall now show that $\{w_1, w_2, \ldots, w_n\}$ is an orthogonal set. For this it is sufficient to show that each $w_i (i \geq 2)$ is orthogonal to its preceding vectors $w_1, w_2, \ldots w_{i-1}$

Let $P(k)$ be the statement: Vector w_k is orthogonal to its preceding vector $w_1, w_2, \ldots, w_{k-1}$
We have,

$$\langle w_2, w_1 \rangle = \langle v_2 - C_{21} w_1, w_1 \rangle$$
$$\Rightarrow \quad \langle w_2, w_1 \rangle = \langle v_2, w_1 \rangle - \langle C_{21} w_1, w_1 \rangle$$
$$\Rightarrow \quad \langle w_2, w_1 \rangle = \langle v_2, w_1 \rangle - C_{21} \langle w_1, w_1 \rangle$$
$$\Rightarrow \quad \langle w_2, w_1 \rangle = \langle v_2, w_1 \rangle - \langle v_2, w_1 \rangle = 0 \qquad \left[\because C_{21} = \frac{\langle v_2, w_1 \rangle}{\langle w_1, w_1 \rangle} \right]$$

$\therefore \quad P(2)$ is true.

Let $P(k)$ be true. Then, each of $w_1, w_2, \ldots, w_{k-1}$ is orthogonal to w_k.
Now,

$$w_{k+1} = v_{k+1} - C_{k+1 1} w_1 - C_{k+1 2} w_2 - \cdots - C_{k+1 k} w_k$$

For any $i = 1, 2, \ldots, k$, we have

$$\langle w_{k+1}, w_i \rangle = \langle v_{k+1} - \sum_{r=1}^{k} C_{k+1 r} w_r, w_i \rangle$$
$$\Rightarrow \quad \langle w_{k+1}, w_i \rangle = \langle v_{k+1}, w_i \rangle - \sum_{r=1}^{k} C_{k+1 r} \langle w_r, w_i \rangle$$

$\Rightarrow \quad \langle w_{k+1}, w_i \rangle = \langle v_{k+1}, w_i \rangle - C_{k+1\ i} \langle w_i, w_i \rangle$

$\Rightarrow \quad \langle w_{k+1}, w_i \rangle = \langle v_{k+1}, w_i \rangle - \langle v_{k+1}, w_i \rangle = 0$

$\Rightarrow \quad w_{k+1}$ is orthogonal to each of the vectors $w_1, w_2, \ldots, w_k$

$\therefore \quad$ P(k+1) is true.

Hence, by induction $\{w_1, w_2, \ldots, w_n\}$ is an orthogonal set of vectors in V. But, every orthogonal set of vectors in an inner product space is linearly independent. So, $\{w_1, w_2, \ldots, w_n\}$ is an orthogonal basis for V.

Let $\quad u_i = \dfrac{w_i}{\|w_i\|}, \, i = 1, 2, \ldots, n$

Then, $\quad \langle u_i, u_i \rangle = \left\langle \dfrac{w_i}{\|w_i\|}, \dfrac{w_i}{\|w_i\|} \right\rangle = \dfrac{1}{\|w_i\|^2} \langle w_i, w_i \rangle = 1 \quad$ for all $i = 1, 2, \ldots, n$.

and, $\quad \langle u_i, u_j \rangle = \left\langle \dfrac{w_i}{\|w_i\|}, \dfrac{w_i}{\|w_j\|} \right\rangle = \dfrac{1}{\|w_i\|\|w_j\|} \langle w_i, w_j \rangle = 0 \quad$ for all $i, j = 1, 2, \ldots, n, \, i \neq j$

Hence, $\{u_1, u_2, \ldots, u_n\}$ is an orthonormal basis for V.

Q.E.D.

REMARK. *In the above theorem, though we have discussed the Gram-Schmidt orthogonalization process for a finite dimensional inner product space. But, the process works equally well for an infinite dimensional vector space.*

ILLUSTRATIVE EXAMPLES

EXAMPLE-1 *Apply the Gram-Schmidt orthogonalization process to the basis $B = \{(1, 0, 1), (1, 0, -1), (0, 3, 4)\}$ of the inner product space R^3 to find an orthogonal and an orhtonormal basis of R^3.*

SOLUTION Let $v_1 = (1, 0, 1)$, $v_2 = (1, 0, -1)$ and $v_3 = (0, 3, 4)$. Further, let $w_1 = v_1 = (1, 0, 1)$.

$$w_2 = v_2 - \dfrac{\langle v_2, w_1 \rangle}{\langle w_1, w_1 \rangle} w_1 = v_2 - 0 = v_2 = (1, 0, -1) \qquad [\because \langle v_2, w_1 \rangle = 0]$$

$$w_3 = v_3 - \dfrac{\langle v_3, w_1 \rangle}{\langle w_1, w_1 \rangle} w_1 - \dfrac{\langle v_3, w_2 \rangle}{\langle w_2, w_2 \rangle} w_2$$

$\Rightarrow \quad w_3 = v_3 - \dfrac{4}{2} v_1 + \dfrac{4}{2} v_2$

$\Rightarrow \quad w_3 = v_3 - 2v_1 + 2v_2$

$\Rightarrow \quad w_3 = (0, 3, 4) + (-2, 0, -2) + (2, 0, -2) = (0, 3, 0)$

Thus, $\{w_1, w_2, w_3\}$ is an orthogonal basis of R^3.

In order to obtain an orthonormal basis of R^3, let us normalize w_1, w_2, w_3
We have,
$$||w_1||^2 = 2, \ ||w_2||^2 = 2 \quad \text{and} \quad ||w_3||^2 = 9$$
Let $\quad u_i = \dfrac{w_i}{||w_i||}; \quad i = 1, 2, 3.$

Then, $\quad u_1 = \left(\dfrac{1}{\sqrt{2}}, 0, \dfrac{1}{\sqrt{2}}\right), \ u_2 = \left(\dfrac{1}{\sqrt{2}}, 0, \dfrac{-1}{\sqrt{2}}\right), \ u_3 = (0, 1, 0)$

form an orthonormal basis for R^3.

EXAMPLE-2 *Let $V = P_3(t)$ be the vector space of all polynomials $f(t)$ of degree less than of equal to 3 with inner product defined by $<f, g> = \int\limits_{-1}^{1} f(t)\,g(t)\,dt$.*

Apply Gram-Schmidt orthogonalization process to find an orthogonal basis with integral coefficients and the an orthonormal basis from the basis $\{1, t, t^2, t^3\}$.

PROOF. Let $f_0 = 1, f_1 = t, f_2 = t^2, f_3 = t^3$ form the given basis. Then,

$$\langle f_0, f_0 \rangle = \int_{-1}^{1} 1\,dt = 2, \ \langle f_1, f_1 \rangle = \int_{-1}^{1} t^2\,dt = \frac{2}{3}, \ \langle f_2, f_2 \rangle = \int_{-1}^{1} t^4\,dt = \frac{2}{5}, \ \langle f_3, f_3 \rangle = \int_{-1}^{1} t^6\,dt = \frac{2}{7}$$

$$\langle f_0, f_1 \rangle = \int_{-1}^{1} t\,dt = 0, \langle f_0, f_2 \rangle = \int_{-1}^{1} t^2\,dt = \frac{2}{3}, \langle f_0, f_3 \rangle = \int_{-1}^{1} t^3\,dt = 0$$

$$\langle f_1, f_2 \rangle = \int_{-1}^{1} t^3\,dt = 0, \langle f_1, f_3 \rangle = \int_{-1}^{1} t^4\,dt = \frac{2}{5}, \langle f_2, f_3 \rangle = \int_{-1}^{1} t^5\,dt = 0$$

Let $g_0 = f_0 = 1$,

$$g_1 = f_1 - \frac{\langle f_1, f_0 \rangle}{\langle f_0, f_0 \rangle} f_0 = t - 0 = t$$

$$g_2 = f_2 - \frac{\langle f_2, f_0 \rangle}{\langle f_0, f_0 \rangle} f_0 - \frac{\langle f_2, f_1 \rangle}{\langle f_1, f_1 \rangle} f_1 = t^2 - \frac{2}{3} \times \frac{1}{2} = t^2 - \frac{1}{3}$$

$$g_3 = f_3 - \frac{\langle f_3, f_0 \rangle}{\langle f_0, f_0 \rangle} f_0 - \frac{\langle f_3, f_1 \rangle}{\langle f_1, f_1 \rangle} f_1 - \frac{\langle f_3, f_2 \rangle}{\langle f_2, f_2 \rangle} f_2 = t^3 - \frac{2}{5} \times \frac{3}{2} \times t - 0 = t^3 - \frac{3}{5} t$$

Thus, $\left\{ g_0 = 1, \ g_1 = t, \ g_2 = t^2 - \dfrac{1}{3}, \ g_3 = t^3 - \dfrac{3}{5}t \right\}$ is an orthogonal basis V. Multiplying g_2 by 3 and g_3 by 5, we obtain $\{\phi_0(t) = 1, \ \phi_1(t) = t, \ \phi_2(t) = 3t^2 - 1, \ \phi_3(t) = 5t^3 - 3t\}$ as an orthogonal basis with integral coefficients.

Now,

$$\|\phi_0(t)\|^2 = <1,1> = \int_{-1}^{1} 1\,dt = 2 \quad \Rightarrow \quad \|\phi_0(t)\| = \sqrt{2}$$

$$\|\phi_1(t)\|^2 = <t,t> = \int_{-1}^{1} t^2\,dt = \frac{2}{3} \quad \Rightarrow \quad \|\phi_1(t)\| = \sqrt{\frac{2}{3}}$$

$$\|\phi_2(t)\|^2 = \langle 3t^2 - 1, 3t^2 - 1 \rangle = \int_{-1}^{1} (3t^2 - 1)^2\,dt = \frac{8}{5} \quad \Rightarrow \quad \|\phi_2(t)\| = 2\sqrt{\frac{2}{5}}$$

$$\|\phi_3(t)\|^2 = \langle 5t^3 - 3t, 5t^3 - 3t \rangle = \int_{-1}^{1} (5t^3 - 3t)^2\,dt = \frac{8}{7} \quad \Rightarrow \quad \|\phi_3(t)\| = 2\sqrt{\frac{2}{7}}$$

Hence, an orthonormal basis of $P_3(t)$ is

$$\left\{ \frac{1}{\sqrt{2}},\ \sqrt{\frac{3}{2}}t,\ \frac{1}{2}\sqrt{\frac{5}{2}}(3t^2 - 1),\ 2\sqrt{\frac{7}{2}}(5t^3 - 3t) \right\}.$$

EXAMPLE-3 *Let S be the subspace, of the inner product space R^4, spanned by the vectors $v_1 = (1, 1, 1, 1)$, $v_2 = (1, 2, 4, 5)$, $v_3 = (1, -3, -4, -2)$ in R^4. Apply the Gram-Schmidt orthogonalization process to find an orthogonal basis and then an orthonormal basis of S.*

SOLUTION We observe that the vectors v_1, v_2, v_3 form a linearly independent set. So, $\{v_1, v_2, v_3\}$ is a basis for S. In order to orthogonalize this basis, let us define:

$w_1 = v_1$

$w_2 = v_2 - \dfrac{\langle v_2, w_1 \rangle}{\langle w_1, w_1 \rangle} w_1 = v_2 - \dfrac{12}{4} v_1 = v_2 - 3v_1 = (-2, -1, 1, 2)$

$w_3 = v_3 - \dfrac{\langle v_3, w_1 \rangle}{\langle w_1, w_1 \rangle} w_1 - \dfrac{\langle v_3, w_2 \rangle}{\langle w_2, w_2 \rangle} w_2 = v_3 + \dfrac{8}{4} v_1 + \dfrac{7}{10} w_2 = \left(\dfrac{8}{5}, \dfrac{-17}{10}, \dfrac{-13}{10}, \dfrac{7}{5} \right)$

Thus, $\{w_1, w_2, w_3\}$ forms an orthogonal basis of S.
Now,

$$\|w_1\|^2 = \langle w_1, w_1 \rangle = 4,\ \|w_2\|^2 = \langle w_2, w_2 \rangle = 10,\ \|w_3\|^2 = \langle w_3, w_3 \rangle = \frac{910}{100}$$

Let $u_i = \dfrac{w_i}{\|w_i\|}$, $i = 1, 2, 3$. Then,

$$\left\{ u_1 = \left(\frac{1}{2}, \frac{1}{2}, \frac{1}{2}, \frac{1}{2} \right),\ u_2 = \left(\frac{-2}{\sqrt{10}}, \frac{-1}{\sqrt{10}}, \frac{1}{\sqrt{10}}, \frac{2}{\sqrt{10}} \right),\ u_3 = \left(\frac{16}{\sqrt{910}}, \frac{-17}{\sqrt{910}}, \frac{-13}{\sqrt{910}}, \frac{14}{\sqrt{910}} \right) \right\}$$

is an orthonormal basis of S.

EXAMPLE-4 Let S be the subspace of R^4 spanned by the vectors $v_1 = (1, 1, 1, 1)$, $v_2 = (1, -1, 2, 2), v_3 = (1, 2, -3, -4)$. Apply the Gram-Schmidt orhtogonalization process to find an orthogonal basis of S and hence, find the projection of $v = (1, 2, -3, 4)$ onto S.

SOLUTION Let $w_1 = v_1 = (1, 1, 1, 1)$.

$$w_2 = v_2 - \frac{\langle v_2, w_1 \rangle}{\langle w_1, w_1 \rangle} w_1 = v_2 - \frac{\langle v_2, v_1 \rangle}{\langle v_1, v_1 \rangle} v_1 = v_2 - v_1 = (0, -2, 1, 1)$$

$$w_3 = v_3 - \frac{\langle v_3, w_1 \rangle}{\langle w_1, w_1 \rangle} w_1 - \frac{\langle v_3, w_2 \rangle}{\langle w_2, w_2 \rangle} w_2 = v_3 + w_1 + \frac{11}{6} w_2 = \left(2, \frac{-2}{3}, \frac{-1}{6}, \frac{-7}{6}\right)$$

Let $w_3' = 6w_3 = (12, -4, -1, -7)$

Clearly, $\{w_1, w_2, w_3'\}$ forms an orthogonal basis of S.

$$\therefore \quad \text{proj}(v, S) = \frac{\langle v, w_1 \rangle}{\langle w_1, w_1 \rangle} w_1 + \frac{\langle v, w_2 \rangle}{\langle w_2, w_2 \rangle} w_2 + \frac{\langle v, w_3' \rangle}{\langle w_3', w_3' \rangle} w_3'$$

$$\Rightarrow \quad \text{proj}(v, S) = w_1 - \frac{1}{2} w_2 - \frac{1}{10} w_3' = \left(\frac{-1}{5}, \frac{12}{5}, \frac{3}{5}, \frac{6}{5}\right)$$

THEOREM-2 Let S be a subspace of a finite dimensional inner product space V. If $B_1 = \{w_1, w_2, \ldots, w_r\}$ is an orthogonal basis of S. Then B_1 can be extended to an orthogonal basis for V.

OR

Every orthogonal list of vectors in a finite dimensional inner product space V can be extended to an orthogonal basis for V.

PROOF. Let V be a finite dimensional inner product space of dimension n. It is given that B_1 is an orthogonal basis of S. Therefore, B_1 is a linearly independent set of vectors in V. So, B_1 can be extendended to a basis $B = \{w_1, w_2, \ldots, w_r, v_{r+1}, \ldots, v_n\}$ of V. Applying the Gram-Schmidt orthogonalization process to B, we first obtain $w_1, w_2, \ldots, w_r$ since B_1 is orthogonal and then we obtain vectors $w_{r+1}, w_{r+2}, \ldots, w_n$ such that $B' = \{w_1, w_2, \ldots, w_r, \ldots, w_n\}$ is an orthogonal basis for V.

Q.E.D.

COROLLARY. In a finite dimensional inner product space V, any orthonormal list of vectors is a part of an orthonormal basis for V.

PROOF. Let $(w_1, w_2, \ldots, w_r)$ be an orthonormal list of vectors in V. Proceeding as in the above theorem this list can be extended to an orthogonal basis $\{w_1, w_2, \ldots, w_r, w_{r+1}, \ldots, w_n\}$ for V. Now, normalize each w_i by defining $u_i = \frac{w_i}{\|w_i\|}$. This gives $\{u_1, u_2, \ldots, u_n\}$ as an orthonormal basis for V.

Q.E.D.

THEOREM-3 *Let S be a subspace of a finite dimensional inner product space V. Then $V = S \oplus S^\perp$.*

PROOF. Since S is a subspace of a finite dimensional inner product space V. Therefore, S itself is a finite dimensional inner product space (its inner product being that of V restricted to S). So, it has an orthonormal basis, say, $B = \{w_1, w_2, \ldots, w_m\}$ (By the Gram-Schmidt orthogonalization process).

Let v be an arbitrary vector in V and let $u = v - \sum_{i=1}^{m} \langle v, w_i \rangle w_i$. Then, for any $j = 1, 2, \ldots, m$, we have

$$\langle u, w_j \rangle = \left\langle v - \sum_{i=1}^{m} \langle v, w_i \rangle w_i, w_j \right\rangle$$

$\Rightarrow \quad \langle u, w_j \rangle = \langle v, w_j \rangle - \left\langle \sum_{i=1}^{m} \langle v, w_i \rangle w_i, w_j \right\rangle$

$\Rightarrow \quad \langle u, w_j \rangle = \langle v, w_j \rangle - \sum_{i=1}^{m} \langle v, w_i \rangle \langle w_i, w_j \rangle$

$\Rightarrow \quad \langle u, w_j \rangle = \langle v, w_j \rangle - \sum_{i=1}^{m} \langle v, w_i \rangle \delta_{ij}$

$\Rightarrow \quad \langle u, w_j \rangle = \langle v, w_j \rangle - \langle v, w_j \rangle = 0$

$\Rightarrow \quad \langle u, w_j \rangle = 0 \quad$ for all $j = 1, 2, \ldots, m$

$\Rightarrow \quad u \in S^\perp$

Clearly, $\sum_{i=1}^{m} \langle v, w_i \rangle w_i \in S \qquad [\because B = \{w_1, w_2, \ldots, w_m\}$ is a basis of $S]$

$\therefore \quad u = v - \sum_{i=1}^{m} \langle v, w_i \rangle w_i$

$\Rightarrow \quad v = u + \sum_{i=1}^{m} \langle v, w_i \rangle w_i = \sum_{i=1}^{m} \langle v, w_i \rangle w_i + u$

$\Rightarrow \quad v \in S + S^\perp \hfill (i)$

Now, let $v \in S \cap S^\perp$. Then,

$\quad v \in S$ and $v \in S^\perp$

$\Rightarrow \quad \langle v, v \rangle = 0$

$\Rightarrow \quad v = 0_V$

$\therefore \quad S \cap S^\perp = \{0_V\} \hfill (ii)$

Hence, $V = S \oplus S^\perp \hfill$ [Using (i) and (ii)]

$\hfill$ Q.E.D.

COROLLARY. If S is any subspace of a finite dimensional inner product space V. Then, $\dim S^\perp = \dim V - \dim S$.

PROOF. We have,
$$V = S \oplus S^\perp$$
$$\Rightarrow \quad \dim V = \dim S + \dim S^\perp$$
$$\Rightarrow \quad \dim S^\perp = \dim V - \dim S$$

Q.E.D.

THEOREM-4 Let S be a subspace of a finite dimensional inner product space V. Then, $\left(S^\perp\right)^\perp = S$.

PROOF. Since S is a subspace of V. Therefore, so is $S^\perp$.

Now,
$$V = S \oplus S^\perp$$
$$\Rightarrow \quad \dim V = \dim S + \dim S^\perp \qquad \text{(i)}$$

Since $S^\perp$ is a subspace of V. Therefore, $\left(S^\perp\right)^\perp$ is also a subspace of V.

$$V = S^\perp \oplus \left(S^\perp\right)^\perp$$
$$\Rightarrow \quad \dim V = \dim S^\perp + \dim \left(S^\perp\right)^\perp \qquad \text{(ii)}$$

From (i) and (ii), we obtain
$$\dim S = \dim \left(S^\perp\right)^\perp \qquad \text{(iii)}$$

Let v be an arbitrary vector in S. Then,
$$\langle v, u \rangle = 0 \quad \text{for all } u \in S^\perp$$
$$\Rightarrow \quad v \in \left(S^\perp\right)^\perp$$

Thus, $v \in S \Rightarrow v \in \left(S^\perp\right)^\perp$

$\therefore \quad S \subset \left(S^\perp\right)^\perp \qquad \text{(iv)}$

Thus, S and $\left(S^\perp\right)^\perp$ are subspace of V such that $\dim S = \dim \left(S^\perp\right)^\perp$.

Hence, $S = \left(S^\perp\right)^\perp$.

Q.E.D.

THEOREM-5 *If S is a subset of a finite dimensional inner product space V. Then, $[S] = S^{\perp\perp}$.*

PROOF. Since $[S]$ is a subspace of V. Therefore,

$$[S]^{\perp} = S^{\perp}$$
$\Rightarrow \quad [S]^{\perp\perp} = S^{\perp\perp}$
$\Rightarrow \quad S^{\perp\perp} = [S] \qquad\qquad [\because \quad [S] \text{ is a subspace of } V \quad \therefore \quad [S]^{\perp\perp} = [S] \text{ By Theorem 4}]$

Q.E.D.

THEOREM-6 *Let S be a subspace of a finite dimensional inner product space V. If $B_1 = \{v_1, v_2, \ldots, v_m\}$ is an orthonormal basis S and $B_2 = \{w_1, w_2, \ldots, w_n\}$ is an orthonormal basis for $S^{\perp}$. Then, $B = B_1 \cup B_2$ is an orthonormal basis for V.*

PROOF. Since B_1 and B_2 are orthonormal bases for S and $S^{\perp}$ respectively. Therefore,

$$\langle v_i, v_j \rangle = \delta_{ij} \quad \text{for } i, j \in \underline{m} \quad \text{and,} \quad \langle w_r, w_s \rangle = \delta_{rs} \quad \text{for } r, s \in \underline{n}$$

Now, $\quad w_s \in S^{\perp}, \quad s \in \underline{n}$
$\Rightarrow \quad \langle u, w_s \rangle = 0 \quad$ for all $u \in S$ and all $s \in \underline{n}$
$\Rightarrow \quad \langle v_i, w_s \rangle = 0 \quad$ for all $i \in \underline{m}$ and for all $s \in \underline{n}$
$\Rightarrow \quad$ Each v_i is orthogonal to each w_s, $i \in \underline{m}$, $s \in \underline{n}$

Thus, $\{v_1, v_2, \ldots, v_m, w_1, w_2, \ldots, w_n\}$ is an orthonormal set of $m+n$ vectors in V. Since an orthonormal set of vectors in an inner product space is linearly independent. Therefore, $\{v_1, v_2, \ldots, v_m, w_1, w_2, \ldots, w_n\}$ is an orthonormal set of $m+n$ linearly independent vectors in V.

Also, $\dim V = \dim S + \dim S^{\perp} = m + n$

Hence, $B = \{v_1, v_2, v_3, \ldots, v_m, w_1, w_2, \ldots, w_n\} = B_1 \cup B_2$ is an orthonormal basis for V.

Q.E.D.

THEOREM-7 *Let S and T be subspaces of a finite dimensional inner product space V. Then,*
(i) $(S+T)^{\perp} = S^{\perp} \cap T^{\perp}$ (ii) $(S \cap T)^{\perp} = S^{\perp} + T^{\perp}$

PROOF. (i) Clearly,

$$S \subseteq S+T \text{ and } T \subseteq S+T$$
$\Rightarrow \quad (S+T)^{\perp} \subseteq S^{\perp}$ and $(S+T)^{\perp} \subseteq T^{\perp}$
$\Rightarrow \quad (S+T)^{\perp} \subseteq S^{\perp} \cap T^{\perp}$ \hfill (i)

Now, we shall show that $S^{\perp} \cap T^{\perp} \subseteq (S+T)^{\perp}$. For this, let v be an arbitrary vector in $S^{\perp} \cap T^{\perp}$. Then,

$$v \in S^\perp \cap T^\perp$$
$\Rightarrow \quad v \in S^\perp$ and $v \in T^\perp$
$\Rightarrow \quad \langle v, u \rangle = 0 \quad$ for all $u \in S$ and $\langle v, w \rangle = 0$ for all $w \in T$
$\Rightarrow \quad \langle v, u \rangle + \langle v, w \rangle = 0 \quad$ for all $u \in S$ and all $w \in T$
$\Rightarrow \quad \langle v, u+w \rangle = 0 \quad$ for all $u+w \in S+T$
$\Rightarrow \quad v \in (S+T)^\perp$
$\therefore \quad S^\perp \cap T^\perp \subseteq (S+T)^\perp$ \hfill (ii)

From (i) and (ii), we obtain
$$(S+T)^\perp = S^\perp \cap T^\perp$$

(ii) Since S and T are subspace of V. Therefore, so are $S^\perp$ and $T^\perp$. Replacing S by $S^\perp$ and T by $T^\perp$ in (i), we get

$$\left(S^\perp + T^\perp\right)^\perp = S^{\perp\perp} \cap T^{\perp\perp}$$

$\Rightarrow \quad \left(S^\perp + T^\perp\right)^\perp = S \cap T \qquad \begin{bmatrix} \because \ S \text{ and } T \text{ are subspaces of finite} \\ \text{dimensional inner product space } V. \\ \therefore \ S^{\perp\perp} = S \text{ and } T^{\perp\perp} = T \end{bmatrix}$

$\Rightarrow \quad \left(S^\perp + T^\perp\right)^{\perp\perp} = (S \cap T)^\perp$

$\Rightarrow \quad S^\perp + T^\perp = (S \cap T)^\perp \qquad \left[\because \ \left(S^\perp + T^\perp\right)^{\perp\perp} = S^\perp + T^\perp\right]$

or, $\quad (S \cap T)^\perp = S^\perp + T^\perp$.

Q.E.D.

EXERCISE 6.7

1. Apply Gram-Schmidt orthogonalisation process to find an orthonormal basis from the basis $\{(2, 0, 1), (3, -1, 5), (0, 4, 2)\}$ of the inner product space R^3.

2. Apply Gram-Schmidt orthogonalisation process to find an orthonormal basis from the basis $\{(1, 0, 0), (1, 1, 0), (1, 1, 1)\}$ of the inner product space R^3.

3. Let V be the inner product space of all polynomials of degree 2 or less in indeterminate t with inner product defined by $\langle f, g \rangle = \int_{-1}^{1} f(t) g(t) dt$. Apply Gram-Schmidt orthogonalization process to obtain an orthonormal basis from the basis $\{1, t, t^2\}$.

4. Let S be the subspace, of inner product space R^4, spanned by the vectors $v_1 = (1, 1, 1, 1)$, $v_2 = (1, 1, 2, 4)$, $v_3 = (1, 2, -4, -3)$. Apply the Gram-Schmidt orthogonalization process to obtain an orthonormal basis of S.

5. Consider the vector space $P(t)$ with inner product $\langle f, g \rangle = \int_0^1 f(t) g(t) dt$. Apply the Gram-Schmidt orthogonalization process to the set $\{1, t, t^2\}$ to obtain an orthogonal set $\{f_0, f_1, f_2\}$ with integer coefficients.

6. Consider the inner product space $P(t)$ of all polynomials with inner product $\langle f, g \rangle = \int_{-1}^1 f(t) g(t) dt$ and S be the subspace of all polynomials of degree 3 or less. Find an orthogonal basis for S by applying the Gram-Schmidt orthogonalization process to $\{1, t, t^2, t^3\}$. Also, find the projection of $f(t) = t^5$ onto S.

7. Let $v = (1, 2, 3, 4, 5) \in R^5$. Find the projection of v onto S, where S is the subspace of R^5 spanned by the vectors.
   (i) $v_1 = (1, 2, 1, 2, 1)$, $v_2 = (1, -1, 2, -1, 1)$
   (ii) $u_1 = (1, 2, 1, 2, 1)$, $u_2 = (1, 0, 1, 5, -1)$

8. Let $\{v_1, v_2, \ldots, v_n\}$ be any basis of an inner product space V. Then there exists an orthonormal basis $\{u_1, u_2, \ldots, u_n\}$ of V such that the change-of-basis matrix from $\{v_1, v_2, \ldots, v_n\}$ to $\{u_1, u_2, \ldots, u_n\}$ is triangular.

9. If $B = \{v_1 = (1, 1, 1, 0), v_2 = (0, 1, 1, 1), v_3 = (0, 0, 1, 1), v_4 = (0, 0, 3, 0)\}$ forms a basis of R^4. Find an orthonormal basis by using Gram-Schmidt orthogonalisation process.

ANSWERS

1. $\left\{ \left(\dfrac{2}{\sqrt{5}}, 0, \dfrac{1}{\sqrt{5}} \right), \left(\dfrac{-7}{\sqrt{270}}, \dfrac{-5}{270}, \dfrac{14}{\sqrt{270}} \right), \left(\dfrac{-1}{3\sqrt{6}}, \dfrac{7}{3\sqrt{6}}, \dfrac{2}{3\sqrt{6}} \right) \right\}$

2. $\{(1, 0, 0), (0, 1, 0), (0, 0, 1)\}$ 3. $\left\{ \dfrac{1}{\sqrt{2}}, \dfrac{\sqrt{3}}{2} t, \dfrac{\sqrt{10}}{4}(3t^2 - 1) \right\}$

4. $\left\{ \left(\dfrac{1}{2}, \dfrac{1}{2}, \dfrac{1}{2}, \dfrac{1}{2} \right), \left(\dfrac{-1}{\sqrt{6}}, \dfrac{-1}{\sqrt{6}}, 0, \dfrac{2}{\sqrt{6}} \right), \left(\dfrac{1}{5\sqrt{2}}, \dfrac{3}{5\sqrt{2}}, \dfrac{-6}{5\sqrt{2}}, \dfrac{2}{5\sqrt{2}} \right) \right\}$

5. $\{1, 2t - 1, 6t^2 - 6t + 1\}$ 6. $\{1, t, 3t^2 - 1, 5t^3 - 3t\}$, $\dfrac{10}{9} t^3 - \dfrac{5}{21} t$

7. (i) $\dfrac{1}{8}(23, 25, 30, 25, 23)$ (ii) $\dfrac{1}{17}(34, 76, 34, 56, 42)$

9. $\left\{\left(\dfrac{1}{\sqrt{3}}, \dfrac{1}{\sqrt{3}}, \dfrac{1}{\sqrt{3}}, 0\right), \left(\dfrac{-2}{\sqrt{15}}, \dfrac{1}{\sqrt{15}}, \dfrac{1}{\sqrt{15}}, \dfrac{3}{\sqrt{15}}\right), \left(\dfrac{1}{\sqrt{15}}, \dfrac{-3}{\sqrt{15}}, \dfrac{2}{\sqrt{15}}, \dfrac{1}{\sqrt{15}}\right),\right.$
$\left.\left(\dfrac{-1}{\sqrt{3}}, 0, \dfrac{1}{\sqrt{3}}, \dfrac{-1}{\sqrt{3}}\right)\right\}$

6.10 INNER PRODUCTS AND MATRICES

MATRIX REPRESENTATION OF AN INNER PRODUCT *Let V be an n-dimensional inner product space with basis $B = \{v_1, v_2, \ldots, v_n\}$ and let $a_{ij} = \langle v_i, v_j \rangle$ for all $i, j \in \underline{n}$. Then, $n \times n$ square matrix $A = [a_{ij}]$ is called the matrix representation of the inner product on V relative to the basis B.*

Clearly,

$$\langle v_i, v_j \rangle = \langle v_j, v_i \rangle \quad \text{for all } i, j \in \underline{n}$$

$\Rightarrow \quad a_{ij} = a_{ji} \quad \text{for all } i, j \in \underline{n}$

$\Rightarrow \quad A = A^T$

$\Rightarrow \quad A$ is a real symmetric matrix.

It is to note here that the matrix representation of an inner product depends on to the inner product on V and the basis B for V.

Now, let B be an orthogonal basis of V. Then,

$$\langle v_i, v_j \rangle = 0 \quad \text{for all } i \neq j$$

$\Rightarrow \quad a_{ij} = 0 \quad \text{for all } i \neq j$

$\Rightarrow \quad A = [a_{ij}]$ is a diagonal matrix.

If B is an orthonormal basis of V, then

$$\langle v_i, v_j \rangle = \delta_{ij} \quad \text{for all } i, j \in \underline{n}$$

$\Rightarrow \quad A = [\delta_{ij}] = I$

Thus, the matrix representation of an inner product relative to an orthonormal basis of V is the identity matrix.

ILLUSTRATIVE EXAMPLES

EXAMPLE-1 *Find the matrix A that represents the usual standard inner product on R^2 relative to the ordered basis $B = \{v_1 = (1, 4), v_2 = (2, -3)\}$.*

SOLUTION Let $A = [a_{ij}]_{2 \times 2}$ be the required matrix. Then,

$a_{11} = \langle v_1, v_1 \rangle = 1 + 16 = 17$, $a_{12} = a_{21} = \langle v_1, v_2 \rangle = 2 - 12 = -10$, $a_{22} = \langle v_2, v_2 \rangle = 4 + 9 = 13$.

$\therefore \quad A = \begin{bmatrix} a_{11} & a_{12} \\ a_{21} & a_{22} \end{bmatrix} = \begin{bmatrix} 17 & -10 \\ -10 & 13 \end{bmatrix}$

EXAMPLE-2 Consider the following inner product on R^2:
$$\langle u, v \rangle = x_1 y_1 - 2x_1 y_2 - 2x_2 y_1 + 5x_2 y_2, \text{ where } u = \langle x_1, x_2 \rangle, v = \langle y_1, y_2 \rangle.$$
Find the matrix representing this inner product relative to basis $B = \{v_1 = (1, -3), v_2 = (6, 2)\}$ of R^2.

SOLUTION Let $A = [a_{ij}]_{2 \times 2}$ be the matrix representation of the given inner product relative to the basis B. Then,

$a_{11} = \langle v_1, v_1 \rangle = 1 + 12 + 45 = 58$ $\qquad [\because \langle u, u \rangle = x_1^2 - 4x_1 x_2 + 5x_2^2]$

$a_{22} = \langle v_2, v_2 \rangle = 36 - 48 + 20 = 8$

$a_{12} = a_{21} = \langle v_1, v_2 \rangle = 6 - 2 \times 1 \times 2 - 2 \times -3 \times 6 + 5 \times (-3) \times 2 = 8$

$$\therefore \quad A = \begin{bmatrix} a_{11} & a_{12} \\ a_{21} & a_{22} \end{bmatrix} = \begin{bmatrix} 58 & 8 \\ 8 & 8 \end{bmatrix}$$

EXAMPLE-3 Consider the inner product space $P_2(t)$ of all polynomials of degree less than or equal to 2 with the inner product

$$\langle f, g \rangle = \int_{-1}^{1} f(t) g(t) dt.$$

Find the matrix A of this inner product relative to the basis $B = \{1, t, t^2\}$.

SOLUTION If $r + s = n$, then

$$\langle t^r, t^s \rangle = \int_{-1}^{1} t^n dt = \left[\frac{t^{n+1}}{n+1} \right]_{-1}^{1} = \begin{cases} 0, & \text{if } n \text{ is odd} \\ \dfrac{2}{n+1}, & \text{if } n \text{ is even} \end{cases}$$

Let $A = [a_{ij}]$ be the matrix of the given inner product relative to the basis B. Then,

$a_{11} = \langle 1, 1 \rangle = 2, \ a_{12} = a_{21} = \langle 1, t \rangle = 0, \ a_{13} = a_{31} = \langle 1, t^2 \rangle = \dfrac{2}{3}$

$a_{22} = \langle t, t \rangle = \dfrac{2}{3}, \ a_{23} = a_{32} = \langle t, t^2 \rangle = 0, \ a_{33} = \langle t^2, t^2 \rangle = \dfrac{2}{5}$

$$\therefore \quad A = \begin{bmatrix} a_{11} & a_{12} & a_{13} \\ a_{21} & a_{22} & a_{23} \\ a_{31} & a_{32} & a_{33} \end{bmatrix} = \begin{bmatrix} 2 & 0 & 2/3 \\ 0 & 2/3 & 0 \\ 2/3 & 0 & 2/5 \end{bmatrix}$$

EXAMPLE-4 Find the matrix representation of standard inner product on R^3 relative to the ordered basis $B = \{(1,3,5), (1,1,0), (1,2,3)\}$.

SOLUTION Let $v_1 = (1,3,5)$, $v_2 = (1,1,0)$ and $v_3 = (1,2,3)$. Let $A = [a_{ij}]$ be the matrix representation of standard inner product relative to the given basis. Then,

$$a_{11} = \langle v_1, v_1 \rangle = 1+9+25 = 35, \ a_{12} = \langle v_1, v_2 \rangle = 1+3+0 = 4 = a_{21}$$
$$a_{22} = \langle v_2, v_2 \rangle = 1+1+0 = 2, \ a_{23} = \langle v_2, v_3 \rangle = 1+2 = 3 = a_{32}$$
$$a_{13} = \langle v_1, v_3 \rangle = 1+6+15 = 22 = a_{31}, \ a_{33} = \langle v_3, v_3 \rangle = 1+4+9 = 14$$

$$\therefore \quad A = \begin{bmatrix} a_{11} & a_{12} & a_{13} \\ a_{21} & a_{22} & a_{23} \\ a_{31} & a_{32} & a_{33} \end{bmatrix} = \begin{bmatrix} 35 & 4 & 22 \\ 4 & 2 & 3 \\ 22 & 3 & 14 \end{bmatrix}$$

THEOREM-1 *Let A be the matrix representation of an inner product relative to basis B of inner product space V. Then, for any vectors $u, v \in V$, prove that*

$$\langle u, v \rangle = [u]^T A [v]$$

where $[u]$ and $[v]$ denote the (column) coordinate vectors relative to the basis B.

PROOF. Let $B = \{w_1, w_2, \ldots, w_n\}$ be the basis of inner product space V and $A = [a_{ij}]$ be the matrix representation of the inner product relative to basis B. Then,

$$a_{ij} = \langle w_i, w_j \rangle$$

Let $u = \sum_{i=1}^{n} \lambda_i w_i$ and $v = \sum_{j=1}^{n} \mu_j w_j$. Then,

$$\langle u, v \rangle = \left\langle \sum_{i=1}^{n} \lambda_i w_i, \sum_{j=1}^{n} \mu_j w_j \right\rangle$$

$$\Rightarrow \quad \langle u, v \rangle = \sum_{i=1}^{n} \sum_{j=1}^{n} \lambda_i \mu_j \langle w_i, w_j \rangle = \sum_{i=1}^{n} \sum_{j=1}^{n} \lambda_i \mu_j a_{ij} \tag{i}$$

Now, $u = \sum_{i=1}^{n} \lambda_i w_i$ and $v = \sum_{j=1}^{n} \mu_j w_j \Rightarrow [u] = \begin{bmatrix} \lambda_1 \\ \lambda_2 \\ \vdots \\ \lambda_n \end{bmatrix}$ and $[v] = \begin{bmatrix} \mu_1 \\ \mu_2 \\ \vdots \\ \mu_n \end{bmatrix}$

$$\therefore \quad [u]^T A [v] = (\lambda_1, \lambda_2, \ldots, \lambda_n) \begin{bmatrix} a_{11} & a_{12} & \cdots & a_{1n} \\ a_{21} & a_{22} & \cdots & a_{2n} \\ \vdots & \vdots & & \vdots \\ a_{n1} & a_{n2} & \cdots & a_{nn} \end{bmatrix} \begin{bmatrix} \mu_1 \\ \mu_2 \\ \vdots \\ \mu_n \end{bmatrix}$$

$$\Rightarrow \quad [u]^T A [v] = \left(\sum_{i=1}^{n} a_{i1} \lambda_i, \sum_{i=1}^{n} a_{i2} \lambda_i, \ldots, \sum_{i=1}^{n} a_{in} \lambda_i \right) \begin{bmatrix} \mu_1 \\ \mu_2 \\ \vdots \\ \mu_n \end{bmatrix}$$

$$\Rightarrow \quad [u]^T A[v] = \sum_{i=1}^{n} a_{i1} \lambda_i \mu_1 + \sum_{i=1}^{n} a_{i2} \lambda_i \mu_2 + \cdots + \sum_{i=1}^{n} a_{in} \lambda_i \mu_n$$

$$\Rightarrow \quad [u]^T A[v] = \sum_{i=1}^{n} \sum_{j=1}^{n} a_{ij} \lambda_i \mu_j = \sum_{i=1}^{n} \sum_{j=1}^{n} \lambda_i \mu_j a_{ij} \qquad \text{(ii)}$$

From (i) and (ii), we get

$$\langle u, v \rangle = [u]^T A[v]$$

Q.E.D.

EXERCISE 6.8

1. Find the matrix representation of the standard inner product on R^2 relative to the basis $B = \{v_1 = (1, -3), v_2 = (6, 2)\}$.

2. Find the matrix A that represents the standard inner product on R^2 relative to each of the following bases of R^2:
   (i) $\{u_1 = (1, 3), u_2 = (2, 5)\}$ (ii) $\{v_1 = (1, 2), v_2 = (4, -2)\}$

3. Find the matrix A that represents the inner product on R^2 defined by

$$\langle u, v \rangle = x_1 y_1 - 2x_1 y_2 - 2x_2 y_1 + 5x_2 y_2$$

   where $u = (x_1, x_2)$, $v = (y_1, y_2)$, relative to the basis $\{v_1 = (1, 4), v_2 = (2, -3)\}$.

4. Find the matrix representation of the standard inner product on R^3 relative to the basis B of R^3 consisting of the vectors $v_1 = (1, 1, 1), v_2 = (1, 2, 1), v_3 = (1, -1, 3)$.

5. Let v be the inner product space of all real polynomials of degree not exceeding 2 with the inner product

$$\langle f, g \rangle = \int_0^1 f(t) g(t) dt.$$

Find the matrix A of the inner product relative to the basis $\{1, t, t^2\}$ of V.

ANSWERS

1. $\begin{bmatrix} 10 & 0 \\ 0 & 40 \end{bmatrix}$

2. (i) $\begin{bmatrix} 10 & 17 \\ 17 & 29 \end{bmatrix}$ (ii) $\begin{bmatrix} 5 & 0 \\ 0 & 20 \end{bmatrix}$

3. $\begin{bmatrix} 65 & -68 \\ -68 & 73 \end{bmatrix}$

4. $\begin{bmatrix} 3 & 4 & 3 \\ 4 & 6 & 2 \\ 3 & 2 & 11 \end{bmatrix}$

5. $\begin{bmatrix} 1 & 1/2 & 1/3 \\ 1/2 & 1/3 & 1/4 \\ 1/3 & 1/4 & 1/5 \end{bmatrix}$

6.11 POSITIVE DEFINITE MATRIX

We have learnt that R^n is an inner product space with the standard inner product. If vectors in R^n are represented by column vectors, that is, by $n \times 1$ column matrices, then $\langle u, v \rangle = u^T v$ defines the standard inner product of R^n.

We use this to define a positive definite matrix as given below.

POSITIVE DEFINITE MATRIX *A real symmetric matrix $A = [a_{ij}]$ is said to be positive definite if*
$$\langle u, Au \rangle = u^T Au > 0 \text{ for every non-zero vector } u \in R^n.$$

or, $\quad \sum_{i=1}^{n} \sum_{j=1}^{n} u_i u_j a_{ij} > 0$, where $u = [u_1, u_2, \ldots, u_n]^T$

THEOREM-1 *A 2×2 real symmetric matrix $A = \begin{bmatrix} a & b \\ b & d \end{bmatrix}$ is positive definite if and only if the diagonal elements a and d are positive and $|A| = ad - b^2 > 0$.*

PROOF. First, Let $A = \begin{bmatrix} a & b \\ b & d \end{bmatrix}$ be a real symmetric positive definite matrix. Then,

$u^T Au > 0 \quad$ for all $u = [x, y]^T$, $u \neq 0$

$\Rightarrow \quad [x\ y] \begin{bmatrix} a & b \\ b & d \end{bmatrix} \begin{bmatrix} x \\ y \end{bmatrix} > 0 \quad$ for all $x, y \in R$ such that at least one is non-zero

$\Rightarrow \quad ax^2 + 2bxy + dy^2 > 0 \quad$ for all $x, y \in R$ such that at least one is non-zero

$\Rightarrow \quad a > 0, d > 0$ and $4b^2 - 4ad < 0 \quad [\because \quad ax^2 + bx + c > 0$ for all $x \Rightarrow a > 0$ and Disc $< 0]$

$\Rightarrow \quad a > 0, d > 0$ and $ad - b^2 > 0$

Conversely, let $A = \begin{bmatrix} a & b \\ b & d \end{bmatrix}$ be a real symmetric matrix such that $a > 0$, $d > 0$ and $|A| = ad - b^2 > 0$. Then, we have to prove that A is a positive definite matrix.

Let $u = [x, y]^T$ be an arbitrary vector in R^2. Then,

$u^T Au = ax^2 + 2bxy + dy^2$

$\Rightarrow \quad u^T Au = a\left(x^2 + \frac{2b}{a}xy + \frac{d}{a}y^2\right) = a\left\{\left(x + \frac{by}{a}\right)^2 + \frac{(ad - b^2)}{a^2}y^2\right\} > 0$ for all $u \neq 0$.

$\Rightarrow \quad$ Matrix A is positive definite.

Q.E.D.

THEOREM-2 *Let A be a real positive definite matrix. Then, R^n is an inner product space with the inner product*
$$\langle u, v \rangle = u^T Av \text{ for all } u, v \in R^n.$$

PROOF. We observe the following properties of the product defined above.

I1: *Linearity*: Let $u, v, w \in R^n$ and $a, b \in R$. Then,

$$\langle au + bv, w \rangle = (au + bv)^T Aw$$
$$= ((au)^T + (bv)^T) Aw$$
$$= (au^T + bv^T) Aw$$
$$= a(u^T Aw) + b(v^T Aw)$$
$$= a\langle u, w \rangle + b\langle v, w \rangle$$

I2: *Symmetry*: For any $u, v \in R^n$, we have

$$\langle u, v \rangle = u^T Av = (u^T Av)^T \qquad [\because \ u^T Av \text{ is a scalar}]$$
$$= v^T A^T (u^T)^T$$
$$= v^T Au \qquad [\because \ A^T = A]$$
$$= \langle v, u \rangle$$

I3: *Positive definiteness*: For any non-zero $u \in R^n$, we have

$$\langle u, u \rangle = u^T Au > 0 \quad [\because \ A \text{ is positive definite} \ \therefore \ X^T AX > 0 \text{ for all } X \in R^n]$$

Also, $\langle O, O \rangle = O^T AO = 0$

$\therefore \quad \langle u, u \rangle \geq 0 \quad$ for all $u \in R^n$

Hence, R^n is an inner product space with the given inner product.

Q.E.D.

THEOREM-3 *Let V be a finite dimensional an inner product space and A be the matrix representation of any basis of V. Then, A is a positive definite matrix.*

PROOF. Let $B = \{v_1, v_2, \ldots, v_n\}$ be any basis for V and $A = [a_{ij}]$ be the matrix representation of the inner product on V relative to basis B. Then,

$$a_{ij} = \langle v_i, v_j \rangle \text{ for all } i, j \in \underline{n}$$

Let X be any non-zero vector in R^n. Then, there exists a non-zero vector $u \in V$ such that X is the coordinate vector of u relative to basis B. Then,

$$X^T AX = [u]^T A[u]$$
$\Rightarrow \qquad X^T AX = \langle u, u \rangle \qquad$ [By Theorem 2]
$\Rightarrow \qquad X^T AX > 0 \qquad\qquad [\because \ u \neq 0_V \ \therefore \ \langle u, u \rangle > 0]$

Thus, $\quad X^T AX > 0 \quad$ for all $O \neq X \in R^n$

Hence, $\quad A$ is positive definite matrix.

Q.E.D.

ILLUSTRATIVE EXAMPLES

EXAMPLE-1 *Show that the symmetric matrix* $A = \begin{bmatrix} 2 & -1 \\ 1 & 3 \end{bmatrix}$ *is positive definite.*

SOLUTION We know that a 2×2 real symmetric matrix is positive definite iff its diagonal entries are positive and if its determinant is positive.

Clearly, diagonal elements of matrix A are positive and $|A| = 7 > 0$. Hence, A is a positive definite matrix.

EXAMPLE-2 *Find the value of λ that make each of the following matrices positive definite:*
(i) $A = \begin{bmatrix} 3 & -6 \\ -6 & \lambda \end{bmatrix}$ (ii) $B = \begin{bmatrix} 16 & 3\lambda \\ 3\lambda & 9 \end{bmatrix}$ (iii) $C = \begin{bmatrix} -5 & 3 \\ 3 & \lambda \end{bmatrix}$

SOLUTION (i) We have,

$$A = \begin{bmatrix} 3 & -6 \\ -6 & \lambda \end{bmatrix}$$

Clearly, A will be positive definite, if

$$\lambda > 0 \text{ and } |A| > 0$$

i.e.   if $\lambda > 0$ and $3\lambda - 36 > 0$

or,   if $\lambda > 0$ and $\lambda > 12$

$\Rightarrow$ $\lambda > 12$

(ii) Matrix B will be positive definite, if

$$|B| > 0 \Rightarrow \begin{vmatrix} 16 & 3\lambda \\ 3\lambda & 9 \end{vmatrix} > 0 \Rightarrow 144 - 9\lambda^2 > 0 \Rightarrow \lambda^2 - 16 < 0 \Rightarrow -4 < \lambda < 4$$

(iii) Given matrix cannot be positive definite for any value of λ as one of the diagonal elements is negative.

EXERCISE 6.9

1. Which of the following symmetric matrices are positive definite?
   (i) $A = \begin{bmatrix} 2 & 5 \\ 5 & 3 \end{bmatrix}$ (ii) $B = \begin{bmatrix} 2 & -3 \\ 3 & 5 \end{bmatrix}$ (iii) $C = \begin{bmatrix} -2 & 1 \\ 1 & 3 \end{bmatrix}$ (iv) $D = \begin{bmatrix} 5 & 4 \\ 4 & 9 \end{bmatrix}$

2. Find the value λ for which the following matrices are positive definite:
   (i) $A = \begin{bmatrix} 3 & -1 \\ -1 & \lambda \end{bmatrix}$ (ii) $B = \begin{bmatrix} 9 & \lambda \\ \lambda & 4 \end{bmatrix}$ (iii) $C = \begin{bmatrix} -3 & 4 \\ 4 & \lambda \end{bmatrix}$ (iv) $\begin{bmatrix} \lambda & 1 \\ 1 & \lambda \end{bmatrix}$

ANSWERS

1. (ii), (iv)   2. (i) $\lambda > \dfrac{1}{3}$ (ii) $-6 < \lambda < 6$ (iii) No value of λ (iv) $\lambda > 1$

6.12 ORTHOGONAL MATRICES

In this section, we shall learn about the relationship between orthonormality of rows (columns) of an orthogonal matrix in R^n with respect to standard inner product.

ORTHOGONAL MATRIX *A square matrix A over R is an orthogonal matrix, if* $AA^T = I = A^T A$.
It is evident from the above definition that A is orthogonal iff A is non-singular and $A^{-1} = A^T$.

Clearly, matrix A given by $A = \begin{bmatrix} \cos\theta & \sin\theta \\ \sin\theta & -\cos\theta \end{bmatrix}$ is orthogonal for all values of θ, because

$$AA^T = \begin{bmatrix} \cos\theta & \sin\theta \\ \sin\theta & -\cos\theta \end{bmatrix} \begin{bmatrix} \cos\theta & \sin\theta \\ \sin\theta & -\cos\theta \end{bmatrix} = \begin{bmatrix} 1 & 0 \\ 0 & 1 \end{bmatrix} = I_2$$

EXAMPLE-1 *Let A be a square real matrix. Prove that:*
  (i) If A is orthogonal, then A is non-singular
  (ii) A is orthogonal iff A^T is orthogonal.
  (iii) If A is orthogonal, then A^{-1} is orthogonal

SOLUTION (i) Let A be an orthogonal matrix. Then,

$$AA^T = I = A^T A$$

$\Rightarrow \quad |AA^T| = 1$

$\Rightarrow \quad |A||A^T| = 1$

$\Rightarrow \quad |A|^2 = 1$ $\qquad [\because \ |A| = |A^T|]$

$\Rightarrow \quad |A| = \pm 1 \Rightarrow A$ is non-singular.

(ii) A is orthogonal

$\Leftrightarrow \quad AA^T = I = A^T A$

$\Leftrightarrow \quad (AA^T)^T = I^T = (A^T A)^T$

$\Leftrightarrow \quad (A^T)^T A^T = I = A^T (A^T)^T$

$\Leftrightarrow \quad A^T$ is orthogonal

(iii) Let A be an orthogonal matrix. Then,

$$AA^T = I = A^T A$$

$\Rightarrow \quad (AA^T)^{-1} = I^{-1} = (A^T A)^{-1}$

$\Rightarrow \quad (A^T)^{-1} A^{-1} = I = A^{-1}(A^T)^{-1}$

$\Rightarrow \quad (A^{-1})^T A^{-1} = I = A^{-1}(A^{-1})^T$ $\qquad [\because \ (A^T)^{-1} = (A^{-1})^T]$

$\Rightarrow \quad A^{-1}$ is orthogonal.

EXAMPLE-2 *If A and B are orthogonal matrices, prove that AB is also an orthogonal matrix.*

SOLUTION It is given that A and B are orthogonal matrices.

$\therefore \qquad AA^T = I = A^T A \quad \text{and} \quad BB^T = I = B^T B$

$\Rightarrow \qquad A^T = A^{-1} \quad \text{and} \quad B^T = B^{-1}$

Now, $\qquad (AB)^T = B^T A^T = B^{-1} A^{-1} = (AB)^{-1}$

$\therefore \qquad (AB)(AB)^T = I$

$\Rightarrow \qquad AB$ is an orthogonal matrix.

EXAMPLE-3 *Let A be an orthogonal matrix. In the inner product space R^n with inner product $\langle u, v \rangle = u^T v$, prove that*

(i) $\langle Au, Av \rangle = \langle u, v \rangle$ *for all* $u, v \in R^n$. (ii) $||Au|| = ||u||$ *for all* $u \in R^n$

PROOF. (i) For any $u, v \in R^n$, we have

$$\langle Au, Av \rangle = (Au)^T (Av) = (u^T A^T)(Av) = u^T (A^T A) v = u^T I v = u^T v = \langle u, v \rangle$$

(ii) For any $u \in R^n$, we have

$||Au||^2 = \langle Au, Au \rangle = \langle u, u \rangle$ [Using (i)]

$\Rightarrow \qquad ||Au||^2 = ||u||^2$

$\Rightarrow \qquad ||Au|| = ||u||$

THEOREM-1 *If A is a square matrix over ring R, then the following statements are equivalent:*

(i) *A is orthogonal*

(ii) *the rows of A form an orthonormal set.*

(iii) *the columns of A form an orthonormal set.*

PROOF. Let A be an $n \times n$ orthogonal matrix having $R_1, R_2, \ldots, R_n$ as its rows and $C_1, C_2, \ldots, C_n$ as its columns, i.e.

$$A = \begin{bmatrix} R_1 \\ R_2 \\ \vdots \\ R_n \end{bmatrix} \quad \text{and} \quad A = [C_1 \ C_2 \ \ldots \ C_n].$$

(i) $\Leftrightarrow$ (ii)

Let A be an orthogonal matrix.

$\Leftrightarrow \quad AA^T = I$

$\Leftrightarrow \quad (AA^T)_{ij} = \delta_{ij} \quad$ for all $i, j \in \underline{n}$

$\Leftrightarrow \quad (i^{\text{th}} \text{ row of } A)(j^{\text{th}} \text{ column of } A^T) = \delta_{ij} \quad$ for all $i, j \in \underline{n}$

$\Leftrightarrow \quad R_i \cdot R_j = \delta_{ij} \quad$ for all $i, j \in \underline{n}$

$\Leftrightarrow \quad R_1, R_2, \ldots, R_n \quad$ form an orthonormal set.

$\Leftrightarrow \quad$ The rows of A form an orthonormal set.

(i) $\Leftrightarrow$ (iii)

Let A be an orthogonal matrix.

$\Leftrightarrow \quad A^T A = I$

$\Leftrightarrow \quad (A^T A)_{ij} = \delta_{ij} \quad$ for all $i, j \in \underline{n}$

$\Leftrightarrow \quad (i^{\text{th}} \text{ row of } A^T)(j^{\text{th}} \text{ column of } A) = \delta_{ij} \quad$ for all $i, j \in$

$\Leftrightarrow \quad (i^{\text{th}} \text{ column of } A)(j^{\text{th}} \text{ column of } A) = \delta_{ij} \quad$ for all $i, j \in$

$\Leftrightarrow \quad C_i \cdot C_j = \delta_{ij}. \quad$ for all $i, j \in \underline{n}$

$\Leftrightarrow \quad C_1, C_2, \ldots, C_n \quad$ form an orthonormal set

$\Leftrightarrow \quad$ The columns of A form an orthonormal set.

Hence, (i) $\Leftrightarrow$ (ii) $\Leftrightarrow$ (iii) i.e. given statements are equivalent.

Q.E.D.

Let us now discuss relationships between orthogonal matrices and orthonormal basis of an inner product space. The following theorems establish the same.

THEOREM-2 *Let V be a finite dimensional inner product space. Let $B_1 = \{u_1, u_2, \ldots, u_n\}$ and $B_2 = \{v_1, v_2, \ldots, v_n\}$ be orthonormal bases for V. Then, prove that the change-of-basis matrix from basis B_1 to basis B_2 is an orthogonal matrix.*

PROOF. Since B_1 is a basis for V. Therefore, for any $i \in \underline{n}$,

$$v_i = \sum_{j=1}^{n} b_{ij} u_j$$

Clearly, $B = [b_{ij}]$ is the change-of-basis matrix from B_1 to B_2. Since B_2 is an orthonormal basis for V.

$\therefore \quad \langle v_i, v_j \rangle = \delta_{ij} \quad$ for all $i, j \in \underline{n}$

$\Rightarrow \quad \left\langle \sum_{r=1}^{n} b_{ir} u_r, \sum_{s=1}^{n} b_{js} u_s \right\rangle = \delta_{ij} \quad$ for all $i, j \in \underline{n}$

$\Rightarrow \quad \sum_{r=1}^{n} \sum_{s=1}^{n} b_{ir} b_{js} \langle u_r, u_s \rangle = \delta_{ij}$ for all $i, j \in \underline{n}$

$\Rightarrow \quad \sum_{r=1}^{n} \sum_{s=1}^{n} b_{ir} b_{js} \delta_{rs} = \delta_{ij}$ for all $i, j \in \underline{n}$

$\Rightarrow \quad \sum_{r=1}^{n} b_{ir} b_{jr} = \delta_{ij}$ for all $i, j \in \underline{n}$

$\Rightarrow \quad \sum_{r=1}^{n} B_{ir} (B^T)_{rj} = \delta_{ij}$ for all $i, j \in \underline{n}$

$\Rightarrow \quad BB^T = I$

$\Rightarrow \quad B$ is an orthogonal matrix.

Q.E.D.

THEOREM-3 Let $\{v_1, v_2, \ldots, v_n\}$ be an orthonormal basis of an inner product space V. Let $A = [a_{ij}]$ be an orthogonal matrix, and let $u_j = \sum_{r=1}^{n} a_{rj} v_r$, $j = 1, 2, \ldots, n$. Then prove that $\{u_1, u_2, \ldots, u_n\}$ is an orthonormal set.

PROOF. For any $i, j \in \underline{n}$, we have

$\langle u_i, u_j \rangle = \left\langle \sum_{r=1}^{n} a_{ri} v_r, \sum_{s=1}^{n} a_{sj} v_s \right\rangle$

$\Rightarrow \quad \langle u_i, u_j \rangle = \sum_{r=1}^{n} \sum_{s=1}^{n} a_{ri} a_{sj} \langle v_r, v_s \rangle$

$\Rightarrow \quad \langle u_i, u_j \rangle = \sum_{r=1}^{n} \sum_{s=1}^{n} a_{ri} a_{sj} \delta_{rs}$

$\Rightarrow \quad \langle u_i, u_j \rangle = \sum_{r=1}^{n} a_{ri} a_{rj}$

$\Rightarrow \quad \langle u_i, u_j \rangle = \sum_{r=1}^{n} (A^T)_{ir} (A)_{rj}$

$\Rightarrow \quad \langle u_i, u_j \rangle = (A^T A)_{ij}$

$\Rightarrow \quad \langle u_i, u_j \rangle = \delta_{ij}$ [$\because A$ is orthogonal matrix $\therefore AA^T = I = A^T A$]

$\Rightarrow \quad \{u_1, u_2, \ldots, u_n\}$ is an orthonormal set

Q.E.D.

ORTHOGONALLY EQUIVALENT MATRICES *Two symmetric matrices of the same order over R are said to be orthogonally equivalent if there exists an orthogonal matrix P such that $B = P^T A P$.*

It can be easily shown that the relation "orthogonally equivalent" is an equivalence relation on the set of all symmetric matrices over R.

ILLUSTRATIVE EXAMPLES

EXAMPLE-1 Find all 2×2 orthogonal matrices of the form $\begin{bmatrix} 1/3 & a \\ b & c \end{bmatrix}$.

SOLUTION Let $A = \begin{bmatrix} 1/3 & a \\ b & c \end{bmatrix}$ be an orthogonal matrix. Then, rows (columns) of A form an orthonormal set in R^2.

$\therefore \quad \frac{1}{9} + a^2 = 1, \quad b^2 + c^2 = 1, \quad \frac{b}{3} + ac = 0$

$\frac{1}{9} + b^2 = 1, \quad a^2 + c^2 = 1, \quad \frac{a}{3} + bc = 0$

Now, $\frac{1}{9} + a^2 = 1, \quad \frac{1}{9} + b^2 = 1, \quad b^2 + c^2 = 1 \quad \text{and} \quad a^2 + c^2 = 1$

$\Rightarrow \quad a = b = \pm \frac{2\sqrt{2}}{3}, \quad c = \pm \frac{1}{3}$

The equations $\frac{a}{3} + bc = 0$ and $\frac{b}{3} + ac = 0$ cannot hold good simultaneously if a, b, c are positive.

Clearly,

$a = b = \frac{2\sqrt{2}}{3} \Rightarrow c = -\frac{1}{3}$ $\left[\text{Putting } a = b = \frac{2\sqrt{2}}{3} \text{ in } \frac{a}{3} + bc = 0 \text{ and } \frac{b}{3} + ac = 0 \right]$

$a = b = -\frac{2\sqrt{2}}{3} \Rightarrow c = \frac{-1}{3}$ $\left[\text{Putting } a = b = \frac{-2\sqrt{2}}{3} \text{ in } \frac{a}{3} + bc = 0 \text{ and } \frac{b}{3} + ac = 0 \right]$

$a = \frac{2\sqrt{2}}{3}, b = \frac{-2\sqrt{2}}{3} \Rightarrow c = \frac{1}{3}$ $\left[\text{Putting } a = -b \text{ in } \frac{a}{3} + bc = 0 \text{ and } \frac{b}{3} + ac = 0 \right]$

$a = \frac{-2\sqrt{2}}{3}, b = \frac{2\sqrt{2}}{3} \Rightarrow c = \frac{1}{3}$ $\left[\text{Putting } a = -b \text{ in } \frac{a}{3} + bc = 0 \text{ and } \frac{b}{3} + ac = 0 \right]$

Hence, all possible values of matrix A are:

$\begin{bmatrix} x & y \\ y & -x \end{bmatrix}, \begin{bmatrix} x & y \\ -y & x \end{bmatrix}, \begin{bmatrix} x & -y \\ y & x \end{bmatrix}, \begin{bmatrix} x & -y \\ -y & -x \end{bmatrix}$, where $x = \frac{1}{3}$ and $y = \frac{2\sqrt{2}}{3}$

EXAMPLE-2 Find an orthogonal matrix whose first row is $(1/3, 2/3, 2/3)$

SOLUTION Let A be the required matrix having its first row as the vector $v_1 = (1/3, 2/3, 2/3)$ and v_2 and v_3 be its second and third rows. Further, let $v_2 = (a, b, c)$ and $v_3 = (x, y, z)$.

Since rows of an orthogonal matrix form an orthonormal set.

$\therefore \quad \langle v_1, v_2 \rangle = 0 \Rightarrow \dfrac{a}{3} + \dfrac{2b}{3} + \dfrac{2c}{3} = 0 \Rightarrow a + 2b + 2c = 0$

and, $\quad \langle v_1, v_1 \rangle = 1 \Rightarrow a^2 + b^2 + c^2 = 1.$

Clearly, $a = \dfrac{2}{3}$, $b = \dfrac{-2}{3}$, $c = \dfrac{1}{3}$ satisfy these equations.

$\therefore \quad v_2 = \left(\dfrac{2}{3}, \dfrac{-2}{3}, \dfrac{1}{3} \right)$

Now, $\langle v_1, v_3 \rangle = 0$, $\langle v_2, v_3 \rangle = 0$ and $\langle v_3, v_3 \rangle = 1$

$\Rightarrow \dfrac{x}{3} + \dfrac{2y}{3} + \dfrac{2z}{3} = 0, \dfrac{2}{3}x - \dfrac{2}{3}y + \dfrac{1}{3}z = 0$ and $x^2 + y^2 + z^2 = 1$

$\Rightarrow x + 2y + 2z = 0, 2x - 2y + z = 0$ and $x^2 + y^2 + z^2 = 1$

$\Rightarrow x + z = 0, x + 2y + 2z = 0$ and $x^2 + y^2 + z^2 = 1$ [Adding first two equations]

$\Rightarrow x = -z, y = -\dfrac{z}{2}$ and $x^2 + y^2 + z^2 = 1$

$\Rightarrow x = -z, y = -\dfrac{z}{2}$ and $z = \pm \dfrac{2}{3}$

$\Rightarrow x = \mp \dfrac{2}{3}, y = \mp \dfrac{1}{3}, z = \pm \dfrac{2}{3}$

$\therefore \quad v_3 = \left(-\dfrac{2}{3}, -\dfrac{1}{3}, \dfrac{2}{3} \right)$ or, $v_3 = \left(\dfrac{2}{3}, \dfrac{1}{3}, -\dfrac{2}{3} \right)$

Hence, $A = \begin{bmatrix} v_1 \\ v_2 \\ v_3 \end{bmatrix} = \begin{bmatrix} 1/3 & 2/3 & 2/3 \\ 2/3 & -2/3 & 1/3 \\ -2/3 & -1/3 & 2/3 \end{bmatrix}$ or, $A = \begin{bmatrix} 1/3 & 2/3 & 2/3 \\ 2/3 & -2/3 & 1/3 \\ 2/3 & 1/3 & -2/3 \end{bmatrix}$

EXAMPLE-3 Find a 3×3 orthogonal matrix A whose first two rows are multiples of $u = (1, 1, 1)$ and $v = (1, -2, 3)$.

SOLUTION Let $w = (x, y, z)$ be the third row of matrix A. Then,

$\langle u, w \rangle = 0, \langle v, w \rangle = 0$

$\Rightarrow \quad x+y+z=0,\ x-2y+3z=0$

$\Rightarrow \quad x+y+z=0,\ 3y-2z=0$ [Subtracting second equation from equation (i)]

$\Rightarrow \quad x = -\dfrac{5}{3}z,\ y = \dfrac{2}{3}z$

$\Rightarrow \quad x = -5,\ y = 2,\ z = 3$ [Taking $z = 3$]

$\therefore \quad w = (-5, 2, 3)$

Normalizing u, v and w, we obtain that the required matrix A is equal to

$$A = \begin{bmatrix} \dfrac{1}{\sqrt{3}} & \dfrac{1}{\sqrt{3}} & \dfrac{1}{\sqrt{3}} \\ \dfrac{1}{\sqrt{14}} & \dfrac{-2}{\sqrt{14}} & \dfrac{3}{\sqrt{14}} \\ \dfrac{-5}{\sqrt{38}} & \dfrac{2}{\sqrt{38}} & \dfrac{3}{\sqrt{38}} \end{bmatrix}$$

EXERCISE 6.10

1. Let $A = \begin{bmatrix} 1 & 2 & 2 \\ 0 & 1 & -1 \\ 4 & -1 & -1 \end{bmatrix}$. Check whether or not

    (i) the rows of A are orthogonal vectors in R^3.

    (ii) the columns of A are not orthogonal vectors in R^3.

    (iii) A is not an orthogonal matrix.

2. Let $A = \begin{bmatrix} 1 & 1 & -1 \\ 1 & 3 & 4 \\ 7 & -5 & 2 \end{bmatrix}$. Determine whether or not

    (i) the rows of A are orthogonal.

    (ii) A is an orthogonal matrix.

    (iii) the columns of A are orthogonal.

3. Show that the relation "is orthogonally equivalent" is an equivalence relation on the set of all real symmetric matrices.

4. Find an orthogonal matrix whose first row is $\left(\dfrac{2}{\sqrt{13}}, \dfrac{3}{\sqrt{13}} \right)$.

5. Find a symmetric orthogonal matrix A whose first row is $\left(\dfrac{1}{3}, \dfrac{2}{3}, \dfrac{2}{3} \right)$.

ANSWERS

1. (i) Yes (ii) No (iii) No

2. (i) Yes (ii) No (iii) No

4. $\begin{bmatrix} \dfrac{2}{\sqrt{13}} & \dfrac{3}{\sqrt{13}} \\ -\dfrac{3}{\sqrt{13}} & \dfrac{2}{\sqrt{13}} \end{bmatrix}$ or, $\begin{bmatrix} \dfrac{2}{\sqrt{13}} & \dfrac{3}{\sqrt{13}} \\ \dfrac{3}{\sqrt{13}} & -\dfrac{2}{\sqrt{13}} \end{bmatrix}$

5. $\begin{bmatrix} 1/3 & 2/3 & 2/3 \\ 2/3 & -2/3 & 1/3 \\ 2/3 & 1/3 & -2/3 \end{bmatrix}$

Chapter 7

Linear Operators on Inner Product Spaces

7.1 INTRODUCTION

In this chapter, we will extend the concept of vector space homomorphisms (linear transformations) on inner product spaces. In Chapter 3, we have learnt about linear transformations and linear operators on ordinary vector spaces. In this chapter, we will discuss these concepts on inner product spaces.

7.2 INNER PRODUCT SPACE MORPHISMS

INNER PRODUCT SPACE HOMOMORPHISM Let V and V' be two inner product spaces. A linear transformation $t : V \to V'$ is said to be an inner product space homomorphism if it preserves the inner products, i.e.,

$$\langle u, v \rangle = \langle t(u), t(v) \rangle \text{ for all } u, v \in V.$$

It should be noted that the inner product on LHS of the above equality is the inner product on V whereas on the RHS it is inner product on V'.

INNER PRODUCT SPACE ISOMORPHISM Let V and V' be two inner product spaces. A linear transformation $t : V \to V'$ is said to be an inner product space isomorphism if

  (i) t is an isomorphism
  (ii) t preserves the inner products, i.e. $\langle u, v \rangle = \langle t(u), t(v) \rangle$ for all $u, v \in V$.

The definitions of monomorphism, epimorphism, automorphism, etc. are similar to what we have learnt in the chapter on linear transformations on vector spaces.

Two inner product spaces V and V' are said to be isomorphic if there exists an isomorphism between them.

THEOREM-1 Let $B = \{u_1, u_2, \ldots, u_n\}$ be an orthonormal basis of an inner product space V. Show that the mapping associating each $v \in V$ to its coordinate vector $[v]_B$ is an inner product space isomorphism.

PROOF. Let v be an arbitrary vector in V. Since B is an orthonormal basis of V, therefore, there exist scalars $\lambda_1, \lambda_2, \ldots, \lambda_n$ such that $v = \sum_{i=1}^{n} \lambda_i u_i$.

Consider the mapping $t: V \to R^n$ defined by

$$t(v) = [v]_B = (\lambda_1, \lambda_2, \ldots, \lambda_n)$$

Clearly, t is well defined as corresponding to each $v \in V$ there exists a unique list $(\lambda_1, \lambda_2, \ldots, \lambda_n)$ of scalars in R^n.

We have to prove that t is an inner product space isomorphism.

t is a linear transformation: Let $u, v \in V$ and $a, b \in R$. Then, $u = \sum_{i=1}^{n} \lambda_i u_i$ and $v = \sum_{i=1}^{n} \mu_i u_i$ for some $\lambda_i, u_i \in R$

$$\therefore \quad t(au + bv) = t\left(\sum_{i=1}^{n} (a\lambda_i + b\mu_i) u_i\right)$$

$\Rightarrow \quad t(au + bv) = (a\lambda_1 + b\mu_1, a\lambda_2 + b\mu_2, \ldots, a\lambda_n + b\mu_n)$

$\Rightarrow \quad t(au + bv) = a(\lambda_1, \lambda_2, \ldots, \lambda_n) + b(\mu_1, \mu_2, \ldots, \mu_n)$

$\Rightarrow \quad t(au + bv) = at(u) + bt(v)$

So, $t: V \to R^n$ is a linear transformation.

t is one-one: For any $u = \sum_{i=1}^{n} \lambda_i u_i, v = \sum_{i=1}^{n} \mu_i u_i \in V$

$$t(u) = t(v)$$

$\Rightarrow \quad t\left(\sum_{i=1}^{n} \lambda_i u_i\right) = t\left(\sum_{i=1}^{n} \mu_i u_i\right)$

$\Rightarrow \quad (\lambda_1, \lambda_2, \ldots, \lambda_n) = (\mu_1, \mu_2, \ldots, \mu_n)$

$\Rightarrow \quad \lambda_i = \mu_i$ for all $i \in \underline{n}$

$\Rightarrow \quad \sum_{i=1}^{n} \lambda_i u_i = \sum_{i=1}^{n} \mu_i u_i$

$\Rightarrow \quad u = v$

$\therefore \quad t: V \to R^n$ is one-one.

t is onto: For any $(\lambda_1, \lambda_2, \ldots, \lambda_n) \in R^n$ there exists a unique vector $u = \sum_{i=1}^{n} \lambda_i u_i \in V$ such that

$$t(u) = t\left(\sum_{i=1}^{n} \lambda_i u_i\right) = (\lambda_1, \lambda_2, \ldots, \lambda_n)$$

So, $t : V \to R^n$ is onto.

t is an inner product morphism: Let $u = \sum_{i=1}^{n} \lambda_i u_i$, $v = \sum_{j=1}^{n} \mu_j u_j \in V$. Then,

$$\langle u, v \rangle = \left\langle \sum_{i=1}^{n} \lambda_i u_i, \sum_{j=1}^{n} \mu_j u_j \right\rangle$$

$\Rightarrow \quad \langle u, v \rangle = \sum_{i=1}^{n} \sum_{j=1}^{n} \lambda_i \mu_j \langle u_i, u_j \rangle$

$\Rightarrow \quad \langle u, v \rangle = \sum_{i=1}^{n} \sum_{j=1}^{n} \lambda_i \mu_j \delta_{ij} = \sum_{i=1}^{n} \lambda_i \mu_i$

and, $\langle t(u), t(v) \rangle = \langle \alpha, \beta \rangle$, where $\alpha = (\lambda_1, \lambda_2, \ldots, \lambda_n)$, $\beta = (\mu_1, \mu_2, \ldots, \mu_n)$

$\Rightarrow \quad \langle t(u), t(v) \rangle = \lambda_1 \mu_1 + \lambda_2 \mu_2 + \cdots + \lambda_n \mu_n = \sum_{i=1}^{n} \lambda_i \mu_i$

$\therefore \quad \langle u, v \rangle = \langle t(u), t(v) \rangle$

Hence, $t : V \to R^n$ is an inner product space isomorphism. Consequently, $V \cong R^n$.

The following is an immediate consequence of the above theorem.

COROLLARY. *Any two finite dimensional inner product spaces V and V' are isomorphic iff $\dim V = \dim V'$.*

THEOREM-2 *Let $B = \{u_1, u_2, \ldots, u_n\}$ and $B' = \{u'_1, u'_2, \ldots, u'_n\}$ be orthonormal bases of inner product spaces V and V' respectively. Let $t : V \to V'$ be the linear map defined by $t(u_i) = u'_i$ for each $i \in \underline{n}$. Show that t is an inner product isomorphism.*

PROOF. It is given that t is a linear map. So, it is sufficient to show that t is a bijection and it is inner product morphism.

t is one-one: Let $u = \sum_{i=1}^{n} \lambda_i u_i$, $v = \sum_{i=1}^{n} \mu_i u_i \in V$. Then,

$$t(u) = t(v)$$

$\Rightarrow \quad t\left(\sum_{i=1}^{n} \lambda_i u_i \right) = t\left(\sum_{i=1}^{n} \mu_i u_i \right)$

$\Rightarrow \quad \sum_{i=1}^{n} \lambda_i t(u_i) = \sum_{i=1}^{n} \mu_i t(u_i)$ \hfill $[\because \ t \text{ is a linear map}]$

$\Rightarrow \quad \sum_{i=1}^{n} \lambda_i u'_i = \sum_{i=1}^{n} \mu_i u'_i$ \hfill $[\because \ t(u_i) = u'_i \text{ for each } i \in \underline{n}]$

$\Rightarrow \quad \lambda_i = \mu_i \quad$ for each $i \in \underline{n}$ $\hfill [\because \ B'$ is a basis of $V']$

$\Rightarrow \quad \sum_{i=1}^{n} \lambda_i u_i = \sum_{i=1}^{n} \mu_i u_i$

$\Rightarrow \quad u = v$

So, $\quad t : V \to V'$ is one-one.

t is onto: Let u' be an arbitrary vector in V'. Then, there exist unique scalars $\lambda_1', \lambda_2', \ldots, \lambda_n' \in R$ such that $u' = \sum_{i=1}^{n} \lambda_i' u_i'$. Therefore, there exists a vector $u = \sum_{i=1}^{n} \lambda_i' u_i \in V$ such that

$$t(u) = t\left(\sum_{i=1}^{n} \lambda_i' u_i\right) = \sum_{i=1}^{n} \lambda_i' t(u_i) = \sum_{i=1}^{n} \lambda_i' u_i' = u'$$

Thus, for each $u' \in V'$, there exists $u \in V$ such that $t(u) = u'$.

So, $t : t : V \to V'$ is onto.

t is an inner product morphism: Let $u, v \in V$. Then,

$$u = \sum_{i=1}^{n} \lambda_i u_i, \ v = \sum_{j=1}^{n} \mu_j u_j$$

$\Rightarrow \quad <u, v> = \left\langle \sum_{i=1}^{n} \lambda_i u_i, \sum_{j=1}^{n} \mu_j u_j \right\rangle$

$\Rightarrow \quad <u, v> = \sum_{i=1}^{n} \sum_{j=1}^{n} \lambda_i \mu_j \langle u_i, u_j \rangle$

$\Rightarrow \quad <u, v> = \sum_{i=1}^{n} \sum_{j=1}^{n} \lambda_i \mu_j \delta_{ij}$ $\hfill [\because B$ is an orthonormal basis]

$\Rightarrow \quad <u, v> = \sum_{i=1}^{n} \lambda_i \mu_i$

and,

$$\langle t(u), t(v) \rangle = \left\langle t\left(\sum_{i=1}^{n} \lambda_i u_i\right), t\left(\sum_{j=1}^{n} \mu_j u_j\right) \right\rangle$$

$\Rightarrow \quad \langle t(u), t(v) \rangle = \left\langle \sum_{i=1}^{n} \lambda_i t(u_i), \sum_{j=1}^{n} \mu_j t(u_j) \right\rangle$

$\Rightarrow \quad \langle t(u), t(v) \rangle = \left\langle \sum_{i=1}^{n} \lambda_i u_i', \sum_{j=1}^{n} \mu_j u_j' \right\rangle$ $\hfill [\because t(u_i) = u_i' \quad$ for each $i \in \underline{n}]$

$\Rightarrow \quad \langle t(u), t(v) \rangle = \sum_{i=1}^{n} \sum_{j=1}^{n} \lambda_i \mu_j \langle u_i', u_j' \rangle$

$$\Rightarrow \quad \langle t(u), t(v) \rangle = \sum_{i=1}^{n} \sum_{j=1}^{n} \lambda_i \, \mu_j \delta_{ij} \qquad [\because B' \text{ is an orthonormal basis}]$$

$$\Rightarrow \quad \langle t(u), t(v) \rangle = \sum_{i=1}^{n} \lambda_i \mu_i$$

$$\therefore \quad <u, v> = \langle t(u), t(v) \rangle$$

So, t is an inner product space morphism.
Hence, t is an inner product space isomorphism.

Q.E.D.

THEOREM-3 *A necessary and sufficient condition for a linear operator t on an inner product space V to be the null operator $(\hat{0})$ is that $\langle t(u), v \rangle = 0$ for all $u, v \in V$.*

PROOF. *Necessary condition:* First, let t be the null operator on V, i.e. $t = \hat{0}$. Then, for all $u, v \in V$

$$\langle t(u), v \rangle = \langle \hat{0}(u), v \rangle = \langle 0_V, v \rangle = 0$$

Sufficient condition: Let t be a linear operator on V such that

$$\langle t(u), v \rangle = 0 \quad \text{for all } u, v \in V$$

$$\Rightarrow \quad \langle t(u), t(u) \rangle = 0 \quad \text{for all } u \in V \qquad [\text{Replacing } v \text{ by } t(u)]$$

$$\Rightarrow \quad t(u) = 0_V \quad \text{for all } u \in V$$

$$\Rightarrow \quad t = \hat{0}.$$

Q.E.D.

REMARK. *It should be noted that $\langle t(u), u \rangle = 0$ for all $u \in V$ need not imply that $t = \hat{0}$. For example, for the linear operator t on R^2 defined by $t(x,y) = (y, -x)$, we have $\langle t(u), u \rangle = \langle (y, -x), (x, y) \rangle = xy - xy = 0$, but $t \neq \hat{0}$.*

7.3 LINEAR FUNCTIONALS AND INNER PRODUCT SPACES

Recall that a linear functional ϕ on a vector space $V(F)$ is a linear transformation $\phi : V \to F$. A linear functional ϕ on an inner product space V is a linear transformation $\phi : V \to R$. In this section, we shall prove a theorem which will be used in proving the existence and uniqueness of the adjoint of a linear operator on an inner product space.

THEOREM-1 *Let ϕ be a linear functional on a finite dimensional inner product space V. Then there exists a unique vector $u \in V$ such that $\phi(v) = <v, u>$ for all $v \in V$.*

PROOF. Let $\{w_1, w_2, \ldots, w_n\}$ be an orthonormal basis for inner product space V. Let

$$u = \phi(w_1)w_1 + \phi(w_2)w_2 + \cdots + \phi(w_n)w_n$$

Clearly, $u \in V$.

Let $f : V \to R$ be a function defined by

$$f(v) = <v, u> \quad \text{for all } v \in V.$$

For any $a, b \in R$ and $v_1, v_2 \in V$, we have

$$\begin{aligned} f(av_1 + bv_2) &= \langle av_1 + bv_2, u \rangle \\ &= a\langle v_1, u \rangle + b\langle v_2, u \rangle \\ &= af(v_1) + bf(v_2) \end{aligned}$$

So, f is a linear functional on V.

For $i = 1, 2, \ldots, n$, we have

$$f(w_i) = \langle w_i, u \rangle$$

$$\Rightarrow \quad f(w_i) = \left\langle w_i, \sum_{j=1}^{n} \phi(w_j) w_j \right\rangle$$

$$\Rightarrow \quad f(w_i) = \sum_{j=1}^{n} \phi(w_j) \langle w_i, w_j \rangle$$

$$\Rightarrow \quad f(w_i) = \sum_{j=1}^{n} \phi(w_j) \delta_{ij} = \phi(w_i)$$

Thus, f and ϕ agree on a basis of V. Therefore, $f = \phi$. Thus, corresponding to a linear functional ϕ on V there exists a vector $u \in V$ such that $\phi(v) = <v, u>$ for all $v \in V$.

Now, we shall prove the uniqueness of vector u.

If possible, let u' be another vector in V for which $\phi(v) = <v, u'>$ for all $v \in V$. Then,

$$\langle v, u \rangle = \langle v, u' \rangle \text{ for all } v \in V.$$

$$\Rightarrow \quad \langle v, u - u' \rangle = 0 \text{ for all } v \in V.$$

$$\Rightarrow \quad \langle u - u', u - u' \rangle = 0 \qquad [\text{Taking } v = u - u']$$

$$\Rightarrow \quad u - u' = 0_V \qquad [\because \langle u, u \rangle = 0 \text{ iff } u = 0_V]$$

$$\Rightarrow \quad u = u'$$

Thus, such a vector is unique.

Q.E.D.

EXAMPLE Let $\phi : R^3 \to R$ be a linear function defined by

$$\phi(x, y, z) = x + 2y - 3z \quad \text{for all } (x, y, z) \in R^3.$$

Find a vector $u \in R^3$ such that $\phi(v) = <v, u>$ for all $v \in V$.

SOLUTION By Theorem 1, we have

$u = \sum_{i=1}^{n} \phi(w_i) w_i$, where $B = \{w_1, w_2, \ldots, w_n\}$ is an orthonormal basis of inner product space V.

Here, $B = \left\{ e_1^{(3)}, e_2^{(3)}, e_3^{(3)} \right\}$ and $\phi(x, y, z) = x + 2y - 3z$.

$\therefore \quad u = \phi(e_1^{(3)}) e_1^{(3)} + \phi(e_2^{(3)}) e_2^{(3)} + \phi(e_3^{(3)}) e_3^{(3)}$

$\therefore \quad u = 1(1,0,0) + 2(0,1,0) + (-3)(0,0,1) = (1, 2, -3)$

Let V^* be the vector space dual to an inner product space V. It is evident from the above theorem that there is one-to-one correspondence between V^* and V i.e., corresponding to each linear functional ϕ in V^* there exists a unique vector $u \in V$ such that $\phi(v) = \langle v, u \rangle$ for all $v \in V$. Thus, each vector u of a finite dimensional inner product space V can be associated to a linear functional (linear form) $\langle -, u \rangle$. The following theorem proves that this correspondence is a vector space isomorphism.

THEOREM-2 *Let V be a finite dimensional inner product space. Then the function associating each vector $u \in V$ to the linear functional (linear form) ϕ_u, defined by $\phi_u(v) = \langle v, u \rangle$ for all $v \in V$, in the dual space V^* is an isomorphism of vector spaces. Hence, $V \cong V^*$.*

PROOF. Let $u, v \in V$ and $a, b \in R$. Then, $au + bv \in V$. So, there exists a unique linear functional $\phi_{au+bv} \in V^*$ such that

$\qquad \phi_{au+bv}(w) = \langle w, au + bv \rangle \qquad$ for all $w \in V$

$\Rightarrow \quad \phi_{au+bv}(w) = \langle w, au \rangle + \langle w, bv \rangle \qquad$ for all $w \in V$

$\Rightarrow \quad \phi_{au+bv}(w) = a\langle w, u \rangle + b\langle w, v \rangle \qquad$ for all $w \in V$

$\Rightarrow \quad \phi_{au+bv}(w) = a\phi_u(w) + b\phi_v(w) \qquad$ for all $w \in V$

$\Rightarrow \quad \phi_{au+bv}(w) = (a\phi_u + b\phi_v)(w) \qquad$ for all $w \in V$

$\Rightarrow \quad \phi_{au+bv} = a\phi_u + b\phi_v$

Now, consider the mapping $\Psi : V \to V^*$ defined by

$\qquad \Psi(u) = \phi_u$ for all $u \in V$.

Ψ is a linear transformation: Let $u, v \in V$ and $a, b \in R$. Then,

$\qquad \Psi(au + bv) = \phi_{au+bv} = a\phi_u + b\phi_v = a\Psi(u) + b\Psi(v)$

$\therefore \quad \Psi$ is a linear transformation.

Ψ is one-one: For any $u, v \in V$

$\qquad \Psi(u) = \Psi(v)$

$\Rightarrow \quad \phi_u = \phi_v$

$\Rightarrow \quad \phi_u(w) = \phi_v(w) \qquad$ for all $w \in V$

$\Rightarrow \quad \langle w, u \rangle = \langle w, v \rangle \qquad$ for all $w \in V$

$\Rightarrow \quad \langle w, u - v \rangle = 0 \qquad$ for all $w \in V$

$\Rightarrow \quad \langle u - v, u - v \rangle = 0 \qquad\qquad\qquad\qquad\qquad\qquad\qquad$ [Taking $w = u - v$]

$\Rightarrow \quad u - v = 0_V$

$\Rightarrow \quad u = v$

$\therefore \quad \Psi$ is one-one.

Thus, $\Psi : V \to V^*$ is a monomorphism such that $\dim V = \dim V^*$. Therefore, Ψ is an isomorphism. Hence, $V \cong V^*$.

Q.E.D.

It follows from the above theorem that each vector u in an inner product space V is identified with the linear form (linear function) $\phi_u = \langle -, u \rangle$. Hence, each vector $u \in V$ can also be considered a linear form on V.

Let $B = \{b_1, b_2, \ldots, b_n\}$ be an orthonormal basis of an inner product space V and $X = \{x_1, x_2, \ldots, x_n\}$ be its dual. Then, each $u \in V$ can be written as $u = \sum_{i=1}^{n} \lambda_i b_i$, where $\lambda_i = x_i(u)$

$\therefore \quad \langle u, b_i \rangle = \left\langle \sum_{j=1}^{n} \lambda_j b_j, b_i \right\rangle = \sum_{j=1}^{n} \lambda_j \langle b_j, b_i \rangle = \sum_{j=1}^{n} \lambda_j \delta_{ji} = \lambda_i$

$\Rightarrow \quad \langle u, b_i \rangle = x_i(u)$ for all $i = 1, 2, \ldots, n$

$\Rightarrow \quad \langle -, b_i \rangle = x_i$ for all $i = 1, 2, \ldots, n$

This means that each basis vector b_i of an orthonormal basis of an inner product space is identified with the corresponding vector x_i in the dual basis. Hence, an orthonormal basis of an inner product space is its own dual.

7.4 ADJOINT OPERATORS

In this section, we will show that corresponding to every linear operator on a finite dimensional inner product space there exists a unique linear operator, known as its adjoint.

Let us begin with its definition.

ADJOINT OF AN OPERATOR *A linear operator t on an inner product space V is said to have an adjoint operator t^* on V if*

$$\langle t(u), v \rangle = \langle u, t^*(v) \rangle \text{ for every } u, v \in V$$

EXAMPLE-1 *Let A be an $n \times n$ matrix over R. Then, there exists a unique linear operator $t_A : R^n \to R^n$ defined by $t_A(u) = Au$ for all $u \in R^n$. Also, R^n is an inner product space with the inner product $\langle u, v \rangle = u^T v$ for all $u, v \in R^n$. Find the adjoint of t_A.*

SOLUTION We have,

$\langle t_A(u), v \rangle = \langle Au, v \rangle = (Au)^T v = u^T A^T v = \langle u, A^T v \rangle$ for all $u, v \in R^n$

$\Rightarrow \quad \langle t_A(u), v \rangle = \langle u, t_{A^T}(v) \rangle$ for all $u, v \in R^n$.

$\Rightarrow \quad t_{A^T}$ is adjoint of operator t_A.

Linear Operators on Inner Product Spaces

REMARK. *In Chapter 3, we have learnt that every matrix can be considered as a linear transformation. Therefore, considering A as an operator, we can say that A^T is adjoint of operator A.*

The following theorem proves the existence and uniqueness of adjoint of a linear operator on a finite dimensional inner product space.

THEOREM-1 *Let t be a linear operator on a finite dimensional inner product space V. Then there exists a unique linear operator t^* on V such that $\langle t(u), v \rangle = \langle u, t^*(v) \rangle$ for all $u, v \in V$.*

PROOF. Let V be an n dimensional inner product space and let v be an arbitrary but fixed vector in V.

Consider the mapping $\phi : V \to R$ given by

$$\phi(u) = \langle t(u), v \rangle \quad \text{for all } u \in R$$

For any $u_1, u_2 \in V$ and $a_1, a_2 \in R$, we have

$$\begin{aligned}
\phi(a_1 u_1 + a_2 u_2) &= \langle t(a_1 u_1 + a_2 u_2), v \rangle \\
&= \langle a_1 t(u_1) + a_2 t(u_2), v \rangle \\
&= \langle a_1 t(u_1), v \rangle + \langle a_2 t(u_2), v \rangle \\
&= a_1 \langle t(u_1), v \rangle + a_2 \langle t(u_2), v \rangle \\
&= a_1 \phi(u_1) + a_2 \phi(u_2)
\end{aligned}$$

$\therefore$ ϕ is a linear functional (linear form) on V.

Therefore, by Theorem 1, there exists a unique vector $v' \in V$ such that

$$\phi(u) = \langle u, v' \rangle \quad \text{for all } u \in V.$$

$\Rightarrow \quad \langle t(u), v \rangle = \langle u, v' \rangle \quad \text{for all } u \in V$

Thus, for each $v \in V$ there exists unique $v' \in V$ such that

$$\langle t(u), v \rangle = \langle u, v' \rangle \text{ for all } u \in V \qquad \text{(i)}$$

This provides a function $t^* : V \to V$ associating v to v', i.e. $t^*(v) = v'$.

$\therefore \quad \langle t(u), v \rangle = \langle u, t^*(v) \rangle \quad \text{for all } u, v \in V.$ \qquad [Replacing v' by $t^*(v)$ in (i)]

Thus, corresponding to every linear operator t on V, there exists a mapping $t^* : V \to V$ such that

$$\langle t(u), v \rangle = \langle u, t^*(v) \rangle \quad \text{for all } u, v \in V \qquad \text{(ii)}$$

We shall now show that t^* is linear map.

For any $v_1, v_2 \in V$ and $a, b \in R$, we have

$$\begin{aligned}
\langle u, t^*(av_1 + bv_2) \rangle &= \langle t(u), av_1 + bv_2 \rangle & \text{for all } u \in V \\
&= \langle t(u), av_1 \rangle + \langle t(u), bv_2 \rangle & \text{for all } u \in V \\
&= a\langle t(u), v_1 \rangle + b\langle t(u), v_2 \rangle & \text{for all } u \in V \\
&= a\langle u, t^*(v_1) \rangle + b\langle u, t^*(v_2) \rangle & \text{for all } u \in V \\
&= \langle u, a\, t^*(v_1) \rangle + \langle u, b\, t^*(v_2) \rangle & \text{for all } u \in V \\
&= \langle u, a\, t^*(v_1) + bt^*(v_2) \rangle & \text{for all } u \in V
\end{aligned}$$

$\therefore \quad t^*(av_1 + bv_2) = at^*(v_1) + bt^*(v_2)$ for all $v_1, v_2 \in V$ and $a, b \in R$

Thus, t^* is a linear operator on V.

Now, we shall prove the uniqueness of t^*.

If possible, let T be another operator satisfying (ii). Then,

$$\langle t(u), v \rangle = \langle u, T^*(v) \rangle \text{ for all } u, v \in V$$

$\Rightarrow \quad \langle u, t^*(v) \rangle = \langle u, T^*(v) \rangle \quad$ for all $u, v \in V$

$\Rightarrow \quad \langle u, t^*(v) \rangle - \langle u, T^*(v) \rangle = 0 \quad$ for all $u, v \in V$

$\Rightarrow \quad \langle u, t^*(v) - T^*(v) \rangle = 0 \quad$ for all $u, v \in V$

$\Rightarrow \quad \langle u, (t^* - T^*)(v) \rangle = 0 \quad$ for all $u, v \in V$

$\Rightarrow \quad \langle (t^* - T^*)(v), (t^* - T^*)(v) \rangle = 0 \quad$ for all $v \in V$ $\qquad$ [Taking $u = (t^* - T^*)(v)$]

$\Rightarrow \quad (t^* - T^*)(v) = 0 \quad$ for all $v \in V$

$\Rightarrow \quad t^* - T^* = \hat{0}$

$\Rightarrow \quad t^* = T^*$

Hence, t^* is unique.

Q.E.D.

REMARK. *In the above theorem it has been proved that each linear operator on a finite dimensional inner product space possesses its adjoint. But, if V is not a finite dimensional inner product space, then some linear operators may possess an adjoint while the others may not. In case, a linear operator possesses an adjoint, then it is unique. The adjoint of t depends on the inner product on V.*

EXAMPLE-2 *Let V be finite dimensional inner product space. Then,*
$\quad$ (i) $I^* = I$ $\qquad\qquad$ (ii) $\hat{0}^* = \hat{0}$

PROOF. (i) For any $u, v \in V$

$$\langle I(u), v \rangle = \langle u, I^*(v) \rangle$$

Also, $\quad \langle I(u), v \rangle = \langle u, v \rangle = \langle u, I(v) \rangle$ $\qquad\qquad$ [$\because \quad I(v) = v$]

$\therefore \qquad \langle u, I^*(v) \rangle = \langle u, I(v) \rangle$ for all $u, v \in V$

Hence, by the uniqueness of the adjoint, we have
$$I^* = I$$

(ii) For any $u, v \in V$
$$\langle \hat{0}(u), v \rangle = \langle u, \hat{0}^*(v) \rangle$$

Also, $\quad \langle \hat{0}(u), v \rangle = \langle 0_V, v \rangle = 0 = \langle u, \hat{0}(v) \rangle$

$\therefore \quad \langle u, \hat{0}^*(v) \rangle = \langle u, \hat{0}(v) \rangle \quad$ for all $u, v \in V$

Hence, by the uniqueness of adjoint, we get
$$\hat{0}^* = \hat{0}.$$

THEOREM-2 *Let V be a finite dimensional inner product space and $t : V \to V$ be a linear transformation. If $A = [a_{ij}]$ is the matrix representing t relative to some basis $B = \{u_1, u_2, \ldots, u_n\}$. Then, $a_{ij} = \langle t(u_j), (u_i) \rangle$ for all $i, j \in \underline{n}$.*

PROOF. Since $A = [a_{ij}]$ is the matrix representing $t : V \to V$ relative to basis B. Then,

$$t(u_j) = \sum_{r=1}^{n} a_{rj} u_r \text{ for all } j \in \underline{n}$$

$\Rightarrow \quad \langle t(u_j), u_i \rangle = \left\langle \sum_{r=1}^{n} a_{rj} u_r, u_i \right\rangle \quad$ for all $i, j \in \underline{n}$

$\Rightarrow \quad \langle t(u_j), u_i \rangle = \sum_{r=1}^{n} a_{rj} \langle u_r, u_i \rangle \quad$ for all $i, j \in \underline{n}$

$\Rightarrow \quad \langle t(u_j), u_i \rangle = \sum_{r=1}^{n} a_{rj} \delta_{ri} \quad$ for all $i, j \in \underline{n}$ $\quad [\because B$ is an orthonormal basis of $V]$

$\Rightarrow \quad \langle t(u_j), u_i \rangle = a_{ij} \quad$ for all $i, j \in \underline{n}$

Q.E.D.

THEOREM-3 *Let V be a finite dimensional inner product space with an orthonormal basis $B = \{u_1, u_2, \ldots, u_n\}$ and $t : V \to V$ be a linear transformation. If A is the matrix representing t relative to basis B, then the matrix representing the adjoint operator t^* relative to the same basis B is A^T.*

PROOF. Let $A = [a_{ij}]$ and $B = [b_{ij}]$ be the matrices represented by t and t^* respectively. Then,

$\quad a_{ij} = \langle t(u_j), u_i \rangle$ and $b_{ij} = \langle t^*(u_j), u_i \rangle \quad$ for all $i, j \in \underline{n}$

$\Rightarrow \quad a_{ij} = \langle u_j, t^*(u_i) \rangle$ and $b_{ij} = \langle t^*(u_j), u_i \rangle \quad$ for all $i, j \in \underline{n}$

$\Rightarrow \quad a_{ji} = \langle u_i, t^*(u_j) \rangle$ and $b_{ij} = \langle t^*(u_j), u_i \rangle \quad$ for all $i, j \in \underline{n}$

$\Rightarrow \quad a_{ji} = \langle t^*(u_j), u_i \rangle$ and $b_{ij} = \langle t^*(u_j), u_i \rangle \quad$ for all $i, j \in \underline{n}$

$\Rightarrow \quad b_{ij} = a_{ji} \quad$ for all $i, j \in \underline{n}$

$\Rightarrow \quad B = A^T$

Q.E.D.

REMARK. It is to note here that no such simple relationship exists between the matrices representing t and t^* if the basis is not orthonormal. Also, this theorem is not valid if V is not a finite dimensional inner product space.

EXAMPLE-3 Find the adjoint of linear transformation $t : R^2 \to R^2$ given by
$$t(x,y) = (x+2y, x-y) \quad \text{for all } (x,y) \in R^2$$

SOLUTION Clearly, $B = \{e_1^{(2)}, e_2^{(2)}\}$ is an orthonormal basis of R^2 such that
$$t(e_1^{(2)}) = t(1,0) = (1,1) = 1e_1^{(2)} + 1e_2^{(2)}$$
$$t(e_2^{(2)}) = t(0,1) = (2,-1) = 2e_1^{(2)} - 1e_2^{(2)}$$

The matrix A that represents t relative to the standard basis B is given by
$$A = \begin{bmatrix} 1 & 2 \\ 1 & -1 \end{bmatrix} \Rightarrow A^T = \begin{bmatrix} 1 & 1 \\ 2 & -1 \end{bmatrix}$$

The adjoint t^* is represented by the transpose of A.

Hence, $\quad t^*(X) = A^T X = \begin{bmatrix} 1 & 1 \\ 2 & -1 \end{bmatrix} \begin{bmatrix} x \\ y \end{bmatrix} = \begin{bmatrix} x+y \\ 2x-y \end{bmatrix}$

$t^*(x, y) = (x+y, 2x-y)$.

EXAMPLE-4 Find the adjoint of $t : R^3 \to R^3$ defined by
$$t(x,y,z) = (x+2y, 3x-4z, y).$$

SOLUTION Clearly, $B = \{e_1^{(3)}, e_2^{(3)}, e_3^{(3)}\}$ is an orthonormal basis of R^3.
Now,
$$t(e_1^{(3)}) = t(1,0,0) = (1,3,0) = 1e_1^{(3)} + 3e_2^{(3)} + 0e_3^{(3)}$$
$$t(e_2^{(3)}) = t(0,1,0) = (2,0,1) = 2e_1^{(3)} + 0e_2^{(3)} + 1e_3^{(3)}$$
$$t(e_3^{(3)}) = t(0,0,1) = (0,-4,0) = 0e_1^{(3)} - 4e_2^{(3)} + 0e_3^{(3)}$$

So, matrix A that represents t relative to standard basis is given by
$$A = \begin{bmatrix} 1 & 2 & 0 \\ 3 & 0 & -4 \\ 0 & 1 & 0 \end{bmatrix} \Rightarrow A^T = \begin{bmatrix} 1 & 3 & 0 \\ 2 & 0 & 1 \\ 0 & -4 & 0 \end{bmatrix}$$

The adjoint t^* is represented by the transpose of A and is given by
$$t^*(X) = A^T X$$

$\Rightarrow \quad t^*(X) = \begin{bmatrix} 1 & 3 & 0 \\ 2 & 0 & 1 \\ 0 & -4 & 0 \end{bmatrix} \begin{bmatrix} x \\ y \\ z \end{bmatrix} = \begin{bmatrix} x+3y \\ 2x+z \\ -4y \end{bmatrix}$

Hence, $\quad t^*(x,y,z) = (x+3y, 2x+z, -4y)$.

SELF-ADJOINT OPERATOR *Let V be a finite dimensional inner product space. A linear operator t on V is said to be self-adjoint or symmetric if $t = t^*$.*

i.e., $\quad \langle t(u), v \rangle = \langle u, t(v) \rangle \quad$ for all $u, v \in V$

Clearly, the identity transformation and the null transformation are self-adjoint (symmetric) transformations.

THEOREM-4 *A necessary and sufficient condition that a self-adjoint operator t on a finite dimensional inner product space V be $\hat{0}$ (null operator) is that $\langle t(u), u \rangle = 0$ for all $u \in V$.*

PROOF. First, let t be a self-adjoint operator on V such that $t = \hat{0}$. Then, for any $u \in V$, we have

$$\langle t(u), u \rangle = \langle \hat{0}(u), u \rangle = \langle 0_V, u \rangle = 0$$

Conversely, let t be a self-adjoint operator on V such that

$$\langle t(u), u \rangle = 0 \quad \text{for all } u \in V \qquad \text{(i)}$$

$\Rightarrow \quad \langle t(u+v), (u+v) \rangle = 0 \quad$ for all $u, v \in V$

$\Rightarrow \quad \langle t(u) + t(v), u+v \rangle = 0$

$\Rightarrow \quad \langle t(u), u \rangle + \langle t(u), v \rangle + \langle t(v), u \rangle + \langle t(v), v \rangle = 0 \quad$ for all $u, v \in V$

$\Rightarrow \quad 0 + \langle t(u), v \rangle + \langle t(v), u \rangle + 0 = 0 \quad$ for all $u, v \in V \qquad$ [Using (i)]

$\Rightarrow \quad \langle t(u), v \rangle + \langle v, t(u) \rangle = 0 \quad$ for all $u, v \in V \qquad$ [$\because t$ is self-adjoint]

$\Rightarrow \quad 2 \langle t(u), v \rangle = 0 \quad$ for all $u, v \in V$

$\Rightarrow \quad \langle t(u), v \rangle = 0 \quad$ for all $u, v \in V$

$\Rightarrow \quad \langle t(u), t(u) \rangle = 0 \quad$ for all $u \in V \qquad$ [Taking $v = t(u)$]

$\Rightarrow \quad t(u) = 0 \quad$ for all $u \in V$

$\Rightarrow \quad t = \hat{0}$

Q.E.D.

THEOREM-5 *A linear operator t on a finite dimensional inner product space V is symmetric iff its matrix $A = [a_{ij}]$ relative to some orthonormal basis is symmetric.*

PROOF. Let $B = \{u_1, u_2, \ldots, u_n\}$ be an orthonormal basis for V and $A = [a_{ij}]$ be the matrix of t relative to basis B. Then,

$$t(b_i) = \sum_{r=1}^{n} a_{ri} b_r \quad \text{for all } i \in \underline{n} \qquad \text{(i)}$$

First, let us assume that t is a symmetric operator on V. Then,

$\langle t(u_i), u_j \rangle = \langle u_i, t(u_j) \rangle$ for all $i, j \in \underline{n}$

$\Rightarrow \langle t(u_i), u_j \rangle = \langle t(u_j), u_i \rangle$ for all $i, j \in \underline{n}$ $\quad [\because \langle u, v \rangle = \langle v, u \rangle]$

$\Rightarrow a_{ji} = a_{ij}$ for all $i, j \in \underline{n}$ $\quad$ [See Theorem 2 on page 613]

$\Rightarrow A = A^T$

$\Rightarrow A$ is a symmetric matrix on V.

Conversely, let $A = [a_{ij}]$ be a symmetric matrix. Then,

$A^T = A$

$\Rightarrow a_{ij} = a_{ji}$ for all $i, j \in \underline{n}$

$\Rightarrow \langle t(u_j), u_i \rangle = \langle t(u_i), u_j \rangle$ for all $i, j \in \underline{n}$ $\quad$ [See Theorem 2 on page 613]

$\Rightarrow \langle t(u_i), u_j \rangle = \langle t(u_j), u_i \rangle$

$\Rightarrow \langle t(u_i), u_j \rangle = \langle u_i, t(u_j) \rangle$ for all $i, j \in \underline{n}$ $\quad$ [By using $\langle u, v \rangle = \langle v, u \rangle$ on RHS]

$\Rightarrow t$ is symmetric on basis B.

$\Rightarrow \langle t(u), v \rangle = \langle u, t(v) \rangle$ for all $u, v \in V$

$\Rightarrow t$ is symmetric on V.

Q.E.D.

SKEW-ADJOINT OPERATOR *Let V be a finite dimensional inner product space. A linear operator t on V is called a skew-adjoint operator or skew-symmetric operator if $t^* = -t$.*

THEOREM-6 *A linear operator t on a finite dimensional inner product space V is a skew-adjoint operator iff its matrix relative to any orthonormal basis of V is skew-symmetric.*

PROOF. Proceed as in the proof of previous theorem.

EXAMPLE-5 *Show that any operator t on an inner product space V is the sum of a self-adjoint operator and a skew-adjoint operator.*

SOLUTION Let t^* be adjoint of t relative to some orthonormal basis of V.

Clearly, $t = \frac{1}{2}(t + t^*) + \frac{1}{2}(t - t^*)$

or, $t = t_1 + t_2$, where $t_1 = \frac{1}{2}(t + t^*)$, $t_2 = \frac{1}{2}(t - t^*)$

$\therefore \quad t_1^* = \left\{ \frac{1}{2}(t + t^*) \right\}^* = \frac{1}{2}(t^* + t^{**}) = \frac{1}{2}(t^* + t) = \frac{1}{2}(t + t^*) = t_1$

and, $t_2^* = \left\{ \frac{1}{2}(t - t^*) \right\}^* = \frac{1}{2}(t^* - t^{**}) = \frac{1}{2}(t^* - t) = -\frac{1}{2}(t - t^*) = -t_2$

Thus, t_1 is self-adjoint and t_2 is skew-adjoint.

Hence, t can be written as the sum of a self-adjoint and a skew-adjoint operator.

THEOREM-7 Let t, t_1, t_2 be linear operators on a finite dimensional inner product space V and let $a \in R$. Then,

(i) $(t_1 + t_2)^* = t_1^* + t_2^*$

(ii) $(at)^* = at^*$

(iii) $(t_1 t_2)^* = t_2^* t_1^*$

(iv) $(t^*)^* = t$

PROOF. (i) Since t_1 and t_2 are linear operators on V. Therefore, so is $t_1 + t_2$.

For any $u, v \in V$

$$\langle (t_1 + t_2)(u), v \rangle = \langle t_1(u) + t_2(u), v \rangle$$

$\Rightarrow \quad \langle (t_1 + t_2)(u), v \rangle = \langle t_1(u), v \rangle + \langle t_2(u), v \rangle$

$\Rightarrow \quad \langle (t_1 + t_2)(u), v \rangle = \langle u, t_1^*(v) \rangle + \langle u, t_2^*(v) \rangle$

$\Rightarrow \quad \langle (t_1 + t_2)(u), v \rangle = \langle u, t_1^*(v) + t_2^*(v) \rangle$

$\Rightarrow \quad \langle (t_1 + t_2)(u), v \rangle = \langle u, (t_1^* + t_2^*)(v) \rangle$

But, $\quad \langle (t_1 + t_2)(u), v \rangle = \langle u, (t_1 + t_2)^*(v) \rangle$

$\therefore \quad \langle u, (t_1 + t_2)^*(v) \rangle = \langle u, (t_1^* + t_2^*)(v) \rangle \quad$ for all $u, v \in V$

Hence, $\quad (t_1 + t_2)^* = t_1^* + t_2^*$

(ii) For any $u, v \in V$

$$\langle (at)(u), v \rangle = \langle at(u), v \rangle = a\langle t(u), v \rangle = a\langle u, t^*(v) \rangle$$

$\Rightarrow \quad \langle (at)(u), v \rangle = \langle u, at^*(v) \rangle = \langle u, (at^*)(v) \rangle$

But, $\quad \langle at(u), v \rangle = \langle u, (at)^*(v) \rangle$

$\therefore \quad \langle u, (at)^*(v) \rangle = \langle u, (at^*)(v) \rangle \quad$ for all $u, v \in V$

$\Rightarrow \quad (at)^* = at^*$

(iii) Since t_1 and t_2 are linear operators on V. Therefore, so is $t_1 t_2$.

For any $u, v \in V$

$$\langle t_1 t_2(u), v \rangle = \langle t_1(t_2(u)), v \rangle = \langle t_2(u), t_1^*(v) \rangle$$

$\Rightarrow \quad \langle t_1 t_2(u), v \rangle = \langle u, t_2^*(t_1^*(v)) \rangle$

$\Rightarrow \quad \langle t_1 t_2(u), v \rangle = \langle u, t_2^* t_1^*(v) \rangle$

But, $\quad \langle t_1 t_2(u), v \rangle = \langle u, (t_1 t_2)^*(v) \rangle$

$\therefore \quad \langle u, (t_1 t_2)^*(v) \rangle = \langle u, t_2^* t_1^*(v) \rangle \quad$ for all $u, v \in V$

$\Rightarrow \quad (t_1 t_2)^* = t_2^* t_1^*$

(iv) For any $u, v \in V$

$$\langle t^*(u), v \rangle = \langle v, t^*(u) \rangle = \langle t(v), u \rangle = \langle u, t(v) \rangle$$

So, by uniqueness of adjoint, we get $(t^*)^* = t$

Q.E.D.

THEOREM-8 *Let V be a finite dimensional inner product space and t be a self-adjoint operator on V. Then,*

$$I_m(t) = [\text{Ker}(t)]^\perp$$

PROOF. Let v be an arbitrary vector in $I_m(t)$. Then, there exists $u \in V$ such that $t(u) = v$.

Let $w \in \text{Ker}(t)$. Then, $t(w) = 0_V$.

We shall now show that $v \in [\text{Ker}(t)]^\perp$. For this, it is sufficient to show that $\langle v, w \rangle = 0$ for all $w \in \text{Ker}(t)$.

Now, $\quad \langle v, w \rangle = \langle t(u), w \rangle$ $\qquad\qquad\qquad\qquad\qquad\qquad\qquad$ [$\because\ t(u) = v$]

$\Rightarrow \qquad \langle v, w \rangle = \langle u, t^*(w) \rangle$

$\Rightarrow \qquad \langle v, w \rangle = \langle u, t(w) \rangle$ $\qquad\qquad\qquad\qquad\qquad$ [$\because t$ is self-adjoint $\therefore t^* = t$]

$\Rightarrow \qquad \langle v, w \rangle = \langle u, 0_V \rangle = 0$

Thus, $\quad \langle v, w \rangle = 0$ for all $w \in \text{Ker}(t)$.

$\therefore \qquad v \in [\text{Ker}(t)]^T$

Thus, $\quad v \in I_m(t) \quad \Rightarrow \quad v \in [\text{Ker}(t)]^T$

$\therefore \qquad I_m(t) \subset [\text{Ker}(t)]^T$ $\hfill$ (i)

By Sylvester's law of nullity, we have

$$\dim(I_m(t)) + \dim(\text{Ker}(t)) = \dim V \qquad\qquad\qquad\qquad\qquad\qquad (ii)$$

Also, we have

$W \oplus W^\perp = V$ for any subspace W of V.

$\therefore \qquad \text{Ker}(t) \oplus (\text{Ker}(t))^\perp = V$

$\Rightarrow \qquad \dim(\text{Ker}(t)) + \dim[\text{Ker}(t)]^\perp = \dim V \hfill$ (iii)

From (ii) and (iii), we get

$\dim(I_m(t)) = \dim[\text{Ker}(t)]^\perp \hfill$ (iv)

From (i) and (iv), we get

$I_m(t) = [\text{Ker}(t)]^\perp$

Q.E.D.

THEOREM-9 *Let t be a self-adjoint linear operator on a finite dimensional inner product space V. If u and v are eigenvectors of t belonging to distinct eigenvalues, then u and v are orthogonal.*

PROOF. Let u and v be eigenvectors of t corresponding to distinct eigenvalues λ_1 and λ_2 respectively. Then,
$$t(u) = \lambda_1 u \quad \text{and} \quad t(v) = \lambda_2 v$$
Now,
$$\lambda_1 \langle u, v \rangle = \langle \lambda_1 u, v \rangle = \langle t(u), v \rangle = \langle u, t^*(v) \rangle$$
$\Rightarrow \quad \lambda_1 \langle u, v \rangle = \langle u, t(v) \rangle \qquad [\because \ t \text{ is self-adjoint} \ \therefore \ t = t^*]$
$\Rightarrow \quad \lambda_1 \langle u, v \rangle = \langle u, \lambda_2 v \rangle$
$\Rightarrow \quad \lambda_1 \langle u, v \rangle = \lambda_2 \langle u, v \rangle$
$\Rightarrow \quad (\lambda_1 - \lambda_2) \langle u, v \rangle = 0$
$\Rightarrow \quad \langle u, v \rangle = 0 \qquad [\because \ \lambda_1 \neq \lambda_2]$

Q.E.D.

THEOREM-10 *Let t be a linear operator on an inner product space V, and let S be a t-invariant subspace of V. Show that $S^\perp$ is invariant under t^*.*

PROOF. Let u be an arbitrary vector in $S^\perp$. In order to prove that $S^\perp$ is invariant under t^*, it is sufficient to show that $t^*(u) \in S^\perp$, i.e. $t^*(u)$ is orthogonal to every vector in S. So, let $v \in S$. Then,
$$t(v) \in S \qquad [\because \ S \text{ is } t\text{-invariant}]$$
Now,
$$\langle v, t^*(u) \rangle = \langle t(v), u \rangle = 0 \qquad [\because \ t(v) \in S \text{ and } u \in S^\perp]$$
$\Rightarrow \quad t^*(u)$ is orthogonal to every vector $v \in S$
$\Rightarrow \quad t^*(u) \in S^\perp$

Thus, $u \in S^\perp \Rightarrow t^*(u) \in S^\perp$
Hence, $S^\perp$ is invariant under t^*.

Q.E.D.

THEOREM-11 *Show that the following three conditions on a linear operator t on an inner product space V are equivalent:*

(i) $t^* t = I$
(ii) $\langle t(u), t(v) \rangle = \langle u, v \rangle$ *for all $u, v \in V$*
(iii) $\|t(u)\| = \|u\|$ *for all $u \in V$.*

PROOF. (i) $\Rightarrow$ (ii)
For any $u, v \in V$, we have
$$\langle t(u), t(v) \rangle = \langle u, t^*(t(v)) \rangle = \langle u, \ t^* t(v) \rangle$$
$$= \langle u, I(v) \rangle \qquad [\because t^* t = I]$$
$$= \langle u, v \rangle$$

This proves that (i) $\Rightarrow$ (ii).

(ii) $\Rightarrow$ (iii)

Let $\quad \langle t(u), t(v) \rangle = \langle u, v \rangle \quad$ for all $u, v \in V$

$\Rightarrow \quad \langle t(u), t(u) \rangle = \langle u, u \rangle \quad$ for all $u \in V$ [Taking $v = u$]

$\Rightarrow \quad ||t(u)||^2 = ||u||^2 \quad$ for all $u \in V$

$\Rightarrow \quad ||t(u)|| = ||u|| \quad$ for all $u \in V$

This proves that (ii) $\Rightarrow$ (iii).

(iii) $\Rightarrow$ (i)

We have, $\quad ||t(u)|| = ||u|| \quad$ for all $u \in V$

$\Rightarrow \quad ||t(u)||^2 = ||u||^2 \quad$ for all $u \in V$

$\Rightarrow \quad \langle t(u), t(u) \rangle = \langle u, u \rangle \quad$ for all $u \in V$

$\Rightarrow \quad \langle u, t^*t(u) \rangle = \langle u, u \rangle \quad$ for all $u \in V$

$\Rightarrow \quad \langle t^*t(u), u \rangle = \langle u, u \rangle \quad$ for all $u \in V$

$\Rightarrow \quad \langle t^*t(u) - u, u \rangle = 0 \quad$ for all $u \in V$

$\Rightarrow \quad \langle (t^*t - I)(u), u \rangle = 0 \quad$ for all $u \in V$

$\Rightarrow \quad t^*t - I = \hat{0} \qquad \begin{bmatrix} \because \quad (t^*t - I)^* = (t^*t)^* - I^* = t^*t - I \\ \therefore \quad t^*t - I \text{ is self-adjoint} \\ \text{Also, see Theorem 4 on page 615} \end{bmatrix}$

$\Rightarrow \quad t^*t = I$

Q.E.D.

THEOREM-12 *Let t be a linear operator on an inner product space V and λ be an eigen-value of t.*

(i) If t is self-adjoint, λ is real.

(ii) If t is skew-adjoint, then λ is purely imaginary.

PROOF. See Chapter 8.

THEOREM-13 *Let t be a self-adjoint linear operator on an inner product space V. Show that: (i) The characteristic polynomial Δt of t is a product of linear polynomials over R. (ii) t has a non-zero eigenvector.*

PROOF. Let A be a matrix representation of t relative to an orthonormal basis of V. Then, $A^T = A$. Let $\Delta(t)$ be the characteristic polynomial of matrix A. Since A has only real eigenvalues (See Theorem 12). Therefore,

$$\Delta(t) = (t - \lambda_1)(t - \lambda_2) \ldots (t - \lambda_n), \text{ where } \lambda_1, \lambda_2, \ldots, \lambda_n \text{ are real.}$$

Hence, $\Delta(t)$ is a product of linear polynomials over R.

As t has a real eigenvalue λ (say). Therefore, there exists a non-zero vector v such that $t(v) = \lambda v$, i.e. t has a non-zero eigenvector.

Q.E.D.

THEOREM-14 *(Principal Axis Theorem) Let t be a self-adjoint operator on a finite dimensional inner product space V. Then there exists an orthonormal basis of V consisting of eigenvectors of t, and in that case t is represented by a diagonal matrix relative to an orthonormal basis such that the diagonal entries are eigenvalues of t and each appears as many times as its multiplicity.*

PROOF. Let V be an n dimensional inner product space. We shall prove the theorem by induction on n.

By Theorem 13, t has a non-zero eigenvector v_1 (say). Let λ_1 be the corresponding eigenvalue. Then, $t(v_1) = \lambda_1 v_1$.

Let $u_1 = \dfrac{v_1}{||v_1||}$. Then, u_1 is also an eigenvector for t and $||u_1|| = 1$.

If $\dim V = 1$, then $\{u_1\}$ is an orthonormal basis for V and u_1 is an eigenvector for t. Thus, the theorem is true if $\dim V = 1$.

Let us now assume that the theorem is true for inner product spaces of dimension less than n. Let S be the one-dimensional subspace spanned by v_1. Then, unit vector $u_1 \in S$.

Let w be any vector in S. Then, $w = au_1$ for some $a \in R$

$\therefore\ t(w) = t(au_1) = at(u_1) = at\left(\dfrac{v_1}{||v_1||}\right) = \dfrac{a}{||v_1||}t(v_1) = \dfrac{a}{||v_1||}\lambda_1 v_1 = \left(\dfrac{a\lambda_1}{||v_1||}\right) = \left(\dfrac{a\lambda_1}{||v_1||}\right)v_1$

$\Rightarrow t(w) \in S$

$\therefore\ S$ is invariant under t.

$\Rightarrow S^\perp$ is invariant under t^* [By Theorem 10]

$\Rightarrow S^\perp$ is invariant under t $[\because t^* = t]$

We know that

$S \oplus S^\perp = \dim V$

$\Rightarrow \quad \dim S + \dim S^\perp = n$

$\Rightarrow \quad \dim S^\perp = n - 1$ $[\because \dim S = 1]$

Let $\hat{t}$ be the restriction of t on $S^\perp$ and $\hat{t}^*$ be restriction of t^* on $S^\perp$. Then, for any $w \in S^\perp$

$\hat{t}^*(w) = t^*(w) = t(w) = \hat{t}(w)$

$\therefore \quad \hat{t}^*$ is self-adjoint on $S^\perp$.

Thus, $\hat{t}^*$ is a self-adjoint operator on $S^\perp$ and $\dim S^\perp = n - 1 < n$.

So, by induction assumption $S^\perp$ has an orthonormal basis $\{u_2, u_3, \ldots, u_n\}$ consisting of eigenvectors of $\hat{t}$.
Now, $\quad \hat{t}(u_i) = t(u_i)$ for $i = 2, 3, \ldots, n$
$\Rightarrow \quad u_2, u_3, \ldots, u_n$ are eigenvectors of t.

Now, $S \oplus S^\perp = V, \{u_1\}$ is a basis for S and $\{u_2, u_3, \ldots, u_n\}$ is an orthonormal basis for $S^\perp$. Also, $\langle u_1, u_i \rangle = 0$ for $i = 2, 3, \ldots n$ because $u_i \in S^\perp$ for $i \geq 2$. Therefore, $\{u_1, u_2, \ldots, u_n\}$ is an orthonormal basis for V and it consists of eigenvectors of t. Hence, by theorem 4 on page 439 of chapter 4, eigen-values of t are exactly the diagonal entries of A and each occurs in the diagonal a number of times equal to the dimension of its eigenspace, i.e. as many times as its multiplicity.

Q.E.D.

EXAMPLE-6 Let t be an invertible linear operator on a finite dimensional inner product space V. Then, $(t^*)^{-1} = (t^{-1})^*$.

PROOF. Let t^{-1} be the inverse of operator t. Then,
$$t\, t^{-1} = I = t^{-1} t$$
$\Rightarrow \quad (t\, t^{-1})^* = I^* = (t^{-1} t)^*$
$\Rightarrow \quad (t^{-1})^* t^* = I = t^* (t^{-1})^* \qquad [\because\ (t_1 t_2)^* = t_2^* t_1^*$ and $I^* = I]$
$\Rightarrow \quad t^*$ is invertible and $(t^*)^{-1} = (t^{-1})^*$

Q.E.D.

EXAMPLE-7 Show that the product of two self-adjoint operators on a finite dimensional inner product space is self-adjoint iff the two operators commute.

SOLUTION Let t_1 and t_2 be two self-adjoint operators on a finite dimensional inner product space V. Then, $t_1^* = t_1$ and $t_2^* = t_2$.

First suppose that t_1 and t_2 commute, i.e. $t_1 t_2 = t_2 t_1$.
We have to prove that $t_1 t_2$ is self-adjoint.
We have,
$\qquad (t_1 t_2)^* = t_2^* t_1^* = t_2 t_1 \qquad\qquad [\because\ t_1^* = t_1, t_2^* = t_2]$
$\Rightarrow \quad (t_1 t_2)^* = t_1 t_2 \qquad\qquad\qquad\qquad [\because\ t_1 t_2 = t_2 t_1]$
$\therefore \quad t_1\, t_2$ is self-adjoint

Conversely, let $t_1\, t_2$ be self-adjoint operator on V. Then,
$\qquad (t_1 t_2)^* = t_1 t_2$
$\Rightarrow \quad t_2^* t_1^* = t_1 t_2$
$\Rightarrow \quad t_2 t_1 = t_1 t_2 \qquad\qquad [\because\ t_1, t_2$ are self-adjoint $\therefore\ t_1^* = t_1, t_2^* = t_2]$
$\Rightarrow \quad t_1$ and t_2 commute.

EXAMPLE-8 *Let t be a self-adjoint operator on an inner product space V and T be any operator on V. Then, T^*tT is a self-adjoint operator on V.*

SOLUTION Since t is self-adjoint operator on V. Therefore, $t^* = t$.

Now,
$$(T^*t\,T)^* = T^*t^*(T^*)^* \qquad [\because\ (t_1\,t_2)^* = t_2^*\,t_1^*]$$
$$\Rightarrow (T^*t\,T)^* = T^*\,t\,T \qquad [\because\ T^{**} = T]$$
$$\therefore\quad T^*\,t\,T \text{ is a self adjoint operator on } V.$$

EXAMPLE-9 *Let t be an operator on an inner product space V and T be an invertible operator on V such that $T^*t\,T$ is self-adjoint. Then t is self-adjoint.*

SOLUTION It is given that

$T^*t\,T$ is self-adjoint operator on V
$$\Rightarrow (T^*t\,T)^* = T^*\,t\,T$$
$$\Rightarrow T^*t^*(T^*)^* = T^*\,t\,T$$
$$\Rightarrow T^*t^*T = T^*\,t\,T$$
$$\Rightarrow (T^*)^{-1}(T^*\,t^*\,T)\,T^{-1} = (T^*)^{-1}(T^*\,t\,T)\,T^{-1} \qquad \begin{bmatrix} \because\ T \text{ is invertible} \\ \therefore\ T^* \text{ is also invertible} \end{bmatrix}$$
$$\Rightarrow ((T^*)^{-1}T^*)\,t^*\,(T\,T^{-1}) = ((T^*)^{-1}T^*)\,t\,(T\,T^{-1})$$
$$\Rightarrow I\,t^*\,I = I\,t\,I$$
$$\Rightarrow t^* = t$$
$$\Rightarrow t \text{ is self-adjoint.}$$

EXAMPLE-10 *Let t be a self-adjoint operator on an inner product space V such that $t^2(v) = 0$, then show that $t(v) = 0$.*

SOLUTION It is given that $t^2(v) = 0$.

$$\therefore\quad \langle t^2(v),\,u\rangle = 0 \quad \text{for each } u \in V$$
$$\Rightarrow \langle t(t(v)),\,u\rangle = 0 \quad \text{for each } u \in V$$
$$\Rightarrow \langle t(v),\,t^*(u)\rangle = 0 \quad \text{for each } u \in V$$
$$\Rightarrow \langle t(v),\,t(u)\rangle = 0 \quad \text{for each } u \in V \qquad [\because\ t^* = t \text{ as } t \text{ is self-adjoint}]$$
$$\Rightarrow \langle t(v),\,t(v)\rangle = 0 \qquad [\text{Taking } u = v]$$
$$\Rightarrow t(v) = 0$$

EXAMPLE-11 *If t is skew-adjoint operator on an inner product space V, then $\langle t(v), v \rangle = 0$ for all $v \in V$.*

SOLUTION For any $v \in V$, we have

$$\langle t(v), v \rangle = \langle v, t^*(v) \rangle = \langle v, -t(v) \rangle \qquad [\because \quad t^* = -t]$$
$$\Rightarrow \quad \langle t(v), v \rangle = -\langle v, t(v) \rangle$$
$$\Rightarrow \quad \langle t(v), v \rangle = -\overline{\langle t(v), v \rangle} \qquad [\because \quad \langle u, v \rangle = \overline{\langle v, u \rangle}]$$
$$\Rightarrow \quad 2\langle t(v), v \rangle = 0$$
$$\Rightarrow \quad \langle t(v), v \rangle = 0$$

7.5 SPECTRAL THEOREM

Let us begin with the following definition.

SPECTRUM OF AN OPERATOR *Let t be a linear operator on a finite dimensional inner product space V. The set of all eigenvalues of t is called the spectrum of t.*

THEOREM-1 *(Spectral Theorem) Let t be a self-adjoint linear operator on a finite dimensional inner product space V. If t has distinct eigenvalues $\lambda_1, \lambda_2, \ldots, \lambda_k$ with the corresponding eigenspaces $V_1, V_2, \ldots, V_k$. Then,*

(i) *$V_1, V_2, \ldots, V_k$ are pairwise orthogonal.*

(ii) *V is the direct sum of $V_1, V_2, \ldots, V_k$, i.e. $V = V_1 \oplus V_2 \oplus \cdots \oplus V_k$*

(iii) *there exists an orthonormal basis of V consisting of eigenvectors of t.*

PROOF. It is given that t is a self-adjoint operator on V. Therefore, $\lambda_1, \lambda_2, \ldots, \lambda_k$ are real.

(i) We have, $V_i = \{v \in V : t(v) = \lambda_i v\}, \quad i \in \underline{k}$.

In order to prove that $V_1, V_2, \ldots, V_k$ are pairwise orthogonal, it is sufficient to show that each $v_i \in V_i$ is orthogonal to each $v_j \in V_j$ for all $i, j \in \underline{k}, i \neq j$.

Let $v_i \in V_i$ and $v_j \in V_j$. Then, $t(v_i) = \lambda_i v_i$ and $t(v_j) = \lambda_j v_j$. Since t is self-adjoint operator on V.

$$\therefore \quad \langle t(v_i), v_j \rangle = \langle v_i, t(v_j) \rangle$$
$$\Rightarrow \quad \langle \lambda_i v_i, v_j \rangle = \langle v_i, \lambda_j v_j \rangle$$
$$\Rightarrow \quad \lambda_i \langle v_i, v_j \rangle = \lambda_j \langle v_i, v_j \rangle$$
$$\Rightarrow \quad \langle v_i, v_j \rangle (\lambda_i - \lambda_j) = 0$$
$$\Rightarrow \quad \langle v_i, v_j \rangle = 0 \qquad [\because \quad \lambda_i \neq \lambda_j]$$

Thus, V_i is orthogonal to V_j for all $i, j \in \underline{k}, i \neq j$.

Hence, $V_1, V_2, \ldots, V_k$ are pairwise orthogonal.

(ii) Since $V_1, V_2, \ldots, V_k$ are pairwise orthogonal, therefore, the sum $V_1 + V_2 + \cdots + V_k$ is direct. Let $S = V_1 \oplus V_2 \oplus V_3 \oplus \cdots \oplus V_k$. Then, S is a non-null subspace.
Clearly, $V = S \oplus S^\perp$ [By Theorem 3 on page 583]
In order to prove that $V = V_1 \oplus V_2 \oplus \cdots \oplus V_k$, it is sufficient to show that $S^\perp = \{0_V\}$. For this, we shall first show that S and $S^\perp$ are invariant under t.
Let $v_1 + v_2 + \cdots + v_k \in S$. Then,

$$t(v_1 + v_2 + \cdots + v_k) = t(v_1) + t(v_2) + \cdots + t(v_k)$$

$\Rightarrow \quad t(v_1 + v_2 + \cdots + v_k) = \lambda_1 v_1 + \lambda_2 v_2 + \cdots + \lambda_k v_k$

$\Rightarrow \quad t(v_1 + v_2 + \cdots + v_k) \in V_1 \oplus V_2 \oplus \cdots \oplus V_k$

$\Rightarrow \quad t(v_1 + v_2 + \cdots + v_k) \in S$

$\therefore \quad S$ is invariant under t.

Let $v \in S^\perp$. Then,

$\langle u, v \rangle = 0 \quad$ for all $u \in S$

$\Rightarrow \quad \langle t(u), v \rangle = 0 \quad$ for all $u \in S \quad [\because S$ is invariant under $t \therefore u \in S \Rightarrow t(u) \in S]$

$\Rightarrow \quad \langle u, t(v) \rangle = 0 \quad$ for all $u \in S \quad\quad\quad\quad\quad\quad [\because t$ is self-adjoint$]$

$\Rightarrow \quad t(v) \in S^\perp$

Thus, $v \in S^\perp \Rightarrow t(v) \in S^\perp$.

So, $S^\perp$ is invariant under t.
Let t' be the restriction of t to $S^\perp$.
We claim that $S^\perp = \{0_V\}$. For, if

$\quad\quad S^\perp \neq \{0_V\}$

$\Rightarrow \quad t'$ has at least one eigenvalue, say λ.

$\Rightarrow \quad$ There exists $u \in S^\perp$ such that $t'(u) = \lambda u$

$\Rightarrow \quad t(u) = \lambda u \quad\quad\quad\quad\quad\quad\quad\quad\quad\quad [\because t'$ is restriction of t to $S^\perp]$

$\Rightarrow \quad u \in$ Some V_i

$\Rightarrow \quad u \in V_1 + V_2 + \cdots + V_k$

$\Rightarrow \quad u \in S$

This is not possible as $u \in S^\perp$ and $S \cap S^\perp = \{0_V\}$.

$\therefore \quad S^\perp = \{0_V\}$.

Hence, $V = S \oplus S^\perp$

$\Rightarrow \quad V = V_1 \oplus V_2 \oplus V_3 \oplus \cdots \oplus V_k$

(iii) Since each V_i is a finite dimensional inner product space. Therefore, each V_i has an orthonormal basis B_i and their union $\bigcup_{i=1}^{k} B_i$ is an orthonormal basis of V (By Theorem 6 on page 585), consisting of eigenvectors of t.

Q.E.D.

SPECTRAL BASIS *Let t be a self-adjoint operator on an inner product space V. Then an orthonormal basis of V consisting of eigenvectors of t is called the spectral basis of V relative to t.*

EXERCISE 7.1

1. Let V be an inner product space and $B = \{u_1, u_2, \ldots, u_n\}$ be an orthonormal basis of V. Then, show that the mapping $t : R^n \to V$ associating each $(\lambda_1, \lambda_2, \ldots, \lambda_n) \in R^n$ to $\sum_{i=1}^{n} \lambda_i u_i$ in V is an inner product space isomorphism.

2. Let V be an inner product space and t be any operator on V. Show that $t + t^*$ is self-adjoint and $t - t^*$ is skew-adjoint.

3. If t_1 and t_2 are self-adjoint linear operators on an inner product space V, then prove that $t_1 t_2 + t_2 t_1$ is self-adjoint and $t_1 t_2 - t_2 t_1$ is skew-adjoint on V.

4. If t_1 and t_2 are skew-adjoint linear operators on an inner product space V, then prove that $t_1 t_2 + t_2 t_1$ is self-adjoint and $t_1 t_2 - t_2 t_1$ is skew-adjoint on V. What happens if one of t_1 and t_2 is self-adjoint and other is skew-adjoint?

5. Let t be a skew-adjoint operator on an inner product space V such that $t^2(v) = 0$, then show that $t(v) = 0$.

6. Let t be a skew-adjoint operator on an inner product space V. Then show that t^2 is self-adjoint and t^3 is skew-adjoint.

7. Let t be a linear operator on an inner product space V such that $t^*t = \hat{0}$. Then, show that $t = \hat{0}$.

8. Let t be a linear operator on a finite dimensional inner product space V. Prove that the image of t^* is the orthogonal complement of the kernel of t, i.e. $I_m(t^*) = [\text{Ker}(t)]^\perp$. Hence, $\text{rank}(t) = \text{rank}(t^*)$.

9. Let V be the vector space of polynomials over R with inner product defined by
$$\langle f, g \rangle = \int_0^1 f(t)g(t)dt.$$
Let D be the derivative operator on V, i.e. $D(f) = \dfrac{df}{dt}$. Show that there is no operator D^* on V such that $\langle D(f), g \rangle = \langle f, D^*(g) \rangle$ for every $f, g \in V$. That is, D has no adjoint.

10. Find the adjoint of linear operator t on R^3 defined by
$t(x,y,z) = (3x + 4y - 5z,\ 2x - 6y + 7z,\ 5x - 9y + z)$ relative to the standard inner product on R^3.

ANSWERS

10. $t^*(x,y,z) = (3x + 2y + 5z,\ 4x - 6y - 9z,\ -5x + 7y + z)$

7.6 ORTHOGONAL OPERATORS

ORTHOGONAL OPERATOR: *A linear operator t on a finite dimensional inner product space V is said to be orthogonal if $tt^* = t^*t = I$.*

It is evident from the above definition that an orthogonal operator on an inner product space V is invertible and $t^* = t^{-1}$, i.e. inverse of t is its adjoint.

THEOREM-1 *Let V be a finite dimensional inner product space and t be a linear operator on V. The following conditions on t are equivalent:*

   (i) t is orthogonal
   (ii) $\langle t(u), t(v) \rangle = \langle u, v \rangle$ for all $u, v \in V$, i.e. t preserves the inner product.
   (iii) $\|t(u)\| = \|u\|$ for all $u \in V$, i.e. t preserves lengths.

PROOF. (i) $\Rightarrow$ (ii)

Let t be an orthogonal operator on V. Then, $tt^* = I = t^*t$.

For any $u, v \in V$

$$\langle t(u), t(v) \rangle = \langle u, t^*(t(v)) \rangle$$
$$\Rightarrow \langle t(u), t(v) \rangle = \langle u, t^*t(v) \rangle = \langle u, I(v) \rangle \qquad [\because t^*t = I]$$
$$\Rightarrow \langle t(u), t(v) \rangle = \langle u, v \rangle$$

Thus, (i) implies (ii).

Let us now prove that (ii) $\Rightarrow$ (iii).

If (ii) holds, then

$$\langle t(u), t(v) \rangle = \langle u, v \rangle \quad \text{for all } u, v \in V$$
$$\Rightarrow \langle t(u), t(u) \rangle = \langle u, u \rangle \quad \text{for all } u \in V$$
$$\Rightarrow \|t(u)\|^2 = \|u\|^2 \quad \text{for all } u \in V$$
$$\Rightarrow \|t(u)\| = \|u\| \quad \text{for all } u \in V.$$

Thus, (ii) implies (iii).

Finally, let us show that (iii) $\Rightarrow$ (i)

If (iii) holds, then

$$\|t(u)\| = \|u\| \text{ for all } u \in V$$
$$\Rightarrow \langle t(u), t(u) \rangle = \langle u, u \rangle \text{ for all } u \in V.$$

For any $u \in V$ we have

$$\langle t^*t(u), u \rangle = \langle u, t^*(t(u)) \rangle = \langle t(u), t(u) \rangle = \langle u, u \rangle = \langle I(u), u \rangle$$
$$\Rightarrow \langle t^*t(u) - I(u), u \rangle = 0 \quad \text{for every } u \in V$$
$$\Rightarrow \langle (t^*t - I)(u), u \rangle = 0 \quad \text{for every } u \in V$$
$$\Rightarrow t^*t - I = \hat{0} \qquad [\because t^*t - I \text{ is self-adjoint}]$$
$$\Rightarrow t^*t = I$$
$$\Rightarrow t \text{ is orthogonal.}$$

Thus, (iii) implies (i).
Hence, given conditions are equivalent.

Q.E.D.

REMARK. *In Section 7.2, we have learnt that an isomorphism from one inner product space into another is a bijective mapping that preserves the three basic operations of an inner product space, viz. vector addition, scalar multiplication and inner products. That is, an invertible linear transformation from one inner product space into another is an isomorphism if it preserves the inner products. It follows from the above theorem that an orthogonal operator on an inner product space may also be characterized as an isomorphism of V into itself.*

THEOREM-2 *Let t be an orthogonal operator on an inner product space V and $(v_1, v_2, \ldots, v_n)$ be an orthonormal list of vectors in V. Then, $(t(v_1), t(v_2), t(v_3), \ldots, t(v_n))$ is also an orthonormal list in V.*

PROOF. Since $(v_1, v_2, \ldots, v_n)$ is an orthonormal list in V. Therefore,

$$\langle v_i, v_j \rangle = \delta_{ij} \quad \text{for all } i, j \in \underline{n}$$

$\Rightarrow \quad \langle t(v_i), t(v_j) \rangle = \delta_{ij} \quad \text{for all } i, j \in \underline{n} \qquad [\because \ t \text{ is orthogonal} \therefore \ \langle t(u), t(v) \rangle = \langle u, v \rangle]$

$\Rightarrow \quad (t(v_1), t(v_2), \ldots, t(v_n))$ is also an orthonormal list in V.

Q.E.D.

THEOREM-3 *Let V be a finite dimensional inner product space. A linear operator t on V is an orthogonal operator iff for every orthonormal basis $B = \{u_1, u_2, \ldots, u_n\}$ of V, $t(B) = \{t(u_1), t(u_2), \ldots, t(u_n)\}$ is also an orthonormal basis.*

PROOF. First, let t be an orthogonal operator on V and B be an orthonormal basis of V. Then, $t(B) = \{t(u_1), t(u_2), \ldots, t(u_n)\}$ is an orthonormal set (see Theorem 2). Also, an orthonormal set is linearly independent. Hence, $t(B)$ is an orthonormal basis of V.

Conversely, let t be a linear operator on V such that for every orthonormal basis $B = \{u_1, u_2, \ldots, u_n\}$ of V, $t(B) = \{t(u_1), t(u_2), \ldots, t(u_n)\}$ is also an orthonormal basis of V. In order to prove that t is an orthogonal operator, it is sufficient to prove that $\langle t(v), t(v) \rangle = \langle v, v \rangle$ for all $v \in V$.
Let $v \in V$. Then,

$$v = \sum_{i=1}^{n} \lambda_i u_i \qquad\qquad [\because \ B \text{ is a basis of } V]$$

$\Rightarrow \quad t(v) = \sum_{i=1}^{n} \lambda_i \, t(u_i) \qquad\qquad [\because \ t \text{ is linear on } V]$

$\therefore \quad \langle t(v), t(v) \rangle = \left\langle \sum_{i=1}^{n} \lambda_i \, t(u_i), \ \sum_{j=1}^{n} \lambda_j \, t(u_j) \right\rangle$

$\Rightarrow \quad \langle t(v), t(v) \rangle = \sum_{i=1}^{n} \sum_{j=1}^{n} \lambda_i \, \lambda_j \, \langle t(u_i), t(u_j) \rangle$

$$\Rightarrow \quad \langle t(v), t(v) \rangle = \sum_{i=1}^{n} \sum_{j=1}^{n} \lambda_i \lambda_j \delta_{ij} \qquad [\because \; t(B) \text{ is an orthonormal basis}]$$

$$\Rightarrow \quad \langle t(v), t(v) \rangle = \sum_{i=1}^{n} \lambda_i^2 \qquad \text{(i)}$$

and,

$$\langle v, v \rangle = \left\langle \sum_{i=1}^{n} \lambda_i \mu_i, \sum_{j=1}^{n} \lambda_j \mu_j \right\rangle = \sum_{i=1}^{n} \sum_{j=1}^{n} \lambda_i \lambda_j \langle u_i, u_j \rangle = \sum_{i=1}^{n} \sum_{j=1}^{n} \lambda_i \lambda_i \delta_{ij} = \sum_{i=1}^{n} \lambda_i^2 \qquad \text{(ii)}$$

From (i) and (ii), we get

$$\therefore \quad \langle t(v), t(v) \rangle = \langle v, v \rangle \quad \text{for all } v \in V$$

Hence, t is an orthogonal operator on V.

Q.E.D.

THEOREM-4 *The composite of two orthogonal operators on an inner product space is an orthogonal operator on it.*

PROOF. Let t_1 and t_2 be two orthogonal operators on an inner product space V. Then, $t_1 \circ t_2$ is also an operator on V. In order to prove that $t_1 \circ t_2$ is orthogonal it is sufficient to show that

$$(t_1 \circ t_2)^* = (t_1 \circ t_2)^{-1}$$

$$(t_1 \circ t_2)^* = t_2^* \circ t_1^* \qquad \text{[Using property of adjoint]}$$

$$\Rightarrow \quad (t_1 \circ t_2)^* = t_2^{-1} \circ t_1^{-1} \qquad [\because \; t_1 \, \& \, t_2 \text{ are orthogonal} \therefore \; t_1^{-1} = t_1, t_2^{-1} = t_2]$$

$$\Rightarrow \quad (t_1 \circ t_2)^* = (t_1 \circ t_2)^{-1}$$

$\therefore \quad t_1 \circ t_2$ is an orthogonal operator on V.

Q.E.D.

THEOREM-5 *Let V be a finite dimensional inner product space. The set of all orthogonal transformations on V (all automorphisms of the inner product space) is a group.*

PROOF. Let $A(V)$ be the set of all automorphisms of the inner product space V. Then,

$$t_1 \circ t_2 \in A(V) \quad \text{for all } t_1, t_2 \in A(V) \qquad \text{[See Theorem 4]}$$

$\Rightarrow \quad A(V)$ is closed under composition of functions as a binary operation.

We observe the following:

Composition of functions is associative on $A(V)$: Since composition of functions is associative. Therefore, for any $f, g, h \in A(V)$

$$f \circ (g \circ h) = (f \circ g) \circ h$$

Existence of identity: The identity map I_V on V is an orthogonal transformation on V. Therefore, $I_V \in A(V)$.

Existence of inverse: Let $t \in A(V)$. Then,

$\quad\quad t \in A(V)$

$\Rightarrow\quad t$ is an orthogonal transformation on V.

$\Rightarrow\quad t$ is an isomorphism from V to itself

$\Rightarrow\quad t^{-1} : V \to V$ exists.

For any $u, v \in V$

$\quad\quad \langle u, v \rangle = \langle tt^{-1}(u), tt^{-1}(v) \rangle$

$\Rightarrow\quad \langle u, v \rangle = \langle t(t^{-1}(u)), t(t^{-1}(v)) \rangle$

$\Rightarrow\quad \langle u, v \rangle = \langle t^{-1}(u), t^{-1}(v) \rangle \quad\quad\quad [\because\ t$ is orthogonal $\therefore\ \langle t(u), t(v) \rangle = \langle u, v \rangle]$

$\therefore\quad t^{-1}$ is an orthogonal transformation on V.

Thus, $t \in A(V) \Rightarrow\ t^{-1} \in A(V)$.

So, every element in $A(V)$ has its inverse.

Hence, $A(V)$ is a group under the composition of functions as a binary operation.

Q.E.D.

REMARK. $A(V)$ *is called the orthogonal group of inner product space V.*

THEOREM-6 *Let t be an orthogonal operator on an inner product space V and S be a subspace of V invariant under t. Show that $S^\perp$ is also invariant under t.*

PROOF. Let t be an orthogonal operator on V. Then, t is invertible and hence a bijection. Therefore, $t(S) = S$, i.e. for any $u \in S$ there exists $v \in S$ such that $t(v) = u$.

Let $w \in S^\perp$. Then for any $u \in S$

$\quad\quad \langle t(w), u \rangle = \langle t(w), t(v) \rangle = \langle w, v \rangle = 0 \quad\quad\quad [\because\ w \in S^\perp$ and $v \in S \therefore\ \langle w, v \rangle = 0]$

$\Rightarrow\quad t(w) \in S^\perp$

$\Rightarrow\quad S^\perp$ is also invariant under t.

Q.E.D.

THEOREM-7 *Let λ be an eigenvalue of an orthogonal linear operator on an inner product space V. Then, $|\lambda| = 1$.*

PROOF. Let v be a non-zero eigenvector of t belonging to eigenvalue λ. Then,

$$t(v) = \lambda v,\ v \neq 0$$

Now,

$\quad\quad t$ is orthogonal

$\Rightarrow\quad \|t(v)\|^2 = \|v\|^2$

$\Rightarrow\quad \|\lambda v\|^2 = \|v^2\|$

$\Rightarrow \quad \lambda^2 ||v||^2 = ||v||^2$

$\Rightarrow \quad \lambda^2 = 1 \Rightarrow |\lambda| = 1.$

THEOREM-8 *Let V be a finite dimensional inner product space. A linear operator t on V is an orthogonal operator iff matrix of t relative to some orthonormal basis of V is orthogonal.*

PROOF. Let $B = \{u_1, u_2, u_3, \ldots, u_n\}$ be an orthonormal basis of V and A be the matrix of t relative to basis B. Then,

$$t(u_i) = \sum_{r=1}^{n} a_{ri} u_r \quad \text{for all } i \in \underline{n}$$

Now,

t is orthogonal

$\Leftrightarrow \quad \langle t(u_i), t(u_j) \rangle = \langle u_i, u_j \rangle \quad \text{for all } i, j \in \underline{n}$

$\Leftrightarrow \quad \left\langle \left(\sum_{r=1}^{n} a_{ri} u_r\right), \left(\sum_{s=1}^{n} a_{sj} u_s\right) \right\rangle = \langle u_i, u_j \rangle \quad \text{for all } i, j \in \underline{n}$

$\Leftrightarrow \quad \left\langle \sum_{r=1}^{n} a_{ri} u_r, \sum_{s=1}^{n} a_{sj} u_s \right\rangle = \delta_{ij} \quad \text{for all } i, j \in \underline{n}$

$\Leftrightarrow \quad \sum_{r=1}^{n} \sum_{s=1}^{n} a_{ri} a_{sj} \langle u_r, u_s \rangle = \delta_{ij} \quad \text{for all } i, j \in \underline{n}$

$\Leftrightarrow \quad \sum_{r=1}^{n} \sum_{s=1}^{n} a_{ri} a_{sj} \delta_{rs} = \delta_{ij} \quad \text{for all } i, j \in \underline{n} \qquad [\because B \text{ is an orthonormal basis of } V]$

$\Leftrightarrow \quad \sum_{r=1}^{n} a_{ri} a_{rj} = \delta_{ij} \quad \text{for all } i, j \in \underline{n}$

$\Rightarrow \quad \sum_{r=1}^{n} (A)^T_{ir} (A)_{rj} = \delta_{ij} \quad \text{for all } i, j \in \underline{n}$

$\Rightarrow \quad (A^T A)_{ij} = \delta_{ij} \quad \text{for all } i, j \in \underline{n}$

$\Rightarrow \quad A^T A = I$

$\Rightarrow \quad A$ is orthogonal matrix.

Q.E.D.

EXERCISE 7.2

1. Prove that the products and inverses of orthogonal matrices are orthogonal.
2. Show that if an orthogonal matrix is triangular, then it is diagonal.
3. Recall that two matrices A and B are orthogonally equivalent if there exists an orthogonal matrix P such that $B = P^T A P$. Show that this relation is an equivalence relation.

7.7 POSITIVE OPERATORS

POSITIVE DEFINITE OPERATOR *A linear operator t on an inner product space V is said to be positive definite, if*

(i) t is self-adjoint, i.e. $t = t^*$.

(ii) $\langle t(u), u \rangle > 0$ for every non-zero vector $u \in V$.

If t is a positive definite operator on an inner product space V, then we write $t > 0$.

If $u = 0_V$, then $\langle t(u), u \rangle = \langle t(0_V), 0_V \rangle = 0$.

Thus, if t is a positive definite operator on an inner product space V, then $\langle t(u), u \rangle \geq 0$ for all $u \in V$ and $\langle t(u), u \rangle = 0 \Rightarrow u = 0_V$.

POSITIVE OPERATOR *A linear operator t on an inner product space V is said to be positive, if*

(i) it is self-adjoint i.e. $t = t^*$.

(ii) $\langle t(u), u \rangle \geq 0$ for all $u \in V$.

Clearly, every positive definite operator is positive. If t is a positive operator, then $\langle t(u), u \rangle = 0$ is possible even if $u \neq 0_V$. Therefore, a positive operator need not be a positive definite operator.

THEOREM-1 *Let V be an inner product space and t be a linear operator on V. Let $f : V \times V \to R$ be defined by*

$$f(u, v) = \langle t(u), v \rangle \quad \text{for all } u, v \in V.$$

Then f is itself an inner product on V iff t is positive definite.

PROOF. Let $u_1, v_1 \in V$ and $a, b \in F$. Then,

$$f(au_1 + bv_1, v) = \langle t(au_1 + bv_1), v \rangle$$

$\Rightarrow \quad f(au_1 + bv_1, v) = \langle at(u_1) + bt(v_1), v \rangle \hfill [\because t \text{ is linear}]$

$\Rightarrow \quad f(au_1 + bv_1, v) = \langle at(u_1), v \rangle + \langle bt(v_1), v \rangle$

$\Rightarrow \quad f(au_1 + bv_1, v) = a \langle t(u_1), v \rangle + b \langle t(v_1), v \rangle$

$\Rightarrow \quad f(au_1 + bv_1, v) = a f(u_1, v) + b f(v_1, v)$

So, f satisfies linearity property.

Now,

$\quad\quad\quad f$ is an inner product on V

$\Leftrightarrow \quad f(u, v) = f(v, u)$ and $f(u, u) > 0 \quad$ for all $u(\neq 0_V) \in V$

$\Leftrightarrow \quad \langle t(u), v \rangle = \langle t(v), u \rangle$ and $\langle t(u), u \rangle > 0$

$\Leftrightarrow \quad \langle t(u), v \rangle = \langle u, t(v) \rangle$ and $\langle t(u), u \rangle > 0$

$\Leftrightarrow$ t is self-adjoint and $\langle t(u), u \rangle > 0$ for all $0_V \neq u \in V$

$\Leftrightarrow$ t is positive definite.

Hence, f is an inner product on V iff t is positive definite.

Q.E.D.

The following theorem proves that on a finite dimensional inner product space every inner product is of the form as proved in the above theorem.

THEOREM-2   *Let V be a finite dimensional inner product space. Let f be any inner product on V. Then there exists a positive linear operator t on V such that $f(u, v) = \langle t(u), v \rangle$ for all $u, v \in V$.*

PROOF.   Let v be a fixed vector in V and $\varphi: V \to F$ be defined by

$$\varphi(u) = f(u, v) \quad \text{for all } u \in V.$$

For any $u_1, u_2 \in V$ and $a, b \in F$, we have

$$\varphi(au_1 + bu_2) = f(au_1 + bu_2, v)$$
$$\Rightarrow \varphi(au_1 + bu_2) = a f(u_1, v) + b f(u_2, v) \qquad [\because f \text{ is an inner product}]$$
$$\Rightarrow \varphi(au_1 + bu_2) = a \varphi(u_1) + b \varphi(u_2)$$

$\therefore$ φ is a linear functional on V.

By Theorem 1 (Section 7.3) there exists a unique vector v' in V such that

$$\varphi(u) = \langle u, v' \rangle \quad \text{for all } u \in V$$
$$\Rightarrow f(u, v) = \langle u, v' \rangle \quad \text{for all } u \in V$$

Thus, corresponding to each $v \in V$ there exists v' in V such that $f(u, v) = \langle u, v' \rangle$ for all $u \in V$. This provides us a mapping $t: V \to V$ defined by $t(v) = v'$.

$\therefore$ $f(u, v) = \langle u, v' \rangle$ for all $u, v \in V$
$\Rightarrow$ $f(u, v) = \langle u, t(v) \rangle$ for all $u, v \in V$

Since f is an inner product on V. Therefore,

$$f(u, v) = \overline{f(v, u)} \quad \text{for all } u, v \in V$$
$$\Rightarrow f(u, v) = \overline{\langle v, t(u) \rangle} \quad \text{for all } u, v \in V \qquad [\text{Using } f(u, v) = \langle u, t(v) \rangle \text{ on RHS}]$$
$$\Rightarrow f(u, v) = \langle t(u), v \rangle \quad \text{for all } u, v \in V$$

Let $u_1, u_2 \in V$ and $a, b \in R$. Then,

$$f(au_1 + bu_2, v) = \langle t(au_1 + bu_2), v \rangle \quad \text{for all } v \in V$$
$\Rightarrow \quad a(f(u_1), v) + b(f(u_2), v) = \langle t(au_1 + bu_2), v \rangle \quad \text{for all } v \in V \quad [\because f \text{ is an inner product}]$
$\Rightarrow \quad a\langle t(u_1), v \rangle + b\langle t(u_2), v \rangle = \langle t(au_1 + bu_2), v \rangle \quad \text{for all } v \in V$
$\Rightarrow \quad \langle at(u_1) + bt(u_2), v \rangle = \langle t(au_1 + bu_2), v \rangle \quad \text{for all } v \in V$
$\Rightarrow \quad t(au_1 + bu_2) = a\, t(u_1) + b\, t(u_2)$

Thus,

$$t(au_1 + bu_2) = a\, t(u_1) + b\, t(u_2) \quad \text{for all } u_1, u_2 \in V \text{ and } a, b \in R$$

So, t is a linear operator on V.

Since f is an inner product on V, therefore, by Theorem 1, t is positive definite.

Uniqueness of t: Let T be another operator on V such that $f(u, v) = \langle T(u), v \rangle$ for all $u, v \in V$. Then,

$$\langle t(u), v \rangle = \langle T(u), v \rangle \quad \text{for all } u, v \in V$$
$\Rightarrow \quad \langle t(u) - T(u), v \rangle = 0 \quad \text{for all } u, v \in V \quad \ldots \text{(i)}$

Let u be fixed. Then, (i) implies that $t(u) - T(u)$ is orthogonal to every vector v in V.

$\therefore \quad t(u) - T(u) = 0_V$
$\Rightarrow \quad t(u) - T(u) = 0_V \quad \text{for all } u \in V$
$\Rightarrow \quad t(u) = T(u) \quad \text{for all } u \in V$
$\Rightarrow \quad t = T$

Hence, t is unique.

Q.E.D.

REMARK. *It follows from the above theorem that corresponding to every inner product f on a finite dimensional inner product space V there exists a unique positive definite linear operator t such that $f(u, v) = \langle t(u), v \rangle$ for all $u, v \in V$.*

THEOREM-3 *Let t be a linear operator on a finite dimensional inner product space V. Then, t is positive definite iff there exists an invertible (non-singular) linear operator T on V such that $t = T^*T$.*

PROOF. First, let t be a linear operator on V such that $t = T^*T$ for some invertible linear operator T on V. Then, we have to prove that t is positive definite.

Now,
$$t = T^*T$$
$\Rightarrow \quad t^* = (T^*T)^* = T^*(T^*)^* \qquad [\because \; (t_1 t_2)^* = t_2^* t_1^*]$

$\Rightarrow \quad t^* = T^*T = t$

$\Rightarrow \quad t$ is self-adjoint operator on V.

Also, for any $v \in V$, we have
$$\langle t(v), v \rangle = \langle T^*T(v), v \rangle = \langle T(v), (T^*)^*(v) \rangle = \langle T(v), T(v) \rangle \geq 0$$

and, $\quad \langle t(v), v \rangle = 0$

$\Rightarrow \quad \langle T(v), T(v) \rangle = 0$ [As shown above]

$\Rightarrow \quad T(v) = 0$

$\Rightarrow \quad v = 0_V$ $\begin{bmatrix} \because & V \text{ is finite dimensional and } T \text{ is invertible} \\ \therefore & T \text{ is non-singular and hence } T(v) = 0_V \text{ iff } v = 0_V \end{bmatrix}$

Thus, t is self-adjoint and $\langle t(v), v \rangle > 0$ for all $v(\neq 0_V) \in V$.

Hence, t is positive definite.

Conversely, let t be a positive definite linear operator on V. Then, by Theorem 1, $f(u, v) = \langle T(u), v \rangle$ is an inner product on V. Let $B_1 = \{u_1, u_2, \ldots, u_n\}$ and $B_2 = \{v_1, v_2, \ldots, v_n\}$ be orthonormal bases for V with respect to inner products $\langle , \rangle$ and f respectively. Then,

$$\langle u_i, u_j \rangle = \delta_{ij} \text{ and } f(v_i, v_j) = \delta_{ij} \quad \text{for all } i, j \in \underline{n}$$

By Theorem 4 on page 242, there exists a unique linear operator T on V such that $T(v_i) = u_i$ for all $i \in \underline{n}$. As T carries a basis of V onto a basis. So, T is invertible.

We have,
$$f(v_i, v_j) = \delta_{ij} = \langle u_i, u_j \rangle = \langle T(v_i), T(v_j) \rangle \quad \text{for all } i, j \in \underline{n}.$$

Now, let u, v be any two vectors in V. Then,

$\qquad u = \sum_{i=1}^{n} \lambda_i v_i$ and $v = \sum_{j=1}^{n} \mu_j v_j$ [$\because \quad B_2$ is a basis for V]

$\therefore \quad \langle T(u), v \rangle = f(u, v)$ [By definition of f]

$\Rightarrow \quad \langle T(u), v \rangle = f\left(\sum_{i=1}^{n} \lambda_i v_i, \sum_{j=1}^{n} \mu_j v_j\right)$

$\Rightarrow \quad \langle T(u), v \rangle = \sum_{i=1}^{n} \sum_{j=1}^{n} \lambda_i \mu_j f(v_i, v_j)$

$\Rightarrow \quad \langle T(u), v \rangle = \sum_{i=1}^{n} \sum_{j=1}^{n} \lambda_i \mu_j \langle T(v_i), T(v_j) \rangle$

$\Rightarrow \quad \langle T(u), v \rangle = \left\langle T\left(\sum_{i=1}^{n} \lambda_i v_i\right), T\left(\sum_{j=1}^{n} \mu_j v_j\right)\right\rangle$ [$\because \quad T$ is linear]

$\Rightarrow \quad \langle T(u), v \rangle = \langle T(u), T(v) \rangle$

$\Rightarrow \quad \langle T(u), v \rangle = \langle T(v), T(u) \rangle$

$\Rightarrow \qquad \langle T(u), v \rangle = \langle v, T^*T(u) \rangle$

$\Rightarrow \qquad \langle T(u), v \rangle = \langle T^*T(u), v \rangle$

Thus, $\quad \langle T(u), v \rangle = \langle T^*T(u), v \rangle \quad$ for all $u, v \in V$

$\Rightarrow \qquad T = T^*T$

Q.E.D.

THEOREM-4 *Let V be an inner product space and t be a linear operator on V. The following conditions on t are equivalent:*

(i) $t = T^2$ for some non-singular self-adjoint linear operator T on V.
(ii) $t = S^*S$ for some non-singular linear operator S on V.
(iii) t is positive definite.

PROOF. (i) $\Rightarrow$ (ii)

Suppose (i) holds, that is, $t = T^2$ for some non-singular self-adjoint linear operator T on V. Then,

$$t = T^2 = TT = T^*T \qquad [\because T^* = T]$$

Thus, (i) implies (ii).

(ii) $\Rightarrow$ (iii)

Now, suppose (ii) holds, that is, $t = S^*S$ for some non-singular linear operator S on V. Then,

$t^* = (S^*S)^* = S^*S^{**} = S^*S = t$

$\Rightarrow \quad t$ is self-adjoint operator on V.

Furthermore for any $v \in V$

$\langle t(v), v \rangle = \langle S^*S(v), v \rangle = \langle S(v), S(v) \rangle \geq 0$

and,

$\langle t(v), v \rangle = 0$

$\Rightarrow \quad \langle S(v), S(v) \rangle = 0 \qquad$ [As shown above]

$\Rightarrow \quad S(v) = 0$

$\Rightarrow \quad v = 0_V \qquad [\because S \text{ is non-singular}]$

$\therefore \quad \langle t(v), v \rangle > 0 \quad$ for all $0_V \neq v \in V$.

So, t is positive definite.

Thus, (ii) implies (iii).

(iii) $\Rightarrow$ (i)

Let t be positive definite, that is, t is self-adjoint and $\langle t(u), u \rangle > 0$ for all $0_V \neq u \in V$.

Now, t is self-adjoint. Therefore, there exists an orthonormal basis $\{v_1, v_2, \ldots, v_n\}$ of V consisting of eigenvectors of t. Let $\lambda_1, \lambda_2, \ldots, \lambda_n$ be the eigenvalues corresponding to eigenvectors $v_1, v_2, \ldots, v_n$ respectively. By Theorem 12 on the page 620, λ_i are real such that

$t(v_i) = \lambda_i v_i \quad$ for all $i \in \underline{n}$

Now,

        t is positive definite

$\Rightarrow\quad \langle t(v_i), v_i \rangle > 0 \quad$ for all $i \in \underline{n}$

$\Rightarrow\quad \langle \lambda_i v_i, v_i \rangle > 0 \quad$ for all $i \in \underline{n}$

$\Rightarrow\quad \lambda_i \langle v_i, v_i \rangle > 0 \quad$ for all $i \in \underline{n}$

$\Rightarrow\quad \lambda_i > 0 \quad\quad\quad\quad$ for all $i \in \underline{n}$ $[\because \langle v_i, v_i \rangle > 0 \quad$ for all $i \in \underline{n}$ as $v_i \neq 0_V]$

Let T be the linear operator defined by
$$T(v_i) = \sqrt{\lambda_i}\, v_i \quad \text{for all } i \in \underline{n}.$$
Then, T is represented by a real diagonal matrix relative to the orthonormal basis $\{v_1, v_2, \ldots, v_n\}$. So, T is self-adjoint.

Also, for each $i \in \underline{n}$, we have
$$T^2(v_i) = T(T(v_i)) = T(\sqrt{\lambda_i}\, v_i) = \sqrt{\lambda_i}\, T(v_i) = \sqrt{\lambda_i}\sqrt{\lambda_i}\, v_i = \lambda_i v_i = t(v_i)$$
Thus, t and T^2 agree on a basis of V. Therefore, $t = T^2$.
Thus, (iii) implies (i).
Hence, all the three statements are equivalent.   Q.E.D.

THEOREM-5 *Let V be an inner product space and t be a linear operator on V. The following conditions on t are equivalent:*

(i) $t = T^2$ for some self-adjoint operator T.
(ii) $t = S^*S$ for some operator S on V.
(iii) t is positive.

PROOF.   Proceed as in Theorem 4.

THEOREM-6 *Let t be a linear operator on a finite dimensional inner product space V. Let $A = [a_{ij}]$ be the matrix of t relative to an ordered orthonormal basis $B = \{v_1, v_2, \ldots, v_n\}$. Then t is positive definite iff the matrix A satisfies the following conditions:*

(i) $A = A^T$
(ii) $\sum\limits_{i=1}^{n} \sum\limits_{j=1}^{n} a_{ij} \lambda_i \lambda_j > 0$, where $\lambda_1, \lambda_2, \ldots, \lambda_n$ are any n scalars not all zero.

PROOF.   Since $A = [a_{ij}]$ is the matrix of t relative to ordered orthonormal basis B. Therefore,
$$a_{ij} = \langle t(v_i), v_j \rangle \quad \text{for all } i, j \in \underline{n} \tag{i}$$
Let v be any vector in V. Then,
$$v = \sum_{i=1}^{n} \lambda_i v_i$$
$$\therefore \quad \langle t(v), v \rangle = \left\langle t\left(\sum_{i=1}^{n} \lambda_i v_i\right), \sum_{j=1}^{n} \lambda_j v_j \right\rangle$$

$\Rightarrow \quad \langle t(v), v \rangle = \left\langle \sum_{i=1}^{n} \lambda_i\, t(v_i), \sum_{j=1}^{n} \lambda_j\, v_j \right\rangle$

$\Rightarrow \quad \langle t(v), v \rangle = \sum_{i=1}^{n} \sum_{j=1}^{n} \lambda_i \lambda_j \langle t(v_i), v_j \rangle$

$\Rightarrow \quad \langle t(v), v \rangle = \sum_{i=1}^{n} \sum_{j=1}^{n} a_{ij}\, \lambda_i\, \lambda_j$ [Using (i)]

First suppose that t is positive definite. Then, t is self-adjoint and $\langle t(v), v \rangle > 0$ for non-zero vector v in V.

Now,

t is self-adjoint.

$\Rightarrow \quad A^T = A$ and, $\langle t(v), v \rangle > 0$ for all non-zero vectors v in V

$\Rightarrow \quad \sum_{i=1}^{n} \sum_{j=1}^{n} a_{ij}\, \lambda_i\, \lambda_j > 0$, where scalars $\lambda_1, \lambda_2, \ldots, \lambda_n$ are not all zero.

Conversely, let A be the matrix of t relative to basis B such that conditions (i) and (ii) are satisfied. Then, we have to prove that t is positive definite.

Now,

$A^T = A$

$\Rightarrow \quad t$ is self-adjoint operator on V and, $\sum_{i=1}^{n} \sum_{j=1}^{n} a_{ij}\, \lambda_i\, \lambda_j > 0$

$\Rightarrow \quad \langle t(v), v \rangle > 0$, if $v \neq 0_V$.

So, t is positive definite.

Q.E.D.

7.7.1 POSITIVE DEFINITE MATRICES

POSITIVE DEFINITE MATRIX A real symmetric matrix A is said to be positive definite if $\langle u, Au \rangle = u^T A u > 0$ for every non-zero vector $u \in R^n$.

Let $A = [a_{ij}]_{n \times n}$ and $u = \begin{bmatrix} \lambda_1 \\ \lambda_2 \\ \vdots \\ \lambda_n \end{bmatrix}$. Then,

$$u^T A u = \sum_{i=1}^{n} \sum_{j=1}^{n} a_{ij}\, \lambda_i\, \lambda_j$$

Thus, a real symmetric matrix $A = [a_{ij}]$ is positive definite if

$$\sum_{i=1}^{n} \sum_{j=1}^{n} a_{ij}\, \lambda_i\, \lambda_j > 0 \quad \text{for } \lambda_1, \lambda_2, \ldots, \lambda_n \text{ not all zero.}$$

POSITIVE MATRIX A real symmetric matrix A is said to be positive if
$$\langle u, Au \rangle = u^T Au \geq 0 \quad \text{for all } u \in R^n.$$

THEOREM-1 *A 2×2 real symmetric matrix $A = \begin{bmatrix} a & b \\ b & d \end{bmatrix}$ is positive definite iff the diagonal entries a and d are positive and $|A| = ad - b^2 > 0$.*

PROOF. See proof of Theorem 1 on page 596.

THEOREM-2 *Let A be a real positive definite matrix. Then the function $\langle u, v \rangle = u^T Av$ is an inner product on R^n.*

PROOF. See proof of Theorem 2 on page 597.

THEOREM-3 *The matrix representation of any inner product on an inner product space V is a positive definite matrix.*

PROOF. See proof of Theorem 3 on page 598. Q.E.D.

EXAMPLE-1 *Prove that every diagonal element of a positive matrix is positive.*

SOLUTION Let $A = [a_{ij}]$ be a positive matrix. Then,

$$\sum_{i=1}^{n} \sum_{j=1}^{n} a_{ij} \lambda_i \lambda_j > 0 \quad \text{for } \lambda_i, \lambda_j \text{ not all zero.}$$

$\Rightarrow \quad a_{ii} > 0 \quad \text{for all } i \in \underline{n} \quad [\text{Taking } \lambda_i = 1 \text{ and } \lambda_1 = \lambda_2 = \cdots = \lambda_{i-1} = \lambda_{i+1} = \cdots = \lambda_n = 0]$

EXERCISE 7.3

1. Show that the sum of two positive definite operators on an inner product space is a positive definite operator.
2. Show that the sum of two positive operators on an inner product space is a positive operator.
3. Determine which of the following matrices are positive (positive definite):
   (i) $\begin{bmatrix} 1 & 1 \\ 1 & 1 \end{bmatrix}$ (ii) $\begin{bmatrix} 0 & 1 \\ -1 & 0 \end{bmatrix}$ (iii) $\begin{bmatrix} 1 & 1 \\ 0 & 1 \end{bmatrix}$ (iv) $\begin{bmatrix} 2 & 1 \\ 1 & 2 \end{bmatrix}$ (v) $\begin{bmatrix} 1 & 2 \\ 2 & 1 \end{bmatrix}$
4. Prove that a diagonal matrix A is positive (positive definite) iff every diagonal entry is non-negative (positive).

ANSWERS

3. (i) & (iv) are positive definite.

Chapter 8

Unitary Spaces

8.1 INTRODUCTION

In Chapter 6, we have learnt about vector spaces over R with inner products defined on them. In this chapter, we shall study Hermitian inner products and other related concepts for vector spaces over the field C of complex numbers. As Hermitian inner product on a vector space over R is same as the inner product on it. So, most of the results on unitary spaces are identical to the corresponding results on inner product spaces.

8.2 UNITARY SPACES

Let us begin with some definitions which will be useful in defining unitary spaces.

LINEAR FORM *Let V be a vector space over the field C of complex numbers. A function $f : V \to C$ is called a linear form if*

$$f(au+bv) = af(u)+bf(v) \quad \text{for all } u, v \in V \text{ and all } a, b \in C$$

CONJUGATE LINEAR FORM *Let V be a vector space over the field C of complex numbers. A function $f : V \to C$ is called a conjugate linear form if*

$$f(au+bv) = \overline{a}f(u)+\overline{b}f(v) \quad \text{for all } u, v \in V \text{ and for all } a, b \in C$$

As we have learnt in earlier chapters that the set of all linear forms $f : V \to C$ is a vector space which is known as the dual space of V and is denoted by V^*. Similarly, the set of all conjugate linear forms $f : V \to C$ is also a vector space over C under the linear combinations defined by $(af_1+bf_2)(v) = af_1(v)+bf_2(v)$ for all $v \in V$ and all $a, b \in C$. This vector space is called adjoint space to V and is denoted by V^Δ.

HERMITIAN INNER PRODUCT *Let V be a vector space over the field C of all complex numbers. A function $\langle , \rangle : V \times V \to C$ assigning each pair of vectors $u, v \in V$ to a complex number denoted by $\langle u, v \rangle$ is known as the Hermitian inner product on V if it satisfies the following axioms:*

HI_1: *Conjugate symmetric property* $\langle u, v\rangle = \overline{\langle v, u\rangle}$ for all $u, v \in V$.
HI_2: *Positive definite property* $\langle u, u\rangle \geq 0$ for all $u \in V$ and $\langle u, u\rangle = 0$ iff $u = 0_V$.
HI_3: *Linearity* $\langle au_1 + bu_2, v\rangle = a\langle u_1, v\rangle + b\langle u_2, v\rangle$ for all $u_1, u_2, v \in V$ and all $a, b \in C$.

UNITARY SPACE *A vector space V over the field C of all complex numbers is called a unitary space, if it is equipped with a Hermitian inner product on it.*

A unitary space is also called a complex inner product space. Note that a unitary space differs from the inner product space only in first axiom.

Let $u, v \in V$ and $a \in C$. By axiom HI_1, we have

$$\langle u, av\rangle = \overline{\langle av, u\rangle} = \overline{a\langle v, u\rangle} = \bar{a}\overline{\langle v, u\rangle} = \bar{a}\langle u, v\rangle$$

That is, we must take the conjugate of a complex number when it is taken out of the second position of a Hermitian inner product.

Let $u, v_1, v_2 \in V$ and $a, b \in C$. By axiom HI_1, we have

$$\begin{aligned}
& \langle u, av_1 + bv_2\rangle = \overline{\langle av_1 + bv_2, u\rangle} \\
\Rightarrow \quad & \langle u, av_1 + bv_2\rangle = \overline{\langle av_1, u\rangle + \langle bv_2, u\rangle} \\
\Rightarrow \quad & \langle u, av_1 + bv_2\rangle = \overline{\langle av_1, u\rangle} + \overline{\langle bv_2, u\rangle} \\
\Rightarrow \quad & \langle u, av_1 + bv_2\rangle = \bar{a}\overline{\langle v_1, u\rangle} + \bar{b}\,\overline{\langle v_2, u\rangle} \\
\Rightarrow \quad & \langle u, av_1 + bv_2\rangle = \bar{a}\langle u, v_1\rangle + \bar{b}\langle u, v_2\rangle
\end{aligned}$$

Thus, the Hermitian inner product is conjugate linear in the second position.

EXAMPLE *Let u, v be two vectors in a unitary space V such that $\langle u, v\rangle = 1 + i$. Find each of the following:*

(i) $\langle (2-i)u, v\rangle$ (ii) $\langle u, (3+2i)v\rangle$ (iii) $\langle (3-2i)u, (2-5i)v\rangle$

SOLUTION We have, $\langle u, v\rangle = 1 + i$. Therefore,

(i) $\langle (2-i)u, v\rangle = (2-i)\langle u, v\rangle = (2-i)(1+i) = 3+i$
(ii) $\langle u, (3+2i)v\rangle = \overline{(3+2i)}\langle u, v\rangle = (3-2i)(1+i) = 5+i$
(iii) $\langle (3-2i)u, (2-5i)v\rangle = (3-2i)(2+5i)\langle u, v\rangle = (16+11i)(1+i) = 5+27i$

By the left linearity of the Hermitian inner product for each $v \in V$, the function $f_v : V \to C$ defined by

$$f_v(u) = \langle u, v\rangle \quad \text{for all } u \in V$$

is a linear form.

By the conjugate linearity of the Hermitian inner product for each $u \in V$, the function $f_u : V \to C$ defined by

$$f_u(v) = \langle u, v\rangle \quad \text{for all } v \in V$$

is a conjugate linear form and hence an element of adjoint space V^{Δ}.

642 • Theory and Problems of Linear Algebra

A form $V \times V \to C$ which is linear in one variable and conjugate linear in other variable is said to be sesquilinear form.

If a_i, $b_i \in C$ and u_i, $v_i \in V$ for $i = 1, 2, \ldots, n$. Then, by combining linearity in first position and conjugate linearity in the second position, we obtain by induction,

$$\left\langle \sum_{i=1}^{n} a_i u_i, \sum_{j=1}^{n} b_j v_j \right\rangle = \sum_{i=1}^{n} \sum_{j=1}^{n} a_i \overline{b_j} \langle u_i, v_j \rangle$$

REMARK-1 *By axiom HI_1, $\langle u, u \rangle = \overline{\langle u, u \rangle}$ for all $u \in V$. Thus, $\langle u, u \rangle$ must be real for all $u \in V$.*

REMARK-2 *By axiom HI_1, we have*

$$\langle 0_V, 0_V \rangle = \langle 0\, u, 0_V \rangle = 0 \langle u, 0_V \rangle = 0$$

$\therefore \quad \langle u, u \rangle > 0 \quad \text{for all } u(\neq 0_V) \in V$

Thus, a function $\langle \, , \rangle : V \times V \to C$ satisfying axioms HI_1, HI_3 and the following axiom:

HI_2' : $\langle u, u \rangle > 0$ *for all non-zero vector* $u \in V$

is the Hermitian inner product on V.

ILLUSTRATIVE EXAMPLES

EXAMPLE-1 *Prove that $C^n = \{(a_1, a_2, \ldots, a_n) : a_i \in C\}$ is a unitary space with the Hermitian inner product defined by*

$$\langle u, v \rangle = \sum_{i=1}^{n} a_i \overline{b_i} \quad \text{for all } u = (a_1, a_2, \ldots, a_n),\ v = (b_1, b_2, \ldots, b_n) \in C^n$$

SOLUTION C^n is a vector space over the field C of all complex numbers (see Example 7 on page 118).

Let us now show that the product defined on C^n satisfies all the axioms of Hermitian inner product.

HI_1: *Conjugate symmetry* For any $u = (a_1, a_2, \ldots, a_n)$, $v = (b_1, b_2, \ldots, b_n)$ in C^n, we have

$$\langle u, v \rangle = \sum_{i=1}^{n} a_i \overline{b_i} = \sum_{i=1}^{n} \overline{b_i} a_i = \sum_{i=1}^{n} \overline{(b_i \overline{a_i})} = \overline{\left(\sum_{i=1}^{n} b_i \overline{a_i} \right)} = \overline{\langle v, u \rangle}$$

HI_2: *Positive definiteness* For any $u = (a_1, a_2, \ldots, a_n) \in C^n$

$$\langle u, u \rangle = \sum_{i=1}^{n} a_i \overline{a_i} = \sum_{i=1}^{n} |a_i|^2 \geq 0$$

Also, $\quad \langle u, u \rangle = \sum_{i=1}^{n} |a_i|^2 = 0$ iff $a_1 = a_2 = \cdots = a_n = 0$, i.e. $u = 0$

HI$_3$: Left linearity For any $u = (a_1, a_2, \ldots, a_n)$, $v = (b_1, b_2, \ldots, b_n)$, $w = (c_1, c_2, \ldots, c_n)$ in C^n and $a, b \in C$, we have

$$\langle au + bv, w \rangle = \sum_{i=1}^{n} (aa_i + bb_i) \overline{c}_i$$

$\Rightarrow \quad \langle au + bv, w \rangle = \sum_{i=1}^{n} (aa_i \overline{c}_i + bb_i \overline{c}_i)$

$\Rightarrow \quad \langle au + bv, w \rangle = \sum_{i=1}^{n} aa_i \overline{c}_i + \sum_{i=1}^{n} bb_i \overline{c}_i$

$\Rightarrow \quad \langle au + bv, w \rangle = a \left(\sum_{i=1}^{n} a_i \overline{c}_i \right) + b \left(\sum_{i=1}^{n} b_i \overline{c}_i \right)$

$\Rightarrow \quad \langle au + bv, w \rangle = a \langle u, w \rangle + b \langle v, w \rangle$

Thus, the given product is Hermitian inner product on C^n. Hence, C^n is a unitary space over C.

REMARK-1 *The Hermitian inner product defined above is called the usual or standard inner product on C^n. The vector space C^n with this Hermitian inner product is called the Complex Euclidean space. Assuming u and v as column vectors, this inner product may be defined as*

$$\langle u, v \rangle = u^T \overline{v}$$

where, as in matrices, $\overline{v}$ means the conjugate of each element of v.

REMARK-2 *In what follows, we will assume that C^n is a unitary space with the above defined inner product unless otherwise stated or implied.*

EXAMPLE-2 *Let V be the vector space of all complex continuous functions on the real interval $[a, b]$. Prove that V is a unitary space with the Hermitian inner product on V defined by*

$$\langle f, g \rangle = \int_{a}^{b} f(t) \overline{g(t)} dt \quad \text{for all } f, g \in V$$

SOLUTION The product defined above satisfies the following axioms:

HI$_1$: Conjugate symmetry For any $f, g \in V$

$$\langle f, g \rangle = \int_{a}^{b} f(t) \overline{g(t)} dt$$

$\Rightarrow \quad \langle f, g \rangle = \int_{a}^{b} \overline{g(t)} f(t) dt$

$\Rightarrow \quad \langle f, g \rangle = \int_a^b \overline{\left\{ g(t) \overline{f(t)} \right\}} dt$

$\Rightarrow \quad \langle f, g \rangle = \int_a^b \overline{g(t)} f(t) dt = \overline{\langle g, f \rangle}$

HI_2: **Positive definiteness** For any $f \in V$

$$\langle f, f \rangle = \int_a^b f(t) \overline{f(t)} dt = \int_a^b |f(t)|^2 dt \geq 0 \left[\because \quad |f(t)|^2 \geq 0 \quad \text{for all } t \in [a, b] \right]$$

Also, $\langle f, f \rangle = 0 \Leftrightarrow \int_a^b |f(t)|^2 dt = 0 \Leftrightarrow f(t) = 0 \quad \text{for all } t \in [a, b]$

HI_3: **Left linearity** For any $f, g, h \in V$ and $\alpha, \beta \in C$

$$\langle \alpha f + \beta g, h \rangle = \int_a^b (\alpha f + \beta g)(t) \overline{h(t)} dt$$

$\Rightarrow \quad \langle \alpha f + \beta g, h \rangle = \int_a^b \{\alpha f(t) + \beta g(t)\} \overline{h(t)} dt$

$\Rightarrow \quad \langle \alpha f + \beta g, h \rangle = \int_a^b \alpha f(t) \overline{h(t)} dt + \int_a^b \beta g(t) \overline{h(t)} dt$

$\Rightarrow \quad \langle \alpha f + \beta g, h \rangle = \alpha \int_a^b f(t) \overline{h(t)} dt + \beta \int_a^b g(t) \overline{h(t)} dt$

$\Rightarrow \quad \langle \alpha f + \beta g, h \rangle = \alpha \langle f, h \rangle + \beta \langle g, h \rangle$

Thus, all axioms of Hermitian inner product are satisfied. Hence, V is a unitary space.

EXAMPLE-3 Let V be the vector space of all $m \times n$ matrices over C, i.e. $V = C^{m \times n}$. V is a unitary space with the Hermitian inner product on V defined by

$$\langle A, B \rangle = tr\left(\overline{B}^T A \right) = \sum_{i=1}^m \sum_{j=1}^n \overline{b_{ij}} \, a_{ij} \quad \text{for all } A = [a_{ij}], B = [b_{ij}] \in V.$$

SOLUTION We observe that the above product satisfies the following axioms:

HI$_1$: *Conjugate symmetry* For any $A = [a_{ij}]$, $B = [b_{ij}] \in V$

$$\langle A, B \rangle = tr\left(\overline{B}^T A\right) = \sum_{i=1}^{m} \sum_{j=1}^{n} \overline{b_{ij}}\, a_{ij}$$

$\Rightarrow \quad \langle A, B \rangle = \sum_{i=1}^{m} \sum_{j=1}^{n} a_{ij}\, \overline{b_{ij}} = \sum_{i=1}^{m} \sum_{j=1}^{n} \overline{(\overline{a_{ij}}\, b_{ij})}$

$\Rightarrow \quad \langle A, B \rangle = \overline{\left(\sum_{i=1}^{m} \sum_{j=1}^{n} \overline{a_{ij}}\, b_{ij}\right)} = \overline{\left(tr(\overline{A}^T B)\right)} = \overline{\langle B, A \rangle}$

HI$_2$: *Positive definiteness* For any $A = [a_{ij}] \in V$

$$\langle A, A \rangle = tr\left(\overline{A}^T A\right) = \sum_{i=1}^{m} \sum_{j=1}^{n} \overline{a_{ij}}\, a_{ij}$$

$\Rightarrow \quad \langle A, A \rangle = \sum_{i=1}^{m} \sum_{j=1}^{n} |a_{ij}|^2 \geq 0$

Also, $\langle A, A \rangle = 0 \Leftrightarrow \sum_{i=1}^{m} \sum_{j=1}^{n} |a_{ij}|^2 = 0 \Leftrightarrow a_{ij} = 0 \quad$ for all $i \in \underline{m}$, $j \in \underline{n}$

$\Rightarrow \quad \langle A, A \rangle \Leftrightarrow A = 0$

HI$_3$: *Left linearity* For any $A = [a_{ij}]$, $B = [b_{ij}]$, $C = [c_{ij}]$ in V and $\alpha, \beta \in C$

$$\langle \alpha A + \beta B, C \rangle = tr\left(\overline{C}^T(\alpha A + \beta B)\right)$$

$\Rightarrow \quad \langle \alpha A + \beta B, C \rangle = \sum_{i=1}^{m} \sum_{j=1}^{n} \overline{c_{ij}}(\alpha\, a_{ij} + \beta\, b_{ij})$

$\Rightarrow \quad \langle \alpha A + \beta B, C \rangle = \sum_{i=1}^{m} \sum_{j=1}^{n} \overline{c_{ij}}\, \alpha\, a_{ij} + \sum_{i=1}^{m} \sum_{j=1}^{n} \overline{c_{ij}}\, \beta\, b_{ij}$

$\Rightarrow \quad \langle \alpha A + \beta B, C \rangle = \alpha \left(\sum_{i=1}^{m} \sum_{j=1}^{n} \overline{c_{ij}}\, a_{ij}\right) + \beta \left(\sum_{i=1}^{m} \sum_{j=1}^{n} \overline{c_{ij}}\, b_{ij}\right)$

$\Rightarrow \quad \langle \alpha A + \beta B, C \rangle = \alpha\, tr\left(\overline{C}^T A\right) + \beta\, tr\left(\overline{C}^T B\right)$

$\Rightarrow \quad \langle \alpha A + \beta B, C \rangle = \alpha \langle A, C \rangle + \beta \langle B, C \rangle$

Thus, the given product satisfies all axioms of Hermitian inner product.

Hence, V is a unitary space.

EXERCISE 8.1

1. Let V be a unitary space and u, v be vectors in V such that $\langle u, v \rangle = 3 + 2i$. Then evaluate each of the following:
   (i) $\langle (2-4i)u, v \rangle$ (ii) $\langle u, (4+3i)v \rangle$ (iii) $\langle (3-6i)u, (5-2i)v \rangle$

2. Let $u = (1+i, 3, 4-i)$ and $v = (3-4i, 1+i, 2i)$ be vectors in unitary space C^3 with standard inner product. Compute each of the following:
   (i) $\langle u, v \rangle$ (ii) $\langle v, u \rangle$ (iii) $\langle u, u \rangle$ (iv) $\langle v, v \rangle$

3. Let $u = (z_1, z_2)$, $v = (w_1, w_2) \in C^2$. Verify that the following is Hermitian inner product on C^2:
$$\langle u, v \rangle = z_1 \overline{w}_1 + (1+i) z_1 \overline{w}_2 + (1-i) z_2 \overline{w}_1 + 3 z_2 \overline{w}_2$$

4. Let $u = (z_1, z_2)$ and $v = (w_1, w_2) \in C^2$. For what values of $a, b, c, d \in C$ is the following an inner product on C^2?
$$\langle u, v \rangle = a z_1 \overline{w}_1 + b z_1 \overline{w}_2 + c z_2 \overline{w}_1 + d z_2 \overline{w}_2$$

5. Let V be a unitary space. For any $u_1, u_2, v_1, v_2 \in V$ and $a_1, b_1, a_2, b_2 \in C$, prove that
$$\langle a_1 u_1 + a_2 u_2, b_1 v_1 + b_2 v_2 \rangle = a_1 \overline{b}_1 \langle u_1, v_1 \rangle + a_1 \overline{b}_2 \langle u_1, v_2 \rangle + a_2 \overline{b}_1 \langle u_2, v_1 \rangle + a_2 \overline{b}_2 \langle u_2, v_2 \rangle$$
More generally, prove that
$$\left\langle \sum_{i=1}^{m} a_i u_i, \sum_{j=1}^{n} b_j v_j \right\rangle = \sum_{i=1}^{m} \sum_{j=1}^{n} a_i \overline{b}_j \langle u_i, v_i \rangle$$

6. Let $u = (z_1, z_2)$, $v = (w_1, w_2) \in C^2$, show that
$$\langle u, v \rangle = 2 z_1 \overline{w}_1 + z_1 \overline{w}_2 + z_2 \overline{w}_1 + z_2 \overline{w}_2$$
defines a Hermitian inner product on C^2.

7. Let V be a vector space over C and let f, g be Hermitian inner products on V. Then show that $f + g$ is a Hermitian inner product on V. Also, show that nf is a Hermitian inner product on V for all $n \in N$.
   [Hint: Proceed as in Example 5 on page 535]

8. Show that the difference of two Hermitian inner products on a vector space $V(C)$ is not necessarily a Hermitian inner product on V.

9. Let $V = C^{n \times n}$ be the vector space of all $n \times n$ matrices over C. Show that V is a unitary space with the Hermitian inner product defined by
$$\langle A, B \rangle = tr(AB^*), \text{ where } B^* \text{ is the conjugate transpose of } B.$$

ANSWERS

1. (i) $14 - 8i$ (ii) $18 - i$ (iii) $129 - 18i$
2. (i) $-4i$ (ii) $4i$ (iii) 28 (iv) 31
4. a and d are positive real numbers, $\overline{c} = b$ and $ad - bc > 0$.

8.3 NORM OF A VECTOR

As with the inner product space, we define norm or length of a vector in a unitary space as follows.

NORM OF A VECTOR *Let V be a unitary space and u be a vector in V. The non-negative square root, of $\langle u,u \rangle$ i.e. $\sqrt{\langle u,u \rangle}$ is defined as the norm or length of vector u and is denoted by $||u||$. That is,*

$$||u|| = \sqrt{\langle u,u \rangle} \quad \Rightarrow \quad ||u||^2 = \langle u,u \rangle$$

THEOREM-1 *Let V be a unitary space. Then, for any $u,v \in V$ and $a,b, \in C$*

$$\langle au+bv, au+bv \rangle = a\bar{a}\langle u,u \rangle + a\bar{b}\langle u,v \rangle + \bar{a}b\langle v,u \rangle + b\bar{b}\langle v,v \rangle$$

i.e. $\quad ||au+bv||^2 = |a|^2||u||^2 + a\bar{b}\langle u,v \rangle + \bar{a}b\langle v,u \rangle + |b|^2||v||^2$

PROOF. Using left linearity in V, we have

$$\langle au+bv, au+bv \rangle = a\langle u, au+bv \rangle + b\langle v, au+bv \rangle$$

$$\Rightarrow \quad \langle au+bv, au+bv \rangle = a\{\langle u,au \rangle + \langle u,bv \rangle\} + b\{\langle v,au \rangle + \langle v,bv \rangle\}$$

$$\Rightarrow \quad \langle au+bv, au+bv \rangle = a\{\bar{a}\langle u,u \rangle + \bar{b}\langle u,v \rangle\} + b\{\bar{a}\langle v,u \rangle + \bar{b}\langle v,v \rangle\}$$

$$\Rightarrow \quad \langle au+bv, au+bv \rangle = a\bar{a}\langle u,u \rangle + a\bar{b}\langle u,v \rangle + \bar{a}b\langle v,u \rangle + b\bar{b}\langle v,v \rangle$$

Q.E.D.

COROLLARY. *Let V be a unitary space. Then, $||au|| = |a|\,||u||$ for all $u \in V$ and $a \in C$.*

PROOF. Substituting $b = 0$ in the above theorem, we get

$$\langle au, au \rangle = a\bar{a}\langle u,u \rangle = |a|^2||u||^2$$

$$\Rightarrow \quad ||au||^2 = |a|^2||u||^2$$

$$\Rightarrow \quad ||au|| = |a|\,||u||$$

Q.E.D.

THEOREM-2 *(Schwarz's inequality) Let V be a unitary space. Then,*

$$|\langle u,v \rangle| \leq ||u||\,||v|| \quad \text{for all } u,v \in V$$

PROOF. If $u = 0_V$, then $\langle u,v \rangle = 0$ and $||u||\,||v|| = 0$ for all $v \in V$. So, the result is true in this case. So, let us assume that $u \neq 0_V$. Following cases arise

Case I *When $\langle u,v \rangle$ is a real number:*

For any non-zero real number a, we have

$\langle au+v, au+v \rangle \geq 0$

$\Rightarrow \quad a\bar{a} \langle u,u \rangle + a\langle u,v \rangle + \bar{a} \langle v,u \rangle + \langle v,v \rangle \geq 0$

$\Rightarrow \quad a^2 \langle u,u \rangle + a\langle u,v \rangle + a\overline{\langle u,v \rangle} + \langle v,v \rangle \geq 0 \qquad [\because \ a \in R \ \therefore \ \bar{a} = a]$

$\Rightarrow \quad a^2 ||u||^2 + 2a\langle u,v \rangle + ||v||^2 \geq 0 \qquad [\because \ \langle u,v \rangle \text{ is real } \therefore \ \overline{\langle u,v \rangle} = \langle u,v \rangle]$

$\Rightarrow \quad a^2 ||u||^2 + 2a\langle u,v \rangle + ||v||^2 \geq 0$ for all $a \in R$

$\Rightarrow \quad 4\{\langle u,v \rangle\}^2 - 4||u||^2 ||v||^2 \leq 0 \qquad$ [Discriminant ≤ 0]

$\Rightarrow \quad \{\langle u,v \rangle\}^2 \leq ||u||^2 ||v||^2$

$\Rightarrow \quad |\langle u,v \rangle| \leq ||u|| \, ||v||$

Case II *When $\langle u,v \rangle$ is not a real number:*

If $\langle u,v \rangle$ is not real, then it is a non-zero complex number.

Let $a = \langle u,v \rangle$. As $a \neq 0$, therefore, $\dfrac{u}{a}$ is defined and $\left\langle \dfrac{u}{a}, v \right\rangle = \dfrac{1}{a}\langle u,v \rangle = \dfrac{\langle u,v \rangle}{\langle u,v \rangle} = 1$.

Thus, $\left\langle \dfrac{u}{a}, v \right\rangle$ is a real number. Therefore, from case I, we get

$\left| \left\langle \dfrac{u}{a}, v \right\rangle \right| \leq \left\| \dfrac{u}{a} \right\| \, ||v||$

$\Rightarrow \quad 1 \leq \dfrac{||u||}{|a|} ||v|| \qquad \left[\because \ \left\langle \dfrac{u}{a}, v \right\rangle = 1\right]$

$\Rightarrow \quad |a| \leq ||u|| \, ||v||$

$\Rightarrow \quad |\langle u,v \rangle| \leq ||u|| \, ||v||$

Hence, in either case, we obtain

$\left| \langle u,v \rangle \right| \leq ||u|| \, ||v||$ for all $u,v \in V$.

Q.E.D.

THEOREM-3 *Let V be a unitary space and $u,v \in V$. Then,*

(i) $2\langle u,v \rangle = ||u+v||^2 + i||u+iv||^2 - (1+i)\left(||u||^2 + ||v||^2\right)$

(ii) $4\langle u,v \rangle = ||u+v||^2 - ||u-v||^2 + i||u+iv||^2 - i||u-iv||^2$

PROOF. (i) For any $u, v \in V$, we have
$$\|u+v\|^2 = \langle u+v, u+v \rangle$$
$$\Rightarrow \|u+v\|^2 = \langle u,u \rangle + \langle u,v \rangle + \langle v,u \rangle + \langle v,v \rangle$$
$$\Rightarrow \|u+v\|^2 = \|u\|^2 + \langle u,v \rangle + \langle v,u \rangle + \|v\|^2 \qquad \text{(i)}$$
and,
$$\|u+iv\|^2 = \langle u+iv, u+iv \rangle$$
$$\Rightarrow \|u+iv\|^2 = \langle u,u \rangle + \langle u,iv \rangle + \langle iv,u \rangle + \langle iv,iv \rangle$$
$$\Rightarrow \|u+iv\|^2 = \langle u,u \rangle - i\langle u,v \rangle + i\langle v,u \rangle + (i)(-i)\langle v,v \rangle$$
$$\Rightarrow \|u+iv\|^2 = \|u\|^2 - i\langle u,v \rangle + i\langle v,u \rangle + \|v\|^2$$
$$\Rightarrow i\|u+iv\|^2 = i\|u\|^2 + \langle u,v \rangle - \langle v,u \rangle + i\|v\|^2 \qquad \text{(ii)}$$

Adding (i) and (ii), we get
$$\|u+v\|^2 + i\|u+iv\|^2 = (1+i)\|u\|^2 + 2\langle u,v \rangle + (1+i)\|v\|^2$$
$$\Rightarrow \|u+v\|^2 + i\|u+iv\|^2 = (1+i)(\|u\|^2 + \|v\|^2) + 2\langle u,v \rangle$$
$$\Rightarrow 2\langle u,v \rangle = \|u+v\|^2 + i\|u+iv\|^2 - (1+i)(\|u\|^2 + \|v\|^2)$$

(ii) For any $u, v \in V$, we have,
$$\|u+v\|^2 = \langle u+v, u+v \rangle$$
$$\Rightarrow \|u+v\|^2 = \langle u,u \rangle + \langle u,v \rangle + \langle v,u \rangle + \langle v,v \rangle$$
$$\Rightarrow \|u+v\|^2 = \|u\|^2 + \|v\|^2 + \langle u,v \rangle + \langle v,u \rangle \qquad \text{(i)}$$
$$\|u-v\|^2 = \langle u-v, u-v \rangle$$
$$\Rightarrow \|u-v\|^2 = \langle u,u \rangle + \langle u,-v \rangle + \langle -v,u \rangle + \langle -v,-v \rangle$$
$$\Rightarrow \|u-v\|^2 = \langle u,u \rangle - \langle u,v \rangle - \langle v,u \rangle + \langle v,v \rangle$$
$$\Rightarrow \|u-v\|^2 = \|u\|^2 + \|v\|^2 - \langle u,v \rangle - \langle v,u \rangle \qquad \text{(ii)}$$

Subtracting (ii) from (i), we get
$$\|u+v\|^2 - \|u-v\|^2 = 2[\langle u,v \rangle + \langle v,u \rangle] \qquad \text{(iii)}$$
$$\|u+iv\|^2 = \langle u+iv, u+iv \rangle$$
$$\Rightarrow \|u+iv\|^2 = \langle u,u \rangle + \langle u,iv \rangle + \langle iv,u \rangle + \langle iv,iv \rangle$$
$$\Rightarrow \|u+iv\|^2 = \langle u,u \rangle - i\langle u,v \rangle + i\langle v,u \rangle + (i)(-i)\langle v,v \rangle$$
$$\Rightarrow \|u+iv\|^2 = \|u\|^2 + \|v\|^2 - i\langle u,v \rangle + i\langle v,u \rangle \qquad \text{(iv)}$$

Replacing i by $-i$, we obtain
$$\|u-iv\|^2 = \|u\|^2 + \|v\|^2 + i\langle u,v \rangle - i\langle v,u \rangle \qquad \text{(v)}$$

Subtracting (v) from (iv), we get
$$\|u+iv\|^2 - \|u-iv\|^2 = -2i\langle u,v \rangle + 2i\langle v,u \rangle$$
$$\Rightarrow i\|u+iv\|^2 - i\|u-iv\|^2 = 2\langle u,v \rangle - 2\langle v,u \rangle \qquad \text{(vi)}$$

Adding (iii) and (vi), we obtain

$$||u+v||^2 - ||u-v||^2 + i||u+iv||^2 - i||u-iv||^2 = 4\langle u,v \rangle$$

Q.E.D.

THEOREM-4 *Let V be a unitary space and $u, v \in V$. Then,*
  (i) $||u+v||^2 + ||u-v||^2 = 2(||u||^2 + ||v||^2)$ *(Parallelogram law)*
  (ii) $||u+v|| \leq ||u|| + ||v||$ *(Triangle in equality)*

PROOF. (i) We have,

$$||u+v||^2 = ||u||^2 + ||v||^2 + \langle u,v \rangle + \langle v,u \rangle \quad \text{[See proof of Theorem 3 (ii)]}$$

and, $\quad ||u-v||^2 = ||u||^2 + ||v||^2 - \langle u,v \rangle - \langle v,u \rangle$

$\therefore \quad ||u+v||^2 + ||u-v||^2 = 2(||u||^2 + ||v||^2)$

(ii) By Schwarz's inequality, we have

$$\left|\langle u,v \rangle\right| \leq ||u|| \, ||v||$$

$\Rightarrow \quad \left|\langle v,u \rangle\right| \leq ||v|| \, ||u||$

$\therefore \quad \left|\langle u,v \rangle\right| + \left|\langle v,u \rangle\right| \leq 2||u|| \, ||v||$

$\Rightarrow \quad \langle u,v \rangle + \langle v,u \rangle \leq 2||u|| \, ||v||$

$\Rightarrow \quad ||u||^2 + ||v||^2 + \langle u,v \rangle + \langle v,u \rangle \leq ||u||^2 + ||v||^2 + 2||u|| \, ||v||$

$\Rightarrow \quad ||u+v||^2 \leq \left(||u|| + ||v||\right)^2$

$\Rightarrow \quad ||u+v|| \leq ||u|| + ||v||$

Q.E.D.

ILLUSTRATIVE EXAMPLES

EXAMPLE-1 *If vectors u, v in a unitary space are linearly dependent, then*

$$\left|\left|\langle u,v \rangle\right|\right| = ||u|| \, ||v||$$

SOLUTION If $v = 0_V$, then $\langle u,v \rangle = 0$ and $||v|| = 0$

$\therefore \quad ||\langle u,v \rangle|| = ||u|| \, ||v||$

Also, if $u = 0_V$, then $\langle u,v \rangle = 0$ and $||u|| = 0$

$\therefore \quad ||\langle u,v \rangle|| = ||u|| \, ||v||$

So, let us assume that u and v are non-zero linearly dependent vectors. Then there exists a scalar $\lambda \in C$ such that $u = \lambda v$.

$\therefore \quad \langle u, v \rangle = \langle \lambda v, v \rangle = \lambda \langle v, v \rangle = \lambda ||v||^2$

$\Rightarrow \quad |\langle u, v \rangle| = |\lambda| \, ||v||^2 = |\lambda| \, ||v|| \, ||v||$

$\Rightarrow \quad |\langle u, v \rangle| = ||u|| \, ||v|| \qquad [\because ||u|| = |\lambda| \, ||v||]$

EXAMPLE-2 *Let u and v be two vectors in a unitary space V such that $|\langle u, v \rangle| = ||u|| \, ||v||$, i.e. the Schwarz's inequality reduces to an equality. Then, u and v are linearly dependent.*

SOLUTION If $u = 0_V$ or $v = 0_V$, then the result holds good trivially as any set containing zero vector is linearly dependent. So, let u and v be non-zero vectors in V satisfying $|\langle u, v \rangle| = ||u|| \, ||v||$.

Consider the vector

$$w = v - \frac{\langle v, u \rangle}{||u||^2} u = v - \lambda u, \text{ where } \lambda = \frac{\langle v, u \rangle}{||u||^2}$$

$\therefore \quad \langle w, w \rangle = \langle v - \lambda u, \ v - \lambda u \rangle$

$\Rightarrow \quad \langle w, w \rangle = \langle v, v \rangle + \langle v, -\lambda u \rangle + \langle -\lambda u, v \rangle + \langle -\lambda u, -\lambda u \rangle$

$\Rightarrow \quad \langle w, w \rangle = \langle v, v \rangle - \overline{\lambda} \langle v, u \rangle - \lambda \langle u, v \rangle + \lambda \overline{\lambda} \langle u, u \rangle$

$\Rightarrow \quad \langle w, w \rangle = \langle v, v \rangle - \frac{\overline{\langle v, u \rangle}}{||u||^2} \langle v, u \rangle - \frac{\langle v, u \rangle}{||u||^2} \overline{\langle v, u \rangle} + |\lambda|^2 \langle u, u \rangle$

$\Rightarrow \quad \langle w, w \rangle = ||v||^2 - 2 \frac{|\langle v, u \rangle|^2}{||u||^2} + \frac{|\langle v, u \rangle|^2}{||u||^2}$

$\Rightarrow \quad \langle w, w \rangle = ||v||^2 - \frac{|\langle v, u \rangle|^2}{||u||^2}$

$\Rightarrow \quad \langle w, w \rangle = ||v||^2 - ||v||^2 \qquad [\because |\langle u, v \rangle| = ||u|| \, ||v||]$

$\Rightarrow \quad \langle w, w \rangle = 0$

$\Rightarrow \quad w = 0_V$

$\Rightarrow \quad v - \lambda u = 0_V$

$\Rightarrow \quad v = \lambda u$

$\Rightarrow \quad v$ is a scalar multiple of u

$\Rightarrow \quad u$ and v are linearly dependent vectors

EXAMPLE-3 *Let u and v be vectors in a unitary space V such that $||u + v|| = ||u|| + ||v||$. Then prove that u and v are linearly dependent vectors. Give an example to show that the converse of this statement is not true.*

SOLUTION We have,

$$||u+v|| = ||u|| + ||v||$$
$$\Rightarrow ||u+v||^2 = (||u|| + ||v||)^2$$
$$\Rightarrow ||u||^2 + ||v||^2 + \langle u,v \rangle + \langle v,u \rangle = ||u||^2 + ||v||^2 + 2||u||\,||v||$$
$$\Rightarrow \langle u,v \rangle + \overline{\langle u,v \rangle} = 2||u||\,||v||$$
$$\Rightarrow 2\,\text{Re}\langle u,v \rangle = 2||u||\,||v||$$
$$\Rightarrow \text{Re}\langle u,v \rangle = ||u||\,||v||$$
$$\Rightarrow ||u||\,||v|| \leq |\langle u,v \rangle| \qquad [\because \text{Re}(Z) \leq |Z|] \quad \text{(i)}$$
But, $\quad |\langle u,v \rangle| \leq ||u||\,||v|| \qquad$ [By Schwarz's inequality] (ii)
$\therefore \quad |\langle u,v \rangle| = ||u||\,||v|| \qquad$ [From (i) and (ii)]
$\Rightarrow \quad u$ and v are linearly dependent vectors [See Example 2]

The converse is not true, because $u = (-1, 0, 1)$ and $v = (2, 0, -2)$ are two vectors in R^3 such that $2u = v$. Therefore, u and v are linearly dependent vectors and, we have

$$||u|| = \sqrt{2},\, ||v|| = 2\sqrt{2},\, ||u+v|| = \sqrt{2}$$
$\therefore \quad ||u+v|| \neq ||u|| + ||v||$

EXERCISE 8.2

1. Let $u = (1-2i,\, 2+3i,\, 3-3i)$ and $v = (1+i,\, 2i,\, 3-4i)$ be two vectors in the unitary space C^3 with the standard inner product. Find $||u||$ and $||v||$.

2. Let u, v be any two vectors in a unitary space V. Prove that

$$\langle u,v \rangle = \frac{1}{4}||u+v||^2 - \frac{1}{4}||u-v||^2 + \frac{1}{4}||u+iv||^2 - \frac{1}{4}||u-iv||^2$$

3. The following results are true in an inner product space V.

    (i) $||u|| = ||v||$ iff $\langle u+v,\, u-v \rangle = 0$

    (ii) $||u+v||^2 = ||u||^2 + ||v||^2$ iff $\langle u,v \rangle = 0$.

    Show by counter examples that these results are not true in a unitary space.

4. Let u and v be vectors in a unitary space V. Then prove that

    (i) $\text{Re}\langle u,v \rangle = \frac{1}{4}||u+v||^2 - \frac{1}{4}||u-v||^2$

    (ii) $\langle u,v \rangle = \text{Re}\langle u,v \rangle + i\,\text{Re}\langle u,iv \rangle$

[Hint: (i) $||u+v||^2 - ||u-v||^2 = 2\langle u,v \rangle + 2\langle v,u \rangle = 2\{\langle u,v \rangle + \overline{\langle u,v \rangle}\} = 4\,\text{Re}\langle u,v \rangle$

(ii) For any complex number $z = a + ib$, we have
$$z = \text{Re}(z) + i\,\text{Re}(-iz)$$
$$\therefore \quad \langle u,v \rangle = \text{Re}\langle u,v \rangle + i\,\text{Re}\left\{-i\langle u,v \rangle\right\}$$
$$\Rightarrow \quad \langle u,v \rangle = \text{Re}\langle u,v \rangle + i\,\text{Re}\langle u,iv \rangle]$$

5. Let $u = (z_1, z_2)$ and $v = (w_1, w_2)$ be any two vectors in C^2. Show that C^2 is a unitary space with the Hermitian inner product defined by
$$\langle u,v \rangle = z_1 \overline{w}_1 + (z_1 + z_2)(\overline{w}_1 + \overline{w}_2).$$
Also, find $\|u\|$, for $u = (3,4) \in C^2$.

ANSWERS

3. $u = (1, 2)$, $v = (i, 2i)$ in C^2.
5. $\sqrt{58}$

8.4 ORTHOGONALITY AND ORTHONORMALITY

The concepts of orthogonality and orthonormality in unitary spaces are analogous to the corresponding concepts in inner product spaces in Chapter 6. For the convenience of the reader, let us refresh the same.

ORTHOGONAL VECTORS *Two vectors u and v in a unitary space V are said to be orthogonal vectors if their Hermitian inner product is zero, i.e.* $\langle u,v \rangle = 0$.

If u is orthogonal to v, then we write $u \perp v$.

Let u and v be vectors in a unitary space V such that

$$u \perp v$$
$$\Rightarrow \quad \langle u,v \rangle = 0$$
$$\Rightarrow \quad \overline{\langle v,u \rangle} = 0 \qquad\qquad \text{[By conjugate symmetry]}$$
$$\Rightarrow \quad \langle v,u \rangle = 0$$
$$\Rightarrow \quad v \perp u$$

Thus, the relation 'is orthogonal to' on a unitary space is a symmetric relation.

Let u be an arbitrary vector in a unitary space V. Then,
$$\langle u, 0_V \rangle = \langle u, 0\,0_V \rangle = 0\langle u, 0_V \rangle = 0$$

Thus, every vector in V is orthogonal to the null vector.

Let $u, v \in V$ such that $u \perp v$ and $a \in C$. Then,
$$\langle au, v \rangle = a\langle u,v \rangle = a\,0 = 0 \qquad\qquad [\because\ u \perp v\ \therefore\ \langle u,v \rangle = 0]$$
$$\Rightarrow \quad au \perp v.$$

Thus, if $u \perp v$, then $au \perp v$ for all $a \in C$.

Now, let a vector $u \in V$ be such that u is orthogonal to every vector in V. Then,

$\langle u, v \rangle = 0$ for all $v \in V$

$\Rightarrow \quad \langle u, u \rangle = 0$ [Taking $v = u$]

$\Rightarrow \quad u = 0_V$

Thus, null vector is the only vector which is orthogonal to every vector in V.

EXAMPLE-1 *Two vectors u and v in a unitary space V are orthogonal iff*
$$\| au + bv \|^2 = \|au\|^2 + \|bv\|^2$$
for all pairs of scalars a and b.

SOLUTION Let u and v be two orthogonal vectors in a unitary space V and $a, b \in C$. Then,

$\|au + bv\|^2 = \langle au + bv,\ au + bv \rangle$

$\Rightarrow \quad \|au + bv\|^2 = \langle au,\ au + bv \rangle + \langle bv,\ au + bv \rangle$

$\Rightarrow \quad \|au + bv\|^2 = \langle au,\ au \rangle + \langle au,\ bv \rangle + \langle bv,\ au \rangle + \langle bv,\ bv \rangle$

$\Rightarrow \quad \|au + bv\|^2 = \langle au,\ au \rangle + a\bar{b}\langle u, v \rangle + b\bar{a}\langle v, u \rangle + \langle bv, bv \rangle$

$\Rightarrow \quad \|au + bv\|^2 = \|au\|^2 + \|bv\|^2 \qquad [\because\ u \perp v\ \therefore\ \langle u, v \rangle = 0$ and $\langle v, u \rangle = 0]$

Conversely, let $u, v \in V$ and $a, b \in C$ such that

$\|au + bv\|^2 = \|au\|^2 + \|bv\|^2$

$\Rightarrow \quad \|au\|^2 + \|bv\|^2 + a\bar{b}\langle u, v \rangle + \bar{a}b\langle v, u \rangle = \|au\|^2 + \|bv\|^2$

$\Rightarrow \quad a\bar{b}\langle u, v \rangle + \bar{a}b\overline{\langle u, v \rangle} = 0$

$\Rightarrow \quad a\bar{b}\langle u, v \rangle + \overline{a\bar{b}\langle u, v \rangle} = 0$

$\Rightarrow \quad a\bar{b}\langle u, v \rangle$ is purely imaginary $\qquad\qquad\qquad\qquad\qquad\qquad\qquad\qquad$ (i)

Taking $a = b$, in (i) we obtain

$a\bar{a} \langle u, v \rangle$ is purely imaginary

$\Rightarrow \quad \text{Re}\ \langle u, v \rangle = 0$

Taking $a = ib$, in (i) we obtain

$\Rightarrow \quad ib\bar{b} \langle u, v \rangle$ is purely imaginary

$\Rightarrow \quad \langle u, v \rangle$ is purely real

$\Rightarrow \quad I_m \langle u, v \rangle = 0$

Hence, $\langle u, v \rangle = 0$. $\qquad\qquad\qquad\qquad [\because\ \text{Re}\ \langle u, v \rangle = 0$ and $I_m \langle u, v \rangle = 0]$

ORTHOGONAL SET *Let V be a unitary space. A set S of non-zero vectors in V is called an orthogonal set if each pair of vectors in S is a pair of orthogonal vectors.*
i.e. $\quad \langle u, v \rangle = 0$ for all $u, v \in S,\ u \neq v$.

THEOREM-1 *Any orthogonal set of non-zero vectors in a unitary space is linearly independent*

PROOF. See proof of Theorem 1 on page 555

THEOREM-2 *Let $S = \{v_1, v_2, \ldots, v_n\}$ be an orthogonal set of non-zero vectors in an inner product space V. If $u \in [S]$, then*

$$u = \sum_{i=1}^{n} \frac{\langle u, v_i \rangle}{\|v_i\|^2} v_i$$

PROOF. See proof of Theorem 3 on page 556.

ORTHOGONAL, BASIS *Let V be a unitary space. An orthogonal set of vectors in V is called an orthogonal basis of V if it is a basis of V.*

An orthogonal set of vectors in a unitary space is linearly independent. Therefore, if V is an n dimensional unitary space and S is an orthogonal set consisting of n non-zero vectors of V, then S forms an orthogonal basis of V.

THEOREM-3 *Let V be a unitary space and $\{v_1, v_2, \ldots, v_n\}$ be an orthogonal basis for V. Then, for any $u \in V$.*

$$u = \sum_{i=1}^{n} \frac{\langle u, v_i \rangle}{\langle v_i, v_i \rangle} v_i$$

PROOF. See proof of Theorem 4 on page 556.

ORTHOGONAL COMPLEMENT OF A VECTOR *Let V be a unitary space and $u \in V$. Then the set of all vectors orthogonal to u is called the orthogonal complement of u and is denoted by $u^\perp$.*

Thus, $u^\perp = \{v : \langle v, u \rangle = 0\}$

The symbol "$u^\perp$" is usually read as "u perpendicular".

THEOREM-4 *Let V be a unitary space and $u \in V$. Then, the orthogonal complement $u^\perp$ of vector u is a subspace of V.*

PROOF. See proof of Theorem 1 on page 561

EXAMPLE-2 *Find an orthogonal basis for $u^\perp$ in C^3, where $u = (1, i, 1+i)$.*

SOLUTION Clearly, $u^\perp = \{v \in C^3 : \langle v, u \rangle = 0\}$. That is, $u^\perp$ consists of all vectors $v = (x, y, z) \in C^3$ such that

$\langle v, u \rangle = 0$

$\Rightarrow \quad x \times 1 + y \times \overline{i} + z \times \overline{(1+i)} = 0$ [Using standard inner product]

$\Rightarrow \quad x - iy + (1-i)z = 0$ \hfill (i)

We observe that $x = 0$, $y = 1 - i$, $z = i$ satisfy equation (i). Therefore, $w_1 = (0, 1 - i, i) \in u^\perp$.

Now, $\langle v, u \rangle = 0$ and $\langle v, w \rangle = 0$

$\Rightarrow \quad x - iy + (1 - i)z = 0$ and $y(1 + i) - iz = 0$

In these two equations z is a free variable. So, let us take $z = 1$ to obtain $y = \dfrac{i}{1+i} = \dfrac{1+i}{2}$

and $x = \dfrac{3}{2}(i - 1)$.

In order to avoid fractions, let us take $w_3 = (3(i - 1), 1 + i, 2)$

The vectors w_1 and w_2 form an orthogonal basis for $u^\perp$.

ORTHOGONAL COMPLEMENT OF A SET *Let S be a subset of a unitary space V. The orthogonal complement of S, denoted by $S^\perp$ (read as "S perpendicular"), is the set of all vectors in V that are orthogonal to every vector in S.*

That is, $S^\perp = \{v \in V : \langle v, u \rangle = 0 \text{ for all } u \in S\}$

THEOREM-5 *Let S be a subset of a unitary space V. Then, $S^\perp$ is a subspace of V.*

PROOF. See proof of Theorem 1 on page 563

EXAMPLE-3 *Let V be the vector space of all $n \times n$ matrices over C, with inner product defined by $\langle A, B \rangle = \text{tr}(AB^*)$. Find the orthogonal complement of the subspace of diagonal matrices.*

SOLUTION Let S be the set of all $n \times n$ diagonal matrices over C. Then,

$S^\perp = \{A \in V : \langle A, B \rangle = 0 \text{ for all } B \in S\}$

or, $S^\perp = \{A \in V : \text{tr}(AB^*) = 0 \text{ for all } B \in S\}$

Now,

$\text{tr}(AB^*) = 0$ for all $B \in S$

$\Rightarrow \quad \sum_{i=1}^{n} (AB^*)_{ii} = 0$ for all $B \in S$

$\Rightarrow \quad \sum_{i=1}^{n} \sum_{r=1}^{n} (A)_{ir}(B^*)_{ri} = 0$ for all $B \in S$

$\Rightarrow \quad \sum_{i=1}^{n} \sum_{r=1}^{n} a_{ir} \overline{b}_{ir} = 0$ for all $b_{ir} \in C$ such that $b_{ir} = 0$ for $i \neq r$

$\Rightarrow \quad \sum_{i=1}^{n} a_{ii} \overline{b}_{ii} = 0$ for all $b_{ii} \in C$

$\Rightarrow \quad a_{ii} = 0$ for all $i = 1, 2, \ldots, n$

Hence, $S^\perp = \{A \in V : (A)_{ii} = 0 \text{ for all } i \in \underline{n}\}$

ORTHOGONAL COMPLEMENT OF AN ORTHOGONAL COMPLEMENT *Let S be a subset of a unitary space V. Then, $S^\perp$ is a subspace of V. The set of all vectors in V which are orthogonal* to each vector in $S^\perp$ is called orthogonal complement of an orthogonal complement and is denoted by $S^{\perp\perp}$. That is,

$$S^{\perp\perp} = \left\{v \in V : \langle v, u \rangle = 0 \text{ for all } u \in S^\perp \right\}.$$

Clearly, $S^{\perp\perp}$ is a subspace of V.

THEOREM-6 Let V be a unitary space and S, S_1, S_2 be subsets of V. Then,

(i) $S \subset S^{\perp\perp}$

(ii) $S_1 \subset S_2 \Rightarrow S_2^\perp \subset S_1^\perp$

(iii) $S^\perp = [S]^\perp$

(iv) $[S] \subseteq S^{\perp\perp}$

(v) $S^\perp = S^{\perp\perp\perp}$

PROOF. Proceed as in the proof of Theorem 2 on page 564.

ORTHONORMAL SET Let V be a unitary space. A set S of vectors in V is called an orthonormal set, if

(i) $\langle u, v \rangle = 0$ for all $u, v \in S$, $u \neq v$

(ii) $\langle u, u \rangle = 1$ for all $u \in S$

That is, an orthogonal set of vectors in an inner product space V is an orthonormal set if each vector is of unit length.

We have, $\langle 0_V, 0_V \rangle = 0 \neq 1$

So, 0_V cannot be a vector in an orthonormal set.

Thus, a list $(v_1, v_2, \ldots, v_n)$ of vectors in a unitary space is an orthonormal list, if $\langle v_i, v_j \rangle = \delta_{ij}$ for all $i, j \in \underline{n}$.

Clearly, every orthonormal set of vectors in a unitary space is an orthogonal set but the converse is not true. Therefore, all properties possessed by an orthogonal set also hold good for an orthonormal set.

ORTHONORMAL BASIS Let V be a unitary space. An orthonormal set of vectors in V which is also a basis for V is called an orthonormal basis.

THEOREM-7 Any orthonormal set of vectors in a unitary space is linearly independent.

PROOF. See proof of Theorem 1 on page 568

COROLLARY-1 An orthonormal set of vectors in a unitary space V is a basis for V iff it spans V.

PROOF. See proof of Corollary 1 on page 568

COROLLARY-2 *A set of n orthonormal vectors in a unitary space V is a basis for V iff* $\dim V = n$.

PROOF. See proof of Corollary 2 on page 569

THEOREM-8 *Let* $S = \{v_1, v_2, \ldots, v_n\}$ *be an orthonormal basis of a unitary space V and* $u \in V$. *Then the coordinates of u relative to S are* $\langle u, v_i \rangle$; $i = 1, 2, \ldots, n$. *Thus,* $u = \sum_{i=1}^{n} \langle u, v_i \rangle v_i$

Moreover, $\quad ||u||^2 = \sum_{i=1}^{n} |\langle u, v_i \rangle|^2$.

PROOF. See proof of Theorem 2 on page 569

EXAMPLE-4 *Let* $S = \{v_1, v_2, \ldots, v_n\}$ *be an orthonormal basis for a unitary space V and* $u, v \in V$. *Then,*

$$\langle u, v \rangle = \sum_{i=1}^{n} \langle u, v_i \rangle \overline{\langle v, v_i \rangle}$$

SOLUTION Since S is an orthonormal basis for V. Therefore,

$$u = \sum_{i=1}^{n} \langle u, v_i \rangle v_i \text{ and } v = \sum_{j=1}^{n} \langle v, v_j \rangle v_j$$

$$\langle u, v \rangle = \left\langle \sum_{i=1}^{n} \langle u, v_i \rangle v_i, \sum_{j=1}^{n} \langle v, v_j \rangle v_j \right\rangle$$

$\Rightarrow \quad \langle u, v \rangle = \sum_{i=1}^{n} \sum_{j=1}^{n} \langle u, v_i \rangle \overline{\langle v, v_j \rangle} \langle v_i, v_j \rangle$

$\Rightarrow \quad \langle u, v \rangle = \sum_{i=1}^{n} \sum_{j=1}^{n} \langle u, v_i \rangle \overline{\langle v, v_j \rangle} \delta_{ij}$

$\Rightarrow \quad \langle u, v \rangle = \sum_{i=1}^{n} \langle u, v_i \rangle \overline{\langle v, v_i \rangle}$

THEOREM-9 *If* $S = \{v_1, v_2, \ldots, v_n\}$ *is an orthonormal set of vectors in a unitary space V and if* $u \in V$, *then the vector* $w = u - \sum_{i=1}^{n} \langle u, v_i \rangle v_i$ *is orthogonal to each of the vectors* $v_1, v_2, \ldots, v_n$, *i.e.* $w \in S^{\perp}$ *and consequently, to the subspace spanned by S.*

PROOF. See proof of Theorem 3 on page 570

THEOREM-10 *(Bessel's inequality)* If $\{v_1, v_2, \ldots, v_n\}$ is an orthonormal set of vectors in a unitary space V and if $v \in V$. Then,

$$\sum_{i=1}^{n} |\langle v, v_i \rangle|^2 \leq \|v\|^2$$

Furthermore, equality holds iff v is in subspace spanned by $\{v_1, v_2, \ldots, v_n\}$.

PROOF. Consider the vector u given by

$$u = v - \sum_{i=1}^{n} \langle v, v_i \rangle v_i$$

$\Rightarrow \quad \|u\|^2 = \langle u, u \rangle$

$\Rightarrow \quad \|u\|^2 = \left\langle v - \sum_{i=1}^{n} \langle v, v_i \rangle v_i, v - \sum_{j=1}^{n} \langle v, v_j \rangle v_j \right\rangle$

$\Rightarrow \quad \|u\|^2 = \langle v, v \rangle - \left\langle v, \sum_{j=1}^{n} \langle v, v_j \rangle v_j \right\rangle - \left\langle \sum_{i=1}^{n} \langle v, v_i \rangle v_i, v \right\rangle + \left\langle \sum_{i=1}^{n} \langle v, v_i \rangle v_i, \sum_{j=1}^{n} \langle v, v_j \rangle v_j \right\rangle$

$\Rightarrow \quad \|u\|^2 = \|v\|^2 - \sum_{j=1}^{n} \overline{\langle v, v_j \rangle} \langle v, v_j \rangle - \sum_{i=1}^{n} \langle v, v_i \rangle \langle v_i, v \rangle + \sum_{i=1}^{n} \sum_{j=1}^{n} \langle v, v_i \rangle \overline{\langle v, v_j \rangle} \langle v_i, v_j \rangle$

$\Rightarrow \quad \|u\|^2 = \|v\|^2 - \sum_{j=1}^{n} |\langle v, v_j \rangle|^2 - \sum_{i=1}^{n} \langle v, v_i \rangle \overline{\langle v, v_i \rangle} + \sum_{i=1}^{n} \sum_{j=1}^{n} \langle v, v_i \rangle \overline{\langle v, v_j \rangle} \delta_{ij}$

$\Rightarrow \quad \|u\|^2 = \|v\|^2 - \sum_{j=1}^{n} |\langle v, v_j \rangle|^2 - \sum_{i=1}^{n} |\langle v, v_i \rangle|^2 + \sum_{i=1}^{n} |\langle v, v_i \rangle|^2$

$\Rightarrow \quad \|u\|^2 = \|v\|^2 - \sum_{i=1}^{n} |\langle v, v_i \rangle|^2 \qquad \qquad \text{(i)}$

$\Rightarrow \quad \|v\|^2 - \sum_{i=1}^{n} |\langle v, v_i \rangle|^2 \geq 0 \qquad \qquad [\because \ \|u\|^2 \geq 0]$

$\Rightarrow \quad \sum_{i=1}^{n} |\langle v, v_i \rangle|^2 \leq \|v\|^2$

This proves the first part of the theorem.

Now, if the equality holds, i.e.

$$\sum_{i=1}^{n} |\langle v, v_i \rangle|^2 = \|v\|^2$$

$\Rightarrow \quad \|u\|^2 = 0 \qquad \qquad \text{[From (i)]}$

$\Rightarrow \quad u = 0_V$

$\Rightarrow \quad v - \sum_{i=1}^{n} \langle v, v_i \rangle v_i = 0_V$

$\Rightarrow \quad v = \sum_{i=1}^{n} \langle v, v_i \rangle v_i$

$\Rightarrow \quad v$ is the linear combination of $v_1, v_2, \ldots, v_n$

$\Rightarrow \quad v \in$ Subspace spanned by $\{v_1, v_2, \ldots, v_n\}$.

Conversely, let v be the linear combination of $v_1, v_2, \ldots, v_n$. Then,

$$v = \sum_{i=1}^{n} \langle v, v_i \rangle v_i \qquad \text{[By Theorem 8]}$$

$\Rightarrow \quad u = 0_V \qquad\qquad\qquad\qquad\qquad\qquad\qquad \left[\because \quad u = v - \sum_{i=1}^{n} \langle v, v_i \rangle v_i \right]$

$\Rightarrow \quad \|u\|^2 = 0$

$\Rightarrow \quad \|v\|^2 = \sum_{i=1}^{n} |\langle v, v_i \rangle|^2$

Q.E.D.

THEOREM-11 *(Another form of Bessel's Inequality)* If $\{v_1, v_2, \ldots, v_n\}$ is an orthogonal set of non-zero vectors in a unitary space V, and $v \in V$, then

$$\sum_{i=1}^{n} \frac{|\langle v, v_i \rangle|^2}{\|v_i\|^2} \leq \|v\|^2$$

PROOF. Let $w_i = \dfrac{v_i}{\|v_i\|}$, $i = 1, 2, \ldots, n$. Then, $\{w_1, w_2, \ldots, w_n\}$ is an orthonormal set of non-zero vectors in V. Proceeding as in Theorem 10, we obtain

$$\sum_{i=1}^{n} |\langle v, w_i \rangle|^2 \leq \|v\|^2 \qquad (i)$$

Now,

$$\langle v, w_i \rangle = \langle v, \frac{v_i}{\|v_i\|} \rangle = \frac{1}{\|v_i\|} \langle v, v_i \rangle \quad \text{for all } i = 1, 2, \ldots, n \qquad (ii)$$

From (i) and (ii), we get

$$\sum_{i=1}^{n} \frac{|\langle v, v_i \rangle|^2}{\|v_i\|^2} \leq \|v\|^2$$

PROJECTION OF A VECTOR *Let V be a unitary space. The projection of a vector v along a non-zero vector w is the scalar multiple cw of w such that $v - cw$ is orthogonal w.*

Now,

cw is the projection of vector v on vector w

$\Rightarrow \quad v - cw$ is orthogonal to w

$$\Rightarrow \quad \langle v-cw, w\rangle = 0$$
$$\Rightarrow \quad \langle v, w\rangle - c\langle w, w\rangle = 0$$
$$\Rightarrow \quad c = \frac{\langle v, w\rangle}{\langle w, w\rangle}$$

The projection of vector v along a non-zero vector w is denoted by $\text{proj}(v, w)$.

Thus, $\quad \text{proj}(v, w) = cw = \dfrac{\langle v, w\rangle}{\langle w, w\rangle} w$

The scalar $c = \dfrac{\langle v, w\rangle}{\langle w, w\rangle}$ is called the component of v along w or the *Fourier Coefficient of v with respect* to w.

EXAMPLE-5 *Find the Fourier coefficient (component) c and projection cw of $v = (3+4i, 2-3i)$ along $w = (5+i, 2i)$ in C^2 with respect to the standard inner product on C^2.*

SOLUTION For any $u = (z_1, z_2)$ and $v = (w_1, w_2)$ in C^2 the standard inner product is defined as
$$\langle u, v\rangle = z_1\overline{w}_1 + z_2\overline{w}_2$$
$\therefore \quad \langle v, w\rangle = (3+4i)\overline{(5+i)} + (2-3i)\overline{(2i)} = (3+4i)(5-i) + (2-3i)(-2i) = 13+13i$
and, $\quad \langle w, w\rangle = (5+i)(5-i) + (2i)\overline{(2i)} = 25+1+4 = 30$

$\therefore \quad$ Fourier coefficient of v with respect to $w = \dfrac{\langle v, w\rangle}{\langle w, w\rangle} = \dfrac{13+13i}{30} = \dfrac{13}{30} + \dfrac{13}{30}i$

Projection of v along $w = cw = \left(\dfrac{13}{30} + \dfrac{13}{30}i\right)(5+i, 2i) = \left(\dfrac{26}{15} + \dfrac{39}{15}i, \dfrac{-13}{15} + \dfrac{1}{15}i\right)$

THEOREM-12 *(Gram-Schmidt orthogonalisation process) Every finite dimensional unitary space has an orthonormal basis.*

PROOF. See proof of Theorem 1 on page 577

EXAMPLE-6 *Find an orthonormal basis of the subspace W of C^3 spanned by the vectors $v_1 = (1, i, 0)$ and $v_2 = (1, 2, 1-i)$.*

SOLUTION Clearly, v_1 and v_2 are linearly independent vectors. So, $B = \{v_1, v_2\}$ forms a basis of W.

Let $w_1 = v_1 = (1, i, 0)$ and $w_2 = v_2 - \dfrac{\langle v_2, w_1\rangle}{\langle w_1, w_1\rangle} w_1$.

Now,
$$\langle v_2, w_1\rangle = 1 \times 1 + 2 \times \overline{i} + (1-i) \times 0 = 1 - 2i$$
$$\langle w_1, w_1\rangle = 1 \times 1 + i \times \overline{i} + 0 \times 0 = 1+1+0 = 2$$

$$\therefore \quad w_2 = v_2 - \frac{\langle v_2, w_1 \rangle}{\langle w_1, w_1 \rangle} w_1 = (1, 2, 1-i) - \frac{1-2i}{2}(1, i, 0) = \left(\frac{1}{2} + i, 1 - \frac{1}{2}i, 1-i\right)$$

Thus, $\{w_1, w_2\}$ is an orthogonal basis of W.

Let us now normalize w_1 and w_2.

We have,
$$\|w_1\|^2 = \langle w_1, w_1 \rangle = 1 \times 1 + i \times \overline{i} + 0 = 2$$

$$\|w_2\|^2 = \langle w_2, w_2 \rangle = \left(\frac{1}{2} + i\right)\overline{\left(\frac{1}{2} + i\right)} + \left(1 - \frac{1}{2}i\right)\overline{\left(1 - \frac{1}{2}i\right)} + (1-i)\overline{(1-i)}$$

$$\Rightarrow \quad \|w_2\|^2 = \left(\frac{1}{4} + 1\right) + \left(1 + \frac{1}{4}\right) + (1+1) = \frac{9}{2}$$

$$\therefore \quad \|w_1\| = \sqrt{2} \text{ and } \|w_2\| = \frac{3}{\sqrt{2}}$$

Hence, $\left\{\dfrac{w_1}{\|w_1\|}, \dfrac{w_2}{\|w_2\|}\right\}$, i.e. $\left\{\left(\dfrac{1}{\sqrt{2}}, \dfrac{i}{\sqrt{2}}, 0\right), \left(\dfrac{1+2i}{3\sqrt{2}}, \dfrac{2-i}{3\sqrt{2}}, \dfrac{2-2i}{3\sqrt{2}}\right)\right\}$ is orthonormal basis of w.

EXAMPLE-7 Let $A = \begin{bmatrix} a & b \\ c & d \end{bmatrix}$, where a, b, c and d are complex numbers such that $|A| \neq 0$. Let $v_1 = (a, b)$, $v_2 = (c, d)$ and suppose that $v_1 \neq 0$. Apply Gram-Schmidt orthogonalisation process to $\{v_1, v_2\}$, using the standard inner product in C^2, to obtain orthogonal basis of C^2.

SOLUTION Let $w_1 = v_1 = (a, b)$.

$$w_2 = v_2 - \frac{\langle v_2, w_1 \rangle}{\langle w_1, w_1 \rangle} w_1$$

$$\Rightarrow \quad w_2 = (c, d) - \frac{c\bar{a} + d\bar{b}}{a\bar{a} + b\bar{b}}(a, b)$$

$$\Rightarrow \quad w_2 = (c, d) - \frac{(c\bar{a} + d\bar{b})}{|a|^2 + |b|^2}(a, b)$$

$$\Rightarrow \quad w_2 = \frac{1}{|a|^2 + |b|^2}\left(c|a|^2 + c|b|^2, d|a|^2 + d|b|^2\right) - \frac{1}{|a|^2 + |b|^2}\left(ca\bar{a} + da\bar{b}, c\bar{a}b + db\bar{b}\right)$$

$$\Rightarrow \quad w_2 = \frac{1}{|a|^2 + |b|^2}\left(ca\bar{a} + cb\bar{b} - ca\bar{a} - da\bar{b}, da\bar{a} + db\bar{b} - c\bar{a}b - db\bar{b}\right)$$

$$\Rightarrow \quad w_2 = \frac{1}{|a|^2+|b|^2}\left(cb\bar{b}-da\bar{b},\, da\bar{a}-c\bar{a}b\right)$$

$$\Rightarrow \quad w_2 = \frac{1}{|a|^2+|b|^2}\left((cb-ad)\bar{b},\, (ad-cb)\bar{a}\right)$$

$$\Rightarrow \quad w_2 = \frac{ad-bc}{|a|^2+|b|^2}\left(-\bar{b},\bar{a}\right) = \frac{|A|}{|a|^2+|b|^2}\left(-\bar{b},\bar{a}\right)$$

Hence, $\{w_1, w_2\}$ is an orthogonal basis of C^2.

EXAMPLE-8 Consider C^3, with standard inner product. Find an orthonormal basis for the subspace spanned by $v_1 = (1, 0, i)$ and $v_2 = (2, 1, 1+i)$.

SOLUTION Let $w_1 = v_1 = (1, 0, i)$.

$$w_2 = v_2 - \frac{\langle v_2, w_1\rangle}{\langle w_1, w_1\rangle}w_1$$

$$\Rightarrow \quad w_2 = (2, 1, 1+i) - \left(\frac{3-i}{2}, 0, \frac{1+3i}{2}\right) \quad \left[\begin{array}{l}\because \langle v_2, w_1\rangle = 2\times 1 + 1\times 0 + (1+i)(-i) = 3-i \\ \langle w_1, w_1\rangle = 1\times 1 + 0 + i\times -i = 2\end{array}\right]$$

$$\Rightarrow \quad w_2 = \left(\frac{1+i}{2}, 1, \frac{1-i}{2}\right)$$

Let us now normalize w_1 and w_2.

We have,

$$\|w_1\|^2 = \langle w_1, w_1\rangle = 1 + 0 + i\times -i = 2$$

$$\|w_2\|^2 = \langle w_2, w_2\rangle = \frac{1+i}{2}\times\frac{1-i}{2} + 1\times 1 + \frac{1-i}{2}\times\frac{1+i}{2} = 2$$

Let $w_1' = \frac{w_1}{\|w_1\|} = \left(\frac{1}{\sqrt{2}}, 0, \frac{i}{\sqrt{2}}\right)$ and $w_2' = \frac{w_2}{\|w_2\|} = \left(\frac{1+i}{2\sqrt{2}}, \frac{1}{\sqrt{2}}, \frac{1-i}{2\sqrt{2}}\right)$

Hence, $\{w_1', w_2'\}$ forms an orthonormal basis for the subspace spanned by $\{v_1, v_2\}$.

THEOREM-13 *Let S be a subspace of a finite dimensional unitary space V. If $B_1 = \{w_1, w_2, \ldots, w_r\}$ is an orthogonal basis of S. Then, B_1 can be extended to an orthogonal basis of V.*

OR

Every orthogonal list of vectors in a finite dimensional unitary space V can be extended to an orthogonal basis of V.

PROOF. Analogous to the proof of Theorem 2 on page 582

COROLLARY. *In a finite dimensional unitary space V, any orthonormal list of vectors is a part of an orthonormal basis for V.*

PROOF. Analogous to the proof of Corollary on page 582

THEOREM-14 *Let S be a subspace of a finite dimensional unitary space V. Then, $V = S \oplus S^\perp$.*

PROOF. Analogous to the proof of Theorem 3 on page 583

COROLLARY. *If S is a subspace of a finite dimensional unitary space V. Then, $\dim S^\perp = \dim V - \dim S$*

PROOF. Analogous to the proof of Corollary on page 584

THEOREM-15 *Let S be a subspace of a finite dimensional unitary space V. Then, $(S^\perp)^\perp = S$.*

PROOF. Analogous to the proof of Theorem 4 on page 584

THEOREM-16 *If S is a subset of a finite dimensional unitary space V. Then, $[S] = S^{\perp\perp}$*

PROOF. Analogous to the proof of Theorem 5 on page 585

THEOREM-17 *Let S be a subspace of a finite dimensional unitary space V. If $B_1 = \{v_1, v_2, \ldots, v_m\}$ is an orthonormal basis for S and $B_2 = \{w_1, w_2, \ldots, w_n\}$ is an orthonormal basis for $S^\perp$. Then, $B_1 \cup B_2$ is an orthonormal basis for V.*

PROOF. Analogous to the proof of Theorem 6 on page 585

THEOREM-18 *If S and T are subspaces of a finite dimensional unitary space V. Then,*
   (i) $(S+T)^\perp = S^\perp \cap T^\perp$ (ii) $(S \cap T)^\perp = S^\perp + T^\perp$

PROOF. Analogous to the proof of Theorem 7 on page 585

EXERCISE 8.3

1. Find the Fourier coefficient c and the projection cw of

   (i) $u = (3+i, 5-2i)$ along $w = (5+i, 1+i)$ in C^2.

   (ii) $u = (1-i, 3i, 1+i)$ along $w = (1, 2-i, 3+2i)$ in C^3.

2. Using Gram-Schmidt orthogonalisation process, find an orthogonal basis and an orthonormal basis for the subspace S of C^3 spanned by $v_1 = (1, i, 1)$ and $v_2 = (1+i, 0, 2)$

3. In C^3, find an orthonormal basis for the subspace spanned by the vector $v = (1, i, 1+i)$ and for its orthogonal complement.

4. Find an orthonormal basis of the subspace S of the unitary space C^4, spanned by $v_1 = (3, -i, 0, 0), v_2 = (2, 2i, 1, 0)$. Also, find an orthonormal basis of C^4 that contains this orthonormal basis of S.

ANSWERS

1. (i) $c = \dfrac{1}{28}(19 - 5i)$ (ii) $c = \dfrac{1}{19}(3 + 6i)$

2. $\left\{ w_1 = \left(\dfrac{1}{\sqrt{3}}, \dfrac{i}{\sqrt{3}}, \dfrac{1}{\sqrt{3}} \right), w_2 = \left(\dfrac{2i}{\sqrt{24}}, \dfrac{1-3i}{\sqrt{24}}, \dfrac{3-i}{\sqrt{24}} \right) \right\}$

8.5 HERMITIAN INNER PRODUCTS AND MATRICES

MATRIX REPRESENTATION OF A HERMITIAN INNER PRODUCT *Let V be an n dimensional unitary space with basis $B = \{v_1, v_2, \ldots, v_n\}$ and let $a_{ij} = \langle v_i, v_j \rangle$ for all $i, j, \in \underline{n}$. Then, $n \times n$ square matrix $A = [a_{ij}]$ is called the matrix representation of the inner product on V relative to the basis B*

We have,

$\langle v_i, v_j \rangle = \overline{\langle v_j, v_i \rangle}$ for all $i, j \in \underline{n}$

$\Rightarrow$ $\quad a_{ij} = \overline{a_{ji}}$ for all $i, j \in \underline{n}$

$\Rightarrow$ $\quad A = (\overline{A})^T$

$\Rightarrow$ $\quad A = A^*$, where $A^* = (\overline{A})^T$

$\Rightarrow$ $\quad A$ is a Hermitian matrix

Clearly, the matrix representation of a Hermitian inner product depends on to the inner product on V and the basis B for V.

Let B be an orthogonal basis for V. Then,

$\langle v_i, v_j \rangle = 0$ for all $i \neq j$

$\Rightarrow$ $\quad a_{ij} = 0$ for all $i \neq j$

$\Rightarrow$ $\quad A = [a_{ij}]$ is a diagonal matrix

If B is an orthonormal basis for V. Then,

$\langle v_i, v_j \rangle = \delta_{ij}$ for all $i, j \in \underline{n}$

$\Rightarrow$ $\quad A = [\delta_{ij}] = I$

Thus, the matrix representation of a Hermitian inner product relative to an orthonormal basis for V is the identity matrix.

EXAMPLE *Find the matrix A that represents the standard inner product on V relative to the ordered basis $B = \{1, i, 1 - i\}$.*

SOLUTION Let $v_1 = 1$, $v_2 = i$ and $v_3 = 1 - i$ and $A = [a_{ij}]$ be the matrix of t relative to basis B. Then, A is 3×3 matrix such that

$$a_{ij} = \langle v_i, v_j \rangle \quad \text{for all } i, j = 1, 2, 3.$$

Now,

$a_{11} = \langle v_1, v_1 \rangle = \langle 1, 1 \rangle = 1$

$a_{12} = \langle v_1, v_2 \rangle = \langle 1, i \rangle = 1 \times \overline{i} = -i$

$a_{13} = \langle v_1, v_3 \rangle = \langle 1, 1-i \rangle = 1 \times \overline{1-i} = 1+i$

$a_{21} = \langle v_2, v_1 \rangle = \overline{\langle v_1, v_2 \rangle} = i$

$a_{22} = \langle v_2, v_2 \rangle = \langle i, i \rangle = i \times \overline{i} = i \times -i = 1$

$a_{23} = \langle v_2, v_3 \rangle = \langle i, 1-i \rangle = i \times \overline{1-i} = i \times (1+i) = -1+i$

$a_{31} = \langle v_3, v_1 \rangle = \overline{\langle v_1, v_3 \rangle} = 1-i$

$a_{32} = \langle v_3, v_2 \rangle = \overline{\langle v_2, v_3 \rangle} = -1-i$

$a_{33} = \langle v_3, v_3 \rangle = \langle 1-i, 1-i \rangle = (1-i)(\overline{1-i}) = (1-i)(1+i) = 2$

$\therefore \quad A = \begin{bmatrix} a_{11} & a_{12} & a_{13} \\ a_{21} & a_{22} & a_{23} \\ a_{31} & a_{32} & a_{33} \end{bmatrix} = \begin{bmatrix} 1 & -i & 1+i \\ i & 1 & -1+i \\ 1-i & -1-i & 2 \end{bmatrix}$

THEOREM-1 *Let $B = \{w_1, w_2, \ldots, w_n\}$ be a basis for a unitary space V and let $A = [a_{ij}]$ be the matrix representing the Hermitian inner product on V. Then, for any $u, v \in V$*

$$\langle u, v \rangle = [u]^T A \overline{[v]}$$

where $[u]$ and $[v]$ are the coordinate column vectors of u and v relative to the basis B.

PROOF. Let $A = [a_{ij}]$ be the matrix representation of the Hermitian inner product relative to basis B. Then,

$$a_{ij} = \langle w_i, w_j \rangle$$

Let $u = \sum_{i=1}^{n} \lambda_i w_i$ and $v = \sum_{j=1}^{n} \mu_j w_j$. Then,

$$\langle u, v \rangle = \left\langle \sum_{i=1}^{n} \lambda_i w_i, \sum_{j=1}^{n} \mu_j w_j \right\rangle$$

$\Rightarrow \quad \langle u, v \rangle = \sum_{i=1}^{n} \sum_{j=1}^{n} \lambda_i \overline{\mu}_j \langle w_i, w_j \rangle$

$\Rightarrow \quad \langle u, v \rangle = \sum_{i=1}^{n} \sum_{j=1}^{n} \lambda_i \overline{\mu}_j a_{ij} = \sum_{i=1}^{n} \sum_{j=1}^{n} a_{ij} \lambda_i \overline{\mu}_j \quad \text{(i)}$

Now, $u = \sum_{i=1}^{n} \lambda_i w_i$ and $v = \sum_{j=1}^{n} \mu_j w_j$

$\Rightarrow \quad [u] = \begin{bmatrix} \lambda_1 \\ \lambda_2 \\ \vdots \\ \lambda_n \end{bmatrix}$ and $[v] = \begin{bmatrix} \mu_1 \\ \mu_2 \\ \vdots \\ \mu_n \end{bmatrix}$

$\therefore \quad [u]^T A \overline{[v]} = [\lambda_1, \lambda_2, \ldots, \lambda_n] \begin{bmatrix} a_{11} & a_{12} & \ldots & a_{1n} \\ a_{21} & a_{22} & \ldots & a_{2n} \\ \vdots & & & \\ a_{n1} & a_{n2} & \ldots & a_{nn} \end{bmatrix} \begin{bmatrix} \overline{\mu}_1 \\ \overline{\mu}_2 \\ \vdots \\ \overline{\mu}_n \end{bmatrix}$

$\Rightarrow \quad [u]^T A \overline{[v]} = \left[\sum_{i=1}^{n} a_{i1} \lambda_i \; \sum_{i=1}^{n} a_{i2} \lambda_i \; \ldots \; \sum_{i=1}^{n} a_{in} \lambda_i \right] \begin{bmatrix} \overline{\mu}_1 \\ \overline{\mu}_2 \\ \vdots \\ \overline{\mu}_n \end{bmatrix}$

$\Rightarrow \quad [u]^T A \overline{[v]} = \sum_{i=1}^{n} a_{i1} \lambda_i \overline{\mu}_1 + \sum_{i=1}^{n} a_{i2} \lambda_i \overline{\mu}_2 + \cdots + \sum_{i=1}^{n} a_{in} \lambda_i \overline{\mu}_n$

$\Rightarrow \quad [u]^T A \overline{[v]} = \sum_{i=1}^{n} \sum_{j=1}^{n} a_{ij} \lambda_i \overline{\mu}_j \qquad\qquad\qquad\qquad\qquad\qquad\qquad\qquad (ii)$

From (i) and (ii), we obtain $\langle u, v \rangle = [u]^T A \overline{[v]}$

Q.E.D.

THEOREM-2 *Let A be a Hermitian matrix such that $X^T A \overline{X}$ is real and positive for every non-zero vector $X \in C^n$. Then, $\langle u, v \rangle = u^T A \overline{v}$ is a Hermitian inner product on C^n.*

PROOF. We observe the following properties:

HI1: *Conjugate Symmetry* Let $u, v \in C^n$. Then,

$$\langle u, v \rangle = u^T A \overline{v} \text{ and } \langle v, u \rangle = v^T A \overline{u}.$$

$\therefore \quad \overline{\langle v, u \rangle} = \overline{(v^T A \overline{u})} = \overline{v}^T \overline{A} u = \left(\overline{v}^T \overline{A} u \right)^T \qquad\qquad [\because \; \overline{v}^T \overline{A} u \text{ is a scalar}]$

$\Rightarrow \quad \overline{\langle v, u \rangle} = u^T (\overline{A})^T (\overline{v}^T)^T = u^T A^* \overline{v} = u^T A \overline{v} \qquad\qquad [\because \; A^* = A]$

$\Rightarrow \quad \overline{\langle v, u \rangle} = \langle u, v \rangle$

HI2: *Positive Definiteness* For any $0 \neq u \in C^n$

$$\langle u, u \rangle = u^T A \overline{u} > 0 \qquad \text{[Given]}$$

HI 3: Linearity For any $u, v, w \in C^n$ and $a, b \in C$

$$\langle au+bv, w\rangle = (au+bv)^T A \overline{w}$$
$$= (au^T + bv^T) A \overline{w}$$
$$= a(u^T A \overline{w}) + b(v^T A \overline{w})$$
$$= a\langle u, w\rangle + b\langle v, w\rangle$$

Hence, $\langle u, v\rangle = u^T A \overline{v}$ is an inner product on C^n.

Q.E.D.

EXERCISE 8.4

1. Find the matrix A that represents the standard Hermitian inner product on V relative to the basis $B = \{1, 1+i, 1-2i\}$
2. Let A be the matrix that represents a Hermitian inner product on V. Then prove that A is Hermitian, and $X^T A X$ is real and positive for any non-zero vector in C^n.

ANSWERS

1. $A = \begin{bmatrix} 1 & 1-i & 1+2i \\ 1+i & 2 & -1+3i \\ 1-2i & -1-3i & 5 \end{bmatrix}$

8.6 UNITARY AND NORMAL MATRICES

In this section, we shall learn about some special type of matrices over the field C of complex numbers. These matrices are analogous to the kinds of matrices over the field R of real numbers as discussed in Sections 6.11 and 6.12.

HERMITIAN MATRIX A square matrix A over C is said to be a *Hermitian matrix*, if its conjugate transpose is equal to A. That is, $\left(\overline{A}\right)^T = A$.

For example, matrix $A = \begin{bmatrix} 3 & 2-3i & i \\ 2+3i & -4 & 1-i \\ -i & 1+i & 2 \end{bmatrix}$ is a Hermitian matrix. Because,

$$\left(\overline{A}\right)^T = \begin{bmatrix} 3 & 2+3i & -i \\ 2-3i & -4 & 1+i \\ i & 1-i & 2 \end{bmatrix}^T = \begin{bmatrix} 3 & 2-3i & i \\ 2+3i & -4 & 1-i \\ -i & 1+i & 2 \end{bmatrix} = A$$

If $A = [a_{ij}]$ is a Hermitian matrix. Then,

$$\left(\overline{A}\right)^T = A$$

$$\Rightarrow \quad a_{ij} = \left(\left(\overline{A}\right)^T\right)_{ij} \quad \text{for all } i, j \in \underline{n}$$

$\Rightarrow \quad a_{ij} = (\overline{A})_{ji}$ for all $i, j \in \underline{n}$

$\Rightarrow \quad a_{ij} = \overline{a_{ji}}$ for all $i, j \in \underline{n}$

$\Rightarrow \quad a_{ii} = \overline{a_{ii}}$ for all $i \in \underline{n}$

$\Rightarrow \quad a_{ii}$ is real for all $i \in \underline{n}$

Thus, diagonal elements of a Hermitian matrix are all real.

Note: $(\overline{A})^T$ is generally denoted by A^*. So, A is Hermitian if $A^* = A$.

SKEW-HERMITIAN MATRIX *A square matrix A over C is skew-Hermitian if its conjugate transpose is equal to $-A$. That is, $(\overline{A})^T = -A$ or, $A^* = -A$.*

For example, $A = \begin{bmatrix} 2i & 1-i & 3+2i \\ -1-i & i & -1+i \\ -3+2i & 1+i & 0 \end{bmatrix}$ is a skew-Hermitian matrix. Because

$(\overline{A})^T = \begin{bmatrix} -2i & 1+i & 3-2i \\ -1+i & -i & -1-i \\ -3-2i & 1-i & 0 \end{bmatrix}^T = \begin{bmatrix} -2i & -1+i & -3-2i \\ 1+i & -i & 1-i \\ 3-2i & -1-i & 0 \end{bmatrix}$

$\Rightarrow \quad (\overline{A})^T = - \begin{bmatrix} 2i & 1-i & 3+2i \\ -1-i & i & -1+i \\ -3+2i & 1+i & 0 \end{bmatrix} = -A$

If $A = [a_{ij}]$ is a skew-Hermitian matrix over C, then

$(\overline{A})^T = -A$

$\Rightarrow \quad ((\overline{A})^T)_{ij} = -(A)_{ij}$ for all $i, j \in \underline{n}$

$\Rightarrow \quad \overline{a_{ji}} = -a_{ij}$ for all $i, j \in \underline{n}$

$\Rightarrow \quad a_{ii} = -\overline{a_{ii}}$ for all $i \in \underline{n}$

$\Rightarrow \quad a_{ii}$ is purely imaginary for all $i \in \underline{n}$

Thus, the diagonal elements of A are purely imaginary.

REMARK. *The definitions of Hermitian and skew-Hermitian matrices are analogous to the definitions of symmetric and skew-symmetric matrices over R.*

UNITARY MATRIX *A square matrix over C is said to be a unitary matrix, if*

$$AA^* = I = A^*A$$

That is, if $A^* = A^{-1}$.

It is evident from the above definition that A is non-singular.

EXAMPLE-1 *The determinant of a unitary matrix has absolute value 1.*

SOLUTION Let A be a unitary matrix. Then,

$$AA^* = I$$
$$\Rightarrow |AA^*| = 1$$
$$\Rightarrow |A||A^*| = 1$$
$$\Rightarrow |A||A| = 1 \qquad \left[\because |A^*| = |\overline{A}^T| = |\overline{A}| = |A|\right]$$
$$\Rightarrow |A|^2 = 1 \Rightarrow |A| = \pm 1$$

EXAMPLE-2 *The transpose of a unitary matrix is a unitary matrix.*

SOLUTION Let A be a unitary matrix. Then,

$$AA^* = I = A^*A$$
$$\Rightarrow A\left(\overline{A}\right)^T = I = \left(\overline{A}\right)^T A$$
$$\Rightarrow \left(A\left(\overline{A}\right)^T\right)^T = I = \left(\left(\overline{A}\right)^T A\right)^T$$
$$\Rightarrow \overline{A}A^T = I = A^T\overline{A}$$
$$\Rightarrow \left(\overline{A^T}\right)^T A^T = I = A^T \left(\overline{A^T}\right)^T$$
$$\Rightarrow \left(A^T\right)^* A^T = I = A^T \left(A^T\right)^*$$
$$\Rightarrow A^T \text{ is a unitary matrix.}$$

THEOREM-1 *The following conditions on a square matrix A over C are equivalent:*

(i) *A is unitary.*

(ii) *The rows of A form an orthonormal set.*

(iii) *The columns of A form an orthonormal set.*

PROOF. Let A be an $n \times n$ unitary matrix over C having $R_1, R_2, \ldots, R_n$ as its rows and $C_1, C_2, \ldots, C_n$ as its columns, i.e.

$$A = \begin{bmatrix} R_1 \\ R_2 \\ \vdots \\ R_n \end{bmatrix} \text{ and, } A = [C_1 \ C_2 \ldots C_n]$$

(i) ⇔ (ii)

Let A be a unity matrix

⇔ $AA^* = I$

⇔ $(AA^*)_{ij} = \delta_{ij}$ for all $i, j \in \underline{n}$

⇔ (ith row of A) (jth column of $\overline{A}^T$) = δ_{ij} for all $i, j \in \underline{n}$

⇔ (ith row of A) (jth row of $\overline{A}$) = δ_{ij} for all $i, j \in \underline{n}$

⇔ $R_i \overline{R}_j = \delta_{ij}$ for all $i, j \in \underline{n}$

⇔ $R_1, R_2, \ldots, R_n$ form an orthonormal set with respect to standard Hermitian inner product on C^n.

Hence, (i) ⇔ (ii)

(i) ⇔ (iii)

Let A be a unitary matrix.

⇔ $A^*A = I$

⇔ $(A^*A)_{ij} = \delta_{ij}$ for all $i, j \in \underline{n}$

⇔ (ith row of A^*) (jth column of A) = δ_{ij} for all $i, j \in \underline{n}$

⇔ (ith column of $\overline{A}$) (jth column of A) = δ_{ij} for all $i, j \in \underline{n}$

⇔ $\overline{C}_i C_j = \delta_{ij}$ for all $i, j \in \underline{n}$

⇔ $C_j \overline{C}_i = \delta_{ji}$ for all $i, j \in \underline{n}$

⇔ $C_1, C_2, \ldots, C_n$ form an orthonormal set with respect to standard Hermitian inner product on C^n.

Thus, (i) ⇔ (iii).

Hence, given statements are equivalent.

Q.E.D.

EXAMPLE-3 *The eigenvalues of a unitary matrix are of unit modulus.*

SOLUTION Let A be a unitary matrix and λ be an eigenvalue of A. Then, there exists a non-zero vector $X \in C^n$ such that

$$AX = \lambda X \qquad \text{(i)}$$

$\Rightarrow \quad \overline{(AX)}^T = \overline{(\lambda X)}^T$

$\Rightarrow \quad \overline{X}^T (\overline{A})^T = \overline{\lambda} \overline{X}^T \qquad \text{(ii)}$

From (i) and (ii), we obtain

$$\left(\overline{X}^T \overline{A}^T\right)(AX) = \left(\overline{\lambda}\overline{X}^T\right)(\lambda X)$$

$$\Rightarrow \quad \overline{X}^T \left(\overline{A}^T A\right) X = \overline{\lambda}\lambda \left(\overline{X}^T X\right)$$

$$\Rightarrow \quad \overline{X}^T I X = |\lambda|^2 \overline{X}^T X \qquad [\because \; A \text{ is unitary} \;\; \therefore \;\; \overline{A}^T A = I]$$

$$\Rightarrow \quad \overline{X}^T X = |\lambda|^2 \overline{X}^T X$$

$$\Rightarrow \quad |\lambda|^2 = 1 \qquad [\because \; \overline{X}^T X = x_1^2 + x_2^2 + \cdots + x_n^2 \neq 0]$$

$$\Rightarrow \quad |\lambda| = 1$$

THEOREM-2 *Let V be a finite dimensional unitary space. Let $B_1 = \{u_1, u_2, \ldots, u_n\}$ and $B_2 = \{v_1, v_2, \ldots, v_n\}$ be orthonormal bases for V. Then the change-of-basis matrix from basis B_1 to basis B_2 is unitary.*

PROOF. Let A be the change-of-basis matrix from basis B_1 to basis B_2.
Since B_2 is an orthonormal basis for V. Therefore,

$$v_i = \sum_{r=1}^{n} b_{ir} u_r \quad \text{for } i = 1, 2, \ldots, n$$

$$\Rightarrow \quad v_j = \sum_{s=1}^{n} b_{js} u_s \quad \text{for } j = 1, 2, \ldots, n$$

Let A be the change of basis matrix from basis B_1 to basis B_2.
Then, $A = B^T$, where $B = [b_{ij}]$.

For any $i \in \underline{n}, j \in \underline{n}$

$$\langle v_i, v_j \rangle = \left\langle \sum_{r=1}^{n} b_{ir} u_r, \sum_{s=1}^{n} b_{js} u_s \right\rangle \qquad [\because \; B_2 \text{ is orthonormal}]$$

$$\Rightarrow \quad \delta_{ij} = \sum_{r=1}^{n} \sum_{s=1}^{n} b_{ir} \overline{b_{js}} \langle u_r, u_s \rangle$$

$$\Rightarrow \quad \delta_{ij} = \sum_{r=1}^{n} \sum_{s=1}^{n} b_{ir} \overline{b_{js}} \delta_{rs} \qquad [\because \; B_1 \text{ is orthonormal}]$$

$$\Rightarrow \quad \delta_{ij} = \sum_{r=1}^{n} b_{ir} \overline{b_{jr}}$$

$$\Rightarrow \quad \delta_{ij} = \sum_{r=1}^{n} (B)_{ir} \left((\overline{B})^T\right)_{jr}, \quad \text{where } B = [b_{ij}]$$

$$\Rightarrow \quad B \left(\overline{B}\right)^T = I$$

$$\Rightarrow \quad \left(B \left(\overline{B}\right)^T\right)^T = I$$

$$\Rightarrow \quad \overline{B} B^T = I$$

$$\Rightarrow \quad \left(\overline{A}^T\right) A = I \qquad [\because \; B^T = A]$$

$\Rightarrow \quad \overline{A}^T A = I$

$\Rightarrow \quad A^* A = I$

$\Rightarrow \quad A$ is unitary matrix

Q.E.D.

THEOREM-3 Let $B_1 = \{u_1, u_2, \ldots, u_n\}$ be an orthonormal basis of a unitary space V. Let $A = [a_{ij}]$ be a unitary matrix, and let

$$v_i = \sum_{j=1}^{n} a_{ji} u_j, \ i = 1, 2, \ldots, n$$

Then $B_2 = \{v_1, v_2, \ldots, v_n\}$ is an orthonormal basis for V.

PROOF. For any $i, j \in \underline{n}$

$$\langle v_i, v_j \rangle = \left\langle \sum_{r=1}^{n} a_{ri} u_r, \sum_{s=1}^{n} a_{sj} u_s \right\rangle$$

$\Rightarrow \quad \langle v_i, v_j \rangle = \sum_{r=1}^{n} \sum_{s=1}^{n} a_{ri} \overline{a_{sj}} \langle u_r, u_s \rangle$

$\Rightarrow \quad \langle v_i, v_j \rangle = \sum_{r=1}^{n} \sum_{s=1}^{n} a_{ri} \overline{a_{sj}} \delta_{rs}$ $\quad [\because B$ is an orthonormal basis$]$

$\Rightarrow \quad \langle v_i, v_j \rangle = \sum_{r=1}^{n} a_{ri} \overline{a_{rj}}$

$\Rightarrow \quad \langle v_i, v_j \rangle = \sum_{r=1}^{n} (A)_{ri} (\overline{A})_{jr}^T$

$\Rightarrow \quad \langle v_i, v_j \rangle = \sum_{r=1}^{n} (\overline{A}^T)_{jr} (A)_{ri}$

$\Rightarrow \quad \langle v_i, v_j \rangle = \left(\overline{A}^T A\right)_{ji}$

$\Rightarrow \quad \langle v_i, v_j \rangle = \delta_{ji} = \delta_{ij}$ $\quad [\because A$ is a unitary matrix $\therefore \overline{A}^T A = I]$

$\Rightarrow \quad B_2 = \{v_1, v_2, \ldots, v_n\}$ is an orthonormal set V.

Since every orthonormal set of n vectors in an n-dimensional vector space is its basis. Hence, B_2 is an orthonormal basis for V.

Q.E.D.

REMARK. In the above theorem A is change-of-basis matrix from basis B_1 to basis B_2. Thus, if the change-of-basis matrix from one basis to another is unitary and one of the bases is orthonormal, then the other basis is also orthonormal.

UNITARILY EQUIVALENT MATRICES Two matrices A and B over C are said to be unitarily equivalent if there exists a unitary matrix P such that $B = P^*AP$.

The relation "unitarily equivalent" is an equivalence relation on the set of all square matrices over C.

EXAMPLE-4 *Find a unitary matrix whose first row is a multiple of* $(1, 1-i)$.

SOLUTION Let A be the required matrix whose first row v_1 is a multiple of $v = (1, 1-i)$ and second row is $v_2 = (a,b)$.

Let $v_1 = \lambda v$, where $\lambda \in R$.

Since rows of a unitary matrix form an orthonormal basis. Therefore, $\{v_1, v_2\}$ is an orthonormal set in C^2.

$\therefore \quad \langle v_1, v_1 \rangle = 1, \langle v_2, v_2 \rangle = 1, \langle v_1, v_2 \rangle = 0.$

Now,

$$\langle v_1, v_1 \rangle = 1$$
$$\Rightarrow \quad \langle \lambda v, \lambda v \rangle = 1$$
$$\Rightarrow \quad \lambda^2 \langle v, v \rangle = 1 \qquad [\because \lambda \in R]$$
$$\Rightarrow \quad \lambda^2 (1 \times 1 + (1-i)(1+i)) = 1$$
$$\Rightarrow \quad \lambda^2 (1+1+1) = 1 \Rightarrow \lambda = \pm \frac{1}{\sqrt{3}}$$

Let us take $\lambda = \frac{1}{\sqrt{3}}$.

$\therefore \quad v_1 = \frac{1}{\sqrt{3}}(1, 1-i) = \left(\frac{1}{\sqrt{3}}, \frac{1-i}{\sqrt{3}} \right)$

Now, $\langle v_2, v_2 \rangle = 1$ and $\langle v_1, v_2 \rangle = 0$

$\Rightarrow \quad |a|^2 + |b|^2 = 1$ and $\frac{1}{\sqrt{3}} \times \bar{a} + \frac{1-i}{\sqrt{3}} \times \bar{b} = 0$

$\Rightarrow \quad |a|^2 + |b|^2 = 1$ and $a + (1+i)b = 0$

$\Rightarrow \quad |a|^2 + |b|^2 = 1$ and $a = -(1+i)b$

$\Rightarrow \quad |a|^2 + |b|^2 = 1$ and $|a| = \sqrt{2}|b|$

$\Rightarrow \quad 3|b|^2 = 1 \Rightarrow |b| = \frac{1}{\sqrt{3}}$

Let $b = \dfrac{1}{\sqrt{3}}$. Then, $a = -\dfrac{1+i}{\sqrt{3}}$

$\therefore \quad v_2 = \left(-\dfrac{1+i}{\sqrt{3}}, \dfrac{1}{\sqrt{3}}\right)$

Hence, $A = \begin{bmatrix} \dfrac{1}{\sqrt{3}} & \dfrac{1-i}{\sqrt{3}} \\ -\dfrac{1+i}{\sqrt{3}} & \dfrac{1}{\sqrt{3}} \end{bmatrix}$

EXAMPLE-5 *Show that unitarily equivalent matrices are similar matrices.*

SOLUTION Let A and B be two $n \times n$ matrices over C such that B is unitarily equivalent to A. Then there exists an $n \times n$ unitary matrix P such that

$$B = P^* A P$$
$\Rightarrow \quad B = (\overline{P})^T A P$
$\Rightarrow \quad B = P^{-1} A P \qquad [\because \; P \text{ is unitary} \;\therefore\; \overline{P}^T = P^{-1}]$
$\Rightarrow \quad B$ is similar to A.

NORMAL MATRIX *A square matrix A over C is said to be normal if it commutes with its conjugate transpose. i.e. if $AA^* = A^*A$*

For example, if $A = \begin{bmatrix} 1 & 1 \\ i & 3+2i \end{bmatrix}$, then $A^* = \begin{bmatrix} 1 & -i \\ 1 & 3-2i \end{bmatrix}$

$\therefore \quad AA^* = \begin{bmatrix} 1 & 1 \\ i & 3+2i \end{bmatrix} \begin{bmatrix} 1 & -i \\ 1 & 3-2i \end{bmatrix} = \begin{bmatrix} 2 & 3-3i \\ 3+3i & 14 \end{bmatrix}$

and, $\quad A^*A = \begin{bmatrix} 1 & -i \\ 1 & 3-2i \end{bmatrix} \begin{bmatrix} 1 & 1 \\ i & 3+2i \end{bmatrix} = \begin{bmatrix} 2 & 3-3i \\ 3+3i & 14 \end{bmatrix}$

$\therefore \quad AA^* = A^*A$

So, A is a normal matrix.

REMARK. *Every unitary matrix is a normal matrix but the converse is not necessarily true.*

EXERCISE 8.5

1. Which of the following matrices are unitary and Hermitian?

    (i) $A = \begin{bmatrix} 1 & i \\ i & 1 \end{bmatrix}$ (ii) $B = \begin{bmatrix} 2 & 1-3i \\ 1+3i & 5 \end{bmatrix}$ (iii) $C = \begin{bmatrix} 1+i & 1-i \\ 1-i & 1+i \end{bmatrix}$

2. Determine which of the following matrices is normal:

    (i) $A = \begin{bmatrix} 1 & i \\ 0 & 1 \end{bmatrix}$ (ii) $B = \begin{bmatrix} 1 & i \\ 1 & 2+i \end{bmatrix}$

3. Find a unitary matrix whose first row is a multiple of $(1, -i, 1-i)$.
4. Prove that the inverse of a unitary matrix is a unitary matrix.
5. Prove that the product of two unitary matrices is a unitary matrix.
6. Prove that the set of all unitary matrices over C forms a group under multiplication of matrices as a binary operation.
7. Show that the relation "Unitarily equivalent" is an equivalence relation on the set of all square matrices over C.
8. Find a unitary matrix which is not orthogonal, and find an orthogonal matrix which is not unitary.

ANSWERS

1. (i) Neither Hermitian nor unitary (ii) Hermitian only (iii) Neither Hermitian nor unitary
2. (ii)
3. $\begin{bmatrix} \dfrac{1}{2} & \dfrac{-i}{2} & \dfrac{1-i}{2} \\ \dfrac{i}{\sqrt{2}} & \dfrac{-1}{\sqrt{2}} & 0 \\ \dfrac{1}{2} & \dfrac{-i}{2} & \dfrac{-1+i}{2} \end{bmatrix}$

Chapter 9

Linear Operators on Unitary Spaces

9.1 INTRODUCTION

Similar to the operators on inner product spaces, we have operators on Unitary Spaces. In this Chapter, we will learn about various operators and their properties on unitary spaces. Note that the definitions of unitary space homomorphisms, monomorphisms, epimorphisms, isomorphisms and automorphisms are exactly identical to the corresponding definitions in inner product spaces. In the previous Chapter, we have defined linear functionals (forms) and conjugate linear functionals (forms) on a vector space over C. Let us first find how these forms are connected or associated to Hermitian products on a unitary space.

9.2 LINEAR FORMS AND UNITARY SPACES

In Chapter 7, we have learnt that corresponding to each linear form $\varphi : V \to R$ there exists a unique vector $u \in V$ such that $\varphi(v) = \langle v, u \rangle$ for all $u \in V$. A similar result hold good in unitary spaces for linear forms and conjugate linear forms as proved below.

THEOREM-1 *Let V be a finite dimensional unitary space and $\phi : V \to C$ be a linear form on V. Then there exists a unique vector $u \in V$ such that $\varphi(v) = \langle v, u \rangle$ for all $v \in V$.*

PROOF. Let $\{w_1, w_2, \ldots, w_n\}$ be an orthonormal basis for the unitary space V. Let

$$u = \overline{\varphi(w_1)}w_1 + \overline{\varphi(w_2)}w_2 + \cdots + \overline{\varphi(w_n)}w_n = \sum_{i=1}^{n} \overline{\varphi(w_i)}w_i$$

Clearly, $u \in V$.

Let $f : V \to C$ be a function defined by
$f(v) = \langle v, u \rangle$ for all $v \in V$.

For any $v_1, v_2 \in V$ and $a, b \in C$

$$f(av_1 + bv_2) = \langle av_1 + bv_2, u \rangle$$
$\Rightarrow \quad f(av_1 + bv_2) = \langle av_1, u \rangle + \langle bv_2, u \rangle$
$\Rightarrow \quad f(av_1 + bv_2) = a\langle v_1, u \rangle + b\langle v_2, u \rangle$
$\Rightarrow \quad f(av_1 + bv_2) = af(v_1) + bf(v_2)$

$\therefore \quad f$ is a linear form on V.

For $i = 1, 2, \ldots, n$, we have
$$f(w_i) = \langle w_i, u \rangle$$
$\Rightarrow \quad f(w_i) = \left\langle w_i, \sum_{j=1}^{n} \overline{\varphi(w_j)}\, w_j \right\rangle$

$\Rightarrow \quad f(w_i) = \sum_{j=1}^{n} \varphi(w_j) \langle w_i, w_j \rangle$

$\Rightarrow \quad f(w_i) = \sum_{j=1}^{n} \varphi(w_j)\, \delta_{ij} = \varphi(w_i)$

Thus, f and φ agree on each basis vector of V. Therefore, $f = \varphi$.

Thus, corresponding to a linear functional (linear form) φ on V there exists a vector $u \in V$ such that $\varphi(v) = \langle v, u \rangle$ for all $v \in V$.

In order to prove the uniqueness of u, suppose u' is another vector in V such that
$$\varphi(v) = \langle v, u' \rangle \quad \text{for all } v \in V$$

Then,

$\langle v, u \rangle = \langle v, u' \rangle \quad \text{for all } v \in V$
$\Rightarrow \quad \langle v, u - u' \rangle = 0 \quad \text{for all } v \in V$
$\Rightarrow \quad \langle u - u', u - u' \rangle \qquad\qquad\qquad\qquad\qquad\qquad$ [Taking $v = u - u'$]
$\Rightarrow \quad u - u' = 0_V$
$\Rightarrow \quad u = u'$

Hence, for each linear form $\varphi : V \to C$ there exists unique vector $u \in V$ such that
$$\varphi(v) = \langle v, u \rangle \quad \text{for all } v \in V.$$

Q.E.D.

EXAMPLE Consider the linear form $\varphi : C^3 \to C$ defined by
$$\varphi(x, y, z) = 2ix + (3 + 2i)y + (2 + i)z$$

Find a vector $u \in C^3$ such that $\varphi(v) = \langle v, u \rangle$ for all $v \in C^3$. Here, the Hermitian inner product on C^3 is the standard inner product.

SOLUTION By Theorem 1, we have

$$u = \sum_{i=1}^{n} \overline{\varphi(w_i)}\, w_i, \text{ where } B = \{w_1, w_2, \ldots, w_n\} \text{ is an orthonormal basis of } V.$$

Here, $B = \{e_1^{(3)}, e_2^{(3)}, e_3^{(3)}\}$ is the standard basis of C^3

$\therefore \qquad u = \overline{\varphi\left(e_1^{(3)}\right)} e_1^{(3)} + \overline{\varphi\left(e_2^{(3)}\right)} e_2^{(3)} + \overline{\varphi\left(e_3^{(3)}\right)} e_3^{(3)}$

$\Rightarrow \qquad u = -2i(1,\ 0,\ 0) + (3 - 2i)(0,\ 1,\ 0) + (2 - i)(0,\ 0,\ 1)$

$\Rightarrow \qquad u = (-2i,\ 3 - 2i,\ 2 - i)$

Hence, required vector $u = (-2i,\ 3 - 2i,\ 2 - i)$.

THEOREM-2 *Let V be a finite dimensional unitary space and $\varphi : V \to C$ be a conjugate linear form on V. Then there exists a unique vector $u \in V$ such that $\varphi(v) = \langle u,\ v \rangle$ for all $v \in V$.*

PROOF. Proceed as in theorem 1 by taking $u = \sum_{i=1}^{n} \varphi(w_i)\, w_i$, where $B = \{w_1, w_2, \ldots, w_n\}$ is an orthonormal basis for V.

In the above discussion, we have learnt that corresponding to each linear form φ on a finite dimensional unitary space V there exists a unique vector $u \in V$ such that $\varphi(v) = \langle v,\ u \rangle$ for all $v \in V$ and corresponding to each conjugate linear form φ on V there exists a unique vector $u \in V$ such that $\varphi(v) = \langle u,\ v \rangle$ for all $v \in V$. Now, we shall show that each basis vector of an orthonormal basis of a unitary space V determines a linear form on it.

Consider an orthonormal ordered basis $B = \{b_1, b_2, \ldots, b_n\}$ of a finite dimensional unitary space V and let v be an arbitrary vector in V. Then,

$$v = \sum_{i=1}^{n} \lambda_i b_i \quad \text{for some } \lambda_i \in C$$

Let $X = \{x_1, x_2, \ldots, x_n\}$ be the basis dual to B. Then for any $j = 1, 2, \ldots, n$

$$\langle v,\ b_j \rangle = \left\langle \sum_{i=1}^{n} \lambda_i b_i,\ b_j \right\rangle = \sum_{i=1}^{n} \lambda_i \langle b_i,\ b_j \rangle = \sum_{i=1}^{n} \lambda_i \delta_{ij} = \lambda_j$$

and, $\qquad x_j(v) = x_j \left(\sum_{i=1}^{n} \lambda_i b_i \right) = \sum_{i=1}^{n} \lambda_i\, x_j(b_i) = \sum_{i=1}^{n} \lambda_i \delta_{ji} = \lambda_j$

$\therefore \qquad x_j(v) = \langle v,\ b_j \rangle \quad \text{for all } v \in V$

Thus, corresponding to each basis vector b_j of an orthonormal ordered basis of a unitary space V there exists a linear form $x_j : V \to C$ such that $x_j(v) = \langle v,\ b_j \rangle$.

680 • *Theory and Problems of Linear Algebra*

REMARK. *An inner product space is linear in both the arguments. Therefore, as we have learnt in chapter 7, that an inner product space is isomorphic to its dual. Also, we have learnt that an inner product space is its own dual, i.e. an inner product space can be identified with its dual. However, this is not true for unitary spaces.*

9.3 ADJOINT OPERATORS

In this section, we shall show that each linear operator on a finite dimensional unitary space determines a unique linear operator on it. This operator is known as the adjoint of the given operator.

THEOREM-1 *Let t be a linear operator on a finite dimensional unitary space V. Then there exists a unique linear operator t^* on V such that*

$$\langle t(u), v \rangle = \langle u, t^*(v) \rangle \text{ for every } u, v \in V$$

PROOF. Let V be an n dimensional unitary space and let v be an arbitrary but fixed vector in V.

Consider the mapping $\varphi : V \to C$ given by

$$\varphi(u) = \langle t(u), v \rangle \quad \text{for all} \quad u \in V$$

For any $u_1, u_2 \in V$ and $a, b \in C$

$$\varphi(au_1 + bu_2) = \langle t(au_1 + bu_2), v \rangle$$
$$\Rightarrow \quad \varphi(au_1 + bu_2) = \langle at(u_1) + bt(u_2), v \rangle \quad\quad [\because \ t \text{ is linear}]$$
$$\Rightarrow \quad \varphi(au_1 + bu_2) = a\langle t(u_1), v \rangle + b\langle t(u_2), v \rangle$$
$$\Rightarrow \quad \varphi(au_1 + bu_2) = a\varphi(u_1) + b\,\phi(u_2)$$

$\therefore$ φ is a linear functional (linear form) on V.

Consequently, by Theorem 1 (Section 9.2), there exists a unique vector $v' \in V$ such that

$$\varphi(u) = \langle u, v' \rangle \quad \text{for all } u \in V$$
$$\Rightarrow \quad \langle t(u), v \rangle = \langle u, v' \rangle \quad \text{for all } u \in V$$

Thus, for each $v \in V$ there exists unique $v' \in V$ such that

$$\langle t(u), v \rangle = \langle u, v' \rangle \quad \text{for all } u \in V \tag{i}$$

This provides a function $t^* : V \to V$ associating v to v' i.e. $t^*(v) = v'$.

$\therefore \quad \langle t(u), v \rangle = \langle u, t^*(v) \rangle \quad \text{for all } u, v \in V \quad\quad$ [Replacing v' by $t^*(v)$ in (i)]

Thus, corresponding to every linear operator t on V, there exists a mapping $t^* : V \to V$ such that

$$\langle t(u), v \rangle = \langle u, t^*(v) \rangle \quad \text{for all } u, v \in V \tag{ii}$$

We shall now show that t^* is linear.
For any $v_1, v_2 \in V$ and $a, b \in C$, we have

$$\langle u, t^*(av_1 + bv_2) \rangle = \langle t(u), av_1 + bv_2 \rangle \quad \text{for all } u \in V$$
$$= \langle t(u), av_1 \rangle + \langle t(u), bv_2 \rangle \quad \text{for all } u \in V$$
$$= \overline{a}\langle t(u), v_1 \rangle + \overline{b}\langle t(u), v_2 \rangle \quad \text{for all } u \in V$$
$$= \overline{a}\langle u, t^*(v_1) \rangle + \overline{b}\langle u, t^*(v_2) \rangle \quad \text{for all } u \in V$$
$$= \langle u, at^*(v_1) \rangle + \langle u, bt^*(v_2) \rangle \quad \text{for all } u \in V$$
$$= \langle u, at^*(v_1) + bt^*(v_2) \rangle \quad \text{for all } u \in V$$

$\therefore \quad t^*(av_1 + bv_2) = at^*(v_1) + bt^*(v_2)$ for all $v_1, v_2 \in V$ and $a, b \in C$

Thus, t^* is a linear operator on V.

In order to prove the uniqueness of t, let T be another operator on V satisfying (ii). Then,

$$\langle t(u), v \rangle = \langle u, T^*(v) \rangle \quad \text{for all } u, v \in V$$
$\Rightarrow \quad \langle u, t^*(v) \rangle = \langle u, T^*(v) \rangle \quad \text{for all } u, v \in V$
$\Rightarrow \quad \langle u, t^*(v) - T^*(v) \rangle = 0 \quad \text{for all } u, v \in V$
$\Rightarrow \quad \langle u, (t^* - T^*)(v) \rangle = 0 \quad \text{for all } u, v \in V$
$\Rightarrow \quad \langle (t^* - T^*)(v), (t^* - T^*)(v) \rangle = 0 \quad \text{for all } v \in V \quad [\text{Taking } u = (t^* - T^*)(v)]$
$\Rightarrow \quad (t^* - T^*)(v) = 0 \quad \text{for all } v \in V$
$\Rightarrow \quad t^* - T^* = \hat{0}$
$\Rightarrow \quad t^* = T^*$

Hence, t^* is unique.

Q.E.D.

REMARK. *In the above theorem it has been proved that each linear operator on a finite dimensional unitary space possesses its adjoint. But, if V is not a finite dimensional unitary space, then some linear operators may possess an adjoint while the others may not. In case, a linear operator possesses an adjoint, then it is unique. The adjoint of t depends on the Hermitian product on V.*

ADJOINT OF AN OPERATOR *Let V be a unitary space and t be a linear operator on V. A linear operator t^* on V is said to be adjoint of t, if*

$$\langle t(u), v \rangle = \langle u, t^*(v) \rangle \text{ for every } u, v \in V$$

EXAMPLE-1 Let V be the unitary space of $n \times 1$ matrices over C with Hermitian inner product $\langle \, , \, \rangle$ defined by $\langle A, B \rangle = B^*A$. If A is an $n \times n$ matrix over C, find the adjoint of the linear operator $t_A : V \to V$ defined by $t_A(X) = AX$.

SOLUTION For any $X, Y \in V$, we have

$$\langle t_A(X), Y \rangle = \langle AX, Y \rangle = Y^*(AX) = (Y^*A)X = (A^*Y)^*X = \langle X, A^*Y \rangle = \langle X, t_{A^*}(Y) \rangle$$

$\therefore \quad (t_A)^* = t_{A^*}$

EXAMPLE-2 Let V be the vector space of all $n \times n$ matrices over C with the inner product $\langle A, B \rangle = tr(B^*A)$. Let M be a fixed $n \times n$ matrix over C. Let t_M be a linear operator on V defined by $t_M(A) = MA$ for all $A \in V$. Find the adjoint of t_M.

SOLUTION For any $A, B \in V$

$\quad\quad\quad \langle t_M(A), B \rangle = \langle MA, B \rangle$

$\Rightarrow \quad \langle t_M(A), B \rangle = tr\,(B^*(MA))$

$\Rightarrow \quad \langle t_M(A), B \rangle = tr\,((MA)B^*)$ $\quad\quad\quad\quad\quad\quad\quad\quad\quad [\because \quad tr\,(AB) = tr\,(BA)]$

$\Rightarrow \quad \langle t_M(A), B \rangle = tr\,(M(AB^*))$

$\Rightarrow \quad \langle t_M(A), B \rangle = tr\,((AB^*)M)$ $\quad\quad\quad\quad\quad\quad\quad\quad\quad [\because \quad tr\,(AB) = tr\,(BA)]$

$\Rightarrow \quad \langle t_M(A), B \rangle = tr\,(A(B^*M))$

$\Rightarrow \quad \langle t_M(A), B \rangle = tr\,(A(M^*B)^*) = tr\,((M^*B)^*A)$ $\quad\quad [\because \quad tr\,(AB) = tr\,(BA)]$

$\Rightarrow \quad \langle t_M(A), B \rangle = \langle A, M^*B \rangle$

$\Rightarrow \quad \langle t_M(A), B \rangle = \langle A, t_{M^*}(B) \rangle$

Hence, $(t_M)^* = t_{M^*}$.

EXAMPLE-3 Let V be the vector space of all polynomials over the field C of all complex numbers, with the inner product given by

$$\langle f, g \rangle = \int_0^1 f(t)\, \overline{g(t)} \, dt \quad \text{for all } f, g \in V.$$

Here, $\overline{g(t)}$ is complex conjugate of $g(t)$, i.e. if $g(t) = \sum a_k\, t^k$, then $\overline{g(t)} = \sum \overline{a_k}\, t^k$. $\overline{g(t)}$ is generally denoted by $\bar{g}(t)$.

Let f be an arbitrary but fixed polynomial in V. Consider the linear operator t_f on V defined by

$$t_f(g) = fg \quad \text{for all } g \in V.$$

Find the adjoint of this operator.

SOLUTION For any $g, h \in V$

$$\langle t_f(g), h \rangle = \langle fg, h \rangle = \int_0^1 (fg)(t)\,\overline{h(t)}\,dt$$

$\Rightarrow \quad \langle t_f(g), h \rangle = \int_0^1 f(t)\,g(t)\,\overline{h(t)}\,dt$

$\Rightarrow \quad \langle t_f(g), h \rangle = \int_0^1 g(t)\,f(t)\,\overline{h(t)}\,dt$

$\Rightarrow \quad \langle t_f(g), h \rangle = \int_0^1 g(t)\,\overline{\{\overline{f}(t)\,h(t)\}}\,dt$

$\Rightarrow \quad \langle t_f(g), h \rangle = \langle g\ \overline{f}h \rangle$
$\Rightarrow \quad \langle t_f(g), h \rangle = \langle g, t_{\overline{f}}(h) \rangle$

Hence, $(t_f)^* = t_{\overline{f}}$.

SELF-ADJOINT OPERATOR *A linear operator t on a finite dimensional unitary space V is said to be self-adjoint if $t^* = t$.*

A self-adjoint linear operator is also known as a Hermitian operator.

Thus, if t is self-adjoint, then

$$\langle t(u), v \rangle = \langle u, t^*(v) \rangle \quad \text{for all } u, v \in V$$
$\Rightarrow \quad \langle t(u), v \rangle = \langle u, t(v) \rangle \quad \text{for all } u, v \in V.$

SKEW-HERMITIAN OPERATOR A linear operator t on a finite dimensional unitary space V is said to be skew-Hermitian if $t^* = -t$.

Thus, if t is skew-Hermitian, then

$$\langle t(u), v \rangle = \langle u, t^*(v) \rangle \quad \text{for all } u, v \in V$$
$\Rightarrow \quad \langle t(u), v \rangle = \langle u, -t(v) \rangle \quad \text{for all } u, v \in V$
$\Rightarrow \quad \langle t(u), v \rangle = -\langle u, t(v) \rangle \quad \text{for all } u, v \in V$

THEOREM-2 Let t, t_1, t_2 be linear operators on a finite dimensional unitary space V and let $a \in \mathbb{C}$. Then,

(i) $(t_1 + t_2)^* = t_1^* + t_2^*$ (ii) $(at)^* = \overline{a}t^*$ (iii) $(t_1 t_2)^* = t_2^* t_1^*$ (iv) $(t^*)^* = t$

PROOF. Analogous to the proof of Theorem 7 on page 617.

THEOREM-3 *The eigenvalues of a self-adjoint (Hermitian) operator on a finite dimensional unitary space are real.*

PROOF. Let t be a self-adjoint operator on a finite dimensional unitary space V and λ be an eigenvalue of t. Then there exists a non-zero vector $v_1 \in V$ such that $t(v_1) = \lambda v_1$.

Now,

$\quad\quad\quad t$ is self-adjoint operator on V

$\Rightarrow \quad\quad \langle t(u), v \rangle = \langle u, t(v) \rangle \quad$ for all $u, v \in V$

$\Rightarrow \quad\quad \langle t(v_1), v_1 \rangle = \langle v_1, t(v_1) \rangle \quad\quad\quad\quad\quad\quad\quad\quad\quad\quad$ [Taking $u = v = v_1$]

$\Rightarrow \quad\quad \langle \lambda v_1, v_1 \rangle = \langle v_1, \lambda v_1 \rangle$

$\Rightarrow \quad\quad \lambda \langle v_1, v_1 \rangle = \overline{\lambda} \langle v_1, v_1 \rangle$

$\Rightarrow \quad\quad \lambda \langle v_1, v_1 \rangle = \overline{\lambda} \langle v_1, v_1 \rangle$

$\Rightarrow \quad\quad \lambda = \overline{\lambda} \quad\quad\quad\quad\quad\quad\quad\quad\quad\quad\quad\quad\quad\quad$ [$\because \langle v_1, v_1 \rangle > 0$ as $v_1 \neq 0_V$]

$\Rightarrow \quad\quad \lambda$ is purely real

Hence, eigenvalues of t are real.

Q.E.D.

THEOREM-4 *The eigenvalues of a skew-Hermitian operator on a finite dimensional unitary space are purely imaginary.*

PROOF. Let t be a skew-Hermitian operator on a finite dimensional unitary space V and λ be an eigenvalue of t. Then, there exists a non-zero vector $v_1 \in V$ such that $t(v_1) = \lambda v_1$.

Now,

$\quad\quad\quad t$ is skew-Hermitian

$\Rightarrow \quad\quad \langle t(u), v \rangle = -\langle u, t(v) \rangle \quad$ for all $u, v \in V$

$\Rightarrow \quad\quad \langle t(v_1), v_1 \rangle = -\langle v_1, t(v_1) \rangle \quad\quad\quad\quad\quad\quad\quad\quad\quad\quad$ [Taking $u = v = v_1$]

$\Rightarrow \quad\quad \langle \lambda v_1, v_1 \rangle = -\langle v_1, \lambda v_1 \rangle$

$\Rightarrow \quad\quad \lambda \langle v_1, v_1 \rangle = -\overline{\lambda} \langle v_1, v_1 \rangle$

$\Rightarrow \quad\quad \lambda = -\overline{\lambda} \quad\quad\quad\quad\quad\quad\quad\quad\quad\quad\quad\quad\quad\quad$ [$\because \langle v_1, v_1 \rangle > 0$ as $v_1 \neq 0_V$]

$\Rightarrow \quad\quad \lambda$ is purely imaginary.

Hence, eigenvalues of t are purely imaginary.

Q.E.D.

THEOREM-5 *Let t be a linear operator on a finite dimensional unitary space V such that $t = T^*T$ for some non-singular operator T on V. If λ is an eigenvalue of t, then λ is real and positive.*

PROOF. It is given that T is non-singular operator on V. Therefore,

$$T(v) \neq 0_V \text{ for any } v(\neq 0_V) \in V$$
$$\Rightarrow \quad \langle T(v), T(v) \rangle > 0 \quad \text{for all } v(\neq 0_V) \in V$$

Since λ is an eigenvalue of t. So, there exists a non-zero vector $v \in V$ such that $t(v) = \lambda v$. Now,

$$\lambda \langle v, v \rangle = \langle \lambda v, v \rangle = \langle t(v), v \rangle$$
$$\Rightarrow \quad \lambda \langle v, v \rangle = \langle T^*T(v), v \rangle$$
$$\Rightarrow \quad \lambda \langle v, v \rangle = \langle T^*(T(v)), v \rangle$$
$$\Rightarrow \quad \lambda \langle v, v \rangle = \langle T(v), (T^*)^*(v) \rangle$$
$$\Rightarrow \quad \lambda \langle v, v \rangle = \langle T(v), T(v) \rangle \qquad [\because \ (T^*)^* = T]$$
$$\Rightarrow \quad \lambda = \frac{\langle T(v), T(v) \rangle}{\langle v, v \rangle} > 0 \qquad [\because \ \langle v, v \rangle > 0 \text{ and } \langle T(v), T(v) \rangle > 0]$$
$$\Rightarrow \quad \lambda \text{ is a positive real number.}$$

Q.E.D.

THEOREM-6 *Let t be a linear operator on a unitary space V, and let S be a t-invariant subspace of V. Then, $S^\perp$ is invariant under t^*.*

PROOF. Let $v \in S$ and $u \in S^\perp$. Then,

$$t(v) \in S \qquad [\because \ S \text{ is } t\text{-invariant}]$$
$$\Rightarrow \quad \langle t(v), u \rangle = 0 \qquad [\because \ u \in S^\perp]$$
$$\Rightarrow \quad \langle v, t^*(u) \rangle = 0 \qquad [\text{By definition of adjoint}]$$
$$\Rightarrow \quad t^*(u) \perp v$$
$$\Rightarrow \quad t^*(u) \in S^\perp$$

Thus, $u \in S^\perp \Rightarrow t^*(u) \in S^\perp$.

Hence, $S^\perp$ is invariant under t^*.

Q.E.D.

THEOREM-7 *Let V be a finite dimensional unitary space and let t be a self-adjoint operator on V. Then eigenvectors corresponding to distinct eigenvalues are orthogonal.*

PROOF. Let λ_1 and λ_2 be distinct eigenvalues of linear operator t on V. Let u and v be eigenvectors corresponding to eigenvalues λ_1 and λ_2 respectively. Then,

$$t(u) = \lambda_1 u \quad \text{and} \quad t(v) = \lambda_2 v$$

Now,

t is self-adjoint operator on V

$\Rightarrow \quad \langle t(u), v \rangle = \langle u, t(v) \rangle$

$\Rightarrow \quad \langle \lambda_1 u, v \rangle = \langle u, \lambda_2 v \rangle$

$\Rightarrow \quad \lambda_1 \langle u, v \rangle = \lambda_2 \langle u, v \rangle \qquad [\because \lambda_1, \lambda_2 \text{ are real (See Theorem 3)}]$

$\Rightarrow \quad (\lambda_1 - \lambda_2) \langle u, v \rangle = 0$

$\Rightarrow \quad \langle u, v \rangle = 0 \qquad [\because \lambda_1 \neq \lambda_2]$

$\Rightarrow \quad u$ is orthogonal to v.

Q.E.D.

THEOREM-8 *Let V be a finite dimensional unitary space and t be a linear operator on V. If $A = [a_{ij}]$ is the matrix representing t relative to some basis $B = \{v_1, v_2, \ldots, v_n\}$. Then,*

$$a_{ij} = \langle t(v_j), v_i \rangle \quad \text{for all } i, j \in \underline{n}$$

PROOF. Since $A = [a_{ij}]$ is the matrix representing t relative to the basis $B = \{v_1, v_2, \ldots, v_n\}$.

$\therefore \quad t(v_j) = \sum\limits_{r=1}^{n} a_{rj} v_r \qquad$ for all $j \in \underline{n}$

$\Rightarrow \quad \langle t(v_j), v_i \rangle = \left\langle \sum\limits_{r=1}^{n} a_{rj} v_r, v_i \right\rangle \qquad$ for all $i, j \in \underline{n}$

$\Rightarrow \quad \langle t(v_j), v_i \rangle = \sum\limits_{r=1}^{n} a_{rj} \langle v_r, v_i \rangle \qquad$ for all $i, j \in \underline{n}$

$\Rightarrow \quad \langle t(v_j), v_i \rangle = \sum\limits_{r=1}^{n} a_{rj} \delta_{ri} = a_{ij} \qquad$ for all $i, j \in \underline{n}$

Q.E.D.

THEOREM-9 *Let V be a finite dimensional unitary space with an orthonormal basis $B = \{v_1, v_2, \ldots, v_n\}$ and let t be a linear operator on V. If P is the matrix representing t relative to basis B, then the matrix representing the adjoint operator t^* relative to the same basis B is conjugate transpose of P, i.e. $\left(\overline{P}\right)^T = P^*$. In other words, if*

$$M_B^B(t) = P, \text{ then } \quad M_B^B(t^*) = P^* \quad \text{or,} \quad M_B^B(t^*) = \left(M_B^B(t)\right)^*$$

PROOF. Let $P = [p_{ij}]$ and $Q = [q_{ij}]$ be the matrices representing t and t^* respectively. Then,

$\qquad p_{ij} = \langle t(v_j), v_i \rangle$ and $q_{ij} = \langle t^*(v_j), v_i \rangle \quad$ for all $i, j \in \underline{n}$

Now, $\quad p_{ij} = \langle t(v_j), v_i \rangle \qquad$ for all $i, j \in \underline{n}$

$\Rightarrow \quad p_{ij} = \langle v_j, t^*(v_i) \rangle \qquad$ for all $i, j \in \underline{n}$

$\Rightarrow \quad \overline{p_{ij}} = \overline{\langle v_j, t^*(v_i) \rangle}$ for all $i, j \in \underline{n}$

$\Rightarrow \quad \overline{p_{ij}} = \langle t^*(v_i), v_j \rangle$ for all $i, j \in \underline{n}$

$\Rightarrow \quad \overline{p_{ij}} = q_{ji}$ for all $i, j \in \underline{n}$

$\Rightarrow \quad q_{ij} = \overline{p_{ji}}$ for all $i, j \in \underline{n}$

$\Rightarrow \quad Q = (\overline{P})^T$

$\Rightarrow \quad Q = P^*$

$\Rightarrow \quad M_B^B(t^*) = P^* = \left(M_B^B(t)\right)^*$

Q.E.D.

REMARK. *The above relationship between the matrices representing t and t^* is true for an orthonormal basis of V. If B is not an orthonormal basis for V, then such a relation may not exist. Also, if V is not a finite dimensional unitary space, then the above theorem may not hold.*

THEOREM-10 *Let t be a linear operator on a unitary space V such that $\langle t(u), v \rangle = 0$ for every $u, v \in V$. Then, $t = \hat{0}$.*

PROOF. We have,

$\langle t(u), v \rangle = 0$ for all $u, v \in V$

$\Rightarrow \quad \langle t(u), t(u) \rangle = 0$ for all $u \in V$ [Replacing v by $t(u)$]

$\Rightarrow \quad t(u) = 0$ for all $u \in V$

$\Rightarrow \quad t = \hat{0}$

Q.E.D.

THEOREM-11 *Let t be a linear operator on a unitary space V such that $\langle t(u), u \rangle = 0$ for all $u \in V$. Then, $t = \hat{0}$.*

PROOF. We have,

$\langle t(u), u \rangle = 0$ for all $u \in V$ (i)

$\Rightarrow \quad \langle t(v+w), v+w \rangle = 0$ for all $v, w \in V$

$\Rightarrow \quad \langle t(v) + t(w), v+w \rangle = 0$ for all $v, w \in V$

$\Rightarrow \quad \langle t(v), v \rangle + \langle t(v), w \rangle + \langle t(w), v \rangle + \langle t(w), w \rangle = 0$ for all $v, w \in V$

$\Rightarrow \quad \langle t(v), w \rangle + \langle t(w), v \rangle = 0$ for all $v, w \in V$ [Using (i)] (ii)

$\Rightarrow \quad \langle t(v), iw \rangle + \langle t(iw), v \rangle = 0$ [Replacing w by iw]

$\Rightarrow \quad -i\langle t(v), w \rangle + i\langle t(w), v \rangle = 0$

$\Rightarrow \quad -\langle t(v), w \rangle + \langle t(w), v \rangle = 0$ (iii)

Adding (ii) and (iii), we get

$$\langle t(w), v \rangle = 0 \quad \text{for all } v, w \in V$$
$$\Rightarrow \quad \langle t(w), t(w) \rangle = 0 \quad \text{for all } w \in V \quad \text{[Replacing } v \text{ by } t(w)]$$
$$\Rightarrow \quad t(w) = 0 \quad \text{for all } w \in V$$
$$\Rightarrow \quad t = \hat{0}$$

Q.E.D.

REMARK. *The above theorem may not be valid in an inner product space. For example, for the linear operator* $t : R^2 \to R^2$ *given by* $t(x, y) = (y, -x)$, *we have* $\langle t(u), u \rangle = 0$ *for every* $u \in R^2$, *but* $t \neq \hat{0}$.

However, it **holds** if t is a self-adjoint operator on an inner product space as proved in the following theorem.

THEOREM-12 *Let t be a self-adjoint operator on a unitary space such that* $\langle t(u), u \rangle = 0$ *for all* $u \in V$. *Then,* $t = \hat{0}$.

PROOF. The result holds for Hermitian inner product on V as shown in the previous theorem. Thus, we have to prove the result for an inner product space.

We have,

$$\langle t(u), u \rangle = 0 \quad \text{for all } u \in V \tag{i}$$
$$\Rightarrow \quad \langle t(v+w), v+w \rangle = 0 \quad \text{for all } u, w \in V \quad \text{[Replacing } u \text{ by } v+w]$$
$$\Rightarrow \quad \langle t(v) + t(w), v+w \rangle = 0 \quad \text{for all } v, w \in V$$
$$\Rightarrow \quad \langle t(v), v \rangle + \langle t(v), w \rangle + \langle t(w), v \rangle + \langle t(w), w \rangle = 0 \quad \text{for all } v, w \in V$$
$$\Rightarrow \quad \langle t(v), w \rangle + \langle t(w), v \rangle = 0 \quad \text{for all } v, w \in V \quad \text{[Using (i)]}$$
$$\Rightarrow \quad \langle t(v), w \rangle + \langle w, t(v) \rangle = 0 \quad [\because t \text{ is self-adjoint}]$$
$$\Rightarrow \quad \langle t(v), w \rangle + \langle t(v), w \rangle = 0 \quad [\langle , \rangle \text{ is an inner product on } V(R)]$$
$$\Rightarrow \quad \langle t(v), w \rangle = 0 \quad \text{for all } v, w \in V$$
$$\Rightarrow \quad \langle t(v), t(v) \rangle = 0 \quad \text{for all } v \in V \quad \text{[Replacing } w \text{ by } t(v)]$$
$$\Rightarrow \quad t(v) = 0_V \quad \text{for all } v \in V$$
$$\Rightarrow \quad t = \hat{0}$$

Q.E.D.

THEOREM-13 *A linear operator t on a finite dimensional unitary space V is Hermitian iff its matrix $A = [a_{ij}]$ relative to some orthonormal basis is Hermitian.*

PROOF. Let $B = \{v_1, v_2, v_3, \ldots, v_n\}$ be an orthonormal basis for V. First, let t be a self-adjoint (Hermitian) operator on V. Then, $t^* = t$. By Theorem 9, we have

$$M_B^B(t^*) = \left(M_B^B(t)\right)^*$$
$$\Rightarrow \quad M_B^B(t) = \left(M_B^B(t)\right)^* \qquad\qquad\qquad\qquad\qquad\qquad [\because\ t^* = t]$$
$$\Rightarrow \quad A = A^*$$
$\Rightarrow \quad A$ is Hermitian

Conversely, let $A = [a_{ij}]$ be a Hermitian matrix. Then,

$$A^* = A, \text{ i.e. } A = \left(\overline{A}\right)^T$$

$\Rightarrow \quad a_{ij} = \overline{a_{ji}}$ for all $i, j \in \underline{n}$

$\Rightarrow \quad \langle t(v_j), v_i \rangle = \overline{\langle t(v_i), v_j \rangle}$ for all $i, j \in \underline{n}$

$\Rightarrow \quad \langle t(v_j), v_i \rangle = \langle v_j, t(v_i) \rangle$ for all $i, j \in \underline{n}$

$\Rightarrow \quad \langle v_j, t^*(v_i) \rangle = \langle v_j, t(v_i) \rangle$ for all $i, j \in \underline{n}$

$\Rightarrow \quad \langle v_j, t^*(v_i) - t(v_i) \rangle = 0$ for all $i, j \in \underline{n}$

$\Rightarrow \quad \langle v_j, (t^* - t)(v_i) \rangle = 0$ for all $i, j \in \underline{n}$

$\Rightarrow \quad t^* - t = \hat{0}$ \hfill [By Theorem 10]

$\Rightarrow \quad t^* = t$

$\Rightarrow \quad t$ is Hermitian operator on V.

\hfill Q.E.D.

THEOREM-14 *A linear operator t on a finite dimensional unitary space V is skew-Hermitian iff its matrix $A = [a_{ij}]$ relative to some orthonormal basis is skew-Hermitian.*

PROOF. Proceed as in the proof of the above theorem.

THEOREM-15 *A necessary and sufficient condition that a linear operator t on a unitary space V be a self-adjoint (Hermitian) is that $\langle t(u), u \rangle$ is real for all $u \in V$.*

PROOF. *Necessary Condition:* Let t be a self-adjoint operator on a unitary space V. i.e. $t^* = t$. Then,

$$\langle t(u), v \rangle = \langle u, t(v) \rangle \quad \text{for all } u, v \in V$$
$\Rightarrow \quad \langle t(u), u \rangle = \langle u, t(u) \rangle \quad \text{for all } u \in V$

$\Rightarrow \quad \langle t(u), u \rangle = \overline{\langle t(u), u \rangle} \quad \text{for all } u \in V$

$\Rightarrow \quad \langle t(u), u \rangle$ is real for all $u \in V$.

Sufficient Condition: Let t be a linear operator on a unitary space V such that $\langle t(u), u \rangle$ is real for all $u \in V$. In order to prove that t is self-adjoint it is sufficient to show that

$$\langle t(u), v \rangle = \langle u, t(v) \rangle \quad \text{for all } u, v \in V$$

For any $u, v \in V$, we have

$\langle t(u+v), u+v \rangle$ is real $\quad [\because \; \langle t(u), u \rangle$ is real for all $u \in V]$

$\Rightarrow \quad \langle t(u)+t(v), u+v \rangle$ is real

$\Rightarrow \quad \langle t(u), u \rangle + \langle t(u), v \rangle + \langle t(v), u \rangle + \langle t(v), v \rangle$ is real

$\Rightarrow \quad \langle t(u), v \rangle + \langle t(v), u \rangle$ is real $\quad [\because \; \langle t(u), u \rangle$ and $\langle t(v), v \rangle$ are real]

$\Rightarrow \quad \langle t(u), v \rangle + \langle t(v), u \rangle = \overline{\langle t(u), v \rangle + \langle t(v), u \rangle}$

$\Rightarrow \quad \langle t(u), v \rangle + \langle t(v), u \rangle = \overline{\langle t(u), v \rangle} + \overline{\langle t(v), u \rangle}$

$\Rightarrow \quad \langle t(u), v \rangle + \langle t(v), u \rangle = \langle v, t(u) \rangle + \langle u, t(v) \rangle \quad$ (i)

$\Rightarrow \quad \langle t(u), iv \rangle + \langle t(iv), u \rangle = \langle iv, t(u) \rangle + \langle u, t(iv) \rangle \quad$ [Replacing v by iv]

$\Rightarrow \quad -i\langle t(u), v \rangle + i\langle t(v), u \rangle = i\langle v, t(u) \rangle - i\langle u, t(v) \rangle$

$\Rightarrow \quad \langle t(u), v \rangle - \langle t(v), u \rangle = -\langle v, t(u) \rangle + \langle u, t(v) \rangle \quad$ (ii)

Adding (i) and (ii), we obtain $\langle t(u), v \rangle = \langle u, t(v) \rangle$

Hence, t is a self-adjoint operator on V.

Q.E.D.

REMARK. *If V is a finite dimensional unitary space, then sufficient part of the above theorem can be proved, by using the existence of t^*, as follows:*

If $\langle t(u), u \rangle$ is real for all $u \in V$, then

$\langle t(u), u \rangle = \overline{\langle t(u), u \rangle} \quad$ for all $u \in V$

$\Rightarrow \quad \langle t(u), u \rangle = \overline{\langle u, t^*(u) \rangle} \quad$ for all $u \in V$

$\Rightarrow \quad \langle t(u), u \rangle = \langle t^*(u), u \rangle \quad$ for all $u \in V$

$\Rightarrow \quad \langle t(u) - t^*(u), u \rangle = 0 \quad$ for all $u \in V$

$\Rightarrow \quad \langle (t - t^*)(u), u \rangle = 0 \quad$ for all $u \in V$

$\Rightarrow \quad t - t^* = \hat{0} \quad$ [By Theorem 12]

$\Rightarrow \quad t = t^*$

$\Rightarrow \quad t$ is a self-adjoint operator on V.

Linear Operators on Unitary Spaces • 691

THEOREM-16 *(Principal Axis Theorem) Let V be a finite dimensional unitary space and t be a self-adjoint operator on V. Then there exists an orthonormal basis B for V such that each vector of which is an eigenvector of t and in such a case t is represented by a diagonal matrix relative to B such that diagonal entries are eigenvalues of t and each appears as many times as its multiplicity.*

PROOF. Proceed as in the proof of Theorem 14 on page 621.

Q.E.D.

ILLUSTRATIVE EXAMPLES

EXAMPLE-1 Let t be a linear operator on C^3 given by

$$t(x, y, z) = \left(x + (1+2i)y, (3-2i)x + y - 2iz, 2ix + 3iy - (4+i)z\right)$$

Find the adjoint of t, if the Hermitian inner product is standard one.

SOLUTION Clearly, $B = \left\{e_1^{(3)} = (1, 0, 0),\ e_2^{(3)} = (0, 1, 0),\ e_3^{(3)} = (0, 0, 1)\right\}$ is the standard orthonormal basis of C^3.

We have,

$$t\left(e_1^{(3)}\right) = t(1, 0, 0) = (1, 3-2i, 2i) = 1\, e_1^{(3)} + (3-2i)\, e_2^{(3)} + 2i\, e_3^{(3)}$$

$$t\left(e_2^{(3)}\right) = t(0, 1, 0) = (1+2i, 1, 3i) = (1+2i)\, e_1^{(3)} + 1\, e_2^{(3)} + 3i\, e_3^{(3)}$$

$$t\left(e_3^{(3)}\right) = t(0, 0, 1) = (0, -2i, -(4+i)) = 0\, e_1^{(3)} + (-2i)\, e_2^{(3)} + (-4-i)\, e_3^{(3)}$$

So, matrix A that represents t relative to the standard basis B is given by

$$A = M_B^B(t) = \begin{bmatrix} 1 & 1+2i & 0 \\ 3-2i & 1 & -2i \\ 2i & 3i & -4-i \end{bmatrix}$$

∴ $M_B^B(t^*) = (M_B^B(t))^*$

⇒ $M_B^B(t^*) = A^* = \left(\overline{A}\right)^T = \begin{bmatrix} 1 & 3+2i & -2i \\ 1-2i & 1 & -3i \\ 0 & 2i & -4+i \end{bmatrix}$

⇒ $t^*(x, y, z) = \left(x + (3+2i)y - 2iz,\ x(1-2i) + y - 3iz,\ 2iy + (-4+i)z\right)$

$$[\because\ M_B^B(t^*(v)) = M_B^B(t^*)\, M_B^B(v)]$$

EXAMPLE-2 *Let* $t : C^2 \to C^2$ *be a linear operator defined by* $t(1,0) = (1,-2)$, $t(0,1) = (i,-1)$. *Find* t^*.

SOLUTION Clearly, $B = \left\{ e_1^{(2)} = (1, 0),\ e_2^{(2)} = (0, 1) \right\}$ is standard orthonormal basis of C^2 such that

$$t(1, 0) = (1, -2) = 1\, e_1^{(2)} + (-2)\, e_2^{(2)}$$

and, $\quad t(0, 1) = (i, -1) = i\, e_1^{(2)} + (-1)\, e_2^{(2)}$

$\therefore \quad M_B^B(t) = \begin{bmatrix} 1 & i \\ -2 & -1 \end{bmatrix}$

$\Rightarrow \quad M_B^B(t^*) = \begin{bmatrix} 1 & -i \\ -2 & -1 \end{bmatrix}^T = \begin{bmatrix} 1 & -2 \\ -i & -1 \end{bmatrix} \qquad [\because\ M_B^B(t^*) = \left(M_B^B(t)\right)^*]$

$\Rightarrow \quad t^*(x, y) = (x - 2y,\ -ix - y) \qquad [\because\ M_B^B(t^*(v)) = M_B^B(t^*)\, M_B^B(v)]$

EXAMPLE-3 *Let* t *be a linear operator on* C^2, *defined by*

$$t(1, 0) = (1 + i, 2),\ t(0, 1) = (i, i)$$

Using the standard inner product on C^2, *find the matrix of* t^* *relative to the standard basis. Also, show that* t *does not commute with* t^*.

SOLUTION Let $B = \left\{ e_1^{(2)} = (1, 0),\ e_2^{(2)} = (0, 1) \right\}$ be the standard orthonormal basis of C^2. Then,

$$t(1, 0) = (1 + i, 2) = (1 + i)\, e_1^{(2)} + 2\, e_2^{(2)}$$

and, $\quad t(0, 1) = (i, i) = i\, e_1^{(2)} + i\, e_2^{(2)}$

$\therefore \quad M_B^B(t) = \begin{bmatrix} 1+i & i \\ 2 & i \end{bmatrix}$

$\Rightarrow \quad M_B^B(t^*) = \begin{bmatrix} 1-i & -i \\ 2 & -i \end{bmatrix}^T = \begin{bmatrix} 1-i & 2 \\ -i & -i \end{bmatrix}$

Now, $\quad M_B^B(t)\, M_B^B(t^*) = \begin{bmatrix} 1+i & i \\ 2 & i \end{bmatrix} \begin{bmatrix} 1-i & 2 \\ -i & -i \end{bmatrix} = \begin{bmatrix} 3 & 3+2i \\ 3-2i & 5 \end{bmatrix}$

and, $\quad M_B^B(t^*)\, M_B^B(t) = \begin{bmatrix} 1-i & 2 \\ -i & -i \end{bmatrix} \begin{bmatrix} 1+i & i \\ 2 & i \end{bmatrix} = \begin{bmatrix} 6 & 1+3i \\ 1-3i & 2 \end{bmatrix}$

Clearly, $\quad M_B^B(t)\, M_B^B(t^*) \neq M_B^B(t^*)\, M_B^B(t)$

$\Rightarrow \quad M_B^B(t o t^*) \neq M_B^B(t^* o t)$

$\Rightarrow \quad t o t^* \neq t^* o t\ \text{or,}\ tt^* \neq t^*t$

EXAMPLE-4 *The product of two self-adjoint linear operators on a finite dimensional unitary space is a self-adjoint operator iff they commute.*

SOLUTION Let t_1 and t_2 be two self-adjoint linear operators on a finite dimensional unitary space V. Then, $t_1^* = t_1$ and $t_2^* = t_2$.

First suppose the $t_1 t_2$ is a self-adjoint operator on V. Then,

$$(t_1 t_2)^* = t_1 t_2$$
$$\Rightarrow \quad t_2^* t_1^* = t_1 t_2 \quad\quad\quad [\because \ (t_1 t_2)^* = t_2^* t_1^*]$$
$$\Rightarrow \quad t_2 t_1 = t_1 t_2$$
$$\Rightarrow \quad t_1 \text{ and } t_2 \text{ commute with each other.}$$

Conversely, let t_1 and t_2 be self-adjoint operators on V such that $t_1 t_2 = t_2 t_1$. Then, we have to prove that $t_1 t_2$ is a self-adjoint operator on V.

Now,

$$(t_1 t_2)^* = t_2^* t_1^* = t_2 t_1 = t_1 t_2 \quad\quad\quad [\because \ t_1 t_2 = t_2 t_1]$$
$$\Rightarrow \quad t_1 t_2 \text{ is a self-adjoint operator on } V.$$

EXAMPLE-5 *If t_1 and t_2 are self-adjoint linear operators on a finite dimensional unitary space V, then*

(i) *prove that $t_1 + t_2$ is a self-adjoint operator on V.*

(ii) *if $t_1 \neq \hat{0}$ and $a \neq 0$, then at_1 is self-adjoint iff a is real.*

SOLUTION (i) It is given that t_1 and t_2 are self-adjoint operators on V.

$$\therefore \quad t_1^* = t_1 \text{ and } t_2^* = t_2$$
Now, $\quad (t_1 + t_2)^* = t_1^* + t_2^* \quad\quad\quad$ [See Theorem 2 on page 683]
$$\Rightarrow \quad (t_1 + t_2)^* = t_1 + t_2 \quad\quad\quad [\because \ t_1^* = t_1 \text{ and } t_2^* = t_2]$$
$$\Rightarrow \quad t_1 + t_2 \text{ is self-adjoint operator on } V.$$

(ii) First, let $a \in R$. Then, $\bar{a} = a$.

$$\therefore \quad (at_1)^* = \bar{a} t_1^* = at_1 \quad\quad\quad [\because \ t_1^* = t_1]$$
$$\Rightarrow \quad at_1 \text{ is self-adjoint operator on } V.$$

Next, let at_1 be a self-adjoint operator on V. Then,

$$(at_1)^* = at_1$$
$$\Rightarrow \quad \bar{a} t_1^* = at_1$$
$$\Rightarrow \quad \bar{a} t_1 = at_1 \quad\quad\quad [\because \ t_1^* = t_1]$$
$$\Rightarrow \quad (a - \bar{a}) t_1 = \hat{0}$$

$\Rightarrow \quad (a-\bar{a})t_1(v) = \hat{0}(v)$ for all $v \in V$

$\Rightarrow \quad (a-\bar{a})t_1(v) = 0_V$ for all $v \in V$

$\Rightarrow \quad a - \bar{a} = 0 \qquad [\because \ t_1 \neq \hat{0} \ \therefore \ t_1(v) \neq 0_V \text{ for some } v \in V]$

$\Rightarrow \quad a = \bar{a}$

$\Rightarrow \quad a$ is purely real.

EXAMPLE-6 *Prove that the eigenvalues of a Hermitian matrix are real.*

SOLUTION Let A be a Hermitian matrix and λ be an eigenvalues of A. Then there exists a non-zero vector $X \in C^n$ such that

$$AX = \lambda X \qquad (i)$$

$\Rightarrow \quad \overline{X}^T A X = \overline{X}^T \lambda X$

$\Rightarrow \quad \overline{\left(\overline{X}^T A X\right)^T} = \overline{\left(\overline{X} \lambda X\right)^T}$

$\Rightarrow \quad (X^T \overline{A}\ \overline{X})^T = \left(X \overline{\lambda} \overline{X}\right)^T$

$\Rightarrow \quad \overline{X}^T \overline{A}^T X = \overline{\lambda} \overline{X}^T X^T$

$\Rightarrow \quad \overline{X}^T A X = \overline{\lambda}\, \overline{X}^T X \qquad [\because \ (\overline{A}^T) = A)]$

$\Rightarrow \quad \overline{X}^T \lambda X = \overline{\lambda}\, \overline{X}^T X \qquad \text{[Using (i)]}$

$\Rightarrow \quad \lambda \overline{X}^T X = \overline{\lambda}\, \overline{X}^T X$

$\Rightarrow \quad (\lambda - \overline{\lambda})\, \overline{X}^T X = 0$

$\Rightarrow \quad \lambda - \overline{\lambda} = 0 \qquad [\because \ X \neq 0 \ \therefore \ \overline{X}^T X \neq 0]$

$\Rightarrow \quad \lambda = \overline{\lambda}$

$\Rightarrow \quad \lambda$ is a real number.

EXAMPLE-7 *Show that every linear operator t on a finite dimensional unitary space can be uniquely expressed as $t = t_1 + i\, t_2$, where t_1 and t_2 are self-adjoint operators on V.*

SOLUTION Let $t_1 = \dfrac{1}{2}(t + t^*)$ and $t_2 = \dfrac{1}{2i}(t - t^*)$. Then, $t = t_1 + i\, t_2$

Now,

$$t_1 = \frac{1}{2}(t + t^*) \text{ and } t_2 = \frac{1}{2i}(t - t^*)$$

$\Rightarrow \quad t_1^* = \left\{\dfrac{1}{2}(t + t^*)\right\}^* \text{ and } t_2^* = \left\{\dfrac{1}{2i}(t - t^*)\right\}^*$

$\Rightarrow \quad t_1^* = \frac{1}{2}(t+t^*)^* \text{ and } t_2^* = -\frac{1}{2i}(t-t^*)^* \quad\quad [\because \ (\lambda t)^* = \overline{\lambda} t^*]$

$\Rightarrow \quad t_1^* = \frac{1}{2}\{t^* + (t^*)^*\} \text{ and } t_2^* = -\frac{1}{2i}\{t^* - (t^*)^*\}$

$\Rightarrow \quad t_1^* = \frac{1}{2}(t^* + t) \text{ and } t_2^* = -\frac{1}{2i}(t^* - t)$

$\Rightarrow \quad t_1^* = \frac{1}{2}(t + t^*) \text{ and } t_2^* = \frac{1}{2i}(t - t^*)$

$\Rightarrow \quad t_1^* = t_1 \text{ and } t_2^* = t_2$

$\Rightarrow \quad t_1$ and t_2 are self-adjoint operators on V.

Thus, t can be expressed in the form $t_1 + i t_2$, where t_1 and t_2 are self-adjoint operators on V.

In order to show the uniqueness of representation of t in the form $t_1 + i t_2$. If possible, let $t = T_1 + iT_2$, where T_1 and T_2 are self-adjoint operators on V. Then,

$\quad\quad t^* = (T_1 + iT_2)^*$

$\Rightarrow \quad t^* = T_1^* + (iT_2)^*$

$\Rightarrow \quad t^* = T_1^* - iT_2^*$

$\Rightarrow \quad t^* = T_1 - iT_2 \quad\quad [\because \ T_1 \ \& \ T_2 \text{ are self-adjoint}]$

$\therefore \quad t = T_1 + iT_2 \text{ and } t^* = T_1 - iT_2$

$\Rightarrow \quad T_1 = \frac{1}{2}(t + t^*) \text{ and } T_2 = \frac{1}{2i}(t - t^*)$

$\Rightarrow \quad T_1 = t_1 \text{ and } T_2 = t_2$

Hence, t can be uniquely expressed as $t = t_1 + i t_2$, where t_1 and t_2 are self-adjoint operators on V.

EXAMPLE-8 *Let t be a linear operator on a finite dimensional unitary space V and u be a vector in V. Then, show that $\phi(v) = \langle u, t(v) \rangle$ for all $v \in V$ is a linear form on V. Also, find a vector $w \in V$ such that $\phi(v) = \langle v, w \rangle$ for all $v \in V$.*

SOLUTION For any $v_1, v_2 \in V$ and $a, b \in C$

$\quad\quad \phi(av_1 + bv_2) = \langle u, t(av_1 + bv_2) \rangle$

$\Rightarrow \quad \phi(av_1 + bv_2) = \langle u, at(v_1) + bt(v_2) \rangle$

$\Rightarrow \quad \phi(av_1 + bv_2) = \overline{\langle at(v_1) + bt(v_2), u \rangle}$

$\Rightarrow \quad \phi(av_1 + bv_2) = \overline{a \langle t(v_1), u \rangle + b \langle t(v_2), u \rangle}$

$\Rightarrow \quad \phi(av_1 + bv_2) = a \overline{\langle u, t(v_1) \rangle} + b \overline{\langle u, t(v_2) \rangle}$

$\Rightarrow \quad \phi(av_1 + bv_2) = a\phi(v_1) + b\phi(v_2)$

Thus, ϕ is a linear form on V. Therefore, by Theorem 1 on page 677 there exists a unique vector $w \in V$ such that $\phi(v) = \langle v, w \rangle$ for all $v \in V$.

$\therefore \quad \langle v, w \rangle = \overline{\langle u, t(v) \rangle} \quad$ for all $v \in V$ $\qquad [\because \ \phi(v) = \overline{\langle u, t(v) \rangle}]$

$\Rightarrow \quad \langle v, w \rangle = \langle t(v), u \rangle \quad$ for all $v \in V$

$\Rightarrow \quad \langle v, w \rangle = \langle v, t^*(u) \rangle \quad$ for all $v \in V$

$\Rightarrow \quad w = t^*(u)$.

Hence, $w = t^*(u)$.

EXERCISE 9.1

1. Let t be an invertible linear operator on a finite dimensional unitary space V. Then, t^* is invertible and $(t^*)^{-1} = (t^{-1})^*$.
2. Prove that the eigenvalues of a skew-Hermitian operator on a unitary space are purely imaginary.
3. Find the adjoint of linear operator $t : C^3 \to C^3$ defined by
$$t(x,y,z) = \Big(2x + (1-i)y, (3+2i)x - 4iz, 2ix + (4-3i)y - 3z\Big)$$
4. If t is a linear operator on a unitary space V and $\lambda \in C$. Then, prove that $(\lambda t)^* = \overline{\lambda} t^*$.
5. Let t be a linear operator on unitary space C^3 with standard inner product such that
$$t(x,y,z) = \Big(ix + (2+3i)y, 3x + (3-i)z, (2-5i)y + iz\Big) \text{ for all } x,y,z, \in C. \text{ Find } t^*(x,y,z).$$
6. Let $\varphi : C^3 \to C$ be a linear form defined by
$$\phi(x,y,z) = ix + (2+3i)y + (1-2i)z$$
Find a vector $u \in C^3$ such that $\varphi(v) = \langle u, v \rangle$ for all $v \in V$.
[Hint: From Theorem 1 on page 677, we have
$u = \sum_{i=1}^{n} \varphi(v_i) v_i$, where $B = \{v_1, v_2, \ldots, v_n\}$ is an orthonormal basis for V.
Taking $B = \{e_1^{(3)}, e_2^{(3)}, e_3^{(3)}\}$, we obtain
$$u = \varphi(e_1^{(3)}) e_1^{(3)} + \varphi(e_2^{(3)}) e_2^{(3)} + \varphi(e_3^{(3)}) e_3^{(3)}$$
$\Rightarrow \quad u = i(1,0,0) + (2+3i)(0,1,0) + (1-2i)(0,0,1) = (i, 2+3i, 1-2i)]$

7. A necessary and sufficient condition that a self-adjoint linear operator t on a unitary space V be $\hat{0}$ is that $\langle t(v), v \rangle = 0$ for all $v \in V$.
8. A necessary and sufficient condition that a linear operator t on a unitary space be $\hat{0}$ is that $\langle t(u), u \rangle = 0$ for all $u \in V$.
9. Show that the product of two self-adjoint operators on a unitary space is a self-adjoint operator iff they commute.

10. Show that every linear operator on a finite dimensional unitary space can be uniquely expressed as the sum of a Hermitian and skew-Hermitian operator.

11. Let V be a unitary space and u be a fixed vector in V. If $\varphi_u : V \to C$ is defined by $\varphi_u(v) = \langle v, u \rangle$ for all $v \in V$, then show that φ_u is a linear form on V. If V is finite dimensional, show that each linear form on V arises in this way from some $u \in V$.

12. Let V be a unitary space and u, v be fixed vectors in V. Show that the mapping $t : V \to V$ defined by $t(w) = \langle u, v \rangle w$ is a linear operator on V. Show that t has an adjoint t^* given by $t^*(w) = \overline{\langle u, v \rangle} w$ for all $w \in V$.

13. Let V be the vector space of all $n \times n$ matrices over C with the inner product defined by $\langle A, B \rangle = tr(AB^*)$. Let P be a fixed invertible matrix in V and let t_P be the linear operator on V defined by $t_P(A) = P^{-1}AP$. Find the adjoint of t_P.

ANSWERS

3. $t^*(x,y,z) = \Big(2x + (3-2i)\,y - 2iz, (1+i)\,x + (4+3i)\,z, 4iy - 3z \Big)$

5. $t^*(x,y,z) = \Big(-ix + 3y, (2-3i)\,x + (2+5i)\,z, (3+i)\,y - iz \Big)$

6. $u = (i, 2+3i, 1-2i)$

9.4 POSITIVE OPERATORS

POSITIVE DEFINITE OPERATORS *A linear operator t on a unitary space V is said to be positive definite, if*

(i) t is self-adjoint i.e. $t = t^*$.

(ii) $\langle t(u), u \rangle > 0$ *for every non-zero vector* $u \in V$

If t is positive definite operator on a unitary space V, then we write $t > 0$.
If $u = 0_V$, then $\langle t(u), u \rangle = \langle t(0_V), 0_V \rangle = 0$
Thus, if t is a positive definite operator on a unitary space V, then

$\langle t(u), u \rangle \geq 0$ for all $u \in V$

and, $\langle t(u), u \rangle = 0 \Rightarrow u = 0_V$.

POSITIVE OPERATOR *A linear operator t on a unitary space V is said to be positive, if*

(i) t is self-adjoint i.e. $t = t^*$.

(ii) $\langle t(u), u \rangle \geq 0$ for all $u \in V$

Clearly, every positive definite operator is positive. If t is a positive operator, then $\langle t(u), u \rangle = 0$ is possible even if $u \neq 0_V$. Therefore, a positive operator need not be a positive definite operator.

THEOREM-1 Let V be a unitary space and t be a linear operator on V. Let $f : V \times V \to C$ be defined by $f(u, v) = \langle t(u), v \rangle$ for all $u, v \in V$. Show that f is a Hermitian inner product on V iff t is positive definite

PROOF. Proceed as in the proof of Theorem 1 on page 632.

THEOREM-2 Let V be a finite dimensional unitary space with inner product $\langle \, , \, \rangle$. Let f be any inner product on V, then there exists a unique positive definite linear operator t on V such that $f(u, v) = \langle t(u), v \rangle$ for all $u, v \in V$.

PROOF. Proceed as in the proof of Theorem 2 on page 633.

THEOREM-3 Let t be a linear operator on a finite dimensional unitary space V. Then t is positive definite iff there exists an invertible (non-singular) linear operator T on V such that $t = T^*T$.

PROOF. Proceed as in the proof of Theorem 3 on page 634.

THEOREM-4 Let V be a unitary space and t be a linear operator on V. The following conditions on t are equivalent:

(i) $t = T^2$ for some non-singular self-adjoint linear operator T on V.

(ii) $t = S^*S$ for some non-singular linear operator S on V.

(iii) t is positive definite.

PROOF. Proceed as in the proof of Theorem 4 on page 636.

THEOREM-5 Let V be a unitary space and t be a linear operator on V. The following conditions on t or equivalent:

(i) $t = T^2$ for some self-adjoint operator T.

(ii) $t = S^*S$ for some operator S on V.

(iii) t is positive

PROOF. Proceed as in the proof of Theorem 5 on page 637.

THEOREM-6 Let t be a linear operator on a finite dimensional unitary space V and let $A = [a_{ij}]$ be the matrix of t relative to an ordered orthonormal basis $B = \{v_1, v_2, \ldots, v_n\}$ of V. Then t is positive definite iff the matrix A satisfies the following conditions:

(i) $A^* = A$, i.e. A is Hermitian matrix.

(ii) $\sum_{i=1}^{n} \sum_{j=1}^{n} a_{ij} \bar{\lambda}_i \lambda_j > 0$, where $\lambda_1, \lambda_2, \ldots, \lambda_n$ are any n scalars not all zero.

PROOF. Proceed as in the proof of Theorem 6 on page 637.

POSITIVE DEFINITE MATRIX A square matrix $A = [a_{ij}]$ over C is said to be positive definite, if

(i) $A = A^*$ i.e. A is Hermitian (ii) $\sum_{i=1}^{n} \sum_{j=1}^{n} a_{ij} \bar{\lambda}_i \lambda_j > 0$ where $\lambda_1, \lambda_2, \lambda_3, \ldots, \lambda_n$ are any n scalars not all zero.

Clearly, a linear operator t on a finite dimensional unitary space is positive definite iff matrix of t relative to an orthonormal basis of V is positive definite.

THEOREM-7 A self-adjoint operator t on a finite dimensional unitary space V is positive iff its eigenvalues are non-negative.

PROOF. Let t be a self-adjoint positive operator on a finite dimensional unitary space V. Let λ be an eigenvalue of t. Then there exists a non-zero vector $v \in V$ such that $t(v) = \lambda v$.

$\therefore \quad \langle t(v), v \rangle = \langle \lambda v, v \rangle = \lambda \langle v, v \rangle = \lambda \|v\|^2$

$\Rightarrow \quad \lambda = \dfrac{\langle t(v), v \rangle}{\|v\|^2} \hfill [\because v \neq 0_V \therefore \|v\|^2 > 0]$

$\Rightarrow \quad \lambda \geq 0 \hfill [\because t \text{ is positive} \therefore \langle t(v), v \rangle \geq 0]$

Conversely, let t be a self-adjoint operator on V such that all its eigenvalues are non-negative. Then, we have to prove that t is positive i.e. $\langle t(u), u \rangle \geq 0$ for all $u \in V$.

Since t is self-adjoint operator on V. Therefore,

There exists an orthonormal basis $B = \{v_1, v_2, \ldots, v_n\}$ consisting of eigenvectors of t. So, for each $i = 1, 2, \ldots, n$ there exists $\lambda_i \geq 0$ such that
$$t(v_i) = \lambda_i v_i$$

Let u be any vector in V. Then, there exist scalars $a_1, a_2, \ldots, a_n$ such that
$$u = a_1 v_1 + a_2 v_2 + \cdots + a_n v_n.$$

$\therefore \quad \langle t(u), u \rangle = \left\langle t\left(\sum_{i=1}^{n} a_i v_i\right), \sum_{j=1}^{n} a_j v_j \right\rangle$

$\Rightarrow \quad \langle t(u), u \rangle = \left\langle \sum_{i=1}^{n} a_i t(v_i), \sum_{j=1}^{n} a_j v_j \right\rangle$

$\Rightarrow \quad \langle t(u), u \rangle = \left\langle \sum_{i=1}^{n} a_i \lambda_i v_i, \sum_{j=1}^{n} a_j v_j \right\rangle$

$\Rightarrow \quad \langle t(u), u \rangle = \sum_{i=1}^{n} \sum_{j=1}^{n} \lambda_i a_i \bar{a}_j \langle v_i, v_j \rangle$

$$\Rightarrow \quad \langle t(u),u \rangle = \sum_{i=1}^{n} \sum_{j=1}^{n} \lambda_i \, a_i \, \bar{a}_j \, \delta_{ij}$$

$$\Rightarrow \quad \langle t(u),u \rangle = \sum_{i=1}^{n} \sum_{j=1}^{n} \lambda_i \, a_i \, \bar{a}_i = \sum_{i=1}^{n} \lambda_i |a_i|^2$$

$$\Rightarrow \quad \langle t(u), u \rangle \geq 0 \qquad [\because \; \lambda_i \geq 0 \text{ for } i=1,2,\ldots,n \text{ and } |a_i|^2 \geq 0]$$

Thus, t is self-adjoint and $\langle t(u),u \rangle \geq 0$ for all $u \in V$.

Hence, t is positive operator on V.

Q.E.D.

9.5 UNITARY OPERATORS

In this section, we will learn about unitary operators on unitary spaces. These operators behave like orthogonal operators on inner product spaces.

UNITARY OPERATOR *A linear operator t on a finite dimensional unitary space V is said to be a unitary operator if $t^* = t^{-1}$ or equivalently $tt^* = t^*t = I_V$.*

It is evident from the above definition that a unitary operator on a unitary space is invertible. The following theorm gives alternative characterization of these operators.

THEOREM-1 *Let V be a finite dimensional unitary space. An operator t on V is a unitary operator iff $\langle t(u),t(v) \rangle = \langle u,v \rangle$ for all $u,v \in V$.*

PROOF. First, suppose that t is a unitary operator on V. Then $t^*t = I_V$. Let $u,v \in V$. Then,

$$\langle t(u),t(v) \rangle = \langle u, t^*(t(v)) \rangle = \langle u, t^*t(v) \rangle = \langle u, I_V(v) \rangle = \langle u,v \rangle$$

Conversely, let t be a linear operator on V such that $\langle t(u),t(v) \rangle = \langle u,v \rangle$ for all $u,v \in V$. Then,

$$\langle t(u),t(v) \rangle = \langle u,v \rangle \text{ for all } u,v \in V$$

$$\Rightarrow \quad \langle u, t^*(t(v)) \rangle = \langle u,v, \rangle \text{ for all } u,v \in V$$

$$\Rightarrow \quad \langle u, t^*t(v) \rangle = \langle u,v \rangle \text{ for all } u,v, \in V$$

$$\Rightarrow \quad \langle u, t^*t(v) \rangle = \langle u, I_V(v) \rangle \text{ for all } u,v \in V$$

$$\Rightarrow \quad \langle u, t^*t(v) \rangle = \langle u, I_V(v) \rangle \text{ for all } v \in V$$

$$\Rightarrow \quad \langle u, (t^*t - I_V)(v) \rangle = 0 \text{ for all } v \in V$$

$$\Rightarrow \quad \langle (t^*t - I_V)(v), v \rangle = 0 \text{ for all } v \in V$$

$$\Rightarrow \quad t^*t - I_V = \hat{0} \qquad \begin{bmatrix} \because \; (t^*t - I_V)^* = (t^*t)^* - I_V^* = t^*t - I_V \\ \therefore \; t^*t - I_V \text{ is self-adjoint} \end{bmatrix}$$

$$\Rightarrow \quad t^*t = I_V$$

Q.E.D.

THEOREM-2 *Let V be a finite dimensional unitary space. An operator t on V is unitary iff $||t(v)|| = ||v||$ for all $v \in V$.*

PROOF. First, let t be a unitary operator on V. Then, $tt^* = I_V$. For any $u, v \in V$

$$\langle t(u), t(v) \rangle = \langle u, t^*(t(v)) \rangle$$
$\Rightarrow \quad \langle t(u), t(v) \rangle = \langle u, t^*t(v) \rangle$
$\Rightarrow \quad \langle t(u), t(v) \rangle = \langle u, I_V(v) \rangle$
$\Rightarrow \quad \langle t(u), t(v) \rangle = \langle u, v \rangle$
$\therefore \quad \langle t(v), t(v) \rangle = \langle v, v \rangle \quad$ for all $v \in V$
$\Rightarrow \quad ||t(v)||^2 = ||v||^2 \quad$ for all $v \in V$
$\Rightarrow \quad ||t(v)|| = ||v|| \quad$ for all $v \in V$

Conversely, let t be a linear operator on V such that $||t(v)|| = v$ for all $v \in V$. i.e.

$$||t(v)||^2 = ||v||^2$$
$\Rightarrow \quad \langle t(v), t(v) \rangle = \langle v, v \rangle \quad$ for all $u \in V$
$\Rightarrow \quad \langle v, t^*(t(v)) \rangle = \langle v, v \rangle \quad$ for all $v \in V$
$\Rightarrow \quad \langle v, t^*t(v) \rangle = \langle v, v \rangle \quad$ for all $v \in V$
$\Rightarrow \quad \langle v, t^*t(v) - v \rangle = 0 \quad$ for all $v \in V$
$\Rightarrow \quad \langle v, (t^*t - I_V)(v) \rangle = 0 \quad$ for all $v \in V$
$\Rightarrow \quad \langle (t^*t - I_V)(v), v \rangle = 0 \quad$ for all $v \in V$
$\Rightarrow \quad t^*t - I_V = \hat{0} \quad \quad [\because t^*t - I_V$ is self-adjoint (See Theorem 12 on page 688)]
$\Rightarrow \quad t^*t = I_V$
$\Rightarrow \quad t$ is a unitary operator on V.

Combining these two theorems, we obtain the following theorem.

Q.E.D.

THEOREM-3 *Let t be a linear operator on a unitary space V. The following conditions on t are equivalent:*

(i) $t^* = t^{-1}$ or, $t^*t = I_V$, i.e. t is unitary
(ii) $\langle t(u), t(v) \rangle = \langle u, v \rangle$ for all $u, v \in V$
(iii) $||t(v)|| = ||v|| \quad$ for all $v \in V$.

PROOF. (i) $\Rightarrow$ (ii)

Suppose (i) holds. Then for every $u, v \in V$

$$\langle t(u), t(v) \rangle = \langle u, t^*t(v) \rangle = \langle u, I_V(v) \rangle = \langle u, v \rangle$$

Thus (i) implies (ii).

(ii) $\Rightarrow$ (iii).

Suppose (ii) holds,

i.e. $\langle t(u), t(v) \rangle = \langle u, v \rangle$ for all $u, v \in V$

$\Rightarrow$ $\langle t(v), t(v) \rangle = \langle v, v \rangle$ for all $v \in V$

$\Rightarrow$ $\|t(v)\|^2 = \|v\|^2$ for all $v \in V$

$\Rightarrow$ $\|t(v)\| = \|v\|$ for all $v \in V$

Hence, (ii) implies (iii).

(iii) $\Rightarrow$ (i)

Suppose (iii) holds,

i.e. $\|t(v)\| = \|v\|$ for all $v \in V$

$\Rightarrow$ $\|t(v)\|^2 = \|v\|^2$ for all $v \in V$

$\Rightarrow$ $\langle t(v), t(v) \rangle = \langle v, v \rangle$ for all $v \in V$

$\Rightarrow$ $\langle v, t^*t(v) \rangle = \langle v, v \rangle$ for all $v \in V$

$\Rightarrow$ $\langle v, (t^*t - I_V)(v) \rangle = 0$ for all $v \in V$

$\Rightarrow$ $\langle (t^*t - I_V)(v), v \rangle = 0$ for all $v \in V$

$\Rightarrow$ $t^*t - I_V = \hat{0}$ [See Theorem 12 on page 688]

$\Rightarrow$ $t^*t = I_V$

$\Rightarrow$ $t^* = t^{-1}$

Hence, (i) $\Leftrightarrow$ (ii) $\Leftrightarrow$ (iii).

Q.E.D.

THEOREM-4 *Let V be a finite dimensional unitary space. A linear operator t on V is unitary iff it takes an orthonormal basis of V onto an orthonormal basis of V.*

PROOF. First suppose that t is a unitary operator on V. Let $B = \{v_1, v_2, \ldots, v_n\}$ be an orthonormal basis for V. We have to prove that $\{t(v_1), t(v_2), \ldots, t(v_n)\}$ is also an orthonormal basis for V. For any $i, j \in \underline{n}$

$\langle t(v_i), t(v_j) \rangle = \langle v_i, v_j \rangle$ [t is unitary]

$\Rightarrow$ $\langle t(v_i), t(v_j) \rangle = \delta_{ij}$ [$\because$ B is an orthonormal basis for V]

$\therefore$ $\{t(v_1), t(v_2), \ldots, t(v_n)\}$ is an orthonormal set

Since t is a bijection. Therefore, $\{t(v_1), t(v_2), \ldots, t(v_n)\}$ is an orthonormal set consisting of n vectors. Also, an orthonormal set is linearly independent. Hence, $\{t(v_1), t(v_2), \ldots, t(v_n)\}$ is an orthonormal basis for V.

Conversely, let t be a linear operator on V such that it takes an orthonormal basis for V onto an orthonormal basis for V. This means that if $\{v_1, v_2, \ldots, v_n\}$ is an orthonormal basis for V, then $\{t(v_1), t(v_2), \ldots, t(v_n)\}$ is also an orthonormal basis for V.

Let u, v be any two vectors in V. Then,

$$u = \sum_{i=1}^{n} \lambda_i v_i \text{ and } v = \sum_{i=1}^{n} \mu_i v_i \text{ for some } \lambda_i, \mu_i \in C$$

$$\therefore \quad \langle u, v \rangle = \left\langle \sum_{i=1}^{n} \lambda_i v_i, \sum_{j=1}^{n} \mu_j v_j \right\rangle = \sum_{i=1}^{n} \sum_{j=1}^{n} \lambda_i \overline{\mu_j} \langle v_i, v_j \rangle = \sum_{i=1}^{n} \sum_{j=1}^{n} \lambda_i \overline{\mu_j} \delta_{ij} = \sum_{i=1}^{n} \lambda_i \overline{\mu_i} \quad \text{(i)}$$

and,

$$\langle t(u), t(v) \rangle = \left\langle t\left(\sum_{i=1}^{n} \lambda_i v_i\right), t\left(\sum_{j=1}^{n} \mu_j v_j\right) \right\rangle = \left\langle \sum_{i=1}^{n} \lambda_i t(v_i), \sum_{j=1}^{n} \mu_j t(v_j) \right\rangle$$

$$\Rightarrow \quad \langle t(u), t(v) \rangle = \sum_{i=1}^{n} \sum_{j=1}^{n} \lambda_i \overline{\mu_j} \langle v_i, v_j \rangle = \sum_{i=1}^{n} \sum_{j=1}^{n} \lambda_i \overline{\mu_j} \delta_{ij} = \sum_{i=1}^{n} \lambda_i \overline{\mu_i} \quad \text{(ii)}$$

From (i) and (ii), we get

$$\langle t(u), t(v) \rangle = \langle u, v \rangle \quad \text{for all } u, v \in V$$

Hence, t is a unitary operator on V.

Q.E.D.

THEOREM-5 *If t is a unitary operator on a unitary space V, and S be a subspace invariant under t. Then, $S^\perp$ is also invariant under t.*

PROOF. It is given that t is unitary. Therefore, it is invertible and hence $t(S) = S$
Consequently for any $v \in S$, there exists $v' \in S$ such that $t(v') = v$.
Let $v \in S^\perp$. Then, for any $u \in S$

$$\langle t(u), v \rangle = \langle t(u), t(v') \rangle$$
$$\Rightarrow \quad \langle t(u), v \rangle = \langle u, v' \rangle \qquad\qquad [\because t \text{ is unitary}]$$
$$\Rightarrow \quad \langle t(u), v \rangle = 0 \qquad\qquad [\because v' \in S \text{ and } u \in S^\perp \quad \therefore \quad \langle u, v' \rangle = 0]$$
$$\Rightarrow \quad t(u) \in S^\perp.$$

Thus, $u \in S^\perp \quad \Rightarrow \quad t(u) \in S^\perp$.

Hence, $S^\perp$ is invariant under t.

Q.E.D.

THEOREM-6 *A linear operator t on a finite dimensional unitary space V is unitary iff matrix of t relative to an orthonormal basis for V is unitary.*

PROOF. Let B be an orthonormal basis of finite dimensional inner product space V and let A be the matrix of t relative to basis B. That is, $A = M_B^B(t)$ Then,

$$M_B^B(t^*) = A^* = (\overline{A})^T \qquad \text{[See Theorem 9 on page 686]}$$

First suppose that t is a unitary operator on V. Then,

$$t^*t = I_V$$

$\Rightarrow \quad M_B^B(t^*t) = M_B^B(I_V)$

$\Rightarrow \quad M_B^B(t^*)\, M_B^B(t) = I \qquad [\because M_B^B(sot) = M_B^B(s)\, M_B^B(t)]$

$\Rightarrow \quad A^*A = I$

$\Rightarrow \quad A$ is a unitary matrix.

Conversely, let A be a unitary matrix. Then,

$$A^*A = I$$

$\Rightarrow \quad M_B^B(t^*)\, M_B^B(t) = M_B^B(I_V)$

$\Rightarrow \quad M_B^B(t^*t) = M_B^B(I_V)$

$\Rightarrow \quad t^*t = I_V$

$\Rightarrow \quad t$ is a unitary operator on V.

Q.E.D.

EXAMPLE *Let V be the vector space of all $n \times n$ matrices over C, with inner product defined by $\langle A, B \rangle = tr\,(AB^*)$ for all $A, B \in V$. For each $M \in V$, let t_M be the linear operator defined by $t_M(A) = MA$. Show that t_M is unitary iff M is a unitary matrix.*

SOLUTION We have, $(t_M)^* = t_{M^*}$ (See Example 2 on page 682).

Now,

t_M is unitary

$\Leftrightarrow \quad t_M(t_M)^* = I_V = (t_M)^* t_M$

$\Leftrightarrow \quad t_M\, t_{M^*} = I_V = t_{M^*}\, t_M \qquad [\because (t_M)^* = t_{M^*}]$

$\Leftrightarrow \quad t_{MM^*} = I_V = t_{M^*M} \qquad [\because t_M\, t_N = t_{MN}]$

$\Leftrightarrow \quad t_{MM^*} = I_V = t_{M^*M}$

$\Leftrightarrow \quad MM^* = I = M^*M$

$\Leftrightarrow \quad M$ is a unitary matrix.

THEOREM-7 *Let A be an $n \times n$ Hermitian matrix. Then there exists a unitary matrix P such that P^*AP is a diagonal matrix.*

PROOF. Let V denote the vector space C^n, with standard inner product and let $B = \{e_1^{(n)}, e_2^{(n)}, \ldots, e_n^{(n)}\}$ be the standard ordered basis for V. Let t be the linear operator on V such that $M_B^B(t) = A$. Then,

$$M_B^B(t^*) = A^*$$

$\Rightarrow \quad M_B^B(t^*) = A$ $\qquad [\because A \text{ is Hermitian} \therefore A^* = A]$

$\Rightarrow \quad M_B^B(t^*) = M_B^B(t)$

$\Rightarrow \quad t^* = t$

$\Rightarrow \quad t$ is a self-adjoint operator on V

$\Rightarrow \quad$ There exists an orthonormal basis C for V such that

$M_C^C(t)$ is a diagonal matrix

Let P be the change-of-basis matrix from B to C. Then, $P = M_C^B(I_V)$ is a unitary matrix. Therefore, $P^* = P^{-1}$.

Now,

$$M_C^C(t) = M_B^C(I_V)\, M_B^B(t)\, M_C^B(I_V)$$

$\Rightarrow \quad M_C^C(t) = P^{-1}AP$

$\Rightarrow \quad M_C^C(t) = P^*AP$

$\Rightarrow \quad P^*AP$ is a diagonal matrix $\qquad [\because M_C^C(t) \text{ is a diagonal matrix}]$

Q.E.D.

EXERCISE 9.2

1. Show that the set of all unitary operators on an inner product space V is a group under the operation of composition.
2. Let V be an inner product space, and suppose $t : V \to V$ (not assumed linear) is surjective and preserves inner products, i.e. $\langle t(u), t(v) \rangle = \langle u, v \rangle$ for every $u, v \in V$. Prove that t is linear and hence unitary.
3. Let S be a subspace of V. For any $v \in V$, let $v = w + w'$, where $w \in S$ and $w' \in S^\perp$. (Such a sum is unique because $V = S \oplus S^\perp$). Let $t : V \to V$ be defined by $t(v) = w - w'$. Show that t is self-adjoint unitary operator on V.
4. Let B and B' be two ordered orthonormal bases for a finite dimensional complex inner product space V. Prove that for each linear operator t on V, the matrix $M_{B'}^{B'}(t)$ is unitarily equivalent to the matrix $M_B^B(t)$.

9.6 NORMAL OPERATORS

NORMAL OPERATOR *A linear operator t on a finite dimensional unitary space V is said to be normal if it commutes with its adjoint, that is, if $tt^* = t^*t$.*

If t is a self-adjoint operator on a unitary space V, then

$$t = t^* \Rightarrow tt^* = tt = t^*t \Rightarrow tt^* = t^*t$$

So, a self-adjoint operator is always a normal operator.

Let t be a unitary operator on a finite dimensional unitary space V. Then,

$$tt^* = I_V = t^*t \Rightarrow tt^* = t^*t$$

So, a unitary space is also a normal operator.

THEOREM-1 *Let V be a finite dimensional unitary space. A linear operator t on V is a unitary operator iff its matrix relative to an orthonormal basis of V is normal.*

PROOF. Let $B = \{v_1, v_2, \ldots, v_n\}$ be an orthonormal basis for V, and let A be the matrix of t relative to basis B, i.e. $A = M_B^B(t)$. Then,

$$M_B^B(t^*) = A^* = (\overline{A})^T.$$

$\therefore \quad t(v_i) = \sum_{r=1}^{n} a_{ri} v_r \text{ and } t^*(v_i) = \sum_{s=1}^{n} \overline{a}_{is} v_s \qquad [\because \ ((\overline{A})^T)_{si} = \overline{a}_{is}]$

Now,

t is normal

$\Leftrightarrow \quad tt^* = t^*t \quad \Leftrightarrow \quad \langle tt^*(v_i), v_j \rangle = \langle t^*t(v_i), v_j \rangle \quad \text{for all } i, j \in \underline{n} \qquad (i)$

For any $i, j \in \underline{n}$

$$\langle tt^*(v_i), v_j \rangle = \langle t(t^*(v_i)), v_j \rangle = \left\langle t\left(\sum_{s=1}^{n} \overline{a}_{is} v_s \right), v_j \right\rangle$$

$\Rightarrow \quad \langle tt^*(v_i), v_j \rangle = \left\langle \sum_{s=1}^{n} \overline{a}_{is} t(v_s), v_j \right\rangle = \sum_{s=1}^{n} \overline{a}_{is} \langle t(v_s), v_j \rangle$

$\Rightarrow \quad \langle tt^*(v_i), v_j \rangle = \sum_{s=1}^{n} \overline{a}_{is} \left\langle \sum_{r=1}^{n} a_{rs} v_r, v_j \right\rangle = \sum_{s=1}^{n} \sum_{r=1}^{n} \overline{a}_{is} a_{rs} \langle v_r, v_j \rangle$

$\Rightarrow \quad \langle tt^*(v_i), v_j \rangle = \sum_{s=1}^{n} \sum_{r=1}^{n} \overline{a}_{is} a_{rs} \delta_{rj} = \sum_{s=1}^{n} \overline{a}_{is} a_{js}$

$\Rightarrow \quad \langle tt^*(v_i), v_j \rangle = \sum_{s=1}^{n} (A)_{js} ((\overline{A})^T)_{si} = (A\overline{A}^T)_{ji}$

and,

$\Rightarrow \quad \langle t^*t(v_i), v_j \rangle = \langle t^*(t(v_i)), v_j \rangle = \left\langle t^*\left(\sum_{r=1}^{n} a_{ri} v_r \right), v_j \right\rangle$

$\Rightarrow \quad \langle t^*t(v_i), v_j \rangle = \left\langle \sum_{r=1}^{n} a_{ri} t^*(v_r), v_j \right\rangle = \sum_{r=1}^{n} a_{ri} \langle t^*(v_r), v_j \rangle$

$\Rightarrow \quad \langle t^*t(v_i), v_j \rangle = \sum_{r=1}^{n} a_{ri} \left\langle \sum_{s=1}^{n} \bar{a}_{rs} v_s, v_j \right\rangle = \sum_{r=1}^{n} \sum_{s=1}^{n} a_{ri} \bar{a}_{rs} \langle v_s, v_j \rangle$

$\Rightarrow \quad \langle t^*t(v_i), v_j \rangle = \sum_{r=1}^{n} \sum_{s=1}^{n} a_{ri} \bar{a}_{rs} \delta_{sj}$

$\Rightarrow \quad \langle t^*t(v_i), v_j \rangle = \sum_{r=1}^{n} \bar{a}_{rj} a_{ri} = \sum_{r=1}^{n} ((\bar{A})^T)_{jr} (A)_{ri} = \left(\bar{A}^T A \right)_{ji}$

Substituting the values of $\langle tt^*(v_i), v_j \rangle$ and $\langle t^*t(v_i), v_j \rangle$ in (i), we find that

$$t \text{ is normal} \Leftrightarrow \left(A\bar{A}^T \right)_{ij} = (\bar{A}^T A)_{ij} \quad \text{for all } i, j \in \underline{n}$$

$\Rightarrow \quad t \text{ is normal} \Leftrightarrow A\bar{A}^T = \bar{A}^T A$

$\Rightarrow \quad t \text{ is normal} \Leftrightarrow A A^* = A^* A$

$\Rightarrow \quad t \text{ is normal} \Leftrightarrow A \text{ is normal.}$

Q.E.D.

THEOREM-2 *Let t be a normal operator on a finite dimensional unitary space V. The necessary and sufficient condition that a vector v be an eigenvector of t with eigenvalue λ is that it be an eigenvector of t^* with eigenvalue $\bar{\lambda}$.*

PROOF. *Necessary Condition:* Let t be a normal operator on V and v be an eigenvector of t with eigenvalue λ. Then, $t(v) = \lambda v$. We have to show that v is also an eigenvector of t^* with eigenvalue $\bar{\lambda}$.

Consider a vector $u = t^*(v) - \bar{\lambda}v$

Now,

$$\langle u, u \rangle = \langle t^*(v) - \bar{\lambda}v, t^*(v) - \bar{\lambda}v \rangle$$

$\Rightarrow \quad \langle u, u \rangle = \langle t^*(v), t^*(v) \rangle + \langle t^*(v), -\bar{\lambda}v \rangle + \langle -\bar{\lambda}v, t^*(v) \rangle + \langle -\bar{\lambda}v, -\bar{\lambda}v \rangle$

$\Rightarrow \quad \langle u, u \rangle = \langle v, tt^*(v) \rangle - \lambda \langle t^*(v), v \rangle - \bar{\lambda} \langle v, t^*(v) \rangle + \lambda\bar{\lambda} \langle v, v \rangle$

$\Rightarrow \quad \langle u, u \rangle = \langle v, t^*t(v) \rangle - \lambda \langle v, t(v) \rangle - \bar{\lambda} \langle t(v), v \rangle + \lambda\bar{\lambda} \langle v, v \rangle \quad [\because t \text{ is normal} \therefore tt^* = t^*t]$

$\Rightarrow \quad \langle u, u \rangle = \langle t(v), t(v) \rangle - \lambda \langle v, \lambda v \rangle - \bar{\lambda} \langle \lambda v, v \rangle + \lambda\bar{\lambda} \langle v, v \rangle$

$\Rightarrow \quad \langle u, u \rangle = \langle \lambda v, \lambda v \rangle - \lambda\bar{\lambda} \langle v, v \rangle - \bar{\lambda}\lambda \langle v, v \rangle + \lambda\bar{\lambda} \langle v, v \rangle$

$\Rightarrow \quad \langle u, u \rangle = \lambda\bar{\lambda} \langle v, v \rangle - \lambda\bar{\lambda} \langle v, v \rangle - \lambda\bar{\lambda} \langle v, v \rangle + \lambda\bar{\lambda} \langle v, v \rangle$

$\Rightarrow \quad \langle u, u \rangle = 0$

$\Rightarrow \quad u = 0_V \Rightarrow \quad t^*(v) = \bar{\lambda}v \Rightarrow v$ is an eigenvector of t^* with eigenvalue $\bar{\lambda}$.

Sufficient Condition: Let t be a normal operator on V such that v is an eigenvector of t^* with eigenvalue μ, i.e. $t^*(v) = \mu v$. We have to prove that v is eigenvector of t with eigenvalue $\bar{\mu}$.

Consider the vector $u = t(v) - \bar{\mu}v$. Then,

$$\langle u, u \rangle = \langle t(v) - \bar{\mu}v, t(v) - \bar{\mu}v \rangle = 0 \qquad \text{[As shown in Necessary Part]}$$

$\Rightarrow \quad u = 0_V \Rightarrow t(v) = \bar{\mu}v \Rightarrow v$ is eigenvector of t with eigenvalue $\bar{\mu}$.

Q.E.D.

THEOREM-3 *Let t be a linear operator on a unitary space V. Then the eigenvectors of t belonging to different eigenvalues of t are orthogonal.*

PROOF. Let u and v be the eigenvectors of t corresponding to two different eigenvalues λ_1 and λ_2 respectively. Then, $\bar{\lambda}_1$ and $\bar{\lambda}_2$ are eigenvectors of t^* corresponding to the vectors u and v respectively.

$\therefore \quad t(u) = \lambda_1 u, \ t(v) = \lambda_2 v, \ t^*(u) = \bar{\lambda}_1 u \text{ and } t^*(v) = \bar{\lambda}_2 v$

Now,

$$\lambda_1 \langle u, v \rangle = \langle \lambda_1 u, v \rangle = \langle t(u), v \rangle = \langle u, t^*(v) \rangle = \langle u, \bar{\lambda}_2 v \rangle = \lambda_2 \langle u, v \rangle$$

$\Rightarrow \quad (\lambda_1 - \lambda_2) \langle u, v \rangle = 0$

$\Rightarrow \quad \langle u, v \rangle = 0 \qquad\qquad\qquad\qquad\qquad\qquad\qquad\qquad\qquad [\because \ \lambda_1 \neq \lambda_2]$

$\Rightarrow \quad u$ is orthogonal to v.

Q.E.D.

COROLLARY-1 *Eigenspaces of a normal transformation are pairwise orthogonal.*

PROOF. Let S_1 and S_2 be the eigenspaces of a normal operator t on an inner product space V corresponding to the distinct eigenvalues λ_1 and λ_2 respectively. Let $u \in S_1$ and $v \in S_2$. Then,

$\qquad t(u) = \lambda_1 u \quad \text{and} \quad t(v) = \lambda_2 v$

$\Rightarrow \quad u$ and v are eigenvectors of t corresponding to two distinct eigenvalues

$\Rightarrow \quad \langle u, v \rangle = 0$

$\Rightarrow \quad u \perp v$

$\Rightarrow \quad S_1$ is orthogonal to S_2.

Q.E.D.

THEOREM-4 *A linear operator t on a finite dimensional unitary space V is normal iff it has diagonal matrix relative to some orthonormal basis for V.*

PROOF. First, let t be a normal operator on a finite dimensional unitary space V. Then we have to prove that matrix of t relative to an orthonormal basis of V is a diagonal matrix. We will prove it by mathematical induction on the dimension n of unitary space V. The theorem is trivially true for $n = 1$.

Let us assume that the theorem is true for all unitary spaces of dimension less than n.

Since V is a vector space over C (the field of complex numbers). Therefore, the characteristic polynomial of t has at least one root λ, that is, at least one eigenvalue λ and hence a non-zero vector v as eigenvector. Therefore, $t(v) = \lambda v$.

Let S be the subspace of V spanned by v and $v_1 = \dfrac{v_1}{\|v\|}$ be a unit vector in S. Let $u \in S$. Then,

$$u = \alpha v \text{ for some } \alpha \in C \qquad [\because \ S = [\{v\}]]$$
$$\Rightarrow \quad t(u) = t(\alpha v) = \alpha t(v) = \alpha(\lambda v) = (\alpha \lambda)v$$
$$\Rightarrow \quad t(u) \in S$$

Thus, $u \in S \Rightarrow t(u) \in S$

So, S is invariant under t.

Since v is an eigenvector of t. Therefore, it is also an eigenvector of t^* and hence S is also invariant under t^*.

Now,
$$S \text{ is invariant under } t^* \ \Rightarrow \ S^\perp \text{ is invariant under } (t^*)^* = t$$

Thus, restriction t_1 of t to $S^\perp$ is a normal operator.

We have, $\quad V = S \oplus S^\perp$
$$\Rightarrow \quad \dim V = \dim S + \dim S^\perp$$
$$\Rightarrow \quad n = 1 + \dim S^\perp \qquad [\because \ \dim S = 1]$$
$$\Rightarrow \quad \dim S^\perp = n - 1$$

By induction, there exists an orthonormal basis $\{v_2, v_3, \ldots, v_n\}$ of $S^\perp$ relative to which matrix of t is a diagonal matrix. But,

$$\langle v_1, v_i \rangle = 0 \quad \text{for } i = 1, 2, \ldots, n \qquad [\because \ v_1 \in S \text{ and } v_i \in S^\perp \text{ for } i = 2, 3, \ldots, n]$$

Therefore, $\{v_1, v_2, \ldots, v_n\}$ is an orthonormal basis for V relative to which matrix of t is a diagonal matrix.

Conversely, suppose that the matrix A of a linear operator t on V relative to an orthonormal basis B of V is a diagonal matrix. Then, we have to prove that t is a normal operator on V.

Now, A is a diagonal matrix

$\Rightarrow \quad A\overline{A}^T = \overline{A}^T A = A\overline{A}$

$\Rightarrow \quad M_B^B(t)\, M_B^B(t^*) = M_B^B(t^*)\, M_B^B(t)$

$\Rightarrow \quad M_B^B(tt^*) = M_B^B(t^*t)$

$\Rightarrow \quad tt^* = t^*t$

$\Rightarrow \quad t$ is a normal operator on V.

Q.E.D.

COROLLARY-1 *A self-adjoint linear operator on a finite dimensional unitary space V has real diagonal matrix relative to some orthonormal basis of V.*

PROOF. Every self-adjoint operator t on a unitary matrix is a normal operator. Therefore, matrix of t relative to an orthonormal basis of V is a diagonal matrix. By Theorem 4 on page 439, the diagonal entries of matrix of t are eigenvalues of t. Since the eigenvalues of a normal transformation are real. Therefore, t has a real diagonal matrix relative to some orthonormal basis of V.

Q.E.D.

COROLLARY-2 *All eigenvalues of a self-adjoint operator on a unitary space are real.*

PROOF. Every self-adjoint operator on a unitary space V is a normal operator. Therefore, if t is a self-adjoint operator on V. Then there exists an orthonormal basis B of V such that $M_B^B(t)$ is a diagonal matrix with diagonal entries as the eigenvalues of t.

i.e. $M_B^B(t) = \text{diag}(\lambda_1, \lambda_2, \ldots, \lambda_n)$, where $\lambda_1, \lambda_2, \ldots, \lambda_n$ are eigenvalues of t.

Let $A = M_B^B(t)$. Then,

$M_B^B(t^*) = (\overline{A})^T$

$\Rightarrow \quad M_B^B(t^*) = \text{diag}(\overline{\lambda}_1, \overline{\lambda}_2, \ldots, \overline{\lambda}_n)$

$\Rightarrow \quad M_B^B(t) = \text{diag}(\overline{\lambda}_1, \overline{\lambda}_2, \ldots, \overline{\lambda}_n)$ $\quad [\because t = t^*]$

$\Rightarrow \quad \text{diag}(\lambda_1, \lambda_2, \ldots, \lambda_n) = \text{diag}(\overline{\lambda}_1, \overline{\lambda}_2, \ldots, \overline{\lambda}_n)$

$\Rightarrow \quad \lambda_i = \overline{\lambda}_i \quad \text{for } i = 1, 2, \ldots, n$

Hence, eigenvalues of t are real.

Q.E.D.

THEOREM-5 *Let V be a finite dimensional unitary space and t be a linear operator on V. Then t is represented by matrices A and A', relative to possibly different orthonormal bases of V iff A' and A are unitarily equivalent.*

PROOF. First, suppose that A and A' are matrices of t relative to orthonormal bases B and C respectively, That is,
$$A = M_B^B(t) \text{ and } A' = M_C^C(t)$$
We have to prove that A and A' are unitarily equivalent matrices. Let P be the change-of-basis matrix from B to C. Then, P is a unitary matrix such that $P = M_C^B(I_V)$.

$\therefore \qquad P^{-1} = M_B^C(I_V)$

Now,
$$P^{-1}AP = M_B^C(I_V) \, M_B^B(t) \, M_C^B(I_V) = M_B^C(I_V) \, M_C^B(t)$$
$\Rightarrow \qquad P^{-1}AP = M_C^C(t) = A'$

$\Rightarrow \qquad P^*AP = A' \qquad\qquad [\because \, P^* = \overline{P}^T = P^{-1}]$

$\Rightarrow \qquad A$ and A' are unitarily equivalent matrices.

Conversely, let A and A' be unitarily equivalent matrices and B be an orthonormal basis of V such that $M_B^B(t) = A$.

Now,

A and A' are unitarily equivalent matrices

$\Rightarrow \qquad$ There exists a unitary matrix P such that $A' = P^*AP$

$\Rightarrow \qquad A' = P^{-1}AP \qquad\qquad [\because \, P^* = (\overline{P})^T = P^{-1}]$

Since P is invertible and B is an orthonormal basis of V. So, there exists exactly one orthonormal basis C of V (See Theorem 2 on page 387.) such that P is the change-of-basis matrix from basis B to basis C, i.e. $P = M_C^B(I_V)$.

$\therefore \qquad P^{-1} = M_B^C(I_V)$

Now,
$$P^*AP = \overline{P}^T AP = P^{-1}AP = M_B^C(I_V) \, M_B^B(t) \, M_C^B(I_V)$$
$\Rightarrow \qquad P^*AP = M_C^C(t)$

$\Rightarrow \qquad P^*AP = A'$

Hence, A and A' are represented by t relative to different orthonormal bases of V.
Let us now prove a theorem which is matrix analog of Theorem 4.

Q.E.D.

THEOREM-6 *A square matrix over C is unitarily equivalent to a diagonal matrix iff it is normal*

OR

Let A be an $n \times n$ matrix over C. Then there exists a unitary matrix P such that P^*AP is a diagonal matrix iff A is a normal matrix.

PROOF. Let $V = C^n$ be the vector space with standard Hermitian inner product defined on V and let B be its standard basis. Let t be a linear operator on V representing matrix A relative to basis B, i.e. $M_B^B(t) = A$. Then, $M_B^B(t^*) = A^*$.

First, let A be a normal matrix. Then, t is a normal operator on V (see Theorem 1 on page 706). Therefore, there exists an orthonormal basis C of V such that $M_C^C(t)$ is a diagonal matrix (See Theorem 4).

Let P be the change-of-basis matrix from B to C. Then,

$$M_C^C(t) = P^{-1}AP$$

By Theorem 6 on page 704, P is a unitary matrix such that $\overline{P}^T = P^{-1}$.

$$\therefore \quad M_C^C(t) = \overline{P}^T AP = P^*AP.$$

Thus, for any $n \times n$ normal matrix A, there exists a unitary matrix P such that P^*AP is a diagonal matrix.

Conversely, let matrix A be unitarily equivalent to a diagonal matrix. Then there exists a unitary matrix P such that P^*AP is a diagonal matrix. Let

$$D = P^*AP$$
$$\Rightarrow \quad \overline{P}^T AP = D$$
$$\Rightarrow \quad A = \left(\overline{P}^T\right)^{-1} DP^{-1} = PD\overline{P}^T \qquad [\because \overline{P}^T = P^{-1}]$$
$$\Rightarrow \quad \overline{A} = \overline{P}\,\overline{D}\,P^T$$
$$\Rightarrow \quad \overline{A}^T = P\,\overline{D}^T\,\overline{P}^T$$
$$\therefore \quad A\overline{A}^T = \left(PD\overline{P}^T\right)\left(P\overline{D}^T\overline{P}^T\right)$$
$$\Rightarrow \quad A\overline{A}^T = PD\left(\overline{P}^T P\right)\overline{D}^T\overline{P}^T$$
$$\Rightarrow \quad A\overline{A}^T = P\left(D\overline{D}^T\right)\overline{P}^T \qquad [\because \overline{P}^T P = I \text{ as } P \text{ is unitary}]$$
and $\quad \overline{A}^T A = \left(P\overline{D}^T\overline{P}^T\right)\left(PD\overline{P}^T\right)$

$\Rightarrow \quad \overline{A}^T A = P\overline{D}^T \left(\overline{P}^T P\right) D\overline{P}^T$

$\Rightarrow \quad \overline{A}^T A = P\overline{D}^T D\overline{P}^T$

$\Rightarrow \quad \overline{A}^T A = P(D\overline{D}^T)\overline{P}^T \qquad \left[\because \text{ D is a diagonal matrix} \therefore D\overline{D}^T = \overline{D}^T D\right]$

$\therefore \quad A\overline{A}^T = \overline{A}^T A$, i.e. $AA^* = A^*A$

$\Rightarrow \quad A$ is a normal matrix.

Q.E.D.

THEOREM-7 *Let t be a normal operator on a unitary space V. If λ is an eigenvalue of t, then the eigenspace $S_\lambda = \{v \in V : t(v) = \lambda v\}$ and its orthogonal complement $S_\lambda^\perp$ both are invariant under t and t^*.*

PROOF. Let $v \in S_\lambda$. Then, $t(v) = \lambda v$.

$\therefore \qquad t(t(v)) = t(\lambda v) = \lambda\, t(v) \qquad\qquad [\because\ t \text{ is linear}]$

$\Rightarrow \qquad t(v) \in S_\lambda$.

Thus, $\quad v \in S_\lambda \Rightarrow t(v) \in S_\lambda$

So, S_λ is invariant under t.

Now,

$\qquad t(t^*(v)) = tt^*(v) = t^*t(v) \qquad\qquad [\because\ t \text{ is normal} \therefore\ tt^* = t^*t]$

$\Rightarrow \qquad t(t^*(v)) = t^*(t(v))$

$\Rightarrow \qquad t(t^*(v)) = t^*(\lambda v)$

$\Rightarrow \qquad t(t^*(v)) = \lambda\, t^*(v)$

$\Rightarrow \qquad t^*(v) \in S_\lambda$

Thus, $\quad v \in S_\lambda \Rightarrow t^*(v) \in S_\lambda$

So, S_λ is invariant under t^*.

Hence, S_λ is invariant under t and t^*.

Now, $\qquad S_\lambda$ is invariant under t.

$\Rightarrow \qquad S_\lambda^\perp$ is invariant under t^*. [By Theorem 6 on page 685]

Again,

$\Rightarrow \qquad S_\lambda$ is invariant under t^*.

$\Rightarrow \qquad S_\lambda^\perp$ is invariant under $(t^*)^* = t$. [By Theorem 6 on page 685]

Q.E.D.

ILLUSTRATIVE EXAMPLES

EXAMPLE-1 *Let t be a normal operator on a unitary space V and λ be a scalar. Show that λt is also a normal operator.*

SOLUTION It is given that t is normal.

$$\therefore \qquad tt^* = t^*t \qquad (i)$$

We have,

$$(\lambda t)^* = \bar{\lambda} t^*$$

$$\therefore \quad (\lambda t)(\lambda t)^* = (\lambda t)\left(\bar{\lambda} t^*\right) = \lambda\bar{\lambda}(tt^*) = |\lambda|^2 \, tt^*$$

and, $\quad (\lambda t)^*(\lambda t) = \left(\bar{\lambda} t^*\right)(\lambda t) = \bar{\lambda}\lambda(t^*t) = |\lambda|^2 \, t^*t = |\lambda|^2 tt^* \qquad$ [Using (i)]

$$\therefore \quad (\lambda t)(\lambda t)^* = (\lambda t)^*(\lambda t)$$

$\Rightarrow \qquad \lambda t$ is a normal operator on V.

EXAMPLE-2 *Let t be a linear operator on a unitary space V. Prove that t is normal iff its real and imaginary parts commute.*

PROOF. Every linear operator on a unitary space can be written in the form $t = t_1 + it_2$, where t_1 and t_2 are self-adjoint operators, i.e. $t_1^* = t_1$ and $t_2^* = t_2$.

First, let t be a normal operator on V. Then,

$$tt^* = t^*t$$

$\Rightarrow \quad (t_1 + it_2)(t_1 + it_2)^* = (t_1 + it_2)^*(t_1 + it_2)$

$\Rightarrow \quad (t_1 + it_2)(t_1^* - it_2^*) = (t_1^* - it_2^*)(t_1 + it_2)$

$\Rightarrow \quad (t_1 + it_2)(t_1 - it_2) = (t_1 - it_2)(t_1 + it_2)$

$\Rightarrow \quad t_1^2 - it_1t_2 + it_2t_1 + t_2^2 = t_1^2 + it_1t_2 - it_2t_1 + t_2^2$

$\Rightarrow \quad 2i(t_1t_2 - t_2t_1) = \hat{0}$

$\Rightarrow \quad t_1t_2 = t_2t_1$

Conversely, let $t_1t_2 = t_2t_1$. Then, we have to show that t is normal. As shown above, we have

$$tt^* = t_1^2 - it_1t_2 + it_2t_1 + t_2^2 = t_1^2 + t_2^2 \qquad [\because \; t_1t_2 = t_2t_1]$$

and $\quad t^*t = t_1^2 + it_1t_2 - it_2t_1 + t_2^2 = t_1^2 + t_2^2 \qquad [\because \; t_1t_2 = t_2t_1]$

$\therefore \qquad tt^* = t^*t$

$\Rightarrow \qquad t$ is a normal operator on V.

EXAMPLE-3 Let t be a linear operator on a finite dimensional unitary space V and $a, b \in C$ such that $|a| = |b| = 1$. Show that $at + bt^*$ is a normal operator on V.

SOLUTION We have,
$$(at + bt^*)(at + bt^*)^* = (at + bt^*)(\bar{a}t^* + \bar{b}t) \qquad [\because \ (at)^* = \bar{a}t^* \text{ and } (bt^*)^* = \bar{b}t]$$
$$= a\bar{a}\, tt^* + a\bar{b}\, t^2 + b\bar{a}\, (t^*)^2 + b\bar{b}\, t^*t$$
$$= |a|^2\, tt^* + a\bar{b}\, t^2 + b\bar{a}\, (t^*)^2 + |b|^2\, t^*t$$
$$= tt^* + a\bar{b}\, t^2 + b\bar{a}\, (t^*)^2 + t^*t \qquad [\because \ |a| = |b| = 1] \text{ (i)}$$

and,
$$(at + bt^*)^*(at + bt^*) = (\bar{a}t^* + \bar{b}t)(at + bt^*)$$
$$= \bar{a}a\, t^*t + \bar{a}b\, (t^*)^2 + \bar{b}a\, t^2 + \bar{b}b\, tt^*$$
$$= |a|^2\, t^*t + \bar{a}b\, (t^*)^2 + \bar{b}a\, t^2 + |b|^2\, tt^*$$
$$= t^*t + \bar{a}b\, (t^*)^2 + \bar{b}a\, t^2 + tt^* \qquad [\because \ |a| = |b| = 1] \text{ (ii)}$$

From (i) and (ii), we get
$$(at + bt^*)(at + bt^*)^* = (at + bt^*)^*(at + bt^*)$$
$\Rightarrow \quad at + bt^*$ is a normal operator on V.

EXAMPLE-4 Let t be a normal operator on a finite dimensional unitary space V. Then, $t(v) = 0$ iff $t^*(v) = 0$.

SOLUTION We have, $tt^* = t^*t$.
Now,
$$\langle t(v), t(v) \rangle = \langle v, t^*(t(v)) \rangle = \langle v, t^*t(v) \rangle$$
$\Rightarrow \quad \langle t(v), t(v) \rangle = \langle v, tt^*(v) \rangle \qquad [\because \ tt^* = t^*t]$
$\Rightarrow \quad \langle t(v), t(v) \rangle = \langle t^*(v), t^*(v) \rangle$
$\therefore \quad \langle t(v), t(v) \rangle = 0$ iff $\langle t^*(v), t^*(v) \rangle = 0$
$\Rightarrow \quad t(v) = 0$ iff $t^*(v) = 0$

EXAMPLE-5 Let t be a normal operator on a finite dimensional unitary space V and $\lambda \in C$. Then, show that $t - \lambda I$ is also a normal operator on V.

SOLUTION We have,
$$(t - \lambda I)(t - \lambda I)^* = (t - \lambda I)(t^* - \bar{\lambda} I)$$
$$= tt^* - \lambda t^* - \bar{\lambda} t + \lambda \bar{\lambda} I$$

$$= t^*t - \bar{\lambda}t - \lambda t^* + \bar{\lambda}\lambda I \qquad [\because \ tt^* = t^*t]$$
$$= (t^* - \bar{\lambda}I)(t - \lambda I)$$
$$= (t - \lambda I)^*(t - \lambda I)$$

$\therefore \quad t - \lambda I$ is a normal operator on V.

EXAMPLE-6 *Let t be a linear operator on a finite dimensional unitary space V. Then any eigenvector of t is also an eigenvector of t^*.*

SOLUTION   Let v be an eigenvector of t with eigenvalues λ. Then,

$$t(v) = \lambda v$$
$$\Rightarrow \quad (t - \lambda I)(v) = 0_V$$
$$\Rightarrow \quad (t - \lambda I)^*(v) = 0_V \qquad \left[\because \ t - \lambda I \text{ is a normal operator (By Example 5)}\right]$$
$$\Rightarrow \quad \left(t^* - \bar{\lambda}I\right)(v) = 0_V$$
$$\Rightarrow \quad t^*(v) - \bar{\lambda}I(v) = 0_V$$
$$\Rightarrow \quad t^*(v) = \bar{\lambda}v$$
$$\Rightarrow \quad v \text{ is an eigenvector of } t \text{ with eigenvalue } \bar{\lambda}.$$

EXAMPLE-7 *Let t be a linear operator on a finite dimensional inner product space V. If $\|t(v)\| = \|t^*(v)\|$ for all $v \in V$, then t is normal.*

SOLUTION   We have,

$$\|t(v)\| = \|t^*(v)\| \quad \text{for all } v \in V$$
$$\Rightarrow \quad \|t(v)\|^2 = \|t^*(v)\|^2 \quad \text{for all } v \in V$$
$$\Rightarrow \quad \langle t(v), t(v) \rangle = \langle t^*(v), t^*(v) \rangle \quad \text{for all } v \in V$$
$$\Rightarrow \quad \langle v, t^*(t(v)) \rangle = \langle v, t(t^*(v)) \rangle \quad \text{for all } v \in V \qquad \text{[By the definition of adjoint]}$$
$$\Rightarrow \quad \langle v, t^*t(v) \rangle = \langle v, tt^*(v) \rangle \quad \text{for all } v \in V$$
$$\Rightarrow \quad \langle v, t^*t(v) - tt^*(v) \rangle = 0 \quad \text{for all } v \in V$$
$$\Rightarrow \quad \langle v, (t^*t - tt^*)(v) \rangle = 0 \quad \text{for all } v \in V$$
$$\Rightarrow \quad t^*t - tt^* = \hat{0} \qquad [\because \ t^*t - tt^* \text{ is self-adjoint}]$$
$$\Rightarrow \quad t^*t = tt^*$$
$$\Rightarrow \quad t \text{ is normal operator on } V.$$

EXAMPLE-8 *If t_1 and t_2 are normal operators on a unitary space V with the property that either commutes with the adjoint of the other, then prove that $t_1 + t_2$ is also a normal operator on V.*

SOLUTION It is given that t_1 and t_2 are normal operators on V such that either commutes with the adjoint of other i.e.

$$t_1 t_2^* = t_2^* t_1 \text{ and } t_2 t_1^* = t_1^* t_2 \qquad \text{(i)}$$

Now,
$$\begin{aligned}
(t_1 + t_2)(t_1 + t_2)^* &= (t_1 + t_2)(t_1^* + t_2^*) \\
&= t_1 t_1^* + t_1 t_2^* + t_2 t_1^* + t_2 t_2^* \\
&= t_1^* t_1 + t_2^* t_1 + t_1^* t_2 + t_2^* t_2 \qquad [\because \ t_1, t_2 \text{ are normal and using (i)}] \\
&= (t_1^* + t_2^*)(t_1 + t_2) \\
&= (t_1 + t_2)^* (t_1 + t_2)
\end{aligned}$$

$\therefore \quad t_1 + t_2$ is a normal operator on V.

EXAMPLE-9 *Let t be a normal operator on a unitary space V and v be a vector in V such that $t^2(v) = 0_V$. Then show that $t(v) = 0_V$. Hence, show that the range and null space of a normal operator are disjoint subspaces of V.*

SOLUTION Let $u = t(v)$. Then,

$$t^2(v) = 0_V \Rightarrow t(t(v)) = 0_V \Rightarrow t(u) = 0_V$$

Since t is a normal operator on V. Therefore,

$$\|t(u)\| = \|t^*(u)\|$$

$\Rightarrow \quad \|t^*(u)\| = 0$

$\Rightarrow \quad t^*(u) = 0_V$

Now, $\quad \langle u, u \rangle = \langle u, t(v) \rangle = \langle t^*(u), v \rangle = \langle 0_V, v \rangle = 0 \qquad [\because u = t(v)]$

$\Rightarrow \quad u = 0_V \Rightarrow t(v) = 0_V \qquad\qquad\qquad\qquad\qquad\qquad\qquad [\because u = t(v)]$

Now, let $v \in I_m(t) \cap \text{Ker}(t)$.

$\Rightarrow \quad v \in I_m(t)$ and $v \in \text{Ker}(t)$

$\Rightarrow \quad v = t(u) \quad$ for some $u \in V$ and $t(v) = 0_V$

Now, $\quad t(u) = v$

$\Rightarrow \quad t(t(u)) = t(v)$

$\Rightarrow \quad t^2(u) = 0_V$

$\Rightarrow \quad t(u) = 0_V \qquad [\because \ t \text{ is a normal operator and } t^2(u) = 0_V]$

$\Rightarrow \quad v = 0_V \qquad [\because \ v = t(u)]$

Thus, $\quad v \in I_m(t) \cap \text{Ker}(t) \Rightarrow v = 0_V$

$\therefore \quad I_m(t) \cap \text{Ker}(t) = \{0_V\}$

Hence, $I_m(t)$ and Ker (t) are disjoint subspaces of V.

EXAMPLE-10 Let t be a linear operator on a finite be dimensional unitary space V. If $B = \{v_1, v_2, \ldots, v_n\}$ is an orthonormal basis for V such that each vector in B is an eigenvector of t. Then, prove that t is a normal operator on V.

SOLUTION For any $i \in \underline{n}$, let λ_i be the eigenvalue of t corresponding to the eigenvector v_i. Then,

$t(v_i) = \lambda_i v_i; \ i = 1, 2, \ldots, n.$

$\Rightarrow \quad M_B^B(t)$ is a diagonal matrix with diagonal elements $\lambda_1, \lambda_2 \ldots, \lambda_n$

$\Rightarrow \quad [M_B^B(t)]^*$ is a diagonal matrix with diagonal elements $\overline{\lambda}_1, \overline{\lambda}_2 \ldots, \overline{\lambda}_n$

$\Rightarrow \quad M_B^B(t^*)$ is a diagonal matrix with diagonal elements $\overline{\lambda}_1, \overline{\lambda}_2 \ldots, \overline{\lambda}_n$

Since product of diagonal matrices is commutative.

$\therefore \quad M_B^B(t) \ M_B^B(t^*) = M_B^B(t^*) \ M_B^B(t)$

$\Rightarrow \quad M_B^B(tt^*) = M_B^B(t^*t)$

$\Rightarrow \quad tt^* = t^*t$

$\Rightarrow \quad t$ is a normal operator on V.

EXAMPLE-11 Let t be a normal operator on a finite dimensional unitary space V and let f be a polynomial with complex coefficients. Then, prove that $f(t)$ is also a normal operator on V.

SOLUTION Let $f(x) = a_0 + a_1 x + a_2 x^2 + \cdots + a_n x^n$ be a polynomial over C. Then,

$f(t) = a_0 I + a_1 t + a_2 t^2 + \cdots + a_n t^n$

For any $m \in M$, we have

$(t^*)^m f(t) = (t^*)^m (a_0 I + a_1 t + a_2 t^2 + \cdots + a_n t^n)$

$= a_0 (t^*)^m I + a_1 (t^*)^m t + a_2 (t^*)^m t^2 + \cdots + a_n (t^*)^m t^n \qquad [\because \ (t^*)^m \text{ is linear}]$

$= a_0 I (t^*)^m + a_1 t (t^*)^m + a_2 t^2 (t^*)^m + \cdots + a_n t^n (t^*)^m \qquad [\because \ t^p (t^*)^q = (t^*)^q t^p \text{ for all } p, q \in M]$

$= (a_0 I + a_1 t + a_2 t^2 + \cdots + a_n t^n)(t^*)^m$

$= f(t)(t^*)^m \qquad \qquad \qquad \qquad \qquad \qquad \qquad \qquad \qquad \qquad \qquad \qquad \text{(i)}$

$$\therefore \quad [f(t)]^* f(t) = \{\bar{a}_0 I + \bar{a}_1 t^* + \bar{a}_2 (t^*)^2 + \cdots + \bar{a}_n (t^*)^n\} f(t)$$
$$= \bar{a}_0 I f(t) + \bar{a}_1 t^* f(t) + \bar{a}_2 (t^*)^2 f(t) + \cdots + \bar{a}_n (t^*)^n f(t)$$
$$= \bar{a}_0 f(t) I + \bar{a}_1 f(t) t^* + \bar{a}_2 f(t) (t^*)^2 + \cdots + \bar{a}_n f(t) (t^*)^n \quad \text{[Using (i)]}$$
$$= f(t) \{\bar{a}_0 I + \bar{a}_1 t^* + \bar{a}_2 (t^*)^2 + \cdots + \bar{a}_n (t^*)^n\}$$
$$= f(t) [f(t)]^*$$

Hence, $f(t)$ is a normal operator on V.

EXAMPLE-12 *Show that the minimal polynomial of a normal operator on a finite dimensional dimensional unitary space has distinct roots.*

SOLUTION Let $f(x)$ be the minimal polynomial of a normal operator t on a finite dimensional unitary space V. Then, $f(x)$ is the monic polynomial of the lowest degree that annihilates t i.e. for which $f(t) = \hat{0}$.

We have to show that $f(x)$ has distinct roots i.e. $f(x)$ is expressible in the form
$$f(x) = (x - \lambda_1)(x - \lambda_2)\ldots(x - \lambda_k),$$
where $\lambda_1, \lambda_2, \ldots, \lambda_k$ all distinct complex numbers.

If possible, let $\lambda_1, \lambda_2, \ldots, \lambda_k$ are not all distinct i.e. some root of $f(x)$ repeats at least twice. Let λ be such root. Then,
$$f(x) = (x - \lambda)^2 g(x) \quad \text{for some polynomial } g(x)$$

Now,
$$f(t) = \hat{0}$$
$$\Rightarrow \quad (t - \lambda I)^2 g(t) = \hat{0}$$
$$\Rightarrow \quad \{(t - \lambda I)^2 g(t)\}(v) = \hat{0}(v) \quad \text{for all } v \in V$$
$$\Rightarrow \quad (t - \lambda I)^2 \{(g(t))(v)\} = 0_V \quad \text{for all } V \in V$$
$$\Rightarrow \quad (t - \lambda I)^2 (\beta) = 0_V \quad \text{for all } \beta, \text{ where } \beta = (g(t))(v)$$
$$\Rightarrow \quad T^2(\beta) = 0_V \text{ for all } \beta, \text{ where } T = t - \lambda I$$
$$\Rightarrow \quad T(\beta) = 0_V \text{ for all } \beta \quad \begin{bmatrix} \because T = t - \lambda I \text{ is a normal operator (See Ex 5),} \\ \text{From Ex 9, } T^2(\beta) = 0_V \Rightarrow T(\beta) = 0_V \end{bmatrix}$$
$$\Rightarrow \quad (t - \lambda I)(\beta) = 0_V \quad \text{for all } \beta$$
$$\Rightarrow \quad (t - \lambda I)(g(t))(v) = 0_V \quad \text{for all } v \in V$$
$$\Rightarrow \quad (t - \lambda I) g(t) = \hat{0}$$
$$\Rightarrow \quad t \text{ annihilates the monic polynomial } \phi(x) = (x - \lambda) g(x)$$
whose degree is less than that of $f(x)$.

This contradicts that $f(x)$ is the minimal polynomial of normal operator t.

Hence, no root λ of $f(x)$ is repeated. Therefore, $f(x)$ has distinct roots.

EXAMPLE-13 Let $A = \begin{bmatrix} 2 & i \\ i & 2 \end{bmatrix}$. Verify that A is normal. Find a unitary matrix P such that P^*AP is a diagonal matrix. Find P^*AP.

SOLUTION The characteristic polynomial $\Delta(\lambda)$ of A is

$$\Delta(\lambda) = |A - \lambda I| = \begin{vmatrix} 2-\lambda & i \\ i & 2-\lambda \end{vmatrix} = \lambda^2 - 4\lambda + 5$$

The eigenvalues of A are given by

$$\Delta(\lambda) = 0 \Rightarrow \lambda^2 - 4\lambda + 5 = 0 \Rightarrow \lambda = 2+i,\ 2-i$$

Eigenvector corresponding to $\lambda_1 = 2+i$: Let $X = \begin{bmatrix} x_1 \\ x_2 \end{bmatrix}$ be an eigenvector corresponding to the eigenvalue $\lambda_1 = 2+i$. Then,

$$(A - \lambda_1 I)X = 0$$

$$\Rightarrow \begin{bmatrix} -i & i \\ i & -i \end{bmatrix} \begin{bmatrix} x_1 \\ x_2 \end{bmatrix} = \begin{bmatrix} 0 \\ 0 \end{bmatrix} \Rightarrow x_1 - x_2 = 0 \Rightarrow x_1 = x_2 = a \text{ (say)}$$

Thus, $S_1 = \left\{ \begin{bmatrix} a \\ a \end{bmatrix} : a \in C \right\}$ is the eigenspace corresponding to the eigenvalue $\lambda_1 = 2+i$.

Clearly, S_1 is spanned by $v_1 = \begin{bmatrix} 1 \\ 1 \end{bmatrix}$. Normalizing v_1, we obtain $u_1 = \begin{bmatrix} \frac{1}{\sqrt{2}} \\ \frac{1}{\sqrt{2}} \end{bmatrix}$.

Eigenvector corresponding to $\lambda_2 = 2-i$: Let $X = \begin{bmatrix} x_1 \\ x_2 \end{bmatrix}$ be an eigenvector corresponding to the eigenvalue $\lambda_2 = 2-i$. Then,

$$(A - \lambda_2 I)X = 0$$

$$\Rightarrow \begin{bmatrix} i & i \\ i & i \end{bmatrix} \begin{bmatrix} x_1 \\ x_2 \end{bmatrix} = \begin{bmatrix} 0 \\ 0 \end{bmatrix}$$

$$\Rightarrow x_1 + x_2 = 0 \Rightarrow x_2 = -x_1 = b \text{ (say)}.$$

Thus, $S_2 = \left\{ \begin{bmatrix} -b \\ b \end{bmatrix} : b \in C \right\}$ is the eigenspace corresponding to the eigenvalue $\lambda_2 = 2-i$.

Clearly, $v_2 = \begin{bmatrix} -1 \\ 1 \end{bmatrix}$ forms a basis for S_2. Normalizing v_2, we get $u_2 = \begin{bmatrix} -\frac{1}{\sqrt{2}} \\ \frac{1}{\sqrt{2}} \end{bmatrix}$.

Clearly, u_1, u_2 are orthonormal vectors.

$$\therefore \quad P = \begin{bmatrix} \dfrac{1}{\sqrt{2}} & -\dfrac{1}{\sqrt{2}} \\ \dfrac{1}{\sqrt{2}} & \dfrac{1}{\sqrt{2}} \end{bmatrix} \Rightarrow P^* = \overline{P}^T = \begin{bmatrix} \dfrac{1}{\sqrt{2}} & \dfrac{1}{\sqrt{2}} \\ -\dfrac{1}{\sqrt{2}} & \dfrac{1}{\sqrt{2}} \end{bmatrix}$$

Clearly, $P^*P = I = PP^*$.

So, P is a unitary matrix.

Now, $\quad P^*AP = \begin{bmatrix} \dfrac{1}{\sqrt{2}} & \dfrac{1}{\sqrt{2}} \\ -\dfrac{1}{\sqrt{2}} & \dfrac{1}{\sqrt{2}} \end{bmatrix} \begin{bmatrix} 2 & i \\ i & 2 \end{bmatrix} \begin{bmatrix} \dfrac{1}{\sqrt{2}} & -\dfrac{1}{\sqrt{2}} \\ \dfrac{1}{\sqrt{2}} & \dfrac{1}{\sqrt{2}} \end{bmatrix} = \begin{bmatrix} 2+i & 0 \\ 0 & 2-i \end{bmatrix}$

EXERCISE 9.3

1. Show that a skew-adjoint operator is normal.
2. Let t be a normal operator on a unitary space V. Prove that:

    (i) t is self-adjoint iff its eigenvalues are real.

    (ii) t is unitary iff its eigenvalues have absolute value 1.

3. If t is a normal operator on a unitary space V, then show that t and t^* have the same kernel and the same image.
4. If t_1 and t_2 are normal operators on a unitary space V such that $t_1 t_2 = t_2 t_1$, show that $t_1 + t_2$ and $t_1 t_2$ are also normal.
5. If t_1 and t_2 are operators on a unitary space V such that t_1 is normal and commutes with t_2. Show that t_1 also commutes with t_2^*.
6. An operator t on a unitary space V is a normal operator iff $||t(v)|| = ||t^*(v)||$ for all $v \in V$.
7. If t is a normal operator on a finite dimensional unitary space V, then for any $m, n \in N$ prove that

    (i) $t(t^*)^n = (t^*)^n t$

    (ii) $t^n t^* = t^* t^n$

    (iii) $(t^*)^n t^m = t^m (t^*)^n$

    [Hint: Use the principle of mathematical induction]

8. Let t_1 and t_2 be normal operators on a finite dimensional unitary space V such that they commute. Show that $t_1 t_2$ is a normal operator on V.

Chapter 10

Bilinear and Quadratic Forms

10.1 INTRODUCTION

In previous chapters, we have learnt about linear forms or linear functionals. In this chapter, we will generalize the notion of linear forms. In fact, we will introduce the notion of a bilinear form on a finite-dimensional vector space. We have studied linear forms on $V(F)$. Here, we will study bilinear forms as mappings from $V \times V$ to F, which are linear forms in each variable. Bilinear forms also give rise to quadratic and Hermitian forms.

10.2 BILINEAR FORMS

BILINEAR FORM *Let V be a vector space of finite dimension over a field F. A bilinear form on V is a mapping $f : V \times V \to F$ such that*

(i) $f(au_1 + bu_2, v) = af(u_1, v) + bf(u_2, v)$

(ii) $f(u, av_1 + bv_2) = af(u, v_1) + bf(u, v_2)$

for all $u, u_1, u_2, v, v_1, v_2 \in V$ and all $a, b \in F$.

Bilinear forms are often called 2–forms.

For a fixed $v \in V$, if we define $\phi_v(u) = f(u, v)$ for all $u \in V$.

Then, for all $u_1, u_2 \in V$ and all $a, b \in F$

$\phi_v(au_1 + bu_2) = f(au_1 + bu_2, v) = af(u_1, v) + bf(u_2, v)$ [From (i)]

$\Rightarrow \quad \phi_v(au_1 + bu_2) = a\, \phi_v(u_1) + b\, \phi_v(u_2)$

$\therefore \quad \phi_v : V \to F$ is a linear form on V.

Thus, condition (i) in the definition of f states that f is linear in the first variable.

For a fixed $u \in V$, let us now define a function $\psi_u : V \to F$ by the rule

$\psi_u(v) = f(u, v) \quad \text{for all } u \in V$

For any $v_1, v_2 \in V$ and $a, b \in F$

$$\phi_u(av_1 + bv_2) = f(u, av_1 + bv_2)$$
$$= af(u, v_1) + bf(u, v_2) \quad \text{[From (ii)]}$$
$$= a\phi_u(v_1) + b\phi_u(v_2)$$

$\therefore \quad \psi_u : V \to F$ is a linear form on V.

Thus, condition (ii) in the definition of f states that f is linear in the second variable.

It follows from the above definition that a bilinear form f on a vector space $V(F)$ is a function from $V \times V \to F$ which is linear as a function of either of its variables when the other is fixed.

ILLUSTRATIVE EXAMPLES

EXAMPLE-1 *Consider the vector space R^2. Show that the mapping $f : R^2 \to R$ given by*

$$f(u, v) = x_1 y_2 - x_2 y_1 \quad \text{for all } u = (x_1, x_2), v = (y_1, y_2)$$

is a bilinear form on R^2.

SOLUTION For any three vectors $u = (x_1, x_2), v = (y_1, y_2), w = (z_1, z_2)$ in R^2 and any $a, b \in R$, we have

(i) $f(au + bv, w) = f((ax_1 + by_1, ax_2 + by_2), (z_1, z_2))$
$= (ax_1 + by_1) z_2 - (ax_2 + by_2) z_1$
$= a(x_1 z_2 - x_2 z_1) + b(y_1 z_2 - y_2 z_1)$
$= af(u, w) + bf(v, w)$

(ii) $f(u, av + bw) = f((x_1, x_2), (ay_1 + bz_1, ay_2 + bz_2))$
$= x_1(ay_2 + bz_2) - x_2(ay_1 + bz_1)$
$= a(x_1 y_2 - x_2 y_1) + b(x_1 z_2 - x_2 z_1)$
$= af(u, v) + bf(u, w)$

Hence, f is a bilinear form on R^2.

EXAMPLE-2 *Show that $f : R^2 \to R$ defined by*

$$f[(x_1, x_2), (y_1, y_2)] = x_1 y_2 + x_2 y_1 \quad \text{for all } (x_1, x_2), (y_1, y_2) \in R^2$$

is a bilinear form on R^2.

SOLUTION Let $u = (x_1, x_2), v = (y_1, y_2), w = (z_1, z_2) \in R^2$ and $a, b \in R$.

Then,

$$f(au+bv, w) = f[(ax_1+by_1, ax_2+by_2), (z_1, z_2)]$$
$$= (ax_1+by_1)z_2 + (ax_2+by_2)z_1$$
$$= a(x_1z_2+x_2z_1) + b(y_1z_2+y_2z_1)$$
$$= af[(x_1,x_2),(z_1,z_2)] + b[(y_1,y_2),(z_1,z_2)]$$
$$= af(u,w) + bf(v,w)$$

and,

$$f(u, av+bw) = f[(x_1,x_2),(ay_1+bz_1, ay_2+bz_2)]$$
$$= x_1(ay_2+bz_2) + x_2(ay_1+bz_1)$$
$$= a(x_1y_2+x_2y_1) + b(x_1z_2+x_2z_1)$$
$$= af[(x_1,x_2),(y_1,y_2)] + bf[(x_1,x_2),(z_1,z_2)]$$
$$= af(u,v) + bf(u,w)$$

Hence, f is a bilinear form on R^2.

EXAMPLE-3 *Show that the mapping* $f : R^2 \to R$ *defined by*

$$f(u,v) = x_1x_2 + y_1y_2 \quad \text{for all } u = (x_1,x_2), v = (y_1,y_2) \text{ in } R^2$$

is not a bilinear form.

SOLUTION Let $u = (x_1, x_2), v = (y_1, y_2), w = (z_1, z_2) \in R^2$ and $a, b \in R$. Then

$$f(au+bv,w) = f[(ax_1+by_1, ax_2+by_2),(z_1,z_2)]$$
$$= (ax_1+by_1)(ax_2+by_2) + z_1z_2$$
$$= a^2x_1x_2 + b^2y_1y_2 + ab(x_1y_2+x_2y_1) + z_1z_2$$

and,

$$af(u,w) + bf(v,w) = a(x_1x_2+z_1z_2) + b(y_1y_2+z_1z_2)$$

Clearly, $f(au+bv,w) \neq af(u,w) + bf(v,w)$

Hence, f is not a bilinear form on R^2.

EXAMPLE-4 *Let* ϕ *and* ψ *be arbitrary linear forms on a vector space* $V(F)$. *Let* $f : V \times V \to F$ *be defined by*

$$f(u,v) = \phi(u)\psi(v) \quad \text{for all } u, v \in V$$

Then, f is a bilinear form on V.

SOLUTION For any $u, u_1, u_2, v, v_1, v_2 \in V$ and any $a, b \in F$, we have
$$\begin{aligned}
f(au_1+bu_2,v) &= \phi(au_1+bu_2)\psi(v) \\
&= \{a\phi(u_1)+b\phi(u_2)\}\psi(v) \qquad [\because \ \phi : V \to F \text{ is a linear form}] \\
&= a\phi(u_1)\psi(v)+b\phi(u_2)\psi(v) \\
&= af(u_1,v)+bf(u_2,v)
\end{aligned}$$

and,
$$\begin{aligned}
f(u,av_1+bv_2) &= \phi(u)\psi(av_1+bv_2) \\
&= \phi(u)\{a\psi(v_1)+b\psi(v_2)\} \\
&= a\phi(u)\psi(v_1)+b\phi(u)\psi(v_2) \\
&= af(u,v_1)+bf(u,v_2)
\end{aligned}$$

Hence, f is a bilinear form on V.

EXAMPLE-5 Let F be a field, A be an arbitrary matrix in $F^n \times F^n$, where $n \in N$. Then a mapping $f : F^n \times F^n \to F$ given by
$$f(X,Y) = X^T AY \quad \text{for all } X, Y \in F^n$$
is a bilinear form on F^n.

SOLUTION For any $X, Y, Z \in F^n$ and $a, b \in F$
$$\begin{aligned}
f(aX+bY, Z) &= (aX+bY)^T AZ \\
&= (aX^T+bY^T)AZ \\
&= a(X^T AZ)+b(Y^T AZ) \\
&= af(X,Z)+bf(Y,Z) \\
f(X, aY+bZ) &= X^T A(aY+bZ) \\
&= a(X^T AY)+b(X^T AZ) \\
&= af(X,Y)+bf(X,Z)
\end{aligned}$$

Hence, f is a bilinear form on F^n.

REMARK. Let $A = [a_{ij}]$ be any $n \times n$ matrix over a field. Then, A may be identified with the following bilinear form f on F^n
$$f(X,Y) = X^T AY = \sum_{i=1}^{n}\sum_{j=1}^{n} x_i\, a_{ij}\, y_j$$
$$= a_{11}\, x_1\, y_1 + a_{12}\, x_1\, y_2 + \cdots + a_{nn}\, x_n\, y_n,$$
where $X = [x_i]$ and $Y = [y_i]$ are column vectors of variables.

This formal expression in the variables x_i, y_i is known as the bilinear polynomial corresponding to the matrix A. In Theorem 3 of Section 10.3, it has been proved that, in a certain sense, every bilinear form is of this type.

EXAMPLE-6 Let F be a field and $V = F^n$ be the vector space of all n tuples over field F. Let $u = (a_1, a_2, \ldots, a_n)$ and $v = (b_1, b_2, \ldots, b_n)$ be any two vectors in V and let $f : V \times V \to F$ be a mapping defined by

$$f(u,v) = a_1 b_1 + a_2 b_2 + \cdots + a_n b_n$$

Show that f is a bilinear form on F^n.

SOLUTION For any $u = (a_1, a_2, \ldots, a_n)$, $v = (b_1, b_2, \ldots, b_n)$ and $w = (c_1, c_2, \ldots, c_n) \in V$ and $a, b \in F$

$$\begin{aligned}
f(au+bv, w) &= f\big((aa_1+bb_1, aa_2+bb_2+\cdots+aa_n+bb_n), (c_1,c_2,\ldots,c_n)\big) \\
&= (aa_1+bb_1)c_1 + (aa_2+bb_2)c_2 + \cdots + (aa_n+bb_n)c_n \\
&= a(a_1c_1+a_2c_2+\cdots+a_nc_n) + b(b_1c_1+b_2c_2+\cdots+b_nc_n) \\
&= af(u,w) + bf(v,w)
\end{aligned}$$

and,

$$\begin{aligned}
f(u, av+bw) &= f\big((a_1,a_2,\ldots,a_n),(ab_1+bc_1, ab_2+bc_2,\ldots, ab_n+bc_n)\big) \\
&= a_1(ab_1+bc_1) + a_2(ab_2+bc_2) + \cdots + a_n(ab_n+bc_n) \\
&= a(a_1b_1+a_2b_2+\cdots+a_nb_n) + b(a_1c_1+a_2c_2+\cdots+a_nc_n) \\
&= af(u,v) + bf(u,w)
\end{aligned}$$

Hence, f is a bilinear form on F^n.

REMARK. If $F = R$, then f is a bilinear form on R^n and $f(u,v) = u \cdot v$ is generally known as the dot product on R^n.

EXAMPLE-7 Let t be a linear operator on a vector space $V(F)$ and f be a bilinear form on V. Show that $\phi : V \times V \to F$ defined as

$$\phi(u,v) = f(t(u), t(v)) \quad \text{for all } u, v \in V$$

is a bilinear form on V.

SOLUTION For any $u, v, w \in V$ and $a, b \in F$

$$\begin{aligned}
\phi(au+bv, w) &= f\big(t(au+bv), t(w)\big) \\
&= f\big(at(u)+bt(v), t(w)\big) && [\because t \text{ is a linear operator on } V] \\
&= af(t(u), t(w)) + bf(t(v), t(w)) && [\because f \text{ is a linear form on } V] \\
&= a\,\phi(u,w) + b\,\phi(v,w)
\end{aligned}$$

and,

$$\begin{aligned}
\phi(u, av+bw) &= f\{t(u), t(av+bw)\} \\
&= f\{t(u), a\,t(v) + b\,t(w)\} && [\because t \text{ is a linear operator}]
\end{aligned}$$

$$= f(t(u),\ a\,t(v)) + f(t(u),\ b\,t(w))$$
$$= af(t(u),\ t(v)) + bf(t(u),\ t(w)) \qquad [\because\ f \text{ is a linear form}]$$
$$= a\,\phi(u,v) + b\,\phi(u,w)$$

Hence, ϕ is a bilinear form on V.

EXAMPLE-8 *Let $V(F)$ be a vector space. Then the zero function $\hat{0} : V \times V \to F$ associating each $(u,v) \in V \times V$ to the zero of F is a bilinear form on V.*

SOLUTION For any $u, v, w \in V$ and $a, b \in F$

$$\hat{0}(au + bv,\ w) = 0 = 0 + 0 = a0 + b0 = a\,\hat{0}(u,w) + b\,\hat{0}(v,w)$$

and $\quad \hat{0}(u,\ av + bw) = 0 = a0 + b0 = a\,\hat{0}(u,v) + b\,\hat{0}(u,w)$

Hence, $\hat{0} : V \times V \to F$ is a bilinear form on V.

EXAMPLE-9 *Let $V(F)$ be a vector space and f be a bilinear form on V. Then the function $-f : V \times V \to F$ defined as*

$$(-f)(u,v) = -f(u,v) \quad \text{for all } u, v \in V$$

is a bilinear form on V.

SOLUTION For any $u, v, w \in V$ and $a, b \in F$

$$(-f)(au + bv,\ w) = -f(au + bv, w)$$
$$= -\{af(u,w) + bf(v,w)\}$$
$$= a\{-f(u,w)\} + b\{-f(v,w)\}$$
$$= a(-f)(u,w) + b(-f)(v,w)$$

and,

$$(-f)(u,\ av + bw) = -f(u, av + bw)$$
$$= -\{af(u,v) + bf(u,w)\}$$
$$= a\{-f(u,v)\} + b\{-f(u,w)\}$$
$$= a\{(-f)(u,v)\} + b\{(-f)(u,w)\}$$

Hence, $-f$ is a bilinear form on V.

EXAMPLE-10 *Let V be the vector space of all $m \times n$ matrices over a field F and A be a fixed $m \times m$ matrix over F. Then show that $f_A : V \to F$ given by*

$$f_A(X,Y) = tr\left(X^T A Y\right) \quad \text{for all } X, Y \in V$$

is a bilinear form on V.

SOLUTION Let X, Y, Z be $m \times n$ matrices over F, i.e. $X, Y, Z \in V$ and $a, b \in F$. Then,

$$\begin{aligned}
f_A(X, aY + bZ) &= tr\{X^T A(aY + bZ)\} \\
&= tr(aX^T AY + bX^T AZ) \\
&= tr(aX^T AY) + tr(bX^T AZ) \\
&= a\, tr(X^T AY) + b\, tr(X^T AZ) \quad \left[\begin{array}{l} \because \; tr(A+B) = tr(A) + tr(B) \\ \text{and,}\; tr(\lambda A) = \lambda tr(A) \end{array}\right] \\
&= a f_A(X, Y) + b f_A(X, Z)
\end{aligned}$$

and,

$$\begin{aligned}
f_A(aX + bY, Z) &= tr\{(aX + bY)^T AZ\} \\
&= tr\{(aX^T + bY^T)AZ\} \\
&= tr\{a(X^T AZ) + b(Y^T AZ)\} \\
&= tr\{a(X^T AZ)\} + tr\{b(Y^T AZ)\} \\
&= a\, tr(X^T AZ) + b\, tr(Y^T AZ) \\
&= a f_A(X, Z) + b f_A(Y, Z)
\end{aligned}$$

Hence, $f_A : V \to F$ is a bilinear form.

EXAMPLE-11 *Let $V(F)$ be a vector space and $a \in F$. If f is a bilinear form on V, show that af is also a bilinear form on V.*

SOLUTION For any $u, v, w \in V$ and $\lambda, \mu \in F$

$$\begin{aligned}
(af)(\lambda u + \mu v, w) &= a f(\lambda u + \mu v, w) & [\because \; (af)(u) = af(u)] \\
&= a\{\lambda f(u, w) + \mu f(v, w)\} & [\because \; f \text{ is bilinear}] \\
&= a\lambda\, f(u, w) + a\mu\, f(v, w) \\
&= \lambda a\, f(u, w) + \mu a\, f(v, w) \\
&= \lambda\, (af)(u, w) + \mu\, (af)(v, w)
\end{aligned}$$

and,

$$\begin{aligned}
(af)(u, \lambda v + \mu w) &= a\, f(u, \lambda v + \mu w) & [\because \; (af)(u) = af(u)] \\
&= a\{\lambda\, f(u, v) + \mu\, f(u, w)\} \\
&= a\lambda\, f(u, v) + a\mu\, f(u, w) \\
&= \lambda a\, f(u, v) + \mu a\, f(u, w) \\
&= \lambda\{(af)(u, v)\} + \mu\{(af)(u, w)\}
\end{aligned}$$

Hence, af is a bilinear form on V.

EXAMPLE-12 Let $V(F)$ be a vector space and f,g be bilinear forms on V. Then, $f+g$ is also a bilinear form on V.

SOLUTION For any $u,v,w \in V$ and $a,b \in F$

$$(f+g)(u,av+bw) = f(u,av+bw)+g(u,av+bw)$$
$$= \{a\,f(u,v)+b\,f(u,w)\}+\{a\,g(u,v)+b\,g(u,w)\}$$
$$= a\{f(u,v)+g(u,v)\}+b\{f(u,w)+g(u,w)\}$$
$$= a(f+g)(u,v)+b(f+g)(u,w)$$

Similarly, we have

$$(f+g)(au+bv,w) = a(f+g)(u,w)+b(f+g)(v,w)$$

Hence, $f+g$ is a bilinear form on V.

EXAMPLE-13 Let $X=(x_1,x_2)$ and $Y=(y_1,y_2)$ be vectors in R^2 and f be a bilinear form on R^2 given by

$$f(X,Y) = x_1y_1+x_1y_2+x_2y_1+x_2y_2$$

Express f in matrix notation.

SOLUTION If $A=[a_{ij}]$ is an $n \times n$ matrix over a field F, then a bilinear form on F^n may be written as

$$f(X,Y) = X^T A Y = \sum_{i=1}^{n}\sum_{j=1}^{n} x_i a_{ij} y_j = a_{11}\,x_1y_1 + a_{12}\,x_1y_2 + a_{13}\,x_1y_3 + \cdots + a_{nn}\,x_ny_n$$

where $X=[x_i]$ and $Y=[y_i]$ are column vectors in F^n.
We have,

$$f(X,Y) = x_1y_1+x_1y_2+x_2y_1+x_2y_2$$

Let $A=[a_{ij}]$, where a_{ij} is the coefficient of x_iy_j. Then,

$$a_{11}=1, \quad a_{12}=1, \quad a_{21}=1, \quad a_{22}=1$$

$$\therefore \quad f(X,Y) = X^T A Y = [x_1\ x_2]\begin{bmatrix}1 & 1 \\ 1 & 1\end{bmatrix}\begin{bmatrix}y_1 \\ y_2\end{bmatrix} \text{ is required matrix notation of } f.$$

EXAMPLE-14 Let $X=(x_1,x_2,x_3)$ and $Y=(y_1,y_2,y_3)$ be vectors in R^3 and f be a bilinear form on R^3 given by

$$f(X,Y) = 3x_1y_1 - 2x_1y_3 + 5x_2y_1 + 7x_2y_2 - 8x_2y_3 + 4x_3y_2 - 6x_3y_3$$

Express f in matrix notation.

SOLUTION Let $A = [a_{ij}]$, where a_{ij} = coefficient of $x_i y_j$. Then,

$$a_{11} = 3, \ a_{12} = 0, \ a_{13} = -2; \ a_{21} = 5, \ a_{22} = 7, \ a_{23} = -8; \ a_{31} = 0, \ a_{32} = 4, \ a_{33} = -6$$

$$\therefore \quad A = \begin{bmatrix} 3 & 0 & -2 \\ 5 & 7 & -8 \\ 0 & 4 & -6 \end{bmatrix}$$

$$\therefore \quad f(X,Y) = X^T A Y = [x_1 \ x_2 \ x_3] \begin{bmatrix} 3 & 0 & -2 \\ 5 & 7 & -8 \\ 0 & 4 & -6 \end{bmatrix} \begin{bmatrix} y_1 \\ y_2 \\ y_3 \end{bmatrix} \text{ is required matrix notation of } f.$$

EXAMPLE-15 Describe explicitly all bilinear forms f on R^3 with the property that $f(u,v) = f(v,u)$ for all $u, v \in R^3$.

SOLUTION We know that any bilinear form on R^n is expressible in the form

$$f(X,Y) = X^T A Y \quad \text{for all } X, Y \in R^n \qquad \text{[See Example 5]}$$

where $A = [a_{ij}]$ is the matrix of f in a basis of R^n.

$\therefore \quad f(u,v) = f(v,u) \quad$ for all $u, v, \in R^3$

$\Rightarrow \quad f(X,Y) = f(Y,X) \quad$ for all $X, Y \in R^3$

$\Rightarrow \quad X^T A Y = Y^T A X \quad$ for all $X, Y \in R^3$

$\Rightarrow \quad [x_1 x_2 x_3] \begin{bmatrix} a_{11} & a_{12} & a_{13} \\ a_{21} & a_{22} & a_{23} \\ a_{31} & a_{32} & a_{33} \end{bmatrix} \begin{bmatrix} y_1 \\ y_2 \\ y_3 \end{bmatrix} = [y_1 y_2 y_3] \begin{bmatrix} a_{11} & a_{12} & a_{13} \\ a_{21} & a_{22} & a_{23} \\ a_{31} & a_{32} & a_{33} \end{bmatrix} \begin{bmatrix} x_1 \\ x_2 \\ x_3 \end{bmatrix}$ for all $x_i, y_i \in R$

$\Rightarrow \quad a_{11} x_1 y_1 + a_{12} x_1 y_2 + a_{13} x_1 y_3 + a_{21} x_2 y_1 + a_{22} x_2 y_2 + a_{23} x_2 y_3 + a_{31} x_3 y_1 + a_{32} x_3 y_2$
$\quad + a_{33} x_3 y_3 = a_{11} x_1 y_1 + a_{12} y_1 x_2 + a_{13} y_1 x_3 + a_{21} y_2 x_1 + a_{22} y_2 x_2 + a_{23} y_2 x_3$
$\quad + a_{31} y_3 x_1 + a_{32} y_3 x_2 + a_{33} y_3 x_3 \quad$ for all $x_i, y_i \in R$

$\Rightarrow \quad (a_{12} - a_{21}) x_1 y_2 + (a_{13} - a_{31}) x_1 y_3 + (a_{23} - a_{32}) x_2 y_3 = 0 \quad$ for all $x_i, y_i \in R$

$\Rightarrow \quad a_{12} - a_{21} = 0, \ a_{13} - a_{31} = 0, \ a_{23} - a_{32} = 0$

$\Rightarrow \quad a_{12} = a_{21}, \ a_{13} = a_{31}, \ a_{23} = a_{32}$

$\therefore \quad f(X,Y) = X^T A Y = \sum_{i=1}^{3} \sum_{j=1}^{3} x_i a_{ij} y_j$

$\Rightarrow \quad f(X, Y) = a_{11} x_1 y_1 + a_{22} x_2 y_2 + a_{33} x_3 y_3 + a_{12}(x_1 y_2 + x_2 y_1) + a_{13}(x_1 y_3 + x_3 y_1)$
$\quad + a_{23}(x_2 y_3 + x_3 y_2),$

$\qquad\qquad\qquad\qquad$ for all $x = (x_1, x_2, x_3), \ y = (y_1, y_2, y_3) \in R^3$

EXAMPLE-16 *Describe a bilinear form f on R^3 which satisfy $f(u,v) = -f(v,u)$ for all $u,v \in R^3$.*

SOLUTION Proceeding as in Example 15, we get

$$a_{11}x_1y_1 + a_{22}x_2y_2 + a_{33}x_3y_3 + (a_{12}+a_{21})x_1y_2 + (a_{13}+a_{31})x_1y_3 + (a_{23}+a_{32})x_2y_3 = 0$$
$$\text{for all } x_i, y_i \in R$$

$\Rightarrow \quad a_{11} = 0, \quad a_{22} = 0, \quad a_{33} = 0, \quad a_{12}+a_{21} = 0, \quad a_{13}+a_{31} = 0, \quad a_{23}+a_{32} = 0$

Hence,

$$f(X,Y) = a_{12}(x_1y_2 - x_2y_1) + a_{13}(x_1y_3 - x_3y_1) + a_{23}(x_2y_3 - x_3y_2)$$
$$\text{for all } X = (x_1, x_2, x_3), \ Y = (y_1, y_2, y_3) \in R^3$$

10.3 SPACE OF BILINEAR FORMS

In the previous section, we have learnt that for any two bilinear forms f, g on a vector space $V(F)$ and any scalar $a \in F$, $f + g$ and af are bilinear forms on V. In this section, we shall show that the set $L(V, V, F)$ of all bilinear forms on a vector space $V(F)$ is a vector space over field F.

THEOREM-1 *Let $V(F)$ be a vector space. Prove that the set $L(V,V,F)$ of all bilinear forms on V is a vector space with the vector addition and scalar multiplication defined as follows:*

$$(f+g)(u,v) = f(u,v) + g(u,v)$$

and $\qquad (kf)(u,v) = kf(u,v),$ *for all $f, g \in L(V,V,F)$ and all $k \in F$.*

PROOF. For the vector addition and scalar multiplication defined in the statement, we will verify vector space axioms V-1 and V-2 (i) to V-2 (iv).

V-1 $L(V,V,F)$ is an abelian group under vector addition:

Associativity: For any $f, g, h \in L(V,V,F)$, and any $(u,v) \in V \times V$, we have

$$\begin{aligned}
((f+g)+h)(u,v) &= (f+g)(u,v) + h(u,v) \\
&= \{f(u,v) + g(u,v)\} + h(u,v) \\
&= f(u,v) + \{g(u,v) + h(u,v)\} \quad [\because \text{ Addition is associative on } F] \\
&= f(u,v) + (g+h)(u,v) \\
&= (f+(g+h))(u,v)
\end{aligned}$$

$\therefore \qquad (f+g)+h = f+(g+h)$

So, vector addition is associative on $L(V,V,F)$.

Commutativity: For any $f, g \in L(V, V, F)$ and any $(u, v) \in V \times V$

$$(f + g)(u, v) = f(u, v) + g(u, v)$$
$$= g(u, v) + f(u, v) \qquad [\because \text{ Addition is commutative on } V]$$
$$= (g + f)(u, v)$$

$\therefore \qquad f + g = g + f$

So, vector addition is commutative on $L(V, V, F)$.

Existence of additive identity: The function $\hat{0}(u, v) = 0$ for all $(u, v) \in V \times V$ is a bilinear form (see Example 8) such that for all $(u, v) \in V \times V$

$$(f + \hat{0})(u, v) = f(u, v) + \hat{0}(u, v) = f(u, v) + 0 = f(u, v)$$

and, $\quad (\hat{0} + f)(u, v) = \hat{0}(u, v) + f(u, v) = 0 + f(u, v) = f(u, v)$

$\therefore \qquad f + \hat{0} = f = \hat{0} + f$

So, $\hat{0}$ is the additive identity.

Existence of additive inverse: Let f be an arbitrary bilinear form on V, i.e. $f \in L(V, V, F)$. Then, $-f$ is also a bilinear form (see Example 9) such that for all $(u, v) \in V \times V$

$$(f + (-f))(u, v) = f(u, v) + (-f)(u, v) = f(u, v) - f(u, v) = 0 = \hat{0}(u, v)$$

and, $\quad ((-f) + f)(u, v) = (-f)(u, v) + f(u, v) = -f(u, v) + f(u, v) = 0 = \hat{0}(u, v)$

Thus, for each $f \in L(V, V, F)$ there exists $-f \in L(V, V, F)$ such that

$$f + (-f) = \hat{0} = (-f) + f$$

So, each $f \in L(V, V, F)$ has its additive inverse $(-f)$.

Hence, $L(V, V, F)$ is an abelian group under the defined vector addition.

V-2 For any $f, g \in L(V, V, F)$ and $a, b \in F$, we have

(i) $\quad (a(f + g))(u, v) = a(f + g)(u, v)$
$$= a\{f(u, v) + g(u, v)\}$$
$$= af(u, v) + ag(u, v)$$
$$= (af)(u, v) + (ag)(u, v) \quad \text{for all } (u, v) \in V \times V$$

$\therefore \qquad a(f + g) = af + ag$

(ii) $\quad ((a + b)f)(u, v) = (a + b)f(u, v)$
$$= af(u, v) + bf(u, v)$$
$$= (af)(u, v) + (bf)(u, v)$$
$$= (af + bf)(u, v) \quad \text{for all } (u, v) \in V \times V$$

$\therefore \qquad (a + b)f = af + bf$

(iii) $((ab)f)(u,v) = (ab)f(u,v)$
$= a\{b(f(u,v))\}$
$= a\{(bf)(u,v)\}$
$= (a(bf))(u,v)$ for all $(u,v) \in V \times V$

$\therefore \quad (ab)f = a(bf)$

(iv) $(1f)(u,v) = 1f(u,v)$
$= f(u,v)$ for all $(u,v) \in V \times V$ $\quad [\because$ 1 is unity in $F]$

$\therefore \quad 1f = f$

Hence, $L(V,V,F)$ is a vector space over field F.

Q.E.D.

THEOREM-2 Let $V(F)$ be a vector space of dimension n. Let $\{x_1, x_2, \ldots, x_n\}$ be any basis of the dual space V^*. Then $\{f_{ij} : i, j = 1, 2, \ldots, n\}$ is a basis of $L(V,V,F)$, where f_{ij} is defined by

$$f_{ij}(u,v) = x_i(u)x_j(v) \quad \text{for all } (u,v) \in V \times V$$

PROOF. Let $B = \{v_1, v_2, \ldots, v_n\}$ be the basis of V dual to the basis $X = \{x_1, x_2, \ldots, x_n\}$ of the dual space V^*. Then,

$$x_i(v_j) = \delta_{ij} \quad \text{for all } i, j \in \underline{n}$$

Let $B' = \{f_{ij} : i, j = 1, 2, \ldots, n\}$. We shall now show that B' is basis for $L(V,V,F)$

B' **spans** $L(V, V, F)$: Let $f \in L(V,V,F)$ be an arbitrary bilinear form on V such that $f(v_i, v_j) = a_{ij}$ for all $i, j \in \underline{n}$.

We claim that $f = \sum_{i=1}^{n} \sum_{j=1}^{n} a_{ij} f_{ij}$. For this, it is sufficient to show that f and $\sum_{i=1}^{n} \sum_{j=1}^{n} a_{ij} f_{ij}$ are same on B

i.e. $f(v_r, v_s) = \sum_{i=1}^{n} \sum_{j=1}^{n} a_{ij} f_{ij}(v_r, v_s) \quad$ for all $r, s \in \underline{n}$

Now,

$$\sum_{i=1}^{n} \sum_{j=1}^{n} a_{ij} f_{ij}(v_r, v_s) = \sum_{i=1}^{n} \sum_{j=1}^{n} a_{ij} x_i(v_r) x_j(v_s)$$
$$= \sum_{i=1}^{n} \sum_{j=1}^{n} a_{ij} \delta_{ir} \delta_{js}$$
$$= \sum_{i=1}^{n} a_{is} \delta_{ir} = a_{rs} = f(v_r, v_s)$$

So, B' spans $L(V,V,F)$.

B' is linearly independent: Let a_{ij}; $i,j \in \underline{n}$ be n^2 scalars such that

$$\sum_{i=1}^{n}\sum_{j=1}^{n} a_{ij}\, f_{ij} = \hat{0}, \text{ here } \hat{0} \text{ is the zero bilinear form.}$$

$\Rightarrow \quad \sum_{i=1}^{n}\sum_{j=1}^{n} a_{ij}\, f_{ij}(v_r, v_s) = \hat{0}(v_r, v_s) \quad \text{for all } r,s \in \underline{n}$

$\Rightarrow \quad \sum_{i=1}^{n}\sum_{j=1}^{n} a_{ij}\, x_i(v_r)\, x_j(v_s) = 0 \quad \text{for all } r,s \in \underline{n}$

$\Rightarrow \quad \sum_{i=1}^{n}\sum_{j=1}^{n} a_{ij}\, \delta_{ir}\, \delta_{js} = 0 \quad \text{for all } r,s \in \underline{n}$

$\Rightarrow \quad \sum_{i=1}^{n} a_{is}\, \delta_{ir} = 0 \quad \text{for all } r,s \in \underline{n}$

$\Rightarrow \quad a_{rs} = 0 \quad \text{for all } r,s \in \underline{n}$

$\therefore \quad B'$ is a linearly independent set.

Hence, B' is a basis for $L(V,V,F)$.

Q.E.D.

COROLLARY. If $V(F)$ is vector space of dimension n, then $L(V,V,F)$ is a vector space of dimension n^2.

PROOF. Let $B = \{v_1, v_2, \ldots, v_n\}$ be a basis for V and $X = \{x_1, x_2, \ldots, x_n\}$ be the basis dual to B. Then, $B' = \{f_{ij} : i,j \in \underline{n}\}$ is a basis for $L(V,V,F)$, where f_{ij} is a bilinear form given by

$$f_{ij}(u,v) = x_i(u) x_j(v) \quad \text{for all } i,j \in \underline{n}$$

$\therefore \quad \text{Dim}(L(V,V,F)) = $ Number of bilinear forms in $B' = n^2$.

Q.E.D.

THEOREM-3 *Let V be a finite dimensional vector space over the field F and let $B = \{v_1, v_2, \ldots, v_n\}$ be an ordered basis for V. If f is a bilinear form on V, then there exists an $m \times n$ matrix A over F such that*

$$f(u,v) = [u]_B^T A [u]_B \quad \text{for all } u,v \in V$$

PROOF. Since $B = \{v_1, v_2, \ldots, v_n\}$ is a basis for V and $u, v \in V$. Therefore, there exist scalars $\lambda_1, \lambda_2, \ldots, \lambda_n$ and $\mu_1, \mu_2, \ldots, \mu_n$ in F such that $u = \sum_{i=1}^{n} \lambda_i v_i$ and $v = \sum_{i=1}^{n} \mu_i v_i$.

Clearly, coordinate vectors (matrices) of u and v relative to basis B are $\begin{bmatrix} \lambda_1 \\ \lambda_2 \\ \vdots \\ \lambda_n \end{bmatrix}$ and $\begin{bmatrix} \mu_1 \\ \mu_2 \\ \vdots \\ \mu_n \end{bmatrix}$

respectively, i.e $[u]_B = \begin{bmatrix} \lambda_1 \\ \lambda_2 \\ \vdots \\ \lambda_n \end{bmatrix}$ and $[v]_B = \begin{bmatrix} \mu_1 \\ \mu_2 \\ \vdots \\ \mu_n \end{bmatrix}$.

Now,

$$f(u,v) = f\left(\sum_{i=1}^{n} \lambda_i v_i, \sum_{j=1}^{n} \mu_j v_j\right)$$

$$\Rightarrow f(u,v) = \sum_{i=1}^{n} \sum_{j=1}^{n} \lambda_i \mu_j f(v_i, v_j)$$

$$\Rightarrow f(u,v) = \sum_{i=1}^{n} \sum_{j=1}^{n} \lambda_i \mu_j a_{ij}, \quad \text{where } a_{ij} = f(v_i, v_j) \text{ for all } i,j \in \underline{n}$$

$$\Rightarrow f(u,v) = \sum_{i=1}^{n} \sum_{j=1}^{n} \lambda_i a_{ij} \mu_j$$

$$\Rightarrow f(u,v) = [u]_B^T A [v]_B$$

Q.E.D.

It follows from the above theorem that if $B = \{v_1, v_2, \ldots, v_n\}$ is an ordered basis of a finite dimensional vector space $V(F)$, then each bilinear form f on V determines an $n \times n$ matrix $A = [a_{ij}]$ over F such that

$$f(u,v) = [u]_B^T A [v]_B \quad \text{for all } u, v \in V$$

where $a_{ij} = f(v_i, v_j)$ for all $i, j \in \underline{n}$

The converse of the above theorem is also true, i.e. if $V(F)$ is a vector space with an ordered basis $B = \{v_1, v_2, \ldots, v_n\}$, then every $n \times n$ matrix $A = [a_{ij}]$ over F determines a unique bilinear form f on V such that $a_{ij} = f(v_i, v_j)$ for all $i, j \in \underline{n}$.

This is stated and proved in the following theorem.

THEOREM-4 *Let $V(F)$ be an n-dimensional vector space with an ordered basis $B = \{v_1, v_2, \ldots, v_n\}$ and let $A = [a_{ij}]$ be an $n \times n$ matrix over F. Then there exists a unique bilinear form f on V such that $f(v_i, v_j) = a_{ij}$ for all $i, j \in \underline{n}$*

PROOF. Since B is a basis for V. Therefore, for any $u,v \in V$ there exist scalars $\lambda_1,\lambda_2,\ldots,\lambda_n$; $\mu_1,\mu_2,\ldots,\mu_n$ in F such that $u = \sum_{i=1}^{n} \lambda_i v_i$ and $v = \sum_{j=1}^{n} \mu_j v_j$

Let us define a mapping $f : V \times V \to F$ given by

$$f(u,v) = \sum_{i=1}^{n} \sum_{j=1}^{n} \lambda_i \, a_{ij} \, \mu_j$$

We shall show that f is a bilinear form on V.

For any $u = \sum_{i=1}^{n} \lambda_i v_i$, $v = \sum_{i=1}^{n} \mu_i v_i$, $w = \sum_{i=1}^{n} \alpha_i v_i \in V$ and $a,b \in F$

$$\begin{aligned}
f(au+bv,w) &= f\left(\sum_{i=1}^{n}(a\lambda_i + b\mu_i)v_i, \sum_{j=1}^{n} \alpha_j v_j\right) \\
&= \sum_{i=1}^{n} \sum_{j=1}^{n} (a\lambda_i + b\mu_i)\alpha_j f(v_i,v_j) \\
&= \sum_{i=1}^{n} \sum_{j=1}^{n} (a\lambda_i + b\mu_i)a_{ij}\alpha_j \\
&= \sum_{i=1}^{n} \sum_{j=1}^{n} (a\lambda_i a_{ij}\alpha_j + b\mu_i a_{ij}\alpha_j) \\
&= \sum_{i=1}^{n} \sum_{j=1}^{n} a\lambda_i a_{ij}\alpha_j + \sum_{i=1}^{n} \sum_{j=1}^{n} b\mu_i a_{ij}\alpha_j \\
&= a\sum_{i=1}^{n} \sum_{j=1}^{n} \lambda_i a_{ij}\alpha_j + b\sum_{i=1}^{n} \sum_{j=1}^{n} \mu_i a_{ij}\alpha_j \\
&= af(u,w) + bf(v,w)
\end{aligned}$$

Similarly, we have

$$f(u,av+bw) = af(u,v) + bf(u,w)$$

Thus, f is a bilinear form on V.

For any $r,s \in \underline{n}$, we have

$$v_r = \sum_{i=1}^{n} \lambda_i v_i, \text{ where } \lambda_r = 1 \text{ and } \lambda_i = 0 \text{ for all } i \neq r$$

$$v_s = \sum_{j=1}^{n} \mu_j v_j, \text{ where } \mu_s = 1 \text{ and } \mu_j = 0 \text{ for all } j \neq s$$

$$\therefore \quad f(v_r,v_s) = \sum_{i=1}^{n} \sum_{j=1}^{n} \lambda_i \, a_{ij} \, \mu_j = a_{rs}$$

Thus, f is a bilinear form on V such that $f(v_i, v_j) = a_{ij}$ for all $i, j \in \underline{n}$.

Bilinear and Quadratic Forms • 737

In order to prove the uniqueness of f. If possible, let g be a bilinear form on V such that $g(v_i, v_j) = a_{ij}$ for all $i, j \in \underline{n}$.

Let $u = \sum_{i=1}^{n} \lambda_i v_i$ and $v = \sum_{j=1}^{n} \mu_i v_i \in V$. Then,

$$g(u,v) = g\left(\sum_{i=1}^{n} \lambda_i v_i, \sum_{j=1}^{n} \mu_j v_j\right)$$

$$\Rightarrow g(u,v) = \sum_{i=1}^{n} \sum_{j=1}^{n} \lambda_i \mu_j g(v_i, v_j)$$

$$\Rightarrow g(u,v) = \sum_{i=1}^{n} \sum_{j=1}^{n} \lambda_i \mu_j a_{ij} = \sum_{i=1}^{n} \sum_{j=1}^{n} \lambda_i a_{ij} \mu_j = f(u,v)$$

Thus, $f(u,v) = g(u,v)$ for all $u, v \in V$

$$\Rightarrow f = g$$

Hence, f is unique.

Q.E.D.

THEOREM-5 Let f be a bilinear form on a vector space $V(F)$ and let $B = \{v_1, v_2, \ldots, v_n\}$ be an ordered basis of V. If $\{x_1, x_2, \ldots, x_n\}$ is basis dual to B, then, $f = X^T A X$, where

$$X = \begin{bmatrix} x_1 \\ x_2 \\ \vdots \\ x_n \end{bmatrix}$$ and $A = [a_{ij}]$ is the matrix of f relative to basis B.

PROOF. If ϕ and ψ are linear forms on a vector space V. Then, $\phi\psi$ defined by $\phi\psi(u, v) = \phi(u) \psi(v)$ is a bilinear form on V (see Example 4 on page 724). Also, if c is a scalar in F, then $\phi\, c\psi$ defined by $\phi c\, \psi(u, v) = \phi(u)\, c\, \psi(v)$ is also a bilinear form on V. Therefore, given n^2 scalars $a_{ij}; i, j \in \underline{n}$

$x_i\, a_{ij}\, x_j$ is a bilinear form on V for all $i, j \in \underline{n}$

Let $u = \sum_{i=1}^{n} \lambda_i v_i$ and $v = \sum_{j=1}^{n} \mu_j v_j$. Then,

$x_i\, a_{ij}\, x_j\, (u,v) = x_i(u)\, a_{ij}\, x_j(v)$

$\Rightarrow x_i\, a_{ij}\, x_j\, (u,v) = x_i\left(\sum_{r=1}^{n} \lambda_r v_r\right) a_{ij}\, x_j \left(\sum_{s=1}^{n} \mu_s v_s\right)$

$\Rightarrow x_i\, a_{ij}\, x_j\, (u,v) = \left\{\sum_{r=1}^{n} \lambda_r\, x_i\, (v_r)\right\} a_{ij} \left\{\sum_{s=1}^{n} \mu_s\, x_j\, (v_s)\right\}$

$$\Rightarrow \quad x_i\, a_{ij}\, x_j\, (u,v) = \left\{\sum_{r=1}^{n} \lambda_r\, \delta_{ir}\right\} a_{ij} \left\{\sum_{s=1}^{n} \mu_s\, \delta_{js}\right\}$$

$$\Rightarrow \quad x_i\, a_{ij}\, x_j\, (u,v) = \lambda_i\, a_{ij}\, \mu_j \quad \text{for each } i \in \underline{n},\ j \in \underline{n}$$

Also,

$$\Rightarrow \quad f(u,v) = f\left(\sum_{i=1}^{n} \lambda_i\, v_i,\ \sum_{j=1}^{n} \mu_j\, v_j\right)$$

$$\Rightarrow \quad f(u,v) = \sum_{i=1}^{n}\sum_{j=1}^{n} \lambda_i\, \mu_j\, f(v_i, v_j)$$

$$\Rightarrow \quad f(u,v) = \sum_{i=1}^{n}\sum_{j=1}^{n} \lambda_i\, \mu_j\, a_{ij} = \sum_{i=1}^{n}\sum_{j=1}^{n} \lambda_i\, a_{ij}\, \mu_j$$

$$\therefore \quad f(u,v) = \sum_{i=1}^{n}\sum_{j=1}^{n} x_i\, a_{ij}\, x_j(u,v) \quad \text{for all } u, v \in V$$

$$\Rightarrow \quad f = \sum_{i=1}^{n}\sum_{j=1}^{n} x_i\, a_{ij}\, x_j$$

$$\Rightarrow \quad f = X^T A X.$$

Q.E.D.

EXERCISE 10.1

1. Let $u = (x_1, x_2)$ and $v = (y_1, y_2)$. Determine which of the following are bilinear forms on R^2:
   (i) $f(u, v) = 2x_1 y_2 - 3x_2 y_1$
   (ii) $f(u, v) = x_1 + y_2$
   (iii) $f(u, v) = 3x_2 y_2$
   (iv) $f(u, v) = 1$
   (v) $f(u, v) = 0$.
   (vi) $f(u, v) = (x_1 - y_1)^2 + x_2 y_2$
   (vii) $f(u, v) = (x_1 + y_1)^2 - (x_1 - y_1)^2$

2. Let $u = (x_1, x_2)$, $v = (y_1, y_2)$. Which of the following functions are bilinear forms on C^2?
   (i) $f(u, v) = 1$
   (ii) $f(u, v) = x_1 + x_2 + y_1 + y_2$
   (iii) $f(u, v) = x_1 \bar{y}_1 + x_2 \bar{y}_2$
   (iv) $f(u, v) = x_1 y_1 + x_2 y_2 + 2x_2 y_1 + 3x_2 \bar{y}_2$

3. Let f be a bilinear form on $V(F)$ and S_1 be a subset of V. Then, prove that $S = \{u \in V : f(u, v) = 0 \text{ for every } v \in S_1\}$ is a subspace of V.

4. Let $V(F)$ be a finite dimensional vector space with basis $B = \{v_1, v_2, \ldots, v_n\}$. Then there exists a basis $B'\{f_{ij} : i, j \in \underline{n}\}$ of $L(V,V,F)$ such that $f_{ij}(v_r, v_s) = \delta_{ir}\, \delta_{js}$ for all $i, j, r, s \in \underline{n}$.

5. Let U and V be vector spaces over field F. A mapping $f : U \times V \to F$ is called a bilinear form on U and V, if

(i) $f(au_1 + bu_2, v) = af(u_1, v) + bf(u_2, v)$
(ii) $f(u, av_1 + bv_2) = af(u, v_1) + bf(u, v_2)$

for every $a, b \in F$ and all $u, u_1, u_2 \in U, v, v_1, v_2 \in V$.

Prove the following:

(i) The set $L(U, V, F)$ of bilinear forms on U and V is a subspace of the vector space of all functions from $U \times V$ in to F.

(ii) If $\{x_1, x_2, \ldots, x_m\}$ is a basis of U^* and $\{y_1, y_2, \ldots, y_n\}$ is a basis of V^*, then $\{f_{ij} : i \in \underline{m}, j \in \underline{n}\}$ is a basis of $L(U, V, F)$, where f_{ij} is defined by

$$f_{ij}(u, v) = x_i(u) y_j(v) \quad \text{for all } u \in U, v \in V.$$

Thus, $\dim L(U, V, F) = \dim(U) \times \dim(V)$

6. Let V be the vector space of all $n \times n$ matrices over C. Show that the function $f : V \times V \to C$ defined by

$$f(A, B) = n \, tr(AB) - tr(A) \, tr(B)$$

defines a bilinear form on V. Is it true that $f(A, B) = f(B, A)$ for all $A, B \in V$?

7. Let V be the real vector space of all real valued continuous functions defined on $[0, \pi]$. Prove that $\phi : V \times V \to R$ defined by

$$\phi(f, g) = \int_0^\pi f(t) g(t) \, dt \quad \text{for all } f, g \in V$$

is a bilinear form on V.

ANSWERS

1. (i), (iii), (v), (vii)   2. (iii)

10.4 BILINEAR FORMS AND MATRICES

Let $V(F)$ be a finite dimensional vector space and let $B = \{v_1, v_2, \ldots, v_n\}$ be an ordered basis for V. Suppose $u, v \in V$ such that $u = \sum_{i=1}^{n} \lambda_i v_i$ and $v = \sum_{j=1}^{n} \mu_j v_j$.

Let f be a bilinear form on V. Then,

$$f(u, v) = f\left(\sum_{i=1}^{n} \lambda_i v_i, \sum_{j=1}^{n} \mu_j v_j\right)$$

$$\Rightarrow \quad f(u, v) = \sum_{i=1}^{n} \sum_{j=1}^{n} \lambda_i f(v_i, v_j) \mu_j$$

It is clear from the above relation that f is completely determined by n^2 scalars $f(v_i, v_j); i, j \in \underline{n}$. These scalars form an $n \times n$ matrix $A = [a_{ij}]$, where $a_{ij} = f(v_i, v_j)$. This matrix is called the matrix of f relative to the basis B or, simply, "the matrix of f in the ordered basis B".

Thus, we may define the matrix of a bilinear form as follows:

MATRIX REPRESENTATION OF A BILINEAR FORM Let $V(F)$ be a finite dimensional vector space, and let $B = \{v_1, v_2, \ldots, v_n\}$ be an ordered basis for V. If f is a bilinear form on V, the matrix of f in the ordered basis B is the $n \times n$ matrix $A = [a_{ij}]$ with entries $a_{ij} = f(v_i, v_j)$.

We denote the matrix of f in the ordered basis B by $M_B^B(f)$.

ILLUSTRATIVE EXAMPLES

EXAMPLE-1 Let f be a bilinear form on R^2 defined by

$$f(X,Y) = 2x_1 y_1 - 3x_1 y_2 + 4x_2 y_2, \text{ where } X = \begin{bmatrix} x_1 \\ x_2 \end{bmatrix} \text{ and } Y = \begin{bmatrix} y_1 \\ y_2 \end{bmatrix}$$

Find the matrix A of f in the ordered basis $B = \{v_1 = (1, 0), v_2 = (1, 1)\}$.

SOLUTION We have, $f(X, Y) = 2x_1 y_1 - 3x_1 y_2 + 4x_2 y_2$. Let $A = [a_{ij}]$ be the matrix of f in the basis B. Then,

$$a_{ij} = f(v_i, v_j) \quad \text{for all } i, j$$

$\therefore \quad a_{11} = f(v_1, v_1) = 2 \times 1 \times 1 - 3 \times 1 \times 0 + 4 \times 0 \times 0 = 2$

$a_{12} = f(v_1, v_2) = 2 \times 1 \times 1 - 3 \times 1 \times 1 + 4 \times 0 \times 1 = -1$

$a_{21} = f(v_2, v_1) = 2 \times 1 \times 0 - 3 \times 1 \times 0 + 4 \times 1 \times 0 = 2$

$a_{22} = f(v_2, v_2) = 2 \times 1 \times 1 - 3 \times 1 \times 1 + 4 \times 1 \times 1 = 2 - 3 + 4 = 3$

Thus, $A = \begin{bmatrix} a_{11} & a_{12} \\ a_{21} & a_{22} \end{bmatrix} = \begin{bmatrix} 2 & -1 \\ 2 & 3 \end{bmatrix}$ is the matrix of f in the basis B, i.e. $M_B^B(f) = \begin{bmatrix} 2 & -1 \\ 2 & 3 \end{bmatrix}$.

EXAMPLE-2 Let V be the vector space of 2×2 matrices over R. Let $M = \begin{bmatrix} 1 & 2 \\ 3 & 5 \end{bmatrix}$, and let $f : V \times V \to R$ be a mapping given by

$$f(A, B) = tr(A^T M B) \quad \text{for all } A, B \in V.$$

(i) Show that f is a bilinear form on V.

(ii) Find the matrix of f in the ordered basis

$$B = \left\{ E_{11} = \begin{bmatrix} 1 & 0 \\ 0 & 0 \end{bmatrix}, E_{12} = \begin{bmatrix} 0 & 1 \\ 0 & 0 \end{bmatrix}, E_{21} = \begin{bmatrix} 0 & 0 \\ 1 & 0 \end{bmatrix}, E_{22} = \begin{bmatrix} 0 & 0 \\ 0 & 1 \end{bmatrix} \right\}$$

SOLUTION (i) See Example 10 on page 727

(ii) We have,

$$f(A, B) = tr(A^T MB) \text{ for all } A, B \in V.$$

Let $P = [p_{ij}]$ be the matrix of f in the basis B. Then,

$p_{11} = f(E_{11}, E_{11}) = tr(E_{11}^T ME_{11}) = tr(E_{11}) = 1$

$p_{12} = f(E_{11}, E_{12}) = tr(E_{11}^T ME_{12}) = tr(E_{11} ME_{12})$ $\quad [\because E_{11}^T = E_{11}]$

$\Rightarrow \quad p_{12} = tr(E_{12}) = 0$

$p_{13} = f(E_{11}, E_{21}) = tr(E_{11}^T ME_{21}) = tr(E_{11} ME_{21}) = tr\left(\begin{bmatrix} 2 & 0 \\ 0 & 0 \end{bmatrix}\right) = 2$

$p_{14} = f(E_{11}, E_{22}) = tr(E_{11}^T ME_{22}) = tr(E_{11} ME_{22})$ $\quad [\because E_{11}^T = E_{11}]$

$\Rightarrow \quad p_{14} = tr\left(\begin{bmatrix} 0 & 2 \\ 0 & 0 \end{bmatrix}\right) = 0$

Similarly, we have

$p_{21} = f(E_{12}, E_{11}) = tr(E_{12}^T ME_{11}) = 0; \quad p_{22} = f(E_{12}, E_{12}) = tr(E_{12}^T ME_{12}) = 1$

$p_{23} = f(E_{12}, E_{21}) = tr(E_{12}^T ME_{21}) = 0; \quad p_{24} = f(E_{12}, E_{22}) = tr(E_{12}^T ME_{22}) = 2$

$p_{31} = f(E_{21}, E_{11}) = tr(E_{21}^T ME_{11}) = 3; \quad p_{32} = f(E_{21}, E_{12}) = tr(E_{21}^T ME_{12}) = 0$

$p_{33} = f(E_{21}, E_{21}) = tr(E_{21}^T ME_{21}) = 5; \quad p_{34} = f(E_{21}, E_{22}) = tr(E_{21}^T ME_{22}) = 0$

$p_{41} = 0, \quad p_{42} = 3, \quad p_{43} = 0, \quad p_{44} = 5.$

Hence, matrix of f in the basis B is $P = \begin{bmatrix} 1 & 0 & 2 & 0 \\ 0 & 1 & 0 & 2 \\ 3 & 0 & 5 & 0 \\ 0 & 3 & 0 & 5 \end{bmatrix}$

THEOREM-1 *Let $V(F)$ be a finite dimensional vector space, and let $B = \{b_1, b_2, \ldots, b_n\}$ be an ordered basis for V. If f and g are bilinear forms on V and $a, b \in F$, then*

$$M_B^B(af + bg) = a M_B^B(f) + b M_B^B(g)$$

PROOF. Let $P = [p_{ij}] = M_B^B(f)$, $Q = [q_{ij}] = M_B^B(g)$. Then, $p_{ij} = f(v_i, v_j)$ and $q_{ij} = g(v_i, v_j)$ for all $i, j \in \underline{n}$.

For any $i, j \in \underline{n}$

$$(af+bg)(v_i,v_j) = (af)(v_i,v_j) + (bg)(v_i,v_j)$$
$$= a\,f(v_i,v_j) + b\,g(v_i,v_j)$$
$$= a\,p_{ij} + b\,q_{ij}$$
$$= (aP+bQ)_{ij}$$

$\therefore \quad M_B^B(af+bg) = aP + bQ = a\,M_B^B(f) + b\,M_B^B(g)$

Q.E.D.

THEOREM-2 *Let $V(F)$ be an n-dimensional vector space and B be an ordered basis of V. Then, $L(V,V,F) \cong F^{n \times n}$.*

OR

Let $V(F)$ be a finite dimensional vector space. For each ordered basis B of V, the function which associates with each bilinear form on V, its matrix in the ordered basis B is an isomorphism of the space $L(V,V,F)$ onto the space $F^{n \times n}$ of $n \times n$ matrices over the field F.

PROOF. Let ψ be a function from $L(V,V,F)$ to $F^{n \times n}$ associating each $f \in L(V,V,F)$ to $M_B^B(f)$ i.e. $\psi(f) = M_B^B(f)$.

We observe the following properties of ψ:

ψ *is a linear transformation*: For any $f, g \in L(V,V,F)$ and any $a, b \in F$

$\psi(af+bg) = M_B^B(af+bg)$
$\qquad = a\,M_B^B(f) + b\,M_B^B(g)$ [By Theorem 1]
$\qquad = a\,\psi(f) + b\,\psi(g)$

$\therefore \quad \psi$ is a linear transformation.

ψ *is one-one*: For any $f, g \in L(V,V,F)$

$\qquad \psi(f) = \psi(g)$
$\Rightarrow \quad M_B^B(f) = M_B^B(g)$
$\Rightarrow \quad f = g$ [By Theorem 4 on page 735]

$\therefore \quad \psi$ is one-one.

ψ *is onto*: Let $A = [a_{ij}]$ be an arbitrary matrix in $F^{n \times n}$. Then, there exists a bilinear form f on V such that

$$M_B^B(f) = A \quad \Rightarrow \quad \psi(f) = A$$

Thus, for each $A \in F^{n \times n}$ there exists $f \in L(V,V,F)$ such that $\psi(f) = A$.
So, ψ is onto.
Thus, $\psi : L(V,V,F) \to F^{n \times n}$ is an isomorphism.
Hence, $L(V,V,F) \cong F^{n \times n}$.

Q.E.D.

10.4.1 CHANGE OF BASIS

As we have seen that the concept of the matrix of a bilinear form in an ordered basis is similar to that of the matrix of a linear operator relative to an ordered basis. We shall now see, how does a matrix representing a bilinear form transform when a new basis is selected?

THEOREM-1 *Let $V(F)$ be a finite dimensional vector space. Let P be a change-of-basis matrix from one basis B to another basis B'. If A and A' are the matrices representing a bilinear form f relative to bases B and B' respectively. Then, $A' = P^T A P$.*

PROOF. Let $u, v \in V$. Since P is the change-of-basis matrix from B to B'. Therefore,
$$P[u]_{B'} = [u]_B \quad \text{and} \quad P[v]_{B'} = [v]_B$$
Now, $\qquad f(u,v) = [u]_B^T A [v]_B$ \hfill [By Theorem 3 on page 734]
$\Rightarrow \qquad f(u,v) = (P[u]_{B'})^T A (P[v]_{B'})$
$\Rightarrow \qquad f(u,v) = [u]_{B'}^T (P^T A P)[v]_{B'}$

Since u and v are arbitrary vectors in V. Therefore, $P^T A P$ is the matrix of f in the basis B'.
Hence, $A' = P^T A P$.

Q.E.D.

Every change-of-basis matrix is invertible. Therefore, $A' = P^T A P$ implies that the matrices A and A' are equivalent. Two matrices are equivalent iff they have the same rank. The above theorem proves that two matrices are equivalent iff they represent the same bilinear form. Thus, we may define the rank of a bilinear form as follows.

RANK OF A BILINEAR FORM *The rank of a bilinear form f on a vector space $V(F)$ is the rank of any matrix representation of f.*

The rank of f is written as rank (f).

THEOREM-2 *Let $V(F)$ be a finite dimensional vector space and f be a bilinear form of rank r on V. Then corresponding to every basis $B = \{v_1, v_2, \ldots, v_n\}$ of V there exists a dual basis $\{x_1, x_2, \ldots, x_n\}$ such that*
$$f = x_1^2 + x_2^2 + \cdots + x_r^2$$

PROOF. Let A be the matrix of f relative to basis B. Then,
$$f = Y^T A Y \qquad \text{[By Theorem 5 on page 737]}$$

where Y is the column matrix of the linear forms in the basis dual to B.

It is given that the rank of bilinear form f is r. Therefore, rank of matrix A is also r. Consequently, A is equivalent to a matrix whose all entries are zero except the first r entries along the main diagonal as unity. That is, A is equivalent to the matrix $D = [d_{ij}]$ such that

$$d_{ii} = 1, \quad i = 1, 2, \ldots, r; \quad d_{ij} = 0 \quad \text{for } i \neq j,\ i = j > r.$$

Since two equivalent matrices represent the same bilinear form. Therefore, matrix D also represents f. Consequently, there exists linear forms $x_1, x_2, \ldots, x_n$ forming a basis of the dual space such that

$$h = X^T D X, \quad \text{where} \quad X = \begin{bmatrix} x_1 \\ x_2 \\ \vdots \\ x_n \end{bmatrix}$$

$$\Rightarrow \quad h = \sum_{i=1}^{n} \sum_{j=1}^{n} x_i\, d_{ij}\, x_j$$

$$\Rightarrow \quad h = x_1^2 + x_2^2 + \cdots + x_r^2$$

Q.E.D.

10.4.2 RANK OF BILINEAR FORM

In the previous subsection, we have defined the rank of a bilinear form on a vector space $V(F)$ as the rank of any matrix which represents the bilinear form in an ordered basis for V. The following theorems provides an alternative definition of the rank of a bilinear form.

THEOREM-1 *Let f be a linear form on a finite dimensional vector space $V(F)$.*

(i) *For any fixed $u \in V$, $L_f(u) : V \to F$ defined by*

$$(L_f(u))(v) = f(u, v) \quad \text{for all } v \in V$$

is a linear form on V.

(ii) *For any fixed $v \in V$, $R_f(v) : V \to F$ defined by*

$$(R_f(v))(u) = f(u, v) \quad \text{for all } u \in V$$

is a linear form on V.

PROOF. (i) For any $v_1, v_2 \in V$ and $a, b \in F$

$$\begin{aligned}(L_f(u))(av_1 + bv_2) &= f(u,\ av_1 + bv_2) \\ &= af(u, v_1) + bf(u, v_2) \\ &= a(L_f(u))(v_1) + b(L_f(u))(v_2)\end{aligned}$$

$\therefore \quad L_f(u) : V \to F$ is a linear form on V.

(ii) For any $u_1, u_2 \in V$ and $a, b \in F$

$$(R_f(v))(au_1 + bu_2) = f(au_1 + bu_2, v)$$
$$= af(u_1, v) + bf(u_2, v)$$
$$= a(R_f(v))(u_1) + b(R_f(v))(u_2)$$

$\therefore \quad R_f(v) : V \to F$ is a linear form on V.

Q.E.D.

THEOREM-2 *Let f be a bilinear form on a finite dimensional vector space $V(F)$. Then,*

(i) *$L_f : V \to V^*$ associating each $u \in V$ to $L_f(u) \in V^*$ is a linear transformation.*

(ii) *$R_f : V \to V^*$ associating each $v \in V$ to $R_f(v)$ in V^* is a linear transformation.*

PROOF. (i) For any $u_1, u_2 \in V$ and $a, b \in F$

$$(L_f(au_1 + bu_2))(v) = f(au_1 + bu_2, v)$$
$$= af(u_1, v) + bf(u_2, v)$$
$$= a(L_f(u_1))(v) + b(L_f(u_2))(v)$$
$$= \big(a\, L_f(u_1) + b\, L_f(u_2)\big)(v) \qquad \text{for all } v \in V$$

$\therefore \quad L_f(au_1 + bu_2) = a\, L_f(u_1) + b\, L_f(u_2)$

Hence, L_f is a linear transformation.

(ii) For any $v_1, v_2 \in V$ and $a, b \in F$

$$(R_f(av_1 + bv_2))(u) = f(u, av_1 + bv_2)$$
$$= af(u, v_1) + bf(u, v_2)$$
$$= a(R_f(v_1))(u) + b(R_f(v_2))(u) \quad \text{for all } u \in V$$
$$= (aR_f(v_1) + bR_f(v_2))(u) \qquad \text{for all } u \in V$$

$\therefore \quad R_f(av_1 + bv_2) = aR_f(v_1) + bR_f(v_2)$

Hence, R_f is a linear transformation.

Q.E.D.

THEOREM-3 *Let f be a bilinear form on the finite dimensional vector space $V(F)$. Let $L_f : V \to V^*$ and $R_f : V \to V^*$ be linear transformations defined by*

$$(L_f(u))(v) = f(u, v) \text{ and } (R_f(v))(u) = f(u, v) \quad \text{for all } u, v \in V.$$

Then, $\quad \text{rank}(L_f) = \text{rank}(R_f)$

PROOF. Let B be an ordered basis for V and A be the matrix of f in basis B. Let $u, v \in V$ such that $[u]_B = X$ and $[v]_B = Y$. Then,
$$f(u, v) = X^T A Y$$

Now, $\quad v \in \text{Ker}(R_f)$
$\Leftrightarrow \quad R_f(v) = \widehat{0}$
$\Leftrightarrow \quad (R_f(v))(u) = \widehat{0}(u) \quad$ for all $u \in V$
$\Leftrightarrow \quad f(u, v) = 0 \quad$ for all $u \in V$
$\Leftrightarrow \quad X^T A Y = 0 \quad$ for all $X \in F^{n \times 1}$
$\Leftrightarrow \quad AY = O$
$\therefore \quad$ Nullity $(R_f) = $ Dimension of the solution space of $AY = O$ \hfill (i)

Similarly,
$\quad u \in \text{Ker}(L_f)$
$\Leftrightarrow \quad L_f(u) = \widehat{0}$
$\Leftrightarrow \quad (L_f(u))(v) = \widehat{0}(v) \quad$ for all $v \in V$
$\Leftrightarrow \quad f(u, v) = 0 \quad$ for all $v \in V$
$\Leftrightarrow \quad X^T A Y = 0 \quad$ for all $y \in F^{n \times 1}$
$\Leftrightarrow \quad X^T A = O$
$\Leftrightarrow \quad A^T X = O$
$\Leftrightarrow \quad$ Nullity $(L_f) = $ Dimension of the solution space of $A^T X = O$ \hfill (ii)

But, A and A^T have the same column rank. Therefore, from (i) and (ii), we get

Nullity $(L_f) = $ Nullity (R_f)
$\Rightarrow \quad$ Rank $(L_f) = $ Rank (R_f)

Q.E.D.

COROLLARY. *Let f be a bilinear form on the finite dimensional vector space $V(F)$ and B be an ordered basis of V'. If A is the matrix of f in basis B. Then,*

$$\text{Rank}(L_f) = \text{Rank}(R_f) = \text{Rank}(A)$$

The above discussion suggests us the following definition of the rank of a bilinear form on a finite dimensional vector space $V(F)$.

RANK OF A BILINEAR FORM *If f is a bilinear form on the finite dimensional vector space $V(F)$, the rank of f is the positive integer $r = \text{rank}(L_f) = \text{rank}(R_f)$.*

ILLUSTRATIVE EXAMPLES

EXAMPLE-1 Let f be a bilinear form on R^2 defined by $f(u, v) = 2x_1y_1 - 3x_1y_2 + 4x_2y_2$, where $u = (x_1, x_2)$, $v = (y_1, y_2)$.

(i) Find the matrix A of f in the basis $B = \{u_1 = (1, 0), u_2 = (1, 1)\}$

(ii) Find the matrix A' of f in the basis $B' = \{v_1 = (2, 1), v_2 = (1, -1)\}$

(iii) Find the change-of-basis matrix P from the basis B to the basis B', and verify that $A' = P^T A P$

SOLUTION (i) Let $A = [a_{ij}]$ be the matrix of f in the basis B. Then, $a_{ij} = f(u_i, u_j)$

$\therefore \quad a_{11} = f(u_1, u_1) = 2 \times 1 \times 1 - 3 \times 1 \times 0 + 4 \times 0 \times 0 = 2$

$a_{12} = f(u_1, u_2) = 2 \times 1 \times 1 - 3 \times 1 \times 1 + 4 \times 0 \times 1 = -1$

$a_{21} = f(u_2, u_1) = 2 \times 1 \times 1 - 3 \times 1 \times 0 + 4 \times 1 \times 0 = 2$

$a_{22} = f(u_2, u_2) = 2 \times 1 \times 1 - 3 \times 1 \times 1 + 4 \times 1 \times 1 = 3$

Hence, $A = \begin{bmatrix} 2 & -1 \\ 2 & 3 \end{bmatrix}$ is the matrix of f in the basis B.

(ii) Let $A' = [\alpha_{ij}]$ be the matrix of f in the basis B'. Then, $\alpha_{ij} = f(v_i, v_j)$

$\therefore \quad \alpha_{11} = f(v_1, v_1) = 2 \times 2 \times 1 - 3 \times 2 \times -1 + 4 \times 1 \times -1 = 6$

$\alpha_{12} = f(v_1, v_2) = 2 \times 2 \times 1 - 3 \times 2 \times -1 + 4 \times 1 \times -1 = 6$

$\alpha_{21} = f(v_2, v_1) = 2 \times 1 \times 2 - 3 \times 1 \times 1 + 4 \times 1 \times -1 = -3$

$\alpha_{22} = f(v_2, v_2) = 2 \times 1 \times 1 - 3 \times 1 \times -1 + 4 \times -1 \times -1 = 9$

Hence, $A' = \begin{bmatrix} 6 & 6 \\ -3 & 9 \end{bmatrix}$ is the matrix of f in the basis B'.

(iii) Let $v_1 = x_1 u_1 + x_2 u_2$ and $v_2 = y_1 u_1 + y_2 u_2$

$\Rightarrow \quad (2, 1) = x_1(1, 0) + x_2(1, 1)$ and $(1, -1) = y_1(1, 0) + y_2(1, 1)$

$\Rightarrow \quad (2, 1) = (x_1 + x_2, x_2)$ and $(1, -1) = (y_1 + y_2, y_2)$

$\Rightarrow \quad x_1 + x_2 = 2, x_2 = 1$ and $y_1 + y_2 = 1, y_2 = -1$

$\Rightarrow \quad x_1 = 1, x_2 = 1, y_1 = 2, y_2 = -1$

$\therefore \quad v_1 = u_1 + u_2$ and $v_2 = 2u_1 - u_2$

So, $P = \begin{bmatrix} 1 & 1 \\ 2 & -1 \end{bmatrix}^T = \begin{bmatrix} 1 & 2 \\ 1 & -1 \end{bmatrix}$

$$\therefore \quad P^T A P = \begin{bmatrix} 1 & 1 \\ 2 & -1 \end{bmatrix} \begin{bmatrix} 2 & -1 \\ 2 & 3 \end{bmatrix} \begin{bmatrix} 1 & 2 \\ 1 & -1 \end{bmatrix} = \begin{bmatrix} 6 & 6 \\ -3 & 9 \end{bmatrix} = A'$$

EXAMPLE-2 Let f be a bilinear form on R^2 defined by

$$f[(x_1, x_2), (y_1, y_2)] = (x_1 + x_2)(y_1 + y_2)$$

for all $(x_1, x_2), (y_1, y_2) \in R^2$.

(i) Find the matrix A of f in the standard basis $B = \left\{ e_1^{(2)} = (1, 0), e_2^{(2)} = (0, 1) \right\}$.

(ii) Find the matrix A' of f in the basis $B' = \{u_1 = (1, -1), u_2 = (1, 1)\}$

(iii) Find the change-of-basis matrix P from basis B to basis B' and verify that $A' = P^T A P$.

SOLUTION (i) Let $A = [a_{ij}]$ be the matrix of f in the basis B. Then, $a_{ij} = f(e_i^{(2)}, e_j^{(2)})$.

$\therefore \quad a_{11} = f\left(e_1^{(2)}, e_1^{(2)}\right) = (1+0)(1+0) = 1; \; a_{12} = f\left(e_1^{(2)}, e_2^{(2)}\right) = (1+0)(0+1) = 1$

$\quad a_{21} = f\left(e_2^{(2)}, e_1^{(2)}\right) = (0+1)(1+0) = 1; \; a_{22} = f\left(e_2^{(2)}, e_2^{(2)}\right) = (0+1)(0+1) = 1$

$\therefore \quad A = \begin{bmatrix} 1 & 1 \\ 1 & 1 \end{bmatrix}$ is the matrix of f in the standard basis of R^2.

(ii) Let $A' = [\alpha_{ij}]$ be the matrix of f in the basis B'. Then, $\alpha_{ij} = f(u_i, u_j)$

$\therefore \quad \alpha_{11} = f(u_1, u_1) = (1-1)(1-1) = 0; \; \alpha_{12} = f(u_1, u_2) = (1-1)(1+1) = 0$

$\quad \alpha_{21} = f(u_2, u_1) = (1+1)(1-1) = 0; \; \alpha_{22} = f(u_2, u_2) = (1+1)(1+1) = 4$

$\therefore \quad A' = \begin{bmatrix} 0 & 0 \\ 0 & 4 \end{bmatrix}$ is the matrix of f in the basis B'.

(iii) We have,

$u_1 = (1, -1) = 1 \times e_1^{(2)} + (-1) e_2^{(2)}$

$u_2 = (1, 1) = 1 \times e_1^{(2)} + 1 \times e_2^{(2)}$

$\therefore \quad P = \begin{bmatrix} 1 & -1 \\ 1 & 1 \end{bmatrix}^T = \begin{bmatrix} 1 & 1 \\ -1 & 1 \end{bmatrix}$

Now,

$$P^T A P = \begin{bmatrix} 1 & -1 \\ 1 & 1 \end{bmatrix} \begin{bmatrix} 1 & 1 \\ 1 & 1 \end{bmatrix} \begin{bmatrix} 1 & 1 \\ -1 & 1 \end{bmatrix} = \begin{bmatrix} 0 & 0 \\ 0 & 4 \end{bmatrix} = A'$$

Bilinear and Quadratic Forms • 749

EXERCISE 10.2

1. Let f be the bilinear form on R^2 defined by
   $f[(x_1, x_2), (y_1, y_2)] = x_1 x_2 + y_1 y_2$ for all $(x_1, x_2), (y_1, y_2) \in R^2$.
   Find the matrix of f in each of the following ordered basis of R^2:
   (i) $B_1 = \{e_1^{(2)} = (1, 0), e_2^{(2)} = (0, 1)\}$ (ii) $B_2 = \{u_1 = (1, 2), u_2 = (3, 4)\}$
   (iii) $B_3 = \{v_1 = (1, 1), v_2 = (0, 1)\}$ (iv) $B_4 = \{w_1 = (1, -1), w_2 = (1, 1)\}$

2. Let f be a bilinear form on R^2 defined by
   $f[(x_1, x_2), (y_1, y_2)] = 2x_1 y_1 - 3x_1 y_2 + x_2 y_2$
   Find the matrix of f in the ordered basis $B = \{u_1 = (1, 0), u_2 = (1, 1)\}$ of R^2.

3. Let f be a bilinear form on R^2 defined by
   $$f[(x_1, x_2), (y_1, y_2)] = 3x_1 y_1 - 2x_1 y_2 + 4x_2 y_1 - x_2 y_2$$
   for all $(x_1, x_2), (y_1, y_2) \in R^2$. Find the

   (i) matrix A of f in the ordered basis $B = \{u_1 = (1, 1), u_2 = (1, 2)\}$
   (ii) matrix A' of f in the ordered basis $B' = \{v_1 = (1, -1), v_2 = (3, 1)\}$
   (iii) change-of-basis matrix P from basis B to basis B', and verify that $B = P^T AP$.

4. Let V be the vector space of all 2×3 matrices over R, and let f be the bilinear form on V defined by
   $$f(X, Y) = \text{tr}(X^T AY) \text{ for all } X, Y \in V, \text{ where } A = \begin{bmatrix} 1 & 2 \\ 3 & 4 \end{bmatrix}.$$
   Find the matrix of F in the ordered basis $B = \{E_{11}, E_{12}, E_{13}, E_{21}, E_{22}, E_{23}\}$, where E_{ij} is the matrix whose only non-zero entry is a 1 in row i and column j.

ANSWERS

1. (i) $\begin{bmatrix} 0 & 0 \\ 0 & 0 \end{bmatrix}$ (ii) $\begin{bmatrix} 4 & 14 \\ 14 & 24 \end{bmatrix}$ (iii) $\begin{bmatrix} 2 & 1 \\ 1 & 0 \end{bmatrix}$ (iv) $\begin{bmatrix} -2 & 0 \\ 0 & -2 \end{bmatrix}$ 2. $\begin{bmatrix} 2 & -1 \\ 2 & 0 \end{bmatrix}$

3. (i) $A = \begin{bmatrix} 4 & 1 \\ 7 & 3 \end{bmatrix}$ (ii) $A' = \begin{bmatrix} 0 & -4 \\ 20 & 32 \end{bmatrix}$ (iii) $P = \begin{bmatrix} 3 & 5 \\ -2 & -2 \end{bmatrix}$

10.5 SYMMETRIC AND SKEW-SYMMETRIC BILINEAR FORMS

In this section, we shall learn about symmetric and skew-symmetric bilinear forms and matrices associated with them. Throughout this section it will be assumed that the characteristic of field F is not equal to 2, i.e. $1 + 1 \neq 0$, where 1 is the unity in F.

SYMMETRIC BILINEAR FORM *A bilinear form f on a vector space $V(F)$ is said to be symmetric, if*
$$f(u, v) = f(v, u) \quad \text{for all } u, v \in V.$$
The bilinear form f on R^2 defined by
$$f[(x_1, x_2), (y_1, y_2)] = x_1 x_2 + y_1 y_2 \text{ for all } (x_1, x_2), (y_1, y_2) \in R^2$$
is symmetric, because
$$f[(x_1, x_2), (y_1, y_2)] = x_1 x_2 + y_1 y_2 = y_1 y_2 + x_1 x_2 = f[(y_1, y_2), (x_1, x_2)]$$

SKEW-SYMMETRIC BILINEAR FORM *A bilinear form f on a vector space $V(F)$ is said to be skew-symmetric, if*
$$f(u, v) = -f(v, u) \quad \text{for all } u, v \in V$$

ALTERNATING BILINEAR FORM *A bilinear form f on a vector space $V(F)$ is said to be an alternating bilinear form, if*
$$f(v, v) = 0 \quad \text{for all } v \in V.$$

THEOREM–1 *Let f be a bilinear form on a vector space $V(F)$ and characteristic of F is different from 2. Then, f is skew-symmetric iff f is alternating.*

PROOF. First, let f be a skew-symmetric bilinear form on V. Then,

$f(u, v) = -f(v, u) \quad$ for all $u, v \in V$
$\Rightarrow \quad f(v, v) = -f(v, v) \quad$ for all $v \in V$
$\Rightarrow \quad f(v, v) + f(v, v) = 0 \quad$ for all $v \in V$
$\Rightarrow \quad f(v, v) = 0 \quad$ for all $v \in V$ $\qquad$ [$\because$ Characteristic $F \neq 2$]
$\Rightarrow \quad f$ is an alternating form on V.

Conversely, let f be an alternating form on $V(F)$. Then,

$f(u+v, u+v) = 0 \quad$ for all $u, v \in V$
$\Rightarrow \quad f(u, u) + f(u, v) + f(v, u) + f(v, v) = 0 \quad$ for all $u, v \in V$
$\Rightarrow \quad f(u, v) + f(v, u) = 0 \quad$ for all $u, v \in V \quad$ [$\because f(u, u) = 0 = f(v, v)$ as f is alternating]
$\Rightarrow \quad f(u, v) = -f(v, u) \quad$ for all $u, v \in V$

Q.E.D.

EXAMPLE-1 *For any $n \times n$ matrix A over a field F, show that the bilinear form $f : F^{n \times 1} \times F^{n \times 1} \to F$ defined by*
$$f(X, Y) = Y^T A X \quad \text{for all } X, Y \in F^{n \times 1}$$
is symmetric or skew-symmetric according as A is symmetric or skew-symmetric.

SOLUTION First, let A be a symmetric matrix over F. Then, $A^T = A$. For any $X, Y \in F^{n \times 1}$

$$f(X, Y) = Y^T A X = (Y^T A X)^T \qquad [Y^T A X \text{ is } 1 \times 1 \text{ matrix, i.e. a scalar in } F]$$

$\Rightarrow \quad f(X, Y) = X^T A^T Y$

$\Rightarrow \quad f(X, Y) = X^T A Y \qquad\qquad\qquad\qquad\qquad [\because A^T = A]$

$\Rightarrow \quad f(X, Y) = f(Y, X)$

$\therefore \quad f$ is symmetric

Now, let A be a skew-symmetric matrix over F. Then, $A^T = -A$.
For any $X, Y \in F^{n \times 1}$

$$f(X, Y) = Y^T A X = (Y^T A X)^T$$

$\Rightarrow \quad f(X, Y) = X^T A^T Y$

$\Rightarrow \quad f(X, Y) = -X^T A Y \qquad\qquad\qquad\qquad [\because A^T = -A]$

$\Rightarrow \quad f(X, Y) = -f(Y, X)$

$\therefore \quad f$ is skew-symmetric

EXAMPLE-2 *Let V be the vector space of all continuous real valued functions defined on R. Let ϕ and ψ be a bilinear forms in V defined by*

$$\phi(f, g) = \int_0^1 \{f(t-1)g(t) + g(t-1)f(t)\} \, dt$$

$$\psi(f, g) = \int_0^1 \{f(t-1)g(t) - g(t-1)f(t)\} \, dt \quad \text{for all } f, g \in V.$$

Show that ϕ is symmetric and ψ is skew-symmetric bilinear form.

SOLUTION We have,

$$\phi(f, g) = \int_0^1 \{f(t-1)g(t) + g(t-1)f(t)\} \, dt$$

$\Rightarrow \quad \phi(g, f) = \int_0^1 \{g(t-1)f(t) + f(t-1)g(t)\} \, dt$

$\Rightarrow \quad \phi(g, f) = \int_0^1 \{f(t-1)g(t) + g(t-1)f(t)\} \, dt$

$\Rightarrow \quad \phi(g, f) = \phi(f, g)$

Now,

$$\psi(f, g) = \int_0^1 \{f(t-1)g(t) - g(t-1)f(t)\} \, dt$$

$$\Rightarrow \quad \psi(g, f) = \int_0^1 \{g(t-1)f(t) - f(t-1)g(t)\}\, dt$$

$$\Rightarrow \quad \psi(g, f) = -\int_0^1 \{f(t-1)g(t) - g(t-1)f(t)\}\, dt$$

$$\Rightarrow \quad \psi(g, f) = -\psi(f, g)$$

Hence, ϕ is symmetric and ψ is skew-symmetric bilinear forms.

THEOREM-2 *Let $V(F)$ be a finite dimensional vector space, and B be an ordered basis for V. Then a bilinear form f on V is symmetric iff its matrix in basis B is symmetric.*

PROOF. Let $B = \{v_1, v_2, \ldots, v_n\}$ and $A = [a_{ij}]$ be matrix of f in basis B. Then,

$$a_{ij} = f(v_i, v_j) \quad \text{for all } i, j \in \underline{n}.$$

First, let f be a symmetric bilinear form on V. Then,

$$f(v_i, v_j) = f(v_j, v_i) \quad \text{for all } i, j \in \underline{n}$$

$\Rightarrow \quad a_{ij} = a_{ji} \quad \text{for all } i, j \in \underline{n}$

$\Rightarrow \quad A = A^T$

$\Rightarrow \quad A$ is symmetric

Conversely, let A be a symmetric matrix. Let $u, v \in V$ and let X, Y be the coordinate matrices of u and v respectively relative to basis B, i.e. $X = [u]_B$ and $Y = [v]_B$. Then,

$$f(u, v) = X^T A Y \text{ and } f(v, u) = Y^T A X$$

Now,

$$f(u, v) = X^T A Y$$

$\Rightarrow \quad f(u, v) = \left(X^T A Y\right)^T \quad \left[\because X^T A Y \text{ is } 1 \times 1 \text{ matrix over } F\right]$

$\Rightarrow \quad f(u, v) = Y^T A^T X$

$\Rightarrow \quad f(u, v) = Y^T A X \quad \left[\because A \text{ is symmetric} \therefore A^T = A\right]$

$\Rightarrow \quad f(u, v) = f(v, u)$

Thus, $f(u, v) = f(v, u)$ for all $u, v \in V$
Hence, f is a symmetric bilinear form on V.

Q.E.D.

THEOREM-3 *Let $V(F)$ be a finite dimensional vector space and B be an ordered basis for V. Then a bilinear form f on V is skew-symmetric iff its matrix in basis B is skew-symmetric.*

PROOF. Proceed as in the proof of Theorem 2.

EXAMPLE-3 Let f be a bilinear form on C^2 defined by

$$f(u, v) = x_1 y_2 + x_2 y_1 - x_1 y_1 \quad \text{for all } u = (x_1, x_2),\, v = (y_1, y_2) \in C^2.$$

Show that f is a symmetric bilinear form on C^2.

SOLUTION We have,

$$f(u, v) = x_1 y_2 + x_2 y_1 - x_1 y_1$$

$\Rightarrow \quad f(v, u) = y_1 x_2 + y_2 x_1 - y_1 x_1 = x_1 y_2 + x_2 y_1 - x_1 y_1 = f(u, v)$

$\therefore \quad f(u, v) = f(v, u) \quad \text{for all } u, v \in V.$

So, f is a symmetric bilinear form on C^2.

Aliter Let X and Y be the coordinate matrices of u and v respectively relative to the standard order basis B of C^2. Then,

$$f(u, v) = x_1 y_2 + x_2 y_1 - x_1 y_1$$

$\Rightarrow \quad f(u, v) = [x_1 \; x_2] \begin{bmatrix} -1 & 1 \\ 1 & 0 \end{bmatrix} \begin{bmatrix} y_1 \\ y_2 \end{bmatrix} = X^T A Y, \text{ where } A = \begin{bmatrix} -1 & 1 \\ 1 & 0 \end{bmatrix}$

Clearly, A is the matrix of f in standard basis of C^2 such that $A^T = A$.

Therefore, matrix of f in basis B is symmetric and hence f is symmetric.

EXAMPLE-4 Let f be a bilinear form on C^2 given by

$$f(u, v) = x_1 y_2 - x_2 y_1 \quad \text{for all } u = (x_1, x_2), v = (y_1, y_2) \in C^2.$$

Show that f is skew-symmetric bilinear form on C^2.

SOLUTION Let X and Y be the coordinate matrices of $u = (x_1, x_2)$ and $v = (y_1, y_2)$ relative to the standard basis B of C^2, i.e. $[u]_B = X$ and $[v]_B = Y$. Then,

$$f(u, v) = x_1 y_2 - x_2 y_1$$

$\Rightarrow \quad f(u, v) = [x_1 \; x_2] \begin{bmatrix} 0 & 1 \\ -1 & 0 \end{bmatrix} \begin{bmatrix} y_1 \\ y_2 \end{bmatrix} = X^T A Y, \text{ where } A = \begin{bmatrix} 0 & 1 \\ -1 & 0 \end{bmatrix}$

Here, $A = \begin{bmatrix} 0 & 1 \\ -1 & 0 \end{bmatrix}$ is the matrix of f in the standard basis of C^2 such that $A^T = -A$

i.e. A is skew-symmetric matrix.

Hence, f is skew-symmetric bilinear form on C^2.

EXAMPLE-5 Let f be a bilinear form on C^2 given by

$$f(u, v) = x_1 y_2 + 2 x_2 y_1 - x_1 y_1 \quad \text{for all } u = (x_1, x_2), v = (y_1, y_2) \in C^2.$$

Show that f is neither symmetric nor skew-symmetric.

SOLUTION Let X and Y be the coordinate matrices of $u = (x_1, x_2)$ and $v = (y_1, y_2)$ relative to the standard basis B of C^2, i.e. $[u]_B = X$ and $[v]_B = Y$. Then,

$$f(u, v) = x_1 y_2 + 2x_2 y_1 - x_1 y_1$$

$$\Rightarrow \quad f(u, v) = [x_1\ x_2] \begin{bmatrix} -1 & 1 \\ 2 & 0 \end{bmatrix} \begin{bmatrix} y_1 \\ y_2 \end{bmatrix} = X^T A Y, \text{ where } A = \begin{bmatrix} -1 & 1 \\ 2 & 0 \end{bmatrix}.$$

Here, A is the matrix of f in the standard basis of C^2 and it is neither symmetric nor skew-symmetric. Hence, f is neither symmetric nor skew-symmetric.

Let A be a square matrix over a field F. Then,

$$(A + A^T)^T = A^T + (A^T)^T = A + A^T$$

and $\quad (A - A^T)^T = A^T - (A^T)^T = A^T - A = -(A - A^T)$

$\Rightarrow \quad A + A^T$ is symmetric and $A - A^T$ is skew-symmetric

Also, $\quad 2A = (A + A^T) + (A - A^T)$

If F is a field of characteristic not equal to 2, then $2A \neq O$

$\therefore \quad 2A = (A + A^T) + (A - A^T)$

$\Rightarrow \quad A = \frac{1}{2}(A + A^T) + \frac{1}{2}(A - A^T)$

Thus, over a field of characteristic not equal to 2, any square matrix is the sum of a symmetric and a skew-symmetric matrix. Note that this is true only for fields of characteristic not equal to 2. Over a field of characteristic 2, say Z_2, a symmetric matrix is also skew-symmetric and the above result is not true.

THEOREM-2 *Every bilinear form on a vector space $V(F)$, where F is a subfield of the field C of all complex numbers, can be uniquely expressed as the sum of a symmetric and skew-symmetric bilinear forms.*

PROOF. Let g and h be mappings from V to F defined by

$$g(u, v) = \frac{1}{2}\{f(u, v) + f(v, u)\} \qquad \text{(i)}$$

$$h(u, v) = \frac{1}{2}\{f(u, v) - f(v, u)\} \qquad \text{(ii)}$$

for all $u, v \in V$.

It can be easily seen that g and h are bilinear forms on V.
We have,
$$g(v, u) = \frac{1}{2}\{f(v, u) + f(u, v)\} = \frac{1}{2}\{f(u, v) + f(v, u)\} = g(u, v)$$
and,
$$h(v, u) = \frac{1}{2}\{f(v, u) - f(u, v)\} = -\frac{1}{2}\{f(u, v) - f(v, u)\} = -h(u, v)$$
$\therefore$ g is symmetric and h is skew-symmetric.

Adding (i) and (ii), we obtain
$$f(u, v) = g(u, v) + h(u, v) \text{ for all } u, v \in V$$
$\Rightarrow \quad f = g + h$

Thus, f can be written as the sum of a symmetric and skew-symmetric bilinear forms.

In order to prove the uniqueness, let us suppose that $f = \phi + \psi$, where ϕ is symmetric and ψ is skew-symmetric. Then,
$$f(u, v) = (\phi + \psi)(u, v)$$
$\Rightarrow \quad f(u, v) = \phi(u, v) + \psi(u, v)$ \hfill (iii)

Also, $\quad f(v, u) = \phi(v, u) + \psi(v, u)$

$\Rightarrow \quad f(v, u) = \phi(u, v) - \psi(u, v)$ \hfill (iv)

From (iii) and (iv), we get
$$\phi(u, v) = \frac{1}{2}\{f(u, v) + f(v, u)\} = g(u, v)$$
and, $\quad \psi(u, v) = \frac{1}{2}\{f(u, v) - f(v, u)\} = h(u, v) \quad$ for all $u, v \in V$

$\therefore \quad \phi = g$ and $\psi = h$

Hence, f is uniquely expressible as the sum of a symmetric and a skew-symmetric bilinear form.

Q.E.D.

THEOREM-3 *Let f be a bilinear form on a vector space $V(F)$ such that characteristic of F is not equal to 2. Then, the following statements are equivalent:*

(i) f is symmetric

(ii) $f(u, v) = \frac{1}{2}\{f(u+v, u+v) - f(u, u) - f(v, v)\}$

(iii) $f(u, v) = \frac{1}{4}\{f(u+v, u+v) - f(u-v, u-v)\}$

Here, characteristic of field F is not equal to 2.

PROOF. First we shall prove that (i) imples (ii).

Let f be a symmetric bilinear form on V. Then,

$$f(u, v) = f(v, u) \quad \text{for all } u, v \in V.$$
$$\therefore \quad f(u+v, u+v) = f(u+v, u) + f(u+v, v)$$
$$\Rightarrow \quad f(u+v, u+v) = f(u, u) + f(v, u) + f(u, v) + f(v, v)$$
$$\Rightarrow \quad f(u+v, u+v) = f(u, u) + 2f(u, v) + f(v, v) \begin{bmatrix} \because \ f \text{ is symmetric and Char } (F) \neq 2 \\ \therefore \ 2f(u, v) \neq 0 \end{bmatrix}$$
$$\therefore \quad f(u, v) = \frac{1}{2}\{f(u+v, u+v) - f(u, u) - f(v, v)\}$$

Hence, (i) implies (ii).

Now, suppose (ii) holds. Then,

$$f(u, v) = \frac{1}{2}\{f(u+v, u+v) - f(u, u) - f(v, v)\} \quad \text{for all } u, v \in V$$
$$\Rightarrow \quad f(v, u) = \frac{1}{2}\{f(v+u, v+u) - f(v, v) - f(u, u)\} \quad \text{for all } u, v \in V$$
$$\Rightarrow \quad f(v, u) = \frac{1}{2}\{f(u+v, u+v) - f(u, u) - f(v, v)\} \quad \text{for all } u, v \in V \ [\because \ u+v = v+u]$$
$$\Rightarrow \quad f(v, u) = f(u, v) \quad \text{for all } u, v \in V$$
$$\Rightarrow \quad f \text{ is a symmetric bilinear form on } V.$$

So, (ii) implies (i)

Hence, (i) $\Leftrightarrow$ (ii).

Now, we shall show that (i) $\Leftrightarrow$ (iii)

Let f be a symmetric bilinear form on V. Then,

$$f(u, v) = f(v, u) \quad \text{for all } u, v \in V.$$
$$\therefore \quad f(u+v, u+v) = f(u, u) + 2 f(u, v) + f(v, v) \qquad [\because \ \text{Chr } (F) \neq 2]$$
$$\text{and,} \quad f(u-v, u-v) = f(u, u) - 2f(u, v) + f(v, v)$$
$$\therefore \quad f(u+v, u+v) - f(u-v, u-v) = 4 f(u, v)$$
$$\Rightarrow \quad f(u, v) = \frac{1}{4}\{f(u+v, u+v) - f(u-v, u-v)\}$$

Thus, (i) implies (iii).

Now, let (iii) hold. Then,

$$f(u, v) = \frac{1}{4}\{f(u+v, u+v) - f(u-v, u-v)\} \quad \text{for all } u, v \in V$$

$\Rightarrow \quad f(v, u) = \frac{1}{4}\{f(v+u, v+u) - f(v-u, v-u)\} \quad \text{for all } u, v \in V$

$\Rightarrow \quad f(v, u) = \frac{1}{4}\{f(u+v, u+v) - f(-(u-v), -(u-v))\} \quad \text{for all } u, v \in V$

$\Rightarrow \quad f(v, u) = \frac{1}{4}\{f(u+v, u+v) - f(u-v, u-v)\} \quad \text{for all } u, v \in V$

$\Rightarrow \quad f(v, u) = f(u, v) \quad \text{for all } u, v \in V.$

$\therefore \quad$ f is a symmetric bilinear form on V.

So, (iii) implies (i)
Thus, (i) $\Leftrightarrow$ (iii)
Hence, (i) $\Leftrightarrow$ (ii) $\Leftrightarrow$ (iii) i.e. all the three statements are equivalent.

Q.E.D.

EXERCISE 10.3

1. Find all skew-symmetric bilinear forms on R^2.
2. Find all symmetric bilinear forms on R^2.
3. For any $n \times n$ matrix A over a field F, show that the bilinear form $f : F^{n \times 1} \times F^{n \times 1} \to F$ defined by

    $$f(X, Y) = X^T AY \quad \text{for all } X, Y \in F^{n \times 1}$$

    is symmetric or skew-symmetric according as A is symmetric or skew-symmetric.
4. Show that the bilinear form f on R^2 defined by

    $$f[(x_1, x_2), (y_1, y_2)] = 2x_1y_1 + 3x_1y_2 + 3x_2y_1 + x_2y_2$$

    for all $(x_1, x_2), (y_1, y_2) \in R^2$, is a symmetric bilinear form.
5. Let f be a bilinear form on R^2 defined by

    $$f(u, v) = 3x_1 y_2 - 3x_2 y_1 \quad \text{for all } u = (x_1, x_2), v = (y_1, y_2) \in R^2.$$

    Show that f is a skew-symmetric bilinear form.
6. Let f be a bilinear form on $R^{2 \times 1}$ such that

    $$f(X, Y) = X^T \begin{bmatrix} 1 & 2 \\ 0 & 3 \end{bmatrix} Y \quad \text{for all } X, Y \in R^{2 \times 1}.$$

    Find the matrix of f in the standard ordered basis of $R^{2 \times 1}$. Is f symmetric?
7. Let f be a bilinear form on R^3 defined by $f(u, v) = -2x_1y_1 + x_2y_2 - x_3y_3 + 4x_1y_2 + 4x_2y_1$ for all $u = (x_1, x_2, x_3), v = (y_1, y_2, y_3) \in R^3$. Show that f is symmetric.

8. Let $S(V, V, F)$ denote the set of all symmetric bilinear forms on a vector space $V(F)$. Show that $S(V, V, F)$ is a subspace of $L(V, V, F)$. If $\dim V = n$, show that $\dim[S(V, V, F)] = \frac{n(n+1)}{2}$.

ANSWERS

1. $f[(x_1, x_2), (y_1, y_2)] = a(x_1 y_2 - x_2 y_1), a \in R$
2. $f[(x_1, x_2), (y_1, y_2)] = a x_1 y_1 + b x_2 y_2 + c(x_1 y_2 + x_2 y_1); a, b, c \in R$
6. No

10.6 QUADRATIC FORMS

Let us begin with the definition of a quadratic form.

QUADRATIC FORM *Let f be a symmetric bilinear form on a vector space $V(F)$. Then a function $q : V \to F$ defined by*

$$q(v) = f(v, v) \quad \text{for all } v \in V$$

is called the quadratic form associated with f.

If F is a field of characteristic not equal to 2, i.e. $1 + 1 \neq 0$, where 1 is unity in F. Then,

$$\begin{aligned} q(u+v) - q(u) - q(v) &= f(u+v, u+v) - f(u, u) - f(v, v) \\ &= f(u, u+v) + f(v, u+v) - f(u, u) - f(v, v) \\ &= f(u, u) + f(u, v) + f(v, u) + f(v, v) - f(u, u) - f(v, u) \\ &= 2f(u, v) \end{aligned}$$

$$\therefore \quad f(u, v) = \frac{1}{2}\{q(u+v) - q(u) - q(v)\} \qquad [\because \text{ Chr }(F) \neq 2 \quad \therefore \quad 2f(u, v) \neq 0]$$

Thus, if V is a vector space over a field of characteristic not equal to 2 and q is a quadratic form associated to a bilinear form f, then f is given by

$$f(u, v) = \frac{1}{2}\{q(u+v) - q(u) - q(v)\} \quad \text{for all } u, v \in V$$

This is known as polarized identity determining the bilinear form f.

If F is a subfield of complex numbers, the symmetric bilinear form f is completely determined by its associated quadratic form, according to the polarized identity

$$f(u, v) = \frac{1}{4}\{q(u+v) - q(u-v)\}$$

EXAMPLE-1 Let $v = R^n$ and let f be the bilinear form defined by $f(u, v) = \sum_{i=1}^{n} x_i y_i$ for all $u = (x_1, x_2, \ldots, x_n)$, $v = (y_1, y_2, \ldots, y_n) \in R^n$. Find the associated quadratic form.

SOLUTION We have,

$$f(u, v) = \sum_{i=1}^{n} x_i y_i \quad \text{for all } u = (x_1, x_2, \ldots, x_n),\ v = (y_1, y_2, \ldots, y_n) \in R^n.$$

Let q be the associated quadratic form. Then

$$q(u) = f(u, u) = \sum_{i=1}^{n} x_i^2 = x_1^2 + x_2^2 + \cdots + x_n^2$$

REMARK-1 Clearly, $q(u) = ||u||^2$ for all $u \in R^n$.

EXAMPLE-2 Let F be a field and n be a positive integer. For some $n \times n$ matrix $A = [a_{ij}]$ over F, $f_A : F^{n \times 1} \times F^{n \times 1}$ given by

$$f_A(X, Y) = X^T A Y \quad \text{for all } X, Y \in F^{n \times 1}.$$

is a bilinear form. Find the quadratic form q_A associated to f_A.

SOLUTION Clearly,

$$q_A(X) = f_A(X, X) \quad \text{for all } X \in F^{n \times 1}$$

$\Rightarrow \quad q_A(X) = X^T A X \quad \text{for all } X \in F^{n \times 1}$

$\Rightarrow \quad q_A(X) = \sum_{i=1}^{n} \sum_{j=1}^{n} a_{ij} x_i x_j$

$\Rightarrow \quad q_A(X) = \sum_{i=1}^{n} a_{ii} x_i^2 + \sum_{\substack{i=1 \\ i \neq j}}^{n} \sum_{j=1}^{n} a_{ij} x_i x_j$

This is the required quadratic form associated to f_A.

Let $A = [a_{ij}]$ be a symmetric matrix over a field F of characteristic not equal to 2. Let f be the bilinear form represented by matrix A. Then,

$$f(X, Y) = X^T A Y \quad \text{for all } X, Y \in F^{n \times 1}.$$

Let q be the quadratic form associated with f. Then,

$$q(X) = f(X, X) = X^T A X$$

$\Rightarrow \quad q(X) = \sum_i \sum_j a_{ij} x_i x_j = \sum_i a_{ii} x_i^2 + 2 \sum_{i<j} a_{ij} x_i x_j$

This suggests us an alternative definition of a quadratic form as given below.

QUADRATIC FORM *A quadratic form in variables* $x_1, x_2, \ldots, x_n$ *is a polynomial such that every term has degree two. That is,*

$$q(x_1, x_2, \ldots, x_n) = \sum_i \alpha_i x_i^2 + \sum_{i<j} \beta_{ij} x_i x_j$$

The quadratic form defined above determines a symmetric matrix $A = [a_{ij}]$, where $a_{ii} = \alpha_i$ and $a_{ij} = a_{ji} = \frac{1}{2}\beta_{ij}$.

If the matrix representation A of q is diagonal, then q has the following diagonal representation

$$q(X) = X^T A X = a_{11} x_1^2 + a_{22} x_2^2 + \cdots + a_{nn} x_n^2$$

What we have learnt so for on quadratic forms on the basis of it, we observe that there is one-to-one correspondence between the set of symmetric bilinear forms on a vector space V (over a field F of characteristic $\neq 2$) and the set of quadratic forms on V. Therefore, results on symmetric bilinear forms can be interpreted as results on quadratic forms and conversely. *Consequently, the rank of a quadratic form is defined to be the rank of associated symmetric bilinear form and the matrix of a quadratic form $q : V \to F$ relative to basis B is defined to be the $n \times n$ matrix of the associated bilinear form $h : V \times V \to F$.*

Thus, the matrix $A = [a_{ij}]$ of the quadratic form $q : V \to F$ relative to an ordered basis $B = \{v_1, v_2, \ldots, v_n\}$ is given by

$$a_{ij} = f(v_i, v_j) = \frac{1}{2}\{q(v_i + v_j) - q(v_i) - q(v_j)\} \quad \text{for all } i, j \in \underline{n}$$

where $f : V \times V \to F$ is the associated symmetric bilinear form.

In particular,

$$a_{ii} = f(v_i, v_i) = q(v_i) \quad \text{for all } i \in \underline{n}.$$

THEOREM–1 *Let V be a finite dimensional vector space over a subfield of the complex numbers, and let f be a symmetric bilinear form on V. Then there exists an ordered basis B for V such that matrix of f in that basis is a diagonal matrix.*

PROOF. In order to prove the theorem, we have to find an ordered basis $B = \{v_1, v_2, \ldots, v_n\}$ for V such that

$$f(v_i, v_j) = 0 \quad \text{for all } i, j \in \underline{n} \text{ and } i \neq j.$$

We shall prove this by induction on n.

If $n = 1$, i.e. V is one-dimensional vector space, then the theorem is obviously true.

Let us assume that the theorem is true for every vector space of dimension $(n-1)$, where $n > 1$.

If $f = \hat{0}$, then theorem is obviously true. So, let us assume that $f \neq \hat{0}$.

If $f(v, v) = 0$ for every $v \in V$, then

$q(v) = 0$ for every $v \in V$, where q is the quadratic form whose associated bilinear form is f

$\Rightarrow \quad q(u+v) = 0$ and $q(u-v) = 0 \quad$ for all $u, v \in V$

$\Rightarrow \quad \frac{1}{4}\{q(u+v) - q(u-v)\} = 0 \quad$ for all $u, v \in V$

$\Rightarrow \quad f(u, v) = 0 \quad$ for all $u, v \in V$

$\Rightarrow \quad f = \hat{0}$, which is a contradiction.

So, there exists a vector $v_1 \in V$ such that $f(v_1, v_1) = q(v_1) \neq 0$.

Let S be the subspace of V spanned by v_1, i.e. $S = [\{v_1\}]$ and T be the set of all vectors $u \in V$ such that $f(v_1, u) = 0$. i.e. $T = \{u \in V : f(v_1, u) = 0\}$

For any $u_1, u_2 \in T$ and $a, b \in F$

$$f(v_1, au_1 + bu_2) = a\, f(v_1, u_1) + b\, f(v_1, u_2) \qquad [\because f \text{ is a bilinear form}]$$
$$\Rightarrow f(v_1, au_1 + bu_2) = a \times 0 + b \times 0 = 0 \; [\because u_1, u_2 \in T \therefore f(v_1, u_1) = 0, f(v_1, u_2) = 0]$$
$$\Rightarrow au_1 + bu_2 \in T$$

Thus, $au_1 + bu_2 \in T$ for all $u_1, u_2 \in T$ and $a, b \in F$.

So, T is a subspace of V.

We shall now show that $V = S \oplus T$.

Let $v \in S \cap T$. Then, $v \in S$ and $v \in T$.

Now,

$v \in S \Rightarrow v = \lambda v_1$ for some $\lambda \in F$.

and, $\quad v \in T$
$\Rightarrow \quad f(v_1, v) = 0$
$\Rightarrow \quad f(v_1, \lambda v_1) = 0 \qquad\qquad\qquad\qquad\qquad\qquad\qquad\qquad [\because v = \lambda v_1]$
$\Rightarrow \quad \lambda f(v_1, v_1) = 0$
$\Rightarrow \quad \lambda = 0 \qquad\qquad\qquad\qquad\qquad\qquad\qquad\qquad\qquad [\because f(v_1, v_1) \neq 0]$
$\therefore \quad v = \lambda v_1 \Rightarrow v = 0_V$

Thus, $v \in S \cap T \Rightarrow v = 0_V$
$\therefore \quad S \cap T = \{0_V\}$

Let v be any vector in V and $w = v - \dfrac{f(v, v_1)}{f(v_1, v_1)} v_1$. Then,

$$f(v_1, w) = f\left(v_1, v - \dfrac{f(v, v_1)}{f(v_1, v_1)} v_1\right)$$
$\Rightarrow \quad f(v_1, w) = f(v_1, v) - \dfrac{f(v, v_1)}{f(v_1, v_1)} f(v_1, v_1)$
$\Rightarrow \quad f(v_1, w) = f(v_1, v) - f(v, v_1)$
$\Rightarrow \quad f(v_1, w) = 0 \qquad\qquad\qquad\qquad\qquad\qquad\qquad\qquad [\because f(v, v_1) = f(v_1, v)]$

$\Rightarrow$ $\quad w \in T$

$\therefore \quad w = v - \dfrac{f(v, v_1)}{f(v_1, v_1)} v_1$

$\Rightarrow \quad v = w + \dfrac{f(v, v_1)}{f(v_1, v_1)} v_1$

$\Rightarrow \quad v \in S + T$ $\qquad\qquad\qquad\qquad\qquad\left[\because \dfrac{f(v, v_1)}{f(v_1, v_1)} v_1 \in S \text{ and } w \in T\right]$

Thus, $\quad v \in V \Rightarrow v \in S + T$

$\therefore \quad V \subset S + T$

But, $\quad S + T \subset V$

$\therefore \quad S + T = V$

Hence, $\quad V = S \oplus T$

$\Rightarrow \quad \dim V = \dim S + \dim T$

$\Rightarrow \quad \dim T = (n-1)$

Let g be the restriction of f to T. Then, g is a symmetric bilinear form on T and $\dim T = n - 1$. So, by induction assumption there exists a basis $\{v_2, v_3, \ldots, v_n\}$ for T such that

$\qquad g(v_i, v_j) = 0 \quad$ for all $i \geq 2, j \geq 2$ and $i \neq j$

$\Rightarrow \quad f(v_i, v_j) = 0 \quad$ for all $i \neq j, i \geq 2, j \geq 2$

Also, $\quad f(v_i, v_j) = 0 \quad$ for $j = 2, 3, \ldots, n$ $\qquad\qquad$ [By definition of T]

$\therefore \quad f(v_i, v_j) = 0 \quad$ for all $i, j \in \underline{n}$ and $i \neq j$

Since $V = S \oplus T$ and $\{v_1\}$, $\{v_2, v_3, \ldots, v_n\}$ are bases for S and T respectively. Therefore, $B = \{v_1, v_2, \ldots, v_n\}$ is a basis for V such that

$$f(v_i, v_j) = 0 \quad \text{for all } i, j \in \underline{n} \text{ and } i \neq j$$

Hence, f is represented by a diagonal matrix, i.e. $M_B^B(f)$ is a diagonal matrix.

Q.E.D.

THEOREM-0 *Every quadratic form over a field F of characteristic $\neq 2$ can be diagonalized.*

PROOF. Let V be a vector space over a field F of characteristic $\neq 2$ and q be a quadratic form on V. Let f be the symmetric bilinear form associated with f. Then,

$$q(v) = f(v, v) \quad \text{for all} \quad v \in V$$

and, $\quad f(u, v) = \dfrac{1}{2}\{q(u+v) - q(u) - q(v)\} \quad$ for all $u, v \in V$.

The matrix of q relative to an ordered basis of V is the matrix of f relative to the same basis. So, it is sufficient to prove that there is an ordered orthonormal basis $B = \{v_1, v_2, v_3, \ldots, v_n\}$ for V such that the matrix of f in basis B is a diagonal matrix. That is,

$$f(v_i, v_j) = 0 \quad \text{for all } i, j \in \underline{n} \text{ and } i \neq j.$$

Proceeding as in the above theorem, we obtain a basis $B = \{v_1, v_2, \ldots, v_n\}$ for V such that

$$f(v_i, v_j) = 0 \quad \text{for all } i, j \in \underline{n} \text{ and } i \neq j.$$

That is, f is represented by a diagonal matrix. Hence, q is represented by a diagonal matrix.

Q.E.D.

ILLUSTRATIVE EXAMPLES

EXAMPLE-1 *Find the symmetric matrix associated with each of the following quadratic forms:*
   (i) $q(x, y) = 8xy + 4y^2$
   (ii) $q'(x, y, z) = x^2 + 2xy + 4xz + 3y^2 + yz + 7z^2$
   (iii) $q''(x, y, z) = 3x^2 + xz - 2yz$
   (iv) $q'''(x, y, z) = 2x^2 - 5y^2 - 7z^2$

SOLUTION The symmetric matrix $A = [a_{ij}]$ representing the quadratic form $q(x_1, x_2, \ldots, x_n)$ has the diagonal entry a_{ii} equal to the coefficient of the square term x_i^2 and non-diagonal entries a_{ij} and a_{ji} each equal to the cross product term $x_i x_j$. Thus, the matrices that represent the given quadratic forms are:

(i) $A = \begin{bmatrix} 0 & 2 \\ 2 & 4 \end{bmatrix}$ (ii) $A' = \begin{bmatrix} 1 & 1 & 2 \\ 1 & 3 & \frac{1}{2} \\ 2 & \frac{1}{2} & 7 \end{bmatrix}$ (iii) $A'' = \begin{bmatrix} 3 & 0 & \frac{1}{2} \\ 0 & 0 & -1 \\ \frac{1}{2} & -1 & 0 \end{bmatrix}$ (iv) $A''' = \begin{bmatrix} 2 & 0 & 0 \\ 0 & -5 & 0 \\ 0 & 0 & 7 \end{bmatrix}$

EXAMPLE-2 *Find the quadratic form $q(X)$ that corresponds to each of the following symmetric matrices:*

(i) $A = \begin{bmatrix} 5 & -3 \\ -3 & 7 \end{bmatrix}$ (ii) $B = \begin{bmatrix} 4 & -5 & 7 \\ -5 & -6 & 8 \\ 7 & 8 & -9 \end{bmatrix}$ (iii) $C = \begin{bmatrix} 2 & 4 & -1 & 5 \\ 4 & -7 & -6 & 8 \\ -1 & -6 & 3 & 9 \\ 5 & 8 & 9 & 1 \end{bmatrix}$

SOLUTION The quadratic form that corresponds to the matrix A is defined by $q(X) = X^T A X$, where $X = [x_i]$ is the column matrix of unknowns.

(i) Let $q(x, y)$ be the quadratic form that corresponds to the matrix A. Then,

$$q(x, y) = [x \, y] \begin{bmatrix} 5 & -3 \\ -3 & 7 \end{bmatrix} \begin{bmatrix} x \\ y \end{bmatrix} = [x \, y] \begin{bmatrix} 5x - 3y \\ -3x + 7y \end{bmatrix} = 5x^2 - 6xy + 7y^2$$

We observe that the coefficient 5 of the square term x^2 and the coefficient 8 of the square term y^2 are diagonal elements of the matrix, and the coefficient -6 of the cross product term xy is the sum of the non-diagonal elements -3 and -3 (or: twice the non-diagonal element -3 as the matrix is symmetric).

(ii) Let $q(x, y, z)$ be the quadratic form that corresponds to the matrix B. Then,
$$q(x, y, z) = 4x^2 - 6y^2 - 9z^2 - 10xy + 14xz + 16yz$$
Here, we use the fact that the coefficients of x^2, y^2, z^2 are respectively the diagonal elements 4, -6, -9 of the matrix and the coefficients of the cross product term $x_i\, x_j$ is twice the non-diagonal elements a_{ij} or a_{ji}.

(iii) Let $q(x_1, x_2, x_3, x_4)$ be the quadratic form that corresponds to the matrix C. Then,
$$q(x_1, x_2, x_3, x_4) = 2x_1^2 - 7x_2^2 + 3x_3^2 + 8x_1x_2 - 2x_1x_3 + 10x_1x_4 - 12x_2x_3 + 16x_2x_4 + 18x_3x_4$$

EXAMPLE-3 *If q is a quadratic form on a vector space $V(F)$, then*
$$q(a\,v) = a^2 q(v) \quad \text{for all } v \in V \text{ and } a \in F.$$

SOLUTION Let f be the symmetric bilinear form whose associated quadratic form is q. Then,

$\quad q(v) = f(v, v) \quad$ for all $v \in V$

$\Rightarrow \quad q(av) = f(av, av) \quad$ for all $v \in V, a \in F$

$\Rightarrow \quad q(av) = a^2 f(v, v) \quad$ for all $v \in V, a \in F$

$\Rightarrow \quad q(av) = a^2\, q(v) \quad$ for all $v \in V, a \in F$

EXAMPLE-4 *Let $q(x, y) = 3x^2 + 2xy - y^2$ be a quadratic form and $x = s - 3t$, $y = 2s + t$.*

(i) *Rewrite $q(x, y)$ in matrix notation, and find the matrix A representing $q(x, y)$.*
(ii) *Rewrite the linear substitution using matrix notation, and find matrix P corresponding to the substitution.*
(iii) *Find $q(s, t)$ using direct substitution.*
(iv) *Find $q(s, t)$ using matrix notation.*

SOLUTION (i) We have,
$$q(x, y) = 3x^2 + 2xy - y^2 = [x\ y]\begin{bmatrix} 3 & 1 \\ 1 & -1 \end{bmatrix}\begin{bmatrix} x \\ y \end{bmatrix} = X^T A X,$$
where $X = \begin{bmatrix} x \\ y \end{bmatrix}$ and $A = \begin{bmatrix} 3 & 1 \\ 1 & -1 \end{bmatrix}$.

Thus, $A = \begin{bmatrix} 3 & 1 \\ 1 & -1 \end{bmatrix}$ is the matrix representing $q(x, y)$.

(ii) We have,
$$x = s - 3t$$
and, $\quad y = 2s + t$

or, $\quad \begin{bmatrix} x \\ y \end{bmatrix} = \begin{bmatrix} 1 & -3 \\ 2 & 1 \end{bmatrix}\begin{bmatrix} s \\ t \end{bmatrix}$

or, $\quad X = PY$, where $X = \begin{bmatrix} x \\ y \end{bmatrix}$, $P = \begin{bmatrix} 1 & -3 \\ 2 & 1 \end{bmatrix}$ and $Y = \begin{bmatrix} s \\ t \end{bmatrix}$

(iii) We have,
$$q(x, y) = 3x^2 + 2xy - y^2, \text{ where } x = s - 3t, y = 2s + t$$
$$\therefore \quad q(s, t) = 3(s-3t)^2 + 2(s-3t)(2s+t) - (2s+t)^2 = 3s^2 - 32st + 20t^2$$

(iv) We have,
$$q(X) = X^T A X \text{ and } X = PY$$
$$\Rightarrow \quad q(s, t) = (PY)^T A\, PY = Y^T (P^T A P) Y$$
$$\Rightarrow \quad q(s, t) = [s\ t] \begin{bmatrix} 1 & 2 \\ -3 & 1 \end{bmatrix} \begin{bmatrix} 3 & 1 \\ 1 & -1 \end{bmatrix} \begin{bmatrix} 1 & -3 \\ 2 & 1 \end{bmatrix} \begin{bmatrix} s \\ t \end{bmatrix}$$
$$\Rightarrow \quad q(s, t) = 3s^2 - 32st + 20t^2.$$

EXERCISE 10.4

1. Find the symmetric matrix A associated with each of the following quadratic forms:
   (i) $q(x, y) = 4xy$ (ii) $q(x, y, z) = 3x^2 + 4xy - y^2 + 8xz - 6yz + z^2$
   (iii) $q(x, y, z) = xy + y^2 + 4xz + z^2$ (iv) $q(x, y, z) = xy + yz$

2. Find the quadratic form $q(X)$ that corresponds to each of the following symmetric matrices:

   (i) $A = \begin{bmatrix} 2 & -6 \\ -6 & 3 \end{bmatrix}$ (ii) $B = \begin{bmatrix} 1 & -4 & 6 \\ -4 & 2 & 8 \\ 6 & 8 & -3 \end{bmatrix}$ (iii) $C = \begin{bmatrix} 3 & 1 & -2 & 5 \\ 1 & 2 & 4 & 1 \\ -2 & 4 & 3 & -2 \\ 5 & 1 & -2 & 1 \end{bmatrix}$

3. Give an example of a quadratic form $q(x, y)$ such that $q(u) = 0$ and $q(v) = 0$ but $q(u+v) \neq 0$.

4. Let $q(x, y) = 2x^2 - 6xy - 3y^2$ and $x = s + 2t, y = 3s - t$.
   (i) Rewrite $q(x,y)$ in matrix notation, and find the matrix A representing the quadratic form.
   (ii) Rewrite the linear substitution using matrix notation, and find the matrix P corresponding to the substitution.
   (iii) Find $q(s, t)$, (i) using direct substitution (ii) matrix notation.

5. Define vector addition and scalar multiplication of quadratic forms on a vector space $V(F)$ by
$$(q+q')(x) = q(x) + q'(x)$$
$$(aq)(x) = aq(x)$$
for all $x \in V$, all quadratic forms q, q' on V, and all $a \in F$.
Show that the set of all quadratic forms on V under the vector addition and scalar multiplication defined above forms a vector space over F.

ANSWERS

1. (i) $A = \begin{bmatrix} 0 & 2 \\ 2 & 0 \end{bmatrix}$ (ii) $B = \begin{bmatrix} 3 & 2 & 4 \\ 2 & -1 & -3 \\ 4 & -3 & 1 \end{bmatrix}$ (iii) $C = \begin{bmatrix} 0 & \frac{1}{2} & 2 \\ \frac{1}{2} & 1 & 0 \\ 2 & 0 & 1 \end{bmatrix}$ (iv) $D = \begin{bmatrix} 0 & \frac{1}{2} & 0 \\ \frac{1}{2} & 0 & \frac{1}{2} \\ 0 & \frac{1}{2} & 0 \end{bmatrix}$

2. (i) $q(x, y) = 2x^2 - 12xy + 3y^2$ (ii) $q(x, y, z) = x^2 - 8xy + 12zx + 2y^2 + 16yz - 3z^2$
   (iii) $q(x_1, x_2, x_3, x_4) = 3x_1^2 + 2x_1 x_2 - 4x_1 x_3 + 10x_1 x_4 + 2x_2^2 + 8x_2 x_3 + 2x_2 x_4$
   $+ 3x_3^2 - 4x_3 x_4 + x_4^2$

3. $q(x, y) = x^2 - y^2$, $u = (1, 1)$, $v = (1, -1)$

4. (i) $A = \begin{bmatrix} 2 & -3 \\ -3 & -3 \end{bmatrix}$ (ii) $\begin{bmatrix} 1 & 2 \\ 3 & -1 \end{bmatrix}$ (iii) $q(s, t) = -43s^2 - 4st + 17t^2$

Index

A
Abelian group, 8
Adjoint operators, 610, 680
Algebra of subspaces, 143
Algorithm, 184, 379
Alternating bilinear form, 750
Alternating function, 462
Annihilators, 325
Artinian modules, 109
Ascending chain condition, 109
Associative binary operation, 5
Augmented matrix, 20

B
Basis, 192
   change of, 378, 743
Bijective map, 3
Bilinear forms, 722
   matrix representation of, 740
   rank of, 743, 744, 746
   space of, 731
Binary operations, 5
   restriction of, 6

C
Characteristic equation, 484
Characteristic polynomial, 477
Co-dimension, 207
Cofactors, 469
Column equivalent matrices, 419
Column matrix, 12
Commutative binary operation, 5
Commutative ring, 10
Conjugate linear form, 640
Coordinates, 229
Cramers rule, 474
Cyclic modules, 105
Cyclic permutation, 444

D
Determinants, 462
   permutations of, 443
   properties of, 447
Diagonal matrix, 12
Diagonalizable linear operator, 500
Diagonalizable matrix, 500
Diagonalization algorithm, 500
Diagonalizing matrices, 497
Diagonalizing real symmetric matrices, 511
Dimension, 192
Disjoint permutations, 444
Dual spaces,
   linear form of, 308
   linear functional of, 308

E

Echelon matrices, 188
Echelon matrix, 13
Element,
   inverse of, 6
Elementary matrices, 419
Elementary operations, 414
Equations in echelon form,
   system of, 21
Equations in triangular form,
   system of, 20
Equivalence relation, 2
Equivalent matrices, 397
Even permutation, 445

F

Fields, 9
Finite dimensional vector space, 192
Finitely generated submodule, 44
Free modules, 82, 86, 96
Functions, 2
   composition of, 4
   inverse of, 4

G

Gram-Schmidt orthogonalization process, 577
Groups, 7

H

Hermitian forms, 722
Hermitian inner products, 640
   matrix representation of, 665
Hermitian matrix, 17, 668
Homomorphic image, 60

I

Identity matrix, 12
Identity relation, 2
Injective map, 3
Inner product space, 531
   homomorphism of, 603
   isomorphism of, 603
   linear functionals and, 607
   morphisms of, 603
Inner products, 588
   matrix representation of, 588
Invertible linear transformation, 296
Invertible matrices, 412
Irreducible module, 41
Isomorphic vector spaces, 251

L

Linear combination, 83, 160
Linear dependence, 84, 178, 188
Linear equations,
   systems of, 19
Linear independence, 83, 178
Linear operator, 236, 294
   characteristic polynomial of, 482
   determinant of, 459
   minimal polynomial of, 525
   polynomials of, 478
Linear spans, 165
Linear transformation, 235, 356
   adjoint of, 347, 348
   basic properties of, 240
   algebra of, 280
   definitions of, 235
   determinant of, 401
   dual of, 347, 348
   eigenvalgues of, 429
   eigenvectors of, 429
   examples of, 235
   images of, 247
   kernel of, 247
   matrices as, 243
   nullity of, 264
   rank of, 264
   trace of, 401
Lower triangular matrix, 13

M

Matrices modules, 96

Matrices, 11, 356, 588
  addition of, 14
  bilinear forms and, 739
  equality of, 14
  equivalency of, 395
  multiplication of, 15
  polynomials of, 477
  similarity of, 395
Matrix multiplication, 15
Matrix polynomial, 16
Matrix,
  adjoint of, 472
  characteristic polynomial of, 479
  column spaces of, 170
  determinant rank of, 469
  eigenvalues of, 434
  eigenvectors of, 434
  images of, 264
  inverse of, 18
  kernal of, 264
  nullity of, 403
  rank of, 22, 402-403, 469
  row spaces of, 170
  trace of, 400
  transpose of, 16
Maximal element, 111
Maximal linearly independent set, 207
Minimal polynomial, 521
Minors, 469
Modules,
  definitions of, 23
  elementary properties of, 38
  examples of, 23
Monoid, 7
Multilinear function, 462
Multilinearity, 462

N
Noetherian modules, 109
Non-singular transformations, 295

Non-trivial combination, 83
Non-trivial linear combination, 178
Normal matrices, 668
Normal matrix, 17, 675
Normal operators, 706
Normal subgroup, 9
Null matrix, 13

O
Operator,
  adjoint of, 610, 681
  spectrum of, 624
Orthogonal basis, 556
Orthogonal diagonalization, 513
  algorithm of, 513
Orthogonal matrices, 595
Orthogonal matrix, 17
Orthogonal operator, 627
Orthogonal set, 555
Orthogonal vectors, 550, 653
Orthogonality, 653
Orthogonally equivalent matrices, 598
Orthonormal basis, 568, 657
Orthonormal set, 567, 657
Orthonormality, 653

P
Polynomials operators, 294
Positive definite matrices, 638
Positive definite matrix, 592, 699
Positive definite operators, 697
Positive operators, 632, 697
Projection mapping, 258
Projections, 333

Q
Quadratic forms, 722, 758
Quotient group, 9
Quotient modules, 54, 56
Quotient spaces, 156

R

R-Algebra, 81
Reflexive relation, 2
Relations,
  Cartesian product of, 1
  inverse of, 1
R-homomorphisms, 58, 77
  kernel of, 60
  sum of, 63
Rings, 9
  characteristic of, 10
R-module,
  rank of, 92
Row canonical form,
  matrix in, 13
Row equivalent matrices, 419
Row matrix, 12

S

Scalar matrix, 12
Scalar multiplication, 280
Self-adjoint operator, 683
Semi-group, 7
Sets,
  Cartesian products of, 1
  linear span of, 165
  orthogonal complement of, 562, 656
Singular linear transformations, 295
Singular matrix, 19, 448
Skew-adjoint operator, 616
Skew-hermitian matrix, 17, 669
Skew-hermitian operator, 683
Skew-symmetric bilinear forms, 749
Skew-symmetric matrix, 17
Spectral theorem, 624
Square matrix, 12
  adjoint of, 19
  eigenvalues of, 487
  eigenvectors of, 487
  positive integral powers of, 16

Submodules, 40
  algebra of, 46
  complement of, 52
  sum of, 48
Subspaces,
  complement of, 151
  direct sum of, 147
  sum of, 146
Surjective map, 3
Symmetric bilinear forms, 749
Symmetric matrix, 17
Symmetric relation, 2

T

Transitive relation, 2
Trivial linear combination, 83, 178

U

Unitarily equivalent matrices, 674
Unitary matrices, 668
Unitary matrix, 17, 669
Unitary module, 26
Unitary operators, 700
Unitary spaces, 640
  linear forms and, 677
  linear operators on, 677
Upper triangular matrix, 13

V

Vector along a subspace,
  projection of, 576
Vector spaces,
  definition of, 115
  elementary properties of, 130
  examples of, 115
Vectors,
  angle between in, 547
  length of, 542
  norm of, 542, 647
  orthogonal compliments of, 561, 655
  projection of, 574, 660